Silicon Carbide and Related Materials 2011

PART 1

Edited by
Robert P. Devaty
Michael Dudley
T. Paul Chow
Philip G. Neudeck

Silicon Carbide and Related Materials 2011

PART 1

Selected, peer reviewed papers from the
14th International Conference on
Silicon Carbide and Related Materials 2011
(ICSCRM 2011),
September 11-16, 2011, Cleveland, Ohio, USA

Edited by

**Robert P. Devaty, Michael Dudley,
T. Paul Chow and Philip G. Neudeck**

TTP

Trans Tech Publications Ltd
Kreuzstrasse 10
CH-8635 Durnten-Zurich
Switzerland
http://www.ttp.net

Volumes 717-720 of
Materials Science Forum *2-vol.-set*
ISSN 0255-5476

Full text available online at *http://www.scientific.net*

Distributed *worldwide by*

Trans Tech Publications Ltd
Kreuzstrasse 10
CH-8635 Durnten-Zurich
Switzerland

Fax: +41 (44) 922 10 33
e-mail: sales@ttp.net

and in the Americas by

Trans Tech Publications Inc.
PO Box 699, May Street
Enfield, NH 03748
USA

Phone: +1 (603) 632-7377
Fax: +1 (603) 632-5611
e-mail: sales-usa@ttp.net

printed in Germany

Preface

The 14th International Conference on Silicon Carbide and Related Materials 2011 (ICSCRM 2011) was held in Cleveland, Ohio USA at the downtown Renaissance Hotel from September 11-16, 2011. 603 registered attendees from 27 countries, including 87 students, participated in this conference to report and to discuss the recent technical progress achieved in this field during the last two years. A total of 379 oral or poster presentations were given during the technical program, 30 of which were invited or plenary. In addition, 34 companies filled the available conference exhibition space, demonstrating continued expansion in industrial interest and economic impact of this technical field.

The International Conference on Silicon Carbide and Related Materials (ICSCRM) is the world's premier technical conference series for microelectronics and MEMS technology based on silicon carbide and related materials. The aim of the conference is to present and discuss recent progress in crystal growth, characterization and control of material properties, as well as other basic research issues concerning silicon carbide (SiC) and other wide-bandgap semiconductors involving III-nitrides and diamond. The latest research results relevant to wafer production processes, device fabrication technologies and device applications were presented at this meeting. This included the latest results in the development and commercialization of advanced devices and circuits used for energy saving, high voltage switching, high frequency, high power amplification and high temperature operation. Work on growth, characterization and device development of epitaxial graphene was also covered. Evolving industrial products and capabilities were also highlighted during an Industrial News session held on Monday evening. The Conference Banquet was held at the world-famous Rock and Roll Hall of Fame and Museum on the scenic Lake Erie shoreline on Thursday evening, which provided a most interesting and entertaining venue for meeting friends and having stimulating discussions.

Other pages of this introductory section list the dedicated individuals and companies that contributed to the success of ICSCRM 2011. The Technical Program Committee, under the leadership of co-chairs D. Kurt Gaskill of the US Naval Research Laboratory and Anant Agarwal of Cree, Inc., carefully selected the contributed and invited papers and put together an interesting scientific program. We thank Lin Cheng of Cree, Inc., for her assistance to the Technical Program Committee in facilitating the organization of abstracts and invited speakers. ICSCRM 2011 marked the first time that ICSCRM manuscript submission and peer-review were carried out entirely online using a web browser based software system (developed by Trans Tech Publications). All the manuscript referees deserve great recognition for their efforts to help guarantee the high scientific level of these Proceedings.

We very much appreciate the generous support of industrial and government sponsors of ICSCRM 2011 that enabled quality conference logistics while keeping conference registration fees affordable. ICSCRM 2011 was locally organized in Cleveland by NASA Glenn Research Center via its conference and event management contract with QinetiQ North America. The QinetiQ team, led by Barbara Kakiris, successfully operated ICSCRM meeting logistics, from meeting space, food, beverage and other service contracts to numerous individual registration fee and sponsor fee financial transactions. We gratefully thank Harsh Naik for assisting the process of editing manuscripts for the proceedings at RPI.

The next conference in the ICSCRM series will be held in Miyazaki, Japan (on Kyushu Island) from Sept. 29 - Oct. 4, 2013. The General Chairman will be Dr. Hajime Okumura of the National Institute of Advanced Industrial Science and Technology (AIST). We wish the organizers of the next conference much success!

Cleveland, March 2012

Philip G. Neudeck
Robert P. Devaty
Michael Dudley
T. Paul Chow

ICSCRM 2011 Major Sponsors

Cree

Keithley Instruments

AIXTRON

Dow Corning Electronics Solutions

General Electric

CuttingEdgeIons.com

Office of Naval Research

NASA Glenn Research Center

Regular Sponsors

Infineon
SemiSouth
Microsemi
OAI

Exhibitors

Nippon Steel Materials Co.
Rigaku Americas
WBG Materials – A Division of II-VI, Inc.
Lasertec Corporation
LPE
Showa Denko K.K.
Tokyo Electron Limited
Cree
AIXTRON SE
Keithley Instruments
KLA-Tencor
TankeBlue Semiconductor Co. Ltd.
Acreo
Power+Energy, Inc.
GeneSiC Semiconductor, Inc.
Yole Developpement
Mersen USA
United Silicon Carbide, Inc.
Dow Corning Corporation
Momentive Performance Materials
Norstel AB
Aymont Technology, Inc.
Evans Analytical Group
Nitride Crystals, Inc.
NOVASiC
Sinmat, Inc.
Shandong Tian Yue Advanced Material Technology Co.
Agilent Technologies, Inc.
Lehighton Electronics, Inc.
Dongguan Tianyu Semiconductor Technology
GTI Technologies
Nanotronics Imaging LLC
Bridgestone
EpiWorld International Co.

ICSCRM International Steering Committee

P. Bergman (Linköping Univ., Sweden)
D. Chaussende (INP – CNRS LMGP, France)
R. P. Devaty (Univ. of Pittsburgh, USA)
P. Friedrichs (SiCED, Germany)
H. J. Kim (Seoul National Univ., Korea)
T. Kimoto (Kyoto Univ., Japan)
M. Krieger (Univ. of Erlangen-Nürnberg, Germany)
P. G. Neudeck (NASA, USA)
R. Nipoti (CNR-IMM, Italy)
H. Okumura (AIST, Japan)
J.W. Palmour (Cree, USA)
R. Stahlbush (Naval Research Laboratory, USA)
S. Yoshida (Saitama Univ., Japan)

ICSCRM Advisory Panel

C. H. Carter (Cree, USA)
W. J. Choyke (Univ. of Pittsburgh, USA)
E. Janzén (Linköping Univ., Sweden)
R. Madar (INPG/CNRS, France)
H. Matsunami (JST, Japan)
G. Pensl (Univ. of Erlangen-Nürnberg, Germany)
D. Stephani (Univ. of Erlangen-Nürnberg, Germany)

ICSCRM 2011 Local Organizing Committee

T. Paul Chow (Renssalaer Polytechnic Institute)
Robert P. Devaty (Univ. of Pittsburgh)
Michael Dudley (SUNY Stony Book)
Kurt Gaskill (US Naval Research Laboratory)
Allen Hefner (National Institutes of Standards and Technology)
Aivars Lelis (US Army Research Laboratory)
Edward Sanchez (Dow Corning Compound Semiconductor)
Phil Neudeck (NASA Glenn Research Center)
John Palmour (Cree, Inc.)
Pirouz Pirouz (Case Western Reserve Univ.)
Stephen Saddow (Univ. of South Florida)
Peter Sandvik (General Electric Global Research)
James Scofield (US Air Force Research Laboratory)
Ilya Zwieback (II-VI, Inc.)

ICSCRM 2011 Chairs

General Chair
Phil Neudeck (NASA Glenn Research Center)

Technical Program Committee Chairs
Anant Agarwal (Cree, Inc.)
Kurt Gaskill (US Naval Research Laboratory)

Technical Proceedings Editors
Robert P. Devaty (Univ. of Pittsburgh)
Michael Dudley (SUNY Stony Book)
T. Paul Chow (Rensselaer Polytechnic Institute)

Poster Session Chairs
Kevin Matocha (SemiSouth)
Stephen Saddow (Univ. of South Florida)

Publicity Chair
James Oliver (Northrop Grumman)

Sponsorship/Exhibits Chairs
Karl Hobart (US Naval Research Laboratory)
John Palmour (Cree, Inc.)
Peter Sandvik (General Electric Global Research)

Financial Aid Chair
H. Scott Coombe (Office of Naval Research)

International Attendee Chair
Pirouz Pirouz (Case Western Reserve University)

Industrial Session Chair
Gary Hunter (NASA Glenn Research Center)

ICSCRM 2011 Technical Program Committee

Technical Program Committee Co-Chairs
Anant Agarwal (Cree, USA)
Kurt Gaskill (Naval Research Laboratory, USA)

Bulk Growth Sub-Committee
Michael Dudley – Chair (SUNY Stony Brook, USA)
Noboru Ohtani (Kwansei Gakuin University, Japan)
Adrian Powell (Cree, USA)
Peter Wellmann (University of Erlangen-Nürnberg, Germany)

Epitaxy Sub-Committee
Kurt Gaskill – Chair (Naval Research Laboratory, USA)
Albert Burk (Cree, USA)
James Oliver (Northrop Grumman, USA)
Bernd Thomas (SiCED, Germany)
Hidekazu Tsuchida (CRIEPI, Japan)

Devices and Circuits Sub-Committee
James Scofield – Chair (Air Force Research Laboratory, USA)
Peter Friedrichs (SiCED, Germany)
Takashi Nakamura (Rohm, Japan)
Sei-Hyung Ryu (Cree, USA)
Victor Veliadis (Northrop Grumman, USA)

SiC Characterization and Theory Sub-Committee
Robert P. Devaty – Chair (University of Pittsburgh)
Adam Gali (Hungarian Academy of Sciences, Hungary)
Tsunenobu Kimoto (Kyoto University, Japan)
Pirouz Pirouz (Case Western Reserve University, USA)
Robert Stahlbush (Naval Research Laboratory, USA)

Processing and Oxide Characterization Sub-Committee
Kevin Matocha – Chair (SemiSouth, USA)
Martin Domeij (TranSiC, Sweden)
Kenji Fukuda (AIST, Japan)
Aivars Lelis (Army Research Laboratory, USA)
John Williams (Auburn University)
Yoshiyuki Yonezawa (Fuji Electric, Japan)

Graphene, GaN, and Related Materials Sub-Committee
T. Paul Chow – Chair (Rensselaer Polytechnic Institute, USA)
Hiroki Hibino (NTT, Japan)
Rachel Myers-Ward (Naval Research Laboratory, USA)
Thomas Seyller (University of Erlangen-Nürnberg, Germany)
Michael Spencer (Cornell University, USA)

ICSCRM 2011 Logistical Support

Conference Implementation: QinetiQ North America
- Barbara Kakiris – Event Manager
- Beth Arent
- Michelle Whitman

Information Technology
- Joshua Glemza (DB Consulting Group, Inc.)
- Joseph Banks (DB Consulting Group, Inc.)
- Anne Wöhlbier (Trans Tech Publications)
- Lea Pazur Katulic (Trans Tech Publications)
- Tanja Lüthi (Trans Tech Publications)
- Phil Neudeck (NASA Glenn Research Center)

Graphics Art
- Katy Kafantaris (Wyle Information Systems LLC)
- Kristin Spear (QinetiQ North America)

Conference Student Assistants Coordinator
- Chris Zorman (Case Western Reserve University)

Conference Student Assistants: Case Western Reserve University
- Andrew Barnes
- Srihari Rajgopal
- Hamidreza Zamani
- Ashley Gan
- Acthara Punya
- Mike LaBarbara
- Cathy Soong
- Jaesung Lee
- Shih-Shian Ho
- Tawinan Cheiwchanchamnangij

Local Tours
- Roger Meredith (NASA Glenn Research Center)
- Michelle Neudeck

Clerical Support
- JoAnne Greene (SGT, Inc.)
- Beverly Smigel (SGT, Inc.)
- Carmen Gonzalez (SGT, Inc.)
- Maryann Saylor (SGT, Inc.)
- Patsy Harris (Case Western Reserve University)

NASA Glenn Management Support
- Karen Crandall
- Jih-Fen Lei
- Mary Zeller
- Ricaurte Chock
- Lawrence Matus
- Gary Seng
- Kathleen Needham

Overview

Part 1

Part 2

Table of Contents

PART 1

Chapter 1: SiC Bulk Growth

1.1 PVT, CVD, Modeling and Theory

1.2 Solution Growth

Chapter 2: SiC Epitaxial Growth

2.1 Homoepitaxial Growth

2.2 Heteroepitaxial and Heteropolytypic Growth

Chapter 3: Physical Properties and Characterization of SiC

3.1 Impurities, Intrinsic Point Defects and Carrier Lifetime

3.2 Extended Defects (See also Section 3.4)

3.3 SiC-SiO_2 Interface (See also Sections 5.1 and 6.4)

3.4 Stress, Structural and Mechanical Properties (See also Section 3.2)

3.5 Fundamental Properties, Nanostructures, Surfaces and Miscellaneous Characterization

Chapter 4: Graphene

4.1 Graphene Growth, Characterization and Theory

4.2 Graphene Processing

4.3 Graphene Devices and Sensors

PART 2

Chapter 5: Processing of SiC

5.1 MOS Processing, SiC-SiO_2 Interface and Other Dielectrics (See also Sections 3.3 and 6.4)

5.2 Doping, Ion Implantation and Contacts

5.3 Cutting, Etching, Micro- and Nano-Processing

Chapter 6: SiC Devices, Circuits and Systems

6.1 Schottky Barrier Diodes

6.2 PiN and Heterojunction Diodes

6.3 JFETs and SITs

6.4 MOSFETs (See also Sections 3.3 and 5.1)

6.5 Bipolar Transistors

6.6 Thyristors and GTOs

6.7 Sensors and Photodetectors

6.8 Systems and Circuits

Chapter 7: Related Materials

7.1 Nanomaterials

7.2 III-Nitrides

7.3 Diamond

7.4 n-ZnO/p-SiC Heterojunctions, Novel Nitride Semiconductors

CHAPTER 1:

SiC Bulk Growth

Materials Science Forum Vols. 717-720 (2012) pp 3-8

doi:10.4028/www.scientific.net/MSF.717-720.3

Status of Large Diameter SiC Single Crystals

Avinash Gupta, Ping Wu[a], Varatharajan Rengarajan, Xueping Xu, Murugesu Yoganathan, Chris Martin, Ejiro Emorhokpor, Andy Souzis, Ilya Zwieback and Tom Anderson

II-VI Incorporated, Wide Bandgap Materials Group.
20 Chapin Road, PO Box 840, Pine Brook, NJ 07058, U.S.A.

[a]pwu@ii-vi.com;

Keywords: bulk crystal growth, sublimation, polytype conversion, crystal defects, dislocations

Abstract. Large-diameter SiC single crystals are grown at II-VI by the sublimation technique. 100mm substrates of semi-insulating 6H SiC and n-type 4H SiC are produced as commercial products; in development, diameter expansion to 125mm has been achieved. Over the last two years, significant improvements have been made in crystal quality. The values of FWHM of x-ray rocking curves are typically 20-40 arc-seconds for 6H SI wafers and 12-30 arc-seconds for 4H n+ wafers. Micropipe density is less than 3 cm^{-2}, and less than 0.1 cm^{-2} in the best substrates. Electrical resistivity of SI substrates is, typically, of 10^{11} $\Omega\cdot cm$ or above. For 4H n+ substrates, the typical dislocation density is about 9×10^3 cm^{-2} and the typical Basal Plane Dislocation density is less than 1×10^3 cm^{-2}.

Introduction

Large-diameter SiC single crystals are grown at II-VI using our Advanced PVT sublimation technique [1]. II-VI manufactures 100 mm diameter commercial substrates of two types: semi-insulating wafers of 6H polytype and low-resistivity, 4° off-cut wafers of 4H polytype. We have previously discussed in some detail the practice of SiC crystal growth at II-VI [2] and reported on properties of the grown crystals, including dominating defects and their evolution during growth [3]. This paper is a short review of the progress achieved at II-VI over the last two years.

Experimental

SiC single crystals with the diameter exceeding 100mm are grown using our Advanced PVT technique, which allows accurate tuning of the thermal field within the growth cell. This is instrumental in achieving and maintaining the desired configuration of the growth interface throughout the entire growth cycle. Semi-Insulating 6H crystals are grown using precise vanadium compensation [4], while conducting 4H crystals are grown in nitrogen-containing ambient. In the past year, diameter expansion to 125mm has been achieved for both crystal types.

Quality of the grown crystals is routinely evaluated using several techniques. Polished wafers are examined macroscopically using crossed polarizers. The technique reveals crystal defects that cause strain in the material, including various defect clusters, inclusions and regions with mixed polytype or sub-grain structure. Examination using polarized light microscopy is used to determine the presence and distribution of the micropipes. Etching in a KOH-based flux is carried out to map the micropipe and dislocation densities. For micropipe density (MPD) mapping, the wafers are etched deeply, followed by polishing to remove shallower etch pits formed on dislocations [5]. The total dislocation density is determined by counting and mapping all etch pits. In addition to the total dislocation density, the density of basal plane dislocations (BPD) is measured on 4° off-axis 4H n+ wafers. BPDs are identified by their elongated etch pits.

Crystal quality of the substrates is evaluated by the x-ray rocking curve scanning; the (0004) and (0006) Bragg reflections are used for the 4H and 6H polytypes, respectively. Linear scans for Ω (sample angle) and FWHM (reflection broadening) are carried out along the orthogonal wafer diameters corresponding to the *m* (-1,1,0,0) and *a* (1,1,-2,0) directions. The results yield the amplitude of lattice curvature (ΔΩ) and the degree of lattice disorder (FWHM). For selected samples, full-wafer rocking curve mapping is performed using a grid of 169 points, with the spot size of 2 x 10 mm^2. Selected wafers are evaluated at the Stony Brook Synchrotron Facility using White Beam X-ray Topography.

Electrical resistivity of the semi-insulating wafers is measured at room temperature using COREMA, a non-contact capacitance-based instrument with the resistivity range of 10^5 to 10^{12} $\Omega \cdot cm$ and mapping capability. Electrical resistivity of the n-type substrates is measured and mapped using a non-contact, eddy current-based instrument Lehighton-1500.

Results and Discussion

Polytype Stability. SiC exists in a large number of polytypes, and it is crucial to maintain the desired polytype, 4H or 6H, during growth. In some cases, we have observed the appearance of pure 6H inclusions in the 4H matrix and 4H inclusions in the 6H matrix. 4H and 6H are isomorphic hexagonal polytypes; therefore presence of such inclusions affects the crystal quality to a relatively small degree. More often, however, polytype inclusions contain also the 15R polytype, especially near the inclusion border.

In the green n-type 4H crystals, 15R regions can be easily detected by their yellow color. Small-size inclusions of 15R (usually shaped as thin layers or lamellae) are often quickly overgrown by the 4H matrix. 15R and 4H are not isomorphic, with 15R being rhombohedral and 4H hexagonal. After conversion to the normal polytype the crystal carries "scars" in the form of dislocation walls emerging due to the lattice mismatch. An example is given in Fig. 1 which shows cross-polarizer images of three wafers (# 7, 10 and 12) sliced from a 4H n+ boule containing a region of mixed polytype (6H and 15R). The wafers in a boule are numbered sequentially with #1 being at the seed end. White triangles delineate the areas of mixed polytype. One can see that during growth the polytype inclusion is "pushed" to the boule periphery: in wafer #12 the mixed-polytype area is a narrow triangle near the edge. One can also see "scars" (indicated by black arrows) left behind by the evolving polytype inclusion.

In the SI crystals, both the 6H matrix and the 4H or 15R inclusions are colorless. Detailed inspection of well polished wafers under cross-polarizer can be used to detect foreign polytypes. Raman microscopy or asymmetrical x-ray diffraction can be used to determine the type of the inclusion. Sketches at the top of Fig. 2 illustrate evolution of a 15R inclusion in a 6H SI crystal, as visible under cross-polarizer. The lines are "scars" remaining in the crystal after 15R was overgrown by the 6H matrix. MPD maps measured before and after the occurance of foreign polytype are shown at the bottom of Fig. 2. The appearance of micropipe clusters is due to the polytype change.

At II-VI, we have made focused efforts to improve polytype stability in both 4H and 6H crystal growth. These efforts included optimization of the growth cell and source morphology, as well as

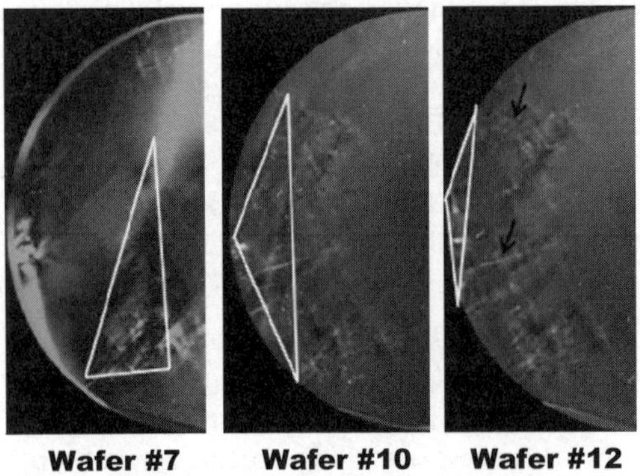

Fig. 1. Effect of polytype change in 4H n+ crystal AC0294 (cross-polarizer images).

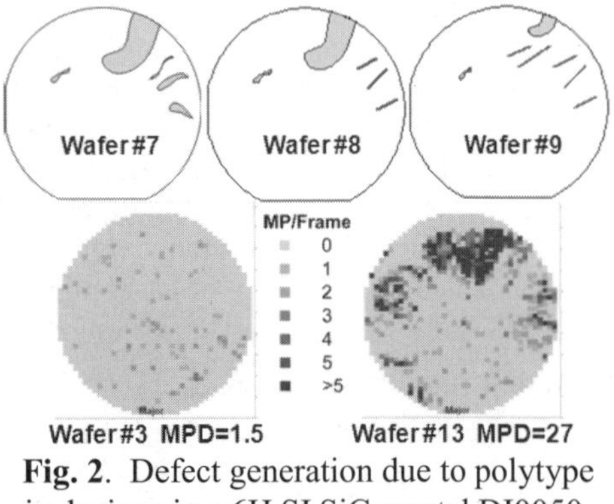

Fig. 2. Defect generation due to polytype inclusions in a 6H SI SiC crystal DI0050.

achieving better temperature and pressure control - all this leading to improved crystal quality. Fig. 3 shows the cross-polarizer images of recently produced, high-quality 100mm 6H SI and 4H n+ substrates.

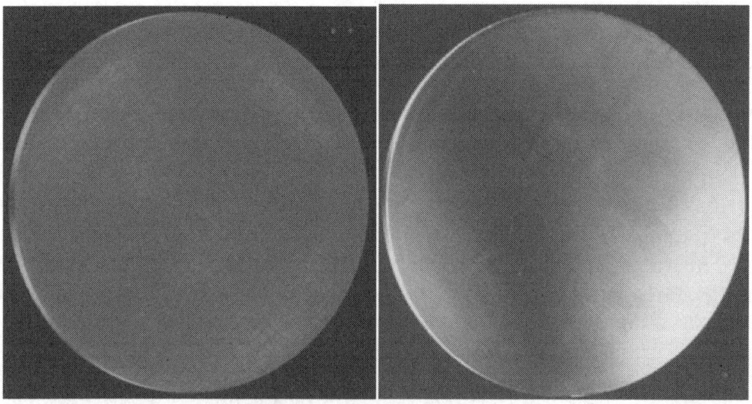

Fig. 3. Cross-polarizer images of recently produced 100mm substrates. Left: 6H SI wafer BR0056-06; Right: 4H n-type wafer DC0088-25.

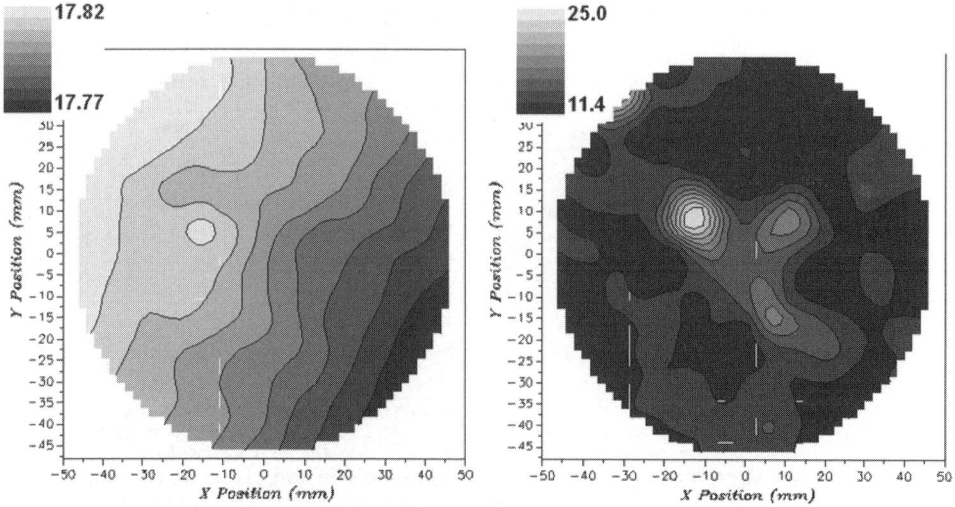

Fig. 4. X-ray rocking curve mapping of a 100mm 4H n+ substrate. Left: Ω map; Right: FWHM.

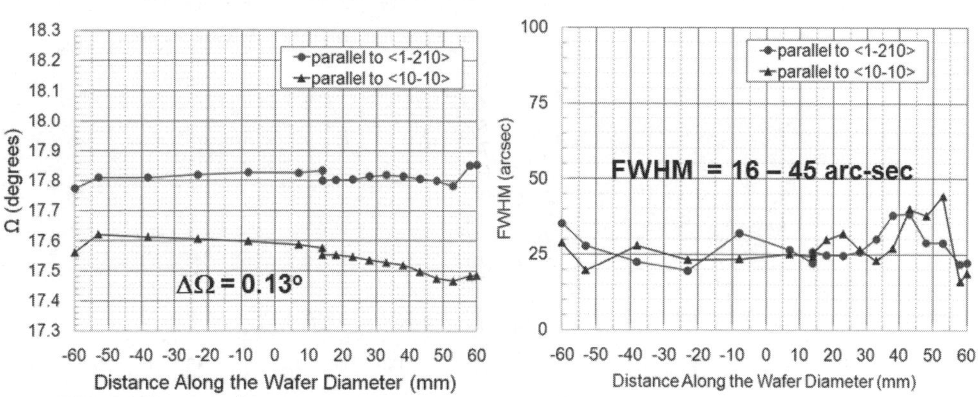

Fig. 5. X-ray rocking curve measurement on a 125mm 6H SI substrate FD1032-11.

Crystal Quality. The median lattice curvature in our 100 mm substrates, both 6H SI and 4H n+, is below 0.1°, and typical FWHM values are between 12 and 40 arc-seconds. Fig. 4 shows Ω and FWHM full-wafer maps measured on a typical 100mm 4H n+ wafer. The lattice curvature of this wafer is below 0.05°. FWHM for most of the wafer is around 13 arc-seconds, and it reaches 25 arc-seconds in a small area near the center. Fig. 5 shows linear x-ray rocking curve scans performed on a 125 mm diameter 6H SI wafer.

For both 100 mm 6H SI substrates and 100 mm 4H n+ substrates, typical MPD values are less than 3 cm^{-2}. Fig. 6 shows MPD maps measured on two recently grown 100 mm crystals, one 6H SI and another 4H n+. For typical 100 mm 4H n+ substrates, the total DD is about 9×10^3 cm^{-2}, with the densities of basal plane dislocations and threading screw dislocations on the order of 1×10^3 cm^{-2}. Examples of total DD and BPD distributions are shown in Fig. 7. Total Dislocation Density (DD) for our 6H SI wafers is about 2×10^4 cm^{-2}, as illustrated in Fig. 8. Compared to the status of two years ago, the above numbers reflect two-three fold reduction. These trends are demonstrated in Fig. 9.

One of our production 100mm 4H n+ wafers was recently evaluated using transmission White Beam X-ray Topography at the Stony Brook Synchrotron Facility. Fig. 10 shows a portion of the image taken with reflection ($\bar{1}\,\bar{1}\,2\,0$). One can see that the BPD loops are clearly resolved. From this image the BPD density is about 3.6×10^3 cm/cm^3. Note that this is the bulk dislocation line density, while counting etch pits on the wafer surface (expressed in cm^{-2}) only measures those dislocation lines that intersect the surface. For a 4° off-cut wafer, the density of BPD etch pits corresponding to the SWBXT results from Fig. 10 would be less than 200 cm^{-2}.

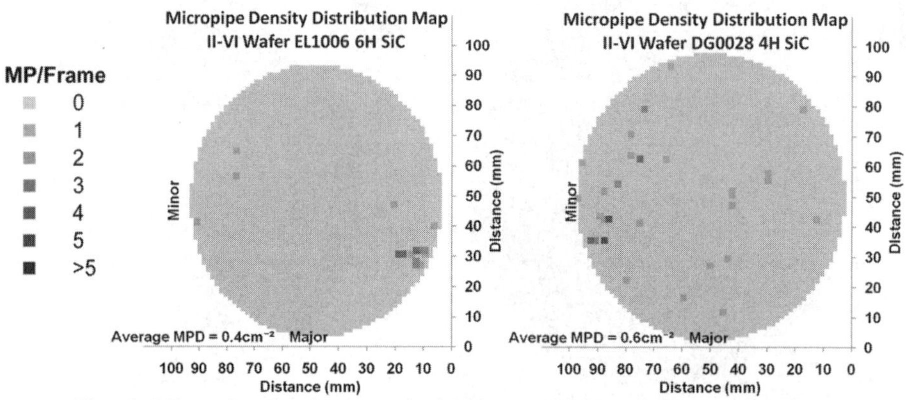

Fig. 6. Micropipe density in typical 100 mm 6H SI and 4H n+ substrates.
Left - MPD map of 100 mm 6H SI wafer EL1006-12 (MPD= 0.4 cm^{-2}).
Right – MPD map of 100 mm 4H n+ wafer DG0028-13 (MPD = 0.6 cm^{-2}).

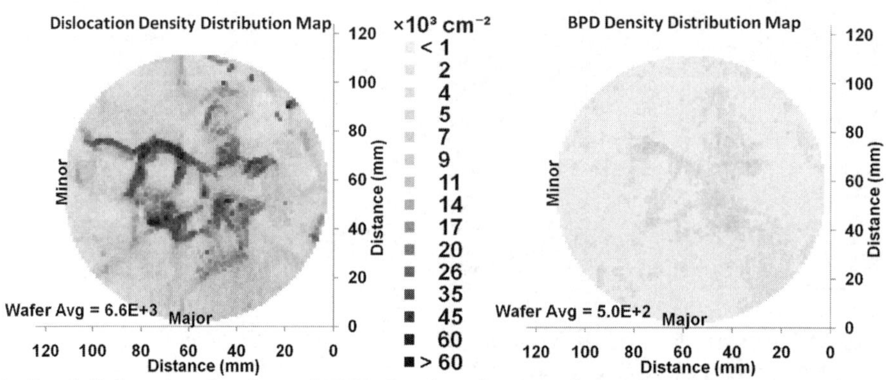

Fig. 7. Total dislocation density and BPD density of a recently produced 4H n+ crystal DF0051.

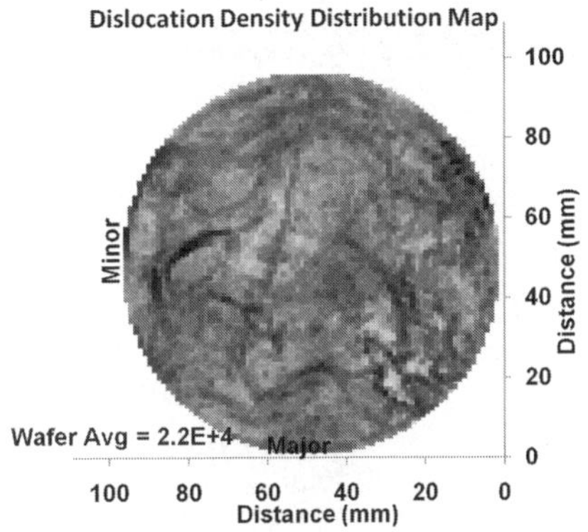

Fig. 8. Dislocation density of recently produced 6H SI crystal DI0058. (The same scale as in Fig. 7 was used.)

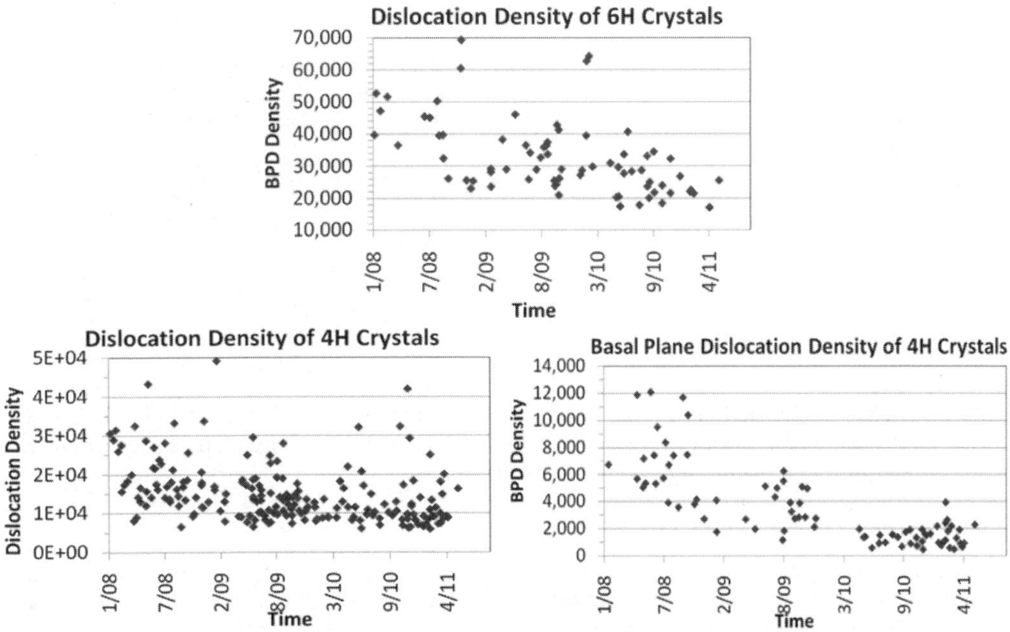

Figure 9. Reduction in dislocation densities over the last few years in both 6H SI (top) and 4H n+ (bottom) substrates produced at II-VI.

Electrical resistivity. For 4H n+ crystals, resisitivity (which is, typically, on the order of 0.020 Ohm-cm) can be adjusted by the nitrogen partial pressure in the growth chamber to meet customer requirements. Growing fully compensated, semi-insulating 6H material is more challenging. We apply special measures to ensure that the levels of boron and nitrogen in the crystal do not exceed 1×10^{16} cm^{-3}, with $(4\text{-}9)\times10^{15}$ cm^{-3} being typical. Vanadium doping is optimized [6] to yield SI material with very high and spatially uniform resistivity: close to 100% of the wafers sliced from a typical 6H SI boule have their resistivity of 10^{11} Ohm-cm or higher. An axial resistivity profile in a typical 6H SI boule and a COREMA resistivity map of a 100 mm 6H SI substrate are given in Fig. 11. The only wafer with low resistivity of about 10^5 Ohm-cm was wafer #1, which is a wafer adjacent to the seed and often containing a portion of the seed.

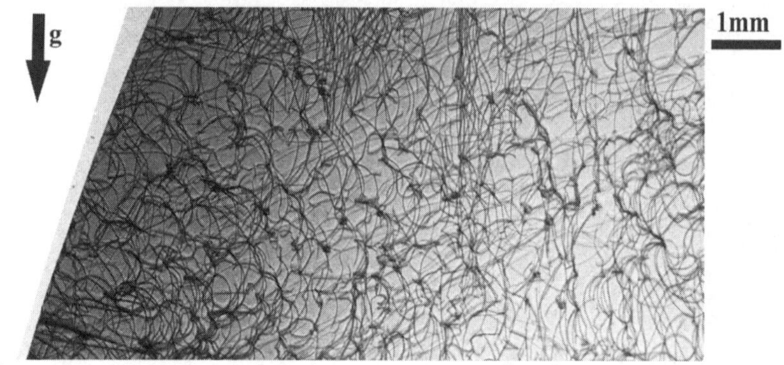

Figure 10. SWBXT transimission measurement done by Dr. Mike Dudley et al. BPD density = 3.6×10^3 cm^{-2}

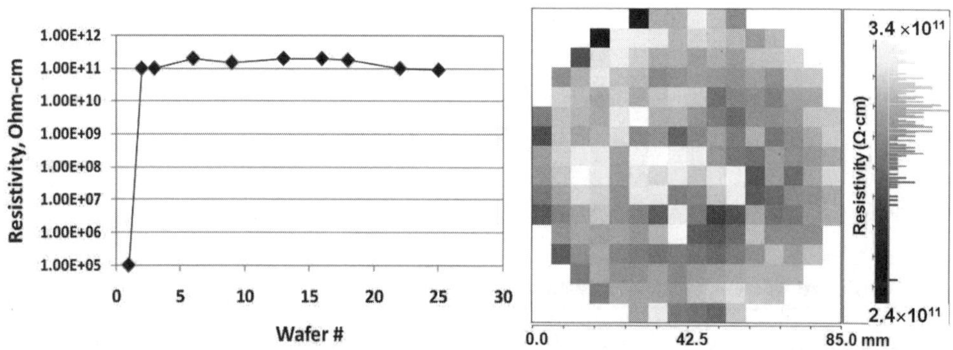

Fig. 11. Left: Axial resistivity profile of typical vanadium-doped 100 mm 6H SI boule EW0022. Each point is an average wafer resistivity. Right: COREMA resistivity map of a 100 mm wafer sliced from this boule (wafer-average $\rho \approx 3\times10^{11}$Ohm-cm).

Conclusions

Large-diameter SiC single crystals grown at II-VI include both 4H n-type and 6H semi-insulating material. Over the last two years, our development efforts have been concentrated on optimization of the growth process parameters and on understanding defect generation and evolution. These efforts resulted in expansion of crystal diameter and significant improvement in the crystal quality.

Acknowledgements

Special thanks to E. Semenas, M. Getkin, M. Nolan, B. Brouhard, P. Flynn, T. Kerr, L. Unick, P. Lemos, E. Loukas, C. Minto, S. Nahra, L. Nguyen and G. Wilson for their practical contributions in growth, fabrication and characterization of SiC substrates.
The authors wish to thank Prof. Michael Dudley and his students Huanhuan Wang, Fangzhen Wu and Shayan Byrappa for x-ray topography measurements and valuable discussions.

References

[1] D. W. Snyder and W. J. Everson, US Patent No. 6,800,136 (2004).
[2] A. Gupta, I. Zwieback, A. Souzis, M. Yoganathan, and T. Anderson: *Mater. Sci. For.*, **600-603** (2009), p. 35.
[3] P. Wu, M. Yoganathan, and I. Zwieback: *J. Cryst. Growth*, **310** (2008), p. 1804.
[4] D. L. Barrett, H. M. Hobgood, J. P. McHugh, and R. H. Hopkins, US Patent No. 5,611,955 (1997).
[5] E. Emorhokpor, T. Kerr, I. Zwieback, W. T. Elkington, M. Dudley, T. A. Anderson, and J. Chen: Compound Semiconductor MANTECH Conference Proceedings, May, 2004, p. 139.
[6] A. K. Gupta, E. Semenas, I. Zwieback, D. L. Barrett, A. E. Souzis: US Patent No. 7,608,524 (2009).

Materials Science Forum Vols. 717-720 (2012) pp 9-12
© (2012) Trans Tech Publications, Switzerland
doi:10.4028/www.scientific.net/MSF.717-720.9

TSD Reduction by RAF (repeated *a*-face) Growth Method

Y. Urakami[1,2,a], I. Gunjishima[1,3,b], S. Yamaguchi[1,3], H. Kondo[1,2], F. Hirose[1,2], A. Adachi[1,4,c] and S. Onda[1,2]

[1]R&D Partnership for Future Power Electronics Technology, Nisshin Branch, 500-1 Minamiyama, Komenoki, Nisshin, Aichi, 470-0111 Japan

[2]Research Laboratories, DENSO CORPORATION, Komenoki, Nisshin, Aichi, 470-0111 Japan

[3]Toyota Central R&D Laboratories Inc., Nagakute, Aichi, 480-1192 Japan

[4]Toyota Motor Corporation, Nishihirose, Toyota, Aichi, 470-0309 Japan

[a]yurakami@rlab.denso.co.jp, [b]gunjishima@mosk.tytlabs.co.jp, [c]ayumu@adachi.tec.toyota.co.jp

Keywords: bulk growth, *a*-face growth, threading screw dislocation, X-ray topography

Abstract. A reduction in threading screw dislocation (TSD) density in 4H-SiC (silicon carbide) crystals is required for SiC power devices. In this study, transformation of TSD by the RAF (repeated *a*-face) growth method [1] is observed by transmission X-ray topography (**g**=0004) of the cross-section of the crystal. Increasing the number of repetitions of *a*-face growth and offsetting *c*-face growth to an angle of several degrees reduce TSDs. TSD density is reduced to 1.3 TSD/cm^2. The RAF growth method is very effective towards growing high quality SiC crystals.

Introduction

SiC power devices are expected as the next-generation energy-saving devices. Some of the problems facing their full-scale practical use are defects in the SiC wafer. Micropipes, which used to be the major defect of SiC wafers, were almost eradicated during the last decade. However, many dislocations remain in the wafers and reducing threading screw dislocations (TSD) is particularly important for SiC power devices [2]. We have continued to develop the RAF growth method devised as a high quality crystal growth technology and *c*-face growth technology using the RAF seed.

In this study, we tried to reduce TSD density by increasing the number of repetitions of *a*-face growth and by offsetting *c*-face growth to an angle of several degrees.

Experiment

Figure 1 shows the RAF growth method. First, the *a*-face seed is sliced from the SiC crystal with many threading dislocations (TDs) parallel to the *c*-axis. In this paper, both the (1-100) face and (11-20) face are called '*a*-face'. In this study, the first *a*-face is the (1-100) face and the first *a*-face growth is in the [1-100] direction. Second, the (11-20) face seed is sliced from the (1-100) face growth crystal and the second *a*-face growth is in the [11-20] direction. Third, *a*-face growth is repeated in the [1-100] and [11-20] directions in order to reduce dislocations along the *c*-axis. Finally, *c*-face growth with an offset angle of several degrees uses a seed sliced from the *a*-face growth crystal.

In this study, we investigated the following two points. As the first point, we investigated how TSD with distortion in the *c*-axis direction changed because of *a*-face growth and repetitions of *a*-face growth. Therefore, the cross-section of another *a*-face perpendicular to the *c*-face of the repeated *a*-face growth was observed by transmission X-ray topograph (**g**=0004). As the second point, we investigated the dependence of the TSD density on the number of repetitions of *a*-face growth. Therefore, *c*-face growth with an offset angle of several degrees used a seed sliced from the result of many (up to 7) repetitions of *a*-face growth and the final crystal was observed by X-ray reflection topography (**g**=11-28). TSD density expressed as small dots was counted.

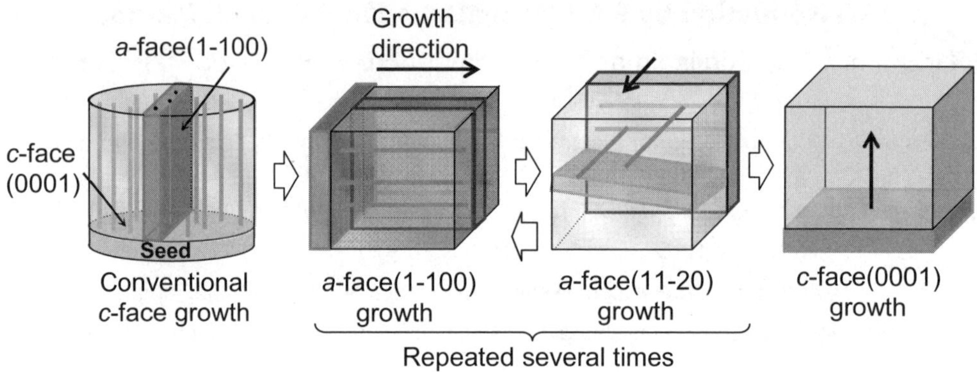

Fig.1. Schematic illustration of RAF (repeated *a*-face) growth method

Results and Discussion

Figure 2 shows a transmission X-ray topographic image (**g**=0004) of a cross-section of the crystal grown on the (1-100) face seed with TSDs. TSDs were transformed into stacking faults (SF) or basal plane edge dislocations (BPD) expressed by various shades of lines. In this case, as some of the lines are angled in the *c*-axis direction, the distortion in the *c*-axis direction remains in the growth crystal. Figure 3 shows a transmission X-ray topographic image (**g**=0004) of a cross-section of the crystal grown on the (11-20) face seed sliced from the first *a*-face ((1-100) face) growth crystal. In comparison with Fig. 2, as some of the lines are angled in the *c*-axis direction, the distortion in the *c*-axis direction is small. As most SFs or BPDs in the first *a*-face growth crystal are parallel to the growth direction, SFs or BPDs that appeared on the surface of the seed sliced from the first *a*-face ([1-100] face) growth crystal are reduced. Therefore, the second *a*-face growth crystal's distortion in the *c*-axis direction is reduced. Figure 4 shows a transmission X-ray topographic image (**g**=0004) of a cross-section of the crystal grown on the (1-100) face seed sliced from the 6th *a*-face growth crystal. Since all of the lines are parallel to the growth direction, the 7th *a*-face growth crystal has small distortion in the *c*-axis direction.

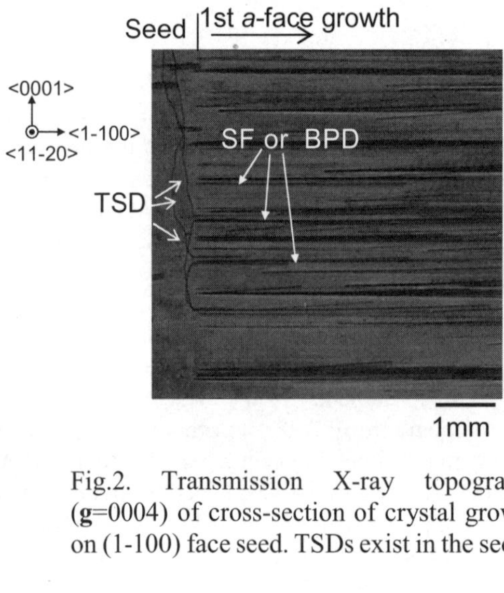

Fig.2. Transmission X-ray topograph (**g**=0004) of cross-section of crystal grown on (1-100) face seed. TSDs exist in the seed.

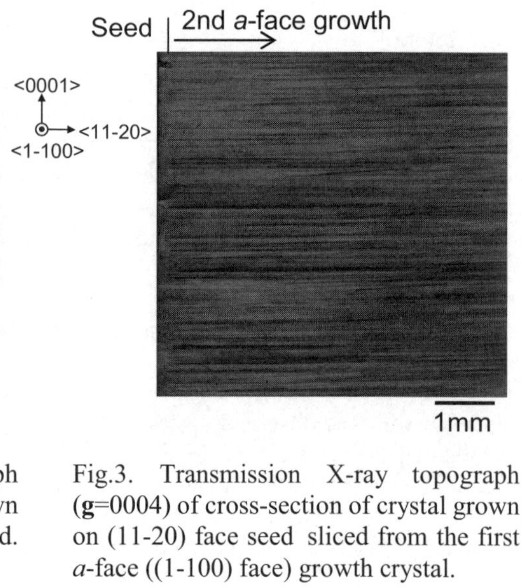

Fig.3. Transmission X-ray topograph (**g**=0004) of cross-section of crystal grown on (11-20) face seed sliced from the first *a*-face ((1-100) face) growth crystal.

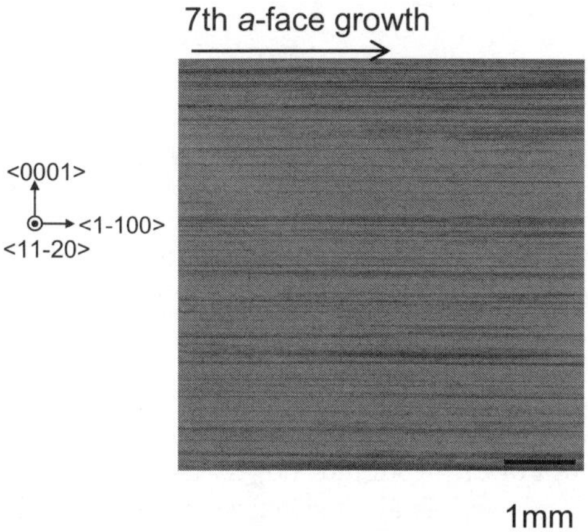

Fig.4. Transmission X-ray topograph (**g**=0004) of cross-section of crystal grown on (11-20) face seed on the (1-100) face seed sliced from the 6th *a*-face growth crystal.

Fig.5. Transmission X-ray topograph (**g**=0004) of cross-section of crystal grown on 8° off-oriented (0001) seed fabricated by RAF growth method (RAF seed).

Figure 5 shows a transmission X-ray topographic image (**g**=0004) of a cross-section of the crystal grown on an 8° off-oriented (0001) seed fabricated by the RAF growth method (RAF seed is without distortion in the c-axis direction). SFs and BPDs in the RAF seed are inherited in the *c*-face, therefore the growth crystal is free from TSDs. As the RAF seed has small distortion in the c-axis direction, SFs and BPDs do not transform into TSD.

Figure 6 shows the reflection X-ray topographic images (**g**=11-28) of (a) a conventional RAF wafer (*c*-face grown crystal after 3-times *a*-face growth) and (b) an improved (*c*-face grown crystal after 7-times *a*-face growth) RAF wafer. The size of each RAF wafer is φ1inch. The number of TSDs is drastically reduced in comparison to the conventional RAF wafer. The TSD density of the improved RAF wafer is 1.3 TSD/cm^2. By increasing the number of repetitions of *a*-face growth, the RAF seed has small distortion in the c-axis direction. Therefore, the TSD density of the improved RAF wafer is reduced.

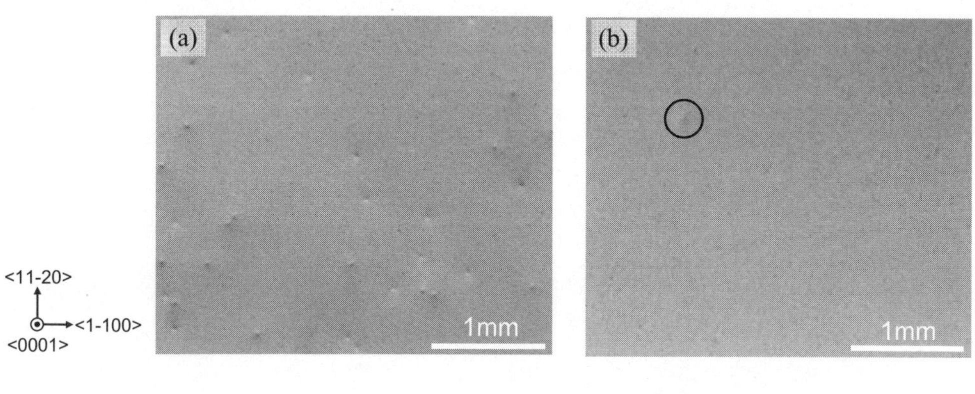

Fig.6. Reflection X-ray topograph (**g**=11-28) of (a) conventional RAF wafer and (b) improved RAF wafer. The small dots are TSDs. In (b), a TSD is enclosed in the circle.

Summary

In this study, we succeeded in reducing TSD density to 1.3 TSD/cm^2 by increasing the number of repetitions of *a*-face growth and by offsetting *c*-face growth to an angle of several degrees. TSDs transform into SFs or BPDs because of *a*-face growth. The distortion of the crystal in the *c*-axis direction is reduced by increasing the number of repetitions of *a*-face growth. SFs or BPDs are inherited in the *c*-face by offsetting *c*-face growth to an angle of several degrees. The RAF growth method is very effective towards growing high quality SiC crystals. We will verify the high quality of the SiC wafer by manufacturing a power device and checking for a reduction in leakage current.

Acknowledgment

This work is supported by the Novel Semiconductor Power Electronics Project for Realizing a Low Carbon Emission Society under the Ministry of Economy, Trade and Industry (METI) and New Energy and Industrial Technology Development Organization (NEDO).

References

[1] D. Nakamura et al., Nature Vol. 430 (2004) 1009.
[2] J. Senzaki et al., Jpn. J. Appl. Phys. 48 (2009) 081404.

Materials Science Forum Vols. 717-720 (2012) pp 13-16

doi:10.4028/www.scientific.net/MSF.717-720.13

Growth of High Quality 4H-SiC Crystals in Controlled Temperature Distributions of Seed Crystals

H. Tsuge, S. Sato, M. Katsuno, T. Fujimoto and W. Ohashi

Nippon Steel Corporation, Advanced Technology Research Laboratories,

20-1 Shintomi, Futtsu, Chiba 293-8511, Japan

TEL: +81-439-80-2282, FAX: +81-439-80-2746,

e-mail: tsuge.hiroshi@nsc.co.jp (corresponding author)

Keywords: 4H-SiC, PVT growth, defect.

Abstract. Large diameter 4H-SiC single crystal wafers with higher quality are required to improve the yields of devices fabricated onto the SiC wafers. For crystal growths with higher quality, it is important to prepare seed crystals with lower defect densities. In particular, the edge part of the seed has to be prepared with considerable care because the crystallinity of the enlarged part of grown crystals depends much upon the surface condition of the seed crystal during radial expansion growth. We found that growth with fewer defect and micropipe densities, specifically at the periphery of the crystal, is possible by utilizing an in-situ etching process for the seed crystal surface. We have also performed numerical calculations of the temperature distribution around the seed surface, and discussed growth conditions which cause the in-situ etching effective to improvement of the crystallinity in enlarged crystals.

Introduction

Silicon Carbide (SiC) is an attractive material for high power and high frequency devices. 4H-SiC single crystal wafers with larger diameters and higher quality are necessary to improve the yields of the SiC devices, and thus recent R&D for SiC wafers emphasizes the importance of 150 mm production of higher crystal quality upon which various types of SiC power devices can be produced with much improved device performance. Up to date, bulk growth of SiC is in general performed by the so-called physical vapor transport (PVT) technique. In order to grow a larger diameter SiC single crystal, the driving force for the radial expansion growth along the radius direction has to be enhanced by controlling relevant growth parameters such as the temperature gradient along the crystal diameter. The crystal quality of the enlarged part of the grown crystal inherits the edge part of seed crystals. Therefore, for crystal growths with higher quality, it is important to prepare the edge of the seed crystal with lower defect densities.

In this study, we have investigated temperature distributions around seed crystals under which in-situ etching occurs exactly at the edge part of the seed crystal, giving rise to an larger diameter crystals of improved quality with lower crystal defect densities during the radial expansion growth.

Experimental Procedure

We have developed our state-of-the-art PVT technique in which various growth conditions such as the growth rate and the temperature distributions are best-optimized for stable growth of 4H-SiC single crystals [1]. A SiC single crystal boule was grown in a graphite crucible containing both a SiC seed crystal and a SiC source material. The crucible was induction-heated to temperatures ranging from 2300K to 2700K under an argon atmosphere. To investigate the shape of the growth front during the growth, we used an interface demarcation technique: during the growth experiment, nitrogen was added intermittently to the inert gas (argon) in the growth cell at a fixed time interval [2]. The grown crystal boules were then cut parallel or perpendicular to the growth direction into the substrates, and the surfaces of the substrates were polished. Substrates were observed by using optical microscopy and X-ray topography.

Results and Discussion

In order to maintain single polytype crystal growth, the thermal distribution of the grown boules has to be convex in shape towards the growth direction. We calculated the temperature distribution inside the crucible, specifically near the seed crystal, and confirmed that the temperature at the front side of the seed crystal became higher compared to the temperature at the front center. For evaluations of the calculations, a single crystal boule was grown with a thick seed crystal. Fig. 1 shows the cross section of the boule obtained. The lines observed in the boule were formed by nitrogen doping during growth, thereby making visible the shape of the growth front at the time of nitrogen doping. As can be seen in Fig. 1, the shapes of growing surfaces are convex in shape, resulting from our careful control of the temperature distribution. The seed crystal was cylindrical in shape before crystal growth, as indicated by the broken line in Fig. 1. The result shown in Fig. 1 suggests that the front side-edge of the seed crystal was sublimed during the initial stage of the crystal growth, since the first two lines near the side-edge of the seed crystal are missing, thus indicating that in-situ etching occurred in this crystal region.

Observations using optical microscopy revealed that almost no micropipe (MP) generation occurred at the front side-edge of the seed crystal where the seed crystal was in-situ etched. In other words, crystal growth with improved crystallinity occurred onto the in-situ etched surface [3], suggesting that the in-situ etched seed surface was cleaned and structurally relaxed surface of the seed crystal was presumably realized prior to the growth.

In order to induce the in-situ etching exactly at the front edge of the seed crystals, the desired temperature distribution for the in-situ etching is necessary. Fig. 2 shows the calculated temperature distributions near the seed surface for the case shown in Fig. 1. The temperature of the front side-edge of the seed crystal is higher by about 10K compared to that for the center surface region. It can be emphasized that such small temperature difference is of particular importance for inducing the in-situ etching phenomenon at an initial stage of PVT growth.

We have performed numerical calculations of the temperature distribution in order to investigate growth conditions under which the in-situ etching occurs exactly at the front side of the seed crystal. We have focused attention on two temperature differences inside the seed crystal at the initial stage of the crystal growth. One is the temperature difference (ΔT_{ce}) which equals T_e-T_c where T_e is the temperature at the front edge surface, and T_c the temperature at the center of the seed surface, as shown in Fig. 3. ΔT_{ce} relates to the degree of in-situ etching. The other one is the temperature difference (ΔT_{bc}) which equals T_c-T_b where T_b is the temperature at the edge of the back-side of the seed crystal. ΔT_{bc} is related to the seed crystal sublimation at the initial stage of the growth. ΔT_{bc} should be positive for a crystal growth with a seed crystal, since the edge part of the seed crystal would sublime in case of negative ΔT_{bc}.

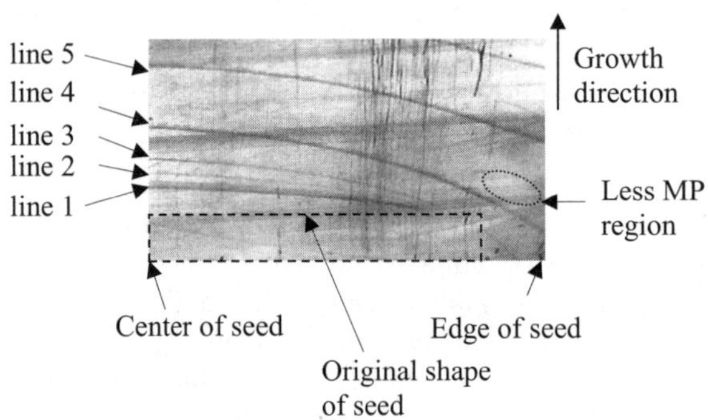

Fig.1 Cross sectional view of the boule grown by using thick seed crystal.

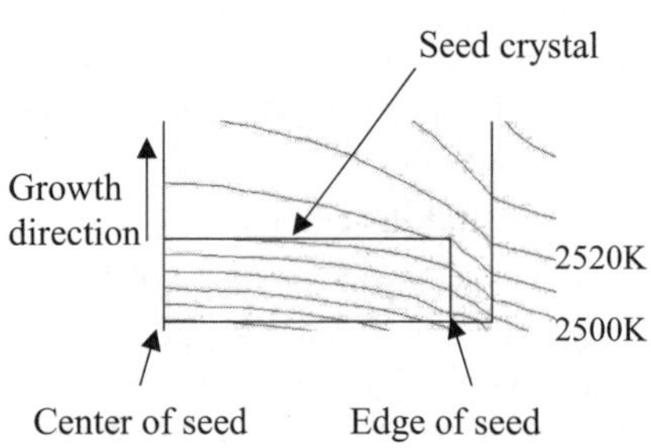

Fig.2 An example of calculated temperature distribution around seed crystal prior to the growth.

We calculated ΔT_{ce} and ΔT_{bc} by changing (a) the distance between the edge of the seed crystal and inner wall of the crucible with a fixed diameter of the seed crystal and (b) the thickness of the seed crystal, as is shown in Fig. 3.

Fig. 4 shows that the gap distance between the edge of the seed crystal and the inner wall of crucible affects ΔT_{ce}: ΔT_{ce} becomes larger with increasing distance, indicating that the effect of the in-situ etching might become greater with larger gap distance. On the other hand, ΔT_{ce} is insensitive to the seed crystal thickness, and the degree of in-situ etching might not change much with the thickness of the seed crystal.

ΔT_{bc} is calculated by varying the seed thickness with the gap distance fixed at 10mm. Fig. 5 shows that ΔT_{bc} is negative when the seed thickness is small. This result suggests that it is necessary to choose a proper seed thickness in order to avoid decomposition of the seed due to sublimation during the in-situ etching process at an initial stage of growth.

Based on these calculated results, a crystal growth experiment was carried out with proper in-situ etching process at the front surface of the seed crystal by using an appropriate seed crystal thickness and a gap distance. We obtained a single crystal boule with 4H polytype by this experiment. The grown crystal boule was cut perpendicular to the growth direction into the substrate and the surfaces of the substrates were polished. The substrate was observed by X-ray topography in order to investigate the distribution of defects.

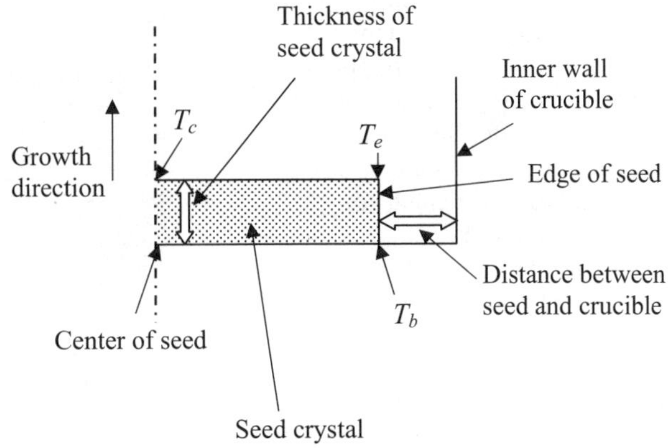

Fig. 3 Model adopted for simulations, in which both the seed thickness and the gap distance are set to be variable parameters.

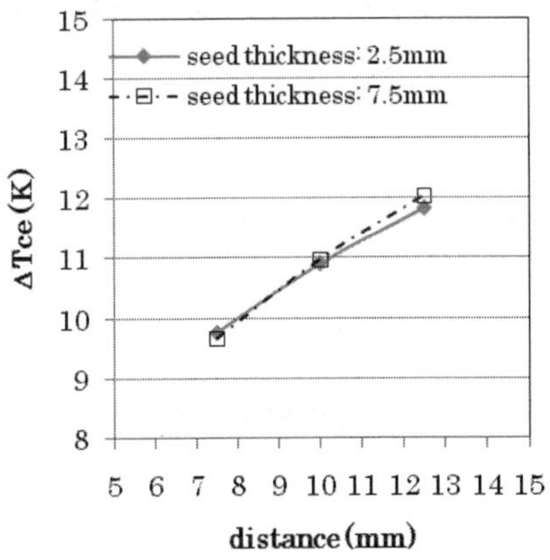

Fig.4 Variation of ΔT_{ce} by changing the distance between the seed edge and the inner wall of the crucible.

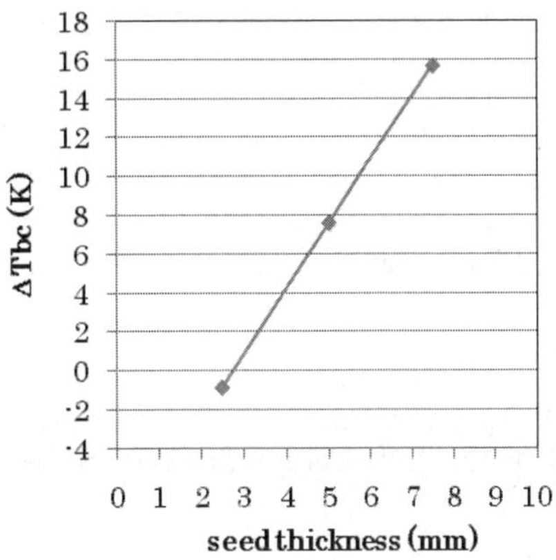

Fig.5 Variation of ΔT_{bc} by changing the seed thickness (for the case that the distance between the seed and inner wall is 10mm).

Fig.6 shows a (1-100) X-ray diffraction image obtained for the substrate we prepared by using transmission X-ray topography (Lang method). Brighter regions correspond to the crystal regions with fewer defects and darker regions to the crystal regions with larger defect densities. We assume that the darker contrast may be caused mainly by basal plane dislocations and threading edge dislocations since these two defects are major dislocations which are measured by etch pit density observations. It can be seen that the grown crystal at the outer part of the seed edge, which is the enlarged part of crystal, is brighter compared to the center part of the grown crystal. Furthermore, just inside the diameter of the seed crystal, shown by the broken line in Fig. 6, the image also exhibits bright contrast. These results indicate that the enlarged part has fewer defects than that for the center part of the grown crystal. We speculate at the moment that such reduction of defect density arises from the effects of in-situ etching at the edge surface of the seed crystal. On the other hand, the effect of the in-situ etching might not be sufficient for the center part of the crystal, since the center part of the crystal showed a dark contrast compared to the outer part.

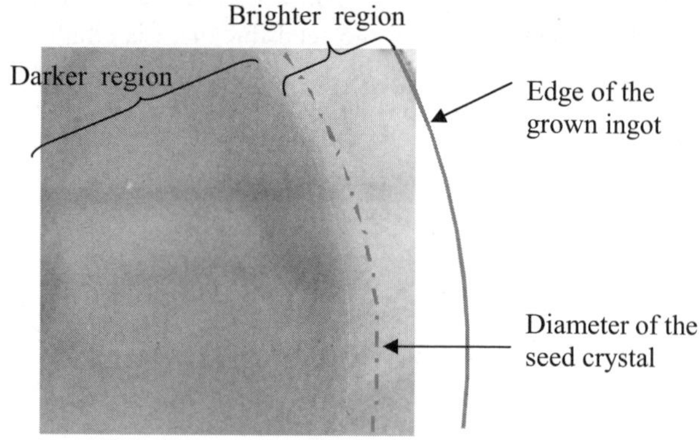

Fig.6 (1-100) transmission X-ray topography of wafer cut out from boule grown under the conditions including in-situ etching process.

It is necessary to realize more appropriate convex temperature distributions near the center surface in order to get more effective in-situ etching near the center part of the grown crystal.

Summary

By utilizing an in-situ etching process at the front side of the seed crystal in the initial stage of crystal growth, an improved quality crystal growth with fewer micropipe densities could be realized. The in-situ etching process is closely related to the temperature distribution around the seed surface. In order to make a desired convex temperature distribution suitable for the in-situ etching, it is necessary to choose both the proper seed thickness and the proper configuration of temperature distribution. Under such growth conditions, we obtained SiC single crystal boules with lower defect densities, specifically at the enlarged part of the grown boule.

These results show that optimized in-situ etching process can realize clean and structurally relaxed surfaces leading to large diameter crystal growths with improved crystallinity.

Acknowledgements

This work is partly supported by NEDO under the grant of "Novel Semiconductor Power Electronics Project Realizing Low Carbon Emission Society".

References

[1] J. Takahashi, M. Kanaya, and Y.F ujiwara: J. Cryst. Growth **135** (1994) p.61.
[2] M. Sleder, L. Kadinski, Yu. Makarov, F. Durst, P. Wellmann, T. Straubinger, D. Hofmann, S. Karpov, and M. Ramm: J. Cryst. Growth **211** (2000) p.333.
[3] T. Kato, S. Nishizawa, and K. Arai: J. Cryst. Growth **233** (2001) p.219.

Materials Science Forum Vols. 717-720 (2012) pp 17-20
© (2012) Trans Tech Publications, Switzerland
doi:10.4028/www.scientific.net/MSF.717-720.17

Process and Crucible Modification for Growth of Highly Doped 4H-SiC Crystal with Larger Diameter

Jong-Hwi Park[1,a], Tae-Kyoung Yang[1,b], Il-Soo Kim[1,c], Won-Jae Lee[1,d], Im-Gyu Yeo[2,e], Tai-Hee Eun[2,f], Seung-Seok Lee[2,g], Jang-Yul Kim[2,h] and Myong-Chuel Chun[3,i]

[1]Electronic Ceramics Center (ECC), Department of Materials and Components Engineering, Dong-Eui University, 995 Eomgwangno, Busanjin-gu, Busan, 614-714, Korea

[2]Research Institute of Industrial Science and Technology, Pohang, Kyungbuk, 790-600, Korea

[3]POSCO center, Daechidong, Gangnamgu, Seoul, 135-777, Korea

[a]popopi19@naver.com, [b]Saenge4u@nate.com, [c]iskim@deu.ac.kr, [d]leewj@deu.ac.kr, [e]igyeo@rist.re.kr, [f]theun@rist.re.kr, [g]winstone@rist.re.kr, [h]jyulkim@rist.re.kr, [i]mcchun@rist.re.kr

Keywords: PVT, Transformation, 4H-SiC, 6H-SiC

Abstract. The present research was focused to investigate various process parameters influencing the large 4H-SiC crystal growth on a 6H-SiC seed by the PVT method. The crucible diameter along the horizontal axial direction and inserted graphite ring were modified to change the growth parameters like the temperature gradient. In the initial stage of growth, foreign polytypes such as 15R/4H were observed on the 6H-SiC seed, indicating the growth temperature to be unstable on the crystal surface. However, from the middle of the growth step, 4H-SiC was successfully formed in the ingot with the modification of growth pressure and a SIMS analysis confirmed the high doping concentration in the grown 4H-SiC crystal.

Introduction

Silicon carbide is one of the most attractive and promising wide band-gap semiconductor materials with excellent physical properties and a huge potential for electronic applications. Up to now, the most successful method for the growth of large SiC crystals with high quality is the seeded sublimation method [1-2]. In particular, 4H-SiC is the preferred polytype for the fabrication of SiC-based high power devices due to its high electron mobility and wide bandgap [3]. Recently, many researchers reported that 4H-SiC single crystals were grown easily on the carbon (000-1) face of a 6H-SiC seed. Among them, Yu. M. Tairov et al. have clearly explained the parameters that favor such polytypic transformation during the growth process of a SiC crystal [4-5]. The process included the transformation of 6H-SiC into 4H-SiC initially and then stabilizing the 4H polytype through the entire growth. Furthermore, while the growth of the 6H-polytype is known to be stable for a wide range of process conditions, 4H crystallization is stable only in a narrow range of growth parameters [6]. Polytypic stability of 4H-SiC during the crystal growth is still an art and few reports explained polytypic stability during the growth process [7], because it is difficult to control polytypes whose growth strongly depends on various parameters such as the growth temperature, radial/axial temperature gradients and the parameters related to crucible properties and design.

Experiments

The present research was focused to investigate various process parameters influenced on the large 4H-SiC crystal growth on a 6H-SiC seed by the PVT method. The crucible design and the growth pressure control were modified to grow a 4H-SiC crystal. The SiC powder source materials were placed on opposite sides in a sealed graphite crucible, which was surrounded by graphite insulator. The growth temperatures were around 2100-2200°C and growth pressures 20, 50 mbar with argon and nitrogen. 2-inch on-axis 6H-SiC with C-face (000-1) was used as the seed material. 4H-SiC crystals with 75mm-diameter and 15~20mm-thickness were typically grown for this study. The crucible diameter along the horizontal axial direction and the inserted graphite ring were modified to change growth parameters like the temperature gradient [8], as shown in Fig. 1 as an example. In the initial stage of growth, foreign polytypes such as 15R/4H were observed on the 6H-SiC seed and the temperature for the crystal growth could be unstable along the growth front. The growth pressure was increased during the growth step to stabilize the polytype formation of SiC crystal. Sliced SiC wafers were polished using a diamond paste to investigate the wafer surface with an optical microscope. X-ray patterns were obtained to identify the polytype of SiC crystal using a high resolution XRD diffractometer (PANalytical-MRD). Nitrogen concentration in the grown SiC crystal was analyzed by SIMS (Secondary Ion Mass Spectrometer). Absorption spectra of SiC crystals were measured by a UV/VIS spectrophotometer.

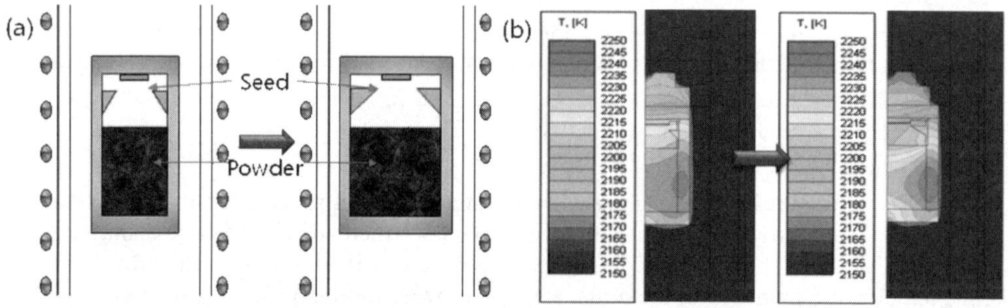

Fig. 1 (a) The schematic diagram for the growth cell and (b) simulation results for the temperature gradient distribution in the modified crucible design.

Results and Discussion

It is well known that the crystal growth of a particular SiC polytype is strongly related to the properties of the seed crystal and adjustment of process parameters. A home-made 6H-SiC ingot with C-face is used as a seed crystal. In the initial stage of SiC crystal growth, various colors indicating different polytypes were observed on the SiC wafer (No.2 in Fig.2). The 6H-SiC polytype region was transformed into 4H polytype as crystal growth proceeds. K. Kakimoto et al. reported that the lower growth pressure increased the probability of polytype instability [9]. Therefore, the growth pressure was increased in the middle of the growth step in this study, and stabilized 4H-SiC polytype was obtained in the crystal ingot.

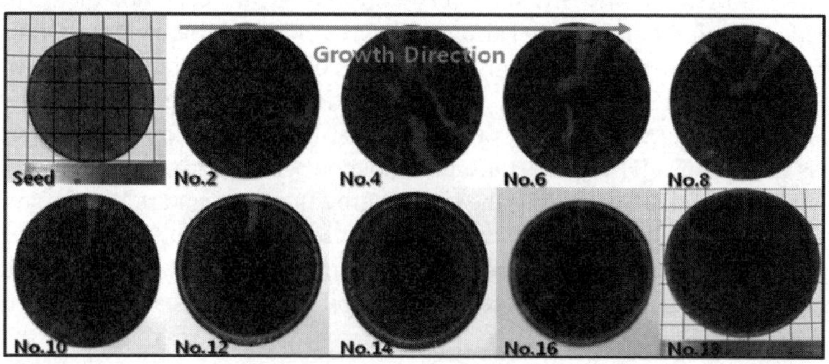

Fig. 2 The optical photographs of SiC wafers made from a SiC boule grown on a 6H-SiC seed.

The polytype of SiC crystals sliced from different positions of the SiC single crystal ingot were analyzed by using XRD patterns, as shown in Fig. 3. Nishino et al. [10] reported that the number of small periodic peaks between the main peaks related to the double diffraction effect could represent the polytype of SiC crystal; 3 peaks for 4H-SiC, 5 peaks for 6H-SiC and 4 peaks for 15R-SiC. In the initial stage of growth, various polytypes such as 6H-, 4H- and 15R-SiC were observed. After the middle period of the growth, various polytypes were not observed in XRD data.

The depth dependence of the concentrations of various impurities in the SiC crystal was analyzed using SIMS(Fig.4). As the formation of 4H-SiC polytype would be difficult for a nitrogen doping level below $2\times10^{19}/cm^3$ [11], a stable 4H-SiC crystal with a nitrogen concentration of $\sim10^{20}/cm^3$ in this work was obtained. The impurity concentration tended to decrease towards the SiC crystal growth surface due to limited source material in the crucible.

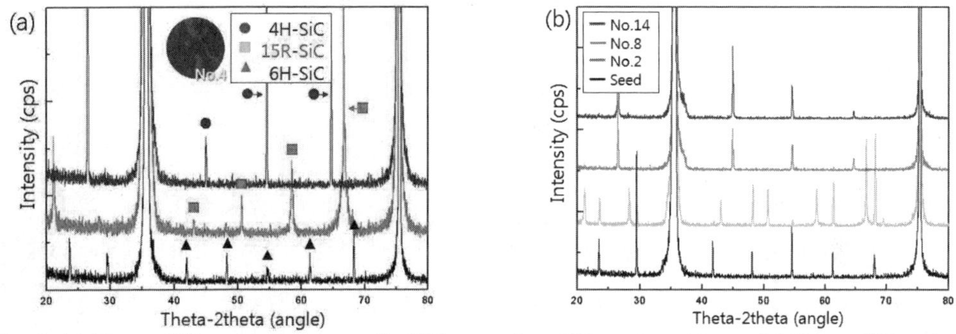

Fig. 3 (a) X-ray diffraction patterns of a SiC crystal at different positions. (b) X-ray diffraction patterns of SiC crystals from the seed to the growth surface.

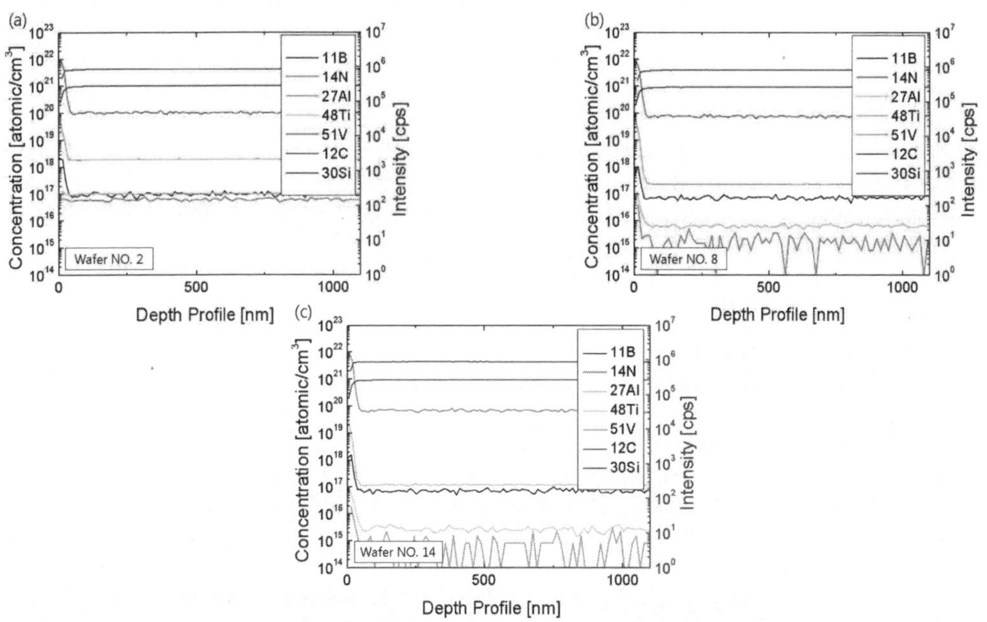

Fig. 4 The impurity concentration at different positions of a SiC bulk crystal determind using SIMS. (a) wafer No.2, (b) wafer No.8, (c)wafer No.14

Fig. 5 shows the absorption and transmittance spectra of SiC wafers sliced from different positions of the SiC bulk crystal. While the typical absorption spectrum of a SiC seed crystal indicated that the SiC polytype was 6H-SiC with fundamental absorption energy of about 3.02eV, the absorption spectrum of the grown SiC crystal exhibited 4H-SiC with fundamental absorption energy of about 3.26eV. A below bandgap absorption (BBGA) relating to the nitrogen donor of about 2.0eV/623nm and 2.67eV/464nm was observed in seed and grown SiC crystals, respectively [12]. The SiC crystal formed in the middle period (No.12) exhibited lower transmittance than SiC in the final period

(No.18). The transmittance at different positions of the SiC grown crystal exhibited different values due to different polytypes and the different doping concentrations due to limited source material in the crucible.

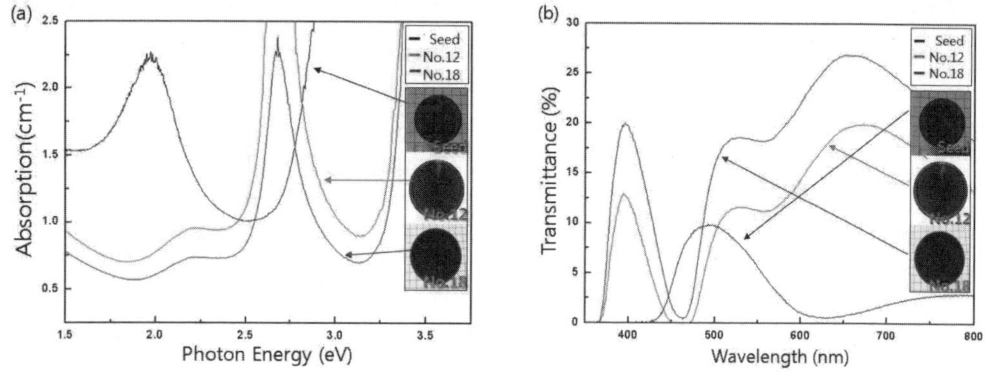

Fig. 5 (a) absorption and (b) transmittance spectra of SiC wafers sliced from different positions of the grown SiC single.

Summary

The present research was focused to investigate various process parameters that influenced 4H-SiC crystal growth on a C-face 6H-SiC seed by the PVT method. The large crystal growth was improved by adopting a modified crucible structure and growth pressure. Increased growth pressures during the growth step improved the stable formation of 4H-SiC polytype. While various polytypes were observed during the initial growth stage, from the middle period of the growth various polytypes were not observed. The high nitrogen concentration at different positions of the SiC grown crystal was confirmed by SIMS. The typical absorption spectra of the SiC seed and grown crystals indicated that the SiC polytypes were 6H-SiC and 4H-SiC with fundamental absorption energies of about 3.02eV and 3.26eV[12], respectively.

References

[1] Yu.M. Tairov and V.F. Tsvetkov, J. Cryst. Growth 52 (1981) 146
[2] Yu.M. Tairov and V.F. Tsvetkov, J. Cryst. Growth 43 (1978) 209
[3] Eugene Y. Tupitsyn, Arulchakkaravarthi Arjunan, Robert T. Bondokov, Robert M. Kennedy and Tangali S. Sudarshan, Mater. Sci. Forum 483-485 (2005) 21-24
[4] Yu.M. Tairov and V.F. Tsvetkov, Crystal Growth Characterization 4 (1982) 111
[5] Yu.A. Vodakov, E.N. Mokhov and A.D. Roenkov, Phys.Stat. Sol. (a) 51 (1979) 209
[6] T.L. Straubinger, M. Bickermann, D. Hofmann, R. Weingartner, P.J. Wellmann and A. Winnacker, Mater. Sci. Forum 333-356 (2001) 25
[7] R. Yakimova, M. Syvajärvi, T. Iakimov, H. Jacobson, R. Raback, A. Vehanen and E. Janzen, J. Cryst. Growth 217 (2000) 255
[8] J.-M. Dedulle et al., Mater. Sci. Forum, vols. 457-460 (2004) 71
[9] K. Kakimoto, B.Gao, T.Shiramomo, S.Nakano, Shi-ichiNishizawa, J. Cryst. Growth 324 (2011) 78–81
[10] S. Nishino, Y. Kojima and J. Saraie, Springer Proceeding in Physics, Amorphous and Crystalline Silicon Carbide III, vol. 56, edited by G. L. Harris, M. G. Spencer, C. Yang (Springer-Verlag, New York, 1992) 15
[11] H.-J. Rost, J. Doerschel, K. Irmscher, M. Roxberg, D. Schulz, D. Siche, J. Cryst. Growth 275 (2005) 451
[12] R. Weingartner, P.J. Wellmann, M. Bickermann, D. Hofmann, T.L. Straubinger and A. Winnacker, Appl. Phys. Lett. 80(1) (2002) 70

Materials Science Forum Vols. 717-720 (2012) pp 21-24

doi:10.4028/www.scientific.net/MSF.717-720.21

Fundamental Study of Sublimation-Recrystallization Phenomena in PVT-growth of SiC Single Crystals

T. Fujimoto[1, a], N. Ohtani[2, b], S. Sato[1], M. Katsuno[1], H. Tsuge[1] and W. Ohashi[1]

[1] Nippon Steel Corporation, Advanced Technology Research Laboratories

20-1 Shintomi, Futtsu, Chiba 293-8511, JAPAN

[2] Kwansei Gakuin University, R&D Center for SiC Materials and Processes

2-1 Gakuen, Sanda, Hyogo 669-1337, JAPAN

[a] fujimoto.tatsuo@nsc.co.jp, [b] ohtani.noboru@kwansei.ac.jp

Keywords: PVT growth, Si-C binary phase diagram, Secondary phase.

Abstract. Sublimation-recrystallization processes occurring during PVT are investigated from the viewpoint of quasi-equilibrium phase transitions of SiC. In addition to the elemental reaction processes of PVT, other phenomena such as silicon droplet formation and in-situ etching are also discussed based upon the Si-C binary phase diagram, and possible mechanisms are proposed.

Introduction

Recent marked development in the field of SiC single crystal growth technology has led to astonishing demonstration of various types of SiC devices with splendid performance at high-temperatures and high-frequencies. Although large diameter SiC single crystals with higher crystal quality are brought to us upon long, continuous research attempts [1], we are still on the way to comprehensive understanding of PVT growth process occurring in crucibles at higher temperatures exceeding around 2273K. PVT is itself a vapor-phase growth process in contrast to other semiconductor crystals such as Si and GaAs, both of which are grown by melt-based (liquid-phase) process. Controlling growth conditions from a viewpoint of quasi-equilibrium phase transition is important for obtaining single crystals of higher quality, and such conditions are believed to be realized much easier in melt-based growth than in vapor-based growth, and from this viewpoint, a solution growth of SiC single crystal has been proposed for improving the crystallinity of SiC single crystals [2]. Although considerable effort has been paid to various new growth attempts, PVT is at the moment a widely-adopted method and embraces advantage as an industrial production process, and therefore, it is important to accurately understand exact aspects of phenomena during PVT.

In this paper, we briefly review typical phenomena occurring in PVT-grown SiC single crystals, and then will show that the phenomena observed in PVT-grown crystals can be explained by examining the Si-C binary phase diagram. A challenge is the so-called "in-situ etching" phenomenon which is often observed in earlier growth stages in PVT [3]. We show that such process can also be understandable as a transition process toward a quasi-equilibrium state.

Experimental Procedure

SiC single crystals with various diameters were grown by PVT using our state-of-the-art growth furnace system. Details of growth conditions were described elsewhere [1]. The crystals obtained were all sliced parallel to the growth direction, and then subjected to mirror-polishing using diamond-based slurry, followed by observations using optical microscopy.

Results and Discussion

Modified Lely method, so-called as PVT and widely-adopted as a standard SiC single crystal growth process as well, is a process consisting of phase transition phenomena of SiC at high temperatures. Fig.1 shows a schematic of the elemental processes occurring in a semi-closed crucible; in source SiC material, SiC powder is decomposed into solid carbon (C(s)) and sublimed gas $v_1(g)$ consisting mainly of Si, Si_2C and SiC_2, whereas, at the surface of a SiC single crystal grown, solidification of SiC (SiC(s)) from sublimed gas occurs and subsequently the composition of sublimed gas varies to $v_2(g)$ which is richer in mole fraction of Si compared to $v_1(g)$'s:

$$SiC(s) \rightarrow C(s) + v_1(g) \quad (1)$$

$$v_1(g) \rightarrow SiC(s) + v_2(g) \quad (2)$$

Note that, for brevity, we assume that the crucible causes no influence on the composition of sublimed gases. It should also be worth noting that if the crucible is sealed up completely, the composition of the gas in the whole inner crucible will change gradually. Such effects of compositional variation are also neglected in the following discussions.

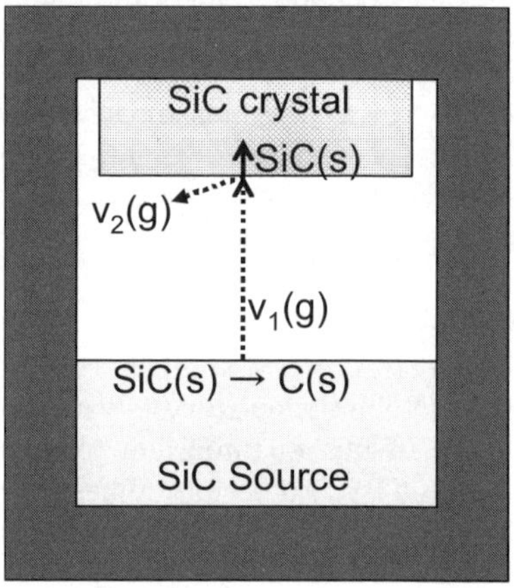

Fig.1 Schematic of PVT processes occurring in crucible for SiC single crystal growth.

Fig.2 shows a phase diagram of the Si-C binary system at a fixed high temperature of around 2500K as a function of both the total gas pressure and the Si-C composition. The diagram shown in Fig.2 is much understandable for discussing the phase transition phenomena described above because, in a dilute environmental circumstance like SiC in PVT, the growth kinetics are strongly dependent on the pressure of the vapor phase.

Fig.3 represents a part of the phase diagram shown in Fig.2. The recrystallization of SiC occurs only when the composition of sublimed gas is in the region ABCD. The sublimed gas with a composition of $v_1(g)$ generated from the source, as shown in Fig.1, becomes energetically unstable. Under a fixed constant pressure of P_0, the composition of the gas shifts toward $v_2(g)$ with solidifying SiC(s) onto the crystal surface, giving rise to the single crystal growth observed in PVT process. The composition of $v_1(g)$ corresponds largely to a peritectic-like reaction point B in Fig.3 at the source powder which is higher in temperature than that of the crystal grown [4]. Fig.3 also provides a possible mechanism for other reactions in PVT. For example, when the pressure decreases below the line BC, the crystal will start decomposition reaction represented as (1). On the other hand, when the gas pressure increases over across the line AD, which is equivalent to overcooling of the crystal, the liquid phase of Si(l) is formed on the crystal. Actually, these reactions are observed as the secondary phase formations such as graphitization in the former and silicon droplets in the latter cases [5-7], respectively.

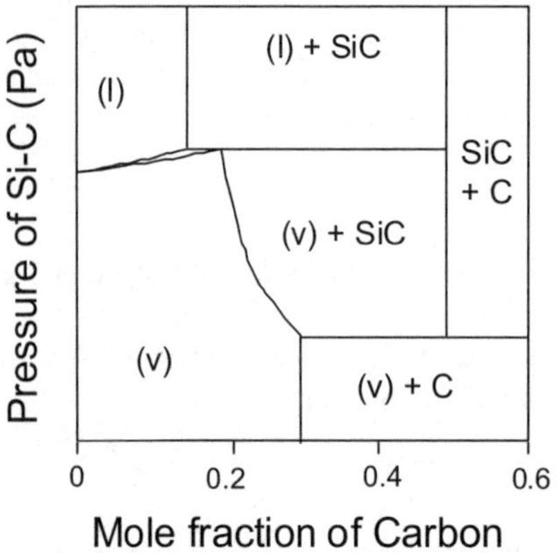

Fig.2 Phase diagram of the Si-C binary system at a fixed temperature of around 2500K

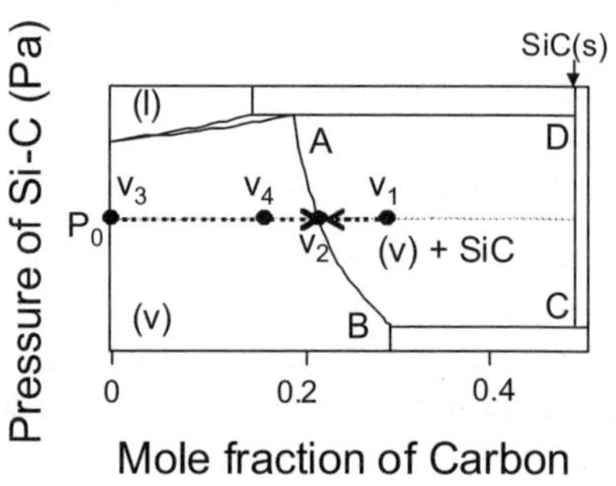

Fig. 3 Phase diagram at the temperature of the crystal surface.

It is known that surface condition of seed crystals is important for crystal quality of the crystal grown, and etching the seed crystal surface has often been performed prior to the growth by reversing the temperature gradient in the crucible [8]. However, as can be understood in Fig. 3, such process has to cause the formation of fine carbon particles on the seed crystal surface, and will induce defect formation such as micropipes during the following growth stages.

Fig.4 shows an example of the in-situ etching observed in a 4H-SiC single crystal of around 50 mm in diameter grown under a nitrogen doping circumstance using a 10 mm-thick, unintentionally doped as-grown 35 mm-diameter 4H-SiC crystal as a seed crystal. As can be seen, the seed crystal is likely to dissipate at earlier growth stages and the seed surface is etched back from the position marked by the white broken line in Fig. 4 to the position of the interface. However, no carbon precipitation occurs at the interface in the following growth stages so that no visible generation of micropipes at the interface was observed.

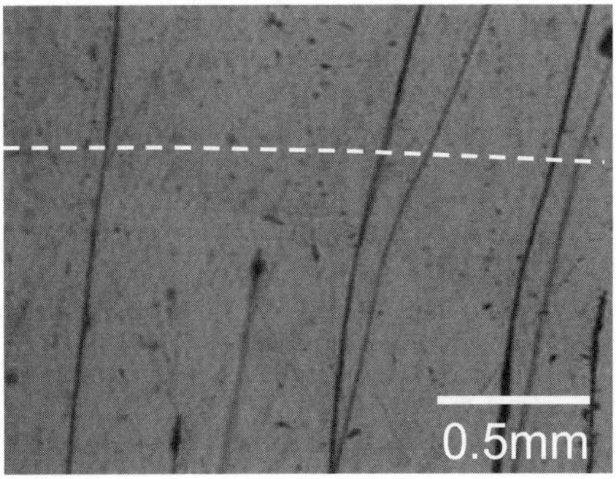

Fig. 4 Cross section of 35 mm diameter 4H crystal grown using 25 mm diameter seed crystal with a thickness of 10 mm. Solid lines are micropipes. For more details, see the main text.

The results described above strongly suggest that, instead of SiC decomposition process (1), another process of SiC sublimation has to exist:

$$SiC(s) \rightarrow v(g) \tag{3}$$

where v(g) represents a vapor with a carbon mole fraction of 0.5.

In order to understand the process (3), we first consider a possible reaction on the supposition that a SiC single crystal is placed in a pure Si vapor gas (= $v_3(g)$ in Fig.3) under a fixed constant pressure of P_0. As shown in Fig.3, such system is energetically unfavorable. SiC starts sublimation, and the vapor subsequently shifts its composition toward $v_2(g)$. This process is exactly the sublimation process of SiC(s) with no generation of carbon precipitates. After the gas composition reaches $v_3(g)$, the system becomes energetically stable and no further visible reactions such as SiC sublimation occur. In other words, an equilibrium state is established:

$$SiC(s) + v_3(g) \rightarrow v_2(g) \tag{4}$$

In short, SiC(s) can be decomposed into the gas phase without carbon precipitations when SiC(s) is placed in a gas circumstance with the composition richer in mole fraction of Si corresponding to the line AB in Fig.3.

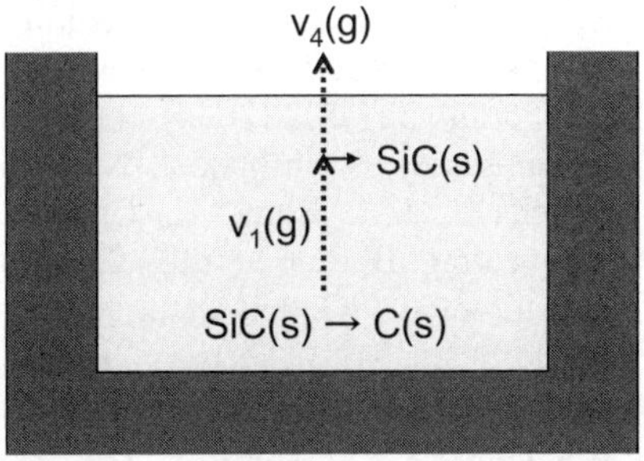

Fig. 5 Possible reaction processes occurring in SiC source material when an appropriate temperature gradient exist in the source material..

Such process written in (4) can take place in actual PVT if the effect of recrystallization process occurring in the source material is considered. In practice, an appropriate temperature gradient along the vertical direction of the crucible is necessary for SiC single crystal growth, and therefore the temperature gradient could also exist in the source powder. As s shown in Fig.5, the higher temperature region of the source materials, for instance, near the lower part of the source material, starts sublimation with a vapor $v_1(g)$ generated. The vapor $v_1(g)$ diffuses upward through gaps between SiC source particles. However, when upper part of the source material is not enough high in temperature to pass away out of the source powder, SiC(s) recrystallization occurs in the source material, and the vapor composition changes from $v_1(g)$ to $v_4(g)$ which is definitely richer in mole fraction of Si compared to $v_1(g)$'s. The composition can be further richer in Si when larger amounts of SiC(s) recrystallization occur in the source material, giving rise to a similar reaction denoted as reaction (4).

Our study described above suggests that elemental processes in PVT can be explained on the basis of quasi-equilibrium phase transition phenomena, and almost all the aspects of the single crystal growth processes can be understood based upon the Si-C binary phase diagram. However, PVT, a vapor phase growth process, inevitably involves risks for unwanted fluctuation of growth conditions compared to the liquid phase based growth process: for instance,

(a) The pressure of sublimed gas can fluctuate much during growth.
(b) The composition of sublimed gas is also variable depending upon growth conditions such as the temperature and the growth velocity.

In these senses, controlling the partial pressure and the composition of the sublimed gas in the crucible is important for SiC single crystal growth with further improved crystal quality, and therefore, quite accurate control of the various growth parameters has to be established in order to maintain quasi-equilibrium processes over the whole growth process.

Summary

Sublimation-recrystallization phenomena occurring in PVT-growth process of SiC single crystals are briefly reviewed. SiC single crystal growth processes in PVT are shown to be explainable by examining the Si-C binary phase diagram. In addition, the second phase formations, such as graphitization and the formation of silicon droplets, are also investigated, and found that both phenomena is a transition process shifting toward an equilibrium state. Furthermore, in-situ etching phenomenon, which is often adopted as a surface cleaning process of seed crystals prior to the growth, is found to be explainable using the Si-C binary phase diagram if the effect of recrystallization process occurring in the region of the source material is included. The in-situ etching phenomenon is a transitioning process to an equilibrium state when SiC(s) is subjected to a Si-rich vapor phase.

References

[1] M. Nakabayashi, T. Fujimoto, M. Katsuno, N. Ohtani, H. Tsuge, H. Yashiro, T. Aigo, T. Hoshino, H. Hirano, and K. Tatsumi, Mater. Sci. Forum, 600-603 (2007) 3-6.
[2] K. Kusunoki, K. Kamei, N. Okada, N. Yashiro, A. Yauchi, T. Ujihara, and K. Nakajima, Mater. Sci. Forum, 527-529 (2006) 119-122.
[3] J. Camsassel, S. Juillaquet, N. Planes, A. Raymond, P. Grosse, G. Basset, C. Faure, M. Couchaud, J.M. Bluet, K. Chourou, M. Anikin, and R. Madar, Materilas Science and Engineering: B, 61-62 (1999) 258-264.
[4] T. Fujimoto, N. Ohtani, *et al.*, to be published.
[5] D. Hofmann, M. Bickermann, R. Eckstein, M. Koelbl, St.G. Mueller, E. Schmitt, A.Weber, A. Winnacker, J. Cryst. Growth, 198/199 (1999) 1005-1010.
[6] R.V.Drachev, G.D.Straty, D.I.Cherednichenko, I.I.Khlebnikov,T.S.Sudarshan, J. Cryst. Growth, 233 (2001) 541-547.
[7] R. C. Glass, D. Henshall, V. F. Tsvetkov, C. H. Carter, Jr., phys. stat. sol. (b), 202 (1997) 149-161.
[8] E.Pernot, P.P.Rejmánková, M.Anikin, M.Pons, C.Faure, P.Grosse, B.Pelissier, C.Moulin, C.Bernard, R.Mader, FED Journal, 11 (2000) 11-15 (in Japanese).

Materials Science Forum Vols. 717-720 (2012) pp 25-28
© (2012) Trans Tech Publications, Switzerland
doi:10.4028/www.scientific.net/MSF.717-720.25

Analysis of growth velocity of SiC growth by the physical vapor transport method

Koichi Kakimoto[1,a], Bing Gao[1], Takuya Shiramomo[1], Satoshi Nakano[1] and Shin-ichi Nishizawa[2]

[1]Research Institute for Applied Mechanics, Kyushu University, Kasuga, Fukuoka 816-8580, Japan

[2]Full National Institute of Advanced Industrial Science and Technology, Tsukuba 305-8568, Japan

[a]kakimoto@riam.kyushu-u.ac.jp

Keywords: Computer simulation; Heat transfer; Substrate; Growth from vapor

Abstract. Crystal growth velocity of SiC in a process of physical vapor transport was studied on the basis of numerical calculation including the effect of compressibility, convection and buoyancy effects, flow coupling between argon gas and species of Si, Si_2C and SiC_2, and the Stefan effect. Calculation in a 2D configuration was performed to clarify the effect of pressure on growth velocity. The results revealed that convection plays a role in the measured values that growers interpret as growth velocity based on a diffusion process of argon gas and species of Si, Si_2C and SiC_2.

Introduction

Silicon carbide (SiC) is one of the candidate semiconductor materials for power electronic devices [1, 2]. Bulk SiC crystals are grown by the physical vapor transport (PVT) method. However, there are still some unsolved problems that limit the commercial utilization of the material in bulk SiC growth. One of them is related to crystal growth velocity, which is important to realize mass production for electric vehicles. There are several papers on growth velocity of SiC grown by the PVT method. Segal et al. [3] reported that their analysis contains a diffusion resistance that includes the clearance between the source and the seed, and the diffusion coefficient, temperature and molar mass of silicon and carbon. However, the origin of the resistance was not clarified since they performed the analysis in a 1D configuration. The buoyancy effect becomes obvious due to a large Grashof number in a larger furnace, and therefore the flow pattern might be changed from the well-known diffusion-dominated flow to a convection-dominated flow. Thus, accurate prediction of the flow pattern is required with a full compressible flow solver, which has an advantage over an incompressible solver using the Boussinesq approximation. The Boussinesq approximation can cause a substantial error when the temperature difference is larger than 151 K [4].

There have been many reports on numerical simulations of SiC sublimation growth [4–15]; however, most of them neglected gas convection [4–6], or neglected the flow coupling between argon gas and species of Si, Si_2C and SiC_2 [7–10] or neglected the buoyancy effect [11,12].

This paper presents an analysis of the growth velocity of SiC based on calculations using a system that includes almost all of the effects in heat and mass transport processes, including the compressibility effect, convection effect, buoyancy effect, flow coupling of argon gas and species of Si, Si_2C and SiC_2, and the Stefan effect, in order to discuss growth velocity and the origin of the diffusion resistance in the gas phase.

Calculations of Heat and Species of Si, Si_2C and SiC_2 Transfer and Growth Velocity

The configuration of the furnace used in the present study is shown in Fig. 1 [16]. The SiC PVT growth system with an axisymmetric configuration consists of SiC powder (1), a graphite crucible (2), an insulation shield (3), a graphite pedestal (4), reflector (5), a wall (6) and induction coils (7). The diameter of the seed was set to 50 mm. Implementation of the present calculation involves the following two steps: first, the temperature distribution of the furnace due to conductive and radiative heat transfer was calculated without gas flow as shown in Fig. 2, and then the flow field and temperature field inside the gas in the furnace were computed using the temperature boundary conditions obtained from the first step.

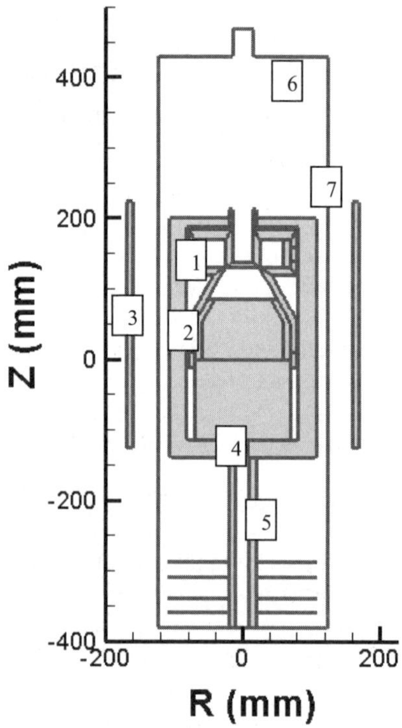

Fig. 1 Furnace configuration.

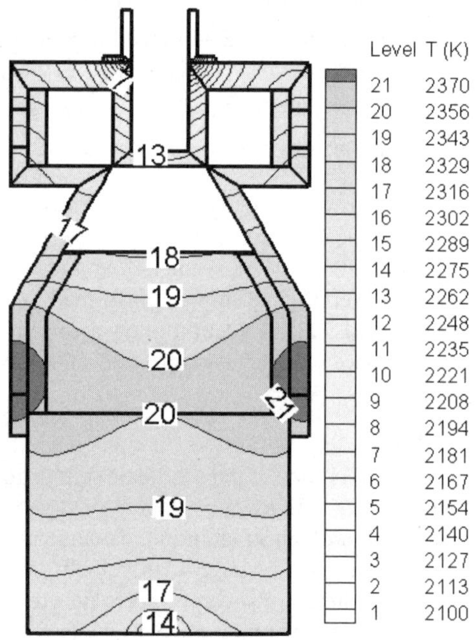

Fig. 2 Temperature distribution in a furnace.

Global modeling of heat transfer in a furnace includes induction heating by an electromagnetic field, conductive heat transfer in all solid components and radiation heat transfer in all enclosures in the furnace. The radiative heat exchange is modeled on the basis of the assumption of diffuse-gray surface radiation [17]. Global simulation of heat transfer shows that the temperatures are about 2329 K at the center of the powder face and about 2270 K at the center of the seed face.

Numerical details of the fully coupled multi-phase flow for SiC were reported elsewhere [18]. The three main species Si, Si_2C and SiC_2 [10,13,15] and argon gas are taken into account in the gas phase for the sublimation and deposition process in the furnace. It is assumed that heterogeneous chemical reactions occur at the surfaces of the crystal and the source and at the container side wall [17]. The boundary conditions of temperature at the surfaces of the source and the seed crystal are set to the values computed from the global heat transfer model. The flow velocities at the source and seed faces are provided according to the definition of Stefan velocity. Calculated distributions of temperature in a furnace and partial pressures of Si, Si_2C and SiC_2 are shown in Fig. 3 (a), (b), (c) and (d), respectively. The evaporation and deposition rates can be markedly affected by the argon gas. Equilibrium pressure is a function of temperature; however, pressure of species of Si, Si_2C and SiC_2 depends on the global flow field, which can be markedly affected by the argon gas. The equilibrium pressure also depends on the furnace pressure, however the effect was neglected in this study. Thus, the effect of argon gas on evaporation and deposition rates is not negligible.

Figure 4 shows calculated results of the pressure dependence of the growth velocity with the experimental results [14] at the center of a seed crystal. The results show that growth velocity increases with decrease in pressure in the furnace. Therefore, the evaporation and deposition rates can be markedly affected by the pressure of argon gas. According to the Hertz–Knudsen equation, evaporation or deposition flux is determined by the difference in equilibrium pressure and pressure of species of Si, Si_2C and SiC_2 near the surface of a source and a seed face. Pressure of species of Si, Si_2C and SiC_2 depends on the global flow field, which can be affected by the argon gas. Thus, the effect of argon gas on evaporation and deposition rates is not negligible. As has been reported [18, 19], there is a vortex in a gas phase with high pressure of argon gas of more than 5 Torr. If species of Si, Si_2C and SiC_2 transport is determined by diffusion and diffusion resistance as reported by Segal et al. [3], growth velocity should be a proportional function of the inverse of pressure in the gas phase. Figure 5 shows the linear relationship between growth velocity and inverse pressure at pressure below 3 Torr, which has the same gradient based on pure diffusion indicated by a triangle in the figure. When the pressure is larger than 3 Torr, the calculated growth velocity deviates from the linear relationship. Thus, the pressure dependence of growth velocity depends on pressure in the furnace. The flow pattern in a gas phase has a vortex [18] that suppresses species of Si, Si_2C and SiC_2 transfer from the source to the seed, and then growth velocity decreases. This suppression of species transfer is the origin of the so-called diffusion resistance [3]. The diffusion resistance has a linear relationship with the distance between the source and the seed, and the resistance therefore becomes large in the case of a large furnace. When a diameter of a crucible is large convective heat and mass transfer become large. Then, the effect of this suppression becomes large. In the case of high pressure, radial distribution of growth rate becomes homogeneous. Consequently, the growth interface becomes more flat than that with low pressure. Thus, analysis of flow coupling of argon gas and species of Si, Si_2C and SiC_2 is important to predict the growth velocity quantitatively, especially in a large furnace.

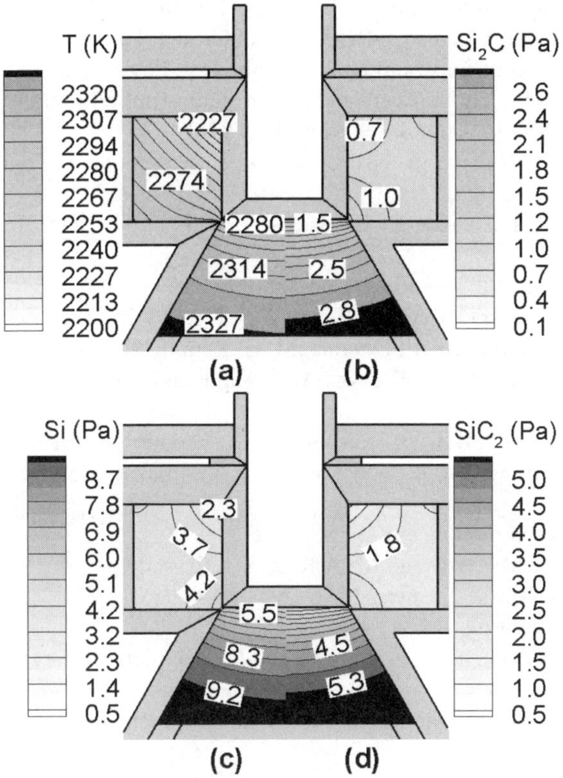

Fig. 3 Distributions of temperature (a), Si_2C (b), Si (c) and SiC_2 (d).

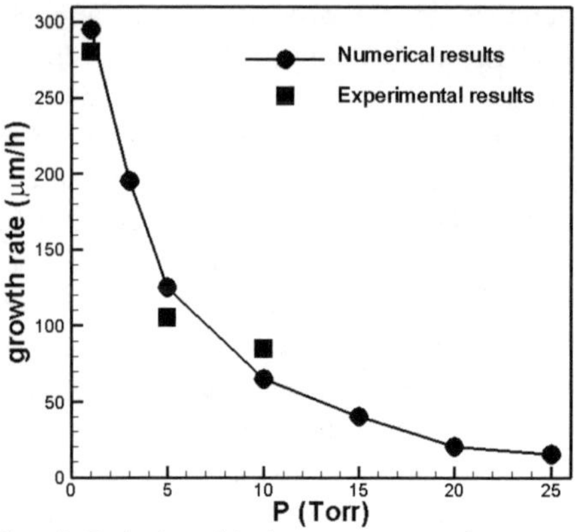

Fig. 4 Relationship between growth rate and furnace pressure.

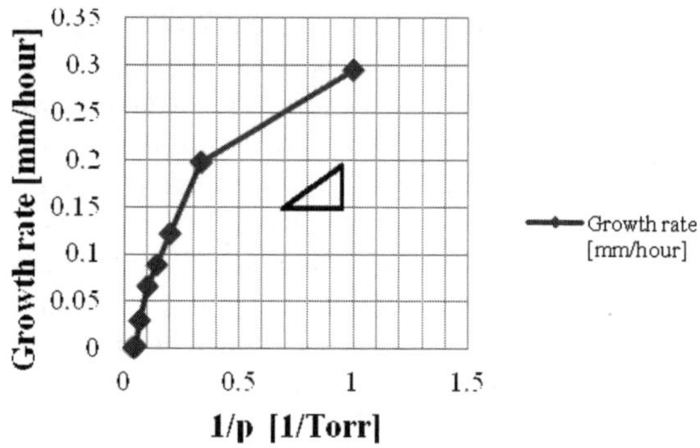

Fig. 5 Relationship between growth rate and inverse pressure.

Summary

The crystal growth velocity of SiC in a process of physical vapor transport is determined by the diffusion process at a low pressure of less than 3 Torr, while suppression of species of Si, Si_2C and SiC_2 transfer by convection reduces the growth rate compared with that of pure diffusion.

References

[1] X. L. Wang, D. Cai and H. Zhang, Int. J. Heat Mass Transfer 50 (2007) 1221.
[2] X. L. Wang, D. Cai and H. Zhang, J. Crystal Growth 305 (2007) 122.
[3] A. S. Segal, A. N. Vorobi and H. Zhang, J. Crystal Growth 305 (2007) 122.
[4] R. A. Stein, P. Lanig and S. Leibenzeder, J. Crystal Growth 131 (1993) 71.
[5] A. A. Maltsev, A. Yu. Maksimov and N. K. Yushin, Inst. Phys. Conf. Ser. 142 (1996) 41.
[6] M. Kanaya, J. Takahashi, Yu. Fujiwara and A. Moritani, Appl. Phys. Lett. 58 (1991) 56.
[7] I. Markov and R. Kaischew, Thin Solid Films 32 (1976) 163.
[8] B. Gao, X. J. Chen, S. Nakano, S. Nishizawa and K. Kakimoto, J. Crystal Growth 312 (2010) 3349.
[9] X. J. Chen, L. J. Liu, H. Tezuka, Y. Usuki and K. Kakimoto, Cryst. Res. Technol. 42, No. 10 (2007) 971.
[10] J. S. Shun, K. H. Chen and Y. Choi, J. Comput. Phys. 106 (1993) 306.
[11] K. Bottcher and D. Schulz, J. Crystal Growth 237–239 (2002) 1196.
[12] M. S. Ramm, E. N. Mokhov, S. E. Demina, M. G. Ramm, A. D. Roenkov, Yu. A. Vodakov, A. S. Segal, A. N. Vorob'ev, S. Yu. Karpov, A. V. Kulik and Yu. N. Makarov, Mater. Sci. Eng. B 61-62 (1999) 107.
[13] A. S. Segal, A. N. Vorob'ev, S. Yu. Karpov, Yu. N. Makarov, E. N. Mokhov, M. G. Ramm, M. S. Ramm, A. D. Roenkov, Yu. A. Vodakov and A.I. Zhmakin, Mater. Sci. Eng. B 61-62 (1999) 40.
[14] S. Nishizawa, private communication.
[15] S. Yu Karpov, Yu. N. Makarov and M. S. Ramm, J. Crystal Growth 169 (1996) 491.
[16] J. Drowart and G. de Maria, J. Chem. Phys. 29 (1958) 1015.
[17] S. K. Lilov, Mater. Sci. Eng. B 21 (1993) 65.
[18] B. Gao, X.J. Chen, S. Nakano, S. Nishizawa and K. Kakimoto, J. Crystal Growth 312 (2010) 3349.
[19] K. Kakimoto, B. Gao, T. Shiramomo, S. Nakano and Shin-ichi Nishizawa, J. Crystal Growth 324 (2011) 78.

Materials Science Forum Vols. 717-720 (2012) pp 29-32

doi:10.4028/www.scientific.net/MSF.717-720.29

Effect of Nitrogen Doping on the Growth of 4H Polytype on the 6H-SiC Seed by PVT Method

E. Tymicki[1,a], K. Grasza[1,2], K. Racka[1], T. Łukasiewicz[1], M. Piersa[1], K. Kościewicz[1,3], D. Teklińska[1,3], R. Diduszko[1], P. Skupiński[2] R. Jakieła[1,2], and J. Krupka[4]

[1]Institute of Electronic Materials Technology, 133 Wólczyńska Str., 01-991 Warsaw, Poland

[2]Institute of Physics Polish Academy of Sciences, Al. Lotników 32/46, 02-668 Warsaw, Poland

[3]Warsaw University of Technology, Faculty of Materials Science and Engineering, 141 Wołoska Str., 02-507 Warsaw, Poland

[4]Institute of Microelectronics and Optoelectronics of Warsaw University of Technology, 75 Koszykowa Str., 00-662 Warsaw, Poland

[a]Emil.Tymicki@itme.edu.pl,

Keywords: nitrogen doping, polytype transformation, 4H-SiC, 6H-SiC, PVT

Abstract. In this work results of nitrogen doping in the amount of 0 vol.%, 3 vol.% and 10 vol.% on the growth of the 4H polytype on the 6H-SiC seed are presented. SiC crystals grown by the PVT method on the (000-1) C-face of 6H seeds using the open seed backside design have been investigated. Structural and electrical properties of the crystals were studied by different experimental methods. The influence of nitrogen dopant on the 4H-SiC polytype stabilization was observed. For SiC with 10 vol.% of nitrogen doping the 4H polytype uniformity was ~95%.

Introduction

In general nitrogen is a basic residual dopant in silicon carbide and it determines the *n*-type of electrical conductivity of SiC [1]. It has also an impact on growth conditions during the growth of crystals by the PVT method [2-4]. Nitrogen doping changes the surface energy and morphology by a reduction of the height and width of crystallization steps. This prevents the 3D-growth on a flat and smooth growth surface as a facet and may confine generation of the foreign polytype inclusions. When the 6H → 4H-SiC transition occurred, a further growth of the 4H polytype with nitrogen doping is more stable. In this work we investigate the aspect of nitrogen doping regarding the kind of polytypes obtained in SiC crystals, as well as their structural quality and electrical properties. As a reference, undoped SiC crystals are also studied. This work is a continuation of our earlier research on the 6H → 4H polytype transformation [5-7].

Experimental

SiC crystals were grown by the PVT method on the (000-1) C-face of 6H-SiC seeds using the open seed backside design. The density of micropipes in these seeds was below 80 cm^{-2}. The temperature measured at the backside of the seeds was in the range of 2150-2300°C. The growth atmosphere was pure argon (for undoped crystals) or it consisted of a mixture of argon and nitrogen. The total pressure was 40 mbar. The initial source to seed distance was 40 mm. The growth rate of SiC crystals was ~0.15 mm/h.

The morphology and the crystalline structure of SiC growth surfaces were studied by optical microscopy, KOH etching and X-ray diffraction (XRD). The concentration of nitrogen atoms in crystals was determined by secondary ion mass spectroscopy (SIMS). The resistivity of samples cut from the crystals was evaluated by the contactless method at microwave frequencies (M-F) at 300 K. The type of electrical conductivity and charge carrier concentrations were estimated from Hall effect measurements.

Results and Discussion

While in the literature are reported many experimental results concerning the growth of 4H polytype on 4H-SiC seeds in the presence of nitrogen dopant [1-4], our studies are focused on the effect of nitrogen doping on the 4H polytype stabilization in SiC crystals grown on 6H-SiC seeds. In general, the authors report the greater amount of 4H polytype in crystals with higher nitrogen concentration [4]. For our undoped and SiC:N (3 vol.%) crystals we observed a mixture of 4H, 6H and 15R polytypes (Fig. 1a, c), while for the crystal with 10 vol.% of nitrogen doping (Fig. 1b) the 4H polytype uniformity is around 95% (the identity of polytypes was determined from XRD diffraction patterns, which are not shown here). Thus, our observations are in agreement with reports in the literature mentioned above.

Generally it is known that 4H-SiC crystals are grown on the carbon face and 6H-SiC crystals on the silicon face of the crystal seed. There is a difference in binding energy values for the C and Si faces. There can be three possible reasons for the 6H → 4H polytype transformation: (i) 4H-SiC nucleation on defects, (ii) bunching of crystallization steps, and (iii) specific thermodynamic conditions [8-11]. It is possible that in growth conditions favorable for the 4H polytype, the 6H → 4H transformation can proceed successfully already in the first phase of the growth process.

One of problems connected with 6H to 4H polytype transformation is the nucleation of 15R polytype, which can result in generation of polytype borders and defects in the form of micropipes. The 15R-SiC crystallizes in different crystallographic structure than 4H- and 6H-SiC, so the polytype boundaries between the 4H-, 6H- and 15R-SiC are not structurally matched. During the crystal growth these boundaries move on the growth surface resulting in degradation of the structural quality of the entire growth surface. In our experiments, with the increase of the nitrogen dopant in argon atmosphere, we observed a stable growth of 4H-SiC and a diminished nucleation of 15R polytype. As a consequence, the structural quality and density of micropipes in 4H-SiC wafers were similar to those in 6H-SiC seeds used for the growth.

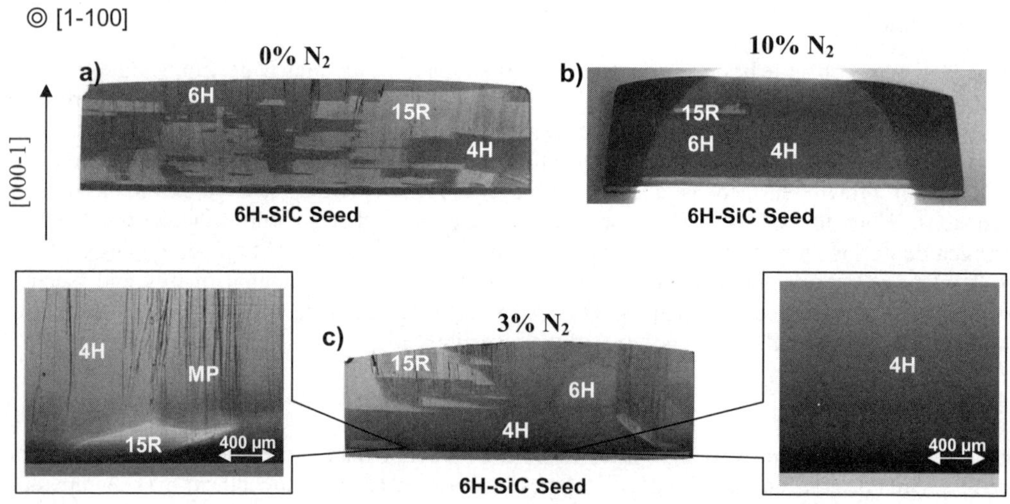

Fig. 1. The cross-sections of 2" SiC crystals grown with different amounts of nitrogen dopant. (4H-, 6H- and 15R-SiC polytypes are: brown, green and yellow areas, respectively). (a) SiC crystal intentionally undoped (coloured areas are due to residual nitrogen coming from SiC source material). Intentionally N-doped SiC crystals: (b) SiC:N (10 vol.%) and (c) SiC:N (3 vol.%). The image to the left of (c) shows micropipes (MP) generated by a 15R polytype inclusion; the image to the right shows no micropipes afer 6H → 4H transformation.

The morphology of 4H-SiC growth surfaces and the main crystalline defects have been studied by optical microscopy (Fig. 2b, d). The bunching of crystallization steps, nucleation of screw dislocations and 3D growth on flat surfaces were observed. In comparison with undoped SiC crystals, for 4H-SiC:N (10 vol. %) crystals a stable growth was observed, which means the growth surfaces were free of foreign polytype inclusions.

In order to evaluate the density of micropipes and dislocations, chemical etching in molten KOH ($T = 420$°C, $t = 20$ min) was used. For undoped SiC crystals the micropipe and dislocation densities of < 150 cm^{-2} and ~6 x 10^4 cm^{-2} (Fig. 3b) were found, while for SiC:N (10 vol.%) crystals they were calculated to be < 80 cm^{-2} and ~2 x 10^4 cm^{-2} (Fig. 3c), respectively. A worse structural quality of undoped crystals was due to the lack of their polytype stability.

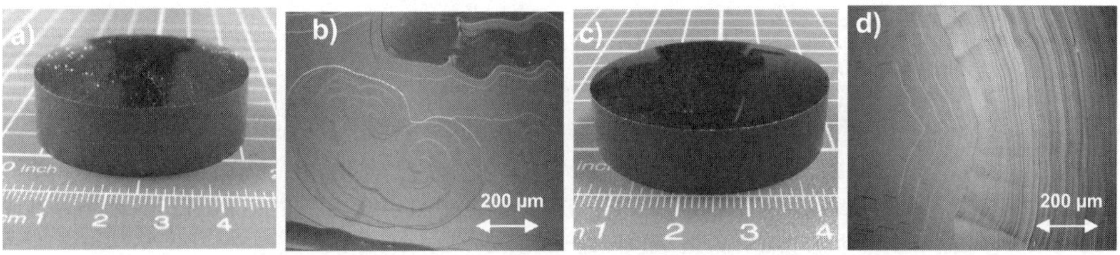

Fig. 2. View and the growth surface morphology of investigated crystals: (a, b) undoped SiC, (c, d) SiC with 10 vol.% of nitrogen doping.

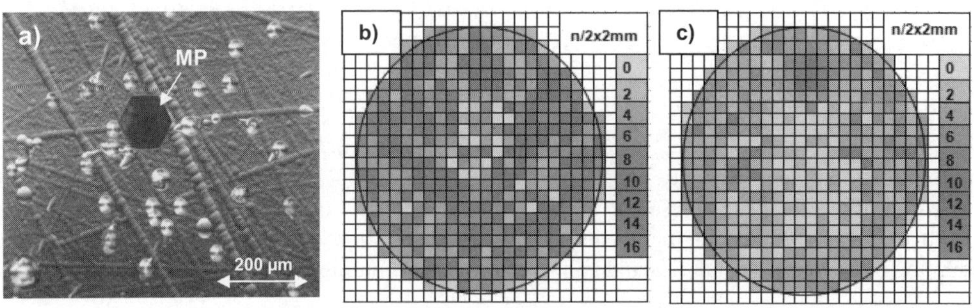

Fig 3. (a) Defects at the Si face of 4H-SiC revealed after etching in molten KOH. (b, c) Micropipe maps of 4H-SiC wafers for undoped SiC and SiC:N (10 vol.%), respectively.

SIMS analysis of 4H-SiC samples cut from undoped and nitrogen doped crystals allowed to detect the nitrogen and common residual impurities of SiC, such as boron and aluminium (Fig. 4). Nitrogen concentration taken from SIMS data was: 2 x 10^{18} cm^{-3} for undoped SiC, 1 x 10^{19} cm^{-3} for SiC:N (3 vol.%) and 2 x 10^{19} cm^{-3} for SiC:N (10 vol.%).

Electrical properties of 4H-SiC samples were also examined at RT by Hall effect measurements and a contactless method at microwave frequencies. The results obtained are shown in Table 1. One can see in this Table that all three SiC samples were of *n*-type electrical conductivity and they were low-resistivity materials. There is a good correlation between resistivity values obtained from both experimental methods. With increasing nitrogen doping of crystals, a rise of the net donor concentration was observed, while the opposite tendency was seen for the electron Hall mobility. By comparison of our electron Hall mobilities with data reported by Iwata *et al.* [12] one can expect in our samples charge compensation effects with the compensation ratio above 0.6.

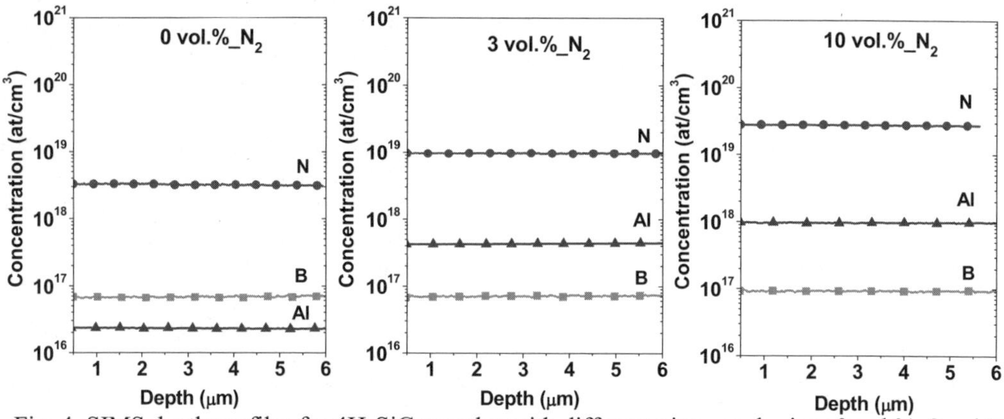

Fig. 4. SIMS depth profiles for 4H-SiC samples with different nitrogen doping: 0 vol.%, 3 vol.% and 10 vol.%.

Table 1. Electrical properties of 4H-SiC samples determined at 300 K by Hall effect measurements and contactless method at microwave (M-F) frequencies.

Sample	Net donor concentration (cm^{-3})	Resistivity (Ω*cm)		Mobility ($cm^2 V^{-1} s^{-1}$)
	Hall effect	Hall effect	M-F	Hall effect
0 % vol. N_2	2.17×10^{18}	0.023	0.027	122.83
3% vol. N_2	7.10×10^{18}	0.014	0.011	61.10
10 % vol. N_2	1.66×10^{19}	0.009	0.009	41.70

Summary

The growth of 4H polytype was the most effective for 10 vol.% of nitrogen doping. In this case, the 4H-SiC polytype uniformity of obtained crystals was ~95%. In 4H-SiC:N (10 vol. %) crystals the density of micropipes was similar to that in 6H-SiC seeds. For undoped SiC crystals an unstable growth of 4H polytype and inclusions of 6H and 15R polytypes were observed, which led to formation of polytype boundaries and an increase of the micropipe density.

Acknowledgements

This work was supported by Polish Ministry of Science and Higher Education under Project: NN 507466837 and the SICMAT Project co-financed by the European Regional Development Fund under the Operational Programme Innovative Economy (Contract No. UDA-POIG.01.03.01-14-155/09).

References

[1] H.J. Rost et al., Mater. Sci. Forum Vol. 433-436 (2003) 91-94
[2] H.J. Rost et al., J. Cryst. Growth Vol. 257 (2003) 75-83
[3] N. Ohtani et al., J. Cryst. Growth Vol. 311 (2009) 1475-1481
[4] H.J. Rost et al., J. Cryst. Growth Vol. 275 (2005) 451-454
[5] K. Grasza et al., Mater. Sci. Forum Vol. 600-603 (2009) 31-34
[6] E. Tymicki et al., Mater. Sci. Forum Vol. 615-617 (2009) 15-18
[7] E. Tymicki et al., Mater. Sci. Forum Vol. 645-648 (2010) 17-20
[8] Y. Zhang et al., J. Electron. Mater. Vol. 39 (2010) 799-804
[9] P. Heuell et al., Surf. Sci. Vol. 331-333 (1995) 965-970
[10] A. Fissel, J. Cryst. Growth Vol. 212 (2000) 438-450
[11] E. Tymicki et al., Cryst. Res. Technol. Vol. 42 (2007) 1232-1236
[12] H. Iwata et al., J. Appl. Phys. Vol. 89 (2001) 6228-6234.

Materials Science Forum Vols. 717-720 (2012) pp 33-36
© (2012) Trans Tech Publications, Switzerland
doi:10.4028/www.scientific.net/MSF.717-720.33

Lateral Growth Expansion of 4H/6H-SiC m-plane Pseudo Fiber Crystals by Hot Wall CVD Epitaxy

Andrew J. Trunek[1, a], Philip G. Neudeck[2, b] Andrew A. Woodworth[3, c], J. A. Powell[4, d], David J. Spry[2, e], Balaji Raghothamachar[5, f] and Michael Dudley[5, g]

[1]OAI, NASA Glenn Research Center, 21000 Brookpark Road, MS 77-1, Cleveland, OH 44135 USA

[2]NASA Glenn Research Center, 21000 Brookpark Road, MS 77-1, Cleveland, OH 44135 USA

[3]NASA Glenn Research Center, 21000 Brookpark Road, MS 106-1, Cleveland, OH 44135 USA

[4]Sest Inc., NASA Glenn Research Center, 21000 Brookpark Road, MS 77-1, Cleveland, OH 44135 USA

[5]Stony Brook University, 310 Engineering, Stony Brook, NY 11794-2275 USA

[a]Andrew.J.Trunek@nasa.gov, [b]Philip.G.Neudeck@nasa.gov, [c]Andrew.A.Woodworth@nasa.gov (NASA Postdoctoral Program Fellow), [d]J.A.Powell@nasa.gov, [e]David.J.Spry@nasa.gov, [f]braghoth@notes.cc.sunysb.edu, [g]mdudley@notes.cc.sunysb.edu,

Keywords: a-plane, a-face, m-plane, m-face, CVD, Hot Wall, Epitaxy

Abstract. Lateral expansion of small mixed polytype 4H/6H-SiC and 6H-SiC slivers was realized by hot wall chemical vapor deposition (HWCVD). Small slivers cut from m-oriented $(1\bar{1}00)$ SiC boule slices containing regions of 4H and 6H-SiC or just single polytype 6H-SiC were exposed to HWCVD conditions using standard silane/propane chemistry for a period of up to eight hours. The slivers exhibited approximately 1500 μm (1.5 mm) of total lateral expansion. Initial analysis by synchrotron white beam x-ray topography (SWBXT) confirms that the lateral growth was homoepitaxial, matching the polytype of the respective underlying region of the seed sliver.

Introduction

Reducing the number of dislocations in SiC substrates is essential for further commercialization of SiC power electronics. Unfortunately, due to the polytypic nature of SiC, dislocations in the seed crystal are essential for maintaining polytype and commercially viable growth rates [1]. There have been some studies of growing SiC in the non-polar $[1\bar{1}00]$ and $[11\bar{2}0]$ directions as a method to reduce dislocations [1,2]. Nakamura et al., explored the possibility of reducing screw dislocations (SD) by developing the physical vapor transport repeated a-face (RAF) growth process which concludes with a final off axis c-face growth [1]. However, these processes are either very difficult or resulted in increased stacking fault formation [2]. The NASA Glenn Research Center has recently proposed a new bulk growth process for producing SiC boules that has the potential for dramatically reducing the number of SD's per wafer, ideally as few as one [3]. Unlike the RAF process, this approach relies on chemical vapor deposition for the lateral homoepitaxial expansion of a SiC fiber without introducing a significant number of additional defects [2, 3]. This paper reports on the lateral growth enlargement of 4H/6H-SiC m-plane slivers up to millimeter diameter crystals via lateral growth in a HWCVD system.

Experimental

Currently, a source of single crystal SiC fibers does not exist, therefore slivers (0.8 mm x 0.5 mm x 15 mm), with a crystal orientation as diagrammed in Fig. 1, were diced from polished m-oriented $(1\bar{1}00)$ SiC boule slices obtained from a commercial source. One m-oriented SiC boule slice contained a thick region of 6H-SiC and small 4H-SiC region; the other boule slice was single crystal 6H-SiC. Both contained numerous micropipes. The slivers act as pseudo fibers, herein referred to as fibers. This work reports on the results from five (Fiber 1) and eight (Fiber 2) hour long growth runs. The dicing results in a fiber with two polished m-plane faces and two saw cut a-plane faces. Prior to growth, fibers were immersed in a 1:1 solution of H_2SO_4 and H_2O_2 for 15 minutes, then blown dry with N_2. Fibers were immediately transferred to an argon filled glove box. All growth experiments proceeded within four hours of cleaning. Growth runs were accomplished in a custom designed, inductively heated, horizontal flow hot wall reactor. The tantalum carbide coated tubular susceptor has a 40mm inside diameter and a length of 140mm. A carbon foam insulation package was used to reduce heat loss and improve temperature uniformity. The fibers were supported in the reactor by an uncoated graphite carrier, which also facilitated in the loading and unloading of the fibers (Fig. 2). Approximately, two millimeters of the fibers' overall length were used for mounting and do not participate in growth. With the exception of the mounted portion of the fiber, the rest of the fiber is uniformly exposed to growth conditions. The carrier gas was hydrogen (H_2) with standard carbon (C_3H_8) and silicon (SiH_4) precursors used for growth. HCl was added to the gas stream to improve pregrowth etching performance and reduce gas phase nucleation during growth. A twelve minute *in-situ* pregrowth etch was performed on the fibers before transitioning directly to growth. Aggressive etching conditions were employed to remove crystalline damage caused by the saw cutting operation. Growth parameters for both fibers are given in Table 1.

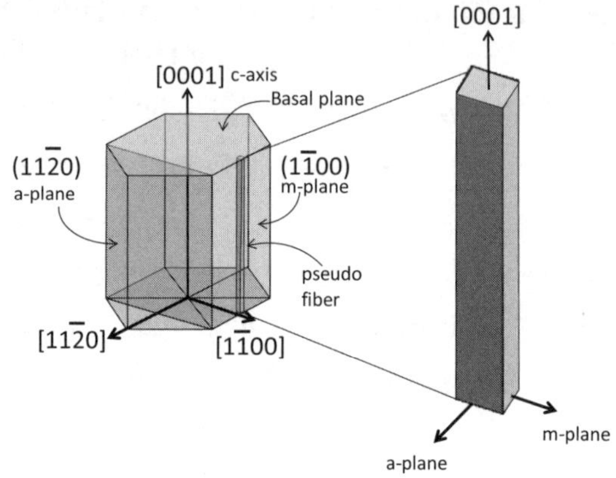

Fig. 1: Schematic diagram showing the SiC unit cell with associated crystallographic planes and the orientation of the seed slivers (pseudo fibers) as cut from m-plane polished boule slices.

Results and Discussion

In the first growth, a fiber from the mixed polytype 4H/6H-SiC m-oriented SiC boule slice was exposed to five hours of growth. Since the fiber is mounted approximately vertically in the gas stream, all surfaces (excluding the 2mm portion used for mounting) are exposed to growth conditions (Fig. 2). The post-growth fiber (Fig. 3) has begun to evolve towards hexagonal crystal symmetry [4]. It is interesting to note that growth progressed on the a-plane side of the pseudo fiber, led by nucleation at the intersection of the m and a–planes. A trough is formed along the a-plane side of the fiber as the growth in the middle trails the corners by a significant amount. Examination of the m-plane face shows that a very small depression exists but is much less than on the a-plane surface. Even though the m-plane surface is ~30% wider than the a-plane surface, the slower

Growth Parameters for Fiber 1 and 2

Growth	*In-situ* HCl etch [min]	Etch Pressure [mbar]	Growth [min]	Growth Pressure [mbar]	Hydrogen [sccm]	Silane[1] [sccm]	Propane[1] [sccm]	HCl[1] [sccm]	Estimated Temperature[2] [°C]
Fiber 1	12	40	300	325	4260	0/4	1.5/1.5	15/20	1600
Fiber 2	12	40	480	325	4260	0/8	1.5/2.25	15/40	1590

1. First number is for etching conditions, second number for growth conditions.
2. No direct observation of temperature by pyrometry was possible. An inferred temperature was calculated based upon the melting points of Si and Pd.

Table 1. Growth parameters for two separate growth runs. Fiber 1 was five hours in length. Fiber 2 was eight hours in length. Note that the intentional Si/C is 2.7 and 3.5 respectively. There is a substantial and unknown quantity of carbon contribution coming from the uncoated graphite sample carrier and carbon foam insulation.

growth rate in the $<1\bar{1}00>$ directions allows the facet time to fill in. The overall growth rate is ~80 μm/hr in the $<11\bar{2}0>$ directions. As expected, the growth rate in the $<1\bar{1}00>$ directions is ~60 μm/hr or 75% of the growth rate in the $<11\bar{2}0>$ directions. Synchrotron white beam x-ray topography (SWBXT) confirms that the growth is homoepitaxial, matching the polytype of the respective underlying region of the fiber.

Fig. 2: Post growth results for the eight hour growth showing three m-plane fibers mounted on top of an uncoated graphite carrier (a). Two other control fibers protrude from the front of the carrier. Yellow 3C-SiC decorates the surface of the

A second 6H-SiC m-plane polished fiber (Fiber 2 in Table 1) was exposed to eight hours of growth at higher super-saturation conditions. Similar to fiber 1, this fiber also has begun to evolve towards hexagonal crystal symmetry (Fig. 4). However, the greatest lateral expansion occurred in the $<1\bar{1}00>$ directions with approximately 1500 μm(1.5 mm) of total lateral expansion. Similar to the five hour growth fiber, a deep trough has developed along the a-plane side of the fiber. A much shallower depression is present on the m-plane face of the fiber. The crystal has several locations where 3C-SiC has nucleated and negatively impacted the growth morphology. It is proposed that carbon particulates coming from the uncoated graphite carrier contributed to the formation of the 3C-SiC. The total overall growth rate is ~180 μm/hr in the $<1\bar{1}00>$ directions. However, the approximate overall growth rate is 30 μm/hr in the $<11\bar{2}0>$ directions. SWBXT confirms the growth was homoepitaxial, matching the polytype of the respective underlying region of the seed sliver. The facets are of high crystalline quality and are mostly free of strain.

Under the higher super-saturation conditions used for fiber 2, the crystal is expanding at the highest rate along the $<1\bar{1}00>$ directions. The fast growth direction is normally understood to be along the $<11\bar{2}0>$ directions. It appears that the evolving crystal shape for these higher super-saturation conditions indicates that the highest nucleation probability is occurring at the edges/corners formed by intersections of the m-planes (Fig. 4). This is not unexpected, as the NASA Glenn Research Center reported similar growth behavior obtained on hexagonal shaped mesas whose sidewalls share the same crystallographic orientation as the sides of the fibers [5].

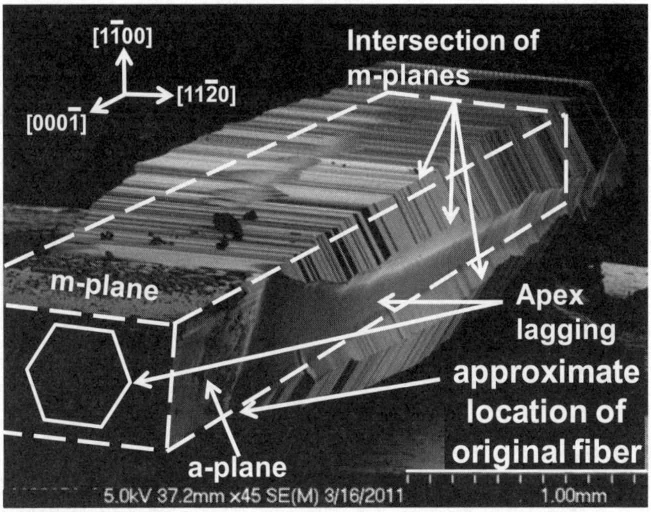

Fig. 3: SEM of m-plane fiber 1 after five hours of growth. The white dashed lines indicate the approximate shape and location of the original fiber. The hexagonal unit cell orientation has been superimposed on the end of the fiber. Growth at the intersection of the m-planes dominates. The portion of fiber visible in the lower left of the SEM was used for mounting and exhibits almost no growth. The fiber has increased in width by ~440 μm along the $<\mathbf{11\bar{2}0}>$ direction and is evolving towards a hexagonal shape.

Both fibers exibit 3C-SiC growth near the tip and a large number of steps are present on both fibers, particularly on the five hour growth fiber. This could be due to the imperfect nature of the starting sliver. However, similar morphology is displayed by the RAF process and a model for such behavior has been proposed by Takahashi et al. [1, 2].

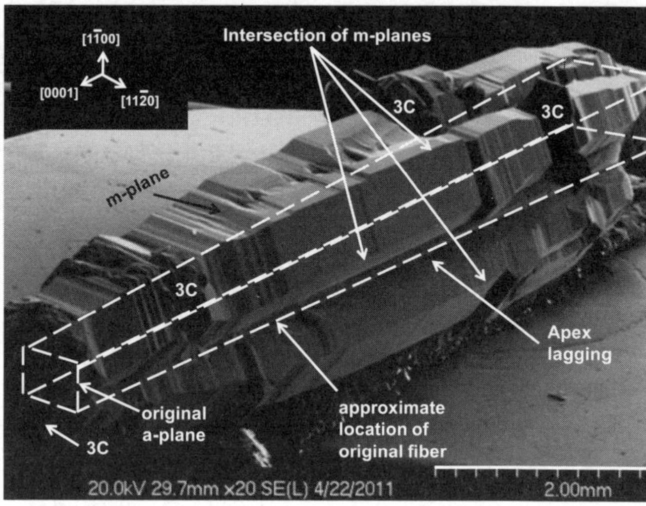

Fig. 4: SEM of 6H-SiC m-plane polished fiber 2 after eight hours of growth at higher growth rate conditions. The white dashed lines indicate the approximate shape and location of the original fiber. Pregrowth etching resulted in a sliver with a tapered shape. Note the deep trough (apex lagging) that has evolved as growth at the intersection of the m-planes has outpaced the growth rate at the center of the fiber. Some regions of 3C-SiC have nucleated near either end of the fiber. The fiber has increased in width by ~1500 μm in the $< \mathbf{1\bar{1}00} >$ direction.

Conclusion

By exposing all surfaces of small slivers diced from m-plane boule slices to uniform growth conditions, homoepitaxial lateral expansion of m-plane oriented fibers has been demonstrated. Growing perpendicular to the (0001) axis without the benefit of SD's to maintain polytype has been demonstrated. Analyses by SWBXT confirms that the underlying substrate polytype was successfully replicated and is of a high crystalline quality with minimal or no strain (Fig. 5).

Acknowledgments
M. Mrdenovich, B. Osborn, T. Hempel, W. John, K. Moses and C. Blaha. Funding for this work is provided by: NASA Vehicle Systems Safety Technologies Project in the Aviation Safety Program, Long Tapered Crystal Project Funding via Space Act Agreement (SAA3-1048) with the US Department of Energy, Vehicle Technologies Program, Susan Rogers, Program Manager.

Fig. 5: SWXBT topograph of fiber 2 showing developing facets.

References

[1] D. Nakamura, I. Gunjishima, S. Yamaguchi, T. Ito, A. Okamoto, H. Kondo, S. Onda, and K. Takatori, "Ultra High-quality silicon carbide crystals", Nature (London), 430 (2004) 1009

[2] J. Takahashi, N. Ohtani, M. Katsuno, and S. Shinoyama, "Sublimation growth of 6H- and 4H-SiC single crystals in the [1$\bar{1}$00] and the [11$\bar{2}$0] directions", J. Crystal Growth, Vol. 181 (1997) 229-240

[3] J. Powell, P. Neudeck, A. Trunek, and D. Spry "Method for the growth of large low-defect single crystals" U.S. Patent 7,449,065 issued November 11, 2008

[4] N. Nordell, S. Karlsson, and A. O. Konstantinov, "Equilibrium crystal shapes for 6H and 4H SiC grown on non-planer substrates", Mater. Sci. and Engineering, B61-62 (1999) 130-134

[5] P. G. Neudeck and J. A. Powell, "Homoepitaxial and heteroepitaxial growth on step-free SiC mesas," in Silicon Carbide: Recent Major Advances, W. J. Choyke, H. Matsunami, and G. Pensl, Eds. Heidelberg, Germany: Springer-Verlag, (2003) 179-205

Materials Science Forum Vols. 717-720 (2012) pp 37-40

doi:10.4028/www.scientific.net/MSF.717-720.37

Synthesis and purification of silicon carbide powders for crystal growth

T. C. Hsiao[a] and S. Tsao

Industrial Technology Research Institute, Rm. 110-4, Bldg. 44, 195, Sec. 4, Chung Hsing Rd., Chuntung, Hsinchu, Taiwan 31040, R.O.C

[a]TaChingHsiao@itri.org.tw

Keywords: silicon carbide powder, synthesis, purification, crystal growth

Abstract. Silicon carbide powders were prepared in a vacuum induction melting furnace (VIM). Silica and silicon were used as sources of silicon, and graphite powder was used as a source of carbon. Pressures of 0.1 and 0.01 atm were selected as the operation conditions, and different silicon carbide powders were prepared. Free carbon and remnant silica were removed by high-temperature baking in air and acid leaching. Low-pressure powders show better crystallinity; moreover, free carbon and silica were rarely found in the product after baking and leaching. The low-pressure grains were prismatic whereas the high-pressure grains were porous. This shows that pressure is a critical parameter in silicon carbide formation, and low-pressure makes the low-temperature synthesis of silicon carbide feasible. Glow discharge mass spectra (GDMS) were used to analyze the impurity content in silicon carbide powders. After baking and leaching, the purity is increased from 3N5 (99.95 wt.%) to 4N5 (99.995 wt.%). Further purification procedures will be combined to meet the quality requirements for crystal growth.

Introduction

Single-crystal silicon carbide is an excellent substrate for high-temperature, high-frequency, and high-power applications such as light-emitting diodes (LEDs), radio frequency devices, and power devices. High-quality single-crystal wafers are needed to manufacture devices that perform well. In the PVT process, the properties of silicon carbide powder, used as the source material, will directly affect the quality of the obtained single crystals. The most popular synthesis process of silicon carbide powder is the Acheson process. However, it is prone to contamination and the purity is unsuitable for crystal growth of silicon carbide. Sol-gel [1], CVD process [2], direct synthesis [3], and arc-discharge process [4] are used to manufacture high-purity silicon carbide powders. In these processes, only the CVD process and direct synthesis of high-purity silicon and carbon can meet the purity requirements for crystal growth. Because of cost constraints, the CVD process is not suitable for mass production. Furthermore, the latter needs high-purity precursors and high synthesis temperatures. This work aims on combining low-purity precursors and low-temperature synthesis conditions. Thus, a lower synthesis cost is expected.

Experiment

The synthesis of high-purity silicon carbide powder was carried out in a vacuum induction melting furnace (VIM). Silicon and silicon dioxide were used as sources of silicon to prepare silicon monoxide. The purity of the silicon source is 99.9 wt.%. Graphite powder with a purity of 99 wt.% was selected as the carbon source. Silicon and carbon were well mixed and loaded into a graphite crucible, which was then placed in the middle of the furnace coil. The furnace power was supplied by an induction generator. The chamber was evacuated by a mechanical and a turbo pump to pressures of 7–70 Torr, and then high-purity argon gas was introduced into the chamber. Before heating, we evacuated the chamber and introduced argon gas twice to decrease the residual oxygen in the chamber. After creating vacuum, we set the power of the induction generator and rapidly increased the temperature of the crucible to 1500 °C. We maintained these conditions for 2 h to let the reaction reach completion. The phase compositions of the products were characterized by powder X-ray diffraction. Product morphology was examined by field emission scanning electron microscopy (FESEM). Glow discharge mass spectroscopy (GDMS) was also performed to evaluate the impurity concentration of the synthesized silicon carbide powder.

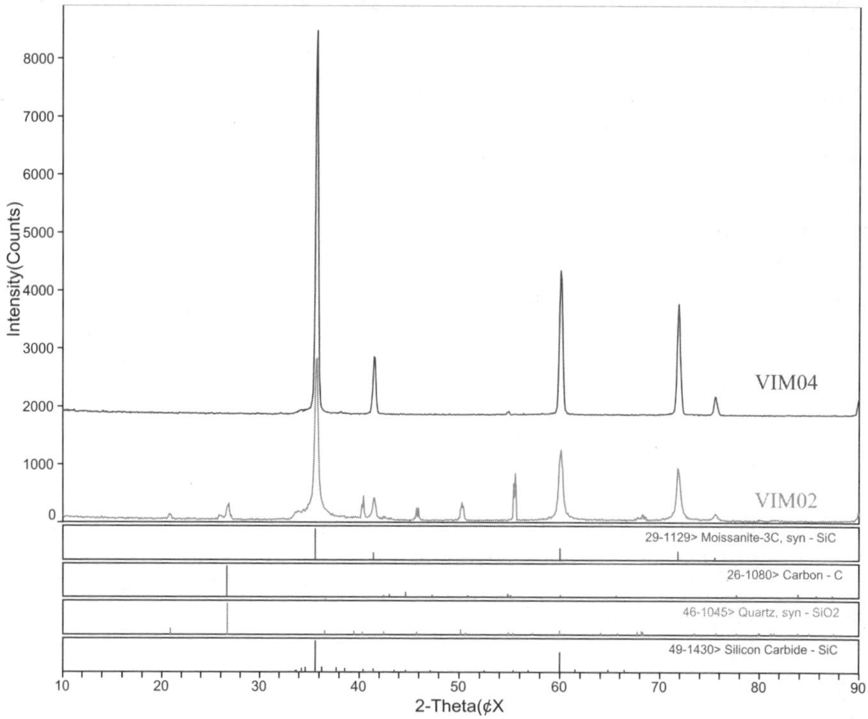

Fig. 1. X-ray diffraction patterns of purified silicon carbide powders.

Fig. 2. SEM images of purified silicon carbide powders: (left) VIM02 and (right) VIM04.

Table 1 Composition of impurities in VIM04 silicon carbide powder by GDMS (ppmw).

	Fe	Al	Cu	Ca	Mg	B	P
Before purification	69.3	163.3	53.3	136.8	24.4	3.9	9.2
After purification	8.8	33.3	2.4	0.7	0.1	3.1	1.5

Results and Discussion

In order to investigate whether the reaction reached completion, i.e., in order to investigate whether any free carbon or silicon remained, powder X-ray diffraction patterns were collected (Fig. 1). In the VIM02 experiment, the background pressure of the reaction was maintained at 70 Torr, and in the VIM04 experiment, the pressure was kept at 7 Torr. Free carbon and silica were found in the pattern of VIM02, which means that the formation of silicon carbide was not complete. No remnant precursor phases were detected in the products of VIM04, and the strongest diffraction peak near $2\theta = 36^o$, which corresponds to 3C-SiC (111), is very sharp. The reaction in VIM04 apparently approached completion, and the crystallinity of the powder was high. Figure 2 shows the SEM images of the silicon carbide powders in experiments VIM02 and VIM04. The particle size was between 1 and 30 μm; however, the particle shapes differed. The surface of the powder in VIM02 is coarse; in contrast, the surface of the powder in VIM04 is well crystalline. Apparently, the lower operation pressures facilitated the formation of silicon carbide at lower temperatures.

The synthesized silicon carbide powder was baked at 950°C for 2 h in air to remove the carbon. Most impurities, especially metals, were removed by leaching with an acid solution of hydrochloric and hydrofluoric acid. After the purification process, the impurity content of the powders was examined by GDMS, and the results are listed in Table 1. Before purification, the purity of the powder was about 3N1; the main impurities were iron, aluminum, copper, calcium, and magnesium. The impurity content clearly decreased after purification, and the purity increased up to 4N5. Boron and phosphorus are the key impurities in silicon carbide that strongly affect the electrical properties of crystals [5-6]. In the present process, no effort was made to remove these impurities.

Because of the low melting point and high evaporation pressure of most chlorides, NaCl was used to synthesize chlorides of the impurities in the precursor. Impurities were expected to be removed by evaporation as the silicon carbide formed. The synthesized powders (VIM05) were also purified by baking and acid leaching. Figure 3 shows a macroscopic view and an SEM image of VIM05. The color of the powder is green, and the crystallinity was slightly worse than that of VIM04. The impurity content is listed in Table 2. The content of most metals was similar except aluminum, and the structure of aluminum in silicon carbide powders needs to be further investigated. It is noteworthy that the content of boron decreased sharply, probably because of the formation of boron chloride during evaporation.

Fig. 3. VIM05: Appearance (left) and SEM image (right) of purified silicon carbide powders.

Table 2. Composition of impurities in VIM05 silicon carbide powder by GDMS (ppmw)

	Fe	Al	Cu	Ca	Mg	B	P
VIM05	4.5	58	1.3	1.4	2.6	0.04	0.72

Summary

The formation of silicon carbide powder by carbon and silicon sources strongly depends on thermodynamic parameters such as temperature, pressure, and gas atmosphere. Single-phase silicon carbide can be synthesized at low temperature and pressure conditions. Kinetics and diffusion between the silicon agents in the carbon powders are the critical mechanisms, which need to be further investigated. The impurities in the silicon carbide powders can be removed by baking, acid leaching, and chloride reaction. Finally, the purity of the silicon carbide powders in our work probably meets the purity requirements for growing single-crystal silicon carbide.

References

[1] G.W. Meng, Z. Cui, L.D. Zhang, and F. Phillipp, J. Cryst. Growth 209 (2000) 801–806.

[2] Z.J. Li, H.J. Li, X.L. Chen, A.L. Meng, K.Z. Li, Y.P. Xu, and L. Dai, Appl. Phys. A 76 (2003)637–640.

[3] P.J.Wellmann, Z.Herro, T.L.Straubinger, and A.Winnacker: Mater. Sci. Forum, 91 (2002) 389-393.

[4] Sheng-Cheng Chiu, Chao-Wei Huang, and Yuan-Yao Li, Synthesis of High-Purity Silicon Carbide Nanowires by a Catalyst-Free Arc-Discharge Method, J. Phys. Chem. C 111 (2007) 10294-10297.

[5] S.A. Reshanov, I.I. Parfenova, and V.P. Rastegaev, Group III -V Impurities in SiC: Lattice Distortions and Solubility, Diamond and Related Materials 10 (2001) 1278-1282

[6] Peter T.B. Shaffer, Solubility of Boron in Alpha Silicon Carbide, Mater. Res. Bull. Vol. 5 (1970) 519-522.

Materials Science Forum Vols. 717-720 (2012) pp 41-44
© (2012) Trans Tech Publications, Switzerland
doi:10.4028/www.scientific.net/MSF.717-720.41

Bulk and surface effects on the polytype stability in SiC crystals

Frédéric Mercier[1,a], Shin-ichi Nishizawa[1,b]

National Institute of Advanced Industrial Science and Technology, 1-1-1 Umezono, Tsukuba, Ibaraki 305-8568, Japan

[a]frederic.mercier@aist.go.jp, [b]s.nishizawa@aist.go.jp

Keywords: Crystal Growth, Polytype Stability, Bulk, Surface, DFT

Abstract. We investigated with ab initio calculations the 3C-, 6H-, 4H- and 2H-SiC polytypes. We discuss the geometry and the energetics of bulk and surface relaxed structures ((0001) Si face and the $(000\bar{1})$ C face surfaces). The polytype stability is discussed regarding the bulk and surface effects.

Introduction

Despite its fascinating fundamental aspect, the existence of polytypism in silicon carbide crystals is still an issue for crystal growers. A lot of work has been devoted to control the polytype, mainly in the sublimation crystal growth process. Among the various process parameters affecting the polytype stability, the nature of the seed, i.e., its polarity and its polytype, is known to affect the bulk crystal polytype [1, 2]. Despite all the experimental evidences, the role of the seed in the polytype stabilization remains unclear.

We performed first-principle calculations in the scheme of the density functional theory (DFT) for 3C-, 6H-, 4H-, and 2H-SiC polytypes. We used the program CASTEP and its implementation of the PW91 generalized gradient approximation (GGA) functional together with ultrasoft pseudopotentials. All the bulk polytypes were simulated in the same hexagonal supercell, having a C_{3v} symmetry and containing 12 SiC bilayers. By this way, the Brillouin zone for the different lattices is sampled with equivalent k-points ($5\times5\times1$). Then, surface calculations were performed using the equilibrium lattice parameters calculated from the respective bulk phases. We used (1×1) supercells containing seven SiC bilayers and a vacuum region of thickness 10 Å. Only the four topmost bilayers of the slab are allowed to relax. Atoms of the bottom surface are covered with H atoms. A Monkhorst-Pack mesh of $9\times9\times1$ was chosen for all structures.

Bulk Calculations

Lattice parameters. The optimized parameters of the relaxed structures are given in Table 1. The calculated values for the a parameter are slightly larger than the experimental values. Also, a decreases and c increases with the bulk hexagonality. As a result, the value c/na differs from the ideal value ($\sqrt{2/3}$) and the difference increases linearly with the hexagonality. This tendancy follows the experimental results [1].

Bulk energy. For each polytype of Table 1, we have calculated the total energy of the relaxed structures. Table 2 gives the bulk energy difference ΔE relative to the 3C structure. Our results reveal that 6H-SiC has the lowest bulk energy. The 2H polytype is clearly less stable than the other structures. However, the energy difference between 3C, 6H and 4H is extremely low, in the range of the uncertainty of the calculation method. Our findings are in agreement with the previous theoretical studies [5, 6, 7, 8, 9] which also reveal a small energy difference between the 3C, 6H and 4H polytypes. Some studies show that 6H is the stable polytype [5, 8], while other studies suggested 4H [6, 7, 9] as the most stable structure. This discrepency in the results shows that the polytype stability cannot be explained only by considering the bulk energies. The surface effects must play a crucial role in the polytype stabilization. For this reason, we investigated them and the results are presented in the following section.

Table 1: Lattice parameters of 3C- and nH-SiC (n=2,4,6). The numbers in parenthesis refers to experimental values [3, 4].

Polytype	Bulk Hexagonality (%)	a (Å)	c/(na)
3C	0	3,086 (3.083)	0,8164 (0.8165)
6H	33	3,084 (3.081)	0,8176 (0.8178)
4H	50	3,083 (3.080)	0,8182 (0.8184)
2H	100	3,081 (3.076)	0,8202 (0.8205)

Table 2: Bulk energy difference ΔE relative to the 3C structure in meV/atom. Comparison with other calculations.

Ref.	3C	6H	4H	2H
This work	0	-0.7	-0.6	2.7
[5]	0	-0.6	-0.4	4.3
[6]	0	-1.5	-1.9	0.9
[7]	0	-1	-1.2	3
[8]	0	-0.6	-0.5	3
[9]	0	0.2	0	5.7

Surface Calculations

Surface relaxation. Due to the existence of dangling bonds on the top surface, the bond condition of top-layer atoms is changed. After optimization, only relaxations perpendicular to the surface are observed, which is consistent with previous theoretical studies [10, 11]. Figure 1 gives the evolution of the surface bond angle. Si-terminated surfaces tends to stay in the sp^3 configuration (ω is close to 109.5, the ideal tetrahedral angle). On the contrary, the angle of the C-terminated surfaces (about 115.9) reveals a geometry close to the planar sp^2 configuration. We also notice a change of the relaxed bond angle with the surface hexagonality. The surface hexagonality defines the hexagonal character of the surface and characterizes the stacking sequences of the different SiC surfaces [11]. For the Si-terminated surfaces, the k-like surfaces (surface hexagonality <0.5) show a larger relaxation than the h-like surfaces. The higher ``rigidity" of h-like surfaces has also been proved by Wang et al. although they observed a smaller relaxation [11]. For the C-terminated surfaces, the h-like surfaces present a larger relaxation than the k-like surfaces. These results show that both the surface atoms (Si- or C-terminated) and the surface stacking structure influence the surface relaxation. As a result, macroscopic features such as the surface energy are affected by the nature of the surface [12].

Stacking selectivity on the (0001) and the (000$\bar{1}$) growth directions. For all the surface structures, we added a new bilayer at the surface of the slab. The additional bilayer can have either a cubic orientation (staggered configuration) or an hexagonal orientation (eclipsed) with respect to the bilayer underneath. For more details on the computation method, we refer the reader to our previous work [12]. Then, from the total energy of the new slab having either a cubic or an hexagonal additional bilayer, we detect if one orientation is preferred. Table 3 gives the results of such calculations. We see that the cubic stacking is preferred on Si-terminated surfaces, while the hexagonal stacking is stable on C-terminated surfaces. The energy differences between the two possible choices is one to two orders

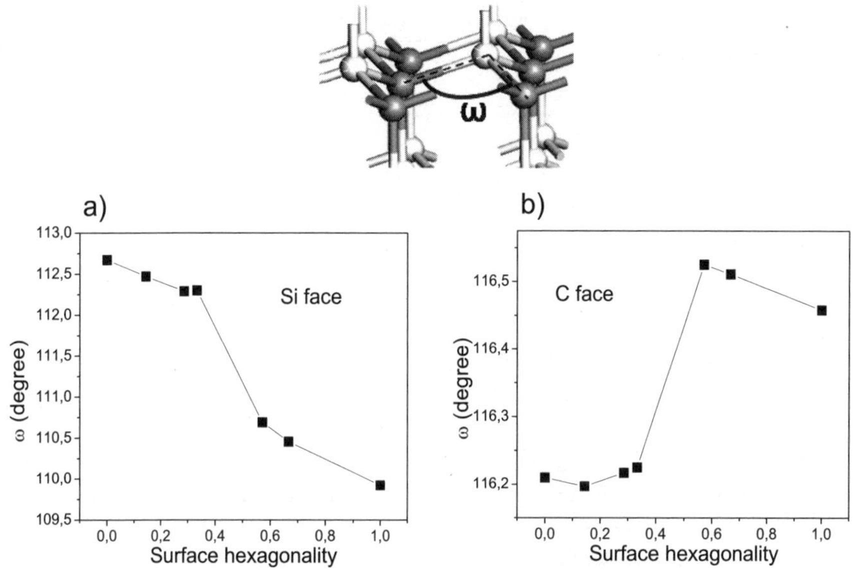

Fig. 1: Relaxed bond angle versus surface hexagonality a) Si-terminated surfaces. b) C-terminated surfaces.

of magnitude larger than those between the cubic and hexagonal bulk polytypes (few units of meV per Si-C pair, see Table 2). Thus, these results suggest that the polytype selectivity is strongly enhanced in the vicinity of the surface. Such energy differences lead to an energy cost of several eV for a full layer of atoms which is significant at growth temperature. Similar conclusions were obtained by Righi et al. on $\sqrt{3} \times \sqrt{3}$ reconstructed (0001) Si-terminated surfaces [13]. Finally, our calculations reveal that the favored surface stacking orientation (h or k) is in agreement with the natural continuity of the bulk stacking sequence. As an example, for the S1-type surfaces the last bilayer has a hexagonal orientation (see Table 3, 6H-S1 and 4H-S1) and the additional stacking must be in a cubic orientation to respect the bulk sequence. For example on (0001) Si surfaces, the probability to have a cubic stacking on 4H-S1 (E_h-E_k=111 meV) is higher than on 4H-S2 (E_h-E_k=88 meV).

Although we consider only one additional bilayer, we can discuss some polytype stability aspects based on this surface selectivity. If we consider a layer-by-layer growth mechanism, which typically occurs on on-axis surfaces, the surface hexagonality and the bulk hexagonality will vary during the growth due to the aforementioned selectivity. On Si-terminated surfaces, the surface hexagonality will tend to decrease during the growth and repeatedly, low hexagonality polytypes such as 6H-SiC or 3C-SiC will be favored. On the contrary, the hexagonal stacking is favored on the C-terminated surfaces. The surface hexagonality and the bulk hexagonality will tend to increase during the growth. As a result, high hexagonality polytypes like 4H-SiC will be favored. These findings are consistent with the experimental facts. In the sublimation process, the use of (0001) Si-terminated substrates (6H or 4H seeds) mainly leads to the occurrence of 6H-SiC polytype while growth of 4H-SiC occurred only on C face seeds [2].

Table 3: Stability of the additional bilayer stacking for all the surface structures investigated. The k and h subscripts denote respectively the cubic and hexagonal orientation of the additional stacking. For the surface sequence, $(k_2k_1h)_\infty$ corresponds to a k_2 site at the surface of the crystal.

Polytype	Surface sequence	Surface hexagonality	E_h-E_k (meV/pair)	
			Si face	C face
3C	$(k)_\infty$	0	88	-57
6H(S3)	$(k_2k_1h)_\infty$	1/7	87	-56
6H(S2)	$(k_1hk_2)_\infty$	2/7	89	-54
4H(S2)	$(kh)_\infty$	1/3	88	-55
6H(S1)	$(hk_2k_1)_\infty$	4/7	113	-44
4H(S1)	$(hk)_\infty$	2/3	111	-45
2H	$(h)_\infty$	1	112	-46

Summary

We have theoretically investigated the SiC polytypes using ab initio pseudopotential calculations in the 3C, 6H, 4H, and 2H structures. We have discussed some aspects of the bulk and the surface effects on the polytype stability. Calculations have revealed the importance of the nature of the surface on the polytype stabilization.

References

[1] Y. Tairov and V. Tsvetkov: Progr. Cryst. Growth and Charact. 7 (1983), p.111

[2] R. A. Stein and P. Lanig: J. Cryst. Growth 131 (1993), p.71

[3] R.F. Adamsky, K.M. Merz: Z. Krist. 111 (1959), p.350

[4] A. Bauer, J. Krausslich, L. Dressler, P. Kuschnerus, J. Wolf, K. Goetz, P. Käckell, J. Fürthmuller and F. Bechstedt: Phys. Rev. B 57 (1998), p.2647

[5] C. Cheng, R. Needs and V. Heine: J. Phys. C: Solid State Phys. 21 (1988), p.1049

[6] P. Käckell, B. Wenzien, and F. Bechstedt: Phys. Rev. B 50 (1994), p.17037

[7] M.J. Rutter and V. Heine: J. Phys.: Condens. Matter 9 (1997), p.8213

[8] C. Raffy, L. Magaud, E. Blanquet, M. Pons and A. Pasturel: Mater. Sci. Forum Vols.353-356 (2001), p.111

[9] T. Ito, T. Kondo, T. Akiyama and K. Nakamura: Phys. Status Solidi C 8 (2011), p.58

[10] M. Sabisch, P. Kruger and J. Pollmann: Phys. Rev. B 55 (1997), p.10561

[11] J. Wang, L. Zhang, Q. Zeng, G. Vignoles and L. Cheng: J. Phys.: Condens. Matter 22 (2010), p.265003.

[12] F. Mercier and S. Nishizawa: J. Cryst. Growth (2011) http://dx.doi.org/10.1016/j.jcrysgro.2011.11.052

[13] M. Righi, C. Pignedoli, G. Borghi, R. Di Felice, C. Bertoni and A. Catellani: Phys. Rev. B 66 (2002), p.045320

Materials Science Forum Vols. 717-720 (2012) pp 45-48
© (2012) Trans Tech Publications, Switzerland
doi:10.4028/www.scientific.net/MSF.717-720.45

Crystallinity evaluation of 4H-SiC single crystal grown by solution growth technique using Si-Ti-C solution

K. Kamei[1,a], K. Kusunoki[1], N. Yashiro[1], N. Okada[1], K.Moriguchi[1], H. Daikoku[2], M. Kado[2], H. Suzuki[2], H. Sakamoto[2] and T. Bessho[2]

[1]Corporate Research & Development Laboratories,Sumitomo Metal Industries Ltd.

Fuso-cho 1-8, Amagasaki, 660-0891 Japan

[2]Higashifuji Technical Center, Toyota Motor Corporation

1200 Mishuku, Susono. Shizuoka, 410-1193, Japan

[a]kamei-kzh@sumitomometals.co.jp

Keywords: Solution growth, 4H-SiC, Ti , EPD, Dislocation, TEM, AFM

Abstract. Crystallinity of 4H-SiC bulk crystals obtained by solution growth technique was characterized mainly by KOH etching of the off-ground and serially ground specimens. Marked reduction of basal plane dislocations, threading edge and screw dislocations during the growth of the on-axis crystal was confirmed. Cross-sectional TEM observation revealed the rapid reduction behavior of threading dislocations microscopically. AFM observation of as-grown morphology showed that screw dislocation dipoles are related to the reduction of threading screw dislocations and single domain formation, which is essential for establishing the high crystallinity.

Introduction

We have long been working on bulk SiC growth from solutions consisting of Si and metallic solvents such as Ti and Cr [1,2]. Through the optimization of growth process conditions, growth of large diameter and long boules are now about to be achieved.

One of the most important advantages of the solution growth is that it gives high crystallinity since it takes place under small deviation from the equilibrium state. We reported good crystallinity of a 6H-SiC bulk crystal [3] and a thin liquid phase epitaxial (LPE) layer of 4H-SiC [4]. However, systematic evaluation of bulky 4H-SiC has not been addressed yet. Thus, in this study we show the etch pit density (EPD) variation during the bulk crystal growth of 4H-SiC. We also characterize the behavior of the basal plane dislocations (BPD), threading edge dislocations (TED) and threading screw dislocations (TSD) by SEM, TEM and AFM in order to discuss their propagation and annihilation mechanisms in the solution growth process.

Experimental Procedure

Growth experiments were carried out using a top seeded solution growth (TSSG) furnace [1], in which a Si-Ti melt was prepared by inductive heating. Seed crystals were on-axis PVT 4H-SiC wafers, for which the typical EPD was in the range of 10^4-10^5cm^{-2}. Fig.1 shows a bulk 4H crystal obtained after 10hr growth at a temperature around 2000℃ using the C-face of the seed.

Fig.1 2” bulk 4H-SiC crystal obtained by solution growth using Si-Ti solvent.

The BPD distribution along the growth direction was examined using a single specimen. As-grown crystals were off-ground at 3.5°, finished by mirror polishing, and etched by the conventional KOH etching technique. The EPD were evaluated from the Si-face since the C-face shows strong durability against KOH etching.

We also conducted another EPD measurement by serial grinding and KOH etching of a Si face on-axis crystal to understand behavior of TED and TSD.

Aiming to observe the propagation behavior of threading dislocations, we have done a cross-sectional TEM (XTEM) observation in the vicinity of an inclusion where dislocations were surely introduced. In addition, we observed surface morphologies by confocal optical microscope, SEM with a low acceleration voltage (5kV) and AFM to reveal the relationship between the growth mechanism and dislocations.

Thin (0.2mm) layers were grown successively and we observed the changes of the overall surface morphology to clarify the macroscopic morphological evolution, focusing on the formation of single domain structure.

Results and Discussion

Basal Plane Dislocation

Fig.2 shows typical variation of BPD density in the off-ground specimen as a function of the distance from the grown-layer/seed interface. The BPD exhibited a marked decrease in the early stage of the growth and almost disappeared at the growth thickness larger than 400μm. Although the BPD density seems to decrease even in the substrate, it is due to the ambiguity of determining the interface between the grown layer and the substrate arising from the inhomogeneous back melting of the substrate at the initial stage of the growth. Fig.3 is an image of a BPD in the vicinity of the grown-layer/seed interface. This image is a plan-view dark field TEM image of a specimen obtained by ion thinning from the other side of the etched surface. We can clearly see the rectilinear morphology of BPD terminated in the pit. As BPDs are lying in basal plane, this image may indicate that the rapid reduction of BPDs is due to their escaping from the side end face of on-axis crystal. Thus, on-axis growth prevents BPDs in seed crystal from propagating toward the grown layer.

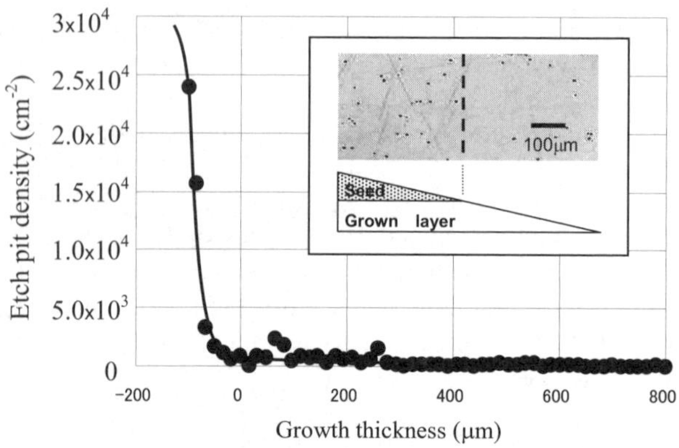

Fig.2 Variations in BPD density as a function of distance from grown layer/seed interface. The inset shows the appearance and geometry of the examined off-ground specimen.

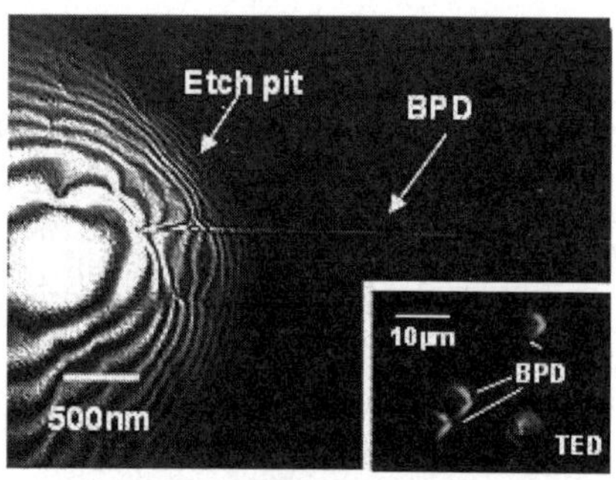

Fig.3 Dark field image of a BPD and its etch pit trace using 10-12 reflection showing rectilinear morphology of a BPD lying in basal plane with typical macroscopic images of BPD and TED etch pits (inset).

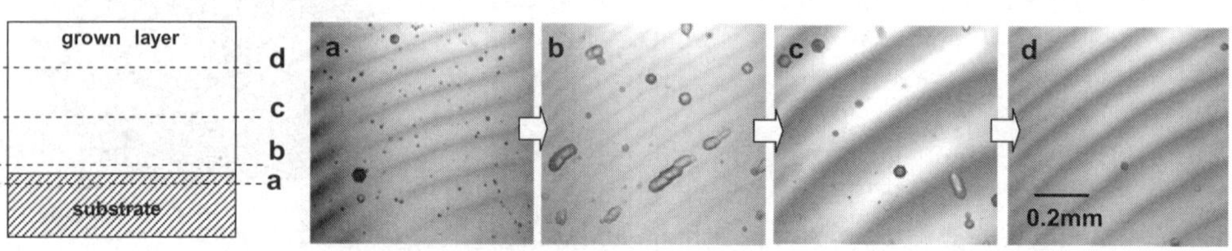

Fig.4 Changes in EPD density and their morphology with the growth thickness of on-axis crystal showing significant improvement in crystallinity.

Threading Edge and Screw Dislocation

Fig.4 shows the changes in etch pit distribution and morphology with growth thickness. We can see the behavior of TED and TSD only, as all figures were obtained for the same area serially thinned parallel to the basal plane of the on-axis crystal. Significant improvement of the crystallinity is clearly seen by the prominent reduction of etch pits.

Fig.5 shows the variations of TED and TSD density as a function of growth thickness. Four strip regions of 0.8mm x 10mm were examined. Both TED and TSD decrease gradually as the crystal grows. TED density decreased drastically to one tenth of that of the seed crystal, while the TSD density dropped steadily to less than half of that of the seed.

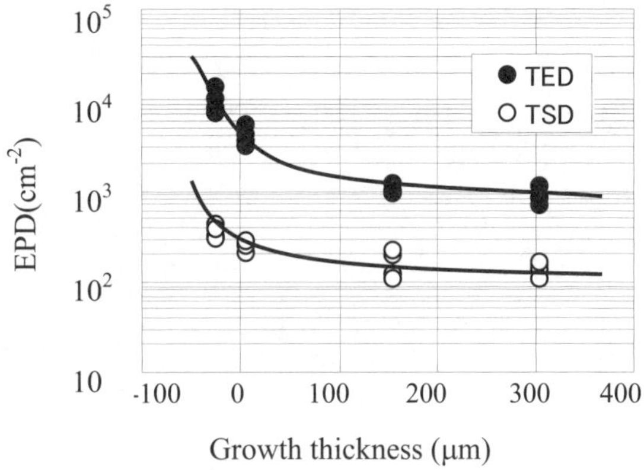

Fig.5 Variations in TED and TSD density as a function of growth thickness, showing their gradual decrease.

Microscopic Behavior of Dislocations

Fig.6a shows the cross-sectional optical image of an area of the grown layer with some inclusions and Fig.6b, c shows the XTEM images of the regions depicted in (a) in the vicinity of the inclusion. The high-density of dislocations initiated around the inclusion (Fig.6b) disappeared in the region less than 1mm above the inclusion (Fig.6c). This behavior indicates that even the newly introduced dislocations disappear quickly with the growth.

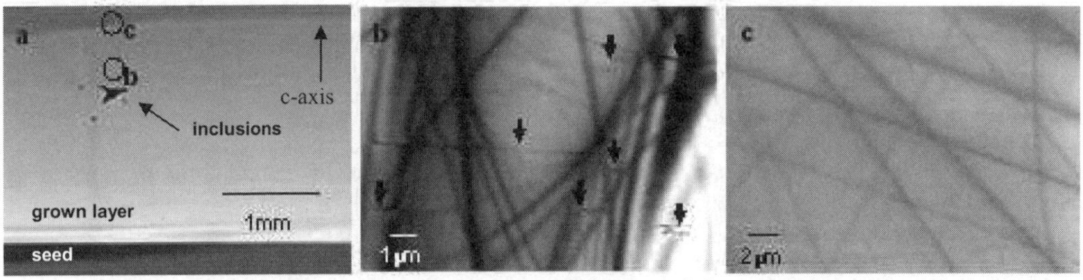

Fig. 6 a: Cross-sectional image showing an area of the growth layer with inclusions. b, c: XTEM images for regions indicated in (a), arrows in (b) indicate introduced dislocations.

Annihilation of TSDs and Single Domain Formation

TSDs are very important defects since they act as a source of spiral growth. They distribute nonuniformly in the grown layer, propagating from the seed crystal with much lower density than BPD and TED. Through observation of many TSD growth centers by AFM and SEM, we have chanced to see a TSD dipole in the as-grown 6H-SiC surface as shown in Fig.7a. This is an interesting image, which may show a possible dislocation

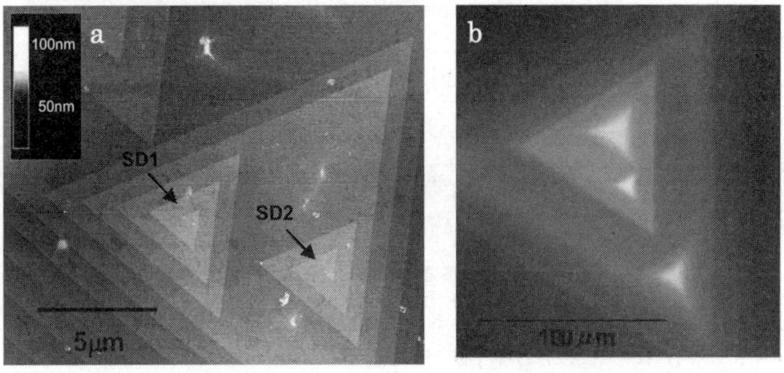

Fig.7 (a) AFM image showing TSD dipole with opposite Burger's vectors at the 6H-SiC as-grown surface. (b) Similar TSD pairs at 4H-SiC as-grown surface.

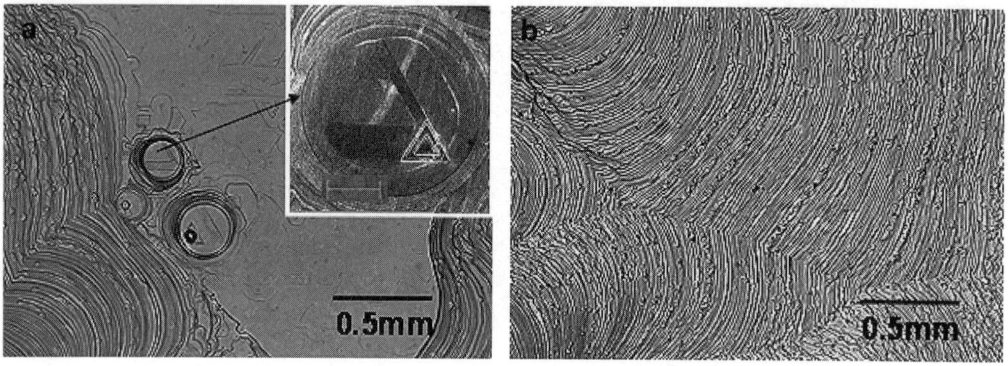

Fig.8 a: Optical image of a huge plateau region with spiral growth center of 0.2mm thick grown layer. The inset is a SEM image of the spiral growth center indicating conjugated SD centers, b: Optical image of the same area after successive 0.2mm growth depicting the annihilation of the spiral growth center.

annihilation process. In this figure, two screw dislocations with opposite Burger's vector (SD1, 2) are about to coalesce to annihilate. As both of them give growth centers, the growth will stall in this region. The same type of TSD pairs can be occasionally seen in 4H-SiC crystals as shown in Fig.7b.

To clarify the role of this TSD dipole on the growth morphology, we have conducted two-stage TSSG growth with repetitive precise observation of as-grown surface morphologies. Fig.8a shows the as-grown surface after 5hr growth with a growth thickness of 0.2mm and we can see the conjugated TSD centers in a huge plateau. Fig.8b shows the same region as Fig.8a after successive 5hr growth with a growth thickness of 0.4mm. It should be noted that the plateau in Fig.8a disappeared totally and the macrosteps initiated at the other TSD center cover the area. Although it is well known that SDs tend to bend and disappear by the step-flow covering, this morphological evolution may indicate that the conjugated TSD centers were a SD dipole and they coalesced to annihilate, resulting in the disappearance of the huge plateau structure. By this annihilation mechanism, hillocks reduce their number and several huge growth centers may prevail and control the solution growth. These phenomena may enhance the formation of the single domain morphology and improve the crystallinity of the grown SiC.

Summary

We have characterized the density variation and the morphology of dislocations in solution grown on-axis 4H-SiC bulk crystal and revealed the marked decrease of BPD and gradual reduction of TED and TSD. XTEM observation of the intentionally introduced threading dislocations showed their rapid decrease during the solution growth. TSD dislocations acting as growth centers with opposite Burgers vectors tend to coalesce to annihilate, resulting in the single domain formation, which is essential for establishing the high crystallinity.

References

[1] K. Kamei, K. Kusunoki, N. Yashiro, N. Okada, T. Tanaka and A. Yauchi: Journal of Crystal Growth Vol. 311 (2009), pp. 855-858

[2] K. Danno, et al., Materials Science Forum Vols 645-648 (2010) pp.13-16

[3] N. Yashiro, K. Kusunoki, K. Kamei and A. Yauchi: Materials Science Forum Vols. 645-648 (2010), pp. 33-36

[4] Ryo Hattori, Kazuhiko Kusunoki, Nobuyoshi Yashiro and Kazuhito Kamei: Materials Science Forum Vols. 600-603 (2009), pp.179-182

Materials Science Forum Vols. 717-720 (2012) pp 49-52

doi:10.4028/www.scientific.net/MSF.717-720.49

SiC Growth by Solvent-Laser Heated Floating Zone

Andrew A. Woodworth[1,a,*], Philip G. Neudeck[2,b], Ali Sayir [1,c], David J. Spry[2,d] Andrew J. Trunek[3,e] and J. Anthony Powell[4,f]

[1]NASA Glenn Research Center, 21000 Brookpark Rd., MS 106-1 Cleveland, OH 44135 USA

[2]NASA Glenn Research Center, 21000 Brookpark Rd., MS 77-1 Cleveland, OH 44135 USA

[3]OAI, NASA Glenn Research Center, 21000 Brookpark Rd., MS 77-1 Cleveland, OH 44135 USA

[4]Sest Inc., NASA Glenn Research Center, 21000 Brookpark Rd., MS 77-1 Cleveland, OH 44135 USA

[a]andrew.a.woodworth@nasa.gov, [b]neudeck@nasa.gov, [c]ali.sayir-1@nasa.gov, [d]david.j.spry@nasa.gov, [e]andrew.j.trunek@nasa.gov, [f]j.a.powell@nasa.gov,
*NASA Postdoctoral Program Fellow and corresponding author

Keywords: laser heated floating zone, LHFZ, fiber, solvent, growth

Abstract: In an effort to grow single crystal SiC fibers for seed crystals the following two growth methods have been coupled in this work: traveling solvent and laser heated floating zone to create the solvent-laser heated floating zone (Solvent-LHFZ) crystal growth method. This paper discusses the results of these initial experiments, which includes: source material, laser heating, and analysis of the first ever Solvent-LHFZ SiC crystals (synchrotron white beam x-ray topography confirmed).

Background

Lateral growth (perpendicular to <0001>) of silicon carbide (SiC) can produce high quality bulk material (defect density <100/cm^2)[1], but obtaining high quality seed crystals for such a growth is still problematic. A possible solution is to isolate a single threading screw dislocation and use it as a source of spiral growth steps to carry out growth on its (0001) surface elongating the crystal vertically in the <0001> direction until it forms a long hexagonal fiber [2]. In order to grow single crystal SiC fibers (seed crystals) a new technique has been created, solvent-laser heated floating zone (solvent-LHFZ), by coupling a solvent system capable of growing SiC [3] with the potential to grow long fibers by laser heated floating zone [4].

Experimental

To investigate this new growth method, a custom system has been constructed (Figure 1) consisting of a CO_2 laser (λ=10.6 μm) and associated optics, vacuum chamber, sample translation system, optical pyrometer and video capture / data collection / experimental control system. Not shown in Figure 1 is the vacuum/gas management system, which is designed to pull a vacuum of $5x10^{-5}$ Torr or better before a growth. During growth ultra high purity Ar gas is brought into the chambers from the top and around the windows and out through the bottom of the chamber in order to keep sensitive areas clean. The pressure in the chamber for these initial experiments is 115 Torr with total Ar flow about 15 l/m.

Figure 1 a) shows the orientation of the laser beams, heated zone and the primary *in-situ* data collection tools. Figure 1 b) shows the orientation of the beams to the seed crystal, heated zone and feed rod. The laser beams enter the chamber on a ~ 5° down angle so they cannot exit the chamber through the opposing window. Before the beam enters the chamber, optics are used to focus the beam into an oval shape (long : short axis ~ 2:1) with the short axis parallel to the vertical axis of the chamber. The long axis is roughly twice the width of the feed rod in order to ensure even heating. The seed crystal is mounted vertically in the upper mount, so that the growth face is down, while the feed rod is mounted vertically in the lower mount. Both mounts can be moved independently of each other with rates of 50-18,000 μm/h (+/- 1 μm/h).

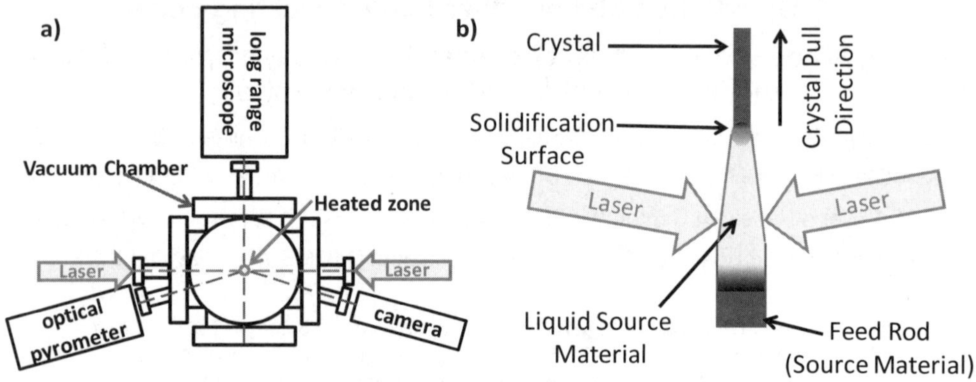

Figure 1. Solvent-LHFZ system. a) Top down view of the configureation of the laser beams and *in-situ* data collection instruments. Note: that all of the components are co-planer. b) Side view of the laser, liquid source material and solidification surface, seed crystal and feed rod.

Feed rods for these experiments consisted of Fe, Si and C (graphite) powders that were mixed by ball milling, formed into a rod by cold isostatic press and then sintered in hydrogen at 1150°C for 1 hour. Under these conditions no dissociation or melting of the rods was visible and density, independent of composition, increased by ~50%. Sintering above 1150°C produced melting or dissociation of the materials depending on composition.

SiC seed crystals used for this experiment are 0.5 mm x 0.5 mm x 15 mm diced from commercially purchased a-face $(11\bar{2}0)$ or m-face $(1\bar{1}00)$ 4H-SiC wafers. The long dimension of the seed is parallel to the c-axis <0001> making one end the C face $(000\bar{1})$, which is also the growth surface. The growth surface varied from on axis to roughly 10° from the $(000\bar{1})$. Seed crystals were cleaned using a H_2SO_4:H_2O_2 (1:1) bath, and then glued (SiC ceramic paste) into an alumina tube used to mount in the growth system. Next, the crystals were re-cleaned by HNO_3:HCl (2:1) dip and then an oxide layer was removed by HF dip (after acids crystals rinsed de-ionized water).

Growths were conducted by first melting the end of the feed rod (Figure 2a. Pre-contact), and then dipping the end of the seed crystal into the melt allowing the liquid to wet the crystal (Figure 2b. wetting). During the entire process the feed rod is continually fed into the heated area, at a rate which maintains approximately the same amount of melted material. The growth ends when the seed/grown crystal is pulled out of the melt. After growth some of the metal solvent remains on the crystal, which is removed using a HF:HNO_3:HCl (1:1:2) bath.

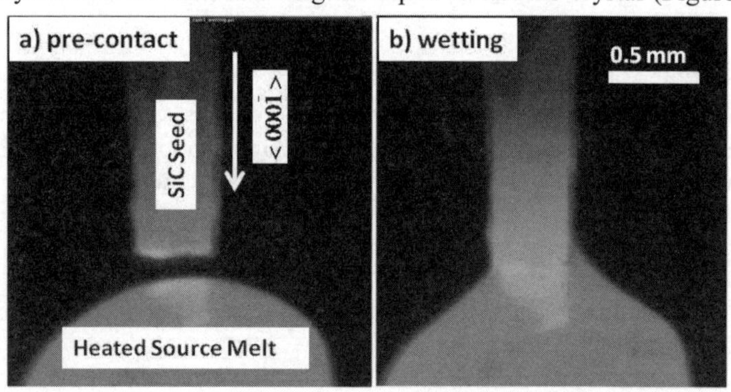

Figure 2. Solvent-LHFZ process. Precontact: Source material is liquid, and the seed approaches the melt. Wetting: Liquid source material wets to the seed crystal.

Results and Discussion

Three compositions were examined: Fe/Si ~ 1.9 with C = 8 at.%, Fe/Si ~ 0.35 with C = 8 at.% and Fe/Si ~ 0.35 with C = 16 at.%. These were designed to examine the compositions where SiC is known to co-exist with Si-Fe-C liquid [5]. Some results of these experiments are found in Table 1. There are two important notes to make about the data given in Table 1. First, due to the possible wide range of emissivity (ε_{Fe}=0.3 [6] to $\varepsilon_{graphite}$=0.9 [7]) and small size (~2 mm^2) heated zone, none of the listed pyrometer-

Table 1. Summary of Results (M.P.= temperature at which the feed rod formed a melt, at.% =atomic %). * Film/Feature growth rates (Figure 3.) Note: Temperatures are not corrected for emissivity.

Fe/Si (atomic ratio)	C (at.%)	M.P. (°C)	Growth Rates (μm/hour) +90 °C	+190 °C	+325 °C
High-Si (Fe/Si~0.35)	8	1170	4	40	135/780*
	16	1195	50	120	N/A
High-Fe (Fe/Si~1.9)	8	N/A	No Growth		

measured temperatures are corrected for emissivity. Second, the increase in growth rates more closely relates to the change in temperature above the melting point (M.P.) of each feed rod composition than change in total temperature. Therefore, results are referred to relative to the feed rod M.P., instead of absolute temperature. Growth times for these experiments were kept under 2 hours and the rate at which the crystal was pulled away from the melted zone was a constant 100 μm/h. This pull rate was chosen for two reasons: First, for these short growth times a maximum movement of 200 μm will only cause a minimal change in position of the growth front relative to the heated zone. Second, small upward motion ensures that the seed crystal is not pulled further into the heated zone by the considerable surface tension forces present in the melt.

Feed rods with high Fe (Fe/Si~1.9) concentration failed to melt easily. Only with the seed crystal in contact with the feed rod during the initial heating stages could a stable melt be formed, and even then the melted material only partially wetted to the seed crystal. Post-experiment analysis failed to show any growth, and in one case severe etching of the seed crystal occurred (>200 μm in less than 20 seconds). We surmise that a possible cause of this difficulty is caused by the high thermal conductivity of the Fe present in the feed rods.

Both high-Si (Fe/Si~0.35) concentration feed rod compositions had similar melting points and wetted, as shown in Figure 2. Table 1 data shows that both high-Si concentration feed rods also exhibited growth rates dependent on temperature and initial C concentration. This suggests two things: First, due to the dramatic increase in growth rate with added C, the growth system is currently C limited. Second, keeping in mind that growth is C limited, the increase in growth rate with temperature suggests that the amount of C transported from the feed rod to seed surface is temperature dependent.

The increase in growth rate with temperature is most apparent in a M.P. +325°C, 10 minute growth experiment. Scanning electron micrographs (SEM) of this growth experiment are shown in Figure 3. Most of the growth surface is covered in a step-flow like morphology 22.5 μm thick at a

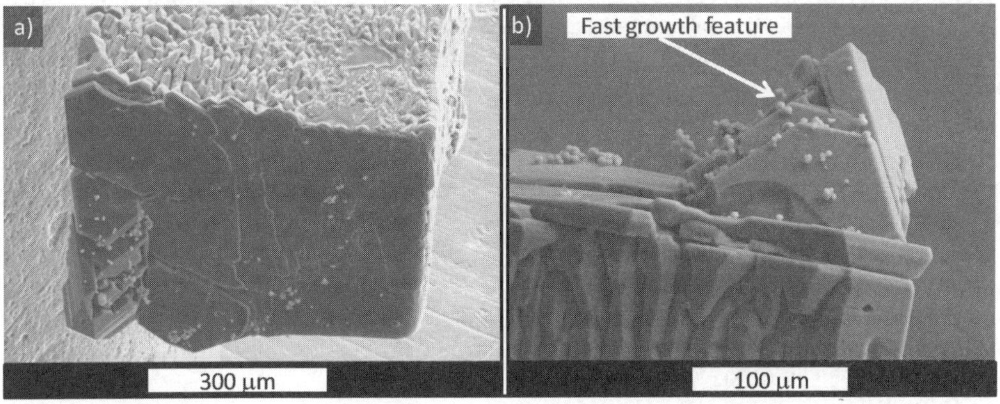

Figure 3. SEM images of the fastest growth (Fe/Si=0.35, C=8 at.%, M.P.+360°C, 10 minutes). a) View of the entire carbon-face $(000\bar{1})$ /growth front. b) edge view of growth surface.

rate of 133 μm/h (33x greater than growth observed at M.P. + 90°C). In one corner a "triangular" feature grew 130 μm tall at a rate of ~780 μm/hour. The polytype of this particular feature has not been ascertained, but several of the initial films have been examined by synchrotron white beam x-ray topography (SWBXT), which has confirmed the replication of the 4H polytype and epitaxial nature of the grown crystals [8].

Overall, two different morphologies were observed: "step flow" (Figure 3a) and "platelet" morphology (Figure 4). These two morphologies are not mutually exclusive to an experiment; both have been observed und a variety of conditions and also in a single growth. While the results suggests the type of growth front is dependent on the seed crystal and not growth conditions, more work remains before this result can be confirmed.

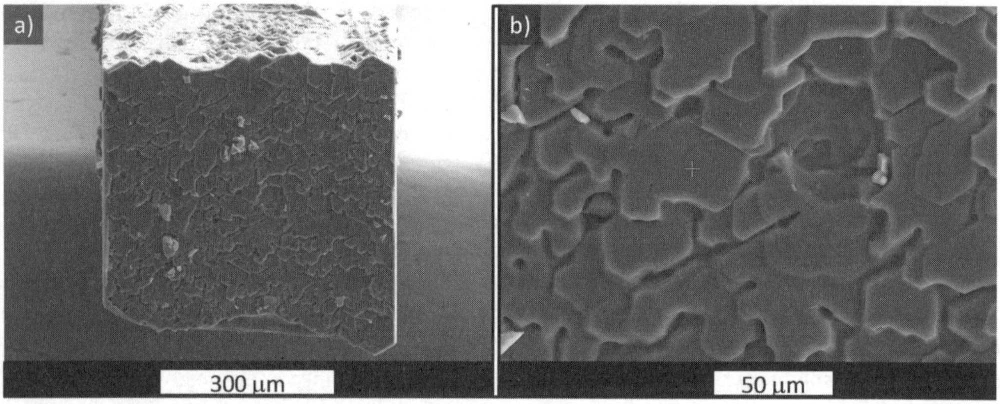

Figure 4: SEM micrograph C-face of a 8μm thick grown film (Fe/Si=0.5, C=8at.%, M.P.+90°C, 2 hours) a) view of the entire grown film. b) close up of the growth fronts.

Conclusion

These initial experimental results show that SiC readily grows under the conditions created by Solvent-LHFZ and that polytype replication from the seed crystal occurs. They also demonstrate the importance of promoting C transport from the feed rod, through the heated zone, to the seed crystal. While these results are encouraging a significant amount of work remains to optimize this new method of growth.

Acknowledgements

NASA Postdoctoral Program, NASA Vehicle Systems Safety Technologies Project in the Aviation Safety Program, US DOE Vehicle Tachnology Program via Space Act Agreement (SAA3-1048) (DOE IA # DE-EE0001093/001) monitored by Susan Rogers, SWBXT by M. Dudley and B. Raghothamachar. G. Ferro and D. Chaussende for the helpful conversation, Technical help from Fredric Dynys, Thomas Sabo, Mark Dieringer, Michelle Mrdonovich-Hill, Beth Osborn and Charles Blaha.

References

[1] D. Nakamura, et al., Nature 430 (2004) 1010
[2] J.A. Powell, et al., U.S. Patent 7,449,065 issued November 11, 2008
[3] J Wollweber, et al., Mat. Sci. Forum 338-342 (2000) 119
[4] C. A. Burrurs, et al., App. Phys. Lett. 26 (1975) 318-320
[5] T Yoshikawa, et al., Int. Conf. on Adv. Structure and Funct. Mater. Des. 2008, J. Phys.: Conf. Series 165 (2009) 012022
[6] H. B. Wahlin et al., Phys. Rev. 74 (1948) 687
[7] M.R. Null et al., J. Appl. Phys. 29 (1958) 1605
[8] M. Dudley and B. Raghothamachar, private communication (2011)

Materials Science Forum Vols. 717-720 (2012) pp 53-56
© (2012) Trans Tech Publications, Switzerland
doi:10.4028/www.scientific.net/MSF.717-720.53

Stable Growth of 4H-SiC Single Polytype by Controlling the Surface Morphology Using a Temperature Gradient in Solution Growth

Yuji Yamamoto[a], Kazuaki Seki[b], Shigeta Kozawa[c], Alexander[d], Shunta Harada[e], and Toru Ujihara[f]

Graduate School of Engineering, Nagoya University, Chikusaku, Nagoya 464-8603, JAPAN

[a]y-ymmt@mercury.numse.nagoya-u.ac.jp, [b]seki@mercury.numse.nagoya-u.ac.jp, [c]kozawa@mercury.numse.nagoya-u.ac.jp, [d]alex@mercury.numse.nagoya-u.ac.jp, [e]harada@numse.nagoya-u.ac.jp, [f]ujihara@numse.nagoya-u.ac.jp

Keywords: solution growth, 4H-SiC, polytype transition, morphology, temperature gradient

Abstract. We introduce a method to grow 4H-SiC single polytype stably by controlling the surface morphology. The polytype transition on on-axis 4H-SiC C-face was investigated from a viewpoint of surface morphology of grown layers. At the area where several hillock-like structures grew adjacently, the polytype transition from 4H-SiC to 6H-SiC or 15R-SiC often occurred. Therefore, we tried a vertical seeded solution growth (VSSG) method to suppress the formation of hillock-like structures. As a result, the hillock-like structure on the grown layer was dramatically reduced. Moreover, the ratio of 4H-SiC polytype to the whole grown surface was increased to be almost 100%.

1. Introduction

Silicon carbide (SiC) is a wide bandgap semiconductor material which provides a tremendeous potential for high power, high temperature, and high frequency devices [1, 2]. Solution method is known as a suitable technique for producing high quality crystals because the growth system is close to thermal equilibrium [3-6]. As is well known, SiC has many polytypes which have different electircal and physical properties [7]. The most common polytypes are 3C-, 4H-, 6H- and 15R-SiC. Among these polytypes, 4H-SiC is the most interesting one for power device applications [8]. It is important to control the growth of 4H-SiC without polytype transition which often occurs during the solution growth using a silicon solvent.

In this study, we performed the solution growth of 4H-SiC on on-axis 4H-SiC C-face and investigated the polytype transition from the viewpoint of surface morphology. To control the surface morphology, we proposed a vertical seeded solution growth (VSSG) method instead of a typical top seeded solution growth (TSSG) method. In the VSSG method, a plate-like seed crystal was vertically dipped into a solvent. As a result, we achieved the growth of 4H-SiC single polytype using the VSSG method.

2. Experimental

SiC single crystals were grown on on-axis 4H-SiC plate-like seed crystals (10 mm x 10 mm x 0.34 mm) by two methods: TSSG method and VSSG method (Fig. 1). A seed crystal was dipped into a high purity (11N) silicon solvent under the temperature gradient. The growth using TSSG method was performed at 1580 °C for 1 h under a high purity Ar gas flow at a growth rate of 7 μm / h. In the case of the VSSG method, the growth time was 2 h and the growth temperatures of the top and bottom parts of the seed crystal were 1680 °C and 1720 °C, respectively. Total growth rate of C-face and Si-face at top parts of the crystal was about 35 μm / h. After the growth, the residual Si solvent on the grown crystal was removed by the mixture of HNO_3 and HF.

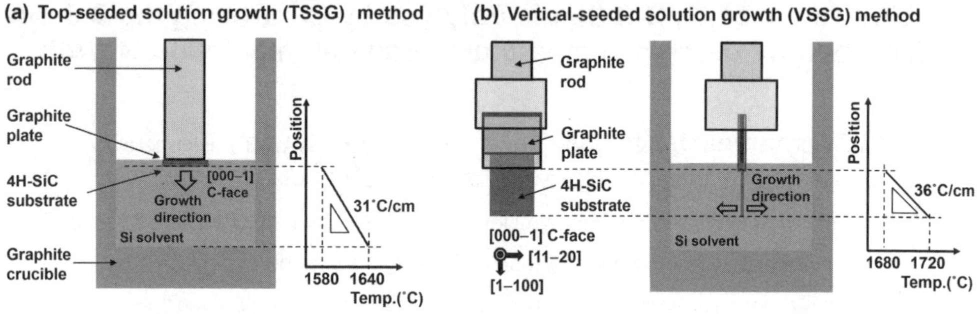

Fig. 1: Schematic illustrations of experimental setups for (a) TSSG method and (b) VSSG method. In the VSSG method, a plate-like seed crystal is vertically dipped into a silicon solvent. The temperatures of the upper side and the lower side of the seed crystal were different.

Surface morphology of the grown crystal was observed using a Nomarski microscope. Polytype identification of the grown layers were performed by micro-Raman spectroscopy (inVia Raman microscope, Renishaw) using an Ar laser beam (λ=532 nm) with a spot size of 10 μm [9]. For the investigation of the polytype spatial distribution of the grown crystal, micro-Raman mappings were carried out with 100 μm intervals.

3. Results and discussion

3.1 Solution growth using TSSG method

Fig. 2 (a) shows the Raman intensity mapping of 4H (776 cm^{-1}), 6H (789 cm^{-1}) and 15R (785 cm^{-1}) TO peaks of the grown crystal on C-face 4H-SiC using the TSSG method. The polytype transitions from 4H-SiC to 6H-SiC or 15R-SiC were observed at several areas. The polytype of grown surface consisted of 92% 4H, 4% 6H and 4% 15R. Fig. 2 (b)-(d) shows the magnified Nomarski images which are three typical morphological types: (1) a train of bunched step structures as shown in Fig. 2 (b); (2) an isolated hillock-like structure surrounded by the bunched steps (indicated by white arrow) as shown in Fig. 2 (c); (3) a cluster of hillock-like structures composed of several hillock-like structures as shown in Fig. 2 (d). The 4H-SiC grew stably at region (b) and (c), where the train of bunched step structures and the isolated hillock-like structure were observed. On the other hand, at region (d), one hillock-like structure (indicated by black arrow) showed polytype transition from 4H-SiC to 6H-SiC. Thus, the polytype transition rarely occurred at such a step train structure, while the polytype transition sometimes occurred at the hillock-like structure.

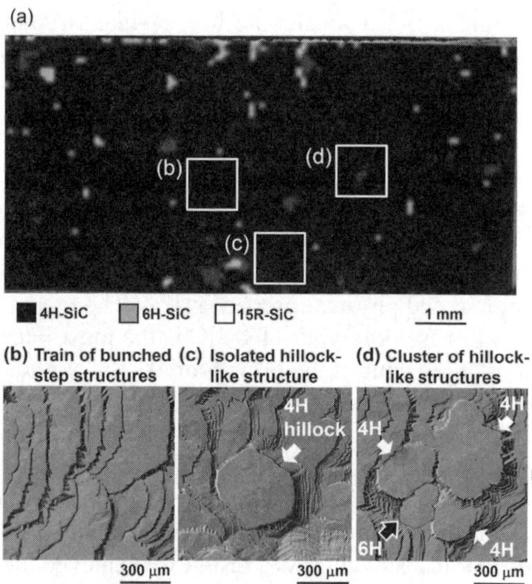

Fig. 2: (a) Raman intensity mapping of 4H, 6H and 15R TO peaks of the crystal grown by TSSG method. Magnified Nomarski images of regions (b)-(d), which are three typical morphological types: (b) a train of bunched step structures, (c) an isolated hillock-like structure, and (d) a cluster of hillock-like structures. One hillock-like structure (indicated by black arrow) showed polyte transition from 4H-SiC to 6H-SiC.

Fig. 3 shows the frequency of the numbers of neighboring hillock-like structures concerning each hillock-like structure. Each hillock was considered individually when counting neighbors. For example, the number of the neighboring hillocks concerning the 6H-SiC hillock indicated by black arrow in Fig. 2 (d) is three. The isolated hillock, of which the number of neighboring hillocks is zero, almost unchanged. On the other hand, the polytype of a hillock adjacent to other hillocks changed "frequently". These results indicate that probability of the polytype transition at the hillock influenced by the neighboring hillocks. Thus, there is a possibility to suppress polytype transition by controlling the surface morphology to reduce the hillock-like structure on the grown surface.

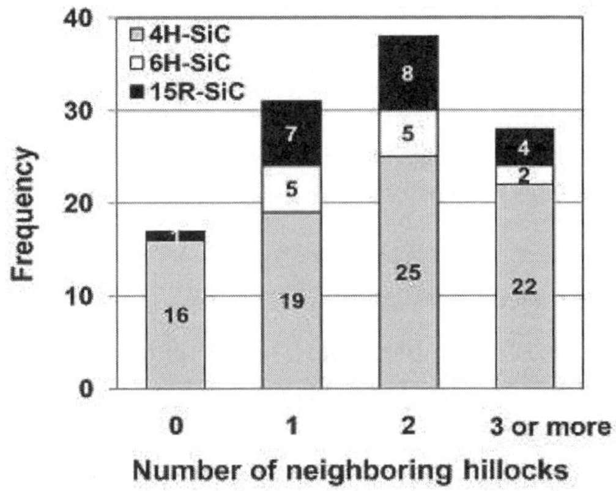

Fig. 3: The frequency of the numbers of neighboring hillock-like structures concerning each hillock-like structure.

3.2 Solution growth using VSSG method

To suppress the formation of hillocks, it seems appropriate to perform a growth method which realizes unidirectional step-flow. We performed the VSSG method which has a possibility to form step-flow from the upper part to the lower part of the seed crystal because the growth rate at upper part is higher than that of the lower part due to the difference of a supersaturation under the temperature gradient as shown in Fig. 1 (b). Fig. 4 (a) shows the cross-sectional image of the grown crystal using the VSSG method as shown in Fig. 1 (b). The thickness of the top and the bottom parts of the grown crystal were 410 μm and 300 μm, respectively. The thickness of the seed crystals was about 340 μm. Thus, the crystal grew at the upper part, while the crystal dissolved at the lower part. The cross-sectional image of the grown crystal indicates that the vicinal surface (about 0.6 deg tilted toward [1−100]) was formed. Fig. 4 (b) shows the morphology of the whole grown surface on C-face. The magnified image of the region A, which shows the typical morphology of the grown surface, is shown in Fig.4 (c). As expected, train of step structures instead of hillock-like structures were formed due to step-flow towards the [1-100] direction. Fig. 4 (d) shows the Raman intensity mapping of 4H-SiC, 6H-SiC and 15R-SiC TO peaks of the whole grown surface. 6H-SiC and 15R-SiC were rarely observed except at the edge of the crystal. Moreover, the ratio of 4H-SiC polytype to the whole grown surface was increased to be almost 100% by using the VSSG method.

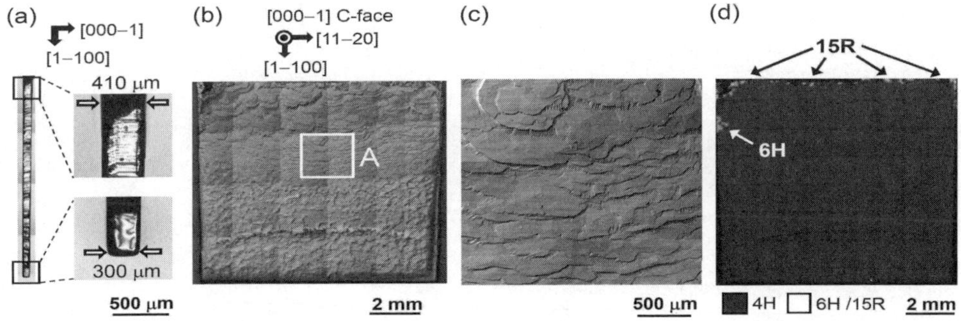

Fig. 4: Nomarski microscope images of (a) the cross-section and (b) the C-face morphology of the crystal grown by VSSG method. (c) The magnified Nomarski image of the region A. (d) Raman intensity mapping of 4H-SiC, 6H-SiC, and 15R-SiC TO peaks of the whole crystal.

We investigated the polytype transition on on-axis 4H-SiC C-face from a viewpoint of surface morphology of grown layers. The polytype transition often occurred at the hillock when the hillock grew adjacent to other hillocks. To suppress hillock formation, we performed the growth using the VSSG method which used a plate-like seed crystal vertically dipped into the silicon solvent under the temperature gradient. As a result, the hillock was rarely formed due to the unidirectional step-flow from the upper part to the lower part of the seed crystal. Consequently, the ratio of 4H-SiC polytype to the whole grown surface was improved to be almost 100%.

Acknowledgements

This work has been performed under Grant-in-Aid for Scientific Research (A) (23246004), Grant-in-Aid for JSPS Fellows (22・8543) from JSPS, Japan-France Integrated Action Program (SAKURA) from JSPS and MAE.

References

[1] P. G. Neudeck: Mater. Sci. Forum Vol. 338-342 (2000), p. 1161

[2] Y. Wang, G. N. Ali, M. K. Mikhov, V. Vaidyanathan, B. J. Skromme, B. Raghothamachar and M. Dudley: J. Appl. Phys. Vol. 97 (2005), p. 013540

[3] T. Ujihara, S. Munetoh, K. Kusunoki, K. Kamei, N. Usami, K. Fujiwara, G. Sazaki and K. Nakajima: Thin Solid Films Vol. 476 (2005), p. 206

[4] K. Kusunoki, K. Kamei, Y. Ueda, S. Naga, Y. Ito, M. Hasebe, T. Ujihara and K. Nakajima: Mater. Sci. Forum Vol. 483-485 (2005), p. 13

[5] K. Kamei, K. Kusunoki, N. Yashiro, N. Okada, T. Tanaka and A. Yauchi: J. Crystal Growth Vol. 311 (2009), p. 855

[6] N. Yashiro, K. Kusunoki, K. Kamei and A. Yauchi: Mater. Sci. Forum Vol. 645-648 (2010), p. 33

[7] A.R. Verma and P. Krishna: Polymorphism and Polytypism in Crystals (Wiley, New York, 1966)

[8] B. J. Baliga: Physics of Semiconductor Power Device (JWS Publishing, 1996)

[9] S. Nakasima and H. Harima: Phys. Stat. Sol. (a) Vol. 162 (1997), p. 39.

Materials Science Forum Vols. 717-720 (2012) pp 57-60

doi:10.4028/www.scientific.net/MSF.717-720.57

Control of Void Formation in 4H-SiC Solution Growth

Takeshi Mitani[1,2,a], Masayuki Okamura[2], Tetsuo Takahashi[1,2], Naoyoshi Komatsu[1], Tomohisa Kato[1,2], and Hajime Okumura[1,2]

[1]R&D Partnership for Future Power Electronics Technology (FUPET), 1-1-1, Umezono, Tsukuba, Ibaraki, Japan

[2]National Institute of Advanced Industrial Science and Technology (AIST), 1-1-1, Umezono, Tsukuba, Ibaraki, Japan

[a] t-mitani@aist.go.jp

Keywords: solution growth, macro-defect, void

Abstract. Void formation in 4H-SiC crystals grown from solution has been investigated by secondary ion mass spectrometry and Raman scattering. It becomes clear that ambient Ar gas fills the voids and the solvent (Si) partially remains. The result indicates that Ar dissolved in the solvent vaporizes and forms bubbles. The trapped bubbles at the crystal growth front are considered to be incorporated in the growing crystal as voids. We also have developed the following methods for suppression of the void formation; (1) dipping seed crystals so that the growth front faces upward, (2) growth under He atmosphere, and (3) the high temperature treatment of the solvent before crystal growth.

Introduction

Crystal growth of SiC from solution has a great advantage for high quality bulk growth, because the defect density can be reduced during growth [1-3]. In addition, intensive research on composition of the solution [4] and convection control [5] has led to high growth rates exceeding 200 μm/h. However, macro-defects, such as voids and incorporation of solvent into the growing crystal, are still serious problems for the solution growth [2]. In this study, we have investigated atomic elements inside voids, and have developed methods to suppress the void formation.

Experimental details

4H-SiC single crystals were grown on 4H-SiC substrates by the solution growth method under Ar atmosphere at 980 kPa and at 1900 °C, which was measured at the bottom of the graphite crucible by a pyrometer. The seed was a 20 mm diameter on-axis 4H-SiC (0001) wafer, and pure Si was used as a solvent. The axial temperature gradient in the solvent was 5~10 °C/cm. Figure 1 shows the experimental settings used in this study. The typical top seeded solution growth (TSSG) method is

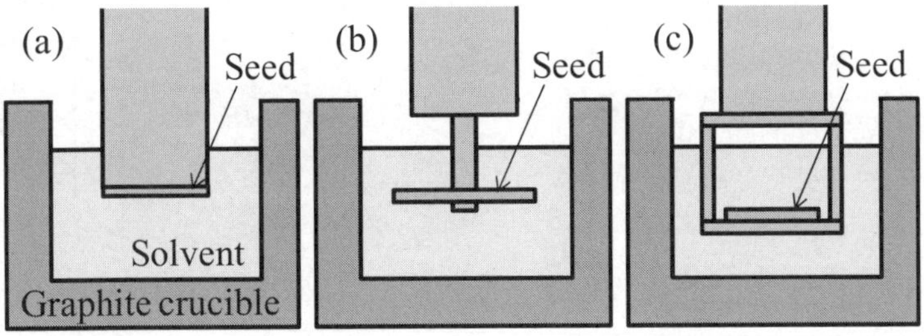

Fig. 1: Growth configurations used in this study. (a) Top seeded solution growth method, (b) configuration for growth on both Si- and C-face in the same time, and (c) configration for dipping seed crystals so that the growth front faces upward.

shown in Fig. 1(a). The setting illustrated in Fig. 1(b) was used for the solution growth on both the Si- and C-face at the same time. The setting in Fig. 1(c) will be discussed later. After growth, residual Si on the seed was removed by HF-HNO_3. The grown crystals were analyzed by Raman scattering for polytype identification and by secondary ion mass spectrometry (SIMS) for identification of atomic elements inside voids.

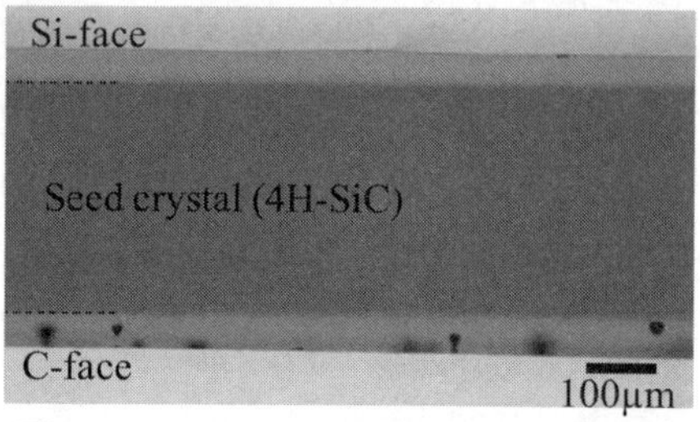

Fig. 2: A cross-sectional optical micrograph of the sample which was grown for 3 hours.

Void formation

Figure 2 shows the cross-sectional optical micrograph of the sample which was grown for 3 hours with the experimental configuration as shown in Fig. 1(b). The growth experiment was carried out with the C-face of the seed crystal facing downward. Voids of several tens μm in diameter are found in the lower grown layer. On the other hand, no void formation is observed in the upper grown layer. The cross-sectional microscopic observations show that most of the voids are hollow and spherical in shape. The experiment was repeated with the Si-face of the seed crystal facing downward. In this case also, void formation is observed only in the lower grown layer. These results show that the generation of voids is independent from the surface polarity of the seed crystal.

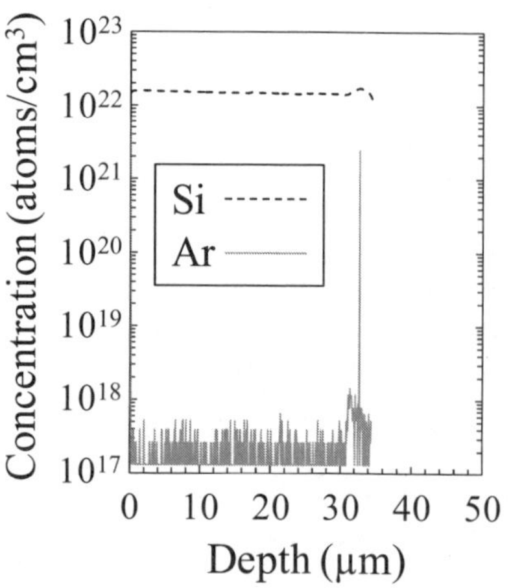

Fig. 3: Depth profiles of Si and Ar obtained by SIMS in the grown layer just above the buried void.

Figure 3 shows the depth profiles of Si and Ar concentration obtained by SIMS. The measurement was performed on the grown layer just above the buried void. The Ar signal instantaneously increased at around 31 μm where it corresponds to the top of the buried void. This result shows that Ar gas fills voids. The Raman measurement shows that Si also remains partially inside the voids as shown in Fig. 4(c). From the Raman intensity map of the transverse optical phonon mode of 4H-SiC at 776 cm^{-1}, there is no significant change in its intensity around the buried void as shown in Fig. 4(d). It shows that no polytype transformation occurs around the buried void. These results indicate that Ar dissolved in the solvent vaporizes and forms bubbles in the solvent. The trapped bubbles at the crystal growth front are considered to be incorporated in the growing crystal as voids.

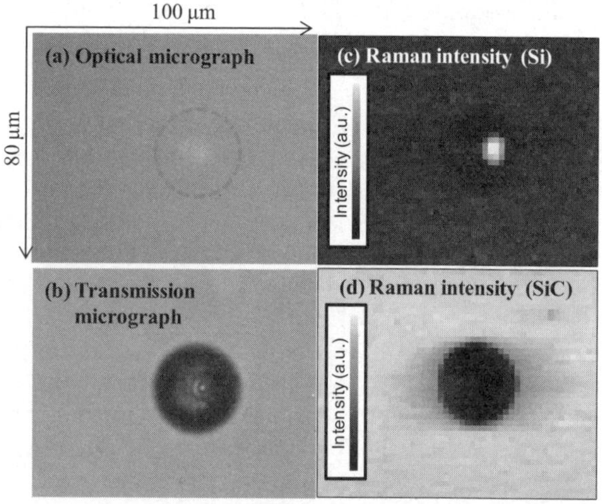

Fig. 4: (a) Optical micrograph of the grown layer. A circle indicates the position of the buried void. (b) Transmission micrograph of the grown layer, (c) intensity map of the Si Raman line at 520 cm^{-1}, (d) Intensity map of the SiC Raman line at 776 cm^{-1}.

Suppression of void formation

We describe three methods for suppression of void formation in the following. The first method is dipping seed crystals so that the growth front faces upward as illustrated in Fig. 1(c). This method suppresses the generation of voids as exemplified by the void-free upper grown layer shown in Fig. 2. In this configuration, the trapped Ar bubbles at the growth front are considered to be desorbed easily by the buoyancy of bubbles and/or convection of the solvent. In order to investigate the effect of forced convection, the variation of the void density on the rotational speed of the seed and crucible was investigated. The rotation ranging from 0 to 30 rpm was applied for both the seed and crucible in counter directions. The growth experiments were performed in the TSSG configuration as illustrated in Fig. 1(a). As shown in Fig. 5, void density decreases with the increase in the rotational speed. With a rotation of 5 rpm, the void density decreases by approximately one-half of that obtained without rotation. However, further decrease in the void density has not been observed with the application of the rotation exceeding 5 rpm. This result indicates that forced convection alone is insufficient to remove the trapped Ar bubbles from the stagnant layer near the seed/solvent interface. With the growth front facing upward, the trapped Ar bubbles are desorbed mainly by the buoyancy of bubbles.

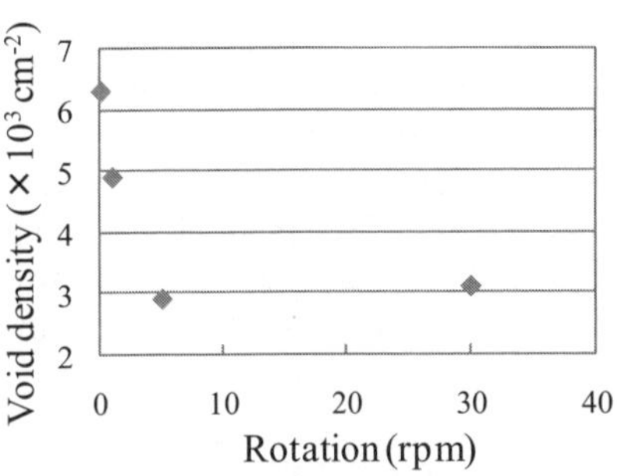

Fig. 5: Void density in grown layers as a function of the rotational speed of the seed and crucible.

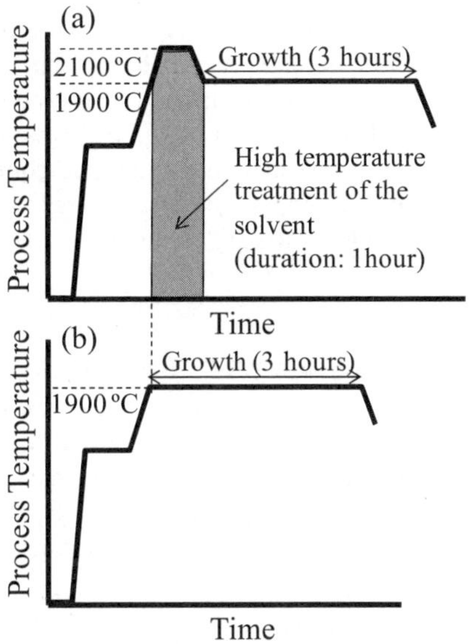

Fig. 6: Temperature regimes of the crystal growth (a) with and (b) without the high temperature treatment.

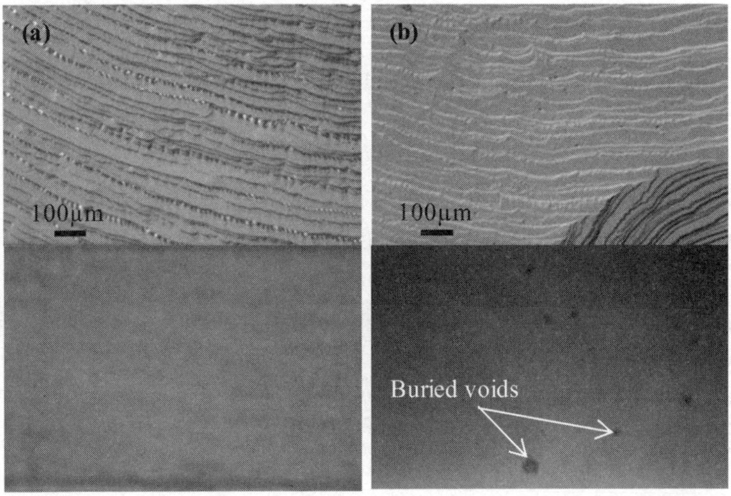

Fig. 7: Optical micrographs of the samples which are grown (a) with and (b) without the high temperature treatment. The upper and lower images are taken at the same position on the samples with reflected and transmitted illumination, respectively.

The second method employed to get rid of voids is growth under He atmosphere. The formation of voids was completely suppressed within our experimental conditions. Though the suppression mechanism is not fully understood, the influence of the small viscosity of He gas on the void formation has been pointed out [6]. Yashiro et al. have reported that solution growth with no void formation can be achieved under the condition that the viscosity of ambient non-oxidizing gas is less than 750 μP (75 μPa·s). While void formation was controlled by growth under He atmosphere, solvent inclusions were still found in the grown crystals.

The third method is the high temperature (HT) treatment of the solvent before crystal growth. As shown in Fig. 6(a), the process temperature is raised to 2100 °C for 1 hour prior to the growth. Figure 7 shows the optical micrographs of the grown layer with and without the HT treatment. The thickness is 80 and 70 μm for the 4H-SiC layers grown with and without HT treatment, respectively. The generation of voids was effectively suppressed with this treatment as shown in Fig. 7. It is worth noting that the void formation can not be suppressed by merely increasing the growth temperature. With the HT treatment, it is supposed that crystal growth proceeds in the solvent whose solubility of ambient gas is less than that of thermal equilibrium. Further studies on the Ar solubility in the solvent as functions of temperature and Ar pressure are needed to give an explanation for the suppression of void formation by the HT treatment.

Summary

Void formation in 4H-SiC crystals grown from solution was investigated. In buried voids, Ar gas is filled and Si remains partially. These results indicate that Ar dissolved in the solvent vaporizes and forms bubbles. The trapped bubbles at the crystal growth front are considered to be incorporated in the growing crystal as voids. We also have developed the following three methods for suppression of the void formation; (1) dipping seed crystals so that the growth front faces upward, (2) growth under He atmosphere, and (3) the high temperature treatment of the solvent before crystal growth.

Acknowledgement

This work is supported by Novel Semiconductor Power Electronics Project Realizing Low Carbon Emission Society under Ministry of Economy, Trade and Industry (METI) and New Energy and Industrial Technology Development Organization (NEDO).

References

[1] S. V. Rendakova, I. P. Nikitina, A. S. Tregubova, and V. A. Dmitriev, J. Electronic Mater. 27 (1998) 292-295.

[2] D. Hofmann and M. Müller, Mater. Sci. Eng. B 61-62 (1999) 29-39.

[3] B. M. Epelbaum and D. Hofmann, J. Crystal Growth 225 (2001) 1-5.

[4] K. Danno, H. Saitoh, A. Seki, H. Daikoku, Y. Fujiwara, T. Ishii, H. Sakamoto, and Y. Kawai, Mater. Sci. Forum 645-648 (2010) 13-16.

[5] K. Kusunoki, K. Kamei, N. Okada, N. Yashiro, A. Yauchi, T. Ujihara, and K. Nakajima, Mater. Sci. Forum 527-529 (2006) 119-122.

[6] Sumitomo Metal Industries, Ltd., N. Yashiro, K. Kamei, and K. Kusunoki, Japanese Patent, 2004-257041, (2004).

Materials Science Forum Vols. 717-720 (2012) pp 61-64

doi:10.4028/www.scientific.net/MSF.717-720.61

Top-Seeded Solution Growth of 4H-SiC Bulk Crystal Using Si-Cr Based Melt

H. Daikoku[1, a], M. Kado[1], H. Sakamoto[1], H. Suzuki[1], T. Bessho[1], K. Kusunoki[2, b], N. Yashiro[2], N. Okada[2], K. Moriguchi[2] and K. Kamei[2]

[1]Higashifuji Technical Center, Toyota Motor Corporation
1200, Mishuku, Susono, Shizuoka, 410-1193, Japan

[2]Corporate Research & Development Laboratories, Sumitomo Metal Industries Ltd.,
Fuso-cho 1-8, Amagasaki, Hyogo, 660-0891, Japan

[a]daikoku@hironori.tec.toyota.co.jp, [b]kusunoki-kzh@sumitomometals.co.jp

Keywords: Solution growth, 4H SiC, Meniscus

Abstract. We have grown high-quality long cylindrical (12 mm thick) 4H-SiC bulk crystals by the meniscus formation technique, which was first applied for the solution growth of bulk SiC. It enabled long-term growth by suppressing parasitic reactions such as polycrystal precipitation around the seed crystal. In addition, we could control the growth angle from −22° to 61° by adjusting the meniscus height. The thickness of the grown cylindrical crystals was 12 mm, which is the largest reported until now, and corresponded to a growth rate of 0.6 mm/h. Smooth morphology growth was maintained on the (000-1) C-face. In cross-sectional transmission optical microscopy images, few solvent inclusions and voids were observed. XRD measurements revealed that the FWHM values of the grown crystals were almost the same as those of the seed crystal.

Introduction

Solution growth of SiC has been studied as the method for obtaining a high-quality wafer, because the growth process occurs under the conditions close to thermal equilibrium [1-2]. The difficulty of obtaining bulk SiC by solution growth is the low growth rate derived from C solubility in the Si solvent. In recent years, some studies reported growing 10 mm thick 6H-SiC using Si-Ti-C melt [3] and 1.8 mm thick 4H-SiC by using Si-Cr based melt with the maximum growth rate of 1 mm/h [4]. These studies demonstrated the possibility of obtaining bulk 4H-SiC by solution growth. However, low cost growth methods to achieve high-quality crystals have not yet been reported.

An important advantage of liquid phase growth is the high controllability of the crystal diameter by applying the Czochralski method, which can produce high-quality crystals, such as Si or GaAs, at low cost [5]. In this study, we investigated the polarity dependence of the stability of surface morphology during 4H-SiC growth and grew long 4H-SiC bulk by the meniscus formation technique. In addition, the crystallinities of the grown bulk crystals are discussed.

Experiments

Solution growth was performed in a top seeded solution growth (TSSG) furnace under an inert gas environment. A graphite crucible was used as both the container and carbon source. An N-type 4H-SiC (0001) Si-face or (000-1) C-face crystal produced by the PVT technique was used as a seed crystal. Silicon-40%Chromium based mixtures was melted in the crucible at 180 kPa. An accelerated crucible rotation technique (ACRT) was used to improve mixing in the melt and to supply carbon to the surface [6]. The maximum crucible and seed rotation rates were 40 and 10 rpm, respectively. The growth temperature was about 2000°C and the growth duration was 10–35 h. The surface morphology was observed by a Nomarski optical microscope, and the crystallinity and polytype were investigated by X-ray diffraction and Raman scattering spectroscopy.

Results and Discussion

Polarity dependence of the growth morphology

As-grown surfaces on the Si-face are shown in Fig. 1(a) and (b) and those on the C-face are shown in Fig. 1(c) and (d). On the Si-face, the distinct facet with hexagonal 3D hillocks was partially observed in Fig. 1(a). Figure 1(b) shows that the entire surface was covered with rough 3D growth. The transition from 2D growth to rough 3D growth seems to occur at the early stage of growth. On the other hand, the surface morphology on the C-face appears to be specular without any 3D hillocks and facets over the entire surface, as shown in Fig. 1(c). Thus, as shown in Fig. 1(d), almost the same morphology was obtained despite the growth progress. This polarity dependence may arise from the large difference between the solid / liquid interfacial energy of the two faces [7]. These results indicate that growth front stability can be maintained on the C-face even during long growth.

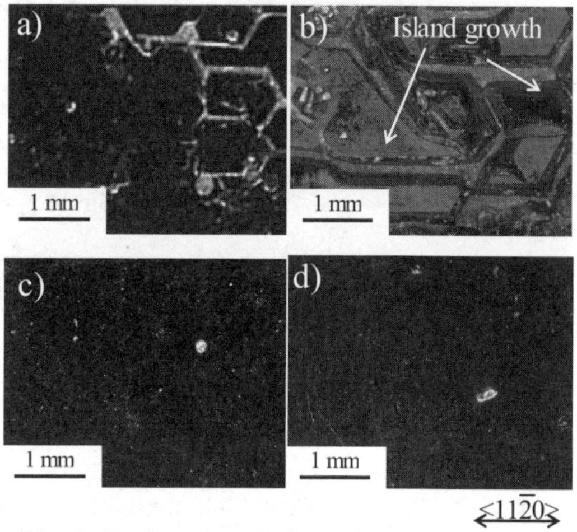

Fig. 1: Surface morphology dependence of bulk crystals on Si polarity (a and b) and C polarity (c and d).
a) and c) 1.5 mm thick: b) and d) 3.0 mm thick.

Meniscus growth

The meniscus formation technique was applied to all growth experiments. The meniscus bridge was formed as shown in Fig. 2: the seed crystal was dipped in the solution and then pulled upward from the solution surface. In this technique, we can control the growth angle defined in Fig. 2 by adjusting the meniscus height. Figure 3 shows the relationship between the meniscus height and the growth angle. The diameter of the grown crystals can be changed by varying the growth angle from −22° to 61°. In addition, parasitic reactions such as polycrystal precipitation around the seed crystal can be suppressed. Figure 4 shows a long cylindrical crystal and an enlarged crystal obtained by using meniscus growth; the seed diameter is 12 mm. The cylindrical crystal in Fig. 4(a) was grown by maintaining the meniscus height at 2.5 mm for 20 h, resulting in a near 0° growth angle. The grown crystal thickness was 12 mm, which is the largest ever reported by solution growth, and corresponded to a growth rate of 0.6 mm/h. The crystal in Fig. 4(b) was enlarged with a 61° growth angle. The diameter of the crystal was enlarged 2.2 times over that of the seed crystal. The crystal in Fig. 4(c) was grown by adjusting the meniscus height from 1.5 mm to a higher value for 35 h. The growth angle gradually decreased from that of the initial growth. This result indicated that a combination of shoulder growth and

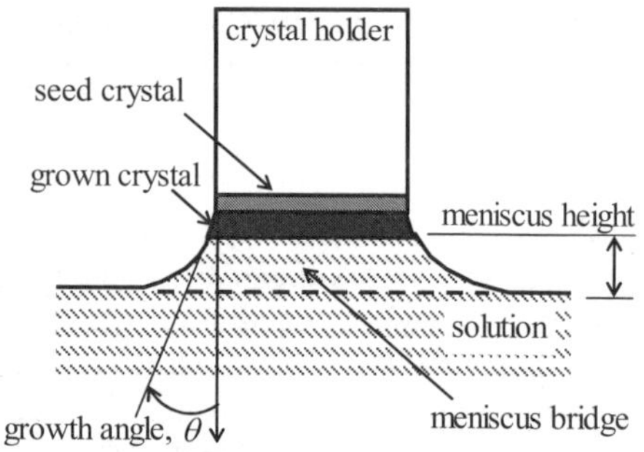

Fig. 2: Schematic representation of the configuration of the meniscus growth method. The distance between the surface of the grown crystal and that of the solution is defined as the meniscus height.

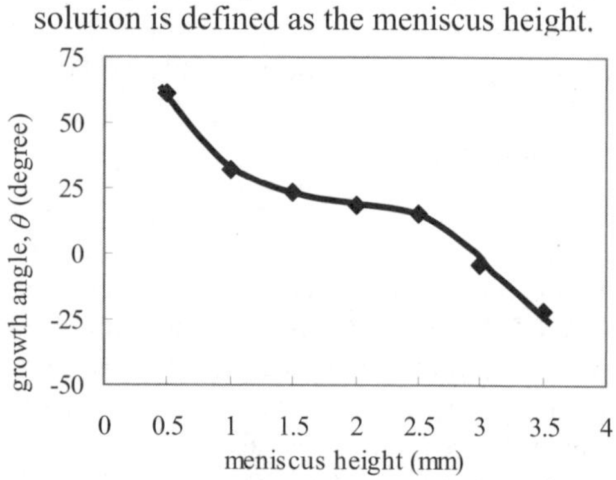

Fig. 3: Dependence of the growth angle on the meniscus height.

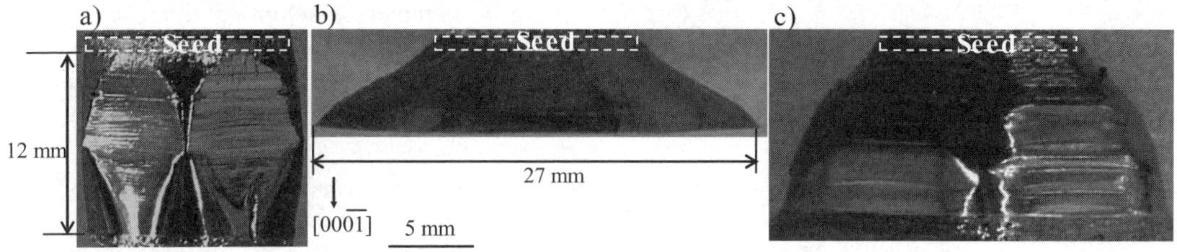

Fig. 4: Side views of 4H-SiC crystals grown by meniscus formation technique: a) growth with a meniscus height of 2.5 mm for 20 h; b) that with a meniscus height of 0.5 mm for 10 h; c) growth by changing the meniscus height from 1.5 mm to a higher value for 35 h.

cylindrical growth can be controlled by adjusting the meniscus height during growth. The side faces of all crystals do not have polycrystals and consist of {1-100} facets. The surface striations generated perpendicularly to the growth direction seemed to be caused by small fluctuations of growth conditions with RF heating and ACRT.

Figure 5(a) shows the surface morphology of the grown cylindrical crystal with a thickness of about 6 mm (0.24 mm/h). In the case of meniscus formation growth, the interface shape is concave to the melt. The peripheral region is about 0.5 mm higher than the central region. The curvature of the interface increases with an increase in the meniscus height. If the curvature is higher than the threshold value, solvent inclusions with giant step bunching will appear and degrade the crystallinity. The temperature of the peripheral region may be lower than that of the central region owing to the radiation cooling around the meniscus bridge. Therefore, the temperature distribution around the meniscus bridge should be chosen so as to minimize the growth front curvature. The surface morphology can be characterized by the occurrence of two regions, as shown in Fig. 5(c). One is the peripheral region, which has flat and large terraces without any spirals detectable by confocal laser scanning microscopy. The other is a concave region containing a central plateau. The major part of the surface consists of a vicinal face and shows smooth morphology. It consists of many steps (terrace < 10 μm) resulting from step bunching. This may indicate that the crystal growth proceeds by 2D nucleation of the edge region and lateral expansion (or step-flow expansion) of these nucleations towards the center.

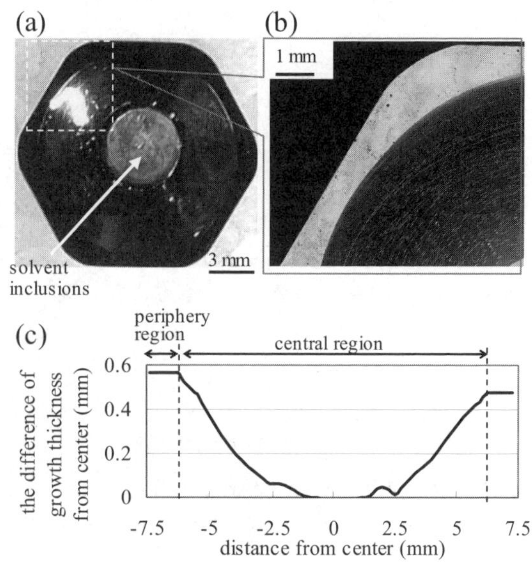

Fig. 5: a) Top views of bulk crystal observed by optical microscopy; (b) observed by confocal laser scanning microscopy; (c) growth distribution depending on the <11-20> lateral position of the crystal.

Crystallinities

Figure 6 shows the cross section of the grown crystal observed using a transmission optical microscope. The growth angle was 19° and the thickness was 3 mm. The crystal contained no cracks and few solvent inclusions were detected. This result indicates that a high-quality bulk crystal without

Fig. 6: Transmission optical microscopy image of the cross section of an enlarged crystal grown by maintaining the meniscus height at 2.0 mm for 10 h.

solvent inclusions can be obtained. The polytype of the grown layer was identified as 4H-SiC by Raman scattering spectroscopy (not shown).

The grown crystal shown in Fig. 4(c), whose initial 6 mm thick layer was sliced into 1 mm thick sections, was subjected to XRD. The excellent crystalline quality was quantified by the full width at half maximum (FWHM) values of the X-ray rocking curve, as shown in Fig. 7. The diffraction plane was 4H-SiC {0004}. The FWHM values ranged from 10 to 15 arcsec. The FWHM values indicated good crystallinity and showed no dependence on the growth thickness. These results indicate that high-quality bulk crystals can be obtained by the meniscus formation technique.

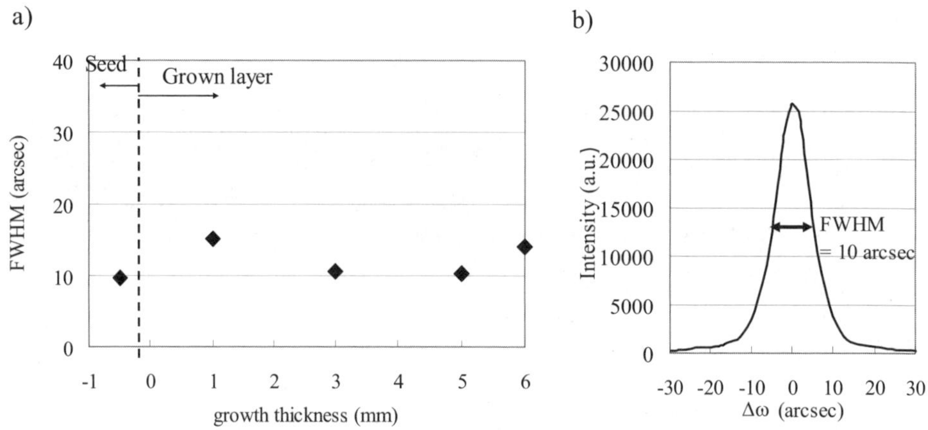

Fig. 7: a) FWHM values depending on the growth thickness for the crystal shown in Fig. 4 (c); b) Selected rocking curve measurements with the FWHM values of 10 arcsec.

Summary

By solution growth using the meniscus formation technique, we have grown high-quality long cylindrical (12 mm thick) 4H-SiC bulk crystals. The growth angle can be controlled from −22° to 61° and growth rates exceeding 0.6 mm/h can be obtained. The growth on the (000-1) C-face can be maintained with a stable smooth surface morphology. In cross-sectional transmission optical microscopy images, no solvent inclusions and voids were observed. XRD measurements revealed that the FWHM values of the grown crystals were almost the same as those of the seed crystal.

References

[1] R. Yakimova, M. Tuominen, A.S. Bakin, J.O. Fornell, A. Vehanen, and E. Janzen: Inst. Phys. Conf. Ser. 142 (1996) 101.

[2] K.Gillessen and W.V. Munch: J. Cryst. Growth 19 (1973) 263.

[3] N. Yashiro et al.: Mater. Sci. Forum 645-648 (2010) 33.

[4] K. Danno et al: Mater. Sci. Forum 645-648 (2010) 13.

[5] W. Zulehner: Mater. Sci. Eng. B 73 1-3 (2000) 7-15

[6] H. J. Scheel and E. O. Schultz-Dubios: J. Cryst. Growth 8 (1971) 304.

[7] R. A. Stein, P. Lanig and S.Leibenzeder: Mater. Sci. Eng. B11 (1992) 69.

Materials Science Forum Vols. 717-720 (2012) pp 65-68

doi:10.4028/www.scientific.net/MSF.717-720.65

Growth of a Thick 2H-SiC Layer in Si-Li Solution under a Continuous CH_4 Flow

Mamoru Imade[1,a,*], Akito Ishikawa[1,b], Yusuke Nakagawa[1,c], Masashi Yoshimura[1,d], Yasuo Kitaoka[1,e], Takatomo Sasaki[1,f], Yusuke Mori[1,g]

[1] Department of Electrical, Electronic and Information Engineering, Osaka University, Yamada-oka 2-1, Suita, 565-0871, Japan

[a]imade@cryst.eei.eng.osaka-u.ac.jp, [b]ishikawa@cryst.eei.eng.osaka-u.ac.jp, [c]nakagawa@cryst.eei.eng.osaka-u.ac.jp, [d]yoshi@eei.eng.osaka-u.ac.jp, [e]kitaoka@cryst.eei.eng.osaka-u.ac.jp, [f]sasaki@eei.eng.osaka-u.ac.jp, [g]mori.yusuke@eei.eng.osaka-u.ac.jp, *corresponding author

Keywords: 2H-SiC, liquid-phase epitaxy, LPE, Si-Li, CH_4

Abstract. The growth of thick 2H-SiC layers in Si-Li solution under a continuous CH_4 flow at atmospheric pressure was investigated. The thickness of the 2H-SiC layer increased linearly with the growth period, reaching as thick as 270-μm with a growth period of 7 hr. The Li concentration in the layers was as low as 1.0×10^{18} atoms/cm^3, which was less than one-hundredth that observed in a previous study.

Introduction

SiC has more than 200 polytypes that are classified according to a stacking sequence along the c-axis. Of the known SiC polytypes, 2H-SiC has the simplest hexagonal structure, giving it some advantageous properties such as the largest band gap (3.33 eV) and highest electron mobility (2.3×10^7 cm/s) [1-4]. Although some researchers have attempted to grow 2H-SiC crystals using chemical vapor deposition [5] or a solvothermal process [6], they have obtained only whiskers or flaky crystals and not bulk 2H-SiC single crystals. Experimental results regarding determination of the the physical properties of 2H-SiC have therefore not been reported. We have previously obtained a 100-μm thick 2H-SiC liquid-phase epitaxy (LPE) layer on a 4H-SiC (0001) substrate by the vapor–liquid–solid (VLS) process in Si-Li solution at a starting CH_4 pressure of 0.9 MPa [7]. In this closed system, however, the thickness of the LPE layer was saturated over the growth period of 12 hr because of consumption of CH_4 gas, preventing the growth of thick 2H-SiC layers by long-term growth. In addition, high concentrations of Li (> 10^{20} atoms/cm^3) were incorporated into the LPE layers. In this study, we attempted to achieve the growth of thick 2H-SiC layers in Si-Li solution under a continuous CH_4 flow (open system) at atmospheric pressure.

Experimental Procedure

The experimental setup used in this study is shown in Fig. 1. A tungsten (W) crucible containing the starting materials (metal Li (purity: 3N), Si (purity: 6N), and an on-axis (0001) 4H-SiC substrate (5 mm × 10 mm) was transferred into a stainless-steel tube in a glove box under an Ar atmosphere. The substrate was set perpendicular to the bottom of the W crucible. The stainless-steel tube was heated to a growth temperature for 1 hr, and CH_4 gas was then introduced at a flow rate of 25 sccm. Ar gas with a flow rate of 0 ~ 75 sccm was used as a carrier gas. The temperature was then maintained at a growth temperature of 900 °C for the growth period (1 ~ 7 hr). After cooling, the substrate was separated from the solidified flux by dissolving the flux in ethanol and water in turn. The LPE layer was characterized by X-ray diffraction (XRD), high-resolution transmission microscopy (HR-TEM), and secondary ion mass spectrometry (SIMS). The yields were calculated as the amounts of metal-Si consumed in the growth of SiC.

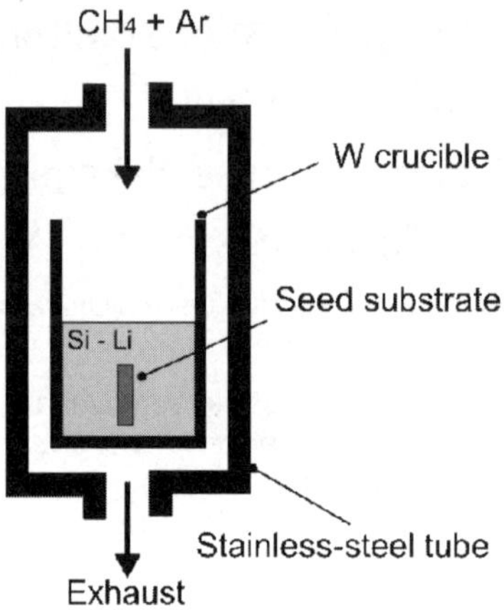

Fig. 1 Schematic illustration of the growth setup used in this study.

Results and Discussion

Polycrystals of SiC were grown in a W crucible for all experiments. Figure 2 shows the dependence of yields of polycrystal (filled circle) and LPE layer (open circle) on the flow rate of Ar carrier gas. We can see in Fig. 2 that the yield of polycrystal linearly decreased with an increasing flow rate of Ar carrier gas. The growth of LPE layer was confirmed at Ar flow rates of 25 and 38 sccm. The change in the yield of polycrystal as a function of the flow rate of Ar carrier gas is likely to be caused by a change in the degree of supersaturation. The solubility of CH_4 gas in the solution is determined by the partial pressure of CH_4, $P[CH_4] = f[CH_4] \cdot P^0 / (f[CH_4] + f[Ar])$, where $f[CH_4]$, $f[Ar]$, and P^0 refer to the flow rate of CH_4, Ar, and total pressure (P^0 = 1 atm), respectively. When $f[Ar]$ increased, $P[CH_4]$ decreased and the concentration of carbon in the solution decreased, resulting in a low degree of supersaturation. However, the yield of the LPE layer increased with the Ar flow rate, reaching a maximum at a flow rate of 38 sccm and decreasing with a higher flow rate. This behavior seems to be explained as follows. In conditions of lower Ar flow rate, high $P[CH_4]$ causes superfluous supersaturation, particularly near the gas–liquid interface, followed by the generation of polycrystals. Once polycrystals are generated, dissolved carbon is preferably consumed for the growth of polycrystals rather than for LPE growth, resulting in a low yield of LPE. When the Ar flow rate increased, dissolved carbon was effectively transported to the seed substrate without contributing to the growth of polycrystals near the gas-liquid interface, which promotes LPE growth. Under the

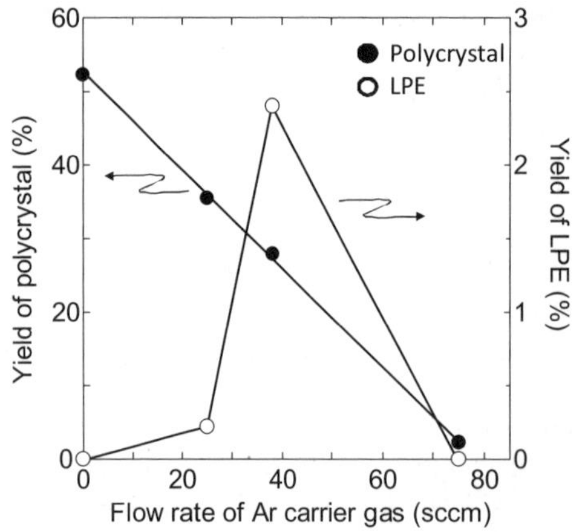

Fig. 2 Dependence of yields of polycrystal and LPE layer on flow rate of Ar carrier gas.

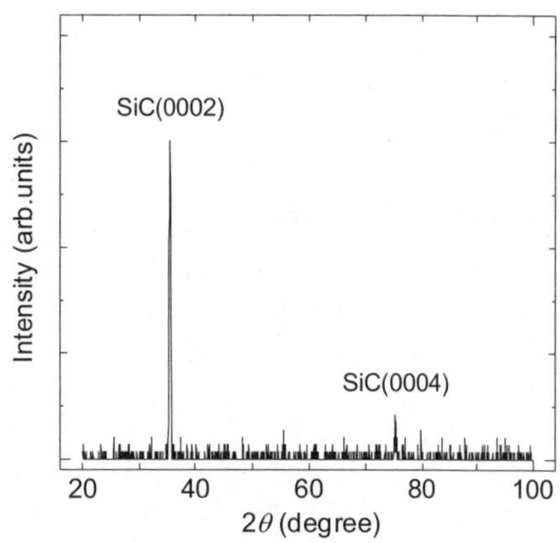

Fig. 3 θ–2θ XRD profile of LPE layer grown at Ar flow rate of 38 sccm.

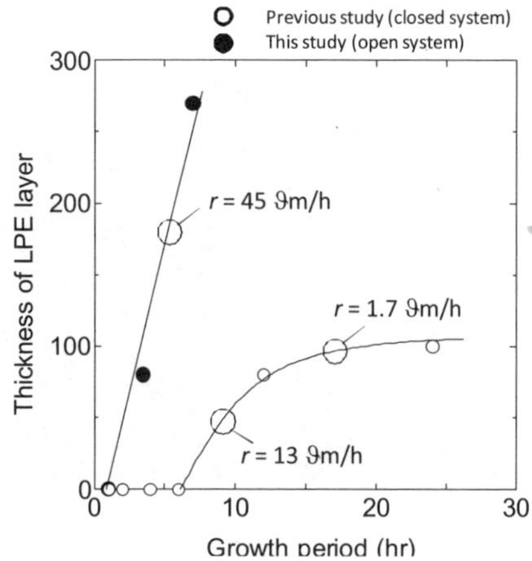

Fig. 4 Thickness of the LPE layer and growth rate as a function of the growth period.

condition of a high Ar flow rate, the degree of supersaturation decreased to a value not sufficient for the growth of SiC crystal, resulting in a low yield of polycrystal and LPE. Figure 3 gives the θ-2θ XRD profile of the LPE layer grown at an Ar flow rate of 38 sccm. In the profile, SiC(0002) and (0004) faces can be clearly resolved, which demonstrates that the LPE layer is a single crystalline layer. However, full width at half maximum values of 2H-SiC(0002) X-ray rocking curves of typical samples were as high as around 1000 arcsec, which must be improved.

Figure 4 shows the thickness of the LPE layer and growth rate (r) as a function of the growth period. In the closed system reported in the previous study, the thickness of the LPE layer was saturated and the growth rate decreased to almost zero over a growth period of 12 hr. In contrast, the thickness of the LPE layer grown in this study (open system) increased linearly with the growth period, reaching as thick as 270-μm with a growth period of 7 hr (growth rate: 45 μm/h). These results indicate that the carbon source continuously dissolved into the solution and was supplied to the substrate during growth.

Figure 5 is the cross-sectional HR-TEM image near the surface of the 270-μm thick LPE layer. The electron beam was irradiated along the <11–20> direction of SiC. As seen in Fig. 5, only a periodic structure with A–B stacking and spacing of 0.5 nm was observed, and the electron diffraction (E.D.) spots corresponded with those of the 2H-SiC polytype, which shows that the LPE layer was single-phase 2H-SiC.

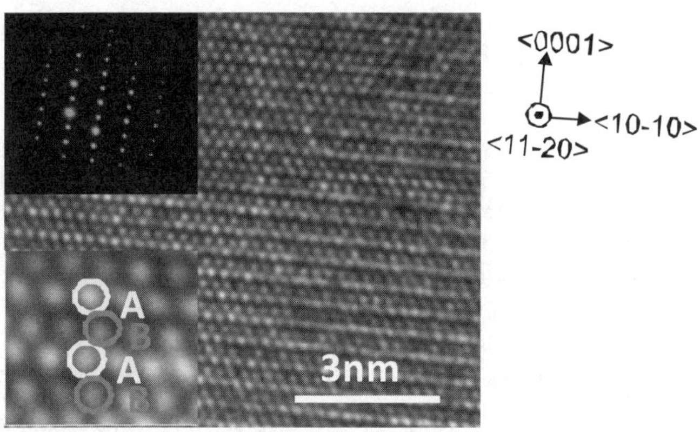

Fig. 5 Cross-sectional HR- TEM images of the LPE layer grown on the 4H-SiC substrate. The inset show the E.D. spots of this sample.

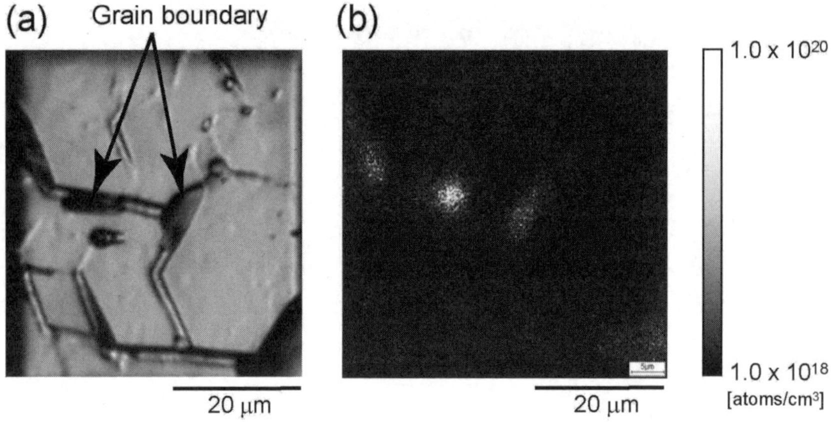

Fig. 6 (a) An optical microscope photograph and (b) an image of the Li concentration distribution of the LPE layer detected by SIMS. In Fig. 6(b), the brightness indicates the Li concentration, with brighter areas containing more Li than darker areas.

An optical microscope photograph and an image of the Li concentration distribution of the LPE layer detected by SIMS are shown in Fig. 6(a) and 6(b), respectively. As seen in Fig. 6(a), the LPE layer was grown with a columnar growth mode and many grain boundaries were observed on the surface. In Fig. 6(b), the brightness indicates the Li concentration, with brighter areas containing more Li than darker areas. In the darker areas, the Li concentrations were as low as 1.0×10^{18} atoms/cm^3, which was less than one-hundredth of that in our previous study (closed system). Brighter areas in Fig. 6(b) corresponded to the grain boundaries in Fig. 6(a), indicating that Li is easily included in grain boundaries. In the future, it will be necessary to promote the 2D growth mode without grain boundaries in order to grow thick 2H-SiC LPE layers with low Li concentrations.

Summary

We investigated the LPE growth of 2H-SiC on a 4H-SiC (0001) substrate by the vapor–liquid–solid (VLS) process in Si-Li solution under a continuous CH_4 flow. The yield of SiC polycrystal generated near the gas–liquid interface linearly decreased with increases in the flow rate of Ar carrier gas. In contrast, the yield of the LPE layer reached a maximum at an Ar flow rate of 38 sccm. The thickness of the LPE layer increased linearly with the growth period, reaching a thickness of 270 μm with a growth period of 7 hr. Although a large amount of Li was incorporated into the grain boundaries, the Li concentration inside the grains was as low as 1.0×10^{18} atoms/cm^3, which was less than one-hundredth of that in our previous study (closed system). We have concluded that the VLS process in Si-Li solution under a continuous CH_4 flow has the potential to grow bulk 2H-SiC crystals with long-term growth.

References

[1] L. Patrick, D.R. Hamilton, and W.J. Choyke, Phys. Rev. 143 (1966) 526-536.
[2] C. Persson and U. Lindefelt, J. Appl. Phys. 82 (1997) 5496-5508.
[3] G. Pennington and N. Goldsman, Phys. Rev. B 64 (2001) 045104-1-10.
[4] H. E. Nilsson and M. Hjelm, J. Appl. Phys. 86 (1997) 6230-6233.
[5] Y. Yao, S.T. Lee, and F.H. Li, Chem. Phys. Lett. 381 (2003) 628-633.
[6] G. Zou, C. Dong, K. Xiong, H. Li, C. Jiang, and Y. Qian, Appl. Phys. Lett. 88 (2006) 071913-1-3.
[7] M. Imade, S. Takeuchi, M. Uemura, M. Yoshimura, Y. Kitaoka, T. Sasaki, Y. Mori, S. Itoh, H. Okuda, and M. Yamazaki, Mater. Sci. Forum 645-648 (2010) 45-48.

Materials Science Forum Vols. 717-720 (2012) pp 69-72
© (2012) Trans Tech Publications, Switzerland
doi:10.4028/www.scientific.net/MSF.717-720.69

Growth rate prediction in SiC solution growth using silicon as solvent

Julien Lefebure[a], Jean-Marc Dedulle[b], Thierry Ouisse[c] and Didier Chaussende[d,*]

LMGP, CNRS UMR 5628, Grenoble INP, Minatec, 3 parvis Louis Neel, BP 257, 38016 Grenoble, France

[a] julien.lefebure@grenoble-inp.fr, [b] jean-marc.dedulle@grenoble-inp.fr, [c] thierry.ouisse@grenoble-inp.fr, [d] didier.chaussende@grenoble-inp.fr, *corresponding author

Keywords: solution growth, bulk, modeling, growth rate.

Abstract. Besides the seeded sublimation process, which is the current industrial crystal growth process for silicon carbide (SiC), solution growth appears as a possible method for high quality bulk crystals. For a further development of the latter technique, a robust numerical model has been implemented with the aim to give quantitative outcomes in addition to qualitative information. Growth rates have been calculated for three different temperature ranges (1700, 1800 and 1900°C) in pure silicon. The computed values were found to be in good agreement with the experimental ones.

Introduction

Top seeded solution growth (TSSG) has been recently reconsidered as a possible process for the growth of large size and high quality single crystalline ingots of silicon carbide [1]. Despite many attempts (see for instance [2]), it is only recently that major achievements have been reported with the real demonstration of a 10 mm thick single crystal [3]. Similarly to sublimation growth, an effective breakthrough could be expected only if the simulation tool was able to open step by step the black box of the process and to gain insight into the intricate relevant phenomena. Through a coupled modeling of heat transfer and fluid dynamics, we have demonstrated in a previous work that the different convective flow contributions have a very important effect on the stabilization/destabilization of the growth front [4]. For example, we have pointed out the importance of controlling both the electromagnetic and the Marangoni convections which can easily give rise to turbulent flows in specific areas of the crucible. After this first phenomenological approach, the present work aims at developing a much more robust and complete numerical model which could be used on a quantitative footing, i.e., able to predict the growth rate.

Results and Discussion

The experimental set-up was a modified Metal Research™ Czochralski puller with medium frequency induction heating (typically 60 kHz). Details about the growth chamber and crucible elements have already been given in ref. [4]. Briefly, pure liquid silicon was used as a solvent and placed in a graphite crucible, which acted both as a container and as the carbon source. We used top seeded solution growth (TSSG) geometry, meaning that the SiC seed was mounted on a graphite rod and dipped into the liquid. The simulations have been conducted by the finite element method with the COMSOL multi-physics® software. Most of the calculation details regarding induction heating, heat transfers and fluid dynamics have been presented in a previous paper which included both the models and data used [4]. Experimental growth rates have been measured by optical microscopy on cross-sections, the formed epilayers having a different color compared to the substrates, due to their higher nitrogen content.

To get rid of thermo-capillary convection and more generally of any turbulent flow, which appeared to be very harmful in the TSSG process of SiC [4], we first designed totally new crucible geometry with a special care paid to the reduction of the temperature gradients. This has practically been achieved by i) increasing the crucible size and insulation thickness with respect to the liquid volume, ii) increasing the induction coil height so as to spread the heating area, iii) closing as much as possible the crucible to avoid any radiative heat sink.

Hereafter will be displayed only the liquid zone as no additional information could be learnt by the full representation of the crucible. Fig. 1 gives an example of the temperature distribution for two different ranges of temperature, namely 1700°C and 1900°C. Note that the axial temperature difference is about 3-5°C and the radial one close to 1°C giving rise to very low temperature gradients, typically about 1°C/cm. Changing the temperature range does not affect the distribution and the isotherms keep exactly the same shape.

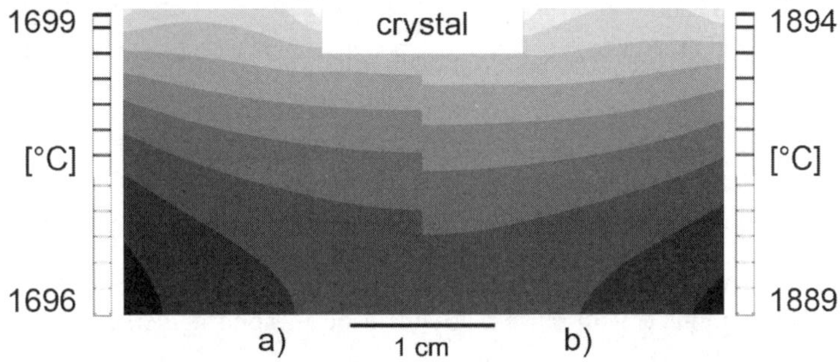

Fig. 1: Computed temperature distribution in the liquid phase for two different temperature ranges: a) 1700°C and b) 1900°C and a crucible rotation rate of 40 rpm.

The carbon concentration (molar concentration C) has been computed using the classical convection-diffusion equation for the steady state (Eq. 1)

$$-D\nabla^2 C + \mathbf{u} \cdot \boldsymbol{\nabla} C = 0\,, \tag{1}$$

D being the diffusion coefficient of carbon in liquid silicon ($1.7 \cdot 10^{-8}$ m²/s) and **u** the fluid velocity vector. A carbon equilibrium concentration (C_{eq}) was set as the boundary condition for both the crucible-liquid and the crystal-liquid interfaces. It was obtained from the carbon molar fraction x_{Ceq}, calculated according to Eq. 2:

$$C_{eq} = \frac{\rho_{Si}}{M_{Si}} \cdot \frac{x_{Ceq}}{1 - x_{Ceq}}, \tag{2}$$

ρ_{Si} being the liquid density (assumed to be equal to the density of pure silicon) and M_{Si} the molar weight of silicon. The temperature dependence of the melt composition, given in ref. [5], has been reminded in Eq. 3:

$$x_{Ceq} = exp(6.249 - 24460/T). \tag{3}$$

A special attention has been paid to the discretization of the liquid area. Triangular meshing has been chosen for the entire computation domain, except at the crystal surface where rectangular meshing has proven to provide a better accuracy and stability to compute the SiC growth rate in the diffusion boundary layer. The computed carbon concentration is presented in Fig. 2a. The carbon flux is indicated by the normalized black arrows. There is a direct correlation between the carbon distribution (shape of isovalues) and the convection patterns (convection loops). As a consequence the supersaturation, calculated as a deviation from the equilibrium concentration (Eq. 4)

$$S = C - C_{eq}, \tag{4}$$

is mainly driven by the convective phenomena in the liquid (Fig. 2b). Black corresponds to supersaturated region (S>0), meaning that crystallization can occur. White corresponds to undersaturated region (S<0), where dissolution of carbon from the crucible should occur. As was expected, the hottest points of the crucible acted as the carbon source and the coldest parts constitute the crystallization zones.

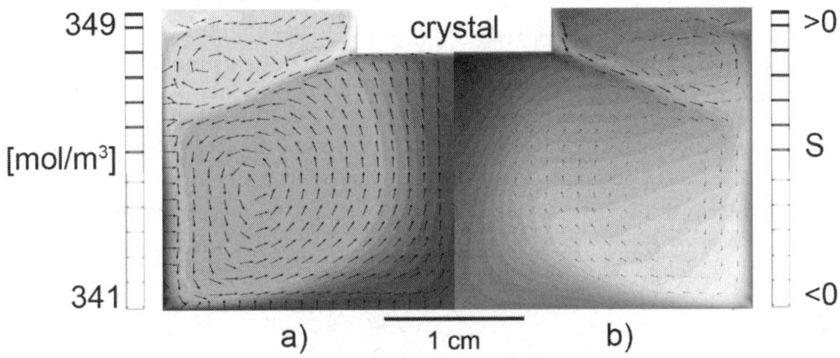

Fig. 2: Modeling of a) the distribution of carbon concentration and b) the distribution of supersaturation in the liquid. Black arrows indicate a) the normalized and b) the real carbon fluxes. Calculations were done for a temperature of 1900°C and a rotation rate of the crystal of 40 rpm.

The growth rate V_g was calculated from the flux of carbon arriving normal to the crystal surface

$$V_g = K(-D\nabla C + C\boldsymbol{u}).\boldsymbol{n}, \tag{5}$$

with $K = 3.6 \cdot 10^9 \cdot M_{SiC}/\rho_{SiC}$. This constant is only used to directly express the growth rate in µm/hr, which is the commonly used experimental unit. **n** is the unit vector perpendicular to the crystal surface. In this case, we assumed that all carbon atoms that reach the surface contribute to the growth without any kinetic limitation.

A profile along the crystal radius of both the calculated and measured growth rate normal to the surface is given in Fig. 3a for an experiment done at 1800°C. The growth rate is almost homogeneous along the radius, except at the edge of the crystal where a drastic increase of the growth rate is observed. In this area, the calculated growth rate is clearly underestimated compared to the measured one. This deviation is not explained yet but the right angle at the edge of the crystal can give rise to numerical errors when computing the liquid parameters (velocity, concentration …) in its close vicinity. The growth rate study as a function of temperature is presented in Fig. 3b. Values are taken on the symmetry axis of the crucible, which corresponds to the center of the crystal. The growth rate increase with temperature is correctly depicted by the model. The activation energies related to the growth rate are 396 kJ/mol and 226 kJ/mol for the calculated and experimental values respectively. Even if the calculated one is a bit overestimated, it keeps the same order of magnitude. The model should be further refined so as to better match the experimental data. The activation energy related to the carbon solubility in pure liquid silicon (from Eq. 3) is 203 kJ/mol. The two experimental activation energies have very close values, much higher than for a transport limited process. The carbon solubility in liquid silicon thus seems to be the limiting parameter for the growth rate.

Conclusion

The present work aimed at developing top seeded solution growth (TSSG) as an alternative to the standard industrial PVT process for the growth of large size and high quality single crystalline SiC ingots. A numerical model has been implemented with the aim to give quantitative outcomes in addition to qualitative information. We have first developed a specific crucible design with very low temperature gradients (1°C/cm). Then, the carbon concentration and the supersaturation distributions have been computed to evidence the dissolution and growth areas. As expected, we found a direct correlation between the concentration fields and the convection patterns. This emphasizes the primary role of the convection loops on the carbon transport from the dissolution to the crystallization zones. Growth rates have been calculated for three different temperatures (1700, 1800 and 1900°C) in pure silicon. In addition, the growth rate profile along the crystal radius has been computed. We obtained a pretty good agreement between the computed and the experimental values. The numerical tool is now ready to assist the future development of the process; the next step consists of increasing the growth rate.

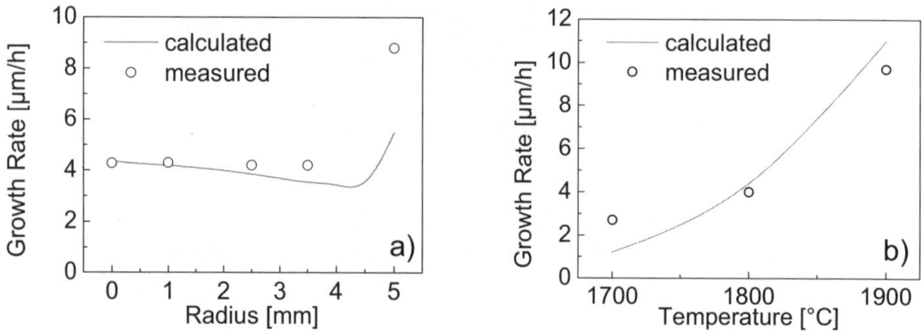

Fig. 3: Correlation between calculated and measured growth rates. a) Growth rate profile along the crystal radius, 0 corresponds to the symmetry axis of the crucible. The experiment was done at 1800°C. b) Plot of the growth rate taken on the symmetry axis of the crucible, for three different temperatures, 1700, 1800 and 1900°C. The crystal rotation rate was 40 rpm.

Acknowledgements

The authors warmly acknowledge Dr. F. Mercier from AIST Tsukuba (Japan) for fruitful discussions. This work has been supported by the French ANR project MINTEX Contract No. ANR-09-BLAN-0189-01.

References

[1] D.H. Hofmann and M.H. Muller, Mater. Sci. Eng. B-Solid State Mater. Adv. Technol. 61-2 (1999) 29-39.

[2] D. Chaussende, P.J. Wellmann, and M. Pons, Journal of Physics D-Applied Physics 40 (2007) 6150-6158.

[3] K. Kamei, K. Kusunoki, N. Yashiro, N. Okada, T. Tanaka, and A. Yauchi, Journal of Crystal Growth 311 (2009) 855-858.

[4] F. Mercier, J.M. Dedulle, D. Chaussende, and M. Pons, Journal of Crystal Growth 312 (2010) 155-163.

[5] F. Durand and J.C. Duby, Journal of Phase Equilibria 20 (1999) 61-63.

CHAPTER 2:

SiC Epitaxial Growth

Materials Science Forum Vols. 717-720 (2012) pp 75-80
© (2012) Trans Tech Publications, Switzerland
doi:10.4028/www.scientific.net/MSF.717-720.75

SiC Epitaxial Layer Growth in a 6×150 mm Warm-Wall Planetary Reactor

A. A. Burk, D. Tsvetkov, D. Barnhardt, M. J. O'Loughlin, L. Garrett, P. Towner, J. Seaman, E. Deyneka, Y. Khlebnikov, and J. Palmour

Cree, Inc., 4600 Silicon Dr., Durham, NC USA

al_burk@cree.com

Keywords: SiC Epitaxy, Warm-Wall Planetary reactors, 150-mm diameter

Abstract. Initial results are presented for SiC-epitaxial growths employing a novel 6x150-mm/10x100-mm Warm-Wall Planetary Vapor-Phase Epitaxial (VPE) Reactor. The increased areal throughput offered by this reactor and 150-mm diameter wafers, is intended to reduce the cost per unit area for SiC epitaxial layers, increasing the market penetration of already successful commercial SiC Schottky and MOSFET devices [1]. Growth rates of 20 micron/hr and short <2 hr fixed-cycle times (including rapid heat-up and cool-down ramps), while maintaining desirable epitaxial layer quality were achieved. No significant change in 150 mm diameter wafer shape is observed upon epitaxial growth consistent with good-quality, low-stress substrates and low (<5°C) cross-wafer epitaxial reactor temperature variation. Specular epitaxial layer morphology was obtained, with morphological defect densities consistent with projected 5x5 mm die yields as high as 80% and surface roughness, Ra, of 0.3 nm. Intrawafer thickness uniformity is good, averaging only 1.6% and within a run wafer-to-wafer thickness variation is 2.7%. N-type background doping densities less that 1E14 cm^{-3} have been measured by CV. Doping uniformity and wafer-to-wafer variation currently average ~12% requiring further improvement. The first 100 μm thick 150-mm diameter epitaxial growths are reported.

Introduction

Initial results are presented for a novel 6x150/10x100-mm Warm Wall Planetary SiC VPE Reactor. The Warm-Wall reactor configuration was first applied to SiC in a 7x2-inch configuration [2] and has been steadily scaled to larger capacity configurations culminating most recently in the 8x100 mm [3] and 10x100-mm [4] configurations. These increases have resulted in a 30x increase in SiC epitaxial layer throughput of up to 20 micron thick layers suitable for 1-2kV SiC power devices. Further scaling from 8x100 mm to 6x150-mm capacity would double the areal throughput yet again. Moreover, increases in growth rate from typical 10 micron/hr would allow even more gains. A much greater impact in throughput (hence device cost reduction) would be realized in the 100 micron 10kV class of power devices as currently they are produced in lower throughput horizontal hot-wall reactors [5]. However, to realize these goals, the requisite layer quality, defect density and uniformity must be achieved.

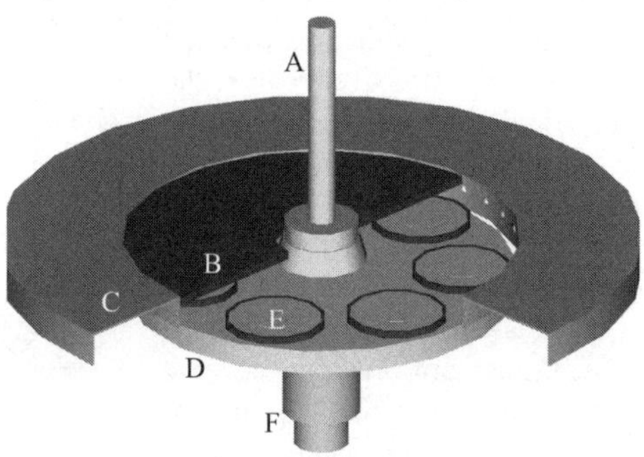

Fig. 1 Generic view of a Multi-Wafer Warm-Wall VPE reactor including, a) water-cooled gas injector, b) warm-wall (ceiling), c) exhaust gas collector, d) susceptor, e) wafer holder, and f) rotating susceptor support.

Warm-Wall Planetary Reactor Design. As shown in Figure 1, the gases in a planetary reactor enter from the center of the reactor and flow outward radially. A unique feature of the design is the use of a gas to levitate and rotate the wafers. Given this geometry, optimized growth conditions are only required over a single wafer diameter. Considering the high 1550-1600°C temperatures required for good SiC crystal quality, and relatively low cracking temperatures of silane 600-700°C, this is particularly helpful for SiC epitaxy. In addition, because of the increase in cross-sectional area with reactor radius, a rapid drop in gas velocity occurs along the flow direction, resulting in a steep drop in reagent concentration and an increase in the boundary layer thickness. For the Si-diffusion limited growth conditions typically employed in SiC epitaxy, this yields a steep but nearly linear drop in growth rate as a function of susceptor radius. The rotation of the wafers about their individual axes effectively averages this linear trend, yielding highly uniform layers. The reactor is also inherently efficient in its use of reagents and carrier gas given the steep depletion of reagents and low reactor volume.

Experimental

4-degree off-oriented, Si-face, 4HN, 150-mm diameter substrates with thicknesses of 500 to 1000 microns were used for these experiments. Short (compared to Semi specifications) 25 mm primary and 15 mm secondary flats were used to minimize epitaxial-area reduction. Prior to epitaxy, the wafers were chemically mechanically polished (CMP) resulting in a surface roughness, Ra, of <0.3 nm. Due to the limited availability of these early 150 mm R&D substrates, only 20 epitaxial runs were performed in the 6x150-mm configuration, with only two containing a full load of six 150-mm diameter wafers. Therefore, to obtain more statistically significant wafer-to-wafer and run-to-run data, additional runs were performed in the same reactor in a 10x100-mm configuration. The epitaxial growths were all performed at reduced pressure and 1550-1600˚C temperatures. Rapid heat and cool down rates enabled total fixed cycle times (excluding epitaxial layer growth) of less than 2 hrs. The growth rate used in this study was ~20 microns/hr.

Results and Discussion

Temperature Uniformity and Wafer Shape. A key requirement for growing highly uniform, flat, device-quality SiC epitaxial layers is to achieve good temperature uniformity over the entire area of each wafer and from wafer to wafer. Figure 2 contains a plot of the wafer and susceptor temperature, at temperatures approaching that of epitaxial growth, as viewed from above by an optical pyrometer. Due to certain geometrical constraints, the radial position of the pyrometer is 25 mm beyond the center of the 150 mm diameter wafers. Inspection of the temperature data indicates that the 6 wafers are ~30°C cooler than the susceptor due to the gas levitation and rotation of the wafers. Importantly, the total wafer-to-wafer and intrawafer temperature range is ~5˚C.

Consistent with the good temperature uniformity is the observation of no significant change in wafer shape after epitaxial layer growth. Figure 3 contains a contour plot of the flatness of a 1000 micron thick 150-mm epitaxial wafer having bow and warp of 12.2 and 21.3 microns respectively.

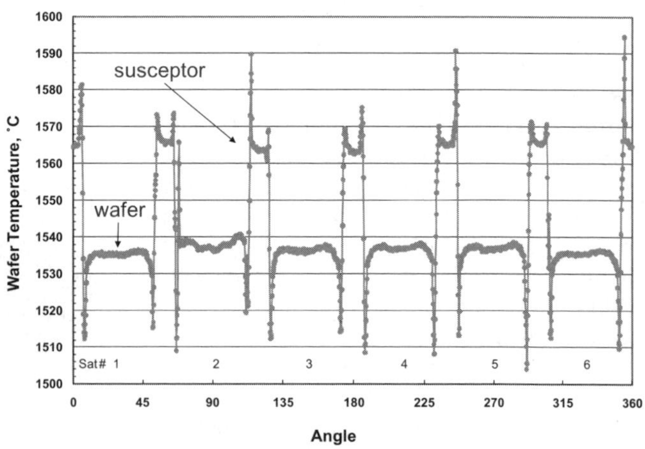

Fig. 2 Top pyrometer temperature as a function of rotation angle for the 6x150 mm, Warm-Wall planetary VPE reactor showing ~5˚C intrawafer and wafer-to-wafer temperature range.

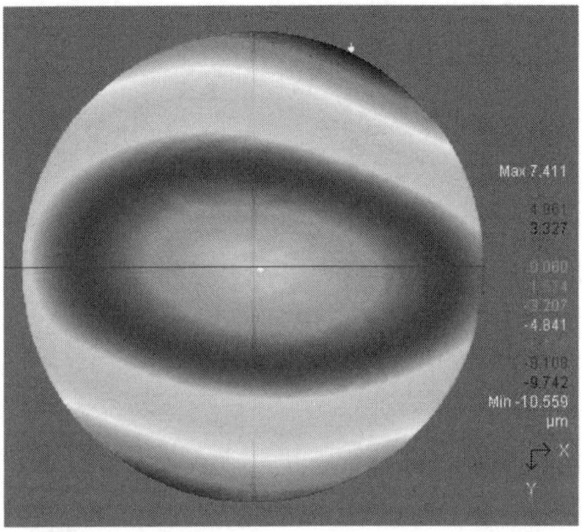

Fig. 3: Tropel shape contours for a 150-mm diameter epitaxial wafer with bow and warp of 12.2 and 21.3 microns, respectively (these values are relatively unchanged from pre-epi values).

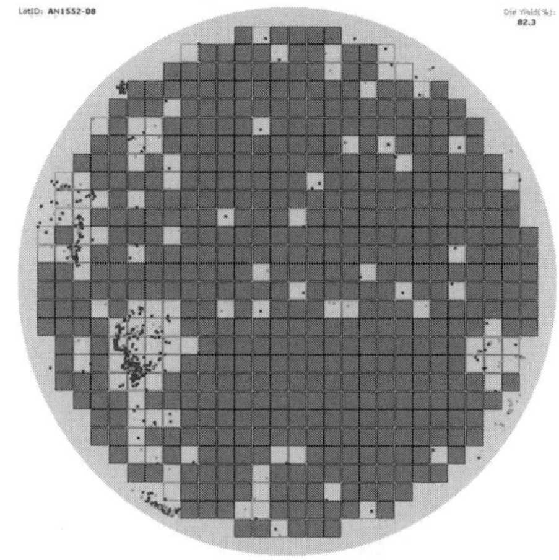

Fig. 4: Candela defect map of a 15 μm thick 150-mm diameter epi wafer. A superimposed 5x5 mm pattern projects a ~80% yield. (defect clusters on the left side of the wafer appeared on adjacent slices so were boule related).

Morphology. Another critical epitaxial material's requirement to enable high device yields is low defect density. In addition, certain device applications may also require low surface roughness. Very promising epitaxial layer morphology and roughness has been observed on layers grown on the initial 150 mm R&D substrates. Figure 4 contains a representative plot of the output of a Candela laser-light scattering measurement. The Candela recipes have been adjusted so that morphological features that are detected on our 100-mm diameter epitaxial wafers correlate well with device failures. The sources of these morphological defects include substrate crystalline defects, surface preparation and the epitaxial growth process. Further investigation will be required to quantitatively determine and reduce the relative contributions of these defects on 150-mm diameter epitaxial wafers and to establish a correlation with device yield. However, as shown in Figure 4, even for 20 micron thick epitaxial layers grown on early R&D wafers with high (but clustered) crystaline defects (total wafer defect density 4.4 cm^{-2}; ~1 cm^{-2} away from crystal defects), have resulted in projected 5x5 mm die yields of ~80%. Further improvements in crystal quality would obviously result in higher projected yields. Additionally, surface roughness of 0.3 nm (Ra), near that of the detection limits on our Zygo interferometer, have been obtained for these epitaxial layers. The combined morphology, defect density and roughness already appear to be suitable for 1-2kV class power electronic device development, however not for production.

Thickness and Doping Uniformity. To achieve desirable distributions of device parameters such as blocking voltage and forward current requires uniform epitaxial layers. Increasing wafer diameter is only economically advantageous from a yield perspective given adequate intrawafer uniformity, intrarun wafer-to-wafer and run-to run variation. Initial intrawafer layer thickness uniformity and intrarun variation are very encouraging in the 6x150 mm configuration. Figure 5 contains a polar plot of the epitaxial layer average and center thickness from a full 6x150-mm growth run as measured by FTIR. Average thickness was ~15 microns, intrawafer thickness uniformity and wafer-to-wafer variation were reasonable, averaging only 1.6% (10 mm edge exclusion) and 2.7% σ/mean respectively.

Intrawafer doping uniformity and intrarun doping variation, however, still require significant improvement. Figure 6 shows the radial dependence of intentional doping near 1E16 cm^{-3} for the same 6 wafers shown in figure 5. In general, there is an undesirable increase in doping toward the

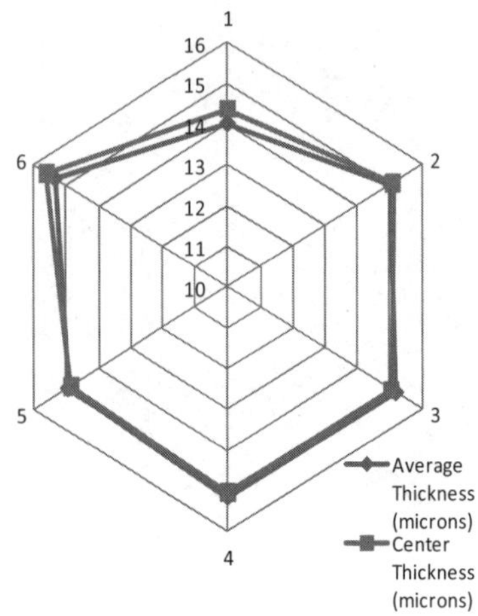

Fig. 5 Polar plot of wafer thickness from a 6x150-mm epitaxial growth run. Mean thickness is 14.9 microns with an average intrawafer uniformity of 1.6% σ/mean and 2.7% σ/mean intrarun variation.

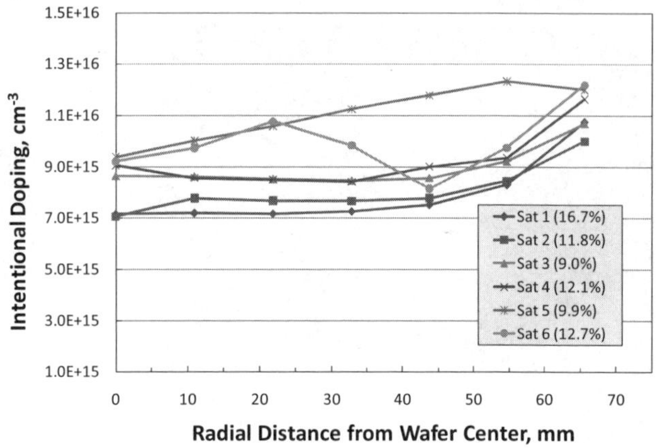

Fig. 6 Wafer-to-wafer doping concentration as a function of wafer radius from the same 6x150-mm epitaxial growth run as in Fig. 5. Mean doping is 9.77E15 cm^{-3} with an average uniformity of 12.8% σ/mean and 12% σ/mean wafer-to-wafer variation.

wafer edge (uniformity averaging 12.8% σ/mean, 9.35 mm edge exclusion). Wafer-to-wafer doping variation is also too high at ~12%. The cause of these greater than desirable doping uniformities and variations will require further study, note however background doping levels are already sufficiently low at <1E14 cm^{-3} for most power-device doping profiles.

Run-to-Run Epitaxial Layer Reproducibility (on 100 mm Wafers). Given the limited number of epi runs performed to date in the 6x150-mm configuration, intrarun and wafer-to wafer layer data is also presented for growths performed in the same reactor using a 10x100-mm susceptor. Given past experience with wafer diameter increases, it is anticipated that with more 150-mm wafers and development, the intrarun and run-to-run results will be similar for 100 mm and 150 mm wafers. Table 1 contains a compilation of results for over three hundred 100-mm wafers. As shown in Table 1 and the histogram contained in Figure 7, the mean and median thickness uniformities are ~1% (7 mm edge exclusion) with a tight distribution; the middle two quartiles range only from 0.89 to 1.075%. Intra-run wafer-to-wafer total thickness range averages only 2%. While run-to-run thicknesses variation is not shown, this adds a negligible contribution to the total thickness range as for each run, slight increases in layer growth times are made to correct for the modest, <5%, drop in growth rate that occurs across an entire epitaxial growth campaign (in between required maintenance).

Table 1: 10x100-mm Configuration Layer Statistics

Attribute	Mean	Median
Thickness Uniformity, % σ/mean	0.99	0.96
Doping Uniformity, % σ/mean	3.1	2.6
IntraRun Thickness Range, %/mean	2.0	1.5
IntraRun Doping Range, %/mean	11.4	7.9
Morphological Defect Density, cm^{-3}	0.84	0.72
Projected 5x5-mm Die Yield, %	89	90

Intrawafer doping uniformities average 3.1% with a 9.35 mm edge exclusion (for 100 mm wafers, sampling indicates similar standard deviation using a smaller 5 mm exclusion for this parameter). The average intrarun wafer-to-wafer doping range is 11% and given the increased sensitivity of doping to many parameters, more outliers are observed than for thickness. However, 90% of this distribution still has a doping range of less than 16%. Conversely, from a yield to doping spec perspective, that ~10% of the runs have one or more wafers outside of a ±8% doping window still leaves room for further improvement. And finally, while zero defect density is the ultimate goal for

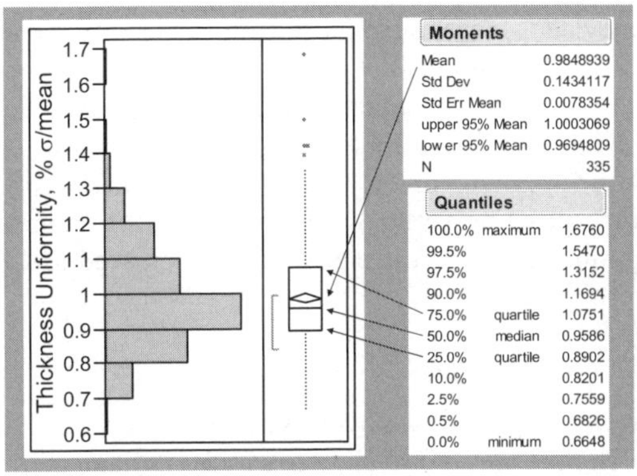

Fig. 7: Histogram, box plot and statistical data for the thickness uniformity (7 mm edge exclusion) for over three hundred 100-mm diameter epitaxial wafers with an average uniformity of 1% σ/mean.

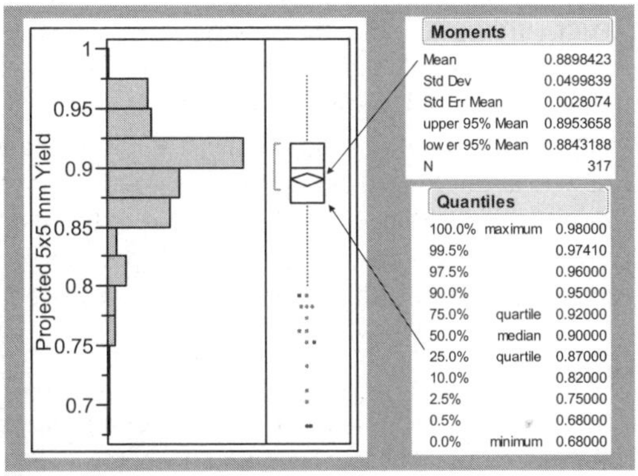

Fig. 8 Histogram, box plot and statistical data for the projected 5x5-mm die yield (by Candela analysis) for over three hundred 100-mm diameter epitaxial wafers with an average yield of 89%.

ever increasing device sizes, current morphological defect densities in the 10x100-mm configuration, including all the contributions from the substrates and epitaxial growth process, are now only ~0.8 defect/cm^2, half of what we reported for the 8x100-mm reactor[3]. As shown in Figure 8, these defect densities are consistent with up to 90% yields of large 5x5 mm power devices. While many more growths and further substrate and epitaxial growth process development will no doubt be required, it is anticipated that ultimately, similar results will be achievable in the 6x150-mm configuration.

Thick Epilayer Demonstration and Lifetime. While the near term goals for the 6x150-mm Warm-Wall reactor are for 10 to 20 micron thick epitaxial layers, suitable for 1-2kV power devices such as Schottkys and MOSFETs, further device developments in 10 kV class power devices could result in significantly increased demand for 10kV and higher devices requiring epi layer thicknesses well over 100 microns. Currently, these very thick layers are preferably grown in Hot-Wall reactors [5]. These reactors produce the highest-quality, lowest background doping density (<5E13 cm^{-3}) and lowest defect density (0.25 cm^{-2}, ~80% 5x5-mm yield) epitaxial material, but, at lower throughput. Given this throughput constraint, and as attempting to grow thick epi layers is an excellent way to access and improve the quality of thinner epitaxial layers, run-to-run layer reproducibility and maintenance cycle durations, we performed one ~110 micron thick growth run in the 10x100-mm (using test-wafer grade substrates) and one in the 6x150-mm configuration.

The morphology of the ~110 micron thick layers were still specular, however, there were now many (dozens) of large point defects visual to the unaided eye. The morphology on the 150 mm diameter wafers were comparable to those on the 100 mm test wafers. Naturally, due to the step flow growth mechanism, the lateral dimensions of many defects scale with layer thickness increasing their ease of detection. In addition, the longer growth times and additional parasitic growth can increase the probability of fall-down and other sources of unwanted particulates. This may help explain the observed increase in total wafer Candela defect densities up to 5-6 cm^{-2}. Given again the clustered nature of the morphological defects induced by crystalline defects, projected 5x5-mm die yields range from 55-62% for the thick epitaxial layers. These defect densities are not yet very promising for 10kV device production, however, the ability to even grow such thick layers by Warm-Wall epitaxy represents a milestone. Note also, away from morphological defects, the surface roughness, Ra, was still relatively smooth at ~1.2 nm.

Another material's property that becomes critical for thick layer bipolar device operation is carrier lifetime. Encouragingly, even at the elevated morphological defect levels of these preliminary samples grown in the 6x150-mm reactor, the post-epi lifetime values for the 100-mm diameter epitaxial wafers were still ~3 microseconds. After subsequent oxidation and annealing was performed (as first developed by Hiyoshi and Kimoto [6]), the lifetime increased to over 6 μs. The ~110 micron thick epi layers grown on the 150-mm diameter wafers initially had lesser lifetimes of ~2 microseconds and also exhibited some discrete, vertically-oriented lines of reduced lifetime. After oxidation was performed on the 150-mm sample its average lifetime increased to 2.8 microseconds (annealing has not been completed yet).

Summary

Initial results were presented for SiC-epitaxial growths employing a novel 6x150-mm, Warm-Wall Planetary Vapor Phase Epitaxial Reactor. The demonstrated combination of 150-mm diameter wafers, growth rates of 20 micron/hr and short <2 hr fixed-cycle times allow a >2x increase in epitaxial areal throughput in comparison to our previous reports [3]. No significant change in wafer shape was observed upon epitaxial growth consistent with good-quality, low-stress substrates and good intrawafer temperature uniformity. Specular epilayer morphology was observed with defect densities of ~4 cm^{-2}, 5x5 mm projected die yields currently as high as 80% and low surface roughness of ~0.3 nm. Even higher projected yields are expected with the reduction of crystalline defects and improvements in surface preparation. Intrawafer thickness uniformity was good, averaging only 1.6% and within a run wafer-to-wafer variation was 2.7%. N-type background doping densities less that 1E14 cm^{-3} were measured by CV. Both doping uniformity and wafer-to-wafer variation currently average ~12% requiring further improvement in the epitaxial growth process. Due to the limited number of and early stage of development for the currently available 150 mm wafers, additional runs were performed in the 6x150-mm reactor using a 10x100-mm susceptor to better determine wafer-to-wafer and run-to run trends. The average intrarun thickness and doping ranges for 100 mm diameter wafers were 2 and 11% respectively. Defect densities were less than 1 cm^{-2} resulting in projected 5x5-mm die yields of 90%. While further improvements are still required in substrate quality and the epitaxial growth process, these early results suggest that 6x150-mm epitaxial material should be of sufficient quality, quantity and throughput to help reduce the cost of 1-2kV power devices

While only a single demonstration, the first 115 μm thick epitaxial layers were grown on 150-mm diameter wafers in the 6x150-mm Warm-Wall reactor. While these epitaxial layers currently exhibit even higher defect densities, 5-6 cm^{-2}, resulting in 5x5-mm projected die yields of only ~55%, they too should significantly improve with increased crystal quality. Post-oxidation lifetimes of 2.8 microseconds were observed on these epitaxial materials. This, albeit very limited data, suggests that in the future there may also be a role for multi-150-mm diameter epitaxial materials in 10kV-class SiC power-electronic devices.

References

[1] http://www.cree.com/press/press_detail.asp?i=1295272745318

[2] A. A. Burk, M. J. O'Loughlin and S. S. Mani: Mater. Sci. Forum Vol.264-268 (1998), p. 83.

[3] A. A. Burk, M. J. O'Loughlin, J. J. Sumakeris, C. Hallin, E. Berkman, V. Balakrishna, J. Young, L. Garrett, K. G. Irvine, A. R. Powell, Y. Khlebnikov, R. T. Leonard, C. Basceri, B. A. Hull and A. K. Agarwal Mater. Sci. Forum Vol. 600-603 (2009) p. 77.

[4] C. Hecht, R .Stein, B. Thomas, L. Wehrhahn-Kilian, J. Rosberg, H. Kitahata, F.Wischmeyer, Mater. Sci. Forum Vol. 645-648 (2010) p. 89.

[5] M. J. O'Loughlin, K.G.Irvine, J. J. Sumakeris, M. H. Armentrout, B. A. Hull and A. A. Burk, Jr., Mater. Res. Soc. Symp. Proc. Vol. 1069, 1069-D04-01 (2008), p. 115.

[6] T. Hiyoshi and T. Kimoto, Appl. Phys. Express 2 (2009) 091101.

Materials Science Forum Vols. 717-720 (2012) pp 81-86
© (2012) Trans Tech Publications, Switzerland
doi:10.4028/www.scientific.net/MSF.717-720.81

Growth of 4H-SiC Epilayers and $Z_{1/2}$ Center Elimination

Tetsuya Miyazawa[1,a]* and Hidekazu Tsuchida[1,b]

[1]Central Research Institute of Electric Power Industry (CRIEPI), 2-6-1 Nagasaka, Yokosuka, Kanagawa 240-0196, Japan

[a]yo-miya@criepi.denken.or.jp, [b]tsuchida@criepi.denken.or.jp

Keywords: Epitaxial growth, Point defect, Carrier lifetime, Deep level transient microscopy, Time-resolved photoluminescence

Abstract. Thick and low-doped epilayers with a low $Z_{1/2}$ center concentration were grown on 8° off-cut 4H-SiC(0001)Si-face substrate. Two post-growth processes, namely, the C^+-implantation/annealing process or the thermal oxidation/Ar annealing process, were applied to the thick epilayers. The dependence of the $Z_{1/2}$ center concentration and the carrier lifetime on process conditions was investigated. Under proper conditions, both processes could eliminate the $Z_{1/2}$ center to a depth of 100 μm or more, and considerably improved the carrier lifetime while maintaining the surface roughness comparable to that of the as-grown sample. The effect of the post-growth processes applied on C-face is also presented.

Introduction

Control of carrier lifetime in thick and low-doped drift layers is critical to realize bipolar power devices with low power dissipation. The $Z_{1/2}$ center, a major intrinsic point defect in as-grown n-type 4H-SiC epilayers, has been recognized as a prominent recombination center that significantly reduces the carrier lifetime [1,2]. Since the concentration of the $Z_{1/2}$ center in epilayers varies with some growth parameters, e.g., growth temperature and C/Si ratio [3,4], it can be minimized by tuning the appropriate growth parameters. However, typical $Z_{1/2}$ center concentrations in epilayers remain on the order of 10^{12} - 10^{13} cm^{-3} and the carrier lifetime is usually limited to a few microseconds at most. To realize very high voltage bipolar devices (>10 kV) with excellent forward characteristics, longer carrier lifetimes, i.e., lower $Z_{1/2}$ center concentrations in the drift layers, are needed.

Two post-growth processes, namely, the C^+-implantation/annealing process [5] and the thermal oxidation/Ar annealing process [6], have been proposed to reduce the $Z_{1/2}$ center concentration and improve the carrier lifetime in n-type 4H-SiC epilayers. Both processes utilize in-diffusion of carbon interstitials which originate from the implanted layer or the oxide/epilayer interface. The in-diffused carbon atoms are considered to eliminate the $Z_{1/2}$ center, which is possibly a carbon vacancy-related defect [7]. After these post-growth processes, the $Z_{1/2}$ center concentrations in the epilayers decreased below the detection limit of deep level transient spectroscopy (DLTS) measurements. The bulk lifetimes of the thick epilayers after the post-growth processes were estimated to be roughly 20 μs or even longer [8,9].

To control the carrier lifetime, the effect of these post-growth processes on very thick drift layers should be quantitatively demonstrated. However, the relationship between these process parameters and the $Z_{1/2}$ center concentration or carrier lifetime has not been clarified yet. In this paper, an epilayer with enough thickness and doping concentration for very high voltage bipolar devices (>10 kV) is grown in a vertical-type reactor. The carrier lifetimes and the depth profiles of $Z_{1/2}$ center concentration in the epilayer before and after the post-growth processes are investigated. The effect of the post-growth processes applied on C-face epilayer is also presented.

Experimental

Thick (~140 μm) 4H-SiC epilayer with a nitrogen doping concentration of ~3×10^{14} cm^{-3} was grown on a 3-inch diameter 4H-SiC(0001) 8° off-axis commercial substrate in a vertical hot wall reactor [10]. The post-growth processes were applied to the diced epi-wafers under different conditions. In the C^+-implantation/annealing process, carbon-ions were implanted at 600 °C to form a 250 nm depth box profile with concentrations of 5×10^{18} cm^{-3} or 5×10^{20} cm^{-3}, and high temperature annealing was performed under an Ar atmosphere at 1500-1800 °C for 30 min with a carbon cap. After annealing, the implanted layer was removed by inductively coupled plasma etching to a depth of 1 μm. In the thermal oxidation/Ar annealing process, dry oxidation was performed once or twice at 1200 or 1300 °C for 5 h. Note that the oxide was removed by HF between the two times oxidation. After removal of the oxide layer, high temperature annealing at 1550 °C was performed for 30 min with a carbon cap under an Ar atmosphere as reported in literature [6].

The carrier lifetimes of the epilayers before and after the post-growth processes under different conditions were measured with time-resolved photoluminescence (TRPL) at room temperature [8]. The 355 nm line of a frequency-tripled Nd:YVO_4 laser was used to inject excess carriers in SiC. The photon density irradiated on epilayer surface was $\sim3\times10^{12}$ cm^{-2}. Note that no intentional surface passivation was done before the TRPL measurement. The surface roughness of the epilayers was measured with atomic force microscope (AFM).

The depth profile of the $Z_{1/2}$ center concentration in the epilayers was investigated by successively polishing epilayers in 25 μm steps with 1 μm diameter diamond slurry. After each step, Ni Schottky contacts with a diameter of 1 mm were formed on the epilayer, and aluminum was deposited on the backside of the substrate to make an ohmic contact. The DLTS spectra were obtained in the temperature range from 200 to 550 K. The reverse bias and the pulse voltage were typically -10 V and 0 V, respectively.

Results and discussion

Epitaxial growth. An epilayer of 140 μm thickness was grown on a 3-inch diameter substrate with a growth rate of 21 μm/h. To obtain a low $Z_{1/2}$ center concentration for the as-grown epilayer, temperature and C/Si ratio were tuned to be 1640 °C and 1.0, respectively. The net doping concentration (N_d-N_a) of the epilayer was determined to be 2-3×10^{14} cm^{-3} from C-V measurement. The typical surface roughness (RMS roughness) was 0.2 nm in a 10×10 μm^2 area from AFM analysis. A relatively low $Z_{1/2}$ center concentration of 1-3×10^{12} cm^{-3} was confirmed by DLTS measurement. The carrier lifetime distribution of the 3-inch diameter epi-wafer, which was obtained by TRPL mapping, is shown in Fig. 1. The carrier lifetime of the as-grown sample was 1.59 μs on the average with very good uniformity (σ=0.07 μs).

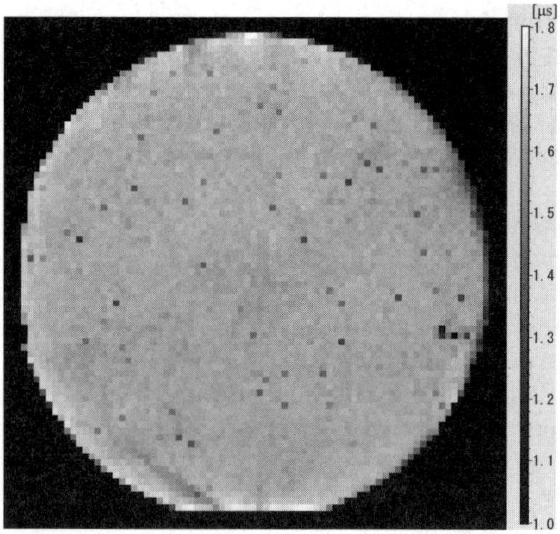

Fig. 1 Carrier lifetime mapping result of a 3-inch diameter 140 μm-thick 4H-SiC epilayer.

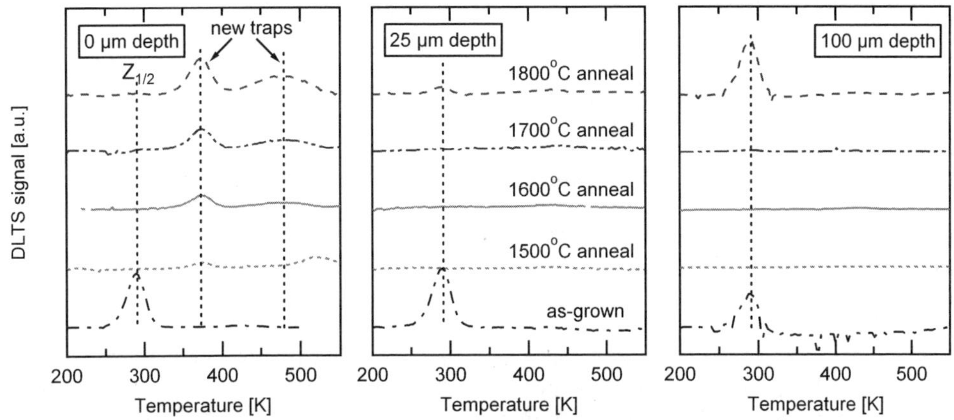

Fig. 2 DLTS spectra of 4H-SiC epilayers before and after C^+-implantation/annealing process at different depths.

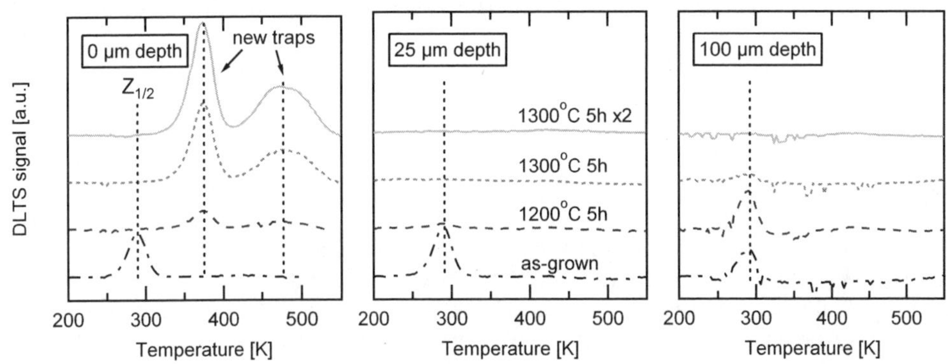

Fig. 3 DLTS spectra of 4H-SiC epilayers before and after thermal oxidation/Ar annealing process at different depths.

$Z_{1/2}$ center elimination. The DLTS spectra before and after the C^+-implantation (5×10^{20} cm^{-3})/annealing processes with different annealing temperatures are shown in Fig. 2. The $Z_{1/2}$ center was the only trap observed in the as-grown sample within the temperature range. After the post-growth processes, no $Z_{1/2}$ peak was observed while two new traps at around 0.9 and 1.0 eV respectively below the conduction band were detected near the surface (before polishing). These two traps disappeared after 25 μm polishing. Samples annealed at 1500-1700 °C showed no $Z_{1/2}$ peak up to 100 μm depth. However, the $Z_{1/2}$ peak was clearly observed at a depth of 100 μm in the sample annealed at 1800 °C.

The DLTS spectra before and after the thermal oxidation/Ar annealing processes with different oxidation conditions are shown in Fig. 3. As the case with the C^+-implantation/annealing process, the $Z_{1/2}$ center was eliminated although the two new traps appeared near the surface. The sample with oxidation at 1300 °C for 5 h twice showed no $Z_{1/2}$ peak up to 100 μm depth, whereas the $Z_{1/2}$ peak comparable to the as-grown sample was observed at a depth of 100 μm in the sample with single 5 h oxidation at 1200 °C.

The depth profiles of the $Z_{1/2}$ center concentration are shown in Fig. 4 for samples in which $Z_{1/2}$ center was detected. In the as-grown epilayer, the $Z_{1/2}$ center concentration decreased slightly with depth, whereas the doping concentration was unchanged. In the sample with C^+-implantation/annealing at 1800 °C, the $Z_{1/2}$ center concentration gradually increased with depth, and surpassed that of the as-grown material at a depth of 100 μm, indicating that excessively high annealing temperature might increase the $Z_{1/2}$ center concentration. In the sample with thermal oxidation at 1200 °C for 5 h and Ar annealing, the $Z_{1/2}$ center concentration increased with depth, and reached the same level as the as-grown material at depths from 50 to 100 μm, indicating the process condition was not sufficient to eliminate $Z_{1/2}$ center in a 100 μm thick epilayer.

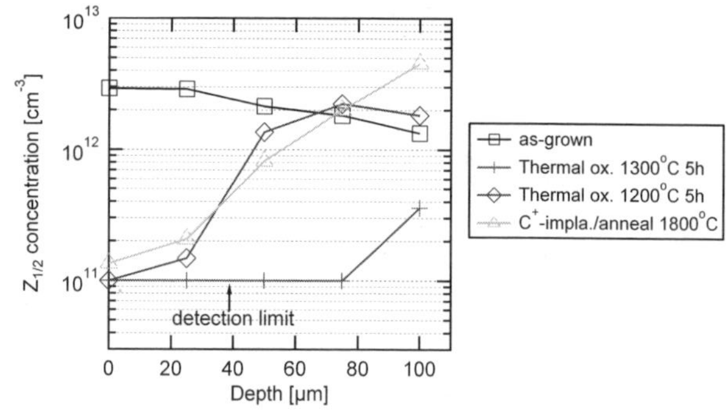

Fig. 4 Depth profiles of $Z_{1/2}$ center concentration in 4H-SiC epilayers. Only samples in which $Z_{1/2}$ center was detected are shown.

After C^+-implantation/annealing at 1500-1700 °C or thermal oxidation at 1300 °C for 5 h twice and Ar annealing, the $Z_{1/2}$ center concentration decreased below the detection limit of DLTS to a depth of 100 μm. In addition, lower dose (5×10^{18} cm^{-3}) C^+-implantation samples annealed at 1500 or 1600 °C were also tested, and the $Z_{1/2}$ center concentration was reduced below the detection limit to a depth of 100 μm, indicating the lower dose C^+-implantation was enough for elimination of $Z_{1/2}$ center in this epilayer.

Carrier lifetime. The carrier lifetimes were extracted from the TRPL decay curves of epilayers before and after the C^+-implantation/annealing process. The TRPL measurement was conducted at one point for each sample. Figure 5(a) shows the effect of the annealing temperature on carrier lifetimes. Note that the TRPL measurements were performed before the successive polishing. All samples with the C^+-implantation/annealing process showed longer lifetimes compared to the as-grown sample. An exceptionally long lifetime was observed in the samples annealed at 1600 °C.

The carrier lifetimes of epilayers before and after the thermal oxidation/Ar annealing process with different oxidation conditions are shown in Fig. 5(b). The sample with thermal oxidation at 1300 °C for 5 h twice exhibited an especially long carrier lifetime. From the TRPL measurements, longer carrier lifetimes were evaluated in samples with lower $Z_{1/2}$ center concentrations. Since the carrier lifetimes obtained from the TRPL measurement include the contribution of carrier recombination at the surface and the interface, the bulk lifetimes should be longer than the values presented in Fig. 5(a) and (b).

Surface roughness. Surface roughness of samples with the C^+-implantation/annealing process is shown in Fig. 6. The high-dose (5×10^{20} cm^{-3}) implantation of carbon ions did not significantly affect the surface roughness of the epilayer. In the subsequent annealing process, RMS roughness increased with the annealing temperature. While the RMS roughness was significantly increased by annealing at 1700 or 1800 °C, it was equivalent to that of the as-grown sample after annealing at 1500 or

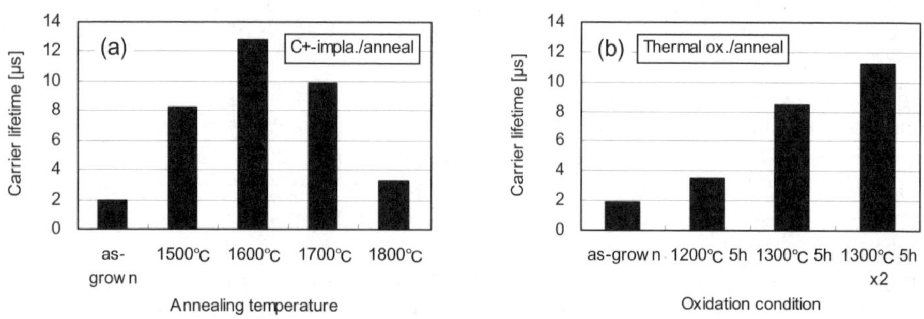

Fig. 5 Carrier lifetimes extracted from TRPL decay curves of 4H-SiC epilayers before and after (a) C^+-implantation/annealing and (b) thermal oxidation/Ar annealing process.

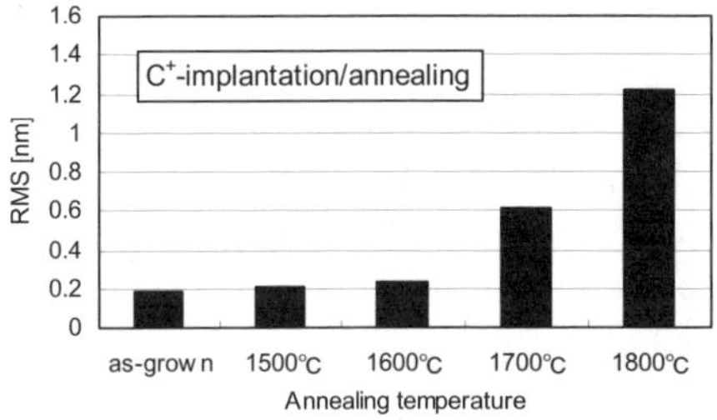

Fig. 6 Surface roughness of 4H-SiC epilayers before and after C^+-implantation/annealing process at different annealing temperatures.

1600 °C. The mid-dose (5×10^{18} cm^{-3}) implantation and annealing at 1500 or 1600 °C did not induce surface roughening either. Significant changes in the surface roughness of epilayers after the thermal oxidation/Ar annealing processes were not observed.

Post-growth process on C-face. Elimination of the $Z_{1/2}$ center in an epilayer grown on the C-face was also tried with the post-growth processes. The epilayer with a thickness above 130 μm was grown on a 4H-SiC(000-1) 8° off-axis substrate. The net doping concentration (N_d-N_a) was 6-7×10^{14} cm^{-3}. The $Z_{1/2}$ center concentration of the as-grown sample was 1.9×10^{14} cm^{-3}, which is about two orders of magnitude higher than that of the Si-face epilayer.

Figure 7 shows the depth profile of the $Z_{1/2}$ center concentration in the C-face epilayer. The high-dose (5×10^{20} cm^{-3}) C^+-implantation and following annealing at 1600 °C for 30 min eliminated the $Z_{1/2}$ center near the surface, however, the $Z_{1/2}$ center concentration at depths from 25 to 100 μm was equivalent to that of the as-grown sample. The thermal oxidation at 1300 °C for 5 h twice and following annealing at 1550 °C for 30 min reduced the $Z_{1/2}$ center concentration only by half near the surface.

The TRPL decay curves of the C-face epilayer before and after the post-growth processes are shown in Fig. 8. The decay curve for the as-grown epilayer shows an extremely short carrier lifetime of 25 ns because of the very high concentration of the $Z_{1/2}$ center. While the thermal oxidation/Ar annealing process did not affect the decay curve, the C^+-implantation/annealing process enhanced the carrier lifetime. Within the conditions tested in this study, C^+-implantation/annealing process was more effective for the reduction of the $Z_{1/2}$ center concentration in the C-face epilayer. However, since the growth conditions were less optimized for C-face and the epilayer used for this study

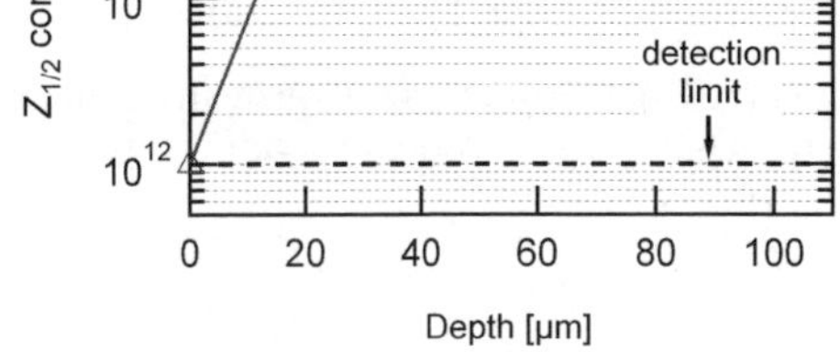

Fig. 7 Depth profile of $Z_{1/2}$ center concentration in 4H-SiC epilayer grown on C-face.

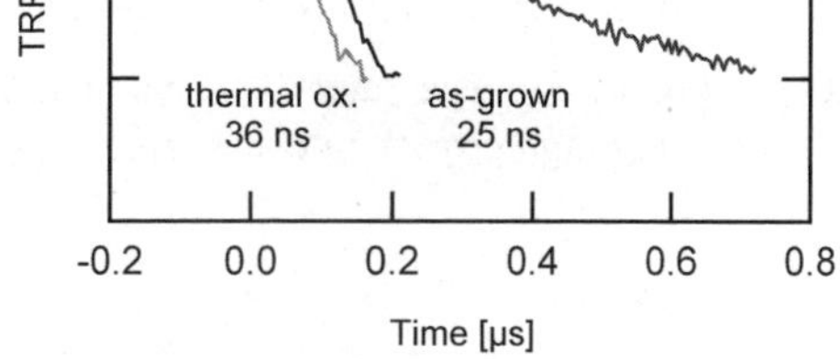

Fig. 8 TRPL decay curves of C-face 4H-SiC epilayers before and after post-growth processes.

included a high concentration of $Z_{1/2}$ center, further investigation for C-face epilayers with initially low $Z_{1/2}$ center concentration will be needed.

Summary

Elimination of $Z_{1/2}$ center concentration in a thick 4H-SiC epilayer was demonstrated by tuning the conditions for both the epitaxial growth and the post-growth processes. Under appropriate conditions, the $Z_{1/2}$ center was reduced below the detection limit (~1×10^{11} cm^{-3}) to a depth of 100 μm or more, and correspondingly the carrier lifetimes of the epilayers were enhanced by the C^+-implantation/annealing process or the thermal oxidation/Ar annealing process for Si-face epilayers. The initial results show that the C^+-implantation/annealing process is more effective for the elimination of $Z_{1/2}$ center for C-face epilayers.

Acknowledgement

This research is partly supported by the Japan Society for the Promotion of Science (JSPS) through its "Funding Program for World-Leading Innovative R&D on Science and Technology (FIRST Program)".

References

[1] J. Zhang, L. Storasta, J. P. Bergman, N. T. Son, and E. Janzén, Electrically active defects in *n*-type 4H-silicon carbide grown in a vertical hot-wall reactor, J. Appl. Phys. 93 (2003) 4708.

[2] P. B. Klein, B. V. Shanabrook, S. W. Huh, A. Y. Polyakov, M. Skowronski, J. J. Sumakeris, and M. J. O'Loughlin, Lifetime-limiting defects in n^- 4H-SiC epilayers, Appl. Phys. Lett. 88 (2006) 052110.

[3] C. W. Litton, D. Johnstone, S. Akarca-Biyikli, K. S. Ramaiah, I. Bhat, T. P. Chow, J. K. Kim, and E. F. Schubert, Effect of C/Si ratio on deep levels in epitaxial 4H-SiC, Appl. Phys. Lett. 88 (2006) 121914.

[4] K. Danno, T. Hori, and T. Kimoto, Impacts of growth parameters on deep levels in *n*-type 4H-SiC, J. Appl. Phys. 101 (2007) 053709.

[5] L. Storasta and H. Tsuchida, Reduction of traps and improvement of carrier lifetime in 4H-SiC epilayers by ion implantation, Appl. Phys. Lett. 90 (2007) 062116.

[6] T. Hiyoshi and T. Kimoto, Elimination of the major deep levels in n- and p-type 4H-SiC by two-step thermal treatment, Appl. Phys. Express 2 (2009) 091101.

[7] L. Storasta, J. P. Bergman, E. Janzén, A. Henry, and J. Lu, Deep levels created by low energy electron irradiation in 4H-SiC, J. Appl. Phys. 96 (2004) 4909.

[8] T. Miyazawa and H. Tsuchida, Evaluation of long carrier lifetimes in thick 4H silicon carbide epitaxial layers, Appl. Phys. Lett. 97 (2010) 202106.

[9] T. Kimoto, T. Hiyoshi, T. Hayashi, and J. Suda, Impacts of recombination at the surface and in the substrate on carrier lifetimes of n-type 4H-SiC epilayers, J. Appl. Phys. 108 (2010) 083721.

[10] M. Ito, L. Storasta, and H. Tsuchida, Development of 4H-SiC epitaxial growth technique achieving high growth rate and large-area uniformity, Appl. Phys. Express 1 (2008) 015001.

Materials Science Forum Vols. 717-720 (2012) pp 87-92

doi:10.4028/www.scientific.net/MSF.717-720.87

Growth and light properties of fluorescent SiC for white LEDs

Mikael Syväjärvi[1,*], Rositza Yakimova[1], M. Iwaya[2], T. Takeuchi[2], I. Akasaki[2] and Satoshi Kamiyama[2]

[1] Department of Physics, Chemistry and Biology, Linköping University, Linköping, Sweden
[2] Department of Materials Science and Engineering, Meijo University, 1-501 Shiogamaguchi, Tenpaku-ku, Nagoya 468-8502, Japan

* Corresponding author. Tel: +46 13285708, Fax: +46 13142337, E-mail: mikael.syvajarvi@ifm.liu.se

Keywords: sublimation epitaxy, fluorescent semiconductor, white LED

Abstract. The LED technology started to developed many years ago with red light emitting diodes. To achieve the blue LED, novel growth technologies and process steps were explored, and made it possible to demonstrate efficient blue LED performance from nitrides. The efficiency was further developed and blue LEDs were commercially introduced in the 1990's. The white LED became possible by the use of the blue LED and a phosphor that converts a part of the blue light to other colors in the visible range to combine into white light. However, even today there are limitations in the phosphor-based white LED technology, in particular for general lighting, and new solutions should be explored to speed the pace when white LEDs will be able to make substantial energy savings. In this paper we overview gallium nitride materials evolution and growth concepts for LEDs. We describe the fluorescent silicon carbide material prepared by a novel growth technology for a new type of white LED in general lighting with pure white light. This paper introduces an interesting research in fundamental growth and optical properties of light emitting silicon carbide.

Development of gallium nitride materials and growth technology

It was found in 1969 that gallium nitride exhibits a direct transition band structure with a bandgap of about 3.39 eV [1]. A couple of years later optical emission was observed [2,3]. The light was generated from needles of single crystal gallium nitride or from spots near the interface of the sapphire substrate that seemed to correlate with grain boundaries. There were major difficulties in the material with high residual donor concentration exceeding 10^{19} cm^{-3} that made fabrication of p-type material impossible. The crystal quality experience cracks in the surfaces and poor uniformity. The efficient luminescent properties were realized when discovering small microcrystallites which exhibited highly efficient light emission and which were prepared by forming n+ regions buried in insulating and n-type gallium nitride [4].

The break-through of the blue LED came when the first p-n junction LED was demonstrated by the group of Isamu Akasaki. In the 1970's there were major problems with gallium nitride material, such as low crystal quality and very high and poorly controlled residual donor concentrations which made it possible to realize p-type conduction. When detecting that the high-quality microcrystals in the poor quality gallium nitride layers were emitting intense light, Akasaki realized that gallium nitride LED with a p-n junction had a great potential if the crystal quality could be controlled. This was possible by using the right growth process, and the choice fell on the MOVPE (metal organic vapor phase epitaxy) growth process even though the method was not commonly used. It has advantages since the reactions of growth species were positive in that atoms were not likely to leave the surface again once they had arrived. The more commonly used growth technologies of MBE (molecular beam epitaxy) and HVPE (hydride vapor phase epitaxy) had either very low growth rate at that time or too high growth rates to provide high crystal quality. The research group built a home-made reactor at Meijo University, Figure 1a. Still, the material quality was poor. Only when the group realized a low temperature buffer layer technology, the high quality material was achieved. The buffer layer technology is a low temperature growth of a thin layer with physical properties similar to both the sapphire substrate used for growth template and the gallium nitride. As a result, near band gap emission was demonstrated, the residual donor concentration was decreased substantially and as well as showing an improved crystal quality [5]. The next step was to achieve controlled p-type doping. The team discovered that the luminescence was increased by irradiating zinc doped gallium nitride with electrons, while still there was not p-type conduction. Akasakis group switched from using zinc as dopant to magnesium in another type of set-up (Figure 1b) applied irradiation on the magnesium-doped films and demonstrated p-type conductive gallium nitride films.

Figure 1. (a) A replica of MOVPE equipment to demonstrate buffer layer technology and high gallium nitride crystal quality; (b) One of the MOCVD equipments that were used to achieve magnesium doping

The conventional nitride technology for white light emitting diodes is simply described by a nitride stack emitting blue light, and a part of this blue light is converted to other colors by a phosphor on top of the nitride layer. There are two main problematic issues of this technology. The first is that the intensity of the remaining blue light is stronger than the other colors from the phosphor due to the low efficiency of the phosphor, and the nitride LED has then a blue tone that is undesired by people using a white light LED on such nitride technology. The second issue is due to the uniformity and stability of the phosphor. When the phosphor is covering the nitride layer, the properties vary from location to location, and cause a slight difference between the light from LEDs made from different areas. As a result the LEDs have to be placed into different bins, which is a drawback in the production and use of the LEDs

.

The blue LED has encountered a profound scientific challenge. When the current to drive the LED is increased to utilize the structure for high illumination purpose, it has been found out that the technology faces a phenomenon in which the efficiency is decreasing with increasing current. Different mechanism have been argued and dismissed, like that the heat that is generated causes the electron and holes to be less confined, or the clustering of indium. The heat theory was dismissed by demonstration of pulsed laser action still showed the droop effect, and the indium clusters which were observed were very likely appearing as an effect of the measurement technique, and not present in the layers before measurement. Still the phenomenon of droop is investigated.

Fluorescent silicon carbide

In 1907, Henry J. Round presented a note in *Electrical World* [6] reporting a "bright glow" from diodes made on carborundum, which was the original denotation of silicon carbide, when exploring crystal detectors which were interesting to demodulate radio-frequency signal in crystal-detector radios. The light was produced by using two electrodes in contact with the material so that a rectifying diode was made when driving a current through the diode, and is the first observation of electroluminescence from a semiconductor.

In the early days of SiC research some studies have been conducted on pn-diodes and donor-to-acceptor pair recombination in single layers but the efficiencies have been low [7-11]. A white light emitting diode structured using a combination of two doped fluorescent SiC layers has been proposed by Satoshi Kamiyama and co-workers at Meijo University. This new white LED does not have a phosphor. Instead the white light comes directly from the material in two broad spectra, covering the visible region, which results in a pure white light, Figure 2. In a simple description the light emitting diode principle is made of (i) a nitride stack which emits UV light into, (ii) a fluorescent silicon carbide which transfers the UV light into a white light, and the white light is emitted out from the LED structure with help of (iii) a moth-eye structure. The moth-eye structure is a non-flat patterned surface, which helps to bring the light out from the fluorescent silicon carbide, since a flat surface is well known to reduce the light extraction due to reflection at the surface from the silicon carbide to air. A great advantage in this LED structure is that the nitride stack that emits near ultraviolet light does not experience the droop effect. The NUV light is absorbed within the volume of the f-SiC substrate, in which visible light is extracted through a moth-eye structure etched on the backside of the f-SiC substrate.

The LED which is produced in this way has a high color-rendering index and is highly suitable for pure white LEDs. The fluorescent SiC structure is built up of two layers, both of which should be sufficienly thick to produce a strong light emission from volume of the fluorescent material. The first is a layer of fluorescent 6H-SiC doped with the optimum concentration of nitrogen and boron, which are acting as donor and acceptor, respectively. The second layer is a 6H-SiC layer doped with nitrogen at a concentration of and aluminum . It was realized for the first time that the material has a very high quantum efficiency [12] and is therefore an ideal material to produce a monolithic white LED chip.

The N and Al doping yields a broad donor-to-acceptor band luminescence, which together with the broad donor-to-acceptor band luminescence in the N-B layer provides most wavelengths in the visible region with a light intensity (wavelength) distribution which does not have any dominating color, i.e. very pure white light. The nitrogen and boron doped SiC layer can emit warm white light with a peak wavelength of around 600 nm. The spectrum from nitrogen and aluminum doped SiC exhibits blue-green emission. By combining these two fluorescent layers and the two broad wavelength light outputs, a pure white light is obtained covering the whole visible spectrum. The luminescence between donors and acceptors is broad due to several energy levels close to the main level resulting in several transition possibilities, as typical in donor to acceptor pair luminescence. The recombination between donor and acceptor makes light output with energies over a continuous range, making a broad peak with an energy which is determined by the doping materials used (N - Al and N - B donor to acceptor pair luminescence have different positions with peaks in the visible region).

As a central part in the white LED based on silicon carbide is the structural quality in the fluorescent silicon carbide material. The quality of this is crucial to provide efficient light. It has been possible to grow silicon carbide since hundreds years, first by the Acheson process in which a mixture of silica, carbon, salt and sawdust were heated in air using an electrical furnace. The yield of single crystal material was low but this material was used for cutting and grinding purposes due to the hardness of silicon carbide. In 1955 Lely presented an important step in growth technology evolution and single crystal material became available. However, the size of these crystals was limited. In 1978 Tairov and Tsvetkov proposed the modified Lely method, in which a starting seed is used, and by this crystal enlargement became possible. This process is based on physical vapor transport (PVT) and has since 1989 been used to commercially produce silicon carbide wafers for transistor applications and as substrate for white light emitting diodes based on nitrides.

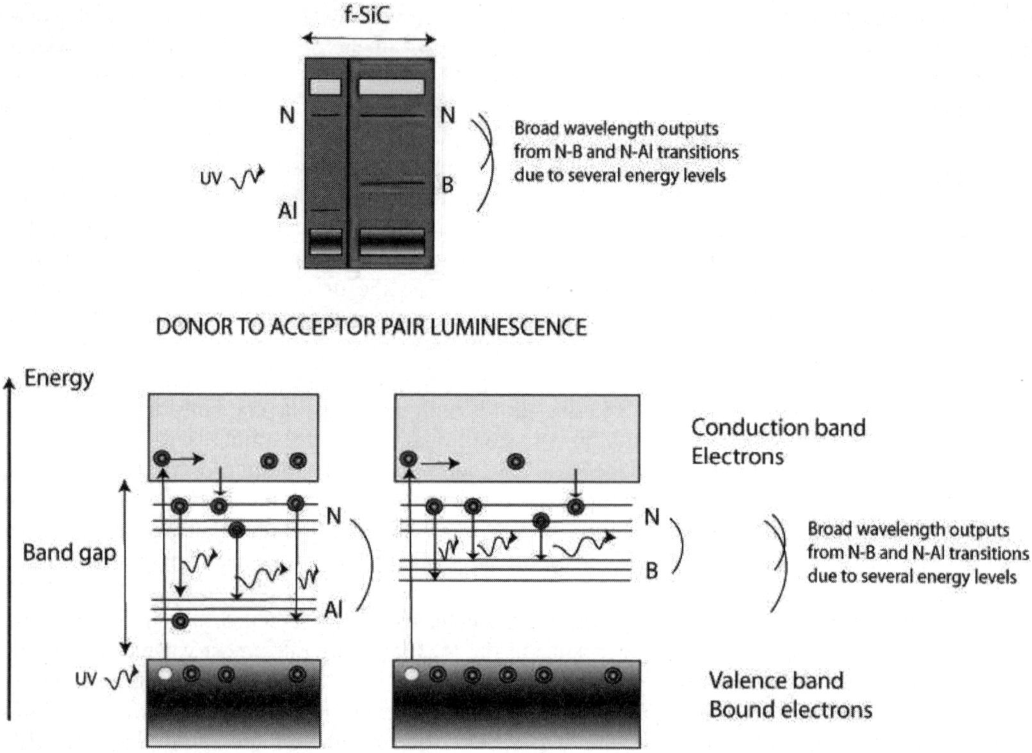

Figure 2. A schematic view of the light output from fluorescent silicon carbide and the optical transition.

In the SiC based LED a very high crystal quality is required. The PVT process used growth temperatures around 2300°C, and the chemical vapor deposition is limited into epitaxial growth at 1500-1600°C. At high growth temperatures it is difficult to maintain low dislocation levels, and dislocations cause non-radiative recombination instead of the desired radiative recombination of electrons. In the last years of 1990's the concept of sublimation epitaxy, which uses a lower growth temperature than PVT (500°C less) while maintaining a high growth rate with high quality of the grown material, was studied. This has since 2007 been further developed into the Fast Sublimation Growth Process (FSGP) [13] which is the perfect growth technology for making the fluorescent silicon carbide in the fluorescent SiC LED since it combines and manages the challenges in having high growth rate, low defect density material, thick layers and uniform doping possibilities at the same time, Figure 3.

Figure 3. The configuration for the Fast Sublimation Growth Process.

The technology of SiC crystal growth via sublimation is a complex process in which a number of parameters have to be controlled. The following sub-processes are of main importance: (i) sublimation of the source material, (ii) mass transport of the vapour species to the growing crystal, (iii) nucleation at the seed/crystal surface, and (iv) feeding of the growing crystal. The crystal growth is driven by the shift along a temperature gradient of the equilibrium between the solid SiC and its vapour. The vapour is produced via incongruent decomposition of SiC source material and reactions with the environment.

Because of the large difference between the silicon and carbon vapor pressures, the Si/C ratio, which determines the stoichiometry of the grown material, is difficult to maintain stable during growth of large crystals. This may affect the polytype stability and cause micropipe formation. Mass transport is predominantly governed by diffusion, which limits the growth rate and may affect crystal perfection. Single crystal SiC nucleates through deposition of the supersaturated vapour species on a SiC seed crystal. Uniformity of the supersaturation is another critical characteristic that is responsible for crystal quality. The surface kinetics and the seed surface conditions also influence the nucleation process and consequently formation of defects.

The driving force for growth is provided by applying a temperature difference which yields a higher temperature at the SiC source compared with the seed. The source can be a piece of SiC or most commonly as powder in boule growth in the conventional PVT method. The crucible has to resist the high temperatures applied and in case of SiC sublimation growth, crucibles made of graphite or TaC are used even though the latter is an expensive solution. Above a certain temperature, the source starts to sublime. The silicon and carbon bearing species are transported to the growing surface in the solid-vapor equilibrium along the temperature gradient. For growth of long boules a long source to seed distance (typically 5-30 mm) is required and interaction of the silicon and carbon species with the graphite walls will occur. The growth is performed in a low-pressure inert gas ambient (typically 5-40 mbar). This requires high growth temperatures to ensure high growth rates.

The epitaxial growth process has several advantages over the conventional PVT method. The distance between the source and the substrate is short, typically 1 millimeter, which has the positive effect that the vapor species do not react with the graphite walls. Instead of a silicon carbide powder as source material, the source is a monolithic silicon carbide plate. When the distance is short, any irregularities of the surface will reflect in the grown material, and hence defects will appear. The solid source is easier to control, while a silicon carbide powder will sublime in a non-uniform way. When the temperature is decreased, the pressure can be decreased to maintain a high growth rate. The lower growth temperature will result in higher quality of the material is the growth conditions can be preserved.

Figure 4 shows the photoluminescence spectra of nitrogen (N) and boron (B) doped and nitrogen (N) and aluminum (Al) doped 6H-SiC epilayers. The pairs of the doped impurities correspond to donor and acceptor, respectively, and these broad light emissions are caused by donor-acceptor pair (DAP) emissions. The N-and-B-doped SiC emits yellow-orange light, while the N-and-Al-doped SiC emits blue-green light as shown in the figure. Therefore, by a combination of these two spectra, a full-range of visible spectrum similar to the sun-light spectrum can be produced. This means that the fluorescent SiC epilayers doped with donor and acceptor impurities are promising phosphor materials for high color rendering index. The CIE Chromaticity coodinates of these two f-SiCs are also measured as shown in Fig. 4. The chromaticity coodinates of x and y in N-and-B-doped SiC are 0.486 and 0.465, respectively, and those in N-and-Al-doped SiC are 0.137 and 0.085, respectively. By a mix of these two epilayers, pure white color can be generated.

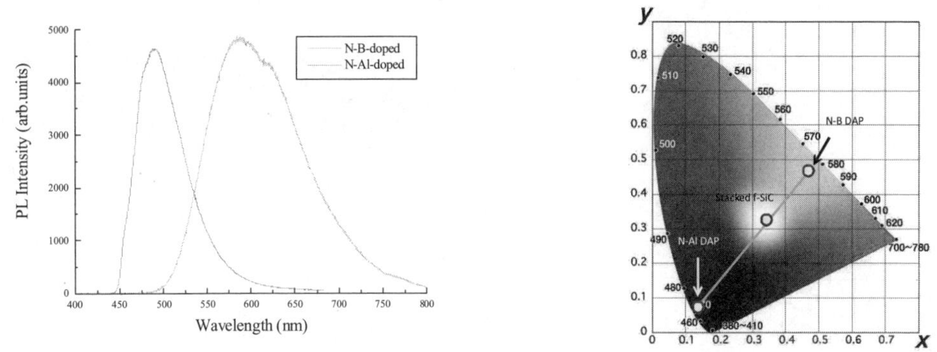

Figure 4. Photoluminescence spectra of fluorescent SiC layers and CIE chromaticity coodinate plot.

Summary

The choice of growth technology is an important decision in development of LEDs since the structural quality has such strong influence on the optical recombination paths. In case of nitride growth the MOCVD is an established method, while the physics demonstrate an effect on the efficiency by the droop effect in blue LEDs. In case of SiC, the PVT method is more difficult to control and the CVD method has difficulties in growth of thick layers that are needed in the new concept of fluorescent SiC. The feature of high structural quality while still maintaining high growth rates is very promising to achieve high optical efficiency of the fluorescent silicon carbide material prepared by the FSGP growth technology

Acknowledgement

This work was supported by Ångpanneföreningen Research Foundation, Richerts Foundation, Swedish Energy Agency, Nordic Energy Research, Swedish Research Council (project No. 2009-5307) and Department of the New Energy and Industrial Technology Development Organization (NEDO).

References

[1] H.P. Maruska and J.J. Tietjen, Appl. Phys. Lett. 15 (1969) 317.
[2] R. Dingle, K.L. Shaklee, R.F. Leheny and R.B. Zetterström, Appl. Phys. Lett 19 (1971) 5.
[3] J.I. Pankove, E.A. Miller, D. Richman and J.E. Berkeyheiser, J. Luminescence 4 (1971) 63.
[4] I. Akasaki and H. Amano, Jap. J. Appl. Phys. 45 (2006) 9001.
[5] I. Akasaki, H. Amano, Y. Koide, K. Hiramatsu and N. Sawaki, J. Cryst. Growth 98 (1989) 209.
[6] H.J. Round, Electrical World 49 (1907) 309.
[7] R.W. Brander, Proc. IEE 116 (1969) 329.
[8] A. Suzuki, M. Ikeda, N. Nagao, H. Matsunami and T. Tanaka, J. Appl. Phys. 47 (1976) 4546.
[9] W.v. Münch and W. Kürzinger, Solid State Electron. 21 (1978) 1129.
[10] S. Nishino, A. Ibaraki, H. Matsunami, and T. Tanaka, Jpn. J. Appl. Phys. 19 (1980) L353.
[11] L. Hoffman, G. Ziegler, D. Theis and C. Weyrich, J. Appl. Phys 53 (1982) 6962.
[12] S. Kamiyama, T. Maeda, Y. Nakamura, M. Iwaya, H. Amano, I. Akasaki, H. Kinoshuta, T. Furusho, M. Yoshimoto, T. Kimoto, J. Suda, A. Henry, I.G. Ivanov, J.P. Bergman, B. Monemar, T. Onuma and S.F. Chichibu, J. Appl. Phys. 99 (2006) 093108.
[13] M. Syväjärvi and R. Yakimova: "Sublimation epitaxial growth of hexagonal and cubic SiC", Elsevier, chapter in encyclopedia - the Comprehensive Semiconductor Science & Technology (SEST), Pallab Bhattacharya, Roberto Fornari and Hiroshi Kamimura (Eds), ISBN 978-0-444-53144-5 (2011).

Materials Science Forum Vols. 717-720 (2012) pp 93-96
© (2012) Trans Tech Publications, Switzerland
doi:10.4028/www.scientific.net/MSF.717-720.93

High Uniformity with Reduced Surface Roughness of Chloride based CVD process on 100mm 4° off-axis 4H-SiC

Hrishikesh Das[1, a], Swapna Sunkari[1,b], Timothy Oldham[1,c], Janna Casady[1,d] and Jeff Casady[1,e]

[1]SemiSouth Laboratories, Inc. 201 Research Blvd., Starkville, MS, 39759, USA

[a]hrishikesh.das@semisouth.com, [b]swapna.sunkari@semisouth.com, [c]timothy.oldham@semisouth.com, [d]janna.casady@semisouth.com, [e]jeff.casady@semisouth.com

Keywords: 6 x 4 inch planetary reactor, epitaxial growth, 4° off-axis, high uniformity, high growth rate, low surface roughness

Abstract. In this work we present the epitaxial growth of 4H-SiC on 100mm 4° off-axis substrates grown in a multi-wafer CVD planetary reactor. Highly uniform epitaxial layers having thickness and doping uniformities of 1.7% and 1.4% respectively were grown in the production reactor with optimized process conditions at 8μm/hr and 30μm/hr growth rates. Process optimizations resulted in 6μm - 15μm thick epitaxial layers having a surface roughness (RMS) of 0.32nm. Epitaxial layers with a thickness of 53μm grown with a 30μm/hr growth process had minimal degradation in surface roughness (RMS of 0.39nm).

Introduction

Homoepitaxial growth on 4H Silicon Carbide (SiC) is the key technology in fabricating low-loss power devices. A remarkable success in the development of the SiC power devices has been observed in the recent years due to the significant advances in the growth of epitaxial layers on good quality substrates. Epitaxial layers grown on 4° off-axis substrates are prone to step-bunching and triangular defects [1]. Step-bunching and surface roughness not only increases the leakage current on Schottky barrier diodes, but also decreases the breakdown voltage. High quality epitaxial layers free of defects with smooth surface morphology are needed to improve device performance [2, 3]. Aigo et al. [4] have demonstrated a surface roughness R_a of 0.2nm on a 10μm thick epilayer grown on 4° off-axis substrates. In our previous work [5] we reported epitaxial growth on 3" substrates with uniformities of 3% and 6% for thickness and doping respectively having a surface roughness (RMS) of 1.2nm. As a continuation of our work, we have been able to achieve better surface roughness and better thickness and doping uniformities on our multi wafer CVD tool, with the combination of an optimized buffer and process conditions. Thickness, doping, wafer-to-wafer homogeneities, surface roughness and characterization of the grown epitaxial layers are presented.

Experimental

The epitaxial growth was conducted in an Aixtron VP2400, a commercial multi-wafer hot-wall CVD planetary reactor. Commercially available 4 inch, n-type, 4° off-axis, Si-face 4H-SiC substrates were used for this work. The epitaxial growth was conducted with a H_2 – SiH_4 – C_3H_8 – HCl chemistry. The growth pressure was varied from 150mbar to 200mbar, while the growth temperature was varied between 1600-1650°C. The C/Si ratio, Cl/Si ratio and H_2 flows were varied to establish the optimal conditions for epitaxial growth.

Fourier Transform Infrared spectroscopy (FTIR), mercury probe Capacitance-Voltage (CV), Atomic Force Microscopy (AFM), X-Ray Diffraction (XRD) and Raman spectroscopy were used to characterize the epitaxial layers. Epitaxial layer thickness was measured using a MKS Filmexpert 2140 on a 17 point grid with 3mm edge exclusion on each wafer. The doping concentration was measured by an automated SSM model 495i Hg probe CV tool by mapping 13 points across the wafer, with an edge exclusion of 3mm. AFM measurements were done on a Dimension Icon AFM system with ScanAsyst. XRD was acquired with a PANalytical X'Pert MRD 6-axis diffractometer equipped with a Copper X-ray tube and sealed proportional detector. Raman spectra were obtained by a LabRam J-Y Spectrometer with a HeNe laser (632.8nm wavelength) and an 1800gr/mm grating.

Results and Discussion

Surface Roughness. Surface roughness of the epitaxial layers was monitored with AFM scans on 10x10μm^2 areas at five locations on each wafer. Process optimization for a smoother surface was done by growing epitaxial layers of 6μm thickness under different conditions keeping a growth rate of 8μm/hr. Fig. 1a shows an epitaxial layer before any optimizations, having sporadic step-bunching and a RMS roughness value of 1.39nm. An optimized etch with H_2 and HCl during heat up and pre-growth was added to the growth process. The grown epitaxial layer was found to be free from any step-bunching and had a RMS roughness of 0.55nm (Fig. 1b). Adding too much HCl to the pre-growth etch regime caused a rougher surface with visible pitting. Therefore optimization of the H_2/HCl flows and the etch time are critical. To further reduce the surface roughness, an optimized buffer layer was added after the pre-etch process. The buffer layer was grown with a much lower growth rate. This further reduced the surface roughness to 0.32nm (Fig. 1c). This optimized growth process was transferred to a 30μm/hr growth regime. Fig. 1d shows a 15μm epitaxial layer grown at 30μm/hr. The surface roughness (RMS) increased only marginally to 0.34nm. A 53μm thick epitaxial layer was grown to check the degradation in surface morphology with increase in thickness. Fig. 1e shows the AFM scan of this epilayer with a surface roughness of 0.39nm. All the substrates used had surface roughness between 0.9nm-1.1nm as seen in Fig. 1f.

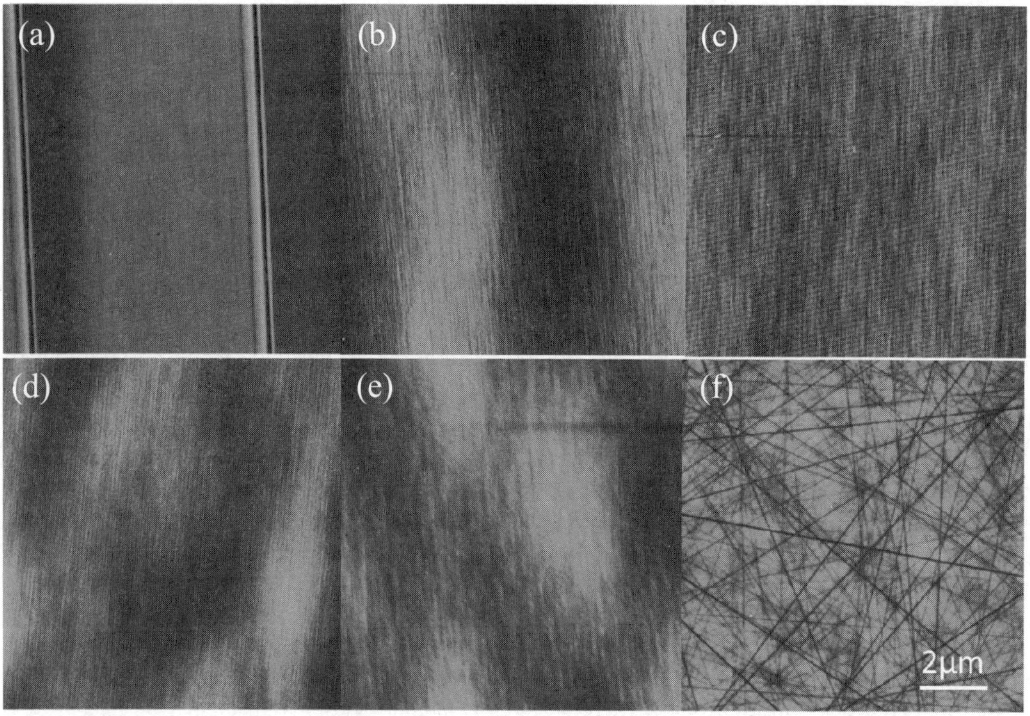

Figure 1. 10x10μm^2 AFM scans of (a) unoptimized epi process (6μm thick, roughness 1.39nm) (b) with H_2/HCl pre-etch epi process (6μm thick, roughness 0.55nm) (c) Pre-etch with optimized buffer epi process (6μm thick, roughness 0.32nm) (d) optimized 30μm/hr growth epi process (15μm thick, roughness 0.34nm) (e) optimized 30μm/hr growth epi process (53μm thick, roughness 0.39nm) (f) typical substrate (roughness 0.9nm-1.1nm)

Uniformity. Another focus of this work was to develop epitaxial growth processes with very good thickness and doping uniformities. The uniformities were optimized with a combination of H_2 flows, Cl/Si ratio and satellite rotation. To maintain similar uniformities at higher growth rates of 30μm/hr, the Cl/Si ratio had to be increased from 1.0 to 3.0. The typical thickness uniformities obtained on 15μm epitaxial layers were 1.75% (σ/mean), and 2.54% (max-min/max+min). The typical doping uniformities on the same epitaxial layers were 1.46% (σ/mean), and 1.96% (max-min/max+min). Fig. 2 shows the normalized intra-wafer thickness and doping profiles.

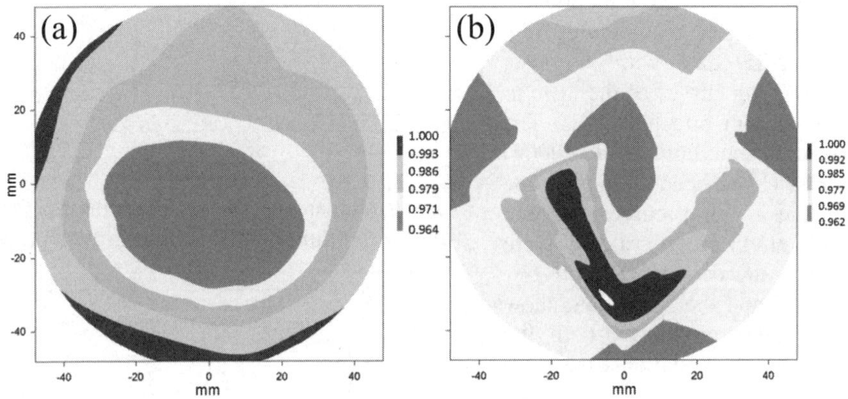

Figure 2. Intra-wafer normalized profiles of (a) thickness and (b) doping uniformity

Fig. 3 shows the normalized wafer to wafer thickness and doping variation on a fully loaded six wafer run. Under typical process conditions, average intra-wafer thickness and doping uniformities of 1.8% and 1.65% were achieved. Good wafer-to-wafer thickness and doping homogeneity of 0.9% and 2.0% respectively was observed.

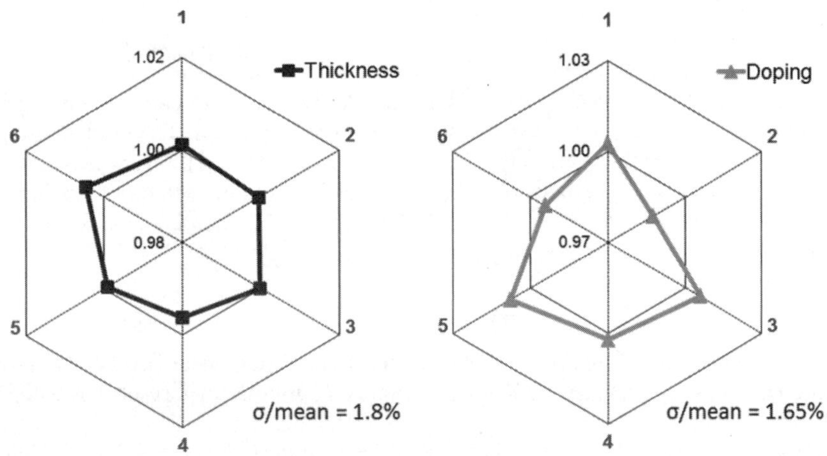

Figure 3. Wafer-to-Wafer thickness (0.9%) and doping (2%) homogeneity for a typical 6x4" run

Run-to-run repeatability of the process was observed for two different product lines (Diodes and FETs) with different epi stacks and specifications. Data collected over 30 product runs (180 wafers) show very consistent repeatibility for both thickness and doping for both diodes and FETs. The variation for FETs was found to be 0.7% and 2.78% for thickness and doping respectively. For Diodes the variations were 1% and 3.79% for thickness and doping respectively.

Epitaxial Quality. Epitaxial layer quality was evaluated using Raman spectra and XRD. Fig. 4 shows the Raman spectra of 15µm epitaxial layers grown with the two growth processes of 8µm/hr and 30µm/hr. The typical peaks of the 4H-SiC polytype are seen at $204cm^{-1}$ (inset), $610cm^{-1}$, 776 cm^{-1}, $796cm^{-1}$ and $964cm^{-1}$ [6,7]. No difference was seen between the epilayers grown with the different growth rates. Fig. 5 shows the XRD data with the inset showing the rocking curves of the 8µm/hr and 30µm/hr growth rate processes. The strong peak corresponding to the (0004) planes of 4H-SiC was seen at 35.57°. The FWHM of the rocking curves for the 8µm/hr and 30µm/hr growth rate epi were 23.0 and 26.6 arcseconds respectively. Both the Raman and the XRD data show the epitaxial layers grown by both the processes are of high crystal quality.

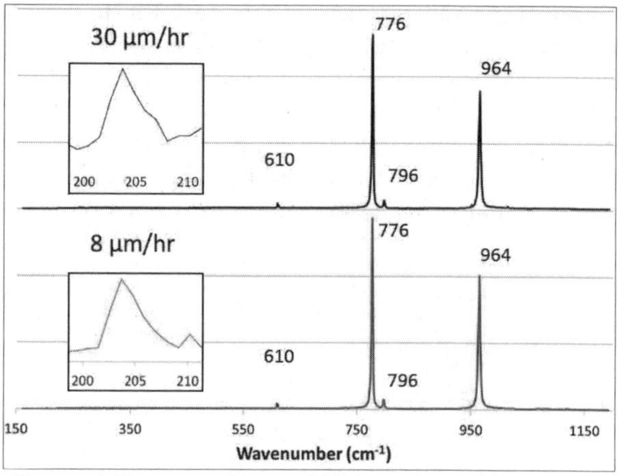

Figure 4. Raman spectra of 15µm epitaxial layers grown with 8µm/hr and 30µm/hr growth processes

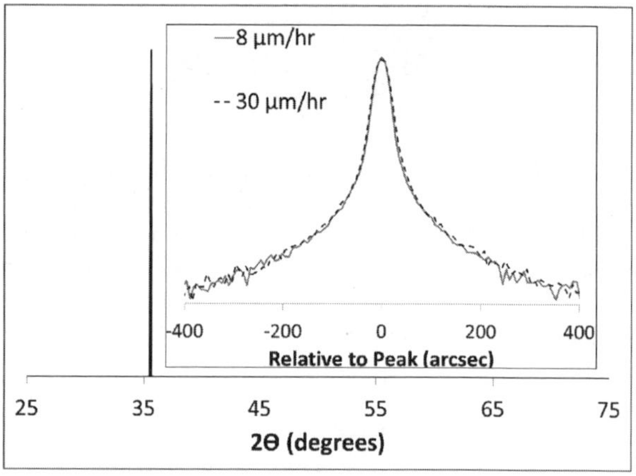

Figure 5. XRD of a 15µm epitaxial layer grown with the 30µm/hr process. The inset shows the rocking curves of epitaxial layers grown with the 8µm/hr and 30µm/hr growth processes having FWHM widths of 23.0 and 26.6 arcsecs respectively

Summary

Highly uniform epitaxial layers with thickness and doping uniformities of 1.75% and 1.46% respectively were grown in a 6x4" planetary reactor. Surface roughness (RMS) was improved from 1.39nm to 0.32nm by a combination of H_2/HCl pre-etch and an optimized buffer. The optimizations were transferred to a 25-30µm/hr growth rate process that maintained similar surface roughness even for a 53µm thick epitaxial layer. The epitaxial layer quality was verified by Raman spectroscopy and XRD measurements.

References

[1] K. Wada, T. Kimoto, K. Nishikawa, H. Matsunami, Journal of Crystal Growth 291 (2006) 370.

[2] S. Leone, H.Pedersen, A.Henry, O.Kordina, E.Janze´n, Journal of Crystal Growth 311 (2009) 3265-3272.

[3] Bernd Thomas, Christian Hecht, and Birgit Kallinger, *Materials Science Forum* Vols. 615-617 (2009), pp. 77.

[4] T.Aigo, A.Tsuge, H. Yashiro, T. Fujimoto, M. Katsuno, M. Nakabayashi, T. Hoshino, and W. Ohashi, *Materials Science Forum*, Vols. 645-648 (2010) 119-122.

[5] Swapna Sunkari, Hrishikesh Das, Carl Hoff, Yaroslav Koshka, Janna Casady, Jeff Casady, *Materials Science Forum* Vols. 615-617 (2009), pp. 423.

[6] Nakashima S, Harima H. Phys Status Solidi A, 162 (1997) pp. 39

[7] Wu Hailei, Sun Guosheng, Yang Ting, Yan Guoguo, Wang Lei, Zhao Wanshun, Liu Xingfang, Zeng Yiping, and Wen Jialiang. Journal of Semiconductors, Vol. 32, (2011), 043005-1

Materials Science Forum Vols. 717-720 (2012) pp 97-100

doi:10.4028/www.scientific.net/MSF.717-720.97

Epitaxial Growth of 4H-SiC using $Si_2(CH_3)_6 + Si_2Cl_6 + C_3H_8 + H_2$ system by atmospheric pressure hot CVD method

H. S. Lee[a, 1], M. J. Kim[b, 1], M. H. Kim[c, 1], S. I. Lee[d, 1], W. J. Lee[e, 1], B. C. Shin[f, 1] and S. Nishino[g, 2, *]

[1]Department of Materials & Components Engineering, Electro Ceramic Center, Dong-eui University, 995 Eomgwangno, Busanjin-gu, Republic of Korea,

[2]Widgap Materials Inc. Kyoto Institute of Technology, Sakyo-ku, Kyoto 606-8585, Japan

[a]haksung83@nate.com, [b]ajacxx@naver.com, [c]hitouyo@naver.com, [d]helloiru@gmail.com, [e]leewj@deu.ac.kr, [f]shinbc@deu.ac.kr, [g, *]s-nishino@widegap-materials.com

Keywords: Epitaxial growth, CVD, 4H-SiC, Hexachlorodisilane, Hexamethyledisilane

Abstract. SiC epitaxial layers grown with chlorinated species were evaluated in this work. The growth of thick and smooth epitaxial layers deposited on 4H-SiC on- and off-axis substrates was demonstrated by using C_3H_8 + HMDS + HCDS. The flow rate of each precursor was changed in the pre-growth and the growth step for SiC epilayer growth. High growth rate of 30 μm/h was achieved at a growth temperature of 1600°C by using HMDS and HCDS as chlorinated precursors.

Introduction

Silicon carbide is an important semiconductor for power device applications. To realize good performance of device operation, high quality and thick epilayer is needed. Therefore, the optimized conditions are necessary to decrease defect. Chemical vapor deposition (CVD) is the primary technique used to produce such high quality SiC material reasonable depositions rate. The most established technique to grow epitaxial layers of SiC is the chemical vapor deposition using silane, MTS, SiH_4 and C_3H_8 or ethylene, C_2H_4 as precursors diluted in hydrogen. The chlorinated precursor is recently focused for high growth rate and high quality epitaxial layer[1]. In the previous studies, the addition of chlorinated species from Si_2Cl_6 in the gas phase eliminated simultaneous Si nucleation which interferes with epitaxy[2-4]. Therefore, we tried to improve the quality of epilayer with controlling ratios of each precursor in the initial growth and the growth step.

Experimental procedure

Epitaxial growth was carried out with a conventional horizontal hot-wall atmospheric-pressure CVD system by using hexachlorodisilane(HCDS: Si_2Cl_6), hexamethyledisilane(HMDS: $Si_2(CH_3)_6$), C_3H_8 as source gases and H_2 as a carrier gas. The diameter and length of reaction tube were 100 mm and 700 mm. The size of pure graphite susceptor, 67 mm width, 20 mm height and 110 mm length was used. The width of the internal passageway was 60 mm and height is 14 mm. The flow rate of H_2 carrier gas was fixed at 10 slm. Nominal gas velocity in the passageway was 41 cm/sec. The growth temperature was 1500 °C-1600 °C and the growth time was typically 1 hour. SiC wafer was set on the graphite tray which was tapered 6° against flow direction. The substrates were (0001) Si face of 4H-SiC 7.5^o off-axis towards the <11-20> direction and almost on-axis(inclined from (0001) by 0.6^o). Fig. 1(a) exhibited the schematic diagram of CVD sytem used in this study. The PL measurements were carried out using He-Cd laser with a wavelength of 325 nm as a light source for excitation at 10K. Free-carrier concentraions was evaluated with Raman spectra of LO-phonon-plasmon-coupled mode.

Results and Discussion

Initial process of epitaxial growth is important to obtain high quality epilayer [5]. In our case, initial process means surface etching at 1600°C for 10 minutes. Two different initial processes were used. One is C_3H_8 addition which supplies C on the surface. The other is HMDS addition which supplies C and Si on the surface. We call those two different processes as "initial process A" and "Initial process B" for convenience. Propane or HMDS was added before the ramp up to growth temperature as shown in Figure 1. Two sets of precursors (HMDS, HCDS) were used at growth step and the propane (C_3H_8) was added when higher C/Si ratio was needed. The Cl/Si ratio was adjusted by flow rate of HCDS, which influenced the growth rate of epilayer. The flow rates of HCDS and C_3H_8 were 0.08-0.4 sccm and 2.5 sccm, respectively. C/Si ratio was varied in the range of 1.8-2.6. Cl/Si ratio was varied in the range of 0.35-1.2. Thus we adjusted C/Si and Cl/Si using C_3H_8. The growth rate definitely depends on the flow rate of HMDS and HCDS. High growth rate of 30μm/h was achieved at flow rates of HCDS: 0.4 sccm, HMDS: 0.6 sccm, H_2:10 slm and growth temperature: 1600°C. The thickness of epilayer was confirmed by a scanning electron microscope (SEM) image.

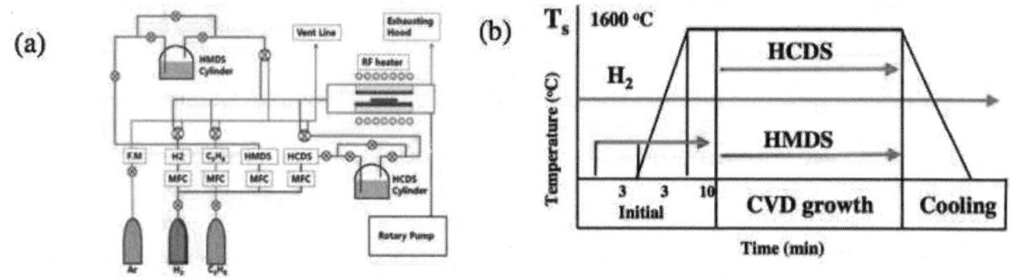

Fig.1(a) Schematic diagram of CVD system used in this work, and (b) temperature profile for epilayer growth.

Surface morphology of the epilayers depends on the position of the graphite tray. Five small substrates were set from up steam to down steam sides. The size of each wafer was 8 mm x 6 mm. When HMDS was used as initial process, surface was rather rough as seen in Figure 2 (d). Position apart from 48 mm from front edge, surface was smooth in both initial processes as shown in Fig. 2 (b,c,e,f).

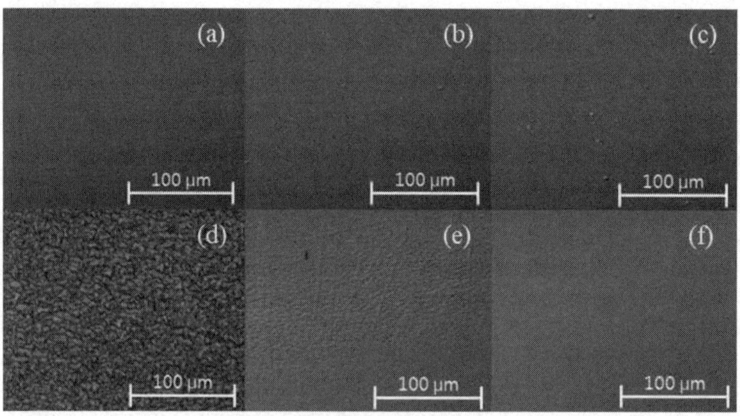

Fig.2 Surface morphologies of epilayers on off-axis substrates depending on position of graphite tray and different initial process. (a,b,c) were grown by initial process A. (d,e,f) were grown by initial process B. (a) and (d) were placed at 32 mm far from edge of the tray. (b) and (e) were placed at 48 mm far from the edge. (c) and (f) were placed at 64 mm far from the edge. Gas flow direction was from left to right in this figure.

Figure 3 shows AFM images of off-axis substrate and epilayer grown on this substrate. The RMS of substrate and epilayers are 8.7 nm and 6.8Å, respectively. This result suggests that the smoothness of the of the epilayer is almost same as the substrate.

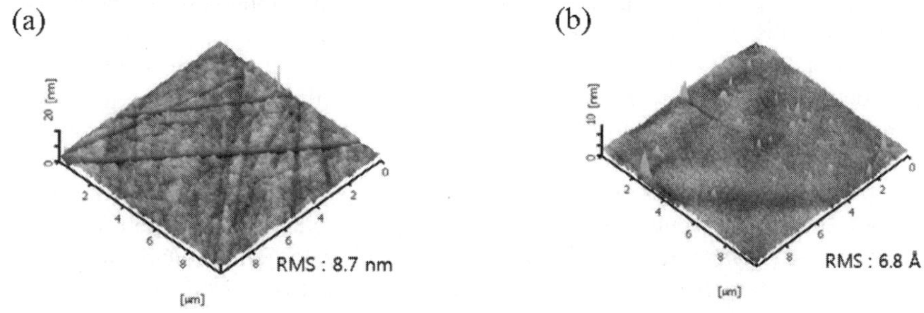

Fig.3 AFM image of 10 μm x 10 μm areas of substrate (a) and epitaxial layers (b).

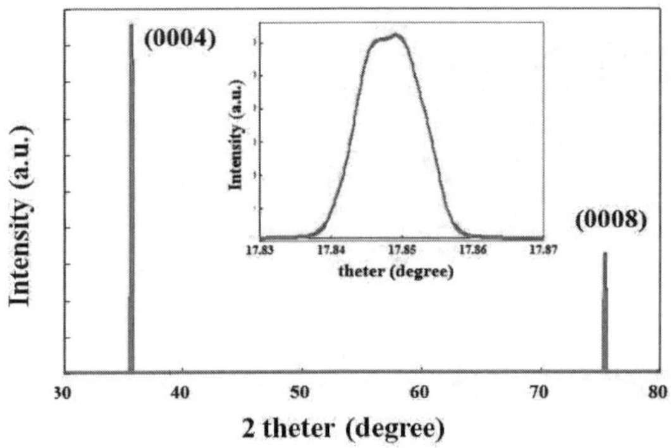

Fig.4 X-ray diffraction of epilayer on off-axis substrate. Rocking curve is in inset.

Figure 4 shows a typical XRD wide scan between 30° to 80° for epilayer. A strong peak at 35.45° is due to the diffraction from the (0004) planes of 4H-SiC. Rocking curves at the (0004) reflection peak of the epilayers were measured. Full width at half maximum (FWHM) of 19.2 arcsec was epilayer on off-axis substrate and 43 arcsec was epilayer on on-axis substrate. FWHM of the substrate was 68 arcsec. This suggests that the crystallinity of the epilayer is better than the substrate.

The PL measurements were carried out using He-Cd laser with a wavelength of 325 nm as a light source for excitation at 10K. Figure 5 shows photoluminescence spectra of epilayer with different initial processes. Those spectra in the photon energy range of 2.8 to 3.2 relates with D-A pair emission from nitrogen donor and aluminum acceptor. Aluminum was not intentionally doped, so small amount of aluminum was contained in the graphite susceptor. Also the broad peak at around 2.3eV (not shown in the figure) was observed and relates to deep level [6]. When we compare those spectra in Figure 5, D-A pair emission was stronger from epilayer prepared by initial process A. Initial process B which supplies simultaneously with C and Si might suppress nitrogen and aluminum doping.

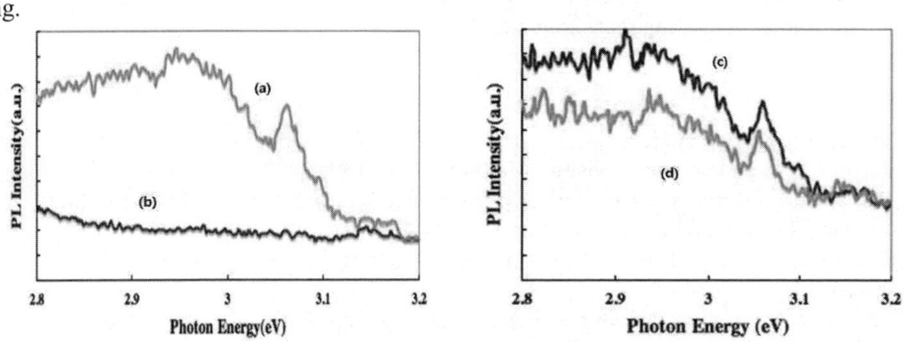

Fig.5 PL spectrum from epilayers with different initial processes. (a) and (b) were grown on off-axis substrate. (a) was grown by initial process A and (b) was grown by initial process B. (c) and (d) were grown on on-axis substrate. (c) was grown by initial process A, ands (d) was grown by initial process B.

Raman spectrum of the epitaxial layer were measured at room temperature in the back-scattering configuration using the Ar^+ laser line at 364 nm as the exciting source. Typical Raman peaks of 4H-SiC polytype at 776 (TO, E_2), 796 (TO, E_1) and 964 (LO, A_1) cm^{-1} corresponding 4H-SiC were observed, as described in Ref.[7]. These results confirmed that the thick 4H-SiC epitaxial layers can be grown on the off-axis 4H-SiC substrates with high crystal quality. Raman spectrum of the coupled mode were obtained as shown in Figure 6. These profiles show typical features of LO-phonon-plasmon coupling. Carrier concentration of the epitaxial layers are obtained by analyzing the line shapes of the axial coupled modes by fitting the theoretical band shape taking the plasmon frequency, phonon-damping rates and effective mass of electrons as adjustable parameters [8]. Curve fitting of LO phonon at around 964 cm^{-1}, nominal carrier concentration of the epilayers was 9 x 10^{17} cm^{-3}. This value is a bit larger than that of the substrates.

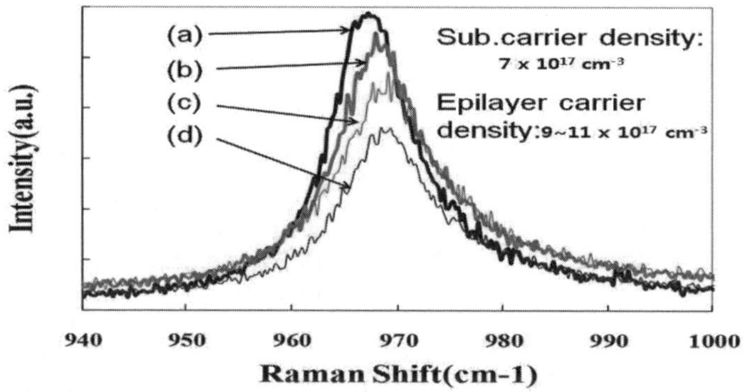

Fig.6 Raman spectra for fitting carrier concentration: (a) 7 x 10^{17} cm^{-3}, (b) 9 x 10^{17} cm^{-3}, (c) 1 x 10^{18} cm^{-3}, (d) 1.1 x 10^{18} cm^{-3}

Conclusions

In this work, the growth of thick and smooth epitaxial layers deposited on 4H-SiC off-axis substrates was demonstrated by using C_3H_8 + HMDS + HCDS. The highest growth rate achieved with a 4H polytype homoepitaxial growth is 30 μm/h using the C_3H_8 + HMDS + HCDS at the Cl/Si ratio of 1.2.

Acknowledgement

The authors would like to express their gratitude to Mr.Jung Gon Kim in Kyoto Institute of technology for Raman scattering measurement.

References

[1] S.Leone, H.Pedersen, A.Henry, O.Kordina and E.Janzen, Mat.Sci.Forum 107 (2009) 600-603.
[2] S.Nishino, T.Miyanagi and Y.Nishio, Mat.Sci.Forum 139 (1998) 264-268.
[3] Y.Masuda, S.Ohshima, C.Jacob and S.Nishino, Mat.Sci.Forum 139 (2001) 355-356.
[4] T.Miyanagi and S.Nishino, Mat.Sci.Forum 199 (2002) 389-393.
[5] T.Kimoto, ZY.Chen, S.Tamura, S.Nakamura, N.Onojima and H.Matsunami, J.Appl.Phys. 40 (2001) 3315.
[6] T.Kimoto, A.Yamashita, A.Itoh and H.Matsunami, J.Appl.Phys. 32 (1933) 1045
[7] S.Nakashima and H.Harimna, Phys.Stat.Sol. A, 162 (1997) 39.
[8] H.Harima, S.Nakashima, and T.Uemura, J.Appl.Phys.78 (1955) 1996

Materials Science Forum Vols. 717-720 (2012) pp 101-104
© (2012) Trans Tech Publications, Switzerland
doi:10.4028/www.scientific.net/MSF.717-720.101

4H-SiC epitaxial growth on 2° off-axis substrates using trichlorosilane (TCS)

T. Aigo[1, a], W.Ito[1], H.Tsuge[1], H.Yashiro[1], M.Katsuno[1], T.Fujimoto[1], and W.Ohashi[1]

[1]Nippon Steel Corporation, Advanced Technology Research Laboratories, 20-1 Shintomi, Futtsu, Chiba 293-8511, Japan

[a]aigoh.takashi@nsc.co.jp

Keywords: Morphology, trichlorosilane, triangle defect, stacking fault, photoluminescence.

Abstract. 4H-SiC epitaxial growth on 2° off-axis substrates using trichlorosilane (TCS) is presented. Good surface morphology was obtained for epilayers with C/Si ratios of 0.6 and 0.8 at a growth temperature of 1600°C. The triangle defect density was reduced to a level below 5 cm^{-2} at 1600°C and below 1 cm^{-2} at 1625°C for a C/Si ratio of 0.8. Photoluminescence (PL) measurements were carried out with band-pass filters of 420 nm, 460 nm, and 480 nm to detect stacking faults. A stacking fault density of below 5 cm^{-2} was achieved at 1600°C and 1625°C with a C/Si ratio of 0.8. The optimal conditions for TCS growth were a C/Si ratio of 0.8 and a growth temperature of 1600°C. The evaluation of stacking faults and etch pit density indicated that the use of 2° off-axis substrates and TCS is effective for reducing basal plane dislocations. Comparing these results to those using silane (SiH_4) with HCl added, it was demonstrated that TCS is much more suitable for obtaining high-quality epilayers on 2° off-axis substrates.

Introduction

Silicon carbide (SiC) is an attractive material for high-power, high-frequency, and high-temperature electronic devices. With the widespread use of SiC-based electronic devices, reduction of the cost of SiC material, which promotes the development of smaller off-axis-angle substrates and the epilayers on them, is necessary. The use of smaller off-axis-angle substrates also has the advantage of reducing basal plane dislocations (BPDs), which are responsible for the stacking faults in SiC electronic devices [1]. Four-degree off-axis substrates have been widely used to reduce the cost and the BPD density and for this reason, epitaxial growth on the substrates has been studied to suppress both step bunching [2] and morphological defects such as triangles and pits [3]. Recently, the surface morphology of epilayers on 2° off-axis substrates has been investigated for the further reduction of the cost and the BPD density [4]. For the epitaxial growth on 2° off-axis substrates, the use of trichlorosilane (TCS) is effective to achieve good surface morphology because radical species produced by TCS are more stable than those by silane (SiH_4) [5]. In this paper, we present the surface morphology, surface roughness, and stacking fault density of epilayers on 2° off-axis substrates grown using TCS with varying growth conditions such as the C/Si ratio and growth temperature. We also compare the results obtained to those using silane (SiH_4) with HCl added.

Experiments

Epitaxial growth was carried out in a horizontal hot-wall reactor. TCS and SiH_4 were used as silicon sources and ethylene (C_2H_4) was used as a carbon source with the carrier gas being hydrogen (H_2). The growth was conducted at reduced pressure of 100mbar. The substrates used in this study were 2-inch, 2° off-axis, Si-face 4H-SiC. The growth temperature was varied from 1600°C to 1650°C and C/Si ratio was varied from 0.6 to 1.0. In the case of epitaxial growth using SiH_4+HCl, the C/Si and Cl/Si ratios were fixed at 0.8 and 3.0, respectively. The surface morphology was inspected using an optical microscope with Nomarski interference contrast. The surface roughness in an area of 10×10μm was measured with atomic force microscopy (AFM). The epilayers obtained were analyzed by Fourier transform infrared spectroscopy (FTIR) for the thickness and by mercury probe C-V measurements for the doping concentration. Etching with molten KOH was performed on the epilayer to reveal dislocations. The BPD density was obtained by manually counting etch pits in about a half-area of the epilayer. To detect stacking faults, photoluminescence (PL) imaging observation was carried out using a highly sensitive CCD camera with band-path filters of 420 nm, 460 nm, and 480 nm. Uniform excitation was performed under Xe-Hg lamp irradiation through a selected UV optical filter. PL with a 0.1-mm spot of He-Cd laser (325 nm) excitation at room temperature was also utilized to confirm the quality of the epilayers grown.

Results

Morphology. Figure 1 shows the surface morphology of epilayers grown on 2° off-axis substrates at 1600°C using TCS with C/Si ratios of 0.6, 0.8, and 1.0. Nomarski micrographs and AFM images are both shown in the figure. The surface morphology of an epilayer grown using SiH_4+HCl at 1600°C is also indicated in the figure. Growth rates were 2.2-3.1 µm /hr. for TCS growth and 3.4 µm /hr. for SiH_4+HCl growth. The total thickness of the epilayers was 6-7 µm. As can be seen in Fig. 1, smooth surfaces are obtained for TCS growth at C/Si ratios of 0.6 and 0.8 and surface roughness, Ra, is comparable to that of epilayers on 4° off-axis substrates [6]. On the other hand, a number of triangle defects are observed on the epilayer grown using SiH_4+HCl. Although the SiH_4+HCl gas system is useful for growth on 4° off-axis substrates [2], the present results demonstrate that TCS is preferable to SiH_4+HCl for growth on 2° off-axis substrates. Surface morphology for TCS growth at 1625°C with C/Si ratios of 0.6 and 0.8 was degraded due to step bunching.

Defects. In the present study, triangle defects were major morphological defects. Defects on 4° off-axis substrates can be largely categorized into two types depending on the shape: one is spatially confined, and the other is extended [3]. The same classification is available for defects on 2° off-axis substrates, and typical shapes of such defects are shown in Fig. 2(a) and 2(b). Figure 3 shows the dependences of the triangle defect density on the C/Si ratio for TCS growth at 1600°C and 1625°C. The density for SiH_4+HCl growth at 1600°C is also indicated. As can be seen in Fig. 3, density is reduced to a level below 5 cm^{-2} at a C/Si ratio of 0.8 for TCS growth at 1600°C. In contrast, the density is over 20 cm^{-2} in the case of the SiH_4+HCl system under the same growth conditions. It is also pointed out that higher growth temperatures lead to lower densities of defects in the TCS growth system.

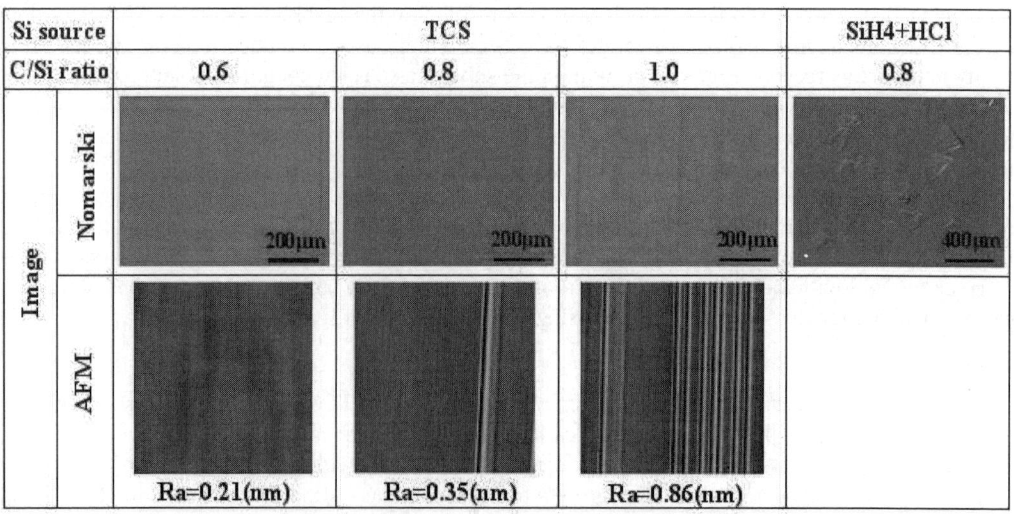

Fig. 1 Surface morphology and surface roughness (Ra) of 4H-SiC epilayers grown on 2° off-axis substrates using TCS and SiH_4+HCl at 1600°C.

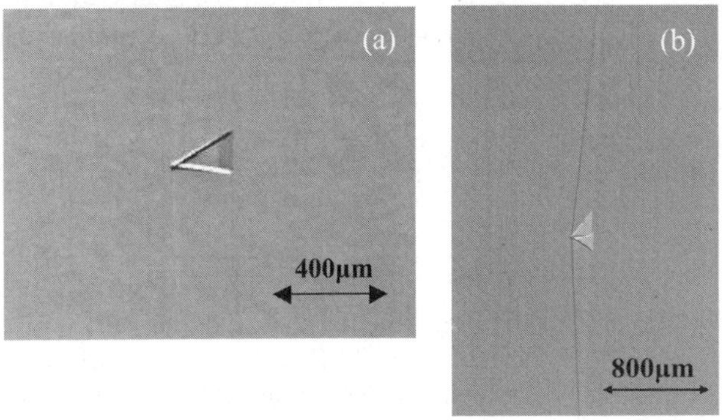

Fig. 2 Typical triangle defects of 4H-SiC epilayers grown on 2° off-axis substrates: spatially confined (a) and extended (b).

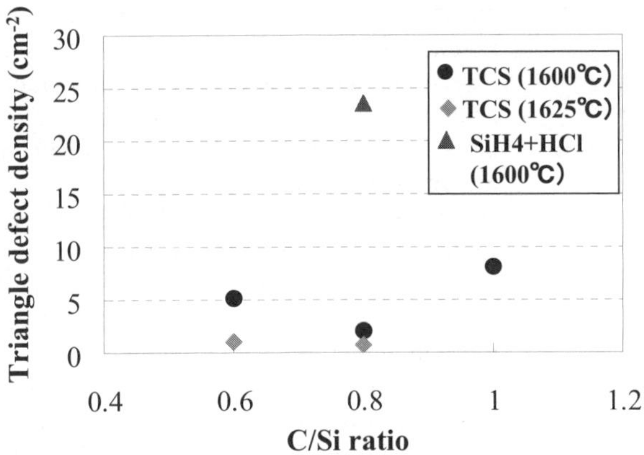

Fig. 3 C/Si ratio dependence of the triangle defect density for 4H-SiC epilayers grown on 2° off-axis substrates using TCS and SiH_4+HCl.

Stacking fault density. Figure 4 shows the dependences of the stacking faults density on the C/Si ratio for TCS growth and SiH_4+HCl growth. In Fig. 4, stacking faults were detected by the PL measurements with band-pass filters of 420 nm, 460 nm, and 480 nm and they are known as Shockley-type stacking faults [7, 8]. In the present study, most of stacking faults were those detected with the filters of 460 nm and 480nm. From Fig. 4, the same dependence as that of the triangle defects is seen for TCS growth at 1600°C. A stacking fault density of below 5 cm^{-2} is achieved at a C/Si ratio of 0.8. Contrary to the results of the triangle defects, a higher growth temperature results in a higher stacking fault density for TCS growth. As for SiH_4+HCl growth, the density is 2-3 times higher than that for TCS growth.

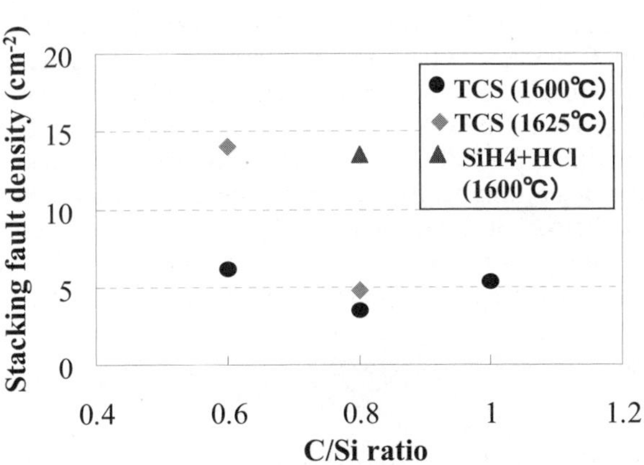

Fig.4 C/Si ratio dependence of the stacking fault density for 4H-SiC epilayers grown on 2°off-axis substrates using TCS and SiH_4+HCl.
The stacking faults are those detected by PL measurements with band-pass filters of 420 nm, 460 nm, and 480nm.

Etch pit density. The BPD density was evaluated using molten KOH etching for the epilayer grown using TCS. The C/Si ratio and the growth temperature for the epilayer were 0.6 and 1625°C, respectively. From the evaluation for an area of 8.1 cm^2 in the epilayer, the BPD density was calculated to be 30-40 cm^{-2}. It is reported that the stacking faults detected by the PL measurements with a band-pass filter of 460nm include BPDs [9]. Figure 4 indicates that the stacking fault density for the C/Si ratio of 0.6 and the growth temperature of 1625°C is larger than that for 0.8 and 1600°C, which are the optimal conditions for TCS growth. Therefore, the BPD density for the optimal conditions is expected to be smaller than 30-40 cm^{-2}. Our previous study has shown that the BPD density could be reduced to less than 50 cm^{-2} by the use of 4° off-axis substrates [6]. The present result confirms that the use of 2° off-axis substrates and TCS has the advantage of reducing BPDs.

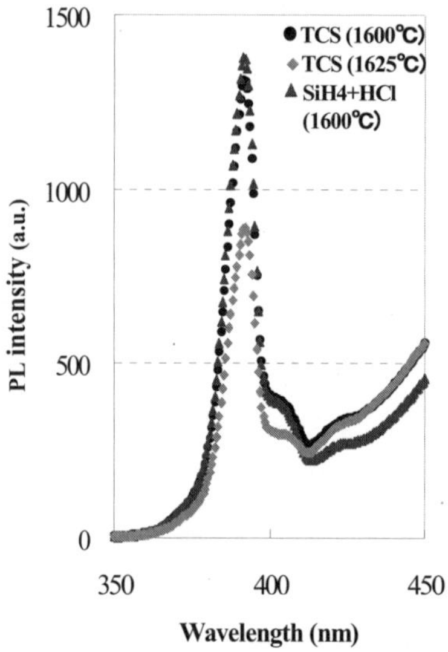

Fig. 5 PL spectra of the epilayers grown on 2° off-axis substrates using TCS and SiH_4+HCl.

Quality of the epilayers grown. PL investigation was carried out and typical spectra for TCS growth and SiH_4+HCl growth with a C/Si ratio of 0.8 are indicated in Fig. 5. The intensity of the band-edge emission is generally decreased by nonradiative recombination through defect levels and surface/interface recombination. Therefore, the intensity of the band-edge emission reflects the crystalline quality of the epilayers. As can be seen from Fig. 5, the intensity for TCS growth is almost the same as that for SiH_4+HCl growth. This result shows that TCS growth promotes step-flow growth and reduces morphological defects including stacking faults although there is no difference in crystalline quality between TCS growth and SiH_4+HCl growth.

Summary

4H-SiC epitaxial growth on 2° off-axis substrates using trichlorosilane (TCS) has been carried out. A smooth surface was obtained for epilayers with C/Si ratios of 0.6 and 0.8 at a growth temperature of 1600°C. The triangle defect density was reduced to a level below 5 cm^{-2} at 1600°C and below 1 cm^{-2} at 1625°C for a C/Si ratio of 0.8. A stacking fault density of below 5 cm^{-2} was achieved at 1600°C and 1625°C for a C/Si ratio of 0.8. The optimal conditions for TCS growth were a C/Si ratio of 0.8 and a growth temperature of 1600°C. The evaluation of stacking faults and etch pit density indicated that the use of 2° off-axis substrates and TCS is effective for the reduction of basal plane dislocations. Comparing these results to those using silane (SiH_4) with HCl added, it has been demonstrated that TCS growth is advantageous for obtaining high-quality epilayers with fewer morphological defects and stacking faults on 2° off-axis substrates.

References

[1] S. Ha et al., Journal of Applied Physics 96 (2004) pp.393-398.
[2] J. Zhang et al., Materials Science Forum 600-603 (2009) pp. 103-106.
[3] S. Leone et al., Journal of Crystal Growth 311 (2009) pp. 3265-3272.
[4] J. Zhang et al., Mater. Res. Soc. Symp. Proc.1069 (2008) 1069-D07-10.
[5] F. La Via et al., Journal of Crystal Growth 311 (2008) pp. 107-113.
[6] T.Aigo et al., Materials Science Forum 645-648 (2010) pp. 119-122.
[7] G. Feng et al., Applied Physics Letters 92 (2008) 221906.
[8] G. Feng et al., Physica B 404 (2009) pp. 4745–4748
[9] R.Hattori et al., Materials Science Forum 615-617 (2009) pp. 129-132.

Materials Science Forum Vols. 717-720 (2012) pp 105-108

doi:10.4028/www.scientific.net/MSF.717-720.105

Development of Vertical 3×2″ LPCVD System for Fast Epitaxial Growth on 4H-SiC

Wanshun Zhao[1, 2, a, *], Guosheng Sun[1, b], Hailei Wu[1, c], Guoguo Yan[1, d], Liu Zheng[1, e], Lin Dong[1, f], Lei Wang[1, g], Xingfang Liu[1, h] and Lijun Yang[2, i]

[1]Novel Semiconductor Material Laboratory, Institute of Semiconductors, Chinese Academy of Sciences, Beijing 100083, China

[2]North China Electric Power University, Beijing 100083, China

[a]zwshuke@semi.ac.cn, [b]gshsun@red.semi.ac.cn, [c]hlwu@semi.ac.cn, [d]ggyan@ semi.ac.cn, [e]liuero@semi.ac.cn, [f]donglin09@semi.ac.cn, [g]wangl@semi.ac.cn, [h]liuxf@semi.ac.cn, [i]yanglj@ncepu.edu.cn

Keywords: 4H-SiC, Low-Pressure Hot-Wall CVD, fast epitaxial growth.

Abstract. A vertical 3×2″ low pressure chemical vapor deposition (LPCVD) system has been developed to realize fast epitaxial growth of 4H-SiC. The epitaxial growth process was optimized and it was found that the growth rate increases with increasing C/Si ratio and tends to saturate when C/Si ratio exceeded 1. Mirror-like thick 4H-SiC homoepitaxial layers are obtained at 1500 ℃ and C/Si ratio of 0.5 with a growth rate of 25 μm/h. The minimum RMS roughness is 0.20 nm and the FWHM of rocking curves of epilayers grown for 1 hour and 2 hours are 32.4 arcsec and 26.2 arcsec, respectively. These results indicate that high-quality thick 4H-SiC epilayers can be grown successfully on the 8° off-cut 4H-SiC substrates.

Introduction

The unique thermal and electrical properties of SiC make it to be an attractive semiconductor for electronic devices designed to operate in extreme conditions, such as high voltage, high temperature, high frequency, and high radiation. In the past few years, various SiC based electronic power devices, like SBD, PIN and MPS, have been reported in Refs. [1, 2, 3]. For industrial applications of SiC power devices, it is a prerequisite to develop 4H-SiC epitaxial reactors with high throughput, which can be realized by enhancing the growth rate or the capacity of the reactors. In recent years, high growth rates of 25–60 μm/h have been achieved in a vertical hot-wall CVD reactor at 1700 ℃ and a horizontal hot-wall reactor at 1530 – 1560 °C, respectively [4,5]. Besides, the addition of chloride in conventional SiH_4+C_2H_4+H_2 system and the usage of trichlorosilane (TCS) as silicon source gas were reported to be able to effectively reduce Si droplets, leading to the improvement on the epilayer quality [6, 7, 8]. However, it is still a challenge to realize high-quality thick 4H-SiC epilayers at fast growth rate.

In this paper, we report recent experimental results on fast epitaxy of 4H-SiC in our vertical 3×2″ low pressure chemical vapor deposition (LPCVD) system. The epitaxial growth rate is up to 30 μm/h at 1530 °C, which is close to the highest growth rate with the comparable growth temperature. The epitaxial surface morphology and the quality of epilayer are also demonstrated.

Experimental procedure

Fast epitaxial growth of 4H-SiC was performed in our home-made vertical LPCVD system with a capacity of 3×2″ . The schematic diagram of the cross section of the reactor is shown in Fig. 1. The graphite susceptor coated with SiC film was inductively heated using RF power and it can rotate during the epitaxial growth. The growth temperature was measured by an infrared pyrometer. TCS, propane and H_2 were used as silicon source, carbon source and carrier gas, respectively. Commercial 8° off-cut 4H-SiC (0001) wafers were used as substrates. The growth temperature range and pressure were 1500-1530 °C and 40 Torr, respectively. The ratio of C/Si was varied from 0.4 to 1.2 by increasing the flow of propane with the fixed TCS flow of 40 sccm. The morphology of epilayers was examined by Nomarski optical microscopy and atomic force microscopy (AFM). The growth quality of epilayers was characterized by high resolution X-ray diffraction.

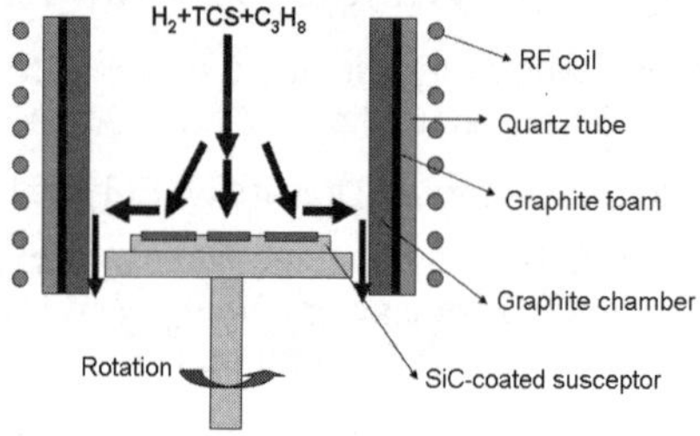

Fig.1 Schematic diagram of the vertical reactor.

Results and Discussions

The dependence of growth rate on the C/Si ratio is shown in Fig. 2. The growth rate increases significantly with increasing the C/Si ratio and tends to saturate under the C-rich condition. This is because the growth rate is limited by the transport of carbon source in Si-rich condition, whereas the rate-limiting step transfers to the supply of silicon source gas in C-rich condition. Excellent surface morphology with a low density of defects was obtained when the C/Si ratio was under 0.5~0.8. With a further increase in the C/Si ratio, the quality of the epilayer began to deteriorate and the density of triangular defects and carrot defects increased. When the C/Si ratio was increased to 1, inclusions of 3C-SiC were observed in large area of epilayers. Similar results have also been reported by other groups [9, 10].

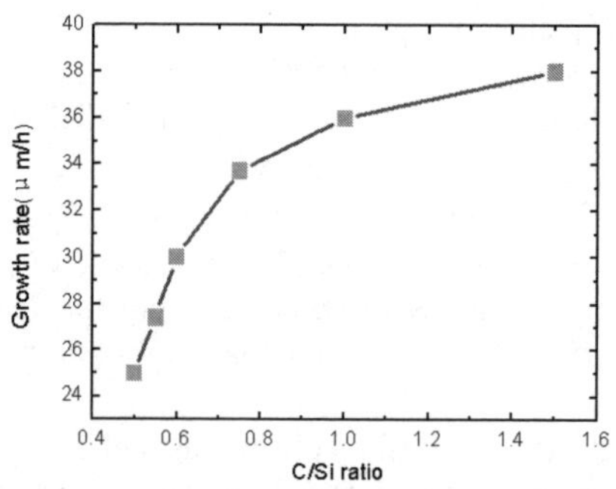

Fig.2 C/Si ratio dependence of growth rate in vertical LPCVD of 4H-SiC (0001).

Figure 3 presents the AFM results of the epilayers which were grown at a rate of 25 μm/h (C/Si ratio: 0.5) for 2 hours. The thick epilayer grown for 2 hours exhibits an RMS roughness of 0.20 nm which is the same as the RMS of substrate. These experimental results indicate that the surface roughness can be well maintained with the increase in growing time.

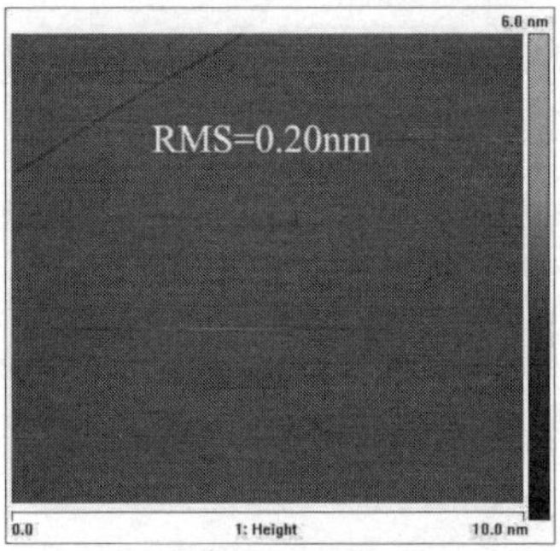

Fig.3 AFM micrographs (10 μm × 10 μm) of the sample grown for 2 hours obtained RMS roughness of 0.20 nm.

To measure the crystalline quality of the 4H-SiC films, the high resolution X-ray diffraction of the epitaxial layers was used. Figure 4 shows the rocking curves at the [0004] refection peak of the epilayers which were grown for 1 hour and 2 hours. The FWHM of these curves are 32.4 arcsec and 26.2 arcsec, respectively, while that of the substrate is 35 arcsec. This suggests that the crystallinity of the epilayer is better than the substrate. And with the growing time increasing, the crystallinity of the epilayer becomes better. It should be noted that this result was achieved under relatively low temperature (1500-1530 °C). This may due to the fact that TCS enables the formation of $SiCl_x$ to effectively suppress the homogeneous nucleation of Si droplets, leading to the good quality of epilayer at such temperature range.

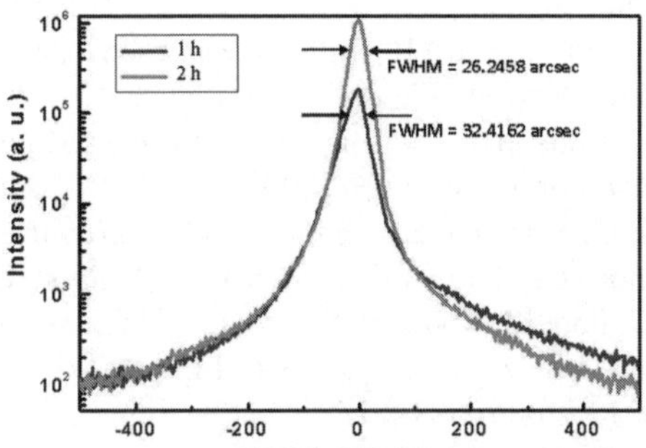

Fig.4 The rocking curves of 4H-SiC epilayers which were grown for 1hour and 2 hours.

The density and the spatial distribution of polytypes of the typical surface morphological defects were investigated. The Nomarski micrograph of the surface on the thick 4H-SiC homoepitaxially grown film is shown in Fig. 5. As shown in the figure, the surface is mirror-like, although several surface morphological defects, including carrot and comet defects, are observed. The density of the typical surface morphological defects is 0.22 $/mm^2$. The investigation on these morphological defects is still underway.

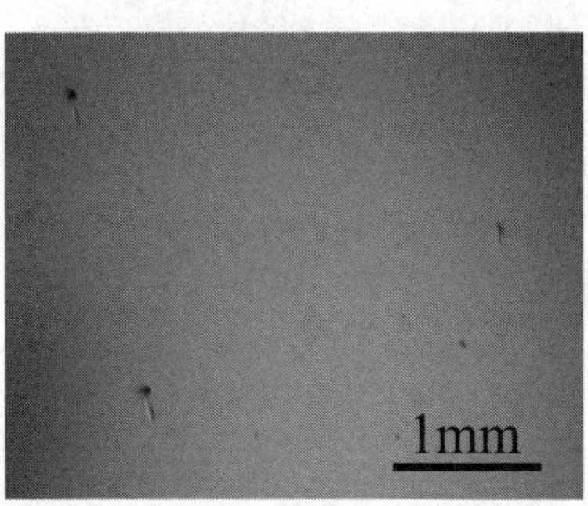

Fig.5 Nomarski micrograph of the surface on the thick 4H-SiC homo-epilayer.

Summary

We have performed fast epitaxial growth of 4H-SiC in our vertical 3×2″ LPCVD system. High-quality thick 4H-SiC homoepitaxial layers with an RMS roughness of 0.20 nm were obtained at 1500 °C and C/Si ratio of 0.5 with a growth rate of 25 μm/h. It was found that the growth rate increases with increasing C/Si ratio and tends to saturate when C/Si ratio exceeds 1. The FWHM of rocking curves of epilayers grown for 1 h and 2 h are 32.4 and 26.2 arcsec, respectively. These results indicate thick 4H-SiC epilayers can be grown successfully on the 8° off-cut 4H-SiC substrates with high crystal quality.

Acknowledgements

This work was supported by the National Natural Science Foundation of China (60876003) and the project of 2011 (2nd) Innovative Research Teams and Leading Talents in Guangdong.

References

[1] Q. Wahab, T. Kimoto, A. Elliso, C. Hallin, M. Tuominen, R. Yakimova, A. Henry, J. P. Bergman, and E. Janzén, A 3 kV Schottky barrier diode in 4H-SiC, Appl. Phys. Lett. 72 (1998) 445-447.

[2] P. Brosselard, N. Camara, J. Hassan, X. Jordá, J. P. Bergman, J. Montserrat, J. Millán, 3.3 kV-10A 4H-SiC PiN Diodes, Mater. Sci. Forum. 600-603 (2009) 991-994.

[3] P. Alexandrov, W. Wright, M. Pan, M. Weiner, L. Jiao, J. H. Zhao, Demonstration of high voltage (600–1300 V), high current (10–140 A), fast recovery 4H-SiC p-i-n/Schottky(MPS) barrier diodes, Solid-State Electron. 47 (2003) 263-269.

[4] K. Fujihira, T. Kimoto, H. Matsunami, Fast epitaxial growth of high-quality 4H-SiC by vertical hot-wall CVD, Mater. Sci. Forum. 433–436 (2003) 161-164.

[5] R.L. Myers, Y. Shishkin, O. Kordina, S.E. Saddow, High growth rates (>30 μm/h) of 4H–SiC epitaxial layers using a horizontal hot-wall CVD reactor, J. Cryst. Growth. 285 (2005) 486-490.

[6] F. L. Via, G. Galvagno, G. Foti, M. Mauceri, S. Leone, G. Pistone, G. Abbondanza, A. Veneroni, M. Masi, G. L. Valente, and D. Crippa, 4H SiC Epitaxial Growth with Chlorine Addition, Chem. Vap. Deposition. 12 (2006) 509–515.

[7] S. Leone, M. Mauceri, G. Pistone, G. Abbondanza, F. Portuese, G. Abagnale, G.L. Valente, D. Crippa, M. Barbera, R. Reitano, G. Foti, and F. La Via, SiC-4H epitaxial layer growth using trichlorosilane (TCS) as silicon preceursor, Mater. Sci. Forum. 527–529 (2006) 179-182.

[8] F. La Via, G. Izzo, M. Mauceri, G. Pistone, G. Condorelli, L. Perdicaro, G. Abbondanza, F. Portuese, G. Galvagno, S. Di Franco, L. Calcagno, G. Foti, G. L. Valente, and D. Crippa, SiC-4H epitaxial layer growth by trichlorosilane (TCS) as silicon preceursor at very high growth rate, Mater. Sci. Forum. 600–603 (2009) 123-126.

[9] T. Kimoto, H. Nishino, S. W. Yoo, and H. Matsunami, Growth mechanism of 6H-SiC in step-controlled epitaxy, J. Appl. Phys. 73 (1993) 726-732.

[10] H. Fujiwara, K. Danno, T. Kimoto, T. Tojo and H. Matsunami, Effects of C/Si ratio in fast epitaxial growth of 4H–SiC(0 0 0 1) by vertical hot-wall chemical vapor deposition, J. Cryst. Growth. 281 (2005) 370–376.

Materials Science Forum Vols. 717-720 (2012) pp 109-112

doi:10.4028/www.scientific.net/MSF.717-720.109

Carrot defect control in chloride-based CVD through optimized ramp up conditions

Milan Yazdanfar [a], Stefano Leone [b], Henrik Pedersen [c], Olof Kordina [d], Anne Henry [e] and Erik Janzén [f]

Department of Physics, Chemistry and Biology, Linköping University, SE-58 183 Linköping, Sweden

[a]milya@ifm.liu.se, [b]leone@ifm.liu.se, [c]henke@ifm.liu.se, [d]olkor@ifm.liu.se, [e]anhen@ifm.liu.se, [f]erija@ifm.liu.se

Keywords: Carrot defects; Chloride-based CVD; Ramp up condition

Abstract: Epitaxial growth of 4H-SiC on 8° off-axis substrates has been performed under different condition during the temperature ramp up in order to study the effect on the carrot defect. The study was done in a hot wall chemical vapor deposition reactor using the single molecule precursor methyltrichlorosilane (MTS). During the temperature ramp up, a small flow of HCl or C_2H_4 was added to the H_2 ambient to study different surface etching conditions. The best result was obtained when HCl was added from 1175 to 1520 °C during the ramp up to growth temperature (1575 °C).

Introduction

Power devices with a large current rating require a large active device area, and thus extended defects such as carrots (Fig. 1) should be avoided. Carrot defects have been shown to appear in many different shapes and structures [1] and with different origins. Most commonly, they are formed at the epilayer/substrate interface where they are pinned to a basal plane dislocation or threading screw dislocation in the substrate, and they consist of two stacking faults intersecting along the stair-rod dislocation [2, 3]. The substrates used for epitaxial growth of SiC influence the defect density in the grown epitaxial layer [4]. Proper surface preparation before the epitaxial growth will reduce the density of the carrot defects. During this process, surface damages will be removed and a smooth surface without scratches will be created. Silicon species evaporate faster than carbon species in vacuum, but in a hydrogen ambient carbon atoms react with the hydrogen to form hydrocarbons leading to a faster removal of carbon than silicon from the surface [5]. Through the addition of gaseous molecules such as HCl or C_2H_4 to the hydrogen ambient, the etching of the SiC surface can be altered. Addition of HCl will increase desorption of silicon from the surface through the formation of $SiCl_x$ species while addition of C_2H_4 will suppress carbon evaporation by the increased partial pressure of hydrocarbons [5]. HCl has been successfully used as etching agent prior to growth for 4° off-axis substrates [6]. In this study epitaxial growth of 4H-SiC on 8° off-axis Si-face substrates has been performed with various temperature ramp up conditions in order to evaluate the impact of the temperature ramp up conditions on the carrot defect density.

Experimental

The growth was done in a horizontal hot wall chemical vapor deposition reactor without rotation. All substrates in this study were approximately 1.5×1.5 cm^2 samples cut from one chemical-mechanical polished 4H-SiC 8° off-axis 3 inch wafer. For all experiments, the growth pressure and growth time were kept constant at 200 mbar and 30 min, respectively. Methyltrichlorosilane (MTS) [7] was used as precursor with a Si/H_2 ratio of 0.15%. The obtained growth rate was about 70 μm/h. All experiments in this study commence with a 10 min temperature stabilization under H_2 purge at 1175 °C followed by a characteristic ramp up to the growth temperature (1575 °C). The ramp up is comprised of a temperature ramp up to a specific temperature (T_{out}, which was in the range of 1500 – 1575 °C), with HCl and/or C_2H_4, added to the H_2 ambient; also studies in pure H_2 were done. At the T_{out} temperature, the etching additive was removed and the sample was heated in a pure H_2 ambient during the remainder of the temperature ramp to the growth temperature. During the characteristic ramp to T_{out}, the flow of HCl was 200 ml/min (Cl/H_2 = 0.4%) and that of C_2H_4 15 ml/min (C/H_2 =0.06%). The morphology of the epilayers was studied using optical microscopy with Nomarski Differential Interference Contrast (NDIC) and in more detail using Atomic Force Microscopy (AFM). Fourier Transform Infrared Reflectance (FTIR) was used to measure the thickness of the epilayers. The quality of the grown epilayers was studied using Low Temperature Photoluminescence (LTPL) at 2K using a frequency double Argon ion laser at 244 nm as excitation source.

Results and Discussion

The carrot defect density is shown in Fig.2 as a function of the T_{out} temperature for two different ambient conditions. Two distinct results are recognized: First there is a well defined minimum of carrot density at T_{out} = 1520°C regardless of ambient; second, the carrot density is slightly lower when HCl ambient is used and $T_{out} \geq 1520$ °C. Typical heating time from 1175 to 1520°C and from 1520 to 1575°C was 4 and 1 minutes respectively. The lower carrot density for HCl ambient compared to ethylene ambient should be explained by different etching conditions during ramp up. With the addition of HCl, the etching of silicon atoms is enhanced through the formation of chlorosilane species, while by adding C_2H_4 carbon etching will be suppressed by the increased partial pressure of hydrocarbons. Also the etching condition during heating from T_{out} to growth temperature is crucial and the best result was obtained for a pure H_2 ambient i.e. when the carbon etching was enhanced. For experiments made with pure H_2 ambient up to 1520 °C, followed by addition of HCl or C_2H_4 up to growth temperature, the carrot density was found to be unaffected by the ramp up conditions, i.e. the same result as with pure H_2 ambient during heating from 1175 to 1575 °C was obtained. It thus can be speculated that modified etching with an etching agent during only one minute at high temperature (1520-1575 °C) is too short time for any effect to be noticed.

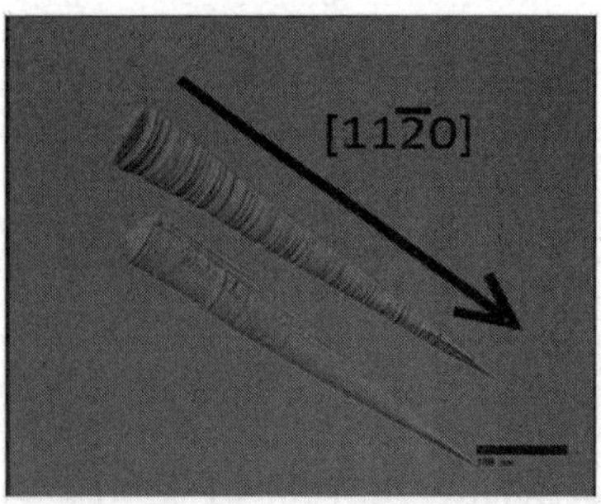

Fig. 1: Optical microscope image at 200 × magnification of 38 μm 4H-SiC thick epitaxial layer grown at 1575 °C. Microscope scale bar is 100 μm. The arrow shows the off-cut direction.

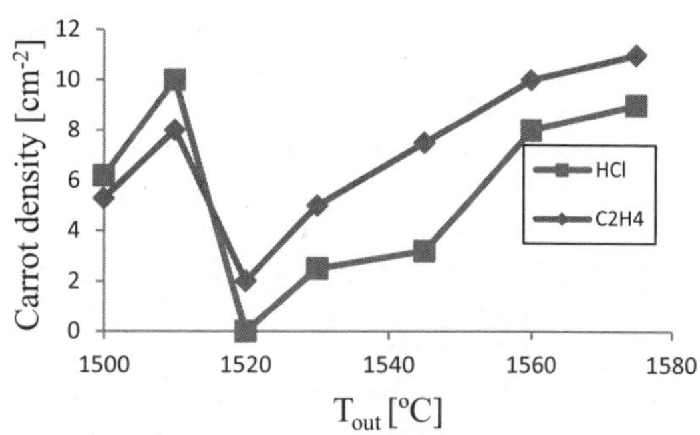

Fig. 2: Carrot - defect density versus T_{out}, the temperature when the ambient HCl or C_2H_4 was removed from the reaction chamber.

Etching first with HCl up to 1520 °C followed by C_2H_4 up to growth temperature, or in C_2H_4 up to 1520°C followed by HCl to growth temperature, resulted in slightly higher carrot densities compared to heating in pure hydrogen ambient. Further, for T_{out} = 1520 °C the dependence of the flow of HCl and the process pressure (Fig. 3, Fig. 4) was also investigated. The best conditions found was 200 ml/min HCl flow and 200 mbar process pressure. In case of higher flow of HCl during rump up, the silicon etching was too high and rather than creating a smooth surface it created a heavily damaged surface. Lower pressure in the cell most likely lead to an increased etch rate due to lower partial pressures of carbon and silicon species which also lead to a damaged substrate surface.

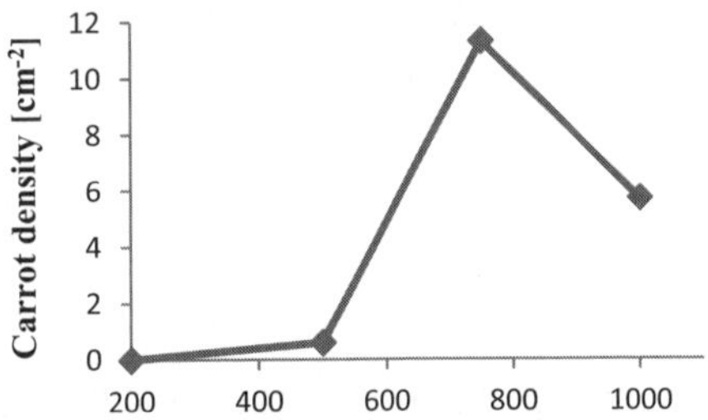

Fig. 3: Carrot - defect density versus the HCl flow during temperature ramp up; all ramp up conditions started after the 10 min stabilization at 1175 to 1520° C. The best result was achieved at a HCl flow of 200 ml/min.

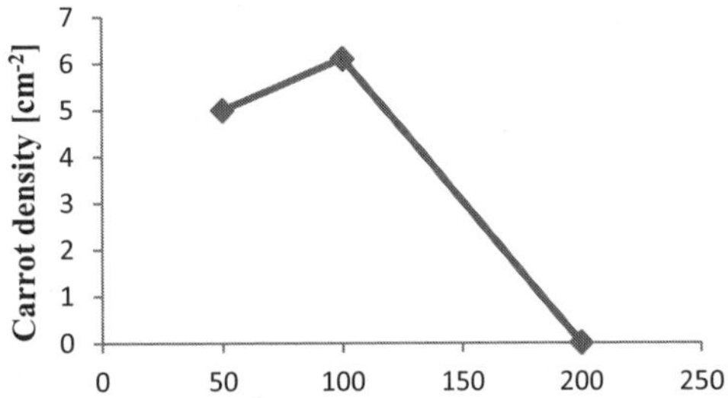

Fig. 4: Carrot - defect density versus the process pressure with a flow of 200 ml/min HCl from 1175 to 1520° C during ramp up condition. The best result was obtained at a pressure of 200 mbar.

All samples were studied with LTPL which showed very intense near bandgap emission, regardless of etching conditions indicating that the epilayers are of high crystalline quality and that the etching conditions does not affect the crystal quality of the grown epilayer. The D_1 line was seen in all samples and the nitrogen-bound exciton (P_{77}) intensity was higher than the free-exciton (I_{76}) intensity. AFM measurements were performed on selected samples to study the surface morphology. Fig.5 and Fig.6 show the AFM image of the substrate and the grown epilayer, respectively. The thickness of the epilayer is 38 μm and the morphology over the sample is uniform. The surface roughness root-mean-square (RMS) value of the epilayers was less than 1.2 nm, while the substrate before the epitaxial growth had a RMS value of 2.3 nm and several scratches from the polishing.

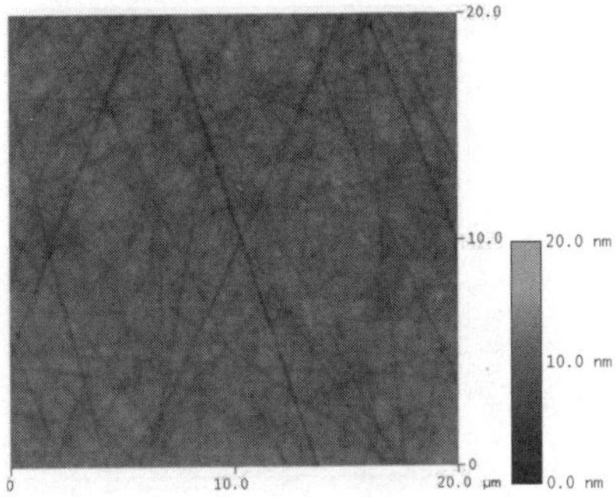

Fig. 5: AFM image of 20×20 μm^2 area before the growth (substrate), with a RMS value of 2.3 nm.

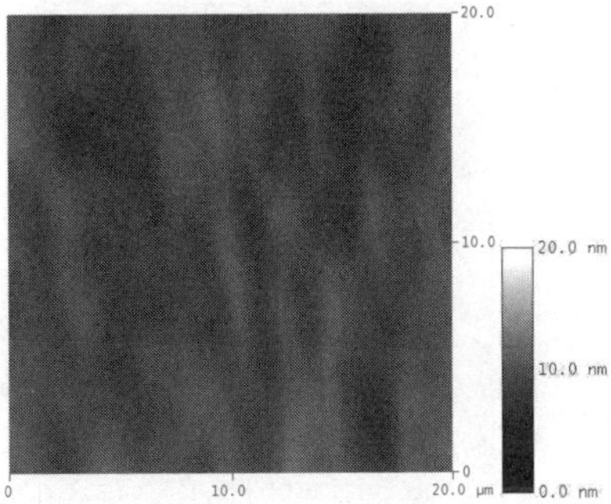

Fig. 6: AFM image of 20×20 μm^2 area of a 38 μm thick epilayer grown at 1575 °C, T_{out} =1520 °C, HCl flow of 200ml/min, pressure of 200 mbar and with a RMS value of 1.1 nm.

Summary

In this study, the in situ etching of the SiC substrate during heating the reactor to growth temperature was studied. By applying an increased silicon etching for the major part of the ramp up followed by a short increased carbon etching, the carrot defect density was significantly lowered. We further noticed that very high HCl flow is increasing the carrot defect density, most likely due to surface damages caused by heavy silicon etching. Also the process pressure was found to have a major impact on the carrot defect density.

Acknowledgement

Authors gratefully acknowledge the financial support from the Swedish Energy Agency, the Swedish Research Council (VR).

References

[1] J. Hassan, A. Henry, P.J. McNally, J.P. Bergman, Journal of Crystal Growth 312 (2010) 1828–1837.

[2] M. Benamara, X. Zhang, M. Skowronski, Applied Physics Letters 86 (2005) 021905.

[3] X. Zhang, S. Ha, M. Benamara, M. Skowronski, Applied Physics Letters 85 (2004) 22.

[4] H. Chen, G. Wang, Y. Chen, X. Jia, M. Dudley, Materials. Research. Society. Symposia. Proceedings 991 (2006) 163.

[5] A.A. Burk, Jr., L.B. Rowland, Applied Physics Letters 68 (1996) 382.

[6] S. Leone, F.C. Beyer, H. Pedersen, O. Kordina, A. Henry, E. Janzén, Materials Research Bulletin 46 (2011) 1272-1275.

[7] H. Pedersen, S. Leone, A. Henry, F.C. Beyer, V. Darakchieva, E. Janzén, Journal of Crystal Growth 307 (2007) 334-340.

Materials Science Forum Vols. 717-720 (2012) pp 113-116
© (2012) Trans Tech Publications, Switzerland
doi:10.4028/www.scientific.net/MSF.717-720.113

Chloride-Based CVD of 4H-SiC at High Growth Rates on Substrates with Different Off-Angles

Stefano Leone[1-#,a*], Henrik Pedersen[1,b], Franziska C. Beyer[1,c], Sven Andersson[1,d], Olof Kordina[1,e], Anne Henry[1,f], Andrea Canino[2,g], Francesco La Via[2,h], and Erik Janzén[1,i]

[1]Department of Physics, Chemistry and Biology, LinköpingUniversity, 58183 Linköping, Sweden

[2] Consiglio Nazionale delle Ricerche IMM, Z. I. VIII Strada 5, 95121 Catania, Italy

[a]leone@ifm.liu.se, [b]henke@ifm.liu.se, [c]fbeyer@ifm.liu.se, [d]svean@ifm.liu.se, [e]olkor@ifm.liu.se, [f]anhen@ifm.liu.se, [g]andrea.canino@imm.cnr.it, [h]francesco.lavia@imm.cnr.it, [i]erija@ifm.liu.se

Keywords: Chloride-based CVD; homoepitaxial growth; substrate off-angle; high growth rate.

Abstract. A review of recently achieved results with the chloride-based CVD on 8° and 4° off-axis and nominally on-axis 4H-SiC wafers is done to clarify the epitaxial growth mechanisms on different off-angle substrates. The process conditions selected for each off-axis angle become even more difficult when running at growth rates of 100 μm/h or more. A fine-tuning of process parameters, mainly temperature, C/Si ratio and *in situ* surface preparation is necessary for each off-angle. Some trends related to the surface properties and the effective C/Si ratio existing on the surface prior to and during the epitaxial growth can be observed.

Introduction

Since the last development in enlarging the diameter of SiC substrates up to 6", and according to the well-known problem with bipolar degradation in case of using substrates with high off-angle (*i.e.* 8°) [1], a reliable epitaxial growth process on low off-angle substrates is required. In facts basal plane dislocations (BPDs) propagates more easily on larger off-angles epitaxial layers. BPDs convert into stacking faults during device operation, which results in a drift of the forward voltage. High growth rates are desired to perform growth of thick epilayers, which are required for very high voltage, power devices [2].

The epitaxial growth of 4H-SiC using a chloride-based CVD process is a well-established technique, which has proven to be suitable for homo- and hetero-epitaxial growth for a very wide range of conditions, achieving growth rates in excess of 100 μm/h [3]. Yet, it is important to understand the different growth mechanisms occurring on different off-angle substrates in order to find the proper growth and surface preparation conditions. Formation of extended defects, such as triangle and carrots, and stepped surfaces are common problems when growing on 8° and 4° off-axis 4H-SiC substrates [4], while 3C-SiC inclusions are usually created when growing on on-axis substrates [5]. Growth parameters have to be fine-tuned according to the substrate off-angle, as it has been demonstrated for the growth with standard precursors [6].

The aim of this work is to understand the effect of the off-angle on the growth mechanism, and which are the most critical growth parameters to be tuned for the deposition on different off-cut substrates, when using different chlorinated precursors.

Experimental procedure

Epitaxial growths were done in a hot-wall CVD reactor.Hydrogen was used as carrier gas (50 l/min) at a reduced pressure of 200 mbar. 4H-SiC samples (2 x 2 cm^2) of different off-angles (8°, 4° and nominally on-axis) were used as substrates. Two different sets of chlorinated precursors were tested: methyltrichlorosilane (CH_3SiCl_3, also called MTS), and silane (SiH_4) plus ethylene (C_2H_4) and hydrogen chloride (HCl). The epitaxial process procedure was as follows: evacuate the reaction chamber to low pressure (10^{-4} mbar), backfill it with hydrogen until process pressure, heat to 1200 °C, then flow another gas (such as HCl or SiH_4) and ramp up to the growth temperature. Growths were usually done for a certain time in order to deposit about 20 µm thick epilayers. The reactor was then cooled down in hydrogen, and evacuated before backfilling it with Argon to atmospheric pressure.

The routine characterizations made on all samples were: morphology analysis with optical microscope; thickness measurement with Fourier Transform Infrared Reflectance (FT-IR); doping measurement by a Capacitance-Voltage (CV) Hg-probe; photoluminescence at 2K (LTPL) using a 244 nm wavelength laser to excite the samples. Spatially resolved micro-photoluminescence at room temperature was used to detect Single Shockley Faults (SSF), which are generated by dissociation of basal plane dislocations (BPD) [7], exhibiting luminescence at 2.93 eV. Further characterizations were performed on selected samples: atomic force microscopy (AFM) in tapping mode was used to get a deeper insight of the surface morphology; etching in molten KOH at 500 °C for 2 minutes was necessary to reveal defects such as BPDs or other dislocations (screw or edge). Minority carrier lifetime was not measured.

Results

Growth optimization. During the optimization process of the growth at very high growth rates with both set or precursors, it turned out that the C/Si ratio had to be set lower when using lower off-angles: C/Si = 1 – 1.2 for 8° off-axis [3]; C/Si = 0.8 for 4° off-axis [8]; C/Si = 0.3 - 0.6 for on-axis [5] unless a specific *in situ* etching was adopted [9]. In all cases, a growth rate of 100 µm/h or more was achieved. The same tendencies were observed at more extreme conditions, such as very low growth temperature (*i.e.* 1300 °C [10]), and concentrated growth conditions (*i.e.* an order of magnitude lower carrier flow and process pressure [11]). The temperature, which is typically 1550-1580 °C for the growth on 8° off-axis substrate, has an important effect mainly on 4° off-axis and on-axis samples. It has to be kept lower than typical, about 1500 – 1535 °C, to avoid step-bunching on 4° off-axis substrates, while a higher temperature of 1600 – 1630 °C can give substantial help to minimize the formation of 3C inclusions during the growth on on-axis substrates.

Growths performed with the same growth parameters but different precursors showed a higher efficiency for MTS. With a Si/H_2 = 0.15 %, substrates with an 8° off-cut, temperature of 1570 °C and pressure of 200 mbar, a growth rate of 70 µm/h was achieved when using MTS, while it was 60 µm/h when using the other set of precursors.

The surface preparation prior to the growth was also different: In the case of 8° or 4° off-axis substrates, HCl etching during the temperature ramp up and for a few minutes prior to the growth was adopted to avoid triangular defects while obtaining a smooth epitaxial layer. For the growth on nominally on-axis wafers, Si-rich etching conditions were needed to achieve homoepitaxial growth.

Characterization. The high quality of the epilayers grown using chloride-based CVD is demonstrated by low temperature photoluminescence measurements at 2 K, which show very strong excitonic 4H luminescence regardless of the off-angle. The background doping measured by CV or LTPL was usually in the high 10^{14} cm^{-3} range, unless very low C/Si ratios were used. Spatially resolved micro-photoluminescence at room temperature, used to detect Single Shockley Faults (SSF), indicated that only the layers grown on on-axis substrates were free of any SSF on the whole sample. KOH etching at 500 °C for 2 minutes was used to calculate the density of BPDs existing in the epilayers. A high amount of BPD (10^4 cm^{-2} range) was found on the material grown on 8° off-axis substrates, while it was less than 10 cm^{-2} on 4° off-axis substrates and zero on on-axis wafers. Electrical measurements indicated very high quality material with leakage current in the pA range.

The only drawback for growth at high growth rates on on-axis substrates is the surface roughness. The growth occurs via island-growth mechanism; the resulting morphology is characterized by very broad and high islands originating from screw dislocations existing in the substrate. Off-axis wafers showed a very low roughness with an RMS value lower than 1 nm due to the dedicated HCl-*in situ* etching process (Fig. 1).

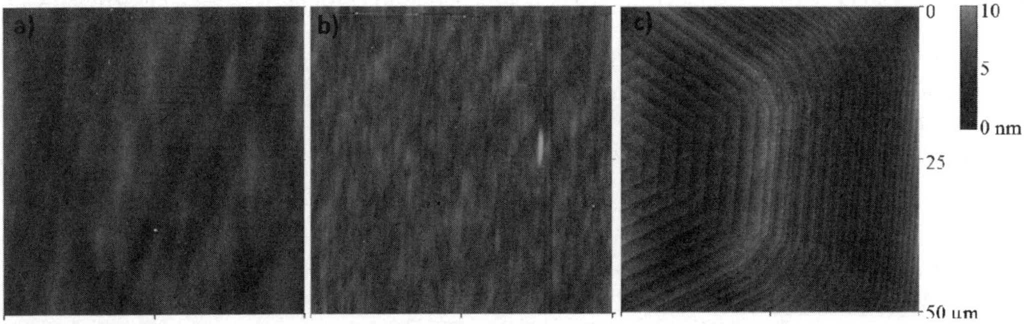

Fig. 1: Atomic Force Microscopy images of 50 µm x 50 µm area of 20 µm thick epitaxial layers grown at 100 µm/h with Cl-CVD on different off-axis substrates: a) 8° off-axis, RMS = 0.58 nm; b) 4° off-axis, RMS = 0.67 nm; c) 0.05° off-axis, RMS = 0.76 nm.

Discussion

In a very simplistic model, the need of more silicon-rich conditions for substrates with lower off-angles could be attributed to the surface properties of 4H-SiC (0001). In Fig.2, a schematic section of an on-axis (top) and off-axis surface (bottom) of 4H-SiC (0001) is displayed. As it can be seen, the on-axis surface exposes almost only silicon atoms which can either be etched by chlorine or evaporate because of the high temperatures and low pressures used during the CVD process. In order to grow stoichiometric SiC, the effective C/Si ratio on the growing surface has to be kept close to 1. This means that in the case of on-axis substrates a Si-rich gas phase has to be used to suppress or reduce silicon etching and evaporation.

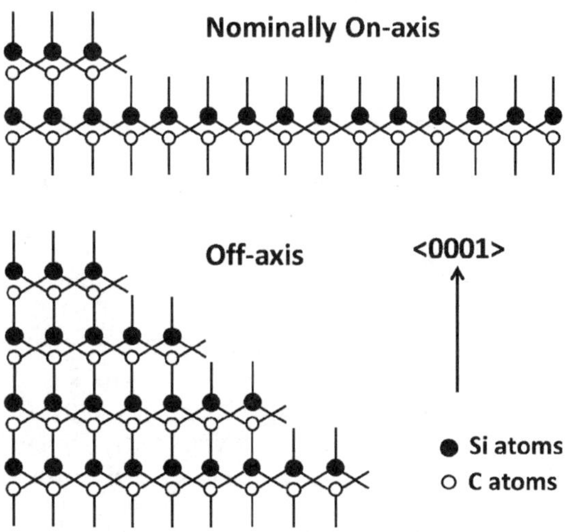

Fig. 2: Schematics of 4H-SiC (0001) crystal structure for nominally on-axis (top) and off-axis (bottom) surfaces.

The off-axis surface has a high density of steps, at the edge of steps the carbon atoms are exposed in addition to the silicon atoms, and every atom has at least two dangling bonds [12]. In this way carbon atoms can be easily removed by the high amount of hydrogen used as a carrier by forming hydrocarbons. Furthermore, the sticking coefficients of the main carbon intermediates are much lower than those of the silicon intermediates [13], this means that carbon atoms will desorb faster from the surface especially at the step edges. Therefore, in the case of off-axis substrates a carbon-rich environment is needed in the gas mixture, especially during the temperature ramp up, to compensate for the lower amount of carbon atoms on the growing surface. But it has been found that carbon-rich environment during the temperature ramp-up can favor the formation of triangular defects [8].

A further confirmation to the hypothesis that on-axis surfaces are silicon-rich comes from experimental evidence [14]. Kojima et al. have found it easier to close micropipes when growth was performed on lower off-angle substrates, especially on-axis. Besides closing of micropipes was facilitated by silicon-rich growth conditions. Therefore it can be assumed that on-axis substrates have a more silicon-rich surface, as speculated in this study.

As discussed elsewhere [9], MTS showed superior efficiency compared to the process with HCl addition, both in terms of growth rate and reduced issues with 3C inclusions.

Summary

In conclusion, for the growth of high quality epitaxial layers, higher off-angle substrates, like in the 8° off-axis case, will need a higher real C/Si ratio reaching the surface, while an on-axis substrate will need a lower C/Si ratio to have more stoichiometric conditions on the growing surface. A fine tuning of the etching and growth conditions in this direction can lead to the growth of high quality epitaxial layers deposited at rates of 100 μm/h and more with the chloride-based CVD process, even on on-axis substrates.

The Swedish Energy Agency and the Swedish Research Council are gratefully acknowledged for financial support.

[#]: Actual address is: Aixtron SE, Kaiserstr. 98, 52134 Herzogenrath, Germany

References

[1] J. P. Bergman, H. Jakobsson, L. Storasta, F. H. C. Carlsson, B. Magnusson, S. Sridhara, G. Pozina, H. Lendenmann, E. Janzén, Materials Science Forum Vol. 389-393 (2002) 9-14.

[2] J. A. Cooper, A. Agarwal, Proceedings of the IEEE Vol. 90 (6) (2002) 956-968.

[3] H. Pedersen, S. Leone, A. Henry, F. C. Beyer, V. Darakchieva and E. Janzén, J. of Crystal Growth vol. 307 (2007) 334-340.

[4] S. Leone, H. Pedersen, A. Henry, O. Kordina, E. Janzén, J. Cryst. Growth 311 (2009) 3265 - 3272.

[5] S. Leone, H. Pedersen, A. Henry, O. Kordina, E. Janzén, J. Cryst. Growth 312 (2010) 24-32.

[6] K. Kościewicz, W. Strupiński, D. Teklinska, K. Mazur, M. Tokarczyk, G. Kowalski, A. R. Olszyna, Materials Science Forum 679-680 (2011) 95-98.

[7]A. Canino, M. Camarda, F. La Via, Mater. Sci. Forum, vol. 679-680, p. 67-70 (2011).

[8] S. Leone , F.C. Beyer, H. Pedersen, O. Kordina, A. Henry, E. Janzén, Mater. Res. Bull. 46 (2011) 1272–1275.

[9] S. Leone, F. C. Beyer, H. Pedersen, O. Kordina, A. Henry, E. Janzén, Crystal Growth & Design 10 (12) (2010) 5334-5340.

[10] S. Leone, F. C. Beyer, H. Pedersen, S. Andersson, O. Kordina, A. Henry, E. Janzén, Thin Solid Films 519 (2011), pp. 3074-3080.

[11] S. Leone, A. Henry, S. Andersson, O. Kordina, E. Janzén, J. Electrochem. Soc. 157 (10) (2010) H969-H976.

[12] T. Yamamoto, T. Kimoto, H. Matsunami, Mater. Sci. Forum 264-268 (1998) 111-114.

[13] Ö. Danielsson, A. Henry, E. Janzén, J. Cryst. Growth 243 (2002) 170-184.

[14] K. Kojima, S. Nishizawa, S. Kuroda, H. Okumura, K. Arai, J. Cryst. Growth 275 (2005) e549–e554

Materials Science Forum Vols. 717-720 (2012) pp 117-120

doi:10.4028/www.scientific.net/MSF.717-720.117

Comparison of 4H silicon carbide epitaxial growths at various growth pressures using dicholorosilane and silane gases

Tawhid Rana[a*], Haizheng Song[b], M.V.S. Chandrashekhar[c], and Tangali Sudarshan[d]

University of South Carolina, SC-29208, USA

[a]rana@email.sc.edu, [b]songha@cec.sc.edu, [c]chandra@cec.sc.edu, [d]sudarsha@cec.sc.edu, *Corresponding author

Keywords: Dichlorosilane, silane, silicon carbide epitaxy, simulation

Abstract. Silicon carbide epitaxial films grown in an inverted chimney reactor using silane-propane-hydrogen and dichlorosilane-propane-hydrogen systems, are compared for growth rates and doping concentrations at various growth pressures. Virtual Reactor, a commercial software, is employed to simulate decomposition of silane and dichlorosilane gases into elemental silicon to explain the growths found by experiment. Parasitic deposition in the gas injector tube using these precursor gases is compared for precursor depletion.

Introduction

The conventional SiC-CVD process uses silane as the Si-precursor, light hydrocarbons, e.g., propane (C_3H_8) or ethylene (C_2H_4), as the C-precursor and hydrogen as the carrier gas. In recent years, chloride precursors such as $SiCl_4$ and/or HCl addition are used to achieve high growth rates and to reduce Si-droplet formation in the 4H or 6H SiC-CVD process. Simulations have been carried out for silane precursor [1, 2] and some chloride precursors [3, 4]. Dichlorosilane (DCS) is recently reported for high growth rate SiC growth [5]. A comparative study is essential to know the advantages of growth with DCS over silane in terms of achieving high growth ratesand high quality epitaxy.

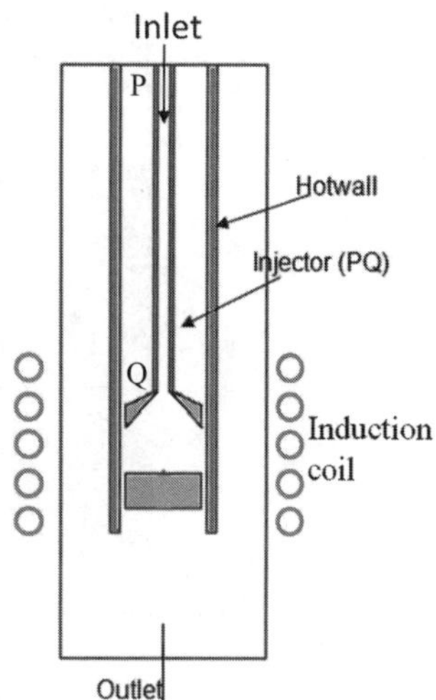

Fig. 1: Furnace geometry

Experimental

Epitaxial growth of SiC is carried out using an inverted chimney vertical hot-wall CVD reactor (Fig. 1). The system is inductively heated by a 20kW generator. Silane/dichlorosilane and propane are used as precursor gases. Propane, silane/dichlorosilane and hydrogen gas flow are kept fixed at 1.5 sccm, 3.28 sccm and 6000 sccm respectively while the growth temperature is fixed at 1550°C. The distance between the injector and sample is kept 33 mm and the growth pressure is varied from 30 to 600 torr. Commercially available 4H-SiC wafers (8° off cut towards **[11$\bar{2}$0],** Si face) are used as the substrate for all of the experiments. Epilayer thickness is measured using FTIR. Simulation is performed using a commercial CVD simulation tool, called Virtual reactor, developed by Semiconductor Technology Research, Inc., for the same conditions using both DCS and silane gases to analyze decomposition of precursor gases into elemental Si.

Results and discussion

Simulation is utilized to analyze the decomposition of silane and DCS precursors into elemental silicon for various growth pressures. Elemental Si with the highest number of Si dangling bonds among the decomposed silicon species, has the highest probability to form Si-Si bond and hence silicon droplets in the gas due to collisions. Partial pressure of the decomposed elemental Si found from simulation, is considered in this paper to be an indication of the degree of gas phase nucleation at different growth pressures. From simulation (for the reactor geometry shown in Fig. 1), Fig. 2 demonstrates that DCS prevents the formation of elemental Si noticeably at low pressures (30 torr) compared to Silane. A higher dissociation bonding energy of Si-Cl (381 KJ/mol) compared to Si-H (318 KJ/mol) [6] is considered to be the reason for this reduced formation of elemental Si for DCS, which, eventually, suppresses gas phase nucleation. This reduction, especially at lower pressures (e.g. 30 torr) increases the availability of silicon containing species for growth and increases the growth rate in the DCS system compared to the silane system (Fig. 3). However, at higher growth pressures, both gases produce a higher amount of elemental Si (Fig. 3, 300-600torr), though, the presence of dissociated Cl restricts the Si droplet formation in the DCS system and gives a higher growth rate even at higher pressures.

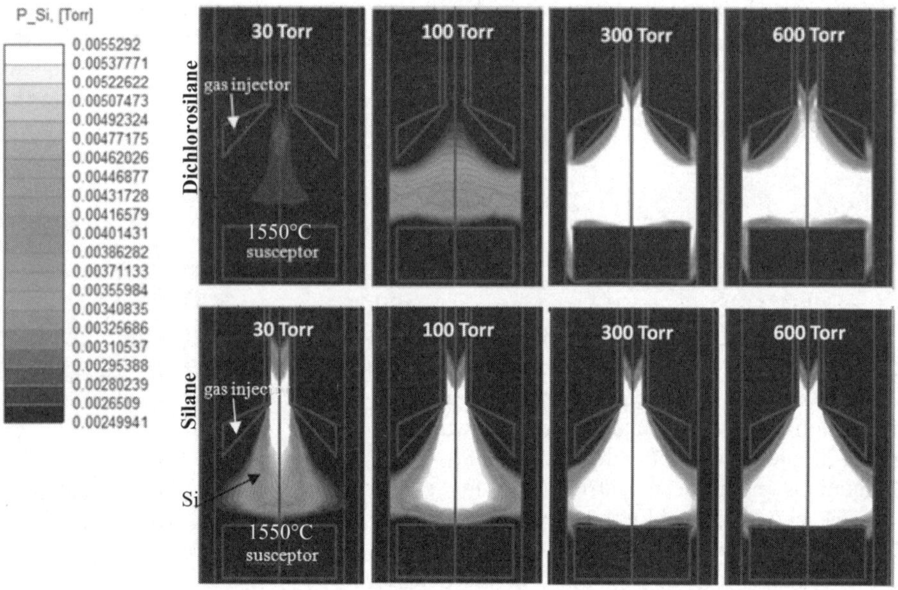

Fig. 2: Simulation of decomposition of DCS (top) and silane (bottom) into elemental Si.

Combining simulation and experimental results (Fig. 2, 3), it can be inferred that growth at 30 Torr in DCS is mostly due to reaction of $SiCl_x$ or SiH_xCl_x species, evidenced by the smaller amount of decomposed Si in this condition at this pressure in Fig. 2. Thus, DCS provides an environment with reduced Si gas phase nucleation, consistent with the assumption that collisions involving $SiCl_x$ or SiH_xCl_x have lower probability to form Si droplets or aerosol. On the other hand, Fig. 2 shows that silane decomposes more easily to form elemental Si even at this lower pre ssure (30 torr) resulting in higher gas phase nucleation and lower gr owth rate

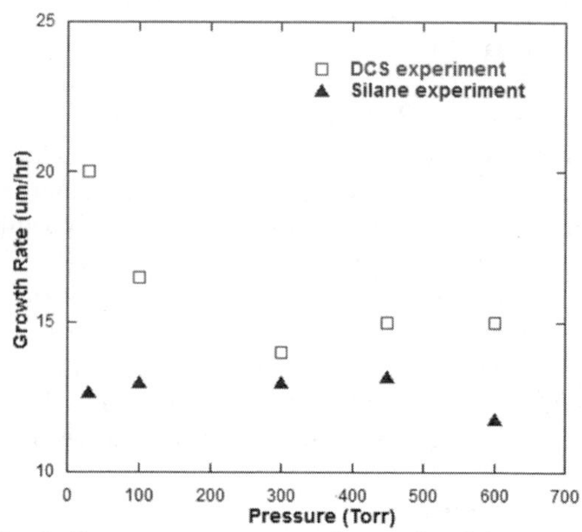

Fig. 3: Pressure versus growth rate for sialne and DCS

(Fig. 3). On the other hand, DCS decomposes into elemental Si in lesser amount (Fig. 2, 30 torr) for the same condition and increases the growth rate (Fig. 3) due to reduced gas phase nucleaiton. Further, in addition to the reduced gas phase nucleation condtion for DCS, it is found that the parasitic deposition region is also reduced for DCS compared to that of silane, which is expected to increase the growth rate by increasing the availability (reduced Si loss) of the precursor for growth. To investigate gas decomposition for these gases, the gas injector tube (shown as 'PQ' in Fig. 1) is axially split into two halves (Fig. 4a and 4b) for silane and DCS growths.

The deposition starts from points ‘A’ and ‘A′’ for silane and DCS gases respectively (Fig. 4). Position ‘A′’ is about 15mm downshifted compared to position ‘A’ for the case of DCS which indicates that the parasitic deposition region is reduced for DCS. A reduced parasitic deposition will increase epi growth rate by reducing the precursor loss. Comparison of these regions indicates that: AB>A′B′, BC>B′C′ and BD=B′D′, and C′D′>CD (Fig. 4a and 4b). A detailed study of the composition of these regions is beyond of the scope of this paper. Regions CD and C′D′ are assumed to be the regions where SiC (3C) growth take place on the tube wall. Elongated C′D′ is a sign of enhanced SiC growth using DCS growth compared to silane (CD). A close look at the positions corresponding to the triangle and circle marks (Fig. 4a and 4b) reveal large, loosely bound granular growths in the tube for silane (Fig. 4c). Growth with this condition is found to be deteriorated with particles on the epi surface, probably due to the exfoliation of these grains from the tube. Parasitic deposition using DCS was found to be smoother at the same position (Fig. 4d). An improved epi surface was observed for this condition. Thus we can deduce that the smoother parasitic deposition in DCS is indicative of improved epi quality due to prevention of particle droplets from the tube wall being delivered on to the growing epi surface.

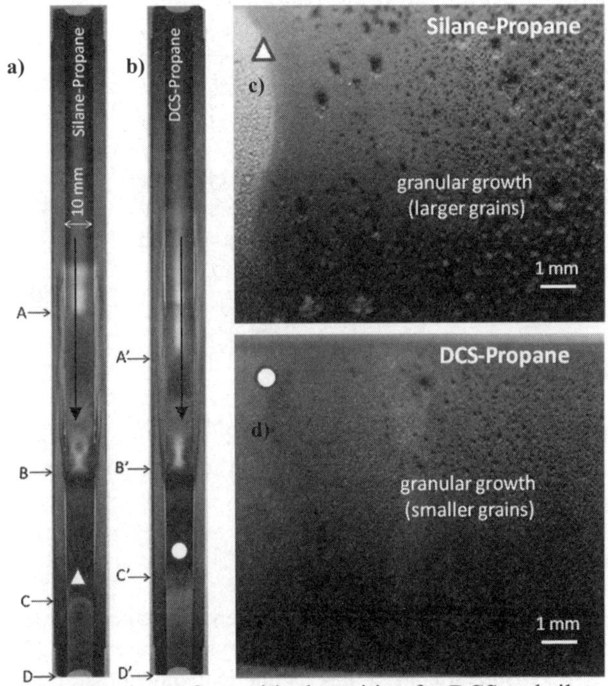

Fig. 4: Comparison of parasitic deposition for DCS and silane in the gas injector tube shown as PQ in Fig.1

A higher n-doping concentration of the epitaxial film is observed at higher pressures (Fig. 5). A higher silicon gas phase nucleation at higher pressure as discussed earlier should deplete the Si precursor and increase the C/Si ratio, which, contrary to the experimental results, should decrease n-doping according to site competition epitaxy [7]. This contradiction can only be explained if carbon depletion is also addressed. We can deduce that as pressure increases, C-depletion also increases relative to the Si depletion, and thus increase the n-doping. This aspect is subject to another concurrent research by the authors’ group.

Summary.

A study is presented for the comparison of growth using DCS and silane at various growth pressures. The higher growth rate achieved using DCS is attributed to the reduction of Si precursor losses, which consist of both gas phase nucleation and parasitic deposition in the reactor. It is also found that DCS improves the epi quality compared to silane by improving the parasitic deposition condition by restricting loose particle formation within the gas injector tube. The behavior of doping at various growth pressures is explained by the effective C/Si ratio which depends on the relative losses of carbon and silicon in the growth chamber rather than the inlet C/Si ratio.

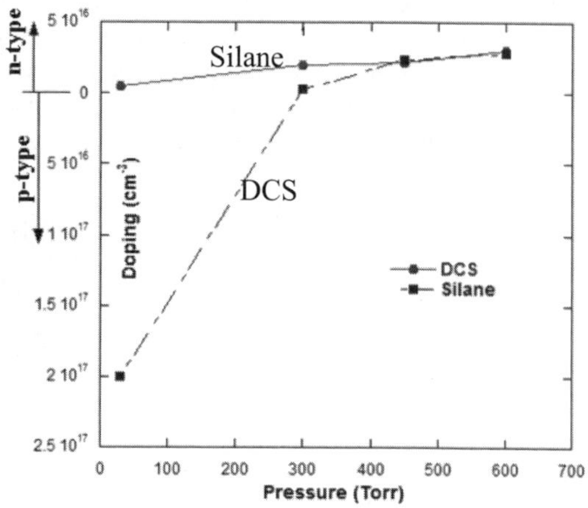

Fig. 5: Doping concentration versus pressure for silane and DCS.

Acknowledgement

This work was supported by the ONR, Grant no. # N000141010530. The authors thank the program manager Dr. H. Scott Coombe for his support of this research.

References

[1] O. Danielsson, A. Henry, E. Janzen, Growth rate predictions of chemical vapor deposited silicon carbide epitaxial layers, Journal of Crystal Growth 243 (2002) 170

[2] S. Nishizawa, M. Pons, Numerical modeling of SiC–CVD in a horizontal hot-wall reactor, Microelectronic Engineering 83 (2006) 100

[3] A. Veneroni and M. Masi, Gas-Phase and Surface Kinetics of Epitaxial Silicon Carbide Growth Involving Chlorine-Containing Species, Chem. Vap. Deposition, 12 (2006) 562

[4] Rong Wang, Ronghui Ma, An integrated model for halide chemical vapor deposition of silicon carbide epitaxial films, Journal of Crystal Growth 310 (2008) 4248

[5] Iftekhar Chowdhury, MVS Chandrashekhar, Paul B Klein, Joshua D. Caldwell, Tangali Sudarshan, High growth rate 4H-SiC epitaxial growth using dichlorosilane in a hotwall CVD reactor, Journal of Crystal Growth, 316 (2011) 60

[6] James E. Huheey, Inorganic chemistry principles of structure and reactivity, Prentice Hall 1972

[7] DJ Larkin, PG Neudeck, JA Powell, LG Matus, Site-competition epitaxy for superior silicon carbide electronics, Applied Physics Letters, 65 (1994), 1659

Materials Science Forum Vols. 717-720 (2012) pp 121-124

doi:10.4028/www.scientific.net/MSF.717-720.121

In-Grown Stacking Faults in SiC-CVD using Dichlorosilane and Propane as precursors

Haizheng Song[a, *], Sabih U. Omar[b], Tawhid Rana[c], M.V.S. Chandrashekhar[d] and Tangali S. Sudarshan[e]

Department of Electrical Engineering, University of South Carolina, Columbia, SC 29208, USA

[a]Songha@engr.sc.edu, [b]Omar@email.sc.edu, [c]Rana@email.sc.edu, [d]Chandra@cec.sc.edu, [e]Sudarsha@engr.sc.edu, *corresponding author

Keywords: epitaxial growth, in-grown stacking faults, dichlorosilane, insulation material, substrate pretreatment.

Abstract. In-grown stacking faults (IGSFs) were studied in 4H-SiC homoepitaxial growth from a SiH_2Cl_2-C_3H_8-H_2 system. Most of the IGSFs start from the epilayer/substrate interfac and exhibit photoluminescence emission peak at 2.58 eV (480nm). The growth parameters, including growth temperature, growth pressure, growth rate, and hydrogen etching, do not significantly influence the formation of IGSFs. Reactor furniture is identified to be a major reason of IGSF formation, especially when the insulation part of the furnace is not completely isolated from the growth zone. Dusting of insulation material greatly increases the formation of IGSFs. When using graphite felt as the insulation material, the IGSF density in the epilayer can be as high as ~10^4 cm^{-2}. Improvement of the insulation material by using graphite foil reduces the density to 30-100 cm^{-2}. Further reduction of IGSF density to less than 10 cm^{-2} is achieved by mild pretreatment of the substrate in molten KOH-NaOH eutectic.

Introduction

In chemical vapor deposition (CVD) epitaxial growth, stacking faults (SFs) can be relatively easily incorporated in 4H-SiC polytype because of the small SF formation energy. The formation energy of single Shockley SF in 4H-SiC is 14.7 mJ/m^2 [1], which is very low compared to that in other semiconductors, such as diamond (279 mJ/m^2) [2], Si (55 mJ/m^2) [1], and GaAs (45 mJ/m^2) [1]. Therefore any disturbance (e.g. particles) to the crystal may generate a SF. When SFs are generated spontaneously during the growth, they are called in-grown SFs (IGSFs) and the IGSFs will extend throughout the epi-growth in the basal plane. The IGSFs are different in structure and optical property from the Shockley SFs generated in degraded PIN diodes [3]. Due to the diversity and complexity of IGSFs, different structures of IGSFs have been reported [4-6]. Although the generation mechanism of IGSFs is still not completely clear, cleanliness of the CVD furnace and surface state of the substrate should have great influence on their formation. It was reported that IGSFs would reduce the Schottky barrier height and breakdown voltage for 4H-SiC Schottky diodes [3] and increase the reverse leakage current in PIN diodes [7]. Therefore, control and reduction of IGSFs is crucial for SiC epitaxial growth.

In previous report [8], SiC epitaxial growth from dichlorosilane (DCS/SiH_2Cl_2) and propane was developed to achieve high growth rate (up to 100 μm/h). However, it was also found that IGSFs could be generated in very high density (~10^4 cm^{-2}) at none or less optimized conditions and DCS did not substantially influence the formation of IGSFs. Therefore, in this paper, we examine the effect of the CVD reactor furniture, which is identified to be a major reason of IGSF formation in the above system. By carefully choosing the furniture material, IGSF density reduces to <100 cm^{-2}.

Experimental

The CVD epitaxial growth was carried out in a homebuilt hot-wall chimney reactor. The substrates used in this work were commercially obtained, 8° off-axis towards [$11\bar{2}0$], highly doped (~10^{18} cm^{-3}, n-type) 4H-SiC wafers. The Si-face was polished by chemical mechanical polishing (CMP) process.

Dichlorosilane (DCS/$SiCl_2H_2$) and propane were used as precursors and hydrogen was the carrier gas. The insulation material in the furnace was examined as a factor that might influence IGSF generation. Some substrates were pretreated by the molten KOH-NaOH eutectic at ~500 °C followed by conventional RCA cleaning prior to being loaded in the furnace. The regular growth condition is at 1550 °C and 80 Torr, resulting in a growth rate of ~12 μm/h. Hydrogen etching of 5 min was performed at growth temperature before growth initiation. Various growth conditions were attempted. After epitaxial growth, the IGSFs in the epilayer were characterized by room temperature photoluminescence (PL) spectroscopy, Nomarski optical microscopy (NOM), and atomic force microscopy (AFM). One sample was subjected to a series of experiments combining reactive ion etching (RIE) process to remove part of the epilayer and eutectic etching to reveal the defects in order to trace an IGSF down close to its starting position.

Results and Discussion

Characterization of IGSFs. When using graphite felt as the insulation material in the furnace, very high IGSF density in the order of 10^3-10^4 cm^{-2} can frequently be found in the epilayers at the regular growth condition (1550 °C, 80 Torr). Room temperature PL spectroscopy (at excitation wavelength of 300 nm) measured a peak with emission wavelength at 480 nm (2.58 eV) only on the epilayer with high IGSF density (as shown in Fig. 1) [9]. This PL peak is close to Fujiwara et al.'s report of 8H-IGSFs (2.56-2.70 eV) [3], but different from Feng et al.'s report (455 nm, i.e. 2.69 eV) [4]. The structure of the IGSFs observed in this study, which is characterized below, is also consistent with 8H-IGSFs [3, 10, 11], so we consider them as 8H-like IGSFs.

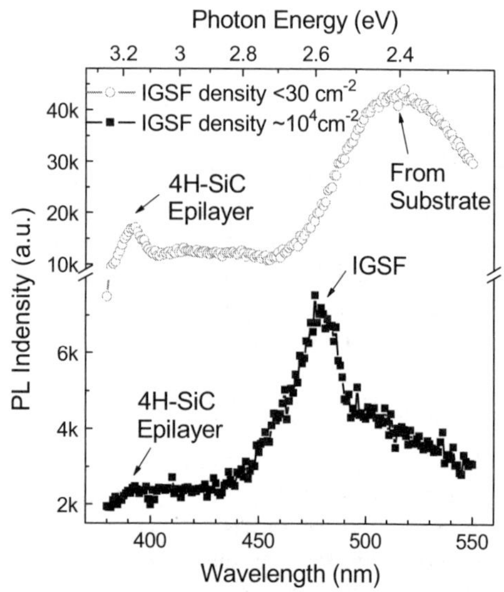

Figure 1: PL spectra of 4H-SiC epilayers with low and high IGSF density.

The IGSFs are 2-dimensional structural defects located on the basal plane. In the off-axis sample, an intersection line between the IGSF and the epi-surface can be observed on the as-grown epilayer by NOM or by AFM (as shown in Fig. 2a and the inset). Because the intersection line is a very shallow groove, it could easily be missed (since a precise microscope focus on the surface is essential to observe the line) unless there is a high density of IGSFs in the epilayer. After KOH or eutectic etching, two shell-like etch pits (two circles in Fig. 2b) connected by an intersection line are characteristic of IGSFs. The two etch pits are intersections between two partials of the IGSF and the epi-surface. One pit has the long-axis along the off-cut direction and the other one is ~30° tilted. Most of the IGSFs in our samples show uniform dimension, and the extended length of these IGSFs implies that they start from the epilayer/substrate interface. It is noted that on the etched surface the intersection lines do not always appear or they are very shallow in some samples; in these cases, IGSFs may be misunderstood to be two individual basal plane dislocations (BPDs).

Particles, screw dislocations or surface depressions on the substrate surface will increase the possibility of IGSF generation, however, most IGSFs observed in this study do not show any features at their starting points (as shown in Fig. 3(b)). A series of experiments combining RIE process to remove part of the epilayer and eutectic etching to reveal the defects were performed on one sample in order to trace the IGSF down close to its starting position near the interface. Fig. 3 shows the RIE process and the NOM images of the relevant marked surfaces. It is confirmed that these IGSFs start from the epi/substrate interface. There is no evidence of a nucleation site for most of this type of IGSF. In Fig. 3(d), the dislocations close to the starting point are two overlapping etch pits naturally

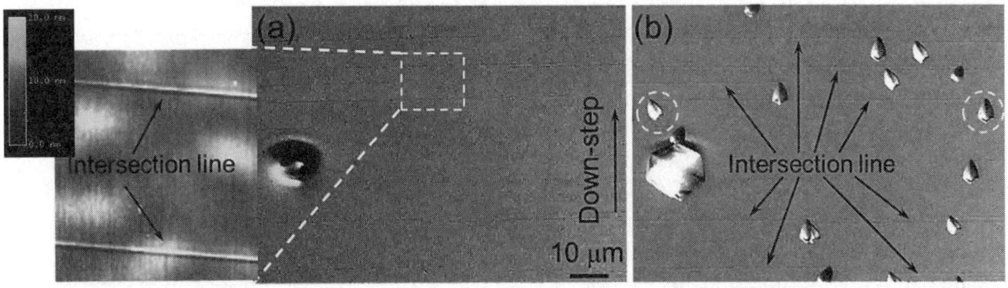

Figure 2: Micrographs of the sample with very high IGSF density (a) in the as-grown epilayer and (b) in the same area after eutectic etching. Inset shows the AFM image (15 μm × 15 μm) of the intersection lines between IGSFs and epi-surface observed on the as-grown epilayer. Almost all of the shell-like pits in (b) are associated with IGSFs. The two shell-like pits in the circles in (b) are two partials in an IGSF. The large feature is a particle related screw dislocation used as the reference.

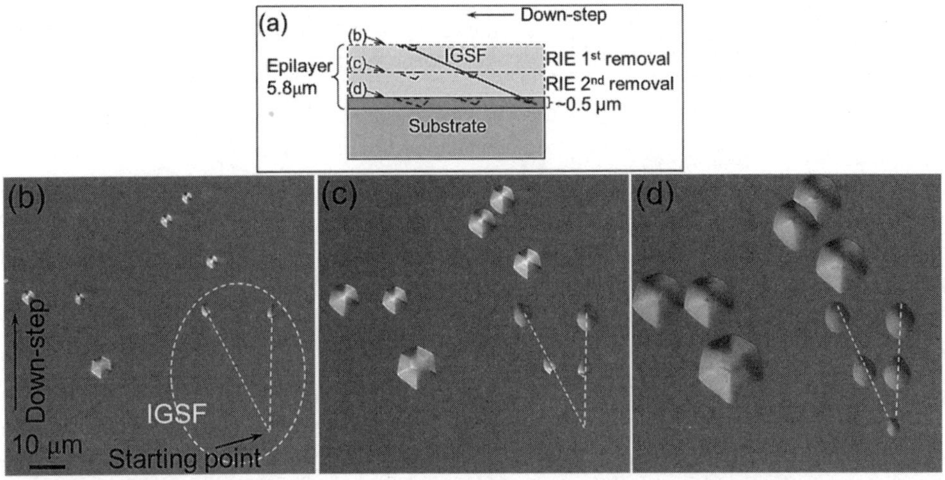

Figure 3: (a) Schematic of RIE and eutectic etching processes to trace IGSF inside the epilayer at different depths. (b)-(d) are micrographs corresponding to the surfaces shown in (a) after eutectic etching. There is no evidence of a nucleation site for this IGSF. Dislocations close to the starting point in (d) are two overlapping etch pits on the two partials. No extra defect is observed near the starting point.

present at the intersection between the IGSF partial and the epi surface; no extra defect is observed near the starting point. Nevertheless, some kind of a nucleation site should exist, which needs to be explored using more precise methods.

Effects of insulation materials and substrate pretreatment on IGSF generation. When using graphite felt as the insulation material in the furnace, the IGSF density could be as high as $\sim10^4$ cm^{-2} in the epilayer. Various growth conditions were attempted in order to reduce IGSF generation, such as different Si-precursor (silane or DCS), growth rate (8-50 μm/h), H_2-etching (unintentional H_2 etching, pure H_2 etching or H_2-C_3H_8 etching), and doping concentration ($\sim10^{15}$ p to $\sim10^{16}$ n); however they did not show effective control on the IGSFs. Since in our furnace, the insulation is not completely isolated from growth chamber, fine particles or fibers released from the graphite felt could drift into the growth zone during sample loading and start of gas flow. They provide very high potential to disturb the step-flow growth resulting in IGSFs formation.

Significant reduction in IGSFs to <100 cm^{-2} is achieved by replacing the soft graphite felts by multilayer graphite foils as the insulation materials (as shown in Fig. 4). The main purpose is to reduce/avoid dusting from any parts of the furnace. It indicates that the cleanliness of the CVD furnace plays an important role in IGSF generation. Considering this, avoidance of IGSFs formation can also be realized by isolating the insulation part completely away from the growth zone.

Further reduction of IGSF density is achieved when the substrates are pretreated by KOH-NaOH eutectic at ~500 °C, 3~45 minutes [12] followed by regular RCA cleaning prior to epitaxial growth (as shown in Fig. 4); this is attributed to improvement of surface state and/or removal of surface damage.

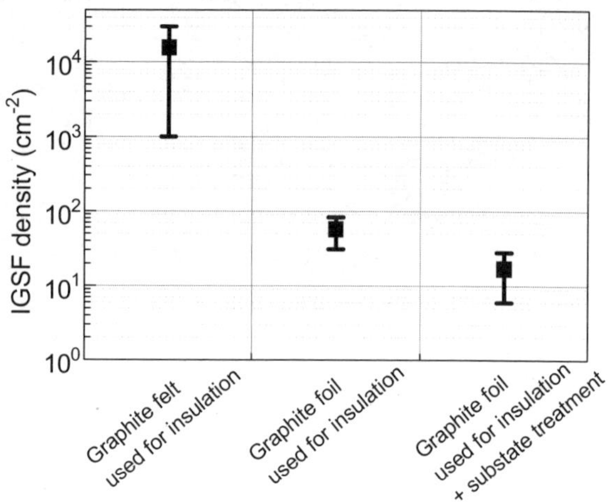

Figure 4: IGSF density at various growth conditions.

Summary

The 8H-like IGSFs in 4H-SiC epitaxial growth is investigated. Selection of reactor furniture material, specifically the insulation material, is very important in controlling IGSF generation, because dusting and degradation of any reactor parts will have unexpected influences on the epigrowth. A suitable substrate pretreatment process further reduces the probability of IGSF formation.

Acknowledgment

This work was supported by the Office of Naval Research, Grant no. # N000141010530. The authors thank the program manager Dr. H. Scott Coombe for his support of this research. The authors also thank Mr. Kevin M. Daniels (USC, Electrical Eng. Dept.) for the helpful suggestions in eutectic etch.

References

[1] M.H. Hong, A.V. Samant, P. Pirouz: Philos. Mag. A Vol. 80 (2000), p. 919
[2] P. Pirouz, D. J.H. Cockayne, N. Sumida, P. Hirsch, A.R. Lang: Proc. R. Soc. A Vol. 386 (1983), p. 241
[3] H. Fujiwara, T. Kimoto, T. Tojo, H. Matsunami: Appl. Phys. Lett. Vol. 87 (2005), p. 051912
[4] G. Feng, J. Suda, T. Kimoto: Appl. Phys. Lett. Vol. 92 (2008), p. 221906
[5] S.I. Maximenko, J.A. Freitas, Jr., P.B. Klein, A. Shrivastava, T.S. Sudarshan: Appl. Phys. Lett. Vol. 94 (2009), p. 092101
[6] J. Hassan, A. Henry, I.G. Ivanov, J.P. Bergman: J. Appl. Phys. Vol. 105 (2009), p. 123513.
[7] J.D. Caldwell, P.B. Klein, M.E. Twigg, R.E. Stahlbush, O.J. Glembocki, K.X. Liu, K.D. Hobart, F.Kub: Appl. Phys. Lett. Vol. 89 (2006), p. 103519
[8] I. Chowdhury, M.V.S. Chandrashekhar, P.B. Klein, J.D. Caldwell, T.S. Sudarshan: J. Cryst. Growth Vol. 316 (2011), p. 60
[9] S.U. Omar, H. Song, T.S. Sudarshan, M.V.S. Chandrashekhar: in this proceeding
[10] S. Izumi, H. Tsuchida, I. Kamata, T. Tawara: Appl. Phys. Lett. Vol. 86 (2005), p. 202108
[11] H. Tsuchida, M. Ito, I. Kamata, M. Nagano: Phys. Status Solidi B Vol. 246 (2009), p. 1553
[12] H. Song, T.S. Sudarshan: in this proceeding

Materials Science Forum Vols. 717-720 (2012) pp 125-128

doi:10.4028/www.scientific.net/MSF.717-720.125

Investigation of Basal Plane Dislocation Reduction/Elimination by Molten KOH-NaOH Eutectic Etching Method

Haizheng Song[a, *] and Tangali S. Sudarshan[b]

Department of Electrical Engineering, University of South Carolina, Columbia, SC 29208, USA

[a]songha@engr.sc.edu, [b]sudarsha@engr.sc.edu, *corresponding author

Keywords: epitaxial growth, BPD reduction/elimination, KOH-NaOH eutectic etching, substrate etching.

Abstract. An optimized molten KOH-NaOH eutectic etching method is developed to reveal defects in highly n-doped SiC substrates and to pre-treat the substrate prior to epitaxial growth. Different from the conventional KOH etching method, by way of eutectic method, the basal plane dislocation (BPD) conversion in the subsequent epitaxial growth is independent of the etch pit size pre-generated on the substrate. Even with a short period (~3 minutes) of pretreatment which does not generate any visible etch pits or degradation of surface morphology on the substrate, an epilayer with low BPD density <22 cm^{-2} is still achieved. This simple and non-destructive method shows high potential to be practically employed as one of the basic pretreatment steps to the substrates in SiC epitaxial growth in order to achieve very low or free BPD density.

Introduction

In recent years, much effort has been put into reducing or eliminating BPDs in SiC epitaxial layers, since the BPDs in the epilayer will convert into Shockley stacking faults and increase the forward voltage drift in bipolar SiC devices [1]. It is well known that BPDs in the epilayer mainly come from the propagation of BPDs in the substrate [2]. One of the effective methods to reduce BPDs is by molten KOH etching of the substrate prior to epitaxial growth [1, 3, and 4]. Although BPD density less than 10 cm^{-2} or even BPD-free epilayers are obtained, the process requires creation of large etch pits ~10 μm on the substrate. These etch pits are all replicated to the epilayer surface as depressions and cause high reverse leakage current [5]. Further, molten KOH etching degrades the epilayer surface morphology even in defect free regions. Therefore, additional polishing is needed to obtain a smooth surface for further application. These limitations restrict extensive application of the KOH etching method. Recently, we developed a "growth-etch-regrowth" method to achieve a BPD-free epilayer with almost no surface degradation [6]. A KOH-NaOH eutectic etching process was used to mildly etch a buffer epilayer followed by the growth of the main epilayer. 100% BPD conversion and no BPD generation were observed in the main epilayer. Although a BPD-free epilayer was obtained, this method doubled the number of steps in epigrowth. In this paper, the KOH-NaOH eutectic etching technique is optimized and applied to replace the conventional KOH etching method as a way of pretreatment of the substrate prior to epitaxial growth. Various pretreatment durations were examined. Using this new approach, high BPD conversion rate is preserved while avoiding surface morphology degradation.

Experiment

The substrates used in this work were 8 mm × 8 mm pieces cut from commercially obtained, 8° off-axis towards $[11\bar{2}0]$, highly doped (~10^{18} cm^{-3}, n-type) 4H-SiC wafers. The Si-face was polished by chemical mechanical polishing (CMP) process. The epitaxial growth was carried out in a homebuilt chimney hot-wall reactor using dichlorosilane (DCS/$SiCl_2H_2$) and propane as precursors and hydrogen as carrier gas. The growth temperature and pressure were 1550 °C and 80 Torr, respectively. The epitaxial layers with thickness of ~6 μm and doping of ~$1x10^{15}$ cm^{-3} n-type were obtained. Sample #1 was regular growth without eutectic treatment of the substrate. For samples #2 to #6, the substrates were etched/treated in an optimized KOH-NaOH eutectic at ~500 °C, 3-45 minutes before epigrowth. The etch pits on the substrate of sample #2 were mapped by Normaski Optical Microscope (NOM) in order to investigate the defect evolution during epitaxial growth. After epitaxial growth, all of the epilayers were performed regular eutectic etching to delineate the defects, except sample #6 epilayer which was first etched using Reactive Ion Etching (RIE) until 0.5 μm thickness of the epilayer was left before final defect delineation. Atomic Force Microscope (AFM, Digital Instruments Dimension 3100, tapping mode) was employed to measure the roughness and growth pits for selected samples. Table 1 summarizes the conditions of the samples.

Table 1. Summary of sample treatment condition by KOH-NaOH eutectic method and BPD density in the epilayer.

Sample No.	Substrate treatment duration [min]	Etch pit size (on the substrate) [μm]	BPD density in the epilayer [cm^{-2}]
# 1	No	--	~200
# 2	45	~4	15
# 3	25	~1.5	22
# 4	12	<1	11
# 5	3	Invisible by AFM	20
# 6*	3	Invisible by AFM	19

BPD density in the substrate is ~$2.8x10^4$ cm^{-2}.

* Sample #6 epilayer was performed Reactive Ion Etching (RIE) until 0.5 μm thickness epilayer was left prior to final defect delineation.

Results and Discussion

By means of the optimal KOH-NaOH eutectic etching, defect delineation shows great controllability, reproducibility and selectivity, not only on low doped SiC epilayers but also on highly n-doped SiC substrates. Fig.1(a) shows the typical defects on a SiC substrate (sample #2) after eutectic etching for 45 minutes. Both threading screw dislocations (TSDs) and threading edge dislocations (TEDs) have hexagonal shapes. The TSDs are a little bigger in size due to the larger Burgers vector. The BPDs are recognized by a distinct oval shape with a randomly oriented long axis. The BPD density on the substrate is around $2.8x10^4$ cm^{-2}. Fig. 1(b) shows that all the etch pits pre-generated on the substrate generate depressions on the as-grown epilayer. This phenomenon is more severe in epigrowth on KOH etched substrates and will increase the reverse leakage current in a SiC PiN diode [5]. However using eutectic etching method, these depressions could be avoided for etch durations < 12 minutes (as shown in Table 1).

Figure 1(c) shows the epilayer after eutectic etching. Compared with the same area of etched substrate (Fig. 1(a)), the evolution of BPDs in the epitaxial growth has been examined. It is found that there are four possible paths for BPD change. (1) BPD conversion to TED (marked as 1, 3, 4, and 5 in Fig. 1(a) converted to 1_TED, etc. in Fig. 1(c)); most of BPD conversions are accompanied by a dislocation shift along the long axis towards the up-step direction (shown as dashed arrows in Fig. 1(c)). (2) BPD disappearance (marked as 2 in Fig. 1(a) and 2_Dep in Fig. 1(c)); some BPDs present on the substrate just disappear leaving depressions on the etched epilayer. (3) BPD propagation (marked as 6 in Fig. 1(a) and a solid arrow and 7_BPD in Fig. 1(c)); very few BPDs are suspected to propagate into the epilayer. However this path is not completely distinguished from path (4), generation of new BPD. As shown in Fig. 1(a) and (c), BPD 6 on the substrate might propagate to 7_BPD on the epilayer, however another possibility is that BPD 6 just disappears (leaving a depression, 6_Dep, on the epilayer) and 7_BPD is newly generated during the epitaxial growth. Due to path (3) and/or (4), a BPD density of $15cm^{-2}$ is achieved in sample #2. The mechanism of different BPD evolution paths needs further investigation, but the high BPD conversion rate may be related to

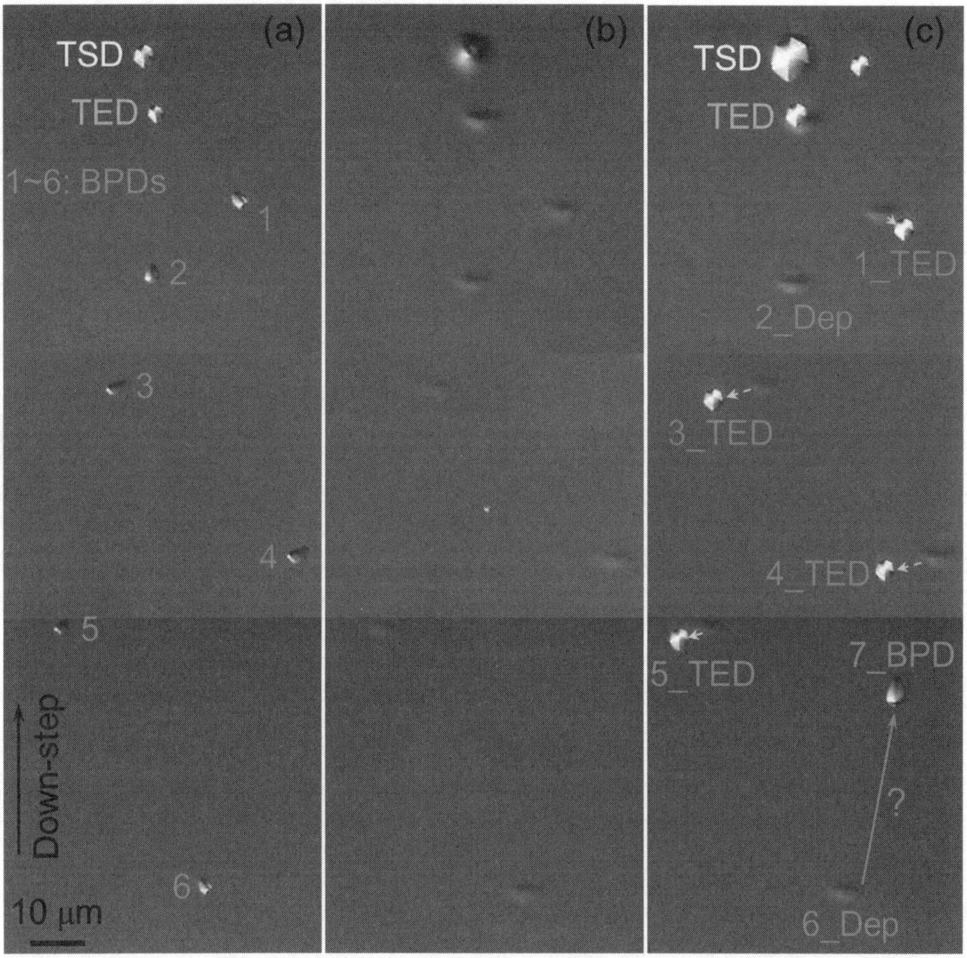

Figure 1: Comparison of microscope images of sample #2 in the same area on (a) the etched substrate, (b) the as-grown epilayer and (c) the epilayer after etching. BPDs 1, 3, 4 and 5 are converted to TEDs with a shift; BPD 2 disappears; BPD 6 might be propagated as 7_BPD, or just disappears and 7_BPD is newly generated. "Dep" stands for "Depression".

the narrow shape of BPD etch pits on the substrate, which enhances the lateral growth to pinch off the step-flow growth inside the etch pits [6, 7].

The substrate pretreatment duration was varied from 45 to 3 minutes in order to examine the effect of the etch pit size on the BPD conversion rate (samples #2 to #5 in Table 1). When the etch pits are less than 2 μm, it is very difficult to map the defects on the substrate. A 3 minutes treatment (sample #5) preserves the substrate surface morphology to the as-received condition; no pits can be detected even by AFM. After regular epitaxial growth on these substrates (samples #2 to #5), these samples were normally etched by eutectic to reveal the defects on the epilayer. The BPD density on the epilayer is summarized in Table 1. Different from the results of BPD conversion in epigrowth on KOH pretreated substrates [4], using the eutectic etching method, the BPD conversion rate is independent of the dimension of etch pits on the substrate. This implies that we can control the etch pits to as small as possible (like sample #5), so that there is no degradation of surface morphology, but the BPD conversion rate is still preserved. The surface roughness of sample #5 was measured by AFM at each stage: before and after 3 minutes eutectic treatment of the substrate, and after epigrowth (as shown in Fig. 2(a), (b) and (c)). The RMS value of sample #5 epilayer is comparable to that of sample #1 epilayer (Fig. 2(d)) which was grown on the untreated substrate. Fig.2 shows that eutectic treatment of the SiC substrate does not bring significant degradation to the substrate surface or to the subsequent epilayer surface.

In order to investigate where the BPDs get converted, sample #6 was performed the same substrate treatment and epigrowth like sample #5, and then the as-grown epilayer was etched by RIE to remove most of the epilayer, leaving only ~0.5 μm thick epilayer. BPD counting was done on the ~0.5 μm epilayer and shown in Table 1. It is found that the BPD density on this ~0.5 μm epilayer is almost the same as that on thicker epilayers. The result implies that all of the BPD conversion occurs within ~0.5 μm of epigrowth, i.e., very close to the epi/substrate interface.

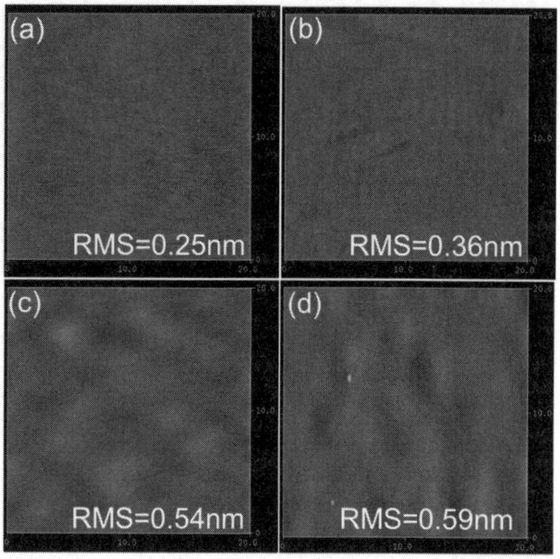

Figure 2: The AFM images (20 μm x 20 μm) and RMS surface roughness of sample #5's (a) substrate before eutectic treatment, (b) substrate after 3-min KOH-NaOH eutectic treatment, and (c) epilayer grown on the treated substrate. (d) The RMS roughness of sample #1's epilayer grown on the untreated substrate. Results show that eutectic treatment of the SiC substrate does not bring significant degradation to the substrate surface or to the subsequent epilayer surface.

Summary

Using KOH-NaOH eutectic treatment of SiC substrates, it is easy to control the etch pit size to as small as possible, so that there is almost no degradation of the surface morphology, but very high BPD conversion rate in the subsequent epitaxial growth is still preserved. No extra polishing process is needed for device fabrication. All the BPD conversion occurs very close to the epi/substrate interface (< 0.5 μm).

Acknowledgment

This work was supported by the Office of Naval Research, Grant no. # N000141010530. The authors thank the program manager Dr. H. Scott Coombe for his support of this research.

References

[1] J.J. Sumakeris, J.P. Bergman, M.K. Das, C. Hallin, B.A. Hull, E. Janzen, H. Lendenmann, M.J. O'Loughlin, M.J. Paisley, S.Y. Ha, M. Skowronski, J. Palmour, C.H. Carter Jr.: Mater. Sci. Forum Vol. 527-529 (2006), p. 141

[2] S. Ha, P. Mieszkowski, M. Skowronski, L.B. Rowland: J. Cryst. Growth Vol. 244 (2002), p. 257

[3] J.J. Sumakeris, U.S. Patents 7,018,554 B2 (2006), and 7,279,115 B1 (2007)

[4] Z. Zhang, T.S. Sudarshan: Appl. Phys. Lett. Vol. 87 (2005), p. 151913

[5] Z. Zhang, A.E. Grekov, P.Sadagopan, S.I. Maximenko, T.S. Sudarshan: Mater. Sci. Forum Vol. 527-529 (2006), p. 371

[6] H. Song, T. Rana, T.S. Sudarshan: J. Cryst. Growth Vol. 320 (2011), p. 95

[7] Z. Zhang, E. Moulton, T.S. Sudarshan: Appl. Phys. Lett. Vol. 89 (2006), p. 081910

Materials Science Forum Vols. 717-720 (2012) pp 129-132
© (2012) Trans Tech Publications, Switzerland
doi:10.4028/www.scientific.net/MSF.717-720.129

Electrical and optical properties of high-purity epilayers grown by the low-temperature chloro-carbon growth method

Siva Kotamraju[1,a], Bharat Krishnan[1,b], Franziska C. Beyer[2,c], Anne Henry[2,d], Olle Kordina[2,e], Erik Janzén[2,f] and Yaroslav Koshka[1,g*]

[1]Mississippi State University, Department of Electrical and Computer Engineering, Mississippi State, Mississippi 39762, United States

[2]Department of Physics, Chemistry and Biology, Linköping University, SE-581 83 Linköping, Sweden

[a]sivakotam@gmail.com, [b]bharat.kris@gmail.com, [c]fbeyer@ifm.liu.se, [d]anhen@ifm.liu.se, [e]olkor@ifm.liu.se, [f]erija@ifm.liu.se, [g]ykoshka@ece.msstate.edu, *corresponding author

Keywords: epitaxial growth, chloro-carbon, deep levels, photoluminescence, DLTS, Schottky.

Abstract. A reduced growth pressure (down to 10 Torr) was employed for the low-temperature chloro-carbon epitaxial growth. More than two times lower H_2 flow rate became possible. The optimal input H_2/Si and C/Si ratios were also lower. A significant reduction of the net free donor concentration resulted from the use of the low pressure, delivering partially compensated epilayers with the net free donor concentration below $7x10^{13}$ cm^{-3}. Deep levels were characterized in the low-temperature epilayers for the first time. No $Z_{1/2}$ or $EH_{6/7}$ centers could be detected by DLTS. No strong D_I photoluminescence signature was observed. The high purity of the obtained epitaxial layers made it possible to use the low-temperature chloro-carbon epitaxial growth to fabricate drift regions of Schottky diodes for the first time. Promising values of the reverse breakdown voltage and the leakage current were obtained from the fabricated devices.

Introduction

Investigation of the possibility to conduct homoepitaxial growth of SiC at lower temperatures has a long history [1,2]. The recently renewed interest in growing at low temperatures owes to the progress with employing chlorinated growth precursors [3,4]. In the previous work, the low-temperature chloro-carbon epitaxial growth method (i.e., a method based on CH_3Cl as the carbon precursor) [5] was further improved by using the chlorinated Si precursor $SiCl_4$ instead of SiH_4 [6]. The homogeneous nucleation was fully eliminated, which also made it possible to establish that the use of chlorinated carbon precursors (CH_3Cl) provides additional benefits beyond supplying Cl for suppressing homogeneous nucleation. Wider C/Si process window and higher growth rate with better epilayer quality were achieved with $CH_3Cl/SiCl_4$ compared to the $C_3H_8/SiCl_4$ growth conducted at the same C/Si and Cl/Si ratios [7]. While the triangular defects (TD) could not be fully eliminated when targeting a growth rate above 3-3.5 μm/h, the low-temperature growth looks promising for commercial applications. Previously, concentrated Cl-based CVD process utilizing low growth pressure has been used at high temperature (1570°C) to grow uniform SiC epilayers [8]. In the present work, the growth pressure of the low-temperature growth was reduced from 140 Torr down to 10 Torr in order to explore the possibility of improving the epilayer quality and evaluate the effect of low pressure on doping.

Experimental

The homoepitaxial growth was conducted in a hot-wall CVD reactor at 1300°C, with H_2 as the carrier gas and $SiCl_4$ and CH_3Cl as the silicon and carbon precursors, respectively. Commercial 8° off-axis 4H-SiC substrates were used. Intentional doping with nitrogen was conducted in some of the experiments to achieve heavier doped n-type epilayers. It should be mentioned that the same epitaxial reactor was also used for vanadium doping experiments and for growing SiC nanowires using metal catalysts. While different growth susceptors and foams were employed for different experiments, some residual contamination of the gas line and the reactor chamber should be expected.

The thickness of the epilayers was measured by reflective Fourier Transform Infrared Spectroscopy (FTIR). Surface morphology was monitored by Nomarski optical microscopy and scanning electron microcopy (SEM). Low-temperature photoluminescence (LTPL) spectroscopy measurements were conducted at 2K using excitation of the 244 nm line of a FreD laser. The luminescence was dispersed by a single monochromator on which a UV sensitive CCD camera was mounted to rapidly detect the PL spectra. Deep levels were studied using deep level transient spectroscopy (DLTS) in a temperature range of 85 to 650K. The steady-state reverse bias was set to $V_r = -4V$. The filling pulse length and height were 10 ms and 4 V, respectively. The capacitance transients were evaluated using a conventional boxcar technique as well as a lock-in amplification simulator to reduce the noise level.

Simple circular Schottky diodes without edge termination were formed by depositing Ni/Ti/Al metal stack by electron beam evaporation on the surface of the low-temperature epilayers serving as the drift regions.

Results and Discussion

Reducing the growth pressure from 140 to 10 Torr made it possible to use more than 2 times lower H_2 flow rate. The optimum values of the input H_2/Si and C/Si ratios were also reduced. Specifically, the optimum C/Si process window required to avoid nucleation of poly-SiC (at too high C/Si) or poly-Si (at too low C/Si) changed from more than 4-5 down to ~1. The growth rate

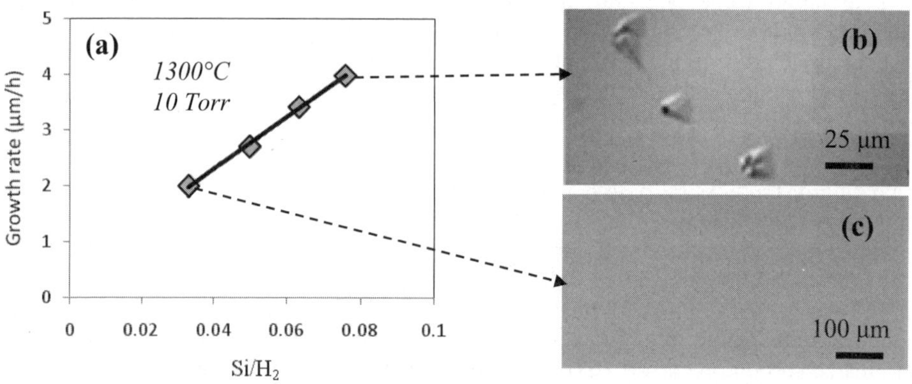

Fig. 1 Growth trends at 1300°C and 10 Torr: (a) the growth rate versus the Si/H_2 ratio, (b),(c) optical micrographs of the epilayer surface morphology for the growth rates of 4 and 2 µm/h respectively.

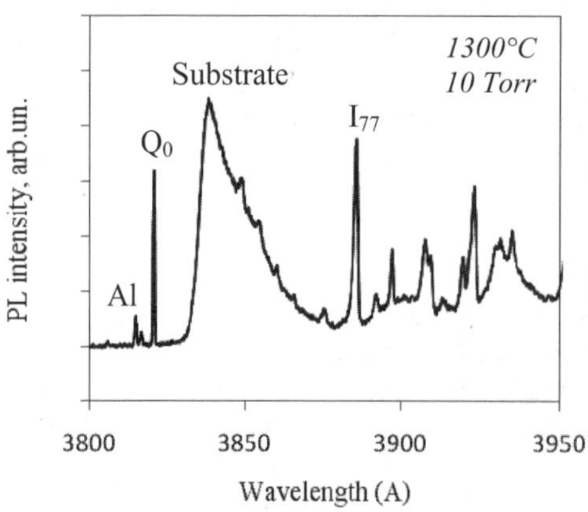

Fig. 2 Low-temperature PL of a 4-µm-thick epilayer with $N_d - N_a = 7 \times 10^{13}$ cm^{-3}.

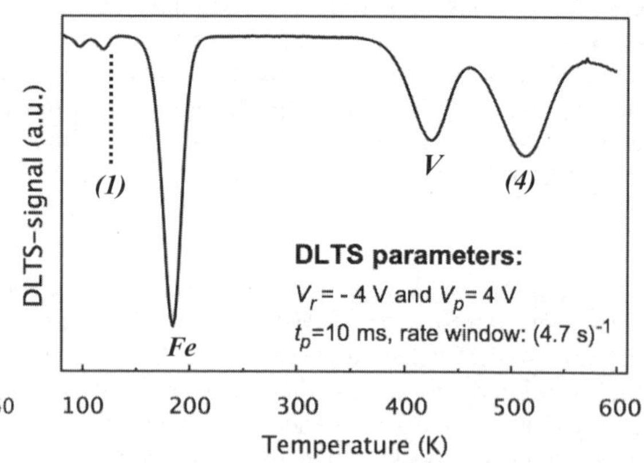

Fig. 3 DLTS of an epilayer intentionally doped with nitrogen to ~2×10^{15} cm^{-3}.

versus the Si/H_2 ratio and the corresponding surface morphology is shown in Fig.1. In comparison, the low-temperature growth at 140 Torr utilized values of Si/H_2 ratio at least an order of magnitude lower [6].

With respect to the pressure effect on the epilayer surface morphology, the low-pressure has yet to bring the desirable widening of the C/Si process window, which is required for eliminating TDs at higher growth rates (Fig.1b). At the investigated growth conditions, the process window boundary corresponding to high values of the C/Si ratio resulted in 3C inclusions and polycrystalline growth. The boundary corresponding to low values of the C/Si ratio resulted in formation of poly-Si islands, similar to the outcome of the growth at higher pressure [7]. The window became too narrow and eventually disappeared at higher growth rates, making it impossible to achieve surface-defect free epilayers at high growth rates. Only modest increase of the growth rate has been achieved after switching to the lower pressure.

A significant reduction of the net free donor concentration resulted from the use of the lower pressure. While the typical value at 140 Torr was $N_d - N_a \geq 10^{16}$ cm^{-3}, the growth at 10 Torr resulted in $N_d - N_a < 7\text{x}10^{13}$ cm^{-3}. Noticeable compensation with Al acceptors was detected by LTPL (Fig.2). While this compensation was contributing to the low value of the net donor concentration, the nitrogen and aluminum concentrations in the best samples were evaluated to be below 2-3x10^{14} cm^{-3} and 1-2x10^{14} cm^{-3} respectively, which is known to be difficult to achieve without a prolonged susceptor pre-baking and conducting much longer growth runs to reduce the background of residual impurities inside the susceptor and insulation foam.

No information about deep levels in epitaxial layers grown by the low-temperature chloro-carbon technique had existed prior to this work. The relatively thin epilayers (4-5 μm-thick) grown at 1300°C and 10 Torr were nevertheless sufficient for DLTS characterization. For this thickness, intentional doping with nitrogen was used to provide sufficiently narrow depletion region under the Schottky contacts as required for DLTS. Large number of circular Schottky contacts per sample were fabricated. Some of the test structures were found to be free from TDs, while the others clearly had one or more TDs inside the Schottky contact area.

The DLTS spectrum from an n-type epilayer with $N_d - N_a \approx 2\text{x}10^{15}$ cm^{-3} measured at a location free from TDs is shown in Fig.3. The values of the activation energy, capture cross-section and concentration of the deep levels for Fig.3 are shown in Table 1. The strongest peaks correspond to Fe (N_T=1.6x10^{14} cm^{-3}) [9], V (N_T=1.1x10^{14} cm^{-3}) [10], and an unidentified defect labeled as (4) in Fig.3 (N_T=7.1x10^{13} cm^{-3}). Since the same reactor (even though with different susceptors and thermal insulation foams) had been used for epitaxial growth of SI substrates with vanadium, as well as for SiC nanowire growth with metal catalysts containing Fe, it could possibly explain the presence of Fe and V in the DLTS spectrum.

Table 1 Deep level parameters from DLTS.

DLTS Peak	E_A, eV	σ , cm^{-2}	N_T, cm^{-3}
(1)	E_C - 0.249	7x10^{-15}	8x10^{12}
Fe	E_C - 0.379	7x10^{-15}	1.6x10^{14}
V	E_C - 0.968	5x10^{-15}	1.1x10^{14}
(4)	E_C - 1.008	8x10^{-17}	7.1x10^{13}

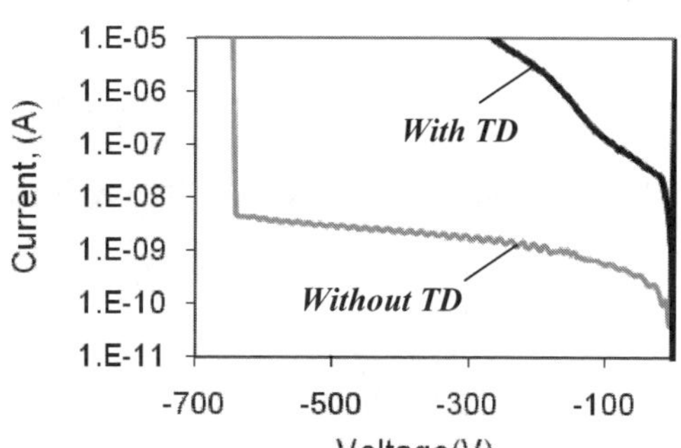

Fig. 4 Reverse bias I-V of two 300 μm-diameter Schottky diodes with a drift region represented by a 10 μm-thick epilayer grown at 1300°C.

No $Z_{1/2}$ or $EH_{6/7}$ centers, which are commonly used as a figure of merit of the epilayer quality, could be detected. Using the DLTS signal noise level at the expected position of $Z_{1/2}$ as a rough estimation for the possible defect concentration, $Z_{1/2}$ concentration was determined to be at least below $5x10^{11}$ cm^{-3}. Furthermore, in most of the similar (but not intentionally doped) samples investigated, no strong D_I luminescence line could be observed by the PL.

Finally, Schottky contacts without edge termination were fabricated on the top of the drift regions represented by the high-purity low-temperature epitaxial layers grown at 10 Torr with N_d - $N_a \approx 7x10^{13}$ cm^{-3} (Fig.4). While the leakage current in the Schottky diodes covering one or more triangular defects (TD) was high and the breakdown voltage V_{brk} was rather low, the maximum and the average values of V_{brk} of 600V and 500V respectively in was observed in the diodes free from TDs with a 5-µm-thick drift region grown at 1300°C. When compared to the results of 2D simulations of the maximum electric field at the Schottky contact edges (assuming that no charges existed at the surface), these experimental values of the breakdown voltage were as good as the theoretical values, even when accounting for the doping dependence of the critical electric field in SiC. While testing of such not-terminated structures cannot be used as a conclusive evidence, this result serves as a preliminary indication that the drift regions grown at 1300°C contain rather low concentrations of lattice defect that are usually responsible for the degradation of the reverse characteristics.

Summary

While the use of the low pressure of 10 Torr has yet to show appreciable improvements in the surface morphology and the growth rate, its effect on the purity of the low-temperature epitaxial layers was spectacular. This improvement makes the low-temperature chloro-carbon growth a promising technique for fabricating low-doped drift regions, even though further improvement of the growth rate is required. Both DLTS measurements and the measurements of the reverse characteristics of the Schottky diodes show a strong promise of high crystalline quality and low defect concentration.

Acknowledgement

The authors are grateful to Dr. Neil Merrett from AFRL for his recommendations and help with Schottky diode process development and device testing.

References

[1] J. Anthony Powell and Hebert A. Will, J. Appl. Phys., 44 (11) (1973), pp. 5177-5178
[2] Atsushi Yamashita, Woo Sik Yoo, Tsunenobu Kimoto and Hirouki Matsunami, Jpn. J. Appl. Phys., 31 (1992), pp. 3655-3661.
[3] Y. Koshka, H.-De Lin, G. Melnychuk, C. Wood, Journal of Crystal Growth, 294 (2006) pp.260–267.
[4] S. Leone, F. Beyer, H. Pedersen, O. Kordina, A. Henry and E. Janzén, Thin Solid Films, 2011, (519), 10, 3074-3080
[5] Y. Koshka, "Method for Epitaxial Growth of Silicon Carbide at Reduced Temperatures," U.S. patent #7404858, July 29, 2008.
[6] S. Kotamraju, B. Krishnan, G. Melnychuk, and Y. Koshka, Journal of Crystal Growth 312 (2010) pp. 645-650.
[7] S. Kotamraju, B. Krishnan, and Y. Koshka, Phys. Status Solidi RRL 4, No. 7, 145-147 (2010).
[8] S. Leone, A. Henry, S. Andersson, O. Kordina, and E. Janzén; *J of Electrochem. Society, 157 (19) H969-H976 (2010).*
[9] F. C. Beyer, C. G. Hemmingsson, A. G ällström, S. Leone, H. Pedersen, A. Henry, and E. Janzén. "Deep levels in iron doped n- and p-type 4H-SiC," submitted to J. Appl. Phys., 2011.
[10] N. Achtziger and W. Witthuhn, Phys. Rev. B 57, 12181, 1998

Materials Science Forum Vols. 717-720 (2012) pp 133-136
© (2012) Trans Tech Publications, Switzerland
doi:10.4028/www.scientific.net/MSF.717-720.133

Use of Vanadium Doping for Compensated and Semi-Insulating SiC Epitaxial Layers for SiC Device Applications

Bharat Krishnan[1,a], Rooban Venkatesh K. G. Thirumalai[1,b], Siva Kotamraju[1,c], J. Neil Merrett [2,d], and Yaroslav Koshka[1,e*]

[1]Mississippi State University, Department of Electrical and Computer Engineering, Mississippi State, Mississippi 39762, United States

[2]Air Force Research Laboratory, Wright-Patterson Air Force Base, Ohio 45433, United States

[a]bharat.kris@gmail.com, [b]rk219@msstate.edu, [c]sivakotam@gmail.com, [d]joseph.merrett@wpafb.af.mil, [e]ykoshka@ece.msstate.edu, *corresponding author

Keywords: semi-insulating, epitaxial growth, chloro-carbon, PiN, drift region.

Abstract. Vanadium doping from $SiCl_4$ source during epitaxial growth with chlorinated C and Si precursors was investigated as a mean of achieving compensated and semi-insulating epitaxial 4H-SiC layers for device applications. Thin epilayers were grown at 1450°C with a growth rate of ~6 μm/h. Experiments at 1600°C resulted in the growth rates ranging from 60 to 90 μm/h producing epilayers with thickness above 30 μm. V concentrations up to about $10^{17}cm^{-3}$ were found safe for achieving defect-free epilayer surface morphology, however certain degradation of the crystalline quality was detected by XRD at V concentrations as low as $3\text{-}5x10^{15}$ cm^{-3}. Controllable compensation of nitrogen donors with V acceptors provided low-doped and semi-insulating epitaxial layers. Mesa isolated PiN diodes with V-acceptor-compensated n^- epilayers used as drift regions showed qualitatively normal forward- and reverse-bias behavior.

Introduction

Semi-insulating (SI) SiC is important for the development and commercialization of RF SiC devices. SI SiC wafers are also preferred and extensively used as substrates for heteroepitaxial growth of GaN for microwave and radio frequency devices [1].

Main approaches to achieving SI properties in SiC include (1) intentional vanadium (V) doping [1,2] and (2) use of deep levels caused by intrinsic defects [3-5]. Other transition metals such as iron (Fe) have also been investigated for producing resistive SiC substrates and epitaxial layers [6,7]. A number of problems, such as doping ihmogeneity and generation of lattice defects, limit the yield of the SI bulk growth, making SI SiC substrates very expensive. For this reason, many applications could benefit from epitaxial growth of SI SiC, to produce SI films of various thicknesses on top of much cheaper n-type substrates.

In our previous work, low-temperature and regular-temperature growth of SiC epitaxial layers with CH_3Cl and $SiCl_4$ chlorinated precursors was investigated [8,9]. Heavily Al doped SiC layers for Ohmic contacts to p-type SiC [10] and p+ anodes for 4H-SiC PiN diodes grown at low-temperature were also demonstrated [11].

In the present work, doping with vanadium (V) acceptors during growth at 1450°C and 1600°C is investigated to obtain controllable compensation as well as semi-insulating behavior in thin (less than 5 μm) and thick (more than 30 μm) epitaxial layers.

Experimental

The previously developed chloro-carbon epitaxial growth technique based on using chloromethane (CH_3Cl) as the carbon precursor was used [8,9]. $SiCl_4$ was used as the silicon precursor. In-situ vanadium doping was carried by H_2 bubbling of vanadium tetrachloride (VCl_4). While the low-temperature chloro-carbon epitaxial growth has been traditionally conducted at 1300°C in our previous work, the early vanadium-doping experiments in this work utilized a higher growth temperature of 1450°C. In addition, experiments at 1600°C were also conducted to achieve thick epitaxial layers better suitable for comprehensive materials characterization.

The thickness of the epilayers was measured by reflective Fourier Transform Infrared Spectroscopy (FTIR). Surface morphology was monitored by Nomarski optical microscopy and scanning electron microcopy (SEM). Photoluminescence (PL) measurements were conducted using an Ar^+ laser line of 300.3 nm wavelength as the PL excitation source. XRD measurements were conducted using a Rigaku SmartLab system with Ge (220) 4-Bounce incident beam monochromator and scintillation counter SC 70. Secondary ion mass spectroscopy (SIMS) was used to establish the atomic concentration of vanadium. The net donor concentration was evaluated by capacitance-voltage (CV) measurements.

Samples with 30 μm-thick drift layers capped with 1.2-1.5 μm thick p+ anode layers were used for pn diode fabrication. Round mesas with diameters of 50, 100, 150, 200 and 300 μm were RIE etched to a height of around 35 μm. Sacrificial thermal oxidation and oxide removal were conducted twice, followed by a dry thermal oxidation at 1100°C for 6.5 hrs. A silicon nitride layer was then deposited by sputtering a Si_3N_4 target with equal flows of Ar and N_2. A 400 Å Ti layer followed by a 1200 Å $Al_{50}Ni_{50}$ layer were sputter deposited for the anode contacts. The backside cathode contact metal was 50 Å of Ti followed by 2000 Å of Ni. The contacts were simultaneously annealed at 1000°C for 2 min in Ar.

Results and Discussion

Featureless epilayer surface was observed by SEM in the epitaxial layers having vanadium concentration [V] = 4-6x10^{16} cm^{-3} (Fig.1a). At higher VCl_4 flows, triangular-shaped defects formed, and their concentration increased with the vanadium supply (Fig.1b,c). Thick epitaxial layers grown

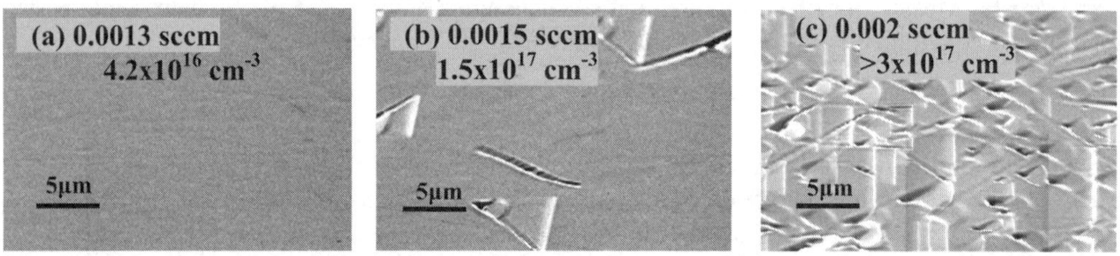

Fig. 1 SEM micrographs (tilted view) of the surface morphology of three epitaxial layers grown at 1450°C with different VCl_4 flow rates. **(a)** Featureless surface morphology is observed at VCl_4 flow below approximately 0.0015 sccm. **(b)** Triangular-shaped defects form at VCl_4 = 0.0015 sccm**.** **(c)** The defect concentration increases at higher VCl_4 flows. A similar degradation was observed in V-doped epilayers growth at 1600°C.

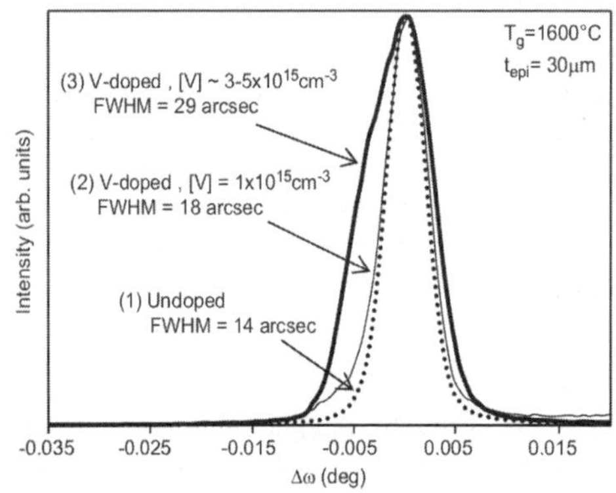

Fig. 2 X-ray rocking curves of V-free and V-doped 30 μm-thick epitaxial layers grown at 1600°C. V incorporation caused degradation of the crystalline quality.

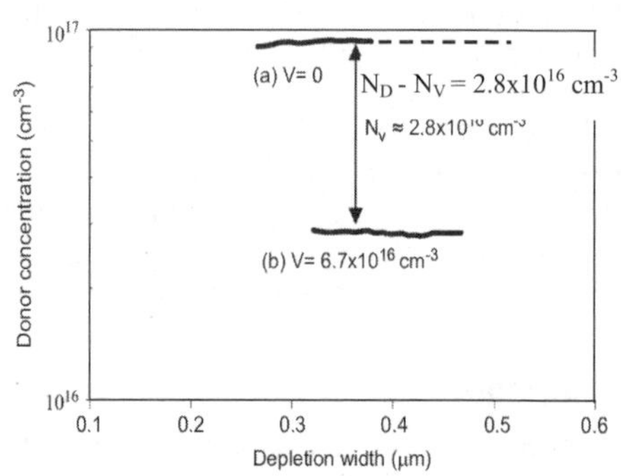

Fig. 3 Net donor concentration (N_d-N_a) in (a) epilayers doped only with nitrogen and (b) N_2-doped epilayers compensated with vanadium. [N] = 9.5x$10^{16}$$cm^{-3}$.

at 1600°C, with the growth rate in the range from 60 to 90 μm/hr, were used to evaluate the effect of V doping on the crystalline quality. Even though thick semi-insulating epitaxial layers had defect-free epilayer surface morphology, X-ray rocking curves showed significant broadening (Fig.2). The asymmetric broadening is possibly due to slightly different lattice spacing related to the epilayer degradation caused by V incorporation.

KOH defect delineation revealed that the concentration of dislocations in the V-doped epilayers was close to that in the substrate. It was concluded that the V-doping did not cause a detectable increase in the dislocation content in the moderately V-doped epilayers. Most of the dislocations in the epilayer were not caused by the epitaxial growth but instead propagated from the substrate.

Electrical activation of the incorporated vanadium was confirmed by the increase in the intensity of the V-related infrared photoluminescence (PL) by almost an order of magnitude in the epitaxial layer doped to [V] = $4x10^{16}$ cm^{-3} as compared to an undoped reference epitaxial layer. Electrical measurements in lower-doped samples showed the values of the net donor concentration consistent with a partial compensation of nitrogen donors with vanadium acceptors (Fig.3). At higher V flow, full compensation and semi-insulating behavior was observed.

The resistivity of a 3-μm-thick fully compensated V-doped epilayers, with V concentrations exceeding $3x10^{16}$ cm^{-3} grown at 1450°C, was found to be on the order of $1–3x10^5$ Ω-cm. In 30 μm-thick samples grown at 1600°C, non-contact capacitance based resistivity measurements revealed the values of resistivity in the range of $2\text{-}4x10^5$ Ω-cm.

In this work, early attempts to investigate the promise of the V-doped epilayers for SiC devices were conducted by using the V-acceptor-compensated n^- epilayers as drift regions in mesa isolated PiN diodes. PiNs with two different 30-μm-thick drift regions were compared: (1) closely V-compensated n- epi with V concentration exceeding $2x10^{15}$ cm^{-3} and $N_D\text{-}N_A \approx 10^{14}$ cm^{-3} and (2) V-free n^- epi with $N_D\text{-}N_A \approx 2x10^{15}$ cm^{-3}. A significant roughness could be observed on the mesa walls even after the sacrificial thermal oxidation. This roughness was likely the cause of the low (< 650 V) and widely varying reverse breakdown values measured for the devices. Temperature dependences of the forward current for the samples (1) and (2) are shown in Fig.4. Note the V-doped diodes had orders of magnitude lower current density than the V-free diodes. Also note that the V-doped diodes showed a positive temperature coefficient as is typical for unipolar conduction. The V-free diode, however, showed an increase in forward current with increasing temperature, as is expected for a p-n diode. Under forward bias, narrow, aligned areas emitting pale white light were observed on one side of the V-doped mesas. These defects are believed to be stacking faults and are suspected to be the cause of the low current density and positive temperature coefficient of the V-doped diodes. Unlike the V-doped diodes, the V-free diodes emitted the characteristic violet light expected from SiC p-n diodes. No stacking faults were observed in the V-free diodes.

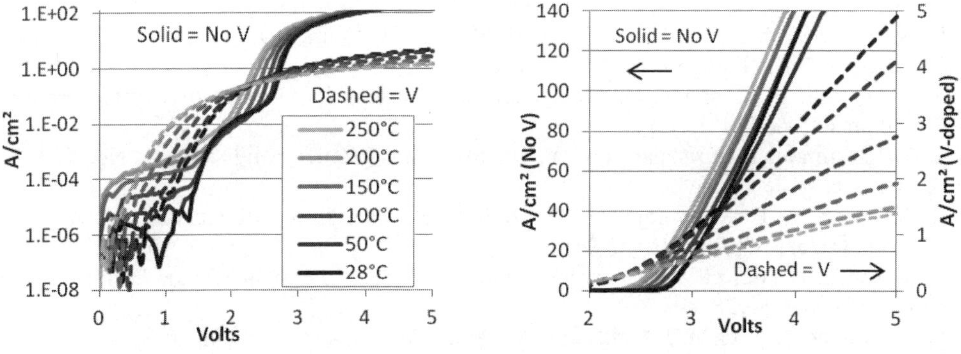

Fig. 4 Forward J-V measurements for SiC PiN diodes with (dashed) and without (solid) V-doping. Note: the data for each diode in the linear plot are plotted on different scales to emphasize the temperature trend. (No V: $N_D\text{-}N_A \approx 2x10^{15} cm^{-3}$; V-doped: $N_D - N_A \approx 10^{14} cm^{-3}$)

Summary

V doped compensated and semi-insulating 4H-SiC epilayers were obtained by chloro-carbon epitaxial growth using in-situ VCl_4 doping. While too high V concentrations (close to $1x10^{17}$ cm^{-3}) caused epilayer surface morphology degradation, the epitaxial process does not require such high V concentrations to achieve SI properties, since the epitaxial growth offers reliable control of shallow donors and acceptors.

Resistivities on the order of 10^5 Ω-cm were achieved at both 1450 and 1600°C, which is what can be expected from compensation of nitrogen donors with insufficiently deep V acceptors. The technique is ready to be extended to achieve V doping in epilayers with Al shallow acceptors. In shallow-acceptor-dominated SiC, V is known to be in the donor state (located at 1.5-1.6eV above the valence band) that are deeper than V acceptors (located at 0.8-1.0eV from the bottom of the conduction band) [12,13], which ensures much higher resistivities in the range of 10^{12} Ω-cm.

Fabrication of mesa-isolated PiN diodes with 30 μm-thick drift region turned out to be too challenging to rely upon reverse breakdown as a figure of merit of the quality of the V-compensated drift region. However, the forward characteristics of the V-doped diodes indicated the formation of stacking faults that were not observed in the V-free diodes.

Acknowledgement

The authors are grateful to Dr. Murugesu Yoganathan for the independent verification of the high resistivity in the SI epilayers using COREMA. The work was partially supported by NASA EPSCoR Grant no. 10-09-021.

References

[1] S.T. Sheppard, K. Doverspike, W. L. Pribble, S. T. Allen, J. W. Palmour, L. T. Kehias, T. J. Jenkins, IEEE Electron Device Letters, Vol. 20, No.4, pp. 161-163, 1999.

[2] W. C. Mitchel, R. Perrin, J. Goldstein, A. Saxler, M. Roth, S. R. Smith, J. S. Solomon, A. O. Evwaraye, J. Appl. Phy., 86, pp.5040-5044, 1999.

[3] W. C. Mitchell, W. D. Mitchell, M. E. Zvanut, G. Landis, Solid-State Electronics 48, 1693, 2003.

[4] A. Ellison, B. Magnusson, C. Hemmingsson, W. Magnusson, T. Iakimov, L. Storasta, A. Henry, N. Henelius, and E. Janzén, Mater. Res. Soc. Symp. 640, H1.2 (2001).

[5] G. Müller, M.F. Brady, W.H. Brixius, R.C. Glass, H.McD. Hobgood, J.R. Jenny, R.T. Leonard, D.P. Malta, A.R. Powell, V.F. Tvestkov, S.T. Allen, J.W. Palmour, and C.H. Carter, Jr., Mat. Sci. Forum 433-436, 39 (2003).

[6] P. G. Baranov, I. V. Ilyin, E. N. Mokhov and V. A. Khramtsov, Semi. Sci. Tech, 16 (2001) 39–43.

[7] H. K. Song, S. Y. Kwon, H. S. Seo, J. H. Moon, J. H. Yim, J. H. Lee, H. J. Kim, and J. K. Jeong, Appl. Phys. Lett., 89, 152112 (2006).

[8] S. Kotamraju, B. Krishnan, G. Melnychuk, and Y. Koshka, Journal of Crystal Growth 312 (2010) pp. 645-650.

[9] S. P. Kotamraju, B. Krishnan, and Y. Koshka, Phys. Status Solidi RRL 3, No. 5, 157-159 (2009).

[10] B. Krishnan, S. P. Kotamraju, G. Melnychuk, H. Das, J. N. Merrett and Y. Koshka, , J. Electron. Mater., 39(1) (2010), p. 34.

[11] B. Krishnan, N. Merrett, G. Melnychuk, and Y. Koshka, Materials Science Forum Vols. 645-648 (2010) pp 925-928.

[12] W. C. Mitchel, R. Perrin, J. Goldstein, A. Saxler, M. Roth, S. R. Smith, J. S. Solomon, A. O. Evwaraye, J. Appl. Phy., 86, pp.5040-5044, 1999.

[13] M. E. Zvanut, V. V. Konovalov, Haiyan Wang, W. C. Mitchel, W. D. Mitchell, and G. Landis, J. Appl. Phy., 96, pp.5484-5489, 2004.

Materials Science Forum Vols. 717-720 (2012) pp 137-140
© (2012) Trans Tech Publications, Switzerland
doi:10.4028/www.scientific.net/MSF.717-720.137

Progress in Growth of Thick Epitaxial Layers on 4 Degree Off-axis 4H SiC Substrates

J. Zhang[1], G. Chung[1], E. Sanchez[1], M.J. Loboda[1], S. Sundaresan[2], R. Singh[2]

[1] Dow Corning Corporation, Midland, MI, USA
[2] GeneSiC Semiconductor, Dulles, VA, USA
j.zhang@dowcorning.com, gil.chung@dowcorning.com, edward.sanchez@dowcorning.com, mark.loboda@dowcorning.com

Keywords: thick epitaxy, BPD density, carrier lifetime, surface defects, high voltage diode

Abstract This paper reports the progress of the thick epitaxy development at Dow Corning. Epiwafers with thickness of 50 – 100 μm have been grown on 4° off-axis 76mm 4H SiC substrates. Smooth surface with RMS roughness below 1nm and defect density down to 2 cm^{-2} are achieved for 80 - 100 μm thick epiwafers. Long carrier lifetime of 2 – 4 μs are routinely obtained, and low BPD density in the range of 50 down to below 10 cm^{-2} is confirmed. High voltage JBS diodes have been successfully fabricated on these wafers with thick epitaxial layers.

Introduction High voltage diodes and transistors fabricated on high quality 4H SiC are suited for new technology markets to control power and reduce energy consumption. Recent progress in the fabrication of 4H-SiC crystals and wafers has resulted in micropipe defect density <0.1/cm^2, screw dislocation densities <2000/cm^2 [1-3], and basal plane defect density <1000/cm^2. CVD epitaxy on these substrates has produced SiC epiwafers with smooth surfaces and low basal plane defect density [4]. In this work, we report the extension of our epitaxy development effort to produce low doped (<1E15/cm^3) epiwafers with thickness of 50-100 um on 3-in 4° off-axis substrates. Thick SiC epitaxy on 8° or 4° off-axis substrates has been reported previously[5, 6, 7, 8, 9, 10]. However, to our knowledge, only O'Loughlin et.al. has reported statistical epi and device data for thick epitaxial growth [5].

Experimental The thick epiwafers are grown on 3-in 4° off-axis 4H SiC substrates in a horizontal hot-wall CVD reactor. Thickness is measured by Fourier Transform Infrared Spectrometry (FTIR) and doping is obtained by mercury probe capacitance voltage (C-V) measurement. Surface defect density is obtained by Laser Light Scattering (LLS) mapping, assuming a 2x2 mm^2 site size, and roughness is measured by Atomic Force Microcscopy (AFM). In addition, carrier lifetime is measured by microwave induced photoconductive decay (micro-PCD). The excitation laser wavelength is 349 nm, and the carrier injection level corresponds to $1x10^{16}$ cm^{-3}. Basal plane dislocations (BPD) are revealed by molten KOH etch at 500 – 550 °C for 5 - 10 minutes. BPD counting is conducted by manual scan along a cross pattern of the wafer with 10mm edge exclusion under a Nomarski optical microscope with a total magnification of 200x. The total inspected area of the cross pattern is 1.33 cm^2.

Results The thick epi development has focused on defect control through improving hot-zone parts design and stability as well as key process parameters. Using an optimized CVD chemistry, epiwafers with thickness up to 100um at intentional n-doping levels around $5x10^{14}$ cm^{-3} have been produced. At this thickness and low doping level, we have obtained a reasonably low defect density with a median value of 5.6 cm^{-2} and best values below 2 cm^{-2}. Figure 1 (a) shows the probability plots of the epi defect density at the different thickness ranges. For the thickness range of 80-100um, 25% of the thick epiwafers display defect densities below 4.2 cm^{-2}. As an example, Figures 1 (b) illustrates the LLS defect site map of a 100 μm thick epitaxy wafer with average n-doping of $4x10^{14}$ cm^{-3}. The defect free site yield of the 100um thick epi wafer is 92%, corresponding to a defect density of 1.9 cm^{-2}. The optimized process has also consistently produced epiwafers with a smooth surface, even at thickness of 80 – 100um, as confirmed by both the LLS topography images and AFM roughness measurements. The RMS roughness of the 80 - 100um thick epi wafers is generally below 1nm, as measured by AFM with a 20x20um^2 scan. However, a deviation from the smooth surface can be sometimes observed at locations close to the wafer edges.

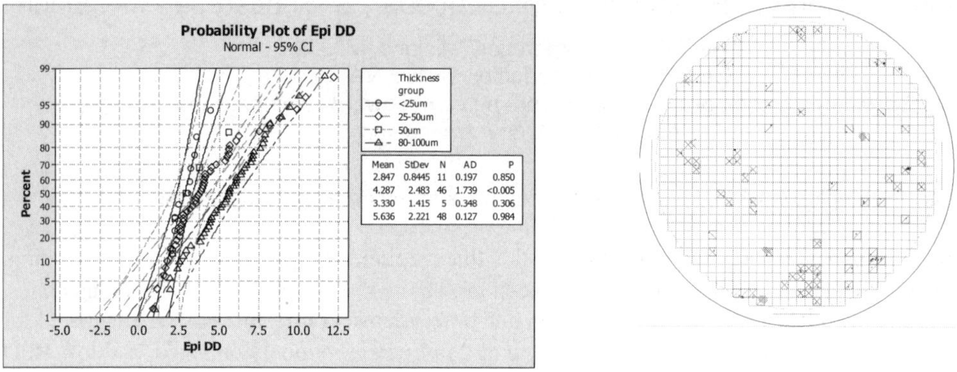

(a) (b)

Figure 1 (a): Probability plots of defect density at different thickness groups. For epiwafers with thickness of 80 – 100um, 50% of the epiwfers have above 80% yield, and 25% of the epiwafers have above 85% yield. **(b):** LLS defect site map of a 100um thick epi wafer with LLS site yield of 92%, corresponding to a defect density of 1.9 cm^{-2}.

In addition to LLS defect density and surface roughness, the epitaxy metrics most pertinent for high power bipolar device applications include carrier lifetime and basal plane dislocation density (BPD) in the thick active epilayer. The carrier lifetime is measured with micro-PCD. The average lifetime lies consistently above 2 μs, as shown in Figure 2. No significant variation in carrier lifetime can be observed in epilayers with thickness from 30 um to 100 um. As an example, the lifetime of a 100 μm thick epilayer ranges from 6 – 8 μs. The lifetime is generally uniform over the wafer.

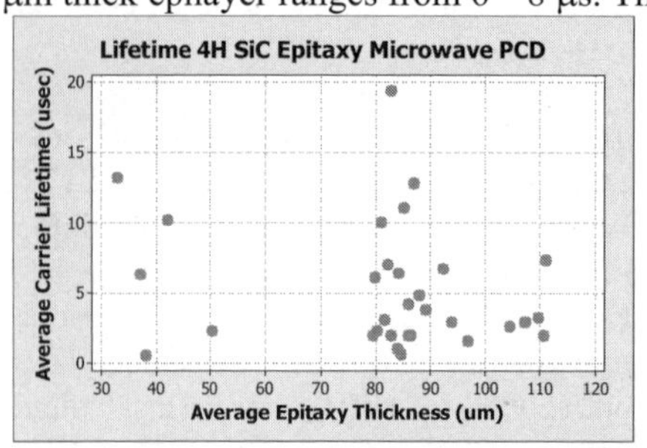

Figure 2 Micro-PCD carrier lifetime of the epi wafers with thickness ranging from 30um to 100um. The average carrier lifetime lies above 2 μs for all epiwafers in this thickness range.

Molten KOH etch has been used to obtain BPD density in a number of thick epilayers. The BPDs, clearly revealed as shell shaped etch pits, are often found associated with morphological defects such as triangles and line shaped defects. Consequently, the total BPD density increases with decreased LLS site yield. Isolated BPDs not associated with any defects are occasionally encountered as well, but with much lower density. Figure 3(a) and (b) are microscope images showing isolated BPDs and BPDs related to a triangular defect or a line defect. Figure 3(c) is a plot of BPD density versus LLS site yield. When an epiwafer has low site yield, the BPDs in the epi are predominantly associated with triangular defects. Considering only isolated BPDs, the average BPD density is below 10 cm^{-2} for 7 of the 12 thick epiwafers. If the BPDs associated with defects are counted, 1 of the 12 thick epiwafers has BPD density < 10 cm^{-2}, and additional 6 of the 12 epiwafers have BPD density below 50 cm^{-2}.

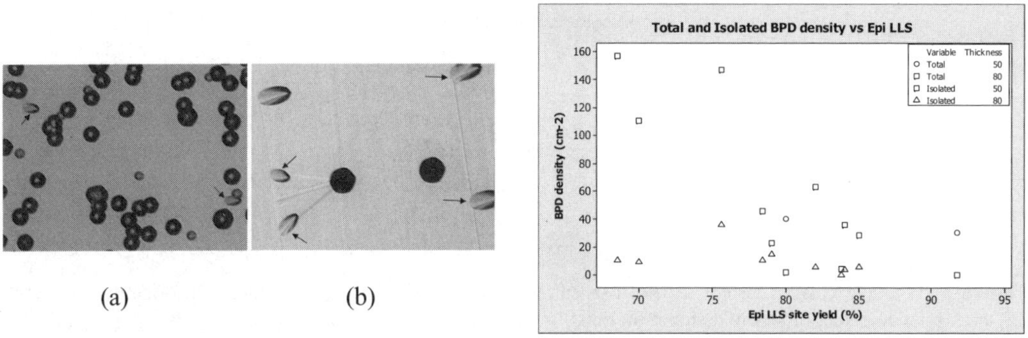

(a) (b) (c)

Figure 3 (a): Nomarski microscope image of KOH etched epi surface. Isolated BPDs are marked with arrows. **(b):** Nomarski microscope image of KOH etched epi surface. BPDs associated with a triangle or a line defect are marked with arrows. **(c):** BPD density dependence on the epiwafer LLS site yield for 50um and 80um epilayer thickness. The total BPD density increases rapidly with decreasing LLS yield, while the isolated BPD density stays low even at low LLS site yield.

High voltage JBS performance JBS diodes with active areas ranging from 0.008 cm^2 to 0.28cm^2 have been processed on the thick epiwafers. Blocking voltages above 90% of the theoretical limit have been achieved on many of the JBS diodes. A blocking voltage of 5kV has been obtained on 0.28cm^2 diodes on 50um thick epitaxy, and 10kV has been demonstrated on 0.18cm^2 diodes on 80um thick epitaxy. Failure analysis on some of the epiwafers after testing suggests that triangular defects are killer defects in the reverse performance. Diodes with a high blocking voltage are free from any triangular defects. Etching in molten KOH was conducted on some of the thick epiwafers after device testing. The diodes with disparate blocking voltages have been inspected under a Nomarski microscope. As shown in Figures 4(a) and (b), the dislocation density does not differ in the diode with blocking voltage of 1kV as compared with that in the diode with blocking voltage of >10 kV. Both diodes have an area of 0.28 cm^2, and not a single BPD is observed in either diode. The substrates selected for the thick epi growth were sent for Synchrotron White Beam X-ray topography (SWXRT) for threading screw dislocation mapping prior to epitaxy. After the JBS diode fabrication and performance tests, an attempt has been made to examine potential correlation between the leakage current at a blocking voltage of 5kV and the screw dislocation density contained in the active diode area for the 50um thick epiwafers. As shown in Figure 5, no correlation can be observed.

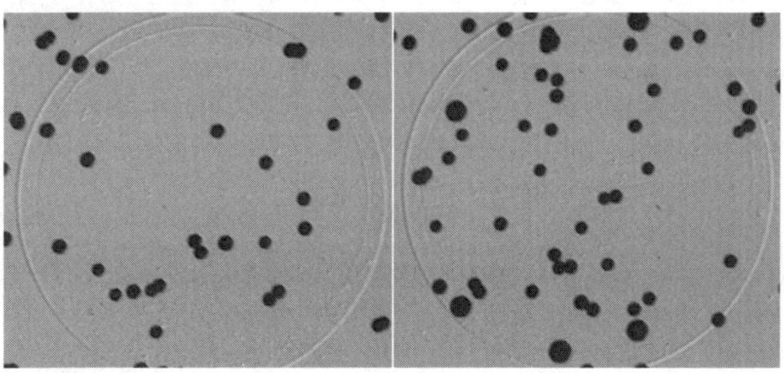

(a) (b)

Figure 4: Nomarski optical microscope images of KOH etched surface of **(a):** a JBS diode with blocking voltage of 1 kV and **(b):** a diode with blocking voltages of >10 kV. The black dots are threading screw (large pits) or threading edge dislocations (small pits) as revealed by KOH etch. No BPDs are observed in either of the diodes.

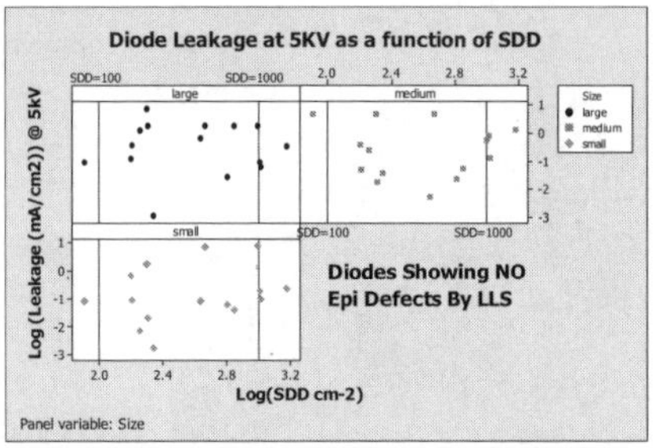

Figure 5: Leakage current versus screw dislocation density (SDD) in logarithmic scale on three different device sizes. No effect of screw dislocation density on leakage current is observed.

Summary 4H SiC epiwafers with thickness of 50 – 100um are grown on 4° off-axis substrates. LLS detected defect density in the range of 2 – 6 cm^{-2} is obtained in the thick epitaxy. Consistent carrier lifetime in the range of 2 – 3 μs has been obtained on these thick epiwafers. Very low BPD density has been confirmed in the thick epiwafers with BPD density down to below 10 cm^{-2}. Epitaxial wafers with thickness of 50-100 um have been used to fabricate diodes. High voltage testing has demonstrated blocking voltages near the theoretical values for 4H-SiC. Blocking voltage of 5kV has been achieved in devices fabricated on 50um thick epitaxial films, and blocking voltage of 10kV has been obtained in devices fabricated on 80um thick films with device areas up to 0.28 cm^2. Failure analysis confirmed triangle defects are killer defects. In addition, the leakage current at the high blocking voltages of the JBS diodes showed no correlation with the screw dislocation density. It is also observed that the primary source of BPD formation originates outside the crystal growth process.

(Authors acknowledge partial support of this work by the U.S. Army Research Laboratory: Contract No - DAAD19-01-C-0067, Dr. Bruce. Geil. The support for XRT measurements from Professor Michael Dudley at Stony Brook University is greatly appreciated as well.)

References

[1] M.J. Loboda, et.al., to be published in the proceedings of the 2011 Compound Semiconductor Manufacturing Technology Conference (CS Mantech), May 2011. Also as internal TIS report #2011-I0000-64083.

[2] N. Zhang, Y. Zhang, B. Raghothamachar, S. Byrappa, G. Choi, E.K. Sanchez, D.M. Hansen, M.J. Loboda, Mater. Sci. Forum 645-648, 2010, p.291

[3] M. Dudley, S. Byrappa, F. Wang, Y. Zu, Y. Zhang, B. Raghothamachar, G. Choi, E.K. Sanchez, D.M. Hansen, R. Drachev, M.J. Loboda, Mater. Res. Soc. Symp. Proc. Vol 1246, 2010, p.29

[4] G. Chung, M.J. Loboda, J. Zhang, J.W. Wan, E.P. Carlson, T.J. Toth, R.E. Stahlbush, M. Skowronski, R. Berechman, S. G. Sundaresan and R. Singh; 4H-SiC Epitaxy With Very Smooth Surface and Low Basal Plane Dislocation on 4 degree Off-axis Wafer, Mater. Sci. Forum 679-680, 2011, p.123

[5] M. J. O'Loughlin, K. G. Irvine, J. J. Sumakeris, M. H. Armentrout, B. A. Hull, C. Hallin, and A. A. Burk, Jr., MRS, Mater. Res. Soc. Symp. Proc. Vol. 1069, D04-01

[6] Bernd Thomas, Christian Hecht, and Birgit Kallinger, Materials Science Forum Vols. 615-617 (2009) pp 77-80

[7] A. Henry, S. Leone, S. Andersson, O. Kordina and E. Janzén, Materials Science Forum Vols. 645-648 (2010) pp 95-98

[8] Iftekhar Chowdhury, MVS Chandrasekhar, Paul B Klein, Joshua D. Caldwell, and Tangali Sudarshan, MRS2010

[9] H. Tsuchida, M. Ito, I. Kamata, M. Nagano, T. Miyazawa and N. Hoshino, Materials Science Forum Vols. 645-648 (2010) pp 77-82

[10] F. La Via, G. Izzo, M. Camarda, G. Abbondanza, and D. Crippa, Materials Science Forum Vols. 615-617 (2009) pp 55-60

Materials Science Forum Vols. 717-720 (2012) pp 141-144

doi:10.4028/www.scientific.net/MSF.717-720.141

Improvement of homoepitaxial layer quality grown on 4H-SiC Si-face substrate lower than 1 degree off angle

Kazutoshi Kojima[1,2,a*], Sachiko Ito[2,b], Akiyo Nagata[2,c], Hajime Okumura[1,2,d]

[1]R & D Partnership for Future Power Electronics Technology, 2-9-5 Toranomon, Minato-ku, Tokyo 105-0001, Japan

[2]Advanced Power Electronics Research Center, National Institute of Advanced Industrial Science and Technology, Central 2 1-1-1 Umezono, Tsukuba, Ibaraki 305-8568, Japan

[a]kazu-kojima@aist.go.jp, [b]sachi-ito@aist.go.jp, [c]akiyo-nagata@aist.go.jp, [d]h-okumura@aist.go.jp

Keywords: 4H-SiC, epitaxy, off angle, vicinal, C/Si ratio, SiH_4, morphology, dislocation, stress

Abstract. In this study, we struck a balance between specular surface morphology and polytype homogeneity on an epitaxial layer grown on 4H-SiC Si-face substrate with off angle less than 1degree by controlling the C/Si ratio with the SiH_4 flow rate. Schottky barrier diodes fabricated on a grown epitaxial layer exhibited a blocking of voltage over 1000 V and an n value of less than 1.1 with a high yield of more than 80%. A substrate with a low off angle was found to have an advantage as regareds the stress that generates the interfacial dislocations at the epitaxial layer/substrate interface during the epitaxial growth process.

Introduction

Recently, reducing the wafer off angle of SiC bulk substrate is strongly required by respects of the wafer manufacture. Indeed, a prototype 6-inch 4H-SiC bulk substrate was presented at ECSCRM 2010[1]. For such wafer, reducing the extraction rate of SiC ingot is important issue to reduce the wafer cost. On-axis or vicinal off-axis growth is very attractive as regards meeting this requirement. With techniques used for epitaxial growth on such substrates, it is important to control both polytype stability and surface morphology. Several groups have reported the homoepitaxial growth of on-axis or vicinal 6H- and 4H-SiC substrate [2-8]. In these reports, the inclusion of another polytype such as 3C-SiC was well suppressed. In terms of the surface morphology, a rough surface caused by macro step bunching or spiral growth still remains a serious problem for the Si-face. At ICSCRM 2009, we reported that the macro step bunching of epitaxial layer surface on 4H-SiC vicinal Si-face substrate could be suppressed by controlling the vicinal off angle [9]. However, it was insufficient to achieve a balance between specular surface morphology and polytype homogeneity. A specular surface without macro step bunching could be obtained by undertaking the growth at a low temperature about 1550°C and with a low C/Si ratio of 0.5. In this case, a 6H or 3C polytype was included in the epitaxial layer. On the other hand, the polytype could be well controlled at a high growth temperature of more than 1600°C. However, the surface becomes rough because macro step bunching occurs at such a high temperature.

In this study, we improved the growth conditions and struk a balance between specular surface and polytype homogeneity. We focused on how to control the surface morphology at temperature higher than 1600°C. We also characterized the surface morphological quality and the propagation of dislocations.

Experiments

Epitaxial growth was carried out using a conventional hot wall type CVD reactor with an H_2-C_3H_8-SiH_4 gas system. The vicinal off-axis substrates were commercial 2-inch wafers except with respect to the off angle. The vicinal off angle of the substrates was from 0.75° to 0.9°. *In-situ* H_2 etching was carried out by adding SiH_4 to prevent the generation of step bunching during this process [10]. For the epitaxial growth, the substrate temperature and the C/Si ratio were used as experimental parameters. The surface quality was characterized using a Nomarski microscope, an atomic force microscope (AFM) and the *I-V* characteristics of Schottky barrier diodes (SBDs). Polytype homogeneity and the propagation of dislocations were investigated with a photoluminescence imaging system.

Results and discussion

To suppress the macro step bunching at high temperatures exceeding 1600°C, we employed the technique reported at ECSCRM 2008 whereby the C/Si ratio was controlled by the SiH_4 flow rate [11]. Figure 1 shows a block diagram of the surface morphology with various growth temperatures and C/Si ratios. The thickness of epitaxial layer was about 5 μm. When the SiH_4 flow rate was fixed at 6.67 sccm and the C/Si ratio was controlled at 0.5, a specular surface morphology without macro step bunching was obtained below 1550°C. However, another polytype was included in this temperature range as reported at ICSCRM 2009[9]. On the other hand, although the macro step bunching could not be suppressed at 1600°C, the polytype could well be controlled. As is well known, growth with a low C/Si ratio is quite effective in suppressing the macro step bunching. Then, to control the C/Si ratio and realize a more Si-rich condition, the C_3H_8 flow rate was fixed at 1.11sccm and the SiH_4 flow rate was increased. As the result, good surface morphology without macro step bunching was obtained at a growth temperature exceeding 1600°C and a C/Si ratio of 0.4 or 0.33. However, macro step bunching was generated again at a C/Si ratio of 0.25. In general, the C/Si ratio is controlled by changing the flow rate of the C source gas with the Si source gas at a fixed flow rate. When the C/Si ratio is changed to a Si rich condition, the C source gas flow rate is reduced. The growth rate is also reduced because the growth rate is limited by the flow rate of the C source gas. Therefore, the influence of H_2 etching for the surface morphology is considered to be enhanced at high temperature. Indeed, the step bunching was also generated during

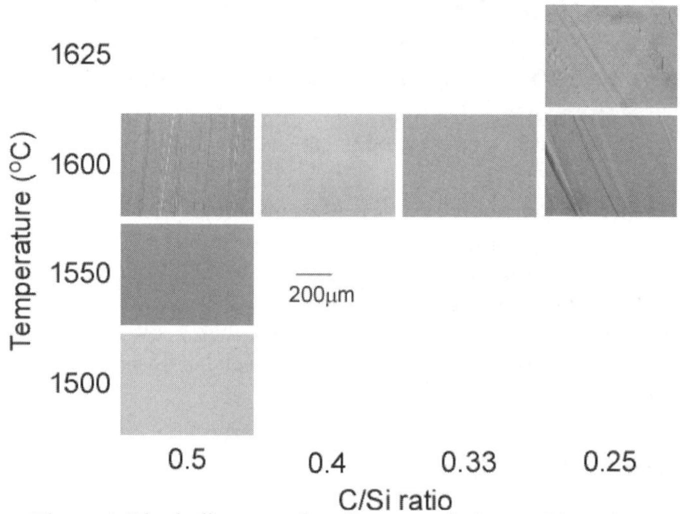

Figure 1 Block diagram of surface morphology with various growth temperature and C/Si ratio. C_3H_8 flow rate is fixed at 1.11sccm.

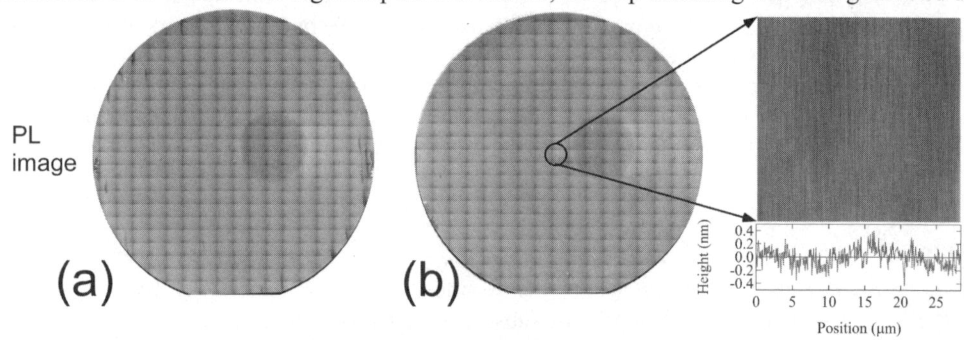

Figure 2 Polytype uniformity of 2-inch wafer. The growth temperature and C_3H_8 flow rate are 1600°C and 1.11 sccm, respectively. The SiH_4 flow rates in Figs. 2(a) and (b) are 6.67 and 8.33 sccm, respectively. AFM image with 30μm square is inset.

the *in-situ* H_2 etching process [11]. However, by increasing the SiH_4 flow rate under C supply limited condition, the C/Si ratio can be changed to a more Si-rich condition but the growth rate remains the same. Indeed, the growth rate of Fig. 1 is limited by C source flow and that growth rate is about 3μm/h for all C/Si ratios. As the result, the generation of macro step bunching was well suppressed at the growth temperature exceeding 1600°C.

Figure 2 shows a photoluminescence image of a 2-inch epitaxial wafer at RT. The growth temperature is 1600°C. The SiH_4 and C_3H_8 flow rates in Figs. 2 (a) are 6.67 and 1.11 sccm, respectively. In this case, the polytype was well controlled at 4H over the entire wafer but the generation of macro step bunching could not be suppressed. However, by increasing the SiH_4 flow rate to 8.33 sccm to reduce the C/Si ratio to 0.4, it is possible to strike a balance between specular surface and polytype control as shown in Figs. 2 (b). AFM measurement revealed that the step height is well controlled below 1 nm for areas in both the 30μm square and 2μm square ranges. These results indicate that controlling the C/Si ratio with the SiH_4 flow rate has the potential to realize a balance between specular surface and polytype homogeneity.

To clarify the surface quality, Ni Schottky barrier diodes (SBDs) were fabricated on the as-grown epilayer surface. The SBD characteristics are sensitive to surface roughness. Figure 3 shows the *I-V* characteristics of SBDs. The growth temperatures in Figs. 3(a) and (b) are 1550°C and 1600°C, respectively. The donor concentrations and epilayer thicknesses in Figs. 3(a) and (b) are 9.3 x 10^{15} cm^{-3} and 9.8 μm and 6.6 x $10^{15}cm^{-3}$ and 6.3 μm, respectively. The diameter of the Ni Schottky metal was 1 mm. Al was used as the backside contact. There were 17 measurement points over the entire 2-inch wafer. A blocking voltage of more than 1000V and a lower leakage current of 1 x 10^{-6} A/cm^2 at 1000V were observed for both samples. For the forward characteristic, an n value of less than 1.1 was also obtained for both samples. However, the yields with good *I-V* characteristics were quite different as shown in this figure. The yield in Figs. 3(a) is clearly worse than that in Figs. 3(b). In Figs. 3(b), the *I-V* characteristics are uniform for all the measurement points and the yield reaches 82%. On the other hand, the *I-V* characteristics in Figs. 3(a) are dispersed and the yield has dropped to 41%. We think these differences are related to the polytype inclusions. For the sample in Fig. 3(a), the growth temperature is 1550°C and another polytype is included. However , the sample in Figs. 3(b) could strike a balance between specular surface and polytype homogeneity as shown in Fig. 2. Then, a higher yield could be obtained.

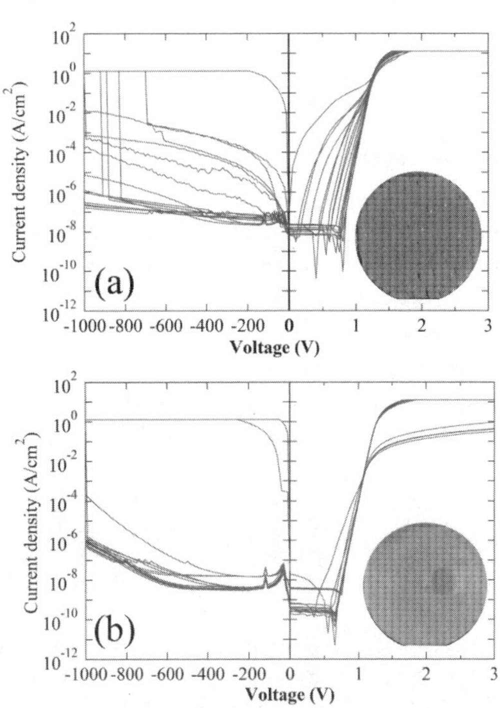

Figure 3 *I-V* characteristics of Schottky barrier diodes. The growth temperature in Figs. 3(a) and (b) are 1550°C and 1600°C, respectively. The PL image of whole wafer is inset.

As regards SiC epitaxy, interfacial dislocations (IDs) are major problem because it prevents the conversion of basal plane dislocations to threading edge dislocations or increases the dislocation density. In particular, IDs cause the generation of dislocation half loop arrays (HLAs) [12]. Figure 4 shows a PL image of dislocations at the center of a 2-inch wafer. Both images are 2.7 mm square. The wafer off angles in Figs. 4(a) and (b) are 8 and 0.76 degrees, respectively. The wavelength of the UV light source is 314nm and the detection wavelength is longer than 750nm. Dislocations were clearly observed. In Figs. 4(a), we see that basal plane dislocations are converted to IDs at the epilayer/substrate interface and this leads the generation of dislocation half loops near surface. On the other hand, basal plane dislocations propagate straight from the substrate into the epitaxial layer and are not converted to IDs in Figs. 4(b). IDs are considered to be generated by the stress induced by, for example, the temperature distribution or carrier concentration difference between the substrate and he epitaxial layer. This result indicates that a SiC wafer with a low off angle has an advantage in term of the stress applied to the wafer during the epitaxial growth process.

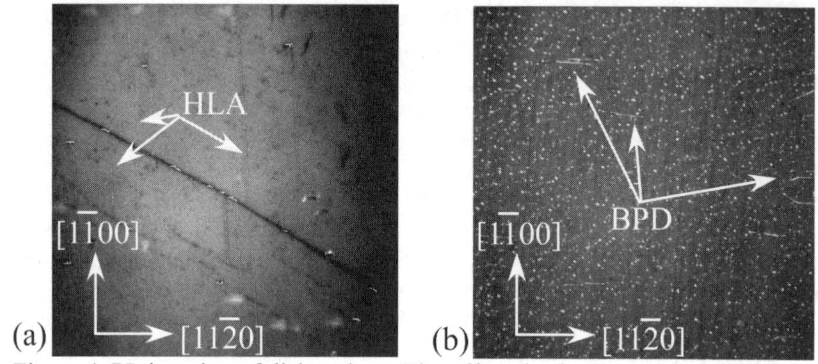

Figure 4 PL imaging of dislocations. The off angles in Figs. 4(a) and 4(b) are 8 and 0.76 degrees, respecively. Both images are 2.7 mm square

Summary

In this study, we improved the growth conditions and struk a balance between the specular surface and polytype homogeneity. We focused on how to control the surface morphology at high temperature exceeding 1600°C. We found that controlling the C/Si ratio with the SiH_4 flow rate is a possible way to strike a balance between a specular surface and polytype homogeneity. This result leads to Schottky barrier diodes with good *I-V* characteristics namely, a blocking voltage of more than 1000V and an n value of less than 1.1 with a yield exceeding 80%. It was also found that a SiC wafer with a low off angle has an advantage as regards the stress applied to the wafer during the epitaxial growth process.

Acknowledgement

This work is supported by Novel Semiconductor Power Electronics Project Realizing Low Carbon Emission Society under Ministry of Economy, Trading Industry (METI) and New Energy and Industrial Technology Development Organization (NEDO).

References

[1] Information on http://www.cree.com.
[2] S. Nakumura, T. Kimoto and H. Matsunami, J. Crystal Growth 256 (2003) 341.
[3] C. Hallin, Q. Wahab, I. Ivanov, J. P. Bergman and E. Janzén, Mater. Sci. Forum 457-460 (2004) 193.
[4] K. Kojima, H. Okumura, S. Kuroda and K. Arai, J. Crystal Growth 269 (2004) 367.
[5] S. Leone, H. Pedersen, A. Henry, O. Kordina and E. Janzén, Mater. Sci. Forum 600-603 (2009) 107.
[6] S. Leone, H. Pedersen, A. Henry, S. Rao, O. Kordina and E. Janzén, Mater. Sci. Forum 615-617 (2009) 93.
[7] K. Kojima, H. Okumura and K. Arai, Mater. Sci. Forum 615-617 (2009) 113.
[8] J. Hassan, J. P. Bergman, A. Henry and E. Janzén, J. Crystal Growth 310 (2008) 4424.
[9] K. Kojima, S. Ito, J. Senzaki and H. Okumura, Mater. Sci. Forum 645-648 (2010) 99.
[10] K. Kojima, S. Kuroda, H. Okumura and K. Arai, Mater. Sci. Forum 556-557 (2007) 85.
[11] K. Kojima, H. Okumura and K. Arai, Mater. Sci. Forum 615-617 (2009) 113.
[12] X. Zhang, M. Skowronski, K. X. Liu, R. E. Stahlbush, J. J. Sumakeris, M. J. Paisley, and M. J. O'Loughlin, J. Appl. Phys. 102 (2007) 093520.

Materials Science Forum Vols. 717-720 (2012) pp 145-148
© (2012) Trans Tech Publications, Switzerland
doi:10.4028/www.scientific.net/MSF.717-720.145

Surface morphology evolution after epitaxial growth on 4°off-axis 4H-SiC substrate

N. Jegenyes[1,a], V. Souliere[1,b*], F. Cauwet[1,c], G. Ferro[1,d]

[1]Université de Lyon, Université Lyon 1, CNRS, UMR 5615, Laboratoire des Multimatériaux et Interfaces 43 Bd du 11 Novembre 1918, 69622 Villeurbanne Cedex, France

[a]njegenyes@gmail.com, [b]veronique.souliere@univ-lyon1.fr*, [c]francois.cauwet@univ-lyon1.fr, [d]gabriel.ferro@univ-lyon1.fr

Keywords: CVD, Surface morphology, step bunching, 4°off-axis

Abstract. Chemical vapour deposition in a cold wall reactor working at atmospheric pressure was used to study the homoeptaxial growth of 4H-SiC on 4°off misoriented substrates from silane and propane precursors. The effect of various growth parameters (temperature 1450-1650°C, C/Si ratio 1-7 and thickness 2.5-10 μm) were studied in order to determine the best conditions for obtaining smooth surfaces after epitaxy. It is shown that the main source of roughness is surface undulation which easily appears during growth, especially at low C/Si ratios or high temperature (up to 1600°C). Temperatures above 1600°C and C/Si ratio of 1 give the best results. When reducing temperature, defects formation is competing with surface undulation.

Introduction

One of the key issues for fabricating 4H-SiC based devices is the epitaxial growth which is mostly performed by chemical vapour deposition (CVD). For the two last decades, 4H commercial wafers were 8°off misoriented in order to promote the so-called "step-controlled epitaxy" [1]. But what was seen as beneficial is now seen as a drawback for two reasons : 1) cutting the SiC boules with such high misorientation induces important material loss for manufacturers, especially when dealing with large diameters of 100 mm or higher; 2) basal plane dislocations (BPDs) coming from the substrates are easily transferred into the epilayer due to this important misorientation [2]. It was shown that BPDs are the nucleation source of Schokley type stacking faults, which are known to degrade device performance [3]. So, it is important to limit this propagation to the growing layers during epitaxy.

Going down to zero misorientation, i.e. (0001) on-axis substrate, was and is still under investigation but the task is not simple [4, 5]. Intermediate solution was proposed by wafer suppliers with 4°off-axis crystals. It was shown that using this lower misorientation compared to 8°off leads to higher conversion of the BPDs into threading edge dislocations [2] which are thought to be less detrimental to devices. Playing with the growth conditions, one can even enhance this conversion up to 100% [6]. However, such intermediate misorientation of 4° leads easily to triangular defect and to higher surface roughness due to enhanced step bunching [6-10]. Growth parameters (temperature, C/Si ratio in the gas phase) play an important role in the appearance of these features. For instance, it was shown that the smoothest layers are rather obtained at low C/Si ratio and low temperatures. However, most of the studies were performed using hot wall CVD reactors working at reduced pressure. To the best of our knowledge, similar study was not achieved yet with a cold wall reactor and/or at atmospheric pressure. This is done in the present work.

Experimental setup

The growths were performed in a home-made, vertical, cold-wall CVD reactor working at atmospheric pressure [3]. The precursor gases were silane (SiH_4) and propane (C_3H_8) diluted in 14 slm of H_2. The precursor fluxes used in this study were 1.5 sccm and 0.5-3.5 sccm respectively for

SiH_4 and C_3H_8. The temperature of the inductively heated graphite susceptor was controlled and monitored by an optical pyrometer. The Si face 4H-SiC(0001) commercial substrates were misoriented 4°off toward [11-20]. Prior to the growth, the seeds were ultrasonically cleaned first in acetone (15 min) and then in methanol (15 min), before being dried under Ar flow and then loaded into the reactor. The growth procedure begins with in-situ desoxidation/cleaning of the seed at 1200°C under H_2 for 5 min. Then the temperature is increased up to the growth temperature (Tg) in 2 min under H_2+C_3H_8. A previous study having shown that the value of the propane flux does not change significantly the surface roughness after this annealing, the propane flux was directly set to the value to be used for the subsequent growth. This gas mixture is kept during 5 min at Tg before adding SiH_4 to start the epitaxial deposition at an average rate of ~3 µm/h. Tg was varied between 1450 and 1650°C and the C/Si ratio in the gas phase was changed from 1 to 7 (by keeping constant the silane flux at 1.5 sccm so that the growth rate did not change significantly within +/- 10 %).

Results and discussion

Effect of temperature. In Fig. 1 are shown the surface morphology of epilayers grown at different temperatures with a fixed C/Si ratio of 1. One can see at low temperatures (1450 and 1500°C) the formation of Si droplets on the surface. Their density decreases with temperature increase up to their complete disappearance above 1500°C. This is due to the increased evaporation rate at higher temperatures. The best surface morphology, as observed using optical microscope, is obtained at 1650°C for which no triangle or carrot-like defects are seen whereas these defects are commonly present at 1600°C. Surface undulation or macrostep formation seems to be present but is hardly detected by optical microscopy. This was better evidenced by atomic force microscopy (AFM) characterization. Regular and parallel undulation can now be seen for all the samples except the one grown at the highest temperature where the features are fainter. Careful investigation of the AFM data shows that these undulations are not consistent with separated macrosteps, even the well defined ones. These undulations correspond rather to hill and valley surface reconstruction.

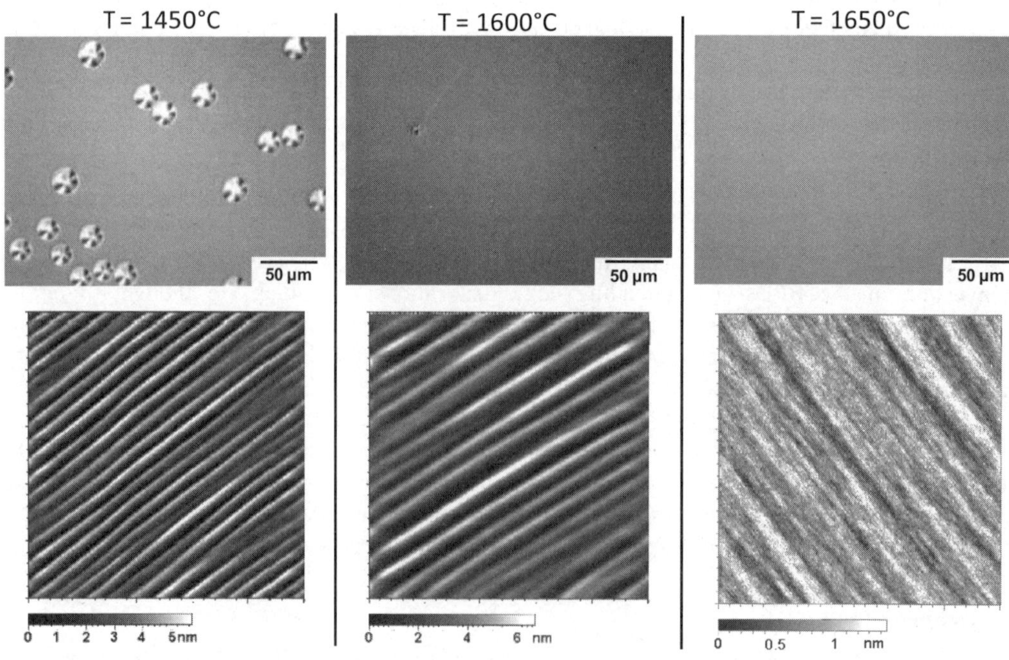

Figure 1 - Optical micrographs and AFM images (5x5 µm) showing the surface morphology evolution of 2.5 µm thick epilayers grown with a C/Si ratio of 1 and at different temperatures.

As shown in Fig. 2 which summarizes the values extracted from the AFM images in Fig. 1, the undulation period significantly increases with temperature while the roughness passes through a maximum at 1600°C. This is due to the fact that at 1650°C, the undulations are getting less clear and they may be decomposed in small and not well defined steps. The resulting roughness is thus very low. As a result, these conditions (1650°C and C/Si = 1) are very interesting for obtaining smooth surfaces after homoepitaxy on 4H-SiC, 4°off oriented substrate.

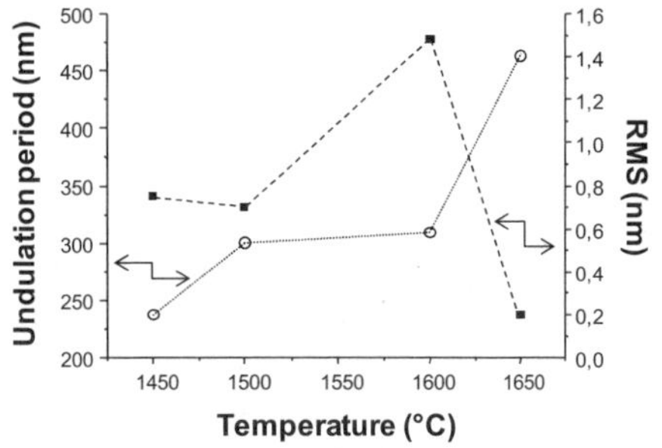

Figure 2 – Evolution of the morphological parameters extracted from AFM images (25x25 μm^2 scan) as a function of growth temperature (2.5 μm thick, C/Si = 1).

Effect of layer thickness. The effect of layer thickness on surface morphology is summarized in figure 3. One can see that increasing the layer thickness leads also to an important increase of the undulation period and of the mean roughness. Furthermore, this thickness dependence gets steeper when increasing the temperature from 1500 to 1600°C. So, if the growth parameters are favoring the formation of surface undulations, increasing thickness will automatically accentuate these features. The best way to keep a smooth surface is thus to avoid the formation of these undulations.

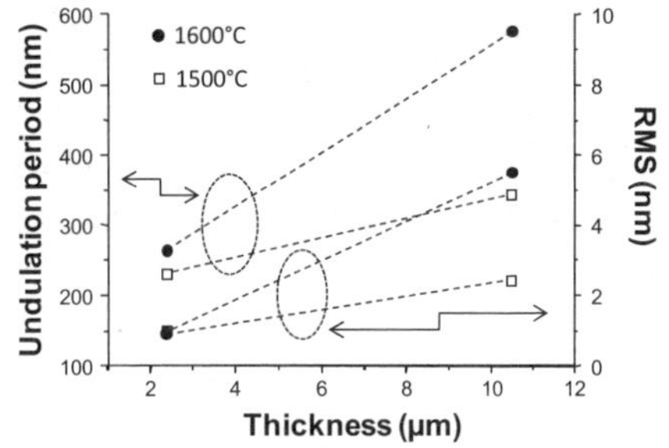

Figure 3 – Evolution of the morphological parameters extracted from AFM images (25x25 μm^2 scan) as a function of layer thickness and growth temperature (C/Si = 1.5)

Effect of C/Si ratio. When increasing the C/Si ratio, it was found that the surface undulation progressively diminishes (Fig. 4) and the mean roughness, as measured by AFM, decreases significantly upon increase of this ratio (Fig. 5a). However, on a macroscopic scale, the layers grown at high C/Si ratio always display a high density of defects, such as triangles or carrots (see Fig. 5b). High C/Si ratio conditions are thus inappropriate for the growth of high quality layers.

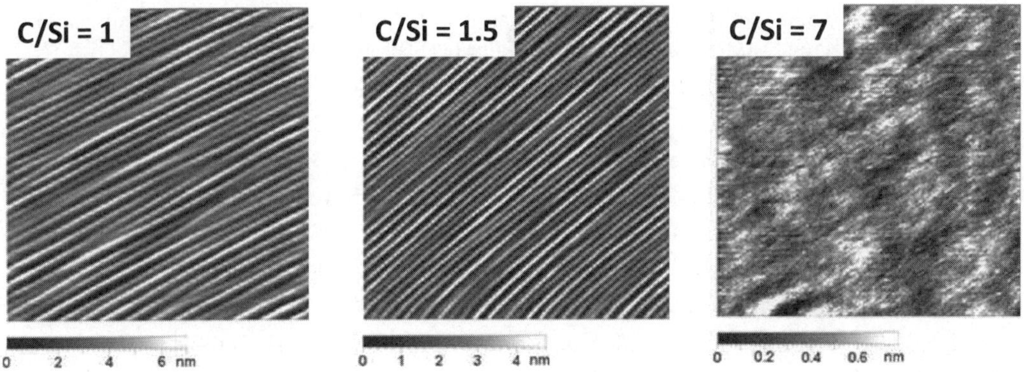

Figure 4 – AFM images (10x10 μm) of samples grown at 1600°C with various C/Si ratio (2.5 μm thick).

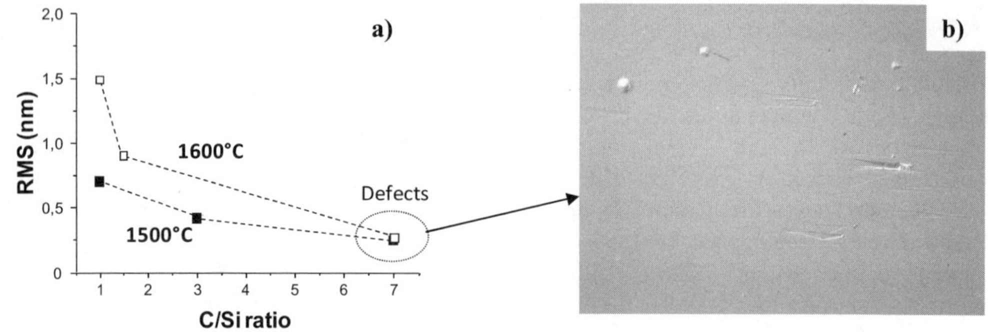

Figure 5 – a) Evolution of the surface roughness as a function of the C/Si ratio at 1500 and 1600°C and for 2.5 µm thick layers (parameters extracted from AFM images, 25x25 μm^2 scan); b) Surface morphology of a sample grown at 1600°C and C/Si ratio of 7 (conditions with a dashed circle on a).

The roughness trends obtained in this study are opposite to the ones found in other works using hot wall reactors and low pressure [6-10]. The reason for this is still unclear even though one can postulate a different surface reaction and/or incorporation chemistry when working at atmospheric pressure and cold wall reactor. For instance, higher growth pressure should reduce the desorption rate of adatoms. The effect of growth rate is under study to gather more information and better understand the mechanism involved.

Conclusion

The main source of roughness after homoepitaxy on 4°off misoriented 4H-SiC substrate is surface undulation which easily appears during growth. Its formation is mostly favored when the adatom diffusion length is high, at low C/Si ratios and high temperature (up to 1600°C). The best surface morphology is obtained for temperatures above 1600°C and C/Si ratio of 1; the surface undulation is smooth and the related roughness very low. If one wants to reduce the temperature, a trade-off has to be found between surface undulation and defect formation at high C/Si ratios.

Acknowledgements : Authors would like to acknowledge the financial support from French ANR in the framework of MINTEX project (ANR-09-BLAN-0189-04).

References

[1] N. Kuroda, K. Shibahara, W. S. Yoo, S. Nishino, and H. Matsunami, Extended Abstracts of the 19th Conference on Solid State Devices and Materials, Tokyo, (1987) 227.

[2] W.Chen, M.A.Capano, J. Appl. Phys. 98 (2005) 114907.

[3] S. Ha, M. Skowronski, H. Lendenmann, J. Appl. Physics 96 (1) (2004) p. 393-398

[4] S. Leone, H. Pedersen, A. Henry, O. Kordina, E. Janzén, J. Crystal Growth 312 (2009) 24–32

[5] S. Nakamura, T. Kimoto, H. Matsunami, J. Crystal Growth 256 (2003) 341–346

[6] R.L. Myers-Ward, B.L. Van Mil et al, Mater. Sci. Forum 615-617 (2009) 105-108

[7] S. Leone, H. Pedersen , A. Henry et al, J. Crystal Growth 311 (2009) 3265–3272

[8] K. Wada, T. Kimoto, K. Nishikawa, H. Matsunami, J. Crystal Growth 291 (2006) 370–374

[9] Y. Ishida, T. Takahashi, H. Okumura et al, Mater. Sci. Forum 600-603 (2009) 473-476

[10] M. Camarda, A. Canino et al, Oral presentation at ICSCRM conference held in Cleveland (USA) in Sept. 2011

Materials Science Forum Vols. 717-720 (2012) pp 149-152

doi:10.4028/www.scientific.net/MSF.717-720.149

Study of the impact of growth and post-growth processes on the surface morphology of 4H silicon carbide films

M. Camarda[a], A. Canino[a], P. Fiorenza[a], C. Bongiorno[a], A. Severino[a], V. Raineri[a], A. La Magna[a] and F. La Via[a]
M. Mauceri[b], G. Abbondanza[b], A. Pecora[b], and D. Crippa[c]

massimo.camarda@imm.cnr.it
[a] Consiglio Nazionale delle Ricerche, Istituto di Microelettronica e Microsistemi CNR-IMM, Z.I. VIII Strada 5 I 95121 Catania, Italy
[b]Epitaxial Technology Center, XVi Strada, Pantano d'Arci, 95121 Catania, Italy
[c]LPE spa, Via falzarego 8. 20021 Baranzate, Italy

Keywords: surface morphology, surface instabilities, step bunching

Abstract. In this paper we study the surface morphology of <11-20> 4° degree off, silicon terminated, 4H Silicon Carbide (4H-SiC) in terms of growth parameters and post growth argon thermal annealing. We find that out-of-equilibrium conditions favor the reduction of the surface roughness. Furthermore, we find preliminary indications that the same growth parameters that lead to the reduction of the surface roughness promote also a reduction of (1,3) and (4,4) stacking faults density.

Introduction

Step-and-terrace structures are widely observed on mis-oriented crystal surface after epitaxial growth, thermal treatment or etching under suitable conditions. Various interesting phenomena related to step-and-terrace structures such as faceting[1], step bunching[2,3] and step meandering are known to occur due to a, far to be fully understood, competition between several mechanisms such as step kinetics, surface diffusion, step stiffness and step–step interactions [3,4,5]. These phenomena are crucial in high-power devices because they lead to electric field crowding, which results in the reduction of blocking voltage and in an increase of the leakage current. Recently it was found that what was previously considered as step bunching is rather a nano-ordering of the step-and-terrace morphology which results in a periodic oscillation of the step density and, thus, to micro undulations of the surface[6]. This nano-ordering might be related to the tendency of the surface to reach an energetically stable configuration rather than an out-of-equilibrium kinetic instability. Aiming to validate this idea and to optimize the growth process in terms of surface roughness we have analyzed several films after growth and post-growth argon annealings in terms of surface morphology. Furthermore, to evaluate the impact of the process on the film quality, we have evaluated the density of stacking faults in the different growth processes.

Experimental results

We have analysed the surface morphology of "as-grown" samples using: AFM, SEM, and TEM varying the growth temperature (ranging from 1620 to 1700 ºC) and C/Si ratio (from 0.8 to 1.5) at fixed growth rate (~16μm/h) and substrate geometry (<11-20> 4° degree off, silicon terminated, 4H Silicon Carbide). We have found that low growth temperatures (below 1650 ºC) and low C/Si ratios (below 1) lead to a clear reduction of the root-mean-square surface roughness (R_q): from ~2 nm to ~0.2 nm (Fig.1). This reduction can be seen also in Fig.2 where the cross sectional TEM analysis of the low temperature and low C/Si ratio film shows almost no evidence of surface reconstruction (note the different marker in the two images). These results support the idea that out-of-equilibrium conditions can hinder the re-organization of the surface morphology.

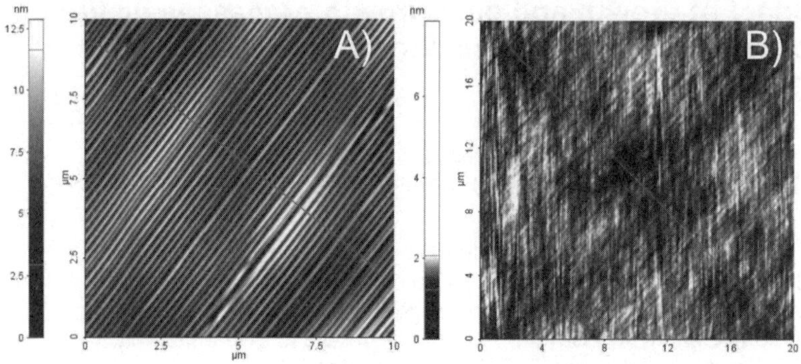

Fig.1 AFM analysis of *A)* high temperature, high C/Si growth *B)* low temperature, low C/Si growth

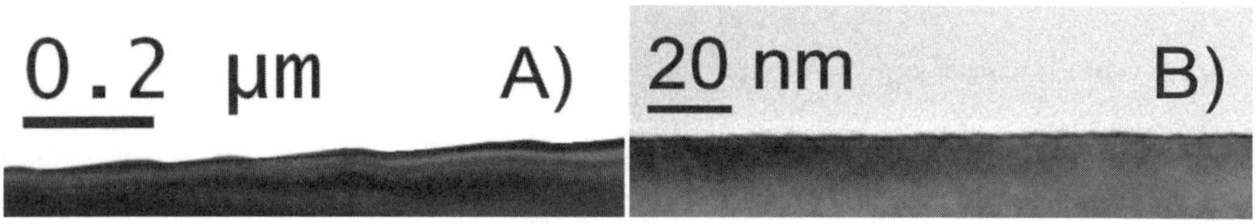

Fig.2 Cross sectional TEM analysis of *A)* high temperature, high C/Si growth *B)* low temperature, low C/Si growth. Note that, even in the later case, the steps are composed not by one but by two or three Si-C bilayers, i.e there is still a residual surface re-organization.

To better evaluate the impact of the growth kinetics on the surface morphology we have annealed the two samples (low and high R_q) for 30 minutes in argon gas ambient at low pressure (20 mbar). This allows obtaining insights on the equilibrium configuration of the surface. In this respect it is important to note that in a compound material such as silicon carbide, the equilibrium configuration is the result of a complex interplay between gas-phase and surface kinetics of both silicon and carbon atoms, so that the final configuration can strongly depends on the specific annealing parameters.

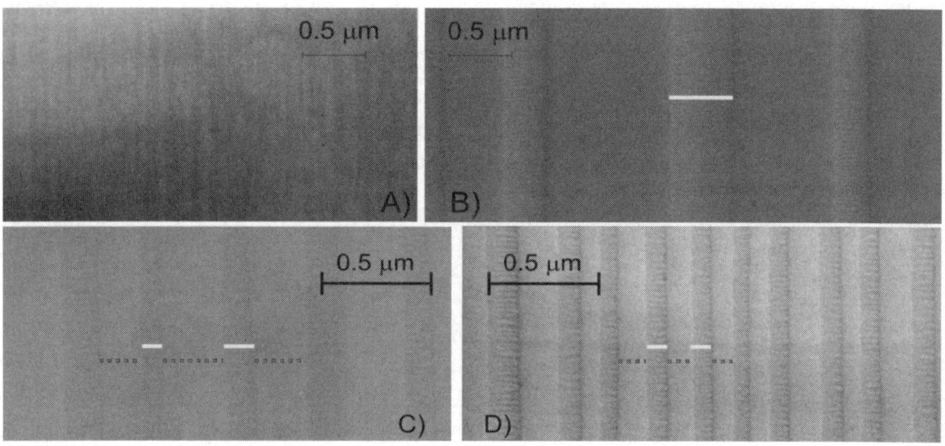

Fig.3 Pre (**A**) and post (**B**) argon annealing of the low R_q film. Pre (**C**) and post (**D**) Argon annealing of the high R_q film. Red and yellow lines indicate the basal plane and the step edge (facet), respectively. In the "as-grown" low R_q film, is not possible to identify the step-and-terrace structures.

As can be seen in fig.3 we have found a strong increase of the R_q value after the thermal treatment, thus confirming the idea that the film surface have an intrinsic (i.e. energy related) tendency towards surface re-organization and thus supporting the idea that out-of-equilibrium conditions can be the key for the reduction of the surface roughness. We also have to note the peculiar lines, perpendicular to the steps, observed after the thermal annealing and in the "as-grown" high R_q film. This feature is related to the geometric local re-arrangement of the steps along the <1-100> direction, i.e. perpendicular to the miscut direction, which generates a zig-zag morphology in the step edges. This particular feature is currently under further investigation. Finally, it should also be noted that the increase in surface roughness after Ar annealing is more pronounced in the low R_q sample. This result, confirmed in other annealing processes, could be an indication of an increased instability of the low R_q film towards surface reorganization. This aspect could be problematic for the fabrication of devices which includes, within the fabrication steps, high temperature thermal treatments. These treatments would lead to higher surface roughness and to a worse device performance for the, *initially*, low R_q films. Work is underway to systematically analyze the surface stability in terms of initial surface morphology and post growth thermal treatments.

Finally, to fully characterize the impact of the growth parameters considered in this study on the film properties we have analyzed the high and low R_q samples using the micro photoluminescence (PL) microscopy to determine the density of stacking faults in the two films. Fig.4 shows an intensity map of the room-temperature PL peaks related to the (1,3) and (4,4) stacking faults (which have an associated peak located 0.2 and 0.44eV below the band gap, respectively [7]).

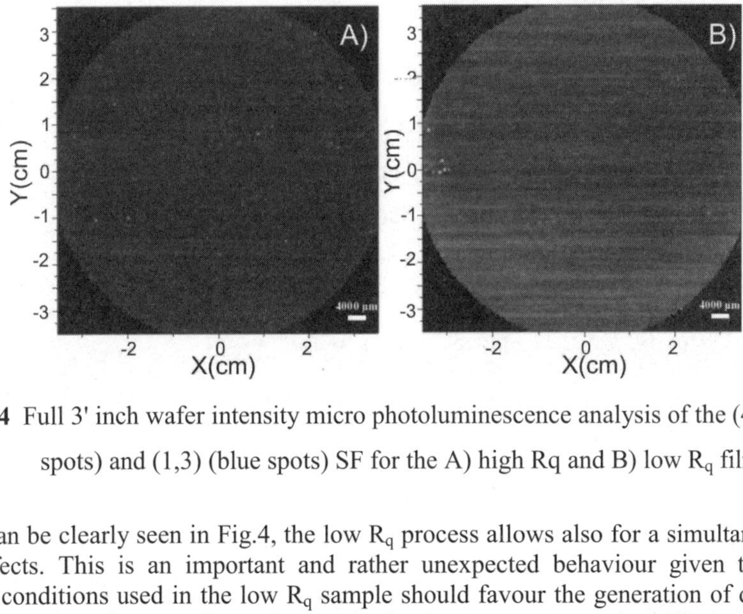

Fig.4 Full 3' inch wafer intensity micro photoluminescence analysis of the (4,4) (green spots) and (1,3) (blue spots) SF for the A) high Rq and B) low R_q film.

As can be clearly seen in Fig.4, the low R_q process allows also for a simultaneous reduction of both defects. This is an important and rather unexpected behaviour given that the out-of-equilibrium conditions used in the low R_q sample should favour the generation of defects. On the other hand, a low surface roughness is the result of narrow terraces (see Fig.1 and 2) which is, indeed, capable of hindering the generation of defects and promote homoepitaxy [8,9,10,11]. The connection between defects generation, surface morphology and growth parameters requires further studies and is currently under investigation.

Summary

In this paper we have studied the impact of growth and post-growth processes on the surface morphology of <11-20> 4° degree off, silicon terminated, 4H-SiC. We found that low temperatures and low C/Si ratios effectivelly reduce the surface roughness. This result, together with the increase of surface roughness observed after high temperature Argon annealing, suggests a connection between high surface roughness (related to a nano-ordering of the step-and-terrace morphology) and equilibrium conditions. We found also preliminary indications that the same growth parameters that lead to the reduction of the surface roughness can also promote a reduction of (1,3) and (4,4) stacking faults density.

References

[1] H. Nakagawa, S. Tanaka and I. Suemune Physical Review Letter, **91**, 226107 (2003)

[2] M. Camarda, A. La Magna, P. Fiorenza, F. Giannazzo and F. La Via Journal of Crystal Growth **310**, 971 (2008)

[3] M. Camarda, A. La Magna, A. Severino and F. La Via, Thin Solid Films **518**, S159 (2010)

[4] H.-C. Jeong and E.-D. Williams Surface Science Report **34** (1999) 171.

[5] M. Camarda, A. La Magna, and F. La Via, Journal of Computational Physics **227**, 1075 (2007).

[6] M. Camarda, A. Severino, P. Fiorenza, V. Raineri, S. Scalese, C. Bongiorno, M. Mauceri, D. Crippa, A. La Magna and F. La Via, Material Science Forum **679**, 358 (2011)

[7] M. Camarda, A. La Magna, P. Delugas and F. La Via. Applied Physics Express **4**, 025802 (2011)

[8] T. Kimoto and H. Matsunami, Journal of Applied Physics **75**, 850 (1994).

[9] M. Camarda, La Magna, A. Canino and F. La Via, Surface Science, **604**, 939 (2010)

[10] M. Camarda, A. La Magna and F. La Via, Surface Science **603**, 2226 (2009)

[11] M. Camarda, A. La Magna and F. La Via, Surface Science **605**, L67 (2011)

Materials Science Forum Vols. 717-720 (2012) pp 153-156

doi:10.4028/www.scientific.net/MSF.717-720.153

Behavior of particles in the growth reactor and their effect on silicon carbide epitaxial growth

Tawhid Rana[a*], Haizheng Song[b], M.V.S. Chandrashekhar[c] and Tangali Sudarshan[d]

[1]University of South Carolina, SC-29208, USA
[a]rana@email.sc.edu, [b]songha@cec.sc.edu, [c]chandra@cec.sc.edu,
[d]sudarsha@cec.sc.edu, *corresponding author

Keywords: Parasitic deposition, gas phase nucleation, particle formation, silicon droplets.

Abstract. Formation of particles and their effect on SiC epitaxial growth in the CVD reactor is investigated. Particle induced defects in the epilayer at different gas decomposition conditions are discussed. A higher number of pits with larger diameters are observed in the epilayer for conditions where gases decompose later in the gas injector tube (i.e. nearer to the substrate). On the other hand, the number and size of these pits reduce for the condition where gas decomposes earlier in the tube. To investigate the effect of particles during the growth, various particles with different size, shape and compositions are intentionally placed on the substrate surface before epitaxial films are grown. Samples are mapped and compared at similar locations in the pre-growth, post growth and post-etch (by molten KOH) conditions. It is found that the nature of particle induced defects depends primarily on size and shape of particles.

Introduction.

One of the greatest challenges in SiC epitaxial growth is to maintain an unadulterated, 'clean' environment by restricting the formation of solid or liquid particles during growth. These particles nucleate mostly in the gas phase as Si droplets and deposit on the gas injector wall as Si and 3C-SiC materials and also get carried to the growth surface by H_2 gas flow. In silicon carbide CVD, the surface is usually kept at around ~1500°C. The gas injector provides a wide temperature zone (~450°C to ~1400°C) suitable for growth of silicon and other materials in the reactor. This favorable condition for Si aerosol formation and particle generation due to the parasitic depositions do not only reduce the growth rate due to precursor loss but also degrade the crystal quality by inducing various defects.

Silicon clusters or particles attaching to the surface during crystal growth may affect the growth adversely [1].Silicon is reported to be used as seed for 3C CVD growth [2] and hence Si material as a form of particles on the SiC substrate during growth is expected to initiate 3C nucleation and generate defects. 3C-SiC deposited at the high temperature zone in the injector tube can also be transported to the growth surface. Use of a hydrocarbon gas (e.g. propane) in the reactor and presence of graphite parts ensure conditions to generate C or C rich particles in the injector tube which may also be carried to the epi surface and introduce defects.

Experimental.

A special technique is applied to study parasitic depositions in the gas injector tube for different growth conditions by using a gas injector tube that is axially split into two halves, which can be assembled for growth (Fig. 1a, b and c, representing one half of each injector tubes for three different temperature profiles). The location of this tube in the reactor is shown as pq in Fig. 1d. The growth surface is inductively heated at 1550°C by a fixed RF coil, whereas the temperature profile within the injector varies, controlled by shaping the eddy current path by introducing some vertical and horizontal gaps in the hotwall. Epilayers are grown (on 8° off cut, 4H-SiC, Si face substrates) for three different gas decomposition conditions evidenced by parasitic depsositions (Fig. 1a, b and c) starting at different places in the tube for three different temperature profiles obtained using the aforesaid method. To investigate the effect of particle on epi growth silicon particles are placed artificially on the substrate and mapped by Nomarski microscope. Epilayer is grown on the substrate and mapped. Finally, molten KOH etching technique is applied to delineate the defects and the etched epilayer is mapped for the same locations.

Results and discussion.

SEM and EDX study reveal that depositions on the injector tube wall are mainly composed of loosely bound silicon particles of different shapes and sizes (Fig. 2; the triangle and circle indicate

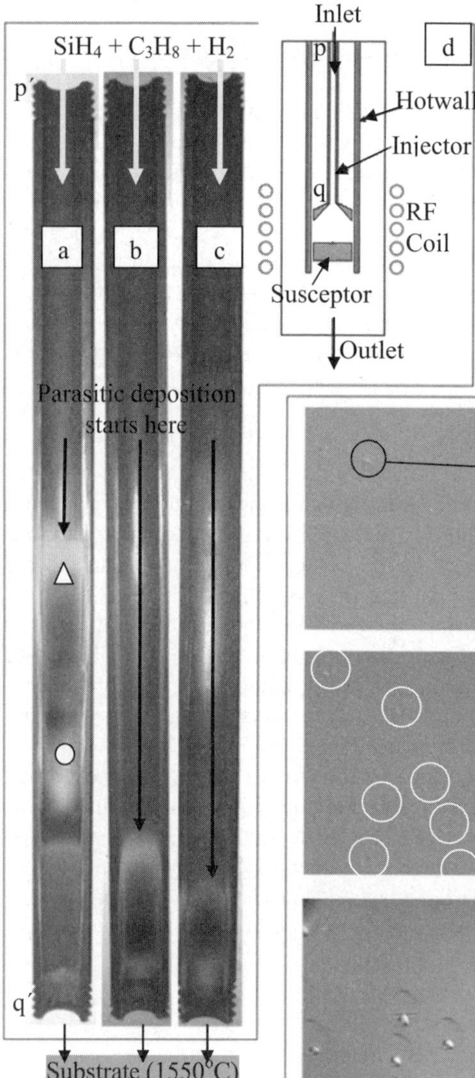

Fig. 1: Axially split gas injector tube (p´q´) showing parasitic depositions for: (a) gas decomposes early in the tube (b) & (c) gas decomposes later in the tube (d) reactor geometry showing gas injector tube (pq)

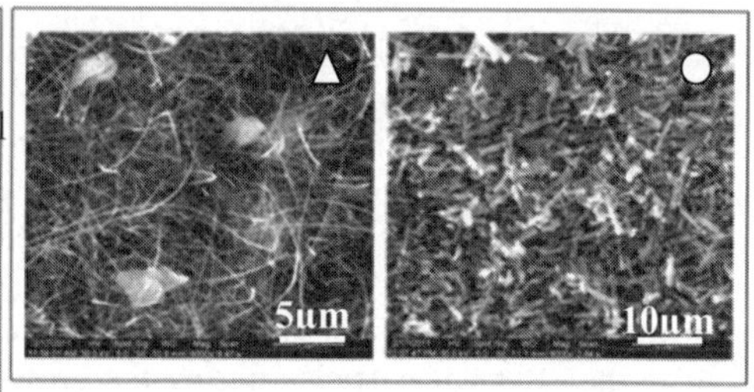

Fig. 2: SEM images of silicon deposition in the injector tube shown in Fig. 1a.

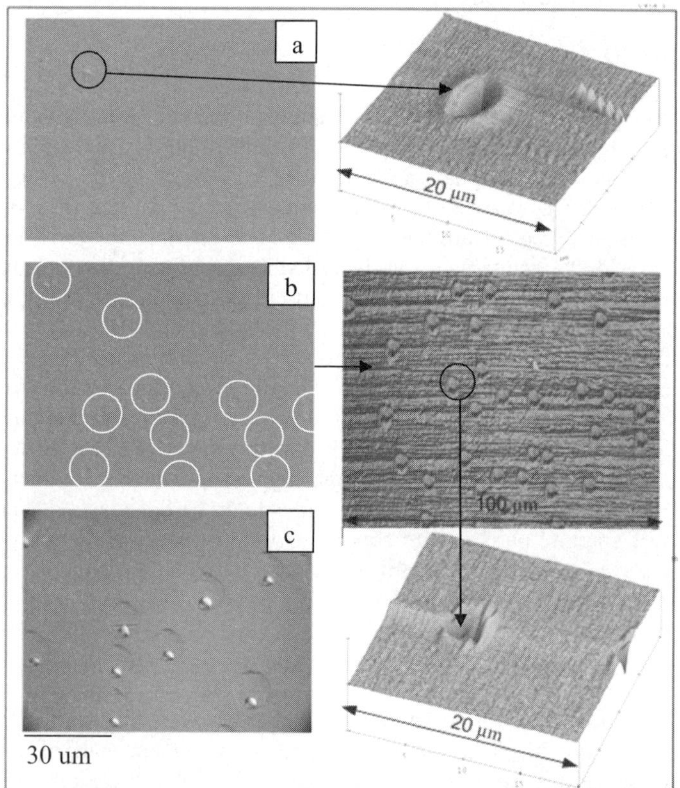

Fig. 3: Epitaxial films grown for conditions in Fig. 1a, b and c. (a) Nomarski and AFM image of a uniform shaped pit where gas decomposes early as Fig. 1a. (b) Nomarski and AFM images of non uniform shaped pits observed in condition Fig. 1b. (c) large size pits with core for growth at condition Fig. 1c

the respective locations in the tube (Fig. 1a)) from where particles can be dislodged and transported to the growth surface during gas flow.

Crystals are grown, analyzed and compared for different gas decomposition conditions (Fig. 1a, b, c), where gases start decomposing at three distinct positions in the injector tube. A 50% increase of growth rate is observed for the case of late gas decomposition in Fig. 1c compared to early gas decomposition in Fig. 1a. The reduced growth rate for gas decomposition condition in Fig. 1a is due to the excessive precursor losses in the form of parasitic deposition. Different size and shape of pits are observed on the epi surface for the above conditions (Fig. 3). Even though the highest growth rate is achieved for Fig. 1c compared to Fig. 1a and 1b, growth pits with large diameter are dominant for this condition (Fig. 3c). The size and density of these pits reduce for the cases where gas decomposes earlier in the injector (Fig. 1a). We can infer that particles travel a longer distance in the hot region and absorb enough energy for the case of Fig.1a to evaporate and disintegrate before reaching the epi surface according to [3].
Effect of particles on the epi growth is investigated by introducing silicon particles artificially on the substrate before growth. The size of the particles or cluster of particles in Fig. 4a vary from ~1μm to 10μm. Epitaxial films are grown with ~3-4μm thickness on this substrate (similarly, growths are also carried out on substrates surface with C and SiC particles). After the growth, pits/bumps with different sizes are observed at the corresponding locations where particles were present (Fig. 4a and Fig. 4b for comparison).

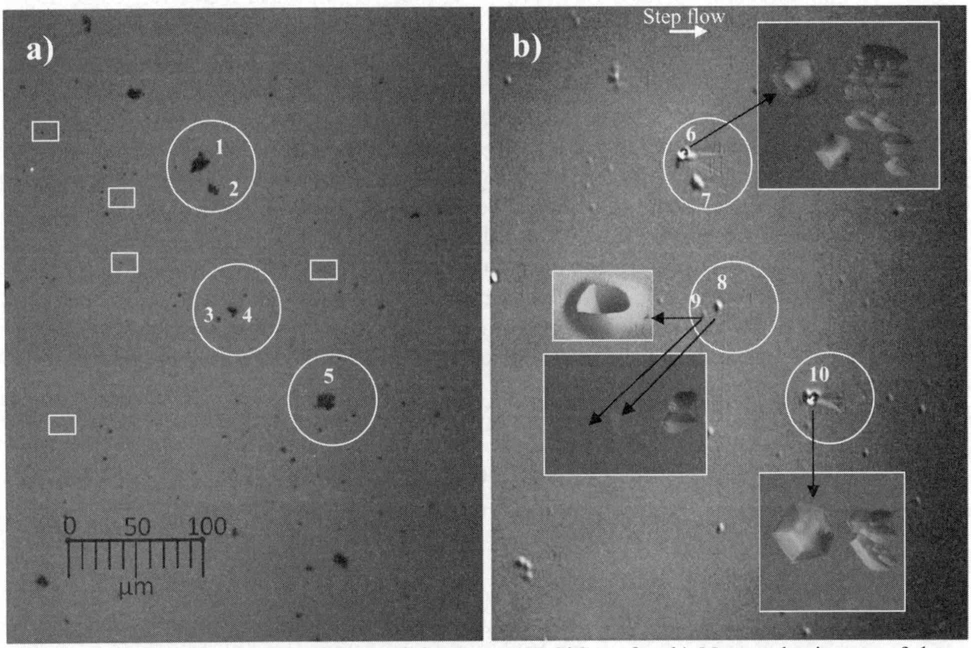

Fig. 4: a) Nomarsky images of Si particles on a 4H-SiC wafer. b) Nomarsky image of the epilaery grown for the same place and magnified images of the positions corresponding the the circles (insets) after etching. AFM image is shown for a pit (#9) found in the epilayer.

Smoother shaped, smaller pits are observed for the smaller particles (<~3μm, Fig. 4a, #3 and alike). Etching by molten KOH is performed for defect delineation and it reveals that these smaller pits do not generate any crystal defects such as TSD or stacking faults but just leave a morphological defect (Fig, 4b, #9). On the other hand, relatively larger pits (usually crater shaped) are observed for the larger particles (~10μm) shown in (Fig. 4b, #6, and #10 corresponding to Fig. 4a, #1 and #5) and

TSDs are revealed at the pit locations after KOH etching. Moreover, pits generated from larger particles (~10um), associated with TSDs, almost always generate stacking faults (SF) which appear as pairs of BPDs originating from the pit (Fig. 4, #6 and #10 insets). Delineation of SF by KOH etching is discussed elsewhere [4]. Fig. 4b shows that pits (Fig. 4b, #6 and #10) generate SF clusters for the largest particles (Fig. 4a, #1, #5). These large particles consist of multiple sharp edges. Medium sized (~5um) particles corresponding to (Fig.4a #2 and #4) generate only a pair of BPDs or a single SFs with or without TSD associated with it (Fig. 4b, #7 and #8). Particles with comparatively very small sizes (<~1um, marked by rectangles in Fig. 4a), often leave no trace in the epilayer (Fig. 4b). However, it is not confirmed whether these very small particles are completely embedded in the crystal (and disappear) or vanish by hydrogen flow prior to the growth.

We can conclude that new crystal and morphological defects, not related to the substrate, are generated due to the intentionally introduced particles with different shapes and sizes. The density of newly generated defects in the epi will depend on the particle density in the substrate. Even though these defects are generated due to artificially introduced particles, a similar condition can also exist for regular epi growth. Firstly, due to the particles introduced by the exfoliation of the parasitic deposition in the reactor and carried to the growth surface by the high gas flow. Secondly, particles may also be generated by the formation of clusters in the gas phase due to supersaturation (especially severe at high growth rate, i.e. high precursor concentration). These in-grown particles can also generate new defects during epi growth similar to the artificially introduced particles discussed earlier. Even though we only report the influence of Si particles in this paper, similar experiments also have been conducted for C and SiC particle. We find that the above discussion is applicable to them as well; these results will be discussed in detail in a future publication.

Summary

Particle formation in the reactor at various growth conditions and their effect on epi growth are discussed. Growth pits are observed in a higher number for the condition where gas decomposes later in the injector tube compared to early decomposition. For the early gas decomposition, even though the particles are formed inside the tube, eventually they re-evaporate in the gas injector tube before reaching the substrate. The effect of particles on growth is investigated- for both morphology and crystal defects by artificially placing silicon particles on the substrate before the growth. It is revealed by molten KOH etching that large size particles (~10um) with irregular contour and sharp edges has the highest chance to induce TSDs and stacking faults in the over grown epilayer.

Acknowledgement

This work was supported by the ONR, Grant no. # N000141010530. The authors thank the program manager Dr. H. Scott Coombe for his support of this research.

References

[1] K. Fujihira, T. Kimoto, H. Matsunami, Growth and characterization of 4H-SiC in vertical hot-wall chemical vapor deposition, Journal of crystal growth 255 (2003) 136

[2] Zorman, Christian A.; Fleischman, Aaron J.; Dewa, Andrew S.; Mehregany, Mehran; Jacob, Chacko; Nishino, Shigehiro; Pirouz, Pirouz, Epitaxial growth of 3C–SiC films on 4 in. diam (100) silicon wafers by atmospheric pressure chemical vapor deposition, J. of Appl. Phys. 78 (1995) 5136

[3] R. Rupp, Yu.N. Makarov, H. Behner, A. Wiedenhofer, Silicon Carbide Epitaxy in a Vertical CVD Reactor: Experimental Results and Numerical Process Simulation, phys. stat. sol. (b) 202 (1997) 281

[4] S. Izumi, H. Tsuchida, I. Kamata, and T. Tawara, Structural analysis and reduction of in-grown stacking faults in 4H-SiC epilayers, App. Phy. Letters 86, (2005) 202108

Materials Science Forum Vols. 717-720 (2012) pp 157-160
© (2012) Trans Tech Publications, Switzerland
doi:10.4028/www.scientific.net/MSF.717-720.157

Influence of Growth Mechanism on Carrier Lifetime in on-axis Homoepitaxial Layers of 4H-SiC

J. Hassan[1,a*], L. Lilja[1,b], I.D. Booker[1,c], J.P. Bergman[1,d] and E. Janzén[1,e]

[1]Department of Physics, Chemistry and Biology, Linköping University, SE-581 83 Linköping, Sweden

[a]jawul@ifm.liu.se, [b]louca@ifm.liu.se, [c]ianbo@ifm.liu.se, [d]peber@ifm.liu.se, [e]erj@ifm.liu.se

Keywords: on-axis, chemical vapor deposition, growth mechanism, carrier lifetime

Abstract. In this report we present homoepitaxial growth of 4H-SiC on Si-face, nominally on-axis substrates with diameters up to 76 mm in a hot-wall chemical vapor deposition reactor. A comparatively low carrier lifetime has been observed in these layers; local variations in carrier lifetime are different from standard epilayers on off-cut substrates. The properties of the layers were studied with focus on charge carrier lifetime and its correlation with starting growth conditions, inhomogeneities of surface morphology and different growth mechanisms.

Introduction

On-axis homoepitaxial layers of 4H-SiC are of prime interest for high power bipolar electronic devices, mainly due to the fact that basal plane dislocations (BPDs) in the substrate cannot replicate into the epilayer and therefore, the forward voltage degradation phenomenon can be controlled. Also, with the increasing diameter of substrates the epitaxial growth technology is being pushed towards lower off-angle substrates to avoid material losses during wafer slicing of on-axis grown boules. Therefore, it is important to develop the growth process for on-axis or near on-axis epitaxy.

We have previously shown and further developed on-axis homoepitaxy on Si-face, 4H-SiC substrates [1]. The layers have been shown to be of very high quality, free of 3C-SiC inclusions, BPDs, typical epi-related surface morphological defects and stacking faults [1]. However, there are a few issues that need to be resolved before the on-axis homoepitaxial growth process can be effectively used for high power device applications. First, the surface morphology is uneven [1-3], which results from different growth mechanisms active in the same layer leading to high surface roughness in the range of micro-meters. Second, there is a relatively low charge carrier lifetime in on-axis epilayers. A high carrier lifetime is required to achieve certain conductivity modulation in high power bipolar electronic devices.

In this report we present on-axis homoepitaxial growth of thick epilayers suitable for very high power bipolar devices and characteristics of the layers with emphasis on carrier lifetime and its correlation with different growth mechanisms.

Experimental

The substrates used in this study were up to 76 mm diameter 4H-SiC, Si-face polished and nominally on-axis (0.08°). The growth was performed in a hot-wall chemical vapor deposition reactor using standard gas phase chemistry (H_2+Ar+SiH_4+C_3H_8) with C/Si ratio 1. Nitrogen gas was used as n-type dopant. The growth temperature and pressure were 1600 $^\circ$C and 100 mbar, respectively. Homoepitaxial layers were grown to a thickness of 100 μm with a growth rate 7 μm/h and intentional n-type doping of 5×10^{14} cm^{-3} - 1×10^{15} cm^{-3}. All the samples were in-situ etched in Si-rich conditions prior to the growth [4]. Surface morphology and roughness were studied using Nomarski optical microscope and atomic force microscope (AFM) in tapping mode, respectively. Doping concentration was measured by CV using a mercury probe station. Charge carrier lifetime was measured using time resolved photoluminescence (TRPL) decay of the near band edge emission of 4H-SiC at room temperature and full wafer mapping was performed with a spatial resolution of 200 μm. Selected area lifetime mapping was done using a high spatial resolution of

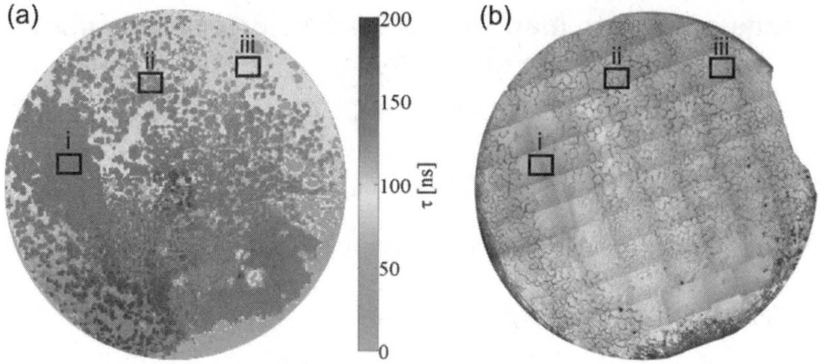

Fig. 1 (a) Time resolved photoluminescence map of a 35 μm thick epilayer grown on a 2-inch wafer. Regions with different carrier lifetime are marked as i, ii and iii. (b) Optical image of the same wafer, where regions with different surface morphology corresponding to different lifetime are marked.

50 μm. The carriers were excited optically using a diode-pumped, frequency-tripled Nd:YAG laser with emission at 355 nm. The 4H polytype identification in the epilayers were made during carrier lifetime mapping and by PL mapping of the wafers at liquid nitrogen temperature under defocused UV laser light. In order to reveal the defects intersecting the epilayer surface, the samples were etched in molten KOH at 480 °C for 4 minutes.

Results and Discussion

Lower carrier lifetimes of about 100-200 ns are observed in 100 μm thick on-axis epilayers compared to about 500 ns in our epilayers grown on 8° off-cut 4H-SiC substrates with comparable thickness. Even though the growth conditions for on- and off-axis epitaxy are comparable, the in-situ etching and growth mechanisms are completely different and could be the reasons for the comparatively low carrier lifetimes in on-axis epilayers. The in-situ etching under Si-rich conditions leads to different gas-phase chemistry (low C/Si ratio) at the beginning of the growth and may enhance point defect incorporation in the interface layers. However, a low carrier lifetime in such thick (100 μm) on-axis epilayer cannot be explained through high recombination rate at the interface layers.

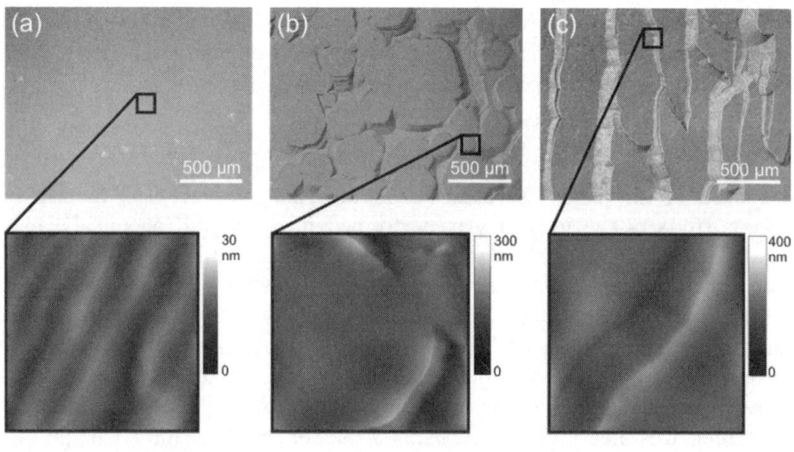

Fig. 2 Optical images taken from the regions marked in Fig.1 showing (a) optically mirror like and (b) columned morphology, followed by spiral growth and (c) step-bunched surface followed by step-flow growth with step height of 300 nm. All the AFM images are (100x100) μm^2.

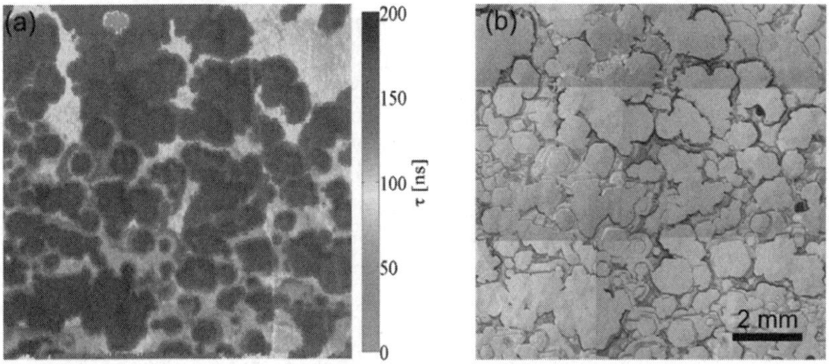

Fig. 3 High spatial resolution carrier lifetime map (a) and optical image of the same region (b), a one-to-one correspondence between locally enhanced lifetime and spiral-mode grown hillocks can clearly be seen.

Using TRPL mapping, we have observed local variations of carrier lifetime in epilayers grown on off-cut substrates in a reactor with non-rotating wafer geometry. The reduction in carrier lifetime was shown to be related to structural defects in the epilayer and variations in the growth conditions along the gas-flow direction [5]. The defects with more pronounced effect on carrier lifetime were either replicated from the substrate like threading screw dislocations and grain boundaries or created during epitaxial growth like the carrot defects and stacking faults. A radial distribution of carrier lifetime, along with defect-related reduction in carrier lifetime, has been observed in the epilayers grown on off-cut substrates in reactor with rotating wafer geometry [6]. Contrasting with the two distributions above, in the case of on-axis epitaxial layers grown in the reactor with non-rotating wafer geometry, a different carrier lifetime distribution has been observed. Along with structural defects related reduction in carrier lifetime, circular regions with relatively high carrier lifetime (150-200 ns) have been observed (Fig. 1a). Comparison of carrier lifetime map with the optical image taken from the same wafer (Fig. 1b) shows an interesting correlation of enhanced carrier lifetime regions with the columnar growth regions and the regions that appear very smooth (area marked as i and ii respectively, in Fig. 1a and b) while a low carrier lifetime (70-80 ns) was found in the regions covered with large steps (area marked as iii in Fig. 1a and b). Optical microscopy analysis and AFM confirms spiral growth in regions with high carrier lifetime (Fig. 2a and b) and step-flow growth in regions with relatively low carrier lifetime (Fig. 2c).

A relatively inhomogeneous surface morphology observed in on-axis epilayers (Fig. 1a) corresponds to different growth mechanisms active in different regions depending on the local off-cut in the substrate [1]. A large local off-cut leads to steps on the surface and therefore, growth is followed by the step-flow. Heavy step bunching is observed in these regions with the step height depending on the local off-cut and epilayer thickness (Fig. 2c). Surface steps as large as 2-3 μm has been observed in 100 μm thick epilayers, however, these steps are not abrupt and extend over a length of more than 100 μm. Also, the terrace width in such regions exceeds 500 μm and under optical microscope the terraces appear mirror like. AFM shows surface coverage with 0.5 to 1 nm steps. Spiral growth has been observed only in the regions where off-cut is less than 0.1°. The surface morphology in such regions appears either extremely smooth with surface roughness (RMS) below 0.5 nm (Fig. 2a) or columned (Fig. 2b) with height and width comparable to the large steps and corresponding terraces, however, extremely smooth surface is observed on the hillock top. No step-bunching is observed in such regions except for the boundaries of hillocks.

Fig. 3a and b are zoom-ins of a selected region of the carrier lifetime map and optical image of the same area, respectively. A one-to-one correspondence is clearly seen between hillocks and regions with relatively enhanced carrier lifetime, however, the areas in between neighboring hillocks show low carrier lifetime. A high surface roughness is known to enhance the surface recombination rate and therefore reduces the apparent carrier lifetime. One might speculate this to

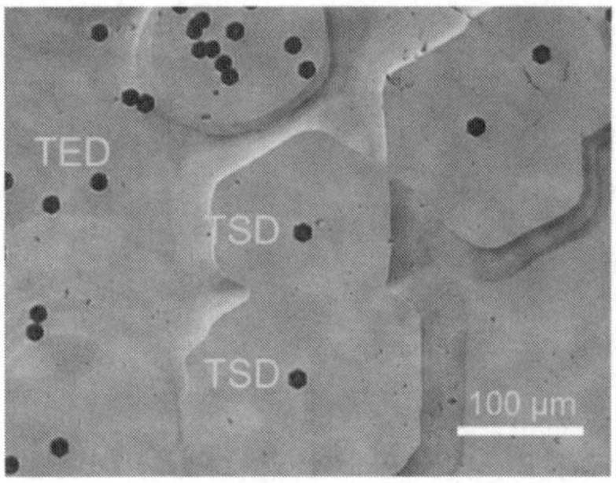

Fig. 4 Optical image taken after etching in KOH at 480 °C for 4 minutes.

be the cause for the observed relatively low carrier lifetime (70-80 ns) in the regions dominated by step-flow growth with relatively high surface roughness. An extremely smooth hillock top could be the reason for high lifetime in these regions. In order to test this, the top 3-4 μm of the epilayer was removed through chemo-mechanical polishing and homogeneous surface morphology was obtained. Comparison of high spatial resolution TRPL map taken before and after polishing show similar carrier lifetime values and features with locally enhanced carrier lifetime in the regions grown by the spiral growth. This clearly shows that the high carrier lifetime in the regions where spiral growth mechanism is dominant and the reduced values in regions dominated by step-flow growth are not related to the observed surface roughness but are properties of the bulk. No additional structural defects have been observed in the regions with reduced lifetime (Fig. 4). Also, LTPL does not show any additional peaks related to stacking faults or residual impurities. Therefore, we believe that the local carrier lifetime variations in on-axis epilayers could be related to intrinsic point defects. Further, the point defects incorporation efficiency is related to different growth mechanisms. The spiral growth leads to low defect density compared to the step-flow growth and results in comparatively high local carrier lifetime.

Summary

Up to 100 μm thick homoepitaxial layers have been grown on Si-face of on-axis 4H-SiC substrates with diameters up to 76 mm. Step height in the regions dominated by step-flow growth and hillock height in the regions dominated by spiral growth increase with the thickness of the epilayer and reach up to 2-3 μm. In-homogeneous distribution of lifetime in on-axis epilayers is related to the growth mechanism with relatively high carrier lifetime in regions dominated by spiral growth and low lifetime in the regions dominated by step flow growth. No additional structural defects are observed in the regions with low lifetime. The step-flow growth results in the incorporation of more point defects compared to spiral growth and leads to local variations in lifetime.

Acknowledgements

This work has benefitted from support by the European and Swedish Research Councils (VR) as well as the Swedish Government Strategic Research Area Grant in Materials Science.

References

[1] J. Hassan, J. P. Bergman, A. Henry, and E. Janzen, Journal of Crystal Growth 310 (2008) 4424.
[2] K. Kosciewicz, W. Strupinski, D. Teklinska, K. Mazur, M. Tokarczyk, G. Kowalski, and A. Olszyna, Materials Science Forum 679-680 (2011) 95.
[3] S. Leone, H. Pedersen, A. Henry, S. Rao, O. Kordina, and E. Janzen, Materials Science Forum 615-617 (2009) 93.
[4] J. Hassan, J. P. Bergman, A. Henry, and E. Janzen, Journal of Crystal Growth 310 (2008) 4430.
[5] J. Hassan and J. P. Bergman, Journal of Applied Physics 105 (2009) 123518.
[6] I. Booker, L. Lilja, J. Hassan, and J. P. Bergman, manuscript in this conference.

Materials Science Forum Vols. 717-720 (2012) pp 161-164

doi:10.4028/www.scientific.net/MSF.717-720.161

The Effect of Growth Conditions on Carrier Lifetime in n-type 4H-SiC Epitaxial Layers

L. Lilja[a,*], J. Hassan[1,b], I.D. Booker[1,c], J.P. Bergman[1,d] and E. Janzén[1,e]

[1]Department of Physics, Chemistry and Biology, Linköping University,

SE-581 83 Linköping, Sweden

[a]louca@ifm.liu.se, [b]jawul@ifm.liu.se, [c]ianbo@ifm.liu.se, [d]peber@ifm.liu.se, [e]erija@ifm.liu.se

Keywords: Epitaxial Growth, Carrier Lifetime, Atomic Force Microscopy, Photoluminescence

Abstract. Carrier lifetime has been studied as a function of C/Si ratio and growth rate during epitaxial growth of n-type 4H-SiC using horizontal hot-wall CVD. Effort has been put on keeping all growth parameters constant with the exception of the parameter that is intended to vary. The carrier lifetime is found to decrease with increasing growth rate and the highest carrier lifetime is found for a C/Si ratio of 1. The surface roughness was correlated with epitaxial growth conditions with AFM analysis.

Introduction

The physical properties of silicon carbide (SiC), such as its wide bandgap, high thermal conductivity and high electric breakdown field strength, makes it promising for high power devices. A long carrier lifetime is necessary for high power devices in order to minimize the forward voltage drop. There have recently been several studies related to the post-growth processing in order to increase the lifetime. This has been done either by surface oxidation [1] or by implantation followed by annealing [2], where the diffusion of mobile C interstitials is assumed to neutralize the $Z_{1/2}$ electron trap, which is proposed as the main lifetime limiting defect [3] and is assumed to be related to the C-vacancy.

In this study, we have instead investigated how growth conditions like growth rate and C/Si ratio can influence carrier lifetime through the incorporation of C-vacancies or other lifetime limiting defects. Furthermore, the influence of growth conditions on surface morphology was also studied.

Experimental

Epitaxial layers of n-type 4H-SiC were grown on 4H-SiC substrates with 8° off-cut towards $[11\bar{2}0]$ in a horizontal hot-wall CVD reactor. The precursor gases were silane (SiH_4) and propane (C_3H_8), together with hydrogen as the carrier gas and nitrogen as the n-type dopant. The growth pressure and temperature were 100 mbar and 1580 °C respectively. All epilayers were grown to a thickness of about 40 μm and with n-type doping of 1×10^{15} cm^{-3}. In order to minimize the influence of variations in growth conditions along and perpendicular to the gas-flow direction, all the samples were grown at the same position inside the susceptor.

Two series of epilayers were grown, all on to pieces cut from the same three inch wafer, in order to reduce the influence of substrate quality and defect distribution. The cutting was done radially, in order to avoid the influence of radial variance in defect distribution. In the first series, the growth rate was changed by varying the precursor flow rates, while keeping all other process parameters constant (pressure, temperature, epi-thickness and C/Si ratio = 1). Growth rates varying from 4.2 μm/h to 18.0 μm/h were investigated. In the second series, the C/Si ratio was varied from

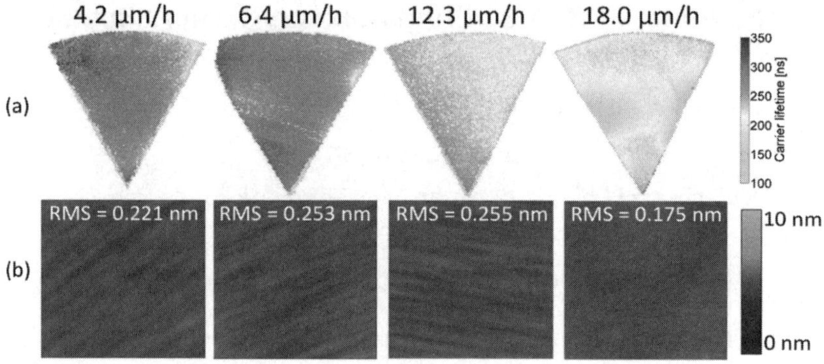

Fig. 1: (a) Carrier lifetime mappings of four ~40 µm thick epitaxial layers grown with different growth rates on 1/8 pieces cut from the same three inch wafer. (b) (10 x 10) μm^2 AFM images taken from the middle of each sample.

0.7 to 1.5, while keeping all other process parameters constant (growth rate, pressure, temperature and epi-thickness). The carrier lifetime were analyzed by mapping the photoluminescence decay of the bandgap emission after excitation with above-bandgap pulsed laser. Optical microscopy with Nomarski diffraction interference contrast was used to study the surface morphology and distribution of surface morphological defects. Surface topography and roughness was analyzed with Dimension 3100 atomic force microscope (AFM) in tapping mode. The thickness of the grown epilayers was measured with Fourier transform infrared reflectance spectroscopy. Low temperature photoluminescence (LTPL) was used to observe the presence of residual impurities in the epilayers. The PL excited by 351 nm Ar+ laser line was dispersed by a single monochromator and was detected by a liquid nitrogen cooled UV sensitive CCD camera to record the spectra.

Results and discussion

Carrier lifetime dependence on epitaxial growth rate: In the first series, the growth rate of the epitaxial layers was increased from 4.2 µm/h to 18.0 µm/h by changing the flow rate of source gases at a fixed C/Si ratio of 1 at the susceptor inlet. The samples grown with a lower growth rate of 4.2 - 6.4 µm/h, showed a significant higher measured carrier lifetime as seen in Fig. 1a. The samples grown at relatively higher growth rates, but otherwise using similar growth conditions, had significantly lower lifetime. An increase in carrier lifetime has been observed with the increasing thickness of the epilayer [4]. The undesired growth rate gradient present along the gas-flow direction may cause small variations in the epilayer thickness. Therefore, in order to reduce the influence of epilayer thickness variations along the gas-flow direction on measured carrier lifetime, the average measured carrier lifetimes over the samples are shown in Fig. 2. We believe that the increased growth rate enhances the possibility for the incorporation of impurities and intrinsic defects such as the $Z_{1/2}$, which could influence the measured lifetime.

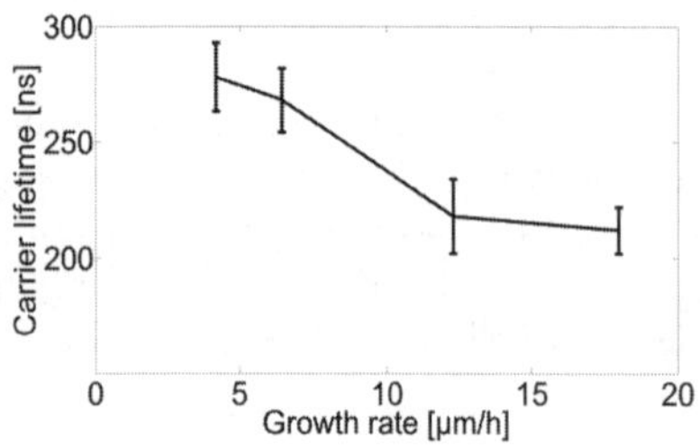

Fig. 2: Plot of measured average carrier lifetime dependence on epitaxial growth rate.

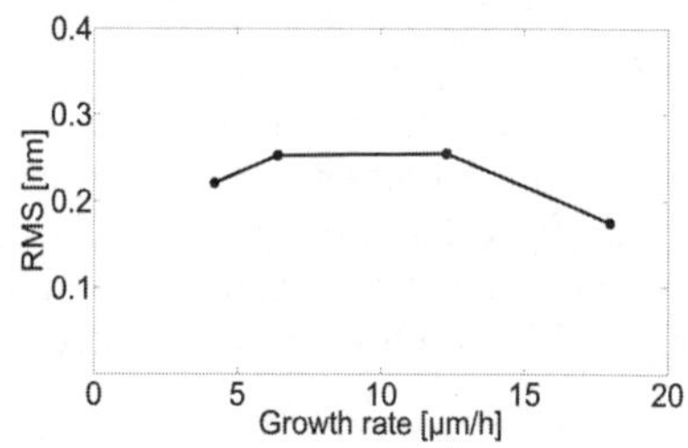

Fig. 3: Plot of RMS dependence on epitaxial growth rate.

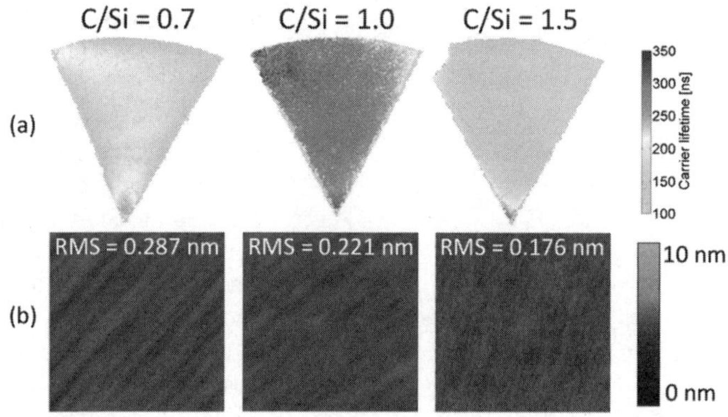

Fig. 4: (a) Distribution of measured carrier lifetimes in three ~40 μm thick epitaxial layers grown with different C/Si ratio on 1/8 pieces cut from the same three inch wafer. (b) (10 x 10) μm^2 AFM images for C/Si ratios of 0.7, 1.0 and 1.5 respectively.

The surface roughness (defined as the root mean square, RMS) usually increases with increasing growth rate and significantly lower the carrier lifetime through high surface recombination rate. No big difference in surface roughness is observed in epilayers grown in the growth rate range of 4.2 to 18 μm/h (Fig. 1b and Fig. 3). However, a few Si-droplets were observed on the surface of the epilayer grown with the highest growth rate (18 μm/h), as expected. Also, no additional structural defects such as stacking faults were observed in the epilayers except for a few carrot defects, that are present in all the samples. Therefore, the observed decrease in the carrier lifetime with increasing growth rate is purely related to different incorporation rates of impurities.

Carrier lifetime dependence on C/Si ratio: In the second series, where the C/Si ratio was varied from 0.7 to 1.5, the growth rate was maintained at 4.2 μm/h since the highest lifetime was observed at this growth rate. The maximum carrier lifetime was observed for a C/Si ratio of 1, as shown in Fig. 4a. A slightly lower lifetime is observed in layer grown at a C/Si ratio of 0.7, however, at a C/Si ratio of 1.5, carrier lifetime reduced considerably. The average measured carrier lifetimes are shown in Fig. 5. If the $Z_{1/2}$ defect incorporation is related to the C supply at the growth surface, we would expect a decreased defect concentration and a positive effect on carrier lifetime. However, we observed a reduced carrier lifetime in the layers with increasing C/Si ratio, which indicates that C incorporation in the crystal during epitaxial growth is also controlled by other factors.

In Fig. 4b, AFM images are shown for the different C/Si ratios. From roughness analysis with AFM, a small decrease in surface roughness (RMS) with increasing C/Si ratio could be observed (Fig. 6). However, the change in surface roughness is within the same order of magnitude and should not affect the carrier lifetime measurements. It is worth mentioning that the RMS is decreasing with increasing C/Si ratio for the epitaxial growth on 8° off-cut substrates, as compared

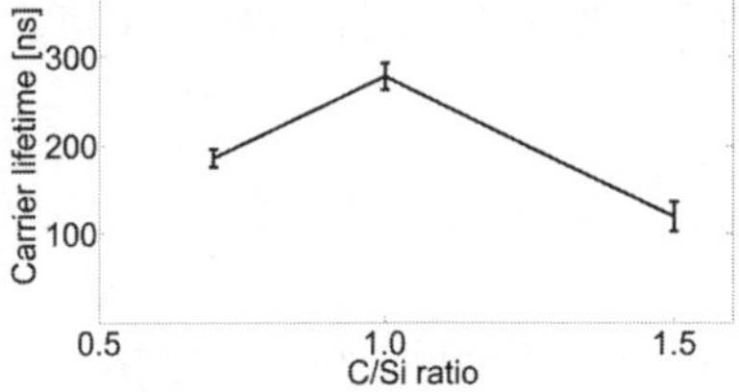

Fig. 5: Plot of measured average carrier lifetime dependence on C/Si ratio.

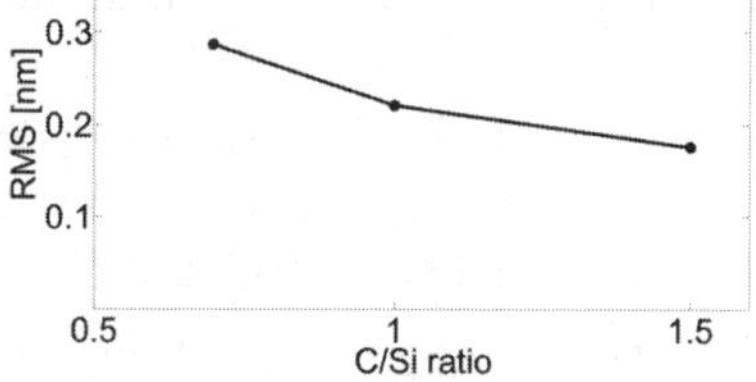

Fig. 6: Plot of RMS dependence on C/Si ratio.

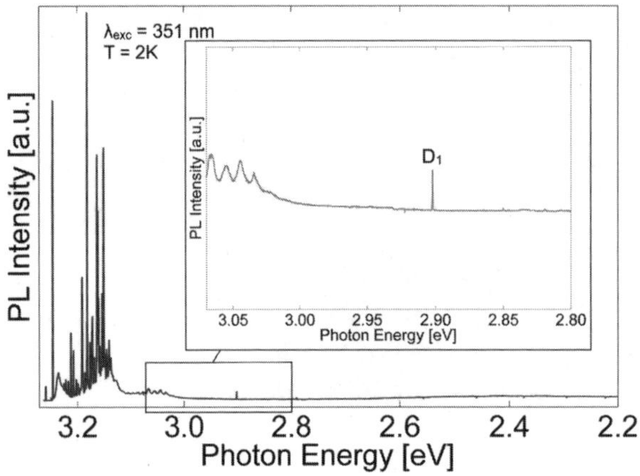

Fig. 7: Low temperature photoluminescence spectrum taken at 2K for a 40 μm thick 4H-SiC epilayer grown with a growth rate of 4.2 μm/h and a C/Si ratio of 1.0.

to the increasing RMS with increasing C/Si ratio on 4° off-cut substrates [5, 6]. The main difference is that macro-steps are not observed on 8° off-axis 4H-SiC. LTPL spectra taken from the samples grown at different growth rates and different C/Si ratios are dominated by near bandgap emissions related to the free exciton and the nitrogen bound exciton (Fig. 7). Weak emission related to the aluminum bound exciton was also observed as well as the so called D_1 defect. The relative intensity of the D_1 emission line was found to decrease with increasing C/Si ratio. Also a weak emission related to titanium was observed in the sample grown with the highest epitaxial growth rate, but was not seen in the other samples grown at relatively low growth rate or with different C/Si ratio.

Summary

A detailed study of the influence of growth rate and C/Si ratio on the measured carrier lifetimes has been presented. A clear dependence of carrier lifetime on growth rate and C/Si ratio has been observed. Any deviation of the C/Si ratio from 1 results in a negative affect on carrier lifetime, however, a C/Si ratio higher than 1 show more pronounced reduction in lifetime. Growth rate depencence of carrier lifetime shows a relatively weak influence on lifetime. Nevertheless, about 30 % reduction in carrier lifetime has been observed in the growth rate range of 4.2 μm/h to 18 μm/h. No big influence of growth rate and C/Si ratio was observed on the surface roughness and epitaxial defects. Therefore, we believe that the observed difference in carrier lifetime is related to the incorporation of intrinsic defects at high growth rates and C/Si ratios other than 1.

Acknowledgements

The Swedish Research Council (VR) is gratefully acknowledged for financial support.

References

[1] T. Hiyoshi, T. Kimoto, Reduction of deep levels and improvement of carrier lifetime in n-type 4H-SiC by thermal oxidation, Appl. Phys. Express 2 (2009) 041101.

[2] L. Storasta et al., Enhanced annealing of the $Z_{1/2}$ defect in 4H-SiC epilayers, J. Appl. Phys. 103 (2008) 013705.

[3] P. B. Klein, Carrier lifetime measurement in n^- 4H-SiC epilayers, J. Appl. Phys. 103 (2008) 033702.

[4] A. Henry et al., Thick epilayer for power devices, Mater. Sci. Forum. 556-557 (2007) 47-52.

[5] T. Kimoto et al., Epitaxial growth of 4H-SiC{0001} and reduction of deep levels, Superlattice. Microst. 40 (2006) 225-232.

[6] W. Chen, M.A. Capano, Growth and characterization of 4H-SiC epilayers on substrates with different off-cut angles, J. Appl. Phys. 98 (2005) 114907.

Materials Science Forum Vols. 717-720 (2012) pp 165-168
© (2012) Trans Tech Publications, Switzerland
doi:10.4028/www.scientific.net/MSF.717-720.165

Study of the lateral growth by VLS mechanism using Al-based melts on patterned SiC substrate

Jean Lorenzzi[1,a], Romain Esteve[2,b], Mihai Lazar[3,c], Dominique Tournier[3,d], Davy Carole[1,e], Gabriel Ferro[1,f*]

[1]Université de Lyon, Université Lyon 1, CNRS, UMR 5615, Laboratoire des Multimatériaux et Interfaces, 43 Bd du 11 Novembre 1918, 69622 Villeurbanne Cedex, France

[2]ACREO AB, Electrum 236, SE-164 40 Kista, Sweden

[3]Laboratoire Ampère, UMR-CNRS 5005, Bât. Léonard de Vinci, 21 avenue Jean Capelle, F 69621 Villeurbanne cedex

[a]lorenzzi@ymail.com, [b]romain.esteve@acreo.se, [c]mihai.lazar@insa-lyon.fr, [d]dominique.tournier@insa-lyon.fr, [e]davy.carole@univ-lyon1.fr, [f]gabriel.ferro@univ-lyon1.fr*

Keywords: VLS growth, Lateral enlargement, Patterned substrate, Homoepitaxy, Aluminum, p-type doping

Abstract. In this work we report on SiC epitaxial growth by vapour-liquid-solid (VLS) mechanism on on-axis 4H-SiC(0001) substrates which were previously patterned to form mesa structures. The liquid phase was set to $Al_{70}Si_{30}$. At 1100°C, it led to very high homoepitaxial lateral growth (140 µm/h) with pronounced spiral growth and in plane anisotropy of growth rate. Upon temperature increase to 1200 °C, this spiral growth was suppressed and the lateral growth was further increased up to 180 µm/h. The in-plane versus out-of-plane anisotropy of growth rate was found to be as high as 60 at this temperature and 46 at 1100°C.

Introduction

Growth on patterned substrate is very interesting for having more fundamental insight of the growth process used. Different parameters can then be evaluated after the growth such as: the lateral growth rates, the growth rate anisotropy or the evolution of crystalline defects. Such knowledge finds frequently some applications in electronic device fabrication when complex structures have to be prepared [1]. For the specific case of SiC, growth on patterned substrate was already studied using chemical vapour deposition (CVD) [2, 3], sublimation [4] and liquid phase epitaxy (LPE) [5] techniques. In the latter case, only liquid Si was used as a melt, which means temperatures higher than 1400°C. Using vapour-liquid-solid (VLS) mechanism, one can grow SiC at lower temperature [6]. Furthermore, depending on the melt composition, VLS mechanism can lead to either 3C-SiC or homoepitaxial layers on 4H-SiC seed, and one can dope either n- or p-type [7]. So, the use of patterned surface for VLS growth can give fundamental insight of the growth from the liquid phase on various unstudied issues like the growth at low temperature, the effect of the melt composition and thus of dopants. This is done in the present work using Al-based melts.

Experimental details

In this work, on-axis 4H-SiC(0001) Si-face substrates were patterned using lithography and dry etching processes. It led to the formation of flat mesas, of 2 µm height above the substrate surface and with different sizes and shapes. The VLS growth experiments were carried out at atmospheric pressure in a home-made vertical water cooled wall reactor specifically dedicated to such experiments. The samples were stuck in the centre of a 20 mm inner diameter graphite crucible, which was subsequently stuck on top of the graphite susceptor. The sources of the $Al_{70}Si_{30}$ alloys were pieces of Si (6N) and Al (4N) which were stacked inside the crucible without pre-alloying. High-purity propane (5% diluted in H_2) was used as carbon source. Initial heating was done under Ar with a ramp of 163.3 K/min and, when the growth temperature was reached, propane was introduced inside the reactor to start the VLS growth. At the end of experiment, propane was removed to stop the growth and the liquid was extracted using an aspiration system before cooling. After cooling, the remaining alloys inside the crucible were dissolved by wet chemical etching in HCl:HF:HNO3 with 1:1:1 volume ratios.

Results and discussion

The surface morphology of sample grown at 1100 °C during 30 min is shown in Figure 1. The morphology is globally rough with the random formation of large hillocks with spiral steps most probably originating from screw dislocations of the substrate. Confirmation that this layer is homoepitaxial was obtained via μ-Raman spectroscopy which did not detect any 3C-SiC signal. In fact, the Raman spectrum is typical of highly p-type doped 4H-SiC material whose formation can be easily explained here by the massive incorporation of Al atoms from the melt into the SiC lattice of the growing layer.

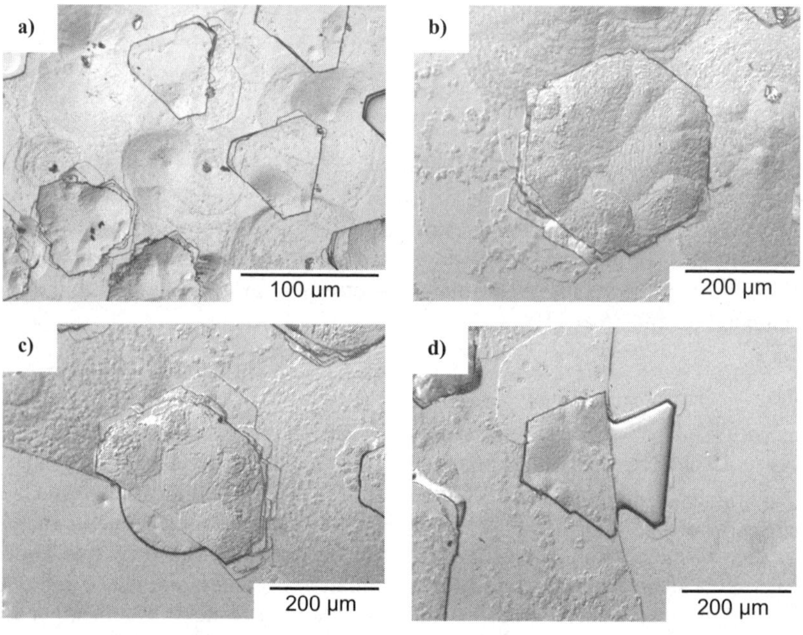

Figure 1 - *Surface morphologies after VLS growth at 1100 °C. The original patterns before growth were triangles and plain circles. The important lateral enlargement after growth is shown more obviously in c) and d) where an area without growth (substrate accidently not wetted by the liquid due to Ar bubble trapping) is crossing some patterns.*

The mesa shape has obviously evolved after growth, as seen from the various examples shown in Figure 1. For examples, the triangles (Figure 1a) and plain circles (Figure 1b) show evidence of faceting: the circles got hexagonal and the triangles got truncated at their tips (and would probably become hexagonal after longer growth run). But the most impressive effect of this lateral growth can be seen on Figures 1c and 1d showing the limit of an area where almost no growth happened accidently due to the trapping of an Ar bubble at the liquid-solid interface. Therefore, one can conveniently see in the same picture the mesas before and after growth. Obviously, the lateral enlargement is very important and can result in the merging of neighboring mesas. From these images, the lateral growth was calculated to be approximately 70 μm, which gives a lateral growth rate as high as ~ 140 μm/h. Note that some mesas were no more visible because completely embedded inside spiral growth. This was probably due to the presence of a screw dislocation within the mesa. In this case, the exact spiral-induced lateral growth rate was difficult to estimate precisely but in some case it was obviously higher than the 140 μm/h mentioned above.

For sample grown at 1200 °C, the procedure was slightly modified in order to limit substrate dissolution by the liquid at such elevated temperature [8]. In a first step, 5 min VLS growth at 1100 °C was performed, followed by a ramp of 3 min up to 1200°C (ramp rate of 33 K/min) while keeping the propane flux inside the reactor. Despite these precautions, the growth duration at 1200 °C was limited to 13 min due to a pronounced reaction between the liquid and the crucible. Fortunately, the VLS growth was sufficient for being analyzed qualitatively and quantitatively (see Figure 2). As expected, increasing temperature to 1200°C led to a considerable decrease of the spiral growth and thus of the epilayer roughness between and on top of the mesas. μ-Raman characterization confirmed the homoepitaxial nature of the deposit and the high p-type doping of the layer obtained in this way. Around the mesas, the lateral growth appears rough and it is systematically composed of two different areas: one of ~ 12 μm wide at the direct contact with the initial mesa, and a second one of ~ 40 μm at the periphery. This is probably the result of the two consecutive growth temperatures. Taking into account this hypothesis, the lateral growth rate at each temperature was calculated to be ~ 145 μm/h at 1100°C and ~180 μm/h at 1200°C. The value at 1100 °C is consistent with the one obtained for sample of Figure 1 while at higher temperature the lateral growth rate has significantly increased. However, the shape faceting is less pronounced at 1200°C than at 1100°: the triangles are less defined (Figure 2a), the circles are still more or less circular (figure 2b) and the hexagons have now rounded edges (Figure 2c).

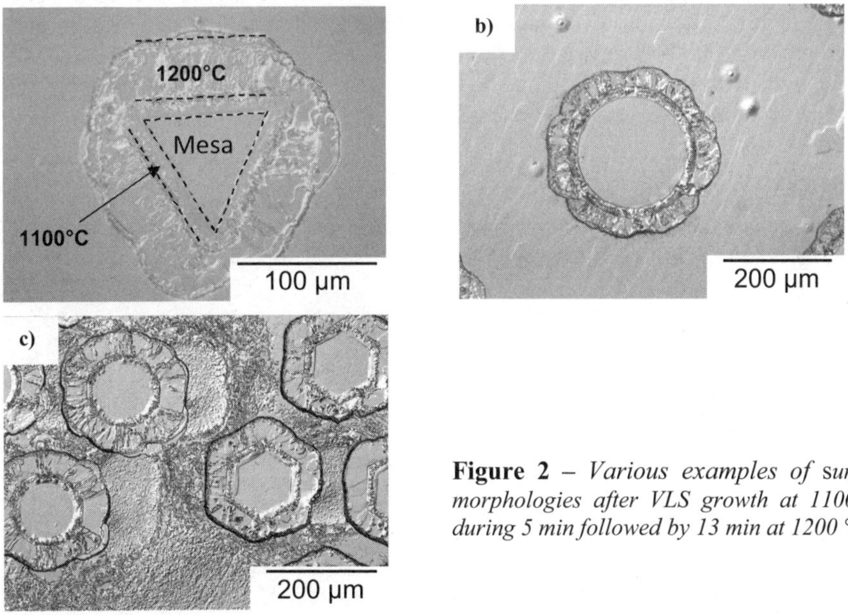

Figure 2 – *Various examples of surface morphologies after VLS growth at 1100 °C during 5 min followed by 13 min at 1200 °C.*

The fact itself that the frontier between the two layers grown at different temperatures can be seen demonstrates that the transition procedure was not perfect for continuous crystalline growth. We believe that it may come from partial dissolution of the grown layer during the temperature increase from 1100 to 1200 °C despite the presence of propane in the gas phase. This hypothesis agrees with the observation on figure 2b of small hexagonal features which are probably substrate dislocations revealed by such dissolution. The increase in C solubility within this temperature range is apparently important so that the liquid phase needs time before being saturated again by the propane flux. One solution would be to slow down the temperature ramp up rate below the actual 33 K/min or to use richer Si-melt in order to lower C solubility due to lower Al content of the liquid phase [6].

Though we did not measure the deposit thickness of each sample, according to our previous experience [9], the growth rate should be < 5 μm/h. Taking 3 μm/h as mean value, this gives an in-plane vs. normal growth rate anisotropy (also noted $\parallel / \perp$) of 46 at 1100°C and 60 at 1200°C.

Similar enhancement of the in-plane growth rate with Al-based melts was already mentioned when growing SiC using the dipping method [10]. The anisotropy of growth rate $\parallel / \perp$ was found to be ~10 for growth at 1650 °C using $Al_{40}Si_{60}$ melt. This is approximately 6 times lower than our results using $Al_{70}Si_{30}$ melt, which are even obtained at much lower temperature. Even if VLS growth promotes the growth anisotropy compare to the dipping technique, both experimental results suggest that this enhanced enlargement is an effect of Al atoms. However, the exact reason is difficult to model.

Let us now compare our values of $\parallel / \perp$ anisotropy with the ones that can be found in the literature for the epitaxial growth of SiC. When pure Si is used for SiC growth from the melt, this anisotropy is as low as ~1 on the temperature range 1450-1600°C [5]. But when CVD technique is used, the typical values of $\parallel / \perp$ anisotropy are in the 10^2 - 10^3 range at 1600°C [2]. Note that when the CVD temperature is decreased down to 1200°C, the anisotropy is apparently reduced down to values below 100, which becomes then in the same order of magnitude as the values obtained for VLS growth with Al-Si melts in this work.

Conclusion

The use of Al-based melts for the SiC epitaxial growth on patterned substrates leads to important $\parallel / \perp$ growth rate anisotropy, as high as 60 due to a very large lateral growth rate, up to 180 µm/h at 120 °C. This lateral enlargement decreases with decreasing temperature, down to 46 at 1100°C. These values are very high compare to other liquid based techniques. The grown material is homoepitaxial with a high p-type doping. Structural and electrical characterizations are under progress in order to estimate the quality of the deposit.

Acknowledgements

The authors acknowledge the European Union (Marie-Curie Research and Training network "MANSiC", Project Nb: MRTN-CT-2006-035735) for financial support.

References

[1] P.O. Vaccaro K. Fujita, IEEE J. Quantum 36 (7), (2000) 849-857

[2] T. Kimoto, H. Matsunami, J. Appl. Phys. 76 (ll) (1994) 7322-7327

[3] P.G. Neudeck, J.A. Powell et al., Mater. Sci. Forum Vols 389-393 (2002) 251-254

[4] G.V. Saparin, S.K. Obyden et al., Scanning 16 (1994) 21.

[5] A.E. Nikolaev, V.A. Ivantsov, S.V. Rendakova, M.N. Blashenkov, V.A. Dmitriev, J. Crystal Growth 166 (1996) 607-611

[6] G. Ferro, M. Soueidan, O. Kim-Hak, F. Cauwet, Y. Monteil, Mater. Sci. Forum Vols. 556-557 (2007) 41-46

[7] C. Jacquier, G. Ferro, M. Zielinski, E.K. Polychroniadis, A. Andreadou, J. Camassel, Y. Monteil, Phys. Stat. Sol. (c) 2, N°4 (2005) 1265-1268

[8] M. Soueidan, G. Ferro, B. Nsouli, F. Cauwet, L. Mollet, C. Jacquier, G. Younes, Y. Monteil, J. Crystal Growth, **293** (2006) 433-437

[9] C. Jacquier, G. Ferro, F. Cauwet, Y. Monteil, Recent Res. Devel. Crystal Growth 4 (2005) 83-103

[10] F. Mercier, O. Kim-Hak, J. Lorenzzi, J-M. Dedulle, G. Ferro, D. Chaussende, Mater. Sci. Forum **645-648** (2010) 67-70

Materials Science Forum Vols. 717-720 (2012) pp 169-172
© (2012) Trans Tech Publications, Switzerland
doi:10.4028/www.scientific.net/MSF.717-720.169

Buried Selective Growth of p-doped SiC by VLS Epitaxy

D. Carole[1,a]*, S. Berckmans[1,b], A. Vo-Ha[1,c], M. Lazar[2,d], D. Tournier[2,e], P. Brosselard[2,f], V. Soulière[1,g], L. Auvray[1,h], G. Ferro[1,i], C. Brylinski[1,j]

[1]Université de Lyon, Université Lyon 1, CNRS, UMR 5615, Laboratoire des Multimatériaux et Interfaces, 43 Bd du 11 Novembre 1918, 69622 Villeurbanne Cedex, France

[2]Laboratoire Ampère, UMR-CNRS 5005,
Bâtiment Léonard de Vinci, 21 avenue Jean Capelle, F 69621 Villeurbanne cedex

[a]davy.carole@univ-lyon1.fr*, [b]stephane.berckmans@hotmail.fr, [c]arthur.vo-ha@univ-lyon1.fr, [d]mihai.lazar@insa-lyon.fr, [e]dominique.tournier@insa-lyon.fr, [f]pierre.brosselard@insa-lyon.fr, [g]veronique.souliere@univ-lyon1.fr, [h]laurent.auvray@univ-lyon1.fr, [i]gabriel.ferro@univ-lyon1.fr, [j]christian.brylinski@univ-lyon1.fr

Keywords: VLS growth, Selective epitaxy, p-type doping, Aluminum

Abstract. Selective epitaxial growth in buried patterns was studied using the vapour-liquid-solid mechanism in Al-Si melt in order to obtain p+-doped SiC localized layers on 4H-SiC substrate. Homogeneous deposition with step bunched morphology was obtained by adding propane at room temperature before growth at 1100°C. Patterns as large as 800 µm and as narrow as 10 µm were completely filled in this way. The deposition kinetics demonstrates that the process is self limited and mainly depends on the initial amount of Si in the liquid. The deposit is highly p-type doped and the p-n junction is demonstrated.

Introduction

All high voltage semiconductor devices for power electronics require peripheral protection, most often based on localized p-type doping. In the case of SiC, diffusion of Al element is not viable, due to slow diffusion processes. That is why implantation of Al ions has been so far the most popular approach, but too many problems still remain unsolved (damage generation, high temperature annealing, dopant activation, limited doping) after more than 15 years of research [1]. Therefore today, localized epitaxy appears as the most attractive approach towards robust and reliable PN junctions with low crystal defect density, as required for peripheral protection. VLS (Vapor-Liquid-Solid) epitaxy has been previously proposed by some of the authors as a way to implement localized high Al doping [2]. The feasibility of such approach has been successfully demonstrated, but the morphology of the deposit was very rough and inhomogeneous. Furthermore, so far, the VLS epitaxy approach has only been performed over the substrate surface and not in buried configuration which may bring additional advantages. On the growth aspect, it may help localizing the liquid in the buried zones while, on the device aspect, it may propose a solution to deep peripheral protection.

Experimental details

In this work, Buried Selective Growth by VLS Epitaxy (BSG-VLS) was investigated using Al-Si melt, as a way to deposit p+ doped SiC inside trenches (see Fig. 1). Vertical trenches with 0.9 µm depth and of various lateral sizes (from 10 to 800 µm) and shapes were formed by standard fluorinated reactive ion etching on commercial n+ doped 4H-SiC(0001), 4°off Si-face substrate. They were filled with Si and Al layers stack by localized patterned evaporation. Targeted Al content in the Al-Si melt was 60-70 at% with a total stacking thickness < 3 µm. The samples were then heated, under Ar or H_2, up to 1100°C to melt the stacking. At this stage, propane (1-3 sccm) was added into the gas phase to start the VLS growth for 5-20 min. Since only the carbon source is supplied by the gas phase, SiC deposition should only occur by VLS mechanism at the places where the liquid is present. The samples were characterized both as grown and after wet chemical etching in order to remove the remaining alloy on the surface.

The deposit morphology was characterized by optical and electronic microscopy. Polytype and doping information was obtained by Raman spectroscopy. The deposit thickness was determined by mechanical profiling.

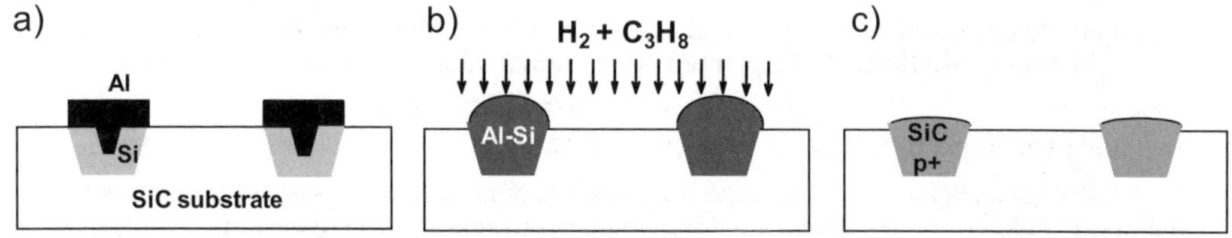

Figure 1 - Schematic description of the BSG-VLS process investigated in this work: a) deposition of Al-Si stack inside the trenches, b) melting of Al-Si and addition of propane to start the VLS growth, c) formation of BSG of p+ doped SiC inside the trenches.

Results and discussion

The first BSG experiments confirmed the previous results on planar configuration [1], i.e. when propane is added in the reactor upon reaching 1100°C, then the liquid segregates and forms droplets (Fig. 2a). The new buried configuration makes no change. However, adding propane earlier, during the initial temperature ramping up, leads to spectacular improvement: the liquid entirely covers the trenches even after the growth (Fig. 2b). That is why propane injection at room temperature before heating has been systematically implemented for the rest of the study. The reason for such improvement is still unclear. One could propose a surfactant effect of carbon favoring the wetting. Some slight SiC growth at low temperature upon alloy melting (T_M = 577°C) may also help improving the wetting.

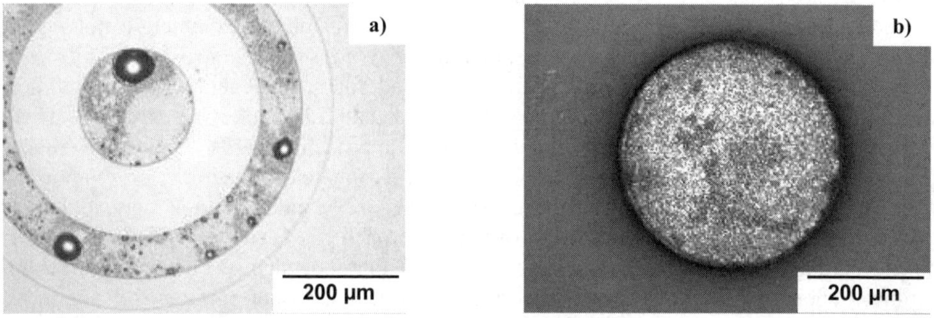

Figure 2 - Optical image of liquid evolution within the pattenrs after BSG-VLS with propane addition a) at 1100°C and b) at room temperature.

After wet chemical etching of the remaining Al-Si alloy on the grown samples, the surface morphology was systematically found to be highly step-bunched (Fig. 3). This is typical for epitaxial growth from a liquid phase. The deposit thickness was found to vary from a pattern to another, especially for the thickest deposits. Raman spectroscopy performed on thick deposit allowed to determine that the grown material was highly p-type doped 4H-SiC (probably with Na $>1x10^{20}at.cm^{-3}$) (see Fig. 4).

Effect of propane partial pressure. Changing propane partial pressure from $1x10^{-4}$ to $6x10^{-4}$ atm did not change significantly the surface morphology (Fig. 3) of the deposit but mainly affected the layer thickness. Surprisingly, the highest propane partial pressure led to the thinnest deposit. It is believed that for too high propane partial pressure, the supersaturation is too high so that a solid crust may form on top of the liquid. This was already mentioned in the case of Ge-Si melts [3] but never for Al-Si ones [4]. On the other hand, the highest growth rate achieved in this work is as high as 24 µm/h for sample of Fig. 3b. This is indeed very high and never achieved in previous works [4, 5]. It is postulated that for too high propane partial pressure, the kinetics of C dissolution from the gas phase into the liquid phase is not fast enough so that it directly reacts with the liquid to form a solid carbon containing compound on top of it. This phase blocks the liquid feeding and thus stops (or drastically slows down) the growth. Raman spectroscopy allowed identifying the presence of Al_4C_3 in such case suggesting the formation of this carbide rather than SiC or even graphite. Furthermore, this is confirmed by the fact that this crust disappears after wet chemical etching.

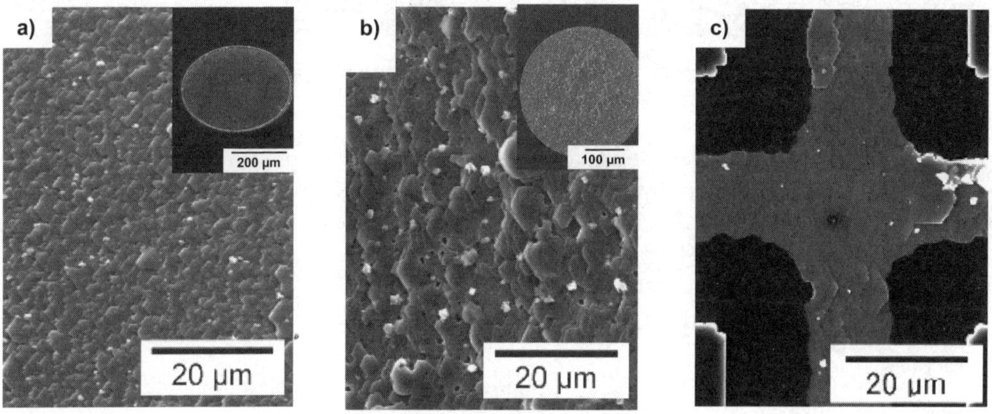

Figure 3 - SEM images taken after chemical etching of the remaining alloy on three BSG performed with different C_3H_8 partial pressures (for 5 min growth): a) $1x10^{-4}$, b) $3x10^{-4}$ and c) $6x10^{-4}$ atm. The deposit thickness is a) 0.9-1 µm, b) 1-2 µm and c) ~0.200 µm. The insets show low-magnification views of the corresponding complete patterns inside which were observed the morphologies.

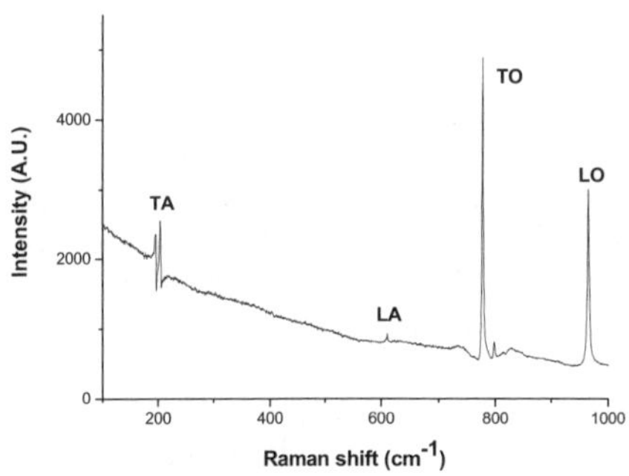

Figure 4 - Typical Raman spectrum collected on the BSG-VLS grown SiC material after chemical etching of the remaining alloy. Note Fano interference close to the TO mode and the increase of the background level at low shift. This is typical of highly p-type doped 4H-SiC.

Effect of growth duration. With intermediate propane partial pressures ($1x10^{-4}$ to $6x10^{-4}$ atm), increasing the growth duration from 5 to 20 min did not increase significantly the deposit thickness. This should due to the fact that the Si content of the liquid diminishes as growth proceeds so that after forming ~1 µm of SiC, the Si content of the liquid may be too low to allow further SiC growth. This is thus an auto-limited growth system which duration depends only on the initial amount of Si in the liquid droplet.

Effect of pattern size and shape. Various sizes and shapes of patterns were studied in order to check the flexibility of the BSG-VLS process. Few more illustrations are given in Fig. 5 where it can be seen that the SiC deposit covers all the patterns and keeps their original shape, even for angular ones. Patterns as large as 800 µm and as narrow as 10 µm were completely filled with SiC deposit in this way. As a result, this growth method can be applied to the fabrication of both large high current densities power devices and interdigitated structures. Extension to bigger and smaller patterns is currently under study.

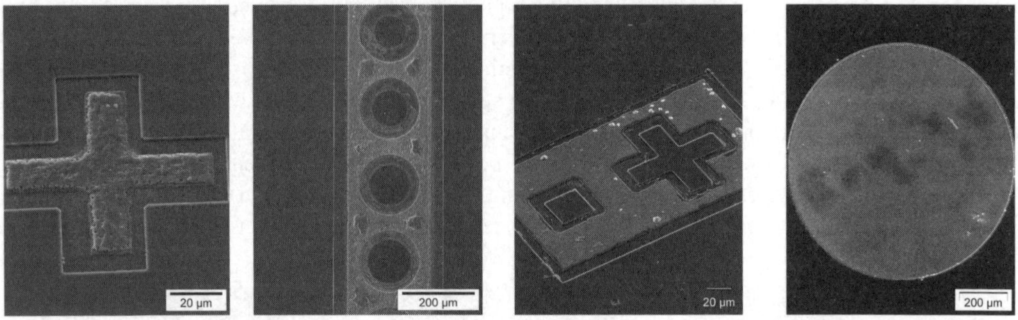

Figure 5 - Few examples of SiC deposit by BSG-BLS epitaxy on different patterns. SEM observations after wet chemical etching.

Finally, preliminary I(V) characterizations were performed to test the p+/n+ junction. The contacts were made without metallization, by putting a tip directly on the grown SiC material (after wet etching of the remaining alloy). As shown in Fig. 6, a rectifier behavior is found with a threshold voltage of ~3V in forward bias. The p-n junction is thus confirmed.

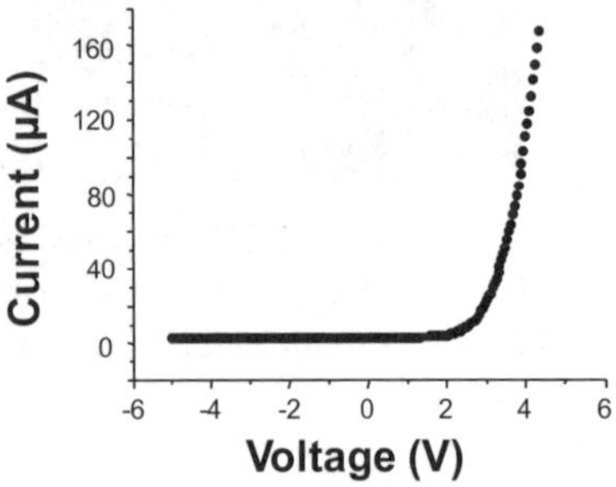

Figure 6 – Typical I-V measurement obtained on a 300 µm circular diode grown by BSG-VLS epitaxy.

Conclusion

The buried selective growth and doping of SiC was successfully achieved using VLS mechanism. It works on various sizes and shapes of patterns. The deposition kinetics demonstrates that the process is self limited and mainly depends on the initial amount of Si in the liquid. Work is under progress to determine the doping level of the deposit and to fabricate a device using this technique.

Acknowledgements

Authors would like to acknowledge the financial support from French ANR in the framework of VHVD-SiC project (ANR-08-BLAN-0191-02).

References

[1] T. Watanabe, S. Aya, et al., Mater. Sci. Forum 645-648 (2010) 705-708

[2] M. Soueidan, G. Ferro, C. Jacquier, P. Godignon, J. Pezoldt, M. Lazar, B. Nsouli, Y. Monteil, Diamond & Related Materials 16 (2007) 37–45

[3] M. Soueidan, G. Ferro, O. Kim-Hak, N. Habka, V. Soulière, B. Nsouli, Crystal Growth & design 8(3) (2008) 1051-1054.

[4] C. Jacquier, G. Ferro, et al., Recent Res. Devel. Crystal Growth, 4 (2005) 83-103

[5] G. Ferro, C. Jacquier, New J. Chem. 28 (2004) 889–896

Materials Science Forum Vols. 717-720 (2012) pp 173-176
© (2012) Trans Tech Publications, Switzerland
doi:10.4028/www.scientific.net/MSF.717-720.173

Reliable Method for Eliminating Stacking Faults on 3C-SiC(001)

Naoki Hatta[1,a], Takamitsu Kawahara[1,b], Kuniaki Yagi[1,c], Hiroyuki Nagasawa[1, d], Sergey Reshanov[2, e], Adolf Schöner[2, f]

[1]SiC Development Center, HOYA Corporation, 1-17-16 Tanashioda, Chuo-ku, Sagamihara, Kanagawa 252-0245 Japan

[2]ACREO AB, Electrum 236, S-16440 Kista, Sweden

[a] Naoki_Hatta@sngw.rdc.hoya.co.jp, [b] Takamitsu_Kawahara@sngw.rdc.hoya.co.jp, [c] Kuniaki_Yagi@sngw.rdc.hoya.co.jp, [d] Hiroyuki_Nagasawa@sngw.rdc.hoya.co.jp, [e] sergey.reshanov@acreo.se, [f] adolf.schoner@acreo.se

Keywords: 3C-SiC, cubic-SiC, stacking fault, homo-epitaxy, simulation, yield

Abstract. A reliable method for reducing the stacking faults (SFs) is demonstrated on the 3C-SiC (001) surface. It is a practical method based on Monte Carlo (MC) simulations of SF propagation during 3C-SiC epitaxial growth, which showed that introducing some discontinuity on the (001) surface enhanced SF reduction. The method is implemented by patterning on the 3C-SiC (001) surface and subsequent homo-epitaxial growth, and this sufficiently reduced the SF density to less than 400 cm^{-1}. A yield of 97.4 % was estimated for a device-ready area of 10 mm^2 by statistical measurements of SF density on the entire epitaxial layer surface.

Introduction

Due to the suppressed trap density at the SiO_2/SiC interface, 3C-SiC is regarded as one of the ideal materials for fabricating high-performance metal-oxide-semiconductor field-effect transistors (MOSFETs). However, degradation of the blocking properties induced by planar defects has been an obstacle for realizing the high-performance MOSFETs using 3C-SiC. For example, stacking faults (SFs) induce current leakage at p-n junction, and the leakage current density strongly depends on the SF density [1]. To obtain MOSFETs for power-switching application, the SF density on the 3C-SiC surface must be reduced to less than 400 cm^{-1} to suppress the leakage current density to less than 1×10^{-4} Acm^{-2} [2]. Although several methods have been proposed for SF reduction [3-5], the SF density has not been reduced to less than 400 cm^{-1}.

Recently, Monte Carlo (MC) simulations of the SF propagation revealed the mechanism of remaining SFs on the 3C-SiC(001) surface [6]. Based on the MC simulations, this paper proposes a practical method for reducing SF density. Remarkable SF reduction was achieved by growing an additional homo-epitaxial layer on a patterned substrate. The yield of the device-ready area was estimated by statistical measurements of SF density on the entire epitaxial layer surface.

Numerical analysis of SF density on the (001) face

Fig. 1 shows the simulated SF density on the 3C-SiC (001) surface as a function of 3C-SiC epitaxial layer thickness. This figure suggests that the conventional SF-reduction methods, using misoriented Si(001) [3], undulant-Si [4], and Switch-Back-Epitaxy [5], become less efficient at reducing SFs with increasing layer thickness due to "spontaneous SF collimation" [6]. The common mechanism that occurs during conventional SF reduction methods is annihilation induced at the junction of counter SFs. Such annihilation rate decreases with increasing SiC thickness because residual SFs are spontaneously aligned in a specific direction as a result of annihilation. Hence, it is impossible to eliminate SFs if annihilation contributes predominantly to SF reduction.

MC simulations were carried out for conventional SF reduction methods by assuming growth in infinite dimensions. This assumption is accurate if the growth area is adequately wide compared to the SiC growth thickness. However, if the effective growth area is comparable to the SiC growth thickness, the SF density will differ from that on an infinite substrate due to the interaction between SFs and the edge of the growth area. Indeed, as seen in the curve labeled "On finite area" for a width of 300-μm, the SF density decreased remarkably for thickness greater than 400 μm. This reduction was promoted by SF termination at the edge of the growth area but not by annihilation. That is, introducing discontinuity on the (001) surface enhanced SF reduction.

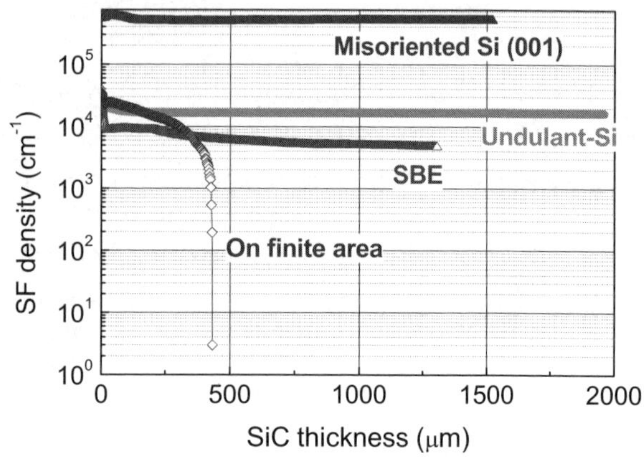

Fig. 1: Change in SF density as a function of SiC thickness simulated for misoriented Si(001), undulant-Si, SBE, and the present method.

Practical method for SF reduction

Based on the aforementioned findings, a practical method for SF reduction was investigated. The method consisted of three sequential processes shown in Fig. 2(a), (b), and (c). First, a continuous 3C-SiC (001) surface was divided into (001) facets bounded by sidewalls parallel to the {110} faces (Fig. 2a). Next, an additional homo-epitaxial layer was grown on the entire surface (Fig. 2b). Although the SFs propagated along the {111} planes during growth, all SFs can be terminated at the sidewall according to the following mechanism. The SF exposed on the lower facet (Lower-SF) could not penetrate into the upper facet because the sidewall acted as stairs. In contrast, the SF exposed on upper facet (Upper-SF) moved toward an edge of the upper facet during growth. Consequently, the Upper-SF reached the sidewall and terminated (Fig. 2c).

To achieve such SF termination, the propagation rate of the Upper-SF toward the edge of the upper facet should be faster than the growth rate of the sidewall in the [110] direction. Hence, the condition defined by $\Delta t_{[001]}/\tan(54.7°) > \Delta t_{[110]}$ must be satisfied during homo-epitaxial growth, where $\Delta t_{[001]}$ and $\Delta t_{[110]}$ are SiC growth rates in the [001] and [110] directions, respectively.

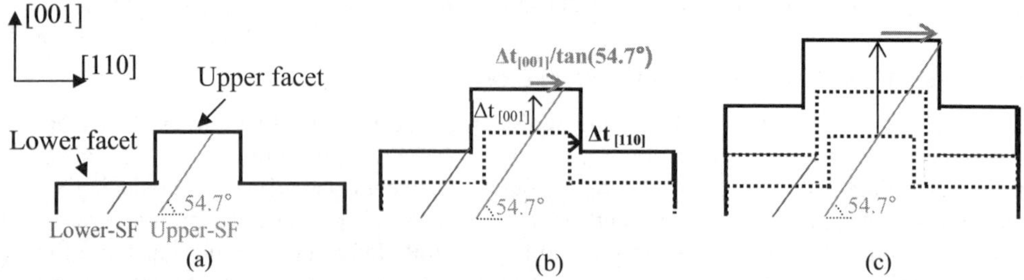

Fig. 2. Sequential processes for reducing SF density: (a) formation of the (001) facet with a sidewall parallel to the {110} faces on a continuous 3C-SiC surface; (b) 3C-SiC homo-epitaxial growth on the entire surface; and (c) SF termination at the sidewall of the (001) facet.

Experimental

To demonstrate the remarkable reduction in SF density, 3C-SiC was grown homo-epitaxialy on the (001) surface divided by sidewalls (Fig. 3). Because the growth conditions on the Si-face were more stable than those on the C-face, an array of rectangular parallelepiped structures aligned in the [-110] direction was fabricated on the (001) surface to unify the polarity of sidewall surface as Si. The structure was fabricated on the 3C-SiC (001) substrate by the following steps. First, a 5-μm-wide striped pattern with a 25-μm period was formed by photolithography, maintaining alignment in the [-110] direction. Then, reactive ion etching was used to form trenches with a depth of 25 μm. The surface was divided into upper and lower (001) facets by (110) and (-1-10) sidewalls.

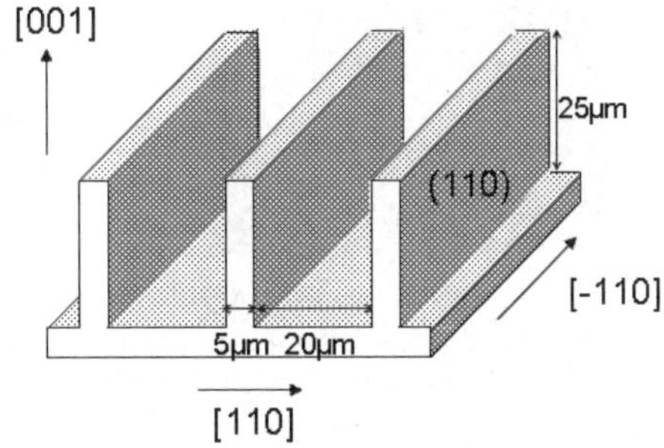

Fig. 3. Schematic structure of the patterned 3C-SiC substrate with (001) facets bounded by sidewalls.

A 21-μm-thick homo-epitaxial layer was grown on the patterned substrate by supplying SiH_2Cl_2 + C_2H_2 as the source gas and H_2 as the carrier gas at 1873 K. The gas flow rates and pressure were optimized for Si-face during growth. As a reference for the present method, the aforementioned homo-epitaxial layer was also grown on an unprocessed 3C-SiC (001) surface.

After homo-epitaxial growth, the SF distribution was observed by optical microscopy on the (001) surface. Prior to observation, the epitaxial layer surface was etched by molten KOH for 5 min at 773 K to visualize the SFs. To obtain a statistical SF distribution, square observation blocks (10 mm^2), including 800 upper-facets with 0.5mm in length, were arbitrary selected on the surface. A cross-sectional image of the cleaved (-110) face was also observed using transmission optical microscopy with a polarizer to determine the SF termination mechanism.

Results and Discussion

Fig. 4 shows a cross-sectional image of the (-110) face of the homo-epitaxial layer on the patterned substrate, in which SFs parallel to the (111) face are visible as inclined bright lines. As shown in Fig. 2, termination of the Upper- and Lower-SFs was clearly observed with consequently generated SF-free region at the vicinity of the (001) surface (relatively dark portion).

Fig. 5 shows SF distributions on the homo-epitaxial layer grown on the 3C-SiC (001) substrate (a) with and (b) without pattering. Few SFs were observed as parallel lines in the [-110] direction on the isolated (001) upper facet in Fig. 5(a). This means that all the remaining SFs on the patterned substrate were parallel to the (1-11) or (-111) planes, exposing a Si-terminated surface. On the other hand, a high density of etch pits in the [-110] direction were observed on the surface of the epitaxial layer on the unprocessed substrate in Fig. 5(b). The dominant SFs correspond to those parallel to the (111) or (-1-11) planes with a C-terminated surface. The difference between the SF distributions in Fig. 5(a) and (b) clearly suggested the vanishing of C-terminated SFs at the sidewall of the patterned substrate, demonstrating the reliability of the present method for SF reduction. Additionally, most of the SFs observed on the upper facet (the Si-terminated SFs) terminated at edges of the facet. This result indicates that the dimensions of the Si-terminated SFs were predominantly determined by the width of the (001) facet due to complete elimination of residual C-terminated SFs on upper (001) facets.

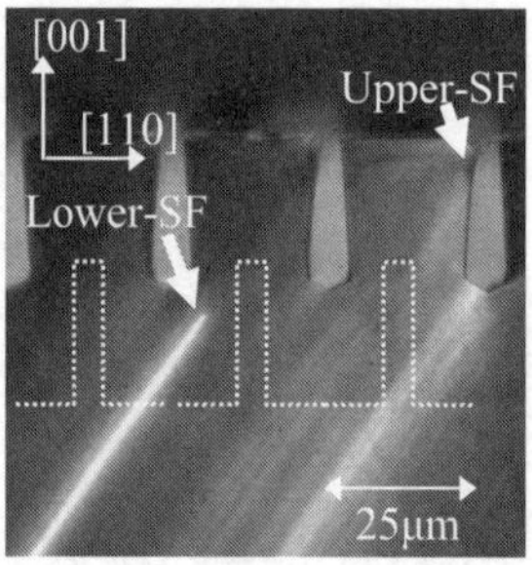

Fig. 4. Transmission optical microscopy cross-sectional image of (-110) of the patterned 3C-SiC substrate using a polarizer.

(a) Patterned region (b)Unprocessed region

Fig. 5. Optical microscopy images of the SF distribution on a homo-epitaxial layer grown on 3C-SiC (001) (a) with and (b) without patterning.

Fig. 6 shows the cumulative frequency of the measured SF density over the entire 3C-SiC surface, which was grown on a patterned substrate. This figure demonstrates that a 10-mm^2 square area with an SF density less than 400 cm^{-1} can be obtained at a probability of 97.4 %, and such stable SF reduction via the present method allows increased yields of MOSFETs compared with the conventional methods. The most remarkable advantage of the present method is that all SFs can be terminated at discontinuities in the lattice, not due to annihilation. Both the density and dimensions of the SFs can be adjusted by geometrical factors such as facet width, trench depth, and homo-epitaxial layer thickness. Thus, an area with low SF density can be obtained with high reproducibility.

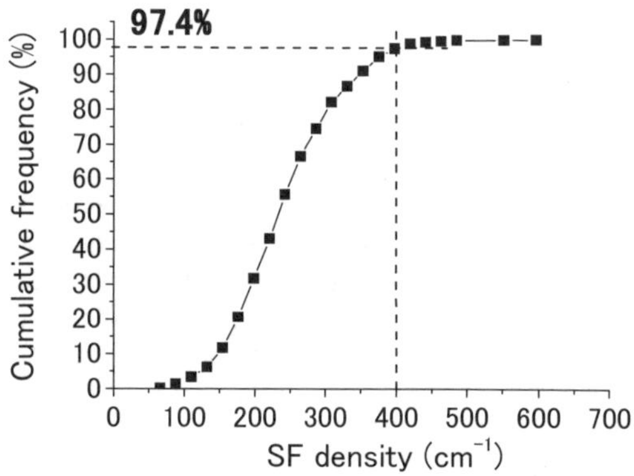

Fig. 6. Cumulative frequency of SF density within an area of 10 mm^2, arbitrarily selected on the patterned 3C-SiC (001) substrate.

Summary

A reliable SF-reduction method involving homo-epitaxial growth on patterned 3C-SiC (001) was investigated. Because all SFs terminated at the sidewalls of the rectangular parallelepiped structure in the [-110] direction, both the density and dimensions of SF can be adjusted by geometrical factors. Thus, an MOSFET-ready area, corresponding to 10-mm^2 square region with an SF density of less than 400 cm^{-1}, was obtained at a yield of 97.4%.

References

[1] T.Kawahara, N.Hatta, K.Yagi, H.Uchida, M.Kobayashi, M.Abe, H.Nagasawa, B.Zippelius, G.Pensl: Mater.Sci.Forum Vol.645-648 (2010) p.339

[2] H. Nagasawa, T. Kawahara, K. Yagi, N. Hatta, H. Uchida, M. Kobayashi, S. Reshanov, R. Esteve, and A. Schöner: Proceedings of HeteroSiC-WASMPE 2011, to be published.

[3] K.Shibahara, S.Nishino, and H.Matsunami: J.Cryst.Growth Vol.78 (1986) p.538

[4] H.Nagasawa, T.Kawahara, K.Yagi: Mater.Sci.Forum Vol. 389-393 (2002) p.319

[5] K.Yagi, T.Kawahara, N.Hatta, H.Nagasawa: Mater. Sci. Forum Vol. 527-529 (2006) p.291

[6] H. Nagasawa, T. Kawahara, K. Yagi, N. Hatta: Mater. Sci. Forum Vol.679-680 (2011), p282

Materials Science Forum Vols. 717-720 (2012) pp 177-180
© (2012) Trans Tech Publications, Switzerland
doi:10.4028/www.scientific.net/MSF.717-720.177

Freestanding 3C-SiC grown by sublimation epitaxy using 3C-SiC templates on silicon

P.Hens[1, 2 *], J.Müller[3], G.Wagner[4], R.Liljedahl[1], R.Yakimova[1], E.Spiecker[3], P.Wellmann[2] and M.Syväjärvi[1]

[1]Department of Physics, Chemistry and Biology, Linköping University, SE-58183 Linköping, Sweden

[2]Materials of Electronics and Energy Technology, University of Erlangen-Nuremberg, Martensstrasse 7, 91058 Erlangen, Germany

[3]Center for Nanoanalysis and Electron Microscopy, University of Erlangen-Nürnberg, Cauerstrasse 6, 91058 Erlangen, Germany

[4]Leibniz Institute for Crystal Growth, Max-Born-Strasse 2, 12489 Berlin, Germany

*corresponding author, email: phihe@ifm.liu.se

Keywords: cubic silicon carbide, heteroepitaxy, sublimation growth, defect reduction

Abstract: In this work a new approach for the production of freestanding cubic silicon carbide (3C-SiC) in (001) orientation is presented which is based on the combination of chemical vapor deposition (CVD) and the fast sublimation growth process (FSGP). Fast homoepitaxial growth of 3C-SiC using sublimation epitaxy on a template created by CVD growth on silicon substrates allows to obtain thick freestanding material with low defect densities. Using standard silicon wafers as substrate material permits a cost efficient process and the applying of wafers with different orientations. The (001) orientation used in this work will potentially allow further heteroepitaxial growth of other cubic semiconductors, like e.g. gallium nitride (GaN).

Introduction: Until now only small bulk crystals [1] or thin epitaxial layers [2] of cubic silicon carbide are available, which strongly limits the potential applications of this material in micro- and optoelectronics. The growth of 3C-SiC on silicon by CVD results in a high concentration of defects, reflected in XRD lines with typical full widths at half maximum (FWHM) values of about 250 arcsec, or even more. In this work we present a new approach to prepare freestanding wafers of 3C-SiC which has the ability to obtain good crystalline qualities regarding increased domain sizes and reduced defect densities.

This technology is based on the combination of chemical vapor deposition and the fast sublimation growth process [3]. Thin layers of silicon carbide are grown heteroepitaxially on silicon (001) substrates by CVD acting as templates for a subsequent sublimation homoepitaxial growth step. The FSGP technology provides a combination of high growth rate and improved crystal quality [4] which is needed for the growth of this semi-bulk material.

Experimental: The new approach uses a solid polycrystalline silicon carbide source and a CVD-grown (001) oriented 3C-SiC template as substrate material. In an inductively heated graphite crucible a spacer with a thickness of one millimeter is applied to create a short source to template distance. To ensure a fast material transport driven by a temperature gradient the growth process was conducted with a vacuum pressure below $5 \cdot 10^{-5}$ mbar and temperatures in the range of 1800 °C.

Since the process temperature exceeds the melting point of the silicon substrate included in the template, the removal of the melt is a crucial step since otherwise the formed droplet will endanger the stability of the thin cubic silicon carbide layer and remaining Si droplets can be incorporated in the growing material degrading the quality of the final cubic silicon carbide layer. This removal of the Si-melt results in an in-situ created free-standing 3C-SiC substrate. The removal of the melt is achieved by adding a graphite plate with a certain low porosity that is infiltrated by the silicon melt.

The samples in this work have been produced using this approach with a growth rate of about 120 µm/h, which is about an order of magnitude higher than typical growth rates of CVD growth systems [2]. Hence this approach is more suitable for the growth of bulk-like 3C-SiC.

Results and discussion: Within this work we have been able to realize the growth of freestanding 3C-SiC samples with a thickness of 20 µm up to 80 µm. During sublimation growth a significant domain size enlargement could be observed. Whereas the typical templates showed domains in the range of 10-20 µm, after 40 µm epitaxial growth domain sizes in the range of 100-150 µm have been achieved by FSGP. In 80 µm thick sublimation grown layers no domain boundaries could be observed by means of light microscopy (Fig. 1a).

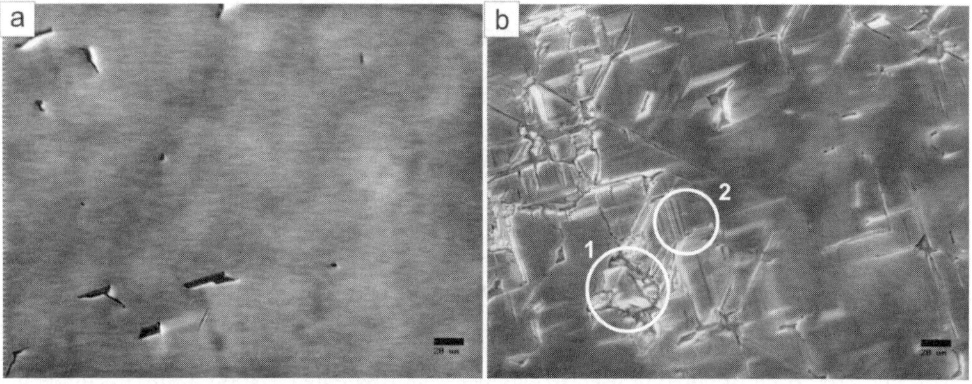

Fig. 1: Areas of different defect densities in plan-view light microscope images of a nearly defect free area (a) compared with highly defective area (b) with inclusions of different orientation (1) and planar defects (2). Scale bars are equivalent to 20 µm.

At this early stage of technology development the samples show some inhomogeneous quality despite their relatively small size in the range of 6-10 mm in diameter. In certain areas a high density of defects can be found. Typical defect structures are inclusions of grains with a different crystalline orientation surrounded by small cracks and planar defects in the layer (Fig. 1b).

These defects originate from localized spots directly at the interface between the CVD grown template and the FSGP grown layer (see Fig. 2). They are very likely to be caused by silicon droplets that formed on the template surface during the heat-up ramp in advance to the start of the growth process. This is supported by the fact that when using a template with a wet chemically removed silicon substrate the defect densities were significantly lower as seen in light microscopy.

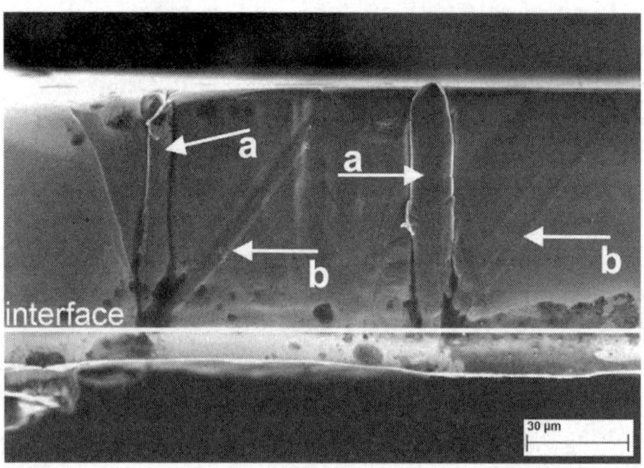

Fig. 2: Cross section view in scanning electron microscopy showing two inlusions (a) and the surrounding planar defects (b). Both emerge from distinctive locations at the interface between the template and the FSGP grown layer.

Studies using transmission electron microscopy (TEM) in cross section geometry revealed a low density of planar defects in the top part of sublimation grown material (see Fig. 3). Especially compared to typical material grown by CVD on silicon [5] the FSGP layers show a very low density of planar defects like stacking faults and microtwins.

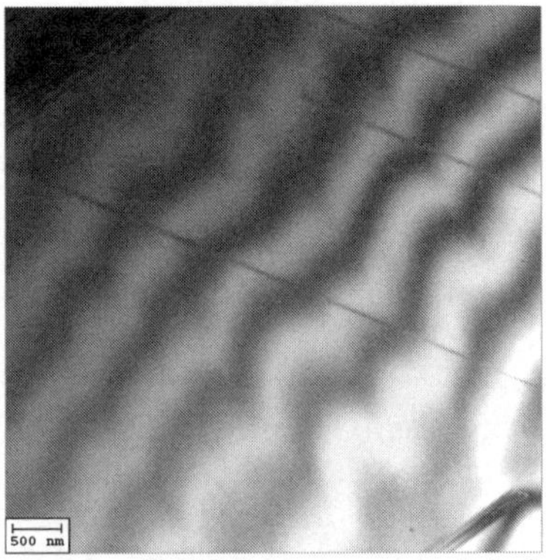

Fig. 3: Cross section TEM image of FSGP grown material showing a low defect density.

Structural investigations by XRD rocking curve measurements on 80 µm thick samples – in an area like in Fig. 1a - show a good crystalline quality with FWHM values for the (002) reflection as low as 78 arcsec measured on a footprint of 1x2 mm^2 (Fig. 4). Areas with a high defect density as shown in Fig. 1b reveal FMHW values up to 1000 arcsec with evidence for several domains.

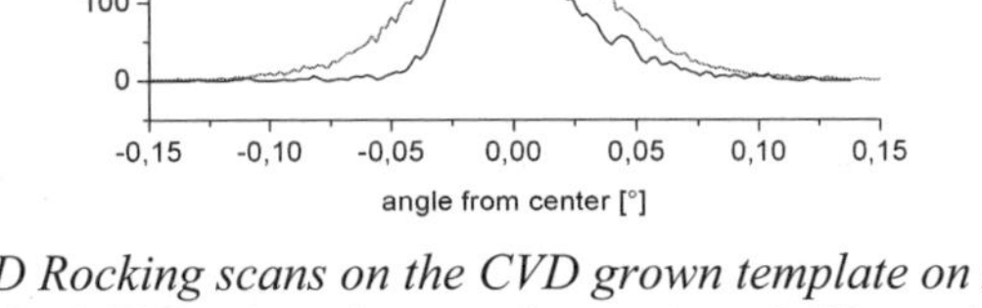

Fig. 4: Comparative XRD Rocking scans on the CVD grown template on silicon (10 µm thick) and on the FSGP grown freestanding material (80 µm thick)

Conclusion:

Within this work we have demonstrated the successful growth of freestanding 3C-SiC in the thickness range of 20 µm to 80 µm and diameters of more than ten millimeters. This was realized using the fast sublimation growth process on templates of cubic silicon carbide on silicon produced by CVD. Structural investigation by XRD rocking curve measurements show a good crystalline quality with FWHM values for the (002) reflection of down to 78 arcsec measured on a footprint of 1x2 mm^2. Further evidence of the high quality of our material was obtained by means of light microscopy and TEM showing the low defect densities.

Acknowledgements: Financial support by the Bundesministerium für Bildung und Forschung (BMBF) under contract number 03SF0393, Swedish Energy Agency and Ångpanneföreningen Research Foundation is gratefully acknowledged.

References:

[1] D. Chaussende, J. Eid, F. Mercier, R. Madar, M. Pons, *Nucleation and Growth of 3C-SiC Single Crystals from the Vapor Phase*, Mat. Sci. Forum, 615-617 (2009) 31-36

[2] M. Zielinski, M. Portrail, T. Chassagne, S. Juillaguet, H. Peyre, *Nitrogen Doping of 3C-SiC thin films grown by CVD in a resistively heated horizontal hot-wall reactor*, J. Cryst. Growth, 310 (2008) 3174-3182

[3] M. Syväjärvi, R. Yakimova, *Sublimation epitaxial growth of hexagonal and cubic SiC*, in: P. Bhattachrya, R. Fornari, H. Kamimura (Eds.), Comprehensive Semiconductor Science & Technology (SEST), Elsevier, 1999

[4] M. Syväjärvi, R. Yakimova, H. Jacobsson, and E. Janzén, *Structural improvement in sublimation epitaxy of 4H-SiC*, J. Appl. Phys. 88 (2000) 1407

[5] A. Severino, M. Camarda, N. Piluso, M. Italia, G. Condorelli, M. Mauceri, G. Abbondanza, F. La Via, *Bow in 6 inch high-quality off-axis (111) 3C-SiC films, Mat. Sci. Forum, 645-648 (2010) 167-170*

Materials Science Forum Vols. 717-720 (2012) pp 181-184
© (2012) Trans Tech Publications, Switzerland
doi:10.4028/www.scientific.net/MSF.717-720.181

Effects of Propane on Low Temperature Growth of 3C-SiC Films on Si (111) by Plasma Assisted CVD

Hideki Shimizu[1, a, *] and Takashi Watanabe[1, b]

[1] Aichi University of Education, 1 Hirosawa, Igaya-cho, Kariya, Aichi 448-8542, Japan

[a]hksimizu@auecc.aichi-edu.ac.jp, [b]however_68@yahoo.co.jp

Keywords: 3C-SiC, Plasma-assisted CVD, XRD, Transmission Electron Microscope, Si (111)

Abstract. To demonstrate the formation of 3C-SiC film on Si (111) at low substrate temperature, the effects of C_3H_8 on the crystalline quality of the 3C-SiC films on Si (111) have been investigated by changing the flow rate of C_3H_8 at the substrate temperature of 950 °C. The crystalline quality has been investigated by transmission electron microscope and X-ray diffraction. 3C-SiC is epitaxially grown on Si(111) and the 3C-SiC films are in either near single crystalline or highly oriented form with stacking faults and twin. It is expected that the film with good crystalline quality may grow at around 2.5 in the ratio of the flow rate of C_3H_8 to SiH_4 and any microstructures of 3C-SiC films on Si (111) can be controlled by accurately controlling the ratio of C/Si.

Introduction

Low temperature growth of 3C-SiC is the most important factor in order to integrate SiC into Si-based device, and makes its application greatly spread [1]. New approaches of low temperature growth of 3C-SiC have been energetically performed by various techniques [2-4]. We also have studied low temperature growth of 3C-SiC on Si by using AC plasma assisted CVD technique [5]. In this work, to demonstrate the formation of 3C-SiC film on Si (111) at low substrate temperature, the effects of C_3H_8 on the microstructures of the films on Si (111) have been investigated by changing the flow rate of C_3H_8 at the substrate temperature of 950 °C. The crystalline quality has been investigated by transmission electron microscope (TEM) and X-ray diffraction (XRD).

Experimental Procedures

3C-SiC films were grown on Si (111) substrates by AC plasma-assisted CVD using SiH_4 and C_3H_8 gases diluted to 0.5% and 1% in H_2 as source materials, respectively. SiC growth processes were carried out after the carbonization process of the Si surface. The effects of C_3H_8 on the crystalline quality of the films on Si (111) have been investigated by changing the flow rate of 1% C_3H_8 from 10 sccm to 150 sccm, while keeping the other conditions constant, such as the substrate temperature (950 °C) and flow rate of H_2 carrier gas (300 sccm). The substrate temperature was measured by an optical pyrometer. Tungsten plates mounted on a water-cooled copper were used as electrodes for exciting the AC plasma. A Si wafer (20x20 mm^2) after cleaning and etching in acetone, deionized water and 2% HF solution, was mounted on the SiC coated graphite susceptor. The plasma was excited by AC power supply (60 Hz) and was located 30 mm away from the substrate. After evacuating the system to 3×10^{-4} Pa, H_2 gas was introduced up to 8×10^3 Pa and the hydrogen plasma was excited at room temperature. C_3H_8 gas diluted in H_2 was passed through the hydrogen plasma. In the two-step carbonization process of the Si substrate, the Si substrate was heated up and held at 600 °C for 15 min and again heated and held at 950 °C for 15 min. 3C-SiC epitaxial layers were continuously grown on the carbonized layer at 950 °C by simultaneously supplying of C_3H_8 and SiH_4 for the deposition time of 20 min. SiH_4 gas and H_2 carrier gas were supplied directly to Si substrate from nozzles placed under the plasma electrodes and were thermally decomposed at the heated substrate, while C_3H_8 gas was supplied through the plasma from nozzles placed over the plasma and was decomposed by the H_2 plasma. The crystalline quality of the films was characterized by using a transmission electron microscope (TEM: JEOL JEM2100) (operation voltage: 200 kV) and an X-ray diffraction (XRD: RIGAKU X-RAY DIFFRACTMETER ATX-G),

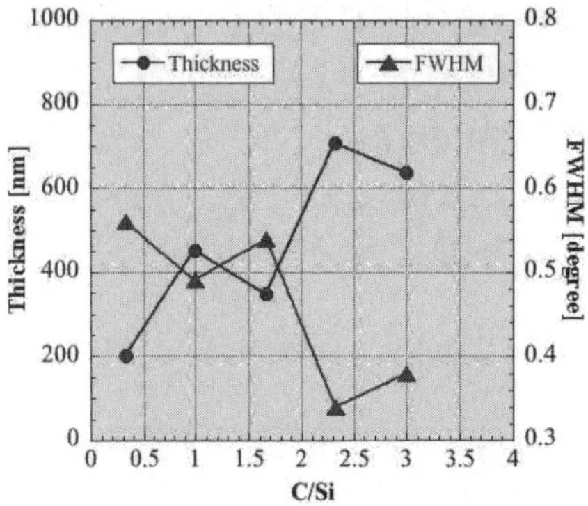

Fig. 1 Changes in the film thickness and the FWHM for the line originated from 3C-SiC (111) plane in XRD spectra as a function of C/Si ratio. The deposition time is 20min.

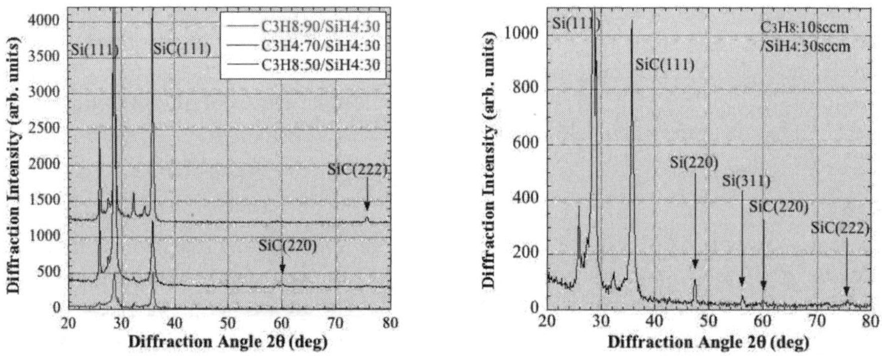

(a) (b)

Fig. 2 XRD spectra of 3C-SiC films deposited at C_3H_8 flow rate of 50sccm, 70sccm and 90sccm (a), and the film deposited at C_3H_8 flow rate of 10sccm (b).

(operation voltage: 50 kV, currents: 300 mA). Specimens for the TEM observation were prepared by mechanical thinning, polishing and argon ion milling.

Result and Discussion

Fig. 1 shows the thickness of the films deposited for 20min and the value of the full width at half maximum [FWHM] for the strongest line originated from 3C-SiC (111) plane in XRD spectra (Fig. 2), as a function of the flow rate of C_3H_8 under SiH_4 flow rate of 30sccm. The thickness of the films monotonously increased with increasing the C_3H_8 flow rate up to around 2.5 in the ratio of the flow rate of C_3H_8 to the flow rate of SiH_4 (C/Si), then slightly decreased with the further increase of the C_3H_8 flow rate. On the other hand the FWHM for the line originated from 3C-SiC (111) plane in XRD spectra monotonously decreased with increasing the C_3H_8 flow rate up to around 2.5 in C/Si, then slightly increased with the further increase of the C_3H_8 flow rate. Under constant flow rate of SiH_4, it is usually supposed that an optimum condition for SiC growth exists at around a point of inflection on the deposition rates as a function of the flow rate of C_3H_8, reflecting that the value of the FWHM in XRD spectra has minimum at around 2.5 in C/Si. Fig. 2 shows XRD spectra of 3C-SiC films deposited at the different flow rate of C_3H_8. For the film deposited at 2.33 in C/Si (C_3H_8 flow rate of 70sccm), only two sharp peaks at diffraction angle (2q) of 35.8° and 75.7° corresponding to 3C-SiC(111) and (222) planes respectively were observed. No other peaks were detected except the peaks at 28.7° due to the underlying Si (111) substrate in XRD spectra, indicating that the 3C-SiC films are in either epitaxial (single crystalline) or highly oriented (textured) form. On the other hand, for the films deposited at much less or more than 2.33 in C/Si, an extra peak was observed at diffraction angle (2q) of 60.4° corresponding to 3C-SiC(220) plane.

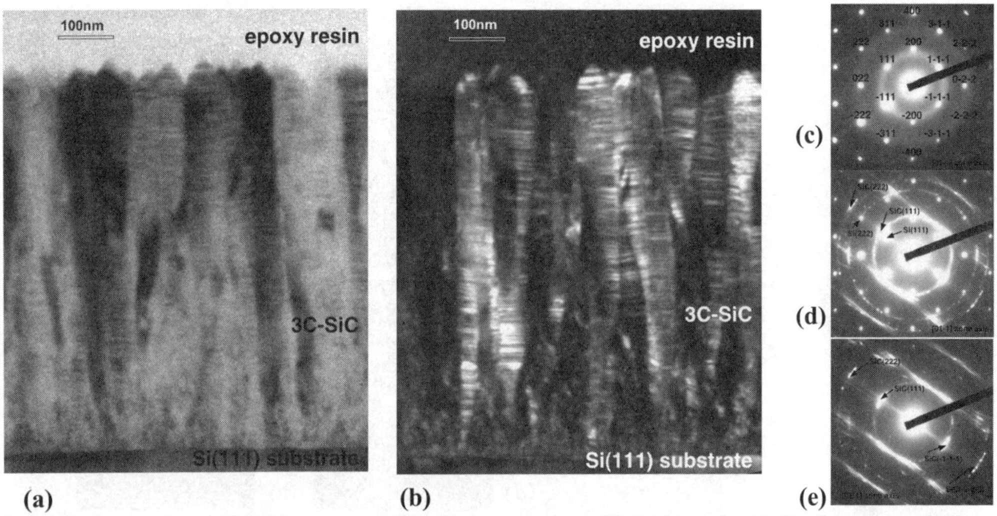

Fig. 3 Cross-sectional TEM images of SiC films grown on Si (111) at the C_3H_8 flow rate of 70 sccm along the zone axis [01-1], (a) bright field (BF) image, (b) dark field (DF) image generated from SiC (111) spot in the selected area electron diffraction (SAED) pattern as shown in (e), (c) SAED pattern on the Si (111) substrate, (d) SAED pattern on the interface between SiC film and Si substrate, (e) SAED pattern on the SiC film.

For the film deposited at the lowest flow rate of C_3H_8 (10sccm, C/Si=0.33), the extra peaks corresponding to Si(220) and Si(311) were observed as shown in Fig. 2b. From the above results, the quality of crystalline structure of the film deposited at the optimum condition was expected to be single crystalline SiC.

Therefore, to precisely evaluate the crystalline quality of 3C-SiC films, the TEM observation was carried out. Fig. 3 shows the cross-sectional TEM images of SiC films grown at the C_3H_8 flow rate of 70 sccm and the SiH_4 flow rate of 30 sccm (C/Si=2.33). From bright field (BF) image and dark field (DF) image by choosing a SiC (111) spot in the selected area electron diffraction (SAED) pattern as shown in Fig. 3(e), 3C-SiC films have columnar structures along the [111] orientation and a lot of stacking faults in (111) plane sequence, giving bright and dark contrasts in the image, and twins, where the streaks are seen along spot to spots and spread as shown in Fig. 3(e). However, the crystal orientations of SiC and Si at the interface can be revealed from the SAED pattern taken at the 3C-SiC/Si(111) interface along the zone axis [01-1]. The SAED spots in the SAED pattern are indexed in Fig. 3d, where the [111] spots of 3C-SiC are in parallel with those of Si. This supports that 3C-SiC[111] is line up with Si[111]. In other words, 3C-SiC is epitaxially grown on Si(111). The films grown at much less than 70 sccm have polycrystalline 3C-SiC included with Si single crystal as shown in Fig. 4, where shows TEM image and SAED patterns taken at the Si substrate, the 3C-SiC/Si(111) interface, the 3C-SiC film and the surface on SiC along to the zone axis [-211]. The homogeneous film labeled as SiC in the TEM image (Fig. 4a) has 3C-SiC polycrystalline structure with Si polycrystalline domain as shown in the SAED pattern (Fig. 4d). The rough part on the homogeneous films labeled as Si in the TEM image (Fig. 4a) has single crystalline Si as shown in Fig. 4(e), where SAED pattern has spots patterns corresponding to the SAED pattern of Si (111)substrate along to zone axis [-211]. At low C/Si ratio, named as poor carbon source condition, the SiC film grows slowly and homogeneously with partially including polycrystalline Si in short term, by the effect of residual carbon sources on the carbonization step. However when the residual carbon source on the carbonization step was consumed to form SiC film, Si crystal growth precedes SiC growth.

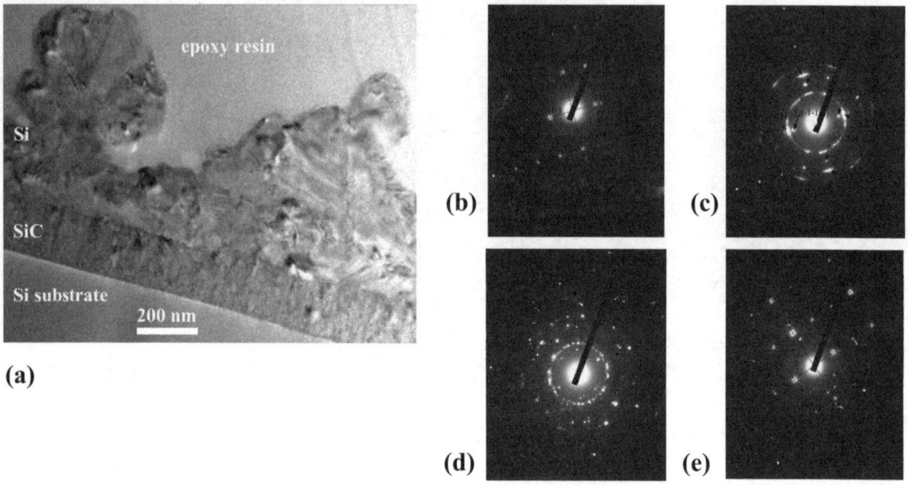

Fig. 4 TEM image (a) and SAED patterns taken at the Si substrate (b), the 3C-SiC/Si(111) interface (c), the 3C-SiC film (d) and the surface on SiC (e) along to the zone axis [-211].

From these results, hydrocarbon radicals, carbon radical, and hydrogen radical supplied by decomposing the propane gas in the plasma region, control the deposition rate and the quality of the 3C-SiC films on Si substrate. It is expected that the film with good crystalline quality may grow at the ratio of the flow rate of C_3H_8 to SiH_4 (C/Si) of 2-3 and any microstructures of 3C-SiC films on Si (111) can be controlled by accurately controlling the ratio of C/S. However, the crystal defects originated from lattice miss match between the SiC and Si, such as the stacking faults and the twins, were not restrained by only the effect of propane. A. Severe et al. reported that the choice of an off-axis (111) substrate resulted in a better quality with lower defect density as compared to standard on-axis growth [6]. So, we will try to investigate the effects of the choice of off-axis Si(111) on low temperature growth of 3C-SiC film by our deposition method.

Summary

In order to demonstrate the formation of 3C-SiC film on Si (111) at low substrate temperature, the effects of C_3H_8 on the crystalline quality of the films on Si (111) have been investigated by changing the flow rate of C_3H_8 at the substrate temperature of 950 °C. 3C-SiC is epitaxially grown on Si(111) and the 3C-SiC films are in either near single crystalline or highly oriented form with stacking faults and twins. It is expected that the film with good crystalline quality may grow at around 2.5 in the ratio of the flow rate of C_3H_8 to SiH_4 and any microstructures of 3C-SiC films on Si (111) can be controlled by accurately controlling the ratio of C/Si.

References

[1] N. Moronuki, M. Kojima and A. Kakuta, Thin Solid Films Vol. 516 (2007) pp.5344-5348
[2] J.J. Sumakeris, K.S. Ailey and R.F. Daivis, Inst. Phys. Conf. Ser., Vols. 137 (1994) p. 75
[3] A.J. Steckl, C. Yuan and J.P. Li, Inst. Phys. Conf. Ser., Vol. 137 (1994) p. 79
[4] C.W. Liu and J.C. Sturn, Inst. Phys. Conf. Ser., Vol. 137 (1994) p. 83
[5] H. Shimizu and A. Kato, Material Science Forum Vols. 556-557 (2007) pp.183-186
[6] A. Severino, C. Bongiorno, N. Piluso, M. Italia, M. Camarda, G. Condorelli, M. S. Di Stefano, B. Cafra, A. La Magna, F. La Via, Thin Solid Films Vol. 518 (2010) pp. 5165-5169

Materials Science Forum Vols. 717-720 (2012) pp 185-188
© (2012) Trans Tech Publications, Switzerland
doi:10.4028/www.scientific.net/MSF.717-720.185

Low Temperature Epitaxy of 3C SiC Using Hexamethyldisilane Precursor on Si<111> Substrates

B. Wagner[1,a], J. Oliver[1], N.B. Singh[1], M. King[1], S. McLaughlin[1], D. Kahler[1], D. Knuteson[1], A. Berghmans[1] and R. Rai[2]

[1] Northrop Grumman Corporation - Electronic Systems, 1212 Winterson Rd. MS 3B10 Linthicum, MD 21090, USA

[2]Nano Materials Characterization Group, Austin TX, USA

[a]Brian.wagner@ngc.com

Keywords: HMDS, CVD, Silicon, Heteroepitaxy, XRD, TEM

Abstract. The CVD growth of SiC thin films using hexamethyldisilane (HMDS) as the singular precursor on Si <111> substrates with an AlN nucleation layer was explored in this study. A statistically designed experiment was used to conclude that growth temperature has the largest impact on crystal quality and surface microstructure. In addition to crystal quality, wafer bow was studied. Crystal quality and growth rate are loosely correlated to wafer bow in our study. SEM surface microstructural analysis of the SiC films shows a changing microstructure with growth temperature consistent changes in measured crystal quality. TEM studies reveal that the films are the 3C polytype having a high density of planar faults.

Introduction

SiC has become a discriminating material for high temperature, high power, and high frequency applications because of its wide band gap, high electron mobility, high electron drift velocity, and high thermal conductivity and stability [1]. The growth of SiC on Si substrates opens the door for integration into large scale, high volume, low cost CMOS processes tailored to Si fabrication. Heteroepitaxial growth of SiC on Si often utilizes the CVD technique for film deposition. A wide array of precursors has be used to deposit CVD SiC on Si substrates including silane /propane [2], monomethylailane [3], tetramethylsilane [4], methyltrichlorosilane [5], and hexamethyldisilane (HMDS) [6]-[8]. In this study, HMDS was chosen as a the precursor that will act as both the Si and C source for the deposition of SiC films below the melting point of Si. Due to the large lattice and thermal expansion coefficient mismatches between the SiC film and the Si substrate, significant amounts of stress can develop during the course of film deposition. Our approach in this study was to use an AlN nucleation layer that has a much closer lattice match to the SiC films to be deposited. AlN films were deposited on 2-inch Si <111> substrates prior to HMDS based low temperature SiC film deposition. The resulting samples were characterized using XRD, SEM, capacitance-based wafer bow, and TEM.

Experimental

The silicon substrates used in this study had a thickness of 250-300 μm, an orientation of (111)±0.5°, prime grade, single-side polished, N-Phos conductivity type, and a 50.8±0.5 mm diameter. The AlN films were deposited in an MOCVD reactor using Trimethylaluminum and Ammonia precursors. The growth process and reactor configuration is similar to that reported by Northrop Grumman in reference [9]. The AlN films in this study were grown to a thickness of 300nm and remained constant throughout this reported work. The SiC thin films were deposited in radial flow, planetary motion CVD reactor capable of handling 5x2-inch, or 5x3-inch wafers. In this particular instance, a liquid HMDS $Si_2(CH_3)_6$ bubbler, using an iso-temp bath with an H2 carrier gas, was used as both the Si and C precursors. Further details of this reactor were reported by Northrop Grumman in reference [10]. For all of the films studied in this report, growth times were held constant at 90 minutes in the film deposition step of the process.

X-ray diffraction studies were performed on a triple-axis D1 diffractometer with Cu-Kα[1] radiation produced by Bede Scientific Instruments. SEM micrographs were collected on an FEI NOVA nanoSEM 200. Wafer bow was measured by the capacitive non-contact Proforma 300 by MTI Advanced Products. TEM samples were examined in a Tecnai F20 TEM/STEM operating at 200keV with point to point resolution of 2.5Å.

Results and Discussion

HMDS was chosen as a precursor for this study since traditional SiC epitaxy methods require temperatures greater than 1500°C, above the melting point of Silicon. Lower deposition temperatures are achievable with HMDS since silicon and carbon atoms are both present in HMDS, and a Si-C bond already exists. This allows for successful deposition of crystalline films in the 1000°C to 1400°C range acceptable for growth on large area, lower cost substrates like silicon.

Table 1 Levels and factors used in this four-factor two-level designed experiment

Run Number	Temperature (°C)	HMDS Flow (sccm)	Growth Pressure (Torr)	Propane Flow (sccm)
1	1150	50	500	100
2	1150	90	500	0
3	1250	50	500	0
4	1250	90	500	100
5	1150	90	100	100
6	1150	50	100	0
7	1250	90	100	0
8	1250	50	100	100

Fig. 1. SEM Micrograph of as-grown AlN surface using a Silicon <111> substrate

The AlN films used as the nucleation layer in this study were grown to a thickness of 300nm using an established MOCVD process that utilizes trimethylaluminum and Ammonia. A representative SEM micrograph of the as-grown AlN surface morphology can be seen in Figure 1. The average X-ray Diffraction omega rocking-curve FWHM value is approximately 2000 arcseconds. The surface of the aluminum nitride film that is grown on the silicon substrate has hexagonal surface defects/voids. The AlN was held constant throughout the entire investigation.

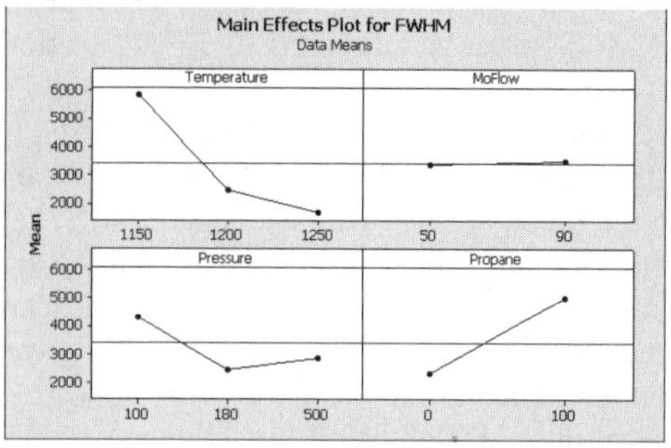

Fig. 2. Main effects plot shows how temperature, pressure, C/Si ratio, and HMDS flow impacted XRD FWHM, an estimation of crystal quality

A four factor – two level statistically designed experiment (DOE) with was performed: (1) Temperature, (2) HMDS (Mo) Flow rate, (3) Growth Pressure, (4) C/Si ratio. Table 1 shows the factors along with the factor levels. Factor (4), C/Si, was manipulated in this experiment by adding propane to the growth atmosphere, thus altering the natural C/Si ratio in HMDS away from 3:1. The effects of the various factors were quantified using on-axis XRD omega rocking curve values to evaluate crystal quality, as well as wafer bow measurements using the contactless capacitance method.

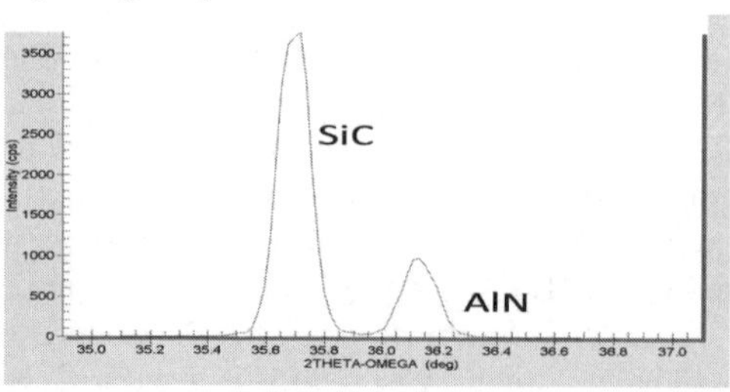

Fig. 3. XRD- 2Theta scan clearly shows SiC and AlN buffer layers, giving indication that the SiC is epitaxailly oriented with the AlN nucleation layer

Figure 2 is a graphical representation of the main effects for all factors measured from the DOE detailed in table 1. Figure 3 is an XRD scan showing the reflections for both the 3C-(111) SiC and the (0002) AlN. From a statistical point of view, temperature was the only significant factor in this set of experiments. Other trends, such as crystal quality also varying with C:Si ratio and growth pressure, appear to be present, but were not determined to be significant in this set of experiments. Wafer bow measurements did not indicate that one individual factor was found to have a statistically significant impact. In this particular experment no higer level interactions between factors were found to be present. A more thorough DOE would be required to elucidate any significant factor interactions. However, this does not cover up the fact that wafer bow did vary widely in this experiment. In general, wafer bow tended to correlate positively with crystal quality, i.e. higher quality films tended to have more wafer bow. It is not possible through this experiement to determine if this is an effect of higher quality films were found to grow at higher temperatures where more stress could be generated due to CTE mismatch, or if this a strain accomodation phenomenon related to film grain boundaries. However, it is entirely likely that one of the two scenarios is the explanation for the correlation between properties, and further experimentation is required to determine the true nature of this relationship. Figure 4 is a comparison between the surface microstructure of two films grown at (A) 1150°C and (B) 1250°C. The (A) sample grown at 1150°C has a rocking curve FWHM of 3863", and the (B) sample grown at 1250°C has a FWHM of 1411". The microstructures shown here are consistent with the idea that higher temperatures encourage grain coalescence, and smoother film surfaces resulting from more thermal energy availability at high temperatures.

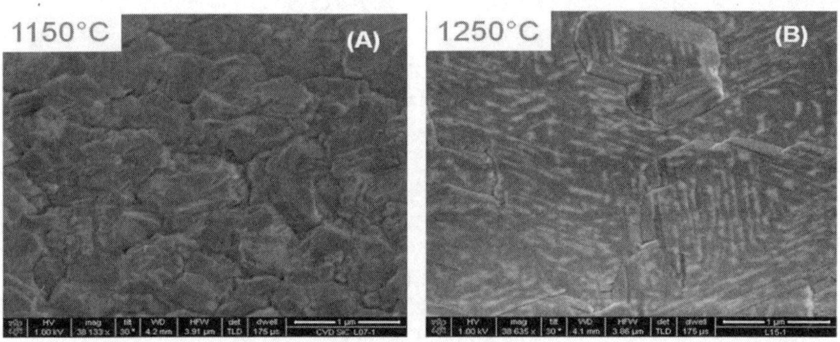

Fig. 4. SEM micrographs of two samples grown at (A) 1150C and (B) 1250C. Higher growth temperatures resulted in better crystal quality, smoother surfaces, and more complete grain coalescence

To gain a more complete understanding of the film microstructure and interfaces, TEM studies were performed. The sample in Figure 4(B), the best quality film from this DOE, was studied in these TEM micrographs. The (111) Si/AlN and the AlN/SiC interfaces were studied, along with defect and polytype identification in the SiC film. As shown by XRD in figure 3, the AlN nucleation layer was confirmed by cross-sectional TEM to be a wurtzitic (0002) oriented film. Figure 5 (A) and (B) are TEM micrographs detailing the interesting interface between AlN nucleation layer and the SiC film. Near the SiC/AlN interface, the structure of the SiC is mainly disordered due to high density of planar faults (complex multiple twinning) and polytype structure cannot be identified.

Figure 6 are High Resolution TEM (HRTEM) micrographs of the HMDS SiC film. Figure 6 (A) clearly shows the characteristic ABC stacking of cubic crystals, along with the [111] direction indicated for reference. Also, the periodicity of the SiC layers is 0.25 nm, consistent with 3C-SiC layer spacing. Figure 6 (B) is another HRTEM micrograph of an area where both faulted and non-faulted SiC can be seen. High densities of planar faults were observed in this film, along with many instances of microtwinning.

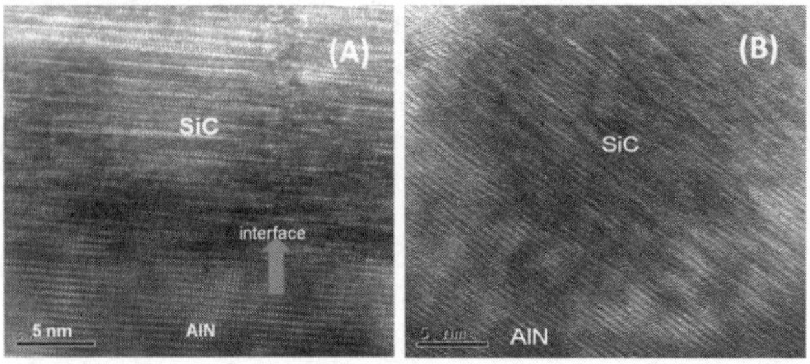

Fig. 5. (A) and (B) AlN/SiC interface diffuse with a high density of planar defects

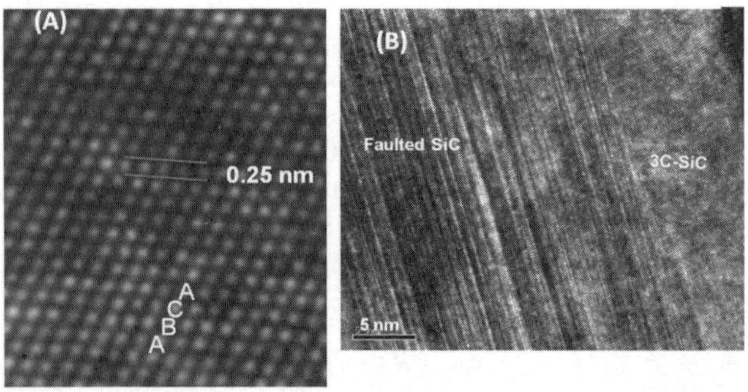

Fig. 6. HRTEM (A) shows the characteristic ABCA cubic stacking in the 3C polytype; (B) Micrograph of planar faults and microtwins in the 3C SiC film.

Summary

CVD SiC films were successfully grown at temperatures below the melting point of silicon using HMDS as a single source precursor for both the Si and C species, utilizing AlN nucleation layers. A partial factorial, two level, four factor DOE was executed to elucidate the relationships between common growth factors, and temperature was found to have the only statistically significant impact on crystal quality, with higher temperatures resulting in lower XRD omega rocking curve values, thus better crystal quality. TEM studies indicate that SiC grown at these temperatures on AlN results in 3C-SiC (111) films that have a high density of planar faults and microtwins.

References

[1] R. F. Davis, J. W. Palmour, J. A. Edmond, IEDM 90 (1990) p. 785

[2] S. Nishino, J. A. Powell, H.A. Will, Appl. Phys. Lett. 42 (1983) p. 460

[3] C. W. Liu, J. C. Sturum, J. Appl. Phys. 82 (1997) p. 4558

[4] A. Figueras, S. Garelick, R. Rodriguez-Clemente, B. Armas, C. Combescure, C. Dupuy, J. Cryst. Growth 110 (1991) p. 528

[5] J. Hoffman, S. Veprek, Thin Solid Films 318 (1998) p. 18

[6] G. Ferro, J. Camassel, S. Juillaguet, C. Balloud, E. K. Polychroniadis, Y. Stoemenos, J. Dazord, H. Peyre, Y. Monteil, S. A. Rushworth, L. M. Smith, Semicond. Sci. Technol. 18 (2003) p.1015

[7] K. Teker, C. Jacob, J. Chung, M. H. Hong, Thin Solid Films 371 (2000) p.53

[8] A. Gupta, D Paramanik, S. Varma, C. Jacob, Bul. Mater. Sci, 27 (2004) p.445

[9] G. H. Jeseen, J. K. Gillespie, G. D. Via, A. Crespo, D. Langley, M. E. Aumer, C.S. Ward, H. G. Henry, D. B. Thomson, D. P. Partlow, IEEE Electron Device Letters 28 (2007) p.354

[10] R. P. Parikh, R. A. Adomaitis, J. D. Oliver, B. H. Ponczak, J. Proc. Ctrl. 17 (2007) p.477

Materials Science Forum Vols. 717-720 (2012) pp 189-192
© (2012) Trans Tech Publications, Switzerland
doi:10.4028/www.scientific.net/MSF.717-720.189

CVD Heteroepitaxial Growth of 3C-SiC on 4H-SiC (0001) Substrates

Xun Li[1, a,*], Stefano Leone[1,b], Sven Andersson[1,c], Olof Kordina[1,d], Anne Henry[1,e], and Erik Janzén[1,f]

[1]Department of Physics, Chemistry and Biology, Linköping University, SE-581 83 Linköping, Sweden

[a]xunli@ifm.liu.se, [b]leone@ifm.liu.se, [c]svean@ifm.liu.se, [d]olkor@ifm.liu.se, [e]anhen@ifm.liu.se, [f]erija@ifm.liu.se

Keywords: heteroepitaxial growth, CVD, 3C-SiC

Abstract. This study has been focused on 3C-SiC epitaxial growth on 4H-SiC (0001) on-axis substrates using the standard CVD chemistry. Several growth parameters were investigated, including growth temperature, in-situ etching process and C/Si ratio. High quality single domain 3C epilayers could be obtained around 1350 °C, with propane present during pre-growth etching and when the C/Si ratio was equal to 1. The best grown layer is 100% 3C-SiC and single domain. The net n-type background doping is around $2x10^{16}$ cm^{-3}. The surface roughness of the layers from AFM analysis is in the 3 to 8 nm range on a 50x50 μm^2 area.

Introduction

3C-SiC has many attractive advantages, such as high electron drift velocity and electron mobility [1]. It has been considered to a promising material for MOSFET devices. Most of the 3C-SiC epilayers, which are grown on silicon wafers, have poor crystalline quality. High quality 3C-SiC could be obtained on hexagonal substrates by VLS [2,3], CVD [4-7] or sublimation epitaxy [8]. However, 3C-SiC epilayers could suffer from serious double position twinning problem, which is a key killer for device performance. A recent study showed that the improved 3C-SiC crystal quality obtained by Cl-based CVD growth enabled large-area Schottky devices with low reverse leakage current and higher barrier height [9]. In this paper, 3C epilayers grown on Si-face nominally on-axis 4H-SiC substrates are reported using the standard, silane and propane, precursors.

Experiments

A horizontal hot-wall CVD reactor equipped with a RF-heated tapered susceptor has been used for this study. The 3C-SiC epilayers were mainly grown on pieces of approximately 2x2 cm^2 n-type doped nominally on-axis Si-face 4H-SiC substrates. No rotation was applied to the substrates. The standard chemistry with propane (C_3H_8) and silane (SiH_4) was used as precursors and a mixture of hydrogen (H_2) and argon (Ar) as carrier gases. The growth runs were done in the following way [10]: clean substrates were loaded in the susceptor on a loading plate. Then the reaction cell was pumped to at least 10^{-6} mbar. The susceptor was heated from room temperature to 1250 °C with carrier gases and the fixed pressure 200 mbar was set during this procedure. After 10 minutes' stabilization at this temperature, the susceptor was heated to the desired temperature. The growth temperature was in the range of 1200 °C to 1500 °C. Different growth parameters, such as C/Si ratio and ramp-up condition were also tested.

The morphology of the samples was assessed by Nomarski differential interference contrast optical microscopy. The layer was qualified as single domain when no double position boundaries (DPBs) were observed and this was confirmed by electron back scattering diffraction (EBSD). The surface roughness was measured by atomic force microscopy (AFM). The thickness of the 3C epilayers was determined using Fourier Transform Infrared (FTIR) reflectance. The quality of the layers was analyzed by low temperature photoluminescence (LTPL) at 2K using 351 nm line of an Ar+ laser as excitation. The net carrier concentration was obtained from Hg probe capacitance-voltage (CV) method at room temperature.

Results and discussion

Growth temperature is one of the most important parameters. 3C-SiC epilayers could be grown between 1200 °C and 1500 °C. The growth rate varied in the range 2 to 6 μm/h with different growth temperatures and also precursor flows. However, temperatures higher than 1375 °C favored polycrystalline growth. At temperatures below 1325 °C, the sample surface was carbonized. High quality 3C epilayers with only single domain could be obtained at temperatures around 1350 °C in this work, thus within a very small temperature window. This differs from the Cl-based chemistry epitaxy for which broader temperature range from 1365 °C to 1425 °C was reported. However, that work was done on 6H-SiC which might favor the formation of single 3C domain on the substrate surface [7]. A special test growth was done when a first layer was grown at the optimal temperature (1350 °C) during 2 h, and then the temperature was raised to 1450 °C for a final growth of 2 h. The single domain morphology was kept and the average growth rate slightly increased.

In-situ etching. Two different ramp-up conditions were tested. One was purging propane through the system together with the carrier gases starting from room temperature. Then the propane flow was changed to the desired growth flow when reaching the deposition temperature and before introducing silane in the reaction chamber. This ramp-up usually took 30 min. The result shows that proper carbon rich pre-growth ambient could effectively reduce the formation of DPBs. In this present study the proper propane flow was identical or very close to the flow used during the growth. If the propane flow was lower than the growth flow, the double boundary twinning problem could not effectively be reduced. In addition, too high flow of propane compared with the growth flow led to fatal surface defects and high density of DPBs. This could be compared to the carbonization step needed when using Si substrate: the formation and the size of DPBs area were found to be influenced by the amount of C precursor flow used during the carbonization layer [11].

Another way was to do the in-situ etching with propane present at the growth temperature before starting the growth. The best etching time found was 5 min when using the same amount of propane as the growth flow. Fig. 1a is an optical microscopy image of 3C-SiC layer grown at 1350 °C with 7 ml/min propane in-situ etching for 5 min at the same temperature. The C/Si ratio was 1, the N/C ratio was 0.004 and the Si/H_2 ratio was 0.015 %, respectively. It has a nice single domain without any DPBs. Fig. 1b shows the surface morphology when using the same process, but without the in-situ etching. The surface is full of DPBs, although the domains seem to try to orient in one particular direction.

Other in-situ surface preparation could be envisaged as pure hydrogen during the temperature ramp-up or silane rich conditions. The former has been tested, however, DPBs were always observed on the 3C-SiC surface. A Si rich environment during the temperature ramp-up was found to be necessary to keep the stability of the 4H polytype [12]. This will be thus in contradiction with the present investigation.

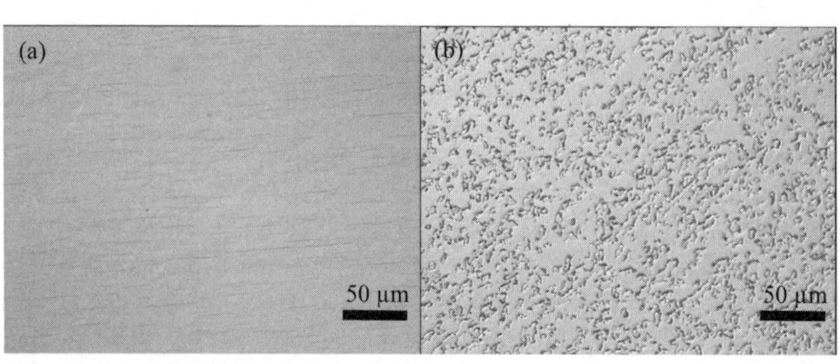

Fig. 1: Optical microscopy images at 400x magnification of 3C-SiC epilayers deposited at 1350 °C on (0001) 4H-SiC substrates with C/Si = 1, N/C = 0.004, Si/H_2= 0.015 % (a) with and (b) without 7 ml/min propane in-situ etching for 5 min. The epilayer thickness is a) 12 μm and b) 10 μm, respectively.

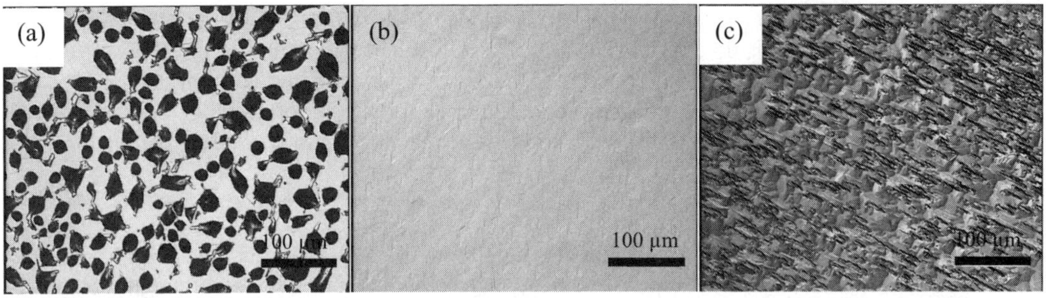

Fig.2 : Optical microscopy images of 3C-SiC epilayers grown at 1350 °C with Si/H_2=0.015% at different C/Si ratios (a) 0.9 (b) 1 (c) 1.1. The thickness of these layers is about 19 µm.

C/Si ratio is another effective growth parameter. Temperatures below 1400 °C and using a C/Si ratio higher than 1.1 (see Fig. 2c) led to polycrystalline material, whereas for a C/Si ratio lower than 0.9, silicon cluster formation was observed on the surface (see Fig. 2a). The most suitable C/Si ratio found equals to 1 at the growth temperature of 1350 °C (Fig. 2b).

Characterization. Most of the samples were grown for 3 to 4 hours. The thickness could be measured by FTIR. As expected the silane flow governs the growth rate. When the Si/H_2 ratio increases from 0.01% to 0.03%, the growth rate almost linearly increases from 2.4 to 4.9 µm/h. For the highest growth rate Si droplets could be observed on the surface of the epilayers. Single domain epilayers could be obtained with a maximum growth rate of about 4 µm/h. With the Cl-based chemistry 10 µm/h growth rate could be achieved maintaining single domain epilayers [7]. When the temperature rises, the growth rate will slightly increase.

EBSD measurement confirmed that the best layer grown was single domain 3C on almost the whole 2x2 cm^2 substrate. In the optical microscopy images, features probably related to stacking faults [9], are, however, observed (see Fig.1.a and Fig.2.b).

After selecting by optical microscopy, the samples with good morphology were analyzed by AFM. Most of the epilayers' surface roughness root-mean-square (RMS) values are in the range of 3 to 8 nm over an area of 50 x 50 μm^2. Fig. 3 is the AFM image of the same sample as shown in Fig. 1a. Its surface roughness is only 0.94 nm over a 10x10 μm^2 area. While the 3C layers grown by Cl-based CVD seem to be smoother, only 0.4 nm on a 10x10 μm^2 area [7].

LTPL technique is a well-known method to examine the quality and doping of these epilayers [13]. The best epilayer is 100 % 3C-SiC according to the LTPL spectrum (Fig. 4). No donor-

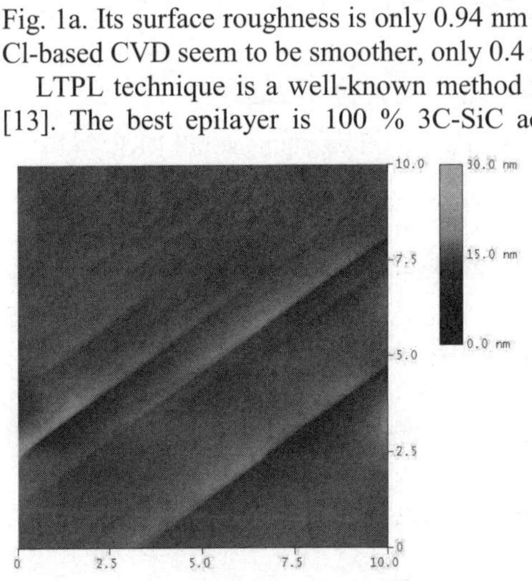

Fig. 3: 10x10 μm^2 AFM image of the same sample as Fig. 1a. Its surface roughness is 0.94 nm.

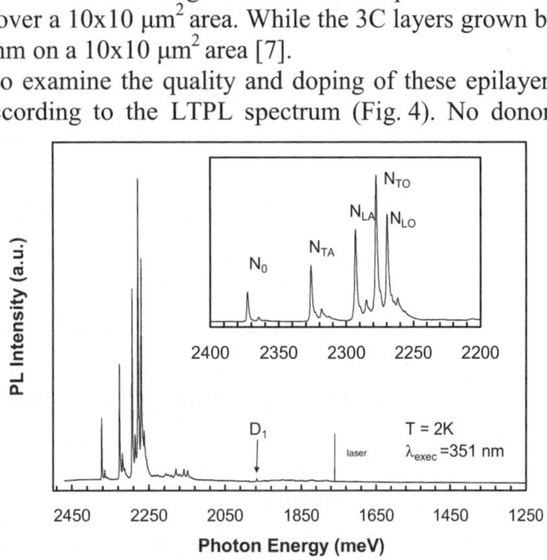

Fig. 4: LTPL spectrum of a 15 µm 3C-SiC epilayer, deposited at 1350 °C with C/Si = 1, N/C = 0.004, Si/H_2= 0.015 %. The insert figure is the near-band gap emission, showing N bound exciton lines.

acceptor pair recombination is observed in the extended spectrum and only weak luminescence associated to the D_I center is detected. From the full width at half maximum of the N-BE transverse acoustic phonon (N_{TA}), the net carrier concentration measured is in the low 10^{16} cm^{-3} range.

CV measurement determined the doping concentration for the nitrogen (N/C = 0.004) doped sample displayed in Fig. 1a to be $5x10^{16}$ cm^{-3}, whereas the undoped one was $2x10^{16}$ cm^{-3}.

Summary

This study shows the results of CVD growth of 3C-SiC on 4H-SiC substrates using standard chemistry. The growth of 3C-SiC epilayers with very smooth surface and only single domain is demonstrated. The quality of the epilayers grown in this work is comparable with that obtained by chloride-based CVD, however, a wide growth window is difficult to obtain.

Acknowledgement

Authors gratefully acknowledge the financial support from the Swedish Energy Agency and the Swedish Research Council (VR). We also want to thank Didier Chaussende from LMGP, Grenoble for EBSD measurements.

References

[1] M. Soueidan and G. Ferro, Adv. Funct. Mater. 16 (2006) 975-979.

[2] M. Soueidan, G. Ferro, O. Kim-Hak, F. Cauwet and B. Nsouli Crystal Growth & Design 8 (2008) 1044-1050.

[3] J. Lorenzzi, M. Lazar, D. Tournier, N. Jegenyes, D. Carole, F. Cauwet, and G. Ferro, Crystal Growth & Design 11 (2011) 2177-2182.

[4] M. V. S. Chandrashekhar, C. I. Thomas, J. Lu and M. G. Spencer, Appl. Phys. Lett. 90 (2007) 173509.

[5] J. Lu, M.V.S. Chandrashekhar, J.J. Parks, D.C. Ralph and M.G. Spencer, App. Phys. Lett. 94 (2009) 162115.

[6] K. Nishino, T. Kimoto, H. Matsunami, Jpn. J. Appl. Phys. 36 (1997) 5202-5207.

[7] S. Leone, F. C. Beyer, A. Henry, O. Kordina, and E. Janzén, Phys. status solidi RRL 4 (2010) 305-307.

[8] A. A. Lebedev, V. V. Zelenin, P. L. Abramov, E. V. Bogdanova, S. P. Lebedev, D. K. Nel'son, B. S. Razbirin, M. P. Shcheglov, A. S. Tregubova, M. Syväjärvi, and R. Yakimova, Semiconductors 41 (2007) 263-265.

[9] J. Eriksson, M. H. Weng, F. Roccaforte, F. Giannazzo, S. Leone, and V. Raineri, Appl. Phys. Lett. 95 (2009) 081907.

[10] A. Henry, J. ul Hassan, J.P. Bergman, C. Hallin and E. Janzén, Chem. Vap. Dep. 12 (2006) 475-482.

[11] S. Roy, M. Portail, T. Chassagne, J. M. Chauveau, P. Vennéguès and M. Zielinski, Appl. Phys. Lett. 95 (2009) 081903.

[12] S. Leone, H. Pedersen, F. C. Beyer, S. Andersson, O. Kordina, A. Henry, A. Canino, F. L. Via, and E. Janzén, submitted to this proceeding (2011).

[13] J. Camassel, S. Juillaguet, M. Zielinski, and C. Balloud, Chem. Vap. Dep. 12 (2006) 549-556.

Materials Science Forum Vols. 717-720 (2012) pp 193-196

doi:10.4028/www.scientific.net/MSF.717-720.193

On stabilization of 3C-SiC using low off-axis 6H-SiC substrates

Valdas Jokubavicius[1, a*], Björn Lundqvist[1,b], Philip Hens[1,c], Rickard Liljedahl[1,d], Rositza Yakimova[1,e], Satoshi Kamiyama[2,f] and Mikael Syväjärvi[1,g]

[1]Department of Physics, Chemistry and Biology, Linköping University, SE-58183, Linköping, Sweden

[2]Department of Materials Science and Engineering, Meijo University, 1-501 Shiogamaguchi, Tenpaku-ku, Nagoya 468-8502, Japan

[a*]valjo@ifm.liu.se, [b]bjolu541@student.liu.se, [c]phihe@ifm.liu.se, [d]ricli@ifm.liu.se, [e]roy@ifm.liu.se, [f]skami@ccmfs.meijo-u.ac.jp, [g]misyv@ifm.liu.se

Keywords: 3C-SiC, sublimation epitaxy, low off-axis 6H-SiC

Abstract. Heteroepitaxial growth of 3C-SiC on 0.8 and 1.2 degree off-oriented 6H-SiC substrates was studied using a sublimation growth process. The 3C-SiC layers were grown at high growth rates with layer thickness up to 300 µm. The formation and the quality of 3C-SiC are influenced by the off-orientation of the substrate, the growth temperature (studied temperature range from 1750 °C to 1850°C), and the growth ambient (vacuum at $5*10^{-5}$ mbar and nitrogen at $5*10^{-1}$ mbar). The largest domains of 3C-SiC and the lowest number of double positioning boundaries were grown using nitrogen ambient and the highest growth temperature. The combined use of low off-axis substrate and high growth rate is a potential method to obtain material with bulk properties.

Introduction

Cubic silicon carbide (3C-SiC) is an excellent candidate for fabrication of advanced metal–oxide–semiconductor field-effect transistors, micro-electro-mechanical systems or even to be used as a substrate for nitrides growth for various optoelectronic applications. Unfortunately, an absence of high quality commercial 3C-SiC substrates makes the development of innovative semiconductor devices difficult and inevitably imposes to grow 3C-SiC heteroepitaxially on substrates like silicon or hexagonal SiC. The advantage of silicon is a low wafer cost. However, large mismatches of lattice constants and thermal expansion coefficients between silicon and 3C-SiC induce a high density of extended defects which deteriorate the quality of the grown 3C-SiC. In contrast, regardless of high substrates cost, the growth of 3C-SiC on hexagonal SiC substrates is highly attractive due to excellent matching between 3C-SiC (111) and hexagonal SiC (0001) crystal planes and thermal expansion coefficients. The most common SiC substrate used for heteroepitaxial growth of 3C-SiC is commercially available on-axis 6H-SiC (0001). The challenge in this case are defects like double positioning boundaries (DPBs) which form when 3C-SiC domains with a different rotation nucleate close to each other as a consequence of the two-dimensional (2D) growth. These should be eliminated in order to unleash the full potential of 3C-SiC grown on hexagonal substrates. Another option for selecting a substrate for heteroepitaxy of 3C-SiC could be off-oriented 6H-SiC (0001). Substrates with larger than 3.5 degrees off orientation are rarely used due to pronounced competition between the step flow growth on the substrate and the 2D nucleation of 3C-SiC on the step terraces which could lead to either repetition of substrate polytype or polytypic mixture in the grown epilayer. However, by lowering the off orientation of the substrate, the step terraces become longer and the probability of spontaneous 3C-SiC nucleation increases [1]. Therefore, the growth of 3C-SiC on low off oriented 6H-SiC (0001) substrates is interesting from a fundamental research point of view as it explores the stability border between these two growth mechanisms. In addition, it could give some insights for improving the growth of 3C-SiC bulk material using on-axis 6H-SiC substrates.

In this study we investigate effects of substrate orientation, growth temperature and growth ambient on reduction of DPBs and stabilization of 3C-SiC on low off-axis 6H-SiC substrates using sublimation epitaxy at high temperatures and high growth rates.

Experimental procedure

Heteroepitaxial layers of 3C-SiC were grown using an epitaxial sublimation growth process [2]. The driving force behind the growth in sublimation epitaxy is a temperature difference between the source and the 6H-SiC substrate. The source material was a polycrystalline SiC plate (with impurities content less than 10^{16} cm^{-3}) which was placed in the bottom of the graphite crucible. The source and the substrate were separated by a graphite spacer, with the thickness of 1 mm, containing a square opening through which SiC species are directly transported from the source to the substrate. In this study, 3C-SiC was grown on substrates having the off-orientation of 0.8 and 1.2 degree from (0001) toward <$11\bar{2}0$> direction. Before the growth, substrates were chemically cleaned by trichlorethylene, acetone and ethanol followed by H_2O:NH_3:H_2O_2 (5:1:1) and H_2O:HCl:H_2O_2 (6:1:1). The growth time was 30 min. The layers were grown at the growth temperatures of 1750°C, 1800°C and 1850°C in two different growth ambient (vacuum ($5*10^{-5}$ mbar) and nitrogen ($5*10^{-1}$ mbar)) for each growth temperature and for different substrate orientation respectively. Experimental details are summarized in Table 1.

Table 1. Experimental details

	Sample ID											
	A1	**A2**	**A3**	**B1**	**B2**	**B3**	**C1**	**C2**	**C3**	**D1**	**D2**	**D3**
Substrate off-orientation [deg]	0.8	0.8	0.8	0.8	0.8	0.8	1.2	1.2	1.2	1.2	1.2	1.2
Growth Temperature [°C]	1750	1800	1850	1750	1800	1850	1750	1800	1850	1750	1800	1850
Growth Pressure [mbar]	$5x10^{-5}$	$5x10^{-5}$	$5x10^{-5}$	$5x10^{-1}$	$5x10^{-1}$	$5x10^{-1}$	$5x10^{-5}$	$5x10^{-5}$	$5x10^{-5}$	$5x10^{-1}$	$5x10^{-1}$	$5x10^{-1}$
Layer thickness [μm]	75	158	300	40	102	240	55	135	300	42	95	200

A temperature ramp up of 20 K/min was applied in all the experiments. The 3C-SiC polytype was identified by low temperature photoluminescence measurements. The surface morphology of the epilayers was analyzed by optical microscopy with Nomarski interference contrast.

Results and discussion

The quality of the substrate is very important for initial 3C-SiC nucleation stage when the growth centers are created. Any deviations in step density or substantial interruptions of the surface, such as scratches or grooves could be intentional 3C-SiC nucleation sites [3, 4]. Therefore, for uniform nucleation of 3C-SiC it is very important to have the lowest possible number of defects and an equal step density all over the substrate surface. It has been reported that during the initial temperature ramping stage (1500-1650°C) in the growth of 3C-SiC using nominally on-axis 6H-SiC (0001) substrates, a homoepitaxy of 6H-SiC governed by a step-flow growth mode takes place before the nucleation of 3C-SiC due to a slight off-orientation of the substrate [5]. In addition, homoepitaxial growth of hexagonal SiC using sublimation epitaxy improves structural quality of the substrate when the growth proceeds via the step-flow growth mode [6]. In this study substrates with 0.8 and 1.2 degree off orientations have been used. Therefore, the step-flow growth on such substrates, before 3C-SiC nucleation, is even more pronounced compared to nominally on-axis 6H-SiC substrates. It implies that the formation of 3C-SiC via 2D nucleation occurs on homoepitaxially grown 6H-SiC surface with improved quality. Therefore, we consider that the density of steps is the main parameter influencing the formation of 3C-SiC unless there are severe substrate surface damages which can not be healed during the initial growth process.

The formation of 3C-SiC on step terraces takes place as soon as the critical supersaturation level needed for 2D nucleation is reached [1]. The off-orientation of the substrate defines the length of step terraces on 6H-SiC (0001) surface. Therefore, the length of step terraces increases by lowering the off-orientation of the substrate and, at the same supersaturation level, 3C-SiC may form on the longer terraces. Two sets of experiments were performed in order to explore the effects of the off-orientation of the substrate on formation of 3C-SiC. In the first one 3C-SiC layers were grown on 0.8 and 1.2 degree substrates in vacuum ($5*10^{-5}$ mbar) at growth temperatures of 1750°C, 1800°C and 1850°C for both off orientations respectively. In the second set all the experiments were replicated as in the first one but in nitrogen ambient ($5*10^{-1}$ mbar). This pressure was selected since it is close to the pressure at which vapor species transport changes from kinetically limited to diffusion limited [2]. At the same growth pressure more 3C-SiC growth centers were created on 0.8 degree substrates compared to 1.2 degree which is depicted in Fig. 1 as 3C-SiC coverage over the substrate surface. By replicating the same growth experiments at in nitrogen atmosphere the coverage of 3C-SiC was reduced (Fig. 1) favoring homoepitaxal growth of 6H-SiC.

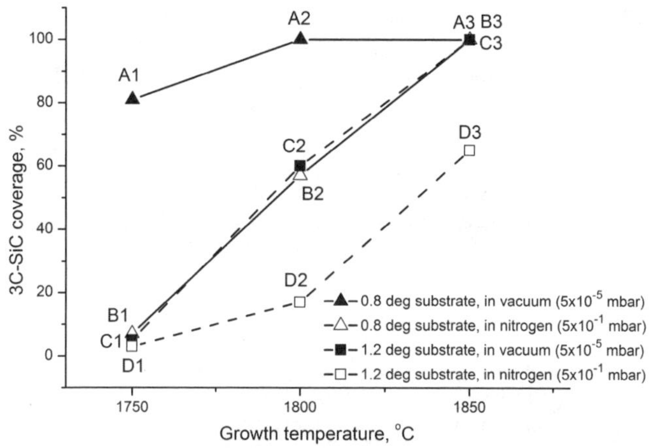

Fig. 1: Surface coverage of 3C-SiC grown on 15x15 mm 6H-SiC substrates

The growth in nitrogen atmosphere is influenced by the diffusion limited mass transfer of SiC bearing species and at the same growth temperature such transfer mechanism is slower compared to kinetically limited growth mechanism which is dominant in the growth at low pressures. Therefore, as a consequence of lower supersaturation, fewer growth centres of 3C-SiC were created and fewer DPBs were observed when using both 0.8 and 1.2 degree substrates (Fig. 1 and 2).

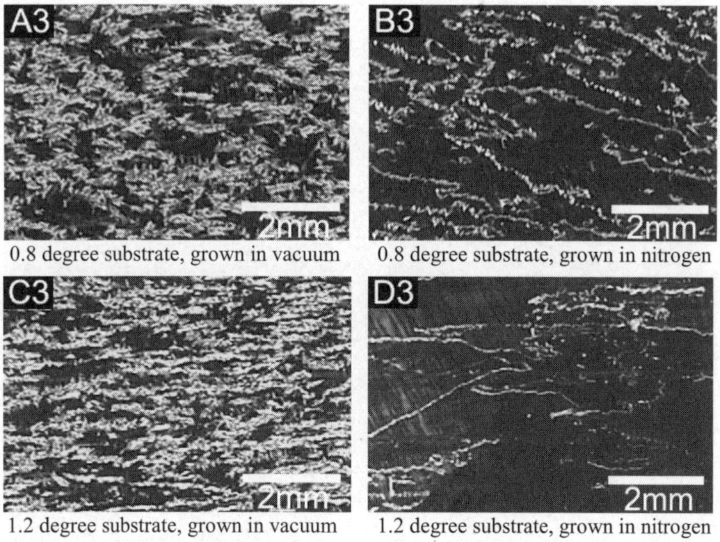

0.8 degree substrate, grown in vacuum — 0.8 degree substrate, grown in nitrogen

1.2 degree substrate, grown in vacuum — 1.2 degree substrate, grown in nitrogen

Fig. 2: Optical micrographs of 3C-SiC domains (growth temperature for all 4 samples– 1850°C)

As demonstrated in this study the number of DPBs can be significantly reduced by changing the growth pressure. Therefore, by controlling the number of 3C-SiC nucleation sites during the initial growth stage in combination with high growth rate and overgrowing domain mechanism [7] the growth of 3C-SiC on low off axis substrates becomes very attractive for further development. In another view, further reduction of the 3C-SiC formation using nitrogen ambient could increase stability of homoepitaxal growth of 6H-SiC using low off-axis substrates.

Summary

We investigated effects of off-orientation of the substrate, the growth temperature and the pressure on heteropepitaxially grown 3C-SiC. A 100% of surface coverage of 3C-SiC was achieved by growing on both 0.8 and 1.2 degree off-oriented substrates at higher growth temperature. The largest density of DPBs was observed in samples grown on 0.8 degree substrates in vacuum. In contrast, the number of DPBs in grown epilayers has been significantly reduced by growing 3C-SiC on 1.2 deg substrates in nitrogen atmosphere ($5*10^{-1}$mbar) and higher growth temperature (1850°C).

Acknowledgement

This work was supported by Ångpanneföreningen Research Foundation, Richerts Foundation, Swedish Energy Agency, Nordic Energy Research, Swedish Research Council (project No. 2009-5307).

References

[1] Y. M. Tairov, V.F Tsvetkov, Prog. Crystal Growth and Charact., Vol. 7 (1983) p. 111-161
[2] M. Syväjärvi et al., J. Cryst. Growth 197 (1999)
[3] T. Kimoto and H. Matsunami, J. Appl. Phys. 75 (1994) p. 850
[4] J.A. Powell et al., Appl. Phys. Lett. Vol. 59 (1991) p. 333
[5] R. Vasiliauskas et al., Mater. Sci. Forum. Vols. 615-617 (2009) p. 189-192
[6] M. Syväjärvi et al., J. Appl. Phys. 88 (2000) p. 1407
[7] M. Syväjärvi et al., Mater. Sci. Forum. Vols. 353-356 (2001) p. 143-146

Materials Science Forum Vols. 717-720 (2012) pp 197-201
© (2012) Trans Tech Publications, Switzerland
doi:10.4028/www.scientific.net/MSF.717-720.197

Si_xC_y thin films deposited at low temperature by DC dual magnetron sputtering: Effect of power supplied to Si and C cathode targets on film physicochemical properties

H. S. Medeiros[1, a], R. S. Pessoa[1, 2, b], J. C. Sagás[1, c], M. A. Fraga[1,3, *], L. V. Santos[1,2, d], H. S. Maciel[1,2, e], M. Massi[1, f] and A. S. da Silva Sobrinho[1, g]

[1]Plasmas and Processes Laboratory, Department of Physics, Technological Institute of Aeronautics, 12228-900, S. J. dos Campos, SP, Brazil

[2]IP&D, University of Vale do Paraíba, 12244-000, S. J. dos Campos, SP, Brazil

[3]Institute for Advanced Studies, Department of Aerospace Science and Technology, 12228-001, S. J. dos Campos, Brazil

[a] henrique_sm_@hotmail.com, [b]rodrigospessoa@gmail.com, [c]jcsagas@pop.com.br, *mafraga@ita.br, [d]santoslv@ita.br, [e]homero@ita.br, [f]massi@ita.br, [g]argemiro@ita.br

Keywords: silicon carbide thin films, DC dual magnetron sputtering, low temperature deposition, film stoichiometry, chemistry analysis.

Abstract. A DC dual magnetron sputtering system with graphite (C) and silicon (Si) targets was used to grow stoichiometric and non-stoichiometric silicon carbide (Si_xC_y) thin films at low temperature. Two independently DC power sources were used to enable the total discharge power be shared, under certain proportions, between the Si and C magnetron cathodes. The motivation was to control the sputtering rate of each target so as to vary the stoichiometric ratio x/y of the deposited films. The species content, thickness and chemical bonds of as-deposited Si_xC_y films were studied by Rutherford backscattering spectroscopy (RBS), profilometry analysis and Fourier transform infrared absorption (FTIR), respectively. Overall, the present work reveals a new reliable plasma sputtering technique for low temperature growth of amorphous Si_xC_y thin films with the capability of tuning the degree of formation of a-SiC, a-Si and a-C bonds in the film bulk.

Introduction

SiC films obtained at low temperature processes are amorphous (a-SiC) or nano-crystallines (nc-SiC) and, thus, can exhibit properties somewhat different from those observed in crystalline films [1]. To better understand these differences it is necessary to characterize the properties of these films, especially those obtained by dual magnetron sputtering (also known as co-sputtering technique [2]), since there are not many references on the subject. It is known that the physical properties of SiC films produced by this technique depend significantly on C/Si atomic ratio in film bulk. In this paper, we studied the chemical composition and bonding of Si_xC_y thin films growth in a graphite/silicon dual target magnetron sputtering system.

Experimental

Deposition system and procedure Si_xC_y films were prepared by using a direct current (DC) dual magnetron sputtering apparatus whose schematic diagram is illustrated in Fig. 1. The system comprises two magnetrons cathodes with normal axis tilted 20° from the central reactor axis. The magnetron targets are graphite and silicon disks (100 mm in diameter each), both of high purity (> 99.9%). The discharge was made in argon (99.999% purity) inserted into the vacuum chamber via a mass flow controller (1179A-MKS). The gas pressure, controlled by a variable conductance throttle valve, was maintained fixed at 5.0 mTorr throughout all the deposition experiments. Before each deposition a turbo pump of 250 l/sec (EXT250-Edwards) backed by a mechanical pump (E2M18-Edwards) evacuated the process chamber to a base pressure of less than 5×10^{-6} Torr.

The magnetron cathodes were powered by two independent variable DC power supplies (MDX-1K Advanced Energy), which allow adjustment of a desired fraction of power in each cathode, always keeping the total discharge power at 200 W. Each sample was deposited on (100) p-type Si wafers (1-10 Ω.cm, 350 μm thick) placed at a substrate-to-target distance (d_{s-t}) fixed at 160 mm. Both targets were pre-sputtered with Ar for 10 min prior to each film deposition to remove contaminations from surfaces. All films were deposited during 120 min without external heating of the substrate, whose temperature, measured by a thermocouple incorporated into the substrate holder, never exceeded 100°C.

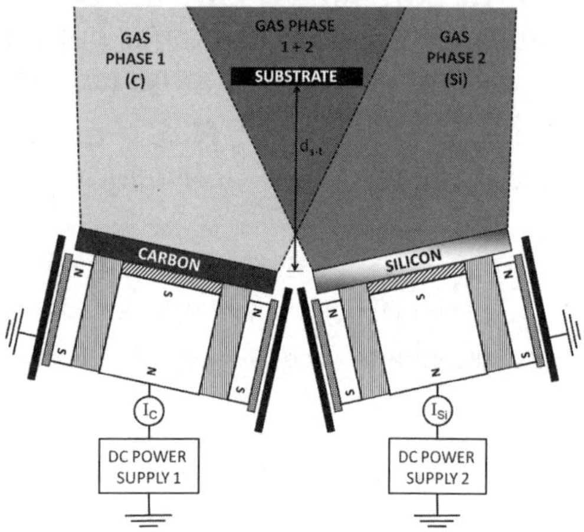

Fig. 1 Schematic diagram of the magnetron co-sputtering deposition set up.

Characterization techniques RBS measurements were carried out using a 2 MeV $^4He^+$ beam from a Pelletron accelerator with a particle detector positioned at 170° to the incident beam and the RBS spectra were analyzed using the RUMP code (RBS spectroscopy analysis package) developed by Cornell University [3]. The thickness of each sample was measured using profilometry (TENCOR Alpha-Step 500 profilometer). Infrared spectroscopy was performed using a Perkin Elmer spectrum 2000 Fourier transform infrared spectrophotometer within the 400- 4000 cm^{-1} range using a (100) silicon wafer as reference.

Results and Discussion

Film composition and deposition rate The relative percentages of Si and C atoms in as-deposited Si_xC_y films were obtained experimentally by RBS characterization. The experimental and simulated curves are shown in Fig. 2. The simulation of RBS spectra allowed calculating the concentrations of C, Si and of contaminants as oxygen (O) and nitrogen (N) in the films.

In the RBS spectrum, the mark of Si, C, N and O denotes the highest energy of backscattering $^4He^+$ obtained from the surface where the collision between $^4He^+$ and these species have occurred as, for example, the marked channel of Si is about 230. For each film deposited under a certain percentage of the total power (fixed at 200W) delivered to C cathode, the C/Si ratio was determined from RBS analyses. By using these data the Fig. 3(a) was plotted in order to evaluate the film stoichiometry dependence on the discharge power supplied to C target. In the same figure a continuous curve is also presented which was obtained from a theoretical estimative based on the following relationship,

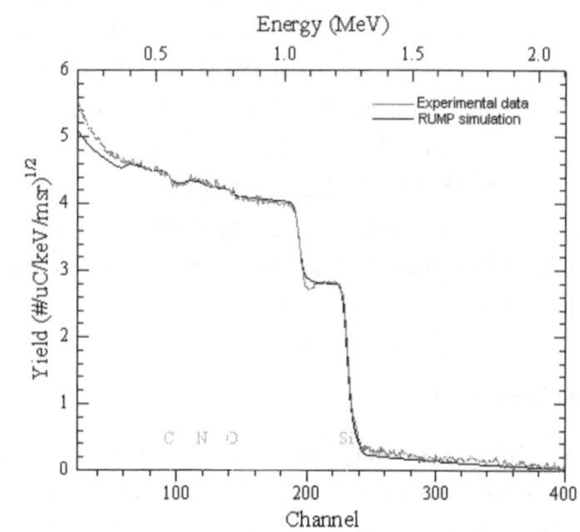

Fig. 2 Typical RBS spectrum of as-deposited Si_xC_y film on Si substrate for condition of 60% power on C target.

$$\frac{R(C)}{R(Si)} = \frac{\frac{J_C}{e}.Y_C.A_{t,C}}{\frac{J_{Si}}{e}.Y_{Si}.A_{t,Si}} = \frac{I_C.Y_C}{I_{Si}.Y_{Si}} \tag{1}$$

where $R(X)$, (X = C or Si), is the sputtering rate of X target in particles/s; I_X and J_X is the current and current density in X target, respectively; Y_X the sputtering yield of X, $A_{t,\,X}$ the target area; and e the fundamental charge. Here, the calculation of Y_X was performed using the theory presented by

Yamamura *et al.* [4, 5]. Furthermore, in calculation of Y_X it was considered that the value of cathode potential (~V_{dc} for magnetron discharge) is the one correspondent to the energy that the ions acquire when crossing, without collisions, the cathode sheath [6].

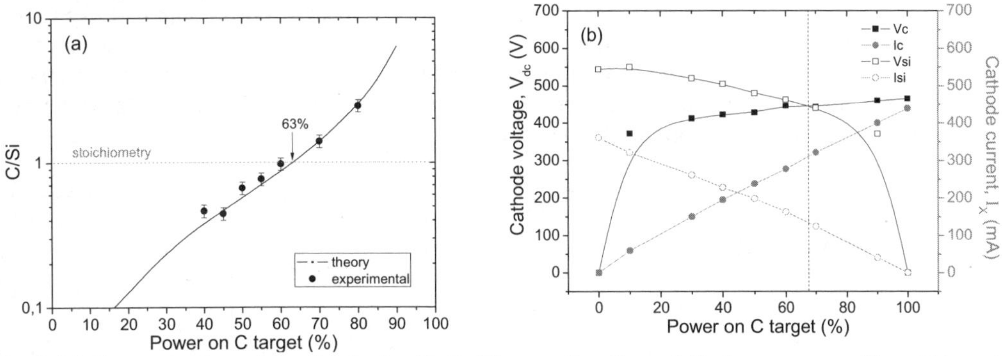

Fig. 3 (a) Concentration of species in the Si_xC_y films as function of discharge power % on C target. (b) Target discharge voltage and current as function of power %.

The reason to use the above ratio, R(C)/R(Si), as equivalent to C/Si ratio in the grown films is due to the assumption that, in non-reactive sputtering, only pure sputtered material is deposited (no reactions with gas species occur) on the collecting area, which includes the substrate and the chamber walls [7]. To solve eq. (1) we must provide the cathode voltage and the cathode current, which can be measured experimentally, for each cathode and each applied power fraction. Accordingly, Fig. 3(b) shows the variation of discharge voltage, V_{dc}, and the current in each magnetron target as a function of power % on C target. Substituting these values in eq. (1), the theoretical curve in Fig. 3(a) is obtained. This figure indicates that with the increase of power % on C target, and consequent decrease in the power applied on Si target, the C/Si ratio increases, as expected. It is also possible to observe a good agreement between theoretical estimative and experimental data, being remarked that stoichiometric film (C/Si = 1), for a total power of 200W, is achieved when about 63% of the discharge power is applied on the C target. Comparing the Fig 3(a) with (b), it is observed that in Fig 3(b) the value of power % on C target where the voltages on both targets are equals is closer to the value of power % where the ratio C/Si = 1. At this stoichiometric ratio, the C cathode current is bigger than the Si cathode current, ie, I_C/I_{Si} about 63%, so maintaining the total discharge power at 200 W. This fact reflects the difference in the sputtering yield (bigger for Si) and secondary electron emission coefficient (bigger for C) of the target materials to determining the targets sputtering rate and consequently the film stoichiometry.

The RBS results also show amounts of oxygen (up to 13%) and nitrogen (up to 3%) in the Si_xC_y films. This is consequence of the low power (200 W) delivered to the discharge plasma so limiting the plasma density, consequently the sputtering rate promoted by the flux of ions coming from the plasma and striking the cathode targets. Increasing the C/Si in film bulk leads to a sharp decrease in the film deposition rate as can be inferred from Fig. 4 that shows an almost linear decay of the film thickness (and respective deposition rate) as a function of power % on C target.

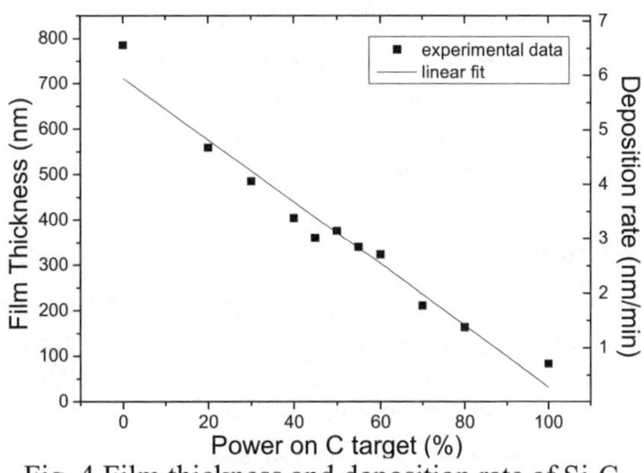

Fig. 4 Film thickness and deposition rate of Si_xC_y thin films as function of power % on C target.

The deposition rate decreases up to 6 times when full power is switched from the Si target to C target, a result that reflects the lower sputtering yield of C compared to Si.

Chemical Bonding As can be seen from RBS and profilometry results, the concentration of Si and C in the Si_xC_y film is largely dependent on the fraction of discharge power supplied to Si and C targets, which in turn determines the flow of Si and C atoms impinging on the substrate surface. The chemical bonds formed in the Si_xC_y films are another important parameter to be investigated. For this, the FTIR characterizations are very useful. Fig. 5 presents the FTIR spectra of the Si_xC_y films for power % on C target ranging from 0 to 100%.

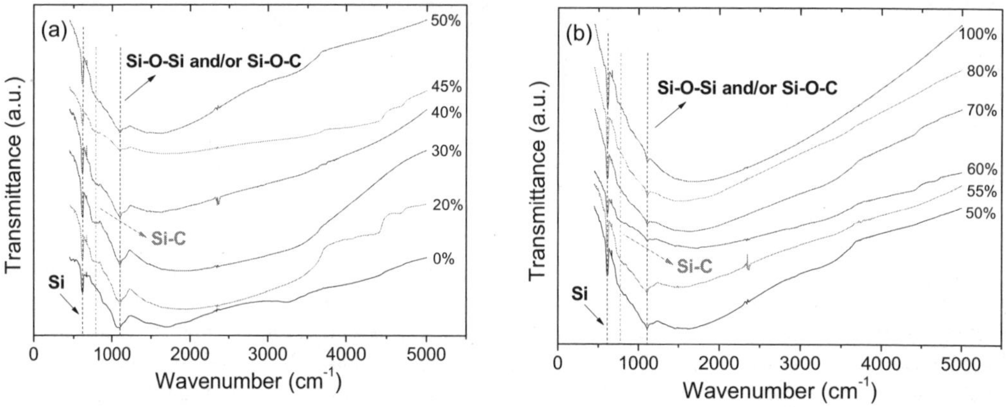

Fig.5 IR spectra of Si_xC_y films for different powers % on C target: (a) 0 to 50% and (b) 50 to 100%.

The absorption peak in the range of 740-820 cm^{-1} is associated with the Si-C stretching vibrational mode observed in diatomic molecules [8-11]. The prominence of a wide absorption band of Si-C suggests the amorphous state of the Si_xC_y films [8, 12]. The FTIR spectra also show the presence of Si-C bonding peak in the films deposited at power fraction on C target varying from 20% to 60% and it practically disappears for greater power fractions. These results are compatible with the C/Si ratio that is greater than 1.0 for C target power fraction above 63% which means that the presence of Si atoms in the film is reduced in comparison with the presence of C atoms. However, the highest peak intensity corresponding to Si-C bonding is observed at 30% of the total applied power on C target (Fig. 5 (a)). The peak located around 1105 cm^{-1} is attributed to Si-O-Si asymmetric stretching of interstitial oxygen in silicon due, probably, to the interstitial oxygen concentration differences within the same wafer from which the substrates were cut for the SiC deposition [13, 14], although, sometimes this peak is assigned to Si-O-C stretching [12]. Increasing the power % applied on C target, the peak corresponding to the Si-O-Si bonding tends to decrease, because the film becomes increasingly rich in carbon and, consequently, the silicon concentration is reduced. However, this peak does not disappear entirely, possibly because the film thickness decreases with the power % on C target increases, which allowing therefore, a possible contribution from the substrate/film interface for this peak.

Conclusions

In this work, we showed an especial configuration of a dual magnetron sputtering reactor as alternative method for deposition of Si_xC_y thin films at low temperature. The system comprises a Si and a C target, each one being powered by independently DC power supplies. The discharge was made in argon and the non-reactive sputtering deposition process was carried out maintaining fixed the total discharge power, the gas pressure, the gas flow rate and the deposition time. In all depositions, the temperature of the substrate holder never exceeded 100°C. The results indicated that a-Si_xC_y thin films with a wide range of C/Si ratio (0.4 to 2.0) can be grown on Si substrate using this dual magnetron sputtering (co-sputtering) technique. The FTIR analyses show the presence of Si-C bonds in the films. Further studies are being conducted in order to induce the formation of nc-SiC in bulk of Si_xC_y film deposited at low temperature. For that, a pathway indicated in the literature [15] would be the use of bias-sputtering technique.

Acknowledgements

The financial support of programs PNPD-CAPES (process number 02765/09-8) and PNM-CNPq (process number 134290/2010-0) is strongly acknowledged. The authors also thank Dr. Marcel Dupret from LAMFI-USP for the RBS measurements and INPE for the Profilometry measurements.

References

[1] G. Foti, Applied Surface Science 184 (2001) 20–26.

[2] G. Golan, A. Axelevitch, B. Gorenstein, Microelectronics Journal 37 (2006) 1538-1542.

[3] L. R. Dolittle, Nucl. Instrum. Methods Phys. Res. B 9 (1985) 344-351.

[4] Y. Yamamura and H. Tawara, Atomic Data and Nuclear Data Tables 62 (1996) 149-253.

[5] M. P. Seah, C. A. Clifford, F. M. Green,I. S. Gilmore, Surf. Interface Anal. 37 (2005) 444-448.

[6] M. A. Lieberman and A. J. Lichtenberg, *Principles of Plasma Discharges and Materials Processing*, 2nd ed., Wiley-Interscience, Hoboken, 2005.

[7] G. Franz, *Low Pressure Plasmas and Microstructuring Technology*, 1st ed, Springer-Verlag Berlin Heidelberg, 2009.

[8] B. P. Swain, R. O. Dusane, Materials Letters 60 (2006) 2915–2919.

[9] K. Mergia, D. Lafatzis, N. Moutis, T. Speliotis, G. Apostolopoulos, F. Cousin, Applied Physics A 92 (2008) 387-395.

[10] H, Zhang, Z, Xu, Optical Materials 20 (2002) 177-181.

[11] Valeri P. Tolstoy, Irina V. Chernyshova, Valeri A. Skryshevsky, *Handbook Of Infrared Spectroscopy of Ultrathin Films*, John Wiley & Sons, Inc., Hoboken, New Jersey, 2003.

[12] V. Jousseaume, N. Rochat, L. Favennec, O. Renault, G, Passemard, Materials Science in Semiconductor Processing 7 (2004) 301-305.

[13] Q. J. Cheng, J. D. Long, Z. Chen, S. Xu, Journal of Physics D 40 (2007) 2304-2307.

[14] M. Kuenle S. Janz, O. Eibl, C. Berthold, V. Presser, K-G. Nickel, Materials Science and Engineering B 159-160 (2009) 355-360.

[15] M. Lattemann, E. Nold, S. Ulrich, H. Leiste, H. Holleck, Surface and Coatings Technology 174-175 (2003) 365-369.

CHAPTER 3:

Physical Properties and Characterization of SiC

Materials Science Forum Vols. 717-720 (2012) pp 205-210
doi:10.4028/www.scientific.net/MSF.717-720.205

Transition Metal Defects in Cubic and Hexagonal Polytypes of SiC: Site Selection, Magnetic and Optical Properties from *ab initio* Calculations

V. Ivády[1,2,a], B. Somogyi[2,b], V. Zólyomi[1,c], A. Gällström[3,d], N.T. Son[3,e], E. Janzén[3,f], and A. Gali[1,2,g,*]

[1]Research Institute for Solid State Physics and Optics, Hungarian Academy of Sciences, PO Box 49, H-1525 Budapest, Hungary

[2]Department of Atomic Physics, Budapest University of Technology and Economics, Budafoki út 8, H-1111 Budapest, Hungary

[3]Department of Physics, Chemistry and Biology, Linköping University, SE-581 83 Linköping, Sweden

[a]ivady.viktor@gmail.com, [b]somogyibalint@gmail.com, [c]zachary@szfki.hu, [d]andga@ifm.liu.se, [e]son@ifm.liu.se, [f]erj@ifm.liu.se, [g]agali@szfki.hu, *corresponding author

Keywords: point defects, hybrid density functional theory, transition metals, electron paramagnetic resonance, photoluminescence, site selection, polytypes

Abstract. Relatively little is known about the transition metal defects in silicon carbide (SiC). In this study we applied highly convergent and sophisticated density functional theory (DFT) based methods to investigate important transition metal impurities including titanium (Ti), vanadium (V), niobium (Nb), chromium (Cr), molybdenum (Mo) and tungsten (W) in cubic 3C and hexagonal 4H and 6H polytypes of SiC. We found two classes among the considered transition metal impurities: Ti, V and Cr clearly prefer the Si-substituting configuration while W, Nb, and Mo may form a complex with a carbon vacancy in hexagonal SiC even under thermal equilibrium with similar concentration. If the metal impurity is implanted into SiC or when many carbon impurities exist during the growth of SiC then complex formation between the Si-substituting metal impurity and the carbon vacancy should be considered. This complex pair configuration exclusively prefers the hexagonal-hexagonal sites in hexagonal polytypes and may be absent in the cubic polytype. We also studied transition metal doped nano 3C-SiC crystals in order to check the effect of the crystal field on the d-orbitals of the metal impurity.

Introduction

The knowledge about transition metal defects in silicon carbide (SiC) is scarce, though some of them (like vanadium) are intentionally introduced into SiC in order to produce semi-insulating samples [1], or can enter SiC unintentionally during or after the growth process [2]. There are unidentified photoluminescence centers that are presumably originated from transition metal defects [3]; however, the number of detected centers does not follow the number of inequivalent substitutional sites in different polytypes, unlike the case of vanadium or chromium [4,5], which makes the defect identification difficult. While *ab initio* calculations do not show the striking accuracy of experimental spectroscopy, still they could predict trends and provide useful information about the relative concentration of defects, their electronic structure, symmetry, spin state and even excitation energies. Combination of data from simulations and experiments can greatly contribute to understanding the nature of point defects in semiconductors. In this study we applied highly convergent and sophisticated density functional theory (DFT) based methods to investigate important transition metal impurities including titanium (Ti), vanadium (V), niobium (Nb), chromium (Cr), molybdenum (Mo) and tungsten (W) in cubic 3C and hexagonal 4H and 6H polytypes of SiC. We call these impurities M-defects in the context. The motivation of this selection was two-fold: (i) experimental data were available for most of these impurities in SiC [6-9] which made a comparison between theory and experiment possible, (ii) the trends could be studied among $4s^2 3d^2$-$3d^4$ M-defects and among $4s^2 3d^4$-$6s^2 5d^4$ M-defects (see Fig. 1).

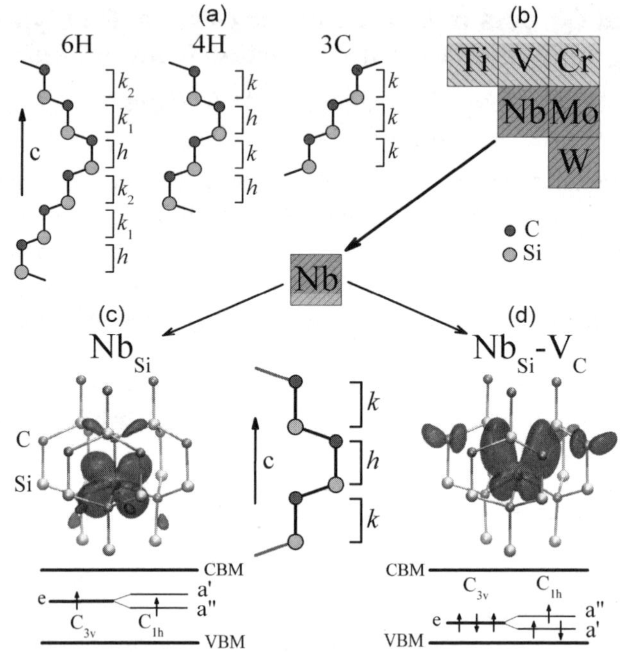

Fig. 1: (a) Stacking sequences and notation of Si-C bilayers in 6H, 4H, and 3C SiC polytypes. (b) The portion of the periodic table with the elements considered in our study. Light grey shaded impurities are preferentially Si-substitutionals in any polytype (c), while dark grey shaded impurities may also form complex with a C-vacancy resulting in an asymmetric split-vacancy configuration in hexagonal polytypes (d). (c) and (d) show the geometry of the neutral niobium impurity at *h* and *h-h* configurations in 4H-SiC, respectively, where the lobes represent the isosurface of the calculated spin density with a value of 0.014 in atomic units. Schematic diagrams of the electronic structure in the band gap are also depicted. Neutral niobium is Jahn-Teller unstable resulting in a reduction in symmetry.

Methodology

We applied DFT with the PBE functional in order to calculate the ground state of the defects (see Ref. [10] for more detail and references). We used 512-atom simple cubic 3C-SiC, 576-atom hexagonal 4H-SiC and 432-atom hexagonal 6H-SiC supercells to accommodate the point defects. In these large supercells we used the Γ-point to sample the Brillouin-zone which is sufficient to obtain a convergent charge density and simultaneously ensures to monitor the degeneracy of the defect states in the band gap. Transition metal elements contain strongly localized d-orbitals that may be poorly described by semilocal DFT functional like PBE because semilocal functionals tend to delocalize these states. We applied two strategies in order to avoid this problem: i) once we obtained a reliable geometry from PBE calculation, we re-calculated the electronic structure by a screened hybrid density functional (HSE06) which was very successful in the quantitative description of native defects in SiC [11]. This functional contains a Fock-exchange term, which tends to localize the defect states and could provide, at least, qualitatively correct electronic structure of transition metal impurities. HSE06 calculations are computationally extremely expensive, so we applied also another strategy. ii) We modeled the defect in nanocrystalline 3C-SiC as well with different sizes of about 0.9-1.4 nm in diameter. Quantum confinement opens the gap in this system and the crystal field also clearly separates the defect states arising from the d-orbitals of transition metal impurities even at the PBE level.

Results and Discussion

Bulk SiC. We first investigated the substitutional and interstitial defects for the selected impurities in cubic and 4H-SiC. We found that the interstitial and carbon substituting M-defects have very high formation energies in all SiC polytypes and will not be considered further. Two configurations have the lowest formation energies: (i) an M impurity replaces Si, M_{Si}, in other words, occupies a Si-vacancy, (ii) M_{Si} forms a complex with a C-vacancy, in other words, an M impurity occupies divacancy. A schematic diagram about the relative stabilities and defect structures is shown in Fig. 1.

It is instructive to discuss the electronic structure of M_{Si} and M_{Si}-C_{vac} defects before providing their calculated geometries and formation energies because the formation energy may depend on the stoichiometry of SiC and the charge state of the defect provided by the actual position of the Fermi-level. In the case of 3C-SiC the five d orbitals split into t_2 and e-levels where the t_2 level will interact with the C sp^3 dangling bonds while the e-level remains atomic-like [12-15]. This e-level will be filled by 0,1,2 electrons when M=Ti,V,Cr, respectively. The position of the e-level tends to lie closer to the valence band edge (E_V) when occupied. In hexagonal polytypes we gave a detailed group theory analysis for the case of Mo_{Si} [7] which is basically valid for the first row M_{Si} defects: in that case the five d orbitals will split as $\{a_1,e(1)\}$ and $e(2)$ orbitals where $\{a_1,e(1)\}$ is similar to the t_2 state in 3C-SiC but a slight mixture between $e(1)$ and $e(2)$ is allowed by symmetry. Generally, high spin states are the ground state of M_{Si} defects because of the Hund-rule is valid for degenerate atomic-like orbitals. For instance, the $e(2)$ level is occupied by two electrons with parallel spins resulting in a S=1 state for neutral Cr, Mo, W and negatively charged V, Nb substituting at the Si-site. It is important to notice that the states originated from strongly localized d orbitals may be not well described by semilocal PBE functional. According to our HSE06 calculation, the position of the fully occupied $e(2)$-level in the spin-up channel (S=1, M_S=1) will shift down with respect to E_V by about 1.0, 0.3, 0.1 eV compared to PBE values for M=Cr,Mo,W, respectively, whereas the gap "opens" by 0.93 eV. The calculated energy difference between cubic and h-sites of M_{Si} defects is within 0.1 eV for any M impurity in any charge state in hexagonal polytypes. This result was obtained both by PBE and HSE06 functionals. This small energy difference (i) indicates that the concentration of M_{Si} defects is about the same, so 2(3) signals are expected for M_{Si} defects in 4H(6H) polytypes, (ii) allows us to discuss the possible charge states of the defects only at the h-site in hexagonal polytypes.

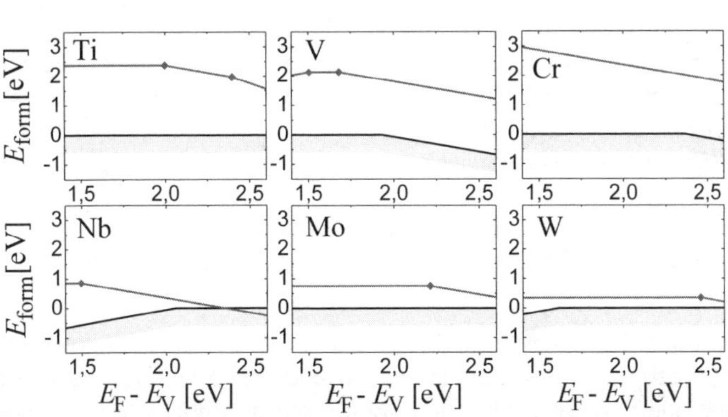

Fig. 2. The relative formation energies between M_{Si} (black line) and M_{Si}-C_{vac} (blue line) defects at h-sites in 4H-SiC as a function of the position of the Fermi-level with respect to the valence band edge (E_F-E_V) calculated by HSE06 functional. We aligned the reference formation energy to that of the neutral M_{Si}-defects. The formation energy of M_{Si}-defects is shown for Si-rich condition. The bottom of the shaded area corresponds to that for C-rich condition.

Next, we describe the M_{Si}-C_{vac} complex. As can be seen in Fig. 1(d) this complex is rather a split-vacancy defect where the M impurity sits in a divacancy. In this compound semiconductor the M impurity is not exactly in the middle of divacancy but stays closer to the Si-vacancy than to the

C-vacancy forming an asymmetric split-vacancy (ASV) configuration. The ASV configuration has not yet been reported so far in any compound semiconductor. The highest symmetry of ASV defects is C_{3v} in 3C-SiC, and at the axial configurations in hexagonal polytypes, whereas the symmetry is reduced to C_{1h} at basal configurations. We analyze the C_{3v} symmetry configuration by invoking the defect-molecule picture and group theory. In a divacancy the silicon and carbon dangling bonds create two a_1 and two e states where the highest energy e state is localized on the silicon dangling bonds (C-vacancy). The s orbital of the M impurity transforms as a_1 while the d orbitals split into one a_1 and two e states. The a_1 states of the divacancy and M impurity can recombine as well as the corresponding e-states resulting in a_1, a_1, e and e bonding states and the corresponding anti-bonding combinations. If we take a M[s^2d^4] impurity with 6 valence electrons and the 6 valence electrons in a divacancy then 12 electrons occupy these states. That means that all the bonding orbitals are fully occupied resulting in a singlet ground state of $a_1^{(2)}a_1^{(2)}e^{(4)}e^{(4)}$. The last e-state and the empty antibonding a_1*, e* states appear in the gap. Briefly, these defects will be either neutral or negatively ionized in hexagonal polytypes under experimental conditions of formation. The formation energies of the most stable h-h ASV configurations are compared with M_{Si} defects in 4H-SiC in Fig. 2.

The results are the same within 0.1 eV in 6H polytypes. Apparently, the formation energies of ASV complexes are comparable to simple M_{Si} defects for M=Nb,Mo,W impurities in hexagonal polytypes. This means that ASV complexes may form with similar concentration as M_{Si} defects in hexagonal polytypes. This is particularly true for Si-rich condition where many C-vacancies are formed. According to our calculations ASV has extremely high stability for M=Nb,Mo,W impurities where the binding energy between M_{Si} and C_{vac} is over 4 eV. This also means that a diffusing carbon vacancy, which is mobile at typical growth temperature, is trapped very effectively by these M impurities. In the case of M=Ti,V,Cr, M_{Si} is definitely preferred. Indeed, 2(3) photoluminescence and deep level transient spectroscopy signals were found associated with these M impurities in 4H(6H)-SiC [6], which can be explained by the number of inequivalent substitutional sites in these polytypes.

Moreover, we found that the h-h configuration of ASV defects is much more stable than any other combinations involving cubic sites in hexagonal polytypes for M=Nb,Cr,Mo,W. The next stable configuration is $M_{Si}(h)$-$C_{vac}(k)$ with energy difference of more than 0.6 eV, at least among these M impurities. The reason of the exclusive site selection is due to the comfortable situation of the $\{d_{xy}, d_{xz}\}$ and $\{d_{yz}, d_{x2}\}$ orbitals which can only maximally overlap with Si dangling bonds of C_{vac} if first neighbor Si atoms of the carbon vacancy are placed just above the first neighbor C-atom of the M impurity [see Fig. 1(d)]. In any other configuration of C_{vac}, the d states have to "twist" in order to form a molecular orbital with Si dangling bonds (c.f., Fig. 3). As a consequence, the corresponding e defect state lies deeper in the gap for the h-h configuration with possessing lower formation energy than that of other configurations. We note that the k-k configuration is, at least, more than 1.0 eV less stable than h-h configuration in hexagonal polytypes. This explains our finding in the cubic polytype: M_{Si} defects are much more stable than ASV defects for any M impurity. The existence of the ASV defect has been recently demonstrated in niobium doped 4H-SiC [2].

Nano 3C-SiC. We studied the most stable M_{Si} and M_{Si}-C_{vac} configurations in nano 3C-SiC with spherical shapes. According to our calculations in bulk SiC, M_{Si}-C_{vac} may not form during the fabrication of doped nano 3C-SiC but it may be created by implantation of the corresponding M ions, thus we also considered this configuration (see Fig. 3).

We used three nanoparticles in size: 0.88 nm, 1.12 nm and 1.40 nm in diameter which contain 35, 71, 147 host atoms, respectively, where a Si atom was in the middle of the cluster. The surface of a nanoparticle was terminated by hydrogen atoms. In the smallest cluster we studied only M_{Si} defects. If the corresponding e-level in the gap of nano 3C-SiC is occupied by two electrons in M_{Si} defects (neutral Cr, Mo, W; negatively charged Cr, V) then the symmetry will be T_d. In the case of

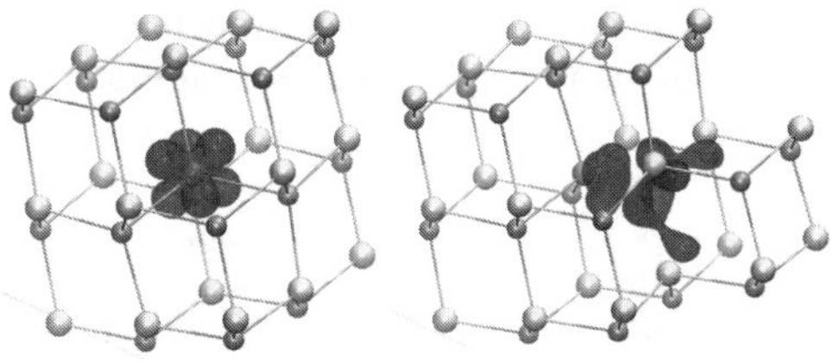

(a) (b)

Fig. 3. (a) M_{Si} and (b) M_{Si}-C_{vac} defects in 3C-SiC cluster cut from nanoparticles where the M impurity is in the middle of the cluster. Typical spin densities are shown by blue lobes given on the example of W impurity.

Table 1. Calculated PBE gap for pristine and selected defective nanocrystals in eV unit

size/conf	pristine	V_{Si}^-	Nb_{Si}^-	Mo_{Si}	W_{Si}	Nb_{Si}^--C_{vac}	Mo_{Si}-C_{vac}	W_{Si}-C_{vac}
0.88 nm	4.45	1.02	0.79	1.51	1.17			
1.12 nm	4.18	0.83	0.50	1.38	1.02	1.29	1.17	1.25
1.40 nm	3.20	0.47	0.23	1.06	0.67	1.01	0.94	1.02

M_{Si}-C_{vac} configuration and the afore-mentioned charge state of the impurity, the symmetry will be C_{3v}. Due to the quantum confinement the band gap of pristine SiC nanoparticles will increase with decreasing their size. The calculated PBE band gaps are shown in Table 1 where we restrict our investigation to the simplest and most promising M defects from an application point of view.

We emphasize that the PBE gaps are not the optical gap that may be measured in experiments. The quasi-particle correction would increase this energy while the electron-hole interaction would compensate this effect. According to our previous work with time-dependent DFT with a hybrid density functional in the kernel [16] the net effect will be an increase of about 1.0 eV with respect to the PBE gap in pristine 3C-SiC nanoparticles. This means that very small nano 3C-SiC particles have a large gap, larger than the band gap of bulk 4H-SiC. In the large gap of nano 3C-SiC, defect states appear in the gap and color the transparent nanoparticles (see Table 1). Beside the *e*-level the empty, antibonding t_2 level is clearly visible in M_{Si} defects (see Fig. 4). In the case of M_{Si}-C_{vac} defects the *e*-level is completely empty. Interestingly, while the symmetry is only C_{3v} still the empty antibonding state is almost triply degenerate. The energy difference between the corresponding states is listed in Table 1. It is expected that an optical transition can occur between these occupied and unoccupied defect states. As can be seen PBE gaps indicate optical transitions in the infrared energy region. This may be expected since strong zero-phonon line transitions were detected for V

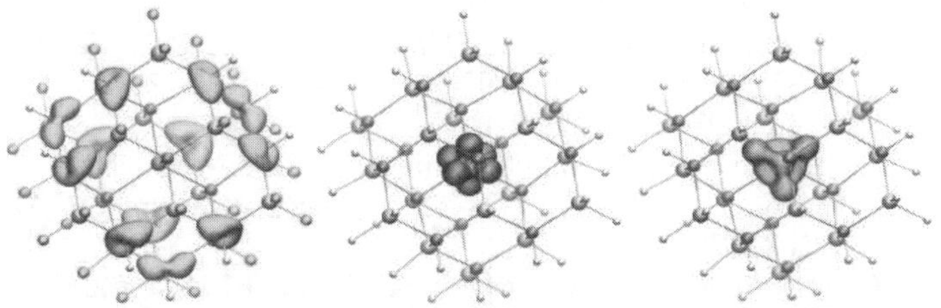

(a) HOMO-1 (b) HOMO (c) LUMO

Fig. 4. Modulus squared of the corresponding wave functions of the W_{Si} defect in 0.88-nm 3C-SiC. The W impurity is in the middle of the cluster. HOMO-1 state corresponds to the highest molecular occupied orbital of pristine nano SiC, which is delocalized. The HOMO state is the occupied *e*-state while LUMO is the empty t_2 state. These states are quite localized states and associated with the *d* orbitals of the W impurity.

[6], Mo [7] and W [2] defects in bulk hexagonal SiC with an energy close to 1.0 eV. We emphasize again that PBE gaps are not quantitative here and there is no rule of thumb even to roughly estimate the excitation energies. Nevertheless, PBE gaps of defect states may be used qualitatively in order

to see the trends in different sizes of SiC nanoparticles. Apparently, the PBE gaps associated with the lowest excitation energies show considerable quantum confinement effect in these small SiC nanoparticles.

From the computation point of view this makes the exact calculation of the excitation spectrum challenging because the excitation spectrum should be calculated in relatively large nanoparticles in order to make any correspondence between the data extracted from nano 3C-SiC and bulk SiC. Nevertheless, it is clear from the present calculations that transition metals indeed act as color centers in SiC nanoparticles, possibly, close to the red and infrared energy region.

Summary

We proposed for some transition metal impurities that asymmetric split-vacancy defects can form in hexagonal polytypes of SiC that are stable only in the *h-h* configuration. Our finding indicates that some electrically and optically active defects may appear in hexagonal polytypes that are absent in cubic polytype, and the number of signals may not follow the number of possible configurations in a given polytype. We found that the transition metal impurities may act as color centers in nano SiC and show quantum confinement in ultrasmall (1-2 nm in diameter) nanoparticles.

Acknowledgment

Support from the Swedish Foundation for Strategic Research, the Swedish Research Council, the Swedish Energy Agency, the Swedish National Infrastructure for Computing Grants No. SNIC 011/04-8 and No. SNIC001-10-223, and the Knut and Alice Wallenberg Foundation is acknowledged.

References

[1] H. M. Hobgood, R. C. Glass, G. Augustine, R. H. Hopkins, J. R. Jenny, M. Skowronski, W. C. Mitchel, and M. Roth, Appl. Phys. Lett. **66** (1995) 1364.

[2] N. T. Son, V. Ivády, A. Gali, A. Gällström, S. Leone, O. Kordina, and E. Janzén, these proceedings.

[3] A. Gällström, B. Magnusson, F. Beyer, A. Gali, N. T. Son, S. Leone, I. Ivanov, A. Henry, and E. Janzén, these proceedings.

[4] J. Schneider, A. Dörnen, S. Leibenzeder, and R. Stein, Appl. Phys. Lett. **56** (1990) 1184.

[5] N. T. Son, A. Ellison, B. Magnusson, M. F. MacMillan, W. M. Chen, B. Monemar, and E. Janzén, J. Appl. Phys. **86** (1999) 4348.

[6] Recent Major Advances in SiC, edited by W. J. Choyke, H. Matsunami, and G. Pensl (Springer-Verlag, Berlin Heidelberg, 2004).

[7] A. Gällström, B. Magnusson, and E. Janzén, Mater. Sci. Forum 615-617 (2009) 405.

[8] F. C. Beyer, C. G. Hemmingsson, A. Gällström, S. Leone, H. Pedersen, A. Henry, and E. Janzén, Appl. Phys. Lett. **98**, (2011) 152104.

[9] N. Achtziger, J. Grillenberger, and W. Witthuhn, Appl. Phys. A **65** (1997) 329.

[10] P. Carlsson, N. T. Son, A. Gali, J. Isoya, N. Morishita, T. Ohshima, B. Magnusson, and E. Janzén, Phys. Rev. B **82** (2010) 235203.

[11] T. Hornos, A. Gali, and B. G. Svensson, Mater. Sci. Forum 679-680 (2011) 261.

[12] M. S. Miao, and W. R. L. Lambrecht, Phys. Rev. B **74** (2006) 235218.

[13] I. I. Parfenova, Semiconductors **38** (2004) 189.

[14] L. V. C. Assali, W. V. M. Machado, and J. F. Justo, Phys. Rev. B **69** (2004) 155212.

[15] V. L. Shaposhnikov, and N. A. Sobolev, J. Phys.: Condens. Matter 16 (2004) 1761.

[16] M. Vörös, P. Deák, T. Frauenheim, and A. Gali, Appl. Phys. Lett. **96** (2010) 051909.

Materials Science Forum Vols. 717-720 (2012) pp 211-216
doi:10.4028/www.scientific.net/MSF.717-720.211

Electronic Configuration of Tungsten in 4H-, 6H-, and 15R-SiC

A. Gällström[1,a,*], B. Magnusson[2,b], F.C. Beyer[1,c], A. Gali[3,4,d], N.T. Son[1,e], S. Leone[1,f], I.G. Ivanov[1,g], A. Henry[1,h], C.G. Hemmingsson[1,i], and E. Janzén[1,j]

[1]Department of Physics, Chemistry and Biology, Linköping University, SE-581 83 Linköping, Sweden

[2]Norstel AB, Ramshällsvägen 15, SE-602 38 Norrköping, Sweden

[3]Department of Atomic Physics, Budapest University of Technology and Economics, Budafoki út 8, H-1111 Budapest, Hungary

[4]Research Institute for Solid State Physics and Optics, Hungarian Academy of Science, PO Box 49, H-1525 Budapest, Hungary

[a]andga@ifm.liu.se, [b]bjorn.magnusson@norstel.com, [c]fbeyer@ifm.liu.se, [d]agali@szfki.hu, [e]son@ifm.liu.se, [f]leone@ifm.liu.se, [g]ivani@ifm.liu.se, [h]anhen@ifm.liu.se, [i]cah@ifm.liu.se, [j]erj@ifm.liu.se
*corresponding author

Keywords: deep level defect, PL, transition metal, Crystal Field Model

Abstract. A commonly observed unidentified photoluminescence center in SiC is UD-1. In this report, the UD-1 center is identified to be tungsten related. The identification is based on (i) a W-doping study, the confirmation of W in the samples was made using deep level transient spectroscopy (DLTS), (ii) the optical activation energy of the absorption of UD-1 in weakly n-type samples corresponds to the activation energy of the deep tungsten center observed using DLTS. The tungsten-related optical centers are reported in 4H-, 6H-, and 15R-SiC. Further, a crystal field model for a tungsten atom occupying a Si-site is suggested. This crystal field model is in agreement with the experimental data available: polarization, temperature dependence and magnetic field splitting.

Introduction

Many of the deep levels observed in optical measurements of silicon carbide (SiC) are still neither identified nor understood. Much effort has been devoted to the understanding of the different defects, since deep level defects play a crucial role in controlling the electronic and optical properties of SiC.

It is well known that transition metal atoms incorporated in semiconductors usually form deep levels in the band gap, this has been used to produce compensated, virtually semi-insulating material (for example, vanadium in 4H-SiC and 6H-SiC [1-2]). Recently, it was reported that the way transition metals are incorporated in SiC, depends on the size of the different atoms [3]. First row transition metals (3d) are suggested to occupy substitional Si-sites in the crystal, whereas the second and third row transition metals (4d and 5d, respectively) can occupy both substitutional Si-sites and divacancies, forming a split-vacancy defect. Further, it has been found that the formation energy for these two different possibilities may be different for hexagonal sites and the quasi-cubic sites. For example, Nb, which is a transition metal from the second row, has been observed to occupy a divacancy, forming an asymmetric split-vacancy defect [4].

The photo luminescence (PL) UD-1 defect is one of several unidentified deep level defects in SiC, and there are some reports published on the properties of UD-1 [5-7]. It has been suggested that UD-1 is an impurity, based on the annealing behavior [7].

The identification of optical centers often involves other experimental techniques, for example magnetic resonance spectroscopy. In this report, UD-1 is suggested to be tungsten-related (W), based on deep level transient spectroscopy (DLTS) and an intentional doping study.

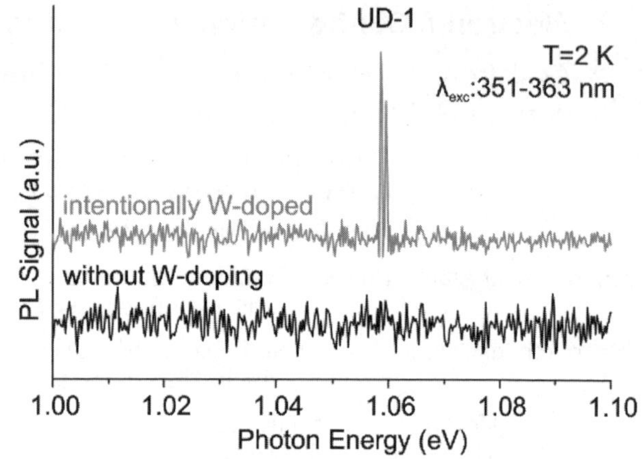

Figure 1 The upper graph shows a PL spectrum from an intentionally W doped 4H-SiC epilayer. The two sharp lines are the UD-1 defect. The lower PL spectrum is from another 4H-SiC epilayer, grown under the same conditions as the sample in the upper spectrum but without W doping, and shows no luminescence from the UD-1 defect

Experimental Details

The doping study samples used in this study are thick epilayers grown with chemical vapor deposition [8]. In the intentional doping study, n-type samples were grown by placing small metallic flakes of W directly on the substrate as well as in the upstream part of the susceptor. Reference samples were also grown, with exactly the same conditions as the doped samples, but before the introduction of W in the growth system to avoid memory effects. Since UD-1 is commonly observed in semi-insulating samples, such samples have also been studied.

The optical experiments were performed using a Bomem DA8 Fourier Transform Infrared Spectrometer. In PL, a quartz beam-splitter and a Germanium detector were used. The samples were excited using the multiline UV emission of an Ar^+-ion laser (351-363 nm). In the absorption measurements, the light source used was a quartz lamp, and the detector an InSb detector. The Zeeman measurements were performed using an optical fiber to access the sample, a setup only allowing the Faraday configuration (magnetic field parallel to the direction of light propagation). The magnetic field was varied between 0 T and 12 T.

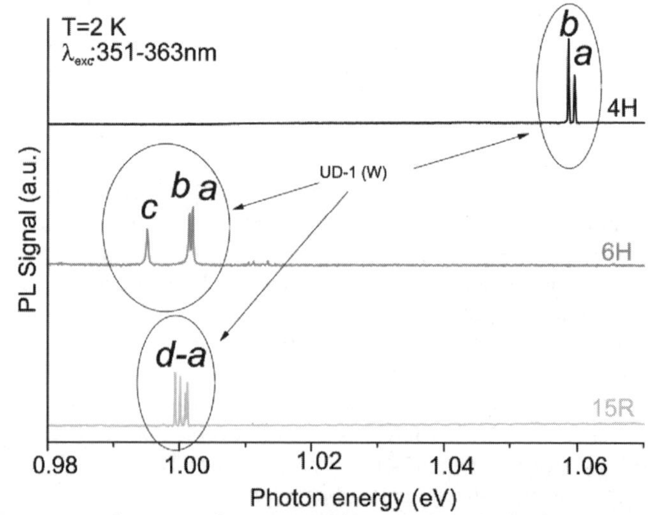

Figure 2 Three PL spectra, the upper from a 4H-SiC sample, middle from a 6H-SiC and the lower from a 15R-SiC sample. All spectra show the W-related UD-1 defect in the three different polytypes

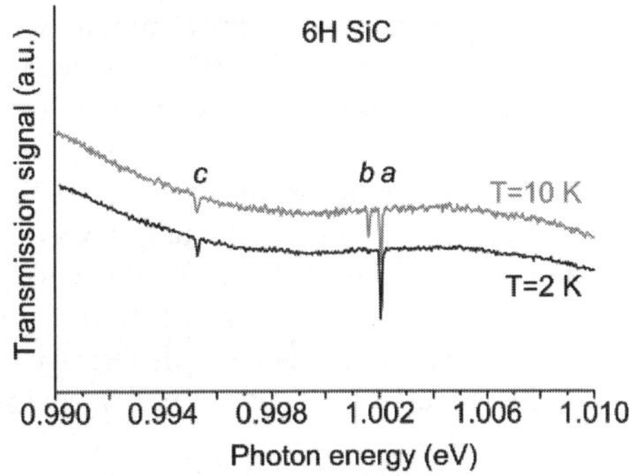

Figure 3 The two spectra show the transmission through a 6H-SiC sample with absorption from the UD-1 defect. The upper spectrum is taken at 10 K and the lower at 2 K.

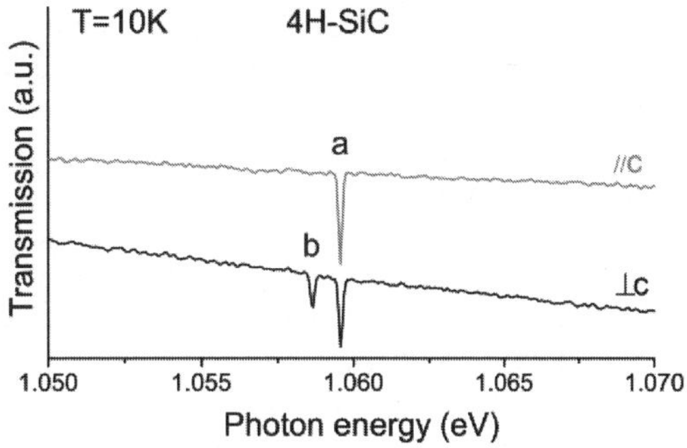

Figure 4 Two transmission spectra from the same 4H-SiC sample, measured at 10 K. The upper showing light polarized parallel and the lower with light perpendicular to the c-axis of the sample, respectively.

Experimental Results and Discussion

The UD-1 defect was observed in the W doped samples but not in the reference samples, see Figure 1. The incorporation of W was confirmed using radiotracer DLTS [9]. The undoped reference samples showed neither the W-related DLTS signal nor UD-1. Earlier published reports on photo-induced absorption showed that the optical activation energy of the absorption of UD-1 is 1.4 eV [6]; this matches well with the thermal activation energy measured for the deep level associated with W in DLTS in 4H-SiC [9].

The fact that the PL lines related to UD-1 are observed in the doped samples, but not in the undoped ones, together with the match of activation energies, suggests that UD-1 is W-related and actually the same W-related defect which is observed in DLTS. But neither optical methods nor electrical techniques usually give any information about the physical structure of the defect. In this case, we know that the number of PL related centers follows the number of inequivalent quasi-cubic sites, and one W-related DLTS signal has been reported in 3C-SiC [10]. This, together with the above mentioned reports on formation energies for second and third row transition metals, indicates that the UD-1 defects are W-atoms occupying a quasi-cubic silicon site. UD-1 has not been observed in 3C-SiC, which may be due to the smaller band-gap of this polytype; thus a level which would be involved in the transition may be in the conduction band, and the recombination is not possible.

In Figure 2, the PL of the UD-1 center, which from now on will be referred to as W-related, can

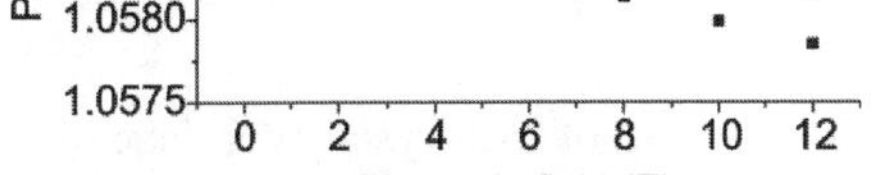

Figure 5 The squares show the peak positions from Zeeman measurements when the magnetic field is parallel to the c-axis of the sample. All measurements were done at 2 K.

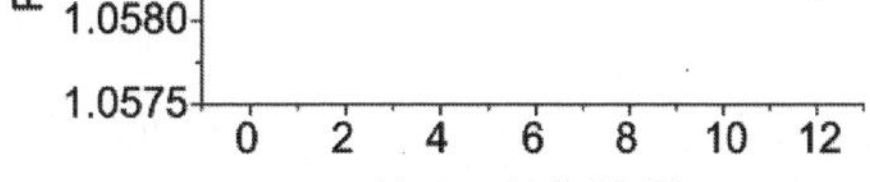

Figure 6 The squares show the peak positions from Zeeman measurements when the magnetic field is perpendicular to the c-axis of the sample. The splitting is non-linear. All measurements were done at 2 K.

be seen in 4H-, 6H, and 15R-SiC. The center in 4H-SiC consists of two spectral lines, from now on, the high energy line will be referred to as *a* and the low energy line as *b*. In 6H-SiC, there are two centers, one similar to the center in 4H-SiC (with the *a* and *b* components), while the other center is only resolved to one line (*c*), however with a shoulder on the high energy side. In 15R-SiC, three centers are observed, one similar to the center in 4H, two similar to the non-split *c*-line in 6H-SiC. The numbers of UD-1 centers do not follow the number of inequivalent sites in the polytypes 4H-, 6H- and 15R-SiC, as for example vanadium [11] and chromium [12] do. Instead, the numbers of UD-1 centers in these polytypes follow the number of inequivalent quasi-cubic sites, 1 in 4H-, 2 in 6H-, and 3 in 15R-SiC. If UD-1 is impurity-related, the impurity should behave differently on quasi-cubic sites and hexagonal sites, in agreement with the reports for second and third row transition metals.

The temperature dependence of the absorption from a W-related defect reveals a clear thermal redistribution of the lines *a* and *b*. At 2 K only *a* is observed while at higher temperatures *b* increases and at 10 K, both *a* and *b* are observed, see Figure 3. In PL no thermal distribution between *a* and *b* is observed. This indicates that, in PL, the transitions *a* and *b* share the same initial state to two different final states and, in absorption it is vice versa.

In tetrahedral symmetry (Td), no preferred symmetry axis is present, which shows that if a strong polarization dependence is present, the symmetry of the defect must be lower. The *b*-component is strongly polarized perpendicular to the c-axis of the crystal, while the *a* component is polarized both perpendicular and parallel to the c-axis, see Figure 4. This shows that the highest possible symmetry of the defect is C_{3V}. The *c* component, on the other hand has no strong polarization dependence.

Zeeman experiments, with the magnetic field either parallel or perpendicular to the c-axis of the sample, were performed. In the case of a magnetic field parallel with the c-axis of the sample both *a* and *b* split into four lines, and the splitting is linearly dependent on the magnetic field magnitude, see Figure 5. PL measurements performed at a fixed magnetic field, but at different temperatures, show that both the excited as well as the ground state for the two transitions split into two states. This shows that all three levels involved in the two transitions *a* and *b* are twofold degenerate.

When the magnetic field is perpendicular to the c-axis of the sample, see Figure 6, the splitting of both *a* and *b* lines are non-linearly dependent on the magnetic field magnitude. This non-linearity of the splitting in this direction shows a mixture of the levels that depends on both the direction and magnitude of the magnetic field.

The highest possible defect symmetry in 4H- and 6H-SiC is C_{3V}. In the single group of C_{3V}, describing an even number of electrons, the transitions between the twofold degenerate Γ_3^{C3v} and one of the two non-degenerate Γ_1^{C3v} or Γ_2^{C3v} together with a Γ_3^{C3v} will give rise to transitions with the same polarization as *a* and *b*. However, since both Γ_1^{C3v} and Γ_2^{C3v} are non-degenerate, they will not split in magnetic fields, contradicting the experiments. Considering the double group of C_{3V}, transitions between Γ_4^{C3v} and $\Gamma_5^{C3v} \oplus \Gamma_6^{C3v}$ and Γ_4^{C3v}, respectively, will give rise to lines with the same polarization dependence as *a* and *b*, see Figure 7. The magnetic field will lower the symmetry and the twofold degeneracy of each level will be broken, in accordance with experiments.

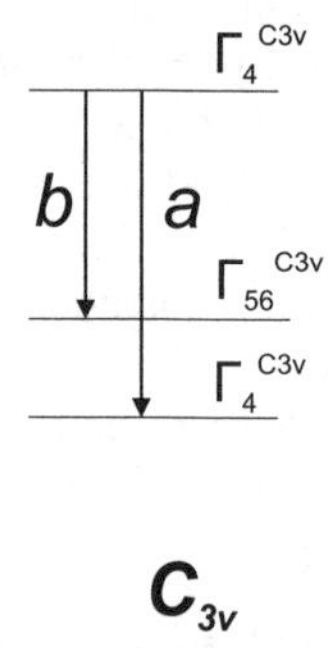

Figure 7 Transition scheme for the UD-1 defect.

Model

The electronic configuration of an impurity occupying a silicon site of the crystal will depend on both the electronic configuration of the impurity per se, as well as on the electron configuration of the silicon vacancy. The electron configuration of a neutral W atom is $[Xe]4f^{14}5d^{4}6s^{2}$, and among these electrons, four will interact with the bonds of the silicon vacancy, leaving $5d^2$ electrons. The neutral charge state of the W_{Si} will then have an even number of electrons, but according to the analysis above, the levels transform as the representations that are exclusive to the C_{3v} double group. Therefore the charge state for which the transitions are observed is odd. The two most

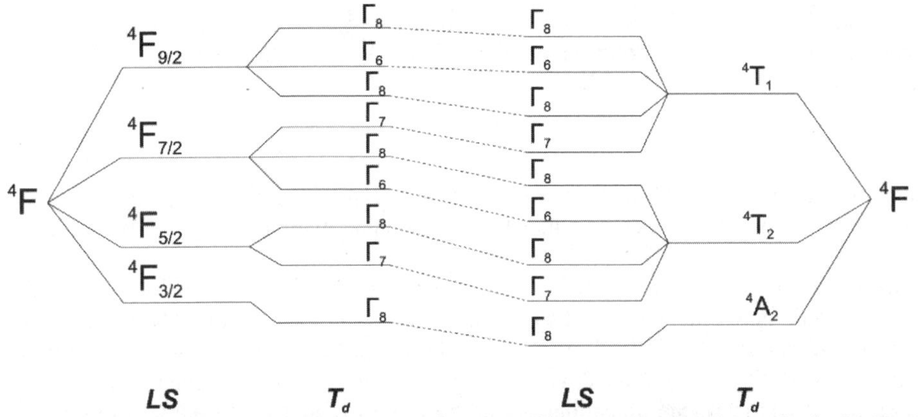

Figure 8 Schematic picture of the states that are of importance for the analysis for T_d. To the left in the figure, the splitting for the free ion W^{+3} due to spin-orbit (LS) interaction can be seen. In the right of the figure, the splitting of ^4F in a T_d crystal field (CF) can be seen.

reasonable charge states are W_{Si}^{+1} and W_{Si}^{-1}. In the case of the charge state W_{Si}^{+1}, one electron is in the 5d orbital, which is similar to Vanadium in SiC (see [13]). The only possible multiplet is ^2D, which splits due to spin-orbit coupling (LS) into $^2D_{3/2}$ and $^2D_{5/2}$ with $^2D_{3/2}$ being the ground state. A crystal field (CF) of T_d symmetry will split these two multiplets further; the $^2D_{5/2}$ will split into a Γ_8^{Td} and Γ_7^{Td}, the latter will have the highest energy, and $^2D_{3/2}$ will transform according to a Γ_8^{Td}. In analogy with other transition metals, the energy splitting between the two Γ_8^{Td} levels should be in the range of 1 eV. When the symmetry is lowered further, to C_{3v}, the Γ_7^{Td} will transform as Γ_4^{C3v}, and the two Γ_8^{Td} levels will each transform as $\Gamma_4^{C3v} \oplus \Gamma_5^{C3v} \oplus \Gamma_6^{C3v}$. This would, however, not lead to the observed polarization dependence or the observed Zeeman splitting of the transitions.

For the negatively charged state, W_{Si}^{-1}, several multiplets can be formed by the three 5d electrons: the doublets ^2D, ^2P, ^2F, ^2G, ^2H, and quartets ^4P and ^4F, the latter being the ground state multiplet. The energy difference between ^4P and ^4F for the free W^{+3} is around 1.75 eV [14].

Since one expects a mixing between the electron states of the W ion with the electron states of the V_{Si}, but still a high localization of the electrons on the W ion, the case is intermediate between two extremes; one being that the CF splitting is much larger than the LS splitting, and the other being the opposite case. Both cases will be briefly discussed, but for a more detailed discussion see [15]. If the CF splitting is much larger than the LS splitting, but still small compared to the splitting of the multiplets, the ^4F will split into three different multiplets: $^4A_2^{Td}$, $^4T_1^{Td}$ and $^4T_2^{Td}$; these will then split due to the LS interaction. The $^4A_2^{Td}$ will not split further and transforms as Γ_8^{Td}, both the $^4T_1^{Td}$ and $^4T_2^{Td}$ will split into Γ_6^{Td}, Γ_7^{Td}, and 2 Γ_8^{Td}. For a schematic see the right hand side of Figure 8.

Starting from the other extreme, when the LS splitting is much larger than the CF splitting, the ^4F multiplets will be further split by the LS coupling. The free ion has $^4F_{3/2}$ as its ground state, and $^5F_{5/2}$ is 0.44 eV higher in energy. The energy distance from the ground state to $^4F_{7/2}$ and $^4F_{9/2}$ is 0.98 eV and 1.24 eV, respectively [14]. In a CF of T_d symmetry, the $^4F_{3/2}$ will not split but transform as Γ_8^{Td}. $^4F_{5/2}$ will split and transform as Γ_7^{Td} and Γ_8^{Td}. $^4F_{7/2}$ will split into Γ_6^{Td}, Γ_7^{Td} and Γ_8^{Td}, and finally the $^4F_{9/2}$ will transform as Γ_6^{Td} and two Γ_8^{Td}, see left hand side of Figure 8.

In both extremes Γ_8^{Td} is the ground state, and Γ_7^{Td} is the closest excited state. This shows that the transitions involved when observing UD-1, in T_d-symmetry, are transitions between Γ_8^{Td} and Γ_7^{Td} also in the intermediate case. This transition is allowed.

When taking more than the closest neighbors in the crystal into account, the symmetry deviates from T_d to C_{3v} symmetry. This will lead to a small splitting, different in size for the different in-equivalent sites, but small compared to the T_d splitting. Since the splitting between *a* and *b* is small, it is reasonable that it is caused by this deviation from T_d. A Γ_8^{Td} level will split into a Γ_4^{C3v} and $\Gamma_5^{C3v} \oplus \Gamma_6^{C3v}$ and Γ_7^{Td} will have the Γ_4^{C3v} representation. As discussed above the transitions between Γ_4^{C3v} and $\Gamma_5^{C3v} \oplus \Gamma_6^{C3v}$ and Γ_4^{C3v}, respectively, agree both with the polarization dependence as well as the Zeeman splitting of them. This charge state, however, leads to the unexpected case of having the charge state W_{Si}^{-3} in n-type 4H-SiC.

Summary

The UD-1 signal has been correlated to W-incorporation in a doping study. The optical activation energy for the absorption of UD-1 [6] corresponds to the thermal activation energy in DLTS for the deep W (W2) [9].

The polarization dependence and magnetic field response of the UD-1 defect show that the involved transitions are between an energetically higher state of Γ_4^{C3v} representation and two closely spaced ground states $\Gamma_5^{C3v} \oplus \Gamma_6^{C3v}$ and Γ_4^{C3v}. The energy difference between the ground state Γ_4^{C3v} and $\Gamma_5^{C3v} \oplus \Gamma_6^{C3v}$ is about 1 meV, and may be explained by a small deviation from T_d to C_{3v}. In the case of T_d symmetry, the transitions would be from a higher Γ_7^{Td} state to a lower Γ_8^{Td} state.

Two different models are discussed, W_{Si}^{+1} and W_{Si}^{-1}. W_{Si}^{-1} can explain the experimental behavior of UD-1 without any ad-hoc hypothesis. However, this charge-state suggests that W_{Si}^{-3} would be present in n-type material. This is unexpected. W_{Si}^{+1} can explain the experimental results, but an adjustment is needed to invert the order of the splitting of the $^2D_{5/2}$ in T_d field, to be the opposite to, for example, Vanadium in SiC. However, the assignment of UD-1 to W_{Si}^{+1} results in more probable charge states, with W_{Si}^{-1} being occupied in n-type material.

References

[1] H.M. Hobgood, R.C. Glass, G. Augustine, R.H. Hopkins, J. Jenny, M. Skowronski, W.C. Mitchel, and M. Roth, Applied Physics Letters 66 (1995), 1364-1366.
[2] T. Kimoto, T. Nakajima, H. Matnsunami, T. Nakata, and M. Inoue, Applied Physics Letters 69 (1996), 1113.
[3] V. Ivády, B. Somogyi, V. Zólyomi, A. Gällström, N.T. Son, E. Janzén, and A. Gali, these Proceedings
[4] N.T. Son, V. Ivády, A. Gali, A. Gällström, S. Leone, O. Kordina, and E. Janzén, these Proceedings
[5] B. Magnusson, and E. Janzén, Materials Science Forum 483-485 (2005), 341
[6] B. Magnusson, A. Ellison, and E. Janzén, Materials Science Forum 389-393 (2002), 505
[7] W.E. Carlos, N.Y. Garces, and E.R. Glaser, Physical Review B 74 (2006), 235201
[8] A. Henry, S. Leone, F.C. Beyer, H. Pedersen, O. Kordina, S. Andersson, and E. Janzén, submitted to Physica B: Condensed Matter (2011)
[9] J. Grillenberger, N. Achtziger, G. Pasold, R. Sielemann, and W. Witthuhn, Materials Science Forum 353-356 (2001), 475.
[10] F.C. Beyer, C.G. Hemmingsson, A. Gällström, S. Leone, H. Pedersen, A. Henry, and E. Janzén Applied Physics Letters 98 (2011), 152104.
[11] J. Reinke, H. Weihrich, S. Greulich-Weber, and J-M Spaeth, Semicond. Sci. Technol. 8 (1993), 1862.
[12] N.T. Son, A. Ellison, M.F. MacMillan, O. Kordina, W.M. Chen, B. Monemar, and E. Janzén, Materials Science Forum 264-268 (1998), 603.
[13] B. Kaufmann, A. Dörnen, and F.S. Ham, Physical Review B 55 (1997), 13009.
[14] S. Fraga, J. Karwowski, and K.M.S Saxena, Handbook of Atomic Data, Elsevier Science Ltd, 1997, pp 245.
[15] A. Gällström, B. Magnusson, F.C. Beyer, A. Gali, N.T. Son, S. Leone, I.G. Ivanov, C.G. Hemmingsson, A. Henry, and E. Janzén, Submitted to Physica B: Condensed Matter (2011)

Materials Science Forum Vols. 717-720 (2012) pp 217-220
© (2012) Trans Tech Publications, Switzerland
doi:10.4028/www.scientific.net/MSF.717-720.217

Identification of Niobium in 4H-SiC by EPR and *ab Initio* Studies

N.T. Son[1, a,*], V. Ivády[2,b], A. Gali[2,3,c], A. Gällström[1,d], S. Leone[1,e], O. Kordina[1,f] and E. Janzén[1,g]

[1]Department of Physics, Chemistry and Biology, Linköping University, SE-581 83 Linköping, Sweden

[2]Research Institute for Solid State Physics and Optics, Hungarian Academy of Sciences, PO Box 49, H-1525 Budapest, Hungary

[3]Department of Atomic Physics, Budapest University of Technology and Economics, Budafoki út 8, H-1111 Budapest, Hungary

[a]son@ifm.liu.se, [b]ivady.viktor@gmail.com, [c]agali@szfki.hu, [d]andga@ifm.liu.se, [e]leone@ifm.liu.se, [f]olk@ifm.liu.se, [g]erj@ifm.liu.se, *corresponding author

Keywords: Transition metal impurity, split-vacancy defect, electron paramagnetic resonance, *ab initio* calculations.

Abstract. In unintentionally Nb-doped 4H-SiC grown by high-temperature chemical vapor deposition (HTCVD), an electron paramagnetic resonance (EPR) center with C_{1h} symmetry and an electron spin S=1/2 was observed. The spectrum shows a hyperfine structure consisting of ten equal-intensity hyperfine (hf) lines which is identified as due to the hf interaction between the electron spin and the nuclear spin of ^{93}Nb. An additional hf structure due to the interaction with two equivalent Si neighbors was also observed. *Ab initio* supercell calculations of Nb in 4H-SiC suggest that Nb may form a complex with a C-vacancy (V_C) resulting in an asymmetric split-vacancy (ASV) defect, Nb_{Si}-V_C. Combining results from EPR and supercell calculations, we assign the observed Nb-related EPR center to the hexagonal-hexagonal configuration of the AVS defect in the neutral charge state, $(Nb_{Si}\text{-}V_C)^0$.

Introduction

Silicon carbide (SiC) single crystals are usually grown at high temperatures and unintentional incorporation of transition metals (TMs) is common. In 3C-SiC with cubic crystal structure, calculations [1] found that 3*d* TMs prefer the Si site to form substitutional impurities. In the 4H and 6H hexagonal polytypes, 3*d* TMs such as Ti [2-4], V [5,6] and Cr [6-8], are also known to substitute for Si and the substitutional centers at different inequivalent-lattice sites in 4H- and 6H-SiC have been identified. However, there is unclear or no corresponding correlation found for 4*d* TM impurities such as Mo and W. Baur and co-workers [9] found only a one EPR spectrum related to the donor state Mo^{5+} ($4d^1$) in 6H-SiC. The EPR of the acceptor state of Mo, Mo^{3+} ($4d^3$), has also been observed for a single lattice site in *n*-type 6H- and 15R-SiC [6]. In 6H-SiC, Irmscher and co-workers [10] assigned an EPR spectrum to the neutral W^{5+} ($5d^1$) substituting for Si. However, also only one EPR spectrum was observed in 6H-SiC. From the reported data, there is no clear analogy in the electronic structure between the 3*d* TMs and others in the 4*d* and 5*d* series in SiC. It has been found that in boron nitride (BN), Mn is energetically more favorable in a divacancy site as compared to the B site to form split-vacancy defects [11]. Recently calculations by Decoster and co-workers [12] suggested that TMs occupy both substitutional and center-bond sites in Ge. In SiC, it is not clear if the 4*d* and 5*d* TMs with larger ionic radii prefer the Si site to form isolated substitutional centers or a divacancy to form split-vacancy defects.

In HTCVD or CVD growth of SiC, tantalum carbide (TaC) is commonly used for coating some parts exposed to high temperatures in the reactor. Recently, niobium carbide (NbC) has been used as an alternative to TaC coated parts. To our knowledge, there has been no report on Nb in SiC so far. In this work, EPR was used to study unintentionally Nb-doped 4H-SiC grown by HTCVD. The electronic structure of substitutional Nb at the Si site, Nb_{Si}, and its complex with the C vacancy in

4H-SiC was studied by *ab initio* supercell calculations. We will show from our EPR and theoretical calculations that in 4H-SiC, Nb prefers to form a complex with a C-vacancy resulting in an asymmetric split-vacancy (ASV) defect, Nb_{Si}-V_C, with Nb being closer to the Si vacancy (V_{Si}) than V_C.

Experiments

The 4H-SiC samples used in this study were grown by HTCVD in a reactor with several parts made of niobium carbide (NbC). The concentration of unintentionally incorporated Nb is $\sim 3\times10^{16}$ cm^{-3} as determined by secondary ion mass spectrometry (SIMS). In addition to Nb, vanadium (V) is also present in the samples as shown from the well-known photoluminescence (PL) signal of V in the near infrared spectral region [5]. All Nb-doped samples show semi-insulating behavior. EPR measurements were performed on X-band (~9.5 GHz) Bruker E500 and E580 spectrometers using a continuous helium-flow cryostat which allows the regulation of the sample temperature from 4 K to room temperature.

Results and Discussion

In all unintentionally Nb-doped 4H-SiC samples, in addition to the broad and isotropic SI-1 signal [13], a new EPR spectrum with a hyperfine (hf) structure consisting of ten equal-intensity lines was observed at temperatures below 100 K. Fig. 1 shows such a spectrum measured at 44 K for the magnetic field **B** parallel to the c-axis of the crystal (**B**||**c**). The observed hf structure is typical for the hf interaction between the electron spin and the nuclear spin I=9/2 of an element having a natural abundance of 100%. There are only a few elements having the maximum possible nuclear spin I=9/2 and a 100% natural abundance: ^{93}Nb, ^{209}Bi, and two indium isotopes, ^{113}In (4.29%) and ^{115}In (95.71%). Bi and In are not common impurities in SiC while the presence of Nb in the sample

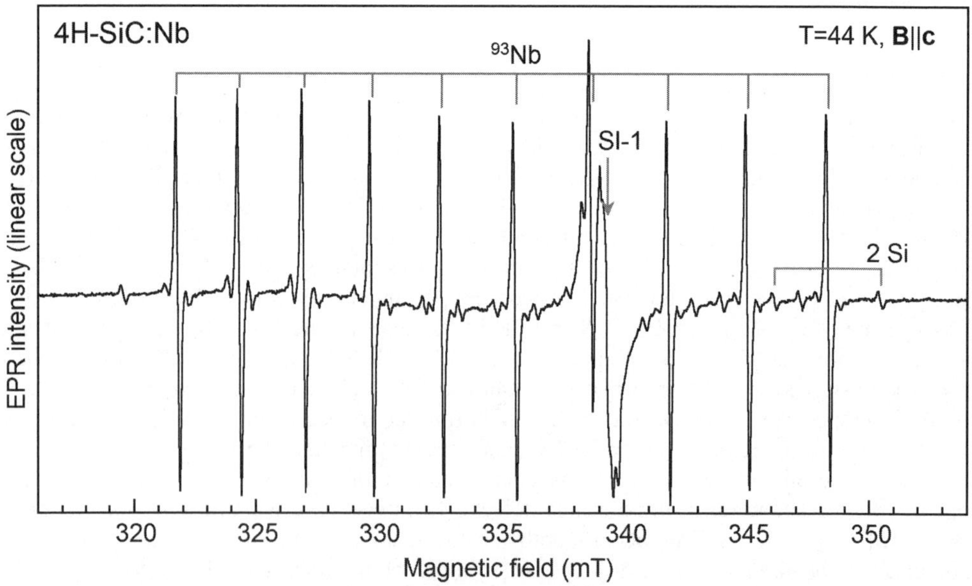

Fig. 1. EPR spectrum of unintentionally Nb-doped 4H-SiC grown by HTCVD measured at 44 K for **B**||**c** showing a ^{93}Nb hf structure consisting of ten lines of equal intensity. The broad line at ~339.26 mT is the isotropic signal of the SI-1 defect (Ref. 13). Each ^{93}Nb hf line is accompanied by a pair of weak hf lines due to the interaction between the electron spin and the nuclear spins of two equivalent Si atoms. The spectrum was measured at a microwave frequency of 9.504 GHz.

Table 1. Spin-Hamiltonian parameters of the Nb-related EPR center in 4H-SiC. The principal A-values of the hf tensors of ^{93}Nb and two Si neighbors are given in the unit of 10^{-4} cm^{-1}. θ and φ are the polar and azimuthal angles, respectively, of the principal axes of the g- and A-tensors measured in degrees. The experimental error is ~0.0005 for g-values and ~0.4×10^{-4} cm^{-1} for A-values.

Parameter	g_{xx}	g_{yy}	g_{zz}	A_{xx}^{Nb}	A_{yy}^{Nb}	A_{zz}^{Nb}	A_{xx}^{2Si}	A_{yy}^{2Si}	A_{zz}^{2Si}
Value	1.9891	2.0180	2.0272	73.7	16.0	30.7	55.1	40.9	41.5
θ	90°	72.81°	17.19°	90°	118.38°	28.38°	90°	90°	0°
φ	0°	90°	270°	0°	90°	90°	0°	90°	90°

has been confirmed by SIMS. Therefore, the observed hf structure can be unambiguously identified to be related to Nb. At this magnetic field direction, each ^{93}Nb hf line is accompanied by two weak hf lines with a hf splitting of ~2.5 mT as indicated in Fig. 1. The intensity ratio between these two hf lines and the main line is ~9-10% which is approximately the natural abundance of two ^{29}Si isotopes (I=1/2, 4.67% natural abundance). This hf structure is therefore identified as due to the hf interaction between the electron spin and the nuclear spin of one ^{29}Si at two equivalent Si neighboring sites.

The angular dependence study shows that the center has an electron spin S=1/2 and C_{1h} symmetry. The hf tensors of ^{93}Nb and two ^{29}Si atoms are also shown to have C_{1h} symmetry. The obtained principal g-values and hf A-values for ^{93}Nb and two ^{29}Si atoms are given in Table 1. From the obtained principal A-values, the spin density was estimated to be ~40.3% on Nb and ~30.3% on two Si atoms. The low symmetry C_{1h} and considerable spin localization on two Si neighbors suggests that the defect may be a complex of Nb_{Si} and a V_C.

We performed *ab initio* supercell calculations of Nb in 4H-SiC. We used the density functional theory (DFT) based Vienna *ab initio* Simulation Package (VASP) [14] with the projector augmented-wave (PAW) method [15] and plane wave basis set. For Nb we utilized small core PAW projectors. We applied a plane-wave cutoff of 420 eV with Γ-point sampling of the Brillouin-zone in 576-atom supercells. In these large supercells the Γ-point calculations provide convergent charge and spin densities, and simultaneously allow us to monitor the degeneracy of the defect levels in the fundamental band gap. We applied both semi-local Perdew-Burke-Ernzerhof (PBE) [16] and non-local HSE06 [17,18] functionals in the calculations where the latter heals the band gap error of the PBE functional [19]. The defects were optimized in the supercell until all forces fell below 10 meV/Å. We calculated the adiabatic charge transition levels as explained in Ref. 20. We found that in Si-rich conditions, the formation energies of Nb_{Si} and the ASV complex, Nb_{Si}-V_C, are comparable and both types of defects can occur. Among all ASV configurations we found the hexagonal-hexagonal (*h-h*) configuration to be much more stable than any other configuration. In the neutral charge state, both the Nb_{Si} and Nb_{Si}-V_C ASV complex are subjected to Jahn-Teller distortion, which reduces the symmetry from C_{3v} to C_{1h}, and have an electron spin S=1/2. However, our calculations show that the distribution of the spin density is very different between these two defects. In Nb_{Si}^0, the spin is localized almost entirely on the *d*-orbital of Nb whereas it is considerably mixed with two Si dangling bonds in $(Nb_{Si}\text{-}V_C)^0$. The calculated distribution of spin density for the ASV defect is in agreement with our EPR observation, suggesting that the Nb-related EPR center is the $(Nb_{Si}\text{-}V_C)^0$ defect. The prediction of the much more stable *h-h* configuration of $(Nb_{Si}\text{-}V_C)^0$ explains why only one EPR spectrum has been detected in the hexagonal 4H-SiC polytype. In our preliminary EPR study of Nb-doped 6H-SiC, we also observed only one EPR spectrum of C_{1h} symmetry and similar hf structures.

Summary

In unintentionally Nb-doped 4H-SiC grown by HTCVD, we have observed an EPR spectrum with spin S=1/2 and C_{1h} symmetry. Based on the obtained hf structure, the spectrum can be unambiguously identified to be related to ^{93}Nb. *Ab initio* supercell calculations of neutral Nb in 4H-SiC suggest that Nb can form a complex with V_C resulting in a Nb_{Si}-V_C ASV defect. Considerable spin localization on two Si dangling bonds predicted for the ASV defect has been detected for the Nb-related EPR center. Comparing with the results from calculations, we assigned the EPR center to the ASV defect in the neutral charge state $(Nb_{Si}\text{-}V_C)^0$. The theoretical prediction that the *h-h* defect is the most stable configuration explains why only one Nb-related EPR spectrum has been observed in the hexagonal 4H-SiC polytype.

Acknowledgements

Support from the Swedish Energy Agency, the Swedish Research Council, the Swedish National Infrastructure for Computing Grants No. SNIC 001-10/223, and the Knut and Alice Wallenberg Foundation is acknowledged.

References

[1] M. S. Miao and W. R. L. Lambrecht, Phys. Rev. B 74 (2006) 235218.

[2] W. J. Choyke and L. Patrick, Silicon Carbide 1973 (University of South Carolina Press, Columbia, 1974), p. 261.

[3] A. W. C. Van Kemenade and S. H. Hagen, Solid State Commun. 14 (1974) 1331.

[4] K. M. Lee, Le Si Dang, G. D. Watkins, and W. J. Choyke, Phys. Rev. B 32 (1985) 2273.

[5] J. Schneider, A. Dörnen, S. Leibenzeder, and R. Stein, Appl. Phys. Lett. 56 (1990) 1184.

[6] J. Baur, M. Kunzer, and J. Schneider, Phys. Stat. Sol. (a) 162 (1997) 153.

[7] P. G. Baranov, V. A. Khramtsov, and E. N. Mokhov, Semicond. Sci. Technol. 9 (1994) 1340.

[8] N. T. Son, A. Ellison, B. Magnusson, M. F. MacMillan, W. M. Chen, B. Monemar, and E. Janzén, J. Appl. Phys. 86 (1999) 4348.

[9] J. Baur, M. Kunzer, K. F. Dombrowski, U. Kaufmann, J. Schneider, P.G. Baranov, and E. N. Mokhov, Inst. Phys. Conf. Ser. 155 (1997) 933.

[10] K. Irmscher, I. Pintilie, L. Pintilie, D. Schulz, Physica B 308-310 (2001) 730.

[11] L. V. C. Assali, W. V. M. Machado, and J. F. Justo, Appl. Phys. Lett. 89 (2006) 072102.

[12] S. Decoster, S. Cottenier, B. De Vries, H. Emmerich, U. Wahl, J. G. Correia, and A. Vantomme, Phys. Rev. Lett. 102 (2009) 065502.

[13] N. T. Son, B. Magnusson, Z. Zolnai, A. Ellison, and E. Janzén, Mater. Sci. Forum 457-460 (2004) 437.

[14] G. Kresse and J. Furthmüller, Phys. Rev. B 54 (1996) 11169.

[15] P. E. Blöchl, Phys. Rev. B 50 (1994) 17953.

[16] J. P. Perdew, K. Burke, and M. Ernzerhof, Phys. Rev. Lett. 77 (1996) 3865.

[17] J. Heyd, G. E. Scuseria, and M. Ernzerhof, J. Chem. Phys. 118 (2003) 8207.

[18] A. V. Krukau, O. A. Vydrov, A. F. Izmaylov, and G. E. Scuseria, J. Chem. Phys. 125 (2006) 224106.

[19] M. Marsman, J. Paier, A. Stroppa, and G. Kresse, J. Phys.: Condens. Matter 20 (2008) 064201.

[20] P. Deák, B. Aradi, T. Frauenheim, E. Janzén, and A. Gali, Phys. Rev. B 81 (2010) 153203.

Materials Science Forum Vols. 717-720 (2012) pp 221-224

doi:10.4028/www.scientific.net/MSF.717-720.221

Mn implantation for new applications of 4H-SiC

M.K. Linnarsson[1, a,*], J. Wong-Leung[2,b], A. Hallén[1,c], S.I. Khartsev[1,d] and A.M. Grishin[1,e]

[1]Royal Institute of Technology, School of Information and Communication Technology P.O Box Electrum 229, SE-164 40, Kista SWEDEN

[2]Centre for Advanced Microscopy and Dept. of Electronic Materials Engineering (RSPE), The Australian National University, Canberra, ACT0200 AUSTRALIA

[a]marga@kth.se, [b]jenny.wongleung@anu.edu.au, [c]ahallen@kth.se, [d]serg@kth.se, [e]grishin@kth.se, *corresponding author

Keywords: Ion implantation, Manganese, 4H-SiC, SIMS, RBS, TEM

Abstract. Structural disorder and lattice recovery of high dose, manganese implanted, semi-insulating, 4H-SiC have been studied by secondary ion mass spectrometry, Rutherford backscattering in channeling directions, visible-to-near infrared optical spectroscopy as well as with transmission electron microscopy. After heat treatment at 1400 and 1600 °C, a substantial rearrangement of manganese is observed in the implanted region. However, the crystal has not been fully recovered. More disorder remains in the $[11\bar{2}3]$ compared to the [0001] channel direction. Stacking faults, voids and 3C inclusions are observed in the implanted region. A Mn containing phase has most likely formed.

Introduction

During the last decade, SiC electronic devices have become commercialized due to energy saving reasons. In addition, SiC:Mn has been recognized as a promising DMS (Diluted Magnetic Semiconductor) candidate for applications in spintronics [1]. However, according to Vodakov et al., a solubility in the range of $1x10^{17}$ Mn atoms/cm^3 is expected in SiC [2]. Hence, the solubility will most likely be exceeded at concentrations applicable for magnetic applications and Mn containing precipitates may form in SiC. Mn implanted 3C-SiC has been extensively studied [3] and recently, a Curie temperature of 245 K has been reported for high concentrations of Mn in 3C-SiC [4]. Curie temperatures well above room temperature have also been measured in Mn-Si-C systems of magnetron sputtered films [5]. According to the Mn-Si-C phase diagram, SiC may coexist with $Mn_{15}Si_{26}$, MnSi, $Mn_5Si_3C_x$, Mn_8Si_2C and Mn_5SiC [6]. In this context, it should be pointed out that Mn_5SiC is well known for its magnetic properties, but has a Curie temperature of 284 K [7]. However, for Mn_5SiC nanostructures, magnetic hysteresis has been measured above 300 K [8].

In this investigation, we have studied heavily Mn implantation doped 4H-SiC. Subsequent heat treatments have been performed. Disorder recovery, rearrangement and incorporation of Mn atoms in the SiC lattice have been traced by Rutherford backscattering in channeling directions (RBS-C), optical spectroscopy in the visible to near IR range as well as transmission electron microscopy (TEM). Secondary ion mass spectrometry (SIMS) has been utilized to obtain Mn concentration versus depth.

Experimental

Manganese implantation in SiC has been performed to create a high Mn concentration well above the solubility limit. Bulk, semi-insulating, 4H-SiC (0001), with an off axis miscut of about 8° toward the $[11\bar{2}0]$ direction has been used as a starting material. The samples have been implanted with 300 keV $^{55}Mn^+$ ions at room temperature to a dose of $1x10^{16}$ cm^{-2}. Subsequent heat treatments were made at 1400 and 1600 °C during 15 minutes in Ar atmosphere, using an RF-heated furnace. After the heat treatment, the samples were exposed to oxygen plasma.

Analysis of the depth distribution of manganese has been made by SIMS utilizing a Cameca ims 4f instrument. The decrease of implantation disorder/damage has been analyzed by Rutherford backscattering spectrometry (RBS) in channeling and non-channeling directions using the tandem accelerator at Uppsala University providing a 2 MeV He^+ beam and a backscattering angle of 170°. The SIMNRA program has been employed for analysis of the RBS spectra [9]. TEM and selected area diffraction were carried out using a Phillips CM300 instrument operated at 300 keV. Optical transmittance spectra were recorded using a Perkin Elmer Lambda 750 spectrometer.

Result and Discussion

Fig. 1 presents room temperature optical transmittance spectra for the virgin 4H-SiC sample, after Mn implantation, as well as Mn implanted samples annealed at 1400 and 1600 °C. The visible range of the transmitted light is indicated in the figure by two vertical dashed lines. In addition, a photo of the four samples is included. They are placed on a ruler graded with cm and mm ticks. On the photo, the backside of the 1400 and 1600 °C samples has been marked with an ink. Virgin semi-insulating 4H-SiC sample (a) is colorless and transparent. After Mn implantation (b), structural disorder makes SiC opaque for visible light. According to TRIM (TRansport of Ions in Matter [10]) calculation, the extension of implantation damage is expected to be less than 0.3 μm, for 300 keV Mn. Only a shallow region close to the sample surface (0.1% of the total thickness) was affected. Although the implantation did not alter the main volume of the SiC samples, it strongly influences both transmitted and reflected light. After heat treatment at 1400 °C the transmittance increases significantly and samples lose color (c). This indicates a substantial reordering of the Mn-implanted region. Further, increase of annealing temperature from 1400 °C to 1600 °C results in a minor increase of the transmittance (d). However, heat treatment does not restore the transmittance to that of the virgin sample. No degradation of the annealed samples' surfaces was observed.

Fig. 2 displays SIMS measurements of the ^{55}Mn concentration versus depth for the as implanted sample and the two samples heat treated during 15 minutes at 1400 and 1600 °C, respectively. The concentration is expressed in atoms/cm^3 and at.% units on the left and right scale, respectively. For the at. % calculation, a total amount of $1x10^{23}$ atoms/cm^3 has been used. No correction of the depth scale is made for the lower atomic densities of the heavily damaged/amorphous, ion implanted, region and the crystalline silicon carbide beneath. In all samples, a manganese concentration above 0.7 at.% is recorded at the depth of 0.18 μm. A substantial rearrangement of Mn-ions has occurred. The full width at half maximum of the concentration profile in Fig. 2 has decreased in the annealed samples compared to the as implanted sample. In the 1600 °C sample, the main part of the

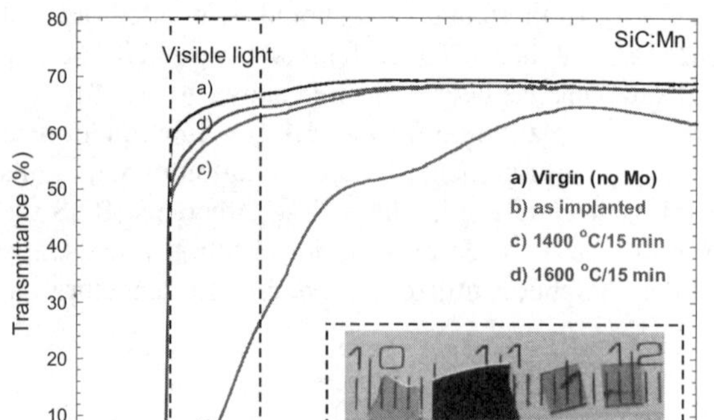

Fig.1. Room temperature optical transmittance spectrum in four 4H-SiC samples: virgin crystal (a), as implanted/300 keV Mn/ dose $1x10^{16}$ atoms/cm^2 (b), and subsequently heat treated at 1400 °C (c) and 1600 °C (d), respectively. The inset shows corresponding photos of the four samples.

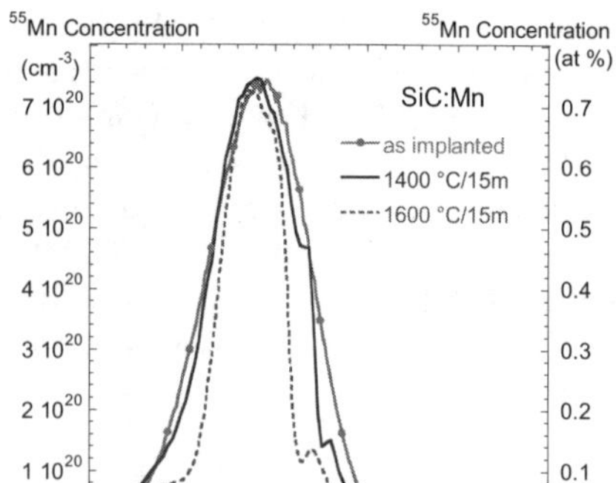

Fig.2. SIMS measurements of ^{55}Mn concentration versus depth in 4H-SiC for one as implanted sample and after heat treatment for 15 minutes at 1400 and 1600 °C, respectively.

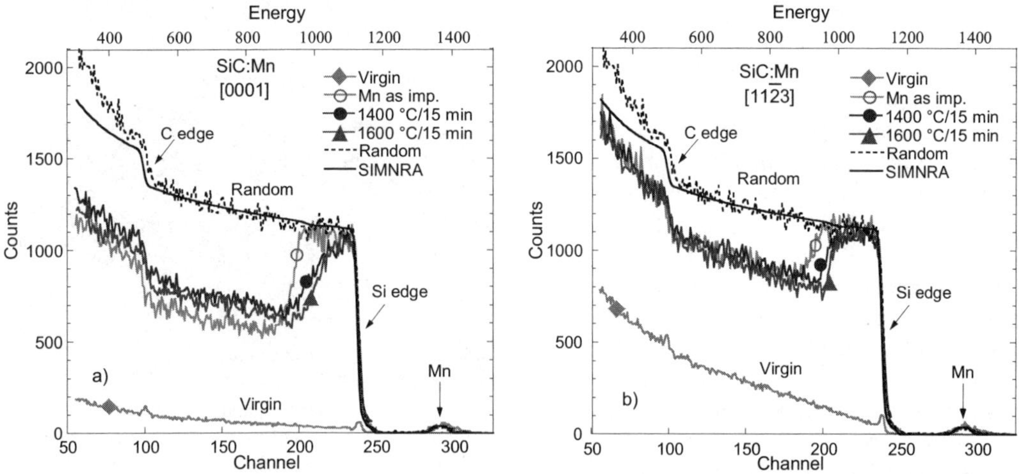

Fig 3. RBS spectra recorded in random and channeling directions [0001] (a) and [11$\bar{2}$3] (b) in four 4H-SiC samples; virgin, as implanted (300 keV Mn, dose 1x10^{16} atoms/cm^2), and heat treated for 15 minutes at 1400 °C and 1600 °C, respectively. Si and C edges as well as the Mn peak position are indicated by arrows. SIMNRA simulations are included.

manganese is found in a 0.1 μm thick slice, between 0.12 and 0.22 μm. On the surface side, Mn diffuses away and a shoulder is formed. The same kind of shoulder does not appear on the opposite - deep side of the implanted peak. Instead, a small Mn peak arises on the deep side. This peak may be related to the detained excess of Mn at defects and/or to the formation of new manganese containing phases. According to the ternary phase diagram (1000 °C), possible candidates that coexist with SiC are Mn_8Si_2C, $Mn_5Si_3C_x$, $Mn_{15}Si_{26}$ and MnSi [6]. However, thermodynamic equilibrium may not be reached during the heat treatment of the implanted material.

RBS-C spectra for 4H-SiC in the [0001] and [11$\bar{2}$3] channeling directions are shown in Fig. 3 a) and b), respectively. Four samples have been measured; virgin, as implanted with Mn and subsequently heat treated at 1400 and 1600 °C. In addition, an RBS spectrum collected in a non-channeling (referred to as random) direction, 8° off, halfway between (1$\bar{1}$00) and (1$\bar{2}$10) has been included in Fig. 3. The edge of the carbon and silicon surfaces and the position of the implanted Mn are indicated by arrows in the figures. SIMNRA [9] has been used for simulation of the expected peak position and yield of manganese in the random RBS spectra. No change in Mn distribution along the measured directions is revealed. The total Mn amount seen in the RBS spectra corresponds to implantation dose, indicating that almost all Mn atoms are still at interstitial sites after heat treatment. Significant decrease in the Si-yield is observed in the undamaged region, below the Mn implant. There is also a clear difference between channeling in the [0001] and [11$\bar{2}$3] directions. Such a discrepancy in yield is expected for non-damaged 4H-SiC caused by different amounts and characteristics of channels in these two directions. Close to the Si surface edge, the yield in random and in channel directions coincide for Mn implanted samples, i.e. this region is heavily damaged. In the 1400 °C annealed samples, there is a decrease in the yield (from the Si part), starting from the inner side of the damage region. This indicates a regrowth and improvement of the crystal quality. A further decrease in the yield is revealed as the temperature increases from 1400 to 1600 °C. In addition, there is a difference in annealing behavior between the [0001] and [11$\bar{2}$3] directions. The recovery in the [0001] direction is more pronounced and extends up to the surface, although the part nearest the surface reaches the random spectra. The discrepancy between these two crystal directions may be related to a rearrangement of the Mn atoms, but no change is observed in the yield in the Mn peak. On the other hand, the SIMS spectra indicate a large Mn rearrangement (Fig. 2). To obtain a good crystal quality after regrowth, the stacking sequence from the bulk must be conserved. The presence of SF (stacking faults) will not appear in the [0001] aligned RBS spectrum but they affect channeling in the [11$\bar{2}$3] direction. There is 17° tilt between the [0001] and [11$\bar{2}$3] directions in the 4H-SiC (1$\bar{1}$00) plane, but this angle will be altered for other polytypes.

TEM analysis of a Mn implanted sample show the presence of voids, previously observed in ion implanted 4H-SiC, 3C inclusions and stacking faults on the c-plane at the end of range (Fig. 4.). An enlargement of a void is included in the figure. The 3C-inclusions have two variants, which are rotational twins on the (111) plane, confirmed by selected area diffraction (not shown). Stacking faults on the c-plane are observed on the deep side of the Mn implantation profile, in agreement with streaks observed in the diffraction pattern (not shown). In addition to SiC related diffraction spots, new spots are detected in the region with high Mn content. This diffraction pattern indicates that a new phase has formed.

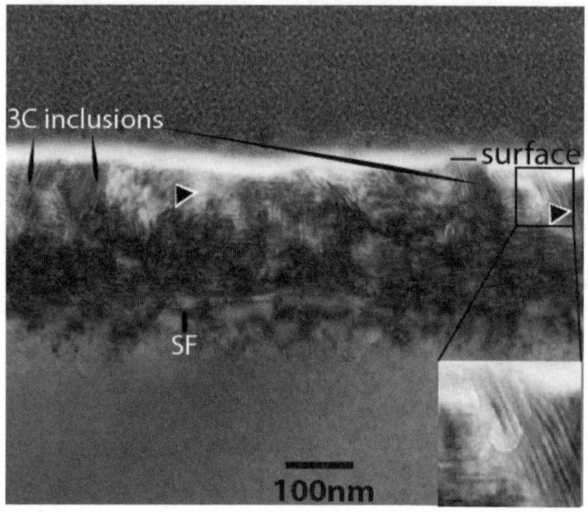

Fig. 4 Bright field TEM micrograph of Mn implanted 4H-SiC after 1600°C anneal showing voids (denoted by arrowheads), stacking faults (SF) and 3C inclusions. The inset shows an enlarged picture of a void.

Summary

In this study we have investigated structural disorder and lattice recovery after ion implantation of 4H-SiC with high Mn concentrations (close to 1 at.%). Strong changes of optical transmittance and color indicate that electronic levels localized in the bandgap appear due to the implantation disorder/damage and are annealed out at temperatures around 1400 °C. However, according to RBS-C and TEM most damage remains. A larger portion of recovery along the [0001] channel direction compared to the [11$\bar{2}$3] channel direction suggests that SF may have formed during the regrowth. These results have been further substantiated by TEM. In addition, SIMS together with TEM indicates that a new Mn containing phase is formed. X-ray diffraction and magnetic measurements are now in progress.

Acknowledgement

The authors acknowledge the Swedish agency for innovation systems (Vinnova), the Australian Research Council for funding, the ANFF and AMMRF facilities (ACT node) and Dr. Animesh Basak for assistance with FIB sample preparation.

References List

[1] M.S. Miao and W.R.L. Lambrecht, Phys. Rev. B68 (2003) 125204.
[2] Yu.A. Vodakov et al., Springer Proc. Phys. 56 (1992) 329-334.
[3] W. Wang et al., New J. Phys. 10 (2008) 055006.
[4] F. Takano et al., AIP Conference Proceedings 1199 (2010) 443-444.
[5] C. Sürgers et al., Phys. Rev. B68 (2003) 174423.
[6] J.C. Schuster, Int. J. of Refractory Metals & Hard Materials 12 (1994) 173-177.
[7] P. Spinat and P. Herpin, Bull. Soc. Fr. Mineral. Crystallogr. 99 (1976) 13-77.
[8] S. Kang et al., J. Appl. Phys. 107 (2010) 104503.
[9] Information on http://www.rzg.mpg.de/~mam/
[10] Information on http://www.srim.org

Materials Science Forum Vols. 717-720 (2012) pp 225-228

doi:10.4028/www.scientific.net/MSF.717-720.225

Diffusion and gettering of transition metals in 4H-SiC

K. Danno[1, a], H. Saitoh[1], A. Seki[1], T. Shirai[1], H. Suzuki[1], T. Bessho[1], Y. Kawai[1] and T. Kimoto[2]

[1]Higashifuji Technical Center, Toyota Motor Corporation, 1200, Mishuku, Susono Shizuoka 410-1193, Japan

[2] Dept. Electronic Science and Engineering, Kyoto University, Kyotodaigakukatsura, Nishikyo, Kyoto 615-8510, Japan

[a]danno@katsunori.tec.toyota.co.jp

Keywords: Transition metals, Diffusion, Gettering

Abstract. Diffusion of transition metals in 4H-SiC has been investigated by secondary ion mass spectrometry using epilayers and substrates implanted with titanium (Ti), chromium (Cr), iron (Fe), or nickel (Ni). In the epilayers, Cr, Fe, and Ni atoms have diffused by argon (Ar) annealing at 1780°C for 30 min. In n^+ substrates, the diffusivity of the metals is smaller than that in the epilayers, and only Ni has diffused by the annealing. By Ar or helium implantation following the implantation of transition metals, diffusion of transition metals can be successfully suppressed.

Introduction

In general, the diffusion coefficients of impurity atoms in silicon carbide (SiC) are much lower than those in silicon (Si) [1]. For example, the diffusion coefficient of nitrogen (N) was reported to be 10^{-20} cm^2/s at 1500°C in SiC [2]. Although there are very limited reports on diffusion of transition metals in SiC [3], the diffusion coefficients are probably higher than those of light elements such as N, as in the case of Si [4,5]. Therefore, it is possible that the metals become mobile under high-temperature conditions. If the metals penetrate device areas, they may degrade device performances. For example, titanium (Ti) [6,7], vanadium [6-8], and chromium (Cr) [6,7] work as deep levels, and can affect electrical properties of SiC devices. In this work, diffusion of major transition metals (Ti, Cr, iron (Fe), and nickel (Ni)) was investigated. The authors also investigated gettering of the transition metals.

Experiments

Epilayers used in this study were grown on 8° off-axis 4H-SiC(0001) n^+ substrates by horizontal hot-wall chemical vapor deposition (CVD) at 1500°C. The thickness and the net donor concentrations of the epilayers were 2-6 μm and (6-8) × 10^{16} cm^{-3}, respectively. The transition metals were introduced into the as-grown epilayers or as-received n^+ substrates by ion implantations at 500°C. The metal concentrations in box profiles were fixed at 1 × 10^{17} cm^{-3}, and the depth was 400 nm for Ti, Cr, and Ni and 200 nm for Fe. The samples were annealed at 1780°C for 30 minutes with graphite-cap technique [9] under argon (Ar) ambient to enhance diffusion of the metals. After the annealing, the caps were removed by oxidation at 1000°C. The depth profiles of the metal concentrations before and after the annealing were investigated by secondary ion mass spectrometry (SIMS). As a trial of intentional gettering of transition metals, argon (Ar) or helium (He) ions were co-implanted after the implantation of the transition metals to introduce gettering sites in the samples.

Results and Discussion

Figure 1 shows depth profiles of (a) Cr and (b) Fe concentrations in the epilayers before and after the Ar annealing. As shown in Fig. 1(a), the depth profile of Cr concentration was changed by the annealing. While the Cr concentration in the implanted region (0-400 nm) decreased, the concentration increased in the region from 0.5 μm to 2.3 μm due to in-diffusion of Cr. Moreover, at

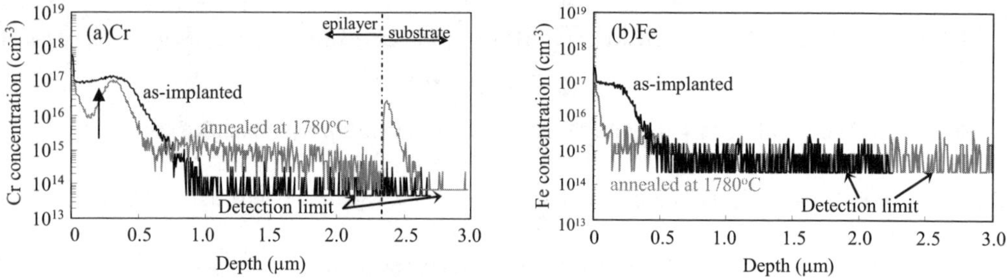

Figure 1: Depth profiles of implanted (a) Cr and (b) Fe concentrations in 4H-SiC epilayers measured by SIMS before and after annealing at 1780°C.

the depth of 2.3 μm (around interface between the epilayer and substrate), the Cr concentration was significantly increased because the Cr might be captured by defects at the interface and/or in the substrate. In part of the implanted region (highlighted by an arrow in Fig. 1(a)), the Cr concentration was almost maintained. In this region, gettering of Cr might occur by structural defects introduced by the Cr implantation. For Fe, most of implanted species diffused away from the implanted region as shown in Fig 1(b). A similar tendency was observed for implanted Ni (not shown). Thus, Fe and Ni showed more rapid diffusion compared to Cr.

Figure 2 shows the depth profiles of concentrations of Ti, Cr, Fe and Ni in the n^+ substrates after the annealing. For Ti, Cr, and Fe, the box profiles were almost maintained after the annealing at 1780°C. For Ni, however, the box profiles were changed by the annealing. At the depth region from 0 to 200 nm, Ar ions were co-implanted for this particular sample. By the co-implantation, Ni diffusion was suppressed (Details will be discussed later.). However, at the deeper region where Ar ions were not implanted (depth range from 200 nm to 400 nm), Ni was diffused. Thus, in the n^+ substrates, diffusivity was decreased for Cr and Fe, indicating that the atoms were trapped by the kinds of defects whose concentrations are high in substrates like basal plane dislocations, impurity atoms, and

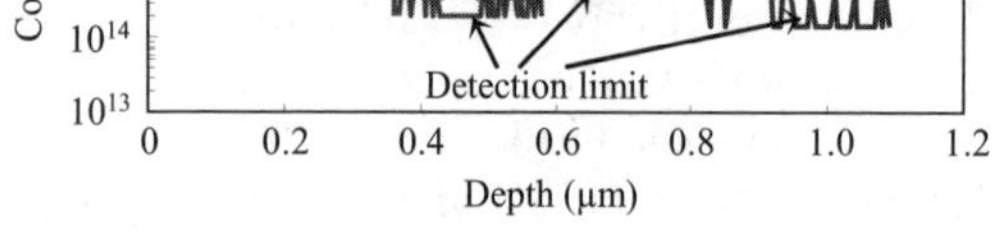

Figure 2: Depth profiles of implanted Ti, Cr, Fe, and Ni atoms in 4H-SiC substrates measured by SIMS after annealing at 1780°C. Depth of implanted atoms is 400 nm for Ti, Cr, Ni, and 200 nm for Fe. For Ni, argon ions were co-implanted in a depth range 0 to 200 nm, in which gettering by Ar implantation occurred (discussed later).

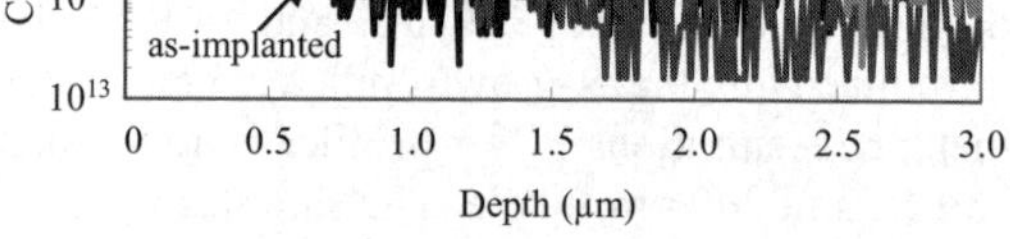

Figure 3: Depth profiles of implanted Cr atoms in as-implanted epilayers and epilayers with annealing at 1300°C, 1500°C, or 1780°C for 30 min. Depth of implanted atoms is approximately 400 nm

Table I. Diffusion of transition metals in SiC epilayers and substrates. The numbers in parentheses indicate the atomic numbers.

Element	Ti (22)	Cr (24)	Fe (26)	Ni (28)
In epilayers	Little	Partly	Highly	Highly
In substrates	Little	Little	Little	Highly

intrinsic defects. To reveal the trapping defects, more systematic investigations using transmission electron microscope, electron paramagnetic resonance, and theoretical calculations, are required.

Figure 3 shows the depth profiles of Cr concentrations of an as-implanted sample, and samples annealed at 1300°C, 1500°C, or 1780°C for 30 min. The depth of the box profile of as-implanted sample was approximately 200 nm for this experiment. The depth profiles of Cr concentration in the epilayer without annealing and in the epilayer annealed at 1300°C were almost the same. After the annealing at 1500°C, however, the depth profile was changed. In the sample annealed at 1780°C, the Cr concentration in the implanted region was decreased compared to that annealed at 1500°C. Thus, the Cr started to diffuse at about 1500°C.

In the epilayers, only a portion of the implanted Cr atoms diffused, while most of the implanted atoms diffused for Fe and Ni as mentioned above. In the substrates, the box profiles of Cr and Fe atoms were almost unchanged and only Ni atoms diffused. These results are summarized in Table I. From these results, the diffusivity seems to be in the order of Ti < Cr < Fe < Ni. It is interesting to note that this order coincides with that of the atomic number (atomic number: Ti; 22, Cr; 24, Fe; 26, Ni; 28). This tendency is also observed in Si [4], in which transition metals show rapid diffusion via interstitial sites [4,5]. Cr, Fe and Ni in SiC may also diffuse interstitially. Higher energy is required for substitutional diffusion compared to interstitial diffusion since the transition metals might be required to substitute both Si and C atoms or to jump from a Si site a transition metal is occupying to a neighboring Si site for substitutional diffusion (The authors assume that transition metals substitute Si sites if they can sit on substitutional sites). For Cr, Fe, and Ni, they might be able to diffuse easily if they occupy interstitial sites. On the other hand, Ti is immobile in SiC probably because Ti substitutes Si sites [10].

As a trial of intentional gettering of transition metals, Ar ions were co-implanted into the epilayers. The implant energy and dose of the Ar ion implantation were 270 keV and 1.1×10^{16} cm^{-2}, respectively. By the implantation, the Ar reached to a depth of 200 nm. As shown in Fig. 4, the Cr concentration was not decreased in the co-implanted region after the annealing at 1780°C. On the

Figure 4: Depth profiles of implanted Cr concentration in 4H-SiC epilayers implanted with Ar ions before and after Ar annealing at 1780°C.

Figure 5: Depth profiles of implanted Ti, Cr, and Fe concentrations in 4H-SiC epilayers implanted with He ions followed by Ar annealing at 1780°C. Depths of implanted atoms are 400 nm for Ti and Cr and 200 nm for Fe. Depth profile of He concentration is measured before annealing.

other hand, Cr concentration was decreased in the deeper region where implanted Ar did not reach. Cr in this region diffused after the annealing, and the Cr reached to the interface between the epilayer and substrate. For Ni, a similar result can be obtained in n^+ substrates as shown in Fig. 2. In SiC, gettering of the metals can be realized only in the implanted region.

He ion implantation was tried to realize gettering in a deeper region than the case of Ar ion implantation. The implant energy and dose of the He ion implantation were 350 keV and 1.6×10^{16} cm^{-2}, respectively. The projected range of implanted He ions was 1.0 μm, which is deeper than the implantation depth of the transition metals. Figure 5 represents the depth profiles of Ti, Cr, Fe and Ni concentrations in epilayers co-implanted with He. Ti, Cr, and Fe concentrations were maintained in the He co-implanted region.

From these results, structural defects and/or intrinsic defects introduced by ion implantation might work as gettering sites of the transition metals. Considering the diffusivity of transition metals in epilayers and substrates, the basal plane dislocation (BPD) is one of candidates for structural defects acting as gettering sites. The reason is that BPD concentration is generally much higher in substrates than that in epilayers. Moreover, the BPDs can be introduced by ion implantation. Therefore, BPD might work as gettering sites. Because a high density of stacking faults can be introduced by ion implantation, they might act as gettering sites. For the threading dislocations, on the other hand, the concentration in epilayers is almost same with that in substrates. Therefore, the threading dislocations can be ruled out. Intrinsic defects such as vacancies, interstitials, and their complexes, might contribute to gettering of transition metals. Since concentrations of Ti, Cr, and Fe atoms were distributed uniformly in the gettering region, the gettering sites should also be distributed in the same way. Intrinsic defects can be introduced uniformly by ion implantation. Therefore, intrinsic defects cannot be ruled out as a candidate for gettering sites.

Summary

In this study, diffusion and gettering of transition metals (Ti, Cr, Fe, and Ni) in 4H-SiC were investigated. Cr, Fe, Ni atoms were revealed to be mobile in 4H-SiC epilayers when annealing temperature was higher than 1500°C. On the other hand, Ti atoms, which substitute Si sites, seemed to be less mobile. The diffusion of the atoms could be suppressed by the Ar or He implantation. The transition metals may be captured by the implantation-related defects.

References

[1] Y. A Vodakov and E. N. Mokhov, Silicon Carbide - 1973 (University of South Carolina Press, Columbia, S. C., USA, 1974) p. 508.

[2] L. J. Kroko and A. G. Milnes, Solid State Electronics 9 (1966) 1125.

[3] G. K. Safaraliev, Y. A. Tairov, V. E. Tsvetkov, and S. S. Shabanov, Sov. Tech. Phys. Lett. 17 (1991) 854.

[4] E. R. Weber, Appl. Phys. A 30 (1983) 1.

[5] S. M. Myers, M. Seibt, and W. Schröter, J. Appl. Phys. 88 (2000) 3795.

[6] T. Dalibor, G. Pensl, H. Matsunami, T. Kimoto, W. J. Choyke, A. Schöner, and N. Nordell, Phys. Stat. Sol. (a) 162 (1997) 199.

[7] N. Achtziger and W. Witthuhn, Appl. Phys. Lett. 71 (1997) 110.

[8] H. M. Hobgood, R. C. Glass, G. Augustine, R. H. Hopkins, J. Jenny, M. Skowronski, W. C. Mitchel, and M. Roth, Appl. Phys. Lett. 66 (1995) 1364.

[9] Y. Negoro, K. Katsumoto, T. Kimoto, and H. Matsunami, Mater. Sci. Forum 457-460 (2004) 933.

[10] K. M. Lee, L. S. Dang, G. D. Watkins, and W. J. Choyke, Phys. Rev. B 32 (1985) 2273.

Materials Science Forum Vols. 717-720 (2012) pp 229-232
© (2012) Trans Tech Publications, Switzerland
doi:10.4028/www.scientific.net/MSF.717-720.229

Chlorine in SiC: Experiment and Theory

G. Alfieri[a,*] and T. Kimoto[b]

Department of Electronic Science and Engineering, Kyoto University, Kyotodaigaku-katsura, Nishikyo, Kyoto, 615-8510, Japan

[a]giovanni@semicon.kuee.kyoto-u.ac.jp, [b]kimoto@kuee.kyoto-u.ac.jp, *corresponding author

Keywords: impurity, defect, chlorine, DLTS, ab initio, DFT.

Abstract. An isochronal annealing study, in the 100-1400 °C temperature range, was carried out on Cl-implanted n- or p-type 4H-SiC epilayers. The electrical characterization of the epilayers shows the rise of seven deep levels and two of these, labeled Ci1 and Ci2, are related to Cl implantation. The role of Cl, on both carrier concentration profile and defects' microscopic structure, is discussed in the light of theoretical results obtained by density functional calculations performed on a cubic SiC supercell.

Introduction

Cl-based chemical vapor deposition (CVD) is a well-established technique for growing thick, high-quality SiC epilayers which are required for the fabrication of high-voltage SiC devices. The use of Cl hinders the formation of Si-droplets, which cause the formation of extended defects in the epilayer [1], and speeds up the growth rate (>100 μm/h). Studies have shown that methyltrichlorosilane (MTS) precursor is best suited for growing homoepitaxial layers at high growth rates [2] and that either N or Al doped SiC epilayers can be obtained [3, 4].

Yet, while H, as part of precursor gases, can be incorporated in SiC and can give rise to electrically active defects, not much is known on the effect of Cl on the electronic properties of SiC. It is known that that Cl incorporation is feasible [5] and that the Cl/Si ratio affects the net-acceptor concentration in 4H-SiC [4]. In addition, the concentration of the $EH_{6/7}$ level is reduced in Cl-CVD grown n-type epilayers relative to traditional CVD growth [1]. However, the nature of the decrease of the net acceptor concentration and the possible existence of Cl-related defects are not known. For this reason, we analyzed the effect of the presence of Cl in implanted n- and p-type 4H-SiC in the light of the results obtained by electrical characterization and density functional calculations.

Results and Discussion

Experimental Results. N or Al-doped 4H-SiC epilayers (doping $7\text{-}8\times10^{15}$ cm^{-3}) were implanted with Cl at 700 keV (dose 5×10^{9} cm^{-2}), which resulted [6] in a mean projected range of 0.5 μm from the surface (Fig. 1). An isochronal annealing series (time step 15 min) was carried out between 100 and 1400 °C in Ar atmosphere. Heat treatments higher than 1200 °C were followed by a dry oxidation process (10 min at 1100 °C), in order to minimize the effect of surface decomposition. The oxide layer was later removed by dipping the sample into HF for 10 min. After each annealing step, Au or Ti Schottky diodes were formed on the surface of n- or p-type 4H-SiC, respectively, by thermal evaporation. For p-type samples, ohmic contacts were formed on the backside of the as-grown samples by depositing a Ti/Al/Ni tri-layer, subsequently sintered at 1000 °C for 10 min.

Electrical characterization was performed by capacitance–voltage measurements (C-V) and by deep level transient spectroscopy (DLTS) [7]. DLTS measurements were carried out in the 200-700 K range, with a reverse bias of ±6 V (to cover the whole implantation profile) and a filling pulse to 0 V for 1 ms. For heat treatments up to 500 °C, the highest DLTS temperature limit coincided with the annealing temperature.

In Fig. 1, the net donor/acceptor concentrations (N_D/N_A), along with the Cl-implantantion and total vacancies profiles are shown. It can be seen that, after implantation, the N_D and N_A profiles are outside the implanted area and the oxidation process has no effect on the N_A/N_D. After high temperature treatment (1400 °C) N_D approaches the implanted region at 0.6 μm, while N_A does not change significantly.

The DLTS spectra of Cl-implanted n- and p-type 4H-SiC are shown in Fig. 2a and 2b, respectively. For n-type 4H-SiC, annealing at 900 °C results in the formation of a broad DLTS peak, which can be identified as the EH4 level [8] (concentration 3×10^{14} cm^{-3}), and of a another wide peak at ~575 K, which we labeled Ci1 (~10^{15} cm^{-3}). Heat treatment at 1400 °C yields the two well-known $Z_{1/2}$ (5×10^{14} cm^{-3}) and $EH_{6/7}$ (7.5×10^{14} cm^{-3}) levels. The latter may have been overlapping with Ci1 and becomes clearly detectable only after annealing of Ci1. For this reason, we considered the spectra at 1400 °C as the final stage from which we subtracted the DLTS signal of the 900 °C annealed sample, considered as the initial stage. The result is shown as a solid line in Fig. 2a which may correspond to the real Ci1 level. By simulating this DLTS signal, as shown in Ref. 7, we obtain an energy level position for the Ci1 at 1.34 eV below the conduction band minimum, E_C, and a capture cross section of ~10^{-16} cm^2.

For p-type (Fig. 2b), two levels labeled Ci2, at 0.51 eV above the valence band edge maximum, E_V, with a capture cross section of 2×10^{-17} cm^2 and concentration of 2×10^{12} cm^{-3}, and UK2 [9] (2.5×10^{12} cm^{-3}) are found after annealing at 900 °C. They are persistent even after 1400 °C when another level, identified as IM7 [10] (~10^{13} cm^{-3}), is detected.

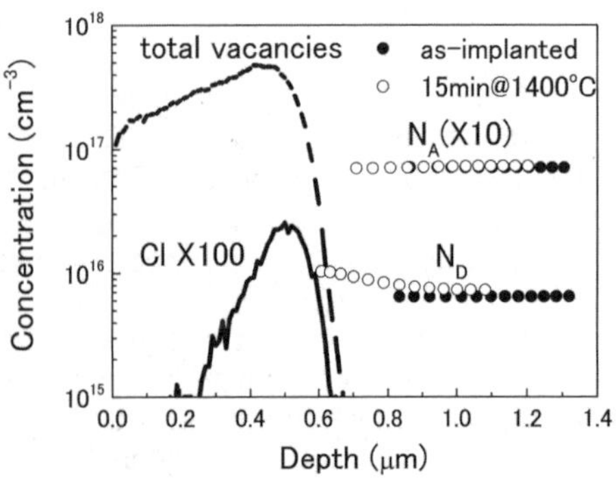

Fig.1 Depth profiles of the net donor and acceptor concentration. The concentration of the total vacancies and Cl implantation profile, obtained by SRIM [6] are also shown.

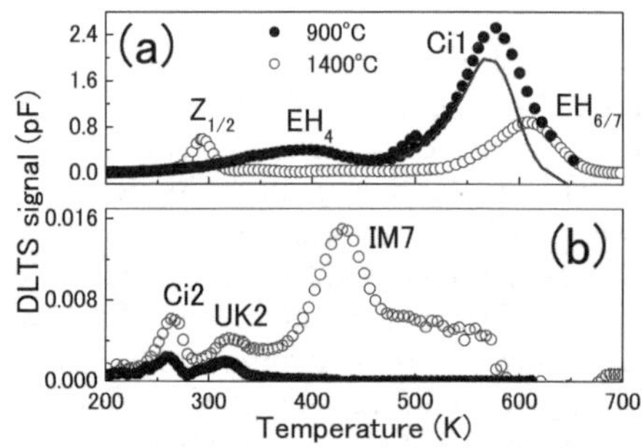

Fig.2 DLTS spectrum of the Cl-implanted (a) n- and (b) p-type 4H-SiC epilayers, after 900 and 1400 °C annealing. The employed period width was 0.2 s.

Theoretical Results. In order to give a more detailed explanation of our experimental results and since neither experimental nor theoretical studies have been performed on the role of Cl in SiC, we carried out spin-unpolarized *ab initio* calculations of isolated (Cl substitutional/interstitial) and complex (Cl-dopant) defects. Calculations were performed by using the SIESTA package [11] on a 64-atom cubic SiC supercell in the local density approximation (LDA). A 90 Ry cutoff and a 3×3×3 Monkhorst-Pack mesh ensured the convergence of the calculations which included the Madelung-type correction, Troullier-Martins norm-conserving pseudopotentials and a double-ζ plus polarization atomic basis set. We point out that the underestimation of the bandgap and the Madelung-type correction can affect the estimation of the total energy which affects the position of the ionization levels. To overcome these problems, hybrid methods can be used but they are too demanding for our computational facilities. For this reason, the calculation of the ionization levels is based on the total energies and their energy position in the bandgap will be referred to E_V.

Regarding substitutional Cl, e.g., Cl_C, Cl_{Si}, it is found that they both act as donors in their neutral state and, after geometry optimization, while Cl_C (Fig. 3a) retains its initial T_d symmetry, Cl_{Si} (Fig. 3b) lowers it from T_d to C_{3v}. To confirm this, we distorted the symmetry of the system to C_{3v}, allowing the atoms to relax. We found that the C_{3v} symmetry is conserved and the total energy is lower than constraining the system to T_d during geometry optimization. The addition of spin-polarization to our calculation showed again that Cl_C retains tetragonal symmetry, while Cl_{Si} lowers it to trigonal. Two interstitial positions were analyzed: Cl_i in T_d (Fig. 3c) and O_h (Fig. 3d) symmetry which are both retained after geometry optimization.

In Fig.4, the formation energy (E_{form}), as a function of the Fermi level (E_F), is shown. E_F varies from 0 to 1.3 eV, or the Kohn-Sham bandgap which, notoriously, is underestimated by LDA with respect to the experimental one (shown by a shaded area in Fig. 4). In Fig. 4a, it is evident that Cl favors a C site rather than a Si site: This is expected, as Cl binds more easily to Si rather than to C [3]. It can also be noted that Cl_C, contrary to Cl_{Si} which gives rise to three positively charged levels, shows only one transition level within the bandgap. Fig. 4b shows that similarly to Cl_{Si}, Cl_i, both in T_d and O_h symmetry, has a very high E_{form} thus is unlikely to occur. For this reason, this defect will not be discussed any further.

We then studied Cl-related complexes, namely Cl-Al and Cl-N and their E_{form} are shown in Fig. 4c. The E_{form} of Cl_CAl_{Si} is negative meaning that it should form spontaneously in p-type SiC and the two considered charge states, (2+/+) and (+/0), are resonant in the conduction band. As expected the E_{form} of the Cl-N complex is ~5 eV higher because N and Cl are both donors and repel each other. It can be noted that the (0/-) is not stable; thus $Cl_{Si}N_C$ displays a negative-U behavior, even though this may be due to the errors introduced by bandgap underestimation and Madelung correction.

In order to overcome the above mentioned limitation of LDA, we apply an *a posteriori* correction, e.g., the scissor operator as defined by Laaksonen et al. [12], which lifts the E_C in order to match the experimental 3C-SiC E_{gap}, and shifts the position of the transition levels proportionally to their distance from E_V. The calculated energy level positions of Cl_C, Cl_{Si}, Cl_CAl_{Si} and $Cl_{Si}N_C$ are shown in Table 1.

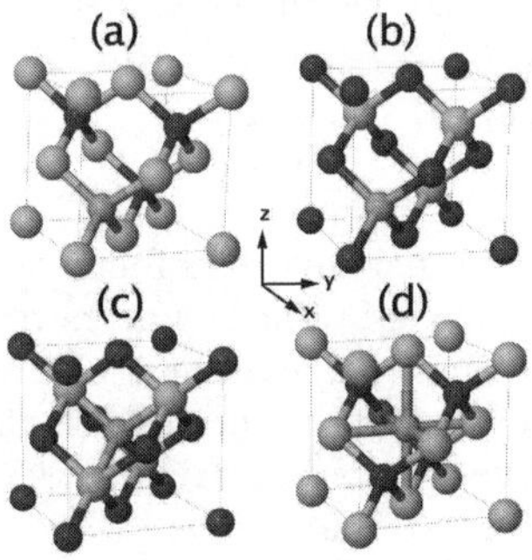

Fig.3. Unrelaxed geometry of (a) Cl_C, (b) Cl_{Si}, (c) Cl_i^{Td} and (d) Cl_i^{Oh}.

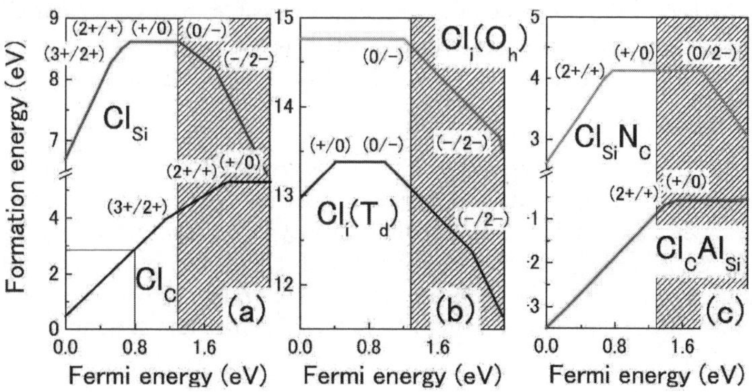

Fig.4. Formation energies of (a) substitutional, (b) interstitial Cl and (c) Cl-Al/N complexes. The shaded area represents the experimental E_{gap}.

Table 1. Energy position, in eV, below E_C for Cl_C, Cl_{Si}, Cl_CAl_{Si} and $Cl_{Si}N_C$.

Charge state	Cl_C	Cl_{Si}	Cl_CAl_{Si}	$Cl_{Si}N_C$
(0/-)		0.20		
(+/0)	0.24	0.40	0.43	1.18
(2+/+)	0.32	1.39	0.52	1.62
(3+/2+)	0.91	1.84		

Discussion. The preference of Cl for the C-site has interesting consequences. First, it can explain the nature of the incorporated Cl in 4H-SiC. In fact, from Fig. 4a, for a n-type doping concentration of 4×10^{14} cm^{-3}, E_{form} becomes 2.94 eV (corresponding to $E_F=E_V+0.81$, see Fig.4a) yielding a $[Cl_C]=2.6\times10^{14}$ cm^{-3} in fair agreement with the concentration of incorporated Cl (10^{14} cm^{-3}) reported in [5]. Second, if the $EH_{6/7}$ level is related to a V_C, its decrease in concentration, for increasing Cl/Si ratio, [1] can be explained in terms of Cl atoms occupying vacant C-sites. Furthermore, the absence of any detrimental effect of the addition of Cl on the n-type conductivity of 4H-SiC [1,3] can also be explained by our results since Cl_C behaves as a donor. Although the $Cl_{Si}N_C$ may give rise to an acceptor in the conduction band that can compensate carriers, this defect would occur at similar concentration as Cl_{Si}, which, however, is expected to have negligible concentration due to its rather high E_{form} (see Fig. 2a). For this reasons, the absence of N_D signal in the implanted area (Fig. 1) cannot be attributed to the implanted Cl but rather to implantation defects, such as silicon or carbon vacancies, affecting N donors [13, 14].

The N_A profile may be due to Al passivation by carbon vacancies or interstitials (e.g., Al_iC_i can be thermally stable up to 1700 °C [15]) and the presence of a compensating center (simple or complex defect) related to Cl can be ruled out because, in the present study, the concentration of implanted Cl is so low that such a complex should not exist in sufficient concentration to compensate carriers. However, a Cl-Al complex would explain, for the case of p-type, why the addition of Cl reduces the N_A as observed in Cl-based grown SiC [4]. Our theoretical results show that the Cl_CAl_{Si} complex forms spontaneously (should be stable up to 2200 °C [4]) and behaves as a deep donor, even if more positive charge states may exist in E_{GAP} in addition to the two considered in this study.

All of the detected levels, with the exception of Ci1 and Ci2, have already been reported in the literature. In n-type, apart from the well known $Z_{1/2}$ and $EH_{6/7}$, for which an extensive literature exists, EH4 has been related to C-clusters [8] while, in p-type, the UK2 was associated to C-displacement [9] and IM7 to a C-related complex [10]. For p-type 4H-SiC, Ci2 is close in energy position to $Cl_{Si}(3+/2+)$ (Table 1). However, as the region probed by DLTS is outside the implanted area and, unless Cl in-diffusion occurs, it can be safely concluded that the Ci2 level should not be related to Cl but rather to implantation defects. Regarding Ci1, which is detected in the n-type region near the implanted Cl profile, the possible involvement of Cl cannot be ruled out. Yet, Ci1 has a concentration of $\sim 10^{15}$ cm^{-3} which is higher than that of Cl at 0.6-0.7 μm (10^{12}-10^{14} cm^{-3}) from the surface. Thus the presence of Cl in the microscopic structure of Ci1 is unlikely.

Summary

The effects of the presence of Cl in SiC were studied by means of DLTS and DFT. It is found that vacancies arising after implantation, rather than Cl, can be held responsible for carrier compensation in n-type SiC, while for p-type SiC the presence of a compensating center, such as the Cl_CAl_{Si} complex, can be put forward. DLTS has shown the presence of seven deep levels, in both n- and p-type SiC, most of them stable up to 1400 °C. Two new levels are found, labeled Ci1 and Ci2, but the involvement of Cl in their microscopic structure may be ruled out.

References

[1] F. La Via et al., Chem. Vap. Deposition 12 (2006) 509.
[2] S. Leone et al., Cryst. Growth Design 10 (2010) 5334.
[3] H. Pedersen et al., J. Cryst. Growth 311 (2009) 1321.
[4] H. Pedersen et al., J. Cryst. Growth 311 (2009) 3364.
[5] H. Pedersen et al., phys. stat. sol. (RRL) 2 (2008) 188.
[6] J.F. Ziegler et al., The stopping and range of ions in solids, Pergamon, New York, 1985.
[7] S. Weiss et al., Solid-State Electronics 31 (1998) 1733.
[8] G. Alfieri et al., J. Appl. Phys. 98 (2005) 043518.
[9] K. Danno et al., J. Appl. Phys. 101 (2007) 103704.
[10] G. Alfieri et al., phys. stat. solidi 246 (2009) 402.
[11] J.M. Soler et al., J. Phys.: Condens. Matter 14 (2002) 2745.
[12] L. Laaksonen et al., J. Phys.: Condens. Matter 21 (2009) 015803.
[13] U. Gerstmann et al., Phys. Rev. B 67 (2002) 205202.
[14] D. Åberg et al., Appl. Phys. Lett. 78 (2001) 2908.
[15] A. Gali et al., Phys. Rev. B 75 (2007) 045211.

Materials Science Forum Vols. 717-720 (2012) pp 233-236

doi:10.4028/www.scientific.net/MSF.717-720.233

Photoluminescence and Raman spectroscopy characterization of boron- and nitrogen-doped 6H silicon carbide

Yiyu Ou[1, a, *], Valdas Jokubavicius[2, b], Chuan Liu[3, c], Rolf W. Berg[4, d], Margareta Linnarsson[5, e], Satoshi Kamiyama[6, f], Zhaoyue Lu[7, g], Rositza Yakimova[8, h], Mikael Syväjärvi[9, i], and Haiyan Ou [10, j]

[1, 7, 10]Department of Photonics Engineering, Technical University of Denmark, DK-2800, Lyngby, Denmark

[2, 8, 9]Department of Physics, Chemistry and Biology, Linköping University, SE-58183, Linköping, Sweden

[3, 4]Department of Chemistry Energy and Materials, Technical University of Denmark, DK-2800, Lyngby, Denmark

[5]School of Information and Communication Technology, Royal Institute of Technology, SE-16440, Kista, Sweden

[6]Faculty of Science and Technology, Meijo University, 1-501 Shiogamaguchi, Tempaku-ku, Nagoya 468-8502, Japan

[a]yiyo@fotonik.dtu.dk, [b]valjo@ifm.liu.se, [c]chliu@kemi.dtu.dk, [d]rwb@kemi.dtu.dk, [e]marga@kth.se, [f]skami@ccmfs.meijo-u.ac.jp, [g]zhalu@fotonik.dtu.dk, [h]roy@ifm.liu.se, [i]misyv@ifm.liu.se, [j]haou@fotonik.dtu.dk, *corresponding author

Keywords: 6H-SiC, photoluminescence, Raman spectroscopy, donor-acceptor-pair emission, sublimation epitaxy

Abstract. Nitrogen-boron doped 6H-SiC epilayers grown on low off-axis 6H-SiC substrates have been characterized by photoluminescence and Raman spectroscopy. The photoluminescence results show that a doping larger than 10^{18} cm^{-3} is favorable to observe the luminescence and addition of nitrogen leads to an increased luminescence. A dopant concentration difference larger than $4x10^{18}$ cm^{-3} is proposed to achieve intense photoluminescence. Raman spectroscopy further confirmed the doping type and concentrations for the samples. The results indicate that N-B doped SiC can serve as a good wavelength converter in white LEDs applications.

Introduction

To date, white light-emitting diodes (LEDs) are the most promising candidates for the illumination market to replace conventional incandescent lamps as energy-saving and environmental friendly light sources. One common method to realize white LEDs is using a UV or blue LED whose light is partially or fully used to optically excite wavelength converters. Therefore, high-efficiency wavelength converters with a high color rendering index (CRI) value and long lifetime are indispensable. Nitrogen (N) and boron (B) doped 6H-silicon carbide (SiC) has been proven as a highly efficient wavelength converter [1]. Combined donor-acceptor-pair (DAP) band luminescence from N-B and nitrogen-aluminium (N-Al) doped 6H-SiC can cover most of the visible spectral range and thus white light with CRI larger than 90 could be produced based on the simulated results [2]. Furthermore, SiC is a well established substrate material for nitride growth and has excellent thermal conductivity.

Experiments and Results

To investigate the route to optimized dopant concentrations of N and B, five samples with various N and B concentrations were studied. The epilayers were grown on 6H-SiC (0001) substrates having 1.4 degree off-orientation in the $[11\bar{2}0]$ direction by the fast sublimation growth process [3] at a

Table 1. Dopant concentrations and normalized PL peak intensities of the samples

Sample	B concentration $[cm^{-3}]$	N concentration $[cm^{-3}]$	Doping type	PL peak intensity [Normalized to sample d]
a	8.0×10^{18}	4.0×10^{16}	P	0.0 %
b	6.9×10^{18}	3.2×10^{18}	P	6.6 %
c	6.9×10^{18}	6.0×10^{18}	P	8.3 %
d	4.4×10^{18}	9.0×10^{18}	N	100 %
e	5.2×10^{18}	9.2×10^{18}	N	77.1 %

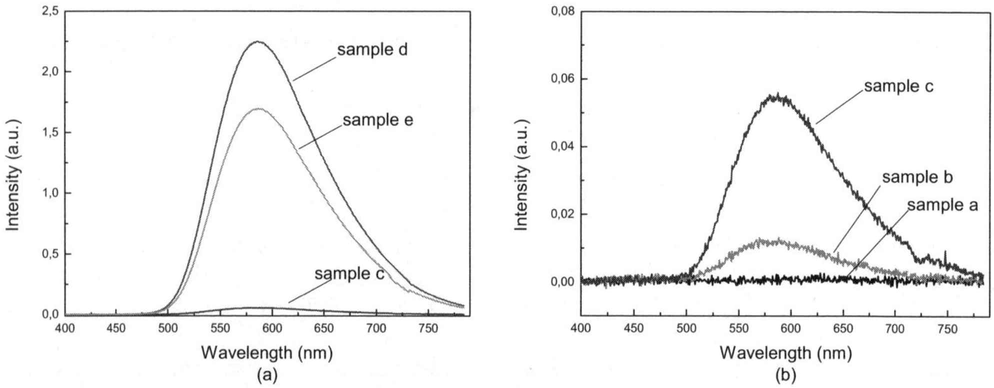

Fig. 1. Measured PL spectra of B-N doped 6H-SiC samples: (a) Comparison of samples a, b, and c; (b) Comparison of samples c, d, and e.

temperature of 1725 °C. The growth process is driven by a temperature gradient created between the source, in the form of a polycrystalline SiC plate, and the substrate. The B was introduced into the epilayers by co-doping from the source and N incorporation was controlled by adjusting N_2 gas pressure during the growth. The atomic dopant concentrations measured by secondary ion mass spectrometry (SIMS) are listed in Table 1, where the N concentration of sample a was close to the background doping level and therefore might not be accurate.

Photoluminescence (PL) measurements were performed by using an Olympus reflected fluorescence system microscope, a 377 nm diode laser as excitation source (focused by a 20X objective), and an Instrument System CAS 140B spectrometer. The integration time was 10 seconds with excitation power density of 0.02 W/cm^2, and the measurements were performed at room temperature. The measured PL spectra are shown in Fig. 1, and the integrated PL peak intensities were normalized to the strongest sample (d) and are listed in Table 1. One can see that the high-level p-type sample (a) exhibits extremely low DAP emission intensity, while the low-level p-type samples (b, c) have relatively stronger DAP emissions but still at a low level. Intense DAP emission was observed in n-type samples (d, e), while the strongest DAP emission occurred in sample d, which has the largest concentration difference of 4.6×10^{18} cm^{-3}. Despite the intensity differences, all the DAP emission spectra show the same peak wavelength at 587 nm and the same full width at half maximum (FWHM) of 120 nm. This result is expected because the spectrum shape of the DAP emission is usually determined by the Coulomb interaction and phonon coupling rather than by the carrier populations [4, 5]. Due to the same dopant materials and experimental conditions, the same value of FWHM was obtained. From the above results, it is clear that both the doping type and the concentration difference affect the DAP emission intensity but do not introduce a change in the peak wavelength and FWHM.

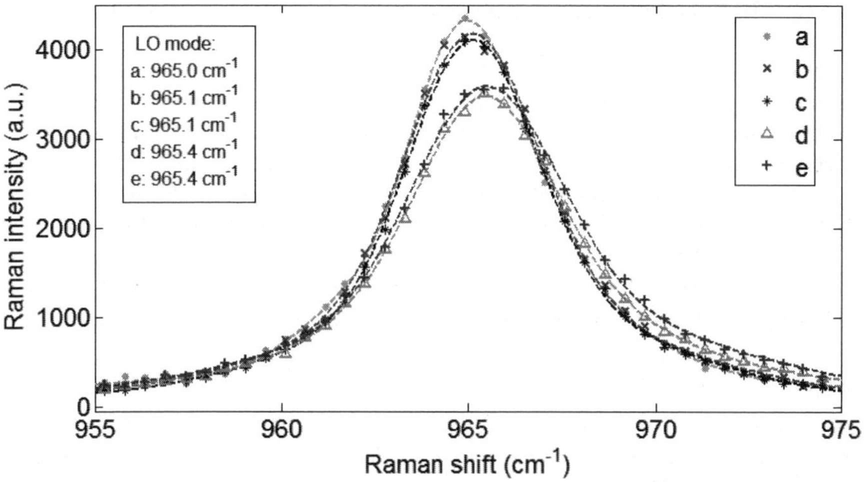

Fig. 2. LO mode in Raman spectra (inset: positions of LO modes of the SiC samples with different dopant concentrations; the peak height decreases in the order of a, b, c, e, d).

Raman scattering spectra of the 6H-SiC samples were acquired in a backscattering configuration using the 514.5 nm line from an Ar ion laser (5 mW). Longitudinal optical (LO) modes of the Raman spectra are shown in Fig. 2, and the mechanisms of the observed Raman shifts in p- and n-type samples are different. The LO mode is much more sensitive to the free electrons than to the free holes due to the large effective mass differences [6].

In p-type samples, very few acceptor states are ionized due to the large ionization energy, and the Raman shift is mainly attributed to the atomic size effect [7]. The B atom is considered to occupy a Si lattice position in SiC. The inter-atomic distance of a Si-C bond is longer than that of a B-C bond due to the smaller atomic radius of B. The biaxial tensile stress will be released which results in a decrease of the phonon oscillation frequency. So the LO mode shifts toward smaller wavenumbers with higher B concentrations.

In n-type samples, the predominant mechanism causing the Raman shift of the LO mode is the coupling interaction between LO phonons and overdamped plasmons, and the LO mode is then redefined as a longitudinal optical phonon-plasmon coupled (LOPC) mode. Although no obvious peak shift has been observed between sample d and e due to the relatively small concentration difference, the peak intensity of the LOPC mode decreases as expected when the N and B concentration difference increases from sample e to d. Furthermore, one can see from Fig. 2 that the LOPC modes of n-type samples occur at significantly higher wavenumbers than the p-type ones.

Discussion and Conclusion

In N-B doped SiC, the acceptor states are almost fully occupied by photo-excited holes due to the large ionization energy, so the hole density at the acceptor states under the excitation condition could be assumed to be the same as the B doping concentration in the sample. On the other hand, the electron density at the donor states under the excitation condition is given by the product of N doping concentration and the electron occupancy probability at the donor states due to the small ionization energy of the donor states. Meanwhile, the electron occupancy probability is proportional to the electron quasi-Fermi level under the excitation condition. So the DAP recombination rate is proportional to the product of the B concentration, N concentration and electron occupancy probability under the excitation condition. Under the same excitation condition, the electron quasi-Fermi level is at a higher energy level in the n-type samples than in the p-type ones, and a larger N and B concentration difference leads to a higher electron quasi-Fermi level. As a result, the electron occupancy probability on donor states increases and so does the DAP recombination rate. This was confirmed by the SIMS and PL measurements results in this work. The detailed interpretation was illustrated in [8].

In conclusion, five samples with various B and N doping concentrations were investigated by photoluminescence and Raman spectroscopy. It is found that n-type doping with both B and N concentrations larger than 10^{18} cm^{-3} is required to achieve intense DAP luminescence, and the concentration difference of the n-type sample needs to be larger than 4 x 10^{18} cm^{-3} as concluded from this study. Doping type and concentrations of the samples were confirmed by the peak shift and intensity change of the LOPC mode in Raman spectra. With adjustment of doping concentrations, intense DAP luminescence peaked at 587 nm with a broad FWHM of 120 nm was obtained. Both peak wavelength and FWHM did not change with varying doping concentration at this doping level, which makes SiC a good candidate for a wavelength converter in white LEDs.

Acknowledgment

The Danish council for strategic research (No. 09-072118), Ångpanneföreningen Research Foundation, Richerts Foundation, Swedish Energy Agency, Nordic Energy Research, and the Swedish Research Council (No. 2009-5307) are acknowledged for financial support.

References

[1]. S. Kamiyama, T. Maeda, Y. Nakamura, M. Iwaya, H. Amano, I. Akasaki, H. Kinoshita, T. Furusho, M. Yoshimoto, T. Kimoto, J. Suda, A. Henry, I. G. Ivanov, J. P. Bergman, B. Monemar, T. Onuma, and S. F. Chichibu: J. Appl. Phys. Vol. 99 (2006) 093108

[2]. Y. Ou, D. Corell, C. Dam-Hansen, P. Petersen, and H. Ou: Opt. Express Vol. 19 (2011) A166-A172

[3]. M. Syväjärvi and R. Yakimova: Sublimation epitaxial growth of hexagonal and cubic SiC, in Pallab Bhattacharya, Roberto Fornari and Hiroshi Kamimura (Eds), Encyclopedia - the Comprehensive Semiconductor Science & Technology (SEST), Elsevier (2011)

[4]. M. Ohishi: Jpn. J. Appl. Phys. Vol. 25 (1986) 1546-1551

[5]. P. Bäume, F. Kubacki, and J. Gutowski: J. Cryst. Growth Vol. 138 (1994) 266-273

[6]. S. Nakashima and H. Harima: Phys. Stat. Sol. (a) Vol. 162 (1997) 39-64

[7]. X. Li, Z. Chen, and E. Shi: Physica B Vol. 405 (2010) 2423-2426

[8]. Y. Ou, V. Jokubavicius, S. Kamiyama, C. Liu, R. W. Berg, M. Linnarsson, R. Yakimova, M. Syväjärvi, and H. Ou: Opt. Mater. Express Vol. 1 (2011) 1439-1446

Materials Science Forum Vols. 717-720 (2012) pp 237-240
© (2012) Trans Tech Publications, Switzerland
doi:10.4028/www.scientific.net/MSF.717-720.237

Simulation of the Temperature Dependence of Hall Carriers in Al Doped 4H-SiC

A. Nath[1,a,*], R. Scaburri[2,b], Mulpuri V. Rao[1,c], and R. Nipoti[2,d]

[1]Electrical and Computer Engineering Department, George Mason University, Fairfax, Virginia 22030-4444, USA

[2]CNR-IMM of Bologna, via Gobetti 101, I-40129 Bologna, Italy

[a]anath@gmu.edu, [b]scaburri@bo.imm.cnr.it, [c]rmulpuri@gmu.edu, [d]nipoti@bo.imm.cnr.it, *corresponding author

Keywords: acceptor doping, dopant ionization energy, simulation, neutrality equation, Hall carriers temperature dependence.

Abstract. A global minimum search weighted and non-weighted least squares algorithm has been employed for a comparative study of various possible models to describe electrical properties of high dose Al implanted 4H-SiC. A wide range of experimental data has been taken from the literature to validate the model. It was found that a single activation energy level, a temperature dependent Hall scattering factor and a degeneracy factor equal to 4 lead to a satisfactory fitting of experimental Hall data up to an Al substitutional density of 10^{20} cm^{-3} in 4H-SiC.

Introduction

The ionization energy of various dopants in SiC varies between a few tens to a few hundreds of meV resulting in a partial ionization of dopants at room temperature and temperature dependent ionization efficiency. As SiC devices operate in a wide temperature range, the temperature dependent impurity ionization efficiency is of great importance in charge carrier transport, especially at low electric fields.

The charge neutrality equation relates the free carrier, dopant and compensation densities to dopant ionization energy and, depending on the SiC polytype and dopant of interest, one (3C-, 2H-SiC), two (4H-SiC), three (6H-SiC) or more energy levels in the forbidden energy gap should be taken into account. In the framework of the Fermi Dirac statistics, the neutrality equation is a transcendental equation that possesses only a numerical solution. However, such an equation can be simplified if the density of free charge carriers is much lower than the density of states of the corresponding allowed band, and the Boltzmann statistics can be adapted in this case to describe the carrier distribution. For a material containing only one dopant species and having b ionization energy levels, the neutrality equation forms a $(b+1)$-th order equation of the free carrier density. Exploiting this approximation, the dopant ionization energy ΔE_A, dopant density N_A, and compensating center density N_{comp} are obtained by a least-squares fit of the experimental Hall data. Unfortunately, in the case of heavily Al doped 6H and 4H-SiC, the obtained total N_A value using this equation is often one order higher than the concentration of Al atoms determined by secondary ion mass spectroscopy [1]. A great deal of attention has been paid to understand this effect due to the importance of Al as a p-type dopant in SiC and for the need to have reliable values of N_A and ΔE_A for simulation and electrical characterization of devices. Realistic Al acceptor concentrations have been obtained by two different approaches: one taking into account an empirically obtained temperature dependence of the Hall scattering factor [1] and the other taking into account the role of excited states on the acceptor dopant ionization [2].

In this study, published Hall carrier data of Al doped 4H-SiC as a function of the temperature [2-4] have been fitted by the solutions of the neutrality charge equation for p-type 4H-SiC following the models of Refs. [1] and [2] or a combination of them. The best fit between experimental and computed data has been performed by using both weighted and non weighted error procedures. The originality of this study lies on the proposal of an algebraic solution of the charge neutrality equation for reducing computation time, a weighted error procedure for a better fitting of the lower temperature carrier density data, and a global minimum searching method for a more accurate fitting.

Fitting of Neutrality Equation for Acceptor Doped 4H-SiC

Experimental data has been obtained from the literature on Al-doped 4H-SiC epitaxial layers [2-4] using Matlab@2007, which allows data collection from digital images with precision. A wide free hole concentration data range (10^{11} cm^{-3} to 10^{20} cm^{-3}) has been selected to validate the adaptability of the model for highly doped 4H-SiC. If the semiconductor is p-type and all acceptors are located in the bottom half of the forbidden band and all compensating donors are in the upper-half of the forbidden band, the free hole concentration p for an arbitrary number of acceptor states is an implicit solution of the charge neutrality equation:

$$p(T)+N_{comp}=\sum_i \frac{N_{A,i}}{1+g_{A,i}\exp\left(\frac{\Delta E_{A,i}}{KT}\right)\cdot\frac{p(T)}{N_V(T)}} \quad , \tag{1}$$

where N_{comp} is the total concentration of compensating donors, i is the index corresponding to different kinds of substitutional sites, $N_{A,i}$ is the concentration of acceptors, $\Delta E_{A,i}$ is the ionization energy of the ith Al acceptor, $g_{A,i,}$ is the temperature dependent acceptor degeneracy factor [2], T is the absolute temperature, K is the Boltzmann constant and $N_V(T)$ is the temperature dependent effective density of states for the valence band [4,5]. Eq. (1) can be transformed to a third order polynomial equation, which is analytically solvable and possesses three real solutions, among which only one is positive and therefore a physically sensible solution [6]. In this study such a solution has been computed as an algebraic expression obtained by the MATLAB Symbolic Toolbox [7], as an alternative to the proposed trigonometric one [6], and has the form:

$$p(T)=N_V(T)\cdot\left[\frac{1}{6}\left(\alpha+12\beta^{\frac{1}{2}}\right)^{\frac{1}{3}}-6\frac{\left(-\frac{1}{3}B-\frac{1}{9}A^2\right)}{\left(\alpha+12\beta^{\frac{1}{2}}\right)^{\frac{1}{3}}}-\frac{1}{3}A\right] \quad , \tag{2}$$

where:

$$\alpha = -36BA + 108C - 8A^3 \quad , \qquad \beta = -12B^3 - 3B^2A^2 - 54BAC + 81C^2 - 12CA^3 \quad , \tag{3}$$

$$A=\frac{N_{comp}}{N_V}+\frac{1}{g_1}+\frac{1}{g_2} \quad , \quad B=\frac{\left(\frac{N_{A,1}-N_{comp}}{g_1}\right)+\left(\frac{N_{A,2}-N_{comp}}{g_2}\right)}{N_V}-\frac{1}{g_1g_2} \quad , \quad C=\frac{\left(N_{A,1}+N_{A,2}-N_{comp}\right)}{\left(g_1g_2\right)N_V} \quad , \tag{4}$$

$$g_1=g_{A,1}\exp\left(\frac{\Delta E_{A,1}}{KT}\right) \quad , \qquad g_2=g_{A,2}\exp\left(\frac{\Delta E_{A,2}}{KT}\right). \tag{5}$$

Eq. 2 was exploited for the calculation of the hole concentration as a function of the fitting parameters $\Delta E_{A,i}$, $N_{A,i}$, and N_{comp} with i=1, 2. A global minimum searching (both weighted and non-weighted) least squares fit algorithm has been employed to eliminate the need for a good initial guess of parameter values, and to avoid the possibility of converging at a local minimum. Typically, the simulations have been performed on a parameter grid spaced 1 meV for activation energy, and refined up-to three significant digits for any particular range of acceptor or compensating center density. All possible combinations of parameters, such as the Hall scattering factor r_H (either a value of 1 or a temperature dependent function [1]), degeneracy factor (a value of 4 or a thermal distribution function for excited states up-to six [2]) and hole density of states effective mass (constant values or a temperature dependent function [5]) have been examined for best fitting.

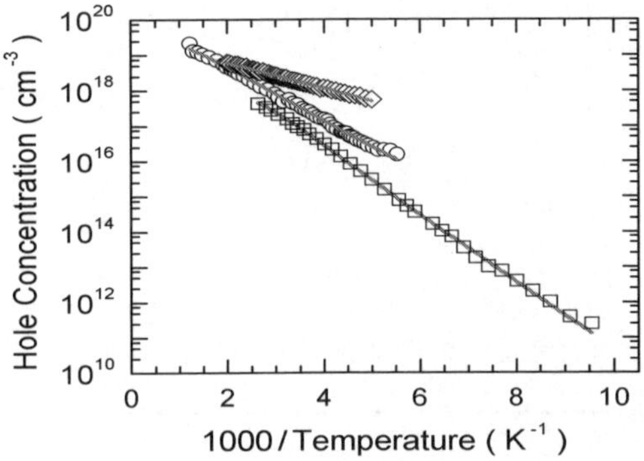

Fig.1. Best fitting of the temperature dependence of free hole concentration in these Al doped 4H-SiC samples: (○) #I4 of Ref. [3], (□) Ref. [2], and (◇) 5.5×10^{19} cm^{-3} of Ref. [4] (labeled #1 in Table 1). The fitting curves correspond to a single energy Al level, $r_H(T)$ [1], $g_{A,1=2} = 4$, and weighted least square error function. Fitting parameters outputs are provided in Table 1.

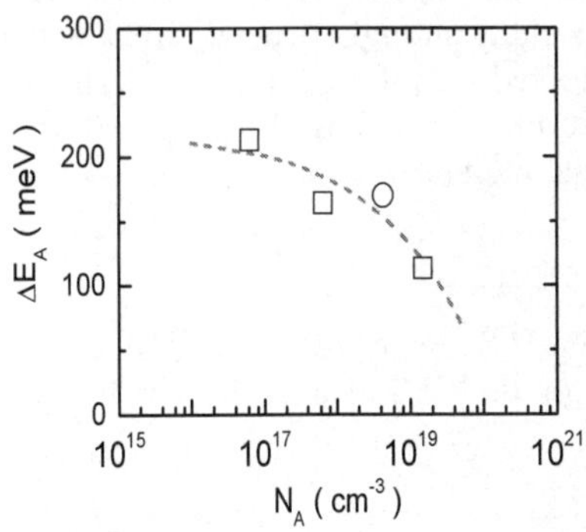

Fig.2. Plot of $\Delta E_{A,1}$ versus $N_{A,1}$ obtained as parameter outputs of the present simulations of Hall carrier data of Al-doped 4H-SiC materials with a negligible [4] (□) and non-negligible [2] (○) compensation ratio. The dashed line is only for guiding the eyes.

Results and Discussion

The fitting quality of the approaches of Refs. [1] and [2] has been confirmed by our simulations, but when a weighted error function and a global minimum searching method is applied the approach of Ref. [1] gives better fits. In particular, it was found that a model consisting of a single activation energy level, a temperature dependent Hall scattering factor [1], a degeneracy factor equal to 4 and a constant effective mass of $2.64m_0$ reproduces all the examined published Hall data up to an Al substitutional density of 10^{20} cm^{-3} in 4H-SiC as shown in Fig. 1.

Table 1 provides a comparison of parameters (Al energy level and substitutional acceptor and compensation center densities) obtained for best fits of Fig. 1 data and other published data using simulations performed by the approach presented here and the approaches of Refs. [1] or [2] for various authors. Results of Table 1 show that the Al energy levels obtained in this study are systematically lower than those obtained in the other studies, while acceptor densities are in some cases higher, in others lower with factors between 1.5 and 3. Concerning N_{comp}, in this study, like in those from the literature, in the case of samples with low compensation ratio, N_{comp} could not be defined, but when the compensation ratio is relevant N_{comp} values larger by a factor between 3 and 9 have been obtained. At the moment the authors believe that the obtained differences in parameter values are mostly due to the use of a weighted fitting procedure.

The $\Delta E_{A,1}$ and $N_{A,1}$ results of Table 1 for this study are displayed in Fig. 2, which confirms the well known trend of decreasing $\Delta E_{A,1}$ for increasing $N_{A,1}$ [4].

Table 1: Comparison of the output parameters: $\Delta E_{A,i}$, $N_{A,i}$ and N_{comp} obtained for best fits of several published Hall carrtier data in the case of epitaxial Al-doped 4H-SiC for simulations performed by the approach of this study and those of Refs. [1] or [2], depending on the authors. A cross (X) indicates a very low compensation ratio for which N_{comp} cannot be obtained.

Data Source	N_{A1} [cm^{-3}]	ΔE_{A1} [meV]	N_{A2} [cm^{-3}]	ΔE_{A1} [meV]	N_{Comp} [cm^{-3}]
Ref. [3] #I4	2.8×10^{19}	143±3	2.8×10^{19}	260	6.8×10^{18}
This study	7.5×10^{19}	116	-	-	2×10^{19}
Ref. [2]	3.2×10^{18}	180	-	-	9×10^{16}
This study	4.1×10^{18}	171	-	-	8.1×10^{17}
Ref. [4] #1	5.5×10^{19}	137	-	-	X
This study	1.5×10^{19}	113	-	-	X
Ref. [4] #2	2.6×10^{18}	201	-	-	X
This study	1×10^{18}	178	-	-	X
Ref. [4] #3	1.8×10^{17}	244	-	-	X
This study	6.3×10^{16}	213	-	-	X

Conclusion

We have fitted the charge neutrality equation, using the Boltzmann approximation, to several sets of temperature dependent experimental Al-doped 4H-SiC hole concentrations taken from the literature. We took into account the possibility of two or one activation energy levels, a temperature dependent Hall scattering factor [1], temperature dependent effective mass, and the influence of excited states on dopant ionization and compared the results we obtained with or without considering these effects. We found that neglecting the role of excited states on dopant ionization and taking into account the correction to experimental data considering a temperature dependent Hall scattering factor with a single energy level and a constant effective mass produces satisfactory and reliable results. However, the results obtained in this study strongly depend on the accuracy of the Hall experimental data as well as on the equations adopted; i.e., if the fitting is weighted or non-weighted and if the algorithm converges for best fit or it searches for a global minimum. Unfortunately, these aspects are frequently neglected in the literature.

Acknowledgement

This paper is based on a work partially supported by the U.S. Army Research Office (contract # W911NF-09-1-0407) and also by the DARPA through the U.S. Naval Research Laboratory (contract # N0017310-2-C006).

References

[1] F. Schmid, M. Krieger, M. Laube, G. Pensl, and G. Wagner, in: W. J. Choyke, H. Matsunami, and G. Pensl (Eds), Silicon Carbide: Recent Major Advances, Springer-Verlag, Berlin-Heidelberg-New York, 2004, pp. 517–536.

[2] H. Matsuura, New J. Phys. 4 (2002) 12.1-12.15.

[3] J. Pernot, S. Contreras, and J. Camassel, J. Appl. Phys. 98 (2005) 023706-9.

[4] A. Koizumi, J. Suda, and T. Kimoto, J. Appl. Phys. 106 (2009) 013716.

[5] G. Wellenhofer and U. Rössler, phys. stat. sol. (b) 202 (1997) 107-123.

[6] G. Rutsch, R. P. Devaty and W. J. Choyke, IEEE proceedings *High Temperature Electronics,1999. HITEN 99. The Third European Conference* on, 1999, pp. 191-194.

[7] MATLAB Symbolic Toolbox, http://www.mathworks.it/help/toolbox/symbolic/index.html

Materials Science Forum Vols. 717-720 (2012) pp 241-246

doi:10.4028/www.scientific.net/MSF.717-720.241

Elimination of Deep Levels in Thick SiC Epilayers by Thermal Oxidation and Proposal of the Analytical Model

Koutarou Kawahara[1, a], Jun Suda[1], Tsunenobu Kimoto[1]

[1] Dept. of Electronic Sci. & Eng., Kyoto University, Nishikyo, Kyoto 615-8510, Japan

[a] kawahara@semicon.kuee.kyoto-u.ac.jp

Keywords: deep level, oxidation, carbon implantation, $Z_{1/2}$, diffusion

Abstract. It has been clarified that the $Z_{1/2}$ center, a well known deep level as a lifetime killer, can be reduced to the concentration below 10^{11} cm^{-3} by thermal oxidation or C^+ implantation plus Ar annealing. In this study, the authors investigate the trap-reduction phenomena systematically (experimentally), and propose a model to analyze the phenomena. Furthermore, prediction of the defect distributions is realized by solving a diffusion equation in accordance with the trap reduction model. This analytical model can explain almost all experimental data: oxidation-temperature dependence, oxidation-time dependence, and initial-$Z_{1/2}$-concentration dependence of the defect reduction. Based on these results, the authors achieved the elimination of the $Z_{1/2}$ center to a depth of ~100 μm in a sample with a relatively high initial-$Z_{1/2}$-concentration of 10^{13} cm^{-3} by thermal oxidation at 1400°C for 16.5 h.

Introduction

Deep levels are generated during epitaxial growth [1-4] and device fabrication steps such as ion implantation [5-10] and reactive ion etching (RIE) [11]. The deep levels work as recombination centers resulting in reduction of the carrier lifetime [12] and as carrier traps leading to reduction of the conductivity. Therefore, deep levels, especially the $Z_{1/2}$ center, a well known deep level as a lifetime killer [13,14], must be controlled. The typical carrier lifetime for commercially available SiC epilayers is only (0.6-1) μs due to a high $Z_{1/2}$ concentration about 10^{13} cm^{-3}, while over 5 μs lifetime is required for 10 kV PiN-diodes.

The origin of the $Z_{1/2}$ center has been extensively investigated [1,3,15,16]. The authors speculate that the $Z_{1/2}$ center is a carbon-vacancy (V_C)-related defect because of the following three experimental results. First, the $Z_{1/2}$ center is generated by low-energy (100-200 keV) electron irradiation, which displaces only C atoms, without subsequent annealing [3,15]. Second, the $Z_{1/2}$ concentration is lower in the samples grown with higher C/Si ratio [3]. Third, the $Z_{1/2}$ is a thermally stable defect [1] showing no change of the concentration up to 1600°C [3], at which temperature interstitials can easily diffuse. Additionally, it has been reported that an energy level (negative U) originating from V_C obtained by ab intio calculation (density functional theory) is very close to an energy level of the $Z_{1/2}$ center (negative U) detected by deep level transient spectroscopy (DLTS) [17].

There are two effective methods to reduce the $Z_{1/2}$ center, (i) C^+ implantation followed by Ar annealing [18] and (ii) thermal oxidation [19]. In the case of the C^+ implantation process, implanted excess carbon atoms should diffuse to the deeper region of a SiC epilayer during post annealing and fill V_C, resulting in reduction of the $Z_{1/2}$ concentration [18]. In the case of thermal oxidation, the $Z_{1/2}$ reduction may be derived from diffusion of interstitials generated at the SiO_2/SiC interface during oxidation [19] as shown in Fig. 1. There are several reports indicating the

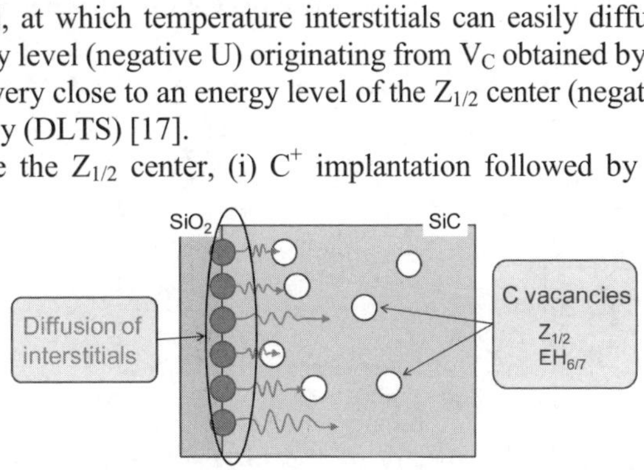

Fig. 1 Cross-sectional diagram of SiC during oxidation. Interstitials generated at the SiO_2/SiC interface diffuse into SiC bulk, and occupy carbon vacancies related to the $Z_{1/2}$ center.

generation of interstitials at the SiO_2/SiC interface, which support the trap reduction model. (i) The experimental oxidation rate of SiC can be well simulated by considering silicon and carbon emission from the oxidation interface [20]. (ii) An increase of carbon concentration at the SiO_2/SiC interface was observed by electron energy loss spectroscopy (EELS) [21]. (iii) It has been predicted from ab initio calculation that carbon clusters are formed at the SiO_2/SiC interface during oxidation [22]. The trap reduction mechanism, however, is not fully understood.

The purposes of this study are to ascertain the trap reduction model shown in Fig.1 and to enable to predict the distributions of deep levels after oxidation in order to realize controlling deep levels by thermal oxidation. For these purposes, the authors compare deep levels in the SiC epilayers after the C^+ implantation process and after thermal oxidation, and investigate the oxidation-temperature dependence, oxidation-time dependence, and initial-$Z_{1/2}$-concentration dependence of the defect reduction in both n-type and p-type epilayers. Based on these data, an analytical model of defect reduction is proposed.

Experiments

The starting materials were n-type (N_d: 10^{14}-10^{15} cm^{-3}) and p-type (N_a: 10^{15}-10^{16} cm^{-3}) 4H-SiC (0001) epilayers. A series of samples was oxidized at different temperatures (1150-1400°C) for various times (1.3-16.5 h), while the other set of samples was implanted with 10-50 keV C^+ ions with a total dose of 1×10^{13} cm^{-2} or 1×10^{14} cm^{-2}, forming a 140-nm-box-profile. The C^+-implanted samples were annealed in Ar ambient at various temperatures (1000-1800°C) for 20 min. For DLTS measurements, Ni and Ti were employed as Schottky contacts on n-type and p-type samples, respectively. The depth profiles of deep-level concentrations up to 10-μm depth were measured by changing the reverse bias voltage up to 100 V in DLTS measurements. In order to monitor deeper regions (over 10 μm), the samples were mechanically polished from the surfaces. It was confirmed that additional deep levels did not appear after the polishing.

Results and Discussion

Defect distributions after trap reduction processes. Fig. 2/Fig. 3 shows the DLTS spectra of the n-type/p-type samples after dry oxidation at 1300°C for 1.3 h, as well as samples after C^+ implantation (dose: 1×10^{14} cm^{-2}) plus Ar annealing at 1800°C/1300°C. Here, signal b_1 is the coefficient of the first sine term in the Fourier series of deep level transient Fourier spectroscopy [23]. After C^+ implantation process (chain lines), the new peaks, ON1 ($E_C - 0.8$ eV), ON2 ($E_C - 1.0$ eV), and ON3 ($E_C - 1.6$ eV) in n-type SiC (Fig. 2), HK0 ($E_V + 0.8$ eV) in p-type SiC (Fig. 3), are observed, while the same four peaks are observed after thermal oxidation. On the other hand, two major deep levels, $Z_{1/2}$ ($E_C - 0.63$ eV) and

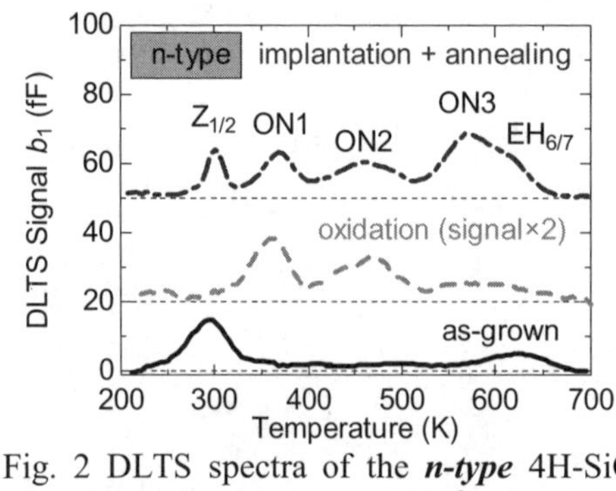

Fig. 2 DLTS spectra of the ***n-type*** 4H-SiC after thermal oxidation at 1300°C for 1.3 h (dashed line) and C^+ implantation plus Ar annealing at 1800°C for 20 min (chain line).

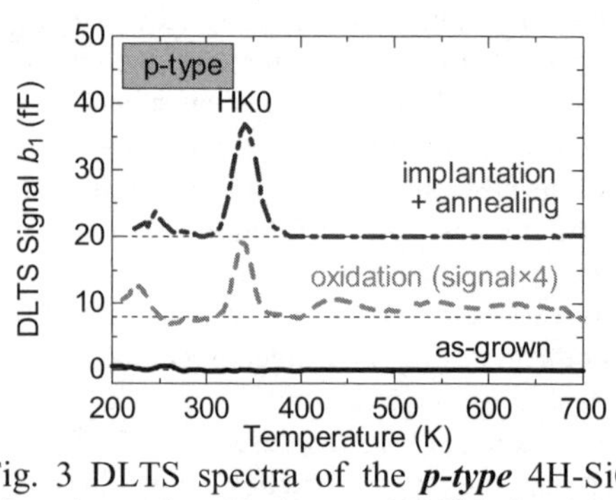

Fig. 3 DLTS spectra of the ***p-type*** 4H-SiC after thermal oxidation at 1300°C for 1.3 h (dashed line) and C^+ implantation plus Ar annealing at 1300°C for 20 min (chain line).

Table 1 Reduced and generated defects in SiC by C^+-implantation process and by thermal oxidation.

C^+ implantation + Ar annealing		Thermal oxidation	
Reduced defects	**Generated defects**	**Reduced defects**	**Generated defects**
$Z_{1/2}$ (n-type)	ON1 (n-type)	$Z_{1/2}$ (n-type)	ON1 (n-type)
$EH_{6/7}$ (n-type)	ON2 (n-type)	$EH_{6/7}$ (n-type)	ON2 (n-type)
	ON3 (n-type)		ON3 (n-type)
	HK0 (p-type)		HK0 (p-type)

$EH_{6/7}$ (E_C – 1.5 eV), are reduced in n-type SiC (Fig. 2) after thermal oxidation. The $Z_{1/2}$ and $EH_{6/7}$ centers are also reduced to below the detection limit by C^+ implantation followed by Ar annealing at 1500°C (not shown), while they are regenerated by high-temperature (over 1700°C) annealing as shown as a chain line in Fig. 2. The reduced and generated defects by the trap reduction processes are summarized in Table 1. The defect behaviors for thermal oxidation clearly agree with those for the C^+-implantation process, which indicates that similar phenomena (such as interstitial diffusion) may occur in the two processes.

Fig. 4 shows the depth profiles of the ON1 center after oxidation at various temperatures for 1.3 h. With increasing oxidation temperature, the ON1 concentration increases and is distributed to a deeper region, suggesting that the ON1 center is related to the interstitials diffused from the SiO_2/SiC interface. The ON2 and HK0 centers showed a similar profile and behavior (not shown). Fig. 5 shows the depth profiles of the $Z_{1/2}$ center after oxidation at various temperatures for 1.3 h. In this particular case, the $Z_{1/2}$ concentration was increased to 1.6×10^{14} cm^{-3} by electron irradiation (150 keV, 1×10^{17} cm^{-2}) in order to investigate the trap reduction in the sample with high initial $Z_{1/2}$ concentration. With increasing oxidation temperature, the $Z_{1/2}$ concentration decreases and is eliminated to a deeper region, suggesting that the $Z_{1/2}$ center is related to the (carbon) vacancies, which are occupied by diffused interstitials during oxidation. The depth of the $Z_{1/2}$ elimination region has been reported to be proportional to a square root of the oxidation time ($t^{1/2}$) [19]. These results clearly suggest that the diffusion phenomena occur during thermal oxidation.

Prediction of defect distributions. As discussed above, the reduction of the $Z_{1/2}$ center by oxidation may be caused by diffusion of interstitials as shown in Fig. 1. Therefore, in order to predict defect distributions after thermal oxidation, the following diffusion equations are solved:

Evolution equations: $$\partial n_I / \partial t = D \cdot \partial^2 n_I / \partial x^2 - \gamma \cdot n_I \cdot n_V, \qquad \partial n_V / \partial t = -\gamma \cdot n_I \cdot n_V, \tag{1}$$

Boundary condition: $$-D\, \partial n_I / \partial x \big|_{x=0} = F_0 \cdot t^{-\alpha}, \tag{2}$$

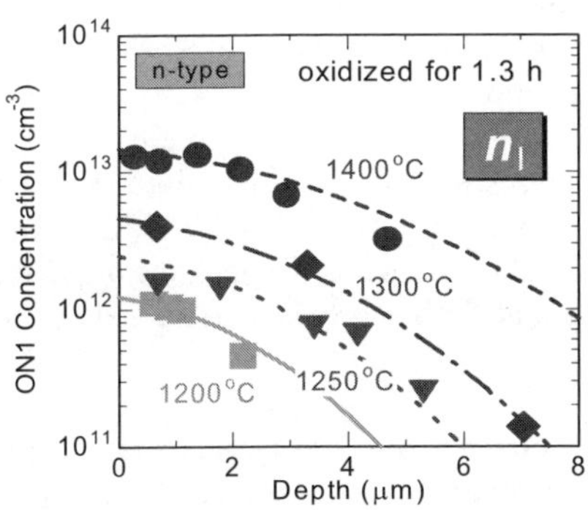

Fig. 4 Depth profiles of ***ON1 center*** after oxidation at various temperatures for 1.3 h. Each symbol indicates the experimental data and each line indicates the calculated $\boldsymbol{n_I}$ curve obtained from Eq. (1)-(3).

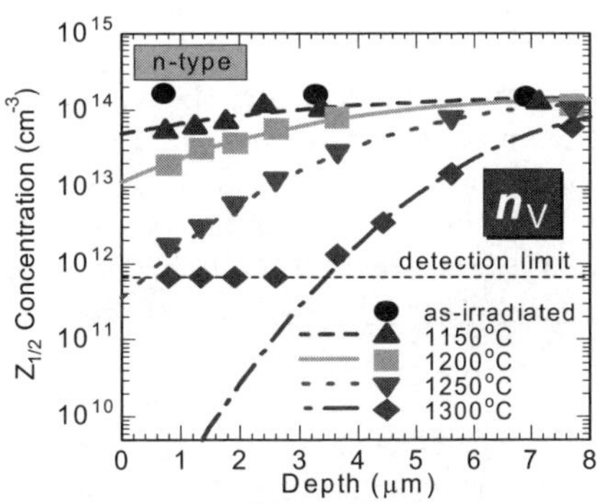

Fig. 5 Depth profiles of ***$Z_{1/2}$ center*** after oxidation at various temperatures for 1.3 h. Each symbol indicates the experimental data and each line indicates the calculated $\boldsymbol{n_V}$ curve obtained from Eqs. (1)-(3).

Fitting parameters: $$X = X_\infty \exp\left(-\frac{E_{aX}}{kT}\right) \quad (X = D,\ F_0,\ \text{or}\ \gamma), \tag{3}$$

where n_I and n_V are the concentrations of interstitials and vacancies, the latter of which may be related to the origin of $Z_{1/2}$ center. D denotes the diffusion coefficient of the interstitials and γ is the recombination coefficient between an interstitial and a vacancy. In this model, vacancies are immobile and decrease through the recombination with diffused interstitials. Equation (2) indicates the boundary condition of interstitial emission at the oxidation interface, which is specified by F_0, the flux of interstitials emitted from the SiO_2/SiC interface when $t = 1$ s (t: oxidation time). The gradual decrease in the flux of emitted interstitials as the oxidation (time) proceeds is considered as a slowdown coefficient α. The flux of interstitials should be in proportion to the oxidation rate. The slowdown of the oxidation reaction at the interface can be estimated from Fig. 6, which shows the dependence of the oxide growth rate on oxidation time at different oxidation temperatures. A result on oxidation at 1100°C reported by Hijikata et al. [20] is also shown in the same figure. When the slope of the plot is described as $-\alpha$, the oxidation rate is proportional to $t^{-\alpha}$. Therefore, interstitial emission can be assumed to decrease in proportion to $t^{-\alpha}$ as shown in equation (2). For simplicity, the oxidation stage is defined as two stages; a high-oxidation-rate stage ($t < 0.8$ h), where α is 1 irrespective of oxidation temperature; and a low-oxidation-rate stage (0.8 h $< t$), where α is 0.23 for (1150-1300)°C oxidation and 0.43 for 1400°C oxidation. As shown in equation (3), each parameter is described as a function of oxidation temperature.

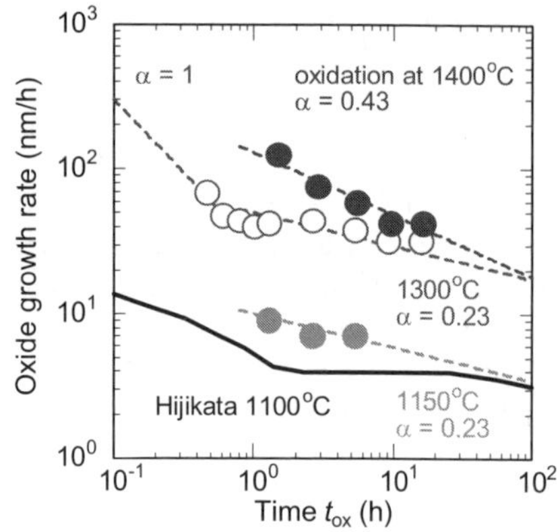

Fig. 6 Dependence of the oxide growth rate on oxidation time at different oxidation temperatures. A result (1100°C) reported by Hijikata et al. [20] is also shown as a black line.

Based on Eqs. (1)-(3), the depth profiles of n_I and n_V can be calculated, which are dependent on six fitting parameters: E_{aD}, D_∞, E_{aF}, $F_{0\infty}$, $E_{a\gamma}$, and γ_∞. As mentioned above, the authors experimentally obtained the depth profiles of $Z_{1/2}$ concentration after oxidation at different temperatures for several samples with different initial $Z_{1/2}$ concentrations (An example is shown in Fig. 5 as symbols). Therefore, the experimental depth profiles of $Z_{1/2}$ concentration were fitted with the n_V profiles calculated from Eqs. (1)-(3). The obtained values of the parameters are summarized in Table 2. The activation energy for the diffusion coefficient of interstitials reducing the $Z_{1/2}$ center was determined as 0.6 eV in this study. On the other hand, the migration barrier for carbon/silicon interstitials in n-type SiC has been reported to be (0.5-0.7) eV/(1.4-1.5) eV as a calculated result by Bockstedte et al. using density functional theory (DFT) in the local density approximation (LDA) [24], and by Gao et al. using molecular dynamics (MD) simulations [25]. The activation energy for the diffusion coefficient of interstitials obtained in this study (0.6 eV) agrees with the reported migration energy for carbon interstitials, which is consistent with the model that carbon interstitials diffuse from the oxidation interface and fill V_C. Using the same parameter values shown in Table 2, the oxidation-time dependence of the $Z_{1/2}$ profiles was predicted assuming that the n_V profile corresponds to the $Z_{1/2}$ profile.

Table 2 Parameter values obtained by fitting results of calculated n_V profiles based on Eqs. (1)-(3) and experimental $Z_{1/2}$ profiles shown in Fig. 5. The top row indicates the quantity "X" in Eq. (3).

	D	F_0	γ
E_{aX}	0.6 eV	1.4 eV	2.1 eV
X_∞	9.7×10^{-9} cm^{2}s^{-1}	4.4×10^{14} cm^{-2}s$^{\alpha-1}$	1.4×10^{-10} cm^{3}s^{-1}

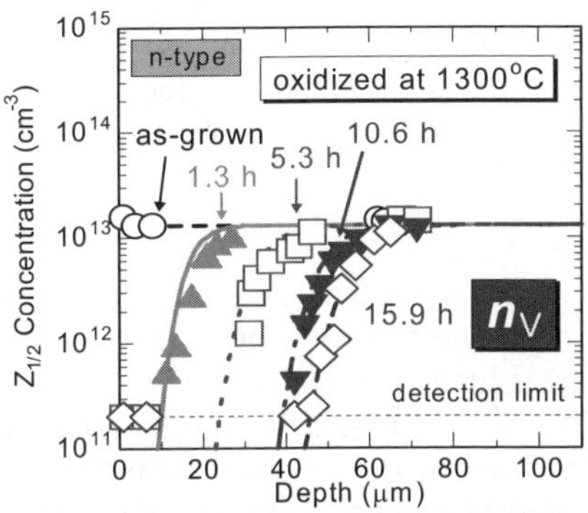

Fig. 7 Depth profiles of $Z_{1/2}$ center (initial $Z_{1/2}$: ***1.3×10^{13} cm^{-3}***) after oxidation at 1300°C for (1.3-15.9) h. Each symbol indicates the experimental data and each line indicates the calculated n_V curve.

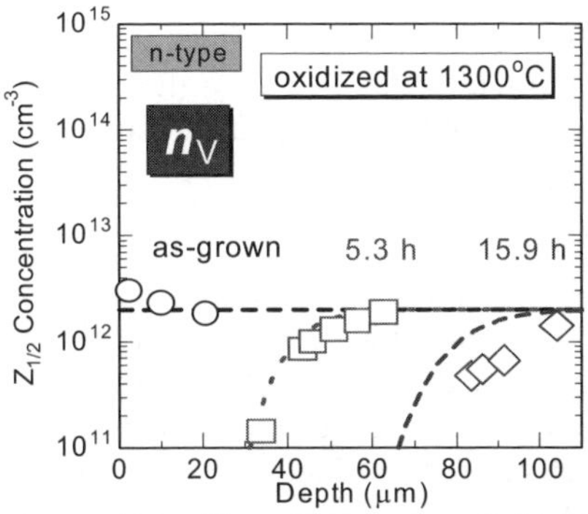

Fig. 8 Depth profiles of $Z_{1/2}$ center (initial $Z_{1/2}$: ***2×10^{12} cm^{-3}***) after oxidation at 1300°C for (1.3-15.9) h. Each symbol indicates the experimental data and each line indicates the calculated n_V curve.

The results are shown as lines in Fig. 7 (initial $Z_{1/2}$ concentration: 1.3×10^{13} cm^{-3}) and Fig. 8 (initial $Z_{1/2}$ concentration: 2×10^{12} cm^{-3}) for 1300°C oxidation, while corresponding experimental data are shown as symbols in the same figures. The depth profiles of $Z_{1/2}$ concentration after various oxidation conditions show very good agreement with the calculated n_V profiles. This agreement indicates the potential in enabling to predict the $Z_{1/2}$ distributions after oxidation at any temperature, for any oxidation time, and any initial $Z_{1/2}$ concentration.

Realization of 100-μm-$Z_{1/2}$-free region. The authors tried to eliminate the $Z_{1/2}$ center to a 100-μm depth, which is required for 10-kV-class bipolar devices. Fig. 9 shows the calculated results of $Z_{1/2}$ (n_V) profiles for various oxidation times and different initial $Z_{1/2}$ concentrations using the parameters obtained in this study. When the initial $Z_{1/2}$ concentration is low, 2×10^{12} cm^{-3}, oxidation at 1300°C for 30 h is enough for $Z_{1/2}$ elimination in the 100 μm-thick SiC epilayer. However, in the case of high initial $Z_{1/2}$ concentration (> 10^{13} cm^{-3}), oxidation over 50 h is required to eliminate the $Z_{1/2}$ center. In order to minimize the oxidation time, the authors propose 1400°C oxidation.

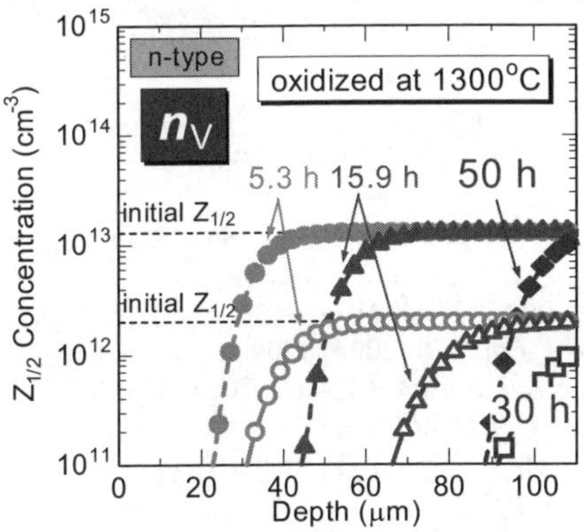

Fig. 9 Calculated results of $Z_{1/2}$ profiles for various oxidation times and different initial $Z_{1/2}$ concentrations using the parameters obtained in this study. It was assumed that n_V corresponds to the $Z_{1/2}$ concentration.

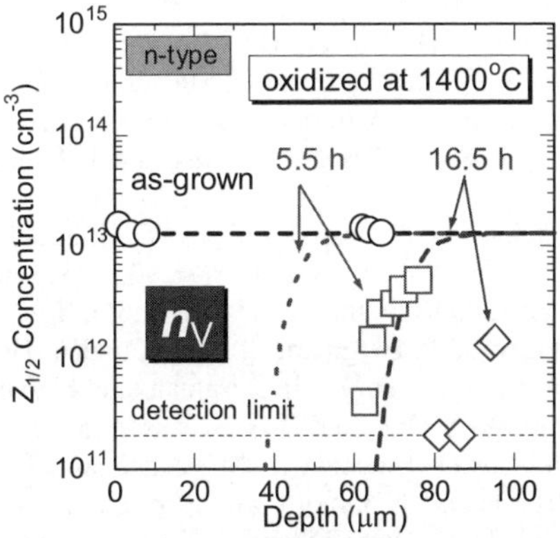

Fig. 10 Depth profiles of $Z_{1/2}$ center after oxidation at 1400°C for 5.5 h (squares), and for 16.5 h (rhombuses). Each line indicates a calculated n_V profile.

Because all the parameters, D, F_0, and γ should increase at high temperature, oxidation at high temperature must be effective in reduction of the $Z_{1/2}$ centers. Fig. 10 shows the depth profiles of the $Z_{1/2}$ center after oxidation at 1400°C for 5.5 h and 16.5 h. Each line indicates the calculated n_V profile with the same parameter values and each symbol indicates experimental data. The $Z_{1/2}$ centers are eliminated to a depth of 90 μm after oxidation for 16.5 h. Although the parameters for calculation must be reconsidered, oxidation at 1400°C is a significantly effective process in accelerating $Z_{1/2}$ reduction.

Summary

It was clarified that deep levels generated by thermal oxidation are the same as those generated by C^+-implantation plus Ar annealing. In addition, the depth profiles of generated/reduced defects represent the distribution of interstitials/vacancies after interstitial diffusion phenomena. These results indicate that the same phenomena, diffusion of interstitials, occur during these processes. Furthermore, the authors proposed an analytical model enabling prediction of the $Z_{1/2}$ distribution after thermal oxidation using a diffusion equation. The results suggested that the initial $Z_{1/2}$ concentration (conditions of epitaxial growth and device processes) must be also considered for eliminating $Z_{1/2}$ and achieving long carrier lifetimes. In order to achieve thick-$Z_{1/2}$-free epilayers for SiC with high initial $Z_{1/2}$ concentration (> 10^{13} cm^{-3}), oxidation at 1400°C is recommended.

Acknowledgements. This work was supported by the Funding Program for World-Leading Innovative R&D on Science and Technology (FIRST Program) and Grant-in-Aid for Scientific Research (21226008) from the Japan Society for the Promotion of Science.

References

[1] T. Dalibor, G. Pensl, H. Matsunami, T. Kimoto, W.J. Choyke, A. Schöner, and N. Nordell, *Phys. Stat. Sol. (a)* **162** (1997) 199.
[2] C. Hemmingsson, N.T. Son, O. Kordina, J.P. Bergman, E. Janzén, J.L. Lindström, S. Savage, and N. Nordell, *J. Appl. Phys.* **81** (1997) 6115.
[3] K. Danno and T. Kimoto, *J. Appl. Phys.* **100** (2006) 113728.
[4] K. Danno and T. Kimoto, *J. Appl. Phys.* **101** (2007) 103704.
[5] T. Dalibor, H. Trageser, G. Pensl, T. Kimoto, H. Matsunami, D. Nizhner, O. Shigiltchoff, and W.J. Choyke, *Mater. Sci. Eng. B* **61-62** (1999) 454.
[6] S. Mitra, M.V. Rao, N. Papanicolaou, K.A. Jones, M. Derenge, O.W. Holland, R.D. Vispute, and S.R. Wilson, *J. Appl. Phys.* **95** (2004) 69.
[7] T. Troffer, M. Schadt, T. Frank, H. Itoh, G. Pensl, J. Heindl, H.P. Strunk, and M. Maier, *Phys. Stat. Sol. (a)* **162** (1997) 277.
[8] D. Åberg, A. Hallén, and B.G. Svensson, *Physica B* **273** (1999) 672.
[9] Y. Negoro, T. Kimoto, and H. Matsunami, *J. Appl. Phys.* **98** (2005) 043709.
[10] K. Kawahara, G. Alfieri, and T. Kimoto, *J. Appl. Phys.* **106** (2009) 013719.
[11] K. Kawahara, M. Krieger, J. Suda, and T. Kimoto, *J. Appl. Phys.* **108** (2010) 023706.
[12] D.V. Lang and C.H. Henry, *Phys. Rev. Lett.* **35** (1975) 1525.
[13] P.B. Klein, B.V. Shanabrook, S.W. Huh, A.Y. Polyakov, M. Skowronski, J.J. Sumakeris, and M.J. O'Loughlin, *Appl. Phys. Lett.* **88** (2006) 052110.
[14] K. Danno, D. Nakamura, and T. Kimoto, *Appl. Phys. Lett.* **90** (2007) 202109.
[15] L. Storasta, J.P. Bergman, E. Janzén, and A. Henry, *J. Appl. Phys.* **96** (2004) 4909.
[16] L. Storasta, A. Henry, J.P. Bergman, and E. Janzén, *Mater. Sci. Forum.* **457-460** (2004) 469.
[17] T. Hornos, A. Gali, and B.G. Svensson, *Mater. Sci. Forum.* **679-680** (2011) 261.
[18] L. Storasta, H. Tsuchida, T. Miyazawa, and T. Ohshima, *J. Appl. Phys.* **103** (2008) 013705.
[19] T. Hiyoshi and T. Kimoto, *Appl. Phys. Express* **2** (2009) 041101.
[20] Y. Hijikata, H. Yaguchi, and S. Yoshida, *Appl. Phys. Express* **2** (2009) 021203.
[21] K.C. Chang, N.T. Nuhfer, L.M. Porter, and Q. Wahab, *Appl. Phys. Lett.* **77** (2000) 2186.
[22] J.M. Knaup, P. Deák, T. Frauenheim, A. Gali, Z. Hajnal, and W.J. Choyke, *Phys. Rev.* **71** (2005) 235321.
[23] S. Weiss and R. Kassing, *Solid-State Electron.* **31** (1988) 1733.
[24] M. Bockstedte, A. Mattausch, and O. Pankratov, *Phys. Rev.* **B68** (2003) 205201.
[25] F. Gao, W.J. Weber. M. Posselt, and V. Belko, *Phys. Rev.* **B69** (2004) 245205.

Materials Science Forum Vols. 717-720 (2012) pp 247-250
© (2012) Trans Tech Publications, Switzerland
doi:10.4028/www.scientific.net/MSF.717-720.247

On the Formation of Intrinsic Defects in 4H-SiC by High Temperature Annealing Steps

B. Zippelius[1,a], J. Suda[1,b] and T. Kimoto[1,c]

[1]Department of Electronic Science and Engineering, Kyoto University, Katsura, Nishikyo, Kyoto 615-8510, JAPAN

[a]bernd@semicon.kuee.kyoto-u.ac.jp, [b]suda@kuee.kyoto-u.ac.jp, [c]kimoto@kuee.kyoto-u.ac.jp

Keywords: high temperature annealing, DLTS, µ-PCD, carrier lifetime, intrinsic defects, $Z_{1/2}$, $EH_{6/7}$

Abstract. In this paper the impact of high temperature annealing on the formation of intrinsic defects in 4H-SiC such as $Z_{1/2}$ and $EH_{6/7}$ is examined. Three epitaxial layers with various initial concentrations of the $Z_{1/2}$- and $EH_{6/7}$-centers (10^{11} – 10^{13} cm^{-3}) are investigated by means of deep level transient spectroscopy (DLTS). It turns out that depending on the initial defect concentration, the high temperature annealing leads to a monotonic increase of the $Z_{1/2}$- and $EH_{6/7}$-concentration in a temperature range from 1600 to 1750°C. For higher temperatures, the resulting defect concentration is independent of the sample's initial values. Therefore, beside the growth conditions such as C/Si ratio, the thermal post-growth processing has a severe impact on the carrier lifetime and must be taken into account during device fabrication.

Introduction

In recent years, huge effort has been put into reducing the concentration of intrinsic defects in silicon carbide. Especially for the widely used 4H polytype the most prominent representatives $Z_{1/2}$ [1] and $EH_{6/7}$ [2] are crucial for controlling the carrier lifetime in SiC devices. There are many hints in the literature that the carbon vacancy is involved in the formation of these defects (e.g. [3]). Consequently, their concentration can be lowered e.g. by C-implantation [4] or oxidation [5] and as a result the carrier lifetime can be greatly improved. However, as the diffusion constants of common dopants in SiC are low, ion implantation and subsequent high temperature annealing steps are required to anneal implantation damage and to electrically activate the implanted ions. Therefore, knowing the impact of thermal treatment on the concentration of intrinsic defects is of great importance and may give further insight to the formation mechanisms of $Z_{1/2}$ and $EH_{6/7}$.

Experimental

Three sets of n-type 4H-SiC epitaxial layers have been investigated (cf. Tab. 1):

- sample set (1), a commercial epitaxial layer (d_{epi} ≈ 50 µm) containing a net doping concentration of $[N]_{net} \approx 1.2 \times 10^{15}$ cm^{-3} and a medium-range initial concentration of $[Z_{1/2}]_{ini} \approx 1.5 \times 10^{13}$ cm^{-3},
- sample set (2), a custom-made (Kyoto University) epitaxial layer (d_{epi} ≈ 125 µm) with a very low net doping concentration of $[N]_{net} \approx (7 \text{ to } 9) \times 10^{13}$ cm^{-3} and a low initial $Z_{1/2}$-concentration of about $[Z_{1/2}]_{ini} \approx 1.0 \times 10^{12}$ cm^{-3} and
- sample set (3), for which the epilayer of set (1) had been oxidized for 5 h at 1300°C prior to the annealing experiments. Subsequently, the oxide was removed by hydrofluoric acid; the initial $Z_{1/2}$-concentration dropped below the detection limit of 2×10^{11} cm^{-3}.

All samples were isochronally annealed for 15 min at temperatures ranging from 1300 to 1850°C under Ar atmosphere using a carbon cap to prevent surface roughening. Ni-Schottky contacts with a thickness of 150 nm were prepared by thermal evaporation and the defect concentration was monitored by Deep Level Transient Spectroscopy (DLTS).

Furthermore, measurements of the carrier lifetime by the differential microwave photoconductance decay (μ-PCD) method have been conducted using a pulsed YLF (yttrium lithium fluoride)-3HG laser with a wavelength of 349 nm to generate excess carriers. The applied photon density corresponds to a generated carrier density of 3.6×10^{16} cm^{-3} at the surface. The conductance decay was monitored by the reflection of microwaves with a frequency of 26 GHz.

Tab. 1 Overview of investigated sample sets with thickness d_{epi} of the epitaxial layer, net doping concentration $[N]_{net}$ and initial concentration $[Z_{1/2}]_{ini}$.

Sample set	d_{epi} (μm)	$[N]_{net}$ (cm^{-3})	$[Z_{1/2}]_{ini}$ (cm^{-3})	Remarks
(1)	50	1.2×10^{15}	1.5×10^{13}	Commercial
(2)	125	$(7 \text{ to } 9) \times 10^{13}$	1.0×10^{12}	Custom-made
(3)	50	1.2×10^{15}	$< 2 \times 10^{11}$	Oxidized set (1)

Results and Discussion

Fig. 1(a) shows the normalized DLTS spectra taken on the non-annealed reference samples. The DLTS signal was normalized to the doping concentration; thus the peak height corresponds to the defect concentration. The non-oxidized samples (solid and dashed lines) show the dominating defects $Z_{1/2}$ (E_C – 0.67 eV) and $EH_{6/7}$ (E_C – 1.5 eV), whereas in the oxidized sample (dash-dotted line) their concentration lies below the detection limit. The normalized DLTS spectra of samples annealed at 1850°C for 15 min are depicted in Fig. 1 (b): The $Z_{1/2}$- and $EH_{6/7}$-signals are nearly independent of the initial concentration.

The ON1- (E_C – 0.9 eV) and ON2a/b-defects (E_C – 1.0 eV) usually emerge in oxidized samples. Their concentrations are nearly independent of the annealing temperature and decrease within 8 μm from the surface by a factor of approximately 5 (not shown). In contrast, the $Z_{1/2}$-concentration in this near-surface region is constant for all samples as detected by DLTS depth profiling.

Fig. 2 shows the observed concentration of (a) $Z_{1/2}$- and (b) $EH_{6/7}$-centers as a function of the reciprocal annealing temperature T_{anneal} taken on samples of sets (1) to (3) with an initial $Z_{1/2}$-concentration from 1×10^{11} cm^{-3} to 1×10^{13} cm^{-3}. The value of T_{anneal} at which a distinct increase of the $Z_{1/2}$- and $EH_{6/7}$-center can be observed lies in the range of 1600 to 1750°C, depending on the

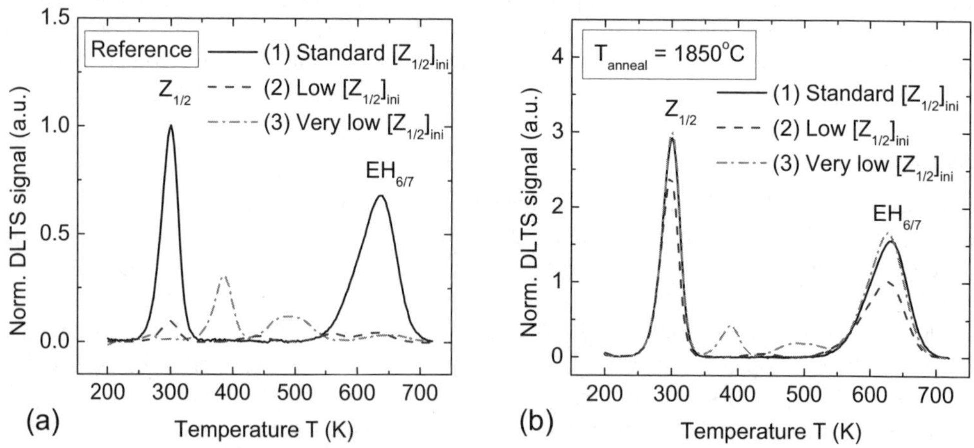

Fig. 1 DLTS spectra taken on samples with (1) standard (as-grown (commercial), black solid line), (2) low (as-grown (Kyoto-University), blue dashed line) and (3) very low (oxidized, red dash-dotted line) initial $Z_{1/2}$-concentration which (a) have not been annealed and (b) have been annealed at 1850°C for 15 min. The annealed samples show nearly identical $Z_{1/2}$-signals, which are even higher than that of sample (1) (Fig. 1(a)). The spectra depict the sine-correlation with a time window of T_W = 204.8 ms. The DLTS measurements were taken with reverse bias of V_R = -10 V, pulse bias of V_P = -0.5 V and filling pulse time of t_P = 1 ms.

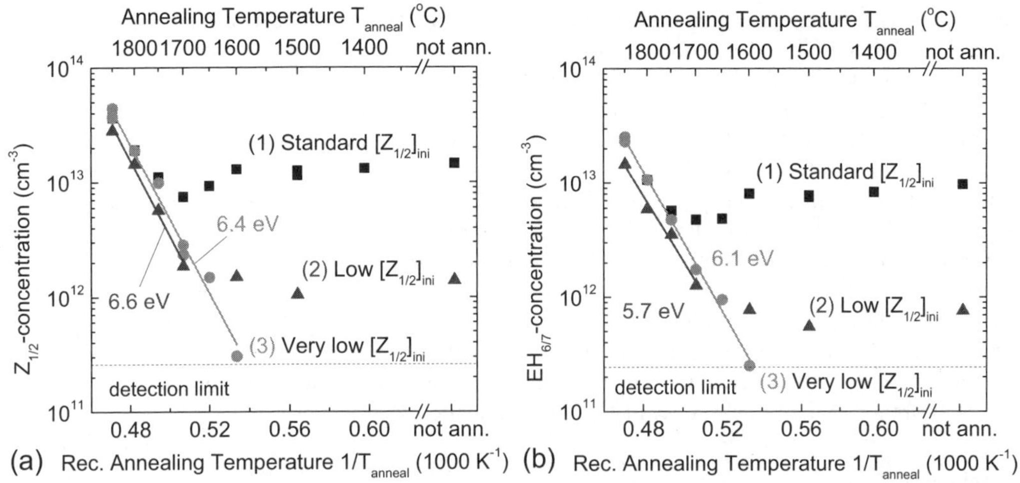

Fig. 2 Concentration of (a) the $Z_{1/2}$- and (b) the $EH_{6/7}$-defect versus the reciprocal annealing temperature. The very right measurement points show the initial concentration of the reference samples which have not been annealed. In the case of the oxidized samples of set (3) the concentrations lie below the detection limit. From the slope of the Arrhenius plot for the sample sets (2) and (3) the formation energy is determined after Eq. (1) as (a) $\Delta H(Z_{1/2}) = 6.4 - 6.6$ eV and (b) $\Delta H(EH_{6/7}) = 5.7 - 6.1$ eV.

initial defect concentration. However, after annealing above these temperatures, the resulting $Z_{1/2}$- and $EH_{6/7}$-concentration are nearly identical, respectively, independent of the initial trap concentrations. This seems to be a unique feature of the generation of these defects, meaning the temperature is another driving force for creating carbon vacancies beside the growth conditions such as the C/Si ratio.

On the one hand, the driving force for the formation of intrinsic defects such as interstitials and vacancies is the irreversibly increasing entropy: with every deviation from a perfect crystal the system's entropy is enlarged. On the other hand, a defined formation energy (also effective enthalpy) ΔH is required in order to form defects. From this simple model the formation energy of a simple point defect under thermal equilibrium can be calculated by [6]

$$N_T = N_0 \exp\left[\frac{S}{k}\right] \exp\left[-\frac{\Delta H}{kT}\right] = A \exp\left[-\frac{\Delta H}{kT}\right], \tag{1}$$

where N_0 is the concentration of lattice sites where the defect may reside and S the formation entropy of the defect.

As the $Z_{1/2}$- and $EH_{6/7}$-concentration for $T_{anneal} > 1700$°C is the same for all sample sets, independent of the initial defect concentration, the system is close to thermal equilibrium and can be described by Eq. (1). From the slope of the Arrhenius plot of the trap concentration of $Z_{1/2}$ and $EH_{6/7}$ versus the inverse annealing temperature the formation energy of these defects can be determined. For all samples, values of approximately 6.4 to 6.6 eV for $Z_{1/2}$ (see Fig. 2(a)) and 5.7 to 6.1 eV for $EH_{6/7}$ are obtained (see Fig. 2(b)).

The theoretical formation energy of the carbon vacancy under stoichiometric conditions varies from 4.2 eV (4H-SiC) [7] to 5.9 eV (3C-SiC) [8], which is slightly lower than the respectively detected formation energy for $Z_{1/2}$ and $EH_{6/7}$ during the annealing experiment. Beside the formation of carbon vacancy-carbon interstitial pairs (Frenkel pairs) from a perfect lattice at high temperatures, other possible mechanisms of forming additional carbon vacancies are the formation of Schottky defects near the surface with subsequent diffusion of carbon vacancies and the diffusion of carbon vacancies from the substrate. At such high temperatures the Fermi-level reaches midgap position and therefore the charge state of V_C is in the transition region from V_C^0 (n-type material) to V_C^{2+}

(intrinsic/p-type material). The migration barrier was calculated by Bockstedte et al. [9] as 3.5 and 5.2 eV for the neutral and the doubly positively charged carbon vacancy, respectively. However, theoretical diffusivity calculations predict a probably too small diffusion length of less than 1 μm, as out-diffusion of V_C was experimentally observed already at relatively low temperatures (T_{anneal} = 1600°C) [10]. To identify the exact process of carbon vacancy generation at high temperature, further investigations are required.

Lifetime measurements show a strong dependence on the sample treatment in agreement with the previous findings. With an epilayer thickness of 50 μm or more the reduction of lifetime due to substrate recombination is strongly suppressed. In oxidized samples (set (3)) which have been annealed at 1400°C the largest lifetime of 2.4 μs was measured at room temperature under high injection condition. In all samples which have been annealed at 1850°C, the lifetime dropped to a value about 0.25 μs. The concentration of the $Z_{1/2}$-center above a value of 10^{13} cm^{-3} is strongly correlated to the inverse carrier lifetime (not shown) which is consistent with previous reports [11]. This is attributed to the Shockley-Read-Hall-recombination via the $Z_{1/2}$-center.

Conclusion

It has been shown that high temperature annealing as it is used for activation of implanted dopants during device fabrication has a great impact on the formation of the $Z_{1/2}$- and $EH_{6/7}$-centers in 4H-SiC. For annealing temperatures $T_{anneal} \geq 1750$°C the detected concentrations of $Z_{1/2}$ and $EH_{6/7}$ are independent of the initial values. Furthermore, a reduction of the carrier lifetime was observed due to the increase in $Z_{1/2}$-concentration, which is in agreement with previous reports. Thus, lifetime enhancing oxidation steps with the result of reduced $Z_{1/2}$- and $EH_{6/7}$-concentrations should be conducted after high temperature treatments during device processing. If this is not possible, an annealing temperature below 1750°C should be chosen with the trade-off of a possibly reduced concentration of activated dopants.

Acknowledgements

This work was supported by the Funding Program for World-Leading Innovative R&D on Science and Technology (FIRST Program) and Grant-in-Aid for Scientific Research (21226008) from the Japan Society for the Promotion of Science.

References

[1] T. Dalibor, C. Peppermüller, G. Pensl, S. Sridhara, R.P. Devaty, W.J. Choyke, A. Itoh, T. Kimoto and H. Matsunami, Inst. Conf. Ser. Vol. 142 (1996) 517

[2] C. Hemmingsson, N.T. Son, O. Kordina, J.P. Bergman, E. Janzén, J.L. Lindström, S. Savage and N. Nordell, J. Appl. Phys. Vol. 81 (1997) 6155

[3] L. Storasta, J.P. Bergman, E. Janzén, A. Henry and J. Lu, J. Appl. Phys. Vol. 96 (2004) 4909

[4] L. Storasta and H. Tsuchida, Appl. Phys. Lett. Vol. 90 (2007) 062116

[5] T. Hiyoshi and T. Kimoto, Appl. Phys. Express Vol. 2 (2009) 041101

[6] K. Danno, T. Hori and T. Kimoto, J. Appl. Phys. Vol. 101 (2007) 53709

[7] L. Torpo, M. Marlo, T.E.M. Staab and R.M. Nieminen, J. Phys.: Condens. Matter Vol. 13 (2001) 6203

[8] C. Wang, J. Bernholc and R. F. Davis, Phys. Rev. B Vol. 38 (1988) 12752

[9] M. Bockstedte, A. Mattausch and O. Pankratov, Phys. Rev. B Vol. 68 (2003) 205201

[10] N. T. Son, B. Magnusson, Z. Zolnai, A. Ellison and E. Janzén, Mater. Sci. Forum Vols. 433-436 (2003) 45

[11] T. Kimoto, K. Danno and J. Suda, phys. stat. sol. (b) Vol. 245 (2008) 1327

Materials Science Forum Vols. 717-720 (2012) pp 251-254
© (2012) Trans Tech Publications, Switzerland
doi:10.4028/www.scientific.net/MSF.717-720.251

$Z_{1/2}$- and EH_6-Center in 4H-SiC: Not Identical Defects?

Bernd Zippelius[1,2,a], Alexander Glas[1,b], Heiko B. Weber[1,c], Gerhard Pensl[1,d], Tsunenobu Kimoto[2,e] and Michael Krieger[1,f,*]

[1]Lehrstuhl für Angewandte Physik, University of Erlangen-Nuremberg, Staudtstr. 7 / A3, D-91058 Erlangen, Germany

[2]Department of Electronic Science and Engineering, Kyoto University, Katsura, Nishikyo, Kyoto 615-8510, Japan

[a]bernd.zippelius@googlemail.com, [b]alex.glas@web.de, [c]heiko.weber@physik.uni-erlangen.de, [d]gerhard.pensl@physik.uni-erlangen.de, [e]kimoto@kuee.kyoto-u.ac.jp, [f]michael.krieger@physik.uni-erlangen.de, *corresponding author

Keywords: Deep level transient spectroscopy, $Z_{1/2}$, $EH_{6/7}$, charge state, intrinsic point defects, Poole-Frenkel effect

Abstract. Deep Level Transient Spectroscopy (DLTS) and Double-correlated DLTS (DDLTS) measurements have been conducted on Schottky contacts fabricated on n-type 4H-SiC epilayers using different contact metals in order to separate the EH_6- and EH_7-centers, which usually appear as a broad double peak in DLTS spectra. The activation energy of EH_6 (E_C - $E_T(EH_6)$ = 1.203 eV) turns out to be independent of the electric field. As a consequence, EH_6 is likely to be acceptor-like according to the missing Poole-Frenkel effect. In this case, it can be excluded that the EH_6-center and the prominent acceptor-like $Z_{1/2}$-center belong to different charge states of the same microscopic defect as theoretically suggested. It is proposed that EH_6 is a complex containing a carbon vacancy and another component available at high concentrations. The activation energy of EH_7 (E_C - $E_T(EH_7)$ = 1.58 eV) has been evaluated indirectly by fitting the DLTS spectra of the $EH_{6/7}$ double peak taking the previously determined parameters of EH_6 into account.

Introduction

At the present, strong activities are pursued to reduce the concentration of electrically active defects, which degrade the performance of SiC power devices. The most prominent defects in 4H-SiC are the $Z_{1/2}$- and the $EH_{6/7}$-centers. The chemical composition and microscopic structure of these defects, which are frequently observed in Deep Level Transient Spectroscopy (DLTS) spectra, are not clarified, yet. However, there are several hints in the literature indicating that a carbon vacancy (V_C) is involved (see e. g. [1]). It has been shown that the $Z_{1/2}$-center is a negative U-center and the observed transition, which determines the DLTS peak, is acceptor-like [2, 3] (note that from DDLTS investigations on n-type SiC, one cannot distinguish between acceptor or double-acceptor).

The $EH_{6/7}$-center represents a broad peak in the DLTS spectra and is assumed to be an overlapping double peak. Recently, Hornos et al. [4] calculated the different charge states of the carbon vacancy in 4H-SiC and attributed - due to similar energy positions in the 4H-SiC bandgap - the acceptor levels of V_C to the $Z_{1/2}$-center and the donor levels of V_C to the $EH_{6/7}$-center. It is the aim of this work to experimentally verify this suggestion.

Experimental

We used n-type 4H-SiC epilayers and conducted DLTS measurements. The charge state of the corresponding defects was determined by Double-correlated DLTS measurements (DDLTS) on the basis of the Poole-Frenkel effect [5]. Our main task was to separate the components of the $EH_{6/7}$ double peak in the DLTS spectra and to reduce the noise level with respect to the small DDLTS signal (for DDLTS, the electron emission of traps in only a small zone in the space charge region is considered).

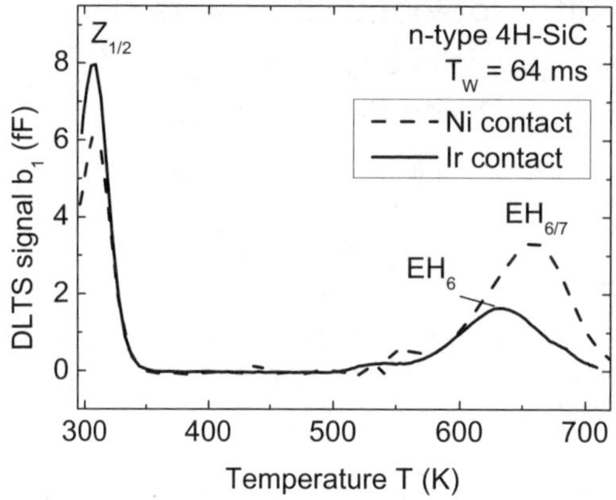

Fig. 1. DLTS spectra (sine correlation b_1) taken with a Ni Schottky contact (Φ_B = 1.80 eV, dashed curve) and with an Ir Schottky contact (Φ_B = 1.35 eV, solid curve). With decreasing barrier height the signal of EH_7 is suppressed and the maximum is shifted towards lower temperatures whereas the position of $Z_{1/2}$ remains unchanged. Using Ir, EH_6 can be investigated separately.

Further a wide range in the doping level was required to be able to determine the activation energy $\Delta E_T = E_C - E_T$ of the defect for a large variation of the applied electric field. Therefore, we put emphasis on the following items:

a) We employed n-type 4H-SiC epilayers with different nitrogen donor concentrations (in the range of (1 ... 20) × 10^{15} cm^{-3}), which allowed to vary the electric field in the range of (25 ... 200) kV/cm.

b) For the Schottky diodes, we used iridium (Ir) resulting in a Schottky barrier height of Φ_B = 1.35 eV as determined by C-V analysis. It is demonstrated in the literature [6, 7] that the barrier height determines the leakage current of the Schottky diode and therefore the position of the Quasi-Fermi-level in the space charge region. As a consequence, depending on the energy position of the defect level in the bandgap, its signal can be suppressed in the DLTS spectra. It turns out that Ir collects the EH_6 signal, but suppresses largely the signal of EH_7.

c) We developed and realized a precise control of the temperature ramp during the DLTS measurement (duration approx. 50 h), which allowed to average a great number of capacitance transients (> 200) and to increase the temperature resolution of the DLTS spectra (ΔT = 1.5 K). This technique resulted in a strongly reduced noise level and improved energy resolution.

Results and Discussion

Fig. 1 compares two DLTS spectra, which were measured on Schottky contacts with different metals. The dashed curve depicts the spectrum measured on a nickel (Ni) Schottky contact (Φ_B = 1.80 eV); a broad peak at approx. T = 650 K consisting of the overlapping signals of EH_6 and EH_7 is revealed. In contrast, the spectrum measured on an Ir Schottky contact (Φ_B = 1.35 eV, solid curve) shows the EH_6 peak with a large suppression of the EH_7 contribution to the signal, resulting in a reduced peak height and a shift of the maximum towards lower temperatures. Using Ir Schottky contacts on samples with different doping concentration, the activation energy $\Delta E_T(EH_6)$ could be determined as a function of the electric field F by means of DDLTS as shown in Fig. 2. The pulse voltage was varied from V_P = (-0.1 ... -4.3) V, while the reverse voltage was kept at V_R = -5 V. It is obvious that $\Delta E_T(EH_6)$ obtained from the Arrhenius analysis is independent of the electric field F and lies closely around

$$E_C - E_T(EH_6) = 1.203 \text{ eV}$$

with a scatter of only about 20 meV due to the improved setup. The electrical capture cross section determined from the measurements is in the range of

$$\sigma(\mathrm{EH}_6) = (1.2\ldots4.7)\times10^{-16}\,\mathrm{cm}^2.$$

Due to the Poole-Frenkel effect, the activation energy ΔE_T of a positively charged electron trap (donor-like) is lowered by

$$\Delta E_{PF} = k\frac{\partial}{\partial T^{-1}}\ln\left[\frac{1+(x-1)\exp(x)}{x^2}+\frac{1}{2}\right] \quad \text{with} \quad x = \frac{e}{kT}\sqrt{\frac{ZeF}{\pi\varepsilon\varepsilon_0}}, \tag{1}$$

where k is the Boltzmann factor and Z is the number of positive elementary charges on the defect center after electron emission [5]. As displayed in Fig. 2, we have fitted eq. (1) to the measured data points for $Z = 0$ (acceptor-like trap, solid curve), $Z = 1$ (donor-like trap, dashed curve) and $Z = 2$ (double-donor-like trap, dotted curve). The best match is achieved for $Z = 0$ meaning that EH_6 is likely to be an acceptor-like or double-acceptor-like trap:

$$EH_6^- \rightarrow EH_6^0 + e^- \quad \text{or} \quad EH_6^{2-} \rightarrow EH_6^- + e^-.$$

This assignment is supported by the small value of the capture cross section, which is typical for a neutral or repulsive trap. It should be noted that a missing Poole-Frenkel effect is not a conclusive evidence of an acceptor-like defect [8]. This will be clarified in further investigations.

Due to the fact that the EH_7-center cannot be observed separately in the DLTS spectra without a distinct contribution of EH_6, an indirect method was employed for the determination of the defect parameters. Fig. 3 shows the DLTS spectra using different sine-correlations, which were taken on Ir Schottky contacts annealed at $T_{ann} \approx 700$ K for several hours. Due to the formation of Ir silicides, this leads to an increased Schottky barrier height ($\Phi_B = 1.45$ eV) and, as a consequence, to a considerable contribution of the EH_7-center to the $EH_{6/7}$ double peak in the DLTS spectrum. In order to obtain the trap parameters $\Delta E_T(EH_7)$, $\sigma(EH_7)$ and the concentration $N(EH_7)$, the DLTS spectra corresponding to four different correlation functions have been simulated and simultaneously fitted to the measured spectra taking the previously determined EH_6 trap parameters into account. A very good agreement between experimental data and simulation could be achieved with the following parameters for the EH_7-center (see Fig. 3, solid curves):

$$E_C - E_T(EH_7) = 1.58 \text{ eV and } \sigma(EH_7) = (1.9\ldots7.1)\times10^{-14}\text{ cm}^2.$$

However, this method is not precise enough for an analysis of the charge state of the EH_7-center.

The chemical nature of $Z_{1/2}$ is most likely the negative-U acceptor state and double-acceptor state of the carbon vacancy. This is supported by many experiments and theoretical calculations reported in the literature (e. g. [9] and references therein). Due to the fact that the concentration ratio of $Z_{1/2}$ and $EH_{6/7}$ is always close to one (e. g. [10]), the formation energies and therefore the origin of both defects seem to be similar. On the other hand, the acceptor-like nature of both defects indicates that $Z_{1/2}$ and at least EH_6 are not identical defects. Combining all facts from the literature and from our experiments, we propose that the EH_6-center is a complex containing a carbon vacancy V_C and another component, which must be present at a high concentration. Suitable candidates could be the nitrogen donor or a second carbon vacancy. The chemical nature of EH_7 and its relation to EH_6 is still an open question: EH_7 could either be the theoretically expected donor-like state of the carbon

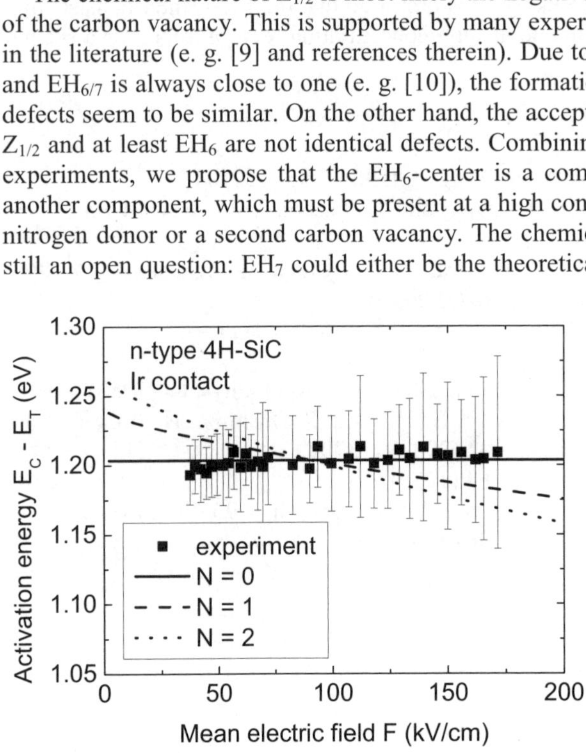

Fig. 2. Activation energy $\Delta E_T = E_C - E_T$ of the EH_6-center obtained by DDLTS measurement on Ir Schottky contacts. The solid, dashed and dotted curves are calculated on the basis of the Poole-Frenkel effect for the corresponding charge states $Z = 0$, 1 and 2 according to eq. (1).

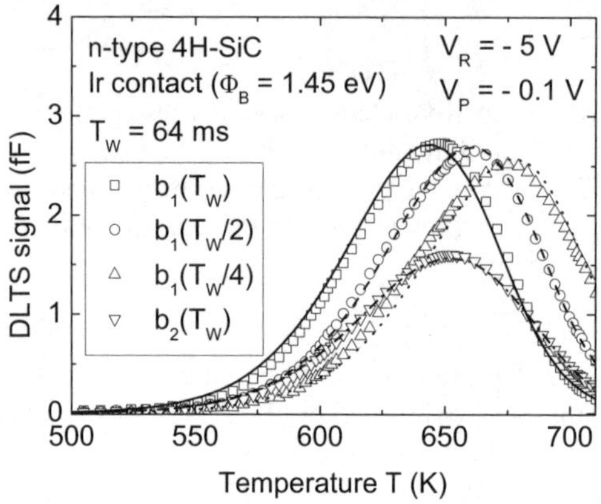

Fig. 3. Experimental (open symbols) and simulated (curves) DLTS spectra of various sine-correlation coefficients taken on annealed Ir Schottky contacts with a barrier height of Φ_B = 1.45 eV. By using the separately determined parameters for the EH_6-center, the fit results in an activation energy of $\Delta E_T(EH_7)$ = 1.58 eV and an electrical capture cross section of $\sigma(EH_7)$ = (1.9 ... 7.1) × 10^{-14} cm^2 for the EH_7-center.

vacancy V_C or a defect complex similar to EH_6 but with different spatial arrangement in the crystal lattice. Then, the deep, donor-like center P_1 ($E_T(P_1) - E_V$ = 1.49 eV) could be a candidate for the donor state of V_C [1].

Summary

The DLTS peak of the $EH_{6/7}$ center has been resolved into its two components by using iridium Schottky contacts. The activation energies and electrical capture cross sections of EH_6 and EH_7 have been determined as $\Delta E_T(EH_6)$ = 1.203 eV, $\sigma(EH_6)$ = (1.2 ... 4.7) × 10^{-16} cm^2, $\Delta E_T(EH_7)$ = 1.58 eV, $\sigma(EH_7)$ = (1.9 ... 7.1) × 10^{-14} cm^2, respectively. A precise DDLTS analysis of EH_6 revealed a missing Poole-Frenkel effect and, thus, very likely an acceptor-like nature of this defect, whereas the charge state of EH_7 remains to be determined. The acceptor-like nature of the $Z_{1/2}$-center and the EH_6-center indicates that both centers should be different defects, although an involvement of the carbon vacancy in both defect centers is very likely. It is proposed that the EH_6-center consists of a carbon vacancy and another component available at high concentration, e. g. nitrogen or a second carbon vacancy.

References

[1] K. Danno, T. Kimoto and H. Matsunami, Appl. Phys. Lett. **86** (2005) 122104.
[2] C. Hemmingsson, N. T. Son, A. Ellison, J. Zhang and E. Janzén, Phys. Rev. B **85** (1998) R10119.
[3] T. Dalibor, C. Peppermüller, G. Pensl, S. Sridhara, R. P. Devaty, W. J. Choyke, A. Itoh, T. Kimoto and H. Matsunami, Inst. Conf. Ser. **142** (1996) 517.
[4] T. Hornos, A. Gali and B. G. Svensson, Mater. Sci. Forum **679-680** (2011) 261.
[5] J. L. Hartke, J. Appl. Phys. **39** (1968) 4871.
[6] Q. Y. Ma, M. T. Schmidt, X. Wu, H. L. Evans and E. S. Yang, J. Appl. Phys. **64** (1988) 2469.
[7] S. A. Reshanov, G. Pensl, K. Danno, T. Kimoto, S. Hishiki, T. Ohshima, H. Itoh, F. Yan, R. P. Devaty and W. J. Choyke, J. Appl. Phys. **102** (2007) 113702.
[8] W. R. Buchwald, N. M. Johnson, J. Appl. Phys. **64** (1988) 958.
[9] L. Storasta, J. P. Bergman, E. Janzén, A. Henry and J. Lu, J. Appl. Phys. **96** (2004) 4909.
[10] K. Danno and T. Kimoto, J. Appl. Phys. **100** (2006) 113728.

Materials Science Forum Vols. 717-720 (2012) pp 255-258
© (2012) Trans Tech Publications, Switzerland
doi:10.4028/www.scientific.net/MSF.717-720.255

Excitation Properties of Silicon Vacancy in Silicon Carbide

A. Gali[1,2,a]

[1]Research Institute for Solid State Physics and Optics, Hungarian Academy of Sciences, PO Box 49, H-1525 Budapest, Hungary

[2]Department of Atomic Physics, Budapest University of Technology and Economics, Budafoki út 8, H-1111 Budapest, Hungary

[a]agali@szfki.hu

Keywords: quantum bit, quantum optics, defect, silicon vacancy, photoluminescence, time-dependent density functional theory, hybrid functional

Abstract. Isolated point defects possessing a high spin ground state and below-band-gap excitation may play a key role in realizing solid state quantum bits in semiconductors which are the basic building blocks of quantum computers. The silicon vacancy in silicon carbide provides these features, making it a feasible candidate in this special and emerging field of science. However, the exact nature of the luminescence of silicon vacancies detected in hexagonal polytypes has not been clarified. This is the first crucial step needed to understand this basic defect in silicon carbide. We report density functional theory based calculations on the silicon vacancy defect. Based on the obtained results we identify the silicon vacancy related photoluminescence signals with the negatively charged defect.

Introduction

The silicon vacancy in silicon carbide (SiC) is a subject of great interest due to its peculiar properties: in the ground state it exhibits a high-spin electronic state and shows luminescence in the near infrared range at around 1.4 eV [1,2]. Photo-electron paramagnetic resonance (EPR) studies indicate that an effective electron spin polarization can occur for the silicon vacancy in SiC [3,4] and its photoluminescence (PL) signal can be detected even at room temperature [5]. Despite recent efforts towards understanding the nature of the silicon vacancy centers [2,6] there is still a debate about the nature of the luminescence. The detailed understanding of the excitation of the silicon vacancy is necessary toward its application in quantum optics and quantum information processes. Recent developments in simulation methods have made it possible to investigate the excited states by *ab initio* calculations where no *a priori* assumptions have to be applied for the defect states. For instance, we have recently shown that the excitation property of the neutral divacancy in SiC or the negatively charged nitrogen-vacancy defect in diamond can be well calculated by time-dependent density functional theory (TD-DFT) [7]. We applied this methodology to the calculation of the excitation spectrum of the silicon vacancy defect in SiC.

Methodology

We investigated the silicon vacancy in a 576-atom hexagonal 4H-SiC clustger by a screened hybrid density functional theory (HSE06) which provided quantitatively accurate results for the thermal ionization energies of point defects in group IV semiconductors (see Refs. [8,9] and references therein). In addition, we applied the TD-DFT method with PBE0 hybrid density functional in the kernel for a nano cubic 3C-SiC cluster with a diameter of about 1.4 nm (see Ref. [7] and references therein). This particular SiC nanocrystal is centered at the interstitial site; thus a silicon vacancy will possess a very slightly distorted T_d symmetry to C_{3v} symmetry due to the crystal field of the nanocrystal, which mimics the hexagonal environment in bulk SiC regarding the symmetry. The relaxation energy due to electronic excitation was calculated by the constrained DFT method [7].

Results and Discussion

First, we studied the silicon vacancy in bulk 4H-SiC. We found that the spin-polarized local density approximation of DFT (DFT-LDA) provided only a single a_1 and a single e level in the fundamental gap, which are occupied by two or three electrons in the neutral and negatively charged state, possessing high spin S=1 and S=3/2 states, respectively. These a_1 and e states are the split t_2 states of the vacancy in the low-symmetry crystal field of the hexagonal crystal. According to our previous group theory analysis [6], another fully occupied a_1 level should exist in this defect which falls in the valance band in DFT-LDA. In the negatively charged state of the silicon vacancy, a generalized gradient approximation of DFT (DFT-PBE) showed the appearance of this low-lying a_1 state just above the valence band in the minority spin channel. The HSE06 functional mimics the quasi-particle correction of DFT levels, which opens the gap by lowering the energy of the valance band edge and increasing the energy of the conduction band edge. In the HSE06 calculation the appearance of the low-lying a_1 state in the gap in the minority spin channel is very clear and is even visible in the spin majority channel just above the valence band edge. In addition, the order of the empty single particle levels is a_1 and e for the split t_2 level in the minority spin channel. The corresponding single particle energies at the cubic and hexagonal sites are equal within 0.05 eV, so the site dependence on the electronic structure of the negatively charged Si-vacancy is not very significant. However, the different amount of splitting of the empty a_1 and e levels may cause different PL spectra at low and elevated temperatures. The situation for the neutral silicon vacancy is not so clear. The low-lying a_1 state falls in the valence band in the HSE06 calculation. In the spin majority channel the e state is fully occupied while the upper a_1 level is empty (see Fig. 1). This gives a $^3A_2$ ground state. (Interestingly, the order of single particle states was reversed in the LDA calculation, which would imply a ^3E ground state.) This is also a high spin S=1 configuration; nevertheless, it cannot be excluded that the many-body singlet state may be the ground state. In 3C-SiC (where the symmetry is T_d) theory proposed that the singlet state of the neutral Si-vacancy may be the true ground state where the multi-determinant nature of the many-body wave function could be directly calculated, although, in a relatively small nanocrystal model [10]. A very recent photo-EPR study [11] indicates that the neutral silicon vacancy has a singlet ground state in 3C-SiC. The corresponding EPR signal could be measured only after photo-excitation with quenching of the S=1 metastable state after system intercrossing in the excitation process. Nevertheless, the crystal field splitting in hexagonal SiC may stabilize the S=1 state as the ground state. Here, we assume that the S=1 state is the ground state in hexagonal SiC and we apply this model in nano SiC studies.

In nano cubic 3C-SiC the gap opens due to quantum confinement; therefore, the vacancy-related states are apparent even in DFT-PBE calculations. The order of single particle states closely follows that obtained in HSE06 calculations in bulk 4H-SiC. The PBE0 hybrid density functional also provided qualitatively the same single particle picture as DFT-PBE did. Next, we calculated the excitation energies by the TD-DFT method based within PBE0 functional both for the neutral and negatively charged silicon vacancy.

In the case of the neutral silicon vacancy, the lowest excitation energy is at about 2.0 eV. This excitation occurs predominantly in the spin majority channel between the occupied e state and the empty a_1 state. In the other excitations the "valence" states mix into the many-body exciton state resulting in higher excitation energies and will not be considered further in the discussion. In the lowest excitation the resulting excited state is ^3E and there is no other feasible spin-conserving excitation route between the localized defect states (see Fig. 1). The ground state was $^3A_2$, so only such photons may be absorbed (infrared absorption) or emitted (PL) in this process that are polarized perpendicular to the symmetry axis (see detailed group theory description in Ref. [6]).

The situation for the negatively charged silicon vacancy is very different. In that case the excitation occurs in the spin minority channel between the occupied lower a_1 level and the empty upper a_1 and e levels (see Fig. 1). The ground state is $^4A_2$ while the symmetries of the corresponding excited states are $^4A_2$ and ^4E, respectively, with calculated excitation energies of 1.568 eV and 1.572 eV. In these processes the selection rules may be applied and it can be shown that the light would be only absorbed with polarization parallel ($E \parallel c$) and perpendicular ($E \perp c$) to the symmetry axis, respectively. The calculated relaxation energy due to excitation is about 0.1 eV, which puts the calculated zero-phonon line at about 1.47 eV.

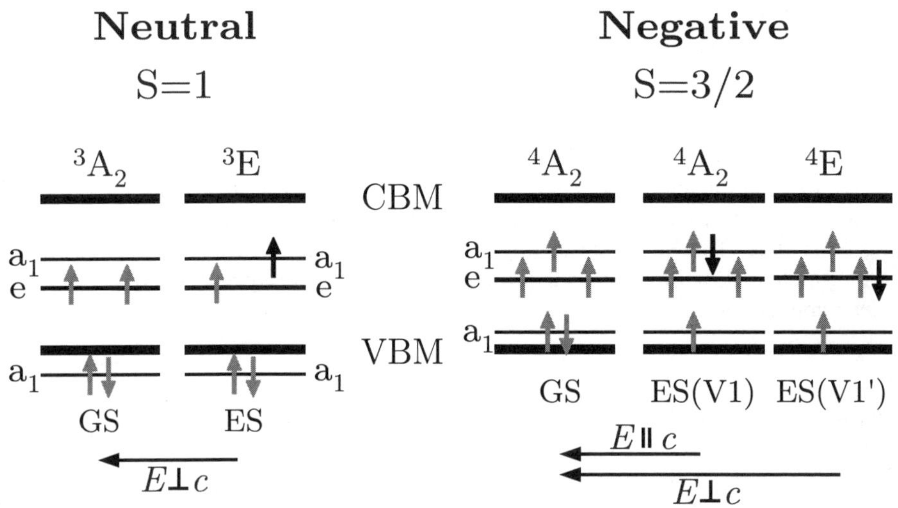

Fig. 1. Schematic figure of single particle levels of the neutrally and negatively charged silicon vacancy in 4H SiC. CBM (VBM) indicates the conduction (valence) band edge. Beside the ground states (GS) the lowest excited states (ES) are shown. The arrows indicate the photon emission processes; the selection rules can be seen above or below the arrows. See text for more explanation.

In 4H-SiC the V1, V1' and V2 zero-phonon lines (ZPL) are associated with the Si-vacancy where the corresponding optically detected magnetic resonance centers (ODMR) of Tv1a, etc. can also be detected. The transition energy of V1 and V1' is at around 1.44 eV while it is at 1.35 eV for V2, where these ZPL signals may be associated with the inequivalent sites of the Si-vacancy in 4H-SiC. According to detailed analysis V1 (and V1') may be associated with the cubic site [6]. The emitted photons of the V1 ZPL are polarized parallel to the symmetry axis while perpendicular to that for the V1' ZPL with about 4.5 meV of energy difference where the V1 ZPL is lower in energy than that of the V1' ZPL. The neutral S=1 Si-vacancy does not show these features because no parallel polarization of the emitted light is expected from that defect, and the calculated excitation energy is significantly larger than the measured one. In contrast, the calculated properties of the negatively charged vacancy can all explain the experimental data, particularly, at the *k*-site which we model in nano 3C-SiC. The estimated ZPL is very close to the measured one, and the V1, V1' ZPLs with the corresponding polarizations can be well explained by the selection rules of optical emission from the $^4A_2$ and ^4E excited states, respectively.

Based on these results we claim that V1 and the closely related other PL centers at different sites are associated with the negatively charged Si-vacancy. The negatively charged Si-vacancy has S=3/2 spin which may be useful in magnetometer applications and the sensitivity might be larger than for the negatively charged nitrogen-vacancy center (NV) in diamond. Since the Si-vacancy has a very localized defect state and spin polarization occurs in photo-excitaton [2], it can be a suitable spin for quantum optics and quantum information applications with close-to-telekom wavelength of luminescence, in contrast to the NV-center in diamond which emits at higher wavelengths (see Ref. [7] and references therein). Silicon carbide can be easily doped and may be easily integrated with existing electro- and optoelectronic devices, and may simultaneously host quantum bits for quantum computing. Progress is needed in controlling of point defects in SiC in order to reduce those defects that may quench the luminescence signals from the Si-vacancy and could provide additional electron spins that may decohere the electron spin of the Si-vacancy. Another important issue is to isotope engineer SiC in order to reduce the hyperfine active ^{29}Si and ^{13}C isotopes around the defect, in order to enhance the coherence time of Si-vacancy spins. Besides, the ultimate control of excitation of the Si-vacancy is needed. For this purpose, the optimization of excitation wave lengths is necessary as well as an investigation of the effects of strain and other perturbations on this defect that may change its luminescence and hyperfine interactions.

Summary

The isolated Si-vacancy was studied in bulk 4H-SiC and nano 3C-SiC. We explored its single particle spectrum in the neutral and negatively charged states. We found that the PL and ODMR lines associated with the Si-vacancy originate from the negative charge state which has a S=3/2 high spin ground state. The PL signal of the negatively charged Si-vacancy is detectable at room temperature, which makes this defect a promising candidate for quantum optics and sensitive magnetometer applications.

References

[1] E. Sörman, N. T. Son, W. M. Chen, O, Kordina, C. Hallin, and E. Janzén, Phys. Rev. B 61 (2000) 2613.
[2] P. G. Baranov, A. P. Bundakova, A. A. Soltamova, S. B. Orlinskii, I. V. Borovykh, R. Zondervan, R. Verberk, and J. Schmidt, Phys. Rev. B 83 (2011) 125203.
[3] N. Mizuochi, S. Yamasaki, H. Takizawa, N. Morishita, T. Ohshima, H. Itoh, and J. Isoya, Phys. Rev. B 66 (2002) 235202.
[4] N. Mizuochi, S. Yamasaki, H. Takizawa, N. Morishita, T. Ohshima, H. Itoh, T. Umeda, and J. Isoya, Phys. Rev. B 72 (2005) 235208.
[5] H. Isono, Michio Tajima, Norihiro Hoshino, and H. Sugimoto, Mater. Sci. Forum 600-603 (2009) 545.
[6] E. Janzén, A. Gali, P. Carlsson, A. Gällström, B. Magnusson, and N. T. Son, Physica B 404 (2009) 4354.
[7] A. Gali, Phys. Stat. Sol. B 248 (2011) 1337.
[8] Peter Deák, Adam Gali, Bálint Aradi, and Thomas Frauenheim, Phys. Stat. Sol. B 248 (2011) 790.
[9] T. Hornos, A. Gali, and B.G. Svensson, Mater. Sci. Forum 679-680 (2011) 261.
[10] P. Deák, J. Miró, A. Gali, L. Udvardi and H. Overhof, Appl. Phys. Lett. 75 (1999) 2103.
[11] Jérémie Lefèvre, Jean-Marc Costantini, Didier Gourier, Stéphane Esnouf, and Guillaume Petite, Phys. Rev. B 83 (2011) 075201.

Materials Science Forum Vols. 717-720 (2012) pp 259-262
© (2012) Trans Tech Publications, Switzerland
doi:10.4028/www.scientific.net/MSF.717-720.259

Investigation of intrinsic carbon-related defects in 4H-SiC by selective-excitation photoluminescence spectroscopy

I.G. Ivanov[1, a,*], A. Gällström[1], R. Coble[2], R.P. Devaty[2], W.J. Choyke[2,b] and E. Janzén[1,c]

[1] Linköping University, Department of Physics, Chemistry and Biology, 581 83 Linköping, Sweden

[2] University of Pittsburgh, Physics & Astronomy Department, 100 Allen Hall 3941 O'Hara Street, Pittsburgh, PA, 15260 USA

[a]ivani@ifm.liu.se, [b]choyke@pitt.edu, [c]erija@ifm.liu.se, *corresponding author

Keywords: carbon aggregates, carbon split interstitial, di-carbon antisite, photoluminescence

Abstract. Emission of carbon-related defects is investigated by means of selectively-excited photoluminescence in high purity 4H-SiC electron-irradiated with very low dose. Two new centers with clearly associated phonon replicas are observed, one of which is tentatively assigned to the carbon split interstitial at the hexagonal site. The temperature dependence of the spectrum is also studied and indicates that at least some of the observed luminescence lines arise from recombination of excitons bound to isoelectronic centers.

Introduction

Defects related to carbon interstitials in SiC have received significant attention in the past few years, because they are commonly created during implantation or electron irradiation. The ability of C interstitials to migrate in the lattice and aggregate at certain lattice sites in different configurations gives rise to a large variety of C-related defects. Several of them have been studied theoretically [1,2] and experimentally [3-6]. The theory is capable of estimating the formation energies of the defects giving us an idea of their abundance in as grown and, perhaps, irradiated material, and their dissociation energy related to the thermal stability of the defects upon annealing. Further output of the calculation is the ground-state energy, but even if no complications with charged states of the defects are involved, the accuracy of the ground state is often not better than a few tenth of an eV, obviously not enough for unambiguous comparison with experimental data from electrical or optical measurements. Fortunately, local-phonon energies can be calculated with much better accuracy [1,2], and since in many cases they are observable as shifts of phonon replicas from the no-phonon line, a rather comprehensive attempt was undertaken in [4,5] to identify some of the C-related defects by comparison of the observed energies of local (defect-specific) phonons in photoluminescence (PL) with the theoretically calculated ones. High-temperature annealing, mapping across the irradiated region, and use of n- and p-type samples were common methods used in these studies to vary and study the relative contributions of different C-related defects. However, the defects were created using relatively high dose of electron irradiation (mostly above 10^{19} cm^{-2}) that inevitably introduces severe lattice damage and large defect concentrations, which may initiate interaction between different defect centers and hinder their individual identification. We introduce a supplementary approach by using a high-purity (virtually intrinsic) sample irradiated with low dose, and using different excitation wavelengths in order to vary and distinguish the emission from different centers.

Experimental

We have studied two high-purity epitaxial layers [both unintentionally doped n-type, with $n \sim 7\ 10^{14}$ cm^{-3} (sample#1, 70 μm thick) and $n \sim 8\ 10^{13}$ cm^{-3} (sample#2, 110 μm thick), irradiated at room temperature with 1 MeV electrons at a fluence of 3 10^{15} cm^{-2}, with subsequent annealing in Argon at 1260°C (sample#1) and 1100°C (sample#2) for 30 min. Sample #2 is the main subject of this work because it can be considered as virtually intrinsic and with the low irradiation dose applied we do not expect significant interaction between the defects or aggregation of larger carbon clusters in significant concentrations. Most of the resulting PL lines are consistent with previous observations in 4H-SiC [4,5], however, with much weaker intensity than in these references (by about 2-3 orders of magnitude judging from comparison of the relative intensities of the PL and the Raman lines). Sample #2 was prepared as a free-standing film with polished side edges used to introduce the laser beam in order to increase its path through the sample and thus the signal. All spectra with below-band-gap excitation were obtained in a geometry of near-backscattering from the polished edge of the sample. Variable excitation was obtained from a frequency-doubled Ti-sapphire laser in the CW-regime.

Results and Discussion

The no-phonon lines and their associated phonon replicas observed in this study are listed in Table 1. The following remarks are based on the data shown in the table. We have listed only lines which are prominent under certain excitation conditions (see below); not all lines are observed under, e.g., above band gap excitation. Most of these lines have been observed also in [4,5,6], except for the two centers with no phonon lines at 4695 Å and 4968 Å. We note also that our assignment of the phonon replicas of the 4718 Å center differs from that of [4].

Apart from visual analysis of the spectra obtained with about 200 different excitation wavelengths, we make use of statistical methods in order to establish correlation between different lines. We consider now the various PL centers in more detail.

The 4695 Å and 4968 Å centers, called here for brevity X- and Y-centers, respectively, have not been reported earlier, to the best of our knowledge. Their emission is very weak under above band gap excitation with a 351 nm laser, as shown in Fig.1a. The dominant emission in this case is due to the alphabet lines (not shown) and the a_0, b_0 and c_0 lines and their phonon replicas (cf. Table 1) previously associated in [5,6] with the dicarbon antisites. However, the X- and Y-centers become dominant under excitation around 3900-4000 Å (3098 – 3178 meV, Fig.1b). In addition, the X-center exhibits a sharp excited state at about 64 meV above the 4695 Å (2640 meV) no-phonon line X_0. The spectrum excited resonantly at this state, 4584.4 Å (2704 meV), is shown in Fig. 1c, and we notice the simultaneous appearance of the peak at 5084 Å (2438 meV) in Fig. 1b,c interpreted here as a phonon replica at 202 meV of the X-center (X_{202}). In previous work [4] a line at the same position as X_{202} was associated with the 189 meV phonon replica of the tricarbon antisite (the 4718 Å center, cf. Table 1). We also observe the 4718 Å no-phonon line in many of our spectra (e.g., Fig.1b and insert), hence further arguments will be given to ascertain that the line at 5084 Å observed in our work is not a replica of the former line. Firstly, confirmation of our assignment comes from the observation of the second order $2X_{202}$, cf. Fig. 1. Secondly, a correlation map built from a series of 55 spectra obtained by scanning the laser across the excitonic band gap (not shown) shows a strong correlation of the X_0 line (4695 Å) with X_{202} (5084 Å), not with the 4718 Å center. In addition, three replicas at 152, 189 and 247 meV were associated in [4] with the 4718 center, but in our series strong correlation is observed only with the 247 meV replica (cf. insert in Fig.1), and the 152 meV replica is not observed at all. Thus we conclude that the 5084 Å line observed in our measurements is indeed a phonon replica of the X-line, although overlap with the 189 meV replica of the 4718 Å center cannot be completely ruled out.

Table 1. The no-phonon PL lines and their replicas observed with different excitation wavelengths.

Notation (position)	Energy (meV)	Local phonons in meV (line position)	Assignment
a_0 (4632 Å)	2675	133 (4874 Å) 180 (4966 Å)	
b_0 (4636 Å)	2674	133 (4878 Å) 178 (4968 Å)	Dicarbon antisite $(C_2)_{Si}$ [6]
c_0 (4643 Å)	2670	132 (4884 Å) 180 (4979 Å)	
X_0 (4695 Å)	2640	202 (5084 Å)	C split interstial (tentative)
4718 (4718 Å)	2627	247 (5208 Å)	Tricarbon antisite $(C_3)_{Si}$ [4]
G_0 (4935 Å)	2512	167 (5287 Å) 170 (5294 Å)	Dicarbon antisite pair [4]
Y_0 (4968 Å)	2495	171 (5332 Å)	–

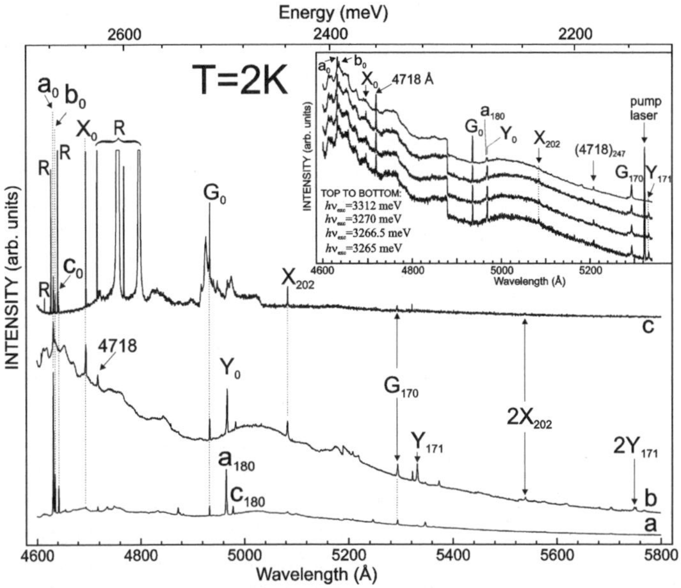

Fig. 1. The PL spectrum of sample #2 with (a) above band gap excitation, (b) excitation at 3126 meV, and (c) resonant excitation at 2704 meV. The insert illustrates the onset of the Y-center (see text). The lines marked 'R' are the Raman bulk modes.

Considering the X-center, the carbon split interstitial (at the hexagonal or cubic carbon site) might be a good candidate based on the reasonable agreement between the observed (202 meV) and calculated local phonon energy (197 [1], or 189-192 [7] meV). Indeed, the higher vibrational energy of the C-C stretch mode compared to the dicarbon antisite is stipulated by the stiffer bond between the two carbon atoms due to less space available to the pair on a carbon site compared to silicon site, and a single vibrational mode above the phonon continuum is calculated in [1]. Also, an excited state is expected due to splitting of the doubly-degenerate *e*-level in the gap originating from the non-bonding π-orbitals on the two C atoms in the lower-symmetry environment of 4H-SiC if compared to 3C-SiC. However, the energy of the experimentally determined excited state – 64 meV – is in strong disagreement with the theoretical calculation for the cubic site (0.7 eV [1]), whereas in the case of a carbon split interstitial at a hexagonal site the two levels are nearly degenerate, leaving the latter configuration a better candidate for the X-center.

Let us consider now the Y-center in more detail. The no-phonon line position Y_0 coincides within the experimental accuracy with the phonon replica b_{178} associated with the b_0 line of the dicarbon antisite [5,6], hence despite the good correlation of this line with its phonon replica Y_{171} and the observation of $2Y_{171}$ another confirmation is desirable that Y_0 is not related to b_0. Indeed, the insert in Fig. 1 displays a series of spectra obtained near the onset of the Y_0-line which show no or even decreasing contribution from b_0 while Y_0 increases more than 10 times from top to bottom in this figure. Again its increase is correlated with the appearance of Y_{171}. No other local phonon modes associated with the Y-center were observed in this study.

Finally, we discuss the G-center associated with a dicarbon antisite pair in [4]. In all our measurements we observe strong correlation of the no-phonon line at 4935 Å with its 170 meV-replica up to the second order, in good agreement with [4]. We also observe the weaker satellite of G_0 at 167 meV discussed in [4]. However, it is worth noting that the formation of antisite pairs in significant concentrations is not likely in our sample due to the low concentration of dicarbon antisites and their low mobility (high dissociation energy). It is worth noting that the structures of the G center (4935 Å) and the Y-center are very similar, both show a single phonon replica at ~170 meV, but we will not attempt identification of the Y-center at this time.

We turn now to consideration of the temperature dependence of the spectrum, some selected spectra of which are displayed in Fig. 2. The most striking feature is the remarkable growth in intensity of the a_0-b_0-c_0 triplet (dicarbon antisite [5]) by a factor of ~2, and the G-center (4935 Å line) by a factor of ~10 in the temperature range 5-40 K. We attribute this growth to the increase in the population of *free* excitons which are freed from the shallow donors and recaptured to the deeper a_0-b_0-c_0- and G-centers. Indeed, if a center possesses a charged state well within the band gap, then a particle of the opposite charge can always be bound by the Coulomb field of the charged

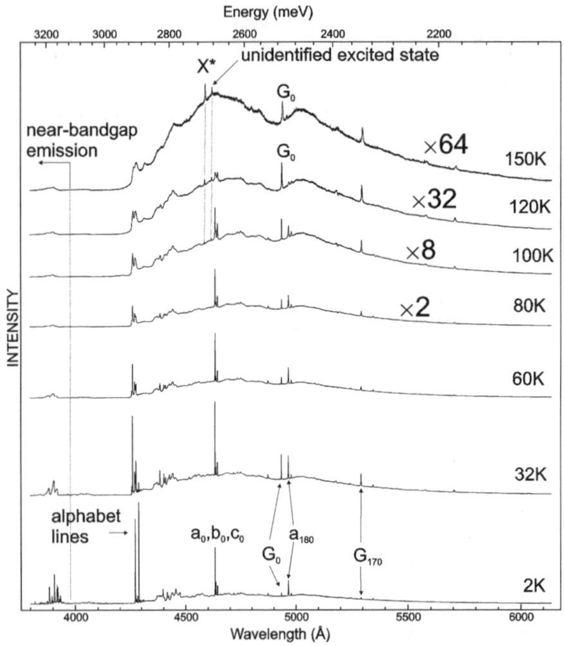

Fig. 2. Temperature dependence of the PL spectrum. All spectra are displayed so as to compare the absolute intensities, but note the scale changes which apply to individual spectra.

center. While excited states for the second particle close to its respective band are expected, the ground state of such a bound exciton need not be shallow, and the recombination lifetime need not be slow as is the case with isoelectronic atoms. Fast recombination rates have been measured in [6] for the center associated with the dicarbon antisite (a_0-b_0-c_0), which was used as an argument against the bound-exciton concept. However, this center possesses two loosely bound electrons in the two carbon π-orbitals which are available for non-radiative Auger recombination, hence short lifetime of an exciton bound to the center is expected. The bound-exciton-recombination picture, suggested originally in Ref. [1], is supported also by the excitation scan across the excitonic band gap; the luminescence of the alphabet lines, the dicarbon antisite and the G_0-center (4935 Å) grows strongly when free excitons (but no free carriers at 2K) are created with the assistance of the lowest energy phonon in 4H-SiC. It should be noted, however, that not all centers exhibit such behavior; on the contrary, the X-center for example decreases in intensity when excited above the phonon-assisted excitonic band gap, which might be an indication that the emission from this center is due to internal electronic transitions. Thus, both the temperature dependence and the scan across the excitonic band gap might provide us with tools to distinguish electronic transition within the center from emission of excitons bound to it.

We notice also that at temperatures above ~80 K two excited states are emerging at 2704 and 2686 meV, the former (X^*) quite obviously being the excited state of the X-center considered earlier. (The other is not yet associated with any of the lines observed at 2K.)

In conclusion, we observe two new luminescence centers in virtually intrinsic 4H-SiC irradiated with a very low dose of electrons. Both centers involve carbon atoms judging from the high energies of the local phonons, and we tentatively propose the carbon split interstitial at a hexagonal site as the origin of one of them (the X-center with no-phonon line at 4695 Å).

Acknowledgment. Financial support from Knut och Alice Wallenbergs Styftelse (Knut and Alice Wallenberg Foundation) and Vetenskapsrådet (VR) in Sweden is gratefully acknowledged. The University of Pittsburgh group thanks the II-VI Foundation for partial support of this research.

References

[1] A. Gali *et al.*, Phys. Rev. B 68 (2003) 125201.
[2] A. Mattausch, M. Bockstedte, and O. Pankratov, Phys. Rev. B 70 (2004) 235211.
[3] G.A. Evans *et al.*, Phys. Rev. B 66 (2002) 035204.
[4] J.W. Steeds and W. Sullivan, Phys. Rev. B 77, 195204 (2008).
[5] F. Yan *et al.*, Mat. Sci. Forum Vols 645-648 (2010) 411-414.
[6] J.W. Steeds, W. Sullivan, S.A. Furkert, and G.A. Evans, Phys. Rev. B 77 (2008) 195203.
[7] M. Bockstedte *et al.*, phys. stat. sol. (b) 245 (2008) 1281-1297.

Materials Science Forum Vols. 717-720 (2012) pp 263-266

doi:10.4028/www.scientific.net/MSF.717-720.263

Local Thermal Expansion and the C-C Stretch Vibration of the Dicarbon Antisite in 4H SiC

R.P. Devaty[1,a,*], Fei Yan[1,b], W.J. Choyke[1,c], A. Gali[2,d], T. Kimoto[3] and T. Ohshima[4]

[1]Dept. of Physics and Astronomy, University of Pittsburgh, Pittsburgh, PA 15260, USA

[2]Research Institute for Solid State Physics and Optics, Hungarian Academy of Sciences, P.O. Box 49, H-1525 Budapest, Hungary

[3]Department of Electronic Sciences & Engineering, Kyoto University, Katsura, Nishikyo, Kyoto, 615-8510, Japan

[4]Japan Atomic Energy Agency, 1233 Watanuki, Takasaki, Gunma 370-1292, Japan

[a]devaty@pitt.edu, [b]feiyan@gmail.com, [c]choyke@pitt.edu, [d]agali@szfki.hu, *corresponding author

Keywords: dicarbon antisite, local vibrational mode, thermal expansion, photoluminescence, point defects

Abstract. The C-C stretch vibration associated with the dicarbon antisite in 4H SiC has been observed out to the fifth harmonic in the low temperature photoluminescence spectrum. The anharmonicity is accounted for reasonably well by fits to the data based on the Morse potential. We combine the observations from experiment, the analytically tractable Morse potential, and results obtained from first principles calculations on this defect to obtain an estimate of the thermal expansion coefficient of the C-C bond. This local thermal expansion coefficient is considerably smaller than the linear thermal expansion coefficient of bulk 4H SiC, in striking contrast with the recent result for the nitrogen-vacancy center in diamond that the local thermal expansion coefficient is larger than the bulk value.

Introduction

While the local thermal expansion associated with a defect in a crystal is of intrinsic scientific interest, it can also play an important role in practical applications. For example, Acosta *et al.* [1] recently proposed that the anomalous temperature dependence of the zero magnetic field splitting parameter D (relevant to electron spin resonances studies) of the nitrogen-vacancy center in diamond is associated with local thermal expansion at the defect about a factor of 4.5 larger than the value of the linear thermal expansion coefficient for bulk diamond. This strong temperature dependence poses a significant challenge to potential applications in ultrasensitive magnetometry.

The dicarbon antisite in 4H SiC is one of the few intrinsic defects in SiC for which a convincing assignment to a low temperature photoluminescence (LTPL) spectrum exists. This defect consists of two carbon atoms replacing a silicon atom, resulting in a cluster of six carbon atoms (See inset to Fig. 1). The two carbon atoms make up a dumb bell, with each of these carbon atoms bonded to two other nearby carbon atoms. The identification [2,3] is based on a comparison of the energies of the local vibrational modes observed as replicas of the "triplet spectrum" with predictions of calculations based on first principles [4,5]. Another argument in support of the presence of a C-C dumb bell in the defect was based on a study of the C-C stretch vibration using a ^{13}C enriched sample [2].

A detailed study of the triplet spectrum [3] reveals a sequence of vibrational replicas of the ~180 meV C-C stretch vibration out to the fifth harmonic, a second local vibrational mode near 132 meV, and combination modes. In this work, we make use of the set of ~180 meV C-C stretch vibration energies, the analytically tractable Morse potential [6], and the results of first principles calculations to estimate the local thermal expansion coefficient. We find that this C-C bond thermal expansion coefficient is significantly smaller than the linear expansion coefficient of bulk 4H SiC.

Experiment

The 4H SiC homoepitaxial layer was grown in a horizontal hot-wall chemical vapor deposition reactor. The n-type epilayer is 20 μm thick and doped around $N_D - N_A \sim 3\times10^{15}$ cm^{-3}. Low dose, uniform electron bombardment was carried out at 1 MeV with fluence 1×10^{15} cm^{-2}. The sample was annealed in Ar for 30 min at 1100°C. LTPL measurements were performed at 2 K using an immersion cryostat. The luminescence was excited using a frequency doubled argon ion laser at 244 nm wavelength and 40 mW power, dispersed in an asymmetric Czerny-Turner spectrometer with approximately 0.5 meV resolution, and detected using a liquid nitrogen cooled CCD array.

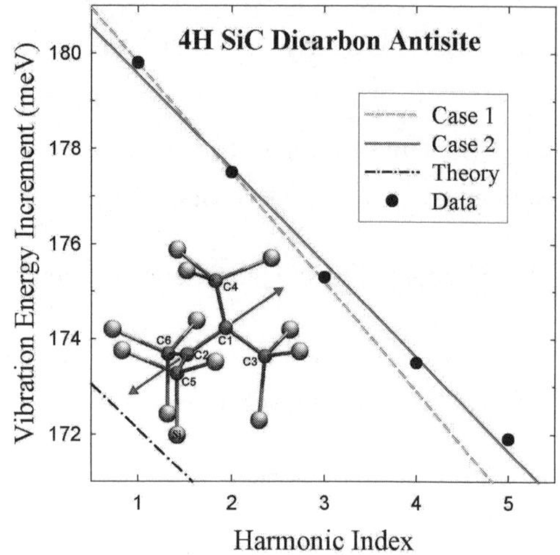

Fig. 1. Morse potential fits to the C-C stretch vibrational replicas plotted as increments. Also shown is the prediction based on first principles calculations. The inset shows the dicarbon antisite schematically. The arrows indicate the motion of the C-C dumb bell for the stretch vibration.

Results, Analysis and Discussion

Fig. 1 shows the sequence of replicas for the C-C stretch vibration plotted as increments $\Delta E_{n,n-1} = E_n - E_{n-1}$, where E_0 refers to the no phonon line a_0 [3] and E_n is the photon energy associated with the nth harmonic observed using LTPL. After reexamination, some of the energy values were slightly adjusted with respect to the listings in Table 1 of Ref. [3]. The Morse potential predicts linear behavior for this sequence of increments, $\Delta E_{n,n-1} = \hbar\omega(1-2n\chi)$, where $\hbar\omega$ is the harmonic energy and the dimensionless parameter χ specifies the anharmonicity. We perform the fit two ways: Case 1: Force the straight line to pass through the first and second harmonics; Case 2: Find the best fit to the five observed replicas. While the Morse potential provides a reasonably good account, the data suggest a slight bowing such that the potential well would be a bit narrower than the Morse form with increasing energy.

The local thermal expansion coefficient is defined by $\alpha = \dfrac{1}{L}\dfrac{\partial L}{\partial T}$, where L is the C-C separation taking into account quantum mechanical averaging (matrix elements) and thermal averaging using the canonical ensemble, and T is the temperature. Calculating the required diagonal matrix elements $\langle m|\hat{x}|m\rangle$, where $\hat{x}$ is the position operator and m an index to label the vibrational state, requires complete knowledge of the Morse potential, which is not provided by the two parameter fit. The corresponding

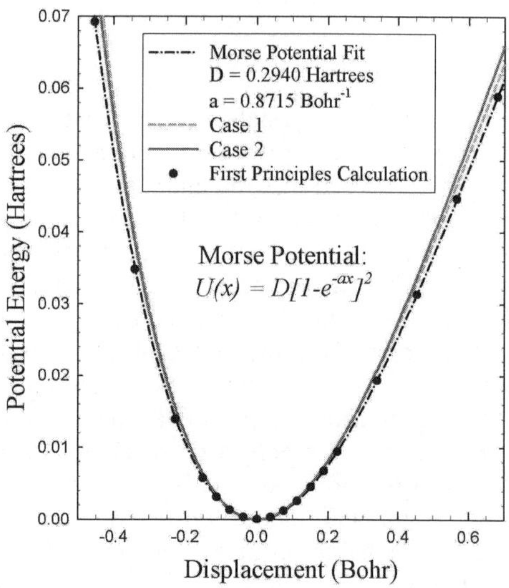

Fig. 2. Morse potential fit to calculated C-C stretch potential energy, used to obtain the reduced mass μ = 10922.31 a.u. Also shown are the Morse potentials obtained using the two fits to the data (Cases 1 and 2) and the value of μ.

Schrödinger equation has three parameters, a reduced mass μ plus the two parameters a and D appearing in the Morse potential energy: $U(x) = D(1-e^{-ax})^2$, where $x = 0$ refers to the configuration of minimum potential energy. The parameter D is determined directly from the fit, $D = \hbar\omega/4\chi$, whereas calculating a from $a = \omega\sqrt{\mu/2D}$ requires a value for μ.

We obtain μ from an analysis of the results of first principles calculations. Details regarding the calculations are provided elsewhere [3]. The important point is that the C-C separation in the dumb bell provides a good configuration coordinate and the resulting "energy surface" is well described by the Morse potential (Fig. 2). Taking the stretch vibrational energy 174 meV calculated in the quasiharmonic approximation using the cluster of six carbon atoms as $\hbar\omega$ leads to μ = 10922.31 a.u. (The atomic unit for mass is the rest mass of an electron.). For an isolated C-C dumb bell the expected value is μ = m(^{12}C)/2. Our value is 99.86% of this value, suggesting only a very small correction due to the participation of the surrounding carbon atoms. Use of this value of μ leads to the Morse potentials for Cases 1 and 2 shown in Fig. 2. Over the range of displacements calculated, Cases 1 and 2 agree with the theoretical potential within 15% and 10%, respectively. The first principles prediction of the energies of the harmonic decrements is shown in Fig. 1. The calculated energies are low, but the slope agrees quite well with experiment. We emphasize that no attempt to "fit" the first principles calculations to the data has been made. The Morse potential parameters are listed in Table 1.

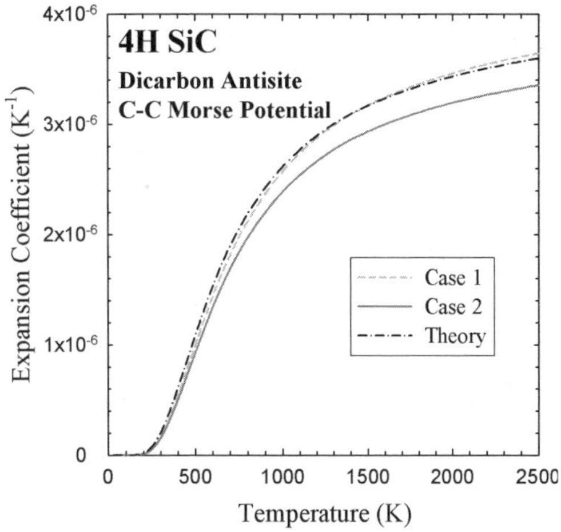

Fig. 3. C-C bond thermal expansion coefficient calculated using three different Morse potentials.

The diagonal matrix elements can be expressed in terms of the digamma function [7]. Fig. 3 shows the C-C bond thermal expansion coefficient obtained using the three Morse potentials. The five vibrational excited states observed experimentally are adequate to calculate the bond thermal expansion up to about 1100 K. Fourteen excited states are adequate for calculations up to 2500 K. For all three cases, the energy of the fourteenth excited state is below 0.09 Hartrees, *i.e.*, just above the energy range shown in Fig. 2. Depending on the chosen potential, the number of bound states ranges from 79 to 92. Fig. 3 demonstrates that the choices do not have a large impact on the result.

Fig. 4 provides a comparison of the calculated C-C dumb bell bond thermal expansion coefficient of the dicarbon antisite in 4H SiC with recent measurements of the uniaxial anisotropic linear thermal expansion coefficient of bulk 4H SiC by Stockmeier *et* al. [8] (α_{11} and α_{33} are the planar and axial components, respectively) as well as older values for SiC compiled by Slack and Bartram [9]. Over most of the temperature range, the bulk values exceed the local bond thermal expansion coefficient by about a factor of two. For very low temperatures the bond thermal expansion coefficient is quenched due to the relatively large thermal energy required to excite the stretch vibrational mode, whereas low energy phonons in the acoustic branches of the bulk phonon dispersion relations can contribute even at low temperature. Since the dicarbon antisite is a small

Table 1. Morse Potential parameters

	$\hbar\omega$ [meV]	χ	D [Hartrees]	a [Bohr^{-1}]	μ [a.u.]
Case 1	182.1	0.006315121	0.2649217	0.9608216	10922.31
Case 2	181.54	0.005453343	0.3058432	0.8914859	10922.31
Theory	174.0	0.005437400	0.2940	0.8715	10922.31

cluster of six carbon atoms, Fig. 4 also shows the linear thermal expansion coefficient for diamond based on the fit by Reeber and Wang [10] and earlier results by Slack and Bartram [9]. The C-C bond expansion is also much less than the expansion coefficient for diamond. Note that the C-C bond in the dicarbon antisite dumb bell differs from the sp^3 tetrahedral covalent bonding in diamond. The dumb bell bond more closely resembles the sp^2 case. However, unlike the double bonding found in organic molecules, which would require the six carbon atoms to be coplanar, the p-orbitals on the two dumb bell atoms are nearly perpendicular, which quenches the bonding. The occupancy of these orbitals determines the charge state of the defect. For the neutral charge state, likely of relevance for the LTPL spectrum, each of the two non-bonding p-orbitals is occupied by one electron.

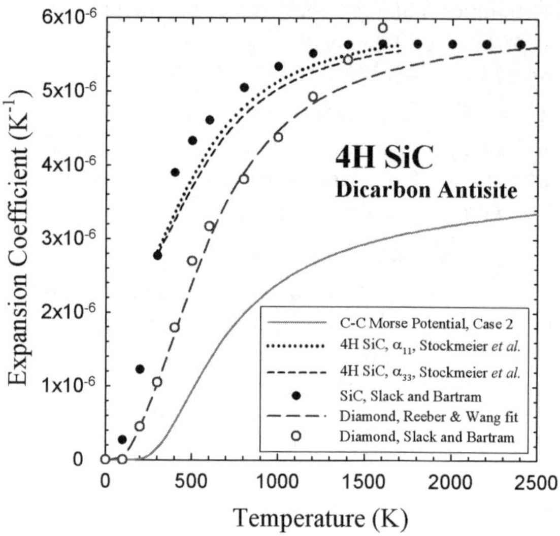

Fig. 4. Comparison of calculated C-C bond thermal expansion coefficient with the bulk linear thermal expansion coefficients of SiC [8,9] and diamond [9,10].

Conclusion

We have combined the observation of five harmonics of the C-C stretch vibration of the dicarbon antisite in 4H SiC, results of first principles calculations, and an analysis based on the analytically tractable Morse potential to estimate the temperature dependence of the linear expansion coefficient associated with the C-C bond in the dumb bell. We find that this local expansion at the defect is *smaller* than the value for bulk 4H SiC, a result in contrast with the dramatic *increase* implied by an analysis of magnetic resonance data on the nitrogen-vacancy complex in diamond.

Acknowledgment

The University of Pittsburgh group acknowledges the II-VI Foundation for support of this research.

References

[1] V.M Acosta, E. Bauch, M.P. Ledbetter, A. Waxman, L.-S. Bouchard and D. Budker: Phys. Rev. Lett. 104 (2010) 070801; 106 (2011) 209901 (*erratum*).

[2] J.W. Steeds, W. Sullivan, S.A. Furkert, G.A. Evans and P.J. Wellman: Phys. Rev. B 77 (2008)195203.

[3] F. Yan, R.P. Devaty, W.J. Choyke, T. Kimoto, T. Ohshima, G. Pensl and A. Gali: Mater. Sci. Forum Vols 645-648 (2010) 411-414.

[4] A. Gali, P. Deák, P. Ordejón, N.T. Son, E. Janzén and W.J. Choyke: Phys. Rev. B 68 (2003) 125201.

[5] A. Mattausch, M. Bockstedte and O. Pankratov: Phys. Rev. B 69 (2004) 045322.

[6] P.M. Morse, Phys. Rev. 34 (1929) 57-64.

[7] E.F. de Lima and J.E.M. Hornos: J. Phys. B: At. Mol. Opt. Phys. 38 (2005) 815-825.

[8] M. Stockmeier, R. Müller, S.A. Sakwe, P.J. Wellmann and A. Magerl: J. Appl. Phys. 105 (2009) 033511.

[9] G.A. Slack and S.F. Bartram, J. Appl. Phys. 46 (1975) 89-98.

[10] R.R. Reeber and K. Wang, J. Electron. Mater. 25 (1996) 63-67.

Materials Science Forum Vols. 717-720 (2012) pp 267-270

doi:10.4028/www.scientific.net/MSF.717-720.267

Defects in an Electron-Irradiated 6H-SiC Diode Studied by Alpha Particle Induced Charge Transient Spectroscopy: Their Impact on the Degraded Charge Collection Efficiency

N. Iwamoto[1,2,a], A. Koizumi[1], S. Onoda[2], T. Makino[2], T. Ohshima[2], K. Kojima[3], S. Koike[1], K. Uchida[1], and S. Nozaki[1]

[1]Department of Electronic Engineering, The University of Electro-Communications, 1-5-1 Chofugaoka, Chofu, Tokyo 182-8585, Japan

[2]Semiconductor Analysis and Radiation Effects Group, Japan Atomic Energy Agency, 1233 Watanuki, Takasaki, Gunma 370-1292, Japan

[3]Advanced Power Electronics Research Center, National Institute of Advanced Industrial Science and Technology, 1-1-1 Umezono, Tsukuba, Ibaraki 305-8568, Japan

[a]iwamoto.naoya@jaea.go.jp

Keywords: alpha particle, charge transient spectroscopy, charge collection, defect, electron irradiation, particle detector

Abstract. Defects in electron-irradiated 6H-SiC diodes have been studied by single alpha particle induced charge transient spectroscopy and deep level transient spectroscopy (DLTS) in order to identify critical defects responsible for the charge collection efficiency (CCE) decreased by high-energy electron irradiation. The defect X_2 detected by the charge transient spectroscopy and the electron trap E_i detected by the DLTS had a similar activation energy of around 0.50 eV. In addition, the annealing at 473K completely removed defects X_2 and E_i, and restored the CCE. The defect X_2 is attributed to the electron trap E_i, and responsible for the decreased CCE.

Introduction

Silicon carbide (SiC) diodes are suitable for application to high-energy particle detectors because of their excellent characteristics, such as small leakage current even under large reverse bias [1, 2]. To develop detectors with high reliability, it is very important to understand the radiation hardness. We previously reported that the charge collection efficiency (CCE) of a SiC diode detector was reduced by the electron irradiation [3]. The decreased CCE was attributed to the increased defect density. The capacitance deep level transient spectroscopy (DLTS) is often employed to study defects in semiconductors, and has also been used to study defects in SiC diodes irradiated with high-energy electrons [4-6]. However, the capacitance of a diode must be large enough for accurate capacitance measurements. A large-area diode with small leakage current is not easy to fabricate with a lightly-doped epilayer, which is required to make a diode detector with high CCE. In addition, the reverse bias voltage applied to a diode in the DLTS is much smaller than that required to collect charges induced by the high-energy particles. Since defects may behave in a different way in the capture and emission of carriers at high electric field, not all defects identified by the DLTS may be responsible for the decreased CCE of a diode detector. In this study we have made an attempt to identify the defects responsible for the decreased CCE of a SiC diode detector fabricated on a lightly-doped epilayer by the single alpha particle induced charge transient spectroscopy [7]. We have also performed the DLTS on SiC diodes fabricated on heavily-doped epilayers, and the results are compared with those obtained by the charge transient spectroscopy.

Experiment

Samples used in this study were 6H-SiC p^+n and n^+p diodes fabricated on n and p-type epilayers grown on n and p-type 6H-SiC substrates, respectively, by hot-wall chemical vapor deposition. Lightly-doped n-type epilayers with a carrier concentration of 2×10^{14} /cm^3 were employed to fabricate p$^+$n diodes as detectors. n- and p-type epilayers with a carrier concentration of 10^{15} /cm^3 were employed to fabricate p$^+$n and n$^+$p diodes for the DLTS. The p$^+$ and n$^+$ regions of the diodes were formed by implantations of aluminum (Al) and phosphorus (P) ions, respectively, and subsequent annealing. Ohmic contacts to p$^+$ regions and p-type substrates were made by Al deposition and sintering. Ohmic contacts to n$^+$ regions and n-type substrates were made by Al deposition without sintering. Some samples were irradiated with 1 MeV electrons at a fluence of 1×10^{15} /cm^2. The collected charges of the lightly-doped p$^+$n diode detectors were measured at room temperature using 5.486 MeV alpha particles and a standard charge collection system, which consists of an Ortec 142A charge sensitive preamplifier, an Ortec 571 spectroscopic amplifier and a Canberra Multiport II multi-channel analyzer. The time-dependent cumulative collected charges of the diode detectors at a reverse bias voltage of 60 V were also measured at temperatures from 180K to 340K in the single alpha particle induced charge transient spectroscopy using the charge sensitive preamplifier and an HP 54520A digital storage oscilloscope. The capacitance transients of the p$^+$n and n$^+$p diodes were measured in the DLTS at temperatures from 100 to 300K using an HP4280A capacitance meter and an HP8112A pulse generator. The diodes were pulsed to 0 V for 1 ms and then reverse-biased at 3 V. The electron-irradiated samples were sequentially annealed at 373 and 473K for 30 min each and characterized by the charge collection measurements, the charge transient spectroscopy and the DLTS after each annealing step.

Results and Discussion

Figure 1 (a) shows the reverse bias voltage-dependent collected charges of the lightly-doped 6H-SiC p$^+$n diode detectors. The collected charge of the unirradiated diode detector increases with the increased reverse bias voltage and becomes saturated at 50 V or higher. The CCE at saturation is estimated to be 93%. Here, the CCE is defined by dividing the experimentally measured collected charge by the ideal value of 122 fC [8]. The detailed analysis of collected charges of the unirradiated diode detector was presented elsewhere [8]. The decreased CCE is found in the electron-irradiated diode detector, and is more pronounced at lower bias voltage. The CCE of the electron-irradiated diode detector at each bias voltage increases after annealing at 373K, and is almost recovered to the initial value by annealing at 473K. Figure 1 (b) shows the time-dependent cumulative collected charges of the diode detectors at a reverse bias voltage of 60 V at temperatures from 250K to 300K. The cumulative collected charge of the unirradiated diode detector increases rapidly and saturates within several nanoseconds at all temperatures, and does not seem to depend on temperature. Since the depletion region width at 60 V is almost the same as the projected range of an alpha particle, 18.2 μm, all carriers (electron-hole pairs) generated by a single alpha particle strike are swept by the electric field and collected within several nanoseconds [8]. The cumulative collected charge of the electron-irradiated diode detector also initially increases rapidly, but the value does not reach that of the unirradiated one. It then continues to increase slowly for several hundred microseconds as shown in Fig. 1 (b). The slower components of cumulative collected charges are charges detrapped from defects in the electron-irradiated diode detector. Since the alpha particle generates a high-density of carriers (>10^{18} /cm^3) in the diode detector, both electron and hole quasi Fermi levels move close to the conduction and valence band edges (E_c and E_v), respectively, even in the depletion region. As a result, defects in the depletion region capture carriers. As free carriers in the depletion region are swept by the electric field and the quasi Fermi levels return close to the midgap, the defects detrap carriers to the

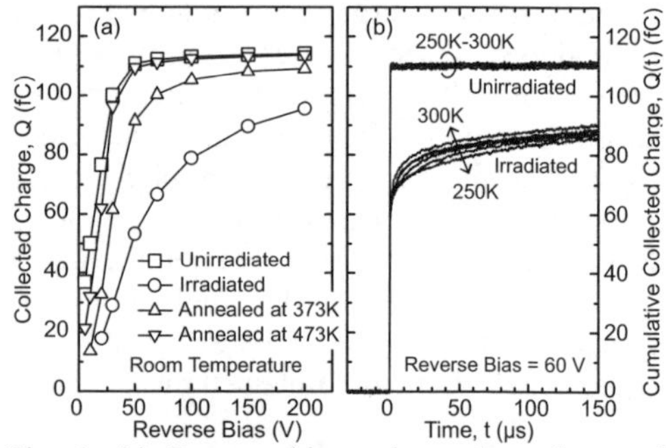

Fig. 1. (a) Reverse bias voltage dependence of collected charges of 6H-SiC p$^+$n diode detectors measured using the standard charge collection system. Degraded CCE by electron irradiation is restored after annealing at 373 and 473K (b) Cumulative collected charge of unirradiated and electron-irradiated 6H-SiC p$^+$n diode detectors.

conduction or valence bands. The detrapped charges are quickly swept by the electric field in the depletion region and collected. The slower components of cumulative collected charges of the electron-irradiated diode detector depend on temperature because the emission rates of the defects vary with temperature.

In order to analyze the slower components of cumulative collected charges, we have employed a rate window analysis in the same manner as the conventional DLTS. Figure 2 shows the single alpha particle induced charge transient spectra of the unirradiated and electron-irradiated diode detectors with and without annealing. The cumulative collected charges are sampled at 5 and 25 μs. A small peak labeled X_0 is observed in the unirradiated diode detector at around 200K. X_0 might be an as-grown or process induced defect. The activation energy of X_0 is estimated to be 0.34 eV from the Arrhenius plot. However, we cannot determine that the activation energy is from E_c or E_v because the detrapped charges could be electrons or holes or both. After the electron irradiation, two peaks labeled X_1 and X_2 emerge at around 200 and 270K, respectively. We could not estimate the activation energy of defect X_1 because of its broad peak. The activation energy of defect X_2 is estimated to be 0.50 eV. After annealing at 373K, peak X_1 becomes more distinct and its activation energy is now estimated to be 0.27 eV, and the intensity of peak X_2 decreases. Defects X_1 and X_2 are further removed by annealing at 473K, and the spectrum measured after annealing at 473K is almost identical to that of the unirradiated one.

It should be noted that the slower components of cumulative collected charges were not measured as the collected charges using the standard charge collection system because the signals of cumulative collected charges were gated with the shaping time, which was set to 0.5 μs in this study, of the spectroscopic amplifier. Therefore, the defects detected here by the charge transient spectroscopy are obviously responsible for the decreased CCE. Since peak X_2 appears near the room temperature, X_2 is more critical to the charge collection at room temperature. As the intensity of peak X_2 is decreased by annealing at 373K and further by annealing at 473K, the CCE are restored as shown in Fig.1 (a) and Fig. 2.

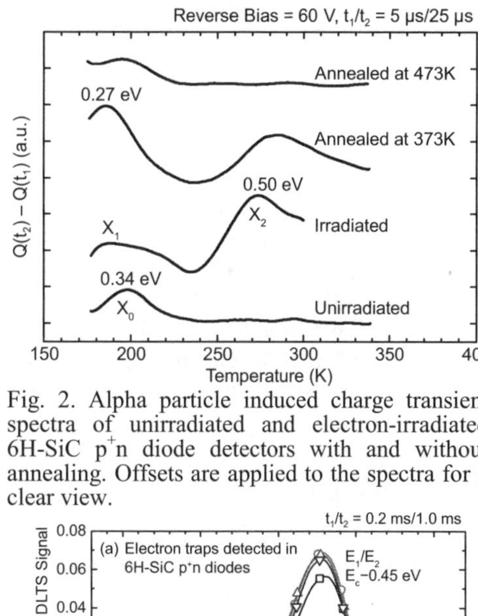

Fig. 2. Alpha particle induced charge transient spectra of unirradiated and electron-irradiated 6H-SiC p$^+$n diode detectors with and without annealing. Offsets are applied to the spectra for a clear view.

We also carried out DLTS on the 6H-SiC p$^+$n and n$^+$p diodes to further understand the defects revealed by the charge transient spectroscopy. Figure 3 shows the DLTS spectra of the unirradiated and the electron-irradiated diodes with and without annealing. The capacitance transients were sampled at 0.2 and 1.0 ms. Accurate capacitance transients of the n$^+$p diodes could not be measured at temperatures below 150K because of their too small capacitances due to the carrier freeze-out. As shown in Fig. 3 (a), an electron trap labeled E_1/E_2 with an energy level of E_c-0.45 eV is observed in the unirradiated p$^+$n diode. After the electron irradiation, the intensity of E_1/E_2 becomes larger, and a new electron trap labeled E_i with an energy level of E_c-0.51 eV emerges. A shoulder also appears on the lower temperature side of E_1/E_2. The intensity of E_i is reduced by annealing at 373K. Defect E_i and the

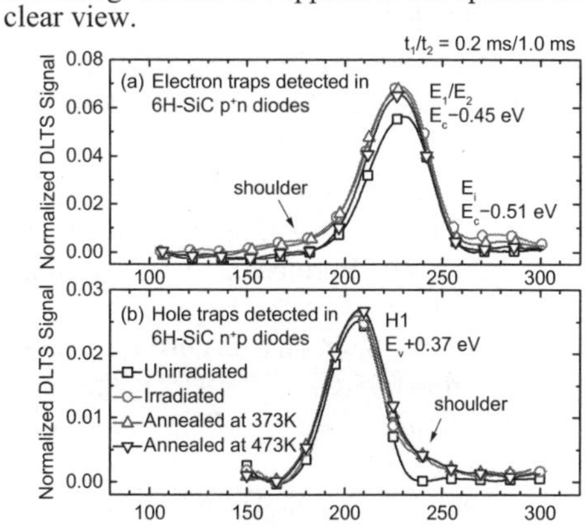

Fig.3 DLTS spectra of unirradiated and electron-irradiated (a) 6H-SiC p$^+$n diodes and (b) 6H-SiC n$^+$p diodes with and without annealing.

shoulder on the lower temperature side of E_1/E_2 are almost annealed out at 473K, while defect E_1/E_2 is still distinct. The energy levels and annealing behaviors of defects E_1/E_2 and E_i are consistent with the results reported in the literatures [4, 5].

In the unirradiated n^+p diode, a hole trap labeled H_1 with an energy level of E_v+0.37 eV is observed as shown in Fig. 3 (b). After the electron irradiation, a shoulder appears on the higher temperature side of peak H_1. Since no significant change is seen in the spectra after annealing at temperature up to 473K, the hole trap H_1 and the shoulder are unlikely to be responsible for the decreased CCE of the electron-irradiated diode detector.

The activation energy and annealing behavior of the electron trap E_i observed by the DLTS are very similar to those of defect X_2 detected by the charge transient spectroscopy. Both activation energies are around 0.50 eV, and both defects are almost completely removed by annealing at 473K. Therefore, we attribute defect E_i to X_2, and conclude that defect E_i is responsible for the decreased CCE of the electron-irradiated diode detector. The activation energy of hole trap H_1 is not much different from that of X_0. The shoulder seen on the lower temperature side of E_1/E_2 and X_1 are removed by annealing at 473K. However, further investigations are required to fully associate defects X_0 and X_1 to the traps identified by the DLTS. Nevertheless, since electron trap E_1/E_2 and all the hole traps are not removed by annealing at 473K, the effect of these traps on the CCE is insignificant. This may be due to their smaller carrier capture cross sections. The traps, which can capture carriers in the depletion region, must have large capture cross sections because carriers in the depletion region are swept out by electric field within several nanoseconds as mentioned earlier. The capture cross section of E_i is indeed larger by two orders than that of E_1/E_2 [4, 5].

Conclusion

Defects in the electron-irradiated 6H-SiC diodes have been studied by the single alpha particle induced charge transient spectroscopy and DLTS. The activation energy and annealing behavior of defect X_2 detected by the charge transient spectroscopy are very similar to those of electron trap E_i detected by the DLTS. After annealing at 473K both defects X_2 and E_i are annealed out and the degraded CCE is restored. Therefore, defect X_2 is attributed to electron trap E_i, and responsible for the decreased CCE of the electron-irradiated diode detector. It is also suggested by this study that the 6H-SiC diode detector degraded by frequent use may be "refreshed" by low-temperature annealing of the detector.

Acknowledgement

This work was supported in part by Grant-in-Aid for JSPS Fellows 217821 and Scientific Research (B) 21360471 from the Ministry of Education, Science, Sports and Culture in Japan.

References

[1] P.J. Sellin, and J. Vaitkus, Nucl. Instr. and Meth. A, 557 (2006) 479-489.

[2] F. Nava, G. Bertuccio, A. Cavallini, and E. Vittone, Meas. Sci. Technol. 19 (2008) 102001-1-25.

[3] N. Iwamoto, S. Onoda, T. Ohshima, K. Kojima, A. Koizumi, K. Uchida, and S. Nozaki, Mater. Sci. Forum 645-648 (2010) 921-924.

[4] C. Hemmingsson, N. T. Son, O. Kordina, E. Janzén, and J. L. Lindström, J. Appl. Phys. 84 (1998) 704-708.

[5] M. Gong, S. Fung, C. D. Beling, and Z. You, J. Appl. Phys. 85 (1999) 7604-7608.

[6] S. Sasaki, K. Kawahara, G. Feng, G. Alfieri, and T. Kimoto, J. Appl. Phys. 109 (2011) 013705-1-6

[7] J. S. Laird, C. Jagadish, D. N. Jamieson, and G. J. F Legge, J. Phys. D: Appl. Phys. 39 (2006) 1342-1351; J. Phys. D: Appl. Phys. 39 (2006) 1352-1362.

[8] N. Iwamoto, S. Onoda, T. Makino, T. Ohshima, K. Kojima, A. Koizumi, K. Uchida, and S. Nozaki, IEEE Trans. Nucl. Sci. 58 (2011) 305-313.

Materials Science Forum Vols. 717-720 (2012) pp 271-274
© (2012) Trans Tech Publications, Switzerland
doi:10.4028/www.scientific.net/MSF.717-720.271

Effects of Sacrificial Oxidation on Characterization of Defects in High-Purity Semi-Insulating 4H-SiC by Discharge Current Transient Spectroscopy

Seiji Nishikawa[*a], Ryota Okada and Hideharu Matsuura[b]

Osaka Electro-Communication University, 18-8 Hatsu-cho, Neyagawa, Osaka, 572-8530, Japan

[a]m10712@oecu.jp, [b]matsuura@isc.osakac.ac.jp

Keywords: High-purity semi-insulating 4H-SiC, Sacrificial oxidation, Discharge Current Transient Spectroscopy.

Abstract. To determine the energy levels of intrinsic defects in high-purity semi-insulating 4H-SiC, we apply discharge current transient spectroscopy (DCTS), a graphical peak analysis method based on the transient reverse current of a Schottky barrier diode, because transient capacitance methods such as deep level transient spectroscopy and isothermal capacitance transient spectroscopy are feasible only in low-resistivity semiconductors. The reverse current consists of the reverse current through the balk and the surface leakage current of the diode. It is elucidated that the sacrificial oxidation could dramatically reduce the surface currents of diodes in the case of high-purity semi-insulating 4H-SiC, suggesting that the densities and emission rates of traps in the bulk of the SiC can be determined from the transient reverse current.

Introduction

We have investigated X-ray detectors operating at room temperature using semi-insulating SiC. Semi-insulating semiconductors are essential for next-generation semiconductor devices. However, there are electrically active intrinsic defects (i.e., traps) in these semiconductors. Because these traps behave as generation centers in the depletion region of the diode and produce a generation current in the reverse-biased diode, they degrade the performance of X-ray detectors. To determine the possibility of semi-insulating SiC being used as a portable X-ray detector operating at room temperature, it is necessary to investigate these traps.

Although transient capacitance methods, e.g., deep level transient spectroscopy (DLTS) [1], are powerful methods to characterize traps in low-resistivity semiconductors, they are not feasible in semi-insulating semiconductors because the measured capacitance of the diode fabricated using semi-insulating semiconductors is determined by the thickness of the diode, not by the depletion region of the junction [2,3]. Thermally stimulated current (TSC) [4] is suitable for characterizing traps in semi-insulating semiconductors. However, it is difficult to analyze experimental TSC data when traps with similar emission rates are included in the semiconductor. Moreover, because the effect of pyroelectric currents and the temperature dependence of the steady-state leakage current must be considered in the TSC analysis, an isothermal measurement is more suitable for characterizing traps than is TSC.

Without any assumptions regarding traps, one of the authors has proposed a graphical peak analysis method to determine the densities and emission rates using the isothermally measured transient current, referred to as discharge current transient spectroscopy (DCTS), and has applied it to high-purity semi-insulating SiC [5,6]. However, these transient currents consisted of currents not only in the bulk, but also on the surface of the diode. The surface leakage current results mainly from surface defects formed by mechanical and chemical polishing of SiC surfaces. To remove those surface defects, the SiC surfaces including the defects are oxidized, and then the oxidized layers are etched using HF, which is referred to as sacrificial oxidation.

In this study, we report on our investigation of the effects of sacrificial oxidation on the characterization of high-purity semi-insulating 4H-SiC

Discharge Current Transient Spectroscopy

DCTS can determine the densities and emission rates of defects in a semi-insulating semiconductor from the transient reverse current $I_{\text{dis}}(t)$ of a Schottky barrier diode at a constant temperature [7,8]. In DCTS, the following evaluation function is defined using the experimental $I_{\text{dis}}(t)$:

$$D(t,e_{\text{ref}}) \equiv \frac{t}{qS}\left[I_{\text{dis}}(t) - I_{\text{S}}(V_{\text{R}})\right]\exp(-e_{\text{ref}}t+1) , \tag{1}$$

where $I_{\text{S}}(V_R)$ is the steady-state leakage current at a reverse bias voltage (V_{R}), q is the electron charge, S is the electrode area, and e_{ref} is the peak-shift parameter [9]. From the time ($t_{\text{peak}i}$) and value ($D(t_{\text{peak}i}, e_{\text{ref}})$) of each peak, we can accurately determine the emission rate ($e_{\text{t}i}$) and sheet density ($N_{\text{t}i}$) of the corresponding defect as:

$$e_{\text{t}i} = \frac{1}{t_{\text{peak}i}} - e_{\text{ref}} \tag{2}$$

and

$$N_{\text{t}i} = \frac{D(t_{\text{peak}i}, e_{\text{ref}})}{1 - e_{\text{ref}}t_{\text{peak}i}} , \tag{3}$$

respectively. When more than one trap with comparable emission rates exist in a semi-insulating semiconductor, the DCTS signal became broader and the individual peaks cannot be distinguished. In this case, the density and emission rate of each trap can be determined from the maximum of the DCTS signal when e_{ref} is changed continuously [7].

Experiment

A 0.3713-mm-thick high-purity semi-insulating 4H-SiC wafer was purchased from Cree Inc., and was cut to a size of 5 mm × 5 mm. The diode without sacrificial oxidation was fabricated as follows: Ni electrodes with a radius of 1.25 mm were evaporated onto both sides of the chip. Since thermal treatment was not carried out, the diodes work as a back-to-back diode. After the current-voltage ($I-V$) characteristics and transient current were measured, the diode with sacrificial oxidation was fabricated. After the Ni electrodes were removed using HCl, each side of the chip surface was oxidized in an O_2 atmosphere at 1273 K for 2 hours. After the sacrificial oxide layer on the chip was removed using HF, Ni electrodes with the same radius were evaporated onto both sides of the chip.

The $I-V$ characteristics of the diodes were measured from 0 to -100 V using a Keithley 236 source-measure unit (SMU236). We waited for 30 s, referred to as the delay time, to measure the reverse current at each bias after the bias was changed by -1 V [6]. After the bias voltage was rapidly changed from 0 V to –100 V, $I_{\text{dis}}(t)$ was measured at V_{R} of -100 V at 393 K. The densities and emission rates of traps in semi-insulating 4H-SiC were determined by DCTS.

Results and Discussion

Figure 1 shows the $I-V$ characteristics obtained devices fabricate on surfaces with and without sacrificial oxidation, denoted by open circles and solid diamonds, respectively. The leakage currents at -100 V were -122.9 and -15.8 pA without and with sacrificial oxidation, respectively. It is clear from Fig. 1 that the sacrificial oxidation could dramatically reduce the current proportional to the bias in the diode without sacrificial oxidation, which might be a surface current.

Figure 2 shows the transient reverse current obtained devices fabricate on surfaces with and without sacrificial oxidation, denoted by open circles and solid diamonds, respectively. The transient reverse currents at 1000 s, which are assumed to be $I_{\text{S}}(V_{\text{R}})$, were -115.5 and -26.0 pA without and with sacrificial oxidation, respectively. Since $I_{\text{S}}(V_{\text{R}})$ was reduced by sacrificial oxidation, we can measure $I_{\text{dis}}(t)$ due to the emission of traps in the bulk over a wide time range.

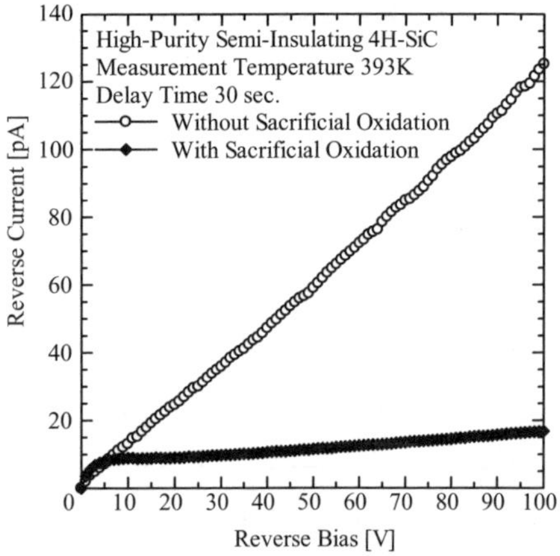

Fig. 1. $I-V$ characteristics for devices fabricated on surfaces with and without sacrificial oxidation.

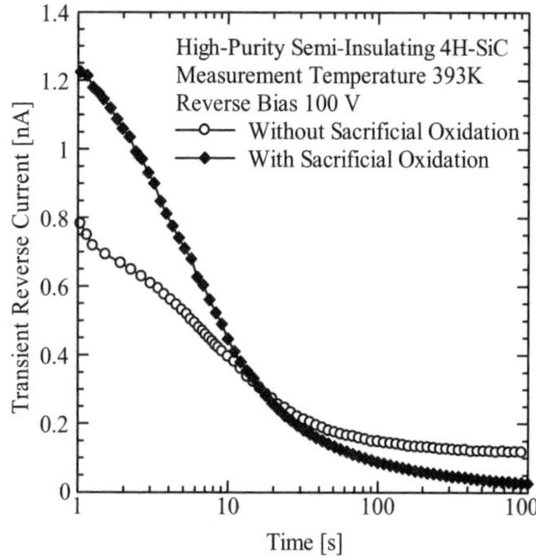

Fig. 2. Transient reverse current for devices fabricated on surfaces with and without sacrificial oxidation.

Figure 3 shows the DCTS signal with $e_{\text{ref}} = 0$ s^{-1}, denoted by the solid line. One peak and one shoulder appear in the figure. From Eqs. (2) and (3), e_{t1} and N_{t1} of the corresponding trap (Trap1) are determined as 5.24 × 10^{-3} s^{-1} and 2.21 × 10^{12} cm^{-2}, respectively. In the figure, the broken line represents the signal simulated with the obtained values using

$$D_{ti}(t, e_{\text{ref}}) = N_{ti} e_{ti} t \exp[-(e_{ti} + e_{\text{ref}})t + 1]. \tag{4}$$

It is clear from Eqs. (3) and (4) that the peak value of the component of the DCTS signal corresponding to each trap can be changed by e_{ref}. Figure 4 shows the DCTS signal with $e_{\text{ref}} = 0.02$ s^{-1}, denoted by the solid line. The peak in Fig. 4 corresponds to the shoulder in Fig. 3. Using Eqs. (2) and (3), e_{t2} and N_{t2} of the corresponding trap (Trap2) are determined as 9.20 × 10^{-2} s^{-1} and 1.46 × 10^{12} cm^{-2}, respectively. In the figure, the broken line represents the signal simulated with the obtained values using Eq. (4). Since the measured signal is larger than the broken line for times < 10 s, the DCTS signal is considered to be affected by other traps. Figure 5 shows the DCTS signal with $e_{\text{ref}} = 0.126$ s^{-1}, denoted by the solid line. Using Eqs. (2) and (3), e_{t3} and N_{t3} of the corresponding trap (Trap3) are determined as 1.27 × 10^{-1} s^{-1} and 1.29 × 10^{12} cm^{-2}, respectively. In the figure, the broken line represents the signal simulated with the obtained values using Eq. (4). Since the broken

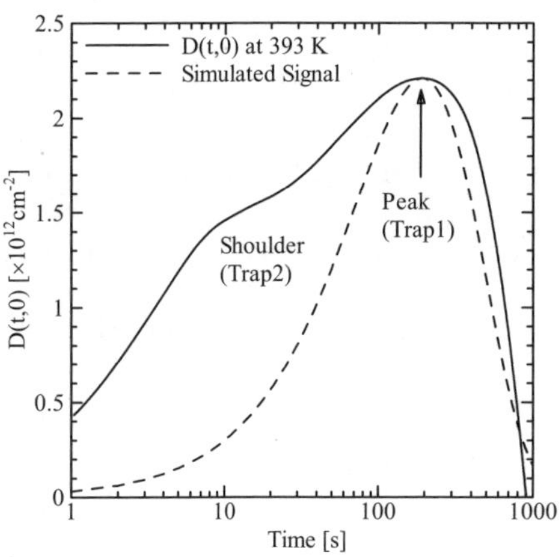

Fig. 3. DCTS signal with $e_{\text{ref}} = 0$ s^{-1}.

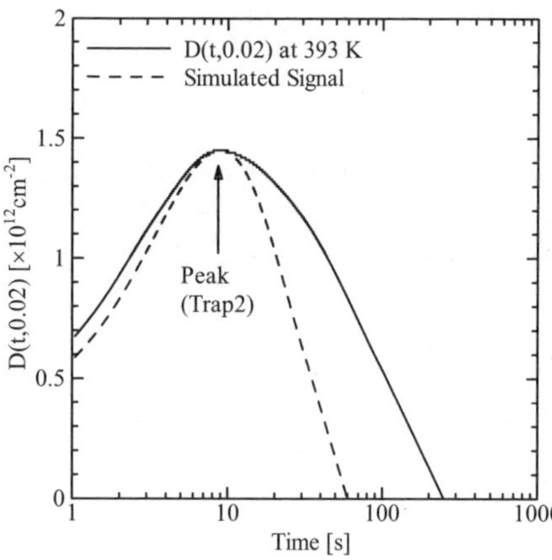

Fig. 4. DCTS signal with $e_{\text{ref}} = 0.02$ s^{-1}.

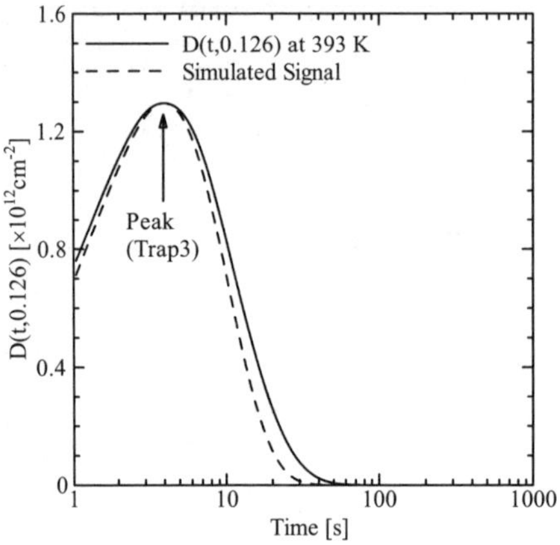

Fig. 5. DCTS signal with $e_{\mathrm{ref}} = 0.126\ \mathrm{s}^{-1}$.

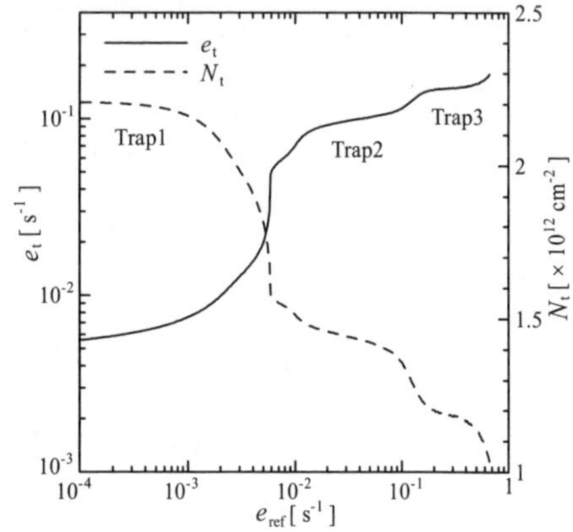

Fig. 6. The e_{ref} dependence of e_{ti} and N_{ti} determined from the maximum of the DCTS signal.

line is fitted with the solid line, the DCTS signal is considered to be affected by other traps with comparable emission rates. Judging from these results, at least three types of traps are included in the high-purity semi-insulating 4H-SiC.

Figure 6 shows the e_{ref} dependence of e_{ti} (solid line) or N_{ti} (broken line) determined from the maximum of the DCTS signal calculated using each e_{ref}. In this analysis [7], the e_{ref} range of almost constant e_{t} and N_{t} corresponds to a trap with a discrete energy level. At least three discrete values of e_{t} or N_{t} clearly appear in the figure, and are close to the obtained values in Figs. 3-5.

Summary

It is elucidated that sacrificial oxidation could dramatically reduce the surface currents of diodes in high-purity semi-insulating 4H-SiC, suggesting that the densities and emission rates of traps in the bulk of the SiC can be determined from the transient reverse current. We applied DCTS to determine those properties of traps, and moreover could characterize three types of traps with comparable emission rates.

References

[1] D. V. Lang, J. Appl. Phys. 45 (1974) 3023-3032.

[2] H. Matsuura, T. Okuno, H. Okushi, and K. Tanaka, J. Appl. Phys. 55 (1984) 1012-1016.

[3] H. Matsuura, N. Minohara, and T. Ohshima, J. Appl. Phys. 104 (2008) 043702 1-6.

[4] R. R. Hearing and E. N. Adams, Phys. Rev. 117 (1960) 451-454.

[5] H. Matsuura, M. Takahashi, Y. Kagawa, S. Tano, and T. Miyake, Mater. Sci. Forum 615-617 (2009) 385-388.

[6] H. Matsuura, Y. Kagawa, M. Takahashi, S. Tano, and T. Miyake, Jpn. J. Appl. Phys. 48 (2009) 056504 1-4.

[7] H. Matsuura, T. Hase, Y. Sekimoto, M. Uchida, and M. Shimizu, J. Appl. Phys. 91 (2002) 2085-2092.

[8] H. Matsuura, M. Yoshimoto, and H. Matsunami, Jpn. J. Appl. Phys. 34 (1995) L185-L187.

[9] H. Matsuura and K. Segawa, Jpn. J. Appl. Phys. 39 (2000) 178-179.

Materials Science Forum Vols. 717-720 (2012) pp 275-278

doi:10.4028/www.scientific.net/MSF.717-720.275

Investigation of Additional States in the Silicon Carbide Surface after Diffusion Welding

János Mizsei[1,a], Oleg Korolkov[2,b,*], Natalja Sleptsuk[2,c], Jana Toompuu[2,d] and Toomas Rang[2,e]

[1]Department of Electron Devices, Technical University of Budapest, Goldmann Gytér 3, H-1521 Budapest, Hungary

[2]Department of Electronics, Tallinn University of Technology, Ehitajate tee 5, Tallinn, 19086, Estonia

[a]mizsei@eet.bme.hu, [b]oleg.korolkov@ttu.ee [c]natalja@elin.ttu.ee, [d]jana@elin.ttu.ee, [e]trang@elin.ttu.ee, *corresponding author

Keywords: Diffusion welding, deep levels, surface states, DLTS measurements, Kelvin probe.

Abstract. This paper is a summary of the experimental study of deep levels in a SiC crystal lattice caused by diffusion welding (DW). Investigations were carried out by DLTS and Kelvin Probe methods. Investigations revealed that DLTS method is not applicable for identification of surface states. Research conducted by the Kelvin Probe method has shown an increase in the density of surface states after the diffusion welding from $2x10^{15}$ cm^{-2} to $3.5x10^{16}$ cm^{-2}.

Introduction

The analysis of current-voltage characteristics of the Al-SiC contacts, which we carried out in our earlier works [1], has led to speculation about the SiC surface contact area condition after the diffusion welding. It has been suggested that as a result of thermal deformation during the diffusion welding, a very thin layer is formed near the silicon carbide surface with a structure close to amorphous. Electron microscopy studies carried out later [2] have confirmed this assumption. The TEM examinations revealed a highly disordered crystal layer of ~ 25 nm thickness that was formed directly in the SiC surface after diffusion welding. A large concentration of defects in the crystal structure should lead to the appearance of numerous additional allowed states, so-called deep centres, in the band gap of the defect zone. To identify and study these deep levels, we used two methods - the method of relaxation spectroscopy, and the Kelvin Probe or vibrating capacitor (VC) method. The results of these mutually complementary studies are presented in this paper.

Relaxation Spectroscopy Study

The most widely used method of determining the parameters of deep levels is currently the method of relaxation spectroscopy of deep levels - Deep Level Transient Spectroscopy (DLTS). DLTS was first introduced by D.V. Lang [3] in 1974 and has evolved into a routine characterisation technique for semiconductor materials and devices. The DLTS method is based on the concept of relaxation of capacitance in a structure with a potential barrier, e.g. a Schottky barrier or an asymmetric p-n junction.

Sample Preparation

For study, the 4H-SiC structure of thickness 0.35 mm, net nitrogen doping density in epi-layer $N_d \sim 1x10^{15}$ cm^{-3} and epi-layer thickness of 6 μm was used. The overall size of the chips was 1.4 x 1.4 mm^2, and the contact diode area was 1 x 1 mm^2. DLTS spectra measurements were performed with a computer-controlled deep-level spectrometer DLS-83D (Semilab, Semiconductor Physics laboratory, Inc., Hungary).

Initially, all the necessary measurements were made on the original chip, and then evaporated aluminium was removed by etching and aluminium foil of 30 μm thickness was welded to the contact surface (the welding process is described in [1]). The same measurements were then repeated on a chip with a diffusion-welded contact.

Comparison of Deep Level Spectra

Due to earlier TEM studies [2] the distortions in the subcontact zone of the crystal structure that arise during the process of diffusion welding can be considered as reliably established fact. As is well known, distortions of crystal structure are accompanied by the formation of various defects, which, in turn, may result in deep levels in the bandgap of the semiconductor. Therefore, it would be logical to expect changes in the spectrum of deep levels after diffusion welding. For this purpose, DLTS temperature and frequency spectra were measured for the samples described above. The appearance of a temperature signature before and after diffusion welding in the form of the specific relation dC/C is shown in Fig. 1.

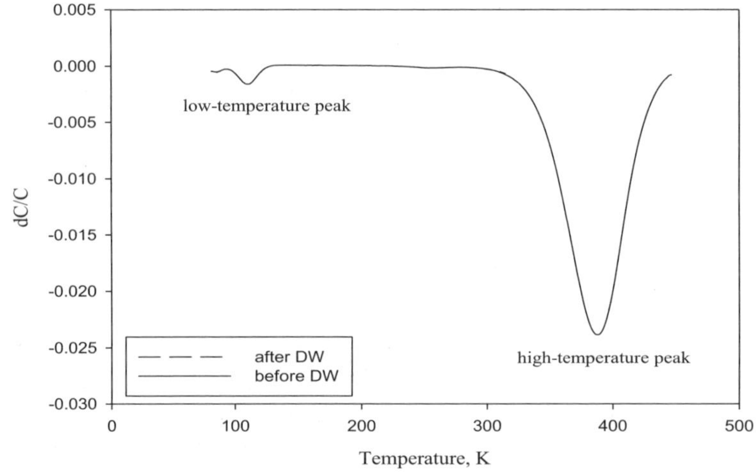

Fig. 1. Temperature scans for Schottky diode before and after DW (U_r = -5 V, f=2000 Hz).

The spectra show that the positions of the peaks before and after diffusion welding coincide both in temperature and frequency scans. A somewhat lower value of dC/C for scans after diffusion welding is associated with an increase in capacity due to lowering of the built-in voltage. The negative amplitude of transient capacitance indicates that these are majority carrier (*id est* electron) traps. The activation energy for the low-temperature peak, obtained from Arrhenius plots, is $E_a = E_c - 0.148$ eV, and for the high-temperature peak $E_a = E_c - 0.46$ eV. Based on the literature data [4], the deep levels with corresponding activation energies are most likely the S centres. These types of defects are characteristic of the epitaxial layer and, therefore, they are not brought in by the diffusion welding. This fact seems to be in disagreement with the fact of the existence of the damaged layer observed in TEM studies. But in reality it is not so. Residual disturbances of the crystalline structure after DW propagate into the bulk SiC for a depth of 25 nm. Of course, it would be wrong to expect the appearance of an exponential relaxation of capacitance in such a narrow zone using the DLTS method. Besides, splitting of energy levels in the interfacial zone of the semiconductor makes it impossible to use the Lang method for fixing the deeper levels in the ultra-thin zone of the semiconductor adjacent to the contact. Hence, one should try to approach the contact area not from the bulk of the semiconductor, but from the surface. One such method available is the vibrating capacitor method.

Vibrating Capacitor (VC) and Surface Photo Voltage (SPV) Method

Basic Principle of Vibrating Capacitor Measurements

The vibrating capacitor (in other words, the Kelvin Probe) is quite an old method and thanks to this is well-developed and adapted for the surface investigations. The basis of this method was first postulated by the renowned Scottish scientist Lord Kelvin, in 1861. Modern implementation of this method is described in detail in [5] and illustrated in Fig. 2.

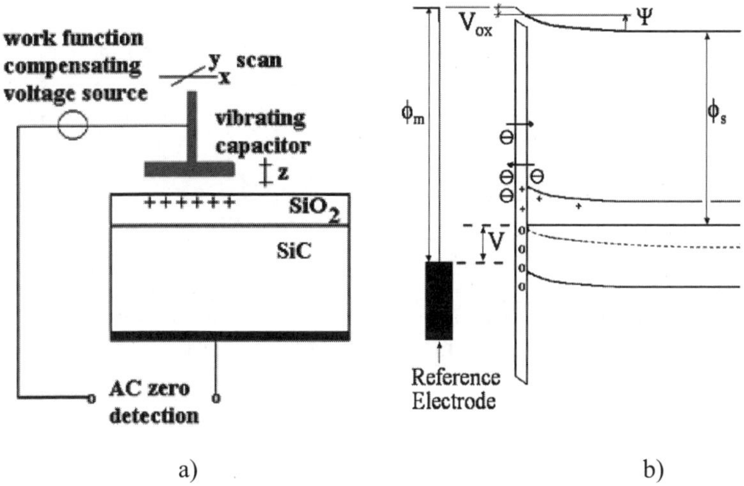

Fig. 2. The principle of the vibrating capacitor method (a) and band diagram of the semiconductor surface (b).

The method is based on the compensation of the electric field between a vibrating (reference) electrode and the surface to be investigated (Fig. 2). The feedback system changes the compensating potential V until the alternating current induced by vibrating electrode is near zero. The compensating voltage then becomes equal to the total surface potential. Assuming zero electric field above the surface, the surface potential depends on the surface charge (Q) according to

$$V = \frac{Qd_{ox}}{\varepsilon_{ox}} + \Psi(Q + Q_{ss}) + \Phi_{ms}, \quad (1)$$

i.e., the total potential V is the sum of the voltage drop in the oxide, ($V_{ox} = Qd_{ox}/\varepsilon_{ox}$), the voltage drop in the semiconductor (Ψ, surface potential barrier), and the work function difference between the reference electrode and the semiconductor (Φ_{ms}) (Fig. 2). In Eq. 1 d_{ox} is the oxide thickness and $\varepsilon_{ox} = \varepsilon\varepsilon_0$ is the oxide permittivity.

Combining the Kelvin probe and the Einstein Photoelectric Effect, the next step of measurement can be realised. In the presence of high intensity light excitation and due to high carrier concentrations, the band bending disappears (the second term in Eq. 1 also disappears). The oxide surface potential V_{FB} can be measured by the photo-capacitive current generated by light pulses:

$$V_{FB} = \frac{Qd_{ox}}{\varepsilon_{ox}} + \Phi_{ms}. \quad (2)$$

Thus, the vibrating capacitor measurement in dark combined with light excitation measurement, gives a surface potential barrier as:

$$\Psi = V - V_{FB}, \quad (3)$$

where the surface potential barrier is the difference between the dark and light values of the measured oxide surface potential.

Material, Sample Preparation and Experimental Results

The surface potential barrier measurements were performed using the Kelvin probe Trek model 320 c type equipped with a 4 x 4 mm reference electrode made of glass, covered by a transparent and conductive tin-oxide thin film. For test measurements the Cree 4H-SiC epi-chips (net doping density in epi-layer $N_d \sim 1x10^{15}$ cm^{-3}) of 5 x 5 mm^2 area were used with a remote metallisation before and after diffusion welding. The results of surface potential measurements in dark and illumination modes (a standard 36 W microscope bulb) after the appropriate mathematical analysis [5] are presented in Table 1.

Table 1. The results of the calculation of surface charges.

DATA			
Initial	E_g, [eV] (bandgap)	3.23	
	T, [K] (temperature)	300	
	n_i, [cm^{-3}] (intrinsic carrier concentration)	10^{-7}	
	L_D, [cm] (intrinsic Debye length)	$9.3x10^4$	
Calculated		**Before DW**	**After DW**
	Ψ, [mV] (surface potential)	-600	-750
	$F(\Psi;\Phi_b)$ (dimensionless function)	$1.1x10^{15}$	$1,99x10^{16}$
	$N=Q/q$, [cm^{-2}] (surface charges density, q – electron charge)	$2x10^{15}$	$3.5x10^{16}$

Conclusion

The DLTS investigations failed to detect the additional states after DW. As seen from the VC results presented in Table 1, after DW an increase in surface potential was detected as well as an increase (more than one order of magnitude) of the surface charge density. Based on the experimental conditions, we have no physical reason to explain the increase in surface potential after DW by increasing the surface barrier. Thus, the increase in the surface potential should be due to the additional charge on the states arising after diffusion welding because of lattice distortions in the surface layer of the semiconductor. Thus, the real surface that separates the metal and the semiconductor is a thin (nanometers), but still three-dimensional buffer area. This area contains a high number of structural defects, which are the main sources of surface states. The name of "surface states" attributed to these states is quite arbitrary, and the density of these states in some cases may exceed the possible density of Tamm states ($\sim 10^{15}$ cm^{-2}).

References

[1] O. Korolkov, T. Rang, N. Kuznetsova and J. Ruut, Mater. Sci. Forum 457-460 (2004) 857-860.

[2] O. Korolkov, N. Kuznetsova, A. Sitnikova, M. Viljus and T. Rang, Mater. Sci. Forum 600-603 (2008) 647-650.

[3] D.V. Lang, J. Appl. Phys. 45 (1974) 3023-3032.

[4] A. Castaldini *et al.*, Appl. Surface Science 187 (2002) 248-252.

[5] J.Mizsei, Solid-State Electronics, 44 (2000) 509-513.

Materials Science Forum Vols. 717-720 (2012) pp 279-284

doi:10.4028/www.scientific.net/MSF.717-720.279

Long Carrier Lifetimes in n-type 4H-SiC Epilayers

P.B. Klein[1, a]

[1]Naval Research Laboratory, Washington DC 20375, USA

[a]paul.klein@nrl.navy.mil

Keywords: 4H-SiC, carrier lifetime, defects, recombination

Abstract. Recent advances in preparing n-type 4H-SiC with long carrier lifetimes have greatly enhanced the possibility of realizing commercially available, very high voltage and high power solid state switching diodes. For the range > several kV, vertical bipolar structures are required with drift layers exhibiting carrier lifetimes ≥ several μsec. Recently, low-doped epilayers with carrier lifetimes in excess of this have been demonstrated, thus approaching a goal that has been pursued for over a decade. Historically, the short lifetimes in early epitaxial layers (a few hundred nsec) were eventually identified with the V_c-related $Z_{1/2}$ lifetime killer. Current strategies to minimize this defect are an essential ingredient in the procedure for obtaining long-lifetime material. In order to optimize the attainable lifetimes, it has been shown that in addition to low $Z_{1/2}$ levels, very thick layers are required to minimize the effects of recombination in the substrate and surface passivation is also necessary to minimize surface recombination (S < 1000 cm/sec).

Introduction

Because of its superior material properties, SiC is considered the material of choice to replace Si for high voltage switching applications. There has been significant progress made in this area over the past several years, as evidenced by the current commercial availability of Schottky barrier diodes operating at voltages up to 1.7kV. For high power applications operating at > several kV, vertical bipolar devices (e.g. p-i-n, IGBT….) are necessary to provide high conductivity in the on-state and still hold off high voltage in the off-state. These devices require a thick, low-doped drift region with long minority carrier lifetime to support effective conductivity modulation and thus minimize dissipative losses. In order to insure device uniformity and to control switching times, even longer lifetimes are required for lifetime control [1]. The 4H polytype is much preferred in these applications because the mobility is much higher, particularly along the c-axis. Consequently, there has been much more effort focused on the material development of 4H-SiC than on 6H-SiC. Much of this effort has been expended toward determining the mechanisms that limit the carrier lifetime and the materials growth/processing techniques that can overcome these limitations.

In general, reduced carrier lifetimes result from fast carrier recombination. Since direct electron-hole recombination is very slow in indirect band gap semiconductors such as SiC, much faster processes that compete for the free carriers determine the measured recombination rate and, consequently, the carrier lifetime. This often occurs through electron-hole recombination at point defects or defect complexes (Shockley-Reed-Hall recombination [2,3]), which are usually assumed to be distributed roughly uniformly throughout the sample. Recombination at structural defects can also produce localized regions with short lifetimes [4]. Since many of these defects propagate from the substrate into the epilayer during growth, the availability of high quality substrates has greatly reduced their effects. However, as carrier lifetimes become longer, the slower recombination rate from structural defects could become a limiting factor. Individual structural defects in the epilayer, such as stacking faults, can reduce the lifetime in the region around the defect [5]. Recombination limiting the carrier lifetime usually occurs at defects in the bulk, on the sample surface, in the substrate or at the epilayer-substrate interface. While the radiative recombination rate is usually considered small in indirect materials, at high injection levels it can also affect the measured lifetime due to the bimolecular recombination kinetics. At even higher injection levels ($>10^{18}cm^{-3}$ [6]) the carrier lifetime can be dominated by Auger recombination.

Some of the major milestones reached in improving the carrier lifetime and leading up to the current state of the art are briefly reviewed and the nature of the carrier recombination mechanisms considered. The recent advances in producing long-lifetime 4H-SiC epilayers, as well as their incorporation into device structures, are discussed.

Defects and Carrier Lifetime

Investigation of the carrier lifetime began [7] shortly after the first demonstration of homoepitaxial growth of 4H SiC in 1993 [8]. With high quality layers available, work focused on characterizing the major defect levels in the material, primarily by DLTS measurements, and on associating these spectral signatures with chemical identities. In particular, the DLTS spectrum in n-type epilayers was found to be dominated by two electron traps: $Z_{1/2}$ and $EH_{6/7}$, which were suspected to be the lifetime killers. Studies of the effect on the defect concentrations of varying growth conditions (C/Si ratio [9], growth temperature [10]) or post-growth annealing [11] led to the conclusion that both defects were probably carbon vacancy (V_c) -related. Further evidence for this was obtained by Storasta et al. [12], showing that these two defects (as well as others) were enhanced by low energy electron irradiation where only the C atoms were displaced. These two defects are currently considered to be V_c-related, although their exact structure has yet to be determined.

Efforts to improve the carrier lifetime focused initially on reducing the defect concentration by optimizing growth conditions. For example, Danno et al. [13] varied the C/Si ratio and were able to attain defect concentrations of (5-6)x10^{11}cm^{-3} and ≈ 2μsec carrier lifetimes. Studies of high temperature annealing [11] indicated that above 1600°C the concentrations of $Z_{1/2}$ and $EH_{6/7}$ defects began to decrease. Jenny et al. [14] employed this behavior to obtain low defect concentrations by annealing high-purity *bulk* samples at temperatures up to 2600°C. Substantial increases in carrier lifetime (from 10nsec to the μsec range) were reported.

From these and other observations [15], it appeared that these two defects were associated with the carrier lifetime. However, by carrying out DLTS measurements in a p-i-n structure under forward bias, it was shown [16,17] that the cross-section for capture of minority holes was large for $Z_{1/2}$ but small for $EH_{6/7}$, indicating that the latter has a minimal effect on the carrier lifetime. Direct correlations observed between the $Z_{1/2}$ concentration and the lifetime [17,18] confirmed that $Z_{1/2}$ was the bulk lifetime killer in n-type 4H-SiC.

Storasta and Tsuchida [19], following up on the previous electron irradiation studies [12], found that by annealing after C$^+$-ion implantation, $Z_{1/2}$ and $EH_{6/7}$ levels in the bulk decreased to below 5x10^{11}cm^{-3} and carrier lifetimes doubled. Their results, shown in Fig. 1, indicated that both implantation and annealing were required to observe the near-disappearance of the $Z_{1/2}$ and $EH_{6/7}$ DLTS signatures. They suggested that carbon interstitials (C_is), created by implantation damage, diffused into the bulk during the anneal and annihilated the V_c-related defects, thus increasing the carrier lifetime. The implantation and anneal parameters were later optimized [20] to reduce deep level concentrations below the detection limit. Following this, Hiyoshi and Kimoto [21] achieved similar results by oxidizing the sample surface at 1300°C. The resulting reduction in defect concentration at depths below the implantation profile was attributed to C_i and Si_i diffusion from

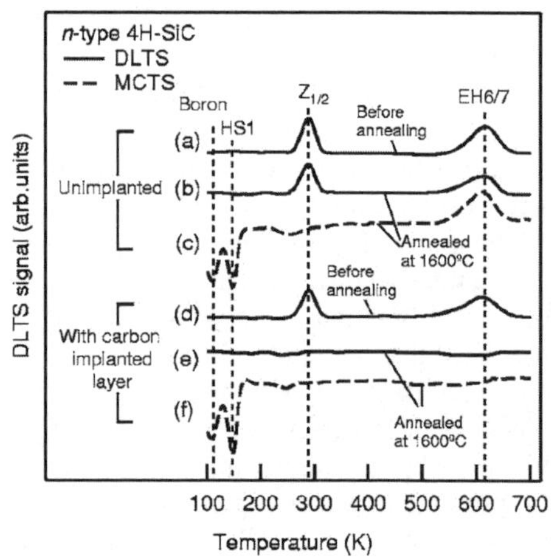

Fig. 1. From Fig. 3 of Ref. 19. DLTS spectra of unimplanted sample (a) before and (b) after annealing; sample with carbon implanted surface layer (d) before and (e) after annealing at 1600 °C. MCTS spectra of (c) unimplanted sample and (f) implanted sample after annealing and removal of the implanted layer.

the damaged oxidized layer into the bulk, leading to V_c annihilation. The carrier lifetime was observed to increase from 0.6 to 1.6μsec. Adding an Ar anneal step to eliminate an oxidation-induced defect (HK0) led to further lifetime enhancement [22]. These two related methods of deep level reduction rely on the same basic mechanism: Damage, induced either by ion implantation or oxidation, creates a high concentration of interstitials in the damaged layer. When diffused further into the bulk at high temperature, these defects annihilate V_c-related defects, thus reducing the concentration of the lifetime killers. This technique forms the basis for the current approach to preparing long-lifetime n-type 4H-SiC.

Limiting Recombination Mechanisms in Low- $Z_{1/2}$ 4H-SiC

The implantation- and oxidation-based procedures enabled the reduction of the bulk defect concentration ($Z_{1/2}$) to levels where it no longer controlled the carrier lifetime [1]. Consequently, it became important to determine what processes dominated recombination in this low-$Z_{1/2}$ material and what conditions would be necessary to achieve very long lifetimes. Kimoto et al. [23] and Klein et al. [24] both considered this problem from very different perspectives.

Kimoto et al. carried out a full calculation of the carrier diffusion equation, solving it through numerical simulation, assuming high injection conditions. From the simulated carrier decays it was evident that when the bulk lifetime is long, recombination of carriers in the substrate (particularly in thinner samples) and recombination at the sample surface dominated the lifetime. Fig. 2, taken from Fig. 3 of Ref. 23, shows simulated decay curves for (a) a 50μm thick sample and (b) a 200μm thick sample for a range of surface recombination velocities (SRV). The bulk lifetime was taken as 5μsec, shown as the dashed decay curve in the figures. The difference between the pure bulk decay (dashed line) and the simulated decay for SRV=0 reflects the carriers lost to recombination in the substrate. As noted by the authors, in thinner epilayers the substrate accounts for a very significant reduction in the carrier lifetime, even when the SRV=0, while for thicker epilayers the reduction in lifetime is not significant until the SRV becomes large. The authors also demonstrated that all measured carrier lifetimes will seriously underestimate the bulk lifetime unless the epilayer is very thick. It was shown that for the measured lifetimes to be within 20% of the actual bulk lifetime, the layer thickness must be at least four times larger than the ambipolar diffusion length. Using samples 50-148μm thick, the authors demonstrated excellent agreement between the simulations and experiment.

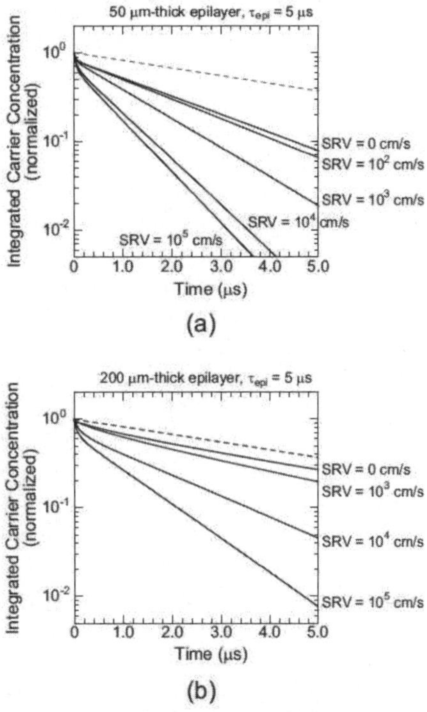

Fig. 2. (From Fig. 3 of Ref. 23). Calculated decay curves of excess carrier concentration in 4H–SiC for different SRVs. In this case, a carrier lifetime of 5.0 μs was used as a bulk lifetime in the (a) 50 μm and (b) 200-μm-thick epilayers. A single exponential decay with a lifetime of 5.0 μs is plotted by a broken line for a reference.

Klein et al. [24] studied the temperature dependence of the lifetime in relatively thin low-$Z_{1/2}$ samples (≈20-35μm) using primarily low-injection conditions, since at low injection the different recombination mechanisms that determine the temperature dependence of the carrier lifetimes can be more easily distinguished. The data were analyzed using an analytical approximation to the solution of the diffusion equation. Under low injection conditions, the band bending at the n^-/n^+ epilayer/substrate interface (≈7kT at 300K) repels minority holes, thus greatly reducing recombination in the substrate or at the interface. In Fig. 3, taken from Ref. 24, the solid symbols represent the experimental temperature dependent recombination rate. The dashed curve reflects the contribution to the recombination rate by carrier diffusion. This follows a power law dependence resulting from the temperature dependent diffusion coefficient [25]. The dotted curves represent the total simulated surface recombination rate that best fits the data, resulting from carrier diffusion and from recombination at the surface and at the substrate interface. The data fits well only for small values of the substrate interface recombination velocity, consistent with the low injection conditions and very long bulk lifetimes (>20μsec). The thermally activated behavior at low temperature results from surface band bending that acts as a capture barrier, reducing the surface recombination rate. At high injection the n^-/n^+ interface is driven into flat-band, and recombination at the substrate becomes large, as in Ref. 23.

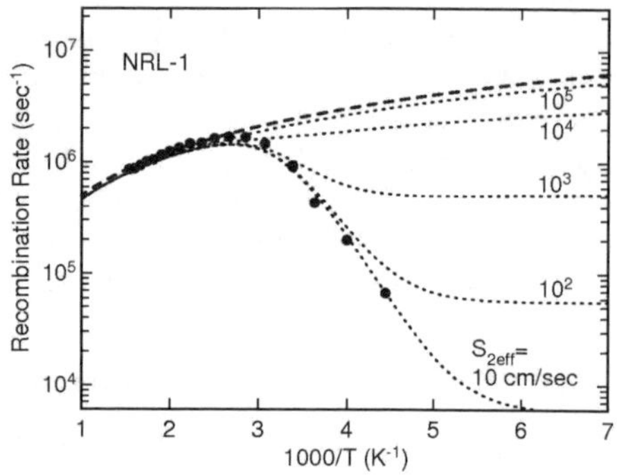

Fig. 3. (From Ref. 24). Temperature dependence of the measured inverse lifetimes (solid symbols) vs. inverse temperature for sample NRL-1. Dotted lines are fits to the data assuming recombination at the surface and substrate interface for varying interface recombination velocities. The dashed line indicates the diffusive component, while the exponential component results from surface capture.

The models employed in both of these approaches were used to give insight into the conditions necessary to obtain long carrier lifetimes. In both cases essentially the same conclusions were obtained: In order to obtain long-lifetimes, it is necessary to 1) reduce the $Z_{1/2}$ concentrations (*i.e.* through oxidation or implantation) well below that obtainable in as-grown material (≈10^{12}cm^{-3}), so that bulk recombination is negligible, 2) employ thick epitaxial layers (*e.g.* >100μm) to limit the effect of the substrate on the lifetime, and 3) limit the surface recombination velocity (by passivation) to < 1000 cm/sec.

Kimoto et al. [23] demonstrated these ideas in ≈150μm thick layers treated with their oxidation procedure, achieving 9.5μsec lifetimes, with actual bulk lifetimes estimated in excess of 30μsec. Effective lifetimes were shown to be further enhanced by surface passivation [26]: Using an oxide passivation annealed in NO, the surface recombination was reduced and an enhancement of the measured lifetime from 9.5μsec to ≈13 μsec was reported. Similarly, Miyazawa et al. [27] used thick C^+-implanted and annealed layers to demonstrate long lifetimes: With layer thicknesses of 265μm and 225μm (free-standing layer) effective carrier lifetimes of ≈19 μsec were demonstrated. Tsuchida et al. [28], employing low-pressure growth to achieve rapid growth rates, observed effective lifetimes approaching 27μsec in a 250μm thick sample after C^+-implantation and annealing. Both deep level reduction approaches have demonstrated long carrier lifetimes, and both groups have reported improved performance in devices fabricated with epilayers that have been treated with the oxidation or implantation procedures.

Application to Device Structures

It is apparent that great strides have been made in attaining long carrier lifetimes. Both lifetime enhancement approaches have been applied to device structures and have demonstrated substantial improvement in device performance. In 2008, Storasta et al. [20] showed proof of principle by fabricating identical p-i-n structures with 90μm drift layers. The devices were C-implanted, but not all were annealed. As shown in Fig. 4, the annealed devices exhibited a much lower forward voltage drop than those that were not annealed (9.5V vs. 4V at 40A/cm^2). This indicated that the enhanced carrier lifetime of the drift layer due to the diffusion of C_is and the annihilation of V_Cs, resulted in much better conductivity modulation. More recently, Nakayama et al. [29] have fabricated HV p-i-n diodes employing thick (120μm) drift layers, where *both* deep level reduction approaches (C-implantation and oxidation) have been compared directly. Both devices exhibited similar low forward voltage drops (≈4V) and exhibited breakdown voltages of 10.9kV and 17.1kV for 100μm and 250μm drift layers, respectively. Miyake et al. [30] have reported a 4H-SiC BJT with record current gain. The authors attribute this to an optimized device geometry, continuous epitaxial growth of the emitter-base junction, surface passivation and deep level reduction ($Z_{1/2}$ and $EH_{6/7}$) via thermal oxidation to improve the lifetime of the p-type base. All of these results tend to confirm that lifetime enhancement can be successfully introduced into real devices and that that the long-lifetime layers can survive the device fabrication process. It will be of interest to see this extended to a device such as an IGBT, where a gate oxide is also involved.

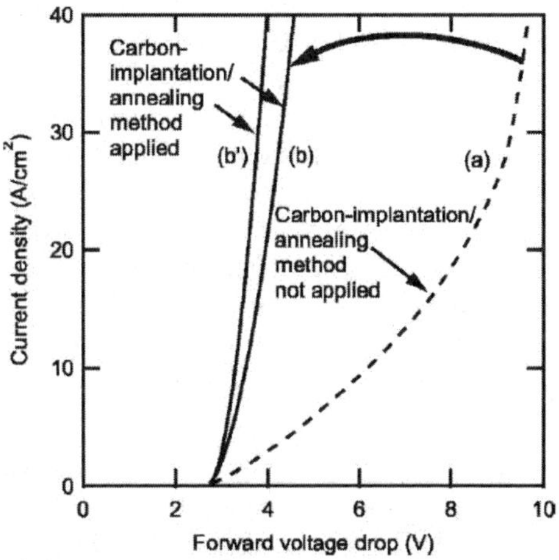

Fig. 4. (From Fig. 6 of Ref. 20.) Current-voltage characteristics of simple *p-i-n* diode structures prepared on the same 90 μm thick 4H–SiC epilayer, (a) without application of the carbon-implantation/annealing method and (b) and (b') with the method applied.

Carrier Lifetimes in p-type 4H-SiC Epilayers

There has also been considerable activity investigating p-type samples, as high voltage devices such as IGBTs employ a p-type base which must also have a long minority carrier lifetime. The as-grown and irradiation-induced defects that are present in p-type samples have been studied in detail [31,32]. Kimoto et al. [18] found that the concentration of hole traps had no significant effect on the carrier lifetime of p-type material, while Matsushita et al. [33] concluded that in p-type samples $Z_{1/2}$ was not controlling the lifetime. Storasta et al. [34] carried out both DLTS and minority carrier transient spectroscopy measurements in electron irradiated p-type material and also found no evidence of any charge state of the $Z_{1/2}$ defect. Thus far, there has been no clear evidence for the existence of this defect in p-type 4H-SiC, even though the electron capture cross section for the stable charge state ($Z^{+/o}$) is large [35]. If present, these defects are probably at very low concentrations. It is therefore unclear which lifetime limiting defects were eliminated from the p-base during oxidation of the BJT [30] noted above. Hayashi et al. studied the effect on the carrier lifetime of injection level and temperature [36] as well as oxidation and passivation [37]. The carrier lifetimes in p-type material showed little change with oxidation or Ar annealing, indicating again that the $Z_{1/2}$ (and HK0) defects do not substantially reduce the lifetime. Significant lifetime enhancement was observed, however, after passivation with a deposited oxide, suggesting that surface recombination plays a significant role. Thus, it seems that no bulk lifetime killer defect, including $Z_{1/2}$, has yet been identified. Consequently, the primary mechanisms controlling the carrier lifetime in p-type samples appear to be recombination at the surface and in the substrate.

Summary

Interest in determining the mechanism limiting the carrier lifetime began soon after the availability of high quality material. This work concluded that the bulk lifetime killer was the $Z_{1/2}$ defect, a V_C-related center. The marked decrease in the concentration of these defects after the annealing of material damaged by C^+-ion implantation was interpreted as the result of the diffusion of C_is into the bulk, thus annihilating the V_c-related centers. This insight led to the development of two related approaches to deep level reduction, involving either implantation or oxidation of the sample surface. The enhanced lifetimes of the resulting low-$Z_{1/2}$ material were no longer controlled by this defect, but were still too short for some device applications. The recombination mechanism in these samples was investigated and found to result from carrier recombination at the surface and the substrate. Consequently, it was concluded that long lifetimes required the reduction of $Z_{1/2}$ levels below that of as-grown material, very thick drift layers and surface passivation. When these criteria were all addressed, long carrier lifetimes (>10μsec) were observed. A marked improvement in device performance has been realized by the incorporation of these materials into real device structures, and represents a milestone in the road toward HV 4H-SiC power devices.

References

[1] K. Danno, D. Nakamura and T. Kimoto, Appl. Phys. Lett 90 (2007) p.202109.
[2] W. Shockley and W.T. Read, Phys. Rev. 87 (1952) p.835.
[3] R.N. Hall, Phys. Rev. 87 (1952) p. 387.
[4] J. Hassan and J.P. Bergman, J. Appl. Phys. 105 (2009) p.123518.
[5] S.I. Maximenko, J.A. Freitas, Jr., P.B. Klein, A. Shrivastava, and T.S. Sudarshan, Appl. Phys. Lett. 94 (2009) p.092101.
[6] V. Grivickas, A. Galeckas and J. Linnros, Lith. J. Phys. 37 (1997) p.473.
[7] O. Kordina, J. P. Bergman, A. Henry, and E. Janzén, Appl. Phys. Lett. 66 (1995) p.189.
[8] T. Kimoto, A. Yamashita, A. Itoh and H. Matsunami, Jpn. J. Appl. Phys. 32 (1993) p.1045.
[9] T. Kimoto, S. Nakazawa, K. Hashimoto and H. Matsunami, Appl. Phys. Lett. 79 (2001) p.2761.
[10] J. Zhang, L. Storasta, J. P. Bergman, N. T. Son, and E. Janzén, J. Appl. Phys. 93 (2003) p.4708.
[11] Y. Negoro, T. Kimoto and H. Matsunami, Appl. Phys. Lett. 85 (2004) p.1716.
[12] L. Storasta, J.P. Bergman, A. Henry, E. Janzén and J. Lu, J. Appl. Phys. 96 (2004) p.4909.
[13] K. Danno, K. Hashimoto, H. Saitoh, T. Kimoto and H. Matsunami, Jpn. J. Appl. Phys. 43 (2004) p.L969.
[14] J.R. Jenny, D.P. Malta, V.F. Tsvetkov, M.K. Das, H. McD. Hobgood, C.H. Carter, Jr., R.J. Kumar, J.M. Borrego, R.J. Gutmann and R. Aavikko, J. Appl. Phys. 100 (2006) p.113710.
[15] T. Tawara, H. Tsuchida, S. Izumi, I. Kamata, and K. Izumi, Mater. Sci. Forum 457-460 (2004) p.565.
[16] S. A. Reshanov, W. Bartsch, B. Zippelius and G. Pensl, Mater. Sci. Forum 615-617 (2009) p.699.
[17] P.B. Klein, B.V. Shanabrook, S.W. Huh, A.Y. Polyakov, M. Skowronski, J.J. Sumakeris and M.J. O'Loughlin, Appl. Phys. Lett. 88 (2006) p.052110.
[18] T. Kimoto, K. Danno and J. Suda, Phys. Status Solidi B 245 (2008) p.1327.
[19] L. Storasta and H. Tsuchida, Appl. Phys. Lett. 90 (2007) p.062116.
[20] L. Storasta, H. Tsuchida and T. Miyazawa, J. Appl. Phys. 103 (2008) p.013705.
[21] T. Hiyoshi and T. Kimoto, Appl. Phys. Express 2 (2009) p.041101.
[22] T. Hiyoshi and T. Kimoto, Appl. Phys. Express 2 (2009) p.091101.
[23] T. Kimoto, T. Hiyoshi, T. Hayashi and J. Suda, J. Appl. Phys. 108 (2010) p.083721.
[24] P.B. Klein, R. Myers-Ward, K.-K. Lew, B. L. VanMil, C.R. Eddy, Jr.,D.K. Gaskill, A. Shrivastava and T.S. Sudarshan, J. Appl. Phys. 108 (2010) p.033713.
[25] P. Grivickas, J. Linnros, and V. Grivickas, J. Mater. Res. 16 (2001) p.524.
[26] T. Kimoto, Y. Nanen, T. Hayashi and J. Suda, Appl. Phys. Express 3 (2010) p.121201.
[27] T. Miyazawa, M. Ito and H. Tsuchida, Appl. Phys. Lett. 97 (2010) p.202106.
[28] H.Tsuchida, M.Ito, I.Kamata, M.Nagano, T.Miyazawa, N. Oshino, Mater. Sci. Forum 645 (2010) p.77.
[29] K.Nakayama, R.Ishii, K.Asano, T.Miyazawa, M.Ito , H.Tsuchida, Mater. Sci. Forum 679 (2011) p.535.
[30] H. Miyake, T. Kimoto and J. Suda, IEEE Electron. Dev. Lett. 32 (2011) p.841.
[31] K. Danno and T. Kimoto, Jpn. J. Appl. Phys. 46 (2006) p. L285.
[32] K. Kawahara, G. Alfieri and T. Kimoto, J. Appl. Phys. 106 (2009) p. 013719.
[33] Y.Matsushita, M.Kato, M.Ichimura, T.Hatayama, and T.Ohshima, Mater. Sci. Forum 645-648 (2010) p. 207.
[34] L. Storasta, I. Kamata, T. Nakamura and H. Tsuchida, Mater. Sci. Forum 527-529 (2006) p.489.
[35] C.G. Hemmingsson, N.T. Son, A. Ellison, J. Zhang and E. Janzén, Phys. Rev. B 58 (1998) p.R10119.
[36] T. Hayashi, K. Asano, J. Suda and T. Kimoto, J. Appl. Phys. 109 (2011) p.014505.
[37] T. Hayashi, K. Asano, J. Suda and T. Kimoto, J. Appl. Phys. 109 (2011) p.114502.

Materials Science Forum Vols. 717-720 (2012) pp 285-288
© (2012) Trans Tech Publications, Switzerland
doi:10.4028/www.scientific.net/MSF.717-720.285

Comparison of Post-Growth Carrier Lifetime Improvement Methods for 4H-SiC Epilayers

I. D. Booker[1,a,*], J. Hassan[1,b], A. Hallén[2,c], E. Ö. Sveinbjörnsson[3,d], O. Kordina[1,e] and J. P. Bergman[1,f]

[1]Department of Physics, Chemistry and Biology, Linköping University, SE-58 183 Linköping, Sweden

[2]Royal Institute of Technology, Lab of Materials and Semiconductor Physics, P. O. Electrum 229, SE-16440 Kista, Sweden

[3]Science Institute, University of Iceland, IS-107 Reykjavik, Iceland

[a]ianbo@ifm.liu.se, [b]jawul@ifm.liu.se, [c]ahallen@kth.se, [d]einars@hi.is, [e]olkor@ifm.liu.se, [f]peber@ifm.liu.se, *corresponding author

Keywords: Carrier lifetime, Implantation, Annealing, Oxidation, Photon counting

Abstract. We compare two methods for post-growth improvement of bulk carrier lifetime in 4H-SiC: dry oxidations and implantations with either ^{12}C or ^{14}N, followed by high temperature anneals in Ar atmosphere. Application of these techniques to samples cut from the same wafer/epilayer yields 2- to 11-fold lifetime increases, with the implantation/annealing technique shown to give greater maximum lifetimes. The maximum lifetimes reached are ~5 μs after ^{12}C implantation at 600 °C and annealing in Ar for 180 minutes at 1500 °C. At higher annealing temperatures the lifetimes decreases, a result which differs from reports in the literature.

Introduction

Future high-voltage bipolar transistors based on 4H-SiC require precise control of the minority carrier lifetime, with several microseconds lifetime necessary for successful conductivity modulation in the blocking layer. Generally, lifetime of as-grown layers is below 1 μs, as measured by time-resolved photoluminescence (TRPL). Apart from extended defects and surface/interface recombination, deep levels in the epitaxial material also lead to involuntary lifetime reduction.

Controlled reduction by irradiation [1] has been demonstrated, as well as improvement by long oxidations [2,3] and high-temperature anneals after implantation with carbon ^{12}C [4]. Both improvement processes are explained using the same model: Carbon interstitials are created at the epilayer surface, either by the lattice-damaging implantation process or the oxide formation. During high temperature annealing these diffuse into the epitaxial layer, and annihilate with the Z_1Z_2 defect, which has been identified as a major lifetime killer in 4H-SIC [5]. Foregoing oxidation or implantation and simply annealing at high temperatures is ineffective, due to this intrinsic defect's high temperature stability up to more than 2000 °C [6]. Although its exact microscopic structure is still debated, this model points to the defect including a carbon vacancy (V_C).

In the following work, we present a comparison of both methods which is not clearly possible from the literature, as work on either method has been performed by different groups using different samples and lifetime analysis techniques, TRPL and microwave photoconductance decay (μ-PCD).

Experimental Details

An n-type 4H-SiC epilayer of 85 μm thickness with a nitrogen doping level of 1×10^{15} cm^{-3} was grown by hot-wall CVD on an 8°-off axis 3 inch 4H-SiC(0001) wafer. The wafer was cut into smaller pieces to which the lifetime improvement processes were applied. Concentrations of the Z_1Z_2 defect could not be measured in these samples but are typically in the 10^{12}-10^{13} cm^{-3} range.

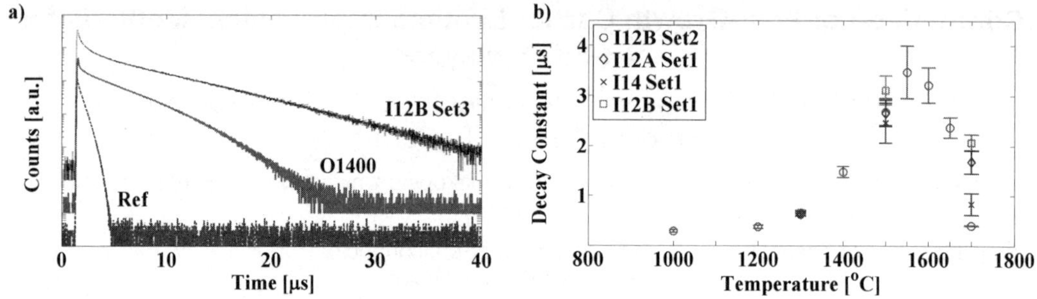

Fig. 1: a) Representative TRPL decay curves: a group I12B Set 3 sample, annealed for 300 minutes total at 1500 °C, a group O1400 sample dry oxidized for 300 minutes at 1400 °C and an as-grown reference sample (Ref). b) Decay constant at room temperature (RT) over annealing temperature for 9 implanted layers of Set 1 (groups I12A, I12B and I14) annealed at 1300 °C, 1500 °C and 1700 °C and a sample of Set 2 (group I12B).

TRPL decays were measured by photon counting at room temperature (RT), using the 355nm line of a frequency-tripled, pulsed Nd:YAG laser, of which about 63% of the photons are absorbed in the epilayer. The resulting near-band edge PL was collected by a photomultiplier tube. Data points in the graphs represent the mean of 49 (for implanted or oxidized) and 9 (as-grown) measurements taken in a ~1.5x2 mm^2 area, along with their standard deviations.

Oxidations. Dry oxidations were performed at temperatures 1150 °C, 1300 °C and 1400 °C for durations of 2.5, 5, 10 and 15 hours. Before oxidation, samples were cleaned by the RCA method. After oxidation the oxide was removed using HF. Several samples were then subjected to high-temperature anneals in Ar atmosphere at 1550 °C and 860 mbar pressure for 30 minutes.

Implantations. Two different implantation species were used, ^{12}C, implanted at either room temperature (sample group designation I12A) or 600 °C (I12B) and ^{14}N implanted at 600 °C (I14). A maximum implantation energy of 140 keV was used to implant an excess density of 1×10^{19} cm^{-3} into the first 300 nm of the layer. The lateral inhomogeneity of the implantation was below 5%.

After implantation, samples were annealed in Ar atmosphere at a pressure of 860 mbar using three different annealing processes – one set of samples (Set 1) was annealed at a specific temperature in the range of 1100 °C to 1800 °C for a duration of 30 minutes. Another set of samples (Set 2) was annealed cumulatively at increasing temperatures starting at 1000 °C and proceeding in 100 °C or 200 °C steps up to 1700 °C for 30 minutes per step. The last set (Set 3) was annealed cumulatively at a temperature of either 1300 °C or 1500 °C, each anneal lasting 30 or 60 minutes up to a total duration of 510 and 300 minutes, respectively. The top implanted layers were only removed after finishing the last anneal of the respective samples by polishing off the top 2-3μm.

Results and Discussion

Three representative decay curves of a reference, an oxidized and an implanted and annealed sample are given in Fig. 1a). The decay curves in the samples generally exhibit a downward bow, evidencing a change in the decay constant as the numbers of carriers in the layer decrease. This bowing is usually not observed in other comparable samples and nearly vanishes for implanted samples with very long lifetime. As it is present in the reference material, it is specific for the sample used and not a product of either the oxidation or the implantation process. For comparison reasons, the values of the decay constants/lifetimes presented in the following are obtained from the high intensity region. The initial fast decay region found in samples with very long lifetime is not included in the fit.

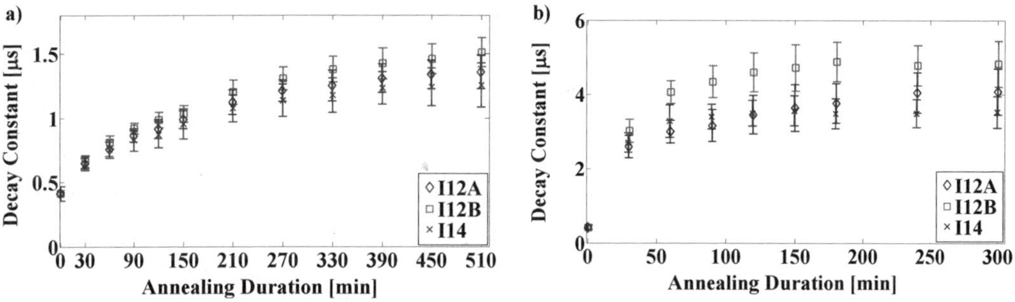

Fig. 2: Set 3 samples, showing the effect of repeated anneals on the decay constant of implanted samples measured at RT. Annealing is done at a) 1300 °C and b) 1500 °C.

The effect of annealing temperature on the resulting carrier lifetime in implanted layers is given in Fig. 1b) for different samples. Minimal changes are already visible after the 30 minute anneal at 1200 ˚C, indicating the beginning in-diffusion of carbon interstitials. The optimal temperature was found to be at 1500-1600 °C, after which lifetimes in the single annealed Set 1 samples and the cumulative Set 2 sample decrease. This is 200-300 °C lower than the optimal temperature found by Storasta, et al. [4], who (using samples similar to our I12B type) showed a factor 2 lifetime increase after annealing at 1600 °C for 30 minutes and a factor 5 increase from 200 ns to ~1 μs after annealing at 1800 °C. Since in the I12B Set 2 sample a large amount of carbon interstitials have already diffused into the bulk before annealing at the highest temperatures, we believe this lifetime decrease to be related to the formation of point defect complexes in the epitaxial layer.

This formation is related to annealing temperature (via an activation energy) and not duration as can be seen in Fig. 2a) and b). Here, the mean lifetimes over annealing duration are plotted for samples annealed at 1300 °C (a) and 1500 °C (b). Samples annealed at 1300 °C have not completely reached their maximal lifetime values after 510 minutes but are seen to converge to values between 1.2 and 1.6 μs. In contrast, samples annealed at 1500 °C reach their maximum lifetimes after an annealing duration of 150 minutes. Judging from Fig. 1b), 2a) and 2b), implantations with ^{12}C at 600 °C (sample type I12B) are more effective than room temperature implanted samples (I12A) or samples implanted with a species different from carbon (I14). The results for I12A and I12B are somewhat surprising, as most of the carbon interstitials originate not from the implantation itself but from implantation induced lattice damage, which is greater when implanting at room temperature.

Lifetime resulting from sample oxidations are given in Fig. 3a), with the thicknesses of the formed oxides given in Fig. 3b). We have found a 1.9-fold and 6.6-fold improvement of the lifetime after 300 minute oxidations at either 1300 °C or 1400 °C, respectively. This is similar to the relative 2.2-fold improvement in carrier lifetime of a 100 μm thick epilayer after thermal oxidation at 1300 °C for 2x300 minutes found by Hiyoshi, et al. [2]. Only relative improvements can be compared here, as this group uses μ-PCD to measure lifetimes.

Similar to the findings in the implanted/annealed samples, the low temperature oxidations have relatively small (1300 °C) or almost no (1150 °C) effect on the lifetime when compared to higher temperature processes (1400 °C). Overall, higher final lifetimes (an 11-fold increase) could be achieved by the implantation/annealing process. Contrary to findings by Kimoto, et al. [7], annealing in Ar after the oxidation had no beneficial effects on our samples but lead to a slight reduction of the fitted lifetimes.

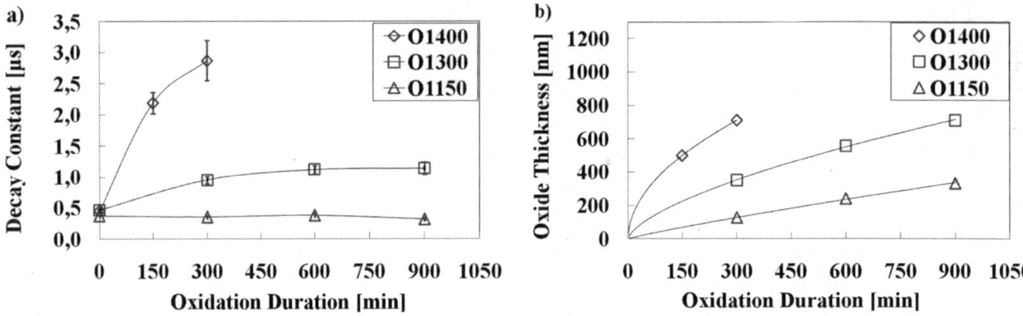

Fig. 3: a) Decay constants measured at RT after oxidations at 1150 °C, 1300°C and 1400 °C. Lines are guides to the eye. b) Oxide thickness over oxidation duration. Lines are a fit to the data points by a power law of the form $y = Ax^B$ (O1150: B = 0.888. O1300: B = 0.647. O1400: B = 0.490).

Conclusion

We have compared two different bulk carrier lifetime enhancing techniques using material cut from the same epilayer/wafer: long oxidations and implantations followed by high-temperature annealing. For oxidations at 1300 °C for 600 minutes we find a 2.5-fold lifetime increase, which is comparable to values found in the literature [2]. Oxidations at 1400 °C yielded a maximum 6.5-fold high-injection lifetime increase and will be investigated in a future paper. Subsequent anneals in Ar atmosphere at 1550 °C were found to have no beneficial impact on carrier lifetime. In terms of maximum lifetimes reached, the implantation/annealing process proved more effective, giving an 11-fold lifetime increase (a mean of 4.85 µs) after annealing at 1550 °C for 180 minutes. Our optimal temperature is 1500-1600 °C in contrast to 1800 °C which was the case in recent annealing studies of Z_1Z_2 [4], with higher temperatures giving steadily reduced lifetimes for both single annealed layers and the 'Set 3' cumulative temperature series. Both processes are similar in that their efficiency improves with high temperatures upwards of 1300 °C.

References

[1] T. Kimoto, K. Danno and J. Suda, Lifetime-killing defects in 4H-SiC epilayers and lifetime control by low-energy electron irradiation, phys. stat. sol. (b) 245 (2008), p. 1327-1336.

[2] T. Hiyoshi and T. Kimoto, Reduction of Deep Levels and Improvement of Carrier Lifetime in n-Type 4H-SiC by Thermal Oxidation, Appl. Phys. Express 2 (2009), p. 041101-1-3.

[3] T. Hayashi, K. Asano, J. Suda and T. Kimoto, Temperature and injection level dependencies and impact of thermal oxidation on carrier lifetimes in p-type and n-type 4H-SiC epilayers, J. Appl. Phys. 109 (2011), p. 014505-1-5.

[4] L. Storasta, H. Tsuchida, T. Miyazawa and T. Ohshima, Enhanced annealing of the Z1/2 defect in 4H-SiC epilayers, J. Appl. Phys. 103 (2008), p. 013705-1-7.

[5] P. B. Klein, Carrier lifetime measurements in n^- 4H-SiC epilayers, J. Appl. Phys. 103 (2008), p. 033702-1-14.

[6] T. Dalibor, G. Pensl, H. Matsunami, T. Kimoto, W. J. Choyke, A. Schöner and N. Nordell, Deep Defect Centers in Silicon Carbide Monitored with Deep Level Transient Spectroscopy, phys. stat. sol. (a) 162 (1997), p. 199-225.

[7] T. Kimoto, T. Hiyoshi, T. Hayashi and J. Suda, Impacts of recombination at the surface and in the substrate on carrier lifetimes of n-type 4H-SiC epilayers, J. Appl. Phys. 108 (2010), p. 083721-1-7.

Materials Science Forum Vols. 717-720 (2012) pp 289-292
doi:10.4028/www.scientific.net/MSF.717-720.289

Radial Variation of Measured Carrier Lifetimes in Epitaxial Layers Grown with Wafer Rotation

J.P. Bergman[1,a], I.D. Booker[1,b], L. Lilja[1,c], J. Hassan[1,d] and E. Janzén[1,e]

[1]Department of Physics, Chemistry and Biology, Linköping University, S-58183 Linköping, Sweden

[a]peber@ifm.liu.se, [b]ianbo@ifm.liu.se, [c]louca@ifm.liu.se, [d]jawul@ifm.liu.se, [e]erj@ifm.liu.se

Keywords: carrier lifetime, wafer rotation

Abstract. In this report we present homoepitaxial growth of 4H-SiC on the Si-face of nominally on-axis substrates with diameter up to 100 mm in a hot-wall chemical vapor deposition reactor. A comparatively low carrier lifetime has been observed in these layers. Also, local variations in carrier lifetime are different from standard off-cut epilayers. The properties of layers were studied with more focus on charge carrier lifetime and its correlation with starting growth conditions, in-homogeneous surface morphology and different growth mechanisms.

Introduction

Silicon carbide, SiC, is a semiconductor material with properties that make it well suited for future more efficient power electronics. The carrier lifetime is an important and critical material property especially for bipolar power devices. There have been a number of investigations of carrier lifetimes [1], and in recent years several studies of post-growth improvements of the carrier lifetime [2,3]. There have, however, been few reports on wafer variation of the carrier lifetime.

At Linköping University we have developed and with time improved a carrier lifetime mapping technique based on photoluminescence detection of the bandgap optical emission. During the last decade we have measured close to one thousand epitaxial wafers with diameters of 35mm, 2", 3" and 4". The epitaxy has mainly been from our own reactors but also from external and commercial epitaxial producers.

In this work we report a physically interesting, and from the device application viewpoint, important observation.

Experimental

A variety of different 4H SiC epitaxial layers has been studied. The epitaxial layers were on substrates from three different commercial suppliers (A, B and C), with epitaxial layers from three different epi-houses (I, II and III). Epi house I corresponds to layers grown with our own Hot Wall epitaxial reactor, whereas epi house II and III are from commercial suppliers. The epi reactor (I) at Linköping University has a non-rotating susceptor while the two commercial suppliers use different reactor models, both having wafer rotation.

Carrier lifetime was measured from the time resolved photoluminescence (TRPL) decay of the near band edge emission at room temperature. The time resolution using this system is approximately 50 ns. Full wafer mapping was performed with a spatial resolution of 200 μm.

Experimental Results

Mapping of full wafers usually shows a quite homogenous distribution of the measured carrier lifetime, as shown in Fig. 1. Figure 1a shows a colormap representing the measured lifetime value, while Fig. 1b shows a line profile of the measured lifetime along the central horizontal and vertical axes, respectively. The epitaxial layer is 100 μm thick and grown on a substrate from supplier A, at Linköping University using a susceptor without rotation. Long scale local variations of carrier lifetime have previously been reported [4] and attributed to variation of growth conditions in the susceptor, such as gas composition and temperature. These variations also influence other parameters such as growth rate and dopant distribution [5,6]. More local variations have also previously been reported and related to structural defects in the substrate replicated into the epitaxial layer. With increasing spatial resolution further structures are revealed, and are present in most epitaxial layers from different suppliers, with different intensities. These features are either related to high dislocation densities, such as grain boundaries, or stripe-like features normally orientated perpendicular to the [11-20] off-cut direction. These are assumed to be related to slip bands or stacking faults along the basal plane.

The lifetime mapping of a 3" wafer from substrate supplier A, with epitaxy from epi producer II, is shown in Fig 2a, together with the horizontal and vertical line lifetime profiles, Fig. 2b. This epilayer is grown in a susceptor with rotating substrate. Local variations attributed to defects replicated from the substrate are still present, but the variations caused by inhomogenous growth conditions, such as temperature variations and gas-phase composition, are no longer present. The distribution of the lifetime is, however, drastically different in epitaxial wafers grown in reactors with rotating substrate. There is, however, an additional effect causing a strong radial dependence of the measured carrier lifetime. This is observed in substrates from four different commercial suppliers (A, B and C), with epitaxial layers from three different commercial epi-houses (I, II and II), even though the shape and variations differ between the wafers. The same radial dependence is observed in epiaxial layers from three different commercial epi-houses (I, II and II), grown on substrates from three different commercial suppliers (A,B and C), even though the shape of the radial variation differs between the wafers.

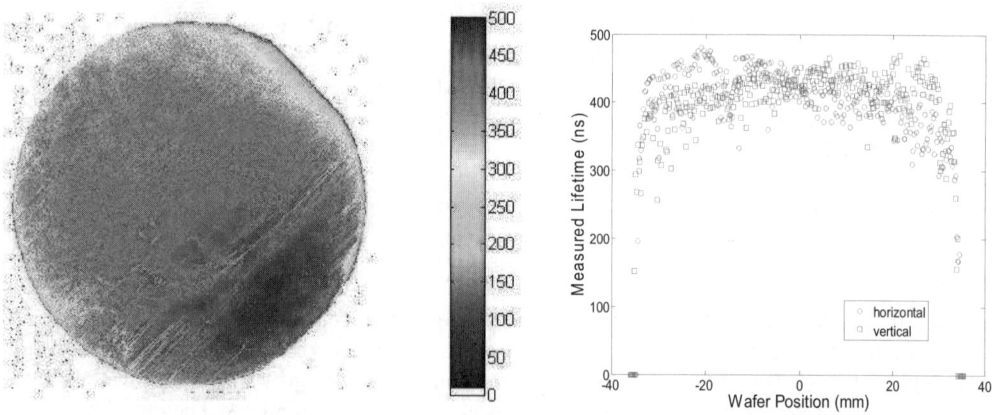

Fig. 1. a) (left) Carrier lifetime mapping of a 100 μm thick epitaxial layer with 3" diameter grown at LiU without wafer rotation. b) (right) Horizontal and vertical line profile of the carrier lifetime.

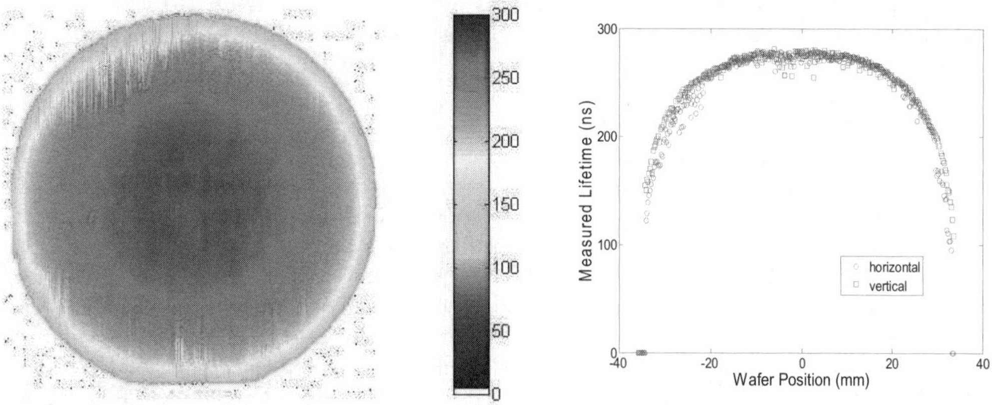

Fig. 2. a) (left) Carrier lifetime mapping of a 100 µm thick epitaxial layer with 3" diameter grown with wafer rotation. b) (right) Horizontal and vertical line profile of the carrier lifetime.

A final example of lifetime mapping from a 4" epitaxial layer from epi-supplier II with substrate from supplier B is shown in Fig. 3, together with the horizontal and vertical lifetime profiles. The epitaxial layer is in this case 15 µm thick, much thinner than the wafers in Fig. 1 and Fig. 2, which influences the relatively lower value of the lifetime. The radial variation is also present in this wafer and even more pronounced than in the studied 4" epitaxial layers.

Discussion

The radial dependence of the measured lifetime is seen in all measured epitaxial wafers from reactors using wafer rotation during growth, and never in any wafer grown in our reactor without wafer rotation. It has been observed in wafers from different epi suppliers and grown on substrates from several different suppliers, and from our experience and opinion not dependent on substrate supplier or individual epi-reactor or growth recipe.

A radial distribution of material properties, such as epi thickness and doping, during growth with wafer rotation has previously been reported [7]. The values were, however, much lower than observed for the measured carrier lifetime, with sigma/mean intra-wafer variation over 3" of 3.3% and 12.3% for thickness and doping, respectively. Similar inhomogeneous variations for 4" wafers

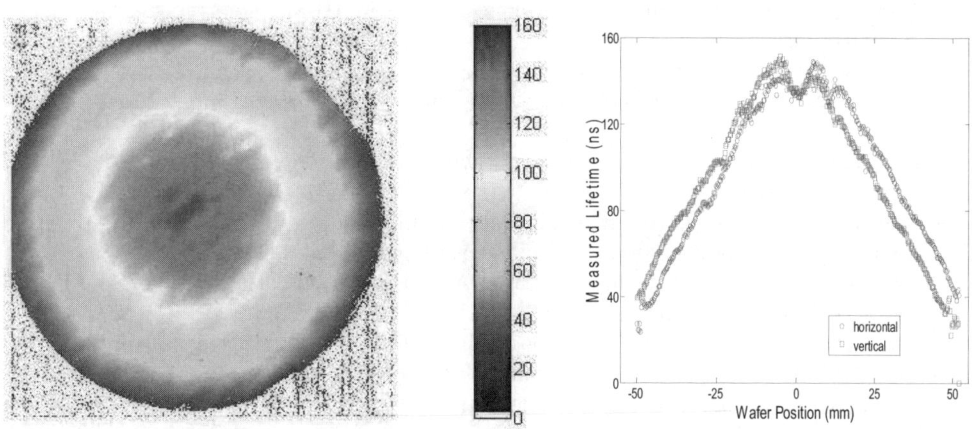

Fig. 3. a) (left) Carrier lifetime mapping of a 15 µm thick epitaxial layer with 100 mm diameter grown with wafer rotation. b) (right) Horizontal and vertical line profile of the carrier lifetime.

have been reported as 1.6% and 6.0% for thickness and doping, respectively [8]. Both of these reactors are comparable to the commercial reactors for epi houses II and III. There is to our knowledge no previous report of inhomogeneities of the carrier lifetime.

We believe instead that this is due to a fundamental mechanism in the epitaxial growth using rotating susceptors. A possibility could be related to radial variations in the gas composition or temperature which change the incorporation of defects acting as recombination centers. Another possibility could be the relative gas flow speed over the surface. A rough estimation of the individual radial wafer rotation rate and the gas flow speed over the substrate gives comparable numbers. This means that the gas flow over the substrate during half of a rotation cycle is close to zero. This influences both the growth rate and the incorporation of mainly intrinsic defects during growth. But whereas the growth rate averages over time as demonstrated by a homogeneous epi-layer thickness, the change in growth conditions and possible increased incorporation of intrinsic defects introduces local lateral regions with lower lifetime, which influences the total measured carrier lifetime. A study showing dependence, even if not very strong, between measured lifetime and growth rate on comparable substrates is published in these proceedings [9].

Further studies of low temperature photoluminescence and DLTS are planned in order to investigate any measurable presence of defects that can influence the observed variations in lifetime.

Summary

Lifetime mapping measurements on full wafers have shown a radial dependence with decreasing lifetime towards the wafer periphery in wafers grown in reactors using wafer rotation. The same effect is not seen in any epitaxial layer grown in a reactor without rotation. The effect is independent of epi or substrate supplier.

Acknowledgements

The Swedish Research Council (VR) is gratefully acknowledged for financial support

References

[1] P.B. Klein, Carrier lifetime measurement in n^- 4H-SiC epilayers, J. Appl. Phys. 103 (2008) 033702.

[2] T. Hiyoshi and T. Kimoto, Reduction of deep levels and improvement of carrier lifetime in n-type 4H-SiC by thermal oxidation, Appl. Phys. Express 2 (2009) 041101.

[3] L. Storasta et al., Enhanced annealing of the $Z_{1/2}$ defect in 4H-SiC epilayers, J. Appl. Phys. 103 (2008) 013705.

[4] E. Janzén, A. Henry, J.P. Bergman, A. Ellison, and B. Magnusson, Material characterization need for SiC-based devices, Materials Science in Semiconductor Processing, v 4, n 1-3 (2001) pp 181-186.

[5] J. Hassan and J.P. Bergman, Influence of structural defects on carrier lifetime in 4*H*-SiC epitaxial layers: Optical lifetime mapping, J. Appl. Phys. 105 (2009) 123513.

[6] I. Booker, J. Hassan, and J.P. Bergman, High-Resolution Time-Resolved Carrier Lifetime and Photoluminescence Mapping of 4H-SiC Epilayers, these Proceedings.

[7] B. Thomas, C. Hecht and B. Kallinger, Large Area Homoepitaxial Growth of Low Doped Thick Epilayers for Power Devices with $V_{BR} > 4$ kV, Mater. Sci. Forum 615-617 (2009) 77.

[8] A.A. Burk et al., SiC Epitaxial Growth on Multiple 100-mm Wafers and its Application to Power-Switching Devices, Mater. Sci. Forum 600-603 (2009), 77.

[9] L. Lilja, J. Hassan, I.D. Booker, J.P. Bergman, and E. Janzén. The effect of Growth Conditions on Carrier Lifetime in n-type 4H-SiC Epitaxial Layers, these Proceedings.

Materials Science Forum Vols. 717-720 (2012) pp 293-296
© (2012) Trans Tech Publications, Switzerland
doi:10.4028/www.scientific.net/MSF.717-720.293

High-Resolution Time-Resolved Carrier Lifetime and Photoluminescence Mapping of 4H-SiC Epilayers

I. D. Booker[1,a,*], J. Hassan[1,b], E. Janzén[1,c] and J. P. Bergman[1,d]

[1]Department of Physics, Chemistry and Biology, Linköping University, SE-58 183 Linköping, Sweden

[a]ianbo@ifm.liu.se, [b]jawul@ifm.liu.se, erj@ifm.liu.se, [d]peber@ifm.liu.se, *corresponding author

Keywords: Full wafer mapping, Carrier lifetime, Photoluminescence, Epitaxial defects.

Abstract. We present a comparison between time-resolved carrier lifetime mappings of several samples and integrated near band edge intensity photoluminescence mappings using a pulsed laser. High-injection conditions and as-grown material are used, which generally allow for the assumption of a single exponential decay. The photoluminescence intensity under these circumstances is proportional to the carrier lifetime and the mappings can be used to detect lifetime-influencing defects in epilayers and give an estimate of the carrier lifetime variation over the wafer. Several examples for the defect detection capability of the system are given.

Introduction

Carrier lifetime is a central parameter for 4H-SiC based power electronics and devices. Apart from the influence of surface and epilayer/substrate interface recombination and point defects acting as recombination centers [1], structural defects [2] such as grain boundaries, dislocation clusters and inclusions also locally degrade lifetime. As extended structural defects are often aligned in the basal plane of off-axis grown epilayers and decrease in length with decreasing epilayer thickness, high resolution is necessary for detection of these defects in thin epilayers.

Extended defects are generally detected by destructive or advanced measurement techniques. KOH etching combined with optical microscopy or scanning electron microscopy for example is able to reveal structural defects, but the analysis makes the epilayer unusable for further processing. Synchrotron white-beam X-ray topography, although non-destructive is available only at certain facilities. An optical carrier lifetime mapping system developed at LiU has been demonstrated to be an effective non-destructive technique to observe the presence of extended defects [2], due to their influence as recombination center. However, the system is limited by measuring speed and hence practical spatial resolution. It is therefore complemented with a high-speed photoluminescence (PL) mapping system for the analysis of lifetime-influencing defects in large diameter 3 or 4 inch wafers.

Experimental Details

All samples are 4H-SiC epilayers grown by hot-wall CVD on 4H-SiC substrates. Samples A and B are 80μm thick and grown on substrates cut from the same boule. Both epilayers are n-type, with doping levels of $1x10^{15}$ cm^{-3}. Sample C is another n-type epilayer of 12 μm thickness grown on a substrate from a different boule than that of samples A and B. Sample D is a 85μm thick epilayer with doping level of $1x10^{15}$ cm^{-3} grown on a substrate cut from yet another boule.

In the time resolved photoluminescence (TRPL) measurement setup, the sample is mounted on a translation stage. A 355nm frequency-tripled Nd:YAG laser is used to inject the excess carriers. The beam has a pulse with of 50ns and is focused to a spot of 100μm diameter, resulting in an injection level of $(1\text{-}5)x10^{17}$ cm^{-3} at 0.13-0.5 mJ/pulse. The photoluminescence signal is collected at room temperature by a photomultiplier tube and a bandpass filter is used to select out the near band edge luminescence (390nm with 15nm full width at half maximum). In the lifetime mapping setup, an oscilloscope is used to record the decay of the measured signal averaged over 200 excitation pulses.

For the photoluminescence intensity mappings, the oscilloscope is replaced by a lock-in amplifier which records only the signal magnitude and not the decay curve shape.

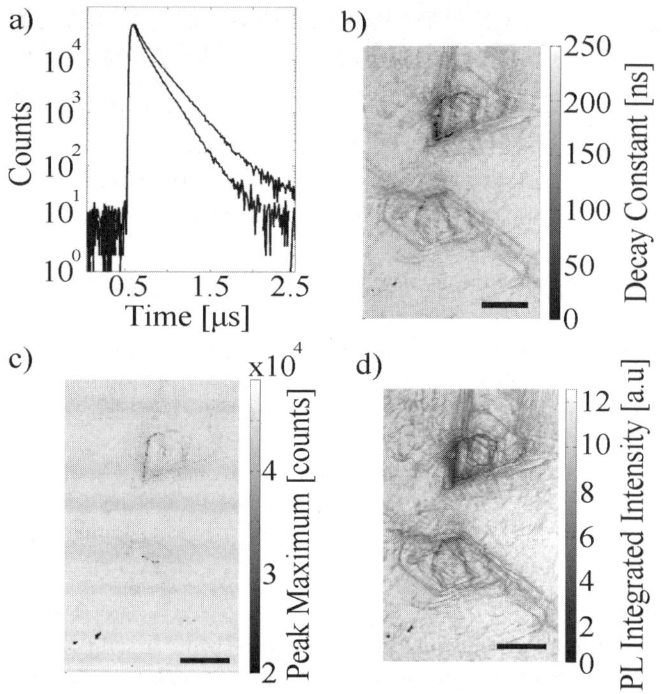

Fig. 1: a) Lifetime decay curves measured at a high and low lifetime regions around the grain boundaries. b) High injection carrier lifetime mapping around grain boundaries in the center of sample B. Step size 60μm. c) Peak maximum value of the measured decay curves. The maximum is decreased only at the grain boundaries. d) PL integrated intensity mapping of the same area, step size 30μm. Notice the good agreement of the relative intensity variations with the relative lifetime variations mapped in a). Scale bars in the three panels (b-d) are equal to 2mm.

Results and Discussion

Carrier lifetime and integrated intensity. In as-grown layers, which generally exhibit carrier lifetimes below ~1μs, the rapidly decaying carrier lifetime signal can be approximated as a single exponential decay. This decay curve is then characterized by the decay constant (lifetime), an initial maximum value and the integrated intensity below the curve. Two examples of decay curves measured using the lifetime mapping setup are given in Fig. 1a). Fig. 1b) and c) show a mapping of the decay constant and the peak maxima. The peak maximum is reasonably constant but decreases at the grain boundary defects due to changes in layer reflectivity and absorption properties. As can be seen in Fig. 1d), the variation in PL integrated intensity, measured with the PL intensity mapping setup, corresponds well to the variation of the lifetimes in b).

Since PL intensity can be mapped several times faster than lifetime decay curves, it can be used to detect lifetime altering defects and give an approximation of carrier lifetimes if correctly calibrated. This approximation however breaks down in high lifetime layers with non-single exponential decays, which occur in layers subjected to oxidations [3] and, albeit less pronounced, also in some as-grown epilayers.

Defect detection. Using PL intensity detection, a full 3 inch wafer can be mapped for lifetime influencing defects using a step size of 60μm in 11 hours. A carrier lifetime mapping under the same conditions would require ~200 hours with our system. An example of a carrier lifetime map at larger step size is given in Fig. 2a), evidencing the presence of downfalls near the wafer center as well as linear defects at the wafer edges. After KOH etching, these linear defects appear to be slip-bands of dislocations [4], propagating along the [11-20] off-cut direction and replicating from the substrate into the epilayer.

The arrays of closely spaced, slightly angled etch pits are visible in a microscope photograph, Fig. 2b). Although the PL intensity mapping in Fig. 2c) does not resolve single dislocations, it reveals the closely spaced linear nature of the dislocation bands.

In thin films of 12μm thickness, such as sample C in Fig. 3a), only 17% of the laser energy is absorbed by the epilayer (compared to 65% absorption in sample D). Lifetime mappings are no longer possible in our system due to the weakness of the signal. PL intensity collection time increases by a factor 2-3 and still allows detection of grain boundaries, in-grown stacking faults and a series of not yet identified linear defects. PL intensity mapping, Fig. 3b) and KOH etching, Fig. 3c), of a thicker sample B exhibiting similar defects prove the propagation of the central grain boundary defects from the substrate into the epilayer. Apart from grain boundaries, a comparison of Fig. 3b) and c) shows the PL mapping system's ability to identify regions of high dislocation density as well as triangular defects originating from downfalls, which most likely are stacking faults.

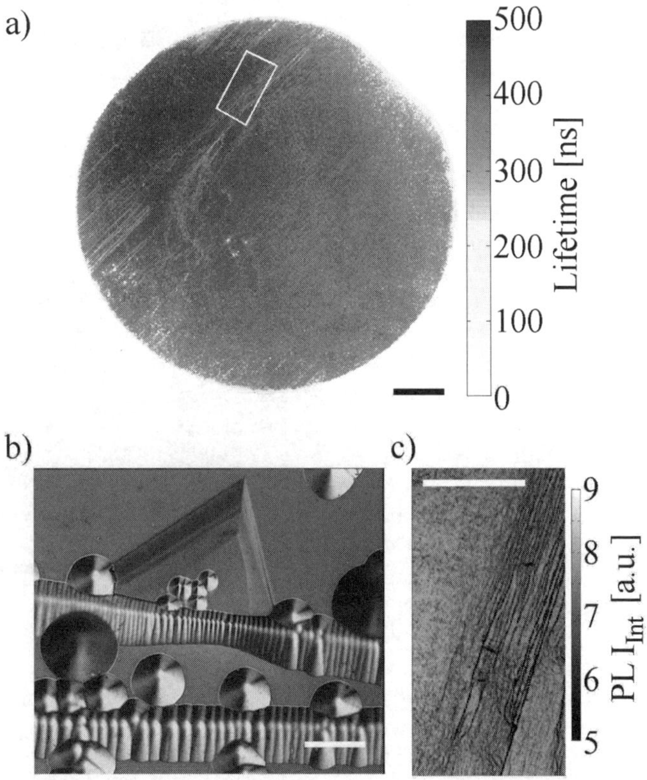

Fig. 2: a) High injection carrier lifetime mapping of sample D (3" wafer) with step size 230μm. The mapping required 16.5 hours. Downfalls (near center) and slip bands (top left quadrant) are visible. The scale bar is equal to 10mm. b) Microscope photograph of KOH etched slip band surface terminations on sample D. These closely spaced dislocations are responsible for local lifetime and PL intensity reductions. The scale bar is equal to 50μm. c) Detail of an integrated PL intensity mapping with step size 60μm (corresponds to the area marked by a white rectangle in a)) of several slip bands before KOH etching. The scale bar is equal to 5mm.

Conclusion

We have presented a comparison between carrier lifetime and faster integrated PL intensity mappings for full wafers. Under the assumption (1) that decay curves are single exponential and (2) the initial decay curve maximum is constant, the PL integrated intensity is proportional to the high injection carrier lifetime. Although the first assumption breaks down in long lifetime layers produced, for example, by oxidations, it holds in a large number of as-grown layers. Assumption number two holds reasonably well except in regions of structural defects, where the decreased initial signal would lead to on overestimation of carrier lifetime in this region. Integrated PL intensity can therefore be used to estimate the relative high injection carrier lifetime variation over a wafer and enables non-destructive detection and mapping of lifetime influencing defects.

Fig. 3: PL intensity mappings of grain boundaries in different samples. a) Mapping of layer C with step size 30µm. In-grown stacking faults (marked by circles), a central large group of grain boundaries, originating from the substrate and as of yet unidentified linear defects (marked by boxes) are visible. The scale bar is equal to 5mm. b) Mapping of layers A (top) and B (bottom) with 30µm step size showing grain boundaries, downfalls and dislocation clusters. The scale bar is equal to 5mm. c) Microscope photograph of the KOH etched sample B, taken in the area marked by a rectangular box in b). The etched grain boundaries (marked G), a carrot (marked C) and downfalls (marked D) are visible. The scale bar is 1mm.

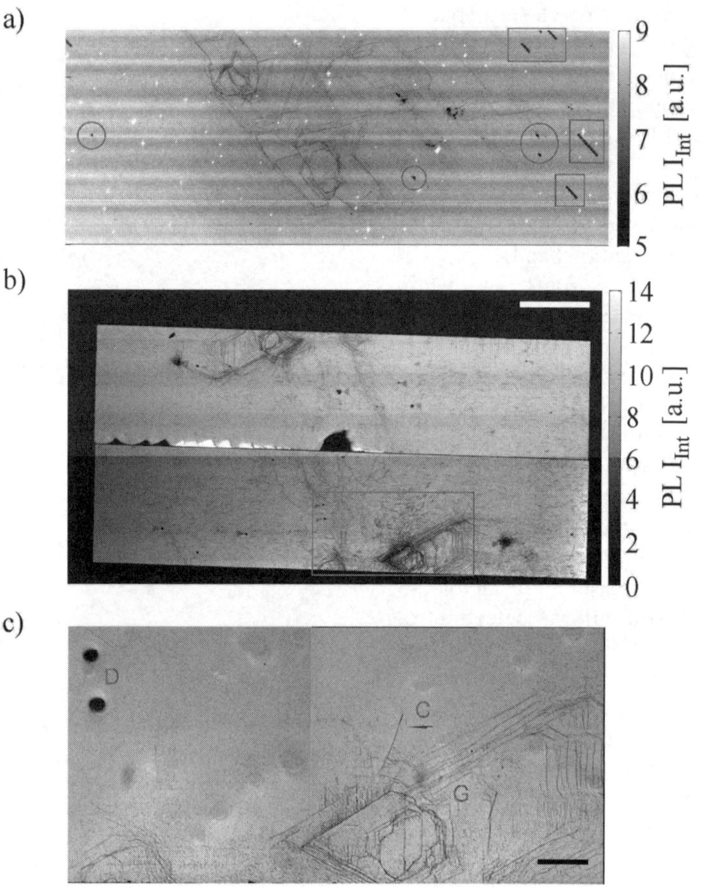

Acknowledgments

The Swedish Research Council (VR) is gratefully acknowledged for financial support

References

[1] P. B. Klein, Carrier lifetime measurements in n⁻4H-SiC epilayers, J. Appl. Phys. 103 (2008) 033702-1-14.

[2] J. Hassan and J. P. Bergman, Influence of structural defects on carrier lifetime in 4H-SiC epitaxial layers: Optical lifetime mapping, J. Appl. Phys. 105 (2009) 123518-1-5.

[3] T. Kimoto, T. Hiyoshi, T. Hayashi and J. Suda, Impact of recombination at the surface and in the substrate on carrier lifetimes of n-type 4H-SiC epilayers, J. Appl. Phys. 108 (2010) 083721-1-7.

[4] S. Ha, M. Skowronski, W. M. Vetter and M. Dudley, Basal plane slip and formation of mixed tilt boundaries in sublimation grown hexagonal polytype silicon carbide single crystals, J. Appl. Phys., 92 (2002) 778-785.

Materials Science Forum Vols. 717-720 (2012) pp 297-300
© (2012) Trans Tech Publications, Switzerland
doi:10.4028/www.scientific.net/MSF.717-720.297

Correlation of Extended Defects on Carrier Lifetime in Thick SiC Epilayers

Nadeemullah A. Mahadik[1,a], Robert E. Stahlbush[1], Joshua D. Caldwell[1], Michael O'Loughlin[2], and Albert Burk[2]

[1]Naval Research Laboratory, 4555 Overlook Ave., Washington, DC 20375

[2]Cree Inc. 4600 Silicon Drive, Durham, NC 27703

[a] Email: mahadik@ccs.nrl.navy.mil

Keywords: Carrier Lifetime, Extended Defects, High power

Abstract. The effect of extended defects on carrier lifetime was investigated in 140 μm thick 4H-SiC epilayers using whole wafer ultraviolet photoluminescence (UVPL) and microwave photoconductive decay (μPCD) mapping. Half-loop arrays (HLA) seen in the UVPL images showed a corresponding lifetime degradation in the same region, even before expansion of the HLAs to form SFs. Lifetime lowering was also seen for a defect comprising of a small 3C-SiC inclusion with a larger misoriented 4H-SiC region. Additionally, formation of slip planes after high temperature annealing was observed, which consequently shows a lifetime reduction in that region.

Introduction

High carrier lifetime (>1 μs) is essential to realize high power 4H-Silicon Carbide (4H-SiC) bipolar devices in order to obtain conductivity modulation in the thick voltage blocking layers [1]. To achieve acceptable low on-state resistance a carrier lifetime of >5 μs is required for 10 kV blocking epilayers and >20 μs for 20 kV blocking epilayers [2]. However, the typical lifetime for SiC epilayers is still about 1 μs, which is a few orders of magnitude lower than silicon. Recently significant work has been done to improve the carrier lifetime in SiC by various oxidation and annealing techniques [3, 4] that showed lifetime as high as 18.5 μs. The lifetime improvements have been attributed to recombination of extrinsically introduced carbon interstitials with carbon vacancies, which in turn lowers the concentration of $Z_{1/2}$ centers. However, during epigrowth and also after the high temperature steps, extended defects are formed in the epilayers, which impact the lifetime adversely. The extended defects appear to be more pronounced in the growth and processing of thicker (>100μm) epilayers due to the longer growths and greater defect propagation. Previously, lifetime degradation resulting from in grown stacking faults, low angle boundaries, micropipes, and carrot defects have been studied in relatively thin epilayers [5, 6]. In this work, we report on the influence of various other extended defects on carrier lifetime such as half loop arrays, slip bands, morphological defects, and single stacking faults in thick epitaxial layers.

Experimental Procedure

Whole wafer non-destructive ultraviolet photoluminescence (UVPL) and microwave photoconductive decay (μPCD) mapping were performed at room temperature on 140 μm thick, $2x10^{14}$ cm^{-3} n-type doped epilayers, commercially grown on 4 degree offcut SiC substrates. For the UVPL imaging, the samples were exposed with a 364 nm, Ar ion laser with penetration depth of 120 μm [7]. The laser power was 0.2 W. Using a computer controlled probe station and CCD detector, 1 mm^2 images in the luminescence range of 600-1000 nm were obtained in a step and repeat fashion. These images were merged to generate the entire wafer map. A Semilab whole wafer lifetime scanner was used to obtain spatially resolved (500 μm step) μPCD transients that provided ambipolar carrier lifetimes at every spot. The penetration depth of the excitation laser is ~50 μm. Lifetime maps were generated to correlate with the extended defects seen in the UVPL maps.

Another set of commercially grown 150μm thick epilayers was also studied to see the effect of high temperature annealing on extended defects and change in lifetime using UVPL and μPCD mapping. The anneals were done in Ar at 1700 °C for 30 mins with a graphite cap.

Results and Discussion

Multiple sets of UVPL and μPCD wafer maps were studied for their correlation of lifetime and extended defects. In the UVPL maps, various extended defects were observed such as polytype inclusions, basal plane dislocations (BPD), slip planes, half loop arrays (HLA), micropipes, in-grown stacking faults, and low angle boundaries. Figure 1 (a)-(d) shows a few of these defects. The carrier lifetime maps show a one-to-one correlation with many of the defects observed in UVPL

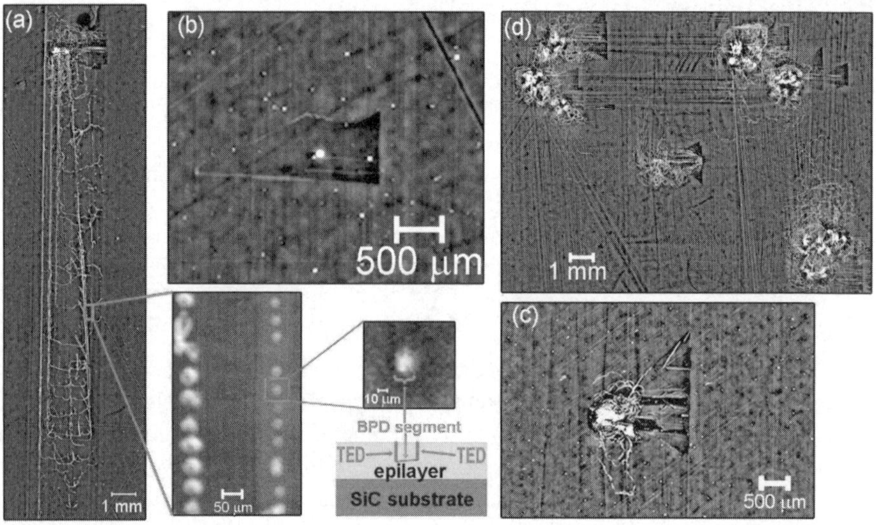

Figure 1: UVPL images from a SiC wafer with 140 μm thick drift layer. (a) cluster of HLAs, a small section has been magnified to show the structure of the HLA along with a cross sectional schematic (b) in-grown stacking fault, (c) morphological defect originating from a 3C-inclusion with a misoriented 4H region, and (d) micropipes with associated stacking faults.

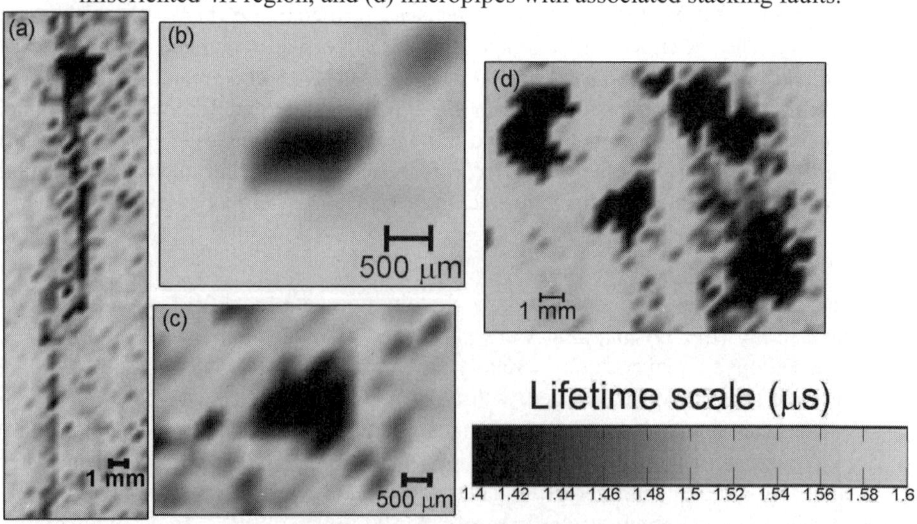

Figure 2: Lifetime map sections from a SiC wafer with 140 μm thick drift layer. (a) cluster of half loop arrays, (b) in-grown stacking fault, (c) morphological defect originating from a 3C-inclusion with a misoriented 4H region, and (d) micropipes with associated stacking faults.

images. Figures 2(a)-(d) show the corresponding lifetime images. In Fig. 1(a), a cluster of seven HLAs is seen along with their interfacial dislocations. Magnified insets show a string of dots, and one of these dots is further magnified to illustrate its structure and compare it with a cross sectional schematic. An HLA is a complex extended defect that contains short BPD segments bounded by two threading edge dislocations. It is formed by the strain-induced glide of a single BPD during the epigrowth [8-10]. During the HLA formation, an interfacial misfit dislocation, often seen to the left side of the HLAs (near to the substrate) is formed to relieve the stain. The HLAs tend to be longer in thicker epilayers and can span several centimeters. The BPD segments in the HLAs will form stacking faults (SF) during device operation, which would continue to expand until it spans from the substrate to the sample surface. From the lifetime map of the HLAs, shown in Fig. 2(a), lifetime degradation is observed at the half loop arrays (HLA) before the stacking faults have expanded no more than 10 μm. Upon superimposing the two maps, the lifetime degradation appears to correlate with the string of BPD segments and not the interfacial dislocation. Figs. 1(b) and 2(b) show the UVPL and lifetime map of an individual in-grown stacking fault, respectively. This defect also spans from the substrate to the epilayer surface, and is thus larger in thick epilayers. Hence, lifetime degradation, as observed in Fig. 2(b), would affect device performance. Another recently reported defect [11] that is a morphological defect originating from a small 3C-SiC particle, which generates a misoriented 4H-SiC region, is seen in Fig. 1(c). This defect also nucleates BPDs in the epilayer. The corresponding lifetime image, shown in Fig. 2(c), shows severe lifetime degradation at this

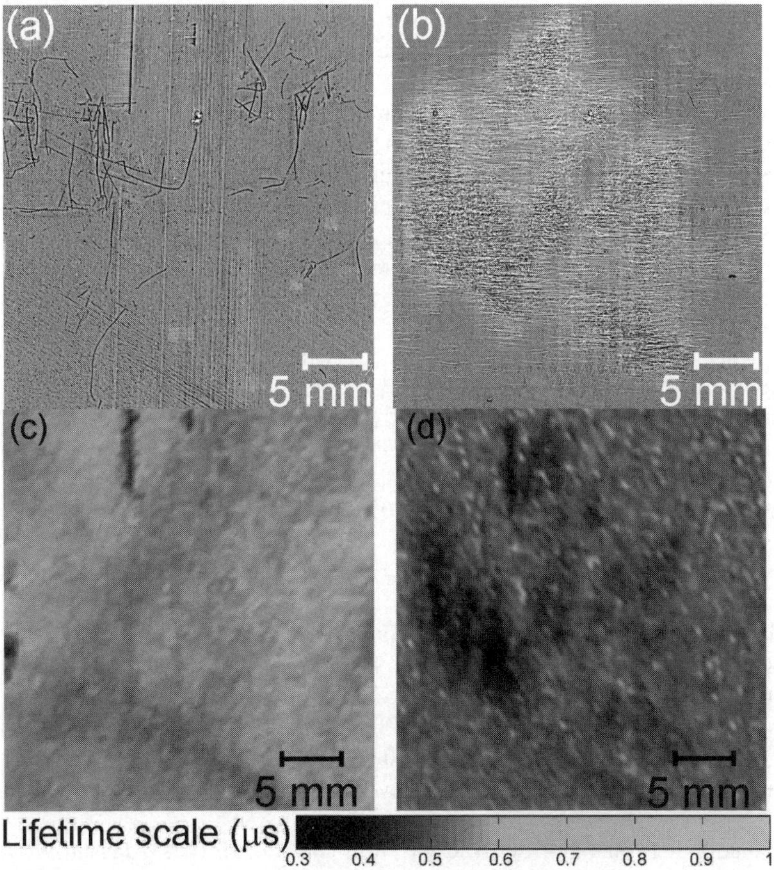

Figure 3: UVPL images of a wafer section (a) before and (b) after high temperature annealing showing generation of slip bands and BPDs. Corresponding lifetime maps of the wafer section (c) before and (d) after annealing showing lifetime degradation from the extended defects.

defect. The various defects shown above depend on various sample and growth conditions. In the 4" wafers studied, 0-30 HLAs were observed. However, these HLAs can span several centimeters and grow upon device operation. The micropipe density was 0-0.05 cm^{-2}, inclusions and in-grown faults were ~0.5 cm^{-2}. There were several areas of the wafers without these defects.

The UVPL and lifetime mapping done on the 150 μm thick samples before and after high temperature annealing showed a ~50% reduction in average carrier lifetime. Earlier studies reported a significant enhancement in carrier lifetime after similar high temperature annealing [3], but these studies performed a carbon implantation into the epilayers prior to the anneal. The anneal in this work also generated slip planes and associated BPDs. Figures 3(a) and 3(b) show UVPL images from our samples before and after the anneal. Figure 3(a) shows a relatively low defect region near the center of the wafer. The dark lines are scratches in the substrate and do not contribute to lifetime in the thick epilayers. After annealing, the same region of the wafer, shown in Fig. 3(b), indicates presence of slip planes and BPDs. These slip planes either relieve strain present in the wafer before annealing or are caused by strain generated during annealing. Both the slip planes and the BPDs are very detrimental to device properties [12]. The lifetime maps shown in Figs. 3(c) and 3(d) show the additional lifetime lowering from the BPDs and slip bands.

Conclusions

Correlation of carrier lifetime with extended defects in thick epilayers was performed using whole wafer UVPL imaging and μPCD mapping. We identified lifetime degradation in 140 μm thick samples with the presence of HLAs even before significant faulting. Lifetime lowering was also observed for the defect comprising of a small 3C-SiC inclusion with a larger misoriented 4H-SiC region. Previously reported lifetime degradation with other extended defects such as in-grown faults, low angle boundaries, micropipes, carrots, and slip planes was also observed. Additionally, creation of extended defects after high temperature (1700 °C/30 min) annealing was observed and contributed to the local lifetime reduction. As epitaxial layer thickness increases, extended defects have a greater impact due to longer growth time and greater defect propagation.

References

[1] H. Lendenmann, N. Johansson, D. Mou, M. Frischholz, B. Astrand, P. Isberg, and C. Ovren, Mater. Sci. Forum 338-342 (2000) 1423

[2] T. Hiyoshi and T. Kimoto, Appl. Phys. Exp. 2 (2009) 041101

[3] T. Miyazawa, M. Ito, and H. Tsuchida, Appl. Phys. Lett. 97 (2010) 202106

[4] T. Kimoto, T. Hiyoshi, T. Hayashi, and J. Suda, J. Appl. Phys. 108 (2010) 083721

[5] J. Hassan and J. P. Bergman, J. Appl. Phys. 105 (2009) 123518

[6] S.I. Maximenko, J.A. Freitas, Y.N. Picard, P.B. Klein, R.L. Myers-Ward, K.K. Lew, P.G. Muzykov, D.K. Gaskill, C.R. Eddy Jr., and T.S. Sudarshan, Mater. Sci. Forum 645-648(2010) 211

[7] R.E. Stahlbush, K.X. Liu, Q. Zhang and J.J. Sumakeris, Mater. Sci. Forum 556-557 (2007) 295

[8] R.E. Stahlbush, B.L. VanMil, K.X. Liu, K.K. Lew, R.L. Myers-Ward, D.K. Gaskill, C.R. Eddy, Jr., X. Zhang and M. Skowronski, Mater. Sci. Forum 600-603 (2009) 317

[9] X. Zhang, M. Skowronski, K.K. Liu, R.E. Stahlbush, J.J. Sumakeris, M.J. Paisley, and M.J. O'Loughlin, J. Appl. Phys. 102 (2007) 093520

[10] N. Zhang, Y. Chen, Y. Zhang, M. Dudley, R. E. Stahlbush, Appl. Phys. Lett., 94 (2009) 122108

[11] N.A. Mahadik, R.E. Stahlbush, S.B. Qadri, O.J. Glembocki, D.A. Alexson, K.D. Hobart, J.D. Caldwell, R.L. Myers-Ward, J.L. Tedesco, C.R. Eddy Jr., and D.K. Gaskill, J. Electron. Mater. 40 (2011) 413

[12] J.P. Bergman, H. Lendenmann, P.A. Nilsson, U. Lindefelt, and P. Skytt, Mater. Sci. Forum, 353-356 (2001) 299

Materials Science Forum Vols. 717-720 (2012) pp 301-304

doi:10.4028/www.scientific.net/MSF.717-720.301

Characterization of Annealed HPSI 4H-SiC for Photoconductive Semiconductor Switches

Cameron Hettler[1,a], W.W. Sullivan III[1,b] and J. Dickens[1,c]

[1]Center for Pulsed Power and Power Electronics, Dept. of Electrical and Computer Engineering, Texas Tech University, Lubbock, TX 79409 USA

[a]cameron.hettler@ieee.org, [b]william.sullivan@ttu.edu, [c]james.dickens@ttu.edu

Keywords: recombination lifetime, microwave photoconductivity decay, photoconductive semiconductor switch.

Abstract. Annealing of high purity semi-insulating (HPSI) 4H-SiC is investigated as a method to improve bulk photoconductive semiconductor switches through recombination lifetime modification. Five samples of HPSI 4H-SiC were annealed at 1810 °C for lengths of time ranging from 3 to 300 minutes. The recombination lifetime of the unannealed and annealed samples was measured using a contactless microwave photoconductivity decay (MPCD) system. The MPCD system consists of a 35 GHz continuous microwave probe and a tripled Nd:YAG pulsed laser. The recombination lifetime was increased from 6 ns, as received, up to 185 ns by annealing for 300 minutes. To experimentally verify switch improvements, identical switches from unannealed and annealed material were fabricated and tested at low voltage. The unannealed device generated a 15 ns pulse with a 2 ns rise-time. The annealed device conducted for upwards of 300 ns with a comparable 2 ns rise-time. The increased recombination lifetime resulted in lower on-state resistance and increased energy transfer.

Introduction and Background

In a variety of pulsed power systems, there is a need for a highly-controllable, long-life, high voltage switch. A few requirements for this device are high voltage (>20 kV), low jitter (< 1 ns), and a compact package with few required external components (gases, pumps, lasers, drive electronics). One option for these applications is a bulk, high-power photoconductive semiconductor switch made from a wide band gap semiconductor, such as silicon carbide. A bulk SiC photoconductive switch offers many potential benefits over traditional gas switches such as negligible jitter, fast rise-time (< 1 ns), electrical isolation, long device lifetime, and overall compactness. Furthermore, a bulk switch structure from high-resistivity material is attractive because it could be significantly lower cost (compared to thick low-doped epitaxy) and could block higher voltages than a junction device.

Previous work in SiC photoconductive semiconductor switches (PCSSs) have been limited by short pulse widths (<20 ns) and high laser power requirements (>1 MW) [1, 2, 3]. The primary reason for this performance limitation is the short recombination lifetime in both vanadium-compensated and high-purity, semi-insulating (HPSI) SiC [4]. Recombination lifetimes in each material have been measured to be 100-200 ps and 5-7 ns, respectively [4,5].

Several groups have demonstrated that the recombination lifetime of HPSI 4H-SiC can be increased by annealing at high temperatures (>1700 °C) which reduces the concentration of carrier trapping defects (primarily Z1/Z2) [4,5,6]. However, the recombination lifetime of vanadium-doped SiC cannot be modified by annealing because the trapping mechanism is through atomic transitions ($V^{4+\rightarrow 5+}$ or $V^{4+\rightarrow 3+}$) rather than crystalline defects [4].

Through removal of shallow defects only (Z1/Z2), it is possible to dramatically increase the recombination lifetime of the HPSI material without changing the overall resistivity of the material. The purpose of this paper is to characterize the annealed HPSI material and to examine the effect of annealing on switch performance.

Annealing and Characterization

Five samples from the same HPSI 4H-SiC wafer were annealed at 1810°C in an induction furnace for various lengths of time ranging from 3 minutes up to 300 minutes. A microwave photoconductivity decay (MPCD) system was used to measure the transient photoconductivity of as-grown and annealed samples. The MPCD system, illustrated in Fig. 1, consists of a continuous wave (CW) 35 GHz Gunn diode as the microwave probe and a pulsed, tripled Nd:YAG laser (355 nm, 14 ns PW) as the exciting light source. When the pulsed laser illuminates the SiC sample, the conductivity of the SiC increases and a portion of the microwaves are reflected backwards to a microwave diode detector. The reflected microwave signal is directly proportional to the transient conductivity of the sample. The MPCD system is a fast, contactless method to measure the recombination lifetime of a semiconductor sample.

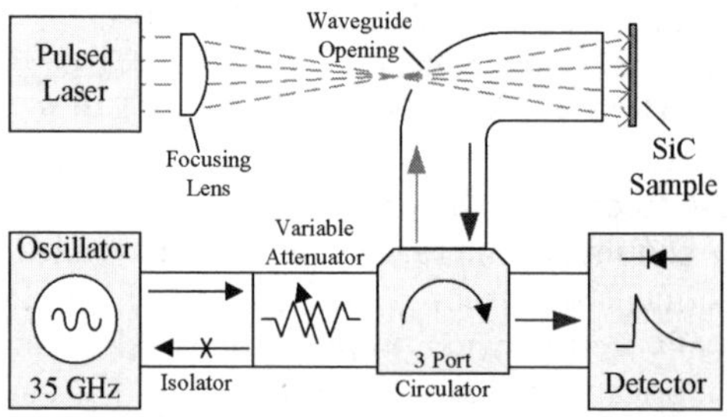

Fig. 1: MPCD system for recombination lifetime measurement [4].

After MPCD tests on the five samples, the recorded photoconductivity transients were curve-fitted to simulations generated from MATLAB. The simulated MPCD waveforms are generated by integrating the recorded laser intensity waveform, as measured by a 5 GHz photodiode. The experimental MPCD results and the corresponding curve-fits are provided in Fig. 2.

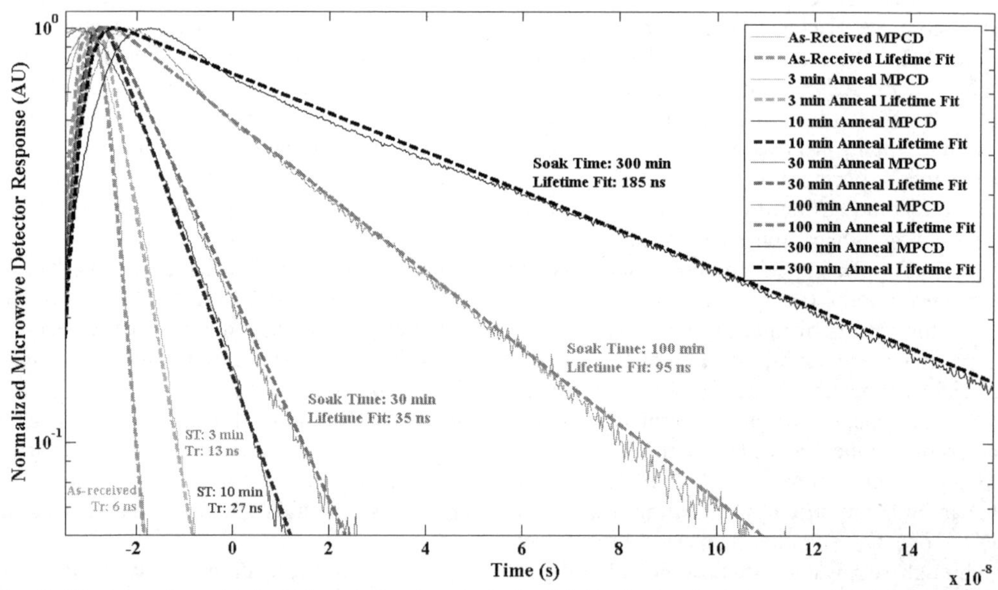

Fig. 2: The experimental (solid) and simulated (dashed) MPCD results for as-received and annealed samples. The 300 minutes anneal increased the recombination lifetime from 6 ns up to 185 ns [6].

The as-received samples all demonstrated recombination lifetimes of 5-7 ns. After 3 minutes at 1810° C, the recombination lifetime increased to 13 ns. The recombination lifetime increased at a nearly linear rate up to 185 ns after a 5 hour anneal.

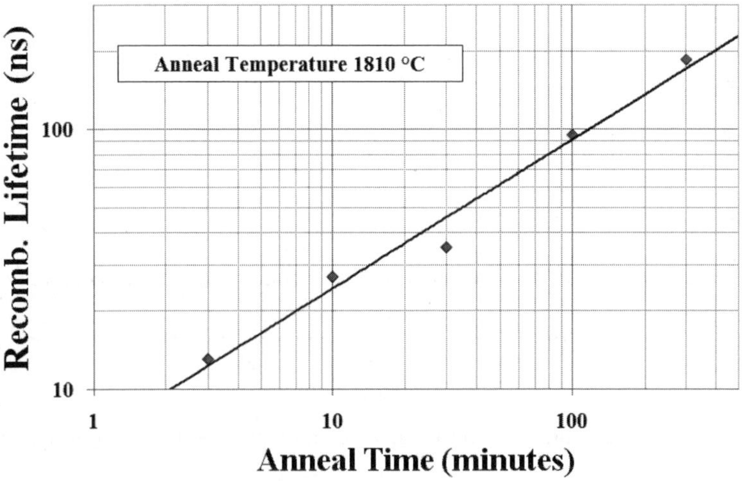

Fig. 3: Measured recombination lifetimes of HPSI 4H-SiC samples annealed at 1810° C.

Experimental Switch Results

Finally, a matching pair of photoconductive switches were fabricated on an as-received sample and an annealed sample for comparison. The switches were 13 x13 mm^2 with the annealed sample heated to 1850 °C for 5 hours. The experimental test setup (Fig. 4) consists of eight parallel 33 nF ceramic capacitors charged to 150 V. The load is a 26 Ω carbon disc resistor and a 0.0495 Ω current viewing resistor. A tunable pulsed laser (365 nm, 8 ns PW) delivers 600 μJ of optical energy into the gridded anode of each photoconductive switch.

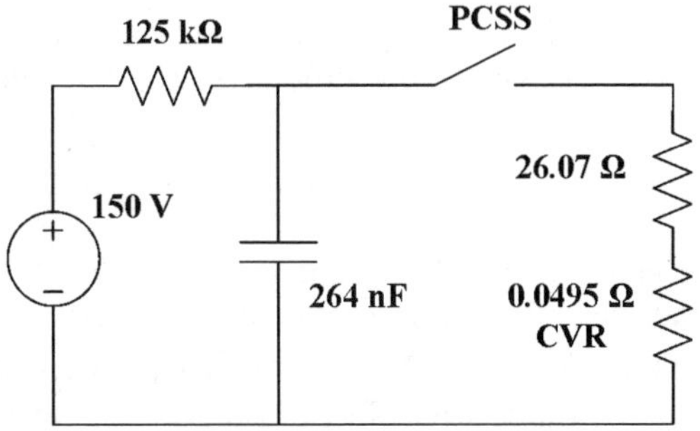

Fig. 4: Experimental setup for switch testing.

In Fig. 5, the as-received switch and annealed switch both demonstrate on-state resistances of less than 0.5 Ω with similar rise-times of approximately 2 ns. As a result of the recombination lifetime increase, the annealed switch conducted for significantly longer than the unannealed device. The annealed device conducted a peak current of 6.1 A with an on-state resistance of less than 50 mΩ. An approximate 300 ns current pulse delivered 128 μJ of energy to the load. The unannealed switch conducted a peak current of 5.9 A with an on-state resistance of 500 mΩ. The 15 ns FWHM pulse from the unannealed PCSS transferred a total of 14 μJ of energy to the load.

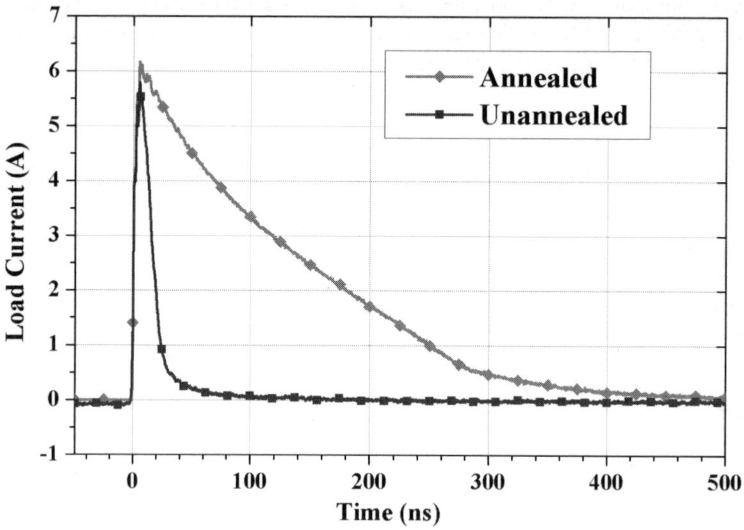

Fig. 5: Low voltage comparison of annealed and unannealed switch performance.

Summary and Conclusions

Recombination lifetime modification has been demonstrated in HPSI 4H-SiC samples for application in photoconductive semiconductor switches. Five samples were annealed at 1810 °C and the increased recombination lifetimes were measured using a contactless MPCD technique. The lifetime improvements were experimentally verified through low voltage tests of identical switches fabricated from annealed and unannealed material. As expected, the unannealed switch conducted for 15 ns because of the measured recombination lifetime of 6 ns. The annealed switch conducted for upwards of 300 ns as a result of the recombination lifetime improvements from high temperature annealing for 5 hours at 1850 °C.

Acknowledgements: This research is funded by the Office of Naval Research. The primary author is supported by the Air Force Office of Scientific Research through the NDSEG fellowship program.

References

[1] J.S. Sullivan and J.R. Stanley, "6H-SiC Photoconductive Switches Triggered at Below Bandgap Wavelengths," IEEE Trans. Ins., 14, 4 (2007) 980-985.

[2] K.S. Kelkar, SiC as a Photoconductive Switch Material for High Power Applications, PhD Dissertation, University of Missouri - Columbia, Columbia, MO, December 2006.

[3] C. Hettler, "Investigation and evaluation of high voltage silicon carbide photoconductive semiconductor switch," M.S. Thesis, Texas Tech University, Lubbock, TX, USA, August 2009.

[4] C. Hettler, C. James, J. Dickens, A. Neuber, "Carrier Lifetime Studies of Semi-Insulating Silicon Carbide for Photoconductive Switch Applications," IEEE Power Modulators and High Voltage Conference, Proceedings of the 2010, May 2010.

[5] J.R. Jenny et al., "Effects of annealing on carrier lifetime in 4H-SiC," Journal of Applied Physics, 100, (2006) 1-6.

[6] C. Hettler, W. Sullivan III, J. Dickens, "Recombination lifetime modification in bulk, semi-insulating 4H-SiC photoconductive switches," presented at 18th International Pulsed Power Conference, Chicago, IL, June 22nd 2011.

Materials Science Forum Vols. 717-720 (2012) pp 305-308
© (2012) Trans Tech Publications, Switzerland
doi:10.4028/www.scientific.net/MSF.717-720.305

Correlation between strain and excess carrier lifetime in a 3C-SiC wafer

Atsushi Yoshida[1,a], Masashi Kato[1,b], Masaya Ichimura[1,c]

[1] Graduate School of Engineering, Nagoya Institute of Technology, Gokiso, Showa, Nagoya, 466-8555 Japan

[a]ciq16607@stn., [b]kato.masashi@, [c]ichimura.masaya@}nitech.ac.jp

Keywords: 3C-SiC, carrier lifetime, strain, μ-PCD method

Abstract. We have obtained excess carrier lifetime maps by the microwave photoconductivity decay (μ-PCD) method in a free-standing n-type 3C-SiC wafer. We compared these lifetime maps with distributions of strains and defects observed by optical microscopy and Raman spectroscopy. We found that the excess carrier lifetimes are short in a strained region in the wafer. This result indicates that structural defects exist around the strained region.

Introduction

3C-SiC is a promising material for low loss and high voltage power devices. Compared with devices fabricated from 4H- and 6H-SiC, 3C-SiC devices are expected to have high electron channel mobility owing to a high quality metal-oxide-semiconductor (MOS) interface. In addition, the electric breakdown field of 3C-SiC MOS field effect transistors (FETs) has been reported to reach 8.5 MV/cm, which is close to values of 4H- and 6H-SiC MOSFETs [1]. However, 3C-SiC devices have never been commercialized because of the presence of defects and strains due to lattice mismatch between Si and 3C-SiC, since 3C-SiC is usually grown epitaxially on Si substrates. Therefore, effects of defects and strains on electrical characteristics should be elucidated. Although some studies of carrier lifetime in 3C-SiC wafers have been reported so far [2-4], correlation between distribution of defects and carrier lifetimes is not clarified. In this work, we evaluated excess carrier lifetime by the microwave photoconductivity decay (μ-PCD) method, and then we compared the lifetime maps with distributions of strains and defects.

Experiments

The sample was a free-standing n-type 3C-SiC wafer. It was epitaxially grown on an undulant Si (100) substrate. The Si (100) substrate has been removed by polishing. A photograph of the sample is shown in Fig. 1. As can be seen in this figure, there are stripes in the sample along the [011] direction. These stripes correspond to positions of depressions and strains discussed below. We defined the grown surface as the epi-side, and the Si substrate-side surface as the sub-side. In the μ-PCD method, we used a 355 nm pulsed YAG laser (the penetration depth is 4.7 μm, the excitation density averaged within the penetration depth from the surface is 1.2×10^{20} cm^{-3} and the spot diameter is 200 μm) as an excitation source and microwaves of 10 GHz as a probe. In the measurements, we used rectangular waveguides in which the TE_{10} mode is the basic mode of the microwave conduction. When the stripes of the sample are parallel to the electric field in the μ-PCD measurements, we call the configuration "parallel". When the strips are perpendicular to the electric field, we call the configuration "perpendicular". From excess carrier decay curves obtained by μ-PCD, we mapped 1/e lifetime, which is defined as the time interval of the excess carrier decay from a peak to 1/e. The strain fields were observed by microscope with crossed polarizers. Defects around a strained region were characterized by micro Raman spectroscopy and observation after molten NaOH etching. Crystalline quality of the sample was evaluated by X-ray diffraction. Surface profiles were obtained by laser microscopy, which measures surface roughness by detection of focus points of a laser beam.

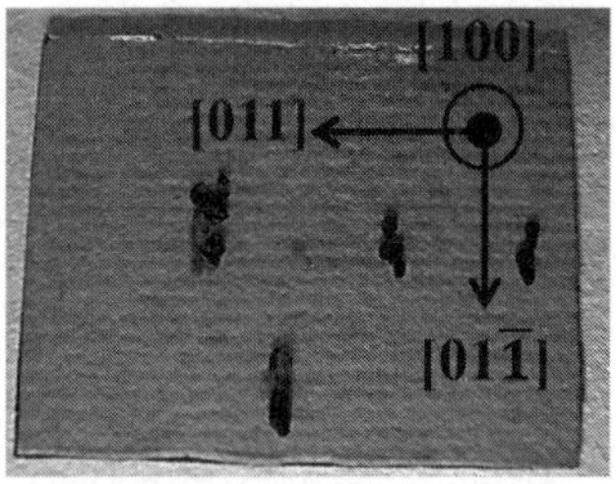

Fig. 1: Sample photograph. Black dots are markings for microscopic measurements.

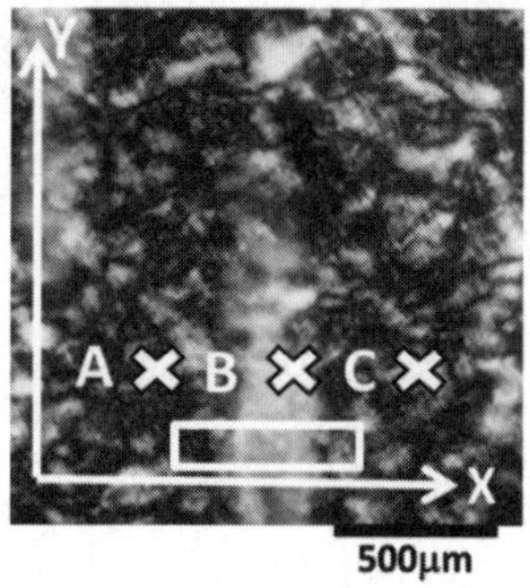

Fig. 2: Optical micrograph with crossed polarizers.

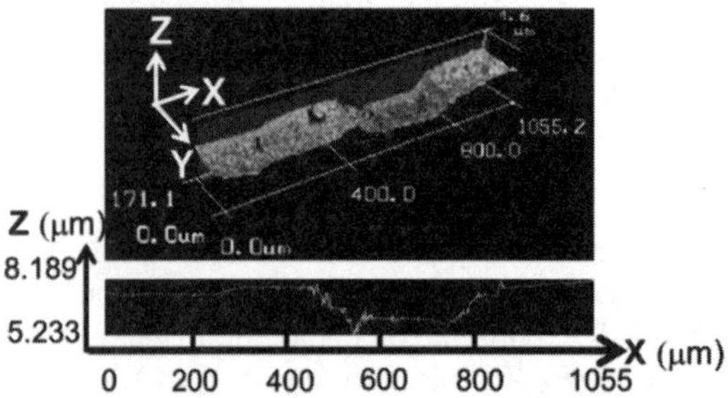

Fig. 3: Profile of the sub-side along the X axis of Fig.2.

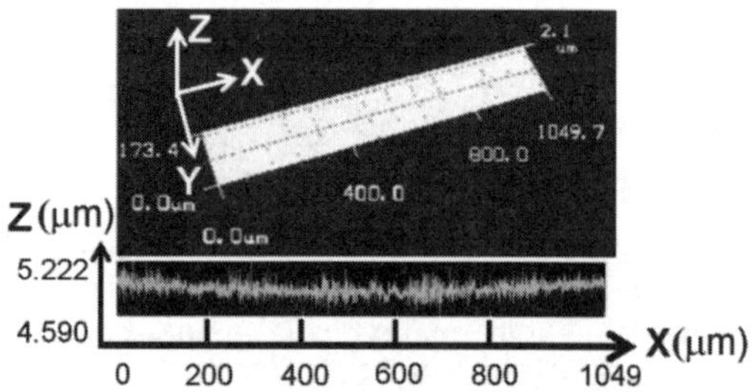

Fig. 4: Profile of the epi-side along the X axis of Fig.2.

Results and Discussion

An optical micrograph of the sample with crossed polarizers is shown in Fig. 2. Contrast due to strains is observed at the middle lower part in this figure. Profiles of the sub- and epi-sides along the X-axis in Fig. 2 are shown in Fig. 3. A depression is observed in Fig. 3, and this depression seems to be located at the strained region in Fig. 2. In contrast, the epi-side is fairly flat as shown in Fig. 4.

The carrier lifetime map for the sub-side measured with the perpendicular configuration is shown in Fig. 5. We found a region with short 1/e lifetimes elongated parallel to the Y axis at X = 500-800 μm. Points A, B and C in this figure correspond to the points in Fig. 2 and the excess carrier decay curves for these points are shown in Fig. 6. These decay curves show initial fast decay components followed by slow decay components (for times > 3 μs), and the time constant for the initial decay at point B corresponding to the strained region is smaller than those for points A and C.

The carrier lifetime map for the epi-side measured with the perpendicular configuration is shown in Fig. 7. The mapped regions are slightly different from Fig. 5, but the axes are scaled so that the coordinates in Figs. 5 and 7 correspond. We found a region with short 1/e lifetimes elongated parallel to the Y axis at X=600-900 μm similarly to Fig. 5. Thus carrier lifetimes in the strained region are short in measurements on both sides. On the other hand, as shown in Figs. 2 and 3, the strained region is depressed on the sub-side but is flat on the epi-side. Therefore, we did not find correspondence between 1/e lifetimes and roughness of the surface.

1/e lifetimes obtained from the epi-side are longer than obtained from the sub-side for most of the regions. For the X-ray rocking curves, the full width at half maximum (FWHM) of the 3C-SiC (200) peak for the epi- and sub-sides are 0.064° and 0.040°, respectively. Generally, the carrier lifetime is long in a high-quality single crystal with small FWHM in the rocking curve, but 1/e lifetimes for the epi-side are larger than those for the sub-side. Therefore, the 1/e lifetime difference between the epi- and sub-sides does not have a general relationship to the crystallinity observed by X-ray diffraction.

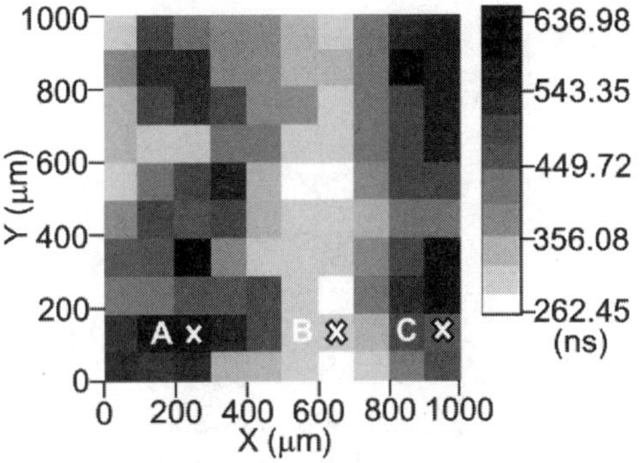

Fig. 5: Carrier lifetime map for the sub-side with the perpendicular configurration.

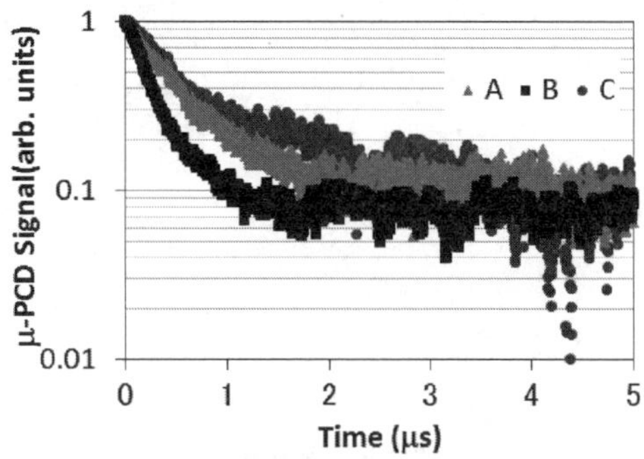

Fig. 6: Excess carrier decay curves for points A, B and C in Fig. 5.

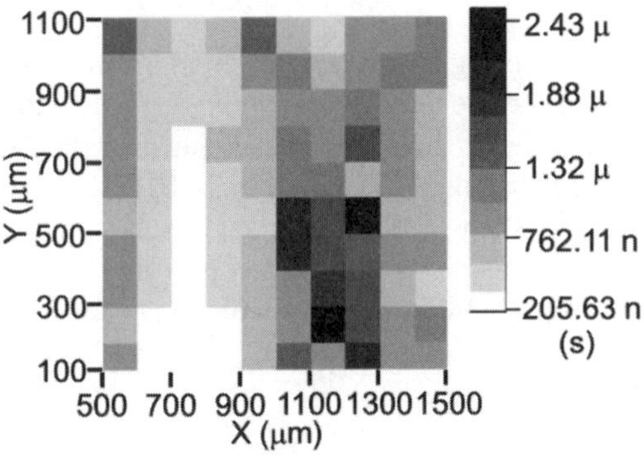

Fig. 7: Carrier lifetime map for the epi-side with the perpendicular configuration.

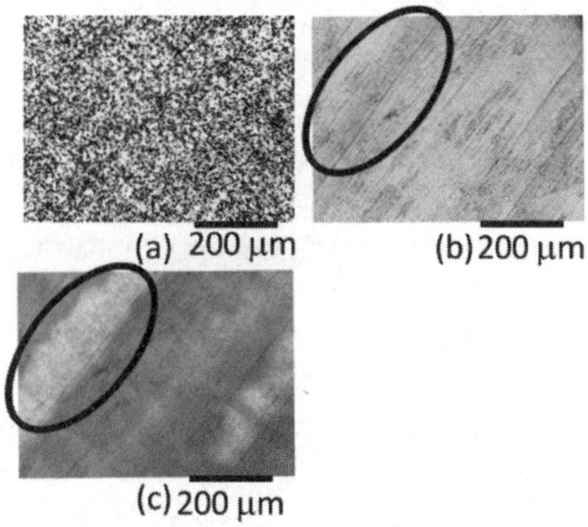

Fig. 8: Optical micrographs after NaOH etching for (a) the sub-side, (b) the epi-side and (c) observed with crossed polarizers. Areas surrounded by ellipsoid indicate a strained region.

After mechanical polishing on the sub-side to remove the depression on the 3C-SiC, the sample was etched by molten NaOH at 550 °C for 10 min. The optical micrographs for each side, from the same region of the sample, are shown in Figs. 8 (a) and (b) along with the micrograph with crossed polarizers (Fig. 8 (c)). In Fig. 8 (a), irregular shaped pits are observed, while line pits are observed in Fig. 8 (b). Contrast due to strains is observed inside the ellipsoid on Fig. 8 (c). We believe that the line pits on the epi-side indicate the presence of stacking faults (SF), and their density seems to be higher in the strained region by comparison of Fig. 8 (b) with the cross polarized image in Fig. 8 (c). This result suggests that the strain regions have a high density of SF.

The Raman spectra inside and outside of the strained regions are shown in Fig. 9. A peak of the TO phonon is observed at 790 cm^{-1}. Because the TO phonon is a forbidden mode for the 3C-SiC (001) orientation in the backscattering arrangement, we assume that the TO peak intensity is related to defect density. The maps of TO/LO peak intensity ratios in the area surrounded by the white line in Fig. 2 are shown in the inset of Fig. 9. The TO intensity becomes larger in the strained region, and thus defect density would be higher in the strained region as observed by etch pits in Fig 8 (b). This result suggests that carrier lifetime decreases due to presence of the defects in the strained region in the 3C-SiC wafer.

A carrier lifetime map for the sub-side measured with the parallel configuration is shown in Fig. 10. Because the sample was rotated by 90°, the X- and Y- axis are also rotated from Figs. 5 and 7. In this figure, we did not find any region with shorter 1/e lifetimes. On the other hand, we were not able to obtain a carrier lifetime map for the epi-side with the parallel configuration because of weak μ-PCD

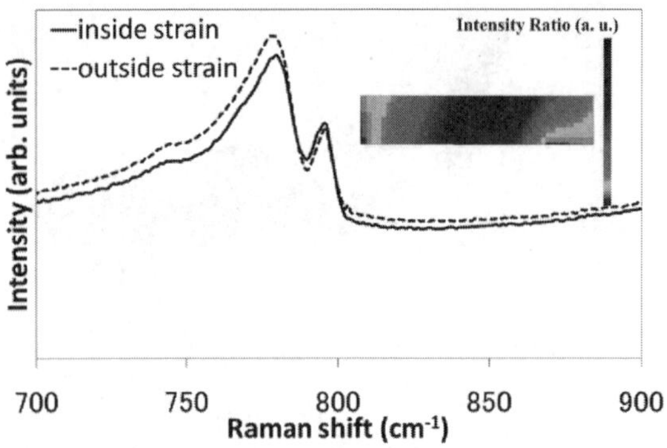

Fig. 9: Raman spectra for regions inside and outside of strain fields. The inset is a Raman map for the TO/LO peak intensity ratio in the area surrounded by the white line in Fig. 2.

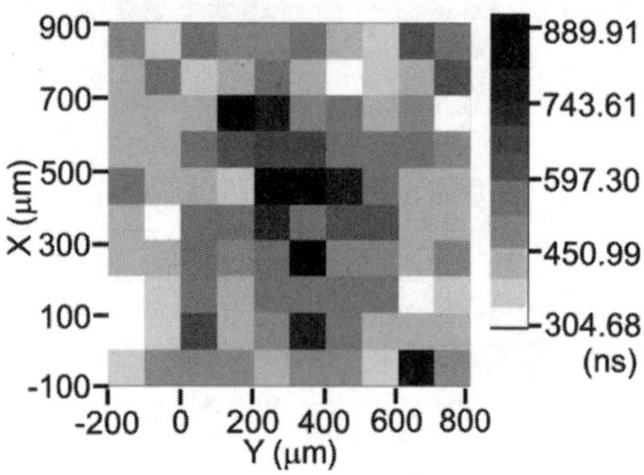

Fig. 10: Carrier lifetime map for the sub-side with the parallel configurration.

signals. Thus μ-PCD signals measured with the parallel configuration are different from those measured with the vertical configuration. The movement of the excited carriers should depend on the direction of the microwave electric field and should be different between measurements with the parallel and vertical configurations. It has been reported that the Hall mobility is larger in the direction parallel to SF than in the direction perpendicular to SF in 3C-SiC grown on undulant Si [5], and thus a carrier scattering mechanism in such 3C-SiC is anisotropic. We can expect that carrier recombination mechanisms in our 3C-SiC are also anisotropic, and the carrier lifetime map depends on the configuration of the sample in the μ-PCD measurements. However, we need further investigation for this configuration dependence of the carrier lifetime.

Conclusions

We have mapped the carrier lifetime in 3C-SiC epitaxially grown on an undulant Si (100) substrate by the μ-PCD method. We found that the carrier lifetime is shorter in strained regions but does not depend on the crystallinity observed by X-ray diffraction. From the Raman spectroscopy and the pit distribution after etching, we also found that defects exist with high concentration in the strained region. These results suggest that the carrier lifetimes decrease in the region with strain and defects.

Acknowledgment

We would like to thank Prof. K. Jarasiunas and Mr. P. Scajev of Vilinius University for their fruitful discussion.

References

[1] T. Ohshima, K. K. Lee, Y. Ishida, K. Kojima, Y. Tanaka, T. Takahashi, M. Yoshikawa, H. Okumura, K. Arai and T. Kamiya, Jpn. J. Appl. Phys. Vol. 42 (2003) p. L625.

[2] V. Grivackas, G. Manolis, K. Gulbinas, K. Jarašiūnas, and M. Kato, Appl. Phys. Lett. Vol.95 (2009) 242110.

[3] M. Ichimura, N. Yamada, H. Tajiri, and E. Arai, J. Appl. Phys. Vol. 84 (1998) p.2727.

[4] P. Ščajev, J. Hassan, K. Jarašiūnas, M. Kato, A. Henry, and J. P. Bergman, J. Electron. Mater. Vol.40 (2010) p.394.

[5] H. Nagasawa, K. Yagi, T. Kawahara, N. Hatta, G. Pensl, W. J. Choyke, T. Yamada, K. M. Itoh and A. Schöner: W. J. Choyke, H. Matsunami, G. Pensl (Eds.), Silicon Carbide, Springer-Verlag, Berlin, Heidelberg (2004) p.207.

Materials Science Forum Vols. 717-720 (2012) pp 309-312
© (2012) Trans Tech Publications, Switzerland
doi:10.4028/www.scientific.net/MSF.717-720.309

Carrier Diffusivity in Highly Excited Bulk SiC, GaN, and Diamond Crystals by Optical Probes

K. Jarašiūnas[1,5,a,*], P. Ščajev[1,b], T. Malinauskas[1,c], M. Kato[2,d], E. Ivakin[3,e], M. Nesladek[4,f], K. Haenen[4,g], Ü. Özgür[5,h], and H. Morkoç[5,i]

[1] Institute of Applied Research, Vilnius University, Vilnius LT-10222 Lithuania

[2] Department Electrical Engineering, Nagoya Institute of Technology, Nagoya 466-8555, Japan

[3] Institute of Physics, National Academy of Sciences, Minsk, 220072, Belarus

[4] Institute for Materials Research, Hasselt University, IMOMEC division IMEC, BE-3590 Diepenbeek, Belgium

[5] Department of Electrical Engineering, Virginia Commonwealth University, Richmond, VA 23228 USA

[a]kestutis.jarasiunas@ff.vu.lt, [b]patrik.scajev@ff.vu.lt, [c]tadas.malinauskas@ff.vu.lt, [d]kato.masashi@nitech.ac.jp, [e]ivakin@dragon.bas-net.by, [f]milos.nesladek@uhasselt.be, [g]ken.haenen@uhasselt.be, [h]umit.ozgur@vcu.edu, [i]hadis.morkoc@vcu.edu *corresponding author

Keywords: gallium nitride, diamond, nonequilibrium carriers, light-induced transient gratings, ambipolar diffusion, degeneracy.

Abstract. Optical monitoring of diffusivity in wide bandgap semiconductors was performed by using a picosecond light-induced transient grating technique. The bandgap renormalization and carrier-carrier scattering manifested themselves at room temperature as a two-fold decrease of the ambipolar diffusion coefficient D_a in cubic SiC and a 5-fold decrease of D_a in diamond at excess carrier density $N > 10^{17}$ cm^{-3}, while for GaN the impact was observed only at T < 100 K. For injections above 10^{19} cm^{-3}, the plasma degeneracy led to enhanced D_a values in SiC and GaN and lessened the diffusivity decrease in diamond.

Introduction

High performance power devices would benefit from knowledge of carrier dynamics at high injection levels up to 10^{19} cm^{-3}. In this realm, the carrier diffusion is a critical parameter for carrier transport in high density plasma when the effects of bandgap renormalization (BGR), many-body interaction, and plasma degeneracy may have considerable impact. Hall effect and time-of-flight (ToF) measurements are commonly used to gauge carrier mobility, where the former requires good ohmic contacts and the latter low conductivity. In contrast, optical monitoring of diffusivity (or mobility) can be easily realized by the time-resolved light-induced transient grating (LITG) technique [1], which is conductivity insensitive and does not require electrical contacts. Its potential has been demonstrated by measurements of minority carrier diffusion (electron mobility) in heavily doped GaAs (up to 2×10^{19} cm^{-3}) [2] as well as bipolar mobility in semi-insulating GaAs and CdTe [3].

In this paper we treat the peculiarities of carrier-density dependent diffusivity, which has been investigated experimentally in wide bandgap semiconductors, namely silicon carbide, gallium nitride, and diamond, and modeled numerically.

Samples and Experimental Techniques

We investigated bulk crystals of cubic SiC, grown by CVD on Si (200 μm thick, E_g=2.36 eV), GaN grown by hydride vapor phase epitaxy (HVPE) (50 and 200 μm thick, E_g =3.45 eV), and bulk diamond (900 and 400 μm, E_g=5.5 eV). The diamond samples were grown by high-pressure high temperature (HPHT) synthesis or by CVD. For carrier generation we used harmonics of a picosecond YAG:Nd^{3+} (or YLF:Nd^{3+}) laser which provided wavelengths required for interband carrier injection (355 nm for 3C-SiC and GaN, and 213 nm for diamond), and 527(351) nm wavelengths for two-photon carrier generation in bulk GaN and diamond. Carrier injection by two coherent pump beams created spatially modulated carrier density $N(x) = N_0 + \Delta N(1+\cos(2\pi x/\Lambda))$ with varying spacing Λ. The experimental scheme is shown in Fig. 1 and utilizes a diffraction optical element as the holographic beam splitter (HBS) [4]. The temporal evolution of the carrier-induced refractive index change, $\Delta n(x,t) \propto \Delta N(t)$, was monitored through the diffraction efficiency of a delayed probe beam at $\lambda = 1053$ nm. The LITG exponential decay with time τ_G is given by $1/\tau_G = 1/\tau_R + 1/\tau_D = 1/\tau_R + 4\pi^2 D/\Lambda^2$, where τ_R, τ_D, and D are the carrier lifetime, diffusive grating decay time, and diffusion coefficient, respectively. This simple relationship was used to determine τ_R and D values from the $1/\tau_G \propto 4\pi^2/\Lambda^2$ plot. The experiments were performed at various carrier densities and temperatures for density and temperature dependences of diffusivity.

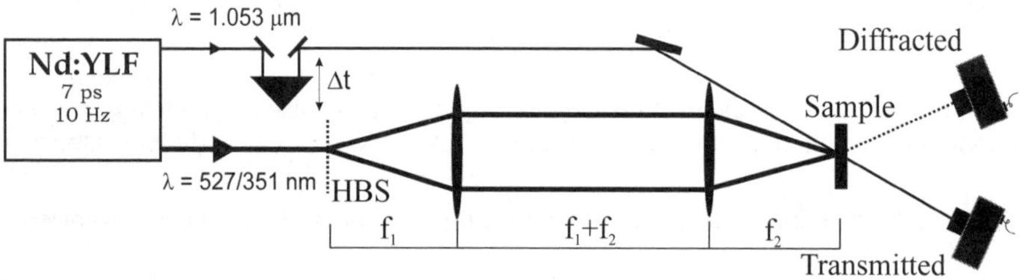

Fig. 1. Optical setup for recording and probing of a light-induced transient grating. Here f_1 and f_2 are the focal lengths of the telescope lenses. An innovative solution to use a holographic beam splitter makes the LITG technique as simple as the well-known "pump-probe" schemes with a single pump beam.

Experimental Data and Discussion

Fig. 2 displays the grating decay kinetics in n-type bulk 3C-SiC (residual density $n_0 \approx 5\times10^{15}$ cm^{-3}) at a carrier density of ~10^{18} cm^{-3} injected into the ~5 μm-thick region, as follows from the *a priori* knowledge of the absorption coefficient $\alpha = 2000$ cm^{-1} at 351 nm. The kinetics with subnanosecond decay times are attributed to the diffusive decay, as carrier lifetimes in the range from 30 to 60 ns were measured in these crystals [5]. Consequently, the values of the room temperature (RT) bipolar diffusion coefficient $D_a = 3.6$ cm^2/s, mobility $\mu_a = 145$ cm^2/Vs, and diffusion length L_D ~5 μm follow. We note that the L_D value in sublimation-grown 3C SiC as well in thin CVD layers on 6H SiC was only 1 μm, thus indicating short carrier lifetime. The temperature dependence of D_a (Fig. 3a) led to the values of bipolar mobility in the 80-800 K range and revealed the dominant carrier scattering mechanisms (by acoustic phonons at T < 250 K and polar optical phonons at T > 250 K). Similarly, the D dependence on the excitation density $D(N)$ was measured in 3C-SiC (Fig. 3b).

In bulk GaN, D values were measured for two distinct cases of carrier injection: that in a very thin surface layer of ~100 nm thickness for single absorption of laser quanta at 355 nm excitation wavelength, or in the entire bulk of 200 μm thick GaN with two-photon absorption of picosecond pulses at 527 nm. Correspondingly, the excess carrier density varied from 10^{19} cm^{-3} down to 10^{15} cm^{-3}, below the doping level of bulk GaN ($n_0 \approx 9\times10^{15}$ cm^{-3}). These two cases are analyzed below.

The extremely high carrier density ΔN resulted in an enhanced recombination rate due to bimolecular recombination ($1/\tau_R \propto B\Delta N$), leading to non-exponential LITG decay (Fig. 2b). Similarly, a faster diffusive decay rate for short period gratings (Λ=2.7 and 3.2 μm) revealed enhanced carrier diffusivity due to plasma degeneracy [6]. The latter effect was more pronounced in the initial stages of the grating decay (similar to the impact of nonlinear recombination), thus making the diffusive decay strongly non-exponential (see Fig. 2b, Λ=2.7 μm). To verify the assumption that carrier degeneracy determines the variation of the diffusion coefficient, we performed numerical modeling of carrier dynamics at high excitation conditions. We solved the two-dimensional carrier continuity equation, which includes carrier generation, linear and nonlinear recombination and diffusion in the degenerate plasma [7]. A strict numerical modeling of $D(N)$ dependence was replaced by an approximation $D(N) = D_0(1+N/N_0)$ with fitting parameters $N_0 = 2.2\times10^{19}$ cm^{-3}, D_0 =1.5 cm^2/s, which correctly described the impact of degeneracy on diffusivity in GaN at RT (see curves in Fig. 2b).

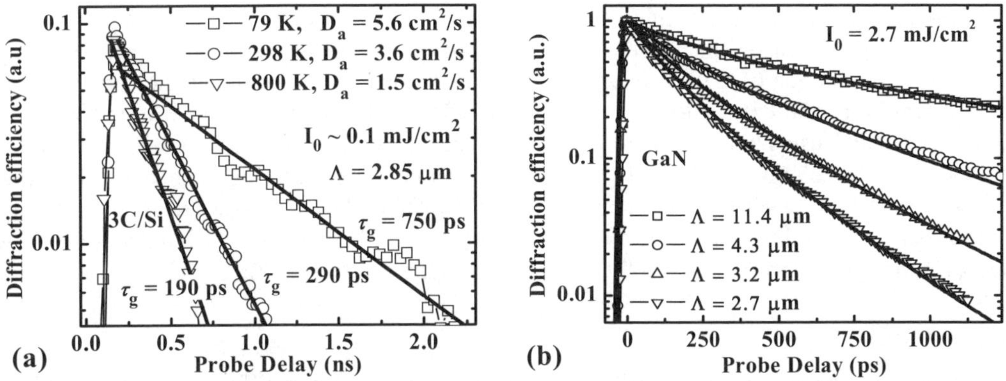

Fig. 2. Light induced diffraction kinetics in 3C/Si (a) and GaN (b) layers under interband carrier injection conditions. Experimental data (points) and numerical modeling (lines) are given on the plots.

In the low excitation regime for bulk GaN, the minority hole mobility and bipolar one were determined by varying the injected bipolar carrier density ($\Delta N_e = \Delta N_p$) below and above $n_0 \approx 10^{16}$ cm^{-3}. The relationship $D_a=[(n+p)D_nD_p]/(nD_n+pD_p)$ describes the cases of minority and bipolar carrier diffusivity, inclusive of the effect of all carriers, both equilibrium (n_0, p_0) and injected (i.e., $n= n_0 +\Delta N_e$ and $p= p_0 +\Delta N_p$). The obtained data for D_a are plotted in Fig. 3b as the low-injection part of the $D(N)$ dependence.

As mentioned previously, electrical and optical techniqies were applied to mobility measurements in high-quality, "nitrogen-free" HPHT and CVD diamond crystals. High electron and hole RT mobility (above 2000 cm^2/Vs, with a corresponding D = 55 cm^2/s value) were determined by electrical ToF measurements [8], but not verified by optical ones. The latter provided only D_a = 9-12 cm^2/s at single photon excitation [9]. This discrepancy was attributed to many-body effects which are known to cause enhanced carrier scattering and bandgap shrinkage resulting in a decrease in D at high injected carrier densities [10,11]. Therefore, $D(N)$ over a wide excess carrier range (10^{15} to 10^{17} cm^{-3}) was obtained under 2-photon carrier injection at 351 nm (see Fig. 3b) and revealed a very notable diffusivity decrease for diamond in this injection range.

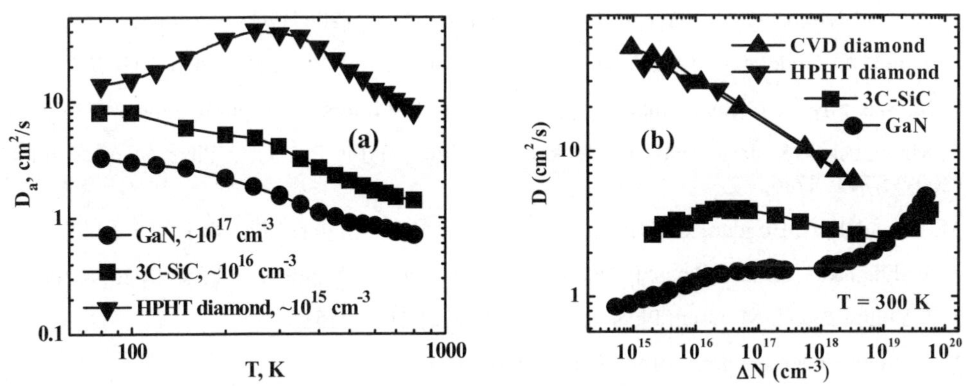

Fig. 3. Dependences of carrier diffusivity in GaN, 3C-SiC, and diamond on temperature (a) and injected carrier density ΔN (b). The dependences D(T) at low injections (a) indicate dominant impact of defects and phonons.

For modeling of hole diffusivity D_h in diamond, we used the equations of Ref. [10]:

$$D_h = D_{h0}\left(\frac{F_{1/2}}{\partial F_{1/2}/\partial x_F} + \frac{\Delta N}{kT}\frac{\partial \Delta E_v}{\partial \Delta N}\right),\ F_{1/2} = \int_0^\infty f_f x^{1/2} dx, \qquad (1)$$

with carrier density dependent Fermi level $x_F = E_F(\Delta N)/kT$, Fermi distribution function $f_f = [\exp(x-x_F)-1]^{-1}$, $x=E/kT$, and using the appropriate carrier energy and temperature dependent scattering rates [12] for calculation of the low-injection diffusivity value D_{h0}. For diamond, the carrier-carrier scattering rate was also taken into account [11]. The effect of band-gap renormalization (BGR) on D_a was calculated using theoretical BGR for the valence band in SiC: $\Delta E_v(\Delta N) = Z\times$ [3 meV $(\Delta N/10^{18})^{1/3}$ + 19 meV $(\Delta N/10^{18})^{1/4}$] [13], Z being a material specific fitting parameter. The value Z = 0.48 for GaN is lower than that for SiC (Z=1.4) probably due to a higher density of valence band states as well as the larger static dielectric constant $\varepsilon_0 = 10.4$ in GaN, which reduces the BGR effect [10] (as opposed to the cases of SiC (ε_0 =9.66) and diamond (ε_0 =5.7)).

Summary

The transient grating technique was utilized to investigate excess carrier diffusivity in wide bandgap semiconductors under high excitation conditions. The diffusivity was found to be reduced by many body effects, which were strongest in diamond, smaller in SiC, and very small in GaN. These peculiarities were explained by carrier scattering and band gap renormalization effects. The increase of diffusivity at even higher injection levels is attributed to plasma degeneracy. The numerical modeling of carrier transport fitted well to the experimental data.

Acknowledgements

The research was supported by FP7 Contract NetTFISiC and CIEE/BAFF foundation (USA).

References

[1] H. J. Eichler, P. Gunter, and D. W. Pohl, Laser-Induced Dynamic Gratings, (Springer, Berlin, 1986) p. 13.

[2] K. Jarašiūnas, R. Aleksiejūnas, T. Malinauskas, V. Gudelis, M. Sudzius, A. Maasdorf, F. Brunner, and M. Weyers, Eur. Phys. J. Appl. Phys. 27 (2004) 181-184.

[3] A. Kadys, Ph. Delaye, G. Roosen, K. Jarašiūnas, Semicond. Sci. Technol. 22 (2007) 1044-1052.

[4] K. Jarašiūnas, R. Aleksiejūnas, T. Malinauskas, V. Gudelis, T. Tamulevičus, A. Guobienė, A. Usikov, V. Dmitriev, H. J. Gerritsen, Rev. Sci. Instrum. 78 (2007) 033901.

[5] P. Ščajev, V. Gudelis, K. Jarašiūnas and P. B. Klein, J. Appl. Phys. 108 (2010) 023705.

[6] S. M. Sze, Physics of Semiconductor Devices (A Wiley-Interscience publication, 1981) p. 29.

[7] T. Malinauskas, K. Jarašiūnas, M. Heuken, F. Scholz, and P. Bruckner, Phys. Status Solidi C 6 (2009) S743–S746.

[8] M. Nesladek, A. Bogdan, W. Deferme, et al, Diam. Relat. Mater. 17 (2008) 1235-1240.

[9] T. Malinauskas, K. Jarašiūnas, E. Ivakin, et al. Phys. Status Solidi A 207 (2010) 2058-2063.

[10] J. F. Young and H. M. Driel, Phys. Rev. B 26 (1982) 2147-2158.

[11] N. H. Fletcher, Proceedings of the IRE (1956) 862-872.

[12] J. Pernot, S. Contreras, and J. Camassel, J. Appl. Phys. 98 (2005) 023706.

[13] C. Persson, U. Lindefelt, B.E. Sernelius, Solid State Electron. 44 (2000) 471-476.

Materials Science Forum Vols. 717-720 (2012) pp 313-318
© (2012) Trans Tech Publications, Switzerland
doi:10.4028/www.scientific.net/MSF.717-720.313

Critical Conditions of Misfit Dislocation Formation in 4H-SiC Epilayers

Xuan Zhang[1,a], Tetsuya Miyazawa[1,b] and Hidekazu Tsuchida[1,c]

[1]Central Research Institute of Electric Power Industry (CRIEPI), 2-6-1 Nagasaka, Yokosuka, Kanagawa 240-0196, Japan

[a]xzhang@criepi.denken.or.jp, [b]yo-miya@criepi.denken.or.jp, [c]tsuchida@criepi.denken.or.jp

Keywords: Misfit Dislocations, Interfacial Dislocations, Basal Plane Dislocations (BPDs), Annealing, Misfit Stress, Thermal Stress, 4H-SiC.

Abstract. Thermal annealing experiments were performed to determine the critical conditions of misfit dislocation formation in 4H-SiC epilayers in a temperature range of 1400-1800 °C. Misfit dislocations were observed to form at a given annealing temperature if the temperature gradient across the epi-wafer exceeded a critical value. It was also found that two types of interfacial dislocations could form under different stress conditions. Their formation mechanisms are discussed.

Introduction

Basal plane dislocations (BPDs) are one of the remaining defects in 4H-SiC epilayers. Besides propagating from the substrates, another mechanism to introduce BPD is stress. Strong evidence is the interfacial dislocations frequently observed in as-grown 4H-SiC epilayers and epilayers processed by high-temperature thermal annealing [1-9]. They are edge-type BPD segments perpendicular to the step-flow direction located at the interface between epilayer and substrate. Several ways of introducing interfacial dislocations have been confirmed, including glide of pre-existing BPDs [3,4,6] and nucleation of BPD half loops at the epilayer surface followed by gliding toward the interface [8]. Both are typical stress-driven dislocation behaviors.

While the characteristics described above are very similar to misfit dislocations, other features of interfacial dislocations are difficult to be explained by the misfit model. For example, they were observed to have a spatial distribution across the epilayer [7,8]. A systematic study on the influence of growth conditions and substrate properties on the formation of interfacial dislocations was performed by Tsuchida *et al* [7]. The results indicated thermal stress during the epitaxial growth played an important role. Direct evidence was obtained by Nagano *et al* [8]. These authors found that interfacial dislocations formed during thermal annealing if a significant radial temperature gradient was exerted on the epi-wafer at high annealing temperatures. Our following study further confirmed the correlation between the distribution of interfacial dislocations and that of thermal stress [9]. Interfacial dislocations were then concluded to be misfit dislocations as a consequence of both misfit stress and thermoelastic stress in the epilayer which have the same sign.

The critical condition of forming misfit dislocations is of particular importance to SiC high-voltage (>10 kV) switching devices like IGBTs, which require thick drift layers as well as multiple stacked layers with different dopings. Formation of misfit dislocations has to be prevented at each interface. Specifically, for each epilayer with its thickness and doping concentration decided by the device design, thermal stresses involved during epitaxial growth and post-growth processing have to be controlled below the critical values at the growth/processing temperatures. The temperature dependence is due to the fact that for a dislocation to glide in a semiconductor material, the shear stress resolved on the dislocation has to exceed the critical resolved shear stress (CRSS) which is strongly temperature-dependent.

In this study, thermal annealing experiments were performed to determine the critical conditions of forming interfacial dislocations in thin (~10 μm) n^- epilayers grown on n^+ substrates in a temperature range of 1400-1800 °C. During the study, two types of interfacial dislocations were observed, which will also be presented in this report. They provide further insight into the formation mechanism of interfacial dislocations.

Experimental

Fig.1. shows the schematics of the heating chamber of the annealer. The crucible and epi-wafer were heated by radiation from the hot wall. During annealing, room temperature argon (Ar) gas was blown through a graphite tube (inner diameter of 1 cm) towards the center of the epi-wafer at a distance of 3 cm above. Such a setup can achieve a temperature distribution with center region of the epi-wafer cooler than its edge. Also, the temperature gradient can be controlled by adjusting the Ar flow rate. Larger thermal gradient can be produced as Ar flow rate increased. In addition, the crucible was made of low thermal conductivity carbon material to help achieve enough temperature gradient. The annealing temperature was in the range of 1100-1800 °C, measured by a pyrometer focusing at the center of the epi-wafer. Radiation emitted from the epi-wafer reached the pyrometer through the graphite tube. The ramp-up rate was 100 °C/min and the ramp-down rate was controlled to be below 35 °C/min. The annealing time was 5 minutes. The epi-wafers tested were of quarter of 3-inch size with ~10 μm unintentionally doped n-type epilayers grown on the Si-face or C-face of commercial n-type conductive 4H-SiC substrates with 8° off-cut towards [11-20]. Grazing incidence synchrotron reflection x-ray topography was performed on the epi-wafers before and after annealing to detect the formation of interfacial dislocations (**g**=-1-128, λ=1.541 Å at SPring-8).

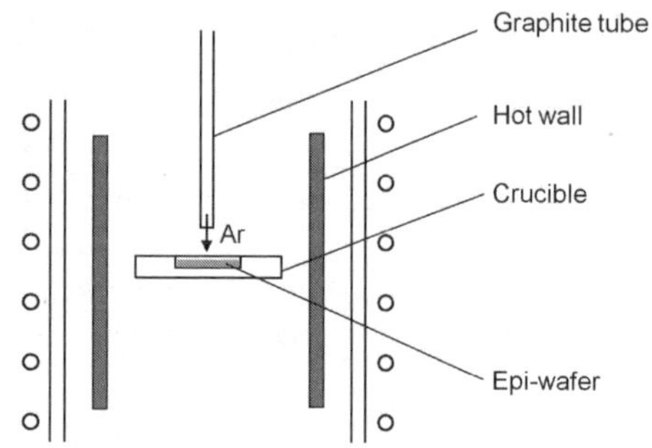

Fig. 1: Schematic drawing of the heating chamber of the annealer.

The critical values of Ar flow rates below which interfacial dislocations could not form were determined for different annealing temperatures. To get the corresponding temperature distribution, the following procedures were performed. First, epi-wafers were implanted with aluminum (Al) ions from the epilayer surface to a depth of 350 nm with a box profile of ~5×10^{18} cm^{-3} at room temperature. Second, activation annealing was performed in the same annealer as shown in Fig. 1, applying the critical Ar flow rate obtained at a certain annealing temperature. Thus the distribution of the sheet resistance (Rs) of the implanted layer was affected by the temperature distribution. Third, metal contacts were fabricated at multiple locations across the epi-wafer and Rs was obtained by the transmission line measurement (TLM). Finally the temperature distribution was determined by converting Rs measured at different locations to temperatures.

Results and discussion

Critical annealing conditions of forming interfacial dislocations. Firstly, the stress distribution produced by a temperature difference between the center of an epi-wafer and its edge is briefly reviewed in order to explain the experimental results clearly. It can be expected that if the center region of an epi-wafer is hotter that its edge, compressive stress will develop in the center region while tensile stress will develop in the region close to the edge. If the center region is cooler than the edge, the stress distribution will be opposite. Considering the off-cut orientation of SiC epi-wafers, the shear stress on the basal plane along the off-cut direction will have a distribution as Fig. 2 shows. In the figure, regions with large shear stresses are encircled by the dashed lines. Positive and negative signs represent shear stresses produced by tensile and compressive normal stresses respectively. It has to be mentioned that the schematic drawings in Fig. 2 only show the approximate distribution of thermal stresses. Detailed analyses can be found in Ref. [9].

As described in the experimental section, the setup of the annealing experiments produced a temperature distribution with center region of the epi-wafer cooler than its edge. According to Fig. 2(b), large compressive

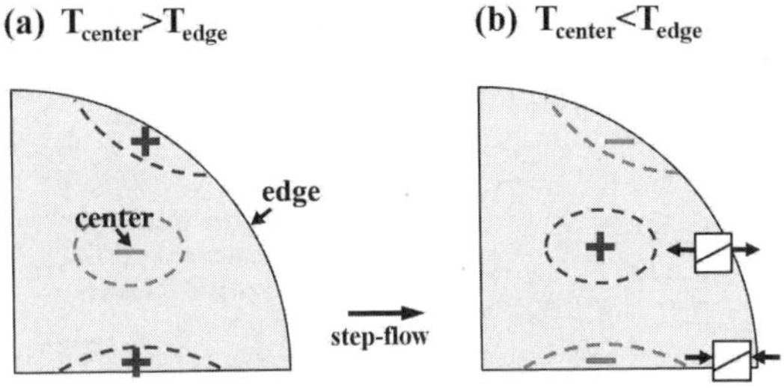

Fig. 2: Regions with large shear stresses on the basal plane of off-cut epi-wafers.

stresses would be produced at the two edges of the epi-wafer perpendicular to the step-flow direction. The X-ray topographs collected from those regions are shown in Fig. 3(a)-(c). The three images were taken from three epi-wafers annealed at 1800 °C (the temperature at the center of the epi-wafer) with Ar flow rate of 1 slm (standard liter per minute), 3 slm and 5 slm, respectively. The positions where the images were collected from the quarters of epi-wafers are illustrated by the schematics at the lower left corners of the topographs. One typical BPD contrast in each epilayer is marked by an arrow. It can be seen that the BPDs in the epilayer annealed with Ar flow rate of 1 slm remained a similar morphology as those in as-grown epilayers. If Ar flow rate increased to 3 slm, interfacial dislocations perpendicular to the step-flow direction started to form. With Ar flow rate further increased to 5 slm, very long interfacial dislocations were observed with their lengths extending several millimeters. As the Ar flow rate increased, larger temperature gradient across the epi-wafer was produced which in turn induced larger thermal stress. The results clearly show that thermal stress is critical to form interfacial dislocations.

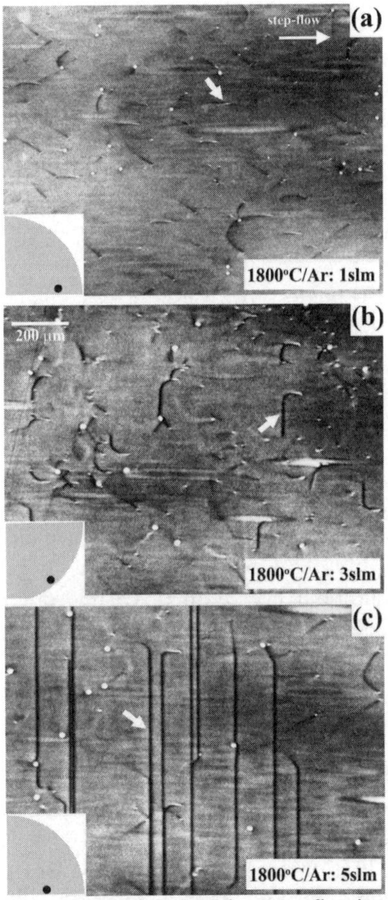

Fig. 3: Grazing incidence reflection X-ray topographs taken from three epi-wafers annealed at 1800 °C with Ar flow rate of (a) 1 slm, (b) 3 slm, and (c) 5 slm, respectively.

In Fig. 3(b) and (c), which were taken from regions with large compressive thermal stress, the interfacial dislocations show black contrast. Later section of this report will explain the details of such contrast. Also the observed BPD behaviors in regions with large tensile thermal stress will be discussed. Here for the purpose of determining the critical annealing conditions of forming misfit dislocations, they will refer to those 'black contrast' interfacial dislocations as shown in Fig. 3(b) and (c). In fact, the interfacial dislocations observed by different research groups [3,4,5,8] have the same characteristics.

The annealing conditions tested are summarized in Fig. 4. At each annealing temperature, different Ar flow rates were tested. The hollow black circles represent the conditions where no misfit dislocations were detected after annealing. The solid red circles represent the conditions where short interfacial dislocations were observed, the maximum length being on the order of 100 μm to a few hundred microns, similar as those shown in Fig. 3(b). The solid black circles represent the conditions where very long misfit dislocations were observed after annealing, similar as those shown in Fig. 3(c). If the critical condition is roughly defined as the situation similar to Fig. 3(b), the critical Ar flow rates were about 5 slm, 4 slm, and 3 slm at temperature of 1500 °C, 1600 °C, and 1800 °C, respectively. More data points need to be collected in order to determine the critical Ar flow rate accurately at each annealing temperature.

The temperature distributions for annealing conditions represented by solid red circles in Fig. 4 were obtained by measuring the distributions of sheet resistance (Rs) of the Al-implanted layers, as introduced in the previous section. The temperature difference between the center and edge of the quarter of 3-inch epi-wafer at 1600 °C was estimated to be ~60 °C. It corresponds to a temperature gradient of about ~5 °C/mm. At 1800 °C, the temperature difference was estimated to be below 10°C. Better temperature uniformity has to be achieved at higher temperatures to prevent the formation of interfacial dislocations because the CRSS for BPDs to glide decreases with temperature.

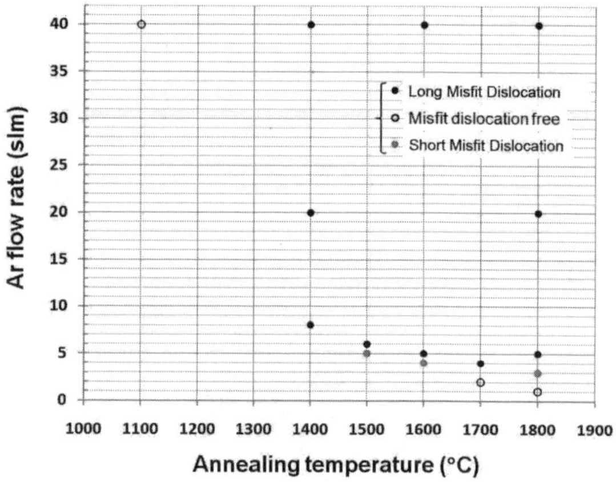

Fig. 4: Temperature dependence of the critical Ar flow rate to produce misfit dislocations in 10 μm 4H-SiC epilayers by thermal annealing.

Response of BPDs in epilayers to thermal stresses. Interfacial dislocations are edge-type BPDs [3]. As explained by Huang *et al.* [10], the contrast of edge-type BPDs in reflection X-ray topographs is determined by the basal plane distortion around the dislocation core. If the extra half plane of an edge BPD points downward, the basal plane becomes like a concave focusing mirror. The higher reflection intensity will produce a darker contrast at the dislocation line, as the cases shown in Fig. 3(b) and (c). Thus the extra half planes associated with those interfacial dislocations are determined to be pointing downward, *i.e.* toward the substrate. Such a configuration seems to indicate a misfit strain of compressive nature in the epilayers. The dislocations formed at interface with their extra half planes in the substrate side could release such compressive strain in the epilayer. A compressive misfit strain coincides with the effect induced by N-doping difference between the epilayer and substrate. N atoms with smaller size substituting C atoms in SiC are expected to reduce the lattice parameter. Thus the low N-doped epilayer grown on highly N-doped substrate will be compressively strained. On the other hand, as stated in the previous section, those interfacial dislocations were found at regions with large compressive thermal stress during annealing. Therefore their formation should be caused by both compressive misfit strain and compressive thermal strain [9].

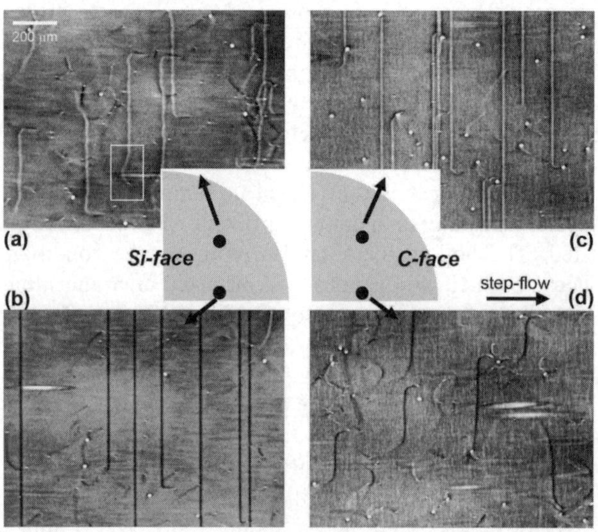

Fig. 5: X-ray topographs taken from the center and edge of one Si-face epi-wafer (a and b) and one C-face epi-wafer (c and d) annealed at 1800 °C with Ar flow rate of 5 slm.

It is interesting to see how BPDs would respond to tensile stress and compare the two situations. Fig. 5 show the X-ray topographs obtained from one Si-face epi-wafer (a and b) and one C-face epi-wafer (c and d), both of which were annealed at 1800 °C with Ar flow rate of 5 slm. As indicated by the schematic drawing, Fig. 5(a) and (c) were taken from the center regions of epi-wafers while (b) and (d) were taken from the regions close to the edges. For both Si-face and C-face epilayers, interfacial dislocations with black contrast (b and d) were found close to the epi-wafer edges with large compressive thermal stresses. However at the centers with large tensile thermal stresses, interfacial dislocations with white contrast (a and c) were observed. The white contrast indicates the extra half planes associated with those dislocations point toward the epilayer [10]. Such a configuration seems to indicate a misfit strain of tensile nature. Although the thermal strain is tensile at the center region, no obvious source of tensile misfit strain has been found in n^- 4H-SiC epilayers. The observation of 'white contrast' interfacial dislocations indicates either another source of tensile misfit strain or a different mechanism contributes to their formation.

From Fig. 5, an obvious difference can be noticed between the morphologies of interfacial dislocaitons with the same contrast (black or white) in epilayers with different polarities. For example, the 'black contrast' interfacial dislocations in the Si-face epilayer (b) appear very straight, however in the C-face epilayer they seem to be a little curved (d). On the contrary, 'white contrast' interfacial dislocations appear very straight in the C-face epilayer (c) while curved in the Si-face epilayer (a). Such difference may be due to the different cores of those dislocations. The black or white contrast indicates the direction of the extra half plane. The core of the dislocation can be determined by the surface polarity pointed by the extra half plane. For the 'black contrast' interfacial dislocations, their extra half planes point toward the substrate side. Thus they will have a C-core in Si-face epilayer (b) and a Si-core in C-face epilayer (d). Applying the same rule, the 'white contrast' interfacial dislocations are determined to have a Si-core in Si-face epilayer (a) and a C-core in C-face epilayer (c). Correlating the core structures with the dislocation morphologies, it can be found that the perfectly straight dislocations in both epilayers are of C-core (b and c) and the curved dislocations are of Si-core (a and c). If taking a closer look, it can be noted that some segments of the curved interfacial dislocations have entered the substrate. One example is marked by a rectangle in Fig. 5(a). Comparing Fig. 5(a) and (c), it seems that given a significant thermal stress, the dislocations with Si-core can enter the substrate while those with C-core cannot. It appears to be consistent with the statement that Si-core partials are more mobile than C-core partials in SiC [11].

Discussion on the formation mechanism of interfacial dislocations. As described in the previous sections, the formation of 'black contrast' interfacial dislocations seems to be the consequence of both compressive misfit strain and compressive thermal strain. The compressive misfit strain coincides with the N-doping difference induced misfit strain in the epilayers. However, the observation of 'white contrast' interfacial dislocations indicates either a different source of misfit strain or the interfacial dislocations are formed by a different mechanism. For the former possibility, it may be due to the different mechanical properties between epilayer and substrate which causes the sign of misfit strain changes with the thermal strain. For the latter possibility, it may be due to the different mobilities of dislocations in epilayer and substrate. It appears that the BPDs could glide more easily in the epilayer than the substrate. Such difference might be caused by the much higher dislocation density in the substrate or the N-doping difference. In fact, the BPD morphologies in 4H-SiC epilayers show a distinct difference between low N-doped and highly N-doped epilayers [12], which indicates N-doping might modify the dislocation properties. Further investigation is needed to clarify the formation mechanism of interfacial dislocations. It has to be noted that all the dislocation behaviors described in this report were still under a relatively low stress situation. Behaviors of BPDs under high level of stress are expected to be different.

Summary

Thermal annealing experiments were performed to determine the critical conditions of misfit dislocation formation in 4H-SiC epilayers in a temperature range of 1400-1800 °C. Misfit dislocations were observed to form at a given annealing temperature if the temperature gradient across the epi-wafer exceeds a critical value. The critical temperature gradient at 1600 °C was estimated to be ~5 °C/mm. At 1800 °C, it was estimated to be below ~1 °C/mm. It was also found that two types of interfacial dislocations, namely 'black contrast' and 'white contrast' interfacial dislocations, could form depending on the signs of thermal stresses. While the N-doping induced misfit strain coincides with the compressive misfit strain to explain the formation of 'black contrast' interfacial dislocations, the observation of 'white contrast' interfacial dislocations indicates either a different source of misfit strain or the interfacial dislocations are formed by a different mechanism. Possibilities include different mechanical properties and different mobilities of dislocations between epilayer and substrate.

Acknowledgements

This research is partly supported by the Japan Society for the Promotion of Science (JSPS) through its "Funding Program for World-Leading Innovative R&D on Science and Technology (FIRST Program)".

References

[1] H. Jacobson, J. Birth, R. Yakimova, M. Syväjärvi, J.P. Bergman, A. Ellison, T.Tuomi, and E. Janzén, J. Appl. Phys. 91 (2002) 6354-6360.

[2] H. Jacobson, J. Birth, C. Hallin, A. Henry, R. Yakimova, T.Tuomi, E. Janzén, and U. Lindefelt, Appl. Phys. Lett. 82 (2003) 3689.

[3] X. Zhang, S. Ha, Y. Hanlumnyang, C.H. Chou, V. Rodriguez, M. Skowronski, J.J. Sumakeris, M.J. Paisley and M.J. O'Loughlin, J. Appl. Phys. 101 (2007) 053517.

[4] X. Zhang, M. Skowronski, K.X. Liu, R.E. Stahlbush, J.J. Sumakeris, M.J. Paisley and M.J. O'Loughlin, J. Appl. Phys. 102 (2007) 093520.

[5] H. Matsuhata, H. Yamaguchi, I. Nagai, T. Ohno, R. Kosugi, and A. Kinoshita, Mater. Sci. Forum 600-603 (2009) 309-312.

[6] R.E. Stahlbush, B.L. VanMil, K.X. Liu, K.K. Lew, R.L. Myers-Ward, D.K. Gaskill, C.R. Eddy, Jr., X. Zhang and M. Skowronski, Mater. Sci. Forum 600-603 (2009) 317-320.

[7] H. Tsuchida, I. Kamata, K. Kojima, K. Momose, M. Odawara, T. Takahashi, Y. Ishida and K. Matsuzawa: Mater. Res. Soc. Symp. Proc. 1069 (2008) 123-128.

[8] M. Nagano, H. Tsuchida, T. Suzuki, T. Hatakeyama, J. Senzaki and K. Fukuda, J. Appl. Phys. 108 (2010) 013511.

[9] X. Zhang, M. Nagano, and H. Tsuchida, Mater. Sci. Forum 679-680 (2011) 306-309.

[10] X.R. Huang, D.R. Black, A.T. Macrander, J. Maj, Y. Chen, and M. Dudley, Appl. Phys. Lett. 91 (2007) 231903.

[11] P. Pirouz and J.W. Yang, Ultramicroscopy 51 (1993) 189-214.

[12] X. Zhang, M. Nagano, and H. Tsuchida, submitted to the proceedings of this conference.

Materials Science Forum Vols. 717-720 (2012) pp 319-322
doi:10.4028/www.scientific.net/MSF.717-720.319

Fourier Transform Analysis of Basal Plane Dislocation Structure in Repeated a-face Grown Crystals

I. Gunjishima[1, 2, a], Y. Urakami[1, 3, b], F. Hirose[1, 3], A. Adachi [1, 4, c], S. Onda[1, 3], and K. Nishikawa[1, 2]

[1]R&D Partnership for Future Power Electronics Technology, Nisshin Branch, 500-1 Minamiyama, Komenoki, Nisshin, Aichi 470-0111, Japan

[2] Toyota Central R&D Laboratories Inc., Nagakute, Aichi 480-1192, Japan

[3]Reserch Laboratories, DENSO CORPORATION, Komenoki, Nisshin, Aichi 470-0111, Japan

[4]Toyota Motor Corporation, Nishihirose, Toyota, Aichi 470-0309, Japan

[a]gunjishima@mosk.tytlabs.co.jp, [b]yurakami@rlab.denso.co.jp, [c]ayumu@adachi.tec.toyota.co.jp

Keywords: bulk growth, basal plane dislocation, X-ray topography, Fourier transformation

Abstract. Morphological features, such as the orientation and linearity of basal plane dislocations (BPDs) in SiC crystals, were analyzed by applying a two-dimensional fast Fourier transform (2D-FFT) to X-ray topographic images of the BPDs. A SiC crystal fabricated by an improved repeated a-face (RAF) method and a SiC crystal fabricated by a conventional RAF method discussed in a previous study were evaluated. In the 2D-FFT images of the improved crystal, streaks along the <1 ,100> directions were observed, indicating that the BPDs were highly oriented along the <11 ,2 0> directions. The degree of orientation of the BPDs, which may reflect their linearity, was calculated, and the improved RAF crystal had a much higher degree of orientation than the conventional RAF crystal.

Introduction

Silicon carbide (SiC) has attracted considerable interest due to its excellent semiconducting properties. However, it contains various types of dislocations that limit device performance. In particular, basal plane dislocations (BPDs) have recently become an issue because micropipes were almost eliminated. BPDs generally have a complex morphology, since they are intricately entangled with other types of dislocations, and this can cause several problems. A twisted BPD can generate multiple threading edge dislocations (TEDs) during epitaxial growth [1]. Furthermore, BPDs tend to decompose into partial dislocations [2] which can degrade device performance, and this decomposition might occur since BPDs are oriented in a crystallographically unstable direction. It is therefore important for BPDs to be linear and crystallographically stabilized to inhibit multiple TED generation and BPD decomposition.

In this study, we performed Fourier transform analysis of X-ray topographic images of BPDs to quantify the BPD morphology, i.e., orientation and linearity, using high quality SiC crystals grown by the repeated *a*-face (RAF) method [3]. A high quality SiC crystal which was fabricated by the improved RAF method was compared with the conventional RAF crystal obtained in the previous study.

Experimental

A SiC crystal was grown by the RAF method (see Fig. 1). In this study, *a*-face growth was repeated 7 times to improve the quality of a-face grown seed compared to that used in conventional study [4]. A plate specimen was cut from the crystal almost parallel to the (0001) face. Transmission X-ray topography using $(1\bar{1}00)$ diffractions was conducted on the specimen produced in this study and a similar specimen from a previous study [4]; then they were analyzed by two-dimensional fast Fourier transform (2D-FFT) (see Fig. 2). As for X-ray topography, three crystallographically equivalent $(1\bar{1}00)$ diffractions, i.e., $(\bar{1}010)$, $(1\bar{1}00)$, and $(01\bar{1}0)$, were used to detect all BPDs. Two out of three BPDs can be imaged, because the dislocation contrast vanishes when the diffraction vector is perpendicular to the Burgers vector of the BPDs as shown in Fig. 3. The three topographic images of the specimen were then digitized. The same square area was extracted from each image and the 2D-FFT process was conducted. 2D-FFT resolves images into sine waves with certain frequencies and amplitudes (see Fig. 2). As shown in Figs. 2 and 3, if the lines in an image are highly oriented, the 2D-FFT image shows streaks perpendicular to the orientation direction of the lines.

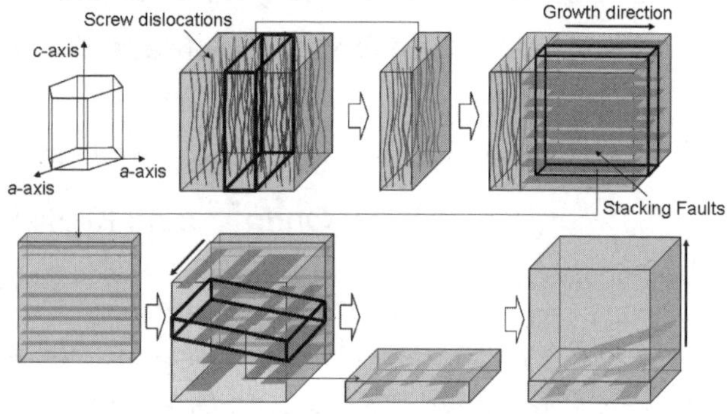

Fig.1 Repeated a-face growth (RAF)

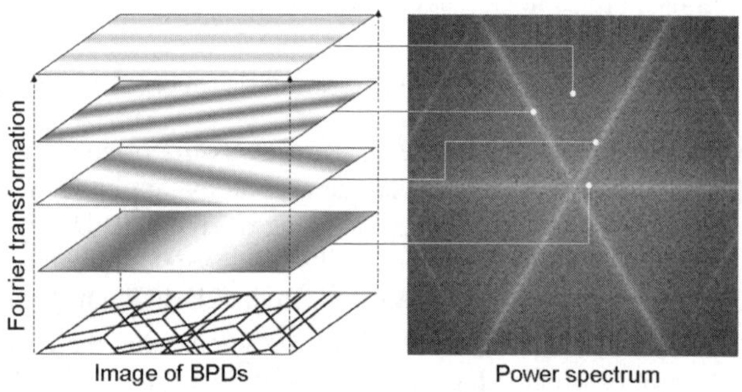

Fig.2 Schematic illustration of 2D-FFT.

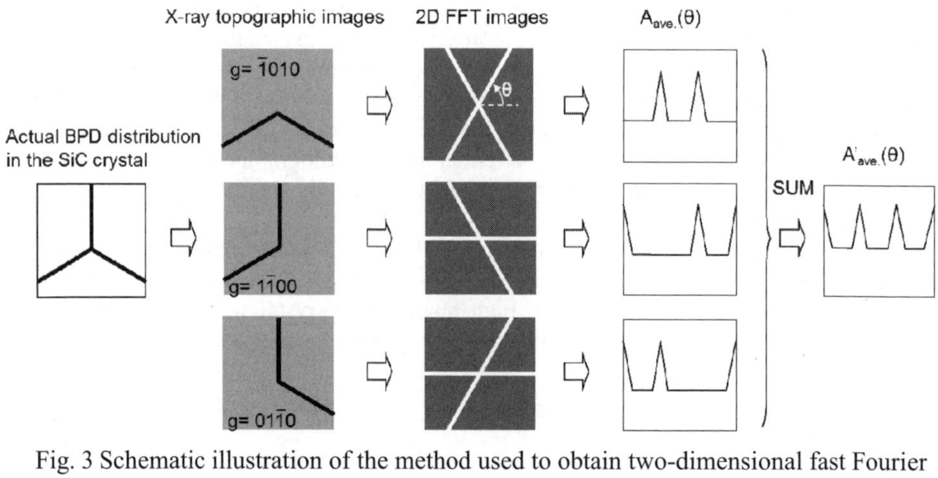

Fig. 3 Schematic illustration of the method used to obtain two-dimensional fast Fourier transform (2D-FFT) images of all BPDs. The center angle θ-dependence of the average amplitude of the Fourier coefficient (A) in the radial direction is shown.

From the 2D-FFT images, functions expressing the degree of orientation were calculated (see Fig. 3) [5]. The amplitude of the Fourier coefficient (A), which reflects the brightness of the 2D-FFT image, was averaged from the center of the image along radial directions (θ) from 0 to 180 degrees [5]. This process was carried for each of the three 2D-FFT images, and the resulting $A_{ave.}(\theta)$ values were summed to obtain $A'_{ave.}(\theta)$, which reflects the actual BPD distribution.

Results and Discussion

X-ray topographic images of (,1010), (1 ,100), and (01 ,10) diffractions of identical areas of the improved RAF crystal and those of the conventional RAF crystal [4] were obtained (not shown). In the improved RAF crystal, there was almost no entanglement of the dislocations, and the BPDs were long and linear. This made the determination of the shapes and lengths of individual BPDs easier compared to conventional RAF crystal. The three topographic images looked quite different, even though they were from identical areas, since a different third of the BPDs was undetected in each image.

Figure 4 shows 2D-FFT images obtained from the X-ray topographic images of (,1 010), (1 ,1 00), and (01 ,10) diffractions of the exact same area of the improved RAF crystal and

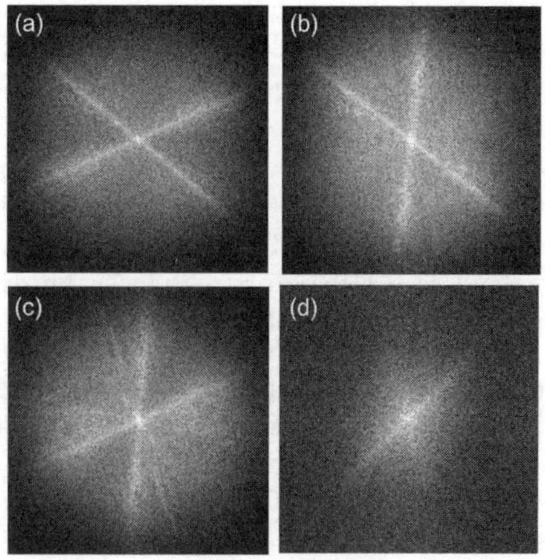

Fig. 4 2D-FFT images of X-ray topographic images of improved RAF crystal from this study using (a) (,1010), (b) (1 ,100), and (c) (01 ,10) diffractions, and (d) that of a conventional RAF crystal using (,1010) diffraction.

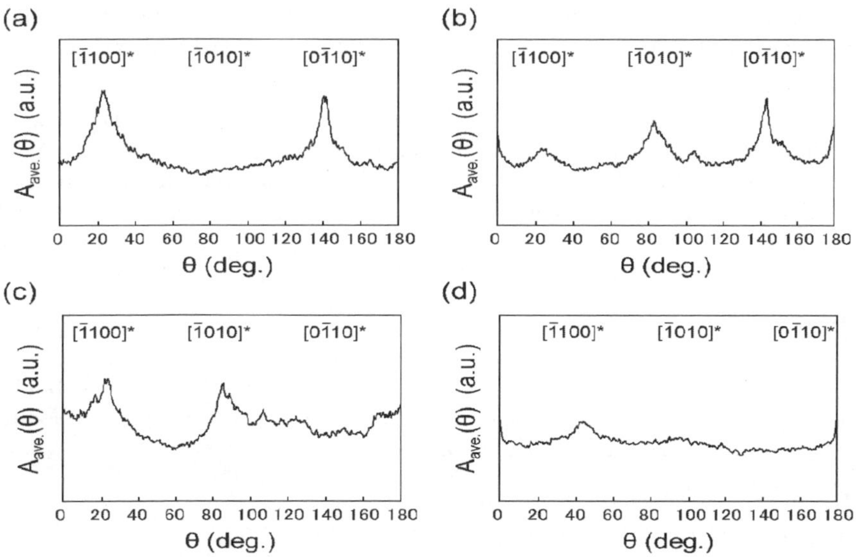

Fig. 5 $A_{ave.}(\theta)$ of X-ray topographic images of the RAF crystal from this study using (a) (,1 010), (b) (1 ,100), and (c) (01 ,10) diffractions, and that of a conventional RAF crystal using (,1010) diffraction. Miller indexes with asterisk (*) indicate that they are in reciprocal space.

(¯,1010) diffraction of the conventional RAF crystal. It can be seen that the images in Figs. 4(a), (b), and (c) exhibit clear streaks. These streaks are oriented along the <1¯,100>* directions in reciprocal space, indicating strong <11¯,20> orientations for the BPDs in real space shown in Fig. 4. In contrast, as seen in Fig. 4(d), the 2D-FFT image obtained from the conventional RAF crystal exhibits no prominent bright streaks, and only a weak variation in contrast with radial angle. These results may imply that the <11¯,20> directions are more stable for BPDs in SiC crystals, and that the crystal produced in this study had a larger portion of such stabilized BPDs than the conventional RAF crystal.

Since BPDs with Burgers vectors of a different two out of the three orientations were detected in each X-ray topographic image, different pairs of streaks oriented along the <1¯,100>* directions appeared in each 2D-FFT image of the crystal produced in this study.

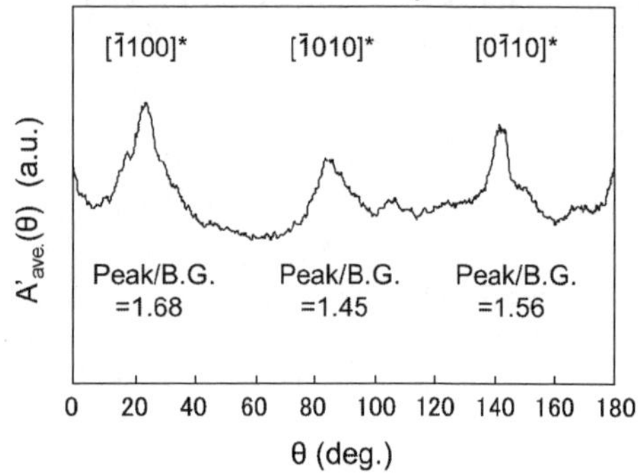

Fig. 6 Sum of $A_{ave.}(\theta)$ from three X-ray topographic images (Figs. 5(a), (b), and (c)) of RAF crystal from this study.

$A_{ave.}(\theta)$ for each of the 2D-FFT images in Fig. 4 is shown in Fig. 5. It can be seen that for the RAF crystal produced in this study, two strong peaks are observed due to the streaks in the 2D-FFT images. The values of θ at these peaks correspond to the <1¯,100>* directions. In contrast, $A_{ave.}(\theta)$ for the previous RAF crystal does not exhibit strong peaks, as shown in Fig. 5(d).

Figure 6 shows $A'_{ave.}(\theta)$, which is the sum of the three $A_{ave.}(\theta)$ curves from Figs. 5(a), (b), and (c). The signal-to-noise ratio at the peaks was 1.68 at [¯,1100]*, 1.45 at [¯,1010]*, and 1.56 at [0¯,110]*. The degree of orientation, which we defined as the mean of the signal-noise ratio, was high, with a value of 1.56. On the other hand, the degree of orientation calculated for the conventional RAF crystal was only 1.16. Owing to the high linearity of the BPDs, the degree of orientation of the improved RAF crystal produced was larger than that of the conventional RAF crystal.

Summary

The orientation and linearity of BPDs in SiC crystals can be quantified using 2D-FFT analysis. The orientation of BPDs can be determined from the appearance and direction of streaks in the 2D-FFT images of the topographic BPD images. The linearity of the BPDs is reflected in the peak strength of $A_{ave.}(\theta)$, which is the dependence of the amplitude of the Fourier coefficient on crystallographic direction. The results of 2D-FFT analysis indicated that improving the RAF method can result in BPDs that are more crystallographically stable and more linear.

Acknowledgements

This study was partially supported by Novel Semiconductor Power Electronics Project Realizing Low Carbon Emission Society of Japan Ministry of Economy, Trade and Industry (METI) and New Energy and Industrial Technology Development Organization (NEDO) through R&D Partnership for Future Power Electronics Technology (FUPET).

References

[1] I. Kamata *et al.* Materials Science Forum 645-648 (2010) 303-306.
[2] R. E. Stahlbush *et al.* Materials Science Forum 645-648 (2010) 271-276.
[3] D. Nakamura *et al.*, Nature 430 (2004) 1009-1012.
[4] D. Nakamura *et al.*, Journal of Crystal Growth 304 (2007) 57-63.
[5] T. Enomae *et al.*, Proceedings of International Papermaking and Environment Conference, Tianjin, P. R. China (May 12-14), Book 2, (2004) pp.355-368.

Materials Science Forum Vols. 717-720 (2012) pp 323-326

doi:10.4028/www.scientific.net/MSF.717-720.323

X-Ray Three-Dimensional Topography Imaging of Basal-Plane and Threading-Edge Dislocations in 4H-SiC

R. Tanuma[1, a], D. Mori[2,b], I. Kamata[1,c] and H. Tsuchida[1,d]

[1]Central Research Institute of Electric Power Industry (CRIEPI)
2-6-1 Nagasaka, Yokosuka-shi, Kanagawa-ken, 240-0196 JAPAN
[2] Fuji Electric Co., Ltd. 1, Fuji-machi, Hino-shi, Tokyo, 191-8502 JAPAN

[a]tanuma@criepi.denken.or.jp, [b]mori-daisuke@fujielectric.co.jp,
[c]kamata@criepi.denken.or.jp [d]tsuchida@criepi.denken.or.jp

Keywords: X-ray, microbeam, topography, three dimensional, dislocation, basal plane dislocation, threading edge dislocation.

Abstract. This paper demonstrates the X-ray three-dimensional (3D) topography of basal-plane dislocations (BPDs) and threading edge dislocations (TEDs) in 4H-SiC for the first time. Stereographic topographs are obtained for BPDs and TEDs, showing the propagation of BPDs from a substrate to an epilayer and the conversion of BPDs into TEDs near the epilayer/substrate interface. Strain analysis is also demonstrated for a TED, providing the image of strains in the order of $\pm10^{-5}$. It is verified that the 3D topography is successfully applicable to BPDs and TEDs.

Introduction

In silicon carbide (SiC) bipolar devices, basal plane dislocations (BPDs) in an epilayer are responsible for the degradation of current-conduction performance with an increase of forward voltage drop [1]. One promising way to decrease BPD density in the epilayer is converting BPDs to threading-edge dislocations (TEDs) near the epilayer/substrate interface. To ensure perfect BPD-TED conversion, however, it is important to have a better understanding of the conversion mechanism [2].

Three-dimensional (3D) topography using a synchrotron X-ray microbeam enables us to investigate the behavior of dislocations even in deep regions [3]. This method has been successfully applied to the characterization of threading screw dislocations (TSDs) with Burgers vector of 1*c* in 4H-SiC [3, 4]. For BPDs and TED, however, 3D topography remains a challenging theme because the magnitude of their 1*a* Burgurs vector is far smaller (about 1/3) than that of the 1*c* vector. This paper demonstrates 3D-topography imaging of BPDs and TEDs for the first time. Strain analysis is also presented for a TED.

Experimental

The sample examined was an 8°-off-cut (0001) Si-face 4H-SiC wafer with a 20 μm-thick epilayer. Prior to 3D-topography analysis, conventional back-reflection X-ray topography was taken with **g** = 1 1 -2 8 (8.046 keV) on the beam line BL08B2 in SPring-8. The wafer was then KOH etched at 500°C for 2 min to reveal the type and location of dislocations.

3D-topography measurements were conducted on the beam line BL24XU [5]. An undulator-radiated X-ray beam (15 keV) was focused on the sample by a bent cylindrical mirror with a focal length of about 1 m. The full widths at half-maximum (FWHM) of the focused beam were typically 1.9 and 0.6 μm in horizontal and vertical directions, respectively, while the beam divergence was approximately 25 μrad in a horizontal direction. The photon flux was about 5×10^6 photons/s on a sample.

Figure 1 shows the setup of microbeam X-ray diffraction for the 3D topography. Measurements were carried out with **g** = 1 1 -2 12 at E = 15 keV. This method uses a special slit with a V-shaped valley (V-slit) in microbeam X-ray diffraction. The V-slit was made of two tungsten-carbide (WC) knives contacting each other on their tips [4]. The knife edges were obliquely cut and polished to make an opening angle of 20°. The V-slit was located 20-30 mm downstream from the sample. The incident X-ray microbeam penetrates the sample, generating a ribbon-shape diffraction beam that impinges on the V-slit. Transmitting the V-slit, a thin X-ray beam reaches a YAP scintillation counter (SC). We thus conducted pinpoint measurements at the point Q where the incident beams crossed the extension of the V-slit transmit beam. Scanning sample positions provides depth-resolved 3D topography. The net spatial resolution of the 3D topography is 1-3 μm depending upon directions and tuning conditions. Since the incident beam divergence (25 μrad) was far larger than the theoretical rocking curve width (~ 1 μrad), the topography will be dominated by kinematical direct image due to off-peak reflections.

Strain analyses [4] were performed for a cross section perpendicular or parallel to a sample surface. We obtained section topographs at step-scanned rocking angles ω. The consecutive topography data were converted to rocking curves, from which effective misorientations (rocking-angle shifts) were calculated by using the Gaussian fitting.

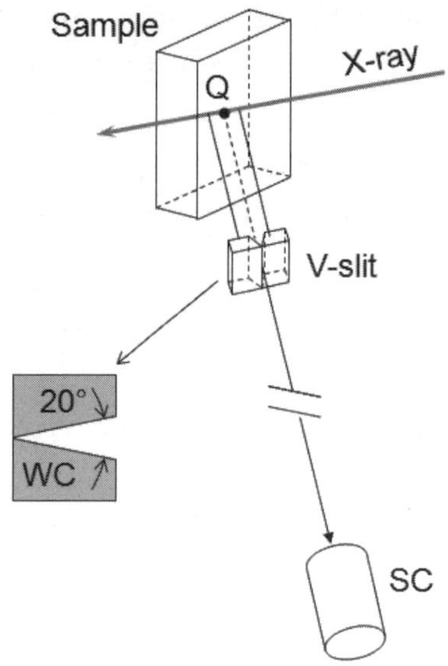

Fig. 1. Setup of microbeam X-ray diffraction for the 3D topography. The V-slit is made of two tungsten-carbide (WC) knives contacting each other on their tips. The knife edges are obliquely cut and polished to make an opening angle of 20°.

Results and Discussion

Figure 2 shows conventional back-reflection synchrotron X-ray topography obtained with **g** = 1 1 -2 8 (8.046 keV) before and after 4H-SiC epitaxial growth for the same position. Comparing the images taken before and after the epitaxial growth, we expect that a BPD (B1) converts to a TED (E1) near the epilayer/substrate interface, and another BPD (B2) propagates from the deep substrate to the epilayer.

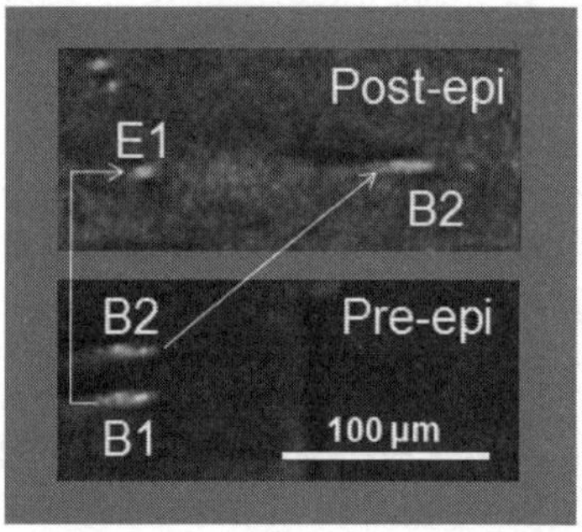

Fig. 2. Back reflection X-ray topographs obtained before and after 4H-SiC epitaxial growth for the same position. We expect that a BPD (B1) converts to a TED (E1), and another BPD (B2) propagates from the deep substrate to the epilayer.

3D-topography measurements were performed for the same position as Fig. 2 in a 30-μm-deep 3D region from the surface. Figure 3 shows stereographic views of the dislocations from different camera angles. The two pits visible in the top view correspond to E1 and B2 (Fig. 3a). Figure 3b clearly shows that B2 propagates from the deep substrate to the epilayer, and B1 converts to E1 near the epilayer/substrate interface (20 μm deep). The horizontal high-intensity region along the surface (arrow in Fig. 3b) will be caused by surface strain (described later). E1 is almost parallel to the surface normal, while B1 and B2 lie on the basal plane at an angle of 8° with the surface (Fig. 3c). Between E1 and B2, we can see a complicated morphology, from which the interaction of the dislocations with other defects, or with the epilayer/substrate interface is anticipated (circled in Figs. 3d, 3e, 3f).

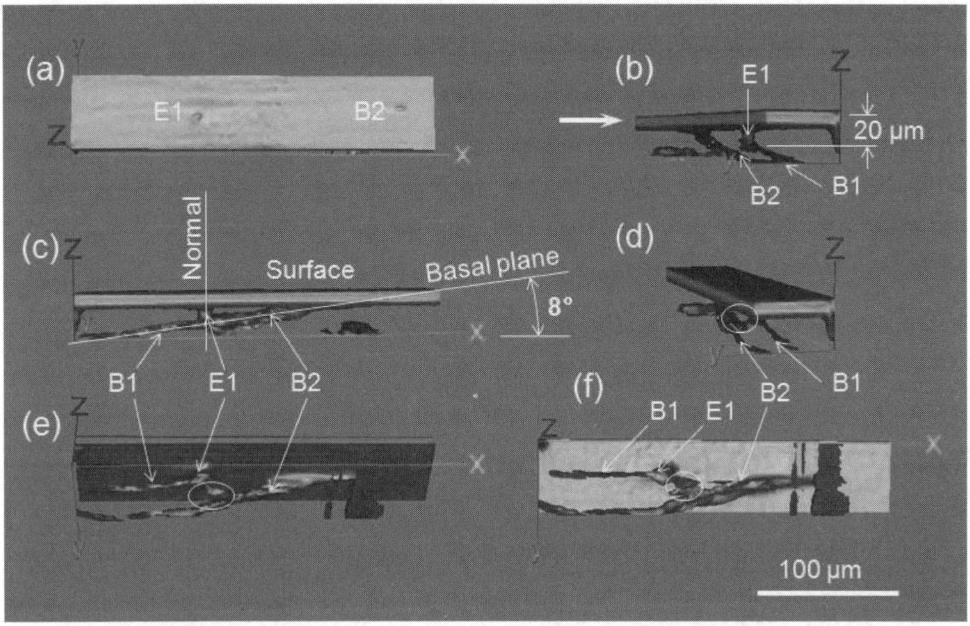

Fig. 3. Stereographic views of BPDs (B1, B2) and a TED (E1). The two pits visible in the top view correspond to E1 and B2 (a). B2 propagates from the deep substrate to the epilayer, and B1 converts to E1 near the epilayer/substrate interface (20 µm deep) (b). The high-intensity region (arrow in b) will be caused by surface strain. B1 and B2 lie on the basal plane at an angle of 8° with the surface, and E1 is almost parallel to the surface normal (c). Based on the complicated morphologies between E1 and B2 (circles in d, e, and f), we anticipate the interaction of dislocations with other defects, or with the epilayer/substrate interface

Strain analysis was applied to E1, providing images of effective misorientations $\Delta\omega$ in Fig. 4. The *z-y* plane is perpendicular to the surface, while the *x-y* plane is parallel to the surface at a depth of $z = -15$ µm (along the lines L). In the *z-y* plane, the sample surface corresponds to the level of $z = 0$, and B1 coming from behind the figure converts to E1, which lies on the *z-y* plane, near the epilayer/substrate interface (20 µm deep). As shown in the *x-y* plane, the Burgers vector **b** is parallel to B1 pointing in the -*x* direction. We can see that the compressive ($\Delta\omega > 0$) and tensile ($\Delta\omega < 0$) sides are explicitly separated. The range of $\Delta\omega$ (±10 µrad) corresponds to strains in the order of $\pm 10^{-5}$. The positive $\Delta\omega$ along the surface indicates compressive superficial strains, showing that the superficial reflection enhancement (arrow in Fig. 3b) is strain originated.

We thus verify that the 3D topography is successfully applicable to BPDs and TEDs. The stereographic imaging of the dislocations enables the detailed analysis of the dislocations near the epilayer/substrate interface, which analysis is difficult in the conventional X-ray topography. Strain imaging also provides new methodology that permits quantitative dislocation analysis.

Summary

X-ray 3D topography was successfully applied to BPDs and TEDs in 4H-SiC. Stereographic imaging enables detailed analysis of dislocations near the epilayer/substrate interface, which is otherwise difficult using conventional X-ray topography. Strain imaging also provides new methodology that permits quantitative dislocation analysis. Including BPDs and TEDs makes 3D topography a powerful tool for defect analysis in SiC-related material research.

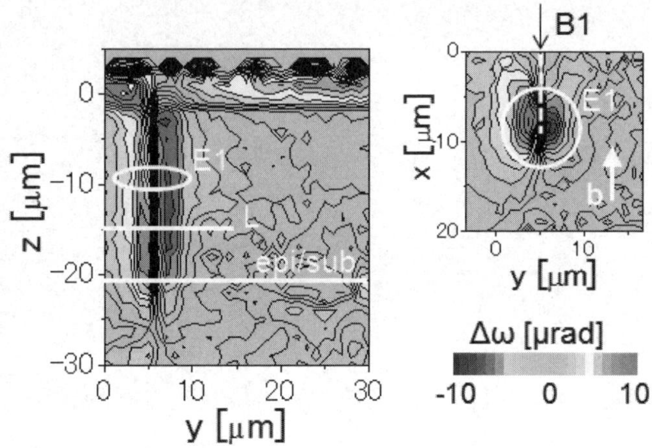

Fig. 4. Results of strain analysis applied to E1. The *z-y* plane is perpendicular to the surface, and the *x-y* plane is parallel to the surface at a depth of z = -15 μm (along the lines L). In the *z-y* plane, the sample surface corresponds to the level of $z = 0$, and B1 coming from behind the figure converts to E1, which lies on the *z-y* plane, near the epilayer/substrate interface (20 μm deep). In the *x-y* plane, the Burgers vector **b** is parallel to B1 pointing in the -*x* direction. The compressive ($\Delta\omega > 0$) and tensile ($\Delta\omega < 0$) sides are explicitly separated. The range of $\Delta\omega$ (±10 μrad) corresponds to strains in the order of $\pm 10^{-5}$. The positive $\Delta\omega$ along the surface indicates compressive superficial strains, showting that the superficial reflection enhancement (arrow in Fig. 3b) is strain originated.

Acknowledgements. We would like to thank Assoc. Prof. Y. Tsusaka of the University of Hyogo for fine tuning the X-ray microbeam. This work was performed as part of research proposal numbers 2010B3237 and 2011A3237 of SPring-8. This research is partly supported by the Japan Society for the Promotion of Science (JSPS) through its "Funding Program for World-Leading Innovative R&D on Science and Technology (FIRST Program)".

References

[1] M. Skowronski, S. Ha, J. Appl. Phys. **99** (2006) 011101.
[2] H. Tsuchida, M. Ito, I. Kamata, M. Nagano, Phys. Status Solidi B **246** (2009) 1553-1568.
[3] R. Tanuma, T. Kubo, F. Togoh, T. Tawara, A. Saito, K. Fukuda, K. Hayashi, Y. Tsusaka, Phys. Status Solidi A **204** (2007) 2706-2713.
[4] R. Tanuma, T. Tamori, Y. Yoneawa, H. Yamaguchi, H. Matsuhata, K. Fukuda, K. Arai, Material Sci. Forum **615-617** (2009) 251-254.
[5] S. Takeda, K. Yokoyama, Y. Tsusaka, Y. Kagoshima, J. Matsui, A. Ogura, J. Synchrotron Radiat. **13** (2006) 373.

Materials Science Forum Vols. 717-720 (2012) pp 327-330
doi:10.4028/www.scientific.net/MSF.717-720.327

Basal Plane Dislocation Multiplication via the Hopping Frank-Read Source Mechanism and Observations of Prismatic Glide in 4H-SiC.

H. Wang[1], S. Byrappa[1], F. Wu[1], B. Raghothamachar[1], M. Dudley[1,a,*], E. K. Sanchez[2] , D. Hansen[2], R. Drachev[2], S. G. Mueller[2] and M. J. Loboda[2]

Department of Materials Sci. and Engr., Stony Brook University, Stony Brook, New York, 11794,

[2]Dow Corning Compound Semiconductor Solutions, Midland, Michigan, 48686, USA

[a]mdudley@notes.cc.sunysb.edu

Keywords: Dislocation, Basal Plane, Slip System, Prismatic Plane, Burgers Vector, Frank-Read Source

Abstract. In this paper, we report on the synchrotron white beam topographic (SWBXT) observation of "hopping" Frank-Read sources in 4H-SiC. A detailed mechanism for this process is presented which involves threading edge dislocations experiencing a double deflection process involving overgrowth by a macrostep (MP) followed by impingement of that macrostep against a step moving in the opposite direction. These processes enable the single-ended Frank-Read sources created by the pinning of the deflected basal plane dislocation segments at the less mobile threading edge dislocation segments to "hop" from one slip plane to other parallel slip planes. We also report on the nucleation of $1/3\langle 11\bar{2}0\rangle\{1\bar{1}00\}$ prismatic dislocation half-loops at the hollow cores of micropipes and their glide under thermal shear stress.

Introduction.

It is well known that the detrimental effect of dislocations on the performance of Silicon Carbide (SiC) electronic devices has made the lowering of their densities a primary goal of SiC crystal growth community [1]. Such "defect engineering" requires knowledge of the detailed configurations of the various kinds of dislocations as well as the origins of these configurations so that strategies might be formulated to enable control to be exerted over the processes involved. There are generally two types of dislocation in 4H-SiC boules: growth dislocations (which self-replicate at the moving growth interface) and deformation-induced dislocations which move under thermal shear stress behind the growth interface. For the latter, minimization of the thermal stress has very effectively lowered the density of this kind of dislocation to the point where the various slip mechanisms can be observed with unprecedented detail [2]. In this paper we report on: (1) the multiplication of $1/3\langle 11\bar{2}0\rangle(0001)$ basal plane dislocations through the operation of single-ended Frank-Read sources which "hop" from one slip plane to other parallel slip planes through a process of double deflection involving overgrowth by a macrostep followed by impingement of that macrostep onto a step moving in the opposite direction; and (2) the nucleation of $1/3\langle 11\bar{2}0\rangle\{1\bar{1}00\}$ prismatic dislocation half-loops at the hollow cores of micropipes and their glide under thermal shear stress.

Results and Discussion.

Fig. 1 shows a Synchrotron White Beam X-ray Topography transmission image recorded from a wafer cut with 4 degree offcut towards $[11\bar{2}0]$ from a PVT-grown boule. Note the many diamond shaped basal plane dislocation spiral loops within the full ellipse. The adjacent dotted rectangular frame highlights an isolated example of a configuration of such loops that is suitable for detailed study. Enlarged images of this region recorded with different g-vectors are shown in Figs. 2(a)-(d). Close examination of this sequence of diamond shaped spiral loops displaced in the vertical direction on this and many other reflections show that it comprises a single dislocation that enters the crystal face at *a* and exits on the opposite face at *b* on figure 2(a).

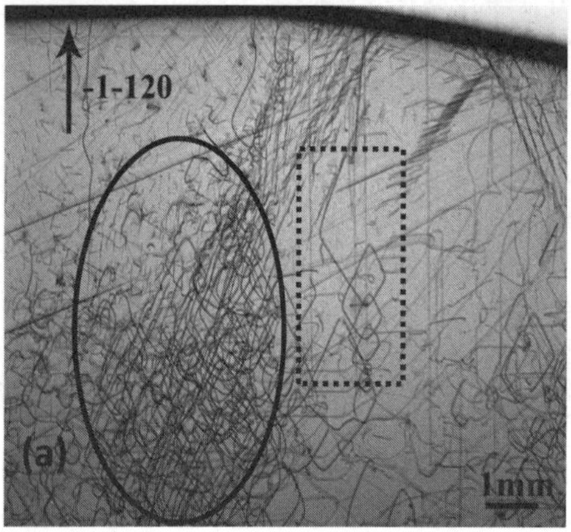

Fig. 1 Synchrotron White Beam X-ray Topography (SWBXT) transmission image ($g=\bar{1}\bar{1}20$) recorded from a wafer cut with 4 degree offcut towards $[11\bar{2}0]$ from a PVT-grown boule.

Note that the vertically displaced spiral loops are out of contrast on Fig. 2(d) ($g=1\bar{1}00$), indicating that the Burgers vector of the spiral loops is 1/3 $[11\bar{2}0]$. The line directions of the loop segments are at 30 degrees to the $[11\bar{2}0]$ direction so that they lie parallel to $[10\bar{1}0]$ and $[01\bar{1}0]$ directions making them 30 degree dislocations. This indicates that $<10\bar{1}0>$ are preferred directions on the (0001) slip plane corresponding to Peierls valleys on that plane. From image 2(a), the loops are clearly superimposed upon each other and yet do not appear to interact strongly indicating that the loops lie on basal slip planes that are significantly displaced along the c-axis from each other. We also observed the jog segments on the loops which are indicated by arrows in figure 2(b) and (c).

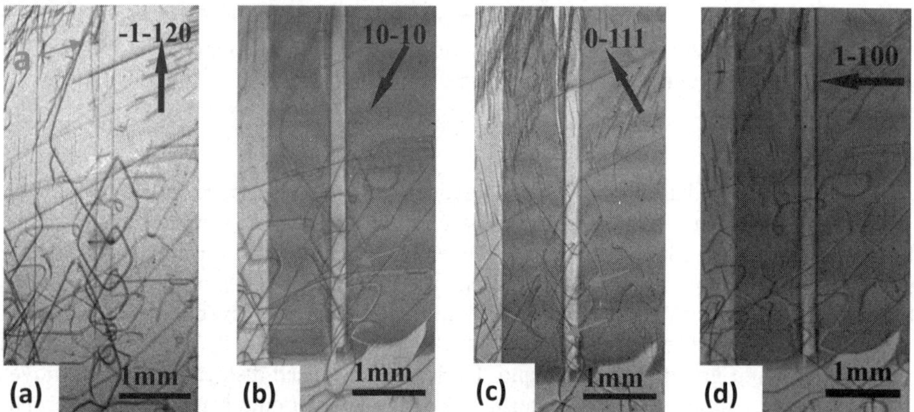

Fig. 2 SWBXT images recorded from the same area of the wafer shows the various contrasts of the diamond shaped loops in different g vectors. The arrows in (b) and (c)indicate the existence of jogs.

Figure 3 (a)-(c) shows dislocations emanating from an isolated micropipe in a 4H-SiC wafer. Basal plane dislocation half-loops are clearly observed. In fact, a significant number of the BPDs in this wafer appear to have nucleated at this micropipe. However, in addition, there are glide dislocations emanating from the micropipe which exhibit a directionality of the slip along close-packed $<11\bar{2}0>$ directions (between the two white arrows on Figs. 3(a)-(b)) and this slip is invisible on Fig. 3(c) suggesting that the slip system is $1/3[1\bar{2}10](10\bar{1}0)$ [3].

The configurations of the loops observed in Figs.1 and 2 have enabled us to postulate a mechanism for their formation. The initial stage of the mechanism involves the macrostep deflection of the surface outcrop of a threading edge dislocation (TED) onto the basal plane to create a segment of basal plane dislocation (BPD) as shown in schematic cross-section in Figure 4(b). This deflection occurs since the macrostep (created by step bunching as unit cell or half unit cell high steps are retarded by various obstacles they encounter on the growth surface, for example, screw dislocation outcrops, foreign particles, other steps, etc.) cannot admit the dislocation into its structure leaving the dislocation with no alternative but to be deflected into the direction of step flow. Once the TED is deflected onto the basal plane, further advancement of the macrostep will replicate the basal plane dislocation in the direction of step flow. If the advancing macrostep encounters a step advancing in the opposite direction, for example a spiral step from a screw dislocation which emerges on an adjacent terrace located in the overall downstep direction, the dislocation will be re-deflected into the threading direction as shown in Figure 4(b).Moreover, the TED segments among BPDs have been confirmed by the existence of jogs as shown above in Fig.2 (b) and (c).

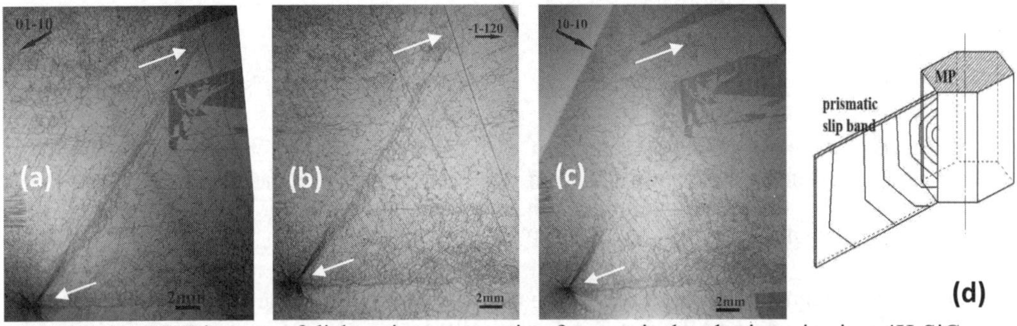

Fig.3 The SWBXT images of dislocations emanating from an isolated micropipe in a 4H-SiC wafer. (a) g=$01\bar{1}0$. (b) g=$\bar{1}\bar{1}20$. (c) g=$10\bar{1}0$ where the slip band is out of contrast.

At a later stage of growth, if the outcrop of the re-deflected TED encounters another advancing macrostep, the whole process can be repeated, as shown Figure 4 (b). Confirmation of the existence of such double deflection process is provided by examination of axial wafers cut parallel to the growth axis. This enables observation of the "hopping" mechanism in cross section, as shown in the SWBXT image presented in Fig.4 (a).

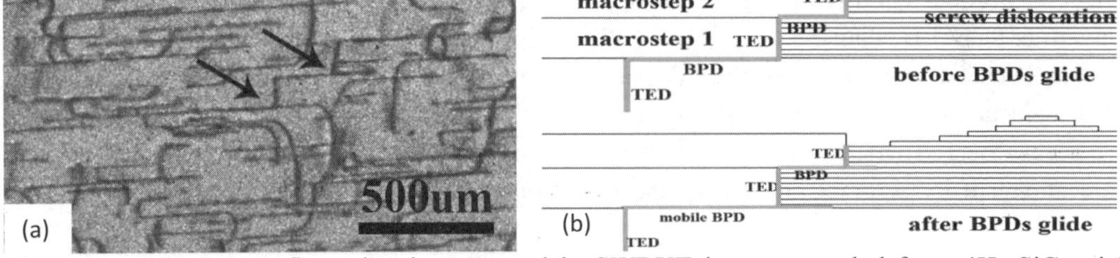

Fig. 4 The hopping configuration is presented in SWBXT image recorded from 4H- SiC axial slice, the accompanying schematic shown in Fig. 4 (b).

During the double deflection between BPDs and TEDs, the segments of mobile BPD will operate as a single-ended Frank-Read source [4] around that pinning point, if at some point the crystal experiences a thermal field that gives rise to a shear stress on the basal plane that exceeds the critical resolved shear stress for their gliding, as shown schematically in perspective in Fig. 5(a)-(d). In this process, the single ended Frank-Read sources created earlier in the process are expected to develop further than the subsequently created sources. In addition, this mechanism is expected to produce a sequence of spiral loops that are displaced with respect to each other in the local down step direction, as shown in Fig. 5(e). This is consistent with observations of the loops in Figs. 1 and 2. This Hopping Frank-read multiplication process is somewhat analogous to the stress induced double cross-slip mechanism that leads to slip band broadening [5] although the processes described here are "growth processes" that occur on the moving crystal growth interface rather than "stress induced" deformation processes occurring behind the growth interface. The vertical separation between the rotating arms of the vertically displaced single ended sources is determined by the length of the TED segment connecting them. Evidently, this is large enough to allow the dislocation arms to pass over each other without experiencing the huge locking force for example encountered when a BPD is pinned by a TSD [6].

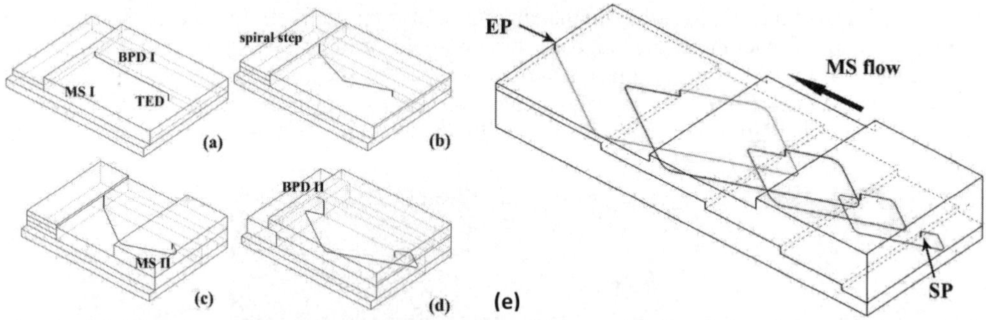

Fig. 5. (a)-(d) Schematic of the formation of a single-ended Frank-Read source through the deflection of a TED into a BPD and back again. The TED segments act as pinning points for the BPD glide; (e) The final configuration of the loops produced by the Hopping Frank-Read Source. SP indicates the starting point, and "EP" end point.

Regarding Fig 3, the nucleation of dislocation half-loops belonging to the prismatic slip system at the hollow core of a micropipe is shown schematically in Fig. 3(d). Slip on this secondary slip system is expected to be less preferable to the primary basal slip system such that when segments of the half loops attain screw orientation they are expected to cross slip onto the basal plane.

In conclusion, we have observed diamond shaped basal plane dislocation spiral loops with the Burgers vector $1/3[11\bar{2}0]$. A formation mechanism has been postulated for these loops termed the "Hopping Frank-read Source" mechanism. This involves the conversion of TEDs into BPDs by macrostep overgrowth creating a single ended Frank-Read source. This BPD can be redeflected into threading orientation via the impingement of the macrostep against a step (for example from a screw dislocation spiral) moving in the opposite direction. The process can then be repeated converting this TED segment again into a BPD. If the re-deflected TEDs segment is long enough, two single-ended Frank-Read sources displaced along the c-axis by the length of the connecting TED segment can operate independently. In addition, slip on the secondary prismatic slip system has been observed emanating from isolated micropipes. A simple model for the nucleation of dislocation half-loops on the prismatic plane at the hollow core of a micropipe is presented.

Acknowledgements. Work supported by Dow Corning. Topography experiments were carried out at the Stony Brook Synchrotron Topography Facility, Beamline X-19C, at the NSLS, BNL (DOE Office of Basic Energy Sciences Contract no. DE-AC02-76CH00016).

References.

[1] St. G. Muller, J. J. Sumakeris, M. F. Brady, R. C. Glass, H. M. Hobgood, J. R. Jenny, R. Leonard, D. P. Malta, M. J. Paisley, A. R. Powell, V. F. Tsvetkov, S. T. Allen, M. K. Das, J. W. Palmour, and C. H. Carter, Jr., Eur. Phys. J. Appl. Phys. 27, 29 (2004)

[2] M. Dudley, N. Zhang, Y. Zhang, B. Raghothamachar, S. Byrappa, G. Choi, E. K. Sanchez, D. M. Hansen, R. Drachev and M. J. Loboda, Mater. Sci. Forum, Vols. 645-648, P 291 (2009)

[3] S. Ha, N. T. Nuhfer, G. S. Rohrer, M. Graff, and M. Skowronski, Journal of electronic materials, Vol. 29, No.7, (2000)

[4] Frank, F. C. & Read, W. T. Phys. Rev. 79, 722–723 (1950)

[5] J. J. Gilman, W. G. Johnston, and G. W. Sears, J. Appl. Phys., V 29, No. 5 (1958)

[6] Y. Chen, M. Dudley, K. X. Liu, and R. E. Stahlbush, Appl. Phys. Lett. 90, 171930 (2007)

Materials Science Forum Vols. 717-720 (2012) pp 331-334

doi:10.4028/www.scientific.net/MSF.717-720.331

Identification of the basal plane component of the Burgers vector of small dislocations in 6H SiC using birefringence microscopy

Le Thi Mai Hoa, T.Ouisse[a,*], M.Seiss and D.Chaussende

LMGP, CNRS UMR 5628, Grenoble INP, Minatec, 3 parvis Louis Néel, BP 257, 38016 Grenoble, France

[a] thierry.ouisse@phelma.grenoble-inp.fr, *corresponding author

Keywords: dislocations, birefringence, residual stress, optical microscopy.

Abstract. Using 6H SiC wafers including regions with a varying residual stress, the birefringence pattern of almost vertical dislocations is measured and modeled. We show that it is possible to identify the basal plane component of "small" dislocations by a quantitative fit of the birefringence pattern. Combining birefringence and etch pit detection after KOH etching shows that most of the vertical dislocations are of a mixed nature, exhibiting both an edge and a screw component.

Introduction

Identification of linear defects in SiC wafers and crystals is often achieved through the use of X-ray topography [1,2] or KOH etching [3]. In our opinion, there is no doubt that X-ray topography is by far the most informative measurement method to this respect. However, white beam topography usually requires a synchrotron radiation facility, which thus ranges this technique in the category of the expensive and sophisticated ones. KOH etching is an easy but destructive technique. Birefringence microscopy is an additional, non-destructive technique which was proposed by Kato *et al.* [4] and Ma *et al.* [5,6] in the case of SiC, and then also developed by others (see, *e.g.*, the refs. collected in [7]). Although birefringence data can only provide a restricted amount of information, the technique is low cost and easy to set up. It is thus interesting to get a clear idea of its advantages and limitations. In a former publication, we showed that a close agreement could be obtained between experimental and simulated birefringence patterns induced by micropipes. Those micropipes were associated to Burgers vectors with a very large modulus. We have now developed a set-up with a microscope of a better quality, which allows us to observe the pattern associated to small or presumably unit dislocations, either in SiC substrates or in HPHT and CVD grown diamond. In this paper, we wish to show that a change of the component of the Burgers vector by only one lattice parameter unit ***a*** in the basal plane induces a noticeable change in the magnitude of the birefringence pattern, so that this component can be identified with reasonable accuracy. We also wish to compare birefringence results with the information obtained from KOH etching and subsequent etch pit detection, so as to shed more light and/or get more confidence about the conclusions that can be drawn so as to know whether SiC vertical dislocations exhibit an edge component or not.

Results and discussion

Birefringence was measured using the rotating polariser method [8], and modelling has already been thoroughly discussed in [7]. The obtained pattern represents an indirect image of the deformation field induced by a dislocation, and is a unique function of the residual and dislocation-induced stress [7]. Here, the plotted quantity is $\sin\delta$, where δ would be the phase shift between two polarised waves crossing the sample and respectively oriented along each of the two principal axes of the optical indicatrix in the horizontal plane [7]. We used relatively old, on-axis commercial 6H SiC wafers so as to have a high density of dislocations to observe (typically around $10^4 cm^{-2}$), and to obtain regions in which the residual stress is locally uniform, but whose mean value substantially changes from one region to the next in the same wafer. Dislocation identification is carried out using the following procedure: the background stress is first accurately determined by adjusting it so as to recover the background value of the birefringence (See Fig.1). Then, the birefringence pattern is adjusted in two successive steps. First, we rotate the direction of the residual, uniaxial stress component with respect to the edge component of the dislocation until the shape of the simulated and experimental patterns become similar. Then, the basal plane component of the Burgers vector being of the form $\mathbf{b}_{\perp}=n\mathbf{a}$ with n integer, we select the value of n which gives the best agreement between experiment and simulation.

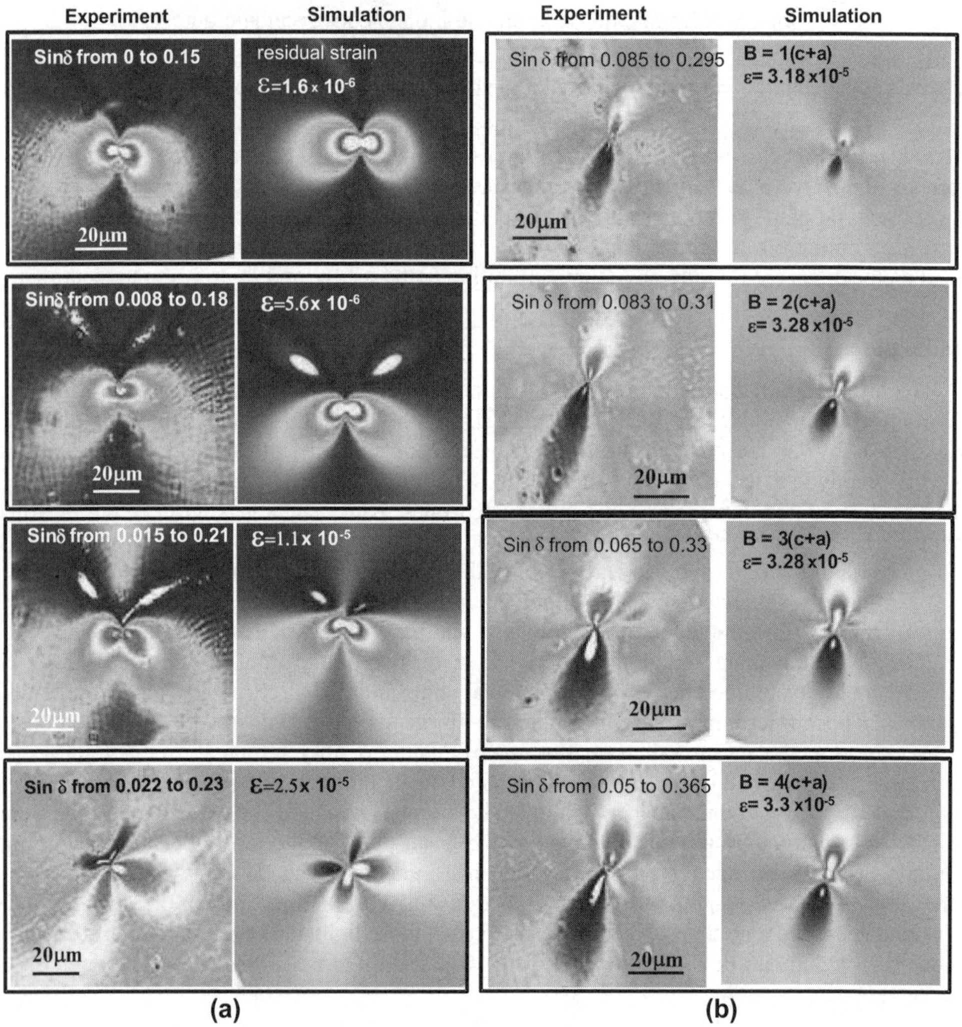

Fig.1: *(a) birefringent pattern induced by dislocations with* $\boldsymbol{b}=3(\boldsymbol{c}+\boldsymbol{a})$ *and various residual strain values (left: experiment and right: simulation) and (b) birefringent pattern induced by dislocations with increasing Burgers vector values; note that the colour scale changes for each dislocation so as to exhibit maximal contrast.*

We note that in the case of a crystal with hexagonal symmetry and almost vertically-oriented **c**-axis, the in-plane component of the Burgers vector associated to a vertically-oriented dislocation is the only one to contribute to the birefringence pattern. Unless the dislocation is not substantially tilted, a Burgers vector equal to $\mathbf{b}=n(\mathbf{c}+\mathbf{a})$ gives the same result as $\mathbf{b}=n\mathbf{a}$. In other words, birefringence is only sensitive to the edge component of vertical dislocations. Besides, since the dislocation-induced pattern eminently depends on the residual stress [7,9], in each region, the birefringence scale has to be finely tuned over a small interval centered at the residual stress value in order to make apparent and recognizable the dislocation-induced pattern. It is therefore worth noticing that for small values of the Burgers vector, a varying residual stress definitely prohibits *automatic* counting of the "small" dislocations, even if our claim is that all of the dislocations with an edge component can be *manually* detected.

To illustrate the validity and interest of the technique, we will develop two aspects: First, Fig.1a shows the influence of the residual stress upon the birefringence pattern. In Fig.1a, dislocations with the same **b**=3(**c**+**a**) Burgers vector have been isolated in regions with a different background stress. There is a close agreement between experiment and simulation, so that it is obvious that changes in the pattern shape are definitely due to changes in the background stress magnitude, and not to the nature of the dislocation. Although this influence has been known for a long time [9], to the best of our knowledge, this is the first time that a complete correspondence such as that of Fig.1a is obtained, with both an experimental and simulated variation of the residual stress.

Fig.1b shows small dislocations down to unit values of $b_\perp$. As an additional proof of the fit quality, we note that only dislocations with extracted values equal to $\mathbf{b}_\perp$=1**a** present no open core, as already noticed by others [1]. We attribute the remaining discrepancy between simulation and experiment to the fact that in contrast with, *e.g.*, diamond, the piezo-optical constants are not as accurately known in the case of SiC [4]. It is also worth noticing that for unit dislocations, the range spanned by the colour scale is extremely narrow and if, in some cases, distinction between $\mathbf{b}_\perp$=1**a** and $\mathbf{b}_\perp$=2**a** may seem difficult (as shown in Fig.1b), it would be definitely impossible to get a reasonable fit of a unit dislocation using $\mathbf{b}_\perp$=3**a**, as the pattern intensity would differ too much from the experimental one. When compiling all our observations, we thus estimate that an improvement in our knowledge of the SiC piezo-optic coefficients should lead to the experimental possibility of a clear distinction between 1**a** and 2**a** edge components.

From the preceding analysis, we conclude that if a unit dislocation remains invisible, this can be due to three possible reasons:
(i) the dislocation has no edge component.
(ii) the residual stress is varying too much in the vicinity of the dislocation and/or the intensity scale is not sufficiently reduced around the background value.
(iii) the tilt of the dislocation line from the vertical axis is too large, such as, *e.g.*, in the case of a basal plane dislocation. In the latter case, the effective thickness over which the strain field induces birefringence, measured parallel to the vertical optical path, is too small, so that the optical phase shift remains undetectable.

To further study points (i) and (iii), we proceeded to KOH etching and microscope observation of the resulting etch pits in wafers already characterized by birefringence. A typical result is given in Fig.2, where both the initial birefringence patterns and the etch pit locations are shown. In Fig.2, it can be seen that to all hexagonal etch pits can be assigned a corresponding birefringent pattern, even when dislocations have no open core (from D1 to D4 in Fig.2). All these dislocations thus exhibit an edge component and are compatible with a Burgers vector of the form **b**=*n*(**c**+**a**). For two wafers among three, we could not find etch pits not corresponding to a previous birefringent pattern, and this seems to be in agreement with recent observations by Dudley *et al.*[2], albeit reported for another polytype (4H SiC). For a third wafer, a substantial fraction (but not the majority) of the hexagonal pits could not be attributed a birefringent pattern, which seems to indicate that those vertical dislocations are pure screw. In Fig.3, it is also worth noting that pear-shaped etch pits (D5 and D6), which are to be attributed to basal plane dislocations [3], do not give rise to any birefringent pattern, as already anticipated in our previous discussion. Our conclusion is therefore mixed: on the one hand, combining birefringence and KOH etching allows one to count all dislocations and to assert whether they own an edge component or not. On the other hand, it seems that depending on the materials used, some wafers contain only mixed vertical dislocations, and others contain both mixed and pure screw vertical dislocations. Since we obtained this result applying the same techniques on different wafers, this might reconcile some previous contradictory views expressed in the literature, so as to know whether vertical dislocations are pure screw or of a mixed nature in SiC (see, *e.g.*, [1] and [10]): as a matter of fact, both pure screw and mixed vertical dislocations can be found, even if mixed dislocations constitute the vast majority in our own observation. This is in qualitative agreement with data obtained by applying X-ray topography to more recent wafers of another polytype (4H SiC) [2].

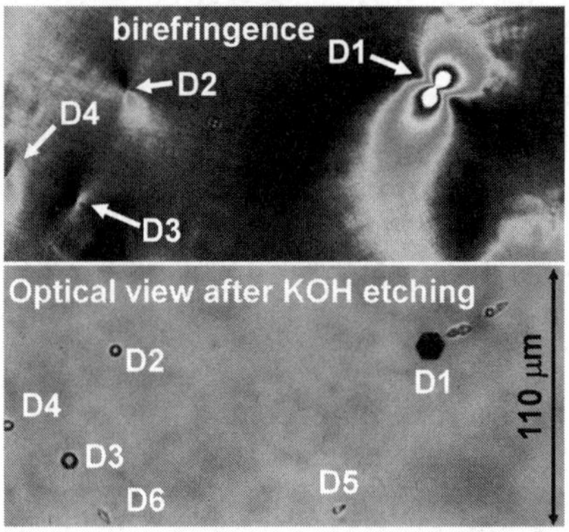

Fig.2: *Comparison between some birefringence patterns and etch pits obtained after KOH etching of a 6H SiC wafer; the upper (birefringence) and lower (etch pits) images are obtained at the same location.*

Conclusion

All vertical dislocations exhibiting an edge component can be detected using birefringence, even the smallest ones. The edge component can be quantitatively identified, and the residual uncertainty should be suppressed by improving our knowledge of the exact piezo-optical coefficients. However, no information can be gained about the screw component, and automatic counting cannot be achieved if the residual stress varies too much across the observed SiC wafer. Combining birefringence and etch pit detection after KOH etching shows that most dislocations are of a mixed nature, even if for some wafers it is also possible to find pure screw dislocations, devoid of any birefringence pattern.

Acknowledgements

This work has been supported by the French ANR (MINTEX project, contract N° ANR-09-BLAN-0189-01).

References:

[1] M.Dudley, X.R.Huang, W.M.Vetter, J. Phys. D: Appl. Phys. 36 (2003) A30-A36.
[2] M.Dudley, H.Wang, F.Wu, S.Byrappa, B.Raghothamachar, G.Choi, E.K.Sanchez, D.Hansen, R.Drachen, S.G.Mueller, M.J.Loboda, Mat. Sci. Forum 679-680 (2011) 269-272
[3] S.A.Sakwe, R.Müller, P.J.Wellmann, J. Cryst. Growth 289 (2006) 520-526
[4] T.Kato, H.Ohsato, A.Okamoto, N.Sugiyama, T.Okuda, Mater. Sci. & Eng. B57 (1999) 147-149
[5] X.Ma, Mat. Sci. & Eng. B 129 (2006) 216-221.
[6] X.Ma, M.Dudley, W.M.Vetter, T.Sudarshan, Jpn J. Appl. Phys. Part 2 42 (2003) L1077-L1079.
[7] T.Ouisse, D.Chaussende, L.Auvray, J. Appl. Cryst. 43 (2010) 122-133.
[8] I.Wood, A.M.Glazer, J. Appl. Cryst. 13 (1980) 217-223.
[9] D.A.Jenkins, J.J.Hren, Philos. Mag. 33 (1976) 173-180.
[10] J.Heindl, W.Dorsch, H.P.Strunk, St.G.Müller, R.Eckstein, D.Hofmann, A.Winnacker, Phys. Rev. Lett. 80 (1998) 740-741

Materials Science Forum Vols. 717-720 (2012) pp 335-338

doi:10.4028/www.scientific.net/MSF.717-720.335

Effects of Nitrogen Doping on the Morphology of Basal Plane Dislocations in 4H-SiC Epilayers

Xuan Zhang[1,a], Masahiro Nagano[1,b] and Hidekazu Tsuchida[1,c]

[1]Central Research Institute of Electric Power Industry (CRIEPI), 2-6-1 Nagasaka, Yokosuka, Kanagawa 240-0196, Japan

[a]xzhang@criepi.denken.or.jp, [b]m-nagano@criepi.denken.or.jp, [c]tsuchida@criepi.denken.or.jp

Keywords: Basal plane dislocations (BPDs), Nitrogen doping, Stacking faults, Si-core partials, C-core partials, 4H-SiC.

Abstract. Morphologies of BPDs in 4H-SiC epilayers with different nitrogen doping concentrations are explained in detail. While BPDs in low-doped epilayers have the typical morphology of gliding dislocations responding to stress, BPDs in highly doped ([N]$\geq 1.0\times10^{18}$ cm^{-3}) epilayers are straight and tilt away from [11-20]. Structures of BPDs were further studied by weak-beam TEM.

Introduction

Basal plane dislocations (BPDs) are one of the remaining defects in 4H-SiC epilayers that need to be addressed before production of high voltage (>10 kV) SiC devices can be realized. Homoepitaxial layers doped with different types of dopants, different levels of concentrations are essential to realize those device structures. However, the effects of doping on the mechanical and electrical properties of dislocations in SiC have not been well studied. This paper reports our observations on the effects of nitrogen doping on the morphology of BPDs in 4H-SiC epilayers.

Experimental

Epilayers of ~10 μm were grown in a vertical hot-wall reactor by chemical vapor deposition (CVD) on the Si-face of commercial 4H-SiC *n*-type conductive wafers 8° off-cut toward [11-20]. The nitrogen doping concentrations ([N]) were in the $6.6\times10^{14} - 2.0\times10^{19}$ cm^{-3} range in the epilayers and in the high 10^{18} cm^{-3} (SIMS results) in the substrates. BPDs in the epilayers were imaged by grazing incidence synchrotron reflection X-ray topography with ***g***=-1-128 and λ=1.541 Å at SPring-8. Their detailed structures were investigated by plan-view transmission electron microscopy (TEM).

Results and discussion

Morphology of perfect BPDs. Fig. 1(a) and (b) show the grazing incidence reflection X-ray topographs with ***g***=-1-128 of two epi-wafers with nitrogen concentrations in the epilayers being 6.6×10^{14} cm^{-3} and 2.0×10^{19} cm^{-3}, respectively. Two BPDs in the epilayers are included in each topograph with their line contrasts roughly along the step-flow direction. It can be seen that while the two BPDs in the low-doped epilayer have the typical 'arc-shape' morphology of gliding dislocations [1,2], the two BPDs in the highly doped epilayer appear quite different. They are straight and tilt away from the step-flow direction. Such unusual morphologies were found in 4H-SiC epilayers with nitrogen concentrations above ~1.0×10^{18} cm^{-3}. This concentration will be used to define the boundary between 'low-doped' and 'highly doped' in this report.

It was also observed that the tilt angle between the dislocation line and step-flow direction was not uniform among all the BPDs in one epilayer. It had a distribution. An example is shown in Fig. 2. The epilayer was doped with [N]=2.0×10^{19} cm^{-3}. Tilt angles were measured for over 100 dislocations collected from different regions. As shown in Fig. 2, for the tilt BPDs, the peaks occur at ~±12° for this specific nitrogen concentration. In addition, about 20% of the population aligns parallel to the

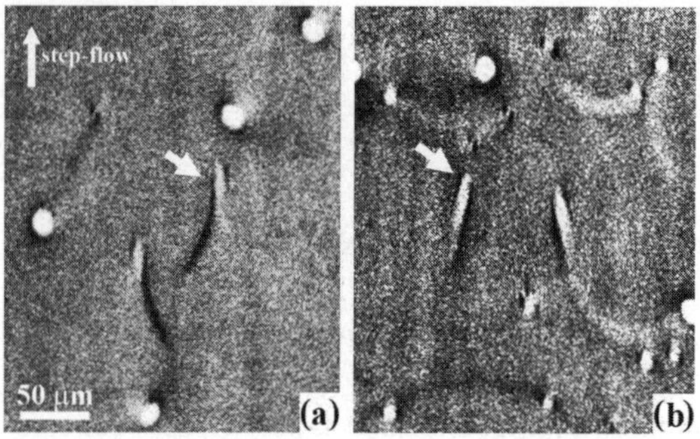

Fig. 1: Grazing incidence (-1-128) reflection X-ray topographs showing the morphologies of BPDs in 4H-SiC epilayers with (a) [N]=6.6×10^{14} cm^{-3}, (b) [N]=2.0×10^{19} cm^{-3}.

step-flow or within ±3°. Here the positive sign is defined as tilting counterclockwise with respect to the step-flow direction. No obvious spatial dependence of the tilt angles was observed. Variation of the tilt angles could be observed in an area of microscopic scale. A series of 4H-SiC epilayers with [N]≥ 1.0×10^{18} cm^{-3} were investigated. Below that, the BPDs tended to show the 'arc-shape' morphology rather than tilt. Dependence of the average tilt angle on the nitrogen concentration is shown in Fig. 3. A steep rise occurs between 1.0×10^{18} cm^{-3} and 2.0×10^{18} cm^{-3}. Above that, the tilt angle only slightly increases as the nitrogen concentration increases. If extrapolating the

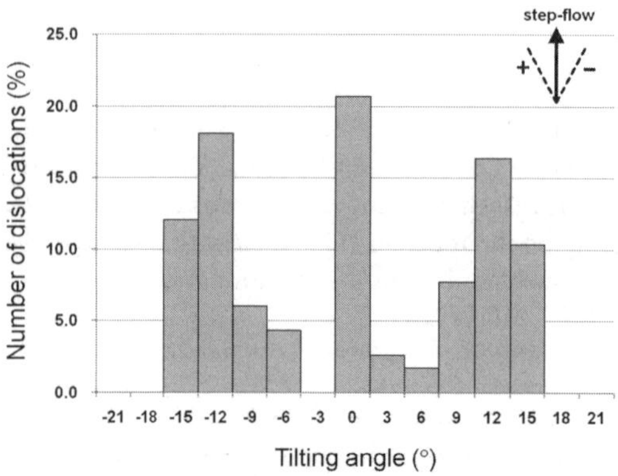

Fig. 2: Distribution of the tilt angles of BPDs in an epilayer doped with [N]=2.0×10^{19} cm^{-3}.

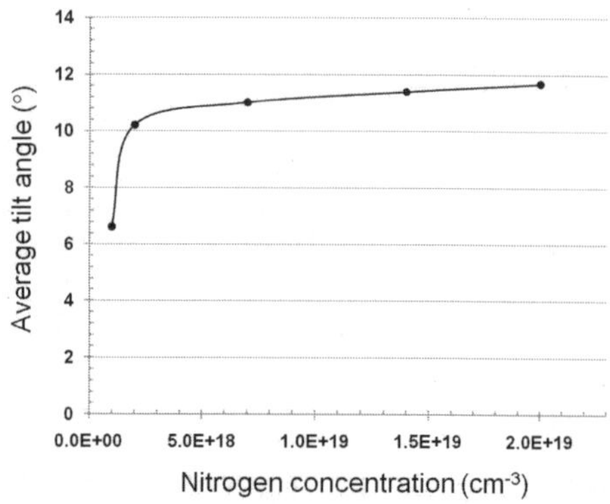

Fig. 3: Dependence of the average tilt angle of BPDs on nitrogen concentration.

dependence curve to [N]=0, the tilt angle also approaches zero, consistent with the fact that <11-20> is the direction of lowest energy for BPDs in SiC if no forces acting on them.

It has been well substantiated that in epilayers grown on substrates off-cut toward [11-20], the BPDs that propagate from the substrates are nearly screw-type with Burgers vectors of ±1/3[11-20]. Using synchrotron reflection X-ray topography, the signs of Burgers vectors could be determined. As examples, one gliding BPD and one tilt BPD, marked by an arrow in Fig. 1(a) and (b) respectively, are selected. Their magnified images are shown in Fig. 4(a) and (b). It can be seen that one similarity between the two is that the contrast of the nearly screw segment is composed of two black lines with white contrast between them, referred to as 'double contrast' [3]. In addition, at the end of the BPD where it intersects the epilayer surface, there is a shift between the double lines. With the step-flow direction pointing upward, the right side of the double lines in Fig. 4(a) shifts upward while in Fig. 4(b) the left side shifts upward. The white dotted lines are drawn to help identify the relative shifts. Such 'double contrasts' of dislocations were simulated by Huang *et al.* based on a 'ray-tracing' method [3]. The relative shift can be explained by the fact that the diffracting planes beside the dislocation core are tilted oppositely in the incident plane. The direction of the relative shift gives information of the strain field around a screw dislocation, which then tells its chirality [3,4]. The BPD in Fig. 4(a) was determined to be left-handed (LH). The one in Fig. 4(b) was right-handed (RH).

In low-doped epilayers, both gliding directions and distances of BPDs can be well explained based on the distribution of stresses and the Burgers vectors of BPDs [5,6]. For example, the LH screw BPD shown in Fig. 4(a) glides toward the right, consistent with a compressive biaxial stress at that location. Reversing either the sign of the stress or that of the Burgers vector of the dislocation will lead to opposite gliding direction, which has been observed in the low-doped epilayers. Correlation between the Burgers vectors of tilt BPDs in highly N-doped epilayers and their tilt directions was also examined. It was found that all the BPDs tilting counterclockwise were LH while all those tilting clockwise were RH. Two examples can be found in Fig. 1(b). Such correlation was confirmed across the entire area of each highly N-doped epilayer. If the tilt directions were determined by stress, such observation would point toward a biaxial tensile stress in the epilayers, which would agree with the N-induced misfit stress in epilayers with [N] higher (*e.g.*, 2.0×10^{19} cm^{-3}) than the substrates. However, the same correlation between the Burgers vectors of BPDs and tilt directions was also observed in the epilayers with [N] lower (*e.g.*, 1.0×10^{18} cm^{-3}) than the substrates. Therefore, N-induced misfit stress alone cannot explain the tilt directions of BPDs in highly N-doped epilayers. On the other hand, an entirely different mechanism is likely.

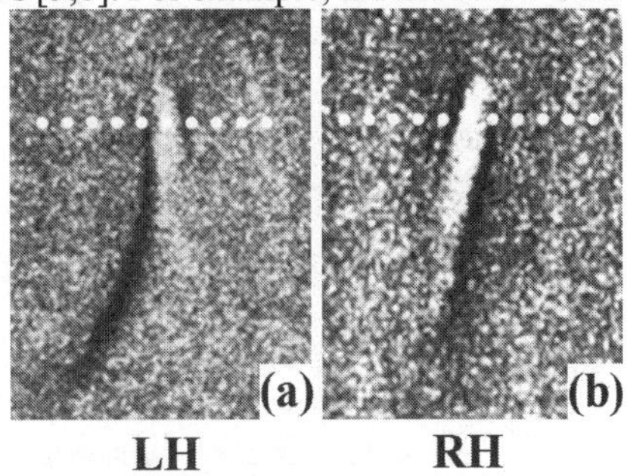

Fig. 4: Magnified images of two BPDs selected from Fig. 1(a) and (b).

Morphology of partial dislocations and stacking faults. To try to elucidate the mechanism accounting for the observed morphologies of BPDs, plan-view TEM was applied to image the two partials and the stacking faults. Fig. 5(a)-(d) were collected from a tilt BPD in the epilayer with [N]=2.0×10^{18} cm^{-3}. Its Burgers vector was determined to be 1/3[11-20] (RH) from its contrast in X-ray topograph. Fig. 5(a) was taken along [0001] zone axis showing the contrast of stacking fault as a dark band. Fig. 5(b)-(d) were recorded under the weak-beam conditions with ***g***=11-20, -2110 and 1-210, respectively. The two partials are in contrast with ***g***=11-20, as shown in Fig. 5(b). Both partials tilt clockwise with respect to [11-20] by 6.6°. Different morphologies of the two partials are recognized instantly. Under the magnification of 10^5, the left one appears to be kinked while the right one appears to be smooth. By ***g·b*** analysis, their Burgers vectors were determined to be $\boldsymbol{b_1}$=1/3[01-10] and $\boldsymbol{b_2}$=1/3[10-10], as labeled in Fig. 5(d) and (c). With the top surface of the plan-view sample being Si-face, the kinked partial was determined to be of Si-core and the smooth partial was determined to be of C-core. Similar observations were obtained on every tilt BPD imaged by TEM. For BPDs in low-doped epilayers, due to the fact that they all aligned parallel to [11-20] when they approached the epilayer surface, TEM samples prepared from those epilayers only included such pure screw-type BPD segments. Both of the partials were smooth along [11-20]. Different morphologies of Si-core and C-core partials were commonly observed in SiC crystals and epilayers, as reported in the literature [7-9].

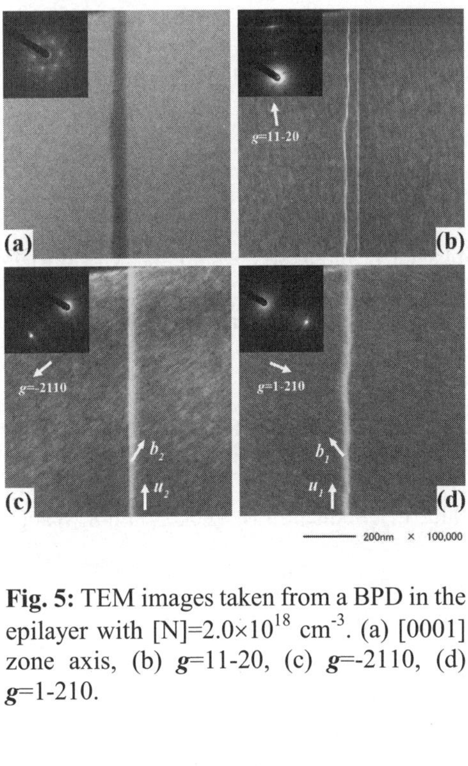

Fig. 5: TEM images taken from a BPD in the epilayer with [N]=2.0×10^{18} cm^{-3}. (a) [0001] zone axis, (b) ***g***=11-20, (c) ***g***=-2110, (d) ***g***=1-210.

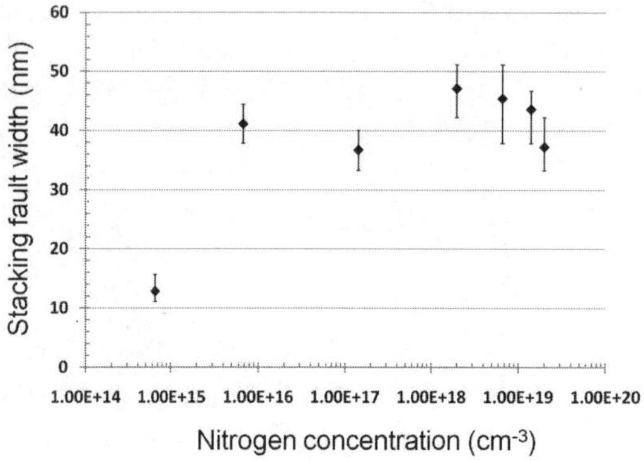

Fig. 6: Dependence of the stacking fault width on the nitrogen concentration in 4H-SiC epilayers.

The widths of stacking faults in different N-doped epilayers were obtained by measuring the distances between the middle of two partials. The results are shown in Fig. 6. Each marker represents the average value of measurement at multiple positions along the stacking fault. The error bar was defined between the maximum and minimum. It can be seen that the stacking fault widths are in the range of 35-50 nm for most of the nitrogen concentrations investigated, in reasonable agreement with the values measured by Hong *et al.* [10]. One exception happens at $[N]=6.6\times10^{14}$ cm^{-3}. The stacking fault is much narrower with a width of ~13 nm. Our attempt of using electron energy loss spectroscopy (EELS) to examine if segregation of nitrogen atoms happened at the stacking faults or partials turned out to be inconclusive.

Summary

Morphologies and structures of BPDs in 4H-SiC epilayers with different nitrogen doping concentrations have been studied by grazing incidence synchrotron reflection X-ray topography and weak-beam TEM. Unlike the 'arc-shape' gliding BPDs in low-doped epilayers, BPDs in highly N-doped ($[N]\geq1.0\times10^{18}$ cm^{-3}) epilayers show totally different morphologies. They are straight and tilt away from [11-20]. A correlation between the Burgers vectors of those BPDs and their tilt directions was found across the entire area of each highly N-doped epilayer. At this stage, it is unclear why those BPDs tilt from [11-20] and what determines the tilt directions and tilt angles. Further investigation is needed.

Acknowledgements

This research is partly supported by the Japan Society for the Promotion of Science (JSPS) through its "Funding Program for World-Leading Innovative R&D on Science and Technology (FIRST Program)".

References

[1] X. Zhang, S. Ha, Y. Hanlumnyang, C.H. Chou, V. Rodriguez, M. Skowronski, J.J. Sumakeris, M.J. Paisley, and M.J. O'Loughlin, J. Appl. Phys. 101 (2007) 053517.

[2] X. Zhang, M. Skowronski, K.X. Liu, R.E. Stahlbush, J.J. Sumakeris, M.J. Paisley, and M.J. O'Loughlin, J. Appl. Phys. 102 (2007) 093520.

[3] X.R. Huang, M. Dudley, W.M. Vetter, W. Huang, W. Si, and C.H. Carter Jr., J. Appl. Cryst. 32 (1999) 516-524.

[4] Y. Chen and M. Dudley, Appl. Phys. Lett. 91 (2007) 141918.

[5] X. Zhang, M. Nagano, and H. Tsuchida, Mater. Sci. Forum 679-680 (2011) 306-309.

[6] X. Zhang, T. Miyazawa, and H. Tsuchida, submitted to the proceedings of this conference.

[7] M.E. Twigg, R.E. Stahlbush, M. Fatemi, S.D. Arthur, J.B. Fedison, J.B. Tucker, and S. Wang, J. Electron. Mater. 33 (2004) 472-476.

[8] X.J. Ning and P. Pirouz, J. Mater. Res. 11 (1996) 884-894.

[9] X.J. Ning, N. Huvey, and P. Pirouz, J. Am. Ceram. Soc. 80 (1997) 1645-1652.

[10] M.H. Hong, A.V. Samant, and P. Pirouz, Phil. Mag. A80 (2000), 919-935.

Materials Science Forum Vols. 717-720 (2012) pp 339-342
© (2012) Trans Tech Publications, Switzerland
doi:10.4028/www.scientific.net/MSF.717-720.339

Analysis of Dislocations Nucleated after Nano Indentation Tests at Room Temperature in 4H-SiC

J.-L. Demenet[1,a,*], M. Amer[1,b], C. Tromas[1,c] and J. Rabier[1,d]

[1]Institut PPRIMME, Département de Physique et Mécanique des Matériaux - UPR 3346 CNRS, Université de Poitiers, ENSMA - SP2MI, BP 30179, F-86962 Chasseneuil Futuroscope Cedex

[a]jean.luc.demenet@univ-poitiers.fr, [b]madyan.amer@etu.univ-poitiers.fr, [c]christophe.tromas@univ-poitiers.fr, [d]jacques.rabier@univ-poitiers.fr,*corresponding author

Keywords: SiC, dislocations, nano indentation, TEM

Abstract. 4H-SiC intrinsic homoepitaxied single crystals have been nano indented at room temperature using a spherical indentor and the related deformation microstructures have been analyzed by TEM. Dislocations are lying on the basal plane but have been found to be perfect, in contrast with observations made at higher temperature. Although such a change in deformation mechanism has been observed in other semiconductors such as silicon and indium antimonide, it was unexpected in a very low stacking fault material such as SiC.

Introduction

The covalent bonding between Si and C atoms makes silicon carbide a material with a high strength and a good candidate for thermomechanical applications. However, as all covalent materials it exhibits a ductile-brittle transition as it has been shown by fracture tests in 4H-SiC which is one of the most common polytypes of this material [1]. Compression tests at constant strain rate led to the same conclusion [2]. Above about 1100°C, the material is ductile and below this temperature it is brittle. The analysis by TEM of samples deformed in the conditions described above shows two types of microstructures corresponding to two deformation mechanisms [2]. At T>1100°C, deformation occurs by nucleation and glide of $a/3<11\bar{2}0>$ perfect dislocations lying in the basal plane. Those dislocations are weakly dissociated into two Shockley partials, the leading and the trailing, with a $a/3<1\bar{1}00>$ Burgers vector. Below 1100°C, only partial dislocations are nucleated, dragging a large stacking fault behind. These leading partial dislocations glide on (0001) slip planes without their trailing companions, whose nucleation seems to be impossible in this domain of temperature. Furthermore, the determination by large angle convergent beam electron diffraction (LACBED) of the core nature of dislocations shows that the observed partial dislocations have a Si(g) core, indicating an easiness of this kind of dislocation to nucleate and to propagate.

Several studies of the behaviour of 4H-SiC in the brittle domain have been performed in the past by different authors. Such studies required the sample to be imbedded in a confining medium in order to reduce fracture and to avoid cracks to propagate. Mussi *et al.* [3] deformed plastically 4H-SiC down to 400°C by using a gaseous confining medium (Paterson press). These authors showed that the plasticity still occurs by partial dislocations at this temperature. However the observation of half-loops of dislocations does not indicate an asymmetry of mobility between Si(g) and C(g) partials. Thanks to a solid confining medium (multi-anvil press), Demenet *et al.* [4] succeeded to deform 4H-SiC down to room temperature. Microstructure consisted of perfect and weakly dissociated dislocations. Nevertheless, the particular experimental conditions and the very high density of dislocations introduced by this harsh method did not allow the authors to know the stress applied to the sample during the experiment.

Another technique used to study the brittle domain, is the indentation test in which the stress tensor possess a hydrostatic component which reduces crack nucleation and propagation. In this paper, we present a detailed analysis of dislocations introduced by nano-indentation at room temperature using a spherical indentor.

Experimental

A homoepitaxied single crystal layer of 4H-SiC was used in this study. The layer was 10µm thick, intrinsic (Ni ~ 6.10^{13} cm^{-3}) and oriented 7° off the (0001) Si-face. Two square networks of 100 indentation impressions, separated each other by 100µm, were made on the Si-face with a CSM Instruments nano indentor. The shape of the tip is spherical with a radius of 0.7 µm. One network was made with a 100 mN maximal load applied to each indentation making a penetration depth of about 420 nm. The other one was made with a 200 mN maximal load applied for each indentation making a penetration depth of about 740 nm. For TEM observations the sample was mechanically thinned from the back face using the Tripod™ technique. Finally, electron transparency was obtained using a Gatan PIPS apparatus. The analysis of dislocations using WBDF and diffraction contrast techniques were performed with a Jeol 200CX microscope whereas a Jeol 2200 FS microscope operating at 200 kV was used for the determination of the core nature of dislocations by the LACBED technique.

Results

Results presented here pertain only to dislocation microstructures analysis.

Fig. 1 is the image of an impression observed using an atomic force microscope (AFM) in the tapping mode and resulting from a 100 mN indentation. Despite the relatively low load applied, cracks are observed emanating from the indent which are aligned approximately along the $<11\bar{2}0>$ and $<1\bar{1}00>$ directions of the basal plane. Fig. 2 shows the TEM image of another indent made under the same conditions. The circular shape of the impression is clearly seen and the correspondence between the two images is very good.

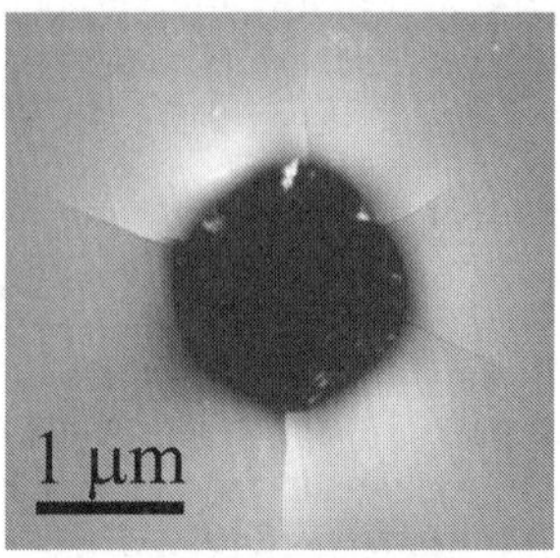

Fig. 1. Image of an indentation impression made under a load of 100 mN and observed by AFM in the tapping mode.

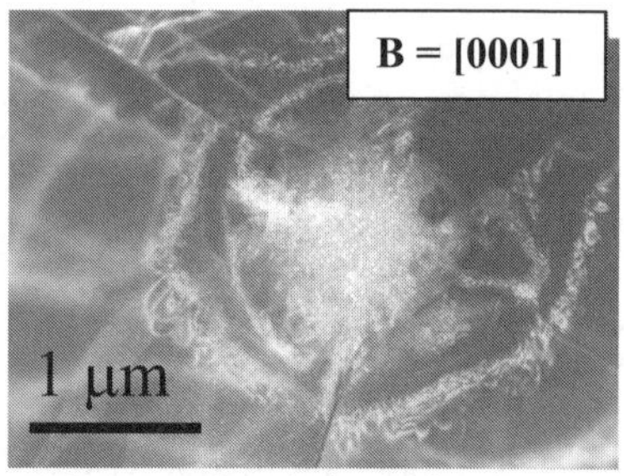

Fig. 2. Image of an indentation impression made under a load of 100 mN and observed by TEM under multi-beam diffraction conditions. Zone axis: **B** = [0001].

With the multi-beam diffraction conditions used in Fig. 2, dislocations are seen all around the indentation impression, forming a rosette with a six-fold symmetry. Despite the spherical shape of the indentor, dislocations seem to start from straight lines forming a hexagon. These lines are along the $<1\bar{1}00>$ directions. Half loops of dislocations did not propagate far from the perturbed region making a fine analysis of dislocations rather difficult. For this reason, the determination of Burgers vectors with the visibility and extinction criterion has been performed around an indentation impression made with a maximum load of 200 mN. The same features as described in Fig. 2 are observed with dislocations somewhat further away from the impression. The **g.b** analysis for one arm of the rosette is reported in Fig. 3. Dislocations are in contrast for $g = \bar{2}110$, $g = \bar{1}2\bar{1}0$ and $g = \bar{1}\bar{1}20$ (Fig. 3a, 3b and 3d respectively) and out of contrast for $g = 10\bar{1}0$ (Fig. 3c). From this, it can be deduced that dislocations have a $b = \pm a/3[\bar{1}2\bar{1}0]$ Burgers vector. In addition, Fig. 3b does not show evidence of dislocation dissociation. If dissociated, dislocations should be dissociated into two partials with $\pm a/3[\bar{1}100]$ and $\pm a/3[01\bar{1}0]$ Burgers vectors, and both partials should be in contrast with $g = \bar{1}2\bar{1}0$.

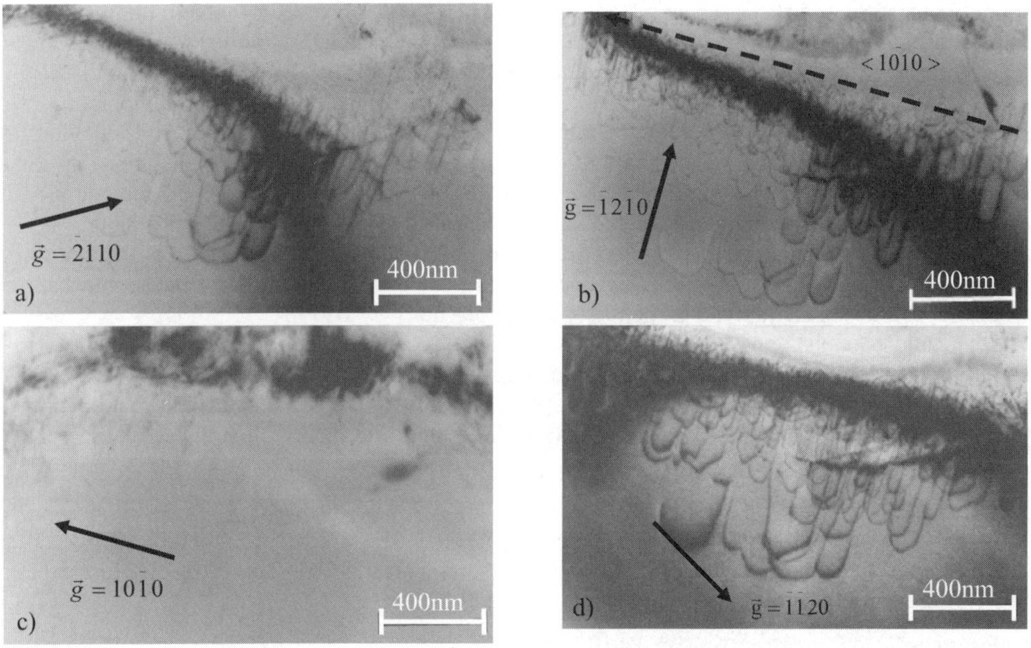

Fig. 3. **g.b** analysis of dislocations around an indentation impression made with a maximum load of 200 mN. Dislocations are in contrast in a), b) and d) and out of contrast in c).

The same contrast analysis, not reported here, has been done for all arms of the rosette around the impression. All dislocations have been found to be perfect, undissociated at the TEM resolution. They have Burgers vectors which correspond to the three slip systems $\pm a/3 < 11\bar{2}0 > \{0001\}$ of the basal plane. For each slip system, half-loops are somewhat elongated in the screw direction corresponding to the propagation direction and the segments constituting the half-loop fronts are composed of Si(g) or C(g) atoms. The sign of Burgers vectors cannot be determined by conventional TEM technique so that it is not possible to know the nature of dislocation fronts. For this reason the LACBED technique has been used to determine the amplitude and sign of Burgers vector. The identification of the Burgers vectors of the dislocations was made by means of the Cherns-Preston rules [5]. In addition, the Bragg line indices and the deviation parameter were identified by comparisons with theoretical patterns drawn by means of the Electron Diffraction software, based on the kinematical theory. Applying the LACBED technique to the zone analyzed in Fig. 3, it has been shown that the Burgers vector of dislocations is $b = a/3[1\bar{2}10]$ [6]. The knowledge of the entry face of electrons in the thin foil in the microscope and application of the FS/RH convention for the Burgers circuit allowed determining the nature of dislocation fronts. Atoms terminating the extra half-planes of these segments are C atoms in the shuffle mode [6]. Despite the fact that dislocations of the six arms around the impression have not been analyzed by LACBED, it is assumed that all dislocations have the same carbon-core front.

Discussion

Nanoindentation of the (0001) Si-face of a 4H-SiC single crystal at room temperature using a spherical indentor has shown that dislocations are emitted on the basal plane. These dislocations are perfect and do not propagate far from the indent. Page *et al.* [7] have performed nanoindentation at room temperature on 6H-SiC using a Berkovich indentor. They observed dislocation loops forming rosettes around indentations, but no quantitative crystallographic analysis of dislocations was made. From our experiments reported in this paper, it is clear that dislocations are perfect and show a change in plasticity mechanism at room temperature as compared to higher temperature. No stacking faults have been observed contrary to what was reported in [4]. This behaviour is analogous to the change in deformation mechanism at low temperature/high stress in some semiconductors with the sphalerite structure. Indeed Rabier and Demenet [8] in silicon and Kedjar *et al.* [9] in indium antimonide have shown that perfect dislocations control deformation in this regime. The low stacking fault energy of SiC (15 $mJ.m^{-2}$ for 4H-SiC [10]) which is thought to be responsible for polytypic transformation does not impede this change in deformation mechanism which appears as a general trend for semi conductor plasticity.

Dislocations lie on the basal plane which is the most favourable plane for dislocation glide in this structure, but the nucleation site of dislocations is a questioning point. Indeed, other planes than the basal one, in which Schmid factor is zero, are necessary to accommodate deformation due to indentor. As observed in Fig. 2 and Fig. 3 (see the dotted line in Fig. 3b), dislocations seem to emerge from a line parallel to $<10\bar{1}0>$. Then dislocations could have been nucleated as dislocation loops in prismatic or pyramidal plane whose intersection with the basal plane is aligned along $<10\bar{1}0>$. This is consistent with Maeda *et al.* [11] and Mussi *et al.* [12] who observed after deformation a moderate activity of dislocations in such planes before they cross-slip to the basal plane. More experiments are needed to justify such a hypothesis.

References

[1] P. Pirouz, M. Zhang, J.L. Demenet and H.M. Hobgood, Transition from brittleness to ductility in SiC, J. Phys.: Condens. Matter 14 (2002) 12929-12945.

[2] P. Pirouz, J.L. Demenet and M.H. Hong, On transition temperatures in the plasticity and fracture of semiconductors, Phil. Mag. A 81 (2001) 1207-1227.

[3] A. Mussi, J. Rabier, L. Thilly and J.L. Demenet, Plasticity and deformation microstructure of 4H-SiC below the brittle-to-ductile transition, phys. stat. sol. (c) 4 (2007) 2929-2933.

[4] J.L. Demenet, X. Milhet and J. Rabier, TEM observations of the coexistence of perfect and dissociated dislocations in SiC under high stress, phys. stat. sol. (c) 202 (2005) 1987-1991.

[5] D. Cherns, A.R. Preston, Convergent beam diffraction studies of interfaces, defects, and multilayers, J. Electr. Microsc. Tech. 13 (1989) 111-122.

[6] J.L. Demenet, M. Amer, C. Tromas and J. Rabier, to be published.

[7] T.F. Page, L. Riester, S.V. Hainsworth, The plasticity response of 6H-SiC and related materials to nanoindentation: slip vs densification, Mater. Res. Soc. Symp. Proc. 522 (1998) 113-118.

[8] J. Rabier and J.L. Demenet, Low temperature, high stress plastic deformation of semiconductors: the silicon case, phys. stat. sol. (b) 222 (2000) 63-74.

[9] B. Kedjar, L. Thilly, J.L. Demenet, J. Rabier, Plasticity of indium antimonide between -176°C and 400°C under hydrostatic pressure. Part II: Microscopic aspects of the deformation, Acta Mater. 58 (2010) 1426-1440.

[10] M.H. Hong, A.V. Samant, and P. Pirouz, Stacking fault energy of 6H-SiC and 4H-SiC single crystals, Phil. Mag. A 80 (2000) 919-935.

[11] K. Maeda, K. Suzuki, S. Fujita, M. Ichihara and S. Hyodo, Defects in plastically deformed 6H-SiC single crystals studied by transmission electron microscopy, Phil. Mag. A 57 (1988) 573-592.

[12] A. Mussi, J.L. Demenet and J. Rabier, TEM study of defects generated in 4H-SiC by microindentations on the prismatic plane, Phil. Mag. Lett. 86 (2006) 561-568.

Materials Science Forum Vols. 717-720 (2012) pp 343-346

doi:10.4028/www.scientific.net/MSF.717-720.343

Synchrotron X-ray Topography Studies of the Propagation and Post-Growth Mutual Interaction of Threading Growth Dislocations with c-component of Burgers Vector in PVT-Grown 4H-SiC

Fangzhen Wu[1], Huanhuan Wang[1], Shayan Byrappa[1], Balaji Raghothamachar[1], Michael Dudley[1,a], Edward K. Sanchez[2,b], Darren Hansen[2], Roman Drachev[2], Stephan G. Mueller[2] and Mark J. Loboda[2]

[1]Department of Materials Science and Engineering, Stony Brook University, Stony Brook, New York, 11794, USA

[2]Dow Corning Compound Semiconductor Solutions, Midland, Michigan, 48686, USA

[a]mdudley@notes.cc.sunysb.edu, [b]edward.sanchez@dowcorning.com

Keywords: Threading Dislocation, Burgers Vector, Annihilation, Vacancy

Abstract. Synchrotron White Beam X-ray Topography (SWBXT) imaging of wafers cut parallel to the growth axis from 4H-SiC boules grown using Physical Vapor Transport has enabled visualization of the evolution of the defect microstructure. Here we present observations of the propagation and post-growth mutual interaction of threading growth dislocations with *c*-component of Burgers vector. Detailed contrast extinction studies reveal the presence of two types of such dislocations: pure *c*-axis screw dislocations and those with Burgers Vector $n_1\mathbf{c}+n_2\mathbf{a}$ [1], where n_1 is equal to 1 and n_2 is equal to 1 or 2. In addition, observations of dislocation propagation show that some of the threading dislocations with *c*-component of Burgers adopt a curved, slightly helical morphology which can drive the dislocations from adjacent nucleation sites together enabling them to respond to the inter-dislocation forces and react. Since all of the dislocations exhibiting such helical configurations have significant screw component, and in view of the fact that such dislocations are typically not observed to glide, it is believed that such morphologies result in large part from the interaction of a non-equilibrium concentration of vacancies with the originally approximately straight dislocation cores behind growth front. Such interactions can lead to complete or partial Burgers vector annihilation. Among the reactions observed are: (a) the reaction between opposite-sign threading screw dislocations with Burgers vectors **c** and –**c** wherein some segments annihilate leaving others in the form of trails of stranded loops comprising closed dislocation dipoles; (b) the reaction between threading dislocations with Burgers vectors of -**c**+**a** and **c**+**a** wherein the opposite **c**-components annihilate leaving behind the two **a**-components; (c) the similar reaction between threading dislocations with Burgers vectors of -**c** and **c**+**a** leaving behind the **a**-component.

Introduction

Recent progress in bulk crystal growth of silicon carbide has enabled the effective elimination of the micropipe, arguably the most harmful defect in this material [2]. However, it is well known that other threading defects in this material can have a detrimental effect on device performance. For example, the impact of closed-core threading screw dislocations has been well documented [3,4]. Recent reports have shown that such threading dislocations are not restricted to those of pure screw character but can comprise all dislocations with *c*-component of Burgers vector, ie. with $b= n_1c+n_2a$ [5,6,7]. In this paper, we report on synchrotron X-ray topographic studies of axial slices taken from 4H-SiC boules wherein the densities of such defects are sufficiently low that detailed interactions can be discerned. Understanding the interactions between such threading dislocations is of great interest as this might provide insight into mechanisms by which their densities may be reduced.

Results and Discussion

Fig. 1 (a) shows an SWBXT image recorded with g=0004 from an axial slice exhibiting low threading dislocation density. Segments of six dislocations are visible although dislocations 3 and 4 appear to have interacted such that parts of their length (between the two arrows on Fig. 1(a)) have apparently annihilated. Figs. 1(b) - (f) show images recorded from the same region with g vectors of 0-110,

1-210, 01-11, 0-111, and 10-1-5. Note that dislocations 2 and 5 are out of contrast on Fig. 1(b)&(c) so that they are pure screw in character, i.e. TSDs, **b**=**c**. The background in Fig. 1(b) consists of horizontal lines which are segments of basal plane dislocations and some short *c*-axis threading segments which are threading edge dislocations; both of these have Burgers vectors of 1/3<11-20> i.e. **b**=**a**, and are invisible (as expected) on Fig. 1(a). Note that line 1 is visible in the 0004, 0-110 and 11-20 reflections, which suggest that it is neither pure screw nor pure edge in character. However, it is invisible in Fig. 1(d) and computation of **g**.**b** shows that its Burgers vector is 1/3[1-213], i.e., **c**+**a**.

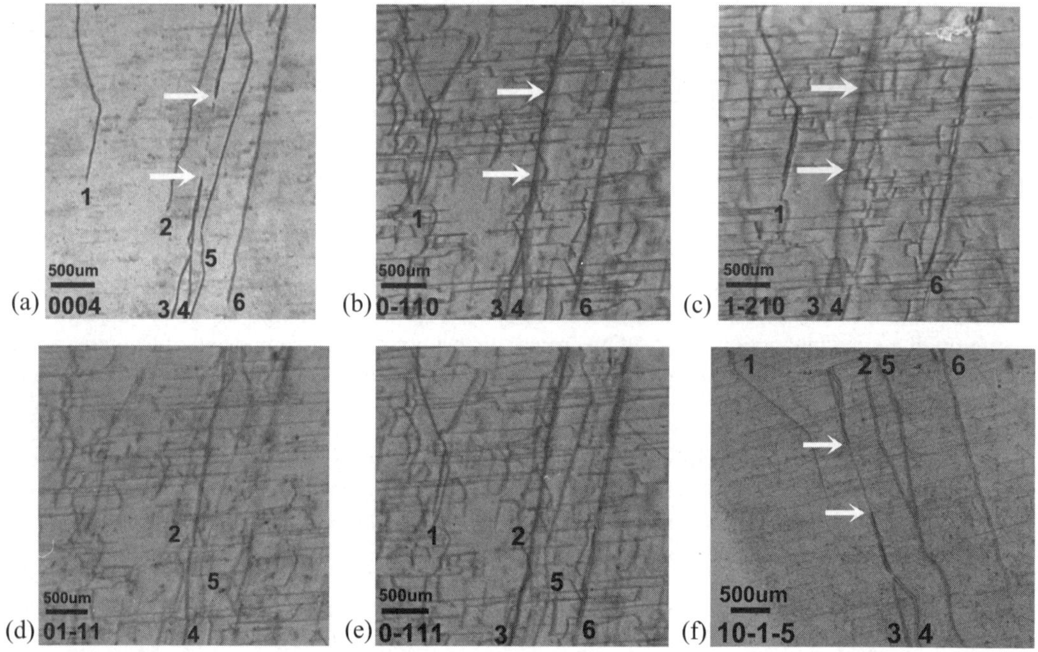

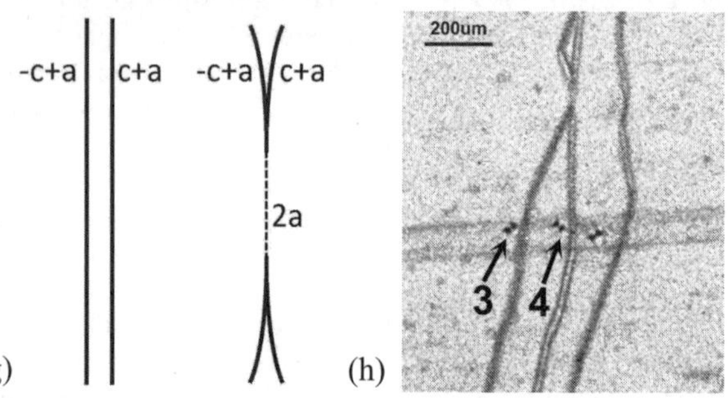

(g) (h)

Figure 1. (a) SWBXT image of 0004 reflection shows dislocation contrast from 1 to 6. Dislocation 3 and 4 contains an annihilation segment, where no contrast is present in this reflection, which is the region between white arrows; (b) 0-110 reflection shows absence of dislocation 2 and 5, but strong contrast for dislocation 3 and 4 with the "annihilation segment" present; (c) 1-210 reflection shows absence of dislocation 2 and 5; (d) 01-11 reflection shows contrast from 2, 4 and 5; (e) 0-111 reflection shows contrast from 1, 2, 3, 5 and 6; (f) 10-1-5 reflection shows sharp difference in contrast between the partially annihilated segment, marked by arrows, and the rest of dislocations 3 and 4; (g) illustration of the reaction between $-c+a$ and $c+a$ dislocation; (h) overlapping 0004 section topograph on projection topograph shows opposite signs of dislocation 3 and 4, since their screw arrangements are mirror images.

By doing similar contrast analysis, the Burgers vectors of dislocations 3, 4 and 6 can be also determined as **c**+**a** type, as summarized in Table 1. In view of the fact that contrast from dislocations 3 and 4 is lighter than that from dislocation 1 in Fig. 1(c), it can be inferred that dislocations 3 and 4 have different *a*-component from dislocation 1. Performing **g**.**b** analysis using the suggested Burgers vectors in Table 1, dislocations 3 and 4 have a smaller **g**.**b** value than dislocation 1 when g is equal to 1-210, which explains the weaker contrast in Fig. 1(c). Careful examination of Figs. 1(d) and (e) shows that dislocations 3, 4 and 6 do not completely disappear in related <01-11> type reflections when **g**.**b**=0. This is due to the fact that these dislocations have mixed character and as such will exhibit very weak but not zero contrast when **g**.**b**=0 [8].

Further evidence for the existence of *c*+*a* dislocations in 4H-SiC can be found from the reaction between dislocation 3 and 4. In Fig. 1(a), a segment of line 3 and 4, marked with arrows, appears to have annihilated and shows no contrast. However, this segment is clearly present on Figs. 1(b)&(c), which suggest that it contains only an *a*-component of Burgers vector so that dislocations 3 and 4 only partially annihilate. As illustrated in Fig. 1(g), dislocation 3 and 4 have opposite sign *c*-components and these *c*-component annihilate while the two *a*-components are left behind. This can be further proved by Fig. 1(f) from which sharp difference in contrast between the partially annihilated segment, marked by arrows, and the rest of dislocations 3 and 4 can be observed. Fig. 1(f) is recorded from 10-1-5 reflection and the fourth index of this reflection is much bigger than others. When performing the **g**.**b** calculation, the *c*-component contributes much more to the final value, which means that the *c*-component shows much stronger contrast than the *a*-component in this reflection. Another interesting thing in Fig. 1(f) is that position of line 2 is shifted compared to its position in Fig. 1(a), because the 10-1-5 reflection is recorded by rotating the crystal through a large angle and differences in the depth position of the dislocations in the axial slice will appear as a change in relative line position. Fig. 1(h) shows a section topograph, which is sensitive to the sense of dislocation, shown overlapped onto an 0004 projection topograph. Arrows point to the contrast from dislocation 3 and 4 in the section topograph. From the sense of the mutual shift between the bimodal image components, it can be seen that dislocation 3 has right-handed screw components while dislocation 4 has left-handed, which confirms that dislocations 3 and 4 have opposite *c*-components which can thus annihilate.

Table 1. **g**.**b** values for dislocations from 1 to 6 in Fig. 1

g.b (**b** \ **g**)	**0004**	**0-110**	**1-210**	**01-11**	**0-111**	**10-1-5**
L1&6:1/3[1-213], *c*+*a*	4	1	2	0	2	-5
L3: 1/3[-1-123], *c*+*a*	4	1	1	0	2	-5
L4: 1/3[-1-12-3], -*c*+*a*	-4	1	1	-2	0	5
L2&5: <0001> , TSD	4	0	0	1	1	-5
Arrow:2/3[-1-120], 2*a*	0	2	2	-2	2	-2

Fig. 2 shows a similar reaction between –*c* and *c*+*a* dislocations. As illustrated in Fig. 2(a), the –*c* TSD annihilate with the opposite sign *c*-component of the *c*+*a* dislocation leaving a single *a*-component behind. Dislocation 7, which is TSD, is visible in Fig. 2(b) and out of contrast in Fig. 2(c). Dislocation 8 is visible in both images, which suggest that it has both *c*- and *a*-components. The annihilated segment, marked by arrows, is invisible in Fig. 2(b) and visible in Fig. 2(c), which indicates that the annihilated segment has only an *a*-component of Burgers vector.

The mechanism involved in these reactions requires a driving force between the pairs of dislocations. In 4H-SiC, the *c*-component of Burgers vector has magnitude of 10.05Å, almost three times as big as that of the *a*-component at 3.08Å in magnitude. Two closed-core threading dislocations with opposite *c*-component experience a strong attractive force. By calculating the square of Burgers vector magnitude of dislocations before and after reaction, it can be found that the *c*-component annihilation also lower the energy of the dislocation. Although the reactions is favorable in energy, the reaction is hard to conduct via slip which is usually considered to be quite difficult because of the large magnitude of the Burgers vectors and the atomically rough slip planes parallel to the *c*-axis leading to high Peierls stress. On the other hand, many threading dislocations with *c*-component of Burgers vectors are observed to adopt a curved, slightly helical morphology. Such morphologies result in large part from the interaction of a non-equilibrium concentration of vacancies with the screw components of the originally approximately straight dislocation cores behind growth front, which can enable the dislocations from adjacent nucleation sites to respond to their mutual attractive forces and come together and react. Fig. 3(a) shows a typical helical morphology of a TSD and when two dislocations like this are close to each other, the interaction between TSDs with opposite signs of Burgers vectors will occur, as shown in Fig. 3(b) where some segments annihilate leaving others in the form of trails of stranded loops comprising closed dislocation dipoles.

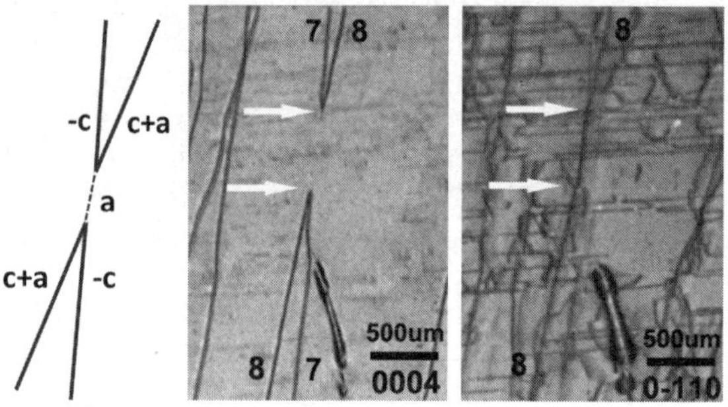

(a) (b) (c)

Figure 2. (a) illustration of the reaction between *-c* and *a+c* dislocation; (b) SWBXT image of 0004 reflection showing the annihilation of *-c* and *c+a* dislocation, named 7 and 8 respectively; (c) 0-110 reflection showing the strong contrast of dislocation 8 as well as "the annihilation segment", and absence of dislocation 7.

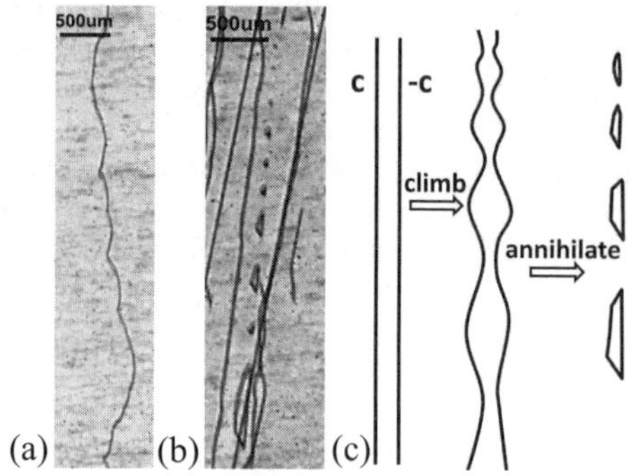

Figure 3. (a) SWBXT image of 0004 reflection showing curved slightly helical morphology of *c* dislocation; (b) 0004 reflection at different region showing annihilation of *c* and *–c* dislocation at some segments leaving others in the form of stranded loops comprising closed dislocation dipoles; (c) illustration of the reaction between *c* and *–c* dislocation.

Summary

We have presented observations of mutual interaction between threading dislocations with *c*-component of Burgers vector. Interaction of a non-equilibrium concentration of vacancies with the *c*-component of Burgers vector enable the dislocations to respond to the attractive forces they experience and come together to react. Such interactions can lead to complete or partial Burgers vector annihilation. Among the reactions observed are: (a) the reaction between opposite-sign threading screw dislocations with Burgers vectors **c** and –**c** wherein some segments annihilate leaving others in the form of trails of stranded loops comprising closed dislocation dipoles; (b) the reaction between threading dislocations with Burgers vectors of **-c+a** and **c+a** wherein the opposite **c**-components annihilate leaving behind the two **a**-components; (c) the similar reaction between threading dislocations with Burgers vectors of **-c** and **c+a** leaving behind the **a**-component.

Acknowledgements

Work was supported by Dow Corning. Topography experiments were carried out at the Stony Brook Synchrotron Topography Facility, Beamline X-19C, at the NSLS (DOE Office of Basic Energy Sciences Contract no. DE-AC02-76CH00016).

References

[1] W. M. Vetter and M. Dudley, Phil. Mag., vol. 86, pp. 1209-1225, (2006).
[2] R.T. Leonard, Y. Khlebnikov, et al., Mater. Sci. Forum, vols. 600-603, pp. 7-10, (2009).
[3] P. G. Neudeck, Mater. Sci. Forum vols. 338-342, pp. 1161-1166, (2000).
[4] H. Lendenmann, F. Dahlquist, et al., Mater. Sci. Forum, vols. 353–356, pp. 727–730, (2001).
[5] M. Dudley, F. Wu, et al., Appl. Phys. Lett., vol. 98, 232110 (2011).
[6] M. Dudley, H. Wang, et al., Mater. Sci. Forum, vols. 679-680, pp. 269-272, (2011).
[7] M. Dudley, S. Byrappa, et al., Mater. Res. Soc. Symp. Proc., vol. 1246, 1246-B02-02 (2010).
[8] B.K. Tanner, X-Ray Diffraction Topography, first ed., Pergamon Pr., pp. 100-102, (1976).

Materials Science Forum Vols. 717-720 (2012) pp 347-350

doi:10.4028/www.scientific.net/MSF.717-720.347

Deflection of Threading Dislocations with Burgers vector c/c+a observed in 4H-SiC PVT –Grown Substrates with associated Stacking faults

Shayan Byrappa[1], Fangzhen Wu[1], Huanhuan Wang[1], Balaji Raghothamachar[1], Gloria Choi[1], Shun Sun[1], Michael Dudley[1,a], Edward K. Sanchez[2,b], Darren Hansen[2], Roman Drachev[2], Stephan G. Mueller[2] and Mark J. Loboda[2]

[1]Department of Materials Science and Engineering, Stony Brook University, Stony Brook, New York, 11794, USA

[2]Dow Corning Compound Semiconductor Solutions, Midland, Michigan, 48686, USA

[a]mdudley@notes.cc.sunysb.edu, [b]edward.sanchez@dowcorning.com

Keywords: Threading Dislocation, Partials, Stacking faults, Burgers Vector **c+a.**

Abstract

A review is presented of Synchrotron White Beam X-ray Topography (SWBXT) studies of stacking faults observed in PVT-Grown 4H-SiC crystals. A detailed analysis of various interesting phenomena were performed and one such observation is the deflection of threading dislocations with Burgers vector c/c+a onto the basal plane and associated stacking faults. Based on the model involving macrostep overgrowth of surface outcrops of threading dislocations, SWBXT image contrast studies of these stacking faults on different reflections and comparison with calculated phase shits for postulated fault vectors, has revealed faults to be of basically four types: (a) Frank faults; (b) Shockley faults; (c) Combined Shockley + Frank faults with fault vector s+c/2; (d) Combined Shockley + Frank faults with fault vector s+c/4.

Introduction

From its inception the Silicon Carbide as a wide band- gap semiconductor has been a promising material for various high-temperature and high- power electronics technologies, but the detrimental influence of defects such as dislocations has hampered its device performance [1]. Over the last two decades SWBXT characterization technique has been an indispensable tool for the SiC crystal growth community as it has played a key role in revealing the detailed configurations of various defects and understanding their origins right from the seed to the devices grown on them. Among the notable contributions are the identification of threading screw dislocations(TSDs) and threading edge dislocations(TEDs) as growth defects [2-4] and basal plane dislocations(BPDs) as deformation induced defects [5]. The latter has promoted the development of stress reduction strategies to minimize the driving force for BPD motion behind the growth interface. The reduction of thermal stress during growth to levels below the critical resolved shear stress of 1MPa which has enabled the lowering of BPD densities by up to three orders of magnitude in both 76 and 100mm wafers [6]. The lower dislocations density wafers have provided an unique opportunity to perceive various interesting phenomena which among them is the observation of deflection of threading dislocations of Burgers vector c/c+a onto the basal plane which produce various kinds of stacking fault configurations. Detailed analysis of these of stacking fault configurations on different reflections and their image contrast studies have enabled us to predict the fault vectors to be of four types. This paper mainly focuses on the formation mechanisms of these four faults and a lucid understanding of their association with threading deflected dislocations is developed, thereby providing the necessary models.

Results and Discussion

The contrast from stacking faults in x-ray topography arises from the phase shift experienced by the x-ray wavefields as they cross the fault plane [7]. This phase shift has been computed to be equal to $\delta=(-2\pi\ g.R)$, where g is the active reciprocal lattice vector for the reflection and R is the fault vector. Contrast is expected to disappear when $\delta=0$ (corresponding to g.R=integer) and is expected to be very weak when $\delta=\pm\pi/6$ (corresponding to g.R=$\pm1/12$, $\pm11/12$) and weak but visible for $\delta=\pm\pi/3$ (corresponding to g.R=$\pm1/6$, $\pm5/6$). Contrast should be very well marked for $\delta=\pm\pi$ and $\pm2\pi/3$ (corresponding to g.R=$\pm1/2$, $\pm1/3$ and $\pm2/3$). This contrast studies from multiple reflections have been carried out to determine the fault vectors.

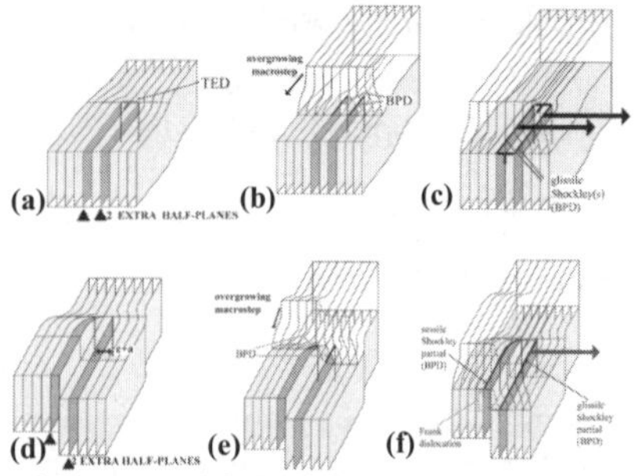

Figure 1: (a)-(c) Overgrowth of the surface outcrop of a TED by a macrostep converting it to a BPD. Once on the basal plane, both Shockley partials are glissile and move in tandem; (d)-(f) Overgrowth of a c + a dislocation with a c-height step. After deflection, one Shockley is sessile and the other glissile.

Two models have been used to explain the formation of these faults. First is the overgrowth of TEDs onto the basal plane[6]. Fig. 1(a)-(c) shows two extra half- planes being overgrown by a macrostep which causes them to get deflected onto the basal plane in the step flow direction. Similarly the second case also involves the overgrowth of dislocations of the type c+a described previously in the papers, [8-10] (i.e. both c and a components are anchored at the dislocation outcrop). Fig. 1(d)-(f) shows the c+a dislocations being overgrown resulting in the deflection of extra half planes associated with 'a' component to deflect onto basal plane but in this case the dislocations are separated by the step height of the 'c' dislocation[6]. In both the cases for overgrowth to happen the stacking position of the underside of overgrowing step should maintain tetrahedral bonding, i.e. stacking sequence rules have to be obeyed[11]. The deflected dislocations under thermal shear stress glides creating large faults, with the leading partial faulting and trailing partial unfaulting the plane in case one and in the latter the partials cannot move in tandem as they are separated by step height due to which depending on direction of force one partial glides creating faults while the other partial gets locked[6]. Based on these two models four fault vectors have been determined which are explained in the following:

Type I: Frank Fault

Frank faults are considered to be "in-grown" stacking faults and are thought to result from the overgrowth of c-axis screw dislocations whose surface growth spiral steps were separated into c/2 increments [12]. Figure 2(a)-(i) correspond to SWBXT images in Transmission geometry used for the analysis of Frank type fault. Figure 2(a)-(c) with triangular shape configuration shows contrast of the fault while the contrast disappeares in all reflections belonging to the [0001]- zone. Detailed analysis of the contrast from different reflections shows the fault vector is ½[0001].

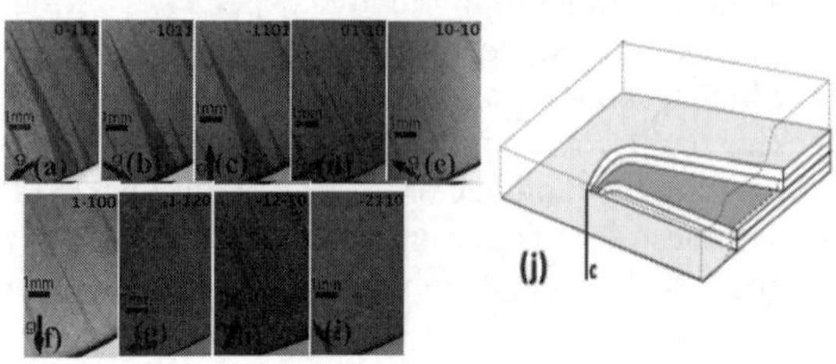

Figure 2: (a)- (i): Transmission SWBXT images of different reflections showing triangular shaped stacking fault corresponding to the Type II Frank fault with R=1/2[0001]. Figure 2(j): Illustration of a Frank type stacking fault resulting from the overgrowth of a c-axis screw dislocation which has separated into two c/2 step heights. The area between the two demisteps corresponds to the Frank fault.

Type II: Shockley faults

Figs. 3(a)- (i) are SWBXT images recorded from a section of a n+ 4H- SiC crystal showing stacking fault configurations which has one straight edge corresponding to the sessile partial and the other being curved corresponding to the glissile partial. In this case the step has an integral number of unit cells high, the resultant expanding fault will be a Shockley fault(since the out-of- plane component is *nc*, where n is an integer). Here we can see the fault contrast in figs. 3(a)-(f) and shows no contrast in figs. 3(g)-(i) from the in- plane reflections. Detailed analysis of the fault contrast revealed the fault vector to be 1/3[01-10] which corresponds to a Shockley fault. Fig. 3(j) is an illustration showing the integral number of step height, where the glissile partial lying below the step has created faults while the sessile partial at the top gets locked.

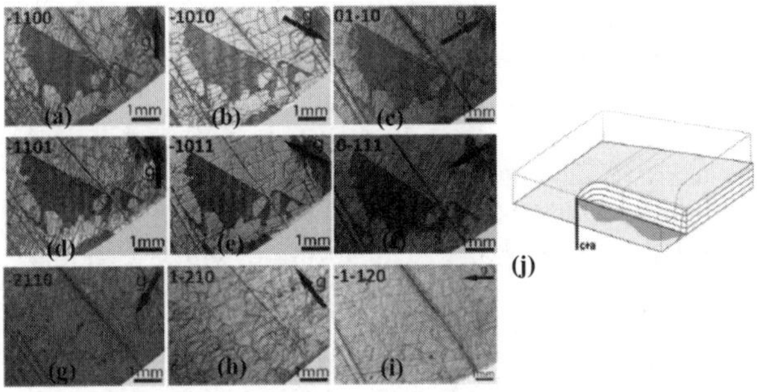

Figure 3: Figs. (a)-(i): SWBXT images recorded from a n+ 4H- SiC crystal cut 4 degrees offcut towards [11-20]. (a)-(i) correspond to various reflections used for analysis of Type I Shockley Fault. Figure 3(j): Illustration for Type I fault vector showing integral number of step height with the glissile partial lying below the step and the sessile at the top which is locked.

Type III: Combination of Shockley and Frank fault with fault vector S+c/2:

These faults are a combination of Shockley and Frank faults resulting from the overgrowth of c-axis screw dislocations whose spiral steps have divided into two c/2- high demisteps and which has several c+a dislocations with c- height spiral step risers protruding onto the terrace between the two demisteps. Figure 4(a)-(i) show SWBXT images of multiple reflections for fault vector analysis by Transmission XRT of a 76mm wafer. Overgrowth of c-axis screw dislocation and c+a dislocations which has protruded the terrace created by two demisteps results in both c/2 Frank fault and Shockley fault associated with c+a to lie on the same basal plane[13].

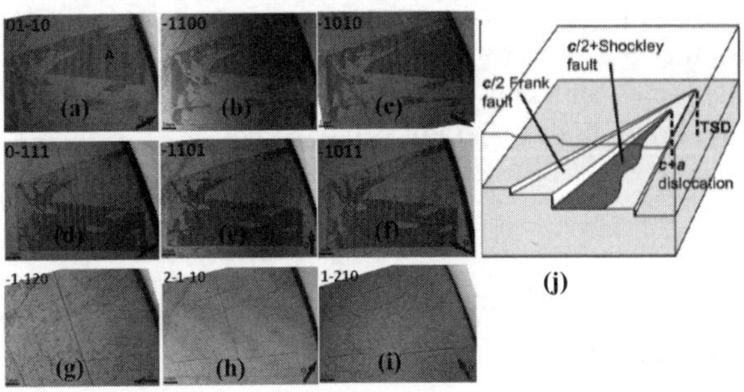

Figure 4(a)-(i): SWBXT images of multiple reflections used for fault contrast analysis. In image4(a) area marked A corresponds to R=1/6[20-23] which is a combination of Shockley and c/2 Frank fault while the area marked B in 4(d) corresponds to a simple Frank fault. Fig. 4(j): Illustration for overgrowth of c dislocation and formation of the c/2 Frank type fault within which a c+a dislocation with one sessile Shockley partial and one glissile Shockley partial converts some part of the fault to c/2 plus a Shockley type.

Type IV: Combination of Shockley plus a Frank fault with Fault vector S+c/4

Also a combination of Shockley and a Frank fault but of the type S+c/4, where the c-axis screw dislocation has divided itself into c/4 and 3c/4 step heights. Dudley et al[14] recently reported that such faults can arise from the deflection of c-axis threading dislocations of Burgers vector c+a onto the basal plane. If the spiral step risers of such dislocations divide into c/4 and 3c/4 increments, overgrowth can be facilitated by the simultaneous dragging of one of the Shockley partials associated with the core structure of the original dislocation by the overgrowing macrostep. This acts like an interfacial Shockley partial and adds the Shockley component to the c/4 Frank component of fault vector. Figure 5(a)-(i) show SWBXT images taken by Transmission geometry used for fault vector analysis, area marked C in fig. 5(d) has been determined to be a fault with R= 1/12[4-403] which is a combination of Shockley and c/4 Frank fault and similarly area marked D in fig. 5(e) corresponds to R=1/12[-4043] which is due to a second c+a dislocation protruding the terrace.

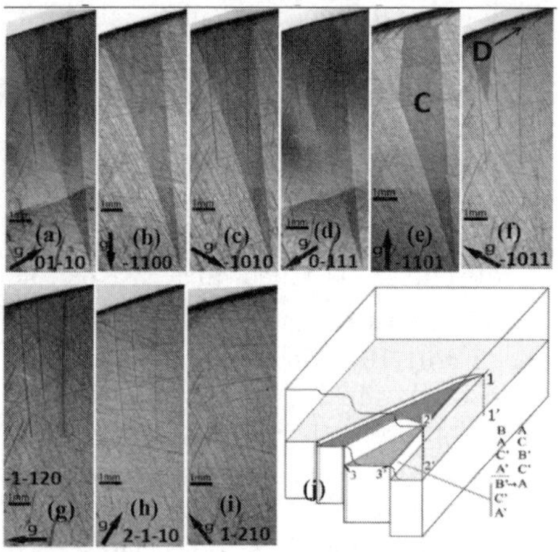

Figure 5(a)-(i): SWBXT images used for fault vector analysis. Area marked C in 5(d) correspond to R=1/12[-4043] while in 5(e) area marked D corresponds to 1/12[-4403]. Figure 5(j): Illustration for overgrowth of c+a dislocation with a c-height step converting it into a Frank dislocation plus two Shockleys with one sessile and the other glissile. Overgrowth of c+a dislocation 11' with c/4 and 3c/4- height steps which has second 22' with c-height spiral step protruding onto the terrace between these two step risers. Stacking sequences are indicated, the "interfacial Shockley" converts A layer into B', allowing overgrowth by the A' layer at the bottom of the macrostep. Following overgrowth the Shockley associated with the deflected 22' dislocation located at 23 glides under stress until it reaches the edge of the step 23' creating the fault of type D shown in fig. 5(e).

Summary

SWBXT characterization technique has been successfully implemented for studying new generation of low density 4H-SiC PVT grown wafers. An interesting observation of deflected threading dislocations with associated stacking faults has been made. Such deflection process generally leads to exit of the threading dislocations through the crystal sidewalls. Overgrowth mechanism in SiC has been used to explain the deflection of c/c+a dislocations onto the basal plane which causes the partials associated with deflected 'a' component to be separated by the vicinal step height. Models developed along with contrast analysis shows the fault vectors are of four types: (a) Type I: Frank fault; (b) Type II: Shockley fault; (c) Combination of Shockley and Frank fault with fault vector S+ c/2; (d) Type IV: Combination of Shockley and Frank fault with fault vector S+c/4.

References

[1] Neudeck PG. J Electron Mater 1995;24:283.
[2] Kamata I, Nagano M, Tsuchida H, Chen Y, Dudley M. J Cryst Growth 2009;311:1416.
[3] Dudley M, Si W, Wang S, Carter C, Glass R, Tsvetkov V. Nuovo Cimento D 1997;19:153.
[4] Dudley M. et al J Phys D Appl Phys 1995;28:A63.
[5] Michael Dudley et al. Materials Science Forum 2008;Volumes 600 - 603:261.
[6] Michael Dudley et al. Materials Science Forum 2010;645 - 648:291.
[7] Authier A, Epelboin Y. Physica Status Solidi a-Applied Research 1977;41:K9.
[8] Ouisse T, Chaussende D, Auvray L. J Appl Crystallogr 2010;43:122.
[9] Presser V, Loges A, Nickel KG. Philosophical Magazine 2008;88:1639.
[10] Heindl J, Dorsch W, Strunk PP, Muller SG, Eckstein R, Hofmann D, Winnacker A. Phys Rev Lett 1998;80:740.
[11] Pirouz P, Yang JW. Ultramicroscopy 1993;51:189.
[12] Tsuchida H, Kamata I, Nagano M. J Cryst Growth 2008;310:757.
[13] Dudley M et al. Appl Phys Lett 2011;98.
[14] Michael Dudley et al. Materials Science Forum 2011;679 - 680:269.

Materials Science Forum Vols. 717-720 (2012) pp 351-354
© (2012) Trans Tech Publications, Switzerland
doi:10.4028/www.scientific.net/MSF.717-720.351

Conversion Mechanism of Threading Screw Dislocation during SiC Solution Growth

Toru Ujihara[1, a, *], Shigeta Kozawa[2, b], Kazuaki Seki[2, c], Alexander[2, d], Yuji Yamamoto[1, e] and Shunta Harada[1, f]

[1]Department of Materials Science and Engineering, Nagoya University, Furo-cho, Chikusa-ku, Nagoya 464-8603, JAPAN

[2]Department of Crystalline Materials Science, Nagoya University, Furo-cho, Chikusa-ku, Nagoya 464-8603, JAPAN

[a]ujihara@numse.nagoya-u.ac.jp, [b]kozawa@mercury.numse.nagoya-u.ac.jp, [c]seki@mercury.numse.nagoya-u.ac.jp, [d]alex@mercury.numse.nagoya-u.ac.jp, [e]y-ymmt@mercury.numse.nagoya-u.ac.jp, [f]harada@numse.nagoya-u.ac.jp, *corresponding author

Keywords: SiC, crystal growth, solution growth, synchrotron X-ray topography, screw dislocation

Abstract. Solution growth is considered to be a powerful method for high quality SiC crystals. This work reports that the conversion process from a threading screw dislocation into a few Frank partial dislocations in basal planes was investigated by synchrotron X-ray topography. This process was effectively assisted by step-flow growth on off-oriented (0001) seed crystals. The Frank partials were not extended into the crystal grown toward the [0001] direction perpendicular to the basal plane. Thus, the conclusion of this study suggests the use of off-oriented seed crystal is important to improve crystal quality.

Introduction

The commercial SiC wafer crystals are prepared by physical vapor transport (PVT) method. The PVT-grown crystals contain many types of dislocations, *e.g.* micropipes (MPs), threading screw dislocations (TSDs), threading edge dislocations (TEDs) and basal plane dislocations (BPDs). For example, the typical value of TSD density in the conventional PVT-grown SiC wafers is in the range of 10^3-10^4 cm^{-2}. The formation of these defects in PVT crystal is caused by unexpected inclusions, polytype precipitations, misoriented 2D nucleation and others (the details are reviewed in the literature[1]). Recently, by taking care of these conditions, the density of MP is less than one per wafer and the density of TSD is also decreased. In addition, as for BPD, the conversion from BPD to TSD in CVD process has been reported [1].

Solution growth is considered to be a powerful method for high quality SiC crystals because some positive effects were reported. Yakimova *et al.* reported that MPs existing in the substrate could be covered by epi-layers during liquid phase epitaxy (LPE) which is one type of solution growth [2]. Kusunoki *et al.* confirmed the BPD density in the LPE crystal was much less than the substrate crystal prepared by PVT method [3]. In our previous study, we have taken the synchrotron X-ray topographic images of the crystal grown on 6H-SiC {0001} on-axis seed crystal by solution method. In the grown layer, no BPD was observed. Moreover, it was revealed that some TSDs were conversed into the other type of defect extended along the step-flow direction [4]. In this study, we performed the step-flow growth on an off-oriented 6H-SiC {0001} seed crystal to assist this conversion process, intentionally. In addition, one of the possible conversion mechanism is proposed referring the formation mechanism of the defects in CVD growth, *e.g.* the carrot defect [5-8].

Experimental details

A SiC crystal was grown by a top-seeded solution method with pure silicon solvent for 3 hr at 1630 °C. A 6H-SiC (0001) Si-face seed crystal (10 mm×10 mm) 1.4° off-oriented towards <11-20> was used. The seed crystal mounted on top of graphite rod was dipped into the solvent in a graphite crucible under a high-purity Ar flow. The details of growth method is described in the literatures [4, 9]. In the present study, we obtained the crystal with the thickness of 24 μm. Synchrotron X-ray topography was carried out under the reflection condition at the high-resolution X-ray diffraction station BL15C in the Photon Factory at the High-Energy Accelerator Research Organization, Japan. Monochromatic X-ray wavelength with the incident angle of 1.9 ° was 1.28 Å and the applied **g** vector was 11-2·12. In this case, the penetration depth was about 3 μm. Chlorine thermal etching was carried out for 30 min at 980 °C in the flow of N_2: 1 slm, O_2: 80 sccm and Cl_2: 80 sccm [10]. Surface morphology was observed by Nomarski microscope.

Results

On the grown surface, the train of bunched steps extended along the [1-100] direction was observed as shown in Fig. 1. The direction of step-flow was <11-20>, which is the off-oriented direction of the seed crystal. The undulation of step structures may be related with the formation of defect, which was explained later. In addition, the comparison with the topographic revealed that some pit structures labeled by arrows indicated the marks that the growth step was pinned by the TSDs.

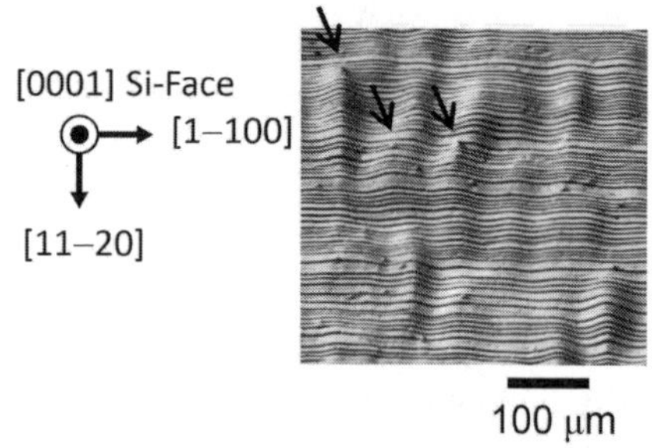

Fig. 1. Nomarski image of the crystal grown on the (0001) 6H-SiC 1.4° off-oriented towards <1120>.

Figure 2 shows the synchrotron X-ray topographic images taken before and after the growth on 6H-SiC off-oriented seed crystal. In the image of the seed crystal, there are mainly the contrasts of two types of defects: bended lines and white dots corresponding to BPDs and TSDs, respectively [11, 12]. On the other hand, after the growth, no BPD image was observed and the number of the TSD spots decreased. Instead of the reduction of the TSD, many dark bended contrasts extended to the step-flow direction appeared. In our previous experiment, we observed that the similar bended contrasts extended from the place where the TSD spot vanished after growth[4]. This fact means that the TSD was converted into the defect corresponding to the bended contrast. Moreover, the bended contrasts always existed along the direction of step-flow as well as the present result. It indicates that

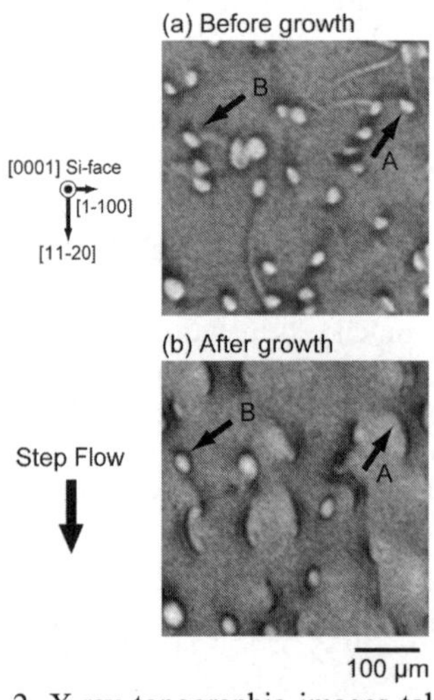

Fig. 2. X-ray topographic images taken before (a) and after (b) growth on the (0001) 6H-SiC seed crystal off-oriented towards <11-20>. Most of TSD contrasts were converted into the bended contrasts (like "A"). A part of TSD still remained (like "B").

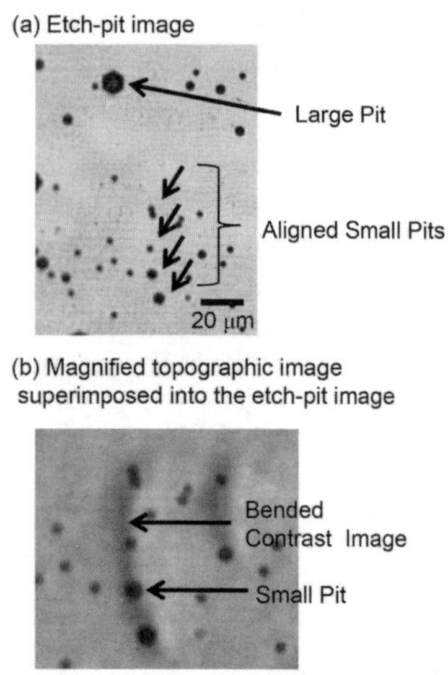

Fig. 3. (a) Etch-pit image after growth and (b) magnified topographic image superimposed into the etch-pit image.

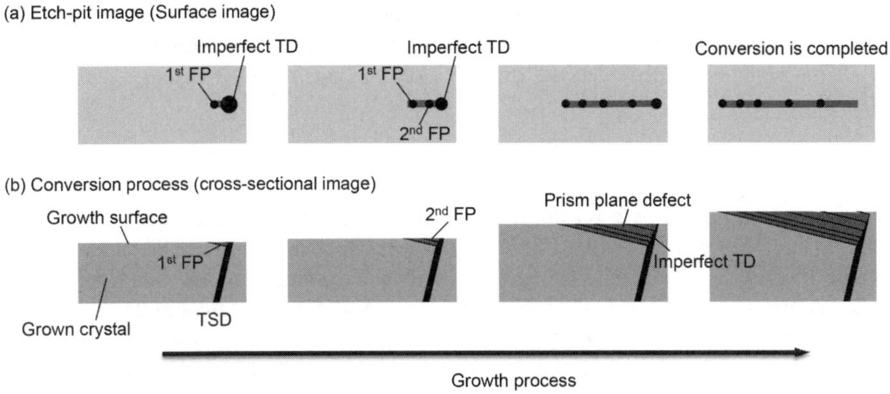

Fig. 4. Schematic drawing of the conversion process: (a) Etch-pit image on the grown surface; (b) Cross-sectional image of the dissociation process of TSD. A TSD is converted into a Frank partial dislocation (FP) and an imperfect threading dislocation (TD), which results in the formation of a prismatic fault from the TD and a stacking fault surrounded by the FP and the prismatic fault. After a few repetitions of this process , the TSD is completely converted into the FPs and eliminated.

the step-flow growth is related with the conversion mechanism. In the present case, about 80% of the TSDs in the seed crystal were converted into the bended contrasts after the growth. This conversion ratio is much higher than that in the case of the use of the on-axis seed. Some features of step structure are different between the off-oriented seed and on the on-axis seed. The height of bunched step of the crystal grown on the off-oriented seed is about 150 nm, which is a few times larger than that on the on-axis seed. In addition, the step-flow direction is random in the case of the on-axis seed unlike the off-oriented seed. These discrepancies, especially the step height may influence the conversion ratio.

In the etched surface image (Fig. 3), the large etch-pits and the small etch pits were observed. The large ones existed at the same positions where the TSD defects were found in the topographic image. In contrast, many small pits often just on the dark bended contrasts. Figure 3(b) shows the magnified topography image superimposed on etch-pit image. It is interesting to note that the four small pits aligned just on the dark bended contrast.

Discussion

These dark contrasts of the present results are similar to the topographic contrast of carrot defect [5-8], which are often observed in a crystal grown on an off-oriented seed crystal by CVD. According to the CVD study, such a contrast is due to a TSD converted into a complex of some defects: (1) a Frank partial dislocations of which Burgers vector has a component of $1/n$ [0001] in a basal plane; (2) a prismatic fault; (3) a stacking fault surrounded by the Frank partial dislocation and the prismatic fault; (4) a threading dislocation (TD). (Here n is 4 for 4H-SiC and 6 for 6H-SiC.) The dark contrasts are the images of the prismatic fault and/or the stacking fault [5-8].

Referring the mechanism of the CVD defect as mentioned above, the bended defect formation of the present result can be interpreted as follows: an original TSD is dissociated into a Frank partial dislocation and an imperfect TD with the Burgers vector smaller than $1c$ as shown in Fig. 4; a prismatic fault is formed along the TD; a few same processes occur repeatedly; the original TSD is completely converted into some Frank partial dislocations when the sum of Burgers vector of the generated Frank partial dislocations is equal to the Burgers vector of the original TSD. The small etch pits are due to the generated Frank partial dislocations as shown in Fig. 4(a). The undulated steps may be connected with this conversion as shown in the Nomarski image (Fig. 1).

According to this mechanism, the TSDs are converted into Frank partial dislocations and stacking faults, which exist on basal plains. These defects cannot be extended to the crystal grown on the (0001) plane. Thus, the repetition of the conversion can result in the decrease of defect density, ultimately. This conclusion provides the direct evidence that the TSD density is reduced in the solution growth.

Summary

We performed the solution growth on the 6H-SiC (0001) 1.4° off-oriented seed crystal. After the growth, 80% of the TSD were converted into the defects, which are observed as bended contrasts in the X-ray topographic image. The bended contrasts always existed along the direction of step-flow. These bended contrasts are similar to the topographic contrast of carrot defect in the crystal grown by CVD. Referring the mechanism of the CVD defect, the bended defect was the TSD converted into a defect complex composed of Frank partial dislocations, a prismatic fault and stacking faults. These defects cannot be extended to the crystal grown on (0001) plane. Thus, the repetition of the conversion can result in the decrease of defect density, ultimately.

Acknowledgements

This study was supported by Novel Semiconductor Power Electronics Project Realizing Low Carbon Emission Society of Ministry of Economy, Trade and Industry (METI) and New Energy and Industrial Technology Development Organization (NEDO) through R&D Partnership for Future Power Electronics Technology (FUPET). The X-ray topography has been performed under the approval of the Photon Factory Program Advisory Committee (Proposal No. 2009G118).

References

[1] M. Dudley, Y. Chen, X. R. Huang and R. Ma: Mater. Sci. Forum vol. 600-603 (2009) 261-266.
[2] R. Yakimova, M. Tuominen, A. S. Bakin, J. O. Fornell, A. Vehanen and E. Janzen: Inst. Phys. Conf. Ser. vol. 142 (1996) 101-104 .
[3] K. Kusunoki, K. Kamei, N. Yashiro and R. Hattori: Mater. Sci. Forum vol. 615-617 (2009) 137-140.
[4] S. Kozawa, K. Seki, Alexander, Y. Yamamoto, T. Ujihara and Y. Takeda: Mater. Sci. Forum vol. 679-680 (2011) 28-31.
[5] M. Benamara, X. Zhang, M. Skowronski, P. Ruterana, G. Nouet, J. J. Sumakeris, M. J. Paisley and M. J. O'Loughlin: Appl. Phys. Lett., vol. 86 (2005) #021905;
[6] X. Zhang, S. Ha, M. Benamara, M. Skowronski, M. J. O'Loughlin and J. J. Sumakeris: Appl. Phys. Lett., vol. 85 (2004) 5209-5211.
[7] H. Tsuchida, M. Ito, I. Kamata and M. Nagano: Mater. Sci. Forum vol. 615-617 (2009) 67-70;.
[8] M. Dudley, F. Wu, H. Wang, S. Byrappa, B. Raghothamachar, G. Choi, S. Sun, E. K. Sanchez, D. Hansen, R. Drachev, S. G. Mueller and M. J. Loboda, Appl. Phys. Lett., vol.98 (2011) #232110.
[9] Alexander, K. Seki, S. Kozawa, Y. Yamamoto, T. Ujihara and Y. Takeda: Mater. Sci. Forum vol. 679-680 (2011) 24-27.
[10] T. Hatayama, T. Shimizu, H. Kouketsu, H. Yano, Y. Uraoka and T. Fuyuki: Jpn. J. Appl. Phys. Vol. 48 (2009) #066516.
[11] T. Ohno, H. Yamaguchi, S. Kuroda, K. Kojima, T. Suzuki and K. Arai: J. Cryst. Growth, vol. 260 (2004), 209-216.
[12] Y. Chen, X. R. Huang, G. Dhanaraj, M. Dudley, E. K. Sanchez and M. F. MacMillan: Mater. Sci. Forum vol. 600-603 (2009) 297-300.

Materials Science Forum Vols. 717-720 (2012) pp 355-358

doi:10.4028/www.scientific.net/MSF.717-720.355

Complex behavior of threading dislocations observed in PVT-grown 4H-SiC single crystals

M. Katsuno[1,2,a], T. Fujimoto[1,2], H. Yashiro[1,2], H. Tsuge[1,2], S. Sato[1,2], H. Hirano[1,2], T. Yano[1,2], and W. Ohashi[1,2]

[1] Futtsu Branch, R&D Partnership for Future Power Electronics Technology (FUPET)

[2] Nippon Steel Corporation, Advanced Technology Research Laboratories

[1,2] 20-1 Shintomi, Futtsu, Chiba 293-8511, Japan

[a] katsuno.masakazu@nsc.co.jp

Keywords: PVT-growth, dislocation, X-ray topography

Abstract. Structures and propagating behaviors of threading dislocations (TDs) in PVT-grown 4H-SiC single crystals were both investigated using Synchrotron monochromatic X-ray topography. Comparative studies by examining images obtained for the crystals with different diffraction geometries of (0004) and (11-20) of 4H-SiC revealed that a large amount of TDs are likely to be mixed in character, i.e., dislocations with Burgers vector components of both <0004> and <11-20>. Closer observations of topography images has revealed that, although TDs lie largely along the c-axis direction, some of the TDs show quite a complex propagating behavior: not extending in a straight line but meandering along the growth direction.

Introduction

Silicon carbide (SiC) substrates are now widely accepted as a promising material for high-power semiconductor devices. SiC Schottky barrier diodes have already been ready for implementation into various power device applications and, up to date, commercial release of SiC power MOSFETs has been announced. However, in the SiC substrates, there still exist various types of dislocations, such as threading screw dislocations (TSD), threading edge dislocations (TEDs), basal plane dislocations (BPDs) and stacking faults (SFs), and the influence on SiC device properties caused by these dislocations has not yet been completely understood.

It is well known that most of the BPDs are converted into TEDs during SiC epitaxial growth. In contrast, threading dislocations such as TSDs and TEDs propagate from substrates into epitaxial layers grown onto the substrates without any conspicuous structural transformation, resulting in deterioration of the reliability of SiC device performance [1]. In particular, as for TSDs, crystallographic defects that affect properties of MOSFETs much compared to TEDs, understanding the behavior of the TSDs during PVT growth is of particular importance for improving the performance of SiC devices. However, there are only a few papers which have so far reported details of the TSD behavior during PVT growth [2, 3].

This paper reports the results of investigation regarding the dislocation structures of TSDs, and then describes detailed observations on the propagating behavior of TSDs during PVT growth of 4H-SiC single crystals. Although most of the TSDs extend along the growth direction, we found from synchrotron X-ray topographs that some of the TSDs show meandering along the growth direction, quite a complex behavior.

Experimental Procedure

4H-SiC single crystals were grown on-axis and off-oriented (000-1) 4H-SiC seed crystals by the physical vapor transport (PVT) growth method. The growth conditions employed in this study were described elsewhere [4], and all the crystals were nitrogen-doped with concentrations in the higher range of 10^{18} cm^{-3}. The growth temperature was set in a range between 2300 to 2400°C. The polytype of the grown crystals was examined by Raman scattering spectroscopy and identified to be 4H-SiC. The crystals grown were then sliced almost parallel to the growth direction to obtain {1-100} wafers with thicknesses of about 0.1-0.3 mm. Finally, both sides of the wafer surface were carefully polished using 1-micron diamond slurry.

Structural investigation using synchrotron monochromatic X-ray topography was performed at the beam line BL15 of SAGA-LS located in Saga, Japan. X-ray topographs were taken with transmission geometries (Laue case), using (0004) and (11-20) diffractions of 4H-SiC. X-ray topographs were recorded onto nuclear plates, Ilford L4, with higher resolutions of several microns.

We have also carried out investigation using BedeScan (Bede Scientific Instruments) X-ray topography by which images with a spatial resolution of around 3 microns can be obtained, much higher compared to other conventional X-ray topography. In this topography, the images obtained were digitally recorded through a CCD camera with a pixel pitch of 2.7 micron.

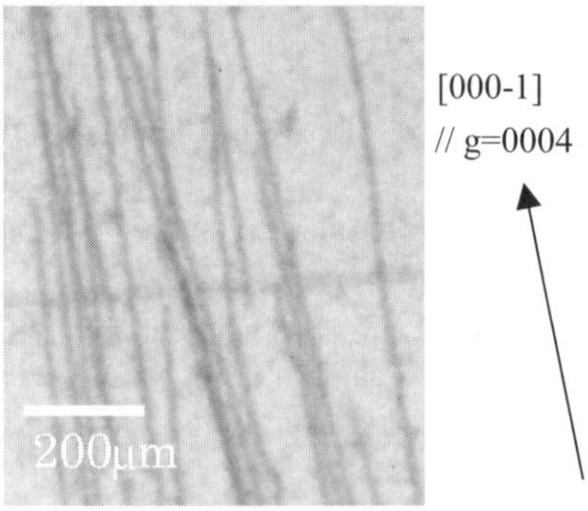

Fig.1. High resolution (0004) X-ray topograph of 4H-SiC (1-100) substrate. Arrows in the inset correspond to the growth direction and g-vector of X-ray diffraction,

Results and Discussion

Fig. 1 shows a high resolution X-ray topograph of a 4H-SiC substrate with (0004) diffraction geometry of 4H-SiC. Threading dislocations (TDs) observed in Fig. 1 accompany a Burgers vector component parallel to the <0004> direction. Almost all the TDs in Fig. 1 are likely to extend in a straight line, but not exactly parallel to the <0001> axis, and therefore, the TDs in Fig. 1 are likely to be mixed dislocations in character.

Fig. 2 shows wider area views for 4H-SiC substrates using high resolution X-ray topography, in

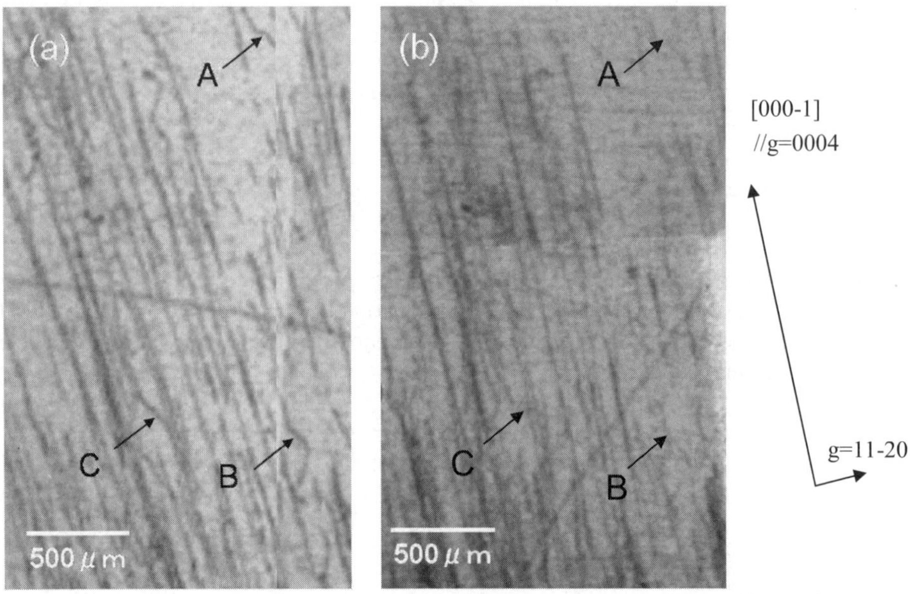

Fig.2. High resolution X-ray topographs of (1-100) substrate, obtained by slicing from as-grown 4H-SiC single crystal sample, with diffraction geometries of (a) (0004) and (b) (11-20) of 4H-SiC, respectively.

which both topographs, obtained under different diffraction geometries of (0004) (Fig.2(a)) and (11-20) (Fig. 2(b)), correspond largely to the same area of the substrate. As can be seen, almost all of the TD lines in Fig. 2(a) are also visible in Fig. 2(b), although some dislocations marked in Fig. 2(a) by arrows A, B and C exhibit very faint contrast in Fig. 2(b). In general, a dislocation line is formed when a glide occurs in a limited area on an easy slip plane, for example 1/3<11-20>(0001) in SiC; for the circular-like dislocation line enclosing such slipped area, screw and edge dislocations are defined as parts of the line with the directions parallel and perpendicular to <11-20>(0001), respectively, and other parts of the line are mixed in character [5]. However, Burger's vectors for all parts of the dislocation lines are in common and have to be 1/3<11-20>(0001), independent of the directions of the dislocation line. Therefore, the results obtained in Fig. 2(a) and (b) suggest that, for the most part, the TDs inherently contains both 1c screw, i.e., <0004>, and in-plane <11-20> Burger's vector components.

Nakamura et al. [2] reported that basal-plane screw dislocations were found in PVT-grown crystals with very low dislocation densities, and discussed possibilities that such dislocations reported in Ref. [2] have Burgers vector components of both long-threading dislocations consisting of c-axis (parallel to <0001>) and a-axis (parallel to <11-20>) components. Our results shown in Fig. 2 are mainly consistent with their conclusions. We also found that, as is seen in Fig. 2(a) and (b), TDs marked by A, B and C show different contrast for (0004) and (11-20) diffraction geometries. Similar phenomena are also observed in Fig. 4 in Ref. [2]. Their results suggest that some TDs are likely to be pure 1c screw dislocations because, in general, the strength of X-ray topograh image contrast has to be proportional to $\mathbf{b}\cdot\mathbf{g}$ where **b** and **g** are the Burgers vector and diffraction vector, respectively. Further investigation has to be performed in order to determine accurate dislocation structures of the mixed dislocations, presumably leading to comprehensive understanding of the origin of dislocations in SiC during PVT growth.

Fig. 3 shows an image obtained using Synchrotron X-ray (0004) diffraction topography. Long, straight line TDs propagating largely parallel to the growth direction are clearly discernible with higher spatial resolution. However, a TD in Fig. 3 exhibits quite a complex extending behavior; it shows meandering largely along the growth direction.

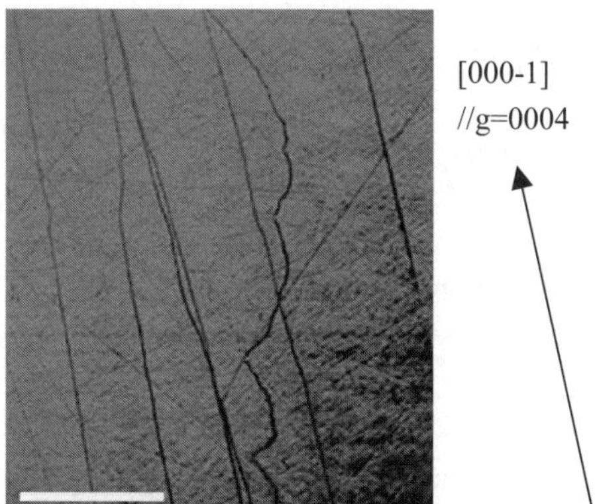

Fig.3. Synchrotron X-ray transimission topograph of 4H-SiC substrate with g=(0004). Dark lines observed in the figure are TDs with a Burger's vector component parallel to <0004>.

Two more examples are also shown in Fig. 4. In particular, in Fig. 4(a), two TDs exhibit meandering in a non-periodic, irregular manner in extending structure. Nakamura et al. [2] also reported that in a 4H-SiC substrate, there existed threading screw dislocations converting to basal-plane dislocations. However, the structurally converted dislocations largely show a form of straight lines, and therefore the TDs we found are in remarkable contrast to the cases reported so far by others. In short, the complex behavior we observed is, to our knowledge, the first to be reported for PVT-grown SiC single crystals. Although there should be a mechanism for explaining such complex behavior of the TDs, it is challenging to discuss it because, as can be understood in both Fig. 4(a) and Fig. 4(b), such TDs coexist with a straight line-like TD in a very narrow crystal region, and therefore at the moment we speculate that the complex behavior observed in Fig. 3 and 4 is presumably related to the dislocation structure.

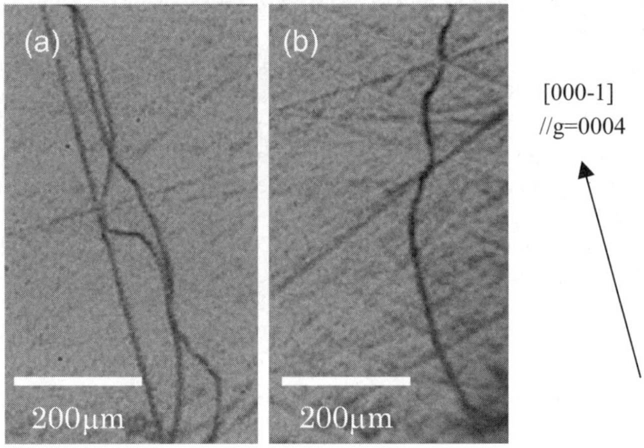

Fig. 4 Synchrotron X-ray transmission topographs of 4H-SiC substrate with a diffraction geometry of g=(0004), showing two other examples of TDs exhibiting complex propagation behavior.

Summary

Intense studies using high resolution X-ray topography with (0004) and (11-20) diffraction geometries suggest that almost all of the threading dislocations in PVT-grown SiC single crystals consist of at least two Burgers vector components parallel to the <0004> and <11-20> directions. Closer observations using synchrotron X-ray topography revealed that a complex behavior of TDs was found to exhibit meandering along the growth direction.

Acknowledgements

This study was supported by Novel Semiconductor Power Electronics Project Realizing Low Carbon Emission Society of Ministry of Economy, Trade and Industry (METI) and New Energy and Industrial Technology Development Organization (NEDO) through R&D Partnership for Future Power Electronics Technology (FUPET).

References

[1] J. Senzaki, A. Shimozato, M. Okamoto, K. Kojima, K. Fukuda, H. Okumura and K. Arai, Jpn. J. Appl. Phys. 48 (2009) 081404.

[2] D. Nakamura, S. Yamaguchi, I. Gunjishima, Y. Hirose, and T. Kimoto, J. Cryst. Growth 304 (2007) 57.

[3] B. Kallinger, S. Polster, P. Berwian, J. Friedrich, G. Muller, A.N. Danilewsky, A. Wehrhahn, and A.-D. Weber, J. Cryst. Grwoth 314 (2011) 21.

[4] J. Takahashi, M. Kanaya, and Y. Fujiwara, J. Cryst. Growth 135 (1994) 61.

[5] Derek Hull "*Introduction to Dislocations*" Pergamon Press, NY (1965).

Materials Science Forum Vols. 717-720 (2012) pp 359-362

doi:10.4028/www.scientific.net/MSF.717-720.359

Correlation between Surface Morphological Defects and Crystallographic Defects in SiC

T. Hatakeyama[1,2 a], K. Ichinoseki[1,2], H. Yamaguchi[1,2], N. Sugiyama[1]
H. Matsuhata[1,2]

[1]R&D Partnership for Future Power Electronics Technology (FUPET), c/o ADPERC, AIST
[2]Advanced Power Electronics Center (ADPERC), National Institute of Advanced Industrial Science and Technology (AIST), Central 2, Umezono 1-1-1, Tsukuba, Ibaraki, 305-8568, Japan

[a] tetsuo-hataketyama@aist.go.jp

Keywords: Defect, dislocation, pit, X-ray topography, SiC, 4H-SiC, Si-face, C-face

Abstract. The origins of certain types of micrometer-scale surface morphological defects on SiC epitaxial layers are clarified using X-ray topography. Two types of surface morphological defects are commonly observed on Si- and C-face epitaxial layers. Relatively large pits (around 4μm×2μm) originate from threading screw dislocations (TSDs). Relatively small pits (around 1.5μm×1μm) originate from threading edge dislocations (TEDs). The shapes and depths of these surface morphological pits depend on the fabrication history of the epitaxial wafers.

Introduction

SiC power devices are regarded as promising candidates for use as high-voltage electrical power switches and diodes owing to the high critical electric field of SiC. However, SiC material has high densities of crystallographic defects (dislocations) such as threading screw dislocations (TSDs), threading edge dislocations (TEDs) and basal plane dislocations (BPDs), whose densities generally exceed tens of thousands per square centimeter. It should be emphasized that understanding and suppressing the effects of these high-density dislocations on device characteristics is a prerequisite for the commercialization of SiC power devices. In fact, it has been reported that dislocations severely degrade the reliability of gate oxides on the Si-face of 4H-SiC [1]. On the other hand, it has been also reported that the dislocation density does not correlate with the reliability of the gate oxide on the C-face of 4H-SiC [2]. Recently, Suzuki *et al.* showed that surface morphological defects that originate from TSDs degrade the reliability of gate oxides [3]. They also showed that TSDs *per se* are not killer defects of metal-oxide-semiconductor field-effect transistors. In fact, gate oxide grown on a commercial wafer without detectable surfaces defects did not exhibit degraded time-dependent dielectric breakdown (TDDB) characteristics. Accordingly, if we are to understand the effect of dislocations on device characteristics, we must investigate the correlation between surface morphological defects and crystallographic defect. In this paper, we classify micrometer-scale surface morphological defects and clarify their origin.

Experimental

The wafers used in this study were off-angled, n-type 4H-SiC with an n-type epitaxial layer on the C- or Si-face. Some of the epitaxial wafers were commercial products, and details of their epitaxial conditions were not known. Surface morphological defects on the epitaxial wafers were observed with a confocal differential interference contrast microscope, scanning electron microscope and atomic force microscope (AFM). Synchrotron X-ray topography images were obtained to allow us to locate the coordinates of crystallographic defects (dislocations) on the wafer. The diffraction geometries were $g = \bar{1}\bar{1}2\bar{8}$, and the X-ray wavelength was λ=0.15 nm [4].

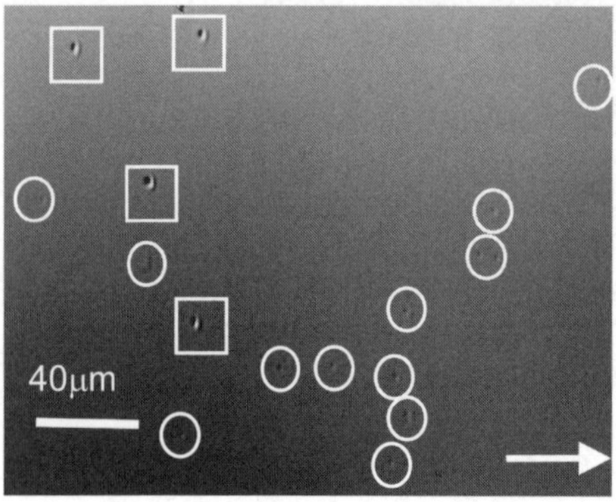

Fig.1 : A confocal differential interference contrast microscopic image of a commercial Si-face SiC epitaxial wafer.

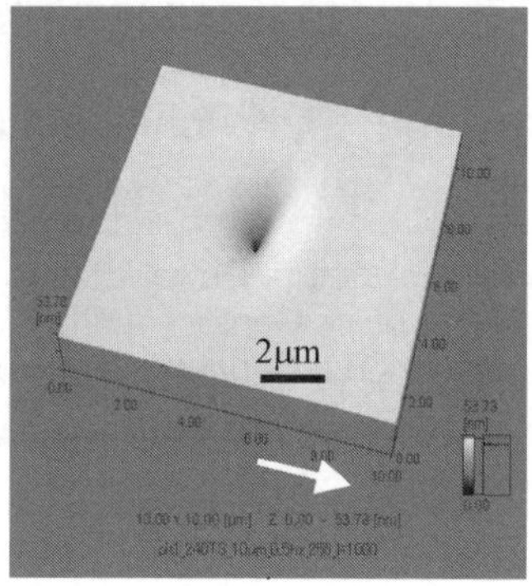

Fig.2 : AFM image of a relatively large pit

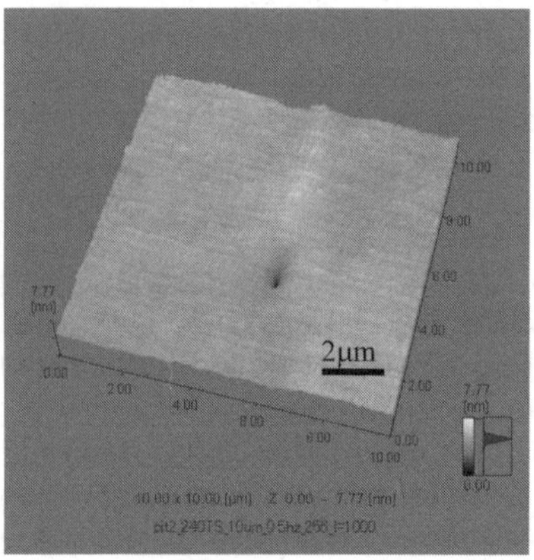

Fig.3 : AFM image of a relatively small pit

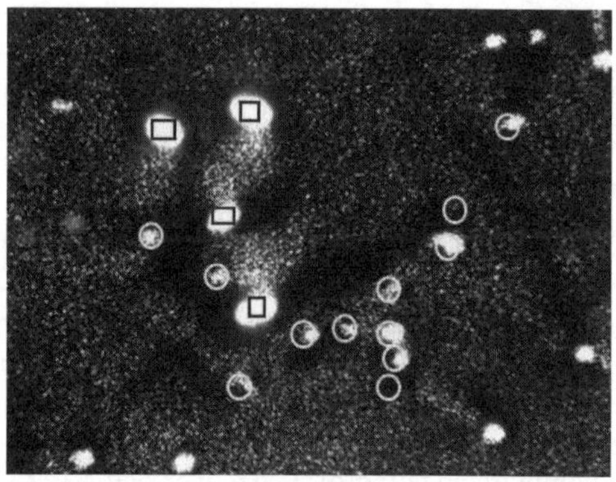

Fig.4 : X-ray topography image at the same location as the microscopic image shown in Fig. 1

Results & Discussions

Fig. 1 shows an example of a conformal differential interference contrast microscope image of a commercial 2-inch epitaxial wafer (8-degree-off, Si-face, 10μm-thick epitaxial layer). Two types of surface morphological defects can be seen: four relatively large pits (white squares) and twelve relatively small pits (white circles). These pits are observed all over the wafer.

Fig. 2 shows an AFM image of a relatively large pit. The size and the depth of the pit are around 4μm and 46nm, respectively. This pit may correspond to the "sharp-apex growth pit" described in ref. [5]. However, the depth of the observed pit is twice that in ref. [5]. Fig. 3 shows an AFM image of a relatively small pit. The size and the depth of the pit are around 1.5μm and 5nm, respectively. This pit may correspond to the "stripe-shaped growth pit" mentioned in ref. [5]. The depth of this pit is almost the same as that in ref. [5]. In this study, X-ray topography imaging is used to correlate surface morphological defects with crystallographic defects, whereas molten KOH etching is used in ref [5]. Fig. 4 shows an X-ray topography image obtained at the same location on the wafer as the microscopic image shown in Fig. 1. In Fig. 4, large white circular patterns (blue squares) and small white irregular patterns (yellow circles) correspond to TSDs and TEDs, respectively. A comparison of Figs. 1 and 4 makes it clear that the TSDs are correlated with relatively large pits, whereas the correlation between TEDs and relatively small pits is somewhat ambiguous. This may be partly due to the limited resolution of the X-ray topography image and partly due to the relatively small stress field around the TEDs.

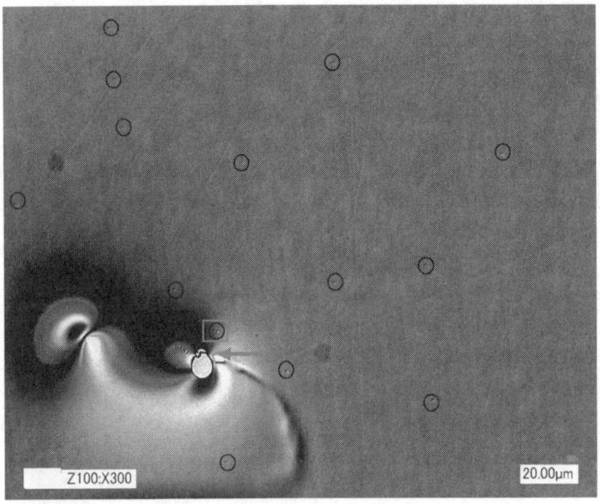

Fig. 5: Microspic image of a commercial 4H-SiC 2-inch C-face epitaxial wafer.

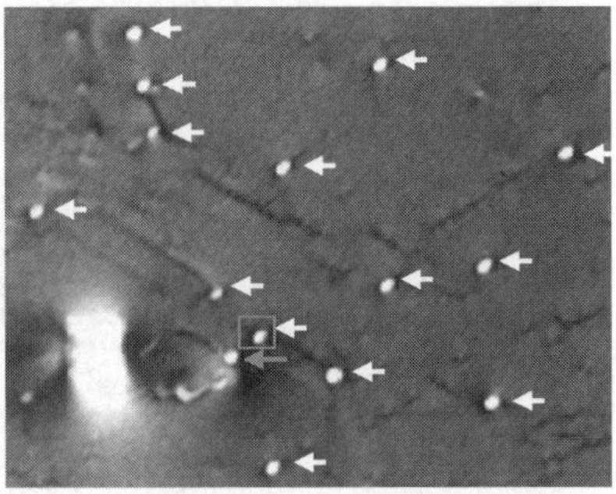

Fig. 6: An X-ray topography image at the same portion of a C-face epitaxial wafer.

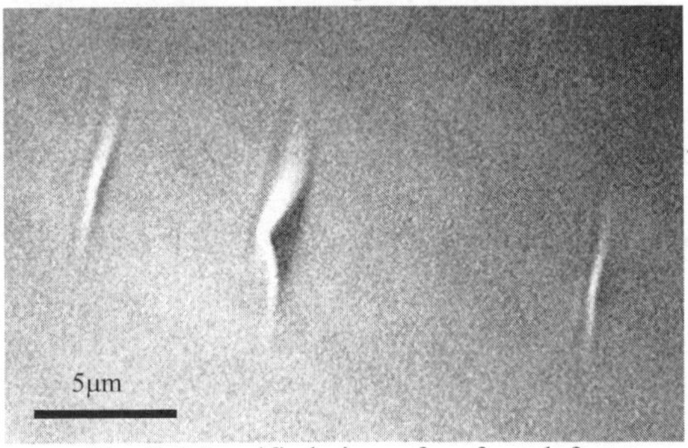

Fig. 7: Magnified view of surface defects around the large spek of dust.

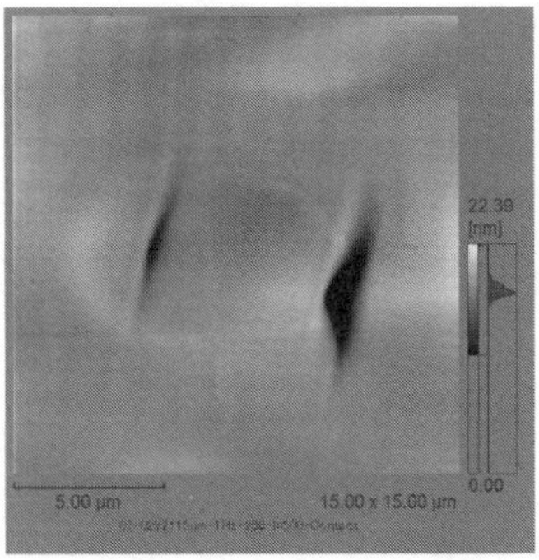

Fig. 8: AFM image of a relatively large pit and a relatively small pit

Fig. 5 shows an example of a differential interference contrast microscopic image of a commercial 4H-SiC 2-inch epitaxial wafer (4-degree-off, C-face, 10μm-thick epitaxial layer). In this case, we also observed two types of surface morphological defects. Relatively large pits are marked by black circles in Fig. 5. However, relatively small pits are observed only in the vicinity of the large speck of dust, which is shown by a red arrow in Fig. 5. Fig. 6 shows an X-ray topography image obtained at the same location on the wafer as the microscopic image shown in Fig. 5. In Fig. 5, large white circular patterns correspond to TSDs. When we compare Figs. 5 and 6, it is clear that the TSDs are correlated with relatively large pits. Fig. 7 is a magnified view of the portion indicated by the red square in Fig. 5. Two relatively small pits can be seen around a relatively large pit. Fig. 8 shows an AFM image of the relatively small pit and the relatively large pit. The two pits are 10nm and 40nm deep, respectively.

Fig. 9 shows an AFM image of a relatively large pit on another commercial Si-face epitaxial wafer. This pit has a hexagonal shape and is 80nm deep. Fig. 10 shows an AFM image of two types of pits on a C-face epitaxial wafer. The large and small pits are around 20nm and 3nm deep, respectively. It appears that the shape and depth of surface morphological pits are dependent on the type and history of the epitaxial wafers.

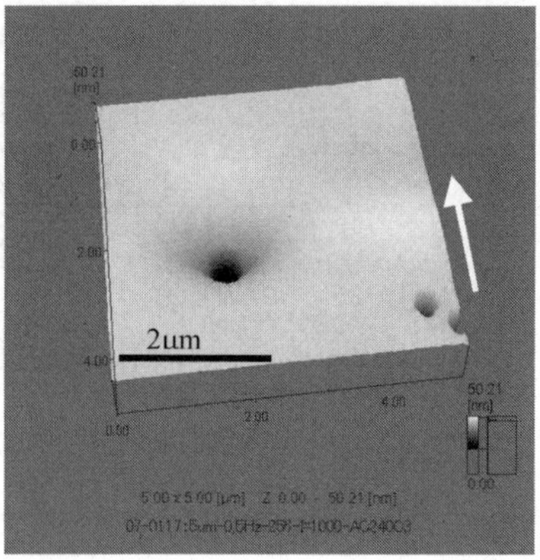

Fig.9 : AFM image of a relatively large pit on another Si-face epitaxial wafer

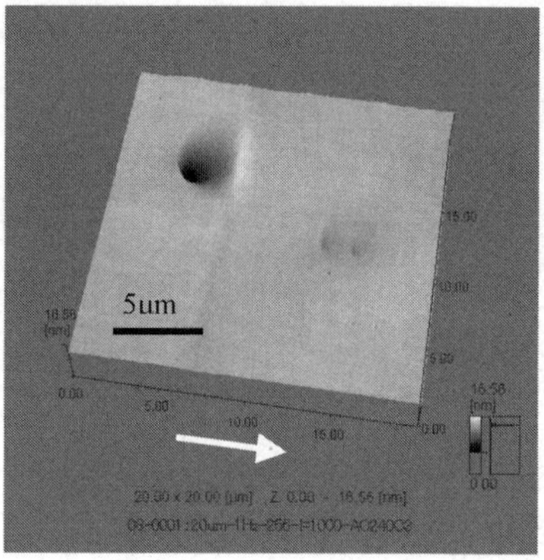

Fig.10 : AFM image of two types of pits on a C-face epitaxial wafer

Summary

The origins of typical surface morphological defects were clarified using X-ray topography. Two types of surface morphological defects are commonly observed on Si- and C-face epitaxial layers. Relatively large pits originate from TSDs. Relatively small pits originate from TEDs. The shape and depth of the surface morphological pits depend on the fabrication history of the epitaxial wafers. Under some epitaxial conditions, dislocations trigger the formation of surface morphological defects.

Acknowledgements

This work is supported by Novel Semiconductor Power Electronics Project Realizing Low Carbon Emission Society under Ministry of Economy, Trade and Industry (METI) and New Energy and Industrial Technology Development Organization (NEDO).

References

[1] J. Senzaki, K. Kojima, T. Kato, A. Shimozato, and K. Fukuda, Appl. Phys. Lett. 89, 022909

[2] T. Hatakeyama, T. Suzuki, J. Senzaki, K. Fukuda, H. Matsuhata, T. Shinohe and K. Arai: Mater. Sci. Forum Vol. 600-603 (2009), p. 783

[3] T. Suzuki, H. Yamaguchi, T. Hatakeyama, H. Matsuhata, J. Senzaki, K. Fukuda, T. Shinohe and H. Okumura, to be published.

[4] H. Yamaguchi, *et al.*, Dislocation Contrast of 4H-SiC in X-Ray Topography under Weak-Beam Condition: Mat. Sci. Forum **600-603** (2009) pp 309-312

[5] Y. N. Picard, K. X. Liu, R. E. Stahlbush, M. E. Twigg, X. Zhang and M. Skowronski, J. Appl. Phys. **103**, 074904 (2008)

Materials Science Forum Vols. 717-720 (2012) pp 363-366

doi:10.4028/www.scientific.net/MSF.717-720.363

Characterization of triangular-defects in 4°off 4H-SiC epitaxial wafers by synchrotron X-ray topography and by transmission electron microscopy

T.Yamashita[1,2,a*], K.Momose[1,2], D.Muto[1,2], Y.Shimodaira[1,2], K.Yamatake[1,2], Y.Miyasaka[1,2], T.Sato[1,2,b], H.Matsuhata[1,3], and M.Kitabatake[1]

[1]R&D Partnership for Future Power Electronics Technology （FUPET）, 2-9-5 Toranomon, Minato-ku, Tokyo 105-0001, Japan

[2]SHOWA DENKO K.K.,1-13-9,Shibadaimon,Minato-ku,Tokyo 105-8518,Japan

[3]Advanced Power Electronics Research Center,National Institute of Advanced Industrial Science and Technology,1-1-1,Umezono,Tsukuba,Ibaraki 305-8568,Japan

[a]t-yamashita@fupet.or.jp,[b]Takayuki_Sato@sdk.co.jp

Keywords: Silicon carbide (SiC); triangular-defect; dislocation; synchrotron X-ray topography; transmission electron microscopy (TEM)

Abstract. We have investigated origins and structures of triangular-shaped defects formed on surfaces of 4-degrees-off 4H-SiC epitaxial wafers. It is well known that some of the triangular-shaped defects are originated by 3C-SiC particles so called "downfalls", but major origins of the defects have been unknown until now. Our TEM observations show that not only anomalous grain formations of SiC in epitaxial films, but also contaminations by foreign materials can be the origins of the defects. Several different types of structure for the isosceles triangular-shaped defects investigated by synchrotron X-ray topography experiments are also reported. It is discussed that there exist different types of Burgers vectors configurations at the interface between epitaxial films and triangular defects.

Introduction

Currently, SiC wafers contain several different types of defects on their epitaxial film. Among those defects, triangle-shaped defects are known as one of the most popular, and one of the most serious device-killer defects [1]. Reduction of the defects densities is required to improve device productions. As one of the origins for the triangular-shaped defects, 3C-SiC particles so called downfalls, which are considered to come from 3C-SiC coat inside a furnace for epitaxial film growth, is well known as one of the origins for the triangular-shaped defect formation [1]. Formation of 3C-SiC inclusions at substrate/epitaxial film interface is also proposed as an origin [2]. However, optical microscopy observations show variety of appearances for the defects [3]. Many of them are not accompanied by the 3C-SiC particles. Also, our optical microscope observations showed similar results. Another model proposed in some articles is lattice defects such as dislocations in substrates surfaces cause the triangular-shaped defects, see for example [4]. Also, by our X-ray topography experiment dislocations seem not to be involved directly for formations of triangular defects. This implies that the formations of the defects are not only by such origins. Systematic investigations for origins and morphologies of those triangular-shaped defects are important and necessary. We report our results on investigations of many triangular-shaped defects. The investigations were carried out by Synchrotron X-ray topography, and by transmission electron microscopy.

Experimentations

Commercially available 4H-SiC wafers with 10 μ m-thick epitaxial film with off-oriented from the exact Si (0001) plane by 4 degrees toward the [112,¯ 0] direction were used. Confocal microscope with differential interference contrast was used to confirm their shapes and coordinates of the triangular-defects. Grazing incidence synchrotron reflection X-ray topography experiments were carried out at BL15C in Photon-Factory to observe dislocations around triangular-defects. Reflections were **g**=1,‾ 1,‾ 28, 2,‾ 118, 12,‾ 18, 1,‾ 018, 01,‾ 18, and 11,‾ 08 at wave length 0.15 nm. Triangular-shaped defects that had neither 3C-SiC particles nor basal-plane dislocations at the origin points of the growths were picked up for investigations. Specimens for transmission electron microscopies (TEM) were prepared by a focused ion beam (FIB) instrument. A typical confocal image of a triangular-defect was shown with TEM specimen position in Fig.1. Cross-sectional specimens at the origin points of the triangular-shaped defects, the tip-points, were picked up. In order to avoid the failure of specimen preparation, thinning of the specimens were stopped at around 1μm in thickness. The thick specimens were observed by high-voltage TEM equipped with energy-filtering system at Kyushu-University. This system enabled us to observe thick specimens clearly. The observations were carried out at the [1,‾ 100] and [2,‾ 110] orientations, **g.b** analysis was carried out in the zero-loss images. EDX analyses were also performed. Thin specimens were also prepared and high-resolution TEM with conventional electron microscope was performed.

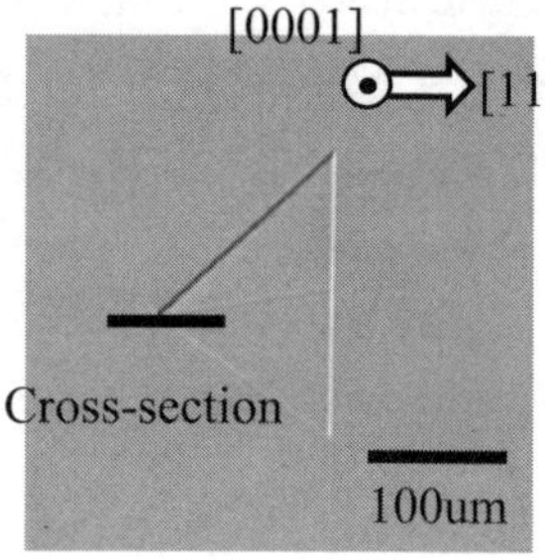

Fig.1 A confocal image of a triangular-defect. Cross-sectional TEM specimen position was also shown.

Results and Discussions

Case 1: Triangular-shaped defect formed by anomalous grain formation.

Fig.2 (a) and (b) show cross-sectional high-voltage TEM images at the tip of a triangular-shaped defect. Formation of a grain was observed. At a lower-right part of the grain, line-shaped contrast, pointed by black arrows and noted by 3C-SiC layer running towards the [11 2,‾ 0] direction along on a (0001) plane, was seen in Fig.2 (a) and (b). This line contrast was thought to be a thin 3C-SiC layer. This thin plate-like layer of 3C-SiC structure grew up to a macro-scale triangular-shaped defect at the right hand side of the Fig. 2 (a). EDX analyses indicated no elements except silicon and carbon were detected at the newly formed grain, the bottom area of the grain, 3C-SiC layer, and starting point of the thin 3C-SiC layer. It was confirmed under the two beam conditions, that there are neither threading dislocations nor basal-plane dislocations coming from the substrate around the grown grain, as shown in Fig.2(c) and (d). Electron diffraction pattern obtained at the grain, shown in Fig.2 (e), was similar to 4H-SiC structure taken at the [2,‾ 110] orientation as shown in Fig. 2 (f). This indicates that anomalous grain has a structure different from the 3C-phase; i.e., particles so called 3C-SiC downfalls. Thus, this grain is originated by neither dislocations nor related lattice defects. Mechanism of the anomalous grain formation observed in this work is not known at this moment.

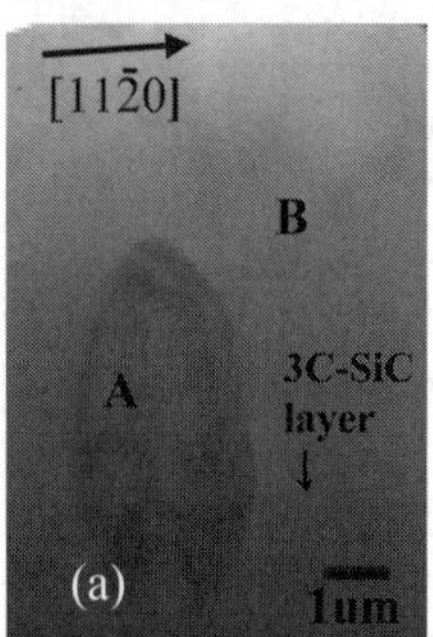

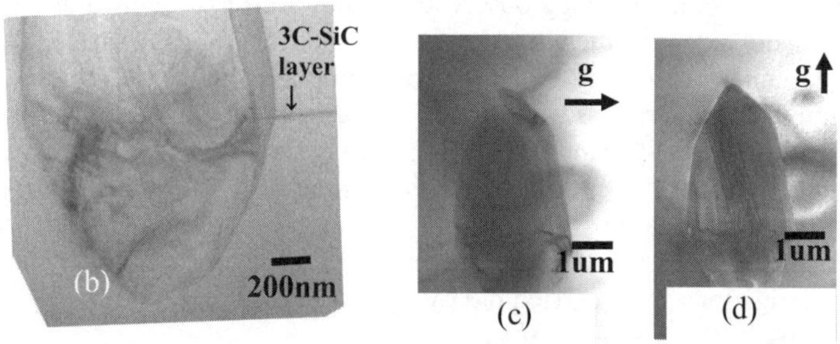

Fig. 2 (a) Cross-sectional TEM image of the new grain observed at [2,‾ 110] orientation; (b) Enlargement from (a) to show the starting point of the thin 3C-SiC layer; (c) and (d) , dark-field images taken from anomalous grain - (c) **g**= 11 $\bar{}$,20,(d), **g**= 0004

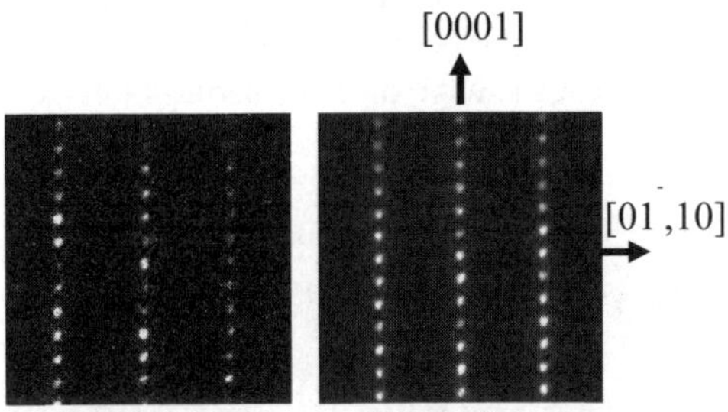

Fig. 2(e) Fig.2 (f)

Fig.2 (e), Electron diffraction pattern obtained from region A in Fig. 2 (a), inside of the grain. Fig.2 (f), Electron diffraction pattern obtained frm region B: normal epitaxial film region. Both corresponded to patterns of 4H-SiC structure with incident beam direction of [,2110].

Case 2: Triangular-shaped defect originated from small foreign materials.

Figure 3 (a) shows cross-sectional TEM image at another tip of a triangular-shaped defect. A plate-like particle with dark contrast, which size is less than one-micron meter, was observed. No dislocations are observed beneath and above of this plate-like particle. Figure 3 (b) shows EDX spectrum from this particle. It indicated that this particle contained Tantalum and Titanium. Apparently this particle is different from the 3C-SiC inclusions such as down-falls in shapes and in compositions. The detected chemical elements imply that this particle is from the furnace. Again dark line contrast running from the dark particle toward the [11 2,¯ 0] direction along on a (0001) plane is observed similarly to the case in Fig. 2 (a). By high-resolution TEM image and nano-area electron diffraction patterns shown in Fig. 3 (a), the formation of 3C-SiC thin layer with 8nm thickness is confirmed. This is recognized as an origin of triangular-shaped defect.

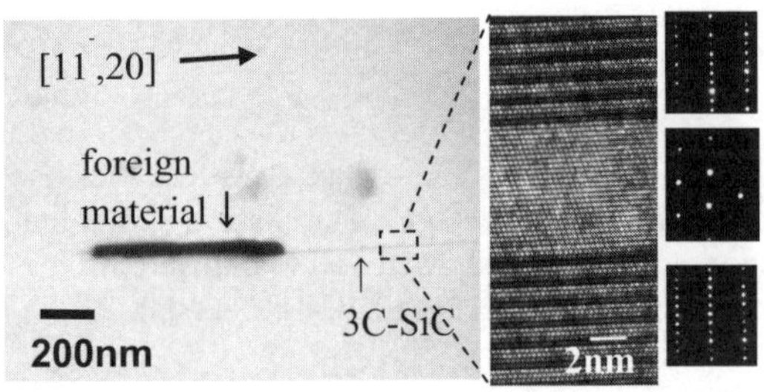

Fig. 3 (a) Cross-sectional TEM image of foreign material and plate-shaped 3C-SiC observed at [2,¯ 110] orientation. High-resolution image of the 3C-SiC dark thin layer and diffraction patterns taken at different area are also shown. *Dark contrasts upper-right of foreign materials are by beam damage.

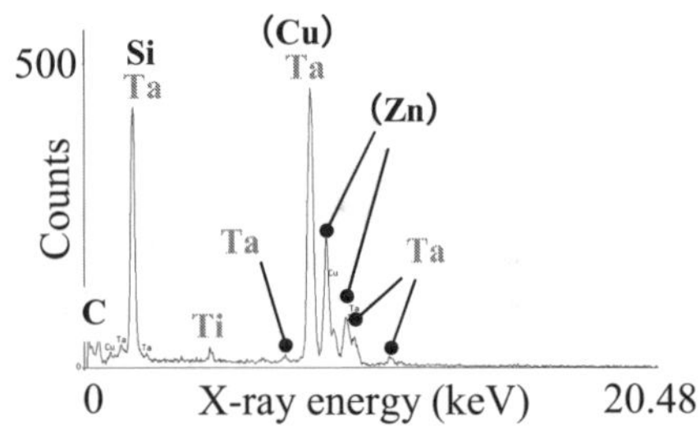

Fig. 3 (b) EDX spectrum taken from the dark particle in Fig. 3 (a). *Cu and Zn signals in the EDX spectrum are from TEM.

Different types of Triangular-shaped defects and 4H epitaxial film interfaces.

By observations of grazing incident synchrotron X-ray topographic images, we have found several different types of contrasts for triangular-shaped defects. Among them we will report two characteristic types in this paper. In Fig. 4 (a), here we call Type-I, clear contrasts at the outlines of the triangular-shapes were observed. The appearances of those contrasts change depending on different diffraction conditions. Particularly, the one of the sides of the triangle becomes stronger, and the other side becomes weaker at **g**= 2,¯ 118, but the appearances in contrast reverse at **g**=1 2,¯ 18. This tendency is the similar to the one that observed by electron microscopy analyses [5]. According to the discussions by Okada, the way in change in contrast indicates that the main dislocations at the interfaces are Shockley type, and one type of Burgers vectors dominate at one side of triangle. This can take places not only for the triangular-shaped defects consist of 3C-SiC layer, but also stacking faults for 1SSF and 3SSF[6]. In Fig.4 (b), here we call Type-II, no clear contrasts of the lattice defects at the outlines of the triangular shapes were observed. The Burgers vectors of dislocations at the interfaces might be cancelled out. For case of 8H-stacking fault for 4SSF[6], burgers vectors of partial dislocations, which locate at the edge of the stacking faults, can be cancelled out. Fig. 4(b) may indicate such strain field, otherwise, only surface morphology with no dislocations.

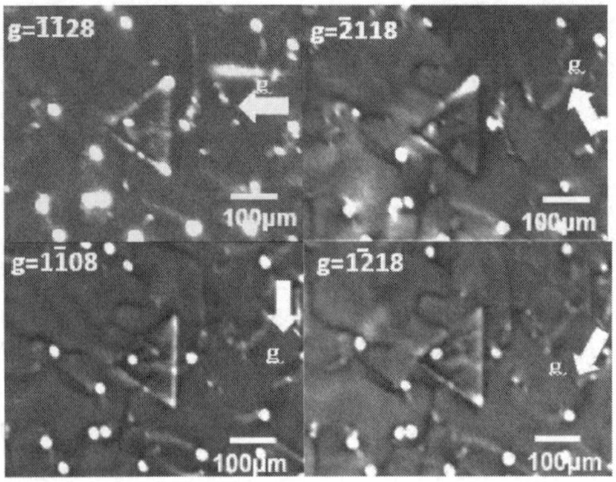

Fig .4 (a)

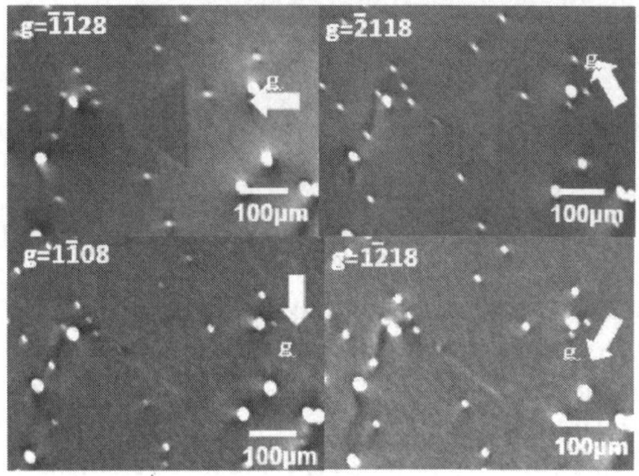

Fig. 4(b)

Fig .4 (a) and Fig. 4(b), 4 series of grazing incident synchrotron X-ray topographies of TYPE-I and TYPE-II. Reflections **g** were shown in each topography.

Summaries

We investigated origins of triangular-defects on 4degrees off 4H-SiC epitaxial wafers that had neither down-falls nor basal-plane dislocations. Contaminations by foreign materials and anomalous grains were observed in the present work. X-ray topography results showed that two different types of microstructure existed at interface between Triangular shaped defects and 4H-SiC crystal.

Acknowledgments

Reseach Laboratory for High-Voltage Electron Microscopy, in Kyushu-University is greatfully acknowledged. This work is supported by Novel Semiconductor Power Electronics Project Realizing Low Carbon Emission Society by METI and NEDO in Japan.

References

[1]T.Hatakeyama, T.Suzuki, K.Ichinoseki, H.Matsuhata, K.Fukuda, T.Shinohe, and K.Arai, Mater. Sci. Forum 645-648(2010)799-804.

[2]R.A.Berechman, M.Skowronski, and Q.Zhang,J.Appl.Phys 105, 074513(2007).

[3]Wein Si, Michael Dudley, Hua-Shuang Kong, Joe Sumakeris, and Calvin Carter,Jr.,J.Electron. Mater.Vol.26(1997)151-159.

[4]T.Okada, T.Kimoto, K.Yamai, H.matsunami, and F.Inoko, Mater.Sci.Eng.A361(2003)67-74.

[5]T.Okada, T.Kimoto, K.Yamai, H.matsunami, and F.Inoko, Mater.Sci.Forum 457-460 (2004) 521-524.

[6]G. Feng, J. Suda, and T. Kimoto, Physica B 404 (2009) 4745-4748.

Materials Science Forum Vols. 717-720 (2012) pp 367-370
© (2012) Trans Tech Publications, Switzerland
doi:10.4028/www.scientific.net/MSF.717-720.367

Variation of Etch Pit Size by Screw Dislocation Tilt in 4H-SiC wafer

Y. Ishikawa[1, a], Y.Yao[1], Y. Sugawara[1], K. Danno[2,b], H. Suzuki[2], Y. Kawai[2] and N. Shibata[1]

[1]Japan Fine Ceramics Center, 2-4-1 Mutsuno, Atsuta, Nagoya 456-8587, Japan

[2]Toyota Motor Corporation, 1200 Mishuku, Susono, Shizuoka 410-1193, Japan

[a]yukari@jfcc.or.jp, [b]danno@katsunori.tec.toyota.co.jp

Keywords: screw dislocation, tilt, 4H-SiC, etch pit, KOH, Na_2O_2, sublimation, wafer, epitaxial

Abstract. The wide size distribution of the hexagonal etch pit of screw dislocations (SD) in 4H-SiC wafer was found in spite of the narrow size distribution of the SD pit in epitaxial film. Calculation on the basis of the strain energy equation indicated that etch pit size depends on the Burgers vector and dislocation tilt. Size variation of SD etch pits in 4H-SiC wafer fabricated by sublimation method is explained to be caused by the dislocation tilt by observing the sizes and the positions of etch pits from the surface of the epitaxial film to the inside of 4H-SiC wafer. The SDs in 4H-SiC wafer fabricated by sublimation method propagate to c-axis direction in macroscopic but changing tilt in microscopic.

Introduction

Etch pit method is powerful to detect dislocations of 4H-SiC because dislocation types can be estimated by etch pit shape and size. The size of hexagonal etch pit on epitaxial 4H-SiC film clearly divided into two levels since the size distribution has two narrow peaks. Generally, it is attributed that the difference of etch pit size is caused by the magnitude of Burgers vector. In fact, large and small hexagonal pits on epitaxial 4H-SiC film corresponding to threading screw dislocations (TSD) and threading edge dislocations (TED), respectively [1,2]. Recently, we succeeded to develop novel etching method that applicable p-n^{++} -type SiC [3]. However, it is found that size distributions of hexagonal etch pits on 4H-SiC wafers by sublimation method were divided into 3-5 levels. One of the examples is shown in Fig.1. Four sharp peaks and tail at larger size range were observed. Etch pits at the wafer were classified by the 1 to 1 correlation between the etch pit at the epitaxial film and that on the wafer. Especially, the large size distribution of TSD pits hampers to discriminate the dislocation types. If we know what changes etch pit size, we can precisely estimate the crystal quality of 4H-SiC substrates.

In this paper, we investigated what changes etch pit size by calculation and experiment in 4H-SiC wafer.

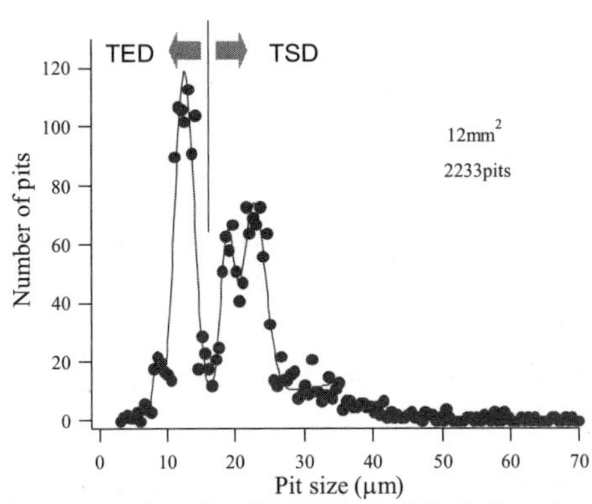

Figure 1. An example of etch pit size distribution on 4H-SiC wafer by sublimation method.

Calculation of Etch Pit Size

We estimated the variation of pit size depending on Burgers vector and dislocation tilt by simple calculation. Because it is well accepted that Burgers vector is main factor to determine the etch pit size and most of threading dislocations in 4H-SiC wafers that we observed by transmission electron microscopy (TEM) were tilted from [0001] axis. The largest tilt was about 60° in the TEM observations of several threading dislocations.

The calculation of etch pit size depending on Burgers vector was performed on the basis of the equation of dislocation strain energy. The dislocation strain energy (E_{strain}) per unit length in a crystal is as follows.

$$E_{strain} = \frac{Gb^2}{4\pi(1-\upsilon)} \ln(\frac{R}{r_0}) \quad \text{Eq. 1}$$

G and ***b*** is shear modulus and Burgers vector, respectively. υ is Poisson's ratio and the value of υ is 0.18 and 0 in an edge dislocation and a screw dislocation, respectively. R and r_0 is a crystal size and a dislocation core radius. The r_0 is nearly equal to b. We hypothesized that pit size is generally defined by the area that over the strain energy threshold ($E_{threshold}$) per unit volume. Therefore pits size can be estimated as R that gives $(\delta E_{strain} / \delta R) / 2\pi R \delta R = E_{threshold}$. Thus, pit size is given following equation.

$$R = \sqrt{\frac{Gb^2}{2(1-\upsilon)E_{threshold}}} \propto \frac{b}{\sqrt{(1-\upsilon)}} \quad \text{Eq.2}$$

Equation 2 indicates that the pit size depends on Burgers vector because $E_{threshold}$ is constant. We calculated the pit sizes of perfect dislocations, which are *a* (***b***=1/3<11-20>), *c* (***b***=<0001>) and *a*+*c* (***b***=1/3<11-20>+<0001>) in 4H-SiC. Calculated pit size ratio is as follows, *a*:*c*:*a*+*c*:2*c*=1:3:3.1:6. We judged these calculated values were almost reliable because measured etch pit size ratio (TSD/TED=*c*/*a*) was from 1.8 to 2.7. This calculation indicates that we can determine edge dislocation (***b***=1/3<11-20>) and simple screw dislocation (***b***=<0001>) by etch pit size. Compounding two *c*-dislocation i.e. *c* -> 2*c* doubles etch pit size. But variation of etch pit size by adding *a*-dislocation to c-*dislocation* i.e. *c* -> *c*+*a* is too small to detect because the size change is within the measurement deviation. It is reasonable as the magnitude of *c*+*a* and *c* is 1.06nm and 1.01nm, respectively. Thus we can't distinguish dislocation types between *c*+*a* and *c* in etch pit size.

The dependence of the etch pit size on dislocation tilt was calculated by taking account of the etching of pit side and the strain energy of screw dislocation. Before calculation, we measured the etch pit structures in the wide size range (28-185μm). It was found that the etch pits are similar six-sided pyramids. Angle between side and [0001] direction in (1-100) cross-section is about 70°. Figure 2 shows the schematic (1-100) cross-section of etch pit. Pit size is estimated by the etching time (t) and etching rate of the side (α) and surface (β). In this calculation, pit size L is defined as the pit length in [11-20] direction. If a dislocation line propagates perpendicular to the surface, pit size is given as follows,

$$L = 2D \tan 70 = 2(\alpha / \sin 70 - \beta) \tan 70 \cdot t \quad \text{Eq. 3.}$$

If the dislocation line tilted θ degree from [0001] axis, etch pit shape is defined by the slowest facet therefore pit size is written as follows,

$$L' = 2D' \tan 70 = 2(\alpha \cdot \cos\theta / \sin(70-\theta) - \beta) \tan 70 \cdot t \quad \text{Eq. 4.}$$

According to the Eq. 4, the pit size increased with the increasing of the angle of dislocation tilt (θ) and the etch pit size of 60° tilt dislocation nearly trebled. For more accurate estimation we have to calculate the variation of strain energy depending on dislocation tilt and the effects of wafer off-cut. There is no space to describe details but it should be noticed that etching of the side is found to be the main factor and similar results were obtained by the pit size calculation including the effects of the strain energy and the wafer off-cut.

Thus the etch pit size is effected by not only the variation of burgers vector but also dislocation tilt.

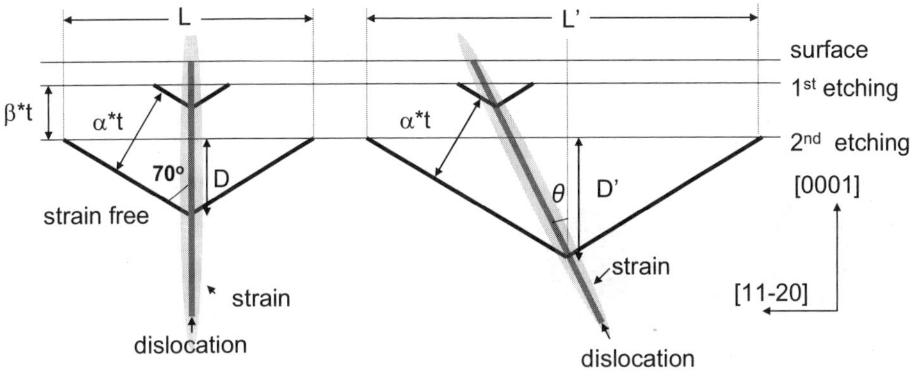

Figure 2. Schematic cross-section of etch pit. The observable size of etch pits, shown as 2nd etching, are formed continuing etching after small facet formation indicated as 1st etching. Diagrams of etch pit formation when the dislocation is parallel to the [0001] (left) and θ° tilt from [0001] (right).

Variation of Size and Position of Etch Pit

The variation of the etch pit size along depth direction from the surface of the epitaxial film to the 100μm inside of the 4H-SiC wafer was investigated by the repeats of etching and polish removing of the surface layer. The wafers were commercially available 8° off 4H-SiC (0001) and fabricated by sublimation method. Epitaxial layer was deposited by cold-wall chemical vapor deposition using CH_3SiH_3 and CH_4 as source gases. Etching was used KOH and Na_2O_2 [3]. Most of etch pits on epitaxial layer were hexagon and the size of hexagonal pits were divided into two levels as shown in Fig.3 (a). The size ratio and the position of hexagonal etch pits were not varied in the range of epitaxial film. It is attributed that dislocations (TSD and TED) in epitaxial film parallel and linearly propagated.

The large hexagonal etch pits (TSDs) in the SiC wafer were identified by the correlation of pit positions between on the epitaxial film and on the SiC wafer as the positions of TSDs were almost unchanged. But the positions of TEDs were shifted in the wafer by the transformation to the basal plane dislocation [4,5] therefore it is hard to prove 1 to 1 correlation. Roughly, small and large hexagonal pits were also TEDs and TSDs in the SiC wafer, respectively. Etch pits estimated TSDs were numbered in Fig.3. The narrow distribution in TSD pit size in the epitaxial film became wide in the wafer. No appearance and disappearance of TSD etch pits were observed from the surface of the epitaxial layer to the surface of the wafer. It means that the large distribution of TSD pit size was not caused by variation of burgers vector but dislocation tilt. In fact, the in-plane positions of etch pit cores were shifted by comparing between the surface of the epitaxal layer (Fig.3 (a))and the surface of the wafer (Fig.3(b)). The shifts were shown in Fig.3 (a) as white bar. The length and direction of the bar indicates distance and direction of dislocation shift, respectively. The calculation described above indicates that shift distance increases with pit size because pit size increases with increasing of the dislocation tilt. However the shift distance of etch pit position has no correlation with the pit size.

In order to investigate this, the pit size variation and the shift of the etch pit core in-plane position were also observed by comparing the surface and the 100μm inside of the wafer. The size of TSD etch pits (1-5) on the surface of the wafer are 2>1,3,5>4 as shown in Fig. 3(b) but it changed to 4>2,3>1>5 at 100μm inside from the surface of the wafer as shown in Fig. 3(c). Thus, pit size has no correlation between the surface and the 100μm inside of wafer. It means that TSDs in 4H-SiC substrate propagate in changing the tilt angle in a short-range (several- several tenth μm). If we accept the dislocation tilt change in a short range we can explain no correlation of dislocation shift with the pit size. In fact shift of etch pit between the surface of the wafer and 100μm inside of the wafer also has no correlation. Besides, we found that SD shifts were limited in ~20μm when we compare the pit positions between the surface of the wafer and 100μm inside of the wafer. In spite of the estimated shift in-plane is 58μm between the surface and 100μm inside of the wafer when the dislocation tilt is 60°. Therefore, we estimated that TSDs largely propagate to c-axis direction in 4H-SiC wafer made by sublimation method. This is well consisted with the observation that is almost straight SD propagation to [0001] direction by X-ray topography [4].

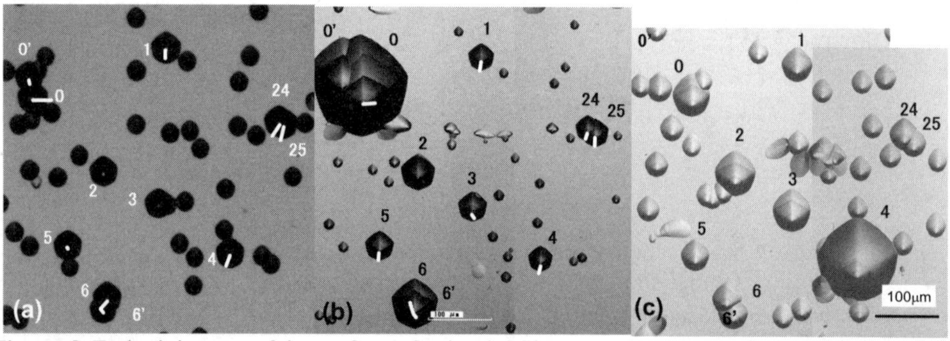

Figure 3. Etch pit images of the surface of epitaxial film (a), nearly the surface of wafer (b) and the 100μm inside of wafer (c). These images were taken at the same position. Numbered hexagonal pits are threading screw dislocations. The white lines in (a) and (b) show the shifts of the pit positions between in (a) and (b), and in (b) and (c), respectively.

Conclusion

It was found that Burgers vector and dislocation tilt are the main factors to decide the etch pit size by calculation. Calculated size ratio of dislocations depending on the Burgers vector is as follows, 1/3<11-20>:<0001>:1/3<11-20>+<0001>:2<0001>=1:3:3.1:6. Edge dislocation (***b***=1/3<11-20>) and simple screw dislocation (***b***=<0001>) can be identified by etch pit size. But identifying a simple screw dislocation (***b***=<0001>) and a complex dislocation (***b***=1/3<11-20>+<0001>) by etch pit size is difficult. Increasing tilt angle of dislocation from [0001] direction increases etch pit size and the etch pit size trebled when the tilt angle become 60°. Size variation of SD etch pits in 4H-SiC wafer fabricated by sublimation method is explained according to the dislocation tilt by observing the sizes and the positions of etch pit cores from the surface of the epitaxial film to the inside of 4H-SiC wafer. The SDs in 4H-SiC wafer fabricated by sublimation method propagated to c-axis direction in macroscopic but changing tilt in microscopic.

References

[1] S.A. Sakwe, R .Müller, P.J. Wellmann: J. Cryst. Growth Vol. 289 (2006) 520-526.

[2] D. Siche, D. Klimm, T. Hölzel, A. Wohlfart: J. Cryst. Growth Vol. 270 (2004) 1-6.

[3] Y. Yao, Y. Ishikawa, Y. Sugawara, H. Saitoh, K. Danno, H. Suzuki, Y. Kawai, and N. Shibata: Mater. Sci. Forum Vol. 679-680 (2011) 294-297.

[4] D. Nakamura, S. Yamaguchi, I. Gunjishima, Y. Hirose, T. Kimoto: J. Cryst. Growth 304 (2007) 57–63

[5] Y. Ishikawa, Y. Yao, and K. Dannno, in preparation for publication.

Materials Science Forum Vols. 717-720 (2012) pp 371-374

doi:10.4028/www.scientific.net/MSF.717-720.371

Critical Density of Nanoscale Pits for Suppressing Variability in Leakage Current of a SiC Schottky Barrier Diode

T. Ishikawa[1, a], T. Katsuno[1, b], Y. Watanabe[1, c], H. Fujiwara[2, d] and T. Endo[3, e]

[1]Toyota Central R&D Labs., 41-1, Yokomichi, Nagakute, Aichi 480-1192, Japan

[2] Toyota Motor Corporation, Nishihirose, Toyota, Aichi 470-0309, Japan

[3]DENSO CORPORATION, Komenoki, Nisshin, Aichi 470-0111, Japan

[a]t-ishi@mosk.tytlabs.co.jp, [b]e1417@mosk.tytlabs.co.jp, [c]y-watanabe@mosk.tytlabs.co.jp, [d]fujiwara@hirokazu.tec.toyota.co.jp, [e]tendo@rlab.denso.co.jp

Keywords: Silicon Carbide (SiC), Schottky barrier diode, Leakage current, Surface morphology, Surface roughness, Device simulation

Abstract. We investigate the influence of SiC surface morphology on increase and variability in reverse leakage current of SiC Schottky barrier diodes using device simulation. It is found that etch pits with only a few tens of nm in depth has a large influence on leakage current and is also shown that leakage current is sensitive to both etch pit shape and density. From these results, we suggest the critical density of nanoscale pit, which is suppressing the variability of leakage current, at various drift layer thickness t_{drift} and doping concentration N_{drift}.

Introduction

Silicon carbide (SiC) is a promising material for power electronics, with its high breakdown voltage, low power loss, high operating temperature and superior thermal conductivity. High breakdown voltage and low on-resistance SiC Schottky barrier diodes (SiC-SBDs) are already commercially available. However large variability is often observed in reverse *I-V* characteristics of SiC-SBDs. Some experimental studies [1, 2] have suggested that this large variability and increase in leakage current are due to non-uniform Schottky contacts with small patches of low barrier height [1] and an increase in SiC surface roughness [2]. Recently, Katsuno et al. have shown that the leakage current is proportional to the crystal dislocation density using molten KOH etch method [3]. They have also shown that a nanoscale pit (pit depth of about 50 nm) originating from crystal dislocation is the main source of the leakage current using emission microscope and AFM [4]. In the present study, we examine the influence of the nanoscale pit on increase and variability in leakage current using device simulation.

Device simulation

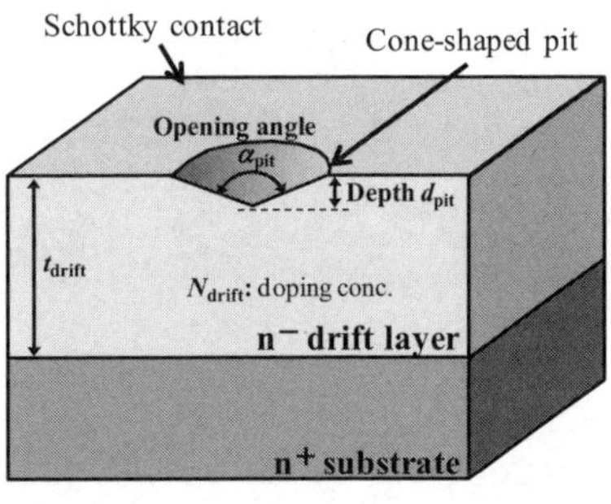

Fig. 1 Schamatic cross section of simulated SiC-SBD with a circular cone-shaped etch pit.

It is well known that circular and hexagonal cone-shaped etch pits appear after molten KOH etching if there are crystal dislocations in SiC [5]. This method is based on defect-selective chemical etching technique in alkaline solution. During device fabrication, SiC wafer is often exposed to the alkaline solution, such as an APM cleaning solution (NH_4OH + H_2O_2 + H_2O) used as a standard cleaning process. Therefore the same type of etch pits can appear on SiC surface before depositing Schottky contact metal. Fig. 1 shows a schematic of the simplified simulation model for the SiC-SBD including a circular cone-shaped etch pit.

Assuming that radius of curvature at the bottom of the etch pit is 5 nm. The shape of pits can be determined by depth d_{pit} and opening angle α_{pit}. Etch pits with opening angle α_{pit} of around 120 - 140° are usually observed [6]. So we investigate etch pits with opening angle α_{pit} of about 120° in this study. Electron tunneling at the Schottky contact is the major source of reverse leakage current. The device simulation to calculate reverse *I-V* characteristics are carried out using the Barrier tunneling model [7], which can express both thermionic field emission [8] and field emission model, assuming that Schottky barrier height Φ_B and barrier tunneling effective mass for electron are 1.2 eV and $0.25m_0$, respectively. The cylindrical coordinate system is used to reduce the simulation time.

Results and discussion

Fig. 2 (a) shows that nanoscale etch pits (d_{pit} = 50 nm, α_{pit} = 120°) on the SiC surface can cause an increase in the reverse leakage current in relatively low reverse voltage condition. This leakage current is due to electron tunneling current at the bottom of the etch pit. This tunneling current is caused by the electric field enhancement at the bottom of nanoscale etch pit. Electric field at metal-semiconductor interface (Schottky junction) near the bottom of pits is four times greater than that of flat surface area, applying reverse voltage V_R = 1200 V (Fig. 2 (b)). Fig. 3 shows the impact of the etch pit shape (depth d_{pit} and opening angle α_{pit}) on leakage current. Though leakage current is sensitive to both depth d_{pit} and opening angle α_{pit}, the remarkable increase in leakage current is not observed in the range of less than 20 nm in depth. Fig. 4 shows that relations between the nanoscale etch pit density (n-EPD) and leakage current at various drift layer doping concentration N_{drift}. For higher n-EPD leakage current is proportional to the n-EPD, so the fluctuation of n-EPD within a SiC wafer causes variability in leakage current. For lower n-EPD, leakage current becomes insensitive to the n-EPD and gradually approaches each dashed line which corresponds to leakage current of SiC-SBDs without etch pits at different N_{drift}. This is because that tunneling current from flat area becomes dominant component of leakage current less than a certain n-EPD. This indicates that variability in leakage current becomes small. We defined this n-EPD as critical density of nanoscale etch pit for suppressing the variability of leakage current. Fig. 5 shows the dependence of this critical n-EPD on drift layer thickness t_{drift} and doping concentration N_{drift} from the simulation results. It is concluded that lower n-EPD is necessary for a high voltage SiC-SBD, which is thicker t_{drift} and lower N_{drift}, to suppress variability in leakage current.

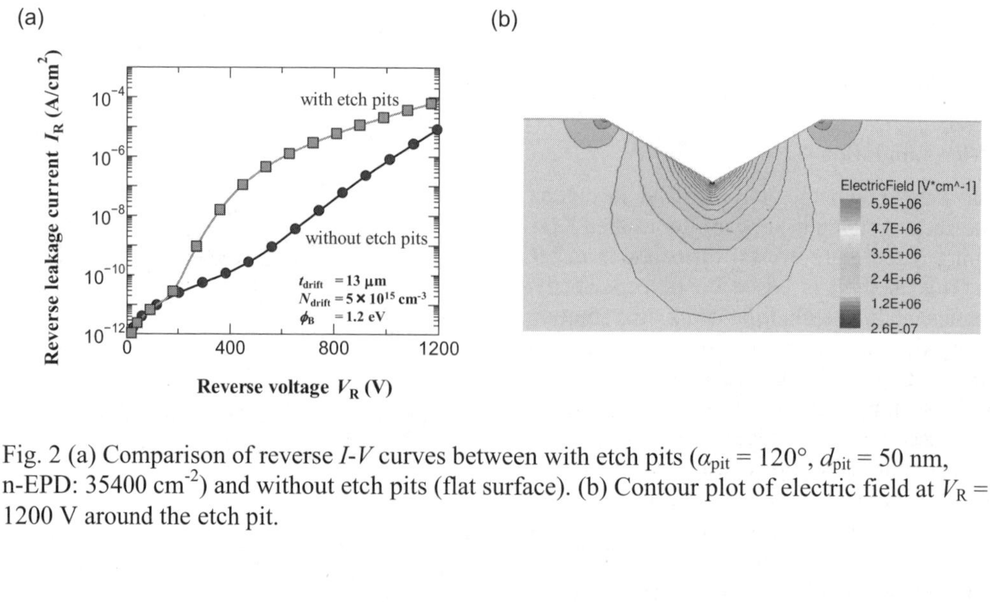

Fig. 2 (a) Comparison of reverse *I-V* curves between with etch pits (α_{pit} = 120°, d_{pit} = 50 nm, n-EPD: 35400 cm^{-2}) and without etch pits (flat surface). (b) Contour plot of electric field at V_R = 1200 V around the etch pit.

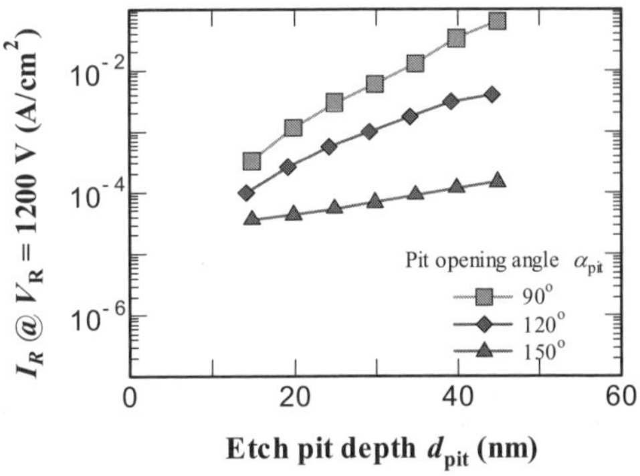

Fig. 3 Dependence of leakage current I_R (at V_R = 1200 V) on opening angle α_{pit} and depth d_{pit} of etch pits. (n-EPD: 1.27×10^6 cm^{-2})

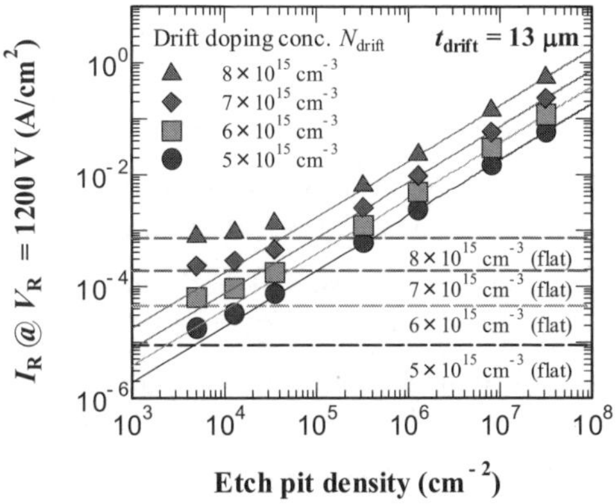

Fig. 4 Relations between leakage current I_R and etch pit (α_{pit} = 120°, d_{pit} = 50 nm) density (t_{drift} = 13 μm). Each dashed line represents leakage current without etch pits at different N_{drift}.

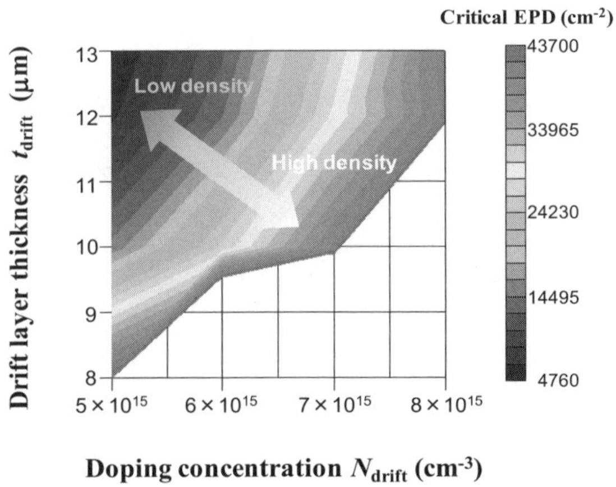

Fig. 5 Critical n-EPD for supressing variability in reverse leakage current at various drift layer conditions (α_{pit} = 120°, d_{pit} = 50 nm).

Summary

We examine the influence of the nanoscale pits on increase and variability in reverse leakage current of SiC-SBDs using device simulation. Nanoscale etch pits (d_{pit} = 50 nm, α_{pit} = 120°) on the SiC surface can cause an increase in the reverse leakage current. It is found that leakage current is sensitive to etch pit shape (depth d_{pit} and opening angle α_{pit}). From these results, we propose the critical density of nanoscale pits on SiC surface for suppressing the variability of leakage current of SiC-SBDs in various drift layer thickness t_{drift} and doping concentration N_{drift}. Lower n-EPD is necessary for a high voltage SiC-SBD, which is thicker t_{drift} and lower N_{drift}, to suppress variability in leakage current.

References

[1] H. Saitoh, T. Kimoto, H. Matsunami, Materials Science Forum **457-460** (2004) pp 997-1000.

[2] T. Kato, A. Kinoshita, K. Wada, T. Nishi, E. Hozomi, H. Taniguchi, K. Fukuda, H. Okumura, Materials Science Forum **645-648** (2010) pp 763-765.

[3] T. Katsuno, Y. Watanabe, H. Fujiwara, M. Konishi, T. Yamamoto, T. Endo, Ext. Abst. 2010 Int. Conf. on Solid State Devices and Materials (2010) pp 906-907.

[4] T. Katsuno, Y. Watanabe, H. Fujiwara, M. Konishi, H. Naruoka, J. Morimoto, T. Morino, T. Endo, Applied Physics Letters **98** (2011) 222111.

[5] T. Sudarshan, Materials Science and Engineering of Bulk Silicon Carbides, in: Z.C. Feng (Ed.), SiC power materials: devices and applications, Springer-Verlag, Berlin, 2004, pp. 47-52.

[6] T. Katsuno, Y. Watanabe, H. Fujiwara, M. Konishi, T. Yamamoto, T. Endo, Materials Science Forum **679-680** (2011) pp 298-301.

[7] Synopsys Sentaurus Device User Gide version D-2010.03 (2010).

[8] T. Hatakeyama, M. Kushibe, T. Watanabe, S. Imai, T. Shinohe, Materials Science Forum **433-436** (2003) pp 831-834.

Materials Science Forum Vols. 717-720 (2012) pp 375-378

doi:10.4028/www.scientific.net/MSF.717-720.375

Surface Morphology of Leakage Current Source of 4H-SiC Schottky Barrier Diode by Atomic Force Microscopy

T. Katsuno[1,a], Y. Watanabe[1,b], T. Ishikawa[1], H. Fujiwara[2,c], M. Konishi[2,d], T. Morino[3,e] and T. Endo[3,f]

[1]Toyota Central R&D Labs. Inc., Yokomichi, Nagakute, Aichi 480-1192, Japan

[2]Toyota Motor Corporation, Nishihirose, Toyota, Aichi 470-0309, Japan

[3]Research Laboratories, DENSO CORPORATION, Komenoki, Nisshin, Aichi 470-0111, Japan

[a]e1417@mosk.tytlabs.co.jp, [b]y-watanabe@mosk.tytlabs.co.jp, [c]fujiwara@hirokazu.tec.toyota.co.jp, [d]masaki@konishi.tec.toyota.co.jp, [e]TOMOO_MORINO@denso.co.jp, [f]tendo@rlab.denso.co.jp

Keywords: 4H-SiC, Schottky barrier diode, surface morphology, defects, threading dislocation

Abstract. The surface morphologies at the leakage current sources of 4H-SiC Schottky barrier diodes (SBDs) were analyzed using atomic force microscopy (AFM). Nanosized circular cone shaped pits (nanopits), which depths distributed from 5 to 70 nm, were observed at the leakage current sources. It is expected that the leakage current of 4H-SiC SBD generates at the nanopits due to the concentration of the electrical field strength. The positions of nanopits corresponded to the positions of threading dislocations (TDs), which were identified from molten potassium hydroxide (KOH) etching.

Introduction

Silicon carbide (SiC) power devices are highly suitable for use in a variety of applications, such as power conversion in electric vehicles, because SiC material shows the excellent physical properties such as a high bandgap, high electron mobility and high thermal conductivity [1-2]. SiC devices have been studied to obtain the high performances, but some of them show the poor properties due to the existences of the surface and crystalline defects [3-5]. In order to know the sensitivities of the defects on the reverse bias characteristics of 4H-SiC SBDs, the relationships between the reverse characteristics of 4H-SiC diodes and the crystalline defects [e.g. threading dislocations (TDs)] are discussed in many literatures [5-9]. However, the sensitivities of TDs on the reverse current of 4H-SiC diodes are not unified. Some reports have noted the influence of TDs on the leakage current [5-7], and contrasting reports have not noted such an influenced by TDs [8-9]. We suggest that the surface morphologies at the leakage current sources of 4H-SiC diode are changed, and this is possible reason for the different sensitivities. In order to confirm the change of surface morphologies at the leakage current sources of 4H-SiC SBDs, the leakage current sources were determined by emission microscopy, and the surface morphologies were then investigated by atomic force microscopy (AFM). Molten potassium hydroxide (KOH) etching was then performed to assist identification of crystalline defects and to determine the presences of TDs at the leakage current sources by laser microscopy.

Experimental

4H-SiC SBDs were fabricated on a 4° off-axis 4H-SiC substrate with 13 μm thick epitaxial layer. Typical TD density of the epitaxial layer used as this experiment is $1{\sim}5\text{x}10^3$ cm^{-2}. The doping concentration of nitrogen was $5.0\text{x}10^{15}$ cm^{-3}. A 100 nm thick, 1 mm diameter Mo layer was deposited on the SiC surface as a Schottky electrode. Al electrode with the thickness of 4 μm was deposited on the Mo electrode. A field limiting ring (FLR) was used as a termination structure, and the breakdown voltage was designed to be approximately 1400 V. 4H-SiC SBD samples with the typical electrical properties were selected in the experiments [10]. In the samples, Schottky barrier height Φ_B and the electric field strength obtained at 1200 V showed 1.13 eV and 1.5 MV/cm, respectively. An emission

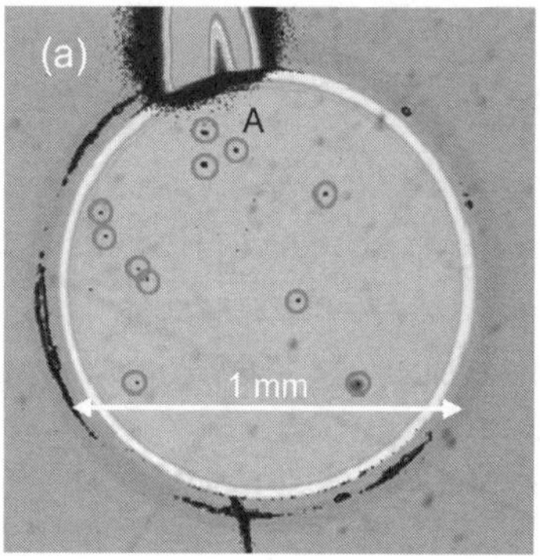

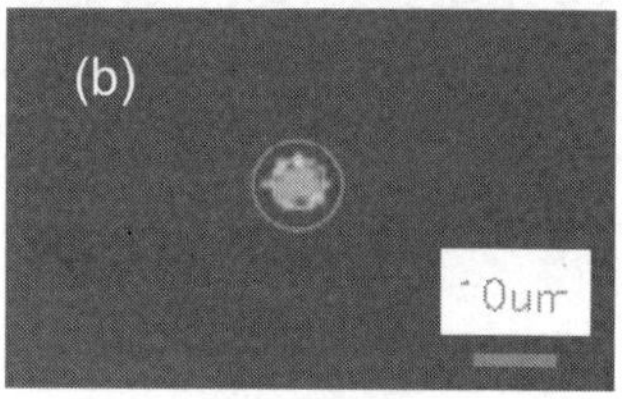

Fig.1 Emission microscope image of 4H-SiC SBD.
(a) View of the entire electrode of 1 mm diameter.
(b) Enlarged view of emission point A as shown in (a).

microscope (Hamamatsu Photonics PHEMOS-1000) was used to determine the leakage current sources of 4H-SiC SBD. The leakage current of 4H-SiC SBD was observed through the partially polished backside electrode using CCD image sensor. A reverse voltage of 1200 V was applied between the surface and backside. The marking positions, which were 10 μm apart from the emission points, were described by focused ion beam (FIB) from surface side to find the emission points easily. The surface morphologies of the emission points were measured by AFM. The analyzed numbers of surface morphologies were 24. After the measurements of surface morphologies, we observed the existences of the crystalline defects or not at the points. The crystalline defects were transformed into etch pits by molten KOH etching, which conditions were 500 °C for 5 min, and the surfaces were then observed by the laser microscope (Keyence: VK-9710).

Results and discussion

Figure 1 shows an emission microscope image of 4H-SiC SBD with the applied voltage of 1200 V. Eleven emission points indicated by the circles in Fig. 1(a) were detected in the 1 mm diameter device. This result indicates that the leakage current observes at the points. In the emission intensity from the detector, the strong emission intensity indicates the high leakage current. Figure 1 (b) shows the enlarged view of the emission point A as shown in Fig. 1(a). The emission intensity of point A seems to be intermediate level among 11 emission points.

Figure 2 shows AFM image of the leakage current source (point A), which analyzed area is 2x2 μm. The different color contrast indicates that the surface morphology of leakage current point A is not flat. To confirm the shape of the leakage current point A, the line profile (X)-(Y) is shown in Fig. 2 (b). Very small circular cone-shaped pit is evident at the leakage current source. It is noted that this is not etch pit. The pit is 45 nm depth and 230 nm wide with a base angle of 137 °. This is nanosized pit, and is therefore called nanopit. It is expected that the nanopit influences the increase of leakage current.

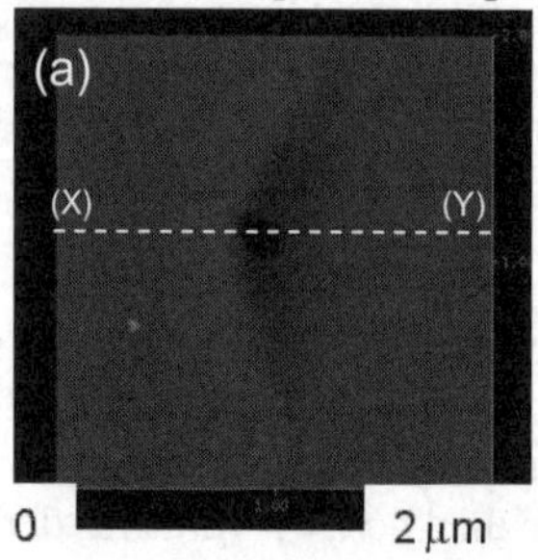

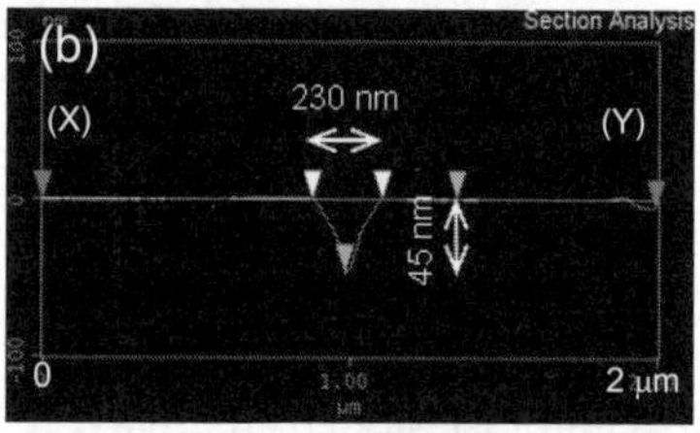

Fig. 2 AFM image of leakage current point A in Fig.1. (a) Surface morphology (b) line profile.

The maximum field strength is applied at the nanopit (equal to small circular cone-shaped structure), which exists at the interface of SiC surface and Mo metal. Therefore, the electrical field concentrates at the nanopit during the reverse applied voltage, and the leakage current generates at the nanopit.

In order to confirm the surface morphologies of the leakage current sources, the different leakage current sources are also measured by AFM. Surprisingly, the nanopits exist at all emission points (total: n=24) in some 4H-SiC SBDs. The dispersion of nanopit depth, width and base angle are concluded in figure 3(a)-(c), respectively. The meshed marking in figure 3 are discussed later. The depths of nanopits have a peak at 15 nm, but distributed from 5 to 70 nm widely. The widths of nanopits have a peak at 250 nm, but also distributed from 160 to 300 nm widely. The base angles of nanopits distributed from 100 to 175 °. According to the simulation results, the leakage currents increase with deep depths and low base angle of nanopits [11]. It looks the correspondences between experimental and simulation results; for example, the nanopit with the strong emission intensity as shown in Fig. 1(a) has deep depth as shown in Fig. 3(a) (the data is not appeared in this paper).

The etch pits formed by the molten KOH etching were observed by laser microscope, in order to confirm the relationship between the positions of the nanopits and the crystalline defects. Eleven etch pits denoted by the circles are observed in Fig. 4(a), and the almost positions of which correspond to the emission points in Fig.1(a). Nanopits to be the leakage current sources are correspondence the positions of TDs. Fig.4(b) shows the enlarged view with the same field as that in Fig.1(b). The crystalline defect is identified as threading edge dislocation (TED) from the shape of the etch pit [12]. Four direction around the etch pit are marked by FIB to find the leakage point easily. The other crystalline defects are identified by the shapes of etch pits, and n=22/24 etch pits are TEDs and the other etch pits (n=2/24) are TSDs. In order to know the differences the nanopit shapes between TEDs and TSDs, the nanopit shapes are rechecked. TSDs and TEDs indicate the mesh and gray fill marks, respectively, as shown in Fig. 3. The information of TSD is a few, but the nanopit sizes of TSD are within these of TED. Therefore, there is no difference the nanopit shapes between TSD and TED. Thus, sensitivities of TSD and TED on the leakage current are almost same. Generally, TSD is sensitive than TED on the leakage current [6-7], but our result is not corresponded with their results. As our proposal, the existence of nanopit is important to increase the leakage current.

Obviously, the creation of nanopit is influenced by the presence of TDs. However, it is unclear why the nanopits are created. As a possibility to create the nanopit, the top of TD at SiC surface can be easily damaged by the thermal conditions due to that TD is crystalline defect, or the nanopit may exist at the bare wafer. TDs do influence the leakage current in some groups [5-7], on the other hand, no influence of TDs on the leakage current are reported [8-9] in many papers. We propose that both results are correct. The former groups had nanopits corresponding to the positions of TDs, which generated the leakage currents. In contrast, the latter group did not have nanopits (or very smooth surface), and no leakage current generated. Thus, leakage current did not depend on TDs.

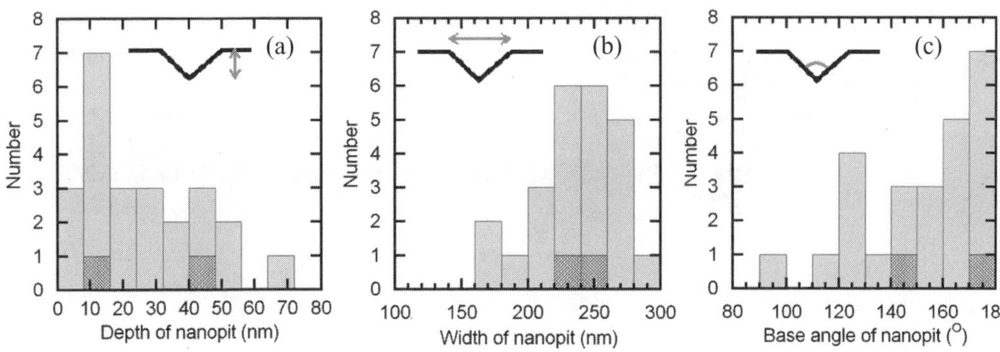

Fig. 3 Distribution of nanopit (a) depth, (b) width and (c) base angle.

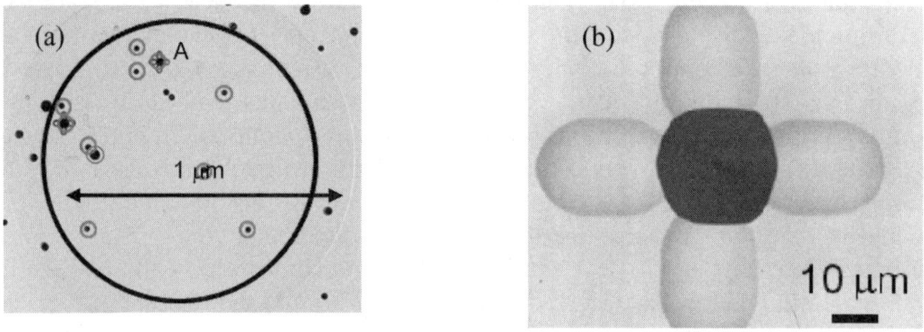

Fig. 4 Laser microscope image of 4H-SiC SBD after molten KOH etching.
(a) View of the entire electrode of 1 mm diameter.
(b) Enlarged view of emission point A as shown in (a).

Summary

The surface morphologies at the leakage current sources of 4H-SiC SBDs were analyzed using AFM. Nanosized circular cone shaped pits (nanopits), which depths distributed from 5 to 70 nm, were observed at the leakage current sources. The leakage currents of 4H-SiC SBDs generated at the nanopits due to the concentration of the electrical field strength. The positions of nanopits corresponded to the positions of TDs, which were identified from molten KOH etching. There is no difference the sensitivities on the leakage current between TSD and TED, because both nanopit shapes are almost same.

References

[1] M. Ruff, H. Mitlehner, and R. Helbig, IEEE Trans. Electron Devices 41 (1994) 1040.
[2] T. Kimoto, N. Miyamoto, and H. Matsunami, IEEE Trans. Electron Devices 46 (1999) 471.
[3] P. G. Neudeck, Mater. Sci. Forum 338-342 (2000) 161.
[4] K.-Y. Lee and M. A. Capano, J. Electron. Mater. 36 (2007) 272.
[5] T. Katsuno, Y. Watanabe, H. Fujiwara, M. Konishi, T. Yamamoto, and T. Endo, Jpn. J. Appl. Phys. 50 (2011) 04DP04.
[6] R. A. Berechman, M. Skowronski, and Q. Zhang, J. Appl. Phys. 105 (2009) 074513.
[7] Q. Wahab, A. Ellison, A. Henry, E. Janzen, C. Hallin, J. Di Persio, and R. Martinez, Appl. Phys. Lett. 76 (2000) 2725.
[8] H. Saitoh, T. Kimoto, and H. Matsunami, Mater. Sci. Forum 457-460 (2004) 997.
[9] D. T. Morisette, and J. A. Cooper, Jr., Mater. Sci. Forum 389-393 (2002) 1133.
[10] T. Katsuno, Y. Watanabe, H. Fujiwara, M. Konishi, H. Naruoka, J. Morimoto, T. Morino, and T. Endo, J. Appl. Phys. 98 (2011) 222111.
[11] T. Ishikawa, T. Katsuno, Y. Watanabe, H. Fujiwara, and T. Endo, submitted to Mater. Sci. Forum (2012).
[12] T. Katsuno, Y. Watanabe, H. Fujiwara, M. Konishi, T. Yamamoto, and T. Endo, Mater. Sci. Forum 679-680 (2011) 298.

Materials Science Forum Vols. 717-720 (2012) pp 379-382

doi:10.4028/www.scientific.net/MSF.717-720.379

Density and Behavior of Etch Pits on C-face 4H-SiC Surface Produced by ClF_3 Gas

Hitoshi Habuka[1, a], Kazuchika Furukawa[1, a], Toshimitsu Kanai[1, a], and Tomohisa Kato[2,b]

[1]Department of Chemical and Energy Engineering, Yokohama National Univ., Japan

[2]Advanced Power Electronics Research Center, National Institutes of Advanced Science and Technology, Japan

[a]habuka1@ynu.ac.jp, [b]t-kato@aist.go.jp

Keywords: 4H-SiC, C-face, chlorine trifluoride gas, etch pit

Abstract. The density and behavior of etch pit produced on the C-face 4H-SiC substrate using chlorine trifluoride gas at various temperatures were evaluated. Because the etch pit density formed at the substrate temperature of 713 K showed the comparable value to the current dislocation level of the Si-face 4H-SiC, the etch pit density obtained by this technique is considered to have a relationship with the crystal quality.

Introduction

For evaluating and improving the SiC crystal quality, the etch pit formation technique applicable to C-face should be developed, similar to S-face [1]. However, conventional KOH etching technique cannot produce the etch pits on the C-face. Thus, for this purpose, the etching technique using chlorine trifluoride (ClF_3) gas [2-4] has been studied and it showed that the etch pits on the C-face 4H-SiC surface could be produced and coincided with the dislocations revealed by the X-ray topograph [4]. However, the etch pit density obtained by this technique has not been evaluated. In this paper, the etch pit density on the C-face 4H-SiC surface formed using chlorine trifluoride gas was studied.

Experimental

The substrate used in this study was the chemical-mechanical-polished (CMP) single-crystalline 4H-SiC having an on-axis (000-1) C-face surface. The substrates used were cut from the same wafer. In order to etch the C-face 4H-SiC surface, the horizontal cold-wall reactor shown in Fig. 1 was used. In this reactor, the 4H-SiC substrate (5 x 5 mm) was placed on the center of the polycrystalline 3C-SiC susceptor.

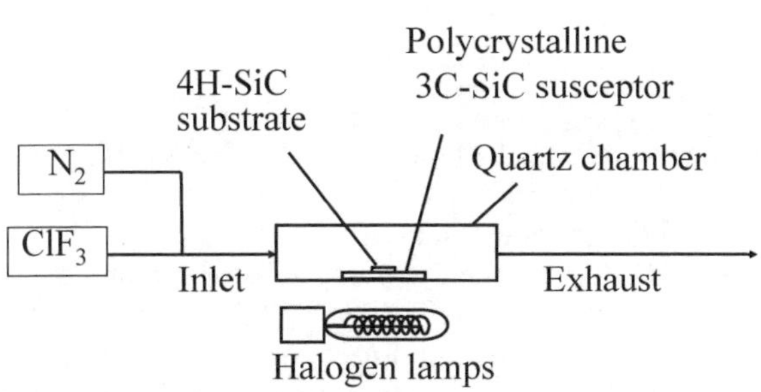

Fig. 1. Horizontal cold-wall reactor for etching 4H-SiC used in this study.

The process for etching the SiC substrate is shown in Fig. 2. During Step (a), nitrogen gas was introduced into the reactor at atmospheric pressure at the flow rate of 1 slm. Electric power to the halogen lamps was fixed to maintain the substrate temperature at 573, 623, 673, 693, 703, 713, 723, 773, and 823 K, in ambient nitrogen at atmospheric pressure. The substrate temperature was measured using thermocouple (R-type). At each fixed temperature in Step (b), the C-face (000-1) 4H-SiC substrate were then etched by the chlorine trifluoride gas (Kanto Denka Kogyo Co., Ltd., Tokyo) without further purification. The concentration and the flow rate of the chlorine trifluoride gas were 100 % and 100 sccm, respectively at atmospheric pressure. In Step (c), the substrate was cooled to room temperature in ambient nitrogen. Every etching used fresh substrates, except of additional etching.

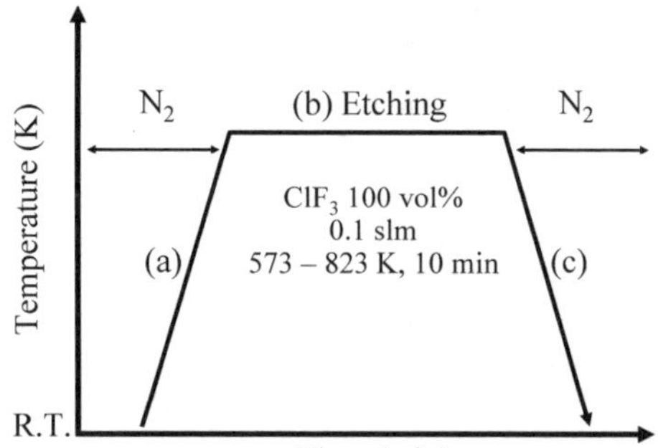

Fig. 2 Process for etching single-crystalline 4H-SiC substrate surface. Step (a): heating up in ambient nitrogen, Step (b): etching using chlorine trifluoride gas and Step (c): cooling to room temperature in ambient nitrogen.

The SiC etching is assumed to follow the overall reaction shown in eq. (1) [2],

$$3SiC + 8ClF_3 \rightarrow 3SiF_4 + 3CF_4 + 4Cl_2. \quad (1)$$

Chlorine trifluoride (ClF_3) gas etches the solid SiC and forms gaseous products, such as silicon tetrafluoride (SiF_4), carbon tetrafluoride (CF_4) and chlorine (Cl_2). Because these products have very low boiling points, no deposition occurred in the downstream region from the reactor throughout the exhaust line.

The average etch rate [2] of the 4H-SiC by the chlorine trifluoride gas was determined by the decrease in the substrate weight. Because the 3C-SiC susceptor and the 4H-SiC substrate were simultaneously etched in the reactor, the etch rate obtained for the 4H-SiC in this reactor is comparable to the average value for the 4H-SiC substrate having the same dimension as that of the 3C-SiC susceptor.

The density of formed etch pits was evaluated using an optical microscope and a scanning electron microscope.

Results and discussion

The etch pit obtained using the chlorine trifluoride gas at 713 K for 10 min at 100% was observed and shown in Fig. 3. The diameter of etch pits was near 10 μm, which was considered to be assigned to screw dislocations, following the previous study [4].

Because the etch pit density (EPD) changed with the substrate temperature [4], the etching was performed at various temperatures around 713 K. The etch pit density obtained in an area of 500 x 500 μm^2 at various substrate temperatures is shown in Fig. 4. At the temperatures below 673 K, the etch pit density was very small. At 683 K, the etch pit density increase to the value near 2×10^4 cm^{-2}. At 713K, the etch pit density showed the maximum value of 4×10^4 cm^{-2}. Although the etch pit density decreased at the temperatures higher than 723 K, its value maintained near 10^4 cm^{-2}.

The etch pit density obtained at 713 K coincided with the values widely accepted as the current dislocation density level of 4H SiC. Thus, the etch pit density of C-face 4H-SiC obtained in this study is acceptable and is expected to show a relationship with the crystal quality. The etch pits obtained in this study were classified to the large circular-shaped and small oval-shaped pits, which ratio were 90 % and 10 %, respectively, at the etching temperature of 713 K. The former is considered to be assigned to the treading screw dislocation, and the latter can be the treading edge dislocation.

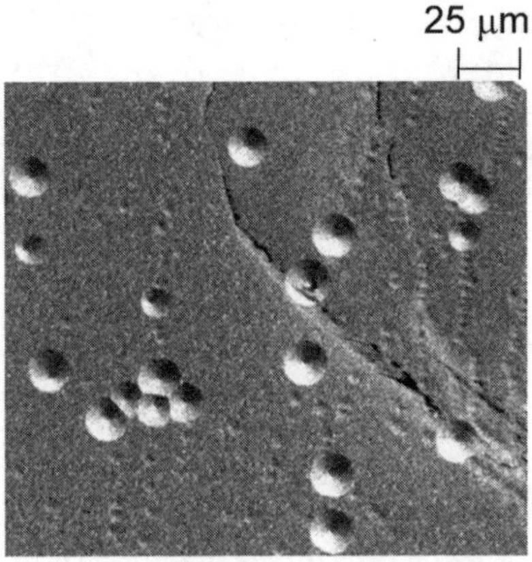

Fig. 3 Etch pits produced on C-face 4H-SiC surface by 100% chlorine trifluoride gas at 713 K for 10 min.

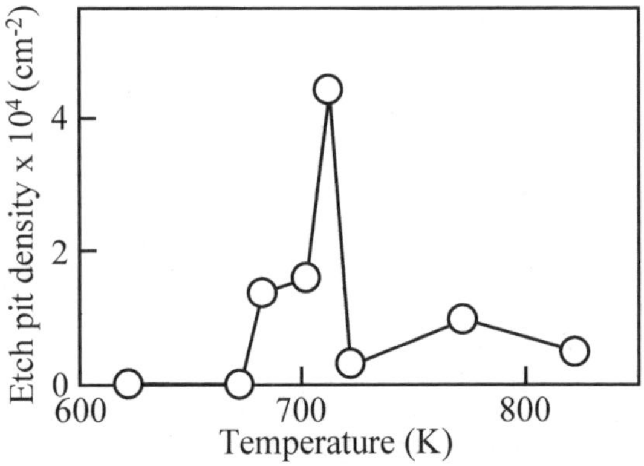

Fig. 4 Etch pit density on C-face 4H-SiC surface produced by chlorine trifluoride gas at various substrate temperatures.

In order to show that the origin of these etch pit was assigned to the crystalline imperfection, the etched surface was further etched using the same etching condition. Fig. 5 (a) shows the nine etch pits observed on the C-face 4H SiC after the etching at 713 K for 10 min using the 100 % chlorine trifluoride gas. Their diameter was about 10 – 15 μm. This substrate was additionally etched at 713 K at the 100 % chlorine trifluoride gas. Fig. 5 (b) shows the surface, 80 μm of which surface was etched off by the additional etching. Fig. 5 (b) shows the large and shallow nine etch pits, which were overlapped with each other. The number and the position of etch pit center in Fig. 5 (b) corresponded to those in Fig. 5 (a), respectively. Thus, over the depth of 80 μm, the origin to cause the etch pit was concluded to exist; the origin could be the dislocation, such as the threading dislocation, existing normal to the substrate surface.

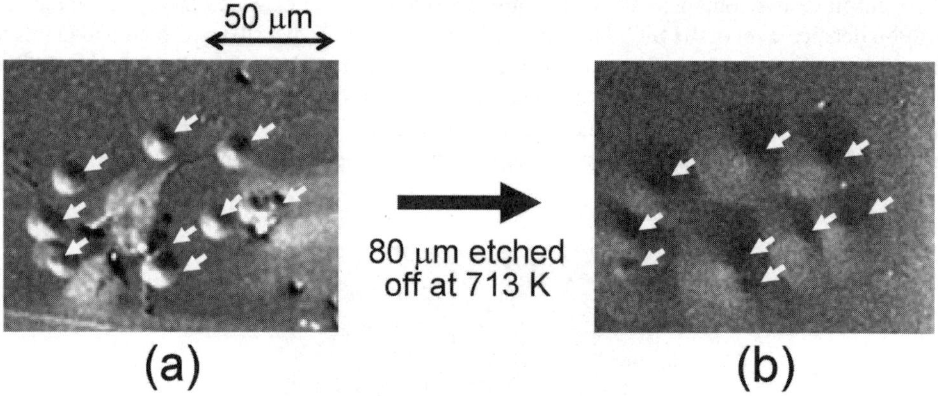

Fig. 5 Comparison of etch pits (a) before and (b) after additional etching. White arrows indicate the etch pits.

The results obtained in this study indicate that the pits caused by the etching using chlorine trifluoride gas at around 713 K has the origin of crystalline imperfection, such as the threading dislocation. However, the interpretation of etch pits with respect to the underlying dislocation type should be further carefully performed by a comparative investigation [4]. Additionally, the density and shape of etch pit by this technique should be further clarified and verified through many characterization. In future, the etching condition applicable to detect the basal plane dislocation [5], existing parallel to the surface, on the C-face 4H-SiC, slightly off-oriented, should be developed.

Conclusions

The density and behavior of etch pit on the C-face 4H-SiC substrate produced using chlorine trifluoride gas at various temperatures were evaluated. At the substrate temperature of 713 K, the etch pit density showed the maximum value of 4×10^4 cm^{-2}. The etch pit density obtained by this technique is considered to show the crystal quality, particularly, the dislocation density.

References

[1] B. Kallinger, S. Polster, P. Berwian, J. Friedrich, G. Mueller, A. N. Danilewsky, A. Wehfhahn and A. D. Weber, J. Cryst. Growth, 314 (2011) 21.

[2] H. Habuka, Y. Katsumi, Y. Miura, K. Tanaka, Y. Fukai, T. Fukae, Y. Gao, T. Kato, H. Okumura and K. Arai, *Mater. Sci. Forum*, Vol. 600-603 (2008) 655-658.

[3] H. Habuka, K. Tanaka, Y. Katsumi, N. Takechi, K. Fukae and T. Kato, *Materi. Sci. Forum,* Vol. 645-648 (2010) 787-790.

[4] H. Habuka, K. Tanaka, Y. Katsumi, N. Takechi, K. Fukae, and T. Kato, *J. Electrochem. Soc.*, 156 (2009). H971-975.

[5] S. Ha, M. Skowronski, W. M. Vetter and M. Dudley, J. Appl. Phys., 92 (2002) 778-785.

Materials Science Forum Vols. 717-720 (2012) pp 383-386
© (2012) Trans Tech Publications, Switzerland
doi:10.4028/www.scientific.net/MSF.717-720.383

Dislocation Formation in Epitaxial film by Propagation of Shallow Dislocations on 4H-SiC substrate

Y. Ishikawa[1, a], K. Sato[1], Y. Okamoto[2,b], N. Hayashi[2], Y. Yao[1], Y. Sugawara[1]

[1]Japan Fine Ceramics Center, 2-4-1 Mutsuno, Atsuta, Nagoya 456-8587, Japan

[2]ACT Corporation, 30 Gokouda-cho, Nishikujyo, Minami, Kyoto 601-8442, Japan

[a]yukari@jfcc.or.jp, [b]okamoto@k-nex.jp

Keywords: dislocation, triangular defect, stacking fault, dislocation array, shallow defect, polish, half-loop

Abstract. Shallow defects, which were induced by mechanical treatment, on 4H-SiC wafers were investigated. The density and the depth distribution of shallow defects on the wafers were dependent on wafer manufacturer. Most of the serious defects, such as the dislocation array (DA), triangular stacking fault (TRSF), and triangular defect (TRD), in epitaxial film were demonstrated to be caused by shallow dislocations on the surface of the wafers. Revised mechanical polishing can reduce the DA, TRSF, and TRD densities in epitaxial film.

Introduction

Decreasing the defect density in epitaxial film is an urgent issue in establishing high performance and in high-yield device fabrication. Most of the dislocations in epitaxial film are formed by the propagation of the surface dislocations of the SiC wafer. The surface dislocations comprise the dislocations in SiC crystal and those formed by mechanical treatment (polish, slice, grinding, etc.). Outstanding advancement of the crystal quality of SiC in recent years has reordered mechanical treatment a key technology. However, reports on defects induced by mechanical treatment are few [1]. The dislocations formed by polishing are generally shallow. However, shallow dislocations remain unclarified because of their difficulty of detection by X-ray topography (XRT), owing to the limitation of resolution, despite the dislocations in SiC crystal being well analyzed by XRT. Knowledge, that is, what types of dislocations are formed in epitaxial film from the surface dislocations and how deep the shallow dislocations are distributed in the SiC substrate, yields solutions for reducing the dislocation density in epitaxial film.

In this paper, we indicate that shallow dislocations on a 4H-SiC wafer form DAs, TRSFs and TRDs by investigating the density variation of these dislocations on epitaxial films depending on the removed thickness by gas-phase etching before epitaxial growth.

Table 1. Sample list

Sample name	Manufacturer	Polishing method	Etch pit density
A	A	MP	9.9e3 cm^{-2}
B	B	CMP	3.2e4 cm^{-2}
C	C	MP	3.1e3 cm^{-2}
D	D	MP by us	8.2e4 cm^{-2}

MP: mechanical polishing, CMP: chemical mechanical polishing

Experimental

Commercially available 0.017-0.025 ohm cm N-doped n-type 4°-off toward [11-20] 4H-SiC(0001) wafers were used, as listed in Table 1. Our revised MP was performed on the wafer from manufacturer D (sample D). The wafers were cut into 10 x14 mm^2 pieces. Epitaxial 4H-SiC film deposition by CVD was started after pretreatment involving etching of the wafer surface by keeping it in H_2 gas flow at high temperature. The etched thickness was varied in the range of 0.1-0.7μm.

Deposited epitaxial films were etched in molten KOH. The shapes and densities of etch pits were observed with a laser microscope. To avoid the effect of cut-induced defects we observed the center area of the wafer.

Results and discussion

On the epitaxial films deposited after 0.1 μm of gas etching, we observed DAs, TRDs and TRSFs. The DA shown in Fig. 1(a) is composed of threading dislocations and basal plane dislocations (BPDs) and is attributed to scratching during the mechanical treatment [1]. The TRD is composed of BPDs and forms a large (over 100 μm) triangular structure, as shown in Fig. 1(b), and is explained as originating from the dust from the chemical vapor deposition system [2]. The TRSF has a simpler and smaller structure than the TRD (Fig. 1(c)). The TRSF is composed of one TD and two BPDs that form a triangle. Sometimes the TRSF is composed only of two BPDs. The TRSF is considered to extend from the basal plane dislocation in the 4H-SiC substrate [3].

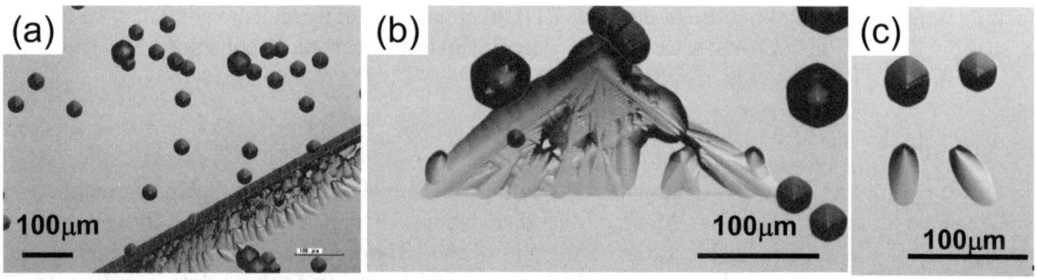

Figure 1 Enlarged etch pit images of (a) dislocation array, (b) triangular defect, and (c) triangular stacking fault. [11-20] direction is downward.

Variations in the densities of TRDs, TRSFs, and DAs on the epitaxial film depending on the removed thickness (0.1−0.7μm) of the wafer surface are summarized in Fig. 2. The densities of TRDs, TRSFs, and DAs decreased with increasing removed thickness. This indicates that shallow dislocations form not only DAs but also TRDs and TRSFs. Figure 3 shows an extreme example where the increase in etched thickness decreases the number of triangular defects (TRD and TRSF) on sample A. There are a great many triangular defects on the epitaxial film deposited after 0.1 μm of gas etching but only a few TRSFs are observed on the film deposited after 0.7 μm gas etching.

The shallow dislocation density on the wafer can be estimated as the densities of DAs, TRDs, and TRSFs on epitaxial film deposited without pre-etching if we accept that DAs, TRDs, and TRSFs originate from shallow defects. In this study, we estimated that the defect density on epitaxial film after 0.1 μm of gas etching approximately indicates the shallow-dislocation density on the wafer. The densities of DAs, TRDs, and TRSFs were dependent on the manufacturer of the wafer, as shown in Fig. 2. In the case of sample A, the defect density was too high to count. The DA density decreases in the order of B>C>D, but the TRD and TRSF densities were B>D>C.

Similar to the defect density, the critical thickness of gas-phase etching that eliminates the DA, TRD, and TRSF varied depending on the substrate manufacturer (Fig. 2). The DA density in samples C and D became 0 cm^{-2} after pre-etchings of 0.7 μm and 0.4 μm, respectively, but the DA density was 9 cm^{-2} for sample B after 0.7 μm of etching. Therefore, defects presumably caused by scratching form DAs in sample D and are the shallowest. The distributions of defects related to DAs in samples become deeper in the order of D<C<B. The TRD densities became 0 cm^{-2} after etching of 0.4 μm and 0.7 μm in samples D and C, respectively. However, TRD densities were 22 and 70 cm^{-2} for samples A and B, despite pre-etching of 0.7μm. Therefore, the defects related to TRDs in samples are distributed deeper in the order of D<C<A, B. It should be noted that dust from CVD parts [2] was a trivial factor in the formation of TRDs in this study because the epitaxial growth was simultaneously performed by putting different types of wafer in the CVD chamber to avoid the effects of differences in the CVD conditions.

On the other hand, the density of TRSFs decreased with increasing etched thickness but did not reach 0 cm^{-2} on any wafer despite the 0.7 μm of etching. This indicates that the remaining defects related to TRSFs are too deep to eliminate by pre-etching of 0.7 μm or that defects originating from the SiC crystal also form TRSFs. In fact, several researchers have pointed out that the TRSF is formed from the BPD. The TRSF density on 0.7 μm-etched wafers was D (36 cm^{-1})>B (26 cm^{-1})>A (14 cm^{-1})>C (9 cm^{-1}). To check the depth of defects related to remaining TRSFs, we polished sample C and removed about several tens μm. The TRSF density in epitaxial film on the repolished sample C was 7 cm^{-2}. This indicates that the origins of the remaining TRSFs were very deep, thus the defects can be attributed to defects in the SiC crystal. Figure 4 shows the relationship between the TRSF density after pre-etching of 0.7 μm and the conventional etch pit (hexagonal and seashell pits) density of wafers. The remaining TRSF density decreased with the etch pit density of wafers. This supports that defects in the SiC crystal cause TRSFs. It means that optimizing the polishing and pretreatment of CVD cannot completely eliminate TRSFs. There is not enough space to discuss what type of dislocation forms TRSFs, but it should be noted that eliminating the screw-type basal plane dislocation parallel to [11-20] in the 4H-SiC wafer is a key to achieving epitaxial film containing no TRSFs.

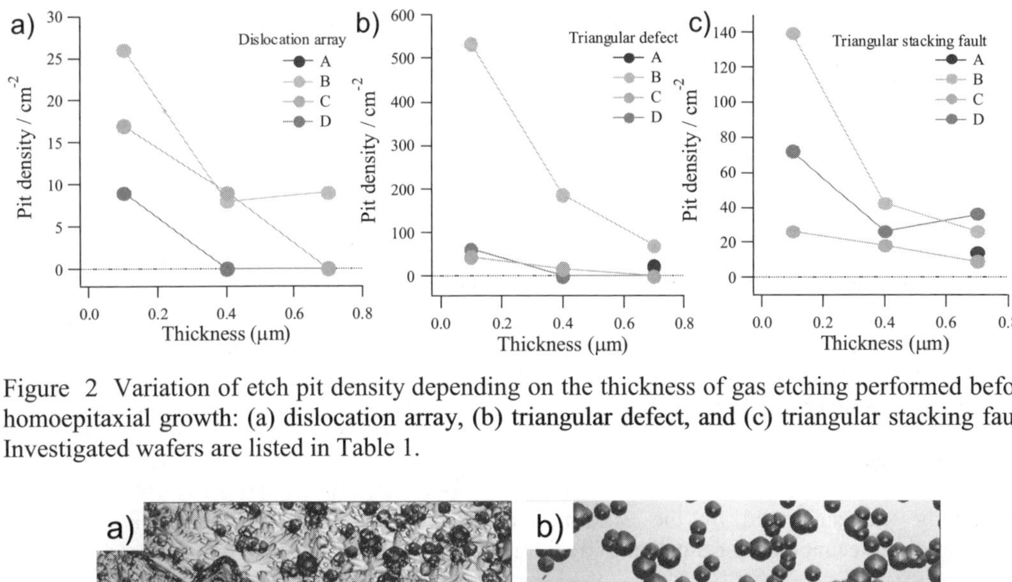

Figure 2 Variation of etch pit density depending on the thickness of gas etching performed before homoepitaxial growth: (a) dislocation array, (b) triangular defect, and (c) triangular stacking fault. Investigated wafers are listed in Table 1.

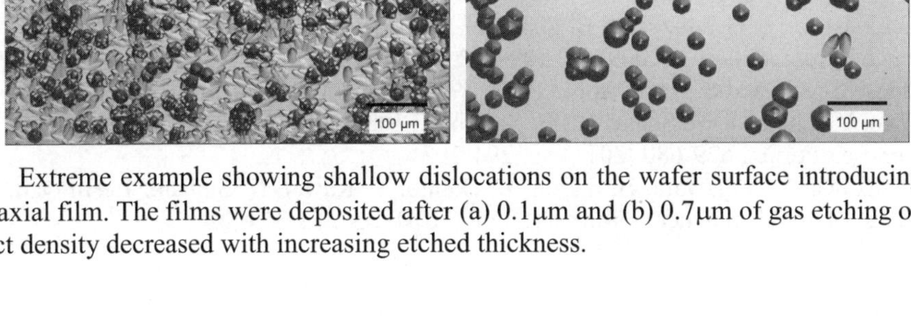

Figure 3 Extreme example showing shallow dislocations on the wafer surface introducing defects into epitaxial film. The films were deposited after (a) 0.1μm and (b) 0.7μm of gas etching on sample A. Defect density decreased with increasing etched thickness.

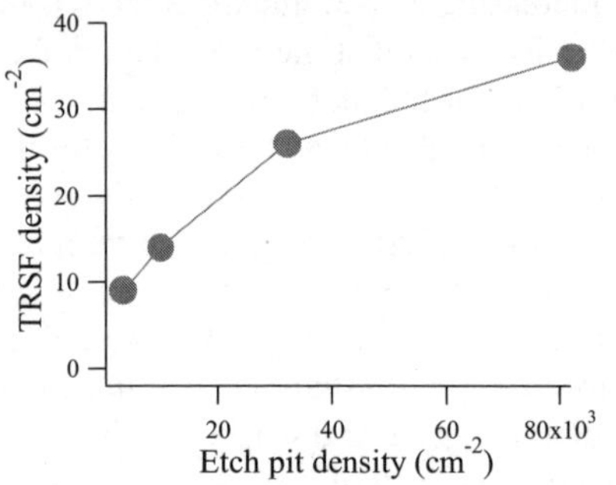

Figure 4 Dependence of TRSF density on etch pit density

To estimate the surface quality of the wafer, we etched wafers in solution of KOH and Na_2O_2 [4]. Round pits and scratches were observed. Scratches were generally caused by polishing and round pits were estimated to be shallow half-loop-type dislocations [5]. Accurate density counting is difficult because the densities of scratches and shallow dislocations vary with depth. However, DA density increased with the scratch density. Similarly, the TRD and TRSF densities increased with increasing round pit density on the wafers. This suggests that TRDs and TRSFs originate from shallow half-loop dislocations [5].

As described above, the shallow-dislocation density depended on the wafer manufacturer. In addition, the lowest shallow dislocation density was not obtained by CMP but by MP. Generally, CMP is performed to obtain a flat surface after MP and is milder than MP. Therefore, we hypothesize that this result derivers from the quality of MP before CMP.

Next, we describe what a good surface treatment is. The surface on which epitaxial film is easily deposited should contains no DA, TRD or TRSF. However shallow defects cannot be avoided, according to the above results, but there is no problem if they are shallow enough to be removed by treatment prior to CVD in low cost. From the viewpoint of defect shallowness, our revised mechanical polish seems to be the most appropriate.

Conclusion

It was demonstrated that serious defects on 4°-off 4H-SiC epitaxial films, that is dislocation arrays, triangular defects, and triangular stacking faults, originate from shallow defects on the 4H-SiC wafer. The dislocation arrays and triangular defects in epitaxial film were eliminated by proper mechanical polishing and 0.4 μm gas-etching before CVD. Triangular defects related to the shallow defects can be eliminated, but these related to crystal defects remained. The density and depth distribution of shallow defects on the wafers were dependent on the wafer manufacturers. The best quality of the surface was obtained by not CMP but MP.

Acknowledgement

This study was supported by the Subsidies for Projects to Promote the Enhancement of Manufacturing Technology for Small and Medium Enterprises (SMEs).

References

[1] N. Zhang, Y. Chen, E. K. Sanchez, D. R. Black, and M. Dudley: Materials Science Forum, 615-617 (2009) pp. 139-142.

[2] A. Shrivastava, P. G. Muzykov, B. Pearman, S. Michael Angel, and T. S. Sudarshan: Materials Science Forum, 600-603 (2009) pp. 109-142.

[3] M. Skowronski, and S. Ha: J. Appl. Phys. 99 (2006) 011101-1 -011101-24.

[4] Y. Yao, Y. Ishikawa, Y. Sugawara, H. Saitoh, K. Danno, H. Suzuki, Y. Kawai, and N. Shibata: Mater. Sci. Forum, 679-680 (2011) pp. 294-297.

[5] Y. Ishikawa, Y. Sugawara, H. Saitoh, K. Danno, Y. Kawai, N. Shibata, T. Hirayama, and Y. Ikuhara: Materials Science Forum, 645-648 (2010) pp. 351-354.

Materials Science Forum Vols. 717-720 (2012) pp 387-390
© (2012) Trans Tech Publications, Switzerland
doi:10.4028/www.scientific.net/MSF.717-720.387

Effect of Stacking Faults Originating from Half Loop Arrays on Electrical Behavior of 10 kV 4H-SiC PiN Diodes

Robert E. Stahlbush[1,a], Qingchun (Jon) Zhang[2], Anant Agarwal[2], Nadeemullah A. Mahadik[1]

[1]Naval Research Laboratory, 4555 Overlook Ave, S.W., Washington, D.C. 20375, USA

[2]Cree, Inc., 3026 E Cornwallis Rd., Research Triangle Park, NC 27709, USA

[a]stahlbush@nrl.navy.mil

Keywords: PiN degradation, half-loop array, stacking fault, basal plane dislocation, BPD

Abstract. The effects of Shockley stacking faults (SSFs) that originate from half loop arrays (HLAs) on the forward voltage and reverse leakage were measured in 10 kV 4H-SiC PiN diodes. The presence of HLAs and basal plane dislocations in each diode in a wafer was determined by ultraviolet photoluminescence imaging of the wafer before device fabrication. The SSFs were expanded by electrical stressing under forward bias of 30 A/cm^2, and contracted by annealing at 550 °C. The electrical stress increased both the forward voltage and reverse leakage. Annealing returned the forward voltage and reverse leakage to nearly their original behavior. The details of SSF expansion and contraction from a HLA and the effects on the electrical behavior of the PiN diodes are discussed.

Introduction

It is well established that Shockley stacking faults (SSFs) that expand from basal plane dislocations (BPDs) in SiC devices can cause an upward drift in the forward voltage [1,2]. This degradation arises in any device that has electron-hole recombination during its operation such as in bipolar devices. It has also been proposed that the stacking faults increase reverse bias leakage [3]. Two aspects of this work lead to a more definitive connection between the SSFs originating from BPDs and increased leakage under reverse bias. First, the wafer was screened using ultraviolet photoluminescence (UVPL) imaging before fabricating 10 kV, 1 A PiN diodes. The UVPL images allow a clear identification of dies with and without BPDs as well as with and without half-loop arrays (HLAs), which are complicated extended defects containing many BPD segments. Second, after expanding the stacking faults during device operation, the SFs were collapsed by annealing and the reverse leakage returned to nearly its original low value before the electrical stress.

Experimental Details

The SiC PiN diode was made on a 125 μm-thick SiC drift layer with $N_d = 5\text{-}6\times10^{14}$ cm^{-3}. Aluminum ion-implantation was used to form the P$^+$N junctions. Multiple-zone junction termination extension (JTE) was implanted with boron around the device periphery. After making alignment marks, but prior to the implantation and metallization, whole wafer UVPL mapping was performed [4], which shows extended defects in the epilayer that can be registered to the device map. Performing the UVPL imaging after creating the alignment marks makes it possible to directly compare the electrical characteristics and BPDs present within the active area of the device. This wafer had a high yield of devices with leakage less than 40 nA at 9 kV. Two diodes were chosen to examine the effect of SSFs on the forward and reverse characteristics of the diodes. One contained half-loop arrays (HLAs), an extended defect that contains many small, ~ 1 μm long, BPD segments [5]. The other was a control without BPDs or HLAs. As shown below, the SSF that originates from these BPD segments expands to form a large stacking fault that spans from the substrate to the anode. The PiN diodes were electrically stressed under forward bias of 1 A DC (~30 A/cm^2) up to 12 hours. Electrical behavior was recovered to pre-stress state of the device upon annealing at 550°C/3hrs, which causes the SSFs to contract [6].

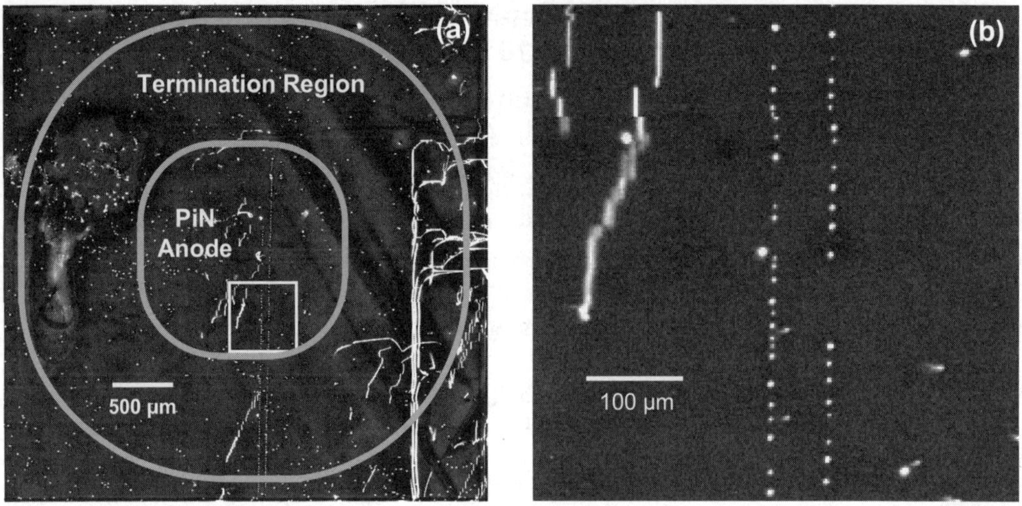

Fig. 1. Ultraviolet photoluminescence (UVPL) images of the area of a PiN diode containing two half-loop arrays (HLAs) taken before device fabrication. The anode and termination regions are shown in (a), and (b) is and a magnified view of the square inside the anode region shown in (a). The two vertical strings of dots are the HLAs.

Results and Discussion

UVPL images of the drift layer of the diode with two HLAs are shown in Fig. 1. Each of these HLAs originates from a single BPD that glides during epitaxial growth [7]. As it glides it leaves a series of short BPD segments and each of these segment turns into a half loop as the epitaxial growth continues [8]. A schematic of the diode area is superimposed on the UVPL image of that area in Fig. 1(a). The square area inside the anode is magnified in Fig. 1(b) to more clearly see the HLAs. Each dot in the two rows of vertical dots is a half loop and the bottom part of each half loop is a BPD segment. Note that these strings of dots extend completely across the anode. The other dots scattered over the die are threading dislocations and do not produce SSFs. There are several vertical lines, which are also dislocations associated with the HLAs, but they do not fault and consequently do not contribute to electrical changes of the PiN diode.

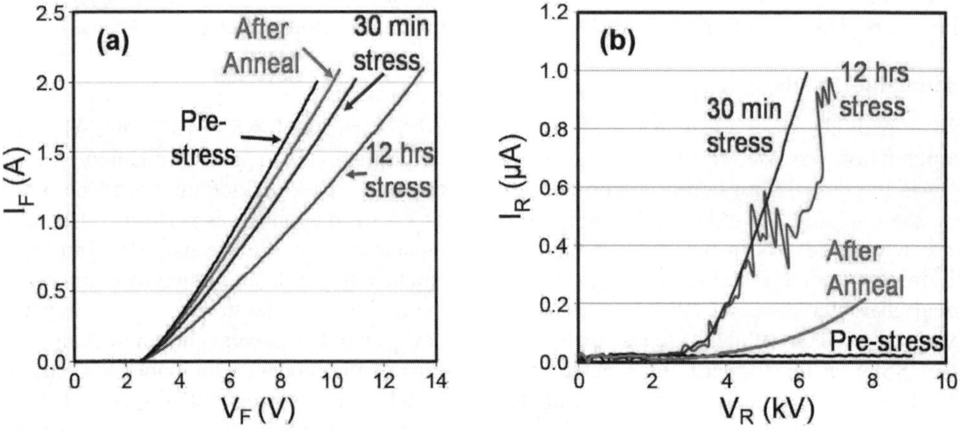

Fig. 2. Pulsed I-V measurements at room temperature of (a) the forward conduction and (b) the reverse leakage of the PiN diode shown in Fig. 1. The anode and termination areas are 3.4 mm^2 and 8.6 mm^2. The forward biased stressing was 30 A/cm^2 and the annealing was at 550 °C.

The changes to the forward and reversed biased behavior of the PiN diodes due to stressing at 30 A/cm^2 show the effects of the expanding and contracting SSF originating from the HLAs. In Fig. 2, the electrical behavior of the diode containing HLAs is shown. The I-V curves were measured after 30 min and 12 hr of stressing. After each stressing the diodes were cooled to room temperature and pulsed I-V curves were taken. Stressing causes the familiar forward voltage upward drift. With annealing the forward voltage returns to nearly its original curve as has been previously observed [6]. The effects of electrical stressing and annealing on the reverse leakage follow the same behavior. It degrades with electrical stress as previously reported [3]. This work also shows that annealing returns the diode to nearly its original leakage curve, which confirms that the SSF introduces a leakage path through the drift layer. As a control, the diode free of BPDs and HLAs was also subjected to the same electrical stress and annealing. Its forward voltage I-V and reverse leakage I-V curves remained unchanged during the stressing and annealing sequence.

The details of the SSF that develop from a HLA explain the reversible behavior of the forward voltage and reverse leakage. The expanding SSF that originates from the BPD segments of the HLA is illustrated in Fig. 3. Frames (a) - (d) are UVPL images from a different wafer in which the SSF expansion was driven by extended UV exposure and recorded as it expanded. Both UV exposure and current injection in PiN diodes introduce excess electrons and holes, and the recombination of these excess carriers drives the SSF expansion [9]. Each dot is from a half loop and the bottom of the half loop is a BPD segment [5]. Frame (a) is at the initial expansion and the two HLAs still appears as two vertical rows of dots. Frame (e) is a cross-sectional schematic with the location of the BPD segments. This frame illustrates that initially they are very localized and unlikely to provide a leakage path through the epi. In frames (b) - (d), the SSFs expand and coalesce. Frames (f) - (h) show their expansion along basal planes as they extend towards the substrate and towards the top of the epi or anode layer in the case of the PiN diodes. This expansion continues until the SSFs fill the basal plane between the substrate and top of the epi. Contraction of the SSF after expansion is illustrated by the UVPL images in Fig. 4. Frames (a) - (d) show the early stage of expansion before the individual SSFs coalesce as shown in Fig.3. As described elsewhere, the conditions are changed to cause the contraction in frames (e) - (h) back to the original HLA [10]. This demonstrates that the SSFs originating from the BPD segments of the HLAs are the source of the reverse leakage displayed in Fig. 2. As shown in Fig. 1, HLAs commonly extend across the active device area under the anode and the fully expanded SSF forms a leakage plane spanning from the bottom to the top of the diode active area. The expanding SSFs are also responsible for the forward voltage increase.

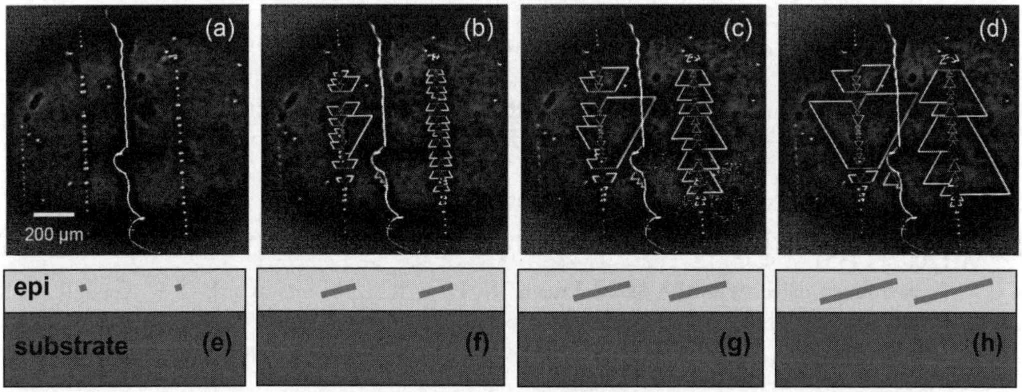

Fig. 3. UVPL images and schematics of HLA expansion. The plan-view UVPL image (a) shows two HLAs before faulting and (b) – (d) show the expanding stacking faults. The cross-sectional schematics (e) – (f) illustrate how the expanding stacking faults eventually completely span the epitaxial layer.

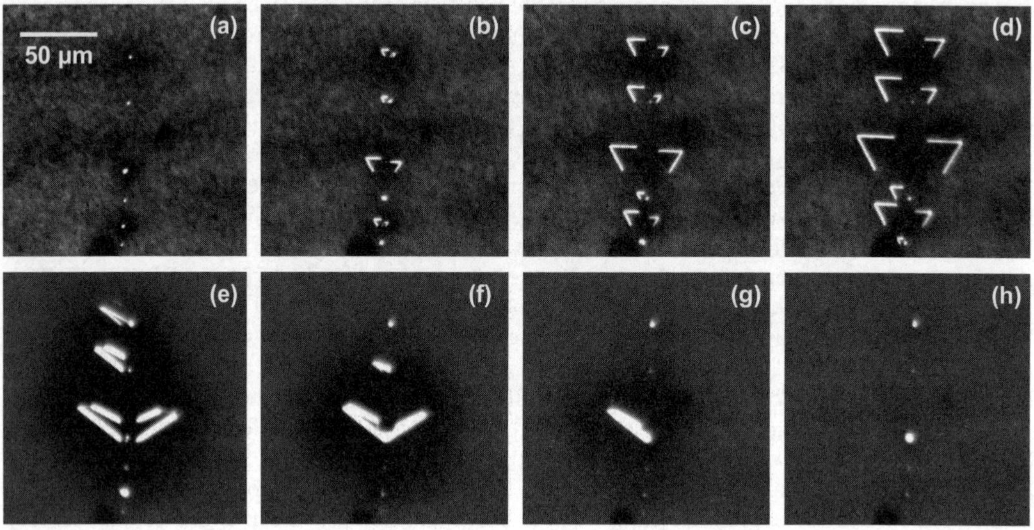

Fig. 4. UVPL images of HLA expansion and contraction. In (a) – (c) stacking faults from the BPD segments in each half-loop are expanding. The same stacking faults are contracting back to the original BPD segments in (d) – (f).

Conclusion

The expansion and contraction of Shockley stacking faults (SSFs) from half-loop arrays (HLAs) in 10 kV PiN diodes demonstrates that SSFs create a leakage path through the PiN drift region. Forward bias stressing expands SSFs across the PiN drift region, which significantly increases reverse leakage and forward voltage. Annealing collapses the SSFs and returns the reverse leakage and forward voltage to nearly their original behavior.

References

[1] H. Lendenmann, F. Dahlquist, N. Johansson, R. Söderholm, P.A. Nilsson, J.P. Bergman, and P. Skytt, Mater. Sci. Forum 353–356 (2001) 727.

[2] J.P. Bergman, H. Lendenmann, P.A. Nilsson, U. Lindefelt, and P. Skytt, Mater. Sci. Forum 353-356 (2001) 299.

[3] A. Agarwal, H. Fatima, S. Haney, and S.-H. Ryu, IEEE Elec. Dev. Lett. 28 (2007) 587.

[4] R. E. Stahlbush, K.X. Liu, Q. Zhang, and J. J. Sumakeris, Mater. Sci. Forum 556-557 (2007) 295.

[5] S. Ha, H.J. Chung, N.T. Nuhfer, M. Skowronski, J. Cryst. Growth 262 (2004) 130.

[6] J. D. Caldwell, R. E. Stahlbush, K. D. Hobart, O. J. Glembocki, and K. X. Liu, Appl. Phys. Lett. 90 (2007) 143519.

[7] R. E. Stahlbush, B. L. VanMil, K. X. Liu, K. K. Lew, R. L. Myers-Ward, D. K. Gaskill, C. R. Eddy, X. Zhang, and M. Skowronski, Mater. Sci. Forum 600-603 (2009) 317.

[8] N. Zhang, Y. Chen, Y. Zhang, M. Dudley, and R. E. Stahlbush, Appl. Phys. Lett. 94 (2009) 122108.

[9] A. Galeckas, J. Linnros, and P. Pirouz, Appl. Phys. Lett. 81 (2002) 883.

[10] N. A. Mahadik, R. E. Stahlbush, J. D. Caldwell, and K. D. Hobart, these Proceedings.

Materials Science Forum Vols. 717-720 (2012) pp 391-394
doi:10.4028/www.scientific.net/MSF.717-720.391

Ultraviolet Photoluminescence Imaging of Stacking Fault Contraction in 4H-SiC Epitaxial Layers

Nadeemullah A. Mahadik[1, a], R. E. Stahlbush[1], J. D. Caldwell[1], and K. D. Hobart[1]

[1]Naval Research Laboratory, 4555 Overlook Ave., Washington, DC 20375, USA

[a] Email: mahadik@ccs.nrl.navy.mil

Keywords: UVPL Imaging, Stacking fault contraction, Basal Plane Dislocations

Abstract: Shockley stacking fault (SSF) contraction in 4H-SiC was investigated, *in-situ*, under varying temperature and ultraviolet (UV) intensity. Contraction of single SSFs at room temperature was observed for the first time under low power UV excitation of 0.04 W/cm^2. At temperatures above 150 °C, complete SSF contraction occurred for UV power at 0.2 W/cm^2. In contrast to expansion, SSF contraction occurred in discrete jumps between pinning sites along existing C-core partials. Luminescence from the pinning sites suggest they may be local concentrations of point defects. Additionally, a change in the line direction of the Si-core partials by ~25° off the <11$\bar{2}$ 0> direction was observed.

Introduction

Silicon carbide (SiC) power devices have forward voltage degradation due to the expansion of Shockley stacking faults (SSF) that originate at basal plane dislocations (BPD) [1,2]. The kinematic constraints that normally restrict dislocation motion to above 900°C are overcome by the recombination enhanced dislocation glide (REDG) mechanism [3]. As long as conditions favor the REDG process, thermodynamic driving forces govern SSF expansion or contraction. Previously, SSF expansion was observed and driving force models addressed this [4-8]. Contraction of multiple SSFs was observed recently using electroluminescence (EL) imaging [9], after high temperature annealing, and also by using EL imaging under concurrent low current bias and heating at 210 °C for 1 hour. Based on the EL studies, and simulations, Caldwell et al. suggested a driving force model that also accounted for SSF contraction [10]. However, details of the SSF contraction and their bounding partials were not addressed. In this work, we investigate the contraction of individual non-interacting SSFs, *in-situ*, under varying temperature (25 °C to 200 °C) and UV laser power (0.04 - 0.2 W/cm^2).

Experimental Procedure

For this work, low n-type doped, 100μm and 20μm thick, 4H-SiC epilayers, commercially grown on 8° offcut 4H-SiC substrates were used. A thermal stage with a Minco heater was used to vary the sample temperature from 25 to 200 °C. UV photoluminescence (UVPL) measurements were carried out using an Ar ion UV laser, with principle lines at 364 nm and 351 nm. A variable UV attenuator was used to control the laser power. UVPL images from the faulted regions were collected in the 600-1000 nm emission range using a CCD camera mounted on a probe station with a 50x objective lens. Using this technique, we obtained real time images to investigate the SSFs and the motion of their bounding partials under a variety of injection and thermal conditions.

Results and Discussion

Real time imaging of individual SSF contraction was found to be different from a time-reversed expansion pattern. After expanding individual SSFs from BPDs, as seen in Fig. 1(a), using higher UV power exposures, the same region was irradiated using a significantly lower UV power of 0.04-0.2 W/cm^2 for ~1700 s. Several images were captured during this process to investigate the movement of the BPD partials and the bounding SSFs. This exposure resulted in the partial contraction of the SSFs as shown in Figs. 1 (b)-(d). These images show direct evidence that SSF contraction is possible even at room temperature. Upon the first low power UV illumination for 10 s [Fig. 1(b)], we observe that the previously expanding Si-core partial changes its line direction ~25° from the original <11$\bar{2}$0> direction.

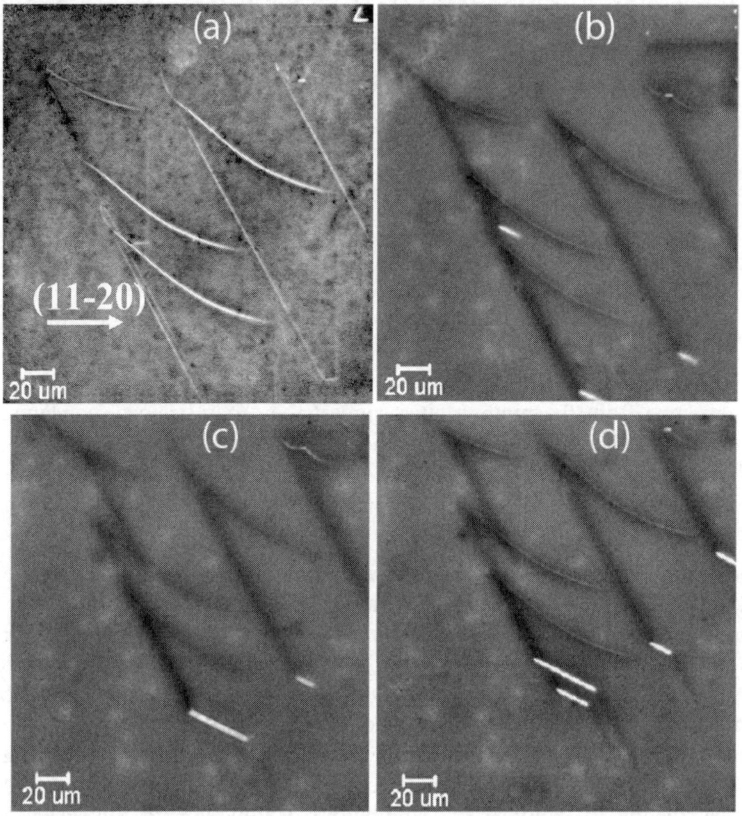

Fig. 1: Room temperature UVPL images of the 20μm thick sample (a) after high power UV exposure with expanded SSFs, (b) Initial image after low power UV exposure of 10 s, (c) SSF contraction after 400 s of low power UV exposure, (d) UVPL image showing partial SSF contraction after 1700 s of low power UV exposure.

This change in line direction was the same for the entire contraction process. In addition, the SSF contraction also occurred in discrete jumps of the Si-core partials, which occur at a random pace during the UV exposures. After a few of these discrete jumps, further SSF contraction did not occur even after long duration (overnight) low power UV exposures.

The sample temperature was increased to 150 °C, and an individual SSF in the 100 μm SiC sample was expanded using higher power UV exposures, shown in Fig. 2(a) after the expansion process. The expanding Si-core partial is seen to lie in the <11$\bar{2}$0> direction during the expansion as expected [11]. This region of the sample was then irradiated with lower UV power at 0.2 W/cm^2, and complete SSF contraction was obtained, as shown in Figs. 2(b)-(e). The contraction behavior was similar to that observed at room temperature, for which discrete jumps of the Si-core partials were observed. In these images, the jumps appear to occur at luminescent centers along the existing C-core partials. These centers, encircled in Figs. 2(c) and 2(d), show a broad luminescence spectrum at 600-800 nm, which is different from the C-core spectrum [12]. It suggests that they are gettering centers for point defects that are known [13] to act as dislocation pinning sites, thereby causing the discrete jumps of the Si partials. Additionally, a change in the line direction of the Si-core partials ~25° off the <11$\bar{2}$0> direction is also seen in these UVPL images for the elevated temperatures. The Burgers vector for the Si-core partial is known to be 30° off the <11$\bar{2}$0> direction

[11]. So the line direction during the contraction nearly aligns with its Burgers vector, which is characteristic of a screw dislocation. In order for the Si core partial to stay mobile, it has to remain below 30° during the SSF contraction process. Beyond that angle it becomes a C-core partial, which is immobile [14]. The change in the line direction seems consistent for

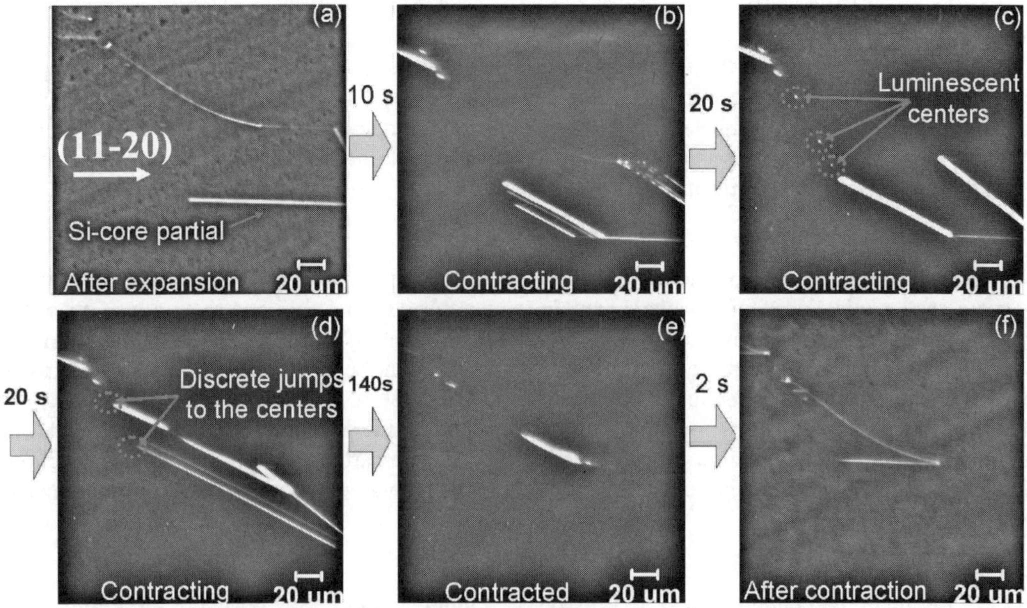

Fig. 2: UVPL images of 100 μm thick 4H-SiC sample at T=150 °C, (a) of SSF in the SiC epilayer after expansion at high UV power (9 W/cm^2), (b) Initial UVPL image after 10s low power (0.2 W/cm^2) UV exposure, (c) after 30s total low power UV exposure showing the luminescent centers along the existing C-core partial, (d) after 50s total low power UV exposure, showing discrete jumps to the luminescent centers, (e) fully contracted SSF after 190 s total low power UV exposure. (f) shows UVPL image to verify the SSF contraction using high UV power exposure.

all contraction conditions, however its cause appears to be quite complex and not clear. After the complete SSF contraction, a UVPL image at higher UV power was taken to verify the contraction, seen in Fig. 2(f). After the contraction, the SSF had completely collapsed to the initial perfect BPD. However, upon a short higher power exposure to obtain the image, the Si-core partial immediately faults to lie in the minimum energy $<11\bar{2}0>$ direction [15] and then expands while in that direction. This image can be compared with Fig. 2(a) and the Si core partial can be seen to have contracted back to the point where the perfect BPD lies.

The effect of variation in the UV power and temperature on the SSF contraction was also studied. Enhancement in the rate at which the Si-core partial jumps causing the SSF contraction is seen with either reduction in UV power or increase in the sample temperature, which is consistent with the contraction model [10]. However, the details of the contraction are not explained by the model. UVPL images revealed that at a particular temperature when the power was maintained for a series of exposures, the SSFs continue to contract and reach a threshold where no further contraction occurs. When the UV power is further reduced, the SSF contraction proceeds at a much faster rate until another threshold is reached or complete SSF contraction occurred. However, for temperatures below 150°C, only partial SSF contraction can be seen even after further lowering of the UV power to 0.04 W/cm^2 for a long duration. This is probably because the density of excess carriers generated by the very low UV power is not sufficient for the REDG process to occur and enable SSF motion. The sample temperature has greater influence on the SSF contraction. Upon increasing the sample temperature to 150°C, complete SSF contraction can be seen even for UV power as high as 0.3 W/cm^2. The UV exposure introduces excess carriers in the SiC epilayers with the carrier concentration proportional to the UV power. Hence, at lower power the carrier density is lower. A previous SSF contraction driving force model [10] suggests that carrier injection at a particular level introduces a quasi-Fermi level for the excess electrons, which is near the conduction band. The SSF has a quantum level ~0.3 eV below the conduction band [16]. The model suggests that if the quasi-Fermi level of the injected carriers lies above the SSF level, SSF expansion will occur and if the quasi-Fermi level is below the SSF level, contraction is favored provided there are sufficient carriers to enable the REDG process. Fermi levels can be lowered towards midgap by either reducing the extrinsic carrier density or increasing the sample temperature. From Fermi statistics, the Fermi level is linearly related to the sample temperature, whereas the carrier density has a logarithmic effect. This is consistent with our observations where by increasing the sample temperature has a more significant effect on SSF contraction.

Conclusions

We have investigated the contraction behavior of individual SSFs in 4H-SiC in real time under varying carrier injection conditions and sample temperature. Partial SSF contraction was observed even at room temperature and UV power of 0.04-0.2 W/cm^2. For complete SSF contraction the sample temperature was increased to 150 °C. During the contraction process, the Si-core partials exhibit discrete jumps between luminescent points on the existing C-core partials. The data suggest that they are gettering centers for point defects that impede the partial dislocation movement. In addition, change in line direction was observed for the Si-core partial ~25° off the $<11\bar{2}0>$ direction. The sample temperature had more significant effect on SSF contraction compared to carrier density, which is consistent with previous SSF driving force models.

References

[1] J.P. Bergman, H. Lendenmann, P.A. Nilsson, U. Lindefelt, and P. Skytt, Mater. Sci. Forum 353-356 (2001) 299

[2] R.E. Stahlbush, J.B. Fedison, S.D. Arthur, L.B. Rowland, J.W. Kretchmer, and S. Wang, Mater. Sci. Forum 389 (2002) 427

[3] A. Galeckas, J. Linnros, and P. Pirouz, Phys. Rev. Lett. 96 (2006) 025502

[4] J.Q. Liu, M. Skowronski, C. Hallin, R. Soderholm, and H. Lendenmann, Appl. Phys. Lett. 80 (2002) 749

[5] P.O.Å. Persson, L. Hultman, H. Jacobson, J. P. Bergman, E. Janzén, J.M. Molina-Aldareguia, W. J. Clegg, and T. Tuomi, Appl. Phys. Lett. 80 (2002) 4852

[6] M. E. Twigg, R. E. Stahlbush, M. Fatemi, S. D. Arthur, J. B. Fedison, J. B. Tucker, and S.Wang, Appl. Phys. Lett. 82 (2003) 2410

[7] W. R. L. Lambrecht and M. S. Miao, Phys. Rev. B, 73 (2006) 155312

[8] S. Ha, M. Skowronski, J.J. Sumakeris, M.J. Paisley, and M.K. Das, Phys. Rev. Lett. 92 (2004) 175504

[9] J.D. Caldwell, R.E. Stahlbush, K.D. Hobart, O.J. Glembocki, and K.X. Liu, Appl. Phys. Lett. 90 (2007) 143519

[10] J.D. Caldwell, R.E. Stahlbush, M.G. Ancona, O.J. Glembocki, and K.D. Hobart, J. Appl. Phys. 108 (2010) 044503

[11] P. Pirouz, J. L. Demenet, and M. H. Hong, Philos. Mag. A 81 (2001) 1207

[12] K.X. Liu, X. Zhang, R.E. Stahlbush, M. Skowronski, and J.D. Caldwell, Mater. Sci. Forum, 600-603 (2009) 345

[13] B. Chen, H. Matsuhata, T. Sekiguchi, T. Ohyanagi, A. Kinoshita, and H. Okumura, Appl. Phys. Lett. 96 (2010) 212110

[14] A. T. Blumenau, C. J. Fall, R. Jones, S. Oberg, T. Frauenheim, and P. R. Briddon, Phys. Rev. B, 68 (2003) 174108

[15] R. E. Stahlbush, B. L. VanMil, R. L. Myers-Ward, K. K. Lew, D. K. Gaskill, and C. R. Eddy Jr., Appl. Phys. Lett. 94 (2009) 041916

[16] M. S. Miao, S. Limpijumnong, and W. R. L. Lambrecht, Appl. Phys. Lett. 79 (2001) 4360

Materials Science Forum Vols. 717-720 (2012) pp 395-398

doi:10.4028/www.scientific.net/MSF.717-720.395

Photoluminescence Study of the Driving Force for Stacking Fault Expansion in 4H-SiC

Rii Hirano[1,2,a], Yuki Sato[3], Michio Tajima[1], Kohei M. Itoh[2], and Koji Maeda[3]

[1]Institute of Space and Astronautical Science, Japan Aerospace Exploration Agency, 3-1-1 Yoshinodai, Chuo-ku, Sagamihara, Kanagawa 252-5210, Japan

[2]School of Fundamental Science and Technology, Keio University, 3-14-1 Hiyoshi, Yokohama 223-8522, Japan

[3]Department of Applied Physics, The University of Tokyo, 7-3-1 Hongo, Bunkyo-ku, Tokyo 113-8656, Japan

[a]hirano@sce.isas.jaxa.jp

Keywords: 4H-SiC, photoluminescence, radiation-enhanced dislocation glide, stacking faults

Abstract. We investigated expansion velocities of Shockley stacking faults (SSFs) in 4H-silicon carbide under laser illumination using photoluminescence methods. The experiments showed that the velocity of SSF expansion or the glide velocity of SSF-bounding 30°-Si(g) partial dislocations (PD) is superlinearly dependent on the excitation intensity. We estimated the sample temperature by analyzing the broadening of band-edge emission and concluded that the lattice heating by laser illumination is not the cause of the enhanced dislocation glide. The superlinear dependence can be accounted for by a photo-induced sign reversal of the effective formation energy of the SSF acting as the driving force of SSF expansion under the illumination.

Introduction

4H-silicon carbide (4H-SiC) is a promising semiconductor for highly efficient power electronic devices due to its high breakdown voltage, wide band gap, and high thermal conductivity. However, after forward current injection, an increase of the forward voltage drop was observed in 4H-SiC pin diodes [1]. This degradation, a critical issue for their application, is caused by the expansion of Shockley stacking faults (SSFs) [2]. The phenomenon of the SSF expansion is induced not only by forward current injection, but also by illumination [3,4] and is caused by enhanced glide of 30°-Si(g) partial dislocations (PD) bounding the SSF. A similar effect, known as the radiation-enhanced dislocation glide (REDG), is observed in many other semiconductors [5]. Comparative studies of REDG in 4H-SiC and the other materials make it possible to elucidate the fundamental mechanism of the REDG, because the 30°-Si(g) PD has been identified well enough for the microscopic mechanism to be discussed in detail [6]. While usually the enhanced dislocation glide velocity is almost proportional to the excitation intensity, the present finding in photoluminescence (PL) experiments is that the velocity of SSF expansion is superlinearly dependent on this intensity. On the basis of the present results, a special characteristic of the driving force for SSF expansion in 4H-SiC is discussed.

Experiment

Sample preparation. The sample used was an 8° off-axis (0001) nitrogen doped 4H-SiC substrate with an epitaxial layer 5 μm in thickness. The nitrogen concentration in the epitaxial layer was 5×10^{15} cm^{-3}. Dislocation sources were intentionally introduced by scratching the sample surface with a diamond scriber.

PL measurement. We performed room-temperature PL mapping using the 266 nm line of a CW quadruple Nd:YVO_4 laser to excite the sample to a penetration depth of 1.2 μm. The excitation laser was focused down to a spot of about 2 μm diameter on the sample surface by a Cassegrain objective (18×). The SSF-related emission at 2.9 eV was selected by band-pass filters and its intensity was mapped on the sample surface. For systematic measurements of the excitation intensity dependence of the velocity of the 30°-Si(g) PDs that bound SSFs, first we scanned a sample area 2-dimensionally with a relatively high excitation intensity of 11 mW to expand the SSFs and simultaneously acquire their PL images. Next, for one of the SSFs, we repeated 1-dimensional sweeping of the sample with the 266 nm spot along the elongated direction of the SSF to induce its expansion and obtain a set of PL intensity profiles under various excitation intensities. In the single sweep, the passage time of the laser spot across a 30°-Si(g) PD, or the effective illumination time, was estimated from an effective carrier diffusion length divided by the sweeping speed of 100 μm/s. From the evolution of these profiles, we deduced the dependence of the 30°-Si(g) PD velocity on excitation intensity. In addition, we also measured PL spectra to estimate the sample temperature rise due to laser irradiation. In this measurement, the 266 nm line of the excitation laser was focused on the sample surface with the same intensity as used in the PL mapping. The PL spectra were obtained in the near band-edge region of 4H-SiC with a spectral resolution of about 3 meV.

Results and Discussion

Figures 1(a-c) show PL maps around the scratch. PL stripes showing SSFs were observed to emanate from the scratch and expand rightward in a <1-100> direction as laser sweeping was repeated. This phenomenon can be explained by the glide of 30°-Si(g) PDs bounding the SSFs from nucleation sites

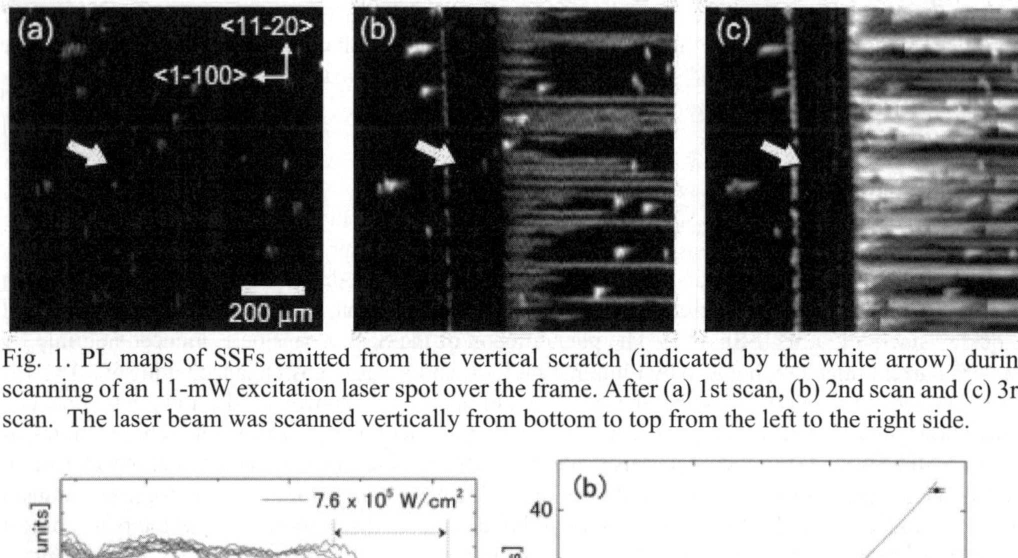

Fig. 1. PL maps of SSFs emitted from the vertical scratch (indicated by the white arrow) during scanning of an 11-mW excitation laser spot over the frame. After (a) 1st scan, (b) 2nd scan and (c) 3rd scan. The laser beam was scanned vertically from bottom to top from the left to the right side.

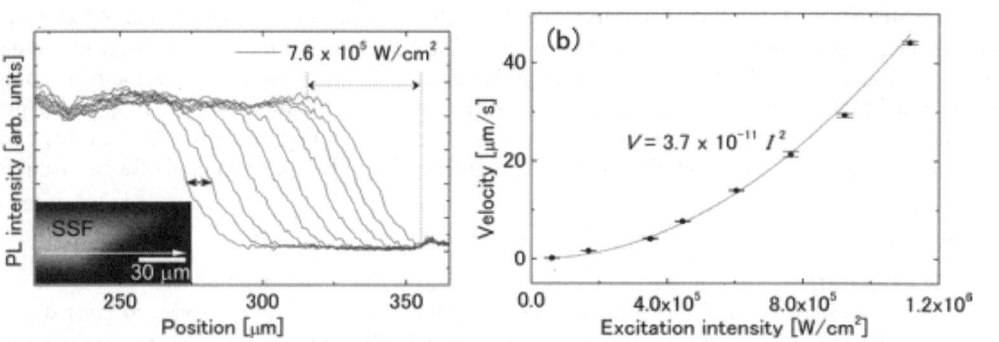

Fig. 2. (a) Evolution of SSF-related PL profiles repeatedly measured along the white arrow in the inset, which shows a PL map of a SSF for which the advancing velocity of a 30°-Si(g) partial dislocation bounding the SSF was measured. The double-headed solid and dotted arrows indicate a length of the SSF expansion per scan and effective carrier diffusion length, respectively. (b) Excitation intensity dependence of the 30°-Si(g) partial dislocation velocity. Dots and a solid line are experimental data and the fitted curve, respectively.

introduced by scratching the sample surface with the diamond scriber. We selected one of the expanded SSFs and obtained sets of PL profiles as shown in Fig. 2(a) under a fixed excitation intensity by 1-dimensional sweep with the laser spot. Figure 2(b) shows the excitation intensity dependence of the SSF expansion velocity or the PD velocity that was evaluated by the average length between the double-headed solid arrows in Fig. 2(a) divided by the effective illumination time of 0.35 s. The effective illumination time was estimated from the effective carrier diffusion length evaluated by the double-headed dotted arrow in Fig. 2(a) divided by the sweeping speed of 100 μm/s. A notable fact in Fig. 2(b) is that shows a superlinear dependence on the excitation intensity.

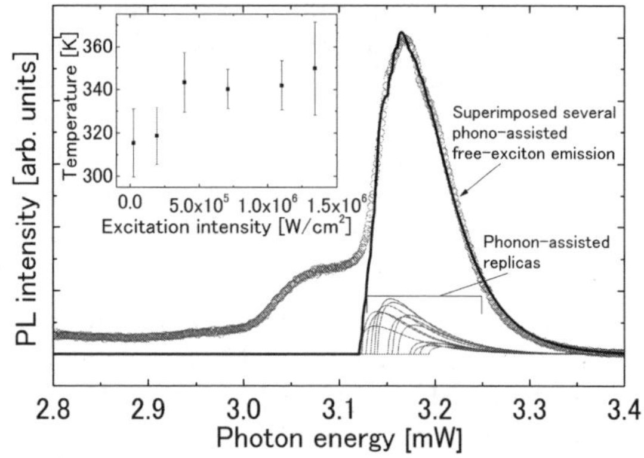

Fig. 3. A PL peak and phonon replicas composing free-exciton emission of 4H-SiC under 1.3×10^6 W/cm^2 excitation intensity. The dots and thick line are experimental results and the sum of the calculated curves shown as thin lines fitted to the experimental data, respectively. The inset shows the excitation intensity dependence of evaluated temperature from curve fitting analyses.

Since the glide of 30°-Si(g) PDs at high temperatures was reported [7,8], we investigated the excitation intensity dependence of sample heating in order to check whether the superlinear relationship in Fig. 2(b) simply arises from the heating by intense laser illumination. A PL spectrum of free-exciton emission of 4H-SiC without SSF [9] under 1.3×10^6 W/cm^2 excitation intensity is shown in Fig. 3. The spectral shape of the radiative recombination involving free excitons is given by a superposition of Maxwell-Boltzmann distributions [10] broadened with Gaussian functions [11] in a form

$$I_{PL}(E) \propto \sum_m \sqrt{E - E_m} \exp\left(\frac{E - E_m}{k_B T}\right) \times \exp\left(-\frac{E - E'_m}{2\sigma^2}\right), \tag{1}$$

where $I_{PL}(E)$ is the luminescence intensity at photon energy E, E_m is the peak threshold energy, E'_m is the peak energy, σ is the Gaussian line width, and T is the temperature. Equation (1) indicates that the full width at half maximum increases linearly with temperature [11]. Therefore, we estimated the sample temperature under laser illumination from a curve fitting using T as an adjustable parameter. The inset of Fig. 3 shows the estimated temperature under a variety of excitation intensities. As seen, the sample temperature actually increased with increase in the excitation intensity, but it did not exceed 375 K. According to direct velocity measurements of 30°-Si(g) PDs by Idrissi *et al.* [12] and indirect evaluation from thermal recovery of device degradation by Caldwell *et al.* [8], the activation energy of 30°-Si(g) PD glide in darkness is 1.35 ± 0.2 eV, from which we could predict that the thermally activated velocity at 375 K would be only 1×10^{-15} m/s, far lower than the velocity observed experimentally. Therefore, we conclude that the heating effect is negligible in the present study.

In ordinary REDG observed in other materials, the driving force for the dislocation glide is usually a shear stress internally present or externally applied [5]. In the case of 4H-SiC, however, SSFs expand in the absence of stress, which was explained by the sign reversal of γ_{SF}^*, the effective formation energy of a SSF, due to the capture of excess electrons into the quantum well associated with a SSF [13]. The difficulty of an electric charge-up could be avoided by taking into account the capture of charge-compensating holes as well [14]. Hence the driving 'stress' may be expressed by

$$-\frac{\gamma_{SF}^*}{b} = \frac{\{\alpha\rho(I) - \gamma_{SF}\}}{b}, \tag{2}$$

where b denotes the magnitude of the Burgers vector of the PD, γ_{SF} the stacking fault energy in thermal equilibrium, and I the excitation intensity. The first term in the numerator in Eq. (2) represents the effect of carrier trapping to a SSF, assumed proportional to ρ, the density of free excitons trapped by the SSF, which effectively lowers the γ_{SF} depending on I with an efficiency factor α. From the fact that the enhanced dislocation velocity in the ordinary REDG is proportional to the stress and the excitation intensity [5], we assume that the SSF expansion velocity V in 4H-SiC should depend linearly on the effective 'stress' under a fixed temperature and hence be given by

$$V \propto \{\alpha\rho(I) - \gamma_{SF}\} I \,. \tag{3}$$

The superlinear (approximately quadratic) relationship in Fig. 2(b) can be accounted for if one considers that $\gamma_{SF} << \alpha\rho(I) \propto I$. This strongly suggests that the effective formation energy of a SSF in 4H-SiC is reduced by an amount proportional to the excitation intensity acting as the driving force for enhanced glide of 30°-Si(g) PDs.

Summary

We performed systematic measurements of PL intensity profiles along the elongated direction of a SSF under various excitation intensities. In contrast to the linear dependence of glide velocity on excitation intensity in other materials [5], the velocity of a 30°-Si(g) PD in 4H-SiC exhibits a superlinear dependence on excitation intensity. Analysis of PL spectra of band-edge emission showed that the sample heating effect under an intense laser illumination is negligible. The superlinear, almost quadratic dependence strongly suggests that the reduction of effective formation energy of a SSF, which acts as the driving force of the SSF expansion, is proportional to the excitation intensity.

Acknowledgements

This work was supported in part by a Grant-in-Aid for the Global Center of Excellence for High-Level Global Cooperation for a Leading-Edge Platform on Access Spaces from the Ministry of Education, Culture, Sports, Science, and Technology in Japan.

References

[1] H. Lendenmann, F. Dahlquist, N. Johansson, R. Söderholm, P. A. Nilsson, J. P. Bergman, and P. Skytt: Mater. Sci. Forum Vol. 353-356 (2001), p. 727.

[2] J. P. Bergman, H. Lenenmann, P. Å. Nilsson, U. Lindefelt, and P. Skytt: Mater. Sci. Forum Vols. 353-356 (2001), p. 299.

[3] A. Galeckas, J. Linnros, and P. Pirouz: Phys. Rev. Lett. Vol. 96 (2006), 025502.

[4] N. Hoshino, M. Tajima, M. Naitoh, E. Okuno, and S. Onda: Mater. Sci. Forum Vols. 600-603 (2009), p. 349.

[5] K. Maeda and S. Takeuchi, Dislocations in Solids, Vol. 10, North-Holland, 1996. p. 444.

[6] M. Skowronski and S. Ha: J. Appl. Phys. Vol. 99 (2006), p. 011101.

[7] T. Miyanagi, H. Tsuchida, I. Kamata,T. Nakamura, K. Nakayama, R. Ishii, and Y. Sugawara: Appl. Phys. Lett. Vol. 89 (2006), p. 062104.

[8] J. D. Caldwell, O. J. Glembocki, R. E. Stahlbush, and K. D. Hobart: J. Electron. Mater. Vol. 37 (2008), p. 699.

[9] Y. Shishkin, R. P. Devaty, and W. J. Choyke: Mater. Sci. Forum Vol. 383-342 (2000), p. 679.

[10] A. Itoh, T. Kimoto, and H. Matsunami: Jpn. J. Appl. Phys. Vol. 35 (1996), p. 4373.

[11] I. Pelant, J. Dian, J. Matoušková, J. Valenta, J. Hála, M. Ambrož, M. Vácha, V. Kohlová, K. Vojtěchovský, and K. Kašlík: J. Appl. Phys. Vol. 73 (2006), p. 3477.

[12] H. Idrissi, G. Regula, M. Lancin, J. Douin, and B. Pichaud: Phys. Stat. Sol. (c) Vol. 6 (2005), p. 1998.

[13] W. R. L. Lambrecht and M. S. Miao: Phys. Rev. B. Vol. 73 (2006), p. 155312.

[14] K. Maeda, Handbook of Reliability and Materials Issues of Semiconductor Optical and Electronic Devices, ed. by O. Ueda and S. Pearton, Springer, 2012, to be published.

Materials Science Forum Vols. 717-720 (2012) pp 399-402
doi:10.4028/www.scientific.net/MSF.717-720.399

Room temperature photoluminescence from 4H-SiC epilayers: Non-destructive estimation of in-grown stacking fault density

Sabih U Omar[a], Haizheng Song[b], Tangali S Sudarshan[c], and MVS Chandrashekhar[d]

Department of Electrical Engineering, University of South Carolina, Columbia, SC 29208, USA

[a]omar@email.sc.edu (*corresponding author), [b]songha@engr.sc.edu, [c]sudarsha@engr.sc.edu, [d]chandra@engr.sc.edu

Keywords: 4H-SiC, stacking fault, photoluminescence, carrier diffusion length

Abstract. Room temperature photoluminescence was obtained by UV excitation of homoepitaxially grown 4H-SiC thin films. A broad band emission from nitrogen donor to deep boron acceptor level transitions was observed along with the band-edge emission of 4H-SiC at 391 nm. The wavelength of the excitation was varied and the change in the relative intensity of the two emission peaks was observed. The variation of the relative intensity was correlated with in-grown stacking fault density in the epilayer. The effect of IGSF on carrier diffusion length can be deduced from this correlation. For epilayers with very high density of in-grown stacking faults, a sharp peak at 480 nm was observed, which corresponds to triple Shockley type stacking faults (3SSF).

Introduction

Silicon carbide is an attractive material for high-power, high frequency, harsh environment electronic applications because of its electronic and mechanical robustness. However, not all polytypes of SiC show the same electrical properties. Inclusion of undesirable polytypes in the form of stacking faults in the SiC crystal may significantly deteriorate material quality. Significant reduction in breakdown voltage has been observed in Schottky barrier diodes fabricated on epilayers with high stacking fault densities [1].

Room temperature photoluminescence (PL) from SiC epilayers is a convenient tool for identifying the in-grown stacking faults (IGSF) in the epilayers. Microphotoluminescence study of the IGSF's show characteristic stacking fault emission lines at wavelengths of 420, 455, 480 and 500 nm along with the band edge emission peak at ~391nm [2,3]. For broad area room temperature measurements, however, broad gaussian peaks are obtained rather than individual emission lines (Fig. 1). Also, for very thin samples (<50μm), the PL spectrum shows significant emission from the substrate along with the epilayer PL. The room temperature PL spectrum of SiC is often dominated by green luminescence from boron-nitrogen donor to deep acceptor recombination [4]. The deep levels are created by boron atoms sharing the same sublattice as a donor atom. Upon illumination by above bandgap wavelengths, the yellow-green luminescence is produced by nitrogen-boron donor-acceptor pair (DAP) recombination of the injected minority carriers [5].

In this study, we investigated stacking faults in 4H-SiC thin films homoepitaxially grown on 8° and 4° off Si-face 4H-SiC substrates using room temperature photoluminescence spectroscopy (RTPL). During RTPL, the area under UV illumination was 8 mm x 4 mm. Since the absorption length of light in a material varies with the wavelength of the incident light, changing the excitation wavelength enables depth profiling of the epilayers. For thin epilayers, excitation with long wavelength can penetrate deep into the sample and even into the substrate. As a result the PL spectrum shows significant emission from the substrate material, too [6]. On the other hand, short wavelengths are confined in the epilayers only, and the PL spectra excited at these wavelengths are expected to reflect characteristics of the epilayer material. In this paper, the difference between the substrate and epilayer emission has been quantified in the form of the intensity (amplitude) ratio of the band edge peak and the nitrogen-boron DAP emission peak. The dependence of this peak ratio

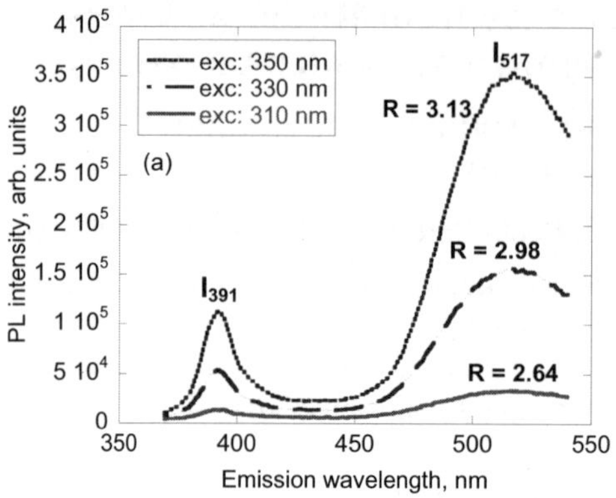

Fig. 1. Typical PL spectra from a 12 μm thick epilayer at three different excitation wavelengths. The band-edge and DAP peaks are denoted as I_{391} and I_{517}, respectively. The ratio of the peak amplitudes I_{517} : I_{391}(R) varies with excitation wavelength (listed in the legend as exc values). The PL spectra from the substrate resemble the epilayer spectrum at 350 nm excitation and do not show any change in the relative intensities of the two peaks (R is constant).

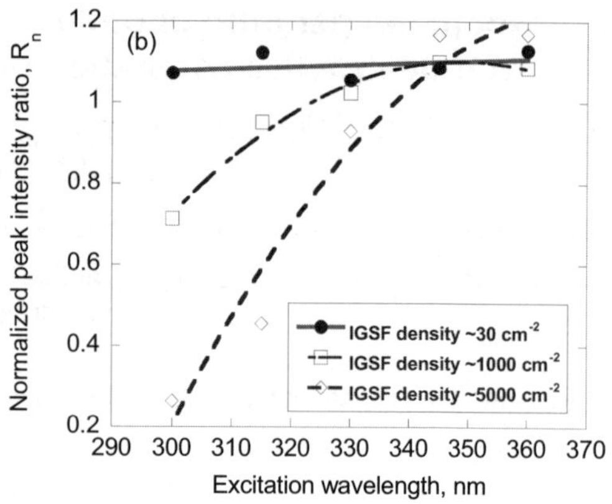

Fig. 2. Peak intensity ratios normalized to that of the substrate ($R_n = R_{epi}/R_{sub}$) obtained from three 12 μm thick epilayers grown on the same substrate with different IGSF densities. The R value for the low IGSF sample follows the substrate profile (R_n~1). The ratio for the high IGSF sample goes down with decreasing absorption length.

(I_{517}:I_{391}) on the excitation wavelength has been correlated to the density of IGSF's in the epilayers. An explanation of this correlation has been proposed in terms of diffusion lengths of the photogenerated carriers in the epilayers.

Experiment

4H-SiC epitaxial films were grown in a vertical hot-wall chimney-style chemical vapor deposition (CVD) reactor. The growths were done on 8 mm x 8 mm samples cut out from commercially grown Cree, Inc. wafers with offcut angles of 4° and 8° along the $[11\bar{2}0]$ direction. The growth experiments were performed at 1550°C at chamber pressures of ~ 50-100 torr. Propane and dichlorosilane were used as carbon and silicon precursors respectively. Typical n-doping concentrations achieved were around 10^{15} cm^{-3}, as estimated by C-V measurements. The thicknesses of the epilayers were in the range of 3~50 μm. In this paper, we will discuss the results on ~12 μm thick epilayers as representative samples, as the observations hold for the ntire thickness range.

Room temperature PL spectra were obtained using a Horiba Jobin Yvon PL spectrometer. The UV source was a 450 W Xe lamp equipped with double grating monochromators for selecting a central excitation wavelength and its bandwidth. The excitation wavelength could be varied from 200 to 900 nm. In our experiments excitation wavelengths of 275~360 nm were used with a bandwidth of 5 nm. The emission was collected by a photomultiplier tube coupled with single photon counting circuitry through an emission monochromator. The room temperature PL spectra of the epilayers are characterized by the band-edge emission peak at 391 nm and a broad DAP peak centered at ~517 nm. PL data were obtained for various excitation wavelengths from 270 nm to 360 nm, which corresponds to a variation of the penetration depth in the range of 2 ~ 50 μm as can be extracted from the fit shown in Fig. 2 of [7]. The DAP peak was present at all wavelengths; the amplitude (height) ratio of the DAP peak to the band-edge peak (I_{517} : I_{391}) varied with excitation wavelength (Fig. 1). The trend in the ratio change with excitation wavelength provides information about the carrier diffusion length (as discussed in the results section), which can be used for identifying in-grown stacking faults.

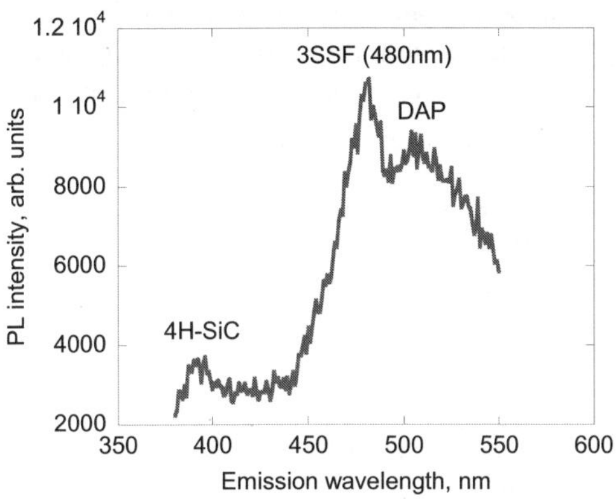

Fig. 3. Emission line at 480 nm from a 3 μm thick epilayer with extremely high IGSF density (> 10,000 cm^{-2}) at an excitation wavelength of 300 nm. The peak at 480 nm has been reported to be associated with triple Shockley type stacking faults (3SSF).

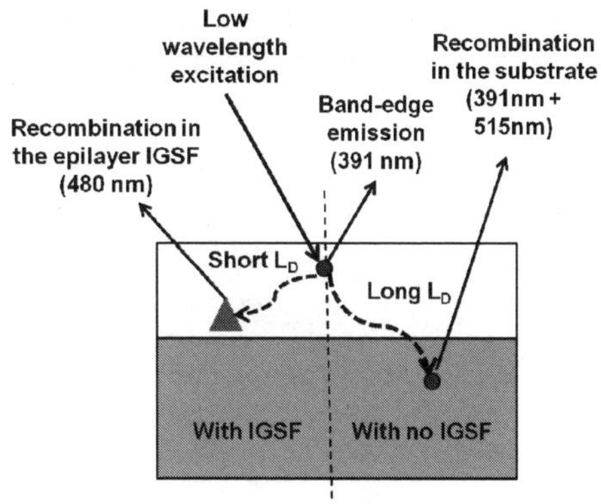

Fig. 4. Epilayers with low IGSF density show long carrier diffusion length. Most of the carriers diffuse through the epilayer to the substrate and recombine there. The resultant PL spectrum resembles that of the substrate for all excitation wavelengths (R_n ~1). In epilayers with high stacking fault density carriers recombine inside the epilayer and the emission does not show substrate-like PL.

The epilayers grown on 8° substrates were etched by molten KOH. The shell-like etch pits that mark the boundaries of the IGSFs reveal their locations in the epilayers. The density of stacking faults was estimated by counting the etch pits using a Nomarski optical microscope. The detailed results are available in [8]. The IGSF density in the 4° off epilayers can also be estimated by counting the number of inverted pyramid type defects on the epi surface, as these defects are known to be associated with IGSF's [7], although these stacking faults may be of a different type than those found in 8° samples.

Results and Discussion

As mentioned earlier, the intensity ratio (R) of the DAP peak (I_{517}) and the band-edge peak (I_{391}) is constant at a given excitation wavelength (λ_{ex}) for a given sample. For a bare substrate, these R values (R_{sub}) remain almost unchanged or decrease slightly with decreased wavelength. The R values are, however, different for different substrates. Fig. 2 shows normalized ratio R_n vs λ_{ex} profiles for three epilayers grown on the same substrate with different stacking fault densities. The R_n values used here are normalized by substrate R_{sub} values to eliminate spectral artifacts ($R_n = R_{epi}/R_{sub}$). As may be noticed in the figure, the R_n values decrease significantly with decreased λ_{ex} for the sample with high IGSF density. On the other hand, the low IGSF sample follows the trend shown by the substrate. Also, the two samples, grown on the same substrate, show similar R values at the long λ_{ex} regime. The overall decreasing trend in the R_n values with reduced absorption lengths in all the epilayers indicates that the boron deep level concentration in the epilayers is less than that in the substrate.

From these observations, it can be said that at longer wavelength the emission from the substrate dominates for both cases. For the low IGSF sample the influence of the substrate persists even at shorter wavelengths, whereas for high IGSF samples, the emission from the epilayers dominates. For epilayers with very high IGSF densities (>10,000 cm^{-2}) a sharp emission peak is observed at 480 nm for short wavelength excitations, as shown in Fig. 3. Feng et al. identified this peak to be associated with triple Shockley stacking faults (3SSF) [2].

The substrate dependence of the PL spectrum can be explained in terms of carrier diffusion lengths, as illustrated in Fig. 4. For the low IGSF sample, the photogenerated carriers have long diffusion lengths, and they can move all the way to the substrate without being recombined in the

IGSF density to Relative Peak intensity Correlation

IGSF density [cm^{-2}]	PL spectrum characteristics at low absorption lengths
<100	Substrate-like R (I_{517}:I_{391}) values
100~5000	Lower R values
>5000	IGSF Peak at 480nm along with lower R-values

Table-1: Summary of IGSF-PL spectrum correlation. For extremely high density of IGSF's the relative intensity of the 480 nm peak can be used as a measure of IGSF density.

epilayers. The carriers recombine at the substrate and give off PL emission. On the other hand, high IGSF density in the epilayers reduces the diffusion length of the photogenerated carriers. In this case, they recombine in the epilayers and show lower DAP emission from boron deep traps present mostly in the substrate. In the presence of very high density of IGSFs in the epilayer, the PL spectrum shows sharp peaks corresponding to that particular IGSF type.

Conclusion

In this study, the room temperature broad area photoluminescence spectrum of 4H-SiC epilayers was observed at different excitation wavelengths. The variation of excitation wavelength enabled depth profiling of the PL emission from the samples. An intense boron-nitrogen DAP emission was observed along with the band-to-band recombination peak. The relative intensity of the DAP peak changed with the variation of excitation wavelength. For samples with low IGSF density, this change resembled profiles observed for the substrate material, indicating a long diffusion length, comparable to thickness of the epilayer. On the other hand, samples with high IGSF density showed a decreasing peak ratio with decreased wavelength. For samples with IGSF density greater than 10,000 cm^{-2}, a sharp SF peak at 480 nm was observed. The peak intensity vs excitation wavelength profiles were compared with IGSF density estimated by molten KOH etching and a good correspondence was found.

References

[1] H. Fujiwara, T. Kimoto, T. Tojo, and H. Matsunami, Appl. Phys. Lett. 87 (2005) 051912.
[2] G. Feng, J. Suda, and T. Kimoto, Appl. Phys. Lett. 92 (2008) 221906.
[3] G. Feng, J. Suda, and T. Kimoto, Appl. Phys. Lett. 94 (2009) 091910.
[4] A. Kakanakova-Georgieva, R. Yakimova, A. Henry, M. K. Linnarsson, M. Syväjärvi, and E. Janzén, J. Appl. Phys. 91 (2002) 2890.
[5] S. G Sridhara, L. L. Clemen, R. P. Devaty, W. J. Choyke, D. J. Larkin, H. S. Kong, T. Troffer, and G. Pensl, J. Appl. Phys. 84, 5 (1998) 2963-2964.
[6] A. Henry, I. G. Ivanov A. Ellison, and E. Janzén, Materials Science and Engineering B61–62 (1999) 234–238.
[7] S. G. Sridhara, T. J. Eperjesi, R. P. Devaty, and W. J. Choyke, Materials Science and Engineering B61–62 (1999) 229–233.
[8] H. Song, S. U. Omar, T. Rana, MVS Chandrashekhar, and T. S. Sudarshan, In-Grown Stacking Faults in SiC-CVD using Dichlorosilane and Propane as Precursors, these Proceedings.
[9] A. Shrivastava, P. Muzykov, and T. S. Sudarshan, J. Appl. Phys. 104 (2008) 093532.

Materials Science Forum Vols. 717-720 (2012) pp 403-406
© (2012) Trans Tech Publications, Switzerland
doi:10.4028/www.scientific.net/MSF.717-720.403

Luminescence Imaging of Extended Defects in SiC via Hyperspectral Imaging

Joshua D. Caldwell[1,a,*], Laurent Lombez[2,b], Amaury Delamarre[2,c], Jean-Francois Guillemoles[2,d], Brice Bourgoin[4,e], Brett A. Hull[3,f], and Marc Verhaegen[4,g]

[1]Naval Research Laboratory, 4555 Overlook Ave, S.W. Washington, D.C. 20375, USA

[2]Chimie ParisTech, Ecole Nationale Superieure, 11, rue Pierre et Marie Curie, 75231 Paris, France

[3]Cree, Inc., E. Corwallis Rd., Research Triangle Park, NC 27709

[4]Photon Etc., 5795 Avenue De Gaspe, #222, Montreal, Quebec H2S 2X3, Canada

[a]joshua.caldwell@nrl.navy.mil, [b]laurent-lombez@chimie-paristech.fr, [c]amaury-delamarre@chimie-paristech.fr, [d]jf-guillemoles@chimie-paristech.fr, [e]bbourgoin@photonetc.com, [f]brett_hull@cree.com, [g]marc.verhaegen@photonetc.com, *corresponding author

Keywords: hyperspectral imaging, photoluminescence, electroluminescence, pin diode, stacking faults

Abstract. Over the past decade, improvements in silicon carbide growth and materials have led to the development of commercialized unipolar devices such as Schottky diodes and MOSFETs. However, much work remains to realize the goal of wide-scale commercialization of both unipolar and bipolar devices, such as pin diodes or IGBTs, for applications requiring high powers, operating at elevated temperatures or in radiation environments or for many fast switching applications. Despite the great strides that have been made in reducing extended and point defect densities during this period, such defects still remain and, with the push to lower off-cut angle substrates, are in many cases seeing increases in prevalence. Thus, spectroscopic and imaging techniques for locating and identifying these defects are in high demand. Luminescence imaging and spectroscopy have both been utilized heavily in such work, yet simultaneously obtaining corresponding spectroscopic and spatial information from such defects is problematic. Here we report on hyperspectral imaging of electroluminescence from SiC pin diodes, whereby a stack of luminescence images are collected over a wide spectral range (400-900 nm), thereby providing the ability to simultaneously image distinct features and identify their corresponding spectral properties. This process is also equally applicable to collecting either photo- or electroluminescence from other materials or devices emitting in either the UV-Vis or NIR spectral range, as well as to reflectance, transmission or other imaging techniques.

Introduction

Extended defects, such as in-grown and recombination induced stacking faults (SFs), threading screw and edge dislocations, half-loop arrays, and low-angle grain boundaries, among others, continue to limit the commercial viability of SiC devices. Recombination-induced SFs have limited the commercialization of SiC bipolar devices to this point, however, significant developments have led to an improved understanding of these defects [1] and growth-based methodologies to eliminate them [2]. Further improvements were also realized upon reducing the off-cut angle of the SiC substrates from 8° to <4°. However, these improvements came with an increase in the prevalence of in-grown SFs (IGSFs) and other extended defects. Therefore, in order for materials development to continue within a commercial environment, methods for identifying the spectral and spatial properties of defects within such materials are required. Currently, such characterization is typically performed through the use of electro- (EL) and photoluminescence spectroscopy or imaging. However, while imaging provides a broad spatial view, enabling the observation of a

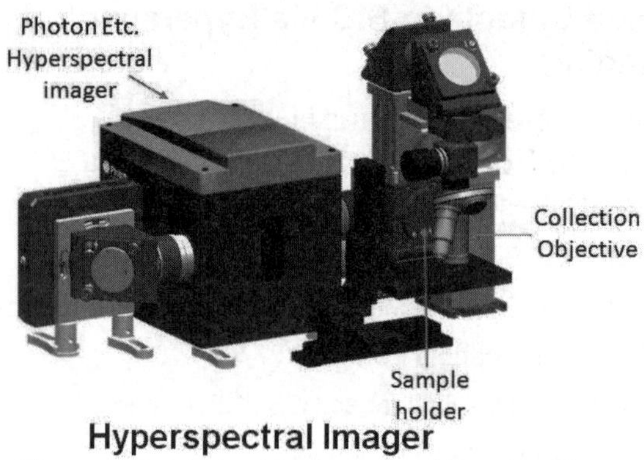

Figure 1: Rendering of hyperspectral imaging filter setup.

large number of defects within the field of view, minimal or no spectral information is obtained [3, 4]. In the case of spectroscopic methods, the spot size used is either quite broad (100's μm to a couple mm's) so that no spatial information is obtained and signatures from small defects can be lost in the background (e.g. threading screw or edge dislocations), or very small as in the case of micro-photoluminescence, albeit at extremely high injection levels, where many spectral features are lost due to quenching or saturation phenomena and spatial information is only obtained through extensive mapping. For electroluminescence (EL) measurements, obtaining spectral information from specific extended defects is quite difficult and is typically acquired with very low spectral resolution [4]. Here, we report on the use of hyperspectral imaging for the simultaneous collection of spectral and spatial information over a user defined field-of-view. This enables the direct correlation between extended defects with their spectral signatures at a defined injection level.

Experimental

A rendering of the hyperspectral imaging setup is provided in Fig. 1. The hyperspectral images were collected using a 20x, 0.40 NA objective. For the images presented here, 28 A/cm^2 was passed through 4H-SiC pin diodes with an Al implanted belt surrounding the 0.93x0.93 mm diode. A gridded top-side contact was used to allow EL to be collected through the top of the device. Details of the diode growth and device fabrication can be found elsewhere [4, 5]. The collected light was focused onto the input of a Photon Etc. hyperspectral filter. The filter provides 2 nm spectral band pass images, with the center wavelength of the pass band tunable from 400-900 nm. During the electrical injection, the filter band pass was varied incrementally in 2 nm steps by rotating the two internal volume Bragg gratings. The exposure time for each image was set to 30s. Following a rectification process, a stack of images, each with their own center wavelength, is created and aligned through computer automation. Once created, any feature or region within the image may be highlighted and the corresponding spectra from that region are provided.

Results and Discussion

Presented in Fig. 2 (a-h) are selected electroluminescence images collected using the hyperspectral imaging setup with each image depicting a different pass-band wavelength as noted. There are several different regions of interest, with some specific sites denoted in Fig. 2. The regions labeled as 1 and 2 highlight different locations within a cluster of SFs, while positions 3 and 4 point to regions containing an IGSF at locations where it penetrates into the p$^+$ layer and within the region where the IGSF passes through the drift layer, respectively. The positions 5 and 6 denote the location of several Si- and C-core partial dislocations (PDs) bounding the various SFs and region 7 denotes the emission from the Al-implanted belt region.

The corresponding spectra collected from regions 1 and 2 are presented in Fig. 3 (a) and from regions 3 and 4 in Fig. 3 (b). Note the emission intensities from the SFs located in the two regions labeled as '1' and '2' are similar when imaged at 420 nm, which is near the expected emission (423 nm) due to the recombination of electrons trapped within the SF quantum well at 0.30 eV below the conduction band [1, 6-8] with holes in the valence band. However, when imaged at 430 nm, region 1 appears brighter. This is consistent with the spectral shift in the SF emission observed in the

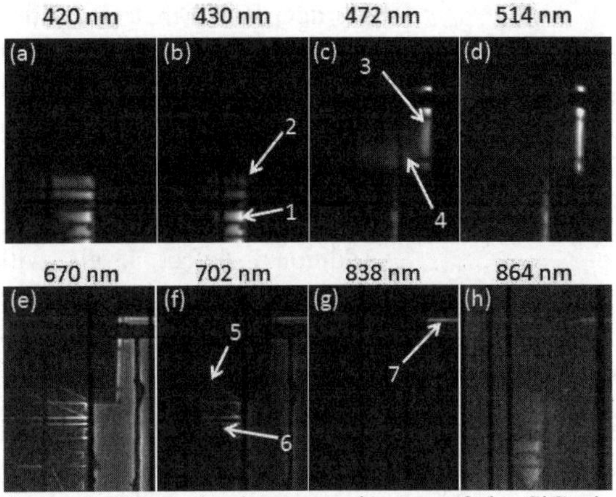

Figure 2: Electroluminescence images of the SiC pin diode discussed in the text collected using the hyperspectral imaging system discussed in the text at 28 A/cm^2. The center wavelength for each image is denoted above the figures.

spectra collected from these two regions [Fig. 3 (a)], where region 1 peaks near 430 nm, while region 2 exhibits the strongest response at 426 nm. These emission lines correspond to energy levels located 0.38 and 0.32 eV below the conduction band for regions 1 and 2, respectively. Currently, the origin of this shift is unclear, but it does appear to be correlated with the regions featuring a higher density of overlapping SFs. Certainly this shift is not indicative of the introduction of double SFs, which are known to induce levels 0.6 eV below the conduction band [9-11]. An emission centered near 864 nm was also observed from the SFs [Fig. 2(h)] in region 1, with this emission presumably due to 2nd order processes, while only a faint emission at 852 nm was observed from regions 1 and 2. In addition, imaging at 670 and 702 nm, the partial dislocations bounding the SFs can be observed, consistent with work by Liu et al. [12]. Furthermore, an emission in the NIR at approximately 838 nm was detected from the region where the Al implant belt was located [region 7, Fig. 2(g)]. This emission corresponds to an energy of 1.48 eV and therefore is clearly not due to the Al acceptor levels themselves (0.19 eV above the valence band) [13], but instead due to some deep level defect induced through the implantation process. By comparison, a boron ring implanted around a diode on the same die featured a broad emission centered near 645 nm (1.92 eV).

Similar variations in spectral position are observed from the IGSF located within the diode near the contact pad in the upper left. The first spectral band is observed from both the IGSF where it passes through the drift layer [region 4, Fig. 2 (c)] and the rectangular region at the right hand side of the defect, where it meets the sample surface [region 3, Fig. 2 (c-f)]. As shown in Fig. 3 (b), both regions 3 and 4 exhibit their strongest emission at a center wavelength of approximately 469 nm (2.64 eV), while region 3 has an additional shoulder that is found to peak near 504 nm (2.46 eV). The emission centered at approximately 2.64 eV is in good agreement with the room temperature emission from the Shockley-type IGSF with (4,4) stacking order in Zhdanov notation [14, 15]. From cathodoluminescence spectroscopy, Maximenko et al. [14] discovered that this IGSF exhibited an excitonic band gap of 2.724 eV and exhibited a right triangular shape, which is consistent with the IGSF observed here. The rectangular feature located where the IGSF reaches the surface of the pin diode has been reported by Caldwell et al. [16, 17] as being due to the penetration of the IGSF into the p^+ layer. The shoulder observed in the spectra from this region illustrates a secondary emission process shifted by approximately 0.18 eV, which is very close to the energy associated with the Al acceptor level (0.19) [13] in the top p^+ layer. Thus, it is likely that this shoulder is due to recombination between electrons trapped within the IGSF with holes located at the acceptor levels, instead of those within the valence band.

Conclusions

Here we have reported on the use of hyperspectral electroluminescence imaging for simultaneously obtaining correlated spatial and structural information from extended defects within SiC pin diodes. Using this technique, the spectral and spatial properties of several defects including recombination induced stacking faults (SFs), partial dislocations, and deep levels within Al and B implanted regions and an in-grown stacking fault (IGSF) were identified and studied. In the case of the SFs, it

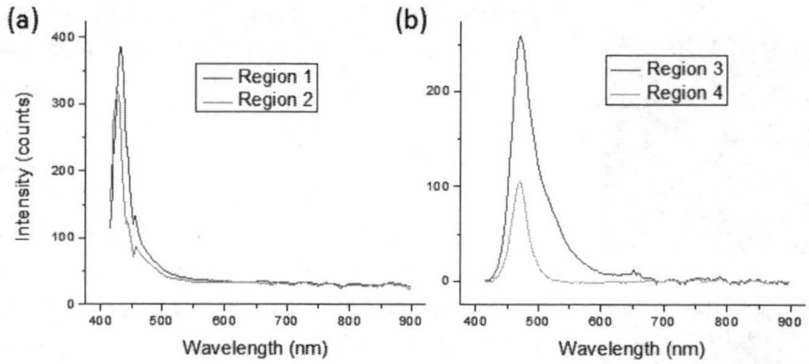

Figure 3: Electroluminescence spectra collected from (a) two regions within the cluster of recombination-induced stacking faults (regions 1 and 2) with varying spectral response and from (b) an in-grown stacking fault at the location where it penetrates into the p-n junction (region 3) and within the body of the fault (region 4).

was determined that in addition to the widely-reported emission near 426 nm, an additional emission band at 30 meV deeper (~430 nm) was also observed in regions with a large number of overlapping SFs. Additional defect levels with emission energies of 1.48 and 1.71 eV were observed in the regions of Al and B implant belts, respectively. Finally, EL spectra from a (4,4) IGSF in Zhdanov notation exhibited two emission lines. The primary line was centered at approximately 2.64 eV, which is consistent with recombination of electrons in the IGSF with holes in the valence band, while the second emission, observed as a shoulder in the spectra collected only where the IGSF meets the diode surface, was centered at 0.18 eV below the primary emission line and is consistent with recombination involving holes at the acceptor levels. These efforts indicate that hyperspectral imaging can provide a unique methodology for simultaneously identifying the presence, spatial extent, location and spectral properties of extended defects in both wide and narrow-band gap semiconductor devices.

Acknowledgements

This work was supported by ONR and DARPA. The authors thank Dr. Rachael Myers-Ward, Dr. Robert Stahlbush and Dr. Orest Glembocki for helpful conversations.

References

[1] J. D. Caldwell, R. E. Stahlbush, O. J. Glembocki, M. G. Ancona and K. D. Hobart, J. Appl. Phys. 108 (2010) 044503.
[2] R. E. Stahlbush, B. L. VanMil, R. L. Myers-Ward, K. K. Lew, D. K. Gaskill and C. R. Eddy, Appl. Phys. Lett. 94 (2009) 041916.
[3] R. E. Stahlbush, K. X. Liu, Q. Zhang and J. J. Sumakeris, Mater. Sci. Forum 556-557 (2007) 295-298.
[4] A. J. Giles, J. D. Caldwell, R. E. Stahlbush, B. A. Hull, N. A. Mahadik, O. Glembocki, K. D. Hobart and K. X. Liu, J. Electron. Mater. 39 (2010) 777-780.
[5] J. D. Caldwell, A. J. Giles, R. E. Stahlbush, M. G. Ancona, O. J. Glembocki, K. D. Hobart, B. A. Hull and K. X. Liu, Mater. Sci. Forum 645-648 (2010) 277-282.
[6] A. Galeckas, J. Linnros and P. Pirouz, Phys. Rev. Lett. 96 (2006) 025502.
[7] K.-B. Park, Y. Ding, J. P. Pelz, J. Grim, M. Skowronski, M. K. Mikhov, Y. Wang and B. J. Skromme, Mater. Sci. Forum *in press* (2006)
[8] M. Skowronski and S. Ha, J. Appl. Phys. 99 (2006) 011101.
[9] R. S. Okojie, M. Xhang, P. Pirouz, S. Tumakha, G. Jessen and L. J. Brillson, Appl. Phys. Lett. 79 (2001) 3056-3058.
[10] H. Iwata, U. Lindefelt, S. Oberg and P. R. Briddon, J. Appl. Phys. 93 (2003) 1577.
[11] H. Iwata, U. Lindefelt, S. Oberg and P. R. Briddon, J. Appl. Phys. 94 (2003) 4972.
[12] K. X. Liu, R. E. Stahlbush, S. I. Maximenko and J. D. Caldwell, Appl. Phys. Lett. 90 (2007) 153503.
[13] M. Ikeda, H. Matsunami and T. Tanaka, Phys. Rev. B 22 (1980) 2842-2854.
[14] S. I. Maximenko, J. A. Freitas, P. B. Klein, A. Shrivastava and T. S. Sudarshan, Appl. Phys. Lett. 94 (2009) 092101.
[15] H. Fujiwara, T. Kimoto, T. Tojo and H. Matsunami, Appl. Phys. Lett. 87 (2005) 051912.
[16] J. D. Caldwell, R. E. Stahlbush, O. J. Glembocki, K. X. Liu, K. D. Hobart and F. Kub, J. Vac. Sci. Technol. B 24 (2006) 2178.
[17] J. D. Caldwell, P. B. Klein, M. E. Twigg, R. E. Stahlbush, O. J. Glembocki, K. X. Liu, K. D. Hobart and F. Kub, Appl. Phys. Lett. 89 (2006) 103519.

Materials Science Forum Vols. 717-720 (2012) pp 407-410
© (2012) Trans Tech Publications, Switzerland
doi:10.4028/www.scientific.net/MSF.717-720.407

Low Temperature Photoluminescence Signature of Stacking Faults in 6H-SiC Epilayers Grown on Low Angle Off-axis Substrates

J. W. Sun[1,a], T. Robert[2,b], V. Jokubavicius[1,c], S. Juillaguet[2,d], R. Yakimova[1,e], M. Syväjärvi[1,f] and J. Camassel[3,g]

[1] Department of Physics, Chemistry and Biology, Linköping University, SE-58183 Linköping, Sweden

[2] Université Montpellier 2, Laboratoire Charles Coulomb UMR 5221, F-34095, Montpellier, France

[3] CNRS, Laboratoire Charles Coulomb UMR 5221, F-34095, Montpellier, France

[a] jiasu@ifm.liu.se, [b] robertt@univ-montp2.fr, [c] valjo@ifm.liu.se, [d] sandrine.juillaguet@univ-montp2.fr, [e] roy@ifm.liu.se, [f] misyv@ifm.liu.se [g] jean.camassel@univ-montp2.fr,

Keywords: Stacking faults, low temperature photoluminescence

Abstract. The radiative recombination spectra of 6H-SiC epilayers grown on low angle (1.4° off-axis) substrates have been investigated by low temperature photoluminescence spectroscopy. Four different types of stacking faults have been identified, together with the presence of 3C-SiC inclusions. From the energies of the momentum-conserving phonons, four excitonic band gap energies Egx have been found to be 2.837, 2.698, 2.600 and 2.525 eV. These photoluminescence features, which give a rapid and non-destructive approach to identify stacking faults in 6H-SiC, provide a direct feedback to improve the material growth.

Introduction

Silicon carbide has been extensively investigated as a promising semiconductor material for high power and high temperature device applications. However, the full commercialization of SiC devices is still limited due to crystallographic imperfections. One of the most harmful defects for the device performance is the appearance of stacking faults (SFs), which can result in the significant deterioration of the forward current-voltage characteristics of bipolar diodes [1]. This attracted much theoretical and experimental research attention and, during the last few years, many photoluminescence, cathodo-luminescence, and transmission electron microscopy studies have been made to investigate the electronic and structural properties of SFs in 4H-SiC [1]. Various types of defects have been identified in this way and, in many cases, the "optical signature" collected from standard photoluminescence investigations allows to identify rapidly and non-destructively the type of SFs present in a given 4H-SiC epitaxial layer [2].

In contrast to the extensive investigation of optical and structural properties of SFs in 4H-SiC, there is much less work focusing on the SFs present in 6H-SiC [3, 4]. First, Takahashi *et al.* [3] observed a high density of SFs in 6H-SiC, which resulted in sizeable anomalies of the resistivity. Independently, Liu *et al.* [4] investigated the microstructure of SFs in 6H-SiC layers grown on (0001) Lely seeds by high-resolution transmission elelctron microscopy. They found a group of SFs with stacking sequences (3, 2), (3, 4), (3, 5), (3, 6), and (3, 2, 2, 2) according to the Zhdanov notation but, to the best of our knowledge, there is still ***no*** detailed investigation of the radiative recombination spectra of stacking faults in 6H-SiC. This point is of interest since, recently, 6H-SiC was proposed as an interesting light emitting material for white LEDs grown on low angle off-axis substrates [5]. Since we have already demonstrated that the homoepitaxial growth of 6H-SiC on a low angle (1.4° off-axis) substrate can be easily accompanied by the creation of 3C-SiC inclusions [6], the main problem in this case is to avoid (or at least control) the presence of 3C SFs.

In this work, low temperature photoluminescence (LTPL) measurements have been done on 6H-SiC homoepitaxial layers grown on 1.4° off-axis 6H-SiC (0001) substrates. From the LTPL signatures we show that, when moving the laser spot from a pure 6H-SiC part to a bulk 3C-SiC inclusion, different PL features appear in between the 6H-SiC matrix signal (at ~ 2.99 eV) and the 3C-SiC bulk-like inclusion lines (at ~ 2.39 eV). Clearly, they are SFs-related emissions with excitonic bandgap energies higher than the 3C-SiC bandgap. Similar to the well-known PL features of stacking faults in 4H-SiC, each of these SF-related emissions can be explained by a quantum confinement effect in which the electrons are confined in the faulted lamella and each series of broad lines is indicative of a specific stacking arrangement.

Experimental Details

Thick, nominally undoped, 6H-SiC homoepitaxial layers have been grown on 1.4° off-axis 6H-SiC (0001) substrates (n-type) by the Fast Sublimation Growth Process (FSGP). For details, see Ref. [7]. The growth was done in a high purity graphite crucible at 1775°C for one hour. The resulting (typical) thickness was ~ 200 μm. Polycrystalline SiC was used as source material and a high purity graphite plate with a triangular opening was used as the spacer in between the source and substrate [6]. Such a sample is shown in the inset of Fig.1(c). The LTPL spectra were collected at 5 K using 30 mW of the 244 nm wavelength of a frequency doubled Ar^+-ion laser as excitation source. A Triax 550 spectrometer from Horiba-Jobin Yvon fitted with a 600 groove/mm gratings and a cooled CCD camera completed the set-up.

Results and Discussion

Most of the epilayers grown on a 1.4° off-axis substrate are made of homoepitaxial 6H-SiC. However, near the border of samples some 3C-SiC polytype can also be identified and, to investigate the conversion of 6H-SiC into 3C-SiC, LTPL spectra were collected when moving from the 6H to 3C parts. A typical example is shown in Fig. 1. Point A is close to the edge of the 6H-SiC deposit. Point C is at the corner of the triangular layer in a 3C-SiC area. Point B is located at some intermediate position when stepping from the 6H-SiC part to the 3C-SiC area. These locations are shown in the inset of Fig. 1(c) and, when moving from points A to C, the PL signatures demonstrate clearly that the epilayer has been converted from 6H to 3C-SiC.

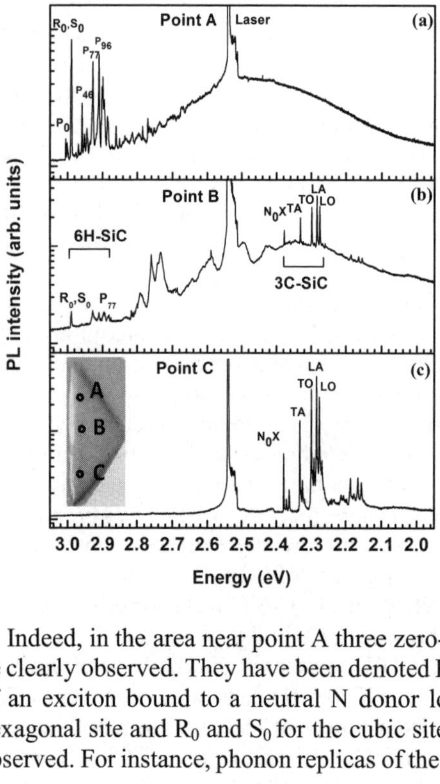

Fig. 1: LTPL spectra collected at 5 K at pure 6H-SiC (a), mixture of 6H- and 3C-SiC polytypes and appearance of SFs (b), and pure 3C-SiC (c). For details see the inset in Fig. 1(c) where the circles A, B, and C indicate the locations where the LTPL measurements have been done. The line at 2.54 eV (488nm) is from the second order of the laser (244nm).

Indeed, in the area near point A three zero-phonon lines (ZPL) of N bound exciton emissions can be clearly observed. They have been denoted P_0, R_0, and S_0 and are associated with the recombination of an exciton bound to a neutral N donor localized at three different sites in 6H-SiC (P_0 for the hexagonal site and R_0 and S_0 for the cubic sites, respectively). A series of phonon replicas were also observed. For instance, phonon replicas of the P_0 line with phonon energies of 46, 77, and 96 meV are denoted as P_{46}, P_{77}, P_{96} in Fig. 1. This is the typical signature of N-bound excitons in 6H-SiC. No contribution from 3C-SiC appears but we notice that, in addition to the sharp N-bound exciton lines, a broad feature manifests at ~2.45 eV. It shows a large red-shift (~0.19 eV) with respect to the usual nitrogen-aluminium (N-Al) donor-acceptor-pair (DAP) spectra in 6H-SiC [8] and can hardly originate from N-Al DAP emission. More detailed works are required to identify the origin of this broad band but we will not focus on this issue here.

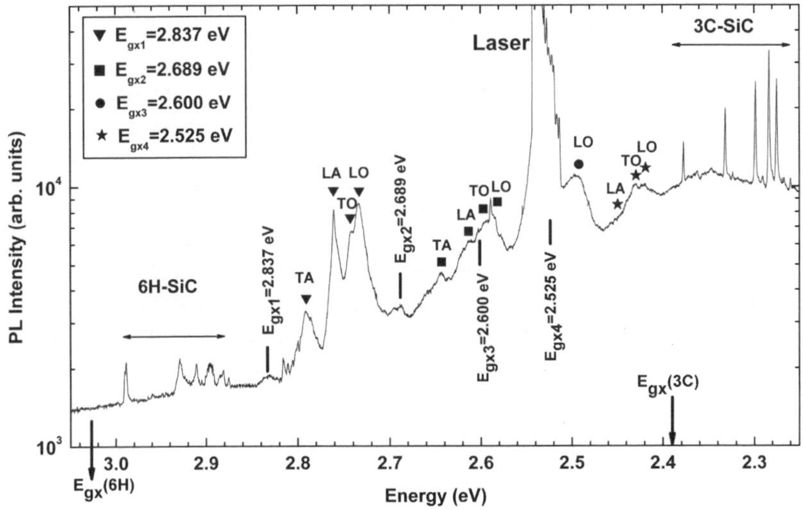

Fig. 2 High resolution LTPL spectrum collected at point B in Fig.1(c). Four series of quantum-well like structures appearing in between the 6H-SiC excitonic emissions (around ~ 2.99 eV) and the bulk-like 3C-SiC excitonic lines (at ~ 2.39 eV) are the specific signatures of stacking faults. Excitonic band gaps of these four stacking faults are also given in the inset of the figure. Excitonic band gaps of 6H-SiC and 3C-SiC are denoted by arrows.

Moving to point C, where the colour of the sample has changed from blue-green to yellow, we find that the LTPL spectrum has changed. It displays now only the near band-edge (NBE) emission spectrum of 3C-SiC [9] without any contribution from 6H-SiC or other defect-related features (including the broad previous band centred at ~2.45 eV). This is shown in Fig. 1(c) and, since all 3C-SiC features (near 2.38 eV) have very narrow line shapes, we conclude that (within the scale of the laser spot) this converted 3C-SiC material has a low defect density.

Finally, let us consider what happens in between the 6H-SiC part (point A) and the 3C-SiC inclusion (point C). This is shown in Fig. 1(b). Clearly, the LTPL spectrum collected in this part of sample shows two different series of features. One is a simple admixture of 3C and 6H polytypes with, in both cases, narrow lines but reduced recombination intensity. This shows that the 3C to 6H transition is not abrupt and that, in this part of the sample, large pieces of 3C and 6H single crystal coexist in which the excitons can trap. The second series of features are made of (large) structures, qualitatively similar to the 3C matrix signal but appearing in between the bulk 6H and 3C-SiC NBE emission lines. This is best shown in Fig. 2 and reminds us of the well-known SFs signature of 3C lamellae in 4H-SiC [2].

Six different series of LTPL features manifest themselves, with different intensities and shapes. Two of them come from excitons bound to residual N donors in the 6H-SiC matrix or the 3C-SiC bulk-like inclusions. They manifest near 2.99 eV for 6H-SiC and 2.38 eV for 3C-SiC, respectively. In between, four series of large, broad structures manifest with fourfold component structures. The components denoted by triangles, for instance, are located at 2.791, 2.761, 2.742 and 2.734 eV, respectively. The corresponding energy differences (which correspond with energy separations between TA (46 meV), LA (76.6 meV), TO (94.5 meV) and LO (104.5 meV) phonon replicas in 3C-SiC) are typical optical signatures of stacking faults [2]. We conclude then that, within the extension of the excitation probe, we could identify four different SFs simultaneously (although some of their phonon replicas are superimposed with the laser line) with excitonic band gap energy E_{gx} equal to 2.837, 2.698, 2.600 and 2.525 eV. Similar SFs signatures have been largely reported in 4H-SiC matrices where they are well described by a type II quantum well (QW) model in which both confinement and built-in electric field effects interplay [2]. Since all SFs excitonic energy gaps in 6H-SiC are above the band gap energy of bulk 3C-SiC, we conclude then that the quantum confinement effect is the dominant contribution (as expected for thin QWs).

It is beyond the scope of this work to discuss all detailed microstructures and electronic properties (energy levels in the QWs) of these SFs. However, we have observed identical SF emissions from other samples grown on such low angle off-axis substrates. We emphasize then that the LTPL signature gives an opportunity to check rapidly whether some SFs do exist or not and (if they exist) if they behave as efficient electron traps or not. Indeed, due to enhanced confinement of the free carries in such faulted regions, stacking faults significantly influence the optoelectronic properties of devices. Most probably, the use of a low off-axis substrate plays an important role in the formation mechanism. Since 3C-SiC can easily form on on-axis 6H-SiC, this is the basic reason why homoepitaxial step-flow growth is usually done on off-axis substrates [6]. In the case of low off-axis substrates, we have shown that the probability of spontaneous nucleation of 3C-SiC is higher, which results in the presence of stacking faults (most probably 3C-SiC lamellae) during growth.

Summary

Thick 6H-SiC layers grown on 1.4° off-axis substrates by FSGP have been investigated by LTPL at 5K. The LTPL spectra show that the homoeptaxial growth on such low off-axis substrates can be easily accompanied by 3C-SiC nucleation. It was found that four different kinds of stacking faults can be identified by LTPL in 6H-SiC with excitonic band gap energies E_{gx} equal to 2.837, 2.698, 2.600 and 2.525 eV. Since all of these band gap energies are above the 3C-SiC band gap energy, the thickness of the SF lamellae in the 6H-SiC matrix should be very thin and the quantum confinement effect should be the dominant term in the binding energy.

Acknowledgements

This work was supported in part by the Ångpanneföreningen Research Foundation, Richerts Foundation, Swedish Energy Agency, Nordic Energy Research, Swedish Research Council (project No. 2009-5307). Partial support from the EU through the Marie Curie Project MANSiC (MRTNCT-2006-035735) is also acknowledged.

References

[1] For a review, see: Silicon Carbide: Recent Major Advances, edited by W.J. Choyke, H. Matsunami, and G. Pensl (Springer, New York, 2004) pages 89-116.

[2] For a recent review, see J. Camassel and S. Juillaguet, Phys. Stat. Sol. (b) 245 (2008) 1337.

[3] J. Takahashi, N. Ohtani, M. Katsuno, and S. Shinoyama, J. Cryst. Growth, 181 (1997) 229.

[4] J. Q. Liu, E. K. Sanchez, and M. Skowronski, Mater. Sci. Forum 389–393 (2002) 435.

[5] S. Kamiyama, M. Iwaya, T. Takeuchi, I. Akasaki, M. Syväjärvi, and R. Yakimova, J. Semiconductors 32 (2011) 13004.

[6] V. Jokubavicius, R. Liljedahl, Y. Y. Ou, H. Y. Ou, S. Kamiyama, R. Yakimova, and M. Syväjärvi, Mater. Sci. Forum 679–680 (2011) 103.

[7] M. Syväjärvi and R. Yakimova: "Sublimation epitaxial growth of hexagonal and cubic SiC", Elsevier, chapter in encyclopedia - The Comprehensive Semiconductor Science & Technology (SEST), Pallab Bhattacharya, Roberto Fornari and Hiroshi Kamimura (Eds), ISBN 978-0-444-53144-5 (2011).

[8] M. Ikeda, H. Matsunami, and T. Tanaka, Phys. Rev. B 22 (1980) 2842.

[9] W. J. Choyke, D. R. Hamilton, and L. Patrick, Phys. Rev. 133 (1964) A1163.

Materials Science Forum Vols. 717-720 (2012) pp 411-414
© (2012) Trans Tech Publications, Switzerland
doi:10.4028/www.scientific.net/MSF.717-720.411

Interaction of 6H-type Stacking Faults with Threading Screw Dislocations in PVT-grown 4H-SiC Single Crystals

S. Sato [a], T. Fujimoto, H. Tsuge, M. Katsuno and W. Ohashi

Nippon Steel Corporation, Advanced Technology Research Laboratories,

20-1 Shintomi, Futtsu, Chiba 293-8511, JAPAN

TEL: +81-439-80-3135, FAX: +81-439-80-2746,

e-mail: [a] sato.shinya@nsc.co.jp

Keywords: silicon carbide, 4H-SiC, PVT growth, stacking fault, threading screw dislocation

Abstract. 6H-type stacking faults (SFs) observed in PVT-grown 4H-SiC single crystals were investigated using Photoluminescence (PL) microscopy at room temperature. Structural analyses using high resolution X-ray topography have revealed that there exist no <000n> (n=4, 8) component in Burger's vectors of the 6H-type SFs we observed, strongly suggesting that the 6H-type SFs are constructed either by insertions of very thin 6H-type foreign polytype inclusions or by successive repetitions of Shockley-type in-plane glides.

Introduction

Silicon carbide (SiC) is a wide band-gap semiconductor material with excellent physical and electronic properties, and much improved performance has been demonstrated for various types of SiC devices fabricated onto SiC wafers grown by PVT technique. An increasing attention has been drawn to various defects, such as threading edge, threading screw and basal plane dislocations as well as various types of stacking faults (SFs) [1, 2], and a number of intense studies have been performed in order to gain a comprehensive understanding on the effect of the dislocations on device performance and reliability.

In the previous paper [3], we reported that 6H-type SFs were identified in 4H-SiC crystals using high resolution TEM, and a possible mechanism for 6H-type SF generation was also proposed; when step-bunching occurs on the growth front of 4H-SiC crystals, 6H polytype embryos are likely to generate on the surface with its terrace area larger enough compared to the surface diffusion length of SiC crystallites, followed by overgrowth of 4H-SiC steps, subsequently giving rise to a thin 6H-type SF [3]. On the other hand, it has been proposed that 6H-type SFs can be constructed by successive repetitions of either Shockley glide or Frank-type insertion/extraction [4], and 6 time-repetition of Shockley partial glides and 4 time-repetition of intrinsic/extrinsic Frank-type insertion can both give rise to an identical structure consisting of 2x(3, 3), i.e., a 2-time repetition structure of single 6H stacking sequence (3, 3) [4].

In this paper, we have first performed the structural analyses of 6H-type SFs by both PL image mappings and high resolution X-ray topography in order to gain suggestive insight of the structural origin for 6H-type SF generation, specifically from the viewpoint of structural comparison of Burgers vector components constructing both the 6H-type SFs and other specific SFs such as single Shockley stacking faults. Once the dislocation structure of the 6H-type SFs is determined, it is of particular interest to investigate the structural interaction with other dislocations, especially threading screw dislocations (TSDs). A possible interpretation of the interaction occurring between the 6H-type SFs and other dislocations is then very briefly discussed.

Experimental Procedure

4H-SiC single crystals with diameters ranging from 50 to 100mm were grown onto the carbon faces of 4H-SiC seed crystals by the PVT technique we have so far developed. Details of the growth conditions were described elsewhere [3, 5]. During the growth, nitrogen doping was carried out by introducing an appriorate amount of nitrogen gas into argon gas ambient. Two different specimens for observations using optical microscopy specimens were prepared by slicing the crystals obtained parallet to the <0001> and <1-100> axis directions, respectively. All the specimens used in this paper were mirror-polished using a diamond-based slurry. SFs in the crystals were also observed using room-temperature Photoluminescence (PL) microscopy in which a UV light source of a 200W class Hg-Xe lamp excited PL signals from the samples, followed by acquisition in a high resolution digital CCD camera through various band-pass filters of selected wavelength windows, such as 420 and 750IR, respectively [6]. The latter filter can pass PL signals with wavelengths larger than 750 nm. High-resolution X-ray topography with a spatial resolution of around 3μm was also adopted in order to determine the dislocation structure by changing the diffraction geometry: (0004) and (0008).

Results and Discussion

Fig. 1(a) shows a mapping of PL signals acquired through a 750IR filter, envisaging that clear images of white-contrasted lines exist with the extending directions perpendicular to the off-orientation direction of the substrate. These defects are exactly identical to those reported by Nakabayashi et al.[3], determined to be 6H-type SFs using high-resolution transmission electron microscopy. A possible mechanism for 6H-type SF generation was proposed in [3]; a very thin 6H polytype layer comprising of several monolayers are likely to be overgrown by 4H-SiC step growth when the step surface becomes larger wit h its terrace area larger enough compared to the surface diffusion length of SiC crystallites [3]. On the other hand, Robert et al.,[4] also identified 6H-type SFs in CVD-grown 4H-SiC epitaxial layers grown on off-axis substrates, and proposed another mechanism for the 6H-type SF generation; the 6H-type SF structure can be constructed by successive repetition of either Shockley glide or Frank-type insertion/extraction [4]. As shown Fig. 2, a sequence of 4 times insertion of an extrinsic Frank-type single Si-C layer leads to an exact 6H structure consisting of 2x(3, 3), and another sequence of 6 times repetitions of Shockley glide can also give rise to an identical structure consisting of 2x(3, 3). Mixed combination of these sequences can also realize the 6H-type SFs as well. In short, there exist different mechanisms for 6H-type SF generation, and therefore at the moment, it is indecisive as for the structural origin of the 6H-type SFs observed in SiC substrates and epitaxial layers.

Fig. 1(b) shows an image obtained for the (0001)-oriented wafer specimen shown in Fig. 1(a) using high-resolution (0008) reflection X-ray topography. The image area corresponds exactly to the area of Fig. 1(a). As can be seen in Fig. 1(b), a lot of thin lines were observed, which are presumably in-plane line defects including the Burgers vectors parallel to the <0001> direction [7]. However, no

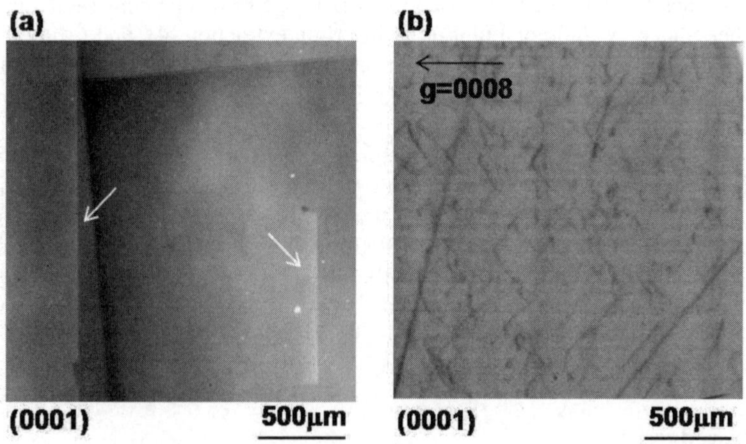

Fig.1 (a) Mapping of PL signal from a 4H-SiC substrate acquired through a 750IR filter indicating the existence of two 6H-type SFs marked by arrows in the figure, and (b) High-resolution (0008) reflection X-ray topograph taken for the area identical to that in Fig. 1(a).

visible contrasts corresponding to SFs observed in Fig. 1(a) were found in Fig. 1(b), suggesting that the 6H-type SFs observed include no Burgers vector components of <0001>. This result is of particular importance for discussing the structural origin of the 6H-type SFs, because, once the 6H-type SFs evolve upon extrinsic Frank-type insertion of single Si-C stacking layer, it has to include a Burgers vector component which is parallel to the <0001> axis direction.

A number of experimental facts can be enumerated as an example for determining the dislocation structure from the results obtained using X-ray topography with various diffraction geometries. Fig. 3 shows a case for single Shockley stacking faults (sSSFs) which are often observed as a typical SF existing in 4H-SiC epitaxial layers grown onto 4H-SiC substrates. The sSSFs, generated in this experiment by irradiation of Hg-Xe light for approximately 120 minutes [6], are well known to emit a PL signal with a peak wavelength of 420 nm [8], and as is shown in Fig. 3(a), the sSSFs are recognized as white-contrasted triangles. High-resolution reflection X ray topography images were taken with (0008) diffraction geometry of 4H-SiC, and the results obtained were shown in Fig. 3(b). The image was obtained for the area identical to that shown in Fig. 3(a). The diffraction image shown in Fig. 3(b) shows no triangle contrasts, and therefore from these results in Fig. 3(a) and (b), it can be conclude that the sSSFs are at least comprised of a Burgers vector component except for <0008>, consistent with results so far reported for sSSF structure stating that sSSFs result from an event of single Shockley glide with Burgers vector lying parallel to the c-face, i.e., (0001) [9].

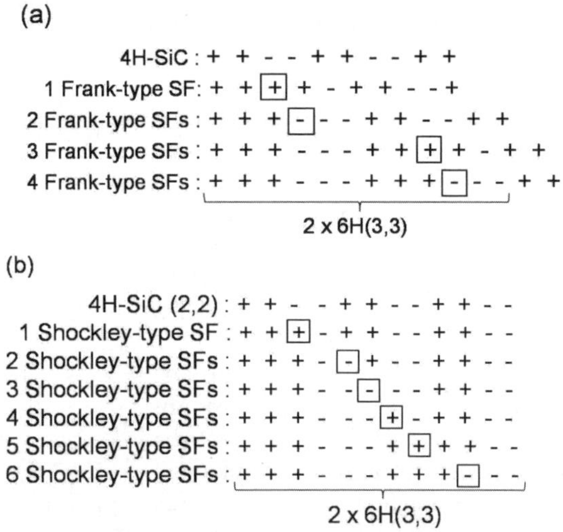

Fig. 2 Possible construction passes for 6H-type SF: (a) 4-times successive insersions of extrinsic single Si-C layer Frank-type dislocations and (b) 6-times successive repetition of single Shockley glide. Further, mixed combination of both (a) and (b) can also give the 6H-type SF structure.

From the discussions described above, we now reach a conclusion that the 6H-type SFs observed in Fig. 1(a) contain at least no Frank-type Burgers vector, and therefore, has to be formed as a result of either 6H-type foreign polytype inclusion or successive repetition of Shockley glides.

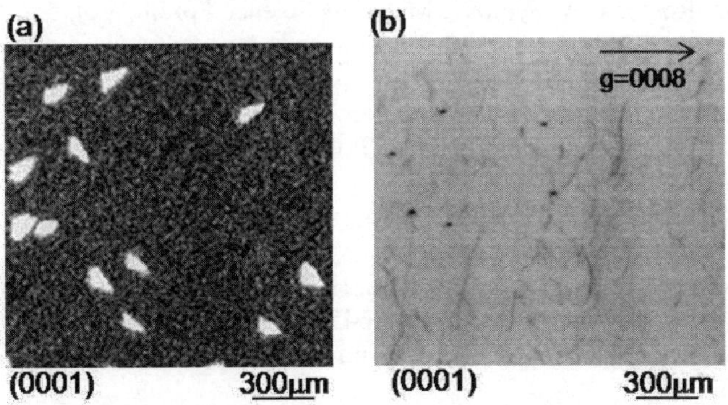

Fig. 3 420 nm PL signal image mapping (Fig. 4(a)), and high -resolution reflection X ray topograph images with (0008) diffraction geometry of 4H-SiC (Fig. 4(b)), both obtained for 4H-SiC epitaxial layer. For detals, see the main text.

Fig. 4 shows a high-resolution (0004) transmission X-ray topograph obtained for the specimen sliced parallel to a $(1\bar{1}00)$ face. A 6H-type SF lies at the position in Fig. 4 marked as "SF", and as can be seen in Fig. 4, threading screw dislocation (TSDs) are observed as a dislocation propagating almost parallel to the growth direction, indicating that the TSDs in

Fig. 4 interact with the 6H-type SF; for instance a TSD marked by TSD-A changes the propagating direction at the 6H-type SF into the in-plane, and then returns back to the original propagating direction parallel to the growth direction.

Earlier studies have reported that both Shockley-type dislocations (basal plane dislocations) [9] and foreign polytype (6H) inclusions [10] can interact with TSDs during the PVT growth, and therefore it can be conjectured that the contrast of 6H-type SF observed in Fig. 4 could be an image decorated by TSDs, although further investigation using Synchrotron X-ray topography is desirable in order to determine fine interacting structures of dislocations.

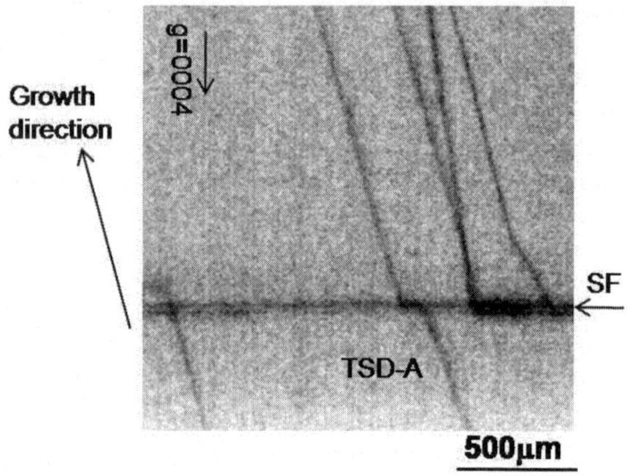

Fig. 4 High-resolution (0004) transmission X-ray topograph obtained for the specimen sliced parallel to $(1\bar{1}00)$. The dark contrasted lines extending parallel to the growth directions are TSDs.

Conclusions

The structural origin of 6H-type stacking faults (SFs) is investigated, and found that it has to be constructed either thin 6H polytype inclusions or successive repetitions of Shockley glides. Such 6H-type SFs can in principle interact with threading screw dislocations.

Acknowledgement

This work is partly supported by NEDO under the grant of "Novel Semiconductor Power Electronics Project Realizing Low Carbon Emission Society ".

References

[1] N. Ohtani, M. Katsuno, M. Nakabayashi, T. Fujimoto, H. Tsuge, H. Yashiro, T. Aigo, H. Hirano, T. Hoshino and K. Tatsumi, Journal of Crystal Growth, Vol. 311 (2009), p. 1475.

[2] D. Sichel, M. Albrechtl, H.-J. Rost and A. Sendzik, Materials Science Forum, Vol. 556-557 (2007), p. 21.

[3] M. Nakabayashi, T. Fujimoto, M. Katsuno, H. Tsuge, T. Aigo, S. Sato, H. Yashiro, T. Hoshino, H. Hirano and W. Ohashi, Materials Science Forum, Vol. 645-648 (2010), p. 9.

[4] T. Robert, M. Marinova, S. Juillaguet, A. Henry, E. K. Polychroniadis and J. Camassell, Materials Scinece Forum, Vol. 645-648 (2010), p. 347.

[5] J. Takahashi, N. Ohtani, M. Katsuno and S. Shinoyama, Journal of Crystal Growth, Vol. 181 (1997), p. 229.

[6] T. Fujimoto, T. Aigo, M. Nakabayashi, S. Sato, M. Katsuno, H. Tsuge, H. Yashiro, H. Hirano, T. Hoshino and W. Ohashi, Materials Science Forum, Vol. 645-648 (2010), p. 319.

[7] D. Nakamura, S. Yamaguchi, I. Gunjishima, Y. Hirose and T. Kimoto, Journal of Crystal Growth, Vol. 304 (2007), p. 57.

[8] G. Feng, J. Suda and T. Kimoto, Applied Physics Letters, Vol. 92 (2008), p. 221906.

[9] N. Ohtani, M. Katsuno, H. Tsuge, T. Fujimoto, M. Nakabayashi, H. Yashiro, M. Sawamura, T. Aigo and T. Hoshino, Microelectronic Engineering, Vol. 83 (2006), p. 142.

[10] N. Ohtani, M. Katsuno, H. Tsuge, T. Fujimoto, M. Nakabayashi, H. Yashiro, M. Sawamura, T. Aigo, T. Hoshino, Journal of Crystal Growth, Vol. 286 (2006), p. 55.

Materials Science Forum Vols. 717-720 (2012) pp 415-418
© (2012) Trans Tech Publications, Switzerland
doi:10.4028/www.scientific.net/MSF.717-720.415

Ab initio calculation of mechanical properties of stacking fault in 3C-SiC: Effect of stress and doping

Yoshitaka Umeno[1,a,*], Kuniaki Yagi[2,b] and Hiroyuki Nagasawa[2,c]

[1]Institute of Industrial Science, Univ. Tokyo, 4-6-1 Komaba Meguro-ku, Tokyo 153-8505, Japan

[2]SiC Development center, HOYA Corp., 1-17-16 Tanashiota, Chuo-ku, Sagamihara, Kanagawa 252-0245, Japan

[a]umeno@iis.u-tokyo.ac.jp, [b]Kuniaki_Yagi@sngw.rdc.hoya.co.jp, [c]Hiroyuki_Nagasawa@sngw.rdc.hoya.co.jp, *corresponding author

Keywords: 3C-SiC, Stacking Fault, Ab Initio Calculation, Density Functional Theory, Stress

Abstract. We carry out *ab initio* density functional theory calculations to investigate the fundamental mechanical properties of stacking faults (SFs) in 3C-SiC, including the effect of stress and doping atoms (substitution of C by N or Si). Stress induced by SF formation is quantitatively evaluated. Extrinsic SFs containing double and triple SiC layers are found to be slightly more stable than the single-layer extrinsic SF, supporting experimental observation. The effect of tensile or compressive stress on SF energies is found to be small. Neglecting the effect of local strain induced by doping, N doping around a SF increases the SF formation energy, while SFs seem to be easily formed in Si-rich SiC.

Introduction

Cubic silicon carbide (3C-SiC) is a promising material for high- and medium-power semiconductor devices due to its prominent electrical properties, including high electron mobility and saturated electron drift velocity. Nevertheless, 3C-SiC has problems of dislocations and planar defects like anti-phase boundaries (APBs) and stacking faults (SFs) that deteriorate the device performance and thus applications to semiconductor devices are still challenging. Efforts have been invested to produce 3C-SiC wafers with reduced density of SFs [1,2]. While such development of fabrication methods is under way, planar defects in 3C-SiC remain serious problems because they cause warpage of wafers [3]. Thus, it is urgent to understand the fundamental mechanical properties of SFs, which may contribute to further refinement of wafer fabrication methods.

Although there have been a number of theoretical studies of SiC and its defects by *ab initio* density functional theory (DFT) [4-6], the mechanical properties of SFs in SiC are not sufficiently revealed. In addition, the effects of impurity doping and adjusted constitution should be investigated as they are practical methods to tune the properties of wafers. In this study we perform *ab initio* DFT calculations to obtain fundamental mechanical properties of SFs in SiC. We first evaluate the energy and induced stress of various SFs of pure SiC (undoped, even constitution of Si and C), and then examine the effect of doping of nitrogen atom and Si-rich constitution on the formation energy of SFs.

Simulation Method and Procedure

For *ab initio* DFT calculations, we employed the Vienna Ab-initio Simulation Package (VASP) [7] to carry out a pseudopotential approach based on the projector augmented wave (PAW) method [8]. By the use of plane wave (PW) basis, not only Hellmann-Feynman forces but also the stress tensor (derivative of total energy with respect to strain) can be easily calculated. We examined both the local density approximation (LDA [9]) and the generalized gradient approximation (GGA [10]) (see next paragraph). Figure 1(a) presents simulation cells for the calculation of pure SFs. Simulation models of intrinsic, single-layer extrinsic, double-layer extrinsic, and triple-layer extrinsic SFs were used, which contained N=10, 10, 14, and 18 atoms, respectively. The cut-off energy of PW (500 eV) and the k-point mesh (11×11×3) were selected to give sufficient accuracy (numerical error down to 0.012

meV/atom, estimated error in SF energy ~0.3 mJ/m^2). Figure 1(b) explains the three sites of doping we examined. Atom sites of carbon indicated by arrows were replaced with N or Si atoms to represent N-doped and Si-rich SFs. In the calculation of SFs with doping, the area of SF (*xy* section of the model) must be large enough to eliminate the contribution from replica cell due to the periodic boundary condition, otherwise neighboring doped atoms interact with each other. Therefore, we used large cells shown Fig. 1(c) for the calculation of N-doped and Si-rich SFs.

Our calculations of the lattice constant of 3C-SiC by GGA and LDA give 4.38Å (+0.45% from the experimental value, 4.36Å) and 4.33Å (-0.65%), respectively, both in good agreement with experiment. The formation energies of the intrinsic SF calculated by LDA and GGA are -0.71 and 10.8 mJ/m^2, respectively. Here we exclude the effect of in-plane and transverse stress induced by SF (details will be described later). Negative energy of the intrinsic SF by LDA is consistent with calculations by Käckell [5,6]. The results by GGA seem more reliable because (i) intrinsic SFs are rarely observed in experiment [11], and (ii) the ANNNI model gives positive and negative energies for the intrinsic and extrinsic SFs, respectively [12]. Thus, we present results by GGA hearafter.

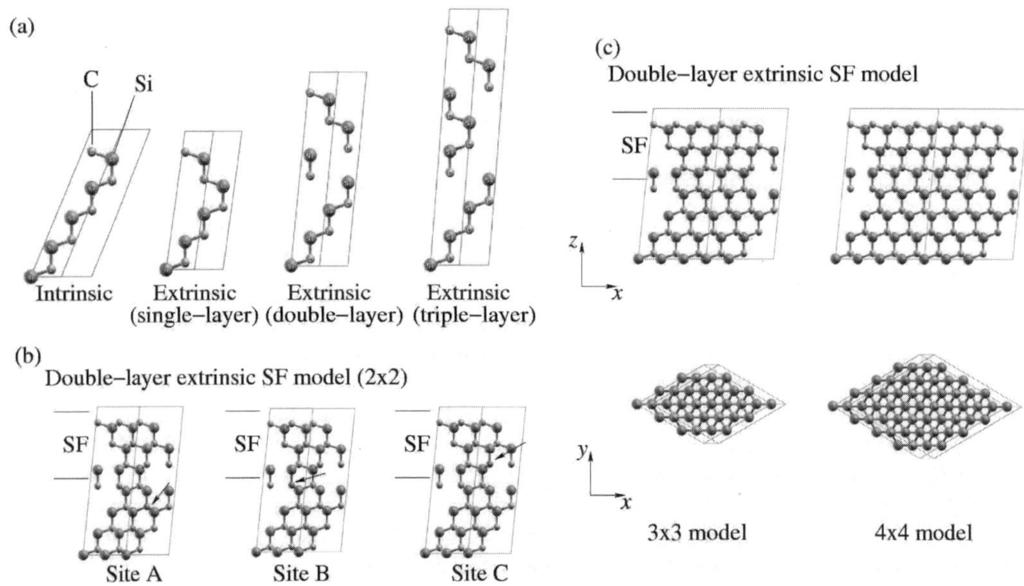

Fig. 1: Simulation cell of SF structure model. (a) Unit cells of SFs with different thickness. (b) Site of doping. The Carbon atom indicated by the arrow was replaced with N or Si. (c) Models with different area of *xy* section used for calculation of N-doped and Si-rich SFs.

Table I: Stacking fault energy (mJ/m^2) obtained by *ab initio* DFT calculations using GGA.

Relaxation	Intrinsic	Extrinsic (single)	Extrinsic (double)	Extrinsic (triple)
No	15.0	-3.44		
Atom only	10.8	-7.83	-11.6	-12.0
Atom & cell		-8.13	-12.2	-12.4

Results and Discussion

Stacking Fault Energy and Induced Stress. Table I lists the calculated formation energies of various SF structures. For the intrinsic and extrinsic SFs, the results without atomic relaxation (atoms are located on the ideal lattice sites) are shown. The energies without relaxation are higher than those with atomic relaxation by about 4 mJ/m^2, indicating that the effect of atomic relaxation around SFs is not negligible. Shown in the last line in the table are the results with "full relaxation", where atom positions and cell size are optimized. The stress (cell) relaxation resulted in slight reduction in the energy (0.3~0.6 mJ/m^2). It should be noted here that the SF energy with the full relaxation is not a well-defined value, because the relaxation of in-plane stress of SFs depends on the cell size normal (perpendicular) to the SF plane. However, the difference in the energy between the atom-only relaxation and the full-relaxation is small because the SF stress is not large (see next section) and because the sizes of the simulation cells are comparable.

The calculated SF energies indicate that the intrinsic SF is unstable, while extrinsic SFs are stable. The double-layer and triple-layer extrinsic SFs are slightly more stable than the single-layer one, which is consistent with experimental observation where SFs with the thickness equal to or larger than two SiC layers often appear [11,13].

Table II lists stress induced by SF, which was calculated with relaxation of atoms around the SF while the cell dimension was determined by the lattice constant of 3C-SiC and fixed. These values therefore present the stresses parallel and normal to the SF plane that arise by the formation of the SF under the strain-fixed condition. Note the unit (N/m) of the in-plane stress, which is because we consider force exerted not on an area but on a plane (zero thickness), similarly to the notation of surface stress. The in-plane stress is almost constant (~0.55 $\pm$ 0.03 N/m) among the different SFs. In contrast, the induced stress normal to the SF plane, which is negative in all the cases, differs depending on the SF thickness. The results indicate that the SF formation causes shrinkage along the SF plane and expansion perpendicular to the plane, which can contribute to warpage in SiC wafers, as is pointed out by Sun *et al.* [3].

Table II: Stress induced by SF formation. Positive values mean tension.

	Intrinsic	External (single)	External (double)	External (triple)
In-plane SF stress f (N/m)	0.523	0.549	0.579	0.542
Normal to SF σ_z (MPa)	-600	-535	-365	-275

Stacking Fault under Stress. Figure 2(a) shows the effect of normal stress perpendicular to the SF plane on the formation energy of extrinsic SFs. Here, normal stress parallel to the SF plane (in-plane stress) is controlled to zero. It is seen that the stress causes almost no effect on the single-layer extrinsic SF, whereas the double-layer and triple-layer SFs are affected by the stress normal to the SF plane. Compression and large tension increases the SF energy. However, the change in the SF energy is modest and even compression of 5 GPa causes only about 1 mJ/m^2 increase. This indicates that local normal stress does not remarkably change the formability of SFs.

(a) (b)

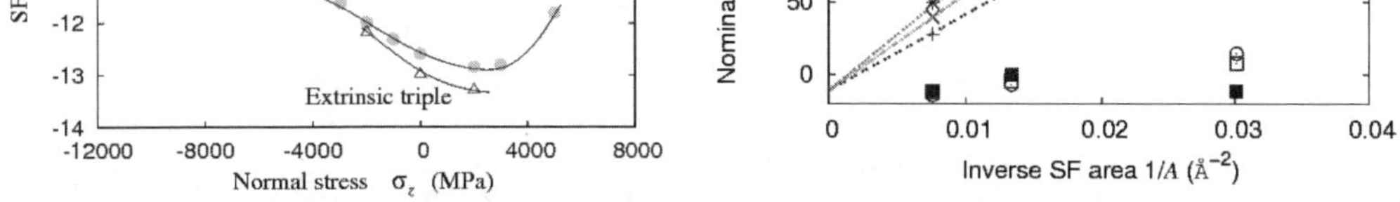

Fig. 2: (a) Energy of extrinsic SFs as a function of stress normal to SF plane (in-plane stress is zero). (b) Nominal SF energy of N-doped and Si-rich double-layer extrinsic SF as a function of inverse area.

Stacking Fault of N-doped and Si-rich SiC. Here we evaluate the effect of doping (substitution of C by N or Si atom) on the SF formation energy. Let E_{dope} be the energy gain (or loss when negative) per doped atom. The "nominal" SF energy, which is calculated by the difference of the total energy with and without SF (both having the doping), can be written as

$$E^*_{SF} = \frac{E_{SF} \cdot A + E_{dope}}{A} = E_{SF} + E_{dope} \cdot \frac{1}{A}. \quad (1)$$

The nominal SF energy (E^*_{SF}) is therefore a linear function of inverse area ($1/A$) provided that the area is sufficiently large and the effect of neighboring doping (due to adjoining replica cells) is negligible. Note that $E^*_{SF} = E_{SF}$ at the limit of $1/A \rightarrow 0$. Figure 2(b) shows the calculated nominal SF energies of N-doped and Si-rich double-layer extrinsic SF models plotted against the inverse area. Here, cell size was kept fixed and only relaxation of atom positions was fulfilled to exclude the effect of stress relaxation and obtain well-defined values (see discussion in the previous section). In the N-doped models, it is seen that the points of $1/A = 0.0075$ and 0.0134 Å^{-2} are on lines of $y = ax + E_{SF}$ ($E_{SF} = -10.9$ mJ/m^2), while those of $1/A = 0.0301$ show large deviations from the lines. This means that the "3×3" models are large enough for the evaluation of SF energies with doping, while the "2×2" ones are insufficient. Calculated values of E_{dope} from the slopes are listed in Table III. The increase in the SF energy by N-doping is in line with experiment that shows 3C becomes more stable than 6H under high nitrogen density [14,15]. The site of the doping affects the energy. The value of abscissa that gives $E^*_{SF} = 0$ corresponds to nitrogen doping of the density of ρ_A=1.4 ~ 2.0x10^{13}/cm^2 on the SF plane. A rough estimation of the corresponding volume density, ρ_V, by simply using the planar density (*e.g.*, $\rho_V = \rho_A{}^{3/2}$) yields the value of ρ_V =5.2 ~ 8.9x10^{19}/cm^3. This is smaller than the solubility limit of nitrogen into SiC, 6x10^{20}/cm^3. It was pointed out, however, that nitrogen does not work as a donor when its density is over 4x10^{19}/cm^3 [16,17]. This indicates that avoiding SF formation solely by nitrogen doping is not practically valid. On the other hand, in the Si-rich models the nominal SF energy (E^*_{SF}) is close to E_{SF} already in the "3×3" models, indicating that the effect of hindering SF formation by the Si-rich condition is very small.

Table III. Effect of doped nitrogen atom on SF energy (double-layer extrinsic SF).

Site of doping	Site A	Site B	Site C
E_{dope} (J/atom)	5.2x10^{-20}	6.6x10^{-20}	7.7x10^{-20}

Summary

To investigate the fundamental mechanical properties of SFs in 3C-SiC, we performed *ab initio* DFT calculations using GGA. Stress induced by SF formation was quantitatively evaluated, from which concomitant strain can be estimated. Calculated SF energies indicated that extrinsic SFs are stable. The double-layer and triple-layer extrinsic SFs were found to be slightly more stable than the single-layer one, which supports experimental observation. Our calculations indicate that N doping around an SF obviously increase the SF formation energy, while SFs seem to be easily formed in Si-rich models. The effect of tensile or compressive stress on SF energies was found to be very small.

References

[1] H. Nagasawa, K. Yagi, T. Kawahara, and N. Hatta, Chem. Vap. Deposition 12 (2006) 502.
[2] H. Nagasawa, M. Abe, K. Yagi, T. Kawahara, and N. Hatta, Phys. Stat. Sol. (b) 245 (2008) 1272.
[3] Y. Sun, S. Izumi, K. Yagi, and H. Nagasawa, Phys. Stat. Sol. (a), submitted.
[4] J.J. Wang *et al.*, Physics Letters A 374 (2010) 2286.
[5] P. Käckell, B. Wenzien, and F. Bechstedt, Phys. Rev. B 50 (1994) 17037.
[6] P. Käckell, J. Furthmüller, and F. Bechstedt, Phys. Rev. B 58 (1998) 1326.
[7] G. Kresse and J. Furthmüller, Phys. Rev. B 54 (1996) 11169.
[8] P.E. Blöchel, Phys. Rev. B 50 (1994) 17953.
[9] D.M. Ceperley and B.J. Alder, Phys. Rev. Lett. 45 (1980) 566.
[10] J.P. Perdew and Y. Wang, Phys. Rev. B 45 (1992) 13244.
[11] N. Hatta, K. Yagi, T. Kawahara and H. Nagasawa, Mater. Sci. Forum 433-436 (2003) 273.
[12] P.J.H. Denteneer and W.v. Haeringen, Solid State Commun. 65 (1988) 115.
[13] K. Yagi and H. Nagasawa, private communication.
[14] G. Ferro, Mater. Sci. Forum 645-648 (2010) 49.
[15] O. Kim-Hak *et al.*, Mater. Sci. Forum 645-648 (2010) 163.
[16] M.V. Rao *et al.*, J. Appl. Phys. 86 (1999) 752.
[17] M. Laube *et al.*, J. Appl. Phys. 92 (2002) 549.

Materials Science Forum Vols. 717-720 (2012) pp 419-422
© (2012) Trans Tech Publications, Switzerland
doi:10.4028/www.scientific.net/MSF.717-720.419

On the Twin Boundary Propagation in (111) 3C-SiC Layers

M. Marinova[1, a], A. Andreadou[1,b], A. Mantzari[1,c] and E.K. Polychroniadis[1,d, *]

[1]Department of Physics, Aristotle University of Thessaloniki, GR54124, Thessaloniki, Greece

[a]marinova@physics.auth.gr, [b]aria@auth.gr, [c]am@auth.gr, [d, *]polychr@auth.gr

Keywords: 3C-SiC, twin boundary propagation, TEM.

Abstract. The present study reports on the propagation of twin boundaries in (111) 3C-SiC by means of conventional (CTEM) and high resolution transmission electron microscopy (HRTEM). The investigated 3C-SiC layers were homoepitaxially grown by Chemical Vapour Deposition (CVD) on layers previously grown by Vapor Liquid Solid (VLS) mechanism on 6H-SiC substrates. At the initial stages of growth the usual twin boundary that occurs is an incoherent $\{\bar{2}11\}$ $\Sigma 3$ one. It transforms to more energetically favorable cases by several ways: (i) The initial $\{\bar{2}11\}$ boundary turns 90°, to a fully coherent (111) interface, forming microtwins; (ii) A step-like interface occurs with facets along the (111) and the $\{\bar{2}11\}$ planes; (iii) It transforms in a fourfold twin complex propagating to the surface.

Introduction

Twin boundaries (TBs) are among the main planar defects in 3C-SiC. The distorted structures at the incoherent as well as most of the coherent twin boundaries influence strongly the electronic properties of a semiconductor, due to local change of stoichiometry at the interface, wrong or dangling bonds, their possible decoration and interaction with impurities and dopants [1]. The appearance of incoherent twin boundaries propagating in the growth direction of the 3C-SiC layers has already been reported to destroy the rectifying behavior of contacts [2]. Thus it is of significant importance to study the type, nucleation and propagation of the different twin boundaries in 3C-SiC. This is the aim of the current study which investigates the evolution of TBs in CVD layers grown on (111) 3C-SiC seeds deposited by the VLS mechanism on 6H-SiC substrates.

Experimental

3C-SiC seeds, grown on Si-face on-axis (0001) 6H-SiC substrates by the VLS mechanism, were used. The VLS growth was performed in Si-Ge melts as described in [3]. The main reason for using such VLS grown layers as seeds was the significant reduction of TBs in the obtained 3C material. The subsequent CVD growth on the (111) VLS seeds was performed in a home-made cold wall reactor working at atmospheric pressure. The growth procedure of the samples is described in detail in Ref. [4]. The microstructure of these samples was studied by Transmission Electron Microscopy. These observations were performed on a JEOL 100CX conventional TEM operating at 100 kV and a JEOL 2011 HRTEM operating at 200 kV.

Results

During the growth of 3C-SiC on (0001) α-SiC substrates (i.e. in the case studied, the cubic VLS layer is heteroepitaxially grown on (0001) 6H-SiC substrates), two equally probable islands are nucleating, having an orientation relationship of 60° rotation along the cubic [111] direction, which coincides with the hexagonal [0001] direction. Both variants have the same free energy and are therefore equally stable, forming a $\Sigma 3$ twin relation, independently of the boundary which separates the two adjacent grains (contact plane). The boundary which is usually formed during growth is an incoherent $\{\bar{2}11\}$ $\Sigma 3$ one, following the growth direction. It is an incoherent one, although the contact plane coincides with the twin plane (and so it is coherent from the crystallographic point of view). The reason for this incoherence is the creation of dangling bonds, something that is explained in the drawing of Fig. 1. At point A the boundary does not stop, but continues, propagating during growth, in three different ways.

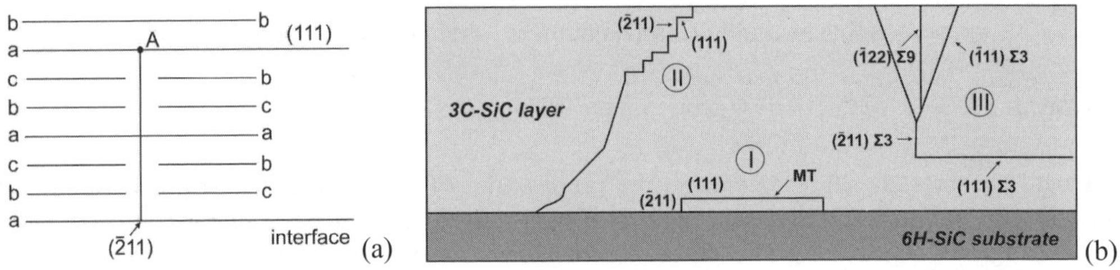

Fig. 1: Schematic illustration of the three different cases the propagation of the $(\bar{2}11)$ boundary after point A in panel (a) during growth.

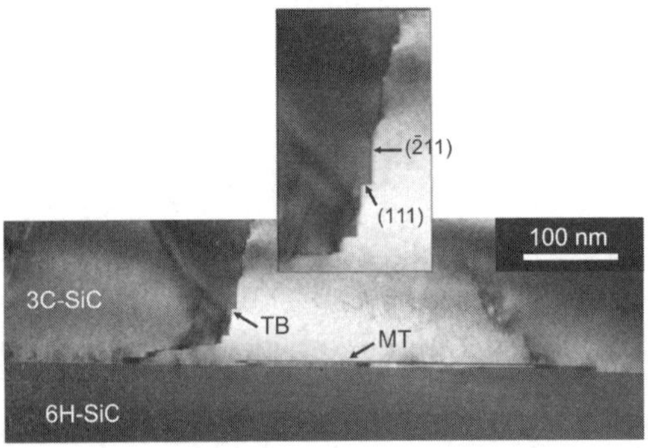

Fig. 2: XTEM micrograph of a step-like twin boundary with facets along the (111) and the $(\bar{2}11)$ planes.

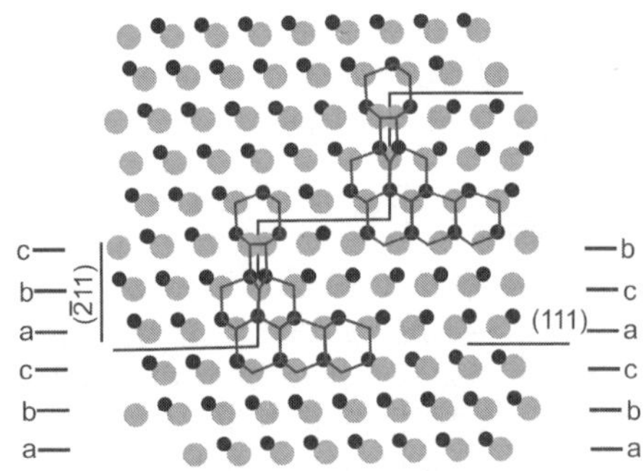

Fig. 3. Atomic model for a step-like propagation of Σ3 interface propagation alternatively along the $(\bar{2}11)$ and (111) planes.

In the first case at the very first stages of the VLS growth, the initial $\{\bar{2}11\}$ boundary "turns" 90°, to a fully coherent (111) interface, forming microtwins (MTs) with thickness of few bilayers lying on the 6H-SiC substrate. This could be triggered by a movement of a partial dislocation on its (111) slip plane due to strain in the first stages of the growth. In the second case, a step-like interface occurs with facets along the (111) and the $\{\bar{2}11\}$ planes. An example of such interface is shown in the TEM image of Fig. 2. In some cases this type of interface tends to close by mutual annihilation and in this way, isolated twin islands are formed leaving the growing crystal twin free. An atomistic model can be proposed for such alternating propagation of the Σ3 interface and it is shown in Fig. 3. This model is simply based on superposition of the symmetric crystals with a common $\{\bar{2}11\}$ interface. It includes dangling and wrong bonds, which indicates that even though it does not include lattice displacements, this boundary has very high energy, is not stable and could appear always pinned between two (111) interfaces separated not at a large distance between each other. Moreover, every third bilayer, the twin crystals from both sides of the $\{\bar{2}11\}$ boundary, have identical sequences e.g. the "a" layer in Fig. 1 (a) and Fig. 3. Indeed such atomic arrangement is observed during 3C to 6H-SiC transformation in which the 6H-SiC unit cell can be considered constituted by two three-bilayer thick twins. This is clearly seen in Fig. 4 where the Inversed Fast Fourier Transformation (IFFT) image given as inset and obtained with the (111) frequencies, clearly shows that there are not lattice displacements at such interfaces. Thus one can assume that only very short segments of such $\{\bar{2}11\}$ $\Sigma3$ interfaces can exist and they will be relaxed by appearance of coherent (111) interfaces forming either microtwins at the 6H-SiC interface, stepped interface or if the thermodynamical considerations permit, transform to 6H-SiC.

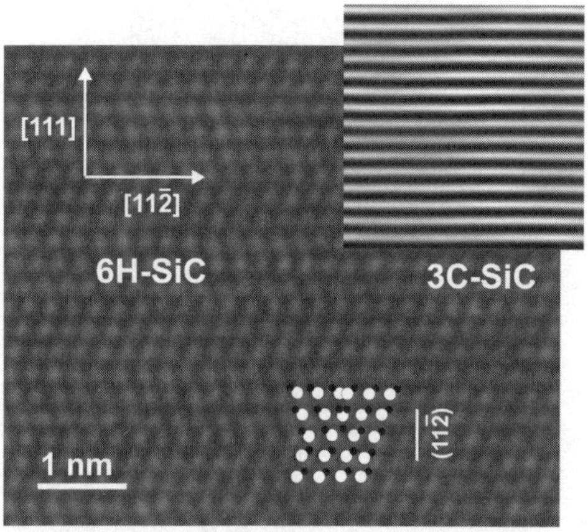

Fig. 4: HRTEM image showing that the 3C/6H-SiC polytypic transformation takes place by three bilayers thin lamellae twinned along the (111) planes and bounded along the (11$\bar{2}$) planes. The IFFT image (inset) from the (111) reflection reveals there is no lattice displacements.

Should however the growth conditions allow, i.e. for specific combinations of temperatures, growth rates or growth mechanisms, the $\{\bar{2}11\}$ $\Sigma3$ interfaces will tend to appear in different atomic structural units where the lattice relaxation will be accommodated through local lattice displacements. Indeed, frequently during CVD growth on VLS layers, strongly depending on the growth conditions, a fourfold twin complex appears which is nucleating in the first stages of the growth [5]. As it is shown in Fig. 5 such manifold twinning does not appear to nucleate from a single point, but from a region on $\{\bar{2}11\}$ interface. As it is seen from Fig. 5 the segment of the $\{\bar{2}11\}$ interface does not propagate perpendicular to the (111) twin boundary and hence the proposed in Fig. 3 atomic model is no longer valid. Moreover, if one examines carefully the regions just above the first (111) interfaces and in the vicinity of the two $(\bar{1}11)$ interfaces as shown in Fig. 5 (insets), it is evident that with the propagation of the $\{\bar{2}11\}$ interface, lattice displacements, which can be up to $0.45a_{<111>}$, are measured. This is indicated in the IFFT images selecting the (111) frequencies. In addition, in the close vicinity of the nucleation area of the fourfold twin, few dislocations appear along the $(\bar{1}11)$ planes as indicated in the IFFT images obtained with the $(\bar{1}11)$ frequencies. One can consider that these dislocations accompanying the introduction of the $(\bar{1}11)$ twin boundary serve to accommodate the strain introduced by the lattice distortions around $\{\bar{2}11\}$ interface. In a similar way the second $(\bar{1}11)$ boundary is introduced and this nucleates the fourfold twin complex. Once formed, either inside the VLS or the CVD layer, these complexes propagate through the entire layer up to the surface.

Summary

The $\{\bar{2}11\}$ twin boundary, which is frequently formed during the growth of 3C-SiC on a 6H-SiC substrate, is a high energy boundary and consequently an unstable one. For its relaxation and with the help of partial dislocations, {111} twin boundaries are always generated. These {111} twin boundaries propagate in three different ways. The (111) one propagates parallel to the interface and by the formation of microtwins parallel to the interface or in a step-like morphology (including $\{\bar{2}11\}$ segments) tends to twin elimination. The $\{\bar{1}11\}$ ones, inclined to the growth direction, usually propagate till the surface, together with the initial $\{\bar{2}11\}$ boundary, which means that in this case there is a multiplication of the twin boundaries. It seems that the (111) boundary pins very short segments of the $\{\bar{2}11\}$ interfaces and if the growth conditions allow relaxation of the $\{\bar{2}11\}$ interfaces, this relaxation would include the formation of the $\{\bar{1}11\}$ boundary.

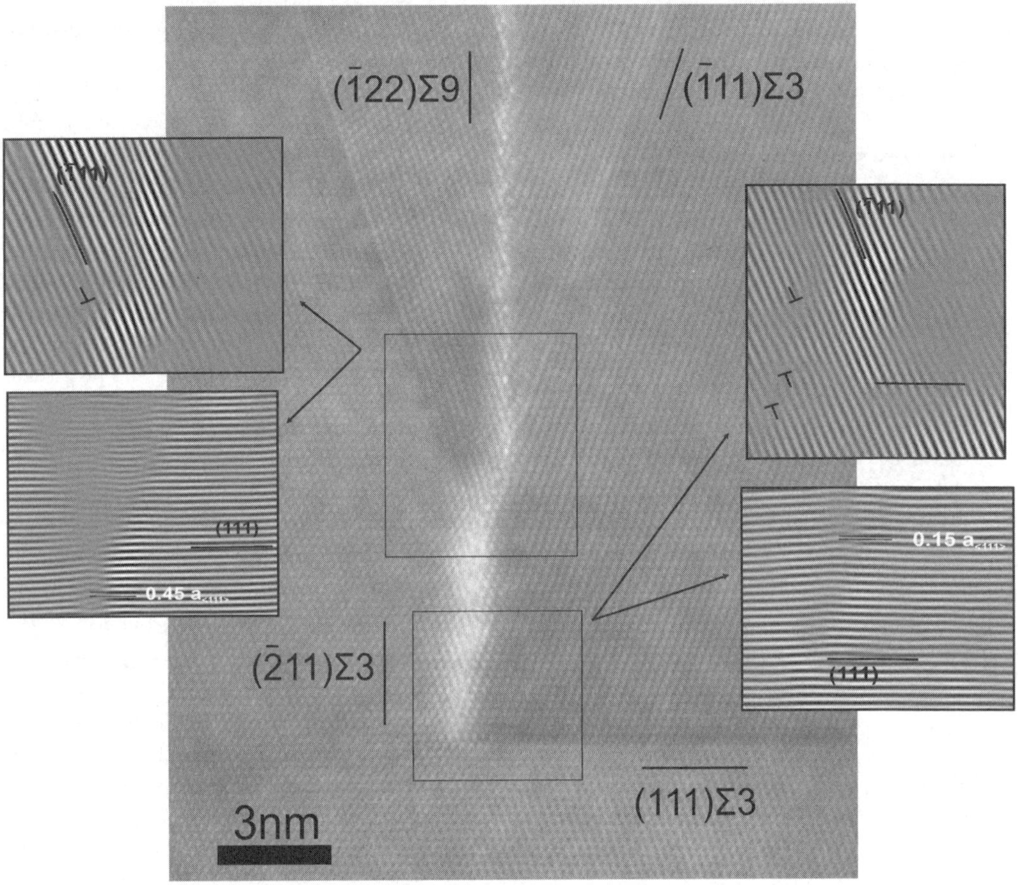

Fig. 5: HRTEM micrograph of a fourfold twin complex. IFFT images from (111) and $(\bar{1}11)$ reflections corresponding to the areas noted on the image are shown as insets. The lattice displacement and the presence of dislocations are clearly revealed.

Acknowledgements

The authors would like to thank N. Jegenyés, J. Lorenzzi and G. Ferro from UCBL, Lyon France, for providing the samples. This work has been supported by the European Commission in the framework of the MANSiC project (Grant No. MRTNCT-2006-035735).

[1] M. Kohyama and R. Yamamoto, Solid State Phenomena Vols. 37-38 (1994) 55-66.

[2] J. Eriksson, M. H. Weng, F. Roccaforte, F. Giannazzo, S. Leone, and V. Raineri, Appl. Phys. Lett. 95 (2009) 081907.

[3] J. Lorenzzi, G. Zoulis, O. Kim-Hak, N. Jegenyés, D. Carole, F. Cauwet, S. Juillaquet, G. Ferro, J. Camassel, Mater. Sci. Forum 645-648 (2010) 171-174.

[4] N. Jegenyés, J. Lorenzzi, V. Soulière, J. Dazord, F. Cauwet, and G. Ferro, Mater. Sci. Forum 645-648 (2010) 127-130.

[5] M. Marinova, N. Jegenyés, A. Andreadou, A. Mantzari, J. Lorenzzi, G. Ferro and E.K. Polychroniadis, AIP Conference Proceedings, 1292 (2010) pp. 95-98.

Materials Science Forum Vols. 717-720 (2012) pp 423-426

doi:10.4028/www.scientific.net/MSF.717-720.423

Defect structures at the silicon/3C-SiC interface

P.Hens[1, 2*], J.Müller[3], E.Spiecker[3] and P.Wellmann[2]

[1] Department of Physics, Chemistry and Biology, Linköping University, SE-58183 Linköping, Sweden

[2] Materials of Electronics and Energy Technology, University of Erlangen-Nuremberg, Martensstrasse 7, 91058 Erlangen, Germany

[3] Center for Nanoanalysis and Electron Microscopy, University of Erlangen-Nuremberg, Cauerstrasse 6, 91058 Erlangen, Germany

*corresponding author, email: phihe@ifm.liu.se

Keywords: cubic silicon carbide, heteroepitaxy, interface, dislocations, roughness, voids

Abstract: In all heteroepitaxial systems the interface between substrate and layer is a crucial point. In this work SEM and TEM studies on the interface between silicon substrate and cubic silicon carbide (3C-SiC) layers obtained by chemical vapor deposition (CVD) are presented. A clear connection between process parameters, like the design of substrate cleaning, and the heating ramp, and resulting defect structures at the substrate-layer interface could be found. Whereas the process step of etching in hot hydrogen for oxide removal is crucial for avoiding the generation of closed voids of type 2, the design of the temperature ramp-up to growth temperature during carbonization influences the interface roughness. Here a fast ramp helps to obtain a flat interface.

Introduction: The heteroepitaxial growth of cubic silicon carbide on silicon substrates leads to the formation of a high defect density at the interface. The large misfit of the lattice parameters (~20 %) and the thermal expansion coefficients (~8 %) makes the material system susceptible for structural defects emerging from the interface. By using field emission scanning electron microscopy (FE-SEM) and transmission electron microscopy (TEM) we investigated the influence of different process parameters of the CVD growth process, like the substrate cleaning and the heating ramp, on the interface roughness, the defect structure and the formation of hollow voids type 1 and filled voids type 2 [1,2] which form during carbonization and the preceding cleaning step, respectively.

Experimental: For the sample preparation two different procedures had been used. Before any processing all samples were cleaned using standard RCA procedure with NH_3, HCl and H_2O_2 followed by dipping the sample/substrate in with H_2O diluted hydrofluoric acid for removal of the oxide layer.

For the standard CVD growth the samples were etched in hot hydrogen directly before carbonization in an Aixtron AIX200/4 cold wall reactor. For this thermal etching step a temperature of 950°C and a pressure of 40 mbar was chosen for 10 minutes. The gas atmosphere during this step consisted of pure hydrogen. The following carbonization step during a continuous ramp to growth temperature was performed in a mixture of propane diluted in hydrogen at 150 mbar [3]. The same pressure was used for the following growth step with propane and monosilane as precursor gases.

In a second procedure carbon was deposited externally on a RCA and HF cleaned substrates either by spin coating of carbon particles dispersed in ethanol or by thermal evaporation of carbon resulting in a thin and closed carbon layer. Subsequently the sandwich structures had been annealed using the CVD reactor under argon or hydrogen atmosphere at reduced pressure or within a lamp heated rapid thermal annealing (RTA) furnace with argon atmosphere at ambient pressure.

SEM imaging was performed in a Zeiss Ultra2 field emission SEM. For transmission electron microscopy a Philips CM30 and a Philips CM300UT were used for bright field and high resolution imaging, respectively.

Results and discussion: A flat interface between the substrate and the silicon carbide nucleation layer, which is formed during the carbonization step, was only found for heating rates of 5 K/s and more (Fig. 1). Lower heating rates caused a severe roughening of the interface between the silicon substrate and the cubic silicon carbide layer. Accompanied by the roughening of the interface a misorientation of the nuclei in the seed layer can by observed causing a decrease in the layer quality. These effects were observed both by SEM and TEM in cross section geometry.

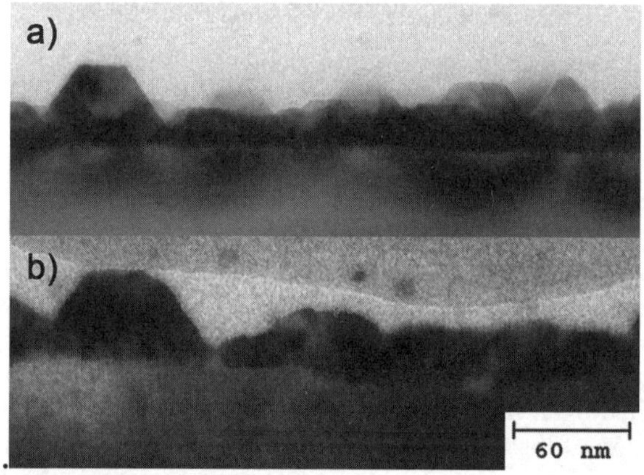

Fig. 1: Interface roughness in TEM cross section images for ramp speeds of (a) 1K/s and (b) 5K/s

HRTEM of the interface (Fig. 2) shows that the large lattice misfit between the 3C-SiC layer and the Si substrate is largely accommodated by dislocations which are located at the interface and are interconnected to stacking faults and twins in the SiC layer. A closer evaluation of the HRTEM image shows that the dislocation density corresponds well to values expected from the misfit as found in the literature. Some residual stress which is also visible in conventional TEM images (e.g. Fig. 1a) may result from the considerable difference in thermal expansion of Si and 3C-SiC.

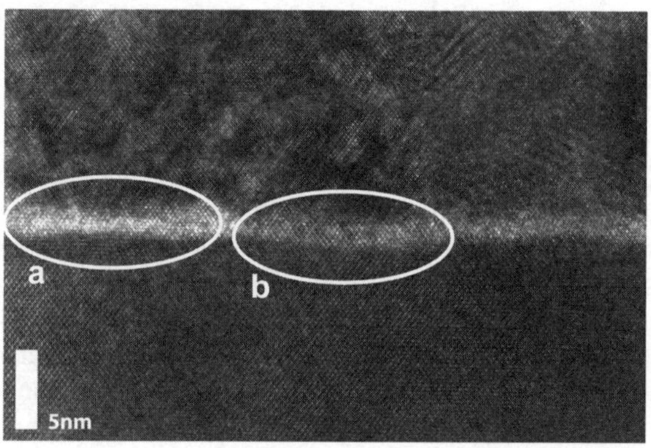

Fig. 2: High resolution cross section TEM image of the silicon/silicon carbide interface with areas of (a) high and (b) low defect density at the interface

Using rapid thermal annealing (RTA) of silicon/carbon sandwich structures grown by thermal evaporation of carbon, the formation of void defects of type 1 and 2 was investigated. Using very high ramping speeds in RTA processing a flat interface was obtained (Fig. 3) as expected from the ramping experiments for the standard carbonization (compare Fig. 1). Below the formed silicon carbide layer hollow voids of type 1 were observed even under the atmospheric pressure conditions used for the rapid thermal annealing. The generation of voids of type 1 was thereby assigned to diffusion processes along the interface during the carbonization step (Fig. 3) and seems not to be related to evaporation [4].

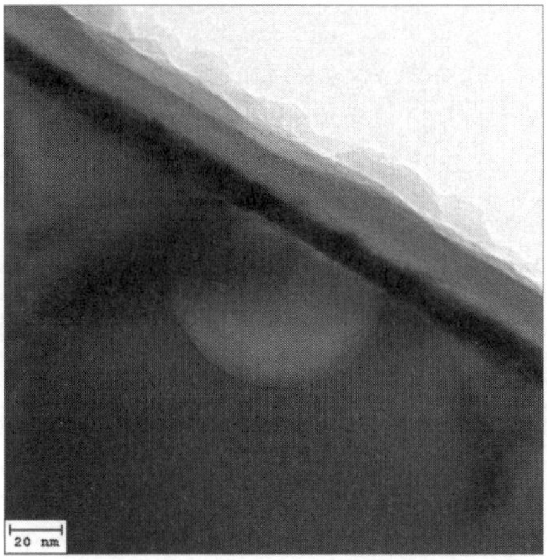

Fig. 3: Hollow void type 1 below a closed silicon carbide seed layer generated by annealing of a silicon/carbon sandwich structure

The formation of type 2 voids, which could also cause a severe roughening of the interface, was found to be related to carbon contaminants on the substrate surface during the etching step for the removal of the native oxide in hot hydrogen (Fig. 4). Intentionally deposited carbon particles on the surface caused the formation of pyramidal pits during annealing at temperatures above 950°C. The composition of the gas atmosphere was found to have no effect since the void type 2 generation was observed under low pressure hydrogen and argon atmosphere as well as inert gas (Ar) at atmospheric pressure.

Fig. 4: Void defects type 2 formed at carbon particles after rapid thermal annealing

Conclusion:

An ex-situ carbonization step using rapid thermal annealing of silicon-carbon sandwich structures allowed studying the influence of particles on the change of the surface structure/morphology. By this the generation of filled voids type 2 could be assigned to the presence of carbon particles contaminating the surface. Besides a roughening due to these voids type 2 an influence of the ramp speed during carbonization on the interface roughness was found for the standard in-situ carbonization method. Furthermore it was shown that fast heating rates can significantly reduce the interface roughness. These results could be confirmed by means of transmission electron microscopy

Acknowledgements: Financial support by the Bundesministerium für Bildung und Forschung (BMBF) under contract number 03SF0393 is gratefully acknowledged

References

[1] P. Hens, J. Müller, L. Fahlbusch, E. Spiecker, P.Wellmann, *Generation of void-like structures during hot-hydrogen etching of Si substrates for 3C-SiC epitaxy*, Mat. Sci. Forum 679-680 (2011) 127-130.

[2] J. Müller, P. Hens, P. Wellmann, E. Spiecker, *Effect of heating rate on the microstructure and quality of CVD-grown 3C-SiC films on Si, Proceedings of the Microscopy Conference* 2011 Vol. 3 (to be published)

[3] P. Hens, G. Wagner, A. Hölzing, R. Hock, P. Wellmann, *Fundamental study of the temperature ramp-up influence for 3C-SiC hetero-epitaxy on silicon (100)*, Mat. Sci. Forum 645-648 (2010) 151-154

[4] L.-O.Björketun, L.Hultman, I.P.Ivanov, Q.Wahab, J.-E.Sundgren, *Interfacial void formation during vapor phase growth of 3C-SiC on Si(001) and Si(111) substrates – Characterization by transmission electron microscopy*, Journal of Crystal Growth, 182 (1997) 379-388

Materials Science Forum Vols. 717-720 (2012) pp 427-432

doi:10.4028/www.scientific.net/MSF.717-720.427

Electrically Detected Magnetic Resonance (EDMR) Studies of $SiC-SiO_2$ Interfaces

T. Umeda[1,a,*], R. Kosugi[2,b], K. Fukuda[2,c], N. Morishita[3,d], T. Oshima[3,e], K. Esaki[4], and J. Isoya[4,g]

[1]Institute of Applied Physics, University of Tsukuba, Tsukuba, 305-8573, Japan

[2]National Institute of Advanced Industrial Science and Technology, Tsukuba, 305-8568, Japan

[3]Japan Atomic Energy Agency, Takasaki, 370-1292, Japan

[4]University of Tsukuba, Tsukuba 305-8550, Japan

[a]umeda@bk.tsukuba.ac.jp, [b]r-kosugi@aist.go.jp, [c]k-fukuda@aist.go.jp, [d]morishita.norio@jaea.go.jp, [e]ohshima.takeshi20@jaea.go.jp, [g]isoya.junichi.fw@u.tsukuba.ac.jp, *corresponding author

Keywords: EDMR, ESR, $SiC-SiO_2$ interface, interface states, MOSFET, hydrogen, nitrogen, shallow donors, channel mobility.

Abstract. We discuss the results of electrically detected magnetic resonance (EDMR) spectroscopy on $SiC-SiO_2$ interfaces interacting with hydrogen and nitrogen. Three types of Si-face 4H-SiC metal-oxide-semiconductor field effect transistors (MOSFETs) were examined by EDMR in the temperature range of 4–300 K. These samples revealed several different characteristics from those of the earlier electron spin resonance (ESR) and EDMR studies on $SiC-SiO_2$ interfaces. The most significant finding was the high-density doping of nitrogen shallow donors into the channel region after the post nitridation anneal. The incorporated nitrogen donors were observed as an "Nh" EDMR signal at 4–20 K. The roles of these nitrogen donors are discussed in connection with the electrical properties of SiC MOSFETs.

ESR and EDMR Studies on $SiC-SiO_2$ Interface Systems

ESR spectroscopy on metal-oxide-semiconductor (MOS) systems was first demonstrated in 1971 [1]. It successfully detected three types of ESR centers, P_a, P_b, and P_c in $Si-SiO_2$ interfaces [1], and currently, we are familiar with the P_b centers [Si dangling bonds (DBs)] which are dominant and intrinsic defects in this interface system. Many other ESR centers in this MOS system have since been reported [2]. In contrast, ESR might not be so successful for studying $SiC-SiO_2$ interfaces, despite them being similar to the $Si-SiO_2$ interface and having a much higher interface-state density. There is a clear difference between Si and SiC; ESR measurements on Si always require a special pre-treatment of the sample surface, owing to undesirable ESR signals from the surface/interface region. In contrast, such a pre-treatment is not necessary for SiC samples, even if their surfaces are oxidized.

Why doesn't ESR work the same way for $SiC-SiO_2$ as it does for $Si-SiO_2$ interfaces? There is as yet no answer, but it may be related to several differences between the two MOS systems.

The first notable difference is the wider band gap of SiC (e.g., 3.24 eV for 4H-SiC) [3]. In principle, this wider band gap makes it possible to involve novel types of interface states in SiC MOS systems that are different from the P_b-like centers in Si. The other notable difference is the presence of carbon atoms in the $SiC-SiO_2$ interface. Thus, carbon-related defects have been proposed as the origin of the SiC MOS interface states [3-6]. For example, ESR studies revealed an amorphous carbon signal of g (g value) ≈ 2.003 in some oxidized 3C/4H/6H-SiC wafers [3,4] (in particular, this signal was strong in *p*-type 3C MOS samples). In addition, theoretical studies predicted a high density of carbon dimers (a pair of threefold coordinated carbons) in the interface [5,6]. Although such carbon dimers have not been experimentally detected so far, similar defects (the HEI5/6 centers) have been identified by conducting ESR on bulk samples [7]. Therefore, ESR may have the potential for detecting the key defects arising from carbon atoms.

There are several important reports on carbon-related interface defects detected by ESR and electrically detected ESR [we shall call this "electrically detected magnetic resonance (EDMR)"]. One famous study reported a carbon DB center (labeled "P_{bC}", their g values are $g_{\parallel} = 2.0023$ and $g_{\perp} = 2.0032$) in oxidized *porous*-SiC [8,9]. This is a really P_b-like center at the SiC-SiO_2 interface, and it exhibited a chemical reaction with hydrogen [9], similarly to the P_b center [2]. Unfortunately, the same ESR signal has not been detected from single-crystal wafer samples, so the areal density of P_{bC} at the interface is not yet known. Accordingly, we cannot conclude if the P_{bC} center accounts for the major part of the high-density interface states.

Another type of carbon-related interface center was found in 4H/6H-SiC MOSFETs by taking advantage of the EDMR technique [10-12]. This center had $g = 2.0027$ and the same signatures as the well-studied silicon vacancy in bulk (the V_{Si}^- center [13]). Accordingly, the $g = 2.0027$ center has been identified as a silicon-vacancy center *in the interface* [12]. Basically, a silicon vacancy has four carbon DBs which generate several energy levels in the mid gap of SiC [13]. The important features of this center are that (1) it was present in some 4H/6H-SiC MOSFETs because it has been detected by EDMR on these devices, (2) its signal intensity was extremely high, suggesting a high concentration of this defect, and (3) it could be annihilated when the MOS interface was nitrided [12]. From these results, this center was proposed as a dominant interface defect [12]. Moreover, after the nitridation of the interface, another type of P_b-like defect was observed by EDMR [11,12]. This was taken to be a sort of DB perpendicular to the interface, with new g values of $g_{\parallel} = 2.0026$ and $g_{\perp} = 2.0010$.

The above ESR/EDMR studies indicate that it is possible to correlate the origin of the SiC-SiO_2 interface states with carbon-related interface defects. Furthermore, their reactions with hydrogen and nitrogen at the interface were also observed. We should, however, point out that these findings are only the beginning, and it will be necessary to continue ESR and EDMR studies on a wider variety of SiC-SiO_2 interfaces. The reason is that we have obtained different EDMR results using different samples.

Low-temperature EDMR Study

We employed the same EDMR technique [10-12] as used in SiC MOSFETs. However, contrary to the earlier ESR/EDMR studies [4,8-12] that were done at room temperature, our EDMR measurements were mainly done at low temperatures (down to 4 K). It is generally known that many ESR centers, related to shallow energy levels or associated with thermal activation processes, become detectable when the temperature goes down (e.g., carbon dimers in the bulk were clearly detected under such conditions [7]). As expected, our study revealed such low-temperature signals.

For the EDMR measurements, we prepared three different MOSFET samples. They are *n*-channel lateral 4*H*-SiC MOSFETs (gate length = 100 μm, gate width = 150 μm) fabricated on *n* or *p* epitaxial layers of Cree 8°-off 4*H*-SiC(0001) Si-face wafers (for the *n* epitaxial layer, [N] $\approx 5\times10^{15}$ /cm^3 and Al implantation was performed with [Al] $\approx 1\times10^{20}$ /cm^3). Source/drain regions were formed by high-dose phosphorus implantation ([P] $\approx 1\times10^{20}$ /cm^3) and were connected to Ni contacts. Gate oxides were grown by dry oxidation (dry SiO_2 thickness = 50 or 60 nm) and were subjected to post-nitridation annealing (NO-N_2O ambient at 1250°C) or post-hydrogen annealing (H_2 ambient at 650°C). We used poly-Si gate electrodes. Finally, three 4*H*-SiC MOSFET samples, labeled "Nitrogen," "Hydrogen," and "Dry" (dry oxidation only), were prepared. Their field-effect channel mobilities (μ_{FE}) varied from 2.5 cm^2/V·s (the Dry sample) to 19.2 cm^2/V·s (the Nitrogen sample). Some of the MOSFETs were subjected to γ-ray irradiation (average energy = 1.2 MeV, dose = 2.3 Mrad) at room temperature.

The EDMR spectra were measured using lateral channel currents (I_{ds}) between the source and drain of each MOSFET in the dark. Only ESR centers interacting with the channel current could be detected in this regime. In principle, the inversion layer of the channel region allows paramagnetic states only in the vicinity of the conduction band edge (E_C). Thus, ESR centers in such a shallow region will be detectable. The currents I_{ds} were 1–100 nA under drain-source biases (V_{ds}) of 1–5 V and gate-substrate biases (V_{gs}) of 1–30 V at 4–300 K. We were not able to measure V_{gs} over 30 V because of large gate leakage currents. At 4 K, only the Nitrogen samples could activate the channel current, and only for these samples could we measure the 4 K EDMR spectra. The reason for this exception will be explained later. To excite the ESR transition, we used a Bruker Bio-Spin TE_{011} microwave cavity and 9.4-GHz microwaves at a power of 32–200 mW. Small changes in I_{ds} (expressed in ppm) due to ESR were monitored by a lock-in amplifier synchronized with a magnetic-field modulation at 1.56 kHz. The readers may consult our papers (Refs. [14] and [15]) for more details on the experimental parameters.

In the following sections, we separately present our EDMR results for the Hydrogen sample and the Nitrogen/Dry samples, from which we will discuss the behaviors of shallow interface defects, hydrogen, and nitrogen in the SiC-SiO_2 interfaces.

Hydrogen-related Interface Defects

EDMR results for the Hydrogen sample were already presented in Ref. [14]. Here, we shall review the important points of those results. (1) In the as-prepared MOSFETs, no EDMR signals were observed at 20–300 K. We emphasize that no interface signals of P_{bC} [8,9], silicon vacancy [10,12], or new DB center [11,12] were observed in our samples. (2) After the γ-ray irradiation, two EDMR signals were detected at 20 K, labeled "P_{H0}" and "P_{H1}" (Fig. 1). The role of γ-ray irradiation is most likely to remove the effect of hydrogen passivation in the Si MOS system [2]. (3) The P_{H1} signal exhibited an isotropic doublet hyperfine splitting (hfs) of 5.4 mT, which is most probably due to a hydrogen nucleus (^{1}H, nuclear spin $I = 1/2$, natural abundance = 99.99%) nearby this defect. In summary, the observed defects should be related to hydrogen.

The P_{H0}/P_{H1} signals disappeared in the temperature range between 20 and 50 K, most probably due to the lifetime broadening. This feature was not seen in the previous interface signals [3,4,8-12]. The g values of P_{H0}/P_{H1} ($g = 2.003$–2.004, slightly anisotropic) were close to those of the silicon-vacancy signal ($g = 2.0027$) [10,12,13] or amorphous carbon signal ($g = 2.0030$) [3,4]. The signal width was found to be fairly anisotropic (peak-to-peak width = 1.3 mT and 0.8 mT for the [0001] and [11-20] directions, respectively), and this indicated an anisotropic structure for the P_{H0}/P_{H1} centers.

Our provisional assignment for the P_{H0}/P_{H1} centers is that they are a family of carbon DBs with/without a nearest hydrogen atom [14]. We imagined that in as-prepared samples, they were passivated by hydrogen atoms and became visible after breaking C-H bonds by γ-ray irradiation. A single ^{1}H hfs (5.4 mT) of the P_{H1} center suggests that a passivated hydrogen is immediately captured at the neighboring site of a carbon DB. In contrast, in the case of the P_{H0} center, a hydrogen atom may diffuse far away from a defect site. These centers may be consistent with the predicted carbon-dimer defects [5,6]. The carbon dimers hold DB-like orbitals on threefold coordinated carbons, which can interact with hydrogen atoms [5].

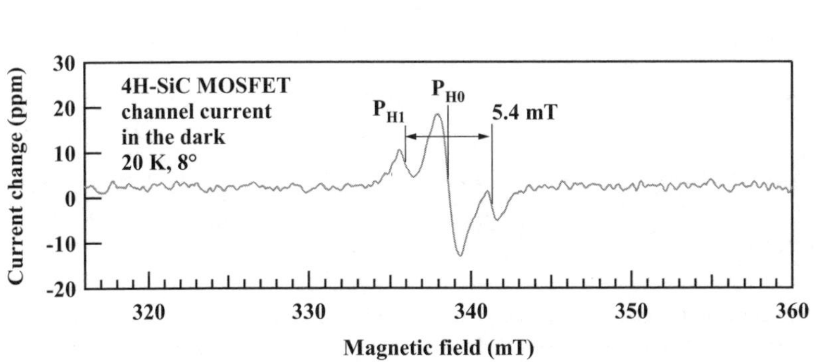

Fig. 1. P_{H0}/P_{H1} EDMR signals in the Hydrogen sample [14]. These signals were observable at 20 K and after γ-ray irradiation.

However, we *do not* consider that the P_{H0}/P_{H1} centers are dominant defects at the interface. Their EDMR signal intensities were smaller than 30 ppm in I_{ds}; they were too small, e.g., compared with the case of the interface silicon-vacancy center (10^2–10^3 ppm) [12]. Furthermore, it should be mentioned that some of the Hydrogen samples did not exhibit these signals even after γ-ray irradiation. Thus, it is difficult to correlate the P_{H0}/P_{H1} centers with the major part of the SiC-SiO_2 interface states. This conclusion seems consistent with the general agreement about SiC-SiO_2 interfaces such that the hydrogen annealing is often ineffective for improving their quality [3,5], and it is strikingly in contrast to the case of Si-SiO_2 interfaces [2].

Nitrogen Shallow Donors in the Channel

We compared the Nitrogen and Dry samples fabricated with almost the same process. EDMR results for these samples will appear in a separate paper in detail [15], and we pick up again notable points of those results. (1) In both the Nitrogen and Dry samples, the room-temperature signals of the P_{bC} center, the interface silicon vacancy, and the new DB center [8-12] were below the detection limit. Therefore, in some types of MOSFETs like ours, these interface defects may not be dominant. (2) On the other hand, at 20 K, EDMR signals became detectable in the Dry sample [Figs. 2(a) and (c)]. One signal was similar to the P_{H0} signal [Fig. 2(c)]. There might also be a very broad resonance in higher magnetic fields [Fig. 2(a)]. If this broad resonance was a powder-pattern signal, its g values were roughly estimated in the range between 1.0 and 1.4. Since this range was out of the normal range of ESR centers in SiC, we can draw no conclusion about this signal at present. However, its signal intensity would be 100 times larger than that of the P_{H0}-like signal because of its large signal width. Thus, this signal might be a candidate for the origin of the interface states. (3) The above EDMR signals were absent in the Nitrogen sample [Figs. 2(b) and (c)]. This behavior is similar to those reported for the room-temperature EDMR signals [10,12]. In summary, nitrogen atoms are effective for removing some interface defects detectable at room temperature and at 20 K as well. (4) Even after γ-ray irradiation, the Nitrogen samples did not show any EDMR signals at 20 K. This behavior was in contrast to the case of hydrogen. Therefore, nitrogen atoms should form stable binding states in the interface region. (5) At 4 K, a very strong sharp signal was observed in the Nitrogen samples (Fig. 3). This signal was unchanged after the γ-ray irradiation. We labeled it "Nh".

The Nh signal was strongly anisotropic. The angular dependence of this signal showed a *c*-axial (C_{3v}) symmetry with $g_{\parallel} = 2.0047$ and $g_{\perp} = 2.0008$, which is similar to the case of nitrogen shallow donors at the carbon *h* site of 4*H*-SiC ($N_C(h)$) [16]. Table 1 summarizes the ESR signatures of the Nh center and two shallow donors, $N_C(h)$ and $N_C(k)$. Nh and $N_C(h)$ showed similar hyperfine splitting of the ^{14}N nucleus (I = 1, natural abundance = 100%) and observable temperatures. Therefore, we concluded that the Nh signal originates from the *h*-site nitrogen shallow donors *in the channel region*. The Nh signal exhibited strong lifetime broadening (Fig. 3) and eventually disappeared above 20 K.

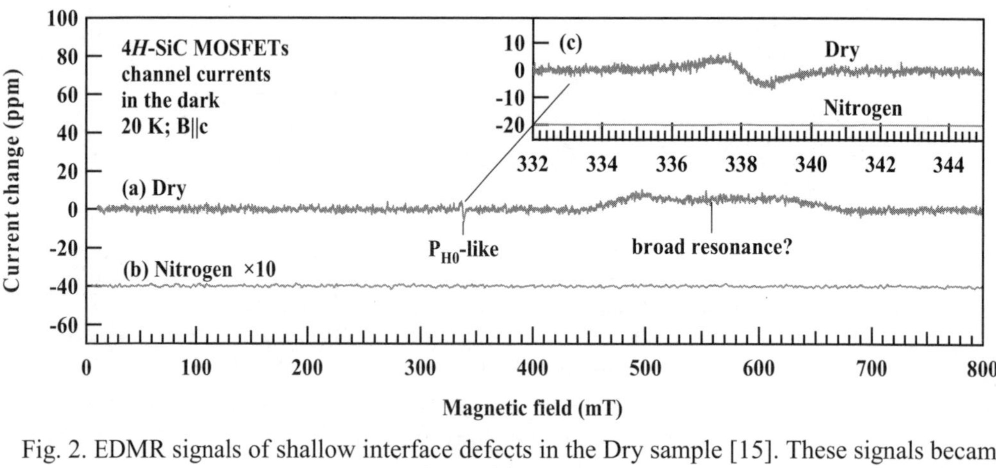

Fig. 2. EDMR signals of shallow interface defects in the Dry sample [15]. These signals became observable at 20 K.

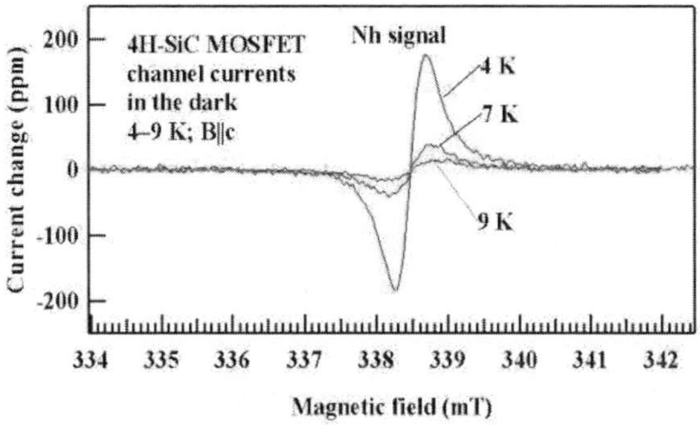

Fig. 3. Nh EDMR signal in the Nitrogen sample, observable below 20 K [15].

Table 1. ESR parameters of nitrogen shallow donors in 4*H*-SiC MOSFET and bulk 4*H*-SiC.

Nitrogen center	Symmetry	$g_{\parallel}$ $g_{\perp}$	^{14}N hfs (mT)	Observable Temp. (K)	Ionization energy to E_C (meV)	Activation energy of lifetime broadening	Ref.
Nh (MOSFET channel)	C_{3v}	2.0047 2.0008	<< 0.4	≤ 20 K	N/A	3.3 meV	[15]
$N_C(h)$ (bulk)	C_{3v}	2.0055 2.0010	0.2	≤ 20 K	52 meV	7.6 meV	[16]
$N_C(k)$ (bulk)	C_{3v}	2.0043 2.0013	3.6	≤ 80 K	92 meV	45.5 meV	[16]

The present results indicate that after post-nitridation annealing, nitrogen atoms can diffuse even into the channel region and act as shallow donors there. In fact, the Nitrogen sample showed an enhanced channel current at 4 K owing to this "nitrogen doping" effect, while the other MOSFET samples could not activate the channel current at the same temperature. Moreover, the very intense Nh signal (= 10^2–10^3 ppm) suggested a substantial level of nitrogen doping.

In fact, we could have measured the "doped" nitrogen atoms by means of x-ray photoemission spectroscopy (XPS) [17]. For the XPS measurements, we prepared "Nitrogen" and "Dry" wafer samples similarly to the corresponding Nitrogen and Dry MOSFET samples. Then, after completely etching the oxide layer with a diluted hydrofluoric acid, we measured the XPS spectra of the two wafer samples. In the Nitrogen and Dry wafer samples, the oxygen peaks due to the oxide layer were completely eliminated after the etching. However, in the Nitrogen wafer sample, nitrogen peaks were still clearly detectable, in addition to silicon and carbon peaks of the substrate. On the basis of this observation, we estimated the nitrogen concentration in the substrate to be on the order of 10^{14} cm^{-2}. Some of the nitrogen atoms may change into shallow donors as a result of the high-temperature annealing during the nitridation. The total nitrogen concentration measured by XPS increased linearly as the nitridation temperature increased in the range between 800 and 1400 °C. Simultaneously, the interface-state density was found to decrease from 2×10^{12} cm^{-2}eV^{-1} to 5×10^{11} cm^{-2}eV^{-1} at $E_C - 0.2$ eV.

Effect of Nitrogen Donors on SiC MOSFET

We consider that nitrogen atoms play two significant roles in the Si-face SiC MOS system. One is to remove some interface defects, as evidenced in both the room-temperature EDMR studies [10-12] and our low-temperature EDMR studies. Another is that they behave as a shallow donor in the channel region. Below, we briefly discuss how the nitrogen shallow donors influence the electrical properties of SiC MOSFETs.

First, nitrogen shallow donors increase the carrier concentration (Q_{ch}) of the channel region. This increase may amount to up to 10^{14} cm^{-2} at maximum. Simultaneously, they cause a drop in the threshold voltage (V_{th}) of the SiC MOSFET via compensation of the acceptor concentration (N_A) in the channel (c.f., $V_{th} \propto N_A^{1/2}$). In fact, a drop in V_{th} has been observed in many nitride SiC MOSFETs [18], and in ours as well [15].

Second, nitrogen shallow donors increase the "Coulomb mobility" [19] of electrons in the channel. Since high-density free carriers suppress the Coulomb scattering mechanism [19], the channel mobility (μ_{FE}) may be improved by the free carriers emitted from the "doped" nitrogen donors. In addition, the increase in Q_{ch} causes a higher I_{ds} current, which appears as a higher μ_{FE} value. Of course, the reduction in the interface states caused by binding nitrogen atoms also contributes to increasing μ_{FE}.

In contrast, the high-density nitrogen donors increase the surface potential fluctuation, causing "surface roughness" scattering [19]. This mechanism degrades μ_{FE}. Consequently, the nitrogen donors may have opposing effects on the channel conductivity, and hence, their presence may result in an intermediate level of performance. This model offers one reason why the nitridation process could not fully alleviate the channel mobility degradation of SiC MOSFETs.

Summary

We studied the SiC-SiO_2 interface regions of three types ("Hydrogen", "Nitrogen", and "Dry") of 4*H*-SiC MOSFETs by means of low-temperature EDMR measurements. The earlier ESR/EDMR studies were done at room temperature, and our study revealed novel behaviors of this interface system at low temperatures. We observed that hydrogen and nitrogen atoms at the interface could passivate or annihilate some of the shallow interface defects. In the case of nitrogen, another significant role was found: nitrogen doping of the channel region. We supposed that the high-density nitrogen donors in the channel region greatly influence the channel conductivity of SiC MOSFETs. We conducted that the post-nitridation technique for SiC MOSFETs must be optimized to make better use of the different behaviors of nitrogen atoms.

References

[1] Y. Nishi: Jpn. J. Appl. Phys. Vol. 10 (1971), p. 52.
[2] P. M. Lenahan and J. F. Conley, Jr.: J. Vac. Sci. Technol. B Vol. 16 (1998), p. 2134.
[3] V. V. Afanas'ev, *et al.*: Mater. Sci. Forum Vol. 483-485 (2005), p. 563, and references there-in.
[4] P. J. MacFarlane and M. E. Zvanut: J. Appl. Phys. Vol. 88 (2000), p. 4122.
[5] S. T. Pantelides *et al.*: Mater. Sci. Forum Vol. 527-529 (2006), p. 935.
[6] S. Wang, *et al.*: Phys. Rev. Lett. Vol. 98 (2007), p. 026101.
[7] T. Umeda, *et al.*: Phys. Rev. B Vol. 79 (2009), p. 115211.
[8] J. L. Cantin, *et al.*: Phys. Rev. Lett. Vol. 92 (2004), p. 015502.
[9] J. L. Cantin, *et al.*: Appl. Phys. Lett. Vol. 88 (2006), p. 092108.
[10] D. J. Meyer, P. M. Lenahan, and A. J. Lelis: Appl. Phys. Lett. Vol. 86 (2005), p. 023503.
[11] M. S. Dautrich, P. M. Lenahan, and A. J. Lelis: Appl. Phys. Lett. Vol. 89 (2006), p. 223502.
[12] C. J. Cochrane, P. M. Lenahan, and A. J. Lelis: J. Appl. Phys. Vol. 109 (2011), p. 014506.
[13] E. Janzén, *et al.*: Mater. Sci. Forum Vol. 615-617 (2009), p. 347, and references there-in.
[14] T. Umeda, *et al.*: Mater. Sci. Forum, Vol. 679-680 (2011), p. 370.
[15] T. Umeda, *et al.*: Appl. Phys. Lett. Vol. 99 (2011), p. 142105.
[16] S. Greulich-Weber: phys. stat. sol. (b) Vol. 210 (1998), p. 415.
[17] R. Kosugi, T. Umeda, and Y. Sakuma: Appl. Phys. Lett. Vol. 99 (2011), p. 182111.
[18] A. Agarwal *et al.*: Mater. Sci. Forum Vol. 600-603 (2009), p. 895.
[19] K. Matocha and V. Tilak: Mater. Sci. Forum Vol. 679-680 (2011), p. 318.

Materials Science Forum Vols. 717-720 (2012) pp 433-436

doi:10.4028/www.scientific.net/MSF.717-720.433

Definitive Identification of an Important 4H SiC MOSFET Interface/Near Interface Trap

C.J. Cochrane[1,a], P.M. Lenahan[2,b] and A.J. Lelis[3,c]

[1,2]The Pennsylvania State University, 212 EES Building, University Park, PA 16802 USA

[3]U.S. Army Research Laboratory, 2800 Powder Mill Road, Adelphi, MD 20783 USA

[a]cjc203@psu.edu, [b]pmlesm@engr.psu.edu, [c]aivars.j.lelis@us.army.mil

Keywords: SiC/SiO_2 interface/near interface traps, magnetic resonance, EDMR, EPR

Abstract. We utilize electrically detected magnetic resonance (EDMR) via spin dependent recombination (SDR) to provide a definitive identification of an interface/near interface defect present in a wide variety of 4H SiC/SiO_2 metal oxide semiconducting field effect transistors (MOSFETs).

Introduction

The performance and reliability of SiC MOSFETs is currently limited by the presence of SiC/SiO_2 interface and near interface traps. Conventional electron paramagnetic resonance (EPR) studies of porous SiC samples have been utilized to argue for carbon dangling bond interface traps [1]. A series of electrically detected magnetic resonance (EDMR) studies provides extensive evidence linking a specific magnetic resonance spectrum with an isotropic $g \approx 2.003$ and poorly resolved side peaks to important SiC/SiO_2 interface/near interface defects in many fully processed SiC MOSFETs [2,3]. The EDMR spectrum has, unfortunately, been only tentatively linked to a specific defect—a silicon vacancy—because it had previously not been possible to quantitatively evaluate the electron nuclear hyperfine interactions at the site. In this work, we provide extremely strong experimental evidence identifying the structure of that defect. The evidence comes from very high resolution and sensitivity "fast passage" (FP) mode [4, 5] electrically detected magnetic resonance (EDMR) or FPEDMR of the ubiquitous EDMR spectrum. The fast passage approach [4, 5] offers multiple advantages in both sensitivity and resolution; however, to the best of our knowledge the work reported herein represents the first application of fast passage to EDMR. Fast passage brings about increased resolution because only the defects that satisfy the requirements will contribute to the overall spectrum. The FPEDMR results provide extremely strong, essentially definitive results linking the spectrum to a negatively charged silicon vacancy. This very strong evidence comes from a comparison of FPEDMR spectra and conventional EPR results reported earlier by several groups studying negatively charged silicon vacancies via both conventional EPR and electron nuclear double resonance (ENDOR) in large volume ($\approx 1 cm^3$) samples which have been subjected to particle bombardment [6]. The negatively charged silicon vacancy has been well characterized via conventional resonance in these large volume samples; it exhibits multiple side peaks which have been convincingly linked to electron nuclear hyperfine interactions with nearby ^{29}Si and ^{13}C nuclei [6].

Experimental

We use the electrically detected form of ESR know as EDMR. EDMR can be used for the detection of deep level defects in a variety of fully processed devices; it can have a sensitivity more than 10 million times greater than conventional EPR [7]. In our measurements, we utilize SDR, a form of EDMR, to observe a spin dependent recombination current from a fully processed 4H SiC MOSFET.

We also utilize the gated diode electrical measurement known as direct current, current vs. voltage (DCIV) [8]. In the DCIV measurement, one forward biases the shorted drain and source to substrate junctions of a MOSFET and then monitors the substrate current while sweeping the gate voltage. The DCIV measurement provides information about both the density and physical location of interface/near interface trapping defects.

In SDR, we expose a MOSFET, biased appropriately for maximum recombination, to a large DC magnetic field and microwave irradiation. The magnetic field is then swept while the spin dependent recombination current is simultaneously monitored. Resonance is achieved when the energy of the applied microwaves equals that of the difference in energy of the electron's spin states. For the simplest case, this energy corresponds to $\Delta E = h\upsilon = g\beta H$, where h is Planck's constant, υ is the radiation frequency, β is the Bohr magneton, H is the magnetic field at resonance, and g is a second rank tensor. The SDR response is virtually identical to the EPR spectrum of the defect under observation. The combination of DCIV and SDR measurements is extremely useful for investigating device reliability and performance problems in MOSFETs.

Our measurements were performed at room temperature on 4H SiC n channel MOSFETs with a gate area of 100 μm x 100μm fabricated by Cree. The source and drain were doped with phosphorus and the gate dielectric is a deposited ONO oxide, which received a 3 hour Ar anneal at 1200°C. The experimental apparatus consisted of a custom built SDR spectrometer which includes a Varian E-line Century four inch magnet and power supply, a Resonance Instruments 8330 X-band bridge, a TE_{102} microwave cavity, and a Lakeshore 475 DSP field controller. SDR currents were measured with a Stanford Instruments SR570 current preamplifier, and a Stanford Instruments SR830 Lock-In Amplifier. The devices were not stressed, so the defects observed were created during device fabrication.

Results

Deep level defects are responsible for electron hole recombination. If the substantial majority of these defects are located at or very near the 4H SiC/SiO_2 interface, the source drain to substrate recombination current (the DCIV current) will peak when the gate is biased so as to yield approximately equal electron and hole densities at the interface. In older generation 4H SiC devices, such a peak could usually be observed. This peak in the DCIV current is generally accompanied by a similar peak in the SDR versus gate bias response. However, such a peak in the DCIV is not often observed in more recently processed devices and the peaks in the SDR response are generally much weaker or entirely absent. We have interpreted this to mean that, in more recent devices, technology improvements in the engineering of the interface region significantly reduce defects precisely at the SiC/SiO_2 interface without substantially eliminating defect density within the SiC substrate region. The devices utilized in this study yield an intermediate response: no peak in the DCIV current measurements, and a weak peak in the SDR versus voltage response. This behavior is illustrated in

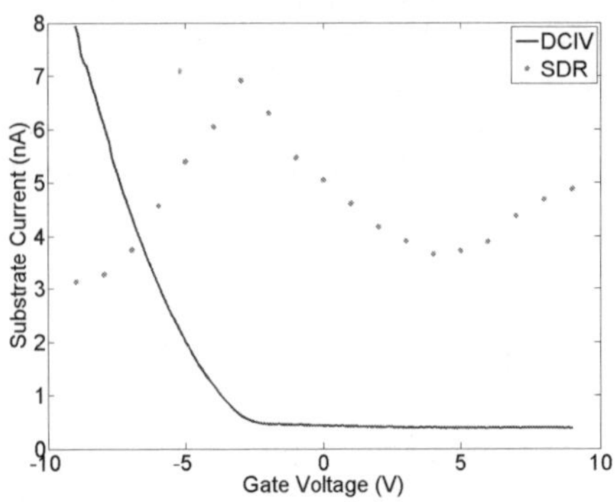

Figure 1: DCIV and SDR versus gate bias.

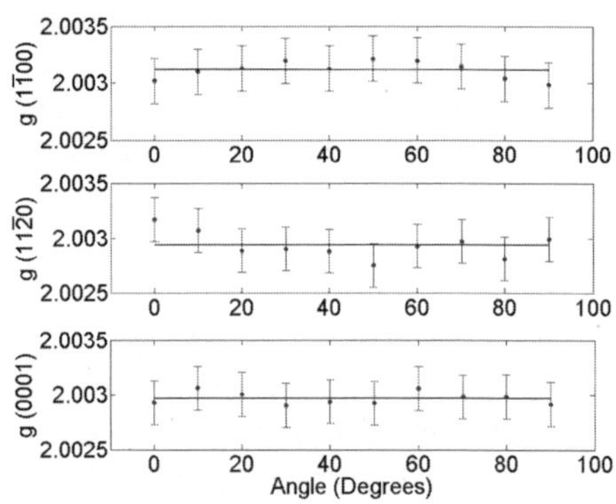

Figure 2: G value versus angle of device with respect to magnetic field for three different axes.

the representative results shown in Fig 1. The weak peak in the in the SDR does indicate that a preponderance of the defects observed via SDR are in the immediate vicinity of the SiC/SiO_2boundary.

One of the most important characteristics of the EDMR spectrum is the relationship between the magnetic field and the orientation of the device in the microwave cavity. In Fig. 2, we plotted the g factor of the center line versus angle of rotation for three different axes using conventional demodulation techniques to observe EDMR. We observed a highly isotropic center line $g \approx 2.0030 \pm 0.0002$. This strongly indicates that the dominating defect is of high symmetry. A problem with conventional demodulation EDMR techniques is that the spectrum exhibits significant broadening. This broadening made it impossible for us to resolve electron nuclear hyperfine interactions in transistors in the conventionally demodulated EDMR spectrum. These interactions provide definitive defect identification. We utilize the fast passage effect to improve resolution in order to detect contributions to EDMR from the ^{13}C hyperfine interactions. In conventional SDR practice, one typically demodulates at the modulation frequency, or the first harmonic. For fast passage EDMR (FPEDMR), we demodulates at the second harmonic of the modulation frequency. A representative FPEDMR trace from a 4H SiC MOSFET is illustrated in Fig. 3 under (a) x1 magnification and (b) x10 magnification. Note the strong central component and the weak side peaks of 13 and 28 Gauss. Fig. 4 compares (b) this spectrum with its relative amplitude measures

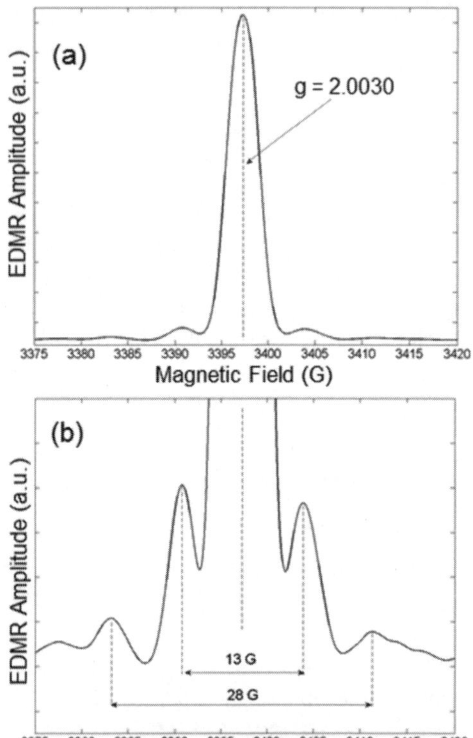

Figure 3: EDMR trace under fast passage conditions at (a) x1 magnification and (b) x10 magnification. Note the side peaks at 13 and 28 Gauss. They are due to hyperfine interactions with ^{13}C nuclei. The magnetic field is parallel to the c axis.

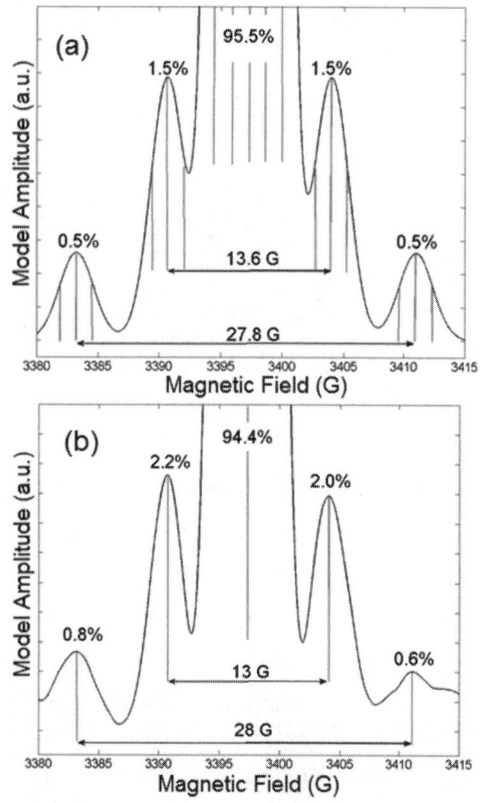

Figure 4: Comparison of (a) a model of the anticipated FPEDMR spectrum for a silicon vacancy and (b) the results of Figure 1(b) with percentages indicating relative amplitudes of the entire spectrum. (The magnetic field is parallel to the c axis.)

with (a) a model for the anticipated resonance response for a silicon vacancy. This model was generated using the relative abundance of ^{29}Si and ^{13}C nuclei (4.67% and 1.08% respectively) and the hyperfine parameters reported in the literature from conventional EPR and ENDOR studies on large volume samples [6].

Note the very close agreement of the two traces. In the conventional EPR studies on large volume samples, the side peak structure corresponding to all the vertical lines of Figure 2a is visible. However, due to broadening of the EDMR spectrum, we are unable to clearly resolve this near side peak structure in the central 9 Gauss, all of which is due to the presence of second nearest neighbor ^{29}Si magnetic nuclei. We are, however, quite clearly able to resolve the much weaker side peaks due to the ^{13}C nuclei of the inner carbon atoms. These lines are much weaker for two reasons. (1) There are only 4 inner carbon atoms but 12 outer silicon atoms at the vacancy. (2) The natural abundance of the magnetic silicon is about 4 times higher than that of magnetic carbon nuclei. (Our inability to resolve the ^{29}Si hyperfine interactions into FPEDMR is likely due to disorder in the near SiC/SiO_2 interface region. This disorder broadens the spectrum so that the closely spaced silicon hyperfine lines are not resolvable but the more widely separated and much weaker carbon lines are sufficiently separated to be definitively resolved.) The extremely close correspondence between the FPEDMR response and the model for the anticipated spectrum of a negatively charged silicon vacancy along with the previously reported gross similarities in spectra provide a very strong argument linking this EDMR spectrum to the silicon vacancy. Since we have previously reported the presence of this EDMR spectrum in a wide range of SiC/SiO_2 devices [2,3], our results provide an essentially definitive identification of an important SiC/SiO_2 MOSFET interface/near interface trap.

Summary

In conclusion, our experimental results and calculated model are in very close agreement. The close correspondence between the hyperfine parameters previously reported for the well-characterized negative silicon vacancy spectrum and our observed SDR response provide definitive evidence linking the electrically active SiC MOSFET defect to this defect. Since this SDR spectrum has been observed in a wide variety of 4H SiC MOSFETs prepared under a wide variety of processing parameters, our study indicates that the silicon vacancy center is an trapping center in the SiC which is of widespread importance in SiC MOS technology.

References

[1] H.J. von Bardeleben and J.L. Cantin, Mat. Sci. Forum 353 (2001) 513-516.

[2] D.J. Meyer, P.M. Lenahan, A.J. Lelis, Appl. Phys. Lett. 86 (2005) 023503.

[3] C.J. Cochrane, P.M. Lenahan, A.J. Lelis, J. Appl. Phys. 109 (2011) 014506.

[4] J.H. Hyde, Phys. Rev. 119 5 (1960) 1483-1492.

[5] M. Weger, The Bell Syst. Tech. J. 39 (1960) 1013-1112.

[6] J. Isoya, T. Umeda, N. Mizuochi, N.T. Son, E. Janzén, T. Ohshima, Phys. Stat. Sol. (b) 245 7 (2008) 1298-1314.

[7] D. Kaplan, I. Solomon, N.F.Mott, J. Phys. Lett. (Paris) 39 4 (1978) L51-L54.

[8] D.J. Fitzgerald, A.S. Grove, Surf. Sci. 9 2 (1968) 347

Materials Science Forum Vols. 717-720 (2012) pp 437-440
© (2012) Trans Tech Publications, Switzerland
doi:10.4028/www.scientific.net/MSF.717-720.437

Comparative Study of Electrical and Microstructural Properties of 4H-SiC MOSFETs

C. Strenger[1, 2, a], V. Häublein[2], T. Erlbacher[2], A. J. Bauer[2], H. Ryssel[1, 2], A. M. Beltran[3, 4], S. Schamm-Chardon[3], V. Mortet[4], E. Bedel-Pereira[4], M. Lefebvre[4], and F. Cristiano[4]

[1]Chair of Electron Devices, Cauerstrasse 6, 91058 Erlangen, Germany

[2]Fraunhofer IISB, Schottkystrasse 10, 91058 Erlangen, Germany

[3] CEMES-CNRS and Univ. de Toulouse, BP 94347, 31055 Toulouse Cedex 4, France

[4]LAAS/CNRS, 7 av. Du Col. Roche, 31077, Toulouse Cedex 4, France

[a]christian.strenger@leb.eei.uni-erlangen.de

Keywords: 4H-SiC MOSFETs, HRTEM, spatially-resolved EELS, SiC/SiO_2 interface, Hall mobility

Abstract. N-channel MOSFETs were manufactured on p-type and on p-implanted, n-type 4H-SiC substrates. The electron mobility in the inversion channel was measured to be correlated with the structural and chemical properties determined by transmission electron microscopy. With regard to what was previously discussed in the literature, interfacial layer formation and carbon distribution across the SiC/SiO_2 interface were considered in relation with the measured Hall electron mobility.

Introduction

Despite the continuous improvement achieved in the development of 4H-SiC MOSFETs with higher performance and stability, the 4H-SiC MOS system still suffers from heavy carrier trapping at the SiO_2/SiC interface. Additionally, the remaining electrons in the inversion layer are seriously affected by Coulomb scattering at the negatively charged trapping states [1,2] resulting in a severe reduction of their mobility. In 2008, A. Agarwal and S. Haney [3] proposed that not only interface states at the SiC/SiO_2 interface are responsible for the electron trapping but also a further component, namely bulk traps in SiC. Furthermore these authors suggested that the density of these bulk traps is significantly increased by ion implantation followed by high-temperature anneals. At the same time several groups provided experimental evidence for a carbon (C) rich transition region on the SiC side of the interface with a C/Si ratio higher than one [4,5]. It was also proposed that the C agglomeration on the SiC side of the interface accounts for the increased bulk trap density in SiC [6] which is boosted by ion implantation and subsequent annealing [3]. Based on the corresponding HRTEM/EELS studies, it was also found that in the fully manufactured MOSFET devices the peak field effect mobility is inversely related to the width of the carbon-rich transition region [5].

In this work we compare n-channel MOSFETs manufactured on p-type epitaxial 4H-SiC layers and on p-implanted, n-type epitaxial layers. The aim is to get more insight on the C distribution and nature across the SiC/SiO_2 interface and to correlate the results with electron mobility measurements. The investigations are based on the combination of structural and compositional analyses carried out by HRTEM and spatially resolved EELS with Hall mobility measurements.

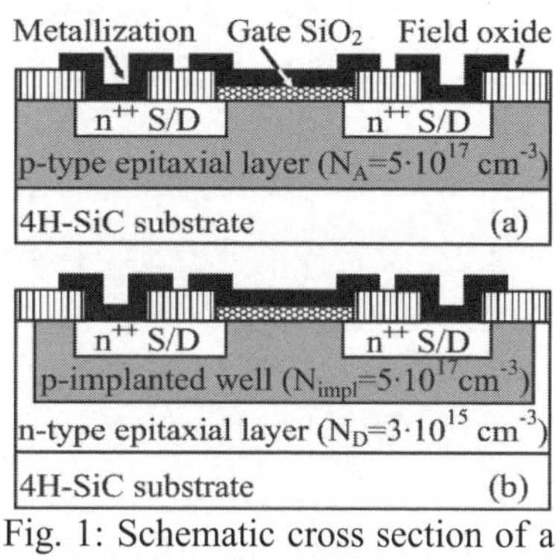

Fig. 1: Schematic cross section of a MOSFET on (a) sample p-epi and (b) on sample p-impl.

Experimental

Two sets of samples with n-channel planar MOSFETs were manufactured on a p-type 4H-SiC epitaxial layer (samples "p-epi", net acceptor concentration $5 \cdot 10^{17}$ cm^{-3}) and on a p-implanted (samples "p-impl") n-type 4H-SiC epitaxial layer (net donor concentration $3 \cdot 10^{15}$ cm^{-3}) from Cree (Fig. 1). On the n-type epitaxial layer multiple Al implantations were carried out to obtain suitable p-type wells (implanted Al concentration about $5\cdot10^{17}$ cm^{-3}) for the fabrication of n-channel MOSFETs. Apart from the p-well implantations the fabrication process for both sets of samples is identical. After capping the Si face substrates with graphite, source and drain implantations with nitrogen as well as p-well implantations were annealed under flowing Ar gas at a temperature of 1973 K for 30 min. 34 nm thick gate oxides were grown by dry oxidation in N_2O atmosphere at 1553 K and subsequently annealed at the same temperature for 30 min under N_2 ambient. Phosphorus-doped polycrystalline silicon was deposited by LPCVD and patterned to form the gate electrodes. For the fabrication of the source, drain, and voltage tap contacts, SiC was alloyed with Ti at 1373 K for 2 min and thereafter a metallization stack containing Ti and Pt was deposited and patterned. The channel length and width used in this study are L/W = 5 μm/100 μm for the lateral and 80 μm/500 μm for the Hall-bar MOSFETs.

MOSFET devices were characterized by I-V and Hall measurements. Transfer characteristics were obtained using a Modular DC Source/Monitor Agilent 4142B. Hall effect measurements were carried out at room temperature with a modified Accent HL 5000 Hall effect setup with a permanent magnet of 0.33 Tesla. The drain voltage was set to 100 mV and the gate voltage was varied from 0 V up to 15 V. A Keithley 2182A nano-voltmeter was used to detect the low level of the Hall voltage and the drain current was measured using a Keithley 2400 source meter. The mobility and the charge carrier density were calculated assuming a Hall scattering factor of 1.

Cross-sectional TEM samples of the lateral MOSFETs have been prepared by focused ion beam (FIB) "lift-out" technique. Atomic structure and elemental analysis were investigated by HRTEM and EELS performed on a field emission TEM, FEI Tecnai™ F20 microscope (spatial resolution 1 nm and frequency 1). A detailed description of the microscope can be found elsewhere [7].

Results

Electrical Characterization. The free charge density in the inversion channel as measured by Hall characterization for samples p-epi and p-impl are plotted in Figure 2(a) and 2(b), respectively. The ideal charge density (dashed lines in Fig. 2) in the inversion channel was calculated by the formula $n_{inv,ideal}=-C_{ox}/q(V_G-V_{th})$ where C_{ox} is the oxide capacitance, q is the electronic charge, V_{th} is the threshold voltage, and V_G is the gate voltage. The threshold voltage was determined using the threshold drain current method with a threshold drain current of 0.1 nA. In Figures 2(a) and 2(b) it can be observed that the slope of the measured free charge density vs gate voltage characteristic is about 50% to 60% lower than the ideal one. The slope being less than C_{ox}/q indicates trapping of free electrons at the interface [8].

The Hall mobilities for both samples are presented in Figure 3. The electron Hall mobility rises with the gate voltage up to 10 cm²/Vs and 25 cm²/Vs for the samples p-epi and p-impl, respectively. This behavior is not intuitive and in contrast to the work of A. Agarwal et al. [3] wherein it was suggested that an Al implantation significantly increases the amount of bulk traps. Further it was proposed that these additional trapping states would than lead to an increased electron trapping, thereby increasing the scattering of free charge carriers and ultimately lower the electron mobility.

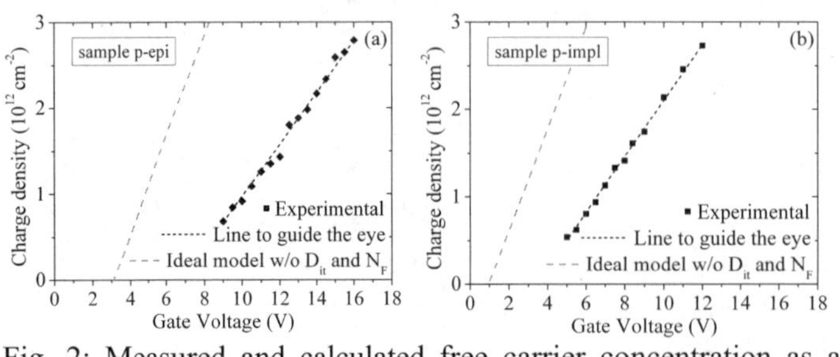

Fig. 2: Measured and calculated free carrier concentration as a function of gate voltage for samples p-epi (a) and p-impl (b). The dotted line should guide the eye only.

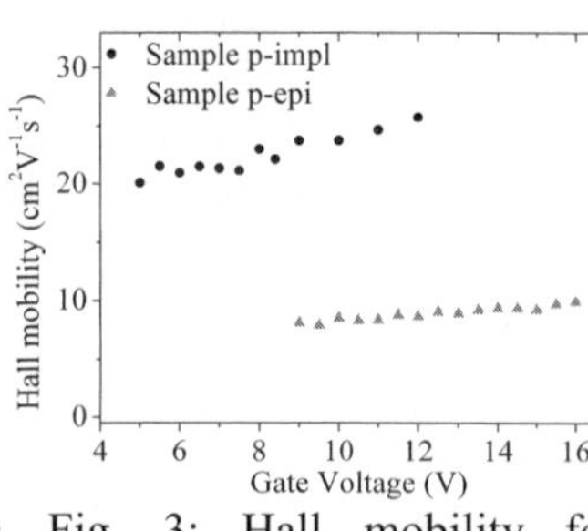

Fig. 3: Hall mobility for samples p-epi and p-impl as a function of the gate voltage.

It should be noted that contrary to the work of A. Agarwal et al. [3], the Al implanted substrate used in this work contains nitrogen. Therefore our mobility results can be interpreted in two ways. First, the nitrogen in the substrate may lead to a passivation of bulk traps in SiC similar to the passivation of interface traps and bulk traps in SiO_2 that was demonstrated by J. Rozen et al. [9]. Then our results indicate that the beneficial effect of nitrogen prevails over the deleterious impact of the Al implantation with respect to the electron mobility. Second, if the passivation of bulk traps in SiC with nitrogen proves to be ineffective, then our results suggest that an Al implantation with a concentration of $5 \cdot 10^{17}$ cm^{-3} cannot increase the bulk trap density so as to significantly affect the electron mobility. Further experiments will be carried out to know more about this feature.

Microscopy Characterization. HRTEM images have been acquired for both samples confirming the good structural quality, in the vicinity of the SiC/SiO_2 interface, since 4H-SiC stacking atomic planes can be observed (Figs. 4(a) and 4(b)). The contrast on the SiO_2 side is the typical one of an amorphous material. There is no added contrast at the transition between the crystalline SiC and the amorphous SiO_2. In particular, no contrast associated to C agglomerations or graphitic clusters can be evidenced. These observations are common to both sets of samples.

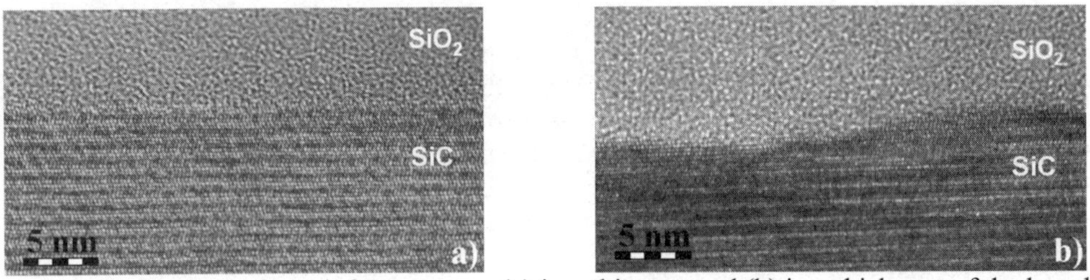

Fig. 4: HRTEM images of p-epi, from an area (a) in a thin part and (b) in a thick part of the lamella.

The structural characterization only reveals a wavy and rather rough interface for the thicker areas (electron beam of the TEM transmitted through a part of the sample lamella with a thickness of about 120 nm) of the prepared sample lamellas (Fig. 4b). Surface roughness can contribute to low carrier mobility at high gate voltages [1], although this feature may be an artifact from sample preparation. These samples were prepared by FIB, which makes them thicker and heterogeneous with respect to SiC and SiO_2 area. This is due to ion-induced damage of the sample and the different hardness of the two materials. Improvements in sample preparation will help to explore this feature in more detail [7].

As the mobility degradation was proposed to be related to the C distribution at the interface, compositional analysis has been performed by STEM-EELS. For this purpose, the C/Si and O/Si ratio profiles have been determined. Concerning the C/Si and O/Si ratio profiles presented in Fig. 5 for the two studied samples we observe, on both sides of the SiC/SiO_2 interface, areas of constant and rather flat C/Si and O/Si profiles up to regions very near the interface. Going from the SiC to the SiO_2 side, just across the interface, the C/Si (and O/Si) elemental profiles are decreasing (increasing) regularly. The width of the change in composition can be associated to the transition layer (TL). On the basis of these results, two points can be stated in relation to published data.

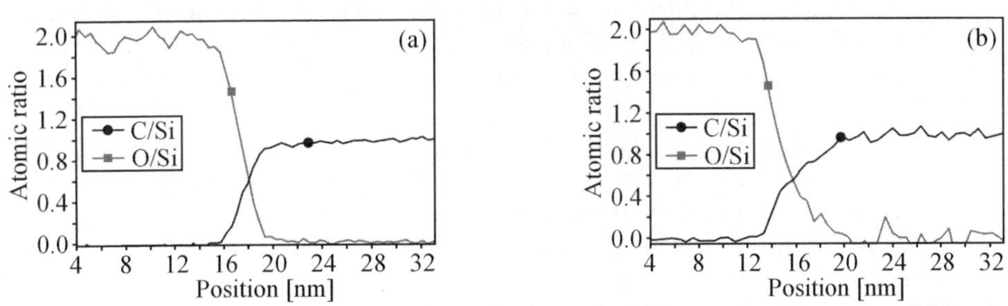

Fig. 5: C/ Si and O/ Si ratio profiles as measured using EELS for samples (a) p-epi and (b) p-impl.

First, as the related C/Si ratio profiles are flat outside the TL, there is no evidence of excess C near the SiC/SiO_2 interface. These results are similar to the one of Pippel et al. performed on MOS samples [10]. These observations are in contradiction with those of Zheleva et al. [4] performed on different MOS structures. They reported the presence of a several nanometers thick transition layer with a C/Si ratio being greater than one up to 4 nm into the SiC and non-zero up to 4 nm into the oxide. Over-stoichiometry of carbon was also reported by Chang et al., particularly for NO-annealed C-face MOSFETs [11,12].

Second, both MOSFETs studied in this work exhibit a thin interface layer (Fig. 5). Therefore, the width of the TL cannot be related to the different measured electron mobilities (about a factor 2) of the two MOSFETs. This is different from reported results of Biggerstaff et al. where the width of the TL is found to be inversely related to the effective channel mobility [5]. It is noticeable that in this latter case the width of the TL is rather large, between 5 and 20 nm.

Summary

The measured electron Hall mobility on the p-implanted samples was found to be about twice as high as the one on the p-epitaxial samples. The mobility difference could not be correlated with the measured TL width at the SiC/SiO_2 interface. Moreover, HRTEM-EELS measurements evidence that the low Hall mobility values compared to the theoretical ones cannot be correlated to a C enrichment near the $4H\text{-}SiC/SiO_2$ interface. One of the possible causes could be a wavy and thus rough interface, if it is induced during the processing of the MOSFETs.

On the basis of these preliminary structural, chemical and electrical investigations, additional Hall measurements have to be carried out in order to further investigate the predominant mechanisms responsible for the degradation of the channel mobility.

Acknowledgments

This work has been performed by the French-German Consortium MobiSiC and supported by the Programme Inter Carnot Fraunhofer (PICF) from BMBF (Grant 01SF0804) and ANR. The support from Robert Bosch GmbH who provided the devices and the Hans L. Merkle-Stiftung is highly appreciated. Authors thank P. Salles and R. Cours for the sample preparation by FIB.

References

[1] S. Potbhare, N. Goldsman, and G. Pennington, J. Appl. Phys. 100 (2006) 044515

[2] N.S. Saks, S.S. Mani, and A.K. Agarwal, Appl. Phys. Lett. 76 (2000) 2250.

[3] A. Agarwal and S. Haney, J. Electron. Mater. 37 (2008) 646-654.

[4] T. Zheleva, A. Lelis, G. Duscher, F. Liu, I. Levin, and M. Das, Appl. Phys. Lett. 93 (2008) 022108, 1-3.

[5] T.L. Biggerstaff, C.L. Reynolds, T. Zheleva, A. Lelis, D. Habersat, S. Haney, S.H. Ryu, A. Agarwal, and G. Duscher, Appl. Phys. Lett. 95 (2009) 032108.

[6] X. Shen and S.T. Pantelides, Appl. Phys. Lett. 98 (2011) 053507, 1-3.

[7] A.M. Beltran, S. Schamm-Chardon, V. Mortet, M. Lefebvre, E. Bedel-Pereira, F. Cristiano, C. Strenger, V. Häublein, and A. J. Bauer (to be published in the Proceedings of HeteroSiC WASMPE 2011).

[8] N. Saks, in: Silicon Carbide: Recent Major Advances, edited by W.J. Choyke, H. Matsunami, and G. Pensl, Springer, New York, 2004, pp. 387-410.

[9] J. Rozen, S. Dhar, M.E. Zvanut, J.R. Williams, and L.C. Feldman, J. Appl. Phys. 105 (2009) 124506.

[10] E. Pippel, J. Woltersdorf, H. Olafsson, and E. Sveinjörsson, J. Appl. Phys. 97 (2005) 034302.

[11] K.C. Chang, L.M. Porter, J. Bentley, C-Y. Lu, J. Cooper Jr, Dhar, L.C. Feldman, and J.R. Williams, J. Appl. Phys. 95 (2004) 8252-8257.

[12] K.C. Chang, Y. Cao, L.M. Porter, J. Bentley, S. Dhar, L.C. Feldman, and J.R. Williams, J. Appl. Phys. 97 (2005) 104920, 1-6

Materials Science Forum Vols. 717-720 (2012) pp 441-444
© (2012) Trans Tech Publications, Switzerland
doi:10.4028/www.scientific.net/MSF.717-720.441

Sub-Bandgap Light-Induced Carrier Generation at Room Temperature in Silicon Carbide MOS Capacitors

Sandeepan DasGupta[1,a,*], Andrew Armstrong[1,b], Robert Kaplar[1,c], Matthew Marinella[1,d], Reinhard Brock[1,e], Mark Smith[1,f], and Stanley Atcitty[1,g]

[1]Sandia National Laboratories, Org. 1748, MS 1084, Albuquerque, NM 87111, USA.

[b]aarmstr@sandia.gov, [c]rjkapla@sandia.gov, [d]mmarine@sandia.gov, [e]rcbrock@sandia.gov, [f]masmit@sandia.gov, [g]satcitt@sandia.gov, [a]sdasgup@sandia.gov, *corresponding author

Keywords: Sub-bandgap, optical generation, metastable defect, silicon vacancy.

Abstract. Carrier generation characteristics in n-type substrate SiC MOS capacitors induced by sub-bandgap energy light are reported. The generation rate is high enough to create an inversion layer in ~20 minutes with monochromatic light (front side illumination) of energy 2.1 eV (intensity ~5×10^{16} $cm^{-2}s^{-1}$) in 4H-SiC for electric fields smaller than 1 MV/cm. Generation and recovery results strongly indicate involvement of a metastable defect whose efficiency as a generation center increases under hole-rich and decreases under electron-rich conditions. The generation dependence on bias history and light energy shows the defect to have properties consistent with the metastable silicon vacancy / carbon vacancy-antisite complex (V_{Si} / V_c - C_{Si}).

Introduction

Silicon carbide (SiC) has traditionally suffered from a large number of defect levels in the energy gap [1] and very high SiC/SiO_2 interface trap densities [2]. Several defect characterization experiments have reported generation lifetimes much smaller than for Silicon at both the SiC/SiO_2 interface [3] and in the SiC substrate [4]. In this study, we report generation characteristics in n-type substrate SiC MOS capacitors induced by sub-bandgap energy light. The generation rate is high enough to create an inversion layer in ~20 minutes due to front-side illumination (where only the edges of the device are exposed to the incident light) with monochromatic light of energy 2.1 eV in 4H-SiC (E_g = 3.2 eV) for -5V bias on a capacitor with 70 nm oxide thickness. It is surprising that such high generation rates can be achieved at room temperature (where $n_i = 10^{-8}$ cm^{-3} for 4H-SiC and the thermal generation rate is negligible) with light having energy much lower than the bandgap.

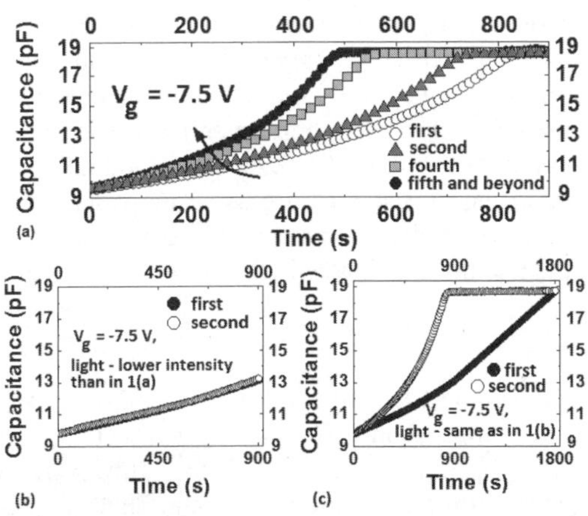

Fig. 1. (a) C-t sweeps at 1 kHz (V_g = -7.5 V) under broad spectrum (predominantly yellow) light at room temperature. (b) C-t sweep on a second device - performed for the same time and bias using much lower intensity light (factor of ~9). (c) C-t sweep on the same device (Fig. 1b) using the same low intensity light (as Fig. 1b) for longer duration.

Sub Bandgap Optical Generation

N-type 4H-SiC MOS capacitors from the wafer line monitor of a state-of-the-art SiC MOSFET manufacturer were used in this study. Similar experiments on 6H-MOS capacitors from a slightly older process showed the same trends. The pulsed capacitor method, which is a conventional technique to study generation lifetime characteristics in MOS systems, was used [5]. Fig. 1a shows C-t measurements at 1 kHz on a device biased in deep depletion at -

7.5 V. The sweeps were performed at room temperature. The probe station microscope lamp (a predominantly yellow light source) was used for illumination. Although the ambient daylight was much more efficient at generation, the microscope light in a closed cabinet was used to ensure constant intensity.

Capacitance values show almost no change over a range of 1 kHz to 1 MHz. The rate of increase of capacitance with time (dC/dt) increases with every sweep and finally reaches saturation. However, when the sweep is performed for the same time and bias using a much lower intensity of light (Fig. 1b), subsequent C-t sweeps show negligible increase in dC/dt. Using the same low intensity light (as in Fig 1b), when the same device is swept for a longer period of time (so that the capacitance reaches values close to the inversion capacitance), subsequent C-t sweeps show a prominent increase in dC/dt (Fig. 1c). Thus, the increase in dC/dt in subsequent sweeps is related not to the bias, time, or intensity of light – but to the value of capacitance reached in the previous sweep.

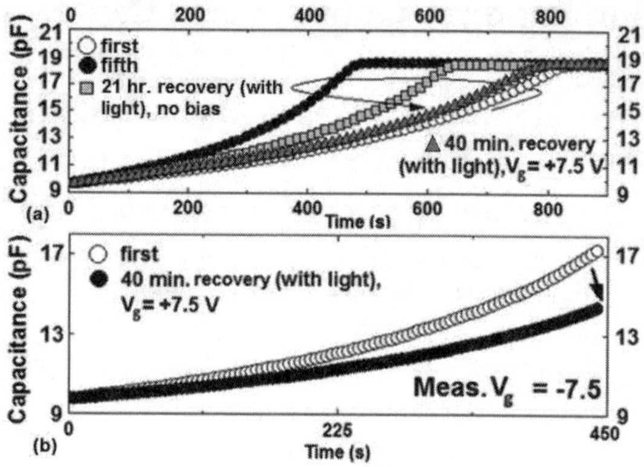

Fig. 2. (a) Room temperature recovery of the device shown in Fig.1 at zero bias (21 hrs.) and at accumulation bias (V_g = +7.5 V) for 40 mins., with light. (b) C-t sweeps for a fresh device and then after a 40 min bias under light at V_g = +7.5 V.

The most important thing to note in Fig. 1 is that every sweep starts from the same capacitance value. This rules out the increase in dC/dt to be due to any trapping or detrapping of bulk oxide charge. While the capacitance is rising, eliminating the light while maintaining the bias causes almost no decay in the capacitance. This clearly shows that the rise in capacitance is not related to the detrapping of only electrons from deep levels. In the absence of optical excitation, traps that have emitted electrons should be quickly filled in an n-type material, since they are empty following emission and free electrons are available to fill up the empty states. Detrapping only holes from deep levels also cannot account for the rise in capacitance because trapped holes should not be stable in an n-type material. Generation of electron-hole pairs does not result in a change in occupancy of a trap, and the absence of empty traps makes it difficult for generated carriers to be captured. So, when the light is withdrawn, a reduction in the number of free electrons and holes can result only from recombination, which is very slow in the presence of a bias strongly separating electrons from holes. Thus, the absence of any significant decay in capacitance following the withdrawal of optical excitation clearly shows that the rise in capacitance is due to the generation of electron hole-pairs.

Recovery of Generation Rate

Figs. 2a and 2b show the recovery characteristics of the device. During recovery, the light was kept at the same level as during stress or measurement to eliminate any bias-independent effects. The device steadily recovered to a lower generation rate (Fig. 2a) in about 21 hours. However, the generation rate was still higher than the pre-stress generation rate. There was a significant increase in recovery when the device was biased in accumulation. Fig. 2a also shows the recovery of the same device at accumulation bias (V_g = +7.5 V) for 40 min with the same intensity light. The accumulation bias was started right after recording the 21 hour recovery in Fig. 2a. Fig. 2b shows the C-t sweeps for another device – first for the fresh device and then after a 40 min bias under light at V_g = +7.5 V. After bias at accumulation, the generation rate is lower than that in the fresh device. Thus, the generation rate increase can be "super recovered". Fig. 3a shows the zero bias recovery characteristics of the device in greater detail. During recovery, the light is kept at the same level as during stress or measurement. The time intervals between successive sweeps from the first to the third recovery measurement were 8, 20 and 60 min, during which the device steadily recovered to lower generation rates (direction shown by large arrow in Fig. 3a). The next sweep was performed after 2 hours and this showed a slight increase in the generation rate. The increase continued and reached saturation in about 18 hours (small arrow in Fig. 3a). The capacitance rate increase under zero bias shows that not only does the hole rich condition of the MOS capacitor enhance the generation rate – but even at zero bias, the illumination causes a slow increase in generation lifetime – the cumulative effect of which over several hours can be significant. The device recovery is stronger when the accumulation type bias is applied under light than without it (Fig. 3b).

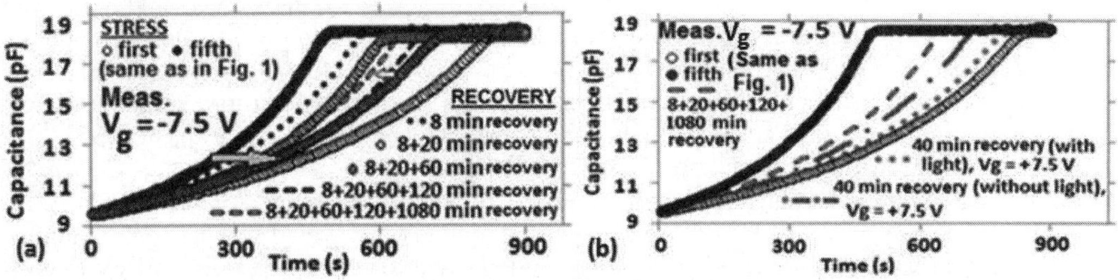

Fig. 3. (a) Zero bias recovery characteristics (large arrow) under illumination for the device shown in Fig.1. After the first three sweeps showing recovery – the capacitance rise rate increases slightly (small arrow) and reaches saturation. (b) Recovery of the same device at accumulation bias (V_g = +7.5 V) for 40 mins., first with and then without the light.

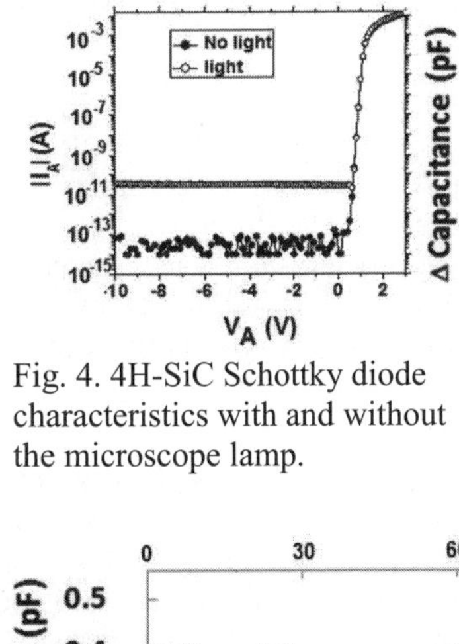

Fig. 4. 4H-SiC Schottky diode characteristics with and without the microscope lamp.

Generation in Schottky Diodes

The results are very similar to previously reported thermal generation characteristics [3] for which the generation rate was proposed to be dominated by interface states. However, Fig. 4 shows an increase in the steady-state reverse bias current of an n-substrate Schottky diode under the same sub-bandgap energy microscope lamp illumination by at least three orders of magnitude, strongly indicating that sub-bandgap generation is related to a bulk SiC defect and not the SiC/SiO_2 interface. The bias history dependence and super recoverable nature of the generation rate strongly indicates that the defect is metastable – switching configurations under electron-rich and hole-rich conditions.

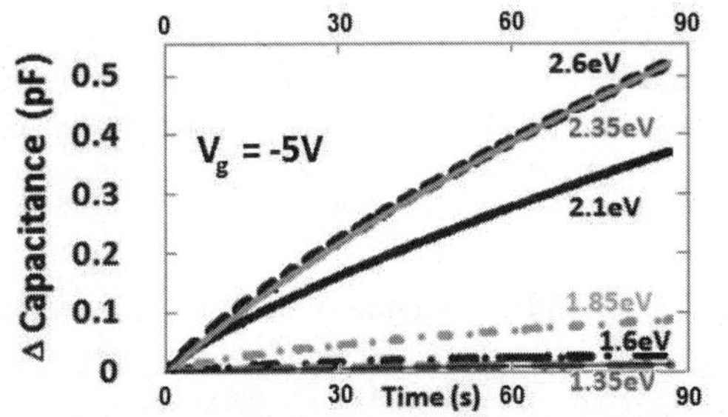

Fig. 5. C-t sweeps using monochromatic light for energies ranging from 1.35 to 2.6 eV. Small sweep times chosen so as to not raise capacitance enough to affect the generation rate for next sweep.

Energy Dependence under Illumination with Monochromatic Light

To study the generation rate as a function of light energy, C-t sweeps were performed using monochromatic light consisting of energies ranging from 0.6 to 3.5 eV. Light from a broadband Xe source was dispersed using a ¼ m monochromator with appropriate mode-sorting filters to achieve 0.05 eV resolution at a photon flux of $\sim 5\times10^{16}$ $cm^{-2}s^{-1}$. Short sweep times were chosen so as to not raise the capacitance enough to affect the generation rate for the next sweep. The generation rate shows the first significant increase (on the order of what is observed with the microscope lamp) above 1.6 eV (Fig. 5). The maximum increase is observed between 1.85 and 2.1 eV. Above 2.35 eV, the generation rate does not increase further for higher energies up to 3.2 eV (sub-band energies above 2.6 eV not shown in for clarity) where band-to-band generation takes over.

TABLE I. Defect levels (energy from E_v) for V_{Si} and its metastable version V_c - C_{Si} as calculated in 4H-SiC at the hexagonal site through density functional theory [8]. Levels closest to midgap are highlighted in boldfont.

Charge State →	(+2/+1)	(+1/0)	(0/-1)	(-1/-2)	(-2/-3)	(-3/-4)
V_{Si}			0.56	**2.01**	2.05	2.81
V_c - C_{Si}	0.83	**1.39**	2.11	2.56	3.12	

Generation rates becoming comparable to the microscope lamp-induced rates above 1.6 eV (half the 4H-SiC bandgap) clearly indicates that the process is dominated by transitions involving two photons. Thus, for different defects, the one with a level closest to midgap will require two photons of minimum energy to generate. The generation characteristics show close similarities to the silicon vacancy (V_{Si}) which has been shown to be metastable [6,7] – transforming to a carbon vacancy-antisite complex V_c - C_{Si} for Fermi level positions at and below midgap. While V_{Si} in 4H-SiC has the closest level (charge state transition from -1 to -2) to midgap at E_v + 2.01eV, V_c - C_{Si} has the closest level (charge state transition from +1 to 0) to midgap at E_v + 1.39 eV [8]. Thus, a configuration change of V_{Si} to V_c - C_{Si} under hole-rich conditions would result in the minimum energy of each of the 2 photons required for a generation event to be reduced from 2.01 eV to 3.2 – 1.39 = 1.81 eV.The saturation of the generation rate around photon energies of 2.35 eV for sub-bandgap energies is likely due to the fact that some of the shallowest levels have electron capture times for levels close to the valence band and hole capture times for levels close to the conduction band too low to generate through a two photon process. V_c - C_{Si} has a level (+2/+1) closest to the valence band at ~0.83eV [8], showing very close consistency with the experimental result of the generation rate saturating at 3.2 – 0.83 ≈ 2.35 eV.

Conclusion

We have demonstrated unexpectedly high generation rates in state-of-the-art SiC MOS capacitors at room temperature for electric fields less than 1 MV/cm due to illumination with light more than 1eV smaller in energy than the SiC bandgap. Generation and recovery results strongly indicate the involvement of a defect exhibiting metastability, whose efficiency as a generation center increases under hole-rich and decreases under electron-rich conditions. The generation rate dependence on bias history and monochromatic light energy shows the defect to have properties consistent with the metastable silicon vacancy (V_{Si}) / carbon vacancy-antisite complex (V_c - C_{Si}).

Acknowledgements

Sandia National Laboratories is a multi-program laboratory managed and operated by Sandia Corporation, a wholly owned subsidiary of Lockheed Martin Corporation, for the U.S. Department of Energy's National Nuclear Security Administration under contract DE-AC0494AL85000. This work was performed under funding from the DOE Energy Storage Program managed by Dr. Imre Gyuk of the DOE Office of Electricity. We thank Drs. Mrinal Das, David Grider and Sarit Dhar of Cree Inc. and Prof. John Williams of Auburn University for providing the devices.

References

[1] V. V. Afanas'ev, F. Ciobanu, S. Dimitrijev, G. Pensl, and A. Stesmans, J. Phys.: Condens. Matter 16 (2004) S1839.
[2] J. A. Cooper, Jr., Phys. Stat. Sol. A, 162 (1997) 305.
[3] M. J. Marinella, D. K. Schroder, T. Isaacs-Smith, A. C. Ahyi, J. R. Williams, G. Y. Chung, J. W. Wan, and M. J. Loboda, Appl. Phys. Lett. 90 (2007) 253508.
[4] Y. Wang, J. A. Cooper, M. R. Melloch, S. T. Sheppard, J. W. Palmour, and L. A. Lipkin, J. Electron. Materials, 25, 5 (1996) 899.
[5] J. S. Kang and D. K. Schroder, Phys. Status Solidi A 89 (1985) 13.
[6] T. Lingner, S. Greulich-Weber, and J.-M. Spaeth, Mater. Sci. Forum 353-356 (2001) 505.
[7] A. Mattausch, M. Bockstedte, and O. Pankratov, Mater. Sci. Forum 353-356 (2001) 323.
[8] Alexander Mattausch, Ph.D. Thesis, University of Erlangen-Nuremberg, (2005).

Materials Science Forum Vols. 717-720 (2012) pp 445-448
© (2012) Trans Tech Publications, Switzerland
doi:10.4028/www.scientific.net/MSF.717-720.445

Oxidation-induced epilayer carbon di-interstitials as a major cause of endemically poor mobilities in 4H-SiC/SiO_2 structures

Xiao Shen[1,a,*] and Sokrates T. Pantelides[1,2,b]

[1]Department of Physics and Astronomy, Vanderbilt University, Nashville, TN 37235, USA

[2]Oak Ridge National Laboratory, Oak Ridge, TN 37831, USA

[a]xiao.shen@vanderbilt.edu, [b]pantelides@vanderbilt.edu, *corresponding author

Keywords: channel traps, mobility degradation, oxidation, defect complex

Abstract. The defects at the interface and in the oxide have been considered as the sources of mobility degradation at the SiC/SiO_2 interface as in the case of the Si/SiO_2 system. By examining available experimental and theoretical results and performing new calculation, we show that thermal oxidation creates immobile carbon di-interstitial defects inside the semiconductor epilayer, which are a major cause of the poor mobility in SiC/SiO_2 structures.

Introduction

Silicon has dominated the microelectronics industry largely due to the superior properties of thermally grown Si/SiO_2 interfaces. The as-grown Si/SiO_2 interface is abrupt with the main defects being Si dangling bonds that can be easily passivated by hydrogen. However, silicon is not ideal for high-power, high-field, or high-temperature devices due to its small band gap and low thermal conductivity. Silicon carbide, especially the 4H polytype, is an ideal candidate for such applications, because it has all the desired bulk properties and its native oxide is also SiO_2. However, compared to the Si/SiO_2 interface, the thermally grown SiC/SiO_2 interface is far inferior, resulting in unacceptably low carrier mobilities at the interface [1].

Both Si and SiC oxidation produce SiO_2, but in the case of SiC a large number of C atoms have to be released. It is generally accepted that the C atoms diffuse out as CO or CO_2 through the oxide. The high density of interfacial defects that limit electron mobilities has been related to residual C atoms in the interface region [2]. Significant efforts have been made to eliminate these defects in the past decade and improvements have been made through post-oxidation annealing (POA) in NO [3,4] or atomic H [5]. Recently, better motilities were achieved by incorporating Na during oxidation or POA using phosphorus [6,7]. But the best mobilities obtained are still only 10-15% of the bulk value.

In 2008, Agarwal and Haney put forward a "conjecture" that thermal oxidation, high-temperature anneal, or ion implantation may introduce defects inside the SiC beneath the interface that may cause significant mobility degradation [8], but no particular defects were singled out. In this paper, we examine theoretical and experimental evidence and show that one such defect, namely a C di-interstitial $(C_i)_2$, is introduced in the epilayer during oxidation, and is most likely to be the source of poor interfacial mobilities in the SiC/SiO_2 system [9]. We make the case for $(C_i)_2$ largely by using the theoretical results already in the literature [10-12], complementing them with new finite-temperature simulations.

Computational Methods

Our new calculation is based on density functional theory. It consists of a quantum molecular dynamical (QMD) simulation of a nitrogen interstitial N_i interacting with a $(C_i)_2$ in 4H-SiC. We employ the PBE [13] version of the exchange-correlation functional, PAW potentials [14], and a plane wave basis as implemented in the VASP code [15]. The calculations are done in a 3×4×1 96 atoms supercell. For QMD simulations, the plane-wave cutoff is 300 eV and the Brillouin zone sampling is done with the Γ point only. The time step for QMD simulations is set to 0.5 fs. The simulation consists of heating to 2200 °C in 2.2 ps and constant temperature QMD for 3.9 ps. The result will be discussed in the next section in context of the experimental information.

Results and Discussion

Whether the SiC epilayer is enriched with a large amount of C during oxidation [16] has been debated, but it is likely that some of the released C atoms do diffuse into SiC, since first-principle calculations have shown that the diffusion barrier for C single interstitials (C_i) is only 0.5~1.4 eV [17]. The notion of C injection into the SiC epilayer is further supported by recent experiments by Hiyoshi and Kimoto [18] that show that the existing $Z_{1/2}$ and $EH_{6/7}$ deep centers in the epilayer, which are likely related to the C vacancy [19], are substantially reduced during thermal oxidation. At the same time, a new defect, the HK0 center [20], appears in a significant concentration.

Due to its small diffusion barrier, isolated C_i is unlikely to exist in SiC after oxidation. However, the clustering of C_i's can generate immobile complexes that can survive at high temperature. The simplest complex is the di-interstitial $(C_i)_2$, which is immobile due to its large size and strong bonding with the surrounding crystal lattice (it is bonded to the lattice by 4 bonds). Unlike the carbon single interstitial, which consists of a carbon dumbbell at a carbon site as $(C_2)_C$, the $(C_i)_2$ consists of a pair of carbon atoms located at the interstitial site in the SiC lattice. It is also very stable and does not dissociate during oxidation (the dissociation energy is 5-6 eV [10,11]).

In thermal oxidation of Si, it is known that Si interstitials (Si_i) are injected into both the SiO_2 layer and the Si [21], but excess Si atoms in Si are generally benign. The Si di-interstitial $(Si_i)_2$ is not very stable, with a relatively small binding energy of 1.8 eV [22]. Therefore although $(Si_i)_2$ complexes may form, they dissociate later during oxidation. In contrast, $(C_i)_2$ in SiC is strongly bonded. Therefore, considerable concentrations of immobile $(C_i)_2$ can remain in the SiC after oxidation. In the remainder of this paper, we examine in detail the properties of $(C_i)_2$ in the context of recent experimental data.

Devynck *et al.* calculated the energy levels of $(C_i)_2$ using a hybrid density functional [12]. Four energy levels in the band gap are reported: (--/-) and (-/0) levels at E_C-0.04 eV and E_C-0.23 eV, and (+/0) and (++/+) level at E_V+0.71 eV and E_V+0.49 eV. We note that the defect is negatively charged in n-type material and positively charged in p-type material, whereby it always causes mobility degradation by Coulomb scattering.

Observations of traps inside SiC have been reported in two sets of experiments. Hiyoshi and Kimoto's experiment [18] supports the notion of C atoms being emitted into the SiC where they annihilate C vacancy-related defects and form the HK0 center. The level of the HK0 center is at E_V+0.78 eV. We propose to assign this level to the theoretical (+/0) level of the $(C_i)_2$ at E_V+0.71 eV. The thermal stability of the HK0 center is also consistent with its identification as $(C_i)_2$. The HK0 center anneals in Ar at 1400 °C [23], which is consistent with the large (5~6 eV) binding energy of $(C_i)_2$. The thermal stability of the HK0 also rules out the other candidate defect, the di-carbon antisite $(C_2)_{Si}$, since it has been reported to anneal out at a lower temperature (1000 °C) [24].

The same group has also reported HK0 centers being created by reactive ion etching (RIE) [23], ion-implantation [25], and electron irradiation [20]. All these processes create carbon interstititals. Notably, in ion-implantation experiments, HK0 is created when Al or Ne is implanted in SiC, but not when N is implanted [25]. This is consistent with the identification of the HK0 defect as $(C_i)_2$. It is known that N can passivate carbon clusters in SiO_2 or at the interface [26]. We performed QMD simulations to probe the interactions between N_i and $(C_i)_2$ inside SiC. The initial configuration is shown in Fig. 1 (left), where an N_i (in split interstitial configuration with a lattice C atom, highlighted by the green dotted circle) is placed next to a $(C_i)_2$ (highlighted by the red dashed circle). Since we are interested in the dynamics in lightly doped SiC during post-implantation annealing [25] or oxidation [27] (as will be discussed in the next paragraph) that happen at temperature ≥1000 °C, the Fermi level is near the middle of the band gap. Under this conduction, $(C_i)_2$ is charge neutral [12] and N_i has charge state -1 [28]. The structure obtained from QMD simulation is shown in Fig. 1 (right). One carbon interstitial is emitted (highlighted by the green dotted circle) and the remaining defect complex restructures into a configuration that can be expressed as $N_C(C_i)_2$ (highlighted by the red dashed circle). Sequential reactions may happen if more nitrogen interstitials are brought in, but are beyond the scope of the present work. The conclusion is that nitrogen interstitials can react with $(C_i)_2$ in SiC and generate a different defect. Therefore, the absence of HK0 in N-implanted SiC can be explained by the implanted nitrogen either modifying any HK0 centers that have formed or simply capturing the C_i's, thus preempting the HK0 formation.

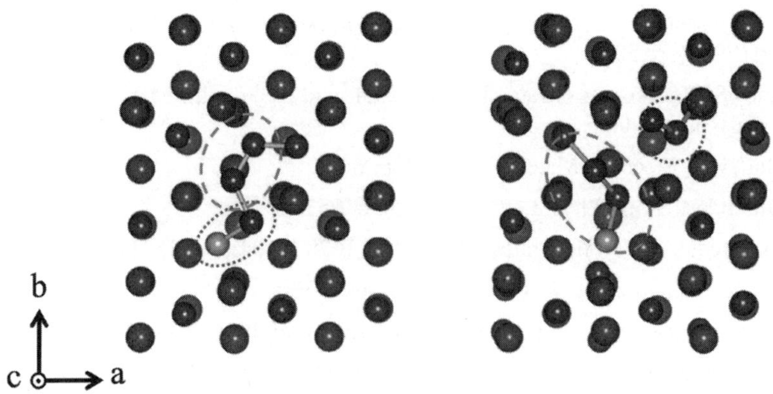

Figure 1. Left: N_i next to a $(C_i)_2$. Right: One carbon atom emitted as a C_i. (Large blue atom: Si, Small brown atom: C, Small light grey atom: N. Only C-C and C-N bonds are shown.)

The second set of experimental observations of defects in SiC is reported by Mooney and co-workers [29]. Using C-V and constant-capacitance DLTS (CCDLTS) measurements they distinguished the traps in n-type SiC MOS capacitors into oxide traps, interface traps, and SiC epilayer traps. The areal densities of SiC epilayer traps in NO-annealed samples are found to be comparable to the densities of remaining oxide and interface traps. Therefore, their contribution to overall trap density should be comparable to other interface or oxide defects, such as the C_O=C_O complexes in the oxide proposed in previous work [26]. The energy levels of epilayer traps are found to be within 0.1 eV below the 4H-SiC conduction band. We propose that these traps correspond to the (--/-) and (-/0) acceptor states of $(C_i)_2$. The authors also found that the shallow traps are not present in samples that had N implantation prior to thermal oxidation [27], which is consistent with our QMD simulation discussed above.

Summary

In summary, we propose that during thermal oxidation, carbon interstitials are injected into the SiC epilayer and form the carbon di-interstitial cluster $(C_i)_2$, which is a major cause of the low channel mobility in thermally grown SiC MOSFETs. The $(C_i)_2$ defect can account for recent experimental observations of SiC epilayer traps and is consistent with the latest pre-oxidation implantation experiments.

We thank A. F. Basile, P. M. Mooney, J. R. Williams and L. C. Feldman for helpful discussions. This work is supported by NSF GOALI grant DMR-0907385 and by the McMinn Endowment at Vanderbilt University. Computing resources are provided by NSF TeraGrid grant TG-DMR100022.

References

[1] V. V. Afanas'ev, M. Bassler, G. Pensl, and M. Schulz: phys. stat. sol. (a) 162 (1997), p. 321.

[2] V. V. Afanas'ev, F. Ciobanu, G. Pensl, and A. Stesmans, Contributions to the Density of Interface States in SiC MOS Structures, in: W. J. Choyke, H. Matsunami and G. Pensl (Eds.), Silicon Carbide: Recently Major Advances, Springer, Berlin, 2003, pp. 343-371.

[3] G. Y. Chung, C. C. Tin, J. R. Williams, K. McDonald, M. Di Ventra, S. T. Pantelides, L. C. Feldman, and R. A. Weller: Appl. Phys. Lett. 76 (2000), p. 1713.

[4] P. Jamet, S. Dimitrijev, and P. Tanner: J. Appl. Phys. 90 (2001), p. 5058.

[5] S. Wang, S. Dhar, S-R. Wang, A.C. Ahyi, A. Franceschetti, J.R. Williams, L.C. Feldman and S.T. Pantelides: Phys. Rev. Lett. 98 (2007), p. 026101.

[6] H. Ö. Ólafsson, G. Gudjónsson, P.-Å. Nilsson, E. O. Sveinbjörnsson, H. Zirath, T. Rödle, and R. Jos: Electron. Lett. 40 (2004), p. 508.

[7] D. Okamoto, H. Yano, S. Kotake, K. Hirata, T. Hatayama, and T. Fuyuki: Mater. Res. Soc. Symp. Proc. 1246 (2010), p. B06-06.

[8] A. Agarwal and S. Haney: J. Electron. Mater. 37 (2008), p. 646.

[9] X. Shen and S. T. Pantelides: Appl. Phys. Lett. 98 (2011), p. 053507.

[10] A. Mattausch, M. Bockstedte, and O. Pankratov: Phys. Rev. B 70 (2004), p. 235211.

[11] A. Gali, N. T. Son, and E. Janzén: Phys. Rev. B 73 (2006), p. 033204.

[12] F. Devynck, A. Alkauskas, P. Broqvist, and A. Pasquarello: AIP Conf. Proc. 1199 (2010), p. 108.

[13] J. P. Perdew, K. Burke, and M. Ernzerhof, Phys. Rev. Lett. 77 (1996), p. 3865.

[14] G. Kresse and D. Joubert, Phys. Rev. B 59 (1999), p. 1758.

[15] G. Kresse and J. Furthmüller, Phys. Rev. B 54 (1996), p. 11169.

[16] T. Zheleva, A. Lelis, G. Duscher, F. Liu, I. Levin, and M. Das: Appl. Phys. Lett. 93 (2008), p. 022108.

[17] M. Bockstedte, A. Mattausch, and O. Pankratov: Phys. Rev. B 68 (2003), p. 205201.

[18] T. Hiyoshi and T. Kimoto: Appl. Phys. Express 2 (2009), p. 091101.

[19] L. Storasta and H. Tsuchida: Appl. Phys. Lett. 90 (2007), p. 062116.

[20] K. Danno and T. Kimoto: J. Appl. Phys. 101 (2007), p.103704.

[21] T. Y. Tan and U. Gösele: Appl. Phys. Lett. 40 (1982), p. 616.

[22] J. Kim, F. Kirchhoff, J. W. Wilkins, and F. S. Khan: Phys. Rev. Lett. 84 (2000), p. 503.

[23] K. Kawahara, M. Krieger, J. Suda, and T. Kimoto: J. Appl. Phys. 108 (2010), p. 023706.

[24] T. Umeda, J. Isoya, N. Morishita, T. Ohshima, E. Janzén, and A. Gali: Phys. Rev. B 79 (2009), p. 115211.

[25] K. Kawahara, G. Alfieri, and T. Kimoto: J. Appl. Phys. 106 (2009), p. 013719.

[26] P. Déak, J. M. Knaup, T. Hornos, C. Thill, A. Gali, and T. Frauenheim: J. Phys. D 40 (2007), p. 6242.

[27] A.F. Basile, S. Dhar, and P. M. Mooney: J. Appl. Phys. 109 (2011), p. 114505.

[28] A. Mattausch, PhD Thesis, Friedrich-Alexander-Universität Erlangen-Nürnberg, Nürnberg, 2005.

[29] A. F. Basile, J. Rozen, X. D. Chen, S. Dhar, J. R. Williams, L. C. Feldman, and P. M. Mooney: ECS Trans. 28 (2010), p. 95.

Materials Science Forum Vols. 717-720 (2012) pp 449-452

doi:10.4028/www.scientific.net/MSF.717-720.449

Silicon Carbide-Silicon Dioxide Transition Layer Mobility

S. Salemi[1], A. Akturk[2], S. Potbhare[2], A. Lelis[3], and N. Goldsman[2]

[1]Department of Reliability Engineering, Univ. of Maryland, College Park, Maryland 20742, USA

[2]Dep. of Electrical and Computer Eng., Univ. of Maryland, College Park, Maryland 20742, USA

[3]U.S. Army Research Laboratory, 2800 Powder Mill Road, Adelphi, Maryland 20783, USA

[1]shahrzadsalemi@gmail.com

Keywords: 4H-SiC/SiO_2 transition layer, defects at SiC/SiO_2 interface, mobility, DFT, DOS

Abstract. We present transition layer electron mobility versus field curves for several 4H-SiC/SiO_2 structures, simulated by a newly developed Monte Carlo simulator that uses density of states calculated by density functional theory (DFT). Our calculations show that among all structures, abrupt SiC/SiO_2 has the highest transition layer mobility.

Mobility of SiC MOSFET Transition Region

Mobility in SiC MOSFET structures is usually calculated using a Matthiessen's methodology which, in general, treats the total average scattering as a linear sum of the average individual scattering mechanisms. In this way, simulations can explore the effects of different types of scattering mechanisms on the overall mobility. In MOSFETs, mobility is often separated into surface and bulk interactions. However, for SiC MOSFETs, the presence of a transition layer makes it advantageous to separate mobility into three regions: surface, bulk and transition layer. In this work we focus on the mobility in the transition region.

The transition region is a type of Silicon-Oxycarbide (SiC_xO_y) structure (an example $Si_{1-x}C_xO_2$ structure formed in SiO_2 with $x=1/3$ is shown in Fig. 1); its exact form and content is still under investigation, and probably depends on the specific fabrication process [1]. In this paper, we present the results for mobility calculations performed for likely $Si_{(1-x)}C_xO_2$, and SiC_xO_y structures comprising the transition layer [1, 2]. The methodology is as follows. We first perform Density Functional Theory (DFT) calculations using "Quantum Espresso" software [3] to determine the density of states of likely relaxed transition layer structures. These densities of states are then input into our Monte Carlo simulator to provide the scattering rates as a function of instantaneous electron energy [4]. We then run the transition layer Monte Carlo simulations as a function of electric field for the given $Si_{(1-x)}C_xO_2$, and SiC_xO_y structures structure and obtain the mobility versus electric field curve. We then repeat the process for the other likely $Si_{(1-x)}C_xO_2$, and SiC_xO_y structures we are investigating. The results for transition layer mobility versus electric field are then compared with other investigations. Best agreement with other investigations provides evidence for determining the more likely transition layer atomic structure. This algorithm is shown in the flowchart of Fig. 2. The density of states for an abrupt structure of SiC/SiO_2, created by putting a 36-atom cluster of α-quartz on a 72-atom cluster of 4H(0001)SiC, is first determined as a reference. Then oxygen and carbon atoms are introduced interstitially in the SiC and SiO_2 sides of the structures, respectively, with different concentrations to make C-O-Si bridges.

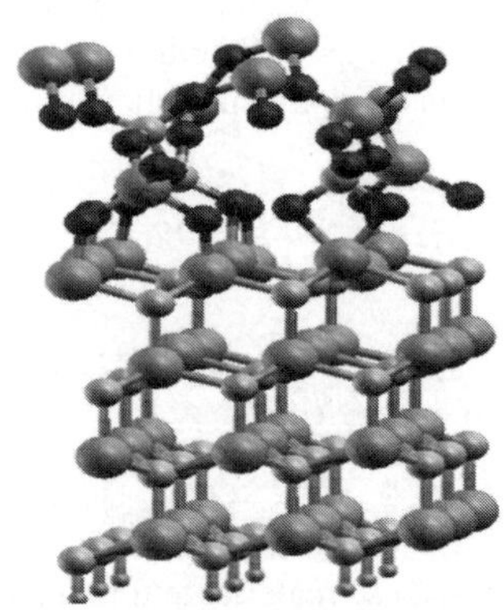

Fig. 1: $Si_{1-x}C_xO_2$ (x=1/3) formed in SiO_2 (top section) plotted as an example structure input to DFT. The structure includes silicon, carbon, oxygen and hydrogen (for passivation).

Monte Carlo Simulations of Transport near Silicon Carbide-Silicon Dioxide Interfaces

To understand the effects of different SiC/SiO_2 transition layers on channel and field-effect mobilities, we feed the results of our density functional theory (DFT) calculations to a Monte Carlo (MC) simulator developed for investigating transport in SiC and at SiC/SiO_2 interfaces. More specifically, the SiC/SiO_2 MC transport simulator requires the following as input: the DFT-calculated densities of states (DOS), physical dimensions of the cell simulated, as well as material and phonon properties of the system considered. We first calculate an average carrier energy dispersion relation based on the DFT-calculated DOS curves, as explained in [4]. This is achieved by relating the energy axis in the DFT-DOS curve to an average crystal momentum. To this end, we evaluate the integral of the DFT-DOS curve starting from the conduction band edge to an energy *E*, and we take this integral to be proportional to the cube of an average momentum for this energy in this quasi-three-dimensional system. Once we obtain an average dispersion relation for electrons, we calculate the possible scattering rates for a given energy. Here we consider acoustic phonon-electron scatterings, optical phonon emission/absorption, polar optical phonon emission/absorption as well as scattering events due to ionized impurities based on the Brooks-Herring formalism. As an example, we below show the scattering rate (based on Fermi's Golden rule) as a function of densities of states for acoustic phonons (Eq. 1), optical phonon absorptions (Eq. 2) and optical phonon emissions (Eq. 3), respectively.

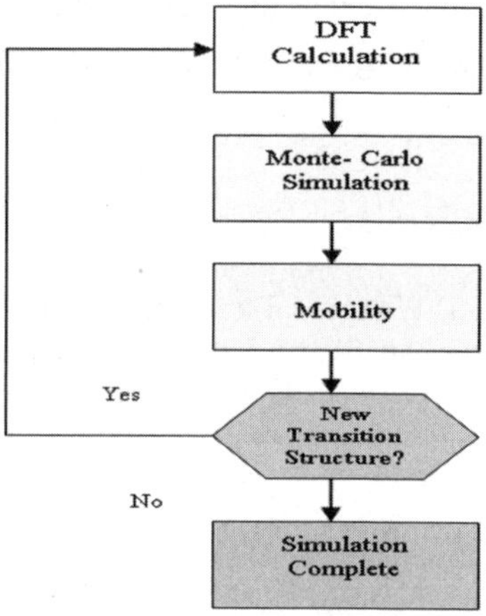

Fig. 2: Algorithm for mobility calculation of different structures.

$$\Gamma_{ac}(E) = \frac{D_{ac}^2 k_B T}{\rho \hbar \upsilon^2 \pi} DOS(E). \tag{1}$$

$$\Gamma_{op}^{ab}(E) = \frac{D_{op}^2 \hbar}{2\rho E_{op} \pi} N(E_{op}) DOS(E + E_{op}). \tag{2}$$

$$\Gamma_{op}^{em}(E) = \frac{D_{op}^2 \hbar}{2\rho E_{op} \pi} [N(E_{op}) + 1] DOS(E - E_{op}). \tag{3}$$

Above, we use material and phonon constants such as D, ρ, υ, and E_{op} which respectively represent deformation potential, density, sound velocity and optical phonon energy. Scattering rates are calculated using these constants, which are taken from our previously calibrated MC 4H-SiC simulator [4], the DFT-calculated DOS curve, the phonon occupation number $N(E_{op})$, as well as the standard constants such as the Boltzmann constant k_B, and temperature T.

We assume that the alpha quartz layer (corresponds to the top section of our unit cell after the exclusion of the vacuum layer) present in our DFT simulations gives rise to energies high above the conduction band edge, that therefore minimally affect the transport characteristics calculated here. Furthermore, the states related to the bulk of the SiC layer result in large mobilities and hence minimally restrict the value of the transition layer mobility. Here the bulk of the SiC layer corresponds to the bottom section of our unit cell after the exclusion of the vacuum layer. This is shown by the bottom three layers of Fig. 1. Moreover, these layers include large and relatively smaller bright spheres that represent silicon and carbon, respectively.

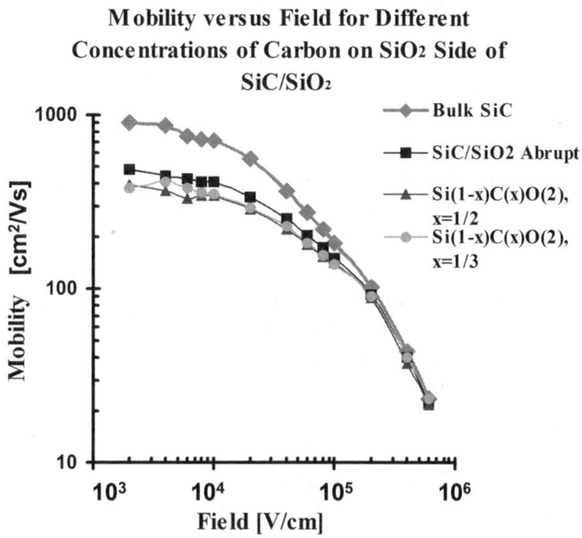

Fig. 3: Mobility versus field curves for the following structures:
-Bulk SiC (red line with diamond)
-SiC/SiO_2 abrupt structure (blue line with square)
-Silicon oxycarbide $Si_{1-x}C_xO_2$ (x=1/3) (yellow line with circle) as shown in Fig. 1
-Silicon oxycarbide $Si_{1-x}C_xO_2$ (x=1/2) (green) formed in SiO_2 part of SiC/SiO_2 (yellow).

Using the semi-classical MC transport calculations, we determine average velocity and mobility for a given structure at a given external field. To achieve this, we first determine the likely time interval for an electron with energy *E* escaping a scattering event. This provides scattering-free drift along the field direction. After this drift, we calculate the likelihood of possible scatterings. In case of no-scattering the electron drifts further along the field direction, and in case of scattering, electron energy and momentum (assuming isotropic scattering) changes according to a randomly chosen scattering event (relative probabilities of scattering events are taken into consideration while selecting a scattering event). We repeat scattering with phonons and ionized impurities as well as drift due to applied field until the electron velocity and energy reach an average.

Key Results Summary

We here compare the field-dependent mobility below the interface for different transition region 4H(0001)-SiC/SiO_2 structures. Figs. 3 and 4 show the resulting electron mobility when the transition layer is on the SiO_2 and SiC side of the interface, respectively. We also compare these mobilities with those calculated for the bulk 4H-SiC and an abrupt interfacial layer. Here the transition region is composed of a silicon-oxycarbide layer such as that shown in Fig. 1. For the transition layer that is on the SiO_2 side of the interface, we consider two structures having a composition that can be described by $Si_{1-x}C_xO_2$ with *x* equal to 1/2 and 1/3. As for the transition layer that is on the SiC side of the interface, the chemical composition can be represented by SiC_xO_y. In the latter structure, we substitute up to nine carbons with oxygens.

As seen in Figs. 3 and 4, the mobility of SiC reduces to less than 50% of the bulk value when there is an abrupt interface between SiC and SiO_2. When a transition layer is present on the SiO_2 side of the interface, the mobility reduces to approximately 42% of the bulk SiC value. (For this transition region, the transition layer mobility is roughly 400 cm^2/Vs.) When the transition layer is on the SiC side of the interface, the electron mobility decreases to between approximately 30% and 8% of the original bulk SiC value (the lowest value is roughly 75 cm^2/Vs). These results for transition layer mobilities are inconsistent with previously reported values extracted using a combination of drift-diffusion device simulation and experiment [5].

In addition to transition region mobilities, we also investigate the impact of different amounts of interface traps on the transport calculations. We calculate the low field mobility as a function of different levels of ionized impurity concentration in these structures. When the ionized impurity concentration is set to 10^{20} cm^{-3} in our quasi three-dimensional interface, the mobility of the abrupt structure, $Si_{1-x}C_xO_2$ with x=1/3 and $Si_{1-x}C_xO_2$ with x=1/2 drops to approximately 50, 40 and 30 cm^2/Vs, respectively. For the same ionized impurity concentration, the calculated mobilities for transition layers formed on the SiC side of the interface vary from 15 to 30 cm^2/Vs. Furthermore, we should also mention that all mobilities are in lower double digits as ionized impurity concentration approaches 10^{20} cm^{-3}, which is consistent with reported field effect mobilities for SiC MOSFETs [5, 6]. Assuming a transition region thickness of 1 nm, the above-mentioned ionized impurity concentration can be translated into an areal concentration of 10^{13} cm^{-2}, which is also consistent with total acceptor type trap levels measured for SiC-SiO_2 interfaces [5, 6]. Additionally, the calculated results along with data in references cited below indicate that the silicon-oxy-carbide structure $Si_{1-x}C_xO_2$ with x=1/3 may be more likely to be found in SiC/SiO_2 interfaces among the possible transition regions discussed here.

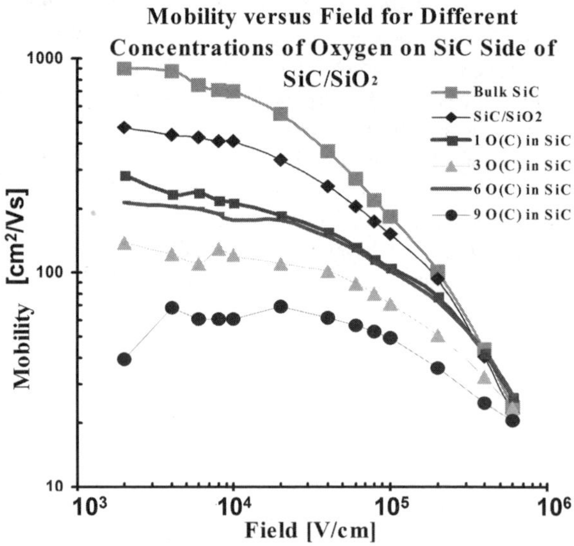

Fig. 4: Mobility versus field for the following structures:
-Bulk SiC (first curve on top, small square)
-SiC/SiO_2 abrupt structure (blue)
-Silicon oxycarbide SiC_xO_y with one (purple), three (yellow), six (green), nine (gray) O_C (oxygen in place of carbon) in the SiC side of SiC/SiO_2.

References

[1] C. Radtke, I. J. R. Baumvol, J. Morais, and F. C. Stedile, Applied Physics Letters 78 (2001) 3601-3603.

[2] C. R. S. da Silva, J. F. Justo, and I. Pereyra, Applied Physics Letters 84 (2004) 4845-4847.

[3] P. Giannozzi et al., Journal of Physics: Condensed Matter 21 (2009) 1-19.

[4] A. Akturk, N. Goldsman, S. Potbhare, and A. Lelis, Journal of Applied Physics 105 (2009) 033703-1-7.

[5] S. Potbhare, A. Akturk, N. Goldsman, A. J. Lelis, S. Dhar, and A. Agarwal, Materials Science Forum, Vols. 645 – 648 (2010) 975-978.

[6] V. Tilak, K. Matocha, and G. Dunne, IEEE Transactions on Electron Devices 56 (2009) 162-169.

Materials Science Forum Vols. 717-720 (2012) pp 453-456
© (2012) Trans Tech Publications, Switzerland
doi:10.4028/www.scientific.net/MSF.717-720.453

Sodium, Rubidium and Cesium in the Gate Oxides of SiC MOSFETs

B. R. Tuttle[1,a,*], S. Dhar[2], S.-H. Ryu[2], X. Zhu[3,4], J. R. Williams[3], L. C. Feldman[1,4] and S. T. Pantelides[1,5]

[1]Department of Physics and Astronomy, Vanderbilt University, Nashville, Tennessee 37235, USA

[2]R&D Division, Cree Inc., Durham, North Carolina, USA

[3]Department of Physics, Auburn University, Auburn, Alabama 36849, USA

[4]Institute of Advanced Materials Devices and Nanotechnology, Rutgers University, New Brunswick, New Jersey, USA

[5]Oak Ridge National Laboratory, Oak Ridge, Tennessee 37831, USA

[a]brt10@psu.edu, *corresponding author

Keywords: Semiconductor Theory, MOSFETs, Shallow Impurities

Abstract. Large group-I elements such as sodium, rubidium and cesium have recently been incorporated in the gate oxide of SiC power MOSFETs. In the case of sodium incorporation, enhanced field effect mobilities have been definitively observed. Based on density functional calculations, we find large group-I elements serve as shallow impurities near the interface. The enhanced mobility, observed in the case of sodium, can be explained in terms of an impurity band model.

Introduction

For high-power and high-temperature applications, MOSFETs based on SiC are a promising alternative to traditional Si based MOSFETs. One limitation has been the relatively low channel-electron mobilities in the SiC devices. Recently, it has been demonstrated that sodium in the gate oxides of SiC MOSFETs improves the channel-electron mobilities [1,2,3]. A major problem with the sodium enhanced MOSFETs is the instability of the I-V characteristics at operating temperatures due to the motion of the sodium ions in the gate oxide. For this reason, larger group-I elements such as rubidium and cesium are currently being investigated to determine if a stable and improved SiC MOSFET can be engineered [4].

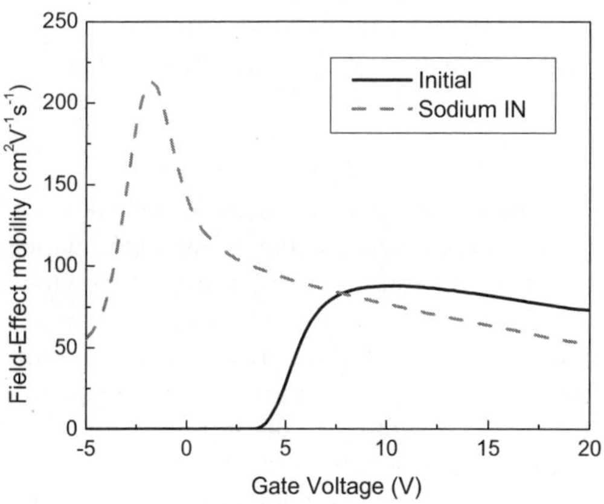

Fig. 1: Field-effect mobility of lateral MOSFET with sodium enhanced oxidation: as-processed (labelled 'initial') and sodium ions drifted to the SiO_2/4H-SiC interface (labelled Sodium IN).

Results

Sodium can be incorporated into gate oxides of SiC MOSFETs by growing the oxide inside an alumina tube containing significant Na contamination [1,2,3]. To verify previously observed sodium enhancement, new sodium incorporated SiC devices have been processed [3]. The drain current versus gate voltage for a drain bias of 0.05 V was recorded. The field effect mobility curve reported in Fig. 1 is extracted from the derivative of the current with respect to the gate voltage. There are two curves for two positions of the sodium ions in the gate oxide. In the case labeled "initial" (dashed curve), the sodium ions are distributed throughout the gate oxide as determined by the growth process. In the case labeled "Sodium IN" (solid), the sodium ions are driven toward the channel-oxide interface with a temperature (175°C) and gate bias (+ 15 V) annealing. The "initial" curve shows a positive turn-on voltage indicating that a large number of interface traps is present. Although not shown, reversing the bias to drive sodium away from the channel-gate oxide interface results in a turn on voltage greater than 20 V, i.e., the device does not turn on. Clearly, sodium incorporation during growth results in a large concentration of interface traps. In contrast, with Na ions driven to the interface region, the "Sodium IN" case results in a negative turn-on voltage and a large initial peak in the field effect mobility. Unfortunately, the beneficial properties of Na ions are unstable during device operation.

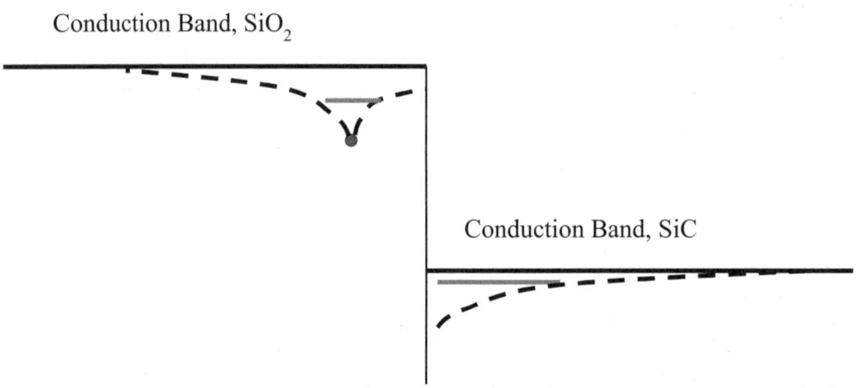

Fig. 2: Band diagram for a sodium ion (solid disk) near the SiC / SiO_2 interface. The ion's Coulombic potential (dashed curves) is shown along with the lowest defect levels (solid horizontal lines) indicating a shallow level in the SiC near the interface.

To explain these observations, we explored the electrochemistry of sodium in the SiC-oxide system using density functional calculations. Sodium does not interact with the oxide, the SiC channel or defects. Instead, these atoms ionize to form isolated positive charges each with a Coulombic potential. Fig. 2 shows a schematic of the defect induced levels near the SiC-oxide interface. In Fig. 2, the dashed curve represents the screened Coulombic potential for a sodium ion in the gate oxide near the semiconductor interface. The theory of Coulombic defects near an interface has been developed previously [5]. As illustrated, a Na ion in the oxide actually results in a shallow donor level in the SiC channel. In Fig. 3, the SiC density of interface states is represented schematically. The dashed peak is the Na ion induced donor levels whereas the solid line represents the background interface states generated during growth. As the concentration of sodium ions increases the defect levels overlap to form a conducting impurity band near the conduction band minimum of the SiC channel. Also, the effective Fermi level in the SiC rises with an increasing concentration of ions at the interface (Fig 3). This behavior explains the lowering of the turn-on voltage as the Na ions are driven toward the interface. The peak in the field effect mobility reported in Fig. 2 can be explained by the peak in the shallow impurity band states due to sodium ions concentrated near the interface as reported in Fig. 3. Finally, the impurity band model presented here also explains differences between Na ions in Si versus SiC MOSFETs, as discussed in Ref. [3].

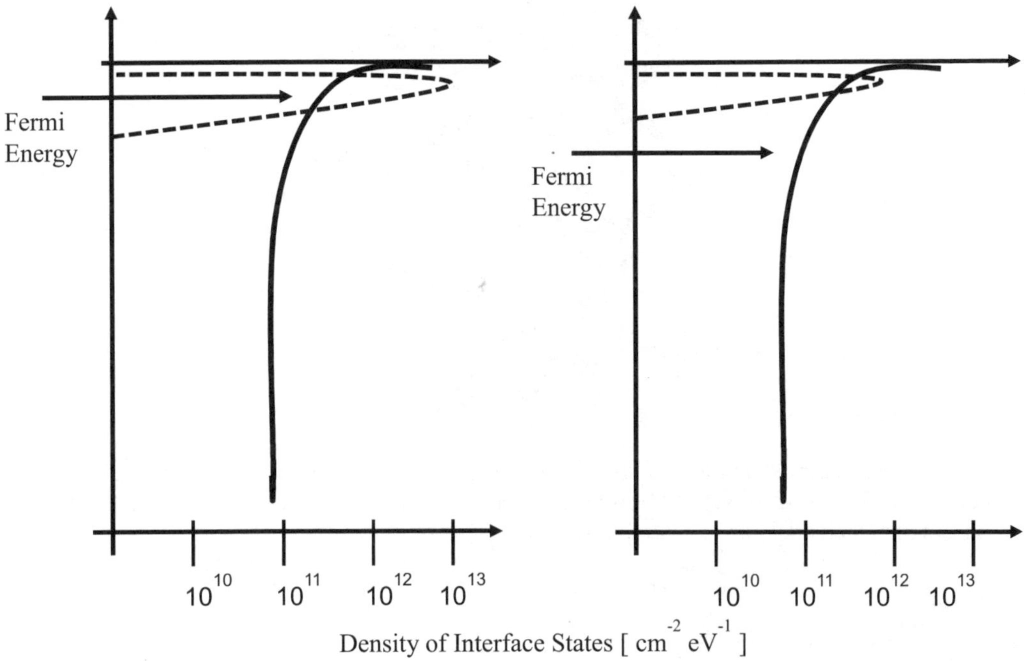

Fig. 3: The densities of interface states are presented schematically for the SiC side of the SiC / SiO_2 interface. The dashed curve represents the Na ion induced impurity band just below the conduction band (CB) at two different Na ion concentrations. The solid curve represents the other interface states in the upper half of the SiC band gap. The left and right curves are for sodium concentrations at the interface that are large and small, respectively.

The successful observation and explanation of sodium enhanced field effect mobilities has encouraged the experimental and theoretical investigation of larger group-I elements in SiC MOSFETs, including rubidium and cesium [4]. From calculations in bulk silica models, all three elements (Na, Rb, and Cs) are found in the present study to favor the open interstitial region of the silica network, i.e. none of them interact strongly with the lattice so they try to move as far from the silica atoms as possible. However, recently it has been observed that Cs can react with O_2 molecules during oxidation to form CsO_2, which will not be electrically active [4]. However, atomic Na, Rb and Cs in the gate oxide near the SiC interface all result in a shallow impurity in the channel of SiC. Once located in an open interstitial region within the oxide, a major difference in the three atoms is the barrier for motion. In the ionic state, the three atoms (Na, Rb, Cs) in the bulk oxide follow the same diffusion path between two neighbor interstitial regions. In Fig. 4, the diffusion pathway is shown for large group-I elements in a silica gate oxide model. While a full determination of the diffusivity is beyond the scope of the present study, the barrier for one diffusion hop has been calculated for each group-I element. We find the diffusion barrier for the sodium ion is 1.2 eV, consistent with observations that sodium is mobile at less than 200°C [2,3]. The barriers for rubidium and cesium are much larger, 3.0 and 3.9 eV, respectively. This indicates that once localized in an open interstitial of the oxide these Rb and Cs ions will be stable to high temperatures. Indeed, observations in bulk SiO_2 find that Rb is stable above 300°C [6]. Of course, given the stability of these elements, incorporating them into the gate oxide without damaging the oxide lattice is a technical challenge. The incorporation of these large elements into gate oxides has been achieved by ion implantation previously [5] and reproduced in our group, showing a pile-up at the interface.

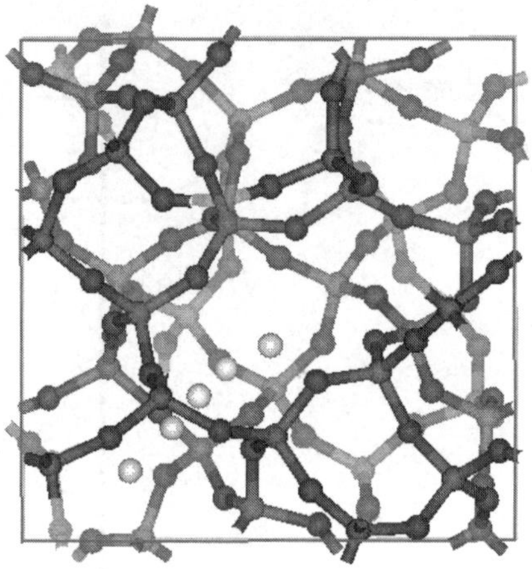

Fig. 4: Shown is a periodic model of silica gate oxide with dark, grey and white balls being oxygen, silicon and sodium, respectively. Multiple sodium ion positions are shown to indicate the diffusion path from one interstitial site to a neighboring site.

Summary

The present experiments [Figs. 1 and 2] confirm that sodium ions near the gate oxide/semiconductor interface of SiC MOSFETs enhance the channel electron mobilities of the device. These observations are semi-quantitatively explained by an impurity band model [Figs. 2 and 3]. All large group-I elements are predicted to form an impurity band if they can be localized in open interstitials in the oxide near the channel interface. The diffusion barriers calculated for rubidium and cesium [Fig. 4] are found to be significantly larger than that for sodium, suggesting a stable device could be designed with these elements if they can be placed near the oxide/channel interface without causing damage.

Acknowledgement

This work has been supported in part by the National Science Foundation via grant DMR-0907385.

References

[1] F. Allerstam, H. O. Olafsson, G. Gudjonsson, D. Dochev, E. O. Sveinbjörnsson, T. Rodle, and R. Jos, J. Appl. Phys. **101** (2007) 124502.

[2] V. Tilak, K. Matocha, G. Dunne, F. Allerstam, and E. O. Sveinbjörnsson, IEEE Trans. Electron Devices **56** (2009) 162.

[3] B. R. Tuttle, S. Dhar, K. Zhu, J. R. Williams, L. C. Feldman and S. T. Pantelides, J. Appl. Phys. **109** (2011) 023702.

[4] A. Chatterjee, H. Piao, K. Matocha, J. Fronheiser, V. Tilak, and V. Smentkowski, J. Appl. Phys. **108** (2010) 083512.

[5] N. O. Lipari, J. Vac. Sci. Technol. **15** (1978) 1412.

[6] H. G. Francois-Saint-Cyr, F. A. Stevie, J. M. McKinley, K. Elshot, and L. Chow, J. Appl. Phys. **94** (2003) 7433.

Materials Science Forum Vols. 717-720 (2012) pp 457-460

doi:10.4028/www.scientific.net/MSF.717-720.457

Total Near Interface Trap Density Calculation of 4H-SiC/SiO$_2$ Structures before and after Nitrogen Passivation

S. Salemi[1,a], A. Akturk[2], S. Potbhare[2,3], A. Lelis[4], N. Goldsman[2]

[1] Department of Reliability Engineering, Univ of Maryland, College Park, MD 20742, USA

[2] Department of Electrical and Computer Engineering, Univ. of Maryland, College Park, MD, 20742, USA

[3] National Institute of Standards and Technology, Gaithersburg, MD 20899-1070

[4] U.S. Army Research Laboratory, 2800 Powder Mill Road, Adelphi, MD 20783, USA

[a] shahrzadsalemi@gmail.com

Keywords: 4H-SiC/SiO$_2$ transition layer, SiC interface states, nitrogen passivation; DFT, DOS

Abstract. We compare the effect of hydrogen, nitrogen, and phosphorus passivation on total near interface trap density and mobility of 4H(0001)-SiC/SiO$_2$ structure. The results show that nitrogen and phosphorus passivation decrease total near interface trap density by pushing the energy levels of interface traps away from the conduction band. The density of states (DOS), including interface states (D_{it}), are calculated for several 4H(0001)-SiC/SiO$_2$ structures using density functional theory (DFT).

Introduction

The density of interface states and its dependence on defects is calculated theoretically. Since channel mobility in SiC MOSFETs is often dominated by Coulomb scattering due to interface traps, we calculated the density of states (DOS) for different 4H-SiC/SiO$_2$ structures using Density Functional Theory (DFT) to investigate the origin of these traps, and especially the effects of hydrogen, nitrogen and phosphorus annealing, which may be used in an effort to reduce the number of interface states near the conduction band.

It has been demonstrated experimentally that annealing SiC/SiO$_2$ MOS structures with nitrogen significantly reduces interface state densities [1,2]. Passivation of the interface with Nitrogen is often attributed to the strong Si-N bond [3,4]. Optical measurements indicate that Si-N bonds exist at the interface and are thus likely to give rise to this passivation [4]. Furthermore, passivation with hydrogen has also been performed. While very limited success at reducing D_{it} with hydrogen passivation has been reported [5], it is generally agreed that nitrogen passivation yields much better results than that of hydrogen.

In this work, we compare resulting interface state densities that arise after hydrogen and nitrogen passivation by using DFT calculations. In addition, we also calculate the resulting D_{it} that would emerge after passivation with phosphorus, since it is in the same column as N and thus should have similar passivation properties. Furthermore, it has been reported that phosphorus anneal gives rise to improved mobility for MOSFET channel electrons [6,7]. Finally, we investigate the effect of nitrogen passivation on silicon-oxycarbide and oxygen vacancy in the oxide side of the SiC/SiO$_2$ structure near the interface.

Calculations and Results

We begin the DFT calculations by constructing a supercell structure of SiC/SiO$_2$, which is formed by putting 36 atoms of α-quartz on 72 atoms of 4H(0001)-SiC. A layer of vacuum has been placed on the top and bottom of the structures to isolate them in the z direction. The atoms at the bottom of the structures are terminated with hydrogen. This resulted in a 1.7 nm thick SiC/SiO$_2$ layer which then received various passivations. We used the DFT solver Quantum Espresso to perform relaxation, band-structure and DOS calculations of the SiC/SiO$_2$ structures, with the Perdew-Burke-Ernzerhof exchange-correlation functional, and a norm-conserving pseudopotential [8]. Figs. 1-3 show the effects of hydrogen, nitrogen, and phosphorus passivation on the DOS of the 4H(0001)-SiC/SiO_2 structure. In all cases, the density of states for SiC/SiO_2 original non-passivated structures (containing a couple of Si dangling bonds, and one strained Si-O bond) are kept as the reference. Besides, the density of states of the structures, before, and after of passivation are aligned with respect to the deep, valence states, far from the gap, and hence unaffected by local changes.

As it can be seen in Fig. 1, hydrogen passivation is not very effective in reduction of the interface trap density in the SiC structure; however, it makes a small shift of the interface traps' energy levels away from the midgap. The wide bandgap of 4H-SiC includes the energy levels of Si–H bonds. This small shift of energy is enough for silicon to push all the near interface traps away from the silicon bandgap because of relatively small bandgap of silicon with respect to SiC [3].

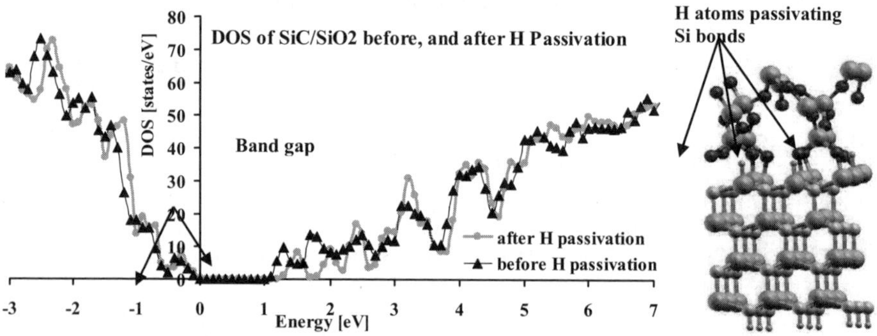

Fig. 1: Density of states of SiC/SiO_2 before (dotted line) and after (triangle-line) hydrogen passivation (left), the atomic structure (right).

In the case of nitrogen passivation, strong $Si \equiv N$ bonds as well as Si-N-O bridges are created, which passivate the interface traps reducing Nit [3,4]. Therefore, as can be seen in Fig. 2, the near interface traps are reduced significantly, which is likely to lead to improved coulomb mobility reduction as well as better reliability.

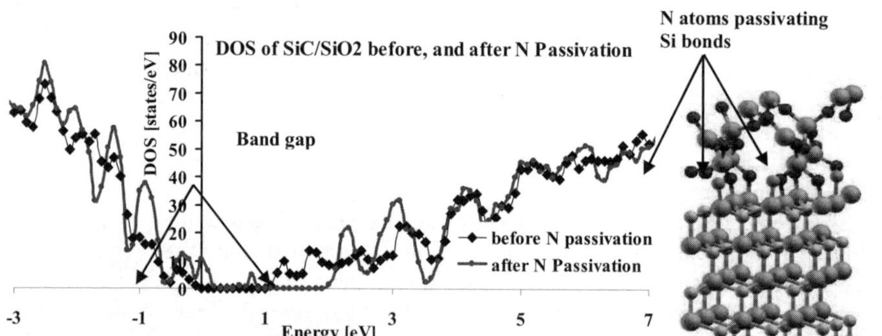

Fig. 2: Density of states of SiC/SiO_2 before (diamond-line) and after (dotted line) nitrogen passivation (left); the atomic structure (right).

It has been reported that phosphorus post oxidation annealing improves inversion layer mobility in SiC MOSFETs [6,7]. Although, the mechanism of phosphorus passivation is not clear yet, our simulation shows that passivation with phosphorus eliminates the near interface traps as well, which may explain the improved mobility. Fig. 3 shows the calculated reduction of interface states that result from phosphorus passivation on the SiC/SiO_2 structure.

To show the trend of D_{it} reduction, the density of interface traps near the conduction band is calculated for the structures before and after hydrogen, nitrogen and phosphorus passivation, relative to one another. As can be seen in Fig. 4, the density of interface traps is reduced after nitrogen and even more after phosphorus passivation.

The impact of nitrogen passivation on the other structures with a transition layer and a defect are also investigated. While there is not agreement on the structure of the transition layer, researchers suggest that a likely structure is to replace the silicon atoms on the oxide side with carbon atoms to make $Si_{1-x}C_xO_2$ with different concentrations [9]. Figs. 5-6 show the effect of nitrogen passivation on $Si_{1-x}C_xO_2$ with x=1/3 and a structure with an oxygen vacancy in the SiO_2 side of SiC/SiO_2. Again, all the dangling, and strained bonds are passivated by nitrogen.

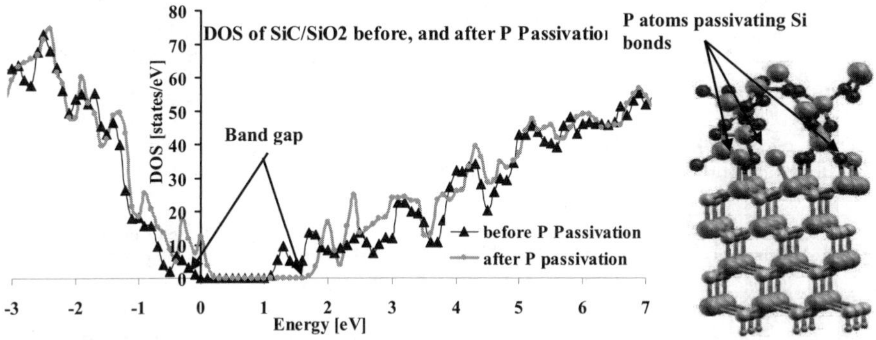

Fig. 3: Density of states of SiC/SiO_2 before (triangle-line) and after (dotted-line) phosphorus passivation (left); the atomic structure (right).

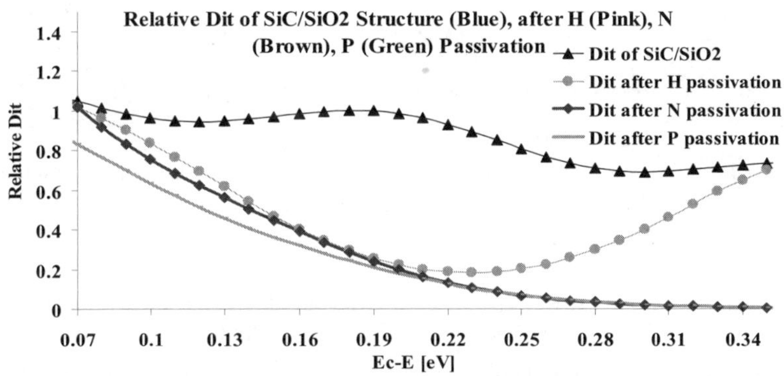

Fig. 4: D_{it} of SiC/SiO2 structure before (triangle-line), and after H (circle-line), N (diamond-line), and Phosphorus (dotted-line) passivation, relative to one another in arbitrary units.

Conclusion

In summary, the impact of hydrogen, nitrogen, and phosphorus passivation on the density of states of 4H-SiC(0001)/SiO_2 is investigated by the density functional theory approach. The calculations show a reduction of the density of states near the conduction band after nitrogen and phosphorus passivation, while hydrogen passivation just makes a small shift of the interface trap density from the conduction band edge. N and P passivations introduce peaks near the valence bands; in Figs. 2-3, the peak is at E_v+0.1eV. There is another peak at 0.8 eV in Fig. 2. In Figs. 5-6 the peaks are near the conduction band again. Furthermore, nitrogen passivation appears to also eliminate near interface states of other structures such as silicon oxycarbide and oxygen vacancy in oxide side of SiC/SiO_2. Therefore, it is likely that the creation of strong bonds between silicon-nitrogen after nitrogen and silicon-phosphorus after phosphorus annealing passivates not only the interface traps due to weak silicon bonds at the interface, but also those interface traps resulting from defects in the oxide side of the 4H-SiC(0001)/SiO_2 structure.

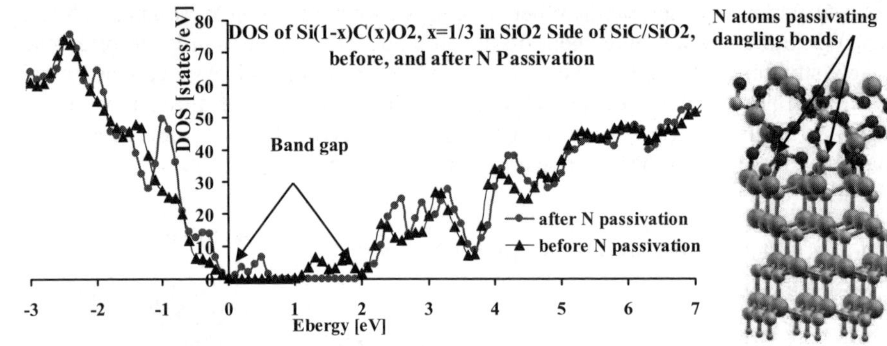

Fig. 5: Density of states of $Si_{1-x}C_xO_2$ (x=1/3) in SiO_2 side of SiC/SiO_2 before (triangle-line) and after (circle-line) nitrogen passivation (left); the atomic structure (right).

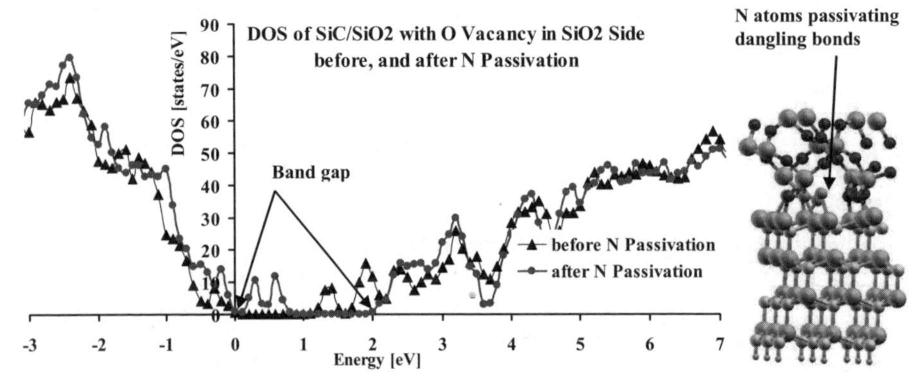

Fig. 6: Density of states of an oxygen vacancy in SiO_2 side of SiC/SiO_2 before (triangle-line) and after (circle-line) N passivation (left); the atomic structure (right).

References

[1] G. Y. Chung, C. C. Tin, J. R. Williams, K. McDonald, M. Di Ventra, S. T. Pantelides, L. C. Feldman, and R. A. Weller, Applied Physics Letters 76 (2000) 1713-1715.

[2] H. Li, S. Dimitrijev, H. Barry Harrison, and D. Sweatman, Applied Physics Letters 70 (1997) 2028-2030.

[3] P. Jamet, S. Dimitrijev, and P. Tanner, Journal of Applied Physics 90 (2001) 5058-5063.

[4] J. W. Chai, J. S. Pan, Z. Zhang, S. J. Wang, Q. Chen, and C. H. A. Huan, Applied Physics Letters 92 (2008) 092119-1-3.

[5] K. Fukuda, S. Suzuki, and T. Tanaka, and K. Arai, Applied Physics Letters 76 (2000) 1585-1587.

[6] D. Okamoto, H. Yano, K. Hirata, T. Hatayama, T. Fuyuki, IEEE Electron Device Letters 31 (2010) 710-712.

[7] D. Okamoto, H. Yano, T. Hatayama, and T. Fuyuki, Applied Physics Letters 96 (2010) 203508-1-3.

[8] P. Giannozzi et al., Journal of Physics: Condensed. Matter 21 (2009) 1-19.

[9] C. Radtke, I. J. R. Baumvol, and J. Morais, and F. C. Stedile, Applied Physics Letters 78 (2001) 3601-3603.

Materials Science Forum Vols. 717-720 (2012) pp 461-464

doi:10.4028/www.scientific.net/MSF.717-720.461

Detection of Mobile Ions in the Presence of Charge Trapping in SiC MOS Devices

Daniel B. Habersat[1,a,*], Aivars Lelis[1,b], and Ronald Green[1,c]

[1]U.S. Army Research Laboratory, 2800 Powder Mill Road, Adelphi, MD 20783, USA

[a]daniel.b.habersat.civ@mail.mil, [b]aivars.j.lelis@us.army.mil, [c] ronald.green39.civ@mail.mil, *corresponding author

Keywords: MOS, charge trapping, mobile ion, bias instability

Abstract. Since power devices such as DMOSFETs will operate at high temperature where mobile ion effects are enhanced, identifying their presence is a key reliability issue for power electronics applications. We have detected the presence of mobile ion contamination in some SiC MOS device sample sets and correlated those results with observed high temperature bias instability. The differing behaviors of these devices to bias stressing as a function of temperature suggests that in some cases mobile ion drift may be counteracting the typical charge trapping effect. Triangular voltage sweep (TVS) data indicates the presence of roughly 1-4x10^{12} cm^{-2} mobile ions in samples where the bias instability significantly decreased with higher temperature, while samples with a nearly flat or positive-trending response showed overall lower ion contaminations of roughly 6-9x10^{11} cm^{-2}. These results, although preliminary, support the theory that mobile ion contamination is the cause of negative bias instability at elevated temperatures in SiC MOS.

Introduction

The purpose of this work is to better understand what role, if any, mobile ions play in altering the electrical characteristics of SiC MOS devices. Numerous groups have discussed the bias-temperature stress (BTS) induced instability of SiC MOS [1-5], with some attributing part of the observed behavior to mobile ion contamination. The typical method to detect mobile ion contamination is to apply a BTS and monitor the change in field-effect mobility [1] or flatband/threshold voltage [2-4], though TVS was also used in [2]. BTS methods will work in cases of gross mobile ion contamination, but it has been postulated [4] that in cases with moderate amounts of mobile ions, the effects may be masked by significant oxide charge trapping (or, conversely, that mobile ions are masking the contribution of charge trapping).

For purposes of this investigation, we have studied devices fabricated via different processing methods and selected devices by the observed behavior of threshold voltage instability due to bias stressing (defined as the difference between threshold voltages following positive and negative bias stress) as a function of temperature. It has been proposed [4] that the mechanism of charge trapping (i.e., channel carriers tunneling into near-interface oxide traps) is responsible for positive instability and that mobile ion drift would cause a negative instability, with the two effects acting in opposition to each other. Accordingly, a reduction in a device's bias instability (e.g., due to a processing change) may not be due to the reduction of charge trapping but instead by the introduction of mobile ions into the oxide. So if we are to effectively evaluate fabrication and processing techniques and accurately characterize the levels of charge trapping, we must also understand the quantities of mobile ions present in the device.

Experiment

The devices used in this study are fabricated on 4H-SiC and consist of lateral MOS capacitor and FET structures, and fully-processed DMOSFETs. Samples were selected from deliverables received over a period of years, each lot being considered state of the art at the time it was produced. In general, devices were fabricated on n-type 4H-SiC wafers with p-type epi and oxidized using standard processes, including post-oxidation NO anneal (oxide thickness

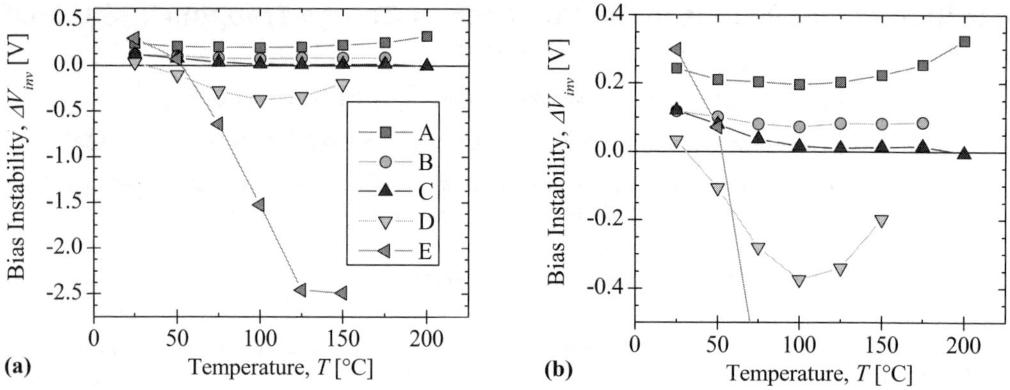

Figure 1. Magnitude of inversion voltage instability due to bias stressing as a function of temperature for all devices studied, showing (a) full range and (b) detailed view.

$t_{ox} \approx$ 500-1000Å). Threshold voltage bias instability was evaluated, calculated using the gate bias needed to achieve a particular drain current [6]. Triangular voltage sweep (TVS) was performed at elevated temperatures to evaluate the presence of mobile ions [7,8].

Bias Instability versus Temperature. Typical instabilities, calculated as the difference between threshold voltage measurements following positive and negative bias stress, are plotted in Figs. 1a and 1b. Samples are ordered by their expected mobile ion contamination, based on the observed threshold instability trends. The two extreme cases are represented by sample A, which exhibited a positive increase in threshold instability at high temperature, and sample E, which produced a significant negative shift following bias stress at temperature.

Triangular Voltage Sweep. TVS measurements were made with a gate sweep from strong inversion to accumulation at a rate < 10 mV/s and at elevated temperature (typically 150 °C). The samples were heated without any bias applied; once at the target temperature, a positive bias was applied to pre-stress the device. Then, the gate bias was slowly ramped from positive to negative and then back to positive again at the same sweep rate but in the reverse direction, producing a triangular waveform. The mobile ion density was estimated by taking the area under the peak of the I_G-V_{GS} curve and scaling by electron charge q, gate area A_G and sweep rate α as appropriate.

Results

A typical I_G-V_{GS} curve measured by TVS is shown in Fig. 2 for sample D. The current has been corrected for capacitive displacement current and bulk conduction through the oxide, and normalized for comparison between data from different α, A_G, and t_{ox}. The strong peak near zero-field is clearly indicative of mobile ion contamination. Asymmetry observed between sweep up and down can be explained by a difference in barrier height on each side of the oxide, which limits the emission rate of ionic charge [9]. The multiple peaks observed in this particular trace may be due to factors such as multiple

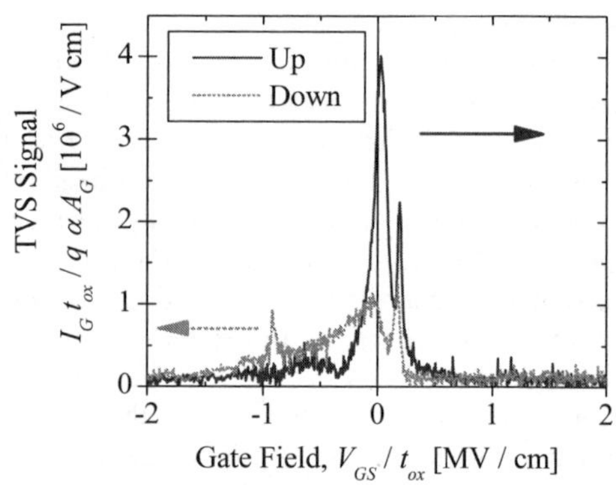

Figure 2. TVS gate leakage versus gate-source bias for sample D, normalized for comparison between different devices. Data are shown for both the upward and downward sweeps.

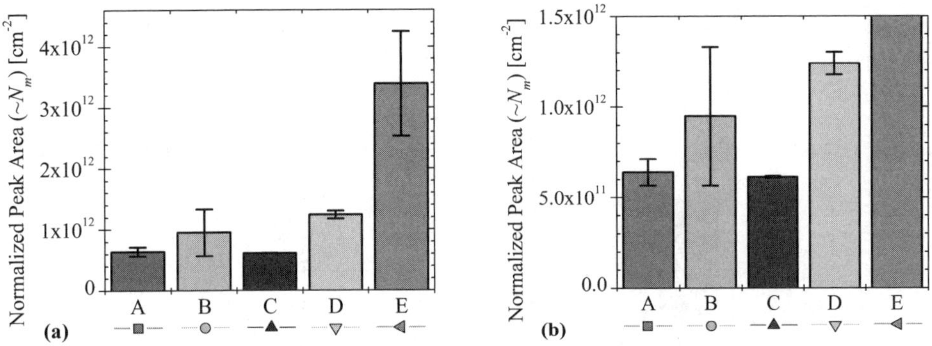

Figure 3. Normalized current peak area from TVS measurements, which should correspond to mobile ion concentration N_m, showing (a) full range and (b) detailed view.

ion species [10], faradaic polarization from the oxide interfaces or high trap density at the SiC/SiO_2 interface [8]. Multiple peaks were observed in the TVS spectra for samples D and E, but the spectra for A, B, and C showed only a single peak.

The normalized peak areas, which should correspond to the number of mobile ions which moved across the oxide during the gate bias sweep, are plotted for each sample in Fig. 3. The bars plot the average peak area for both the upward and downward sweeps for each sample, while the error bars correspond to the values measured for each sweep direction. Ideally, the normalized peak area will be independent of sweep rate & direction [8] and thus the error bars should be small. This data allows us to compare the mobile ion concentration as detected by TVS with the high-temperature threshold instability behavior from Fig. 1. Of immediate note is that sample E and, to a lesser extent, sample D, which had negative instability at high temperature, show significantly more mobile ions than the other samples and support our hypothesis of mobile ion contamination. Additionally, samples A and C show the same mobile ion concentration but have different ΔV_T vs. T behavior. In this case the improvement in ΔV_T appears to have come not from a balancing of mobile ions but instead from a reduction in charge trapping. Since both samples came from the same source but sample C is of a newer vintage, we can infer that their device processing has substantially improved. When Okayama et al. used TVS to study SiC DMOSFETs [2], they detected lower concentrations of $\sim 1\text{x}10^{11}$ cm^{-2} but suggested much more may be present due to contamination from field oxide [11].

For reference, the bias instability due to 10^{12} cm^{-2} mobile ions moving completely across a 500 Å oxide would shift I_D-V_{GS} characteristics by ≈2.3 V. This represents the "worst-case" scenario, which the back-and-forth instability measurements would likely not fully see due to temperature and stress time. The mobile ion concentrations measured in these samples are sufficiently large to account for the observed negative threshold instability or to potentially mask the effects of charge trapping.

Chou wrote about the need for quasi-equilibrium during a TVS measurement [8], and identified it by three parameters that should be independent of temperature, bias ramp rate, and sweep range: normalized peak area, peak current, and gate bias corresponding to peak current. Typical measurement conditions used to establish quasi-equilibrium for Si/SiO_2 are 200-300 °C, ≤ 10 mV/s, and ±1 MV/cm; however, it is by no means certain that the same holds for SiC/SiO_2. The first evaluation parameter Chou referred to is plotted in Fig. 3; peak current and corresponding gate bias are shown in Fig. 4. The data here shows a fair amount of variation, both sample-to-sample and comparing different sweep directions (though some asymmetry is expected). In particular, the peak area should be independent of sweep direction under the right conditions. We did not always see good agreement between the two values for our devices, suggesting that perhaps the devices were not in perfect quasi-equilibrium. This underscores the need to make multiple measurements of TVS under varying conditions, to confirm that the assumptions that the analysis is based upon are correct.

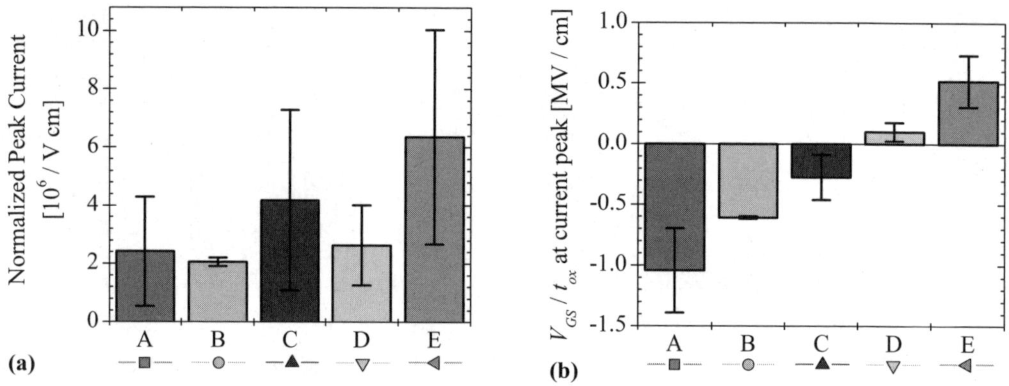

Figure 4. Additional quasi-equilibrium parameters from TVS measurements, showing (a) normalized peak current and (b) gate bias corresponding to the current peak.

Conclusion

Using the TVS technique on SiC lateral test structures and DMOSFETs, we detected mobile ion concentrations of ~6-9x10^{11} cm^{-2} for devices with positive threshold instability, and ~1-4x10^{12} cm^{-2} for devices with negative threshold instability. The contamination levels detected are sufficiently large to explain the large negative instabilities observed, and there is a rough correlation between mobile ion densities and the different behaviors observed in ΔV_T vs. T. These results support the hypothesis that mobile ion drift and oxide charge trapping are counteracting with each other at high temperature in SiC MOS. However, the TVS data collected so far was not able to distinguish well between samples with similar ΔV_T vs. T behavior (such as samples A-C). By analyzing additional samples and refining the measurement system and technique, we hope to increase confidence in the results and more clearly differentiate between similarly behaved devices.

It is important to remember that SiC/SiO_2 is a much more complex material system than Si/SiO_2 and appropriate care must be taken to ensure that measurement conditions are appropriate (in particular, the need for quasi-equilibrium during the TVS measurement). TVS appears to be a useful tool for quantifying mobile ion concentrations in SiC devices and, when combined with other techniques, could be invaluable to evaluate the effects of processing improvements on device performance.

References

[1] Z. Chen et al., J. Electron. Mat, 39 (2010) 526-529.

[2] T. Okayama, S. D. Arthur, J. L. Garrett, and M. V. Rao, Solid State Electron. 52 (2008) 164-170.

[3] M. J. Marinella et al., Appl. Phys. Lett. 90 (2007) 253508.

[4] A. J. Lelis, D. Habersat, R. Green and N. Goldsman, Mat. Sci. Forum 600-603 (2009) 807-810.

[5] K. Matocha, and V. Tilak, Mat. Sci. Forum 679-680 (2011) 318-325.

[6] A. J. Lelis et al., IEEE Trans. Electron. Dev. 55 (2008) 1835-1840.

[7] M. Kuhn and D. J. Silversmith, J. Electrochem. Soc. 118 (1971) 966-970.

[8] N. J. Chou, J. Electrochem. Soc. 118 (1971) 601-609.

[9] E. H. Nicollian, and J. R. Brews, MOS (Metal Oxide Semiconductor) Physics and Technology, John Wiley & Sons, USA, first ed. (1982) 442-443.

[10] D. K. Schroder, Semiconductor Material and Device Characterization, John Wiley & Sons, USA, third ed. (2006) 340-342.

[11] J. R. Trost et al., Proceedings of the 11th ISPSD Conference, (1999) 189-192.

Materials Science Forum Vols. 717-720 (2012) pp 465-468

doi:10.4028/www.scientific.net/MSF.717-720.465

Two-way Tunneling Model of Oxide Trap Charging and Discharging in SiC MOSFETs

A. Lelis [1,a], D. Habersat [1,b], R. Green [1,c], and N. Goldsman [2,d]

[1] U.S. Army Research Laboratory, Adelphi, MD 20783, USA

[2] University of Maryland, College Park, MD 20742, USA

[a] aivars.j.lelis@us.army.mil, [b] daniel.b.habersat.civ@mail.mil, [c] ronald.green39.civ@mail.mil, [d] neil@ece.umd.edu

Keywords: MOSFET, oxide trap, threshold-voltage instability, tunneling model.

Abstract. A two-way tunneling model describing simultaneous oxide trap charging and discharging in SiC MOSFETs is presented, along with a comparison with experimental results. This model can successfully account for the variation in threshold-voltage instability observed as a function of bias-stress time, bias-stress magnitude, and measurement time.

Introduction

A number of experimental studies in recent years have demonstrated the threshold-voltage instability effect in SiC MOSFETs—for example, see [1-5]. This effect has been attributed to the direct tunneling of electrons into and out of near-interfacial oxide traps [1], and the significant increase in this instability under high-temperature bias stress conditions has been attributed to the activation of additional oxide traps over time [5].

This work reports the development of a simultaneous two-way tunneling model, based on an existing one-way tunneling model [6], to simulate the time-dependent and field-dependent charging and discharging of the near-interfacial oxide traps in response to an applied gate-bias stress. It is important that a two-way model be used, since, depending on the electric field in the oxide and the distribution of charged and uncharged traps at a given spatial depth, electrons are likely to be both tunneling in and out simultaneously during a given period of time. This model is applied to both the gate-bias stress periods, as well as during the measurement periods, when the gate-to-source voltage, V_{GS}, is being ramped.

Model Details

The basic algorithm for the tunneling model is as follows. First, the total number of active oxide traps is defined, along with their spatial distribution. In the simulations presented here, an average oxide trap density of 1×10^{12} cm^{-2} was assumed over the first 50 Å from the SiC interface, with an exponential decay of the trap density into the oxide. The interface traps were taken to vary from mid 10^{13} cm^{-2} values near the band edges down to mid 10^{11} cm^{-2} values at mid-gap [7]. The oxide traps, which are assumed to be either positively charged or neutral based on the likely physical nature of the trap [5], were initially set to a zero-gate-bias steady-state condition to ensure a realistic initial charge. Then, for a given gate-bias stress and measurement simulation, a series of applied gate-bias stresses and stress times is determined. With the application of each new gate bias, the surface potential is found self-consistently with the assumed interface trap and oxide trap occupations. Next, the electric potential in the oxide is determined using numerical methods, from which the electric field is calculated. This is followed by the tunneling-calculation loop. With every time step, the change in oxide-trap occupancy is calculated. The new distribution of charged oxide traps can be found from the present distribution using a similar procedure to that for calculating the time-dependent occupation of interface traps [8]:

$$n_{OT}(x,t+\Delta t) = n_{OT}(x,t) + \Delta n_{OT}(x,t) \ . \qquad (1)$$

The incremental change in the charged oxide trap distribution is found by taking the difference between the electrons tunneling in and those tunneling out (equivalent to holes tunneling in),

$$\Delta n_{OT}(x,t) = \Delta t \cdot p_{tun\_h}(x) \cdot \left[n_{OT\_total}(x) - n_{OT}(x,t)\right] - \Delta t \cdot p_{tun\_e}(x) \cdot n_{OT}(x,t) \ , \quad (2)$$

wherein p_{tun_e} and p_{tun_h} are the tunneling probabilities for electrons and holes, respectively, and n_{OT_total} is the total number of oxide traps present whereas n_{OT} is the number of oxide traps that are positively charged. The probability of electrons or holes tunneling in is calculated using the WKB approximation [6], wherein the tunneling barrier height,

$$\beta(x) = \sqrt{\left(\frac{2 \cdot m_t^*}{\hbar^2}\right) \cdot \left[E_t - q \cdot \mathcal{E}_{OX} \cdot x\right]} \ , \quad (3)$$

is a function of effective tunneling mass, m^*_t, oxide trap energy level, E_t, and localized oxide field, $\mathcal{E}_{OX}$; and wherein q is the electronic charge and *h-bar* is Planck's constant divided by $2 \cdot \pi$. The barrier height will in general be different for electrons and holes, due to differences in the effective tunneling mass, oxide trap energy level, and the sign of the electronic charge. The energy level of the oxide trap was assumed to vary, depending on the charge state [9], with the positively charged oxide trap having an energy level just below the valence band of the SiC, based on values in the literature [10].

The rate of change of the oxide trap density with respect to time is found to be sharply peaked spatially, with a full-width half-max value of around 2 Å. This peak moves logarithmically with time into the oxide, at an approximate rate of 2 Å per decade of time, which leads to the notion of a tunneling front, wherein most trap states in its wake have changed charge state [6]. For a two-way tunneling model, trap states in the wake of the tunneling front have generally achieved a new steady-state value, with the number of electrons tunneling in equal to the number tunneling back out. The initial time steps are set to the initial tunneling transition time of about 0.1 ps [6]. As the tunneling front proceeds deeper into the oxide, the time step is increased since the change in oxide trap occupation described by (2) continually decreases due to the exponential decay in tunneling probability with oxide depth. The surface potential, electric field, etc., are re-calculated several times per decade of time. This process proceeds until the total simulated stress time is reached, and is repeated for all the gate-bias stress steps, including those simulating a measurement.

Model Results and Discussion

Fig. 1 shows the simulated change in the charged oxide trap distribution, normalized by the total number of oxide traps present, versus oxide depth following a long-term 10^6-s negative bias stress of –15 V (when electrons have mostly tunneled out leaving behind uncovered positive charge which

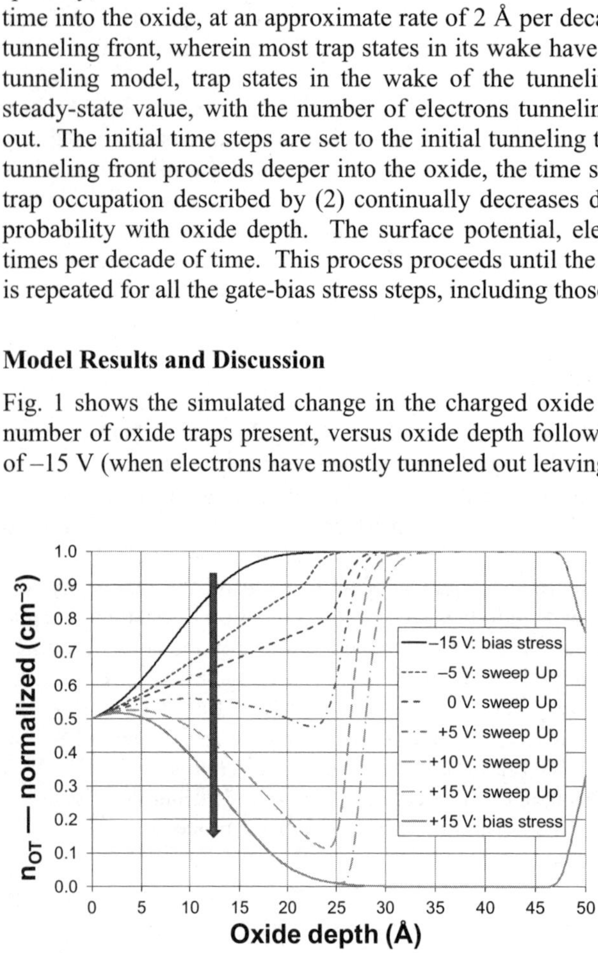

Fig. 1.

Variations in the normalized positively charged oxide trap distribution versus oxide depth as a function of gate bias following:
(1) an initial –15 V bias stress,
(2) a 1-s measurement sweeping up in gate bias, and
(3) a final +15 V 10^6-s bias stress.

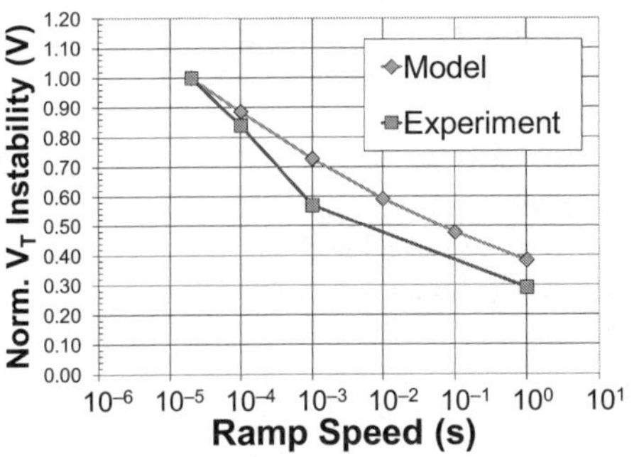

Fig. 2.

Plot of the back-and-forth threshold-voltage instability (normalized by the V_T shift at 20 μs) versus measurement ramp speed, comparing model calculations and experimental results.

causes a negative shift of the threshold voltage); during selected points of a 1-s measurement during which the gate bias is swept up from –5 to +15 V: V_{GS} = –5, 0, +5, +10, and + 15 V; and following a long-term 10^6-s positive bias stress of +15 V (when electrons have generally tunneled back in, neutralizing the oxide traps, which causes the threshold voltage to shift back positively). Clearly, some of the electrons tunnel back in during the measurement itself, causing a significant change to occur in the charged oxide trap distribution over the first 25 Å during the simulated1-s measurement. A slower measurement would result in oxide traps even further away from the SiC interface changing charge state during the measurement, whereas a faster measurement would result in oxide traps only very close to the interface changing charge state. This effect explains the experimentally observed dependence of the threshold-voltage instability on measurement time [1], which is matched by the simulation as shown in Fig. 2. The faster the measurement, the less time for oxide traps to change charge state, thus revealing more of the effect of the gate-bias stress itself. It is interesting to note that the oxide traps at an oxide depth of 50 Å have not yet fully changed charge state for the 10^6-s bias stress simulations shown in Fig. 1. This is once again due to the time dependence of the trap response, with traps further into the oxide taking longer to respond.

The model was also used to simulate the variation in threshold-voltage instability for different gate-bias stress magnitudes. Fig. 3 shows these results, as a function of stress time for simulated 1-s measurements, with significant differences due to the magnitude of the positive-bias stress, but no real difference seen under negative-bias stress. The asymmetrical nature of the response with the polarity of the bias stress occurs due to the differences in the assumed values of the oxide trap energy level and effective tunneling mass for the electrons and holes. These model values were determined from the literature [10] and by comparing with experimental results. The total instability modeled is of course dependent on the total number of oxide traps assumed.

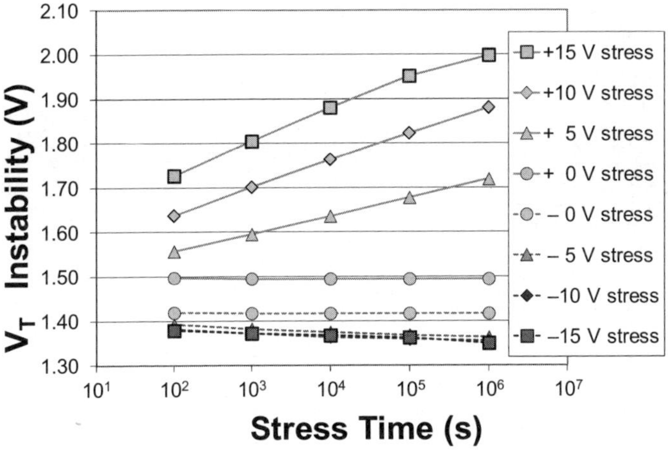

Fig. 3.

Plot of the calculated threshold-voltage instability versus the bias-stress, for both magnitude and polarity.

However, further improvements to the model are still needed since it does not yet account for the variation in subthreshold swing with measurement direction [11]. Although the present model does include the charging of interface traps with change in surface potential, and does show an increase in the stretch-out of the subthreshold slope with increasing interface trap density, it does not predict a variation in the subthreshold swing with measurement direction. To address this, the model should better account for the time response of deep-level interface traps [8]. In addition, since it is likely that at least some of the oxide trap states lie within the wide bandgap of SiC, future modeling work should also consider the possibility of a two-step mechanism involving interface traps acting as an intermediary.

Summary

In summary, a well-established tunneling model based on the WKB approximation was extended to consider a simultaneous two-way process of electrons tunneling both into and out of near-interfacial oxide traps. The concept of a tunneling front, wherein oxide traps in its wake have assumed a new steady-state value whereas trap states further into the oxide have not yet been affected, was used to explain the strong time dependence of the oxide trap occupation for both the gate-bias stress and the measurement. An outline of the basic tunneling algorithm was presented. The model results of the threshold-voltage instability as a function of measurement time, and as a function of bias-stress magnitude and polarity, compare very well with experimental data. However, further improvement of the model is still required since it does not yet account for the variation in stretch-out of the subthreshold *I-V* characteristics with measurement direction.

References

[1] Lelis, et al., IEEE Trans. Elec. Dev., vol. 55:8 (2008), p. 1835.
[2] Tadjer, et al., Mater. Sci. For. vols. 600-603 (2009), p. 1147.
[3] Grieb, et al., Mater. Sci. Forum, vols. 645-648 (2010), p. 681.
[4] Yano, et al., Mater. Sci. Forum, vols. 679-680 (2011), p. 603.
[5] Lelis, R. Green, and D. Habersat, Mater. Sci. Forum, vols. 679-680 (2011), p. 599.
[6] McLean, Harry Diamond Lab., Adelphi, MD, HDL-TR-1765 (1976).
[7] Potbhare, et al., IEEE Trans. Elec. Dev., vol. 55:8 (2008), p. 2029.
[8] Potbhare, et al., IEEE Trans. Elec. Dev., vol. 55:8 (2008), p. 2061.
[9] Nicklaw, et al., IEEE Trans. Nucl. Sci., vol. 49:6 (2002), p. 2667.
[10] Lelis, et al., IEEE Trans. Nucl. Sci., vol. 36:6 (1989), p. 1808.
[11] Lelis, et al., IEEE IIRW Final Report (2008), p. 72.

Materials Science Forum Vols. 717-720 (2012) pp 469-472

doi:10.4028/www.scientific.net/MSF.717-720.469

Peak Degradation of Heavy-Ion Induced Transient Currents in 6H-SiC MOS Capacitors

T. Makino[1,a], N. Iwamoto[1,2], S. Onoda[1], T. Ohshima[1], K. Kojima[3], and S. Nozaki[2]

[1]Semiconductor Analysis and Radiation Effects Group, Japan Atomic Energy Agency, 1233 Watanuki, Takasaki, Gunma, 370-1292 Japan

[2]Department of Electronic Engineering, The University of Electro-Communications, 1-5-1 Chofugaoka, Chofu, Tokyo, 182-8585 Japan

[3] National Institute of Advanced Industrial Science and Technology (AIST), 1-1-1 Umezono, Tsukuba, Ibaraki, 305-8568 Japan

[a]makino.takahiro@jaea.go.jp

Keywords: heavy ion irradiation, MOS capacitor, single event transient current

Abstract. The degradation of the peak value of transient currents induced by heavy-ions in Metal-Oxide-Semiconductor (MOS) capacitors fabricated on n-type and p-type 6H-SiC is observed. The capacitance of the MOS capacitors measured during ion irradiation suggests that the depletion layer width decreases as the number of incident ions increases and also reaches a saturation value. This saturation coincides with the saturation of the peak value of the transient currents so that the degradation of the peak current is interpreted as a decrease in the depletion layer width.

Introduction

Silicon carbide (SiC) is regarded as a promising candidate for electronic devices with high radiation tolerance (rad-hard devices). Ohshima et al. [1] reported that SiC Metal-Oxide-Semiconductor (MOS) Field Effect Transistors (FETs) showed higher gamma-ray radiation resistance than Si MOSFETs. It was also reported that SiC Static Induction Transistors (SITs) [2] and Metal-Semiconductor FETs (MESFETs) [3] were operated up to 10 MGy without significant degradation of their electrical characteristics. These results indicate that SiC has superior radiation tolerance from the point of view of Total Ionizing Dose effects (TIDs). For the development of rad-hard SiC devices, it is also necessary to understand the response of their performance under dense charge generated by an incident ion, which is known as Single Event Effects (SEEs). Permanent malfunctions are induced for power devices by SEEs, and the probability of SEEs increases with increasing electric field in devices. Previously, SEEs were thought to be relevant for space applications only. However, terrestrial silicon (Si) power devices were reported to be damaged by SEEs due to cosmic rays (high energy neutrons from space). Since the electric field in SiC power devices must be higher than Si power devices, it is very important to clarify SEEs in SiC devices. In a previous study of the ion incidence effects, Nava et al. reported charge generated in SiC Schottky diodes by He ions and its response to damage creation for particle detector applications [4]. Charge generated in SiC pn diodes by heavy ions was also reported [5,6]. In the case of MOS devices, Iwamoto et al. [7] reported that the ion-induced charges were not affected by gamma-ray irradiation in SiC MOS capacitors. It was also reported by Ohshima et al. [8] that the peak amplitude of ion induced transient currents in n-type 6H-SiC MOS capacitors decreased as the number of incident ions increased. The peak amplitude was recovered to the initial value by applying a positive bias of +1 V. Although such a unique behavior in n-type SiC MOS capacitors was experimentally observed, the origin is not yet fully understood.

In this study, we detect transient current induced in n-type as well as p-type 6H-SiC MOS capacitors by 18 MeV-oxygen ions, and also investigate the change in the capacitance of the MOS capacitors during ion irradiation.

Experiments

The samples used in this study were MOS capacitors (200 μm in diameter) fabricated on either a 4.9 μm thick n-type or a 5.5 μm thick p-type 6H-SiC epitaxial layer on n-type and p-type 6H-SiC substrates (3.5 deg. off, Si face, CREE), respectively. The carrier concentrations of the n- and p-type epitaxial layers were 3.9 x 10^{15} /cm^3 and 4.3 x 10^{15} /cm^3, respectively. The MOS capacitors were fabricated by the following process. First, the field oxide at a thickness of 100 nm was grown on the sample surface by pyrogenic oxidation (H_2 : O_2 = 1 : 1) at 1100 °C. Next, the field oxide with a diameter of 200 μm was removed by conventional photolithography and wet etching techniques using buffered hydrofluoric acid (HF : NH_4F = 1 : 5). The gate oxide was grown on the SiC surface by pyrogenic oxidation at 1100 °C for 60 min. After the oxidation, the samples were cooled down to 900 °C from 1100 °C in 60 min under an Ar atmosphere. Re-oxidation was then carried out at 900 °C for 15 min under pyrogenic conditions. Both the top and bottom contacts were made by Al evaporation without annealing. The thickness of the gate oxide for n- and p-type MOS capacitors was estimated to be 68 nm and 87 nm, respectively, from capacitance–voltage (C-V) measurements at 1 MHz.

Transient currents induced by 18 MeV oxygen ions were measured using a Transient Ion Beam Induced Current (TIBIC) system in conjunction with a 3 MV tandem accelerator at JAEA, Takasaki [9]. In this study, a single ion hitting system was applied to minimize the damage creation [9]. During the TIBIC measurements, a negative bias of -15 V and a positive bias of 15 V were applied to the top contact of the n-type and p-type MOS capacitors, respectively in order to form a depletion region. In addition, the capacitance of the MOS capacitors was also measured during ion irradiation.

Results and Discussion

Figure 1 shows the TIBIC peak current obtained from (a) an n-type and (b) a p-type MOS capacitor as a function of the number of incident oxygen ions. As shown in Fig. 1 (a), the peak current of the n-type MOS capacitor decreases with increasing number of incident ions up to 700 ions, and then the saturation of the peak value is observed. Well after saturation had been reached (the 1000th incident ion), a positive bias of +1 V was applied to the sample for 1 min, and the TIBIC measurements were continued under the same conditions (under an applied bias of -15 V). By briefly applying the positive bias, the peak amplitude of the TIBIC signal was recovered to the initial value. This cycle of peak recovery to its initial value by applying a bias of +1 V and the subsequent reduction of the peak value to 0.013 mA (about 18% of the initial value) was repeatable. The collected charge, which was estimated from

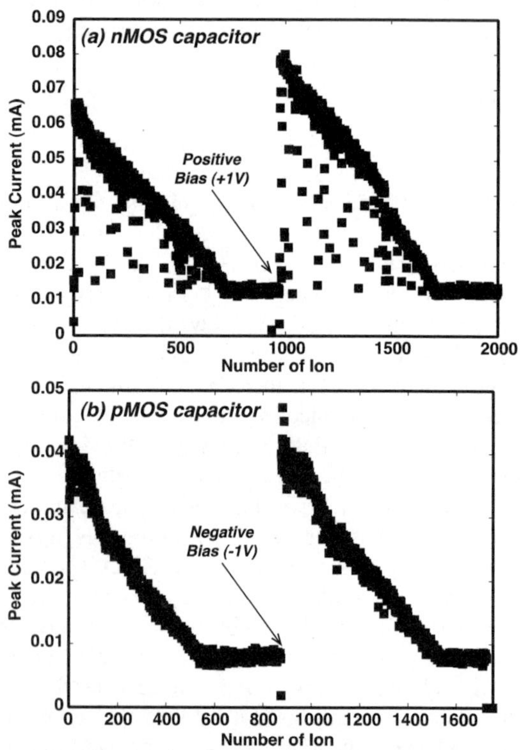

Fig. 1: TIBIC signal peak as a function of number of incident ions in the case of (a) negative biased n-type and (b) positive biased p-type MOS capacitor.

the integrated TIBIC signals, also decreased as the number of incident ions increased up to 700 ions. Beyond this the value saturated (not shown here). The collected charge value also regained its initial value after the application of a positive bias. This behavior (decreasing TIBIC peaks and their recovery by applying a positive bias) agrees with those reported in Ref. [8]. Since the recovery of TIBIC signals is observed by applying a positive bias, the behavior obtained in this study cannot be explained in terms of the damage creation in SiO_2 and/or SiC due to ion incidence.

For the positively biased p-type MOS capacitor, the same cycle as obtained for the n-type MOS capacitor is observed as shown in Fig. 1 (b) although the decrease in the peak current occurs more rapidly than that observed for the n-type MOS capacitor, and the saturation value of the peak value is 0.008 mA. Since the difference in spreading behavior of ion induced-carriers in the gate region of p-type and n-type capacitors is not known, it is difficult to explain a faster decrease in TIBIC peak of p-type MOS capacitors. In order to quantitatively understand the phenomena, further investigation is necessary.

Figure 2 shows the variation in the normalized capacitance of (a) n-type and (b) p-type MOS capacitors during ion irradiation. The biases of -15 V and +15 V are applied to n-type and p-type MOS capacitors, respectively. The TIBIC peak current degradation shown in Fig. 1 is superimposed in Fig. 2 by using right axes. The closed triangles show the variation of the capacitance normalized by the maximum capacitances of the n- and p-type MOS capacitors. The capacitances of both n- and p-type MOS capacitors increase with increasing number of incident ions and saturate at the values 0.23 and 0.41, respectively. This result indicates that the depletion layer decreased to 1 μm from 2.3 μm for n-type and to 0.4 μm from 1.3 μm for p-type capacitors, respectively. Since the saturation of the capacitance and the TIBIC peak are observed at the same number of incident ions, we can conclude that the decrease in peak current comes from the decrease in the depth for the ion induced charges (depletion layer). In other words, the minority carriers may result in the formation of an inversion layer. A possible explanation is the increase of trapped minority carriers near the SiO_2/SiC interface. Because the quality of the SiO_2/SiC interface is not as high as the SiO_2/Si interface and defects exist near the SiO_2/SiC interface, despite considerable efforts to develop an oxidation process for high quality SiO_2/SiC [10]. Since defects near the interface act as carrier traps, the obtained results suggest that the electron and/or hole traps also affect the variation of TIBIC signals observed for SiC MOS capacitors, in addition to charge in the inversion layers.

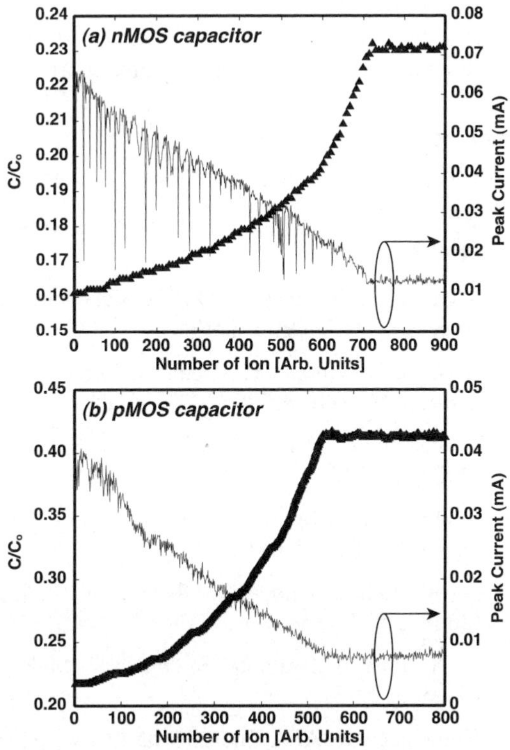

Fig. 2: Capacitances of (a) n-type and (b) p-type MOS capacitors during ion irradiation. TIBIC peak degradations shown in Fig. 1 are super-imposed.

Figure 3 shows the capacitance of the n-type MOS capacitor biased at -15 V as a function of elapsed time. The capacitance was also monitored after ion irradiation. For the "Beam ON" region (between 0 and 140 sec), the capacitance variation shown in Fig. 2 (a) is re-plotted. The capacitance in the "Beam OFF" region means the capacitance variation of the n-type MOS capacitor biased at -15 V after irradiation. As shown in Fig. 3, the capacitance slowly decayed from the saturation value of 0.23 in the "Beam OFF" region. The capacitance decay occurs by the release of charge from the SiO_2/SiC interface. The decay of capacitance observed in the "Beam OFF" region is very slow. The decay time may have the information of charge release time from the SiO_2/SiC interface. Further investigations are necessary to clarify the contribution of such traps to TIBIC variation for SiC MOS capacitors.

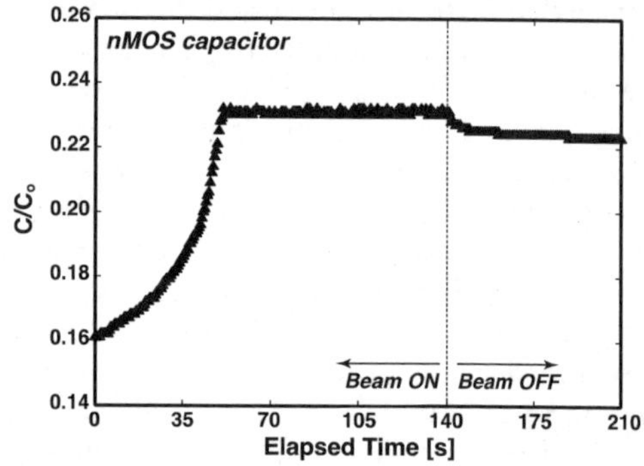

Fig. 3: Capacitances of n-type MOS capacitor biased at -15 V as a function of elapsed time.

Conclusions

The response of heavy-ion induced transient currents in MOS capacitors fabricated on n-type and p-type 6H-SiC epitaxial layers was observed. The peak values of the transient currents decrease with increasing number of incident ions. The decreased transient current peak was recovered to the initial value by applying a positive bias of 1 V for 1 min. The variation of the capacitance for 6H-SiC MOS capacitors was also measured during ion irradiation. The capacitance increased with increasing number of incident ions. Therefore, we conclude that the decrease in the transient current during ion irradiation is due to the decrease in the depletion layer width.

Acknowledgement

A part of this work was supported in part by a Grant-in-Aid for Scientific Research (B) No. 21360471 from the Ministry of Education, Culture, Sports, Science and Technology.

References

[1] T. Ohshima, H. Itoh, and M. Yoshikawa, J. Appl. Phys. 90 (2001) p.3038.

[2] Y. Tanaka, S. Onoda, A. Takatsuka, T. Ohshima, and T. Yatsuo, Mater. Sci. Forum 645-648 (2010) p.941.

[3] S. Onoda, N. Iwamoto, S. Ono, S. Katakami, M. Arai, K. Kawano, and T. Ohshima, IEEE Trans. Nucl. Sci. 56 (2009) p. 3218.

[4] F. Nava, E. Vittone, P. Vanni, G. Verzellesi, P.G. Fuochi, C. Lanzieri, and M. Glaser, Nucl. Instrum. Methods A 505 (2003) p. 645.

[5] T. Ohshima N. Iwamoto, S. Onoda, T. Kamiya, and K. Kawano, Nucl. Instrum. Methods B 267 (2009) p. 2189.

[6] S. Onoda, N. Iwamoto, M. Murakami, T. Ohshima, K. Kojima, and K. Kawano, Mater. Sci. Forum 615-617 (2009) p. 861.

[7] N. Iwamoto, S. Onoda, T. Ohshima, K. Kojima, and K. Kawano, Mater. Sci. Forum 615-617 (2009) p.517.

[8] T. Ohshima, N. Iwamoto, S. Onoda, T. Makino, S. Nozaki, and K. Kojima, Mater. Sci. Forum 679-680 (2011) p. 362.

[9] Home page of JAEA Takasaki, http://www.taka.jaea.go.jp/index_e.html

[10] K. Danno and T. Kimoto, J. Appl. Phys. 100 (2006) 113728.

Materials Science Forum Vols. 717-720 (2012) pp 473-476

doi:10.4028/www.scientific.net/MSF.717-720.473

Low Frequency Noise in 4H-SiC Lateral JFET Structures

H.K.Chan[1,a*], R.C. Stevens[1,b], J.P. Goss[1,c], N.G.Wright[1,d] and A.B. Horsfall[1,e]

[1]School of Electrical, Electronic and Computing Engineering, Newcastle University, England

[a]h.k.chan1@ncl.ac.uk, [b]rupert.stevens@ncl.ac.uk, [c]jonathan.goss@ncl.ac.uk, [d]nick.wright@ncl.ac.uk, [e]alton.horsfall@ncl.ac.uk

Keywords: Low frequency noise, low-level lateral JFET, spectral noise density

Abstract: Low frequency noise in 4H-SiC low-level signal-lateral JFETs was systematically investigated. In contrast to previous studies, which were based upon high power vertical structures, this work investigates the low-frequency noise behavior of low-level signal-lateral devices that are more relevant to the realization of small signal amplifiers. The JFETs studied share an identical cross section, but different gate lengths and widths. For high temperature operation between 300K and 700K at a gate source voltage (V_{GS}) of 0V, the normalized power spectral density (NPSD) of the JFETs is inversely proportional to frequency (f), referred to as $1/f$ noise. The NPSD increases monotonically up to a critical temperature, where it starts to decline. Two unique noise origins, fluctuations from bulk and SiO_2-SiC interface traps, were observed across all devices investigated. Low frequency noise for devices with a 50μm gate width is localized at the SiO_2-SiC interface, whereas for wider devices the noise is seen to be of bulk/substrate origin, following Hooge's model.

Introduction

Junction field effect transistors (JFETs) have long been favored for front-end amplifiers, preamplifiers and oscillator circuitry due to their low intrinsic noise [1-3]. Low frequency noise plays an important role in defining the operability of JFET based circuitry. Whilst low frequency noise may only extend up to hundreds of Hz, in applications such as oscillators this can be up-converted to high frequencies, thus distorting the output signal. In recent years, 4H-SiC has demonstrated an enhanced performance over 6H-SiC in high frequency operation because of the superior carrier mobility and saturation velocity, enabling the construction of microwave devices with greater power density [1]. In the past, reports have explored low frequency noise in power devices, which may not be generally valid since the signal to noise ratio is then extremely large, and the semiconductor noise may not be significant. In this paper we present an investigation of low frequency noise in an array of twelve different 4H-SiC epitaxial n-channel, normally on, low-level signal planar JFETs under different operating conditions. Since the investigation involved multiple devices with different current capability, direct comparison of their measured power spectral density (PSD) is not feasible, as the parameter is heavily influenced by the drain source current, I_{DS}. The discrepancies between large and small devices can be eradicated by normalizing the PSD against I_{DS}^2, thus introducing a universal parameter which is independent of the I-V characteristic.

Experimental

A schematic cross-section of the 4H-SiC JFET studied in this work is illustrated in Fig. 1. The devices under investigation were categorized into 4 sets with gate widths (W) of 200μm, 150μm, 100μm and 50μm. Each gate width has 3 gate lengths (L) of 9μm, 15μm and 21μm. Each device has 3 terminals on the top surface of the substrate with no back-contact to simplify system integration and device characterization. The

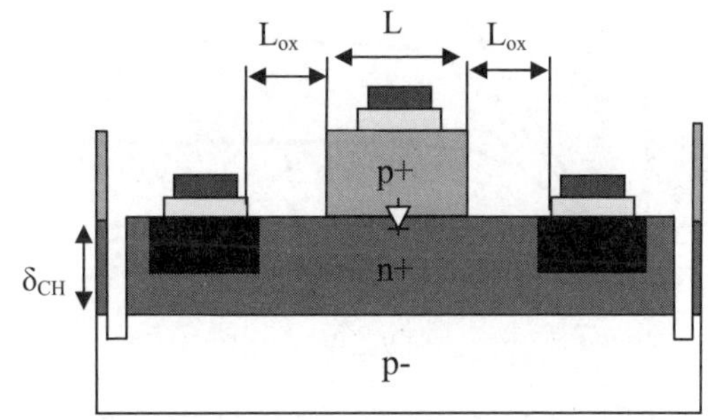

Fig. 1 – Schematic cross-section of epitaxial JFET.

doping concentration in the channel is $10^{17}cm^{-3}$ and the channel depth is 0.3μm. The p+ epitaxial layer is partially etched to form the gate region while exposing the channel epilayer to facilitate contact formation. A 25 nm oxide-layer was grown in an oxidation furnace at 1150°C for 6 hours, followed by formation of Ti(5nm)/Ni(100nm) and Ti(5nm)/Al(60nm)/Ti(5nm)/Ni(40nm) contacts to the n-type and p-type regions, respectively. The contacts were annealed at 1050°C and 950°C respectively before a gold layer was deposited to enable wire bonding. L_{OX}, shown in Fig. 1, is the length of the passivation layer used to isolate the device terminals. The oxide dimensions play an important role in defining the number of traps in the oxide, important for the intrinsic noise [2]. For the noise measurements a Keithley 2611 with a 1 Hz low pass filter was used to bias the gate. The drain noise current was amplified using a SR570 low noise current amplifier (with built in voltage source to provide drain source bias) before being analyzed with a SR760 FFT analyzer.

Results and Discussion

Fig. 2 shows the normalized drain-source current (I_{DS}/W) for two devices for gate geometries (W/L) of 200μm/21μm and 50μm/21μm as a function of the gate source voltage, V_{GS}. The characteristics are expected to follow the relationship in Eq. 1,

$$I_{DS} = \frac{Wq\mu N_D}{L}\left\{V_{DS} - \frac{2}{3\sqrt{\Psi_P}}\left[(\Psi_{bi} + V_{DS} - V_{GS})^{\frac{3}{2}} - (\Psi_{bi} - V_{GS})^{\frac{3}{2}}\right]\right\} \tag{1}$$

where q the elementary charge, μ the carrier drift mobility, N_D the n-channel doping concentration, Ψ_P the pinch-off voltage, Ψ_{bi} the built in voltage of the p-n junction, and V_{DS} the drain source voltage. Deviation of the graphs of I_{DS}/W vs. V_{GS} is believed to be caused by differences in the turn-on voltage (V_{TO}) from doping and thickness variations in the wafer. The contact metallization developed by Newcastle University has shown excellent temperature stability, low resistivity and fairly good ohmic contacts [4].

The results plotted in Fig. 3 show drain-source resistance, R_{DS}, for the devices with gate geometries (W/L) of 200μm/21μm and 50μm/21μm as a function of V_{GS}. Values of V_{TO} are calculated from plots of $\sqrt{I_{DS}}$ vs. V_{GS} to be −3.2V and −3.7V, respectively. Variation in channel depth (δ_{CH}) and doping concentration (N_D) can account for the difference between these values. Measurements of low frequency noise at temperatures in the range of 300K to 700K were performed with V_{DS} biased in the linear ohmic region and $V_{GS} = 0$V. This is to ensure that the conduction channel is fully open for both devices (as R_{DS} approaches a constant value above 0V in Fig. 3) and to simplify the calculation of the number of free electrons.

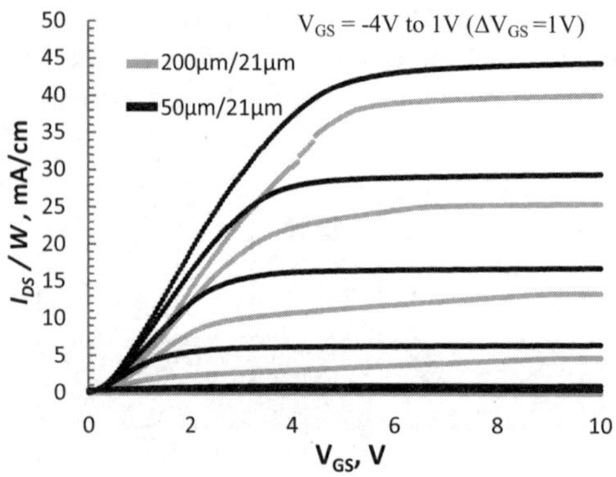

Fig. 2 – I-V characteristics for devices with gate geometries (W/L) of 200μm/21μm and 50μm/21μm at 300K.

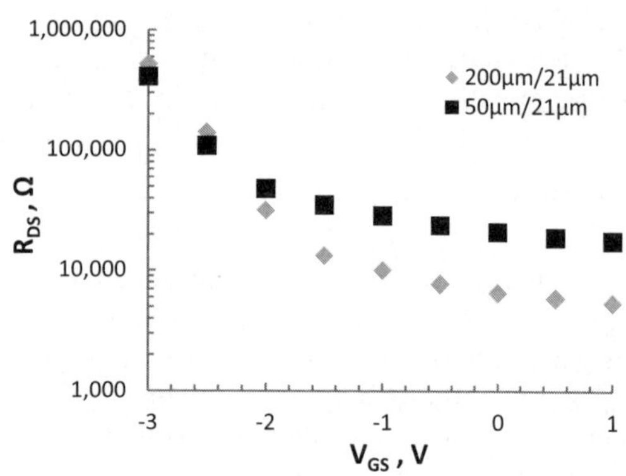

Fig. 3 – R_{DS} vs. V_{GS} for devices with gate geometries (W/L) of 200μm/21μm and 50μm/21μm at 300K.

Fig. 4 shows the NPSD for JFETs with gate geometry (W/L) of 200µm/21µm at different frequencies as a function of temperature. This is a common pattern found to be applicable to all devices fabricated on the same wafer. The results in Fig. 4 illustrate that NPSD increases proportionally with temperature until a critical point (which varies between different devices), where it begins to decay and increase again at 650K [1]. The increase in NPSD at 650K has not been previously reported and we suspect this will rise further beyond 700K. However, due to limitation of our instruments no test can be conducted above 650K. The temperature dependence suggests multiple mechanisms govern the noise behavior in SiC JFETs for high temperature operation.

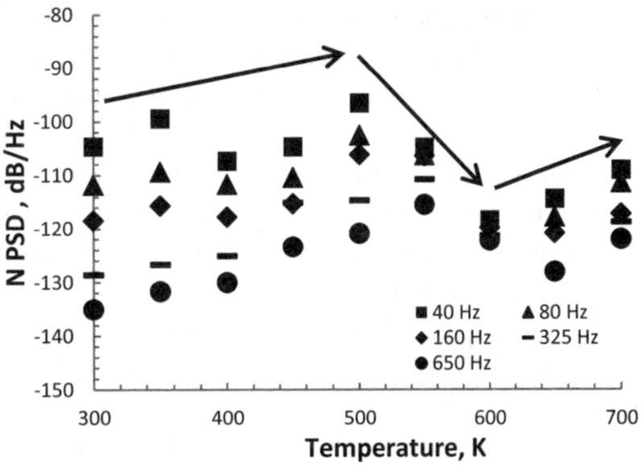

Fig. 4 – NPSD (S_I/I^2) vs. temperature for different frequencies in devices with gate geometry (W/L) of 200µm/21µm.

Conventionally, semiconductor noise is described in terms of fluctuation in the number of charge carriers in the channel according to Hooge's formula [5]:

$$\frac{S_I}{I^2} = \frac{\alpha}{N_{CH}\, f^{\lambda}} \tag{2}$$

$$N_{CH} = N_D W \delta_{CH}(2L_{ox} + L) \tag{3}$$

where α is the quality factor of the substrate shown to increase monotonically with crystal detect concentration [1,5], f the frequency, and λ defines the slope of noise spectral density, typically 0.8—1.5 [1, 3]. N_{CH} is the total number of free electron in the conduction channel, calculated using Eq. 3.

As the temperature rises, more free carriers will be generated [6] thus effectively increases N_{CH}, so that based on Eq. 2 S_I/I^2 should decrease accordingly. However, that it is not the case, as shown in Fig. 4, implies that α increases significantly with temperature. The decrease in NPSD with temperature may also be influenced by a reduction in carrier mobility at elevated temperatures [1].

Previous work has shown that low-frequency noise in 4H-SiC and 6H-SiC JFETs is caused by either free carrier fluctuation at the SiC-SiO_2 interface (traps) or in the channel (bulk/substrate). The results show that the SiC-SiO_2 interface is the dominant factor for devices with passivation layers [1-3]. However, to-date little attention has been paid to the origin of noise in devices with different dimensions and hence SiO_2-SiC interface areas. Fig. 5 shows the NPSD as a function of frequency for devices with gate W/L of 200µm/21µm and 50µm/21µm at different operating temperatures. It can clearly be seen that for the temperature range studied the NPSD is proportional to f^{-1} to a reasonable degree. Theoretically, devices with larger SiC-SiO_2 interface areas are expected to exhibit higher NPSD values because free carriers in the conduction channel have a greater tendency to be involved in generation-recombination mechanisms with the oxide traps. However, this is at odds with our observations as the 200µm/21µm device which has a four times larger SiC-SiO_2 interface area is 5 dB/Hz lower in NPSD in comparison to 50µm/21µm device.

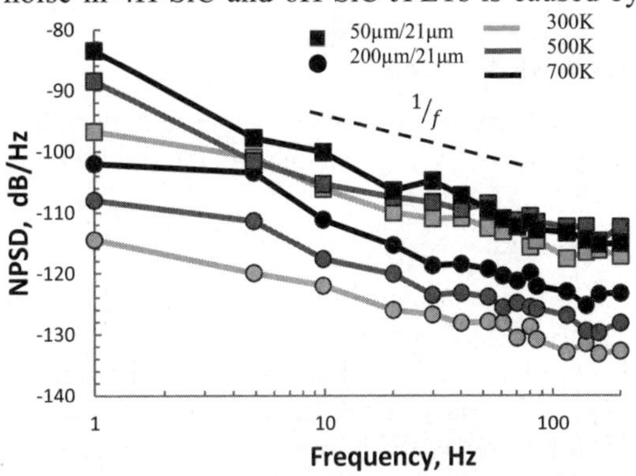

Fig. 5 – NPSD (S_I/I^2) versus frequency for devices with W/L ratio of 200µm/21µm and 50µm/21µm at 300K, 500K and 700K

To validate the origin of the noise plotted in Fig. 5, the NPSD was measured across different devices as a function of gate bias. As V_{GS} becomes more negative, the effective channel depth δ_{CH} becomes narrower until it is totally depleted, which can be observed as a higher R_{DS}. When the dominant noise source is bulk in origin, NPSD is proportional to R_{DS} as it is described by $l/N_{CH}\mu eA$ (where l is the length of the conduction channel, A the cross section area of the channel, μ the carrier mobility and e the elementary charge). The NPSDs were acquired by fixing V_{DS} at 1.5V and changing V_{GS} between −1.5V and 1.0V in 0.5V steps to ensure all the JFETs are working in the linear ohmic region.

Fig. 6 plots NPSD against R_{DS} for devices with various oxide dimension at different frequencies. The results show that NPSD increases monotonically with R_{DS} for 200μm/9μm and 200μm/21μm devices with almost identical magnitude. This implies that the noise in both devices is generated in the bulk/substrate as described by Hooge's formula. In contrast, the 50μm/21μm device shows R_{DS}^2 behavior and is 5 dB/Hz higher in magnitude. This phenomenon suggests this device that has distinct noise source which may originate from the SiO_2-SiC interface [1].

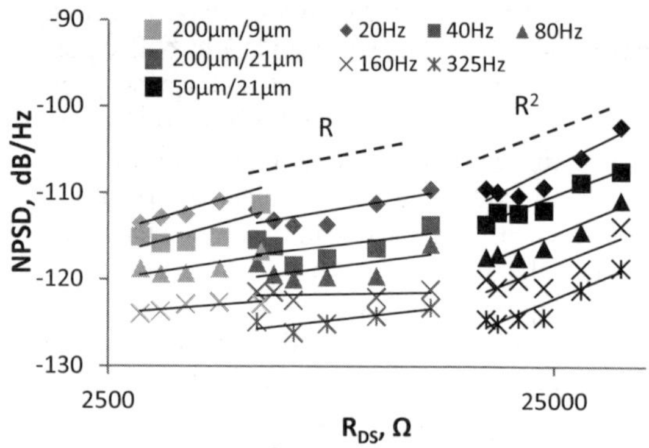

Fig. 6 –NPSD (S_I/I^2) versus R_{DS} for devices with gate W/L of 200μm/9μm, 200μm/21μm and 50μm/21μm at 20 Hz, 40 Hz, 80 Hz, 160 Hz, and 325 Hz.

Conclusions

Two distinct low-frequency noise-sources were observed for lateral 4H-SiC JFETs, originating from carrier fluctuations in both the bulk/substrate and SiO_2-SiC interface. We believe that both noise sources are present in the devices simultaneously. However, which mechanism is dominant, the overall noise source is critically dependent on device size and hence R_{DS}. As R_{DS} increases, the noise source switches from originating in the bulk/substrate to the SiO_2-SiC interface. This is analogous to two noisy resistors connected in parallel: when one resistance becomes substantially bigger than the other, it has less significant effect to the overall resistance.

Acknowledgements

The lead author is financially supported by EPSRC and BAE Systems through the Dorothy Hodgkin Postgraduate Awards scheme and the research via EPSRC grant number EP/D068827/1.

References

[1] M.E. Levinshtein, S.L. Rumyantsev, J.W. Palmour, J. Appl. Phys. 81 (1997) 1758-1762
[2] Y. Fu, H.Wong, J.J. Liou, Electron Devices & Solid State Circuits IEEE Conf. (2005) 39-42
[3] P. Flatresse, T. Ouisse, Solid-State Electronics 38 (1995) 971-975
[4] P. Nikitina, K.V. Vassilevski et al, J. Appl. Phys. 97 (2005) 083709
[5] F.N. Hooge, IEEE Trans. on Elec. Devices 41 (1994) 1926-1935
[6] M.E. Levinshtein, S.L. Rumyantsev et al, "Properties of adv.semicon.material", John Wiley & Sons, New York, 2001, pp. 93-112

Materials Science Forum Vols. 717-720 (2012) pp 477-480

doi:10.4028/www.scientific.net/MSF.717-720.477

Influence of Threading Dislocations on Lifetime of Gate Thermal Oxide

K. Yamamoto [a], M. Nagaya, H. Watanabe, E. Okuno,

T. Yamamoto and S. Onda

Material R&D Div., DENSO CORPORATION, Minamiyama, Komenoki-cho, Nisshin, Japan

[a]KENSAKU_YAMAMOTO@denso.co.jp

Keywords: MOS, TDDB, Threading Dislocation, TEM, KOH, TED, TSD

Abstract. The influence of the threading dislocations on the lifetime of the gate thermal oxide was investigated. The MOS diodes were fabricated and their lifetimes were measured by TDDB (Time Dependent Dielectric Breakdown). The types of defects under the breakdown points were investigated by molten KOH and TEM (Transmission Electron Microscopy). As the results, it was clarified that the threading edge dislocations (TED) degraded the lifetime by one and threading screw dislocations (TSD) by two orders of magnitude.

Introduction

The reliability of gate oxides is a fundamental issue for realizing commercial SiC MOSFETs. Many reports mentioned that crystal defects shorten the lifetime of the gate oxide [1]. The epi defects, the basal plane dislocations and TSD are considered as killer defects. However, because of the high TSD density of commercial SiC wafers, the exact relationship between other kinds of dislocations with lifetime has not been revealed. On the other hand, RAF (Repeated a-Face [2]) wafers, that we developed earlier and used in this study, have low TSD density, so it is easy to evaluate the relationship between other kinds of dislocations and lifetime. Using a RAF wafer, we clarified the relationship between the lifetime of the gate oxide and crystal defects.

Experimental

We prepared two 8°off-axis 4H-SiC (0001) wafers with epitaxial layers doped with nitrogen density of 5×10^{15} cm^{-3}. One of the wafers was grown by the RAF method, by which it is possible to produce wafers with low threading dislocation density. The other was a commercially available wafer (high grade) which was prepared for reference. The gate oxide was formed by thermal oxidation (1080 ℃, 300 min) in wet ambient. The thickness of the oxide was 39 nm. The electrodes were formed with phosphorus doped poly-silicon with diameters of 50, 100, 200 and 400 μm. Above the electrode, an inter-layer dielectric was deposited. After the contact hole was formed, the metal electrode was formed. A cross-sectional view of the sample is shown in Fig. 1.

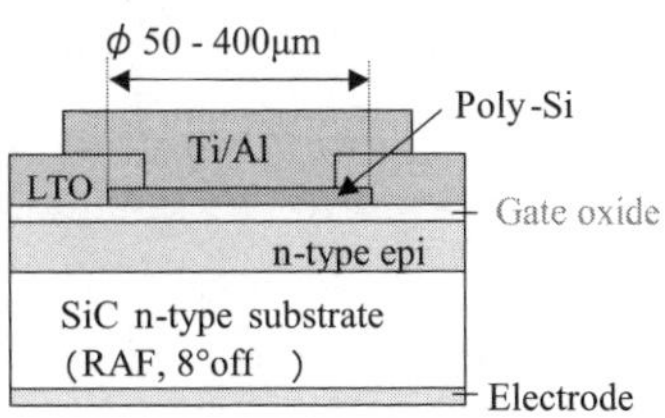

Fig. 1. Cross sectional view of the fabricated MOS diode

The lifetime of the gate oxide was evaluated by TDDB measurement at stress electric field of 8.5 MV/cm and temperature of 250 ℃. The breakdown positions were defined by the photo-emission method. The dislocation types under the breakdown positions were investigated by molten KOH etching analysis and cross sectional TEM observation.

Results and Discussion

Gate Oxide Lifetime Measured by TDDB. Fig. 2 shows a Weibull plot of the results in which electrodes of various diameters were used. The electrode with diameter of 50 μm exhibited a wear-out breakdown mode (m=5.1). The larger the electrode diameter, the shorter the lifetime was. Additionally, the lifetimes were divided into three groups, named modes A, B and C. A was classified as intrinsic breakdown. Modes B and C exhibited shorter lifetimes than mode A. Fig. 3 shows a lifetime comparison between the RAF wafers with diode diameter of 50 μm and the commercially available wafers. In the case of the RAF wafer, all samples exhibited the intrinsic breakdown mode A. On the other hand, in the case of the commercial wafer, the lifetime was short and divided into all three modes. We infer that this fast degradation is caused by higher threading dislocation density than in the case of the RAF wafer.

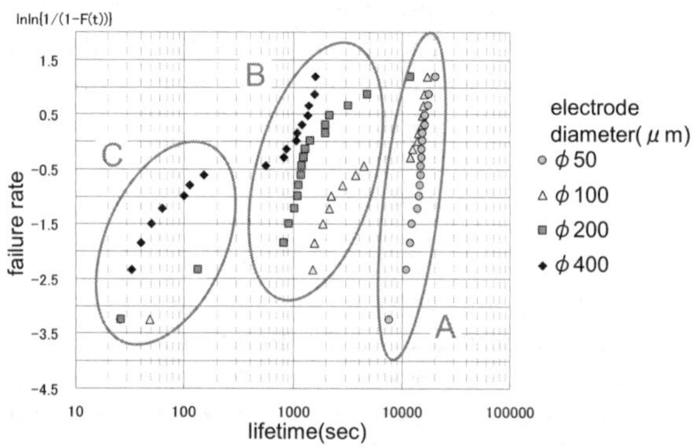

Fig. 2. Results of TDDB measurements. The electrode diameters were 50, 100, 200, 400 μm. Stressed electric field was 8.5 MV/cm. The ambient temperature was 250 ℃.

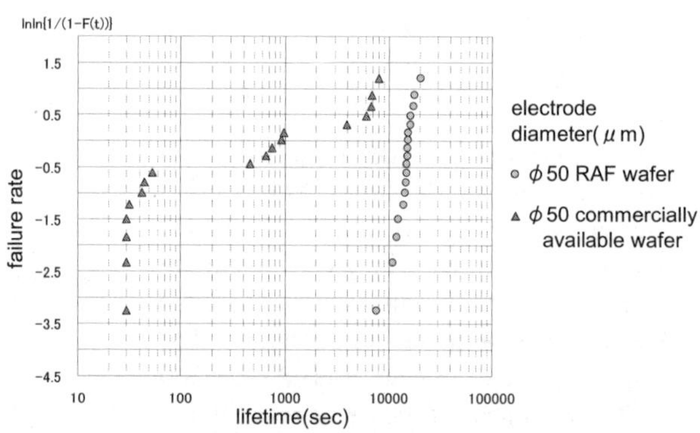

Fig. 3. Results of TDDB measurements. Lifetime comparison between RAF wafers of ϕ 50 μm diodes and commercially available wafers.

Clarification of the Origin of Modes B and C. In order to clarify the relationship between lifetime and crystal defects, we performed photo-emission observation, molten KOH etching and TEM observation in the case of the RAF wafer sample. Before photo-emission observation, the metal electrode was removed. The emission from the breakdown position was detected through the Poly-Si layer. In photo-emission observation, only one emission spot was observed in each sample. The samples were etched with molten KOH, in which the Poly-Si layer was removed. After the etching, etched pits were observed on the SiC surface by optical microscope. At least one etched pit was found in all diode areas. One of the etched pits in each sample coincided with the emission spot (Fig. 4). The pit shapes of breakdown positions were divided into four groups, and had a strong correlation with lifetime (Fig. 5 (a), (b)). The modes A, B and C of Fig. 5 (a) correspond to those of Fig. 2. In the case of mode A the shapes of etched pits were hemispherical. In mode B the shapes were hexagonal shallow or conic, and in mode C they were hexagonal deep or hemispherical shapes.

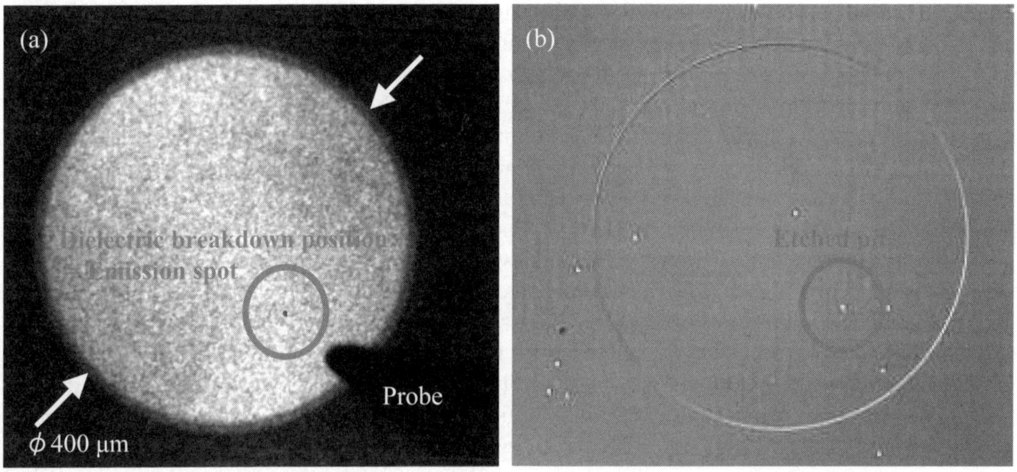

Fig. 4. Results of (a) photo-emission and (b) molten KOH etching analysis.

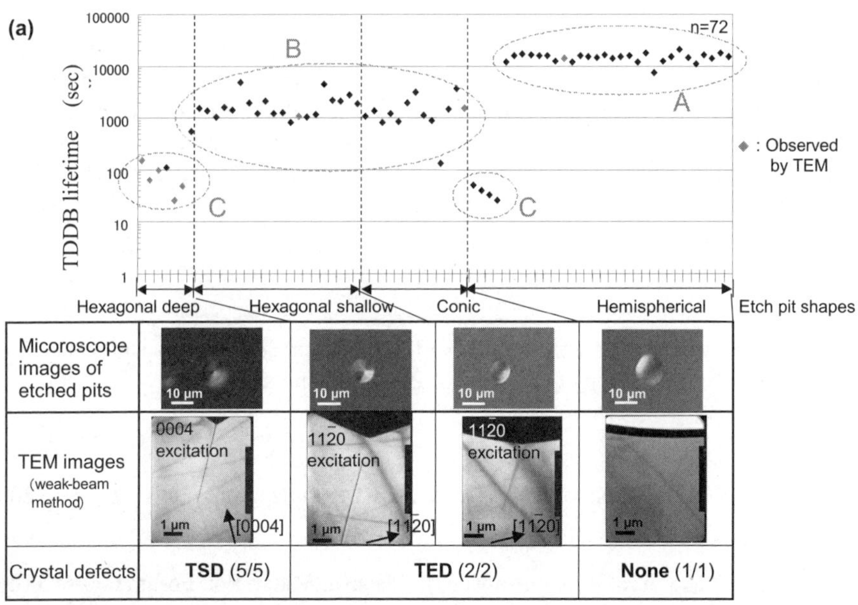

Fig. 5. Relationship between TDDB lifetime and crystal defect. (a) TDDB lifetimes of MOS diodes. (b) Optical microscope images of etched pit shapes at the breakdown points. (c) Cross sectional TEM images under the pits.

TEM observations were performed at 1000 keV acceleration voltage. Owing to the high acceleration voltage, the electron transmission of a thick sample, for example 1 μm, is possible. Therefore, the observation of the dislocation is easy. Additionally, the identification of dislocation type is possible from the information of the dislocation strain direction which can be obtained from the relationship between the electron beam direction and the detected image. The results are shown in Fig. 5 (c). In mode A, no defects were observed directly under the hemispherical KOH pits. Because there was only one etched pit under the electrode, we assumed that the etched pit was formed from surge current and/or surge voltage when the gate oxide was intrinsically broken by TDDB measurement. On the other hand, threading dislocations were observed in modes B and C. In mode B, the pit shapes were hexagonal shallow or conic, and both dislocation types were TED, in which the strain direction is perpendicular to the dislocation. In mode C, the pit shapes were mainly deep hexagonal, and their dislocation types were TSD, in which the strain direction is parallel to dislocation. The hemispherical pits were also formed in mode C. Although the origin of the hemispherical pit in mode C is not revealed, we assumed that the origin was a kind of process defect.

From Fig. 5, it is clear that TED degraded the lifetime by one and TSD by two orders of magnitude. However, the mechanism by which threading dislocations degrade the gate oxide lifetime was not revealed. To explain the degradation, we assumed two models, the shape effect and the oxide quality degradation models. In the shape effect model, the high oxidation rate on a threading dislocation makes the SiC/oxide interface pit-shaped, and the shape causes the electric field enhancement which causes the electron injection from the top of the pit to the oxide [3]. In the oxide quality degradation model, the thermal oxide inherits the strain and/or the impurities which are originally included in the threading dislocation. We need to identify the lifetime degradation mechanism hereafter. In any case, we assume that the degradation is caused by thermal oxidation of the SiC crystal with threading dislocations. So, the deposited film will have the possibility to inhibit the lifetime degradation, when applied as the gate film. We will verify the possibility hereafter.

Conclusion

Using a RAF wafer, we clarified the relationship between the lifetime of the gate oxide and crystal defects. We fabricated MOS diodes and measured their lifetimes by TDDB measurement. The breakdown positions were defined by the photo-emission method. Finally, we classified the defects by TEM. As the results, it was revealed that a TED decreases the lifetime as does a TSD in a certain condition. In our test condition, the lifetime of the gate oxide area, in which a TED was included, was shorter by one order of magnitude than the intrinsic breakdown. In addition, a TSD led to a decrease of two orders of magnitude.

Acknowledgment

TEM observations were carried out at the HVEM Laboratory of Nagoya University through High Performance Electron Microscopy in Support of Bio and Nanoscience.

References

[1] J. Senzaki, K. Kojima, T. Kato, A. Shimozato & K. Fukuda, Appl. Phys. Lett. 89 (2006) 022909

[2] D. Nakamura et al.: Nature 430 (2004) p. 1009.

[3] Y. Nakano, Y. Nakamura, A. Kamisawa & H. Takasu, Mater. Sci. Forum 600-603 (2009) p. 377

Materials Science Forum Vols. 717-720 (2012) pp 481-484

doi:10.4028/www.scientific.net/MSF.717-720.481

Doping-Induced Lattice Mismatch and Misorientation in 4H-SiC Crystals

S. Sasaki[1, a], J. Suda[1, b], and T. Kimoto[1, c]

[1]Dept. of Electronic Sci. & Eng., Kyoto University, Nishikyo, Kyoto 615-8510, Japan

[a]sasaki@semicon.kuee.kyoto-u.ac.jp, [b]suda@kuee.kyoto-u.ac.jp, [c]kimoto@kuee.kyoto-u.ac.jp

Keywords: lattice mismatch, misorientation, HRXRD, reciprocal space mapping

Abstract. The *c*- and *a*-lattice constants of nitrogen-doped 4H-SiC were measured in the wide temperature range (RT - 1100°C). The samples used in this study were heavily doped substrates and lightly-doped free-standing epilayers. The lattice constants at room temperature are almost identical for all the samples. However, the lattice contraction by heavy nitrogen doping was clearly observed at high temperatures, which indicates that the thermal expansion coefficients are dependent on the nitrogen concentration. The lattice mismatch ($\Delta d/d$) between a lightly-doped free-standing epilayer ($N_d = 6x10^{14}$ cm^{-3}) and a heavily-doped substrate ($N_d = 2x10^{19}$ cm^{-3}) was calculated as $1.7x10^{-4}$ at 1100°C. The authors also investigated lattice constants of high-dose N^+, P^+, and Al^+-implanted 4H-SiC. Reciprocal space mapping (RSM) was utilized to investigate the lattice mismatch and misorientation. The RSM images show the *c*-lattice expansion and *c*-axis tilt of the ion-implanted layers, irrespective of ion species. The authors conclude that the lattice expansion is not caused by heavy doping itself, but by secondary defects formed after the ion-implantation and activation-annealing process.

Introduction

Although crystal quality of 4H-SiC has been much improved in recent years, further reduction of various defects is essential to improve the device performance and reliability. A basal plane dislocation (BPD) is one of the most detrimental defects in 4H-SiC bipolar devices, since it acts as a nucleation site of stacking-fault generation upon carrier injection. Zhang et al. reported that BPDs can glide during epitaxial growth, forming interfacial dislocations (IDs) at the epilayer/substrate interface [1]. IDs were also observed at the bottom of the ion-implanted region [2]. The formation of IDs may indicate the existence of doping-induced lattice mismatch in 4H-SiC. However, to date, only limited data on doping-induced lattice mismatch are available [3,4,5]. Since high-temperature process is inevitable in the SiC device fabrication, study of lattice mismatch at elevated temperatures is also of great importance for understanding the misfit strain during the high-temperature process. In this work, the authors attempt to clarify the influence of doping on lattice parameters of 4H-SiC crystals.

Experiments

High-resolution x-ray diffraction (HRXRD) measurements were conducted in the wide temperature range from room temperature to 1100°C. The samples were nitrogen-doped (8°-off or on-axis) 4H-SiC(0001) substrates and free-standing epilayers (100-147 μm thick), having different nitrogen concentrations, ranging from $6x10^{14}$ to $2x10^{19}$ cm^{-3}. The *c*-lattice constants were determined from (0008) reflection, while *a*-lattice constants from (11-20) reflection.

Ion-implanted layers were also prepared to investigate the lattice parameters of ion-implanted 4H-SiC. The starting materials were lightly-doped p-type 4H-SiC epilayers grown on heavily-doped n-type 8° off-axis 4H-SiC(0001) substrates. The epilayer thickness and doping concentrations were 50 μm and $9.6x10^{14}$ cm^{-3}, respectively. N, P, Al, or (C+Si) ions were implanted at 600°C from the sample surface to a depth of 330 ~ 520 nm with a box profile of $1x10^{19}$-$1x10^{20}$ cm^{-3}. After the ion-implantation, activation annealing was performed in Ar ambient at 1800°C for 10 min. Lattice parameters were measured at room temperature with reciprocal space mapping (RSM) by using (0008) reflection.

Results and discussion

Lattice constants of nitrogen doped 4H-SiC Figs. 1 (a) and (b) show the temperature dependence of *c*- and *a*-lattice constants of a heavily-doped (8x10^{18} cm^{-3}) substrate and a lightly-doped (1x10^{15} cm^{-3}) epilayer. As shown in Fig. 1, the *c*- or *a*-lattice constants of these samples at room temperature are almost identical within the measurement accuracy ($\Delta d/d \sim 10^{-5}$). However, the *c*- and *a*-lattice constants at high temperatures are clearly different for each sample, indicating that the thermal expansion coefficients of 4H-SiC crystals are dependent on the doping concentration. The lattice mismatch between the lightly-doped epilayer and the heavily-doped substrate does increase as temperature increases. The lattice contraction by nitrogen doping can be attributed to the size difference between the nitrogen and the carbon atoms. Nitrogen atoms have a smaller covalent radius than carbon atoms, which leads to the contraction of lattice parameters by heavy nitrogen doping. Note that the *c*/*a* ratio is almost constant at each temperature, which means isotropic thermal expansion of *c*- and *a*-axes of 4H-SiC crystals.

Fig. 2 shows the *c*-lattice constants of investigated samples with various nitrogen concentrations. In this figure, the *c*-lattice constants at room temperature and 1100°C are shown as a function of nitrogen concentration. From the results, the lattice mismatch between the n$^-$ epilayer (N_d = 6x10^{14} cm^{-3}) and the n$^+$ substrate (N_d = 2x10^{19} cm^{-3}) is calculated as 1.7x10^{-4} at 1100°C. The lattice mismatch would reach about (2-3)x10^{-4} at 1600°C (typical growth temperature). The same value would be expected for the *a*-lattice mismatch because of the isotropic thermal expansion. The authors estimated the critical thickness of nitrogen-doped 4H-SiC layers. In case of lattice mismatch of (2-3)x10^{-4}, which is expected at 1600°C, the critical thickness is estimated as approximately 600 μm from the People and Bean model [6]. However, there are very few reports on the observation of misfit dislocations in 4H-SiC. Experimental evidence is necessary to verify the calculated critical thickness. At least, 100 μm-thick 4H-SiC epilayers would not exceed critical thickness in the typical device structure (n$^-$/n$^+$), but the increasing lattice mismatch at elevated temperature may partly contribute to the nucleation of interfacial dislocations, which can be formed by thermal strain and misfit strain [7].

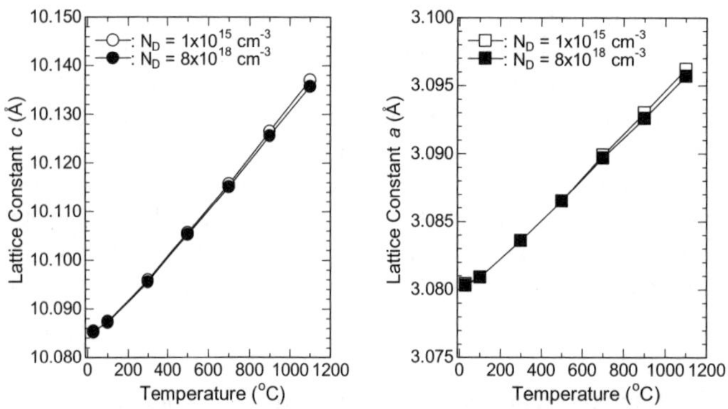

(a) (b)

Fig. 1. Temperature dependence of (a) *c*- and (b) *a*-lattice constants taken from a 4H-SiC substrate with a nitrogen concentration of 8x10^{18} cm^{-3}, and an epilayer doped to 1x10^{15} cm^{-3}.

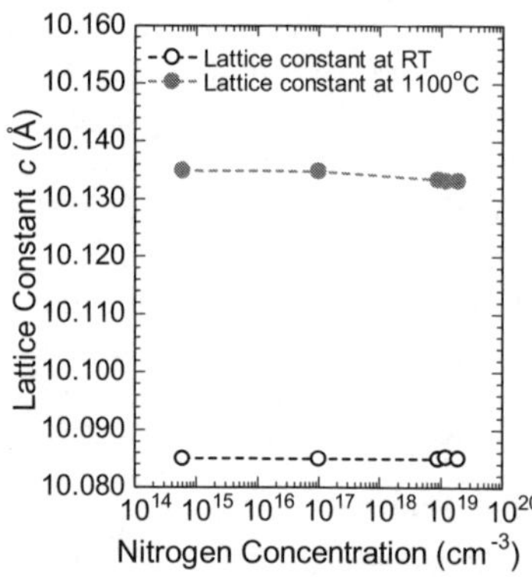

Fig. 2. *c*-lattice constants in 4H-SiC at RT and 1100°C as a function of nitrogen concentration.

Lattice parameters of ion-implanted 4H-SiC Figs. 3 (a) and (b) show the RSM images for (0008) reflection of the P$^+$-implanted ((a) 1x10^{19} cm^{-3} and (b) 5x10^{19} cm^{-3}) 4H-SiC. The sample, which was P$^+$-implanted with 1x10^{19} cm^{-3}, does not exhibit clear peak splitting in the RSM image (Fig. 3 (a)). On the other hand, Fig. 3 (b) reveals two distinct reflection peaks, originating from the epilayer and the implanted layer, respectively. The RSM image shows the increase in the *c*-lattice constant (decrease in q_z) of the ion-implanted layer compared with that of the epilayer. Substitutional P atoms should not cause the lattice expansion in 4H-SiC crystals, since P atoms have a smaller covalent radius than Si atoms. Therefore, the c-lattice expansion observed in the P^+-implanted layer can not be explained by the size difference. Fig. 3 (b) also shows that the two peaks are located at the slightly different q_x values, meaning that the basal planes of the P^+-implanted layer are tilting with respect to the basal planes of the epilayer. The c-lattice mismatch between the P^+-implanted layer and epilayer can be estimated as about 9×10^{-4}, and the tilt angle as about 0.012°. These values increased as implantation dose increases.

Figs. 4 (a) and (b) depict the RSM images for (0008) reflection of the Al^+-implanted ((a) 1×10^{19} cm^{-3} and (b) 5×10^{19} cm^{-3}) samples. The c-lattice expansion and c-axis tilt were also observed in the high-dose Al^+-implanted sample. The tilt direction is the same as that observed in the P^+-implanted layers. N^+-implanted samples also show the similar results (not shown). These results indicate that the c-lattice expansion and c-axis tilt are caused by heavy ion-implantation, irrespective of ion species. Note that all the implanted samples with a relatively low concentration of 1×10^{19} cm^{-3}, did not cause detectable peak splitting in RSM images. In addition, a similar result was observed on the RSM image of the (C^++Si^+)-coimplanted sample. Since (C^++Si^+)-coimplantation also causes the c-lattice expansion and c-axis tilt, ion-implantation-induced damage may be responsible for the lattice parameter change.

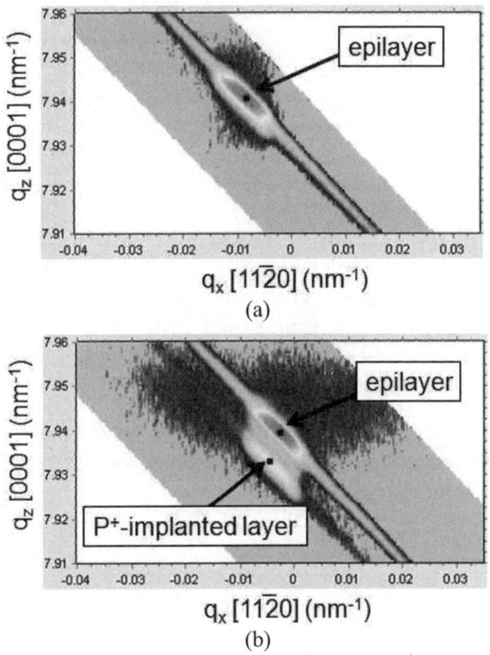

Fig. 3. (0008) RSM images of 4H-SiC, P^+-implanted with (a) 1×10^{19} cm^{-3} and (b) 5×10^{19} cm^{-3} (annealing: 1800°C, 10 min).

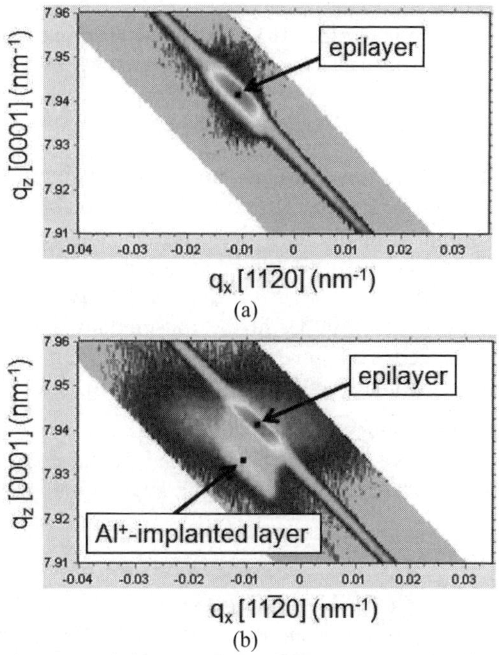

Fig. 4. (0008) RSM images of 4H-SiC, Al^+-implanted with (a) 1×10^{19} cm^{-3} and (b) 5×10^{19} cm^{-3} (annealing: 1800°C, 10 min).

Fig. 5 depicts the c-lattice mismatch, obtained from (0008) RSM measurements on ion-implanted 4H-SiC, as a function of the implanted ion concentration. In the figure, the result taken from the (C^++Si^+) co-implanted sample is also included. In the case of (C^++Si^+) co-implantation, a half dose was employed for each ion species to generate a box profile of 5×10^{19} cm^{-3}. Fig. 5 shows that heavier and larger ions tend to cause larger lattice mismatch at a given concentration of implanted atoms. Since heavier and larger ions generate higher amount of interstitial atoms and (vacancies) in the crystals, it is likely that interstitial atoms generated by ion bombardment play an important role. Note that the c-lattice mismatch observed in the ion-implanted 4H-SiC is much higher than the mismatch expected from the nitrogen-doped epilayer/substrate (as seen in Fig. 2).

It has been reported that excess interstitials generated by ion bombardment create secondary defects after high-temperature annealing [8,9]. These secondary defects are formed on the basal planes. The observed lattice mismatch might be caused by the extra planes inserted parallel to (0001) planes. As a result of the generation of extra planes, the crystallographic tilt may be introduced to accommodate the vertical lattice mismatch, as schematically shown in Fig. 6. The tilt direction observed in this study is consistent with the model shown in the figure.

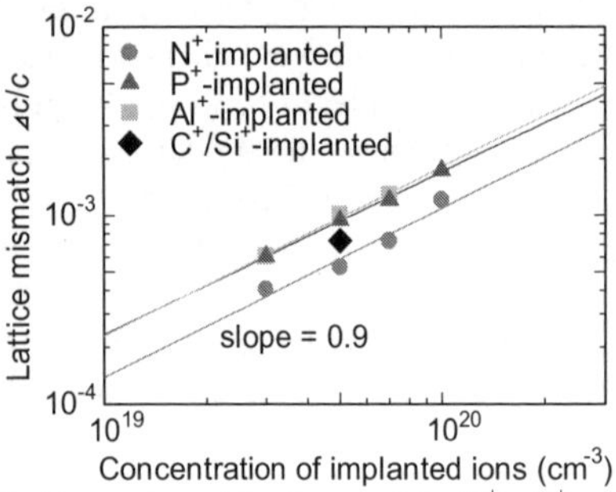

Fig. 5 Lattice mismatch observed in N^+-, P^+-, Al^+-, or (C^++Si^+)-implanted 4H-SiC is plotted as a function of the implanted ion concentration.

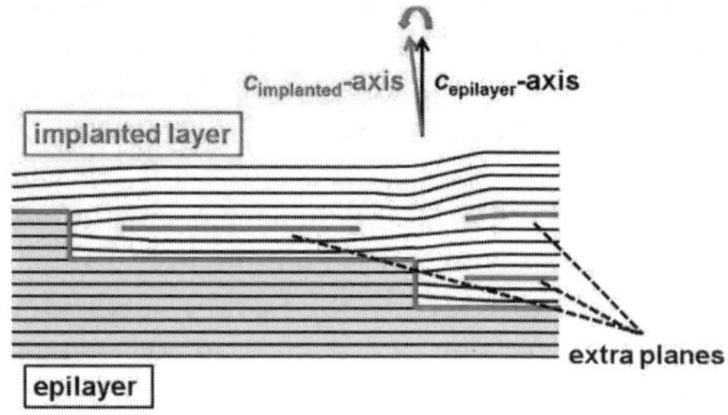

Fig. 6. Schematic illustration of the crystallographic tilt caused by extrinsic stacking faults on the basal planes of 4H-SiC.

Summary

HRXRD measurements showed no distinguishable difference in the *c*- and *a*-lattice constants at room temperature. However, the lattice contraction by heavy nitrogen doping was clearly observed at high temperatures. These results indicate that the thermal expansion coefficients of 4H-SiC are dependent on the doping concentration. The lattice mismatch between a lightly-doped free-standing epilayer (N_d = 6x10^{14} cm^{-3}) and a heavily-doped substrate (N_d = 2x10^{19} cm^{-3}) was calculated as 1.7x10^{-4} at 1100°C. In the ion-implanted layers, *c*-lattice expansion and *c*-axis tilt were caused, irrespective of ion species. The observed lattice mismatch and *c*-axis tilt may not be caused by heavy doping itself, but by implantation-induced secondary defects.

Acknowledgements: This work was supported by the Funding Program for World-Leading Innovative R&D on Science and Technology (FIRST Program) and a Grant-in-Aid for Scientific Research (21226008) from the Japan Society for the Promotion of Science.

References

[1] X. Zhang, M. Skowronski, K. X. Liu, R. E. Stahlbush, J. J. Sumakeris, M. J. Paisley, and M. J. O'Loughlin, J. Appl. Phys. Vol. 102 (2007), p. 093520.

[2] M. Nagano, H. Tsuchida, T. Suzuki, T. Hatakeyama, J. Senzaki, and K. Fukuda, J. Appl. Phys. Vol. 108 (2010), p. 013511.

[3] H. Jacobson , J. Birch, C. Hallin, R. Yakimova, T. Tuomi, E. Janzén, and U. Lindefelt, Appl. Phys. Lett. Vol. 82 (2003), p. 3689.

[4] R. S. Okojie, T. Holzheu, X. Huang, and M. Dudley, Appl. Phys. Lett. Vol. 83 (2003), p. 1971.

[5] M. Stockmeier, R. Müller, S. A. Sakwe, P. J. Wellmann, and A. Magerl, J. Appl. Phys. Vol. 105 (2009), p. 033511.

[6] R. People and J. C. Bean, Appl. Phys. Lett. Vol. 47 (1985), p. 322.

[7] X. Zhang, M. Nagano, and H. Tsuchida, Mater. Sci. Forum Vol. 679-680 (2011), p. 306.

[8] T. Ohno and N. Kobayashi, J. Appl. Phys. Vol. 89 (2001), p. 933.

[9] P. O. Å. Persson, L. Hultman, M. S. Janson, and A. Hallén, J. Appl. Phys. Vol. 100 (2006), p. 053521.

Materials Science Forum Vols. 717-720 (2012) pp 485-488

doi:10.4028/www.scientific.net/MSF.717-720.485

Strain build-up, swelling and stacking fault formation in implanted 4H-SiC

J-F. BARBOT[1,a,*], S. LECLERC[2,b], C. TROMAS[1,c], V. AUDURIER[1,d], A. DECLEMY[1,e], M. TEXIER[3,f], M-F. BEAUFORT [1,g]

[1]Institut Pprime, CNRS - Université de Poitiers - ENSMA - UPR 3346 Département Physique et Mécanique des Matériaux, SP2MI, Bd M. et P. Curie - BP 30179, 86962 FUTUROSCOPE CHASSENEUIL Cedex, France

[2]CIMAP, CEA-CNRS-ENSICAEN, BP 5133, 14070 Caen Cedex 05, France

[3]IM2NP Faculté des Sciences et Techniques, campus de Saint Jérôme, av. Escadrille Normandie Niemen, F-13397 Marseille Cedex, France

[a]jean.francois.barbot@univ-poitiers.fr, [b]sleclerc.ganil.fr, [c]christophe.tromas@univ-poitiers.fr, [d]valerie.audurier@univ-poitiers.fr, [e]alain.declemy@univ-poitiers.fr, [f]michael.texier@univ-cezanne.fr, [g]marie.france.beaufort@univ-poitiers.fr

Keywords: SiC, elastic strain, swelling, cavities, stacking fault

Abstract. Ion implantation into 4H-SiC induces a local gradient of strain which increases with the nuclear energy losses. With the increase of temperature of implantation the strain tends to become uniform in the whole implanted area requiring the migration of particles. In case of helium implantation, defects are more stabilized and their evolutions observed post thermal annealing are consistent with the evolution of surface swelling. The local modifications imputed to the ion process lead to the formation and the pile-up of stacking faults in the highly damaged region.

Introduction

Ion implantation is a technique currently used to create localized doped areas in the fabrication of SiC-based devices. However, the well-known drawback of this method is the accumulation of damage along the ion path. Post thermal annealing or implantations at elevated temperatures allow reducing the defect concentration [1]. These defects induce elastic strain all along the ion path that can be determined through X-ray diffraction (XRD) measurements [2,3]. The defect formation and the as-resulting strain gradient induce dimensional changes (defect-induced swelling) that must be more and more controlled (or integrated) with the device miniaturization toward the nanoscale. In this study we compare the strain build-up induced by nitrogen (N) and helium (He) implantation as a function of the implantation temperature. N atoms are commonly used as dopants and He atoms are used to locally induce damage in the formation of SiC-On-Insulator (SiCOI) structures. Moreover with the increase of temperature more complex defects can be formed; it is thus fundamental to understand both their formation and nature to improve the device performances.

Experimental details

Samples of n-type $(0001)_{Si}$ 4H-SiC single crystals (8° off-axis toward the $[11\bar{2}0]$ direction) of resistivity 0.02 $\Omega.cm$ supplied by Cree Research Inc. were implanted with nitrogen ions at different temperatures. Nitrogen implantations were performed at fixed incident energy of 180 keV and given fluence of $1x10^{15}$ cm^{-2} that gives a peak nitrogen concentration of about 0.1 at.% at a mean projected range R_p close to 290 nm ($\Delta R_p = 50nm$) and a maximal deposited energy of $2x10^{21}$ $keV.cm^{-3}$ or 0.25 displacement per atom (dpa), according to SRIM calculations (the damage peak is located around $R_D = 250nm$) [4]. Helium implantations were performed at 160 keV to $1x10^{16}$ cm^{-2} ($R_p = 590nm$, 0.15 dpa). Additional helium implantations were performed under different conditions and then annealed (up to 1700°C) in order to compare the as-resulting strain and also to understand the contribution of defects in surface swelling. The strain gradient was studied through XRD measurements conducted on a two-circle goniometer provided by a 5 kW RIGAKU RU-200 generator. The surface swelling (step height) was measured by atomic force microscopy (AFM) and the defects were studied by using cross-sectional transmission electron microscopy (TEM) carried out on a Jeol 200CX or a FEI Titan 80-300 microscope.

Results and discussion

Strain build-up

In Figure 1 are plotted the XRD curves of the $1x10^{15}$ $N^{+}cm^{-2}$ implanted samples at different temperatures. All curves show a main sharp Bragg diffraction peak located at zero strain resulting from the diffraction of the non-implanted bulk 4H-SiC, and a large tail of scattered intensity that is characteristic of a dilatation gradient of the lattice in the implanted area [2,3]. At RT the scattered intensity consists of an asymmetrical peak (see the arrow figure 1) followed by an interference fringe pattern resulting from the coherent diffraction of two damaged zones of same level of strain [2]. This asymmetrical peak is ascribed to the diffraction of the near surface region. The normal strain profile deduced from the XRD simulation showed that it fits the nitrogen profile around R_p [5] as reported for He implantation [6]. When performing implantations at temperatures higher than RT the shape of curves evolves. The near surface peak is strongly attenuated and the last fringe is shifted towards the main Bragg peak. The maximum of strain given by the position of this last fringe away the Bragg peak (see the dotted lines in Fig. 1) thus strongly decreases with the implantation temperature showing the enhancement of dynamical annealing with temperature. At 500°C the scattered intensity consists of a well defined main satellite peak which means that the whole implanted region is roughly homogeneously strained (the strain gradient becomes blurred). The last effect is much less marked for helium implantations for which a surface peak and a strain gradient are still observed at 800°C ($\varepsilon_s = 0.2\%$), see figure 2.

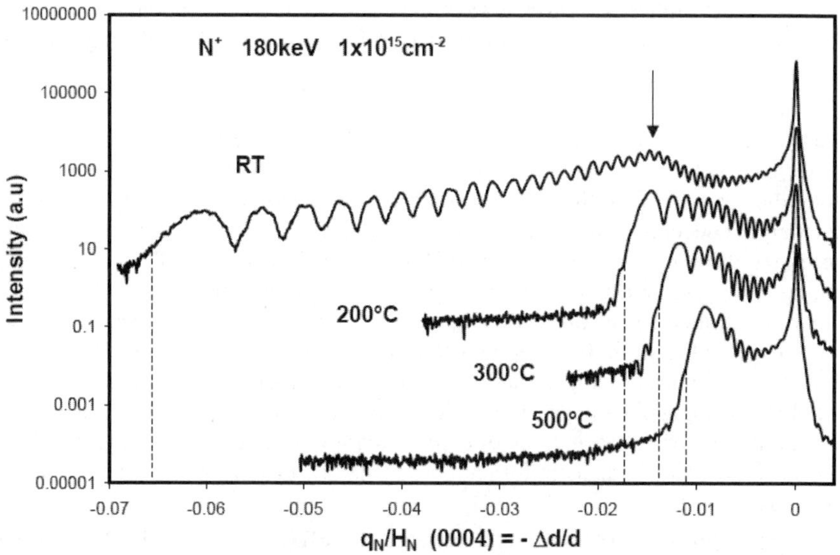

Fig. 1: X-ray scattered intensity distribution along the surface normal direction close to the (0004) reflection in N-implanted 4H-SiC at different temperatures of implantation to a fixed energy of 180 keV and a given fluence of $1x10^{15}$ cm^{-2} ($\varepsilon = (\Delta d/d)$ is the normal strain). The arrow shows the near surface peak clearly observed in the RT implanted sample only. The dotted lines give the values of maximal strain $\varepsilon_{\max}$.

For the two implanted ions, N and He, a strong enhancement of the dynamical annealing occurs at low temperatures (lower than 200°C, see figs. 1 and 2 [7]). As shown in table I which summarizes the evolution with the implantation temperature T of the maximal strain normalized to the value at RT, the drop of strain is more significant in case of N ions which suggests an enhancement of dynamical annealing with defect density. Similarly, in He implanted samples, the maximal strain decrease is more abrupt than the near surface strain decrease (see fig. 2 and Ref. [7]). Moreover, the homogenization of strain with temperature in the implanted region of N implanted sample suggests, in addition to recombination, a mechanism of defect migration as previously proposed [5]. A defect migration has been reported in case of He implantation at high fluence for which the accumulation of interstitials has been pointed out [7]. He atoms are also known to stabilize vacancy-type defects which may also be responsible for a drop of maximum strain lower than observed for N implantation as shown in table I.

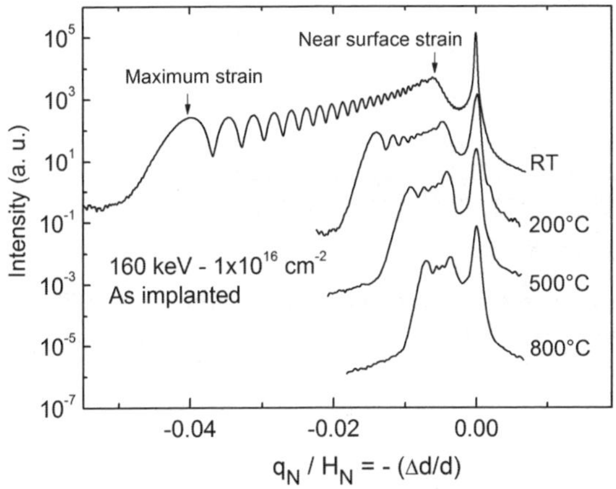

Fig. 2: XRD curves obtained close to the (0004) reflection in He-implanted SiC.

Table 1: Comparison of variations of the ratio $\varepsilon^{T}_{max}/\varepsilon^{RT}_{max}$ *between the maximum strain* ε_{max} *and the value at RT as function of the temperature of implantation T for N et He implantations.*

T(°C)	RT	200	300	500	800
N *0.25 dpa*	1 *6.4%*	0.26	0.21	0.16	-
He *0.15 dpa*	1 *4.3%*	0.35	-	0.23	0.18

Surface swelling

The formation volume of a Frenkel pair being positive it contributes to the surface swelling under implantation. The significant reduction of the Frenkel pair concentration with annealing temperature leads to strain relief and therefore to the decrease of the elastic strain magnitude. This effect is clearly seen on the step height variations in the low fluence He implanted sample (RT-$1x10^{16}cm^{-2}$), fig. 3. Upon annealing the elastic strain progressively relaxes according to the different migration energies of defects leading to the continuous decrease of swelling with temperature. The residual step height above 1000°C results from the formation of large two-dimensional clusters of bubbles [6].

Under more severe conditions of He implantation (400°C, $5x10^{16}cm^{-2}$) the step height remains stable with annealing temperature. The point defect recombination is hidden by the growth of helium-vacancy complexes into spherical bubbles during post-implantation annealing. The recombination is also less efficient due to the migration of interstitials toward the highly damaged region [7] and their agglomeration into clusters.

Under harsh conditions of implantation (700°C, $1x10^{17}$ cm^{-2}) the value of step height remains constant for annealing temperatures up to 1000-1100°C and then sharply rises. This significant increase is concomitant with the growth of extended defects, see fig. 3. Beyond 1000°C, vacancies become mobile resulting in the huge growth of cavities. At 1700°C large cavities are observed among a huge density of stacking faults. The formation of stacking faults is the consequence of the strain gradient which enhances the defect clustering. A more detailed study [8] showed that drifted interstitials form Franck dislocation loops disrupting the 4H sequence and therefore promoting the emission of Shockley partials leaving stacking faults in their wake. The formation of stacking faults has also been reported in phosphorus implanted 4H-SiC [9] which confirms that the non-homogeneous distribution of damage induced by ion implantation is responsible for the stacking fault formation.

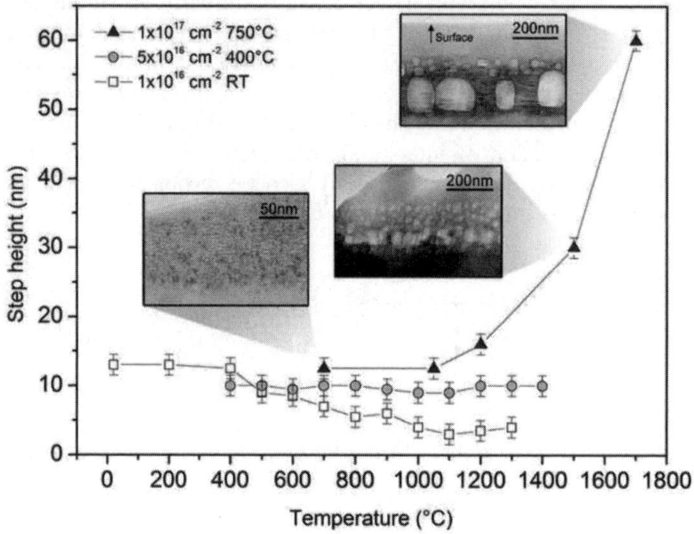

Fig. 3: Evolution of the step height in helium implanted 4H-SiC under three different conditions of implantation. TEM images of the samples implanted at high temperature and fluence show the microstructures observed at temperatures of annealing of 750°C, 1500 and 1700°C.

Conclusions

Ion implantation into 4H-SiC results in the formation of an elastic strain gradient whose magnitude increases with nuclear energy losses. With increasing temperature of implantation the strain decreases with a profile which tends to become homogenous. This requires the recombination and the migration of defects and possibly the migration of the implanted ions. However, in He implanted sample the homogenization of the profile is limited due to the interaction of helium with defects. The evolution of the as-resulting swelling with annealing temperature strongly depends on the implantation conditions, i.e., on the as-formed defects. In particular, the local modifications of the microstructure caused by the implantation under harsh conditions lead to the formation of stacking faults.

References

[1] W. Jiang, W.J. Weber, S. Thevuthasan, D.E. McCready, J. of Nucl. Mat. 257 (1998) 295-302.

[2] S. Leclerc, A. Declémy, M.F. Beaufort, C. Tromas, J.F. Barbot, J. Appl. Phys. 98 (2005) 113506-10

[3] A. Debelle *et al.*, J. Phys. D :Appl Phys. 43 (2010) 455408

[4] J.F. Ziegler, J.P. Biersack and U. Littmark, http:/www.srim.org

[5] M. Amigou, M.F. Beaufort, A. Declémy, S. Leclerc, J.F. Barbot, Mat. Sci. Forum 679-680 (2011) 185-188

[6] S. Leclerc, M.F. Beaufort, A. Declémy, J.F. Barbot, Appl. Phys. Lett. 93 (2008) 122101-3.

[7] S. Leclerc, M.F. Beaufort, A. Declémy, J.F. Barbot, J. of Nucl. Mat. 397 (2010) 132-134.

[8] J. F. Barbot, M.F. Beaufort, M. Texier, C. Tromas, J. of Nucl. Mat. 257 (2011) 162-165.

[9] J. Wong-Leung, M.K. Linnarsson, B.G. Svensson, D.J.H. cokayne, Phys. Rev. B71 (2005) 165210

Materials Science Forum Vols. 717-720 (2012) pp 489-492
© (2012) Trans Tech Publications, Switzerland
doi:10.4028/www.scientific.net/MSF.717-720.489

High Resolution X-ray Diffraction (HRXRD) Studies of the Initial Stages of PVT-Growth of 4H-SiC Crystals

N. Ohtani[1,2,a,*], M. Katsuno[1,3,b], T. Fujimoto[1,3,c], S. Sato[1,3,d], H. Tsuge[1,3,e], W. Ohashi[1,3,f], H. Matsuhata[1,g] and M. Kitabatake[1,h]

[1]R&D Partnership for Future Power Electronics Technology (FUPET)
2-9-5 Toranomon, Minato-ku, Tokyo 105-0001, Japan

[2]Kwansei Gakuin University, R&D Center for SiC Materials and Processes
2-1 Gakuen, Sanda, Hyogo 669-1337, Japan

[3]Nippon Steel Corporation, Advanced Technology Research Laboratories
20-1 Shintomi, Futtsu, Chiba 293-8511, Japan

[a]ohtani.noboru@kwansei.ac.jp, [b]katsuno.masakazu@nsc.co.jp, [c]fujimoto.tatsuo@nsc.co.jp, [d]sato.shinya@nsc.co.jp, [e]tsuge.hiroshi@nsc.co.jp, [f]ohashi.wataru@nsc.co.jp, [g]h.matsuhata@aist.go.jp, [h]kitabatake.makoto@jp.panasonic.com, *corresponding author

Keywords: physical vapor transport (PVT) growth, high resolution x-ray diffraction (HRXRD), lattice plane bending.

Abstract. Defect formation during the early stages of physical vapor transport (PVT) growth of 4H-SiC was investigated using high resolution x-ray diffraction (HRXRD). Characteristic lattice bending behaviors were revealed in the nearby seed crystal regions of grown crystals. The lattice bending was localized in close proximity to the seed/grown crystal interface, and the (0001) basal planes bended convexly toward the growth direction, indicative of the insertion of extra-half planes pointing toward the growth direction during the initial stages of crystal growth. This paper discusses the possible mechanisms of the observed lattice bending and sheds light on the defect formation processes during PVT-growth of 4H-SiC single crystals.

Introduction

Over the last decade, silicon carbide (SiC) has been under intensive investigation as a promising material for high-power semiconductor devices, and this has led to the implementation of widely-accepted SiC Schottky barrier diodes across various applications as well as the recent announcement of the commercial release of SiC power MOSFETs. Nevertheless, SiC semiconductor technology still suffers from reliability issues, particularly originating from the crystallographic defects that exist in SiC epitaxial wafers. Commercially available SiC wafers contain dislocations, typically in the range of 10^3-10^4 cm^{-2}, and most of them are introduced during crystal growth. Hence, understanding the growth kinetics and defect formation during SiC bulk crystal growth is essential to obtain SiC crystals with a lower defect density.

This paper reports studies on defect formation during the early stages of SiC bulk crystal growth through an analysis of lattice plane bending in the near seed crystal regions of grown crystals. We observed characteristic behaviors of the peak shift of x-ray rocking curves across the crystals, which give a useful insight into understanding defect formation during the initial stages of SiC bulk crystal growth.

Experimental

4H-SiC single crystals were grown on well- (<0.25°) and 4° off-oriented $(000\bar{1})$C seed crystals by the physical vapor transport (PVT) growth method. The growth conditions employed in this study were the same as those previously described [1] except the growth crucible size, and the crystals were nitrogen-doped in the high 10^{18} cm^{-3} range. The typical growth temperature was about 2300–2400°C.

The polytype of the grown crystals was examined by Raman scattering spectroscopy and found to be identical to that of 4H-SiC seed crystals. The grown crystals were sliced almost perpendicularly or parallel to the growth direction into {0001} or {11$\bar{2}$0} wafers; results were similar among crystals prepared for each wafer orientation. The thickness of the wafers was about 1 mm and the surface was carefully polished on both sides and then lightly etched in molten KOH prior to x-ray rocking curve measurements to prevent the lattice plane bending from being affected by surface damaged layers.

X-ray rocking curve measurements were performed using a double-axis diffractometer, where double crystal x-ray optics for Cu $K_{\alpha 1}$ radiation is employed. The x-ray beam was incidental on the (0001)Si or (11$\bar{2}$0) surface of crystals, and the footprint size of the incident x-ray beam on the sample surface was about 1 mm × 1 mm. For measurements on the {0001} wafers, we examined three wafers sliced from the near-seed regions of grown crystals, and the rocking curve measurements of these wafers gave the information on the crystal parts of grown SiC crystals at about 0.5, 3.1 and 4.4 mm above the seed crystal.

Results and Discussion

Defect formation at the initial stages of crystal growth is one of the most important topics in the bulk crystal growth of SiC since most threading dislocations in SiC crystals are introduced during the early stages of crystal growth. Sanchez *et al.* examined, using defect selective etching and transmission electron microscopy (TEM), the formation of threading dislocations at the initial stages of PVT-growth of 6H-SiC [2]. They observed the formation of basal plane stacking faults, with densities showing a good correlation with densities of threading dislocations, suggesting that basal plane defects and threading dislocations are closely related in terms of their formation. From this perspective, we examined basal plane bending at the near-seed regions of grown crystals, which provides a useful insight into the understanding of defect formation at the initial stages of crystal growth.

Figure 1 shows the plot of the relative 0008 x-ray rocking curve peak position as a function of the beam position across the {0001} wafer along the [11$\bar{2}$0] direction; the red, blue and black circles represent the data from the crystal portions at 0.5, 3.1 and 4.4 mm above the seed crystal, respectively. As seen in the figure, the rocking curve for the crystal at 4.4 mm above the seed crystal (circles) showed a positive peak shift as the x-ray beam scanned from the beam incidence to the exit direction. This indicates that the basal plane bent slightly in the crystal and was concave toward the [000 $\bar{1}$]C growth direction.

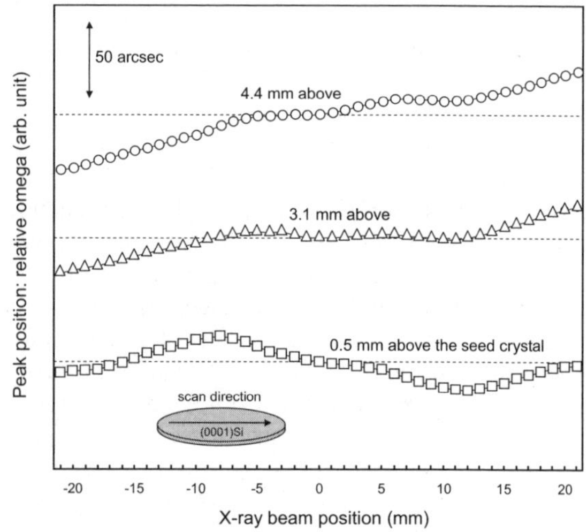

Fig. 1 Relative 0008 x-ray rocking curve peak shift as a function of x-ray beam position across the crystal at 0.5, 3.1 and 4.4 mm above the seed crystal. The inset schematically shows the x-ray beam scan direction.

The observed basal plane bending (about 1.6 arcsec/mm) of the crystal is reasonably understood in terms of the temperature distribution in SiC crystals during the PVT-growth process [3,4]. In the PVT-growth process, crystal growth is primarily driven by the temperature gradient along the *c*-axis, and thus large shear stresses are imposed on the crystal during growth. At typical PVT-growth temperatures (> 2400ºC), the plastic deformation of SiC crystals occurs and shear stresses are largely relaxed by activating basal plane dislocation glides within the crystal. In the PVT-growth process, the temperature decrease toward the seed crystal is maintained during growth, resulting in negative shear stresses within the grown crystal. Therefore, during the plastic deformation, basal plane dislocations that have an extra-half-plane pointing toward the seed crystal are introduced and cause the basal plane to bend concavely toward the crystal growth direction when the crystal is cooled to room temperature.

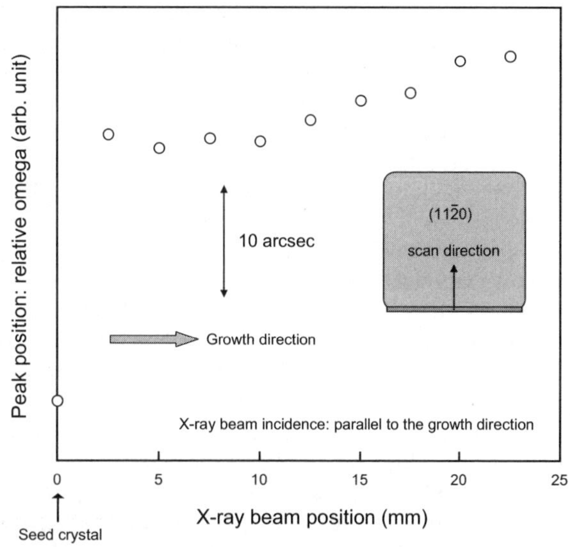

Fig. 2 Relative 11$\bar{2}$0 x-ray rocking curve peak position as a function of x-ray beam position. The x-ray beam scanned along the growth direction and was directed from the seed to the grown crystal. The inset schematically shows the x-ray beam scan direction.

As seen in the figure, the peak shift decreased when the crystal region closer to the seed crystal was measured. Almost no peak shift was observed over the central area of crystals at 3.1 mm above the seed crystal. This tendency was enhanced when the region of measurement was even closer to the seed crystal; the rocking curves from crystals at about 0.5 mm above the seed crystal showed a negative peak shift (about -1.8 arcsec/mm) at the center of the crystals, indicating that this part of the crystal was convex toward the [000$\bar{1}$]C growth direction. The observed behaviors of the basal plane bending cannot be simply explained by the temperature distribution in grown crystals, and different mechanisms need to be involved.

One possible mechanism is the difference in the coefficient of thermal expansion (CTE) between SiC and crucible materials (graphite). Generally, graphite shows a larger CTE compared to SiC, and thus could give rise to a convex shape bending of the basal plane toward the [000$\bar{1}$]C growth direction, resulting in a negative peak shift in the rocking curve measurements. It was found, however, that the observed lattice bending was quite localized and extended only a few mm or less above the seed crystals. Figure 2 shows a plot of the relative 11$\bar{2}$0 x-ray rocking curve peak shift as a function of the beam position along the growth direction. The x-ray beam was directed along the growth direction, from the seed to the grown crystal. As seen in the figure, the peak position almost abruptly changed in close proximity to the seed crystal (~13 arcsec/mm), suggesting that the causes of the observed lattice plane bending were more significantly related to the initial growth processes than the thermoelasticity.

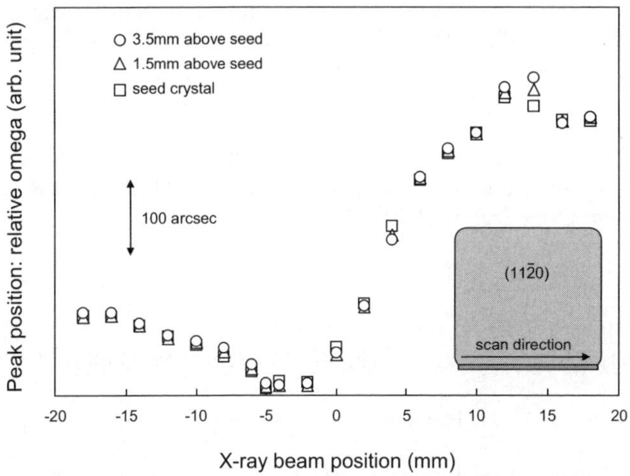

Fig. 3 Relative 11$\bar{2}$0 x-ray rocking curve peak shift along the normal to the growth direction. The x-ray beam was incident along the normal to the growth direction. The inset schematically shows the x-ray beam scan direction.

Figure 3 shows plots of the relative 11$\bar{2}$0 x-ray rocking curve peak shift along the normal to the growth direction. The x-ray beam was incidental and scanned perpendicularly to the growth direction within the seed crystal (squares), about 1.5 (triangles) and 3.5 mm (circles) above the seed crystal. As seen in the figure, the bending or variation of the $(11\bar{2}0)$ lattice plane along the normal to the growth direction was almost the same as that of the seed crystal, and the deviation from that of the seed crystal was fairly small (≤5 arcsec/mm) compared to the scan along the growth direction, indicating that the lattice plane bending in the initial parts of grown crystals was anisotropic: large along the growth direction and small along the normal to the growth direction.

The negative peak shift in the 0008 rocking curves measured from the near seed regions of grown crystals implies that BPDs having an extra-half plane pointing toward the growth direction were preferably introduced during the initial stages of crystal growth to those with an extra-half plane of the opposite direction. The mechanism of this preference is assumed to be closely related to the nucleation process of growth nuclei and their coalescence on the seed crystal surface. PVT-growth of SiC generally proceeds *via* Volmer-Weber growth mode at the initial stages of growth, where three-dimensional (3D) adatom clusters or growth nuclei are formed on the seed crystal surface. These small 3D growth nuclei are likely to show different lattice spacings from those of bulk crystals, particularly in close proximity to the nucleus surface. Surface relaxation occurs at the nucleus surface, and in SiC, the lattice spacings of prism planes are narrower at the $(11\bar{2}0)$ or $(1\bar{1}00)$ surface [5], which implies that the growth nuclei are contracted in the lateral directions increasingly as they grow from the (0001) seed crystal surface. When they coalesce, therefore, they tend to incorporate between them basal plane defects with an extra-half plane toward the growth direction, causing the (0001) basal plane to bend convexly toward the growth direction.

Summary

The localized lattice bendings in the near seed crystal regions of grown crystals of 4H-SiC were revealed through HRXRD studies of 4H-SiC crystals grown by the PVT growth method. The 0008 x-ray rocking curves showed a negative peak shift in close proximity to the seed/grown crystal interface as the x-ray beam scanned across the (0001) wafer, indicative of the basal plane bending convexly toward the growth direction. The $11\bar{2}0$ rocking curves with the x-ray beam incidence along the crystal growth direction exhibited an abrupt change in the rocking curve peak position at the seed/grown crystal interface, while that with the beam incidence normal to the growth direction hardly changed across the interface. The results indicate that basal plane defects with an extra-half plane pointing toward the growth direction are preferably introduced during the initial stages of PVT-growth of 4H-SiC, to those with an extra-half plane of the opposite direction. The paper discusses the mechanisms of the observed localized lattice bendings and relates them to the growth kinetics at the initial nucleation stages of PVT-growth of SiC.

Acknowledgements

This work was partially supported by Novel Semiconductor Power Electronics Project Realizing Low Carbon Emission Society under Ministry of Economy, Trade and Industry (METI) and New Energy and Industrial Technology Development Organization (NEDO).

References

[1] J. Takahashi, M. Kanaya, Y. Fujiwara, J. Cryst. Growth 135 (1994) 61.

[2] E. K. Sanchez, J. Q. Lin, M. De. Graef, M. Skowronski, W. M. Vetter, M. Dudley, J. Appl. Phys. 91 (2002) 1143.

[3] S. Ha, G. S. Rohrer, M. Skowronski, V. D. Heydemann, D. W. Snyder, Mater. Sci. Forum 338–342 (2000) 67.

[4] N. Ohtani, M. Katsuno, T. Fujimoto, M. Nakabayashi, H. Tsuge, H. Yashiro, T. Aigo, H. Hirano, T. Hoshino, W. Ohashi, Jpn. J. Appl. Phys. 48 (2009) 065503.

[5] S. P. Mehandru, A. B. Anderson, Phys. Rev. B 42 (1990) 9040.

Materials Science Forum Vols. 717-720 (2012) pp 493-496
© (2012) Trans Tech Publications, Switzerland
doi:10.4028/www.scientific.net/MSF.717-720.493

On the stability of 3C-SiC single crystals at high temperatures

D. Dompoint[1, a], I. G. Galben-Sandulache[2,b], A. Boulle[1,c], D. Chaussende[2,d,*], D. Eyidi[3,e], J. -L. Demenet[3,f], M. -F. Beaufort[3,g] and J. Rabier[3,h]

[1] Science des Procédés Céramiques et de Traitements de Surface (SPCTS), CNRS UMR 6638, Centre Européen de la Céramique, 12,rue Atlantis 87068 LIMOGES,France

[2] Laboratoire des Matériaux et du Génie Physique (LMGP), CNRS UMR 5628, Grenoble INP, Minatec, 3 parvis Louis Néel, BP 257, 38016 Grenoble Cedex 01, France

[3] Institut Pprime, CNRS UPR 3346, Département Physique et Mécanique des Matériaux, SP2MI-Téléport 2-Bd Marie et Pierre Curie, BP 30179, 86962 Futuroscope-Chasseneuil Cedex,France

[a]deborah.dompoint@etu.unilim.fr, [b]irina.galben@grenoble-inp.fr, [c]alexandre.boulle@unilim.fr, [d]didier.chaussende@grenoble-inp.fr, [e]dominique.eyidi@univ-poitiers.fr,

[f]jean.luc.demenet@univ-poitiers.fr, [g]marie.france.beaufort@univ-poitiers.fr,

[h]jacques.rabier@univ-poitiers.fr,
*corresponding author

Keywords: 3C-SiC, 3C-6H transition, stacking faults, dislocations.

Abstract. The 3C-6H polytypic transition in 3C-SiC single crystals is studied by means of diffuse X-ray scattering (DXS) coupled with transmission electron microscopy (TEM). TEM reveals that the partially transformed SiC crystals contain regions of significantly transformed SiC (characterized by a high density of stacking faults) co-existing with regions of pure 3C-SiC. The simulation of the diffuse intensity allows to determine both the volume fraction of transformed material and the transformation level within these regions. It is further shown that the evolution with time and temperature of the transition implies the multiplication and glide of partial dislocations, the kinetics of which are quantified by means of DXS.

Introduction

The question of the relative stability of the different polytypes is a longstanding issue in the studies of SiC materials. In the case of 3C-SiC this point is of special importance since this polytype is believed to be unstable and transforms into 6H-SiC at the temperatures usually required to grow SiC single crystals from the vapor phase [1,2]. As a consequence, the crystals grown are in general of poor crystalline quality, characterized by a high density of stacking faults (SFs) and/or 6H inclusions. The improvement of the overall crystalline quality of 3C-SiC single crystals hence requires a significant optimization of the growth conditions. Along this line, a better understanding of the stability of 3C-SiC at high temperatures, for instance in terms of the 3C-6H transformation kinetics and the mechanisms involved, would provide an invaluable help. This is the topic of the present article.

Because of the rarity of 3C-SiC single crystals, the 3C-6H transition in SiC has only been scarcely studied (see Ref. [1] and references therein) and the transformation kinetics are yet unknown. Up until recently even the transformation mechanism was not clearly identified. A previous study using diffuse X-ray scattering (DXS) revealed that the transformation implies the glide of partial dislocations [3], probably according to a mechanism first described by Pirouz (Ref. [4] and references therein). In the present work, we pursue these DXS experiments, combined with transmission electron microscopy (TEM), in order to analyse the 3C-6H polytypic transition on both a qualitative (i.e. what are the mechanisms involved) and quantitative (i.e. the transformation kinetics) standpoint.

Experimental details

In this work we used (001)-oriented 3C-SiC single crystals produced by Hoya Corp. by chemical vapor deposition on "undulant" (001) Si substrates [5]. It is known that the SF distribution in these samples is anisotropic [5]: the (-111) and (1-11) planes exhibit a high SF density, whereas the (111) and (-1-11) planes exhibit a low SF density. In the following we shall only consider the planes exhibiting a high SF density, since the 3C-6H transition mainly takes place in the direction normal to these planes. In order to determine the transformation kinetics, a first set of crystals have been annealed at 2000°C for a time ranging between 1 h and 8 h. A second set of crystals annealed 5 h for temperatures ranging between 1700 and 2100°C allows us to assess the activation energy of the transformation. All crystals have been characterized by DXS on a high-resolution laboratory diffractometer. This technique allows to determine, in a non-destructive way, the mechanism involved in the transformation, the level of transformation (i.e. disorder), the nature of the polytypes and their relative amount. Full details concerning the DXS technique can be found in Ref. [6]. Plane-view and cross-section samples have also been prepared from a crystal annealed at 1800°C for TEM observations.

Results and discussion

Microstructure of the annealed crystals. A plane-view of a typical microstructure formed in the annealed crystals is shown in Fig. 1(a). There are regions extending parallel to the {111} planes, of several micrometers wide, characterized by a high density of SFs. The volume fraction of these regions is denoted x_t and the level of transformation in these regions is denoted τ ($\tau = 0$ for the 3C phase and $\tau = 1$ for the 6H phase). Between these regions, the 3C-SiC (with volume fraction $1 - x_t$) is unaffected by the transformation. A zoom on an isolated SF terminating within the crystal, Fig. 1(b), shows that it is bounded by a dislocation whose Burgers vector was determined to be 1/6[2-1-1], which correspond to the Burgers vector of partial dislocations in 3C-SiC. Such partial dislocations result from the dissociation of a perfect basal plane dislocation according to 1/2<1-10> → 1/6<2-1-1> + 1/6<1-21>. A basal plane dislocation undergoing that type of dissociation is clearly visible in Fig. 1(c) (see arrowheads). Since the two partials are characterized by different mobilities [4], the expansion of one of the partial forms a loop that produces a SF in the entire lattice plane. Double cross-slipping then allows successive lattice planes to be faulted (according to a mechanism proposed in [4]) so as to, starting from the 3C stacking, form the final 6H sequence.

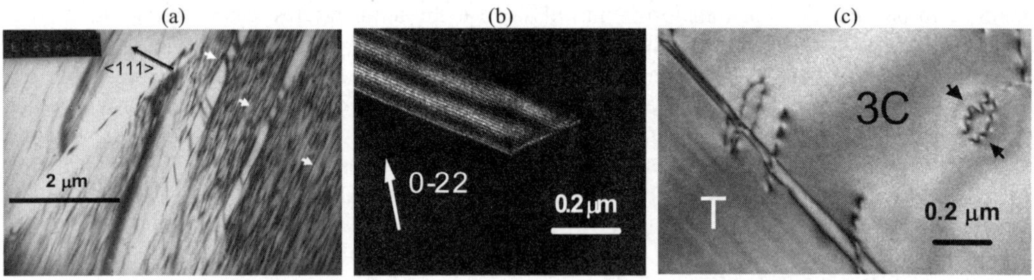

Fig. 1: TEM observations of a 3C-SiC crystal annealed at 1800°C. (a) overview showing transformed, stacking-fault-rich bands (arrowheads) between pure 3C-SiC areas. (b) weak-beam image of a typical stacking fault and a bounding dislocation. (c) Bright-field image of short partial dislocations and long dislocations lying at the interface between transformed areas (T) and 3C-SiC. Two dissociated partials in pure 3C-SiC are also visible (arrowheads).

Transformation kinetics. All DXS experiments reveal that, upon high temperature annealing, 3C-SiC systematically transforms into the 6H polytype [3,6,7]. The presence of SFs produce extended diffuse X-ray scattering, the simulation of which allows to determine the volume fraction of transformed material x_t and the transformation level within the partially transformed regions τ. The model used for the simulation is built on the following findings obtained by TEM: (i) the SiC crystals contain regions of heavily transformed SiC co-existing with regions of perfect 3C-SiC, so that the overall scattered intensity is written as a volume average of the intensities scattered from each type of region, and (ii) the transformed regions contain SFs which are bounded by partial dislocations which result from the dissociation of basal plane dislocations (for details see [6]).

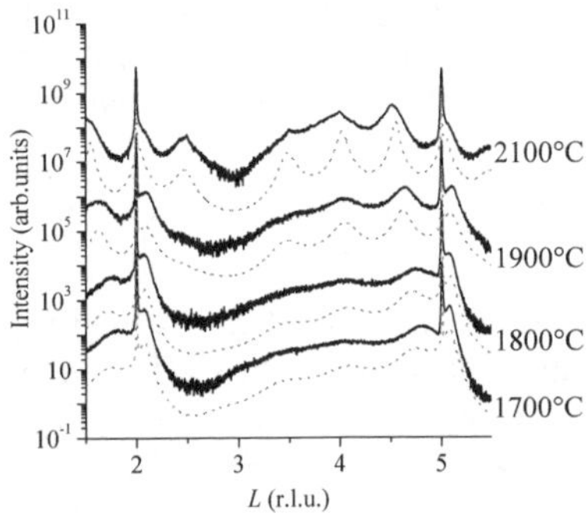

Fig. 2: DXS curves (black line) and simulation (red dotted line) corresponding to SiC crystals annealed 5h for increasing temperatures.

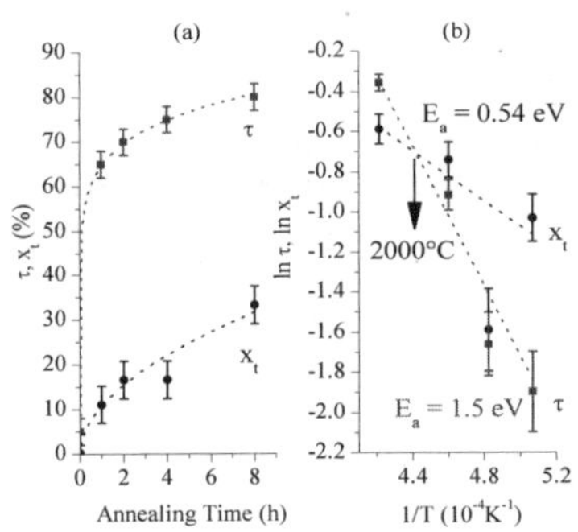

Fig. 3: (a) kinetics and (b) Arrhénius plot corresponding to the the transformation level τ and the volume fraction x_t.

The DXS curves (together with the simulations) recorded along the $[10L]_h$ row corresponding to the second set of crystals is shown in Fig. 2 (the subscript 'h' stands for hexagonal). The index L corresponds to the Miller index in a three-layer hexagonal unit cell, *i.e.* the Bragg peaks occurring at $L = 2$ and 5 correspond to the (002) and (-113) Bragg peaks of 3C or, alternatively, to the (104) and (10 10) peaks of 6H [6]. Upon increasing temperature the diffuse intensity located between the Bragg peaks increases and forms broad pseudo-peaks (corresponding to the Bragg peaks of the forthcoming 6H phase) indicating the progress of the transformation.

The volume fraction x_t and the transformation level τ deduced from the simulations for all investigated crystals are displayed in Fig. 3. Fig. 3(a) represents the evolution with time of τ and x_t at a temperature of 2000°C. Fig. 3(b) represents the evolution of τ and x_t as a function of temperature in the form of an Arrhénius plot. Let us first consider the evolution with annealing time. The first striking observation is that the transformation level and the volume fraction are characterized by two very distinct kinetics. Whereas the transformation level exhibits a fast kinetic and reaches a value of 80 % after 8 hours, the volume fraction exhibits a much slower kinetic and only reaches 33 % after 8 hours. This observation suggests that these two evolutions imply two distinct mechanisms. The increase of the transformation level necessarily implies the creation of new SFs. As mentioned earlier, new SFs can be created in SiC by the multiplication of partial dislocations using the double cross-slip mechanism [4]. On the contrary, the increase of the volume fraction does not necessarily imply the creation of new SFs: the glide of partial dislocations in their planes produces an extension of the SFs that increases the volume fraction of transformed material. However, in order to fault significant regions of the crystal, these dislocations have to glide over several tens of micrometers, a path along which they might be pinned by structural defects. Moreover, two dislocations, gliding in two different glide planes (say (-111) and (1-11)) can not cross each other (since they correspond to two different orientations of the upcoming 6H phase) so that the motion of one of the dislocation is stopped. These arguments might explain why the corresponding kinetic is much slower and does not reach very high values (contrarily to τ).

More can be learned from the activation energies corresponding to each mechanism, Fig. 3(b). The increase of τ is characterized by an activation energy E_a = 1.5 eV. Since the increase of τ necessarily implies the creation of new SFs, this activation energy simply correspond to the energy required to activate double cross-slip. This value is in excellent agreement with the value (E_a = 1.54 eV) deduced from *ab initio* calculations [8]. The increase of x_t is characterized by an activation energy approximately three time smaller (E_a = 0.54 eV). This demonstrates that it indeed implies a distinct mechanism than τ, and in particular a mechanism that requires a much smaller activation energy. This value might hence correspond to the activation energy for the glide of partial dislocations in 3C-SiC. Although this energy is not clearly determined in the literature, it is fair to think that the corresponding energy is smaller than for the double cross-slip motion which implies a significant dislocation core reconfiguration [4].

Conclusions

The 3C-6H polytypic transition in SiC has been studied by diffuse X-ray scattering and transmission electron microscopy. TEM revealed that the transformation does not affect the whole crystal simultaneously. Instead, there are regions with a high density of SFs co-existing with perfect 3C regions. The SF are bounded by partial dislocations which result from the dissociation of basal plane dislocations. DXS revealed that the 3C-6H polytypic transition in SiC actually involves two different mechanisms: the first one corresponds to the nucleation of new stacking faults through the double cross-slipping of partial dislocations, and the second one corresponds to the expansion of the partial dislocations. The former is a fast process and characterized by an activation energy of E_a = 1.5 eV, whereas the latter is a slow process and characterized by an activation energy of E_a = 0.54 eV.

Acknowledgements

The authors would like to warmly thank the Region Limousin (France) for its financial support. LMGP thank the MINTEX-French ANR program (contract No. ANR-09-BLAN-0189-01) and the MANSiC-Marie Curie Research and Training Network (contract No. MRTN-CT-2006-035735) for their financial support.

References

[1] N. W. Jepps, T. F. Page, Polytypic transformations in silicon carbide, Proc. Cryst. Growth Charact. 7 (1983) 259-307.

[2] D. Chaussende, F. Mercier, A. Boulle, F. Conchon, M. Soueidan, G. Ferro, A. Mantzari, A. Andreadou, E.K. Polychroniadis, C. Balloud, S. Juillaguet, J. Camassel, M. Pons, Prospects for 3C-SiC bulk crystal growth, J. Cryst. Growth 310 (2008) 976-981.

[3] A. Boulle, J. Aube, I. Galben-Sandulache, D. Chaussende, The 3C-6H polytypic transition in SiC as revealed by diffuse X-ray scattering, Appl. Phys. Lett. 94 (2009) 201904 1-3.

[4] P. Pirouz, Polytypic transformations in SiC, Sol. State Phenomena 56 (1997) 107-132.

[5] H. Nagasawa, K. Yagi, T. Kawahara, 3C-SiC hetero-epitaxial growth on undulant Si(0 0 1) substrate, J. Cryst. Growth 237-239 (2002) 1244-1249.

[6] A. Boulle, D. Dompoint, I. Galben-Sandulache, D. Chaussende, Quantitative analysis of diffuse X-ray scattering in partially transformed 3C-SiC single crystals, J. Appl. Cryst. 43 (2010) 867-875.

[7] D. Dompoint, A. Boulle, I. Galben-Sandulache, D. Chaussende, L. T. M. Hoa, T. Ouisse, D. Eyidi, J. -L. Demenet, M. -F. Beaufort, J. Rabier, Kinetics of the 3C-6H polytypic transition in 3C-SiC single crystals: a diffuse X-ray scattering study, J. Appl. Phys. 110 (2011) 053508.

[8] P. Käckell, J. Furthmüller, F. Bechstedt, Polytypic transformations in SiC: an ab initio study, Phys. Rev. B 60 (1999) 13261-13264.

Materials Science Forum Vols. 717-720 (2012) pp 497-500
© (2012) Trans Tech Publications, Switzerland
doi:10.4028/www.scientific.net/MSF.717-720.497

XRD Characterization for Al- and N-Doped 3C-SiC on Si (100) Substrate after Pulsed Excimer Laser Anneal

Kung-Yen Lee [1,a,*], Yan-Hao Huang [1,b], Chih-Fang Huang [2,c], Cheng Yueh Chung [1,d], Shun Chiu Lin [1,e], and Feng Zhao [3,f]

[1] Department of Engineering Science and Ocean Engineering, National Taiwan University, Taipei 10617, Taiwan, ROC

[2] Institute of Electronics Engineering, National Tsing Hua University, Hsinchu 30013, Taiwan, ROC

[3] School of Engineering and Computer Science, Washington State University, Vancouver, WA 98683, USA

[a] kylee@ntu.edu.tw, [b] r99525061@ntu.edu.tw, [c] cfhuang@ee.nthu.edu.tw, [d] r99525026@ntu.edu.tw, [e] sclin@ntumems.net, [f] feng.zhao@wsu.edu,

Keywords: XRD, 3C-SiC, Pulsed excimer laser anneal, AFM

Abstract. This paper characterizes Al-, N-doped and undoped 3C-SiC samples after pulsed excimer laser anneal by using X-ray diffraction (XRD) and atomic force microscopy (AFM). In order to protect surface morphology, the low energy density of 0.2444 J/cm^2 per shot was applied to Al- and N-doped samples. The results show damage caused by implantation can be recovered by the anneal, depending on the total energy applied to the surface. The peaks of the Bragg diffraction spectra of Al- and N-doped samples were barely moved and moved, respectively. As the energy density per shot was reduced to 0.0667 J/cm^2 and the total energy density of 40.0 J/cm^2 was applied to the undoped samples, the peaks of the Bragg diffraction spectra of the undoped samples were shifted.

Introduction

Silicon carbide (SiC) is a promising semiconductor material due to its material properties such as wide band gap, high thermal conductivity and high breakdown field. Among various polytypes of SiC, an advantage of cubic SiC (3C-SiC) is that 3C-SiC epilayer can be grown on Si substrates. However, the quality of 3C-SiC epilayer is influenced by the mistmatch in lattice parameter and thermal expansion coefficient between 3C-SiC and Si materials. In addition, incorporation of impurities into SiC materials requires the elevated temperature ion implantation and post-implant anneal to activate dopants and to remove damage. The post-implant anneal performed at extremely high temperature might degrade surface morphology, leading to poor performance of SiC power devices. The highest anneal temperature is also limited by the melting point of Si substrates. Therefore, the pulsed excimer laser can be used for a post-implant anneal at lower temperature for SiC because the high energy pulses can accumulate enough energy in a very short time into the surface region to activate dopants [1]. However, the surface of SiC may be burned if the total applied energy is too high. In this paper, a laser pulse with the lower energy density at a repetition rate was chosen to achieve the desired energy on SiC surface which can recovery damage caused by implantation and activate dopants without degrading surface morphology [1].

Experiments

Table 1 Energies and doses of Al and N ion implantation

Al		N	
Energy [KeV]	Dose [cm^{-2}]	Energy [KeV]	Dose [cm^{-2}]
20	4e14	30	7.8e14
60	6e15	120	1.3e15
120	1e13	240	1.3e15
180	1e13	360	1.3e15

A 5.5 um 3C-SiC epilayer grown on a Si (100) substrate was used for aluminum and nitrogen ion implantation at 650°C with multiple energies/doses which are listed in Table 1. The doping concentrations were designed for the p+ and N+ regions of DMOSFET. The highest Al doping concentration is about 1x10^{20} cm^{-3} simulated by Trim and uniformly distribute about 0.1 um.

The pulsed excimer laser manufactured by Coherent Inc. with an energy density of 0.2444 J/cm^2 per shot used for the post-implant anneal was then performed at a repetition rate of 10 Hz. Square areas of 9 mm^2 were exposed to 50, 100 and 200 shots of laser pulses at room temperature, as listed in Table 2. For N-doped 3C-SiC samples, the doping concentration is about 1x10^{20} cm^{-3} simulated by Trim and uniformly distribute about 0.5 um. The pulsed excimer laser with an energy density of 0.2444 J/cm^2 per shot was carried out at a repetition rate of 10 Hz as well. The energy densities on the 3C-SiC samples were 12.2 (50 shots) and 48.8 J/cm^2 (200 shots), as listed in Table 2. For comparison, the pulsed excimer laser energy densities of 0.0667 and 0.2444 J/cm^2 per shot were performed for the undoped 3C-SiC samples. The total energy densities applied to the undoped 3C-SiC samples, x, z, and y were 22.2, 40.0 and 44.4 J/cm^2, respectively, as listed in Table 3. An undoped 3C-SiC sample, w, without the pulsed excimer laser anneal was used for reference. Atomic force microscopy (AFM) manufactured by ITRI (OBJ-204C) and X-ray diffraction (XRD) manufactured by Philips (X'Pert PRO) operated at 45 kV and 40 mA were used to characterize surface morphology and SiC crystalline.

Table 2 Applied energy densities of the pulsed excimer laser anneals for Al- and N-doped 3C-SiC samples

Al-doped sample code	N-doped sample code	Total energy density [J/cm2]	Shot
b-1	nb-1	12.2	50
b-2		24.4	100
b-3	nb-3	48.8	200

Table 3 Applied energy densities of the pulsed excimer laser anneals for the undoped 3C-SiC samples

Undoped sample code	Total energy density [J/cm2]	Shot	Energy density per shot [J/cm2]
w	0	0	0
x	22.2	150	0.0667
		50	0.2444
y	44.4	300	0.0667
		100	0.2444
z	40.0	600	0.0667

Results and Discussion

Fig. 1 (a), (b), and (c) show the Bragg diffraction spectra of the Al-doped 3C-SiC epilayers on Si (100) substrates with the total energy densities of 12.2, 24.4 and 48.8 J/cm^2, respectively.

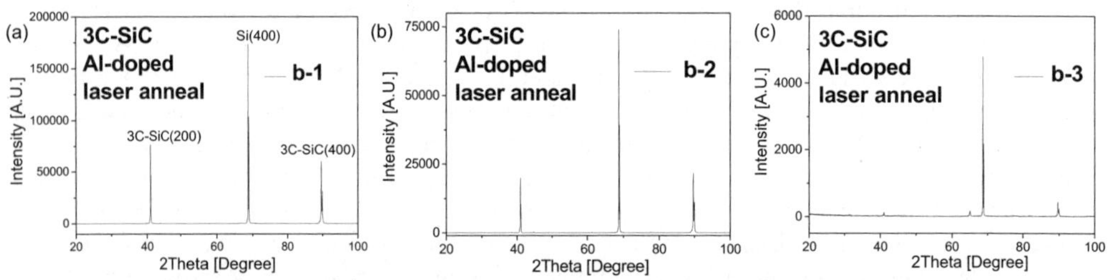

Fig. 1 Bragg diffraction spectra of the Al-doped 3C-SiC epilayers on Si (100) substrates with laser energy densities of (a) 12.2, (b) 24.4 and (c) 48.8 J/cm^2. Spectra of 3C-SiC (200), Si (400) and 3C-SiC (400) planes are at 2θ angles of ~41°, ~70° and ~89°, respectively.

The peaks with high intensities in Fig. 1 (a) and (b) are only at 2θ angles of ~41°, ~70° and ~89° which are corresponding to 3C-SiC (200), Si (400) and 3C-SiC (400) planes, respectively. There are no other peaks corresponding to different 3C-SiC or Si planes. It indicates the 3C-SiC epilayer is strong orientation aligned to Si [2]. Nevertheless, as the total energy density increases to 48.8 J/cm^2, the intensities of the peaks at 2θ angles of ~41° and ~89° are significantly reduced, particularly at the 2θ angle of ~41°, as shown in Fig 1 (c) [3].

Fig. 2 (a) and (b) show the magnifications of Fig. 1 (a), (b) and (c) at 2θ angles of ~41° and ~89°. It is obvious that the damage (the shoulders of the b-1 curves) caused by implantation still exist after annealing at 12.2 J/cm^2, and partially recovery at 24.4 J/cm^2 (the b-2 curves) [4]. However, the intensities of the sample with a laser energy density of 48.8 J/cm^2 (the b-3 curves) are very weak so that damage or recovery is hard to be determined.

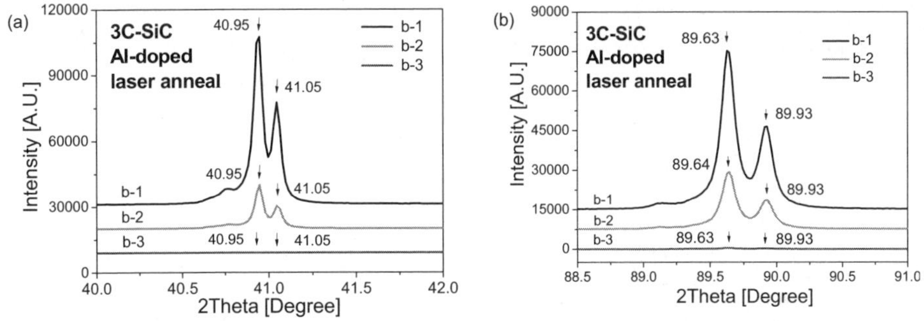

Fig. 2 The magnifications of Fig. 1 at 2 θ angles of (a) ~41° and (b) ~89°.

The Bragg diffraction spectra at 2 θ angles of ~41° and ~89° of the undoped 3C-SiC sample, w, without the pulsed excimer laser anneal are shown in Fig 3 (a) and (b), respectively. Comparing Fig. 2 with Fig. 3, there are no shoulders in Fig. 3 and the peaks are barely shifted. It means that only very limited implanted Al atoms might move onto the substitutional sites so that the dimension of lattice does not change [5]. The surface roughness RMS (Root-Mean-Square) of an undoped sample without the pulsed excimer laser anneal and Al-doped samples with the pulsed excimer laser anneal are 3 and 4 nm, respectively. It indicates that the pulsed excimer laser anneal can recovery damage caused by implantation without degrading surface morphology.

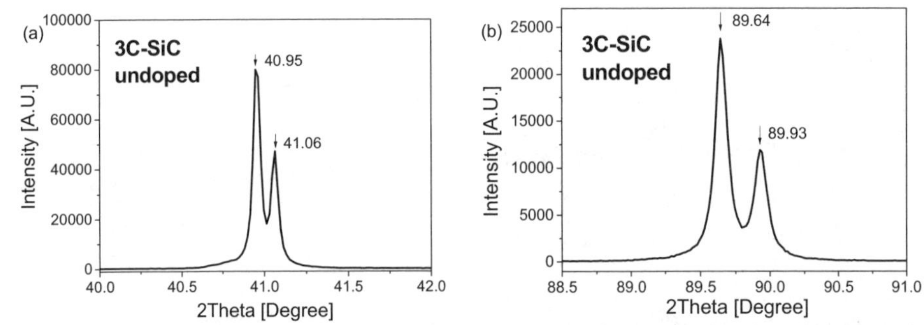

Fig. 3 Bragg diffraction spectra of the undoped 3C-SiC samples without pulsed excimer laser anneals at 2 θ angles of (a) ~41° and (b) ~89°.

In order to investigate the influence of the pulsed excimer laser anneal energies on Bragg diffraction spectra of the undoped 3C-SiC samples at 2 θ angles of ~41° and ~89°, three undoped 3C-SiC samples were annealed with the total energy densities of 22.2, 40.0 and 44.4 J/cm^2 which are listed in Table 3. Fig. 4 (a) and (b) show Bragg diffraction spectra of undoped 3C-SiC samples annealed with different energies at 2 θ angles of ~41° and ~89°. Comparing Fig. 3 with Fig. 4, there are no shoulders induced by the elevated temperature ion implantation and laser anneal at 2 θ angles of ~41° and ~89° in both figures. The peaks of the undpoed 3C-SiC samples with laser energy densities of 22.2 and 44.4 J/cm^2 almost do not shift. However, the laser energy density of 40.0 J/cm^2 (the top curve, z) causes the peaks dramatically to shift at both 2 θ angles of ~41° and ~89°. The difference in the anneal process between the samples z and y are the lower energy density per shot and the relatively longer annealing time (600 shots) on the sample z, which might cause the dimension change of the lattice [5].

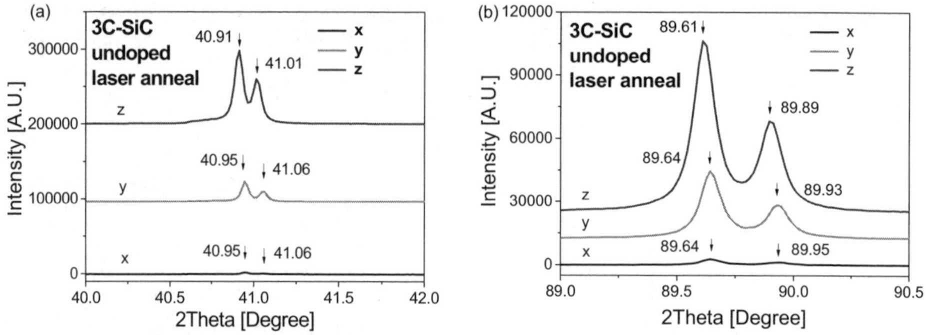

Fig. 4 Bragg diffraction spectra of the undoped 3C-SiC samples with laser energy densities of 22.2 (curve x), 40.0 (curve z) and 44.4 J/cm^2 (curve y) at 2 θ angles of ~41° and ~89°.

Fig. 5 (a) and (b) show Bragg diffraction spectra of N-doped 3C-SiC epilayer on Si (100) substrates with the total laser energy densities of 12.2 and 48.8 J/cm^2 at 2 θ angles of ~41° and ~89°. The intensities of the peaks at both 2 θ angles of ~41° and ~89° are very week, particularly at the 2 θ angle of ~41°. Comparing Fig. 3 with Fig. 6, the peaks dramatically shift and there are other peaks in Fig. 6 (b). The surface roughness RMS of both the undoped samples and N-doped samples after the pulsed excimer laser anneal are 4 nm and 5 nm, respectively. It indicates the laser anneal in this work did not degrade surface morphology.

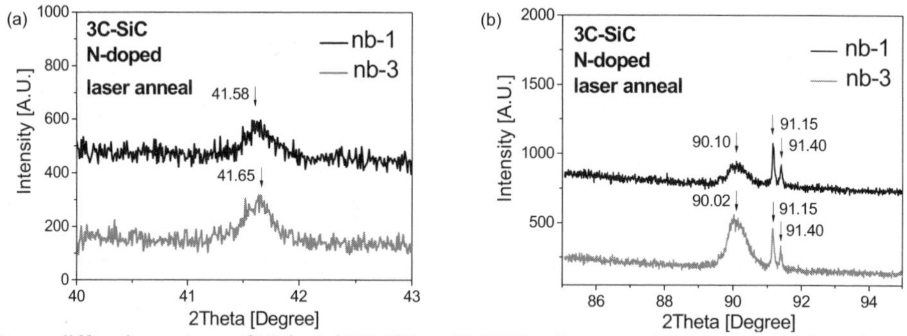

Fig. 5 Bragg diffraction spectra of N-doped 3C-SiC on Si (100) substrates with laser energy densities of 12.2 and 48.8 J/cm^2 at 2 θ angles of (a) ~41° and (b) ~89°

Summary

This paper demonstrates damage induced by the high temperature Al ion implantation can be recovered by the pulsed excimer laser anneal with a low energy density while the total energy applied to the surface is high enough. The peaks of Al-doped samples with the pulsed excimer laser anneal barely shift. However, the peaks of N-doped samples with the pulsed excimer laser anneal dramatically shift, though the intensities are very weak. As the energy density per shot is reduced to 0.0667 J/cm^2, having the relatively longer annealing time, the peaks of the undoped sample with the total energy density of 40.0 J/cm^2 also dramatically shift comparing with the other two undoped samples with different energy densities during the relatively shorter annealing time.

References

[1] C. Boutopoulos, et al., Appl. Surf. Sci., 253 (2007) 7912-7916
[2] R. Analone et al., J. Appl. Phys., 105 (2009) 084910
[3] Y. Dong and P. Molian, Appl. Phys. Lett. 84 (2004) 10-12
[4] D. J. Brink, H. Peyre, and J. Camassel, J. Appl. Phys., 105 (2009) 063527
[5] K. B. Mulpuri, et al., Solid-state Electronics 50 (2006) 1035-1040

Materials Science Forum Vols. 717-720 (2012) pp 501-504
© (2012) Trans Tech Publications, Switzerland
doi:10.4028/www.scientific.net/MSF.717-720.501

Origin of the Warpage of 3C-SiC Wafer: Effect of Nonuniform Intrinsic Stress

Yu Sun[1], Satoshi Izumi[1,*], Shinsuke Sakai[1], Kuniaki Yagi[2] and Hiroyuki Nagasawa[2]

[1]Department of Mechanical Engineering, School of Engineering, The University of Tokyo, 7-3-1 Hongo, Bunkyo-ku, Tokyo 113-8656, Japan

[2]SiC development center, HOYA Corporation, 1-17-16 Tanashioda, Chuo-ku, Sagamihara-shi, Kanagawa 252-0245 Japan

*izumi@fml.t.u-tokyo.ac.jp

Keywords: 3C-SiC, Epitaxy, Warpage, Bilayer formula, Intrinsic stress, Stacking fault

Abstract. Technique of bulk-like 3C-SiC film (up to 300 μm) growth on undulant-Si substrate is known to be very effective to reduce stacking fault density as well as that of other planar defects. However, freestanding 3C-SiC wafer shows anisotropic warpage involving large convex curvature in the direction perpendicular to the ridge of undulation ([110] direction), and slight concave curvature in parallel direction ([$\bar{1}$10] direction), i.e. saddle shape. In this paper the origin of the warpage of the 3C-SiC wafer is investigated. *Ex-situ* curvature measurements and stress calculation reveal that large compressive intrinsic stress is generated during high-temperature growth process (1623 K) in both parallel and perpendicular directions. In order to investigate the intrinsic stress distribution along the [001] direction, a reactive ion etching (RIE) is conducted for the 3C-SiC on Si substrate to observe the dependence of the SiC/Si system curvature as a function of 3C-SiC thickness. This observation shows that the intrinsic stress component perpendicular to the ridge of undulation presents nonuniform distribution in [001] direction. The remarkable change in the intrinsic stress is observed in the 50 μm-thick region from SiC/Si interface. A finite element method simulation using the obtained intrinsic stress distribution clearly explains that the anisotropic warpage of SiC wafer is induced by the intrinsic stress distribution in quantitative manner. Microstructure change induced by stacking fault reduction process (stacking fault collision) would be the cause of the intrinsic stress variation.

Introduction

Among the various SiC polytypes, the cubic SiC (3C-SiC) owns unique properties of high electron mobility and high-saturation drift velocity [1]. More significantly, 3C-SiC can be grown epitaxially on single-crystal Si, which reduces the fabrication cost of substrates with large area [2,3].

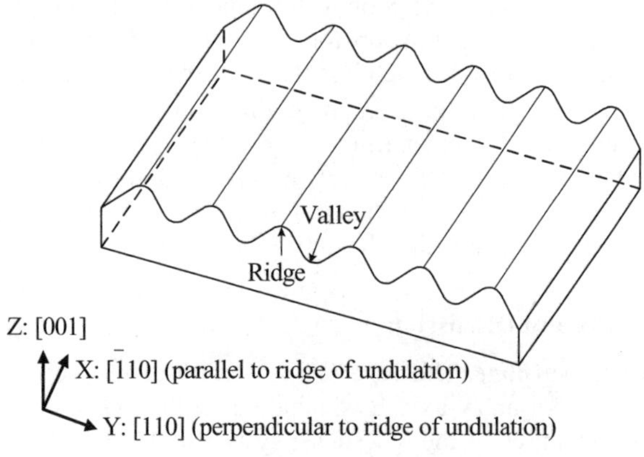

Figure 1: Schematic of the undulant-Si substrate.

However, mainly owning to the unallowable defect density in the epitaxial layers and large film stress (wafer warpage), which result from the large lattice mismatch (~20%) and difference of thermal expansion coefficient (~8%) between the Si substrate and 3C-SiC film, the extensive developments of 3C-SiC/Si have been somewhat prevented [4,5]. The wafer contains high density of defects is still applicable to the basic functions whereas the warpage could cause problems even in those cases. Recently, the technique for growing the thick 3C-SiC layer (up to 300 μm) on an undulant-Si substrate (shown as Fig. 1) was performed and effectively reduced densities of stacking fault and other planar defects [6]. Nevertheless, after removing the Si substrate, 3C-SiC freestanding layer shows anisotropic warpage involving large convex curvature in the [110] direction and slight concave curvature in the $[\bar{1}10]$ direction, i.e. saddle shape.

Basically, the wafer warpage is determined by the residual stress in the hetero-structure system. The residual stress is generally divided into two parts, a thermal stress induced by cooling process and an intrinsic stress induced by deposition process. The former in 3C-SiC/Si systems is always tensile (concave) since thermal expansion coefficient of SiC is larger than that of Si, while the latter, which would induce the SiC wafer warpage, is more complicated and its details are unknown.

Several researches have been conducted to elucidate the mechanism of the intrinsic stress generation for very thin SiC wafer [7,8,9,10]. As for thickness of 300 μm SiC wafer, there must be microstructure change and intrinsic stress distribution along the thickness direction. From that point of view, it is indispensable to discuss the warpage of thick SiC wafer.The present paper investigates the origin of significant anisotropic shape of 3C-SiC wafer referring the experimental curvature measurement and numerical stress calculation approaches.

Experimental

Vapor phase epitaxy (VPE) of 3C-SiC. The 3C-SiC layer was grown on the 725 μm–thick, 3 inches Si substrate by a vapor phase epitaxy (VPE) using cold-wall type low-pressure reactor. In order to reduce the anti-phase boundaries and the stacking faults, undulant-Si substrate was used, which was fabricated by scratching the surface of Si(001) in the $[\bar{1}10]$ direction with a 15 μm diameter diamond slurry with a load of 0.1 kg$\cdot$cm^{-2} to create continuous undulations, given by Nagasawa *et al.* in details [6]. The growth procedure of 3C-SiC in the [001] direction is described below. First, the system was heating from room temperature to 1623 K under a continuous flow of 10 sccm C_2H_2 and 100 sccm H_2 to carbonize the surface of the undulant-Si substrate. Then, while holding the temperature at 1623K, the 3C-SiC layer was grown on the undulant-Si by using 50 sccm SiH_2Cl_2 and 10 sccm C_2H_2 as precursors and 100 sccm H_2 as carrier gas. The system pressure during the carbonization and growth process was kept at 100 Pa. Under these conditions, the growth rate was approximately 40 μm$\cdot$h^{-1}. The 3C-SiC layer with thickness of 310 μm has been obtained.

Curvature measurement and residual stress calculation. The warpage of wafer was measured by interferometer using 633 nm laser. The sampling interval along in x- and y-direction was 2 mm. After the measurement, the Stoney formula is usually utilized to convert the measured curvature into residual stress in a film-substrate system. The Stoney formula is derived based on the assumption that the film thickness is much less than the substrate thickness, while in the present sample, the thickness of SiC layer (310 μm) is not negligible compare with thickness of Si substrate (725 μm) thus violates the assumption. To determine the residual stress in the present system, the bilayer formula is applied [11].In order to take account the plastic deformation effect of Si substrate, the curvature of the 3C-SiC/Si system corresponding to each thickness was calculated by excluding the curvature of Si substrate after polishing the entire SiC layer off.

Results and Discussion

Wafer warpage. The 3C-SiC/Si system presents a concave shape (negative) and similar curvatures along x-axis and y-axis. By excluding the thermal stress induced by cooling after the growth process, the curvatures of the 3C-SiC/Si system at 1623 K are estimated as shown in Fig. 2 The negative values show the curvatures under RT which include both of the intrinsic stress and thermal stress. By considering the thermal expansion coefficient difference of 3C-SiC and Si, we calculated the theoretical thermal stress and excluded it to obtain the intrinsic stress shown as the positive values in Fig. 2. It was found that compressive intrinsic stress was generated at the growth temperature (1623 K). After removing the Si substrate, the SiC layer appears to be a saddle-shape, i.e. slight concave (negative) along direction parallel to the ridge of undulation (x-axis in Fig. 1) while notable convex (positive) along perpendicular direction (y-axis in Fig. 1). The change in curvatures along the x- and y-axis as function of freestanding SiC thicknesses measured at room temperature is shown in Fig. 3. It indicates that there is a nonuniform distribution of intrinsic stress along the [001] direction. The *ex-situ* result was verified by *in-situ* warpage measurement system (kSA Mini-MOS) up to 1273 K.

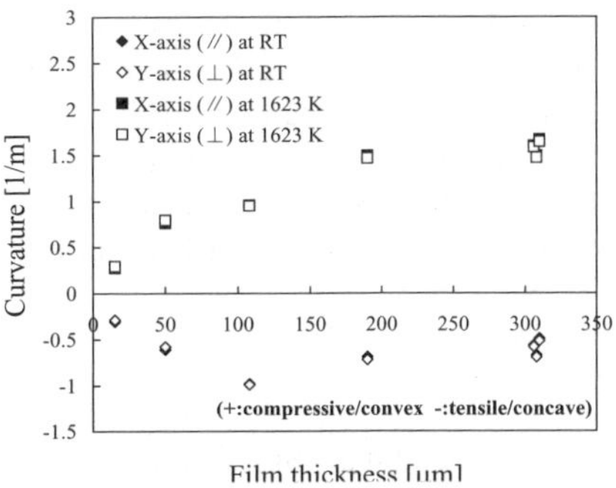

Figure 2: The curvatures of the 3C-SiC/Si system at room temperature and 1623 K.

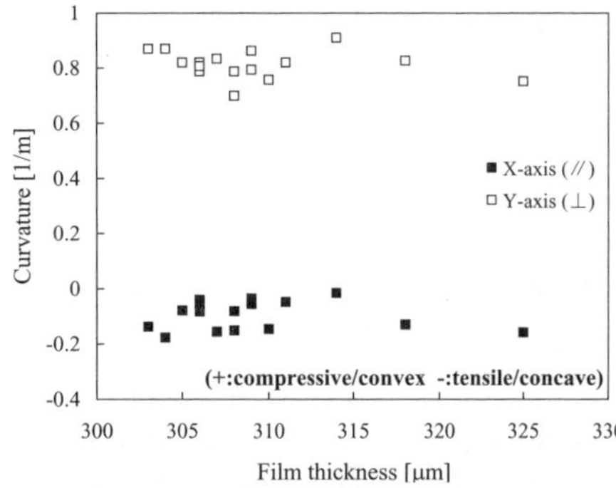

Figure 3: The curvatures of SiC wafer after removing Si substrate along the x-axis and y-axis at room temperature.

The convex shape has been observed under high temperature and the curvature was increasing with temperature. Therefore, although the *in-situ* measurement did not reach the growth temperature of 1623 K due to technology problems, we predicted the tendency and verified our *ex-situ* result.

Intrinsic stress distribution. In order to investigate the distribution of intrinsic stress, SiC layer on Si substrate was gradually etched by RIE. The variation in curvatures is shown in Fig. 4. Although the wafer appears to be a concave shape along both the x-axis and y-axis during repeating etching process, the difference is evident and the warpage along y-axis is larger than that of x-axis. After removing the total SiC layer, the Si substrate shows a convex shape because of the plastic deformation. With the purpose of knowing the intrinsic stress of SiC, the plastic effect of Si substrate should be taken into account.

By means of the bilayer formula, the corresponding residual stress in the wafer system during RIE process has been estimated. By excluding the thermal stress and the plastic deformation of Si substrate, the dependence of intrinsic stress on SiC

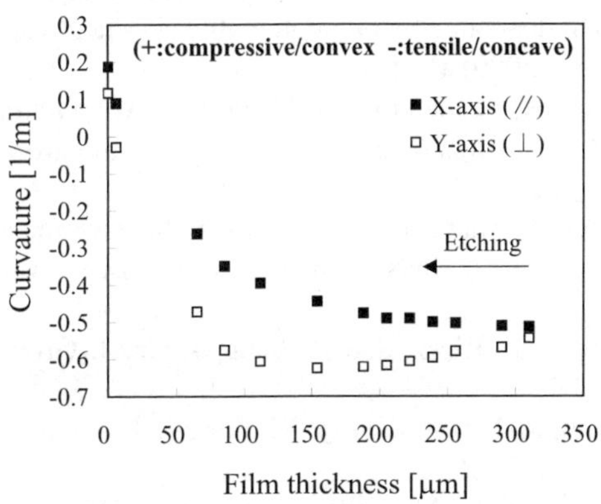

Figure 4: The variation of the curvatures with gradual SiC etching.

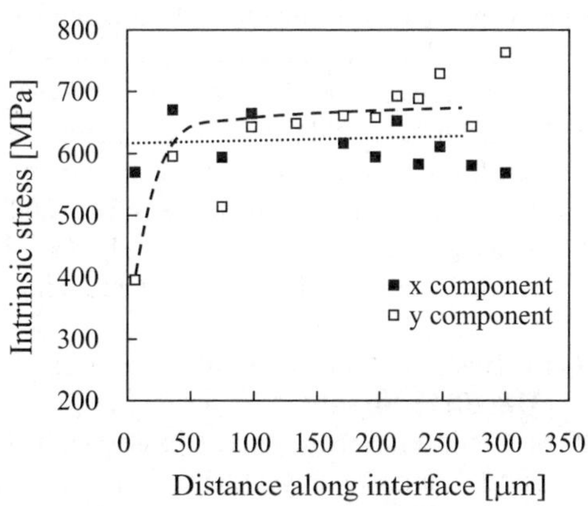

Figure 5: The distribution of intrinsic stress from interface along the SiC film thickness.

thickness has shown that both of x- and y-component of intrinsic stress are compressive and y-component (perpendicular to ridge of undulation) varies nonlinearly, especially below thickness of 50 μm.

Fig. 5 shows the change in intrinsic stress as function of SiC thickness. This shows nearly constant intrinsic stress in the x-component (parallel to ridge of undulation), while that of y-component (perpendicular to ridge of undulation) changes remarkably at thinner than 50 μm.

In order to know the effect of the nonuniform intrinsic stress distribution on SiC anisotropic warpage, a Finite Element Method (FEM) simulation was conducted. The general-purpose FEM software, ANSYS with an axial symmetric 2D-model using 8-node solid elements (PLANE183) was applied. Various fitting curves of intrinsic stress (y-axis) were performed because of large scattering as Fig. 5. The resulting curvature is around 0.8 m^{-1} to 1.5 m^{-1}, which is very close to experimental values (Fig. 3). As for x-axis, curvature is thought to be very small as compared with y-axis. We speculate that the high stacking fault density near the interface could induce the intrinsic stress distribution as well as warpage of SiC wafer. However, relationship between the intrinsic stress and microstructures such as stacking fault distribution and collision is not known. Further studies on microstructures are necessary.

Conclusions

The analysis on the intrinsic stress of 3C-SiC layer grown on the undulant-Si substrate has been performed to study the mechanism of 3C-SiC wafer warpage. RIE of SiC enabled us to obtain the dependence of SiC/Si system warpage on the remained SiC thickness, from which we have estimated the intrinsic stress distribution of SiC layer in the [001] direction. It was observed that the compressive intrinsic stress σ_x (parallel to ridge of undulation) kept almost constant while σ_y (perpendicular to ridge of undulation) showed nonuniform distribution and sharper increase near the interface particularly, which was speculated to be induced by the high density of stacking fault and the its collision structures within the corresponding part.

References

[1]S. Nishino, H. Suhara, H. Ono, and H. Matsunami, Epitaxial growth and electric characteristics of cubic SiC on silicon, J. Appl. Phys. **61** (2009) 4889–4893.

[2] M. Kitabatake, SiC/Si heteroepitaxial growth, Thin Solid Films **369** (2000) 257–264.

[3]M. Bosi, G. Attolini, B.E. Watts, F. Rossi, C. Ferrari, F. Riesz, and L. Jiang, Wafer curvature analysis in 3C-SiC layers grown on (001) and (111) Si substrates, J. Cryst. Growth **318** (2010) 401-405.

[4] S. Veprek, T. Kunstmann, D. Volm, and BK Meyer, Relaxation of interfacial stress and improved quality of heteroepitaxial 3C--SiC films on (100) Si deposited by organometallic chemical vapor deposition at 1200 °C, J. Vac. Sci. & Technol. A **15** (2009) 10–17.

[5] M. Zielinski, S. Ndiaye, T. Chassagne, S. Juillaguet, R. Lewandowska, M. Portail, A. Leycuras, and J. Camassel, Strain and wafer curvature of 3C-SiC films on silicon: influence of the growth conditions, Phys. Status Solidi A **204** (2007) 981–986.

[6] H. Nagasawa, M. Abe, K. Yagi, T. Kawahara, and N. Hatta, Fabrication of high performance 3C-SiC vertical MOSFETs by reducing planar defects, Phys. Status Solidi B **245** (2008) 1272–1280.

[7] M. Bosi, G. Attolini, B. Watts, F. Rossi, C. Ferrari, F. Riesz, and L. Jiang, J. Cryst. Growth **318**(2011) 401–405.

[8] M. Zielinski, S. Ndiaye, T. Chassagne, S. Juillaguet, R. Lewandowska, M. Portail, A. Leycuras, and J. Camassel, Phys. Status Solidi A 204(2007), 981–986.

[9] G. Ferro, T. Chassagne, A. Leycuras, F. Cauwet, and Y. Monteil, Chem. Vapor Deposition 12(2006), 483–488.

[10] D. Choi, R. Shinavski, W. Steffier, and S. Spearing, J. Appl. Phys. 97(2005), 074904.

[11] L. Freund, J. Floro, and E. Chason, Extensions of the Stoney formula for substrate curvature to configurations with thin substrates or large deformations, Appl. Phys. Lett. **74** (1999) 1987.

Materials Science Forum Vols. 717-720 (2012) pp 505-508
© (2012) Trans Tech Publications, Switzerland
doi:10.4028/www.scientific.net/MSF.717-720.505

Raman scattering and X-ray absorption from CVD grown 3C-SiC on Si

Zhe Chuan Feng[1,a,*], Cheng Chen[2,b], Qiang Xu[1,3,c], Suwan P. Mendis[4,d], Ling-Yun Jang[5,e], Chin-Che Tin[4,f], Kung-Yen Lee[6,g,#], Chee Wee Liu[1,7,h], Zhengyun Wu[3,i], and Zhi Ren Qiu[2,j,+]

[1]Institute of Photonics & Optoelectronics and Department of Electrical Engineering, National Taiwan University, Taipei, 106-17 Taiwan

[2]State Key Laboratory of Optoelectronic Materials and Technologies and School of Physics and Engineering, Sun Yat-Sen University, Guangzhou, 510275 China

[3]Department of Physics, Xiamen University, Xiamen 361005, China

[4]Department of Physics, Auburn University, Auburn, AL 36849, USA

[5]National Synchrotron Radiation Research Center, Hsinchu, 30076 Taiwan

[6]Department of Engineering Science and Ocean Engineering, National Taiwan University, Taipei, 106-17 Taiwan

[7]National Nano Device Laboratories, Science-based Industrial Park, Hsinchu, 30078 Taiwan

[a]fengzc@cc.ee.ntu.edu.tw, [b]ynu2008@163.com, [c]xuqiang-1322@163.com, [d]mendisp@auburn.edu, [e]lyjang@nsrrc.org.tw, [f]tinchin@auburn.edu, [g]kylee@ntu.edu.tw, [h]cheeweeliu@gmail.com, [i]zhywu@xmu.edu.cn, [j]stsqzr@mail.sysu.edu.cn

[*,#,+]corresponding authors

Keywords: 3C-SiC, Raman scattering, synchrotron radiation, X-ray absorption, bond length

Abstract. FTIR, Visible and UV Raman scattering, as well as synchrotron radiation X-ray absorption, in combination, have been employed to investigate a series of CVD grown 3C-SiC/Si (100). Significant results on the optical and atomic bonding properties are obtained from these comparative studies.

Introduction

Cubic silicon carbide (3C-SiC), with large energy gap (2.2 eV at room temperature), high electric breakdown field, high saturation drift velocity, moderately high electron mobility, chemical inertness and temperature stability, is a promising material for electronic and optical devices working under high temperature, high radiation flux and other severe environments [1,2]. 3C-SiC has a lattice constant of 4.359 Å at room temperature (RT), about 20% lower than that of Si, 5.430 Å at RT [2]. This large lattice difference and an 8% mismatch in thermal expansion coefficients with Si made it difficult to grow epitaxial 3C-SiC layers on Si wafers. Due to the pioneering work of employing the buffer technology, large size of 3C-SiC films on Si wafers has been achieved by way of chemical vapor deposition (CVD) [3], which has caused a great deal of effort and progress [4,5].

Although the current development and device applications of SiC are more focused on 6H- and 4H-SiC materials [4,5], the 3C-SiC/Si system is still very attractive due to the following reasons. (i) The high cost and difficulty in obtaining large area of 6H-/4H-SiC materials limit their wide commercial use. One way to overcome these limits is to use the 3C-SiC films on Si substrates. It is possible to obtain larger area SiC wafers on Si with low cost. (ii) 3C-SiC has been shown to provide an excellent microwave device with high (~400°C) temperature operation and having a high output power density of 67 kW/cm^2, which is about 30 times larger than the devices made from GaAs. (iii) There is a small difference of lattice constants and thermal expansion coefficients between SiC and GaN, and therefore, SiC can serve as a promising substrate for the epitaxial growth of cubic and hexagonal GaN, cubic InGaN and InGaN/GaN multiple quantum wells. (iv) 3C-SiC is possibly a convenient intermediate layer material for GaN epitaxy, for example, on silicon-on-insulator (SOI) structure. (v) It is also attractive to realize the combined devices of Si and wide bandgap SiC and GaN through the well-established Si-based integrated technology [2]. Therefore, great efforts have been continually made on the hetero-epitaxy and applications of 3C-SiC/Si materials and structures/devices in recent years [6-11]. In this paper, we have employed Raman scattering and synchrotron radiation (SR) extended X-ray absorption fine-structure (EXAFS), both being powerful non-destructive techniques, to perform a comparative study on a series of CVD grown 3C-SiC/Si.

Experimental

Materials Growth. The experimental samples involved in this paper were prepared by chemical vapor deposition (CVD). This CVD procedure produces an initial SiC layer from a carbonization, which was done by way of an initial temperature ramp on the Si substrate in the presence of a hydrocarbon [1]. This thin (~ several hundreds of Å) SiC mixed alloy layer served as a buffer for a subsequent CVD growth of SiC. After a certain thickness of CVD film growth, good and high quality single crystalline cubic SiC films on Si with size larger than 1" dimension were obtained. The 3C-SiC films were found to be single crystalline with the (100) surface [12]. Three 3C-SiC samples (C702, 713 and 716) were involved in this study. SiC films are naturally n-type with nitrogen (N) doped in the CVD process. The film thicknesses were calculated,from FTIR spectra measured as described in [2], being 3.06 (C702), 5.60 (C716) and 5.40 (C713) μm, respectively.

Characterization Techniques. Raman spectra were measured in the backscattering geometry by two J.Y. micro-Raman systems under excitation of 325 and 532 nm at room temperature (RT), respectively. Synchrotron radiation (SR) beam-line 16A at the National Synchrotron Radiation Research Center (NSRRC) in Hsinchu, Taiwan, was used. The absorption spectra were recorded by detecting the fluorescence from the sample as a function of incident photo energy. Here we focus on the extended X-ray absorption fine structure oscillations region from about tens of eV above the absorption edge to up to 2700-eV photon energy, including the absorption K edge of a Si atom (1839 eV). EXAFS can probe the physical and chemical structure of matter at the atomic scale, and provide information needed to simulate the nearest and second nearest bond lengths of the X-ray absorbing atom [13,14]. Raman parameters such as intensity, width, peak frequency and polarization provide fruitful information on the crystal quality and properties [15,16].

Results and Discussion

Raman Scattering. Figure 1 shows Raman spectra of three 3C-SiC/Si (100), under the 514.5 nm excitation from an Ar^+ laser. All spectra were normalized to the Raman mode from Si at ~520 cm^{-1} (not shown), which is dominant because of the large penetration depth of 514.5 nm light in SiC [2]. The 1st order Raman transverse optical (TO) and longitudinal optical (LO) phonon modes [2,12] from 3C-SiC at ~795 cm^{-1} and 970 cm^{-1}, respectively, are presented. The changes in peaks at TO and LO modes and width for different samples are observed for different film thicknesses. For the thinnest film, the TO mode is stronger than LO, and for two thicker films, LO is stronger than TO, which are similar to previous observation [2,12].

Comparative UV excited Raman spectra of these three 3C-SiC films are exhibited in Figure 2. Under the UV excitation of 325 nm from a HeCd laser with a shallow penetrating depth in SiC [2], the Raman mode from Si at ~520 cm^{-1} is weakly seen from the ~3 μm thick cubic SiC film, while almost not observed from the other two films thicker than 5 μm. Raman TO(Γ) and LO(Γ) modes [2,12] from 3C-SiC at ~795 cm^{-1} and 970 cm^{-1}, respectively, are observed as well.

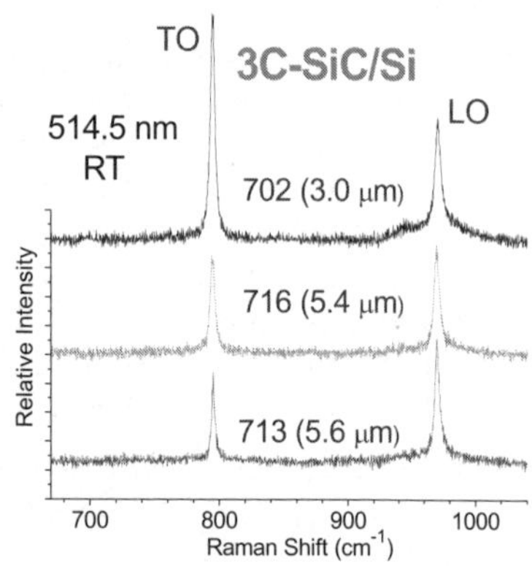

Fig. 1. Raman spectra of three 3C-SiC/Si (100) samples, under the 514.5 nm excitation, showing single crystalline 3C-SiC characteristic TO and LO modes at the Γ-point.

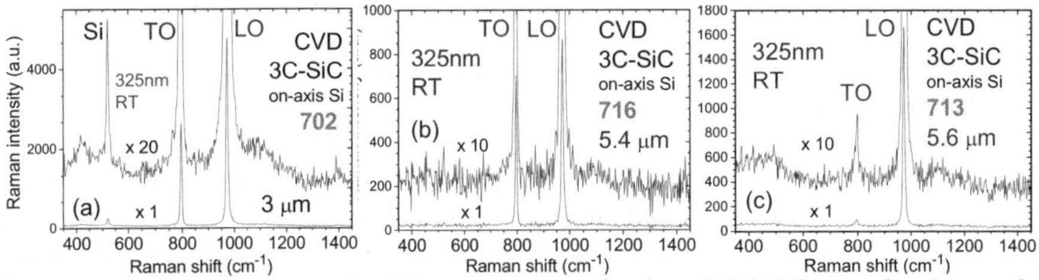

Fig. 2. UV excited Raman spectra, under 325 nm excitation, for three 3C-SiC/Si samples. Due to the shallow penetrating depth of UV 325 nm light, the Si Raman mode is not seen for 3C-SiC films thicker than 5 μm.

Extended X-ray Absorption Fine-Structure (EXAFS). The EXAFS signal was extracted from the raw x-ray absorption spectrum as shown in Figure 3(a). Transforming the spectra from energy scale to k-space scale with k^3-weighted multiplication is shown in Figure 3(b), because EXAFS depends on the wave-nature of the photoelectron. The fitting program IFEFFIT then transforms the raw data into Fourier transform spectra as shown in Figure 3(c). Peaks in Fourier transform spectra indicate reflective waves from the atoms around the (Si) atom which has absorbed an X-ray.

We can simulate the EXAFS data for the nearest and next nearest neighbors to the absorbing Si atom to obtain the nearest Si-C and next nearest Si-Si bond lengths, respectively [17,18 and references therein]. Figure 4 exhibits the simulation fitting results on Fourier transform spectra of three 3C-SiC/Si samples: (a) C702, (b) C716 and (c) C713. The fitted parameters are listed in Table 1.

The first peak of Figs. 4 (a)-(c) is associated with the reflected wave from the nearest atomic bonding of Si-C; the second peak is from the next nearest atomic bonding of Si-Si, according to the 3C-SiC structure. Fitting results show that bond lengths of Si-C for C702, C716 and C713 samples are 1.879, 1.883 and 1.883 Å; bond lengths of Si-Si for C702, C716 and C713 are 3.070, 3.075 and 3.074 Å, respectively. From Raman and EXAFS results, the change of Si-C bond length may be related to the dopants substituting for carbon atoms or stress-induced distortion in the SiC lattice; the small change in Si-Si bond length may be from the influence of doping or defects or distortion near the absorbing atom.

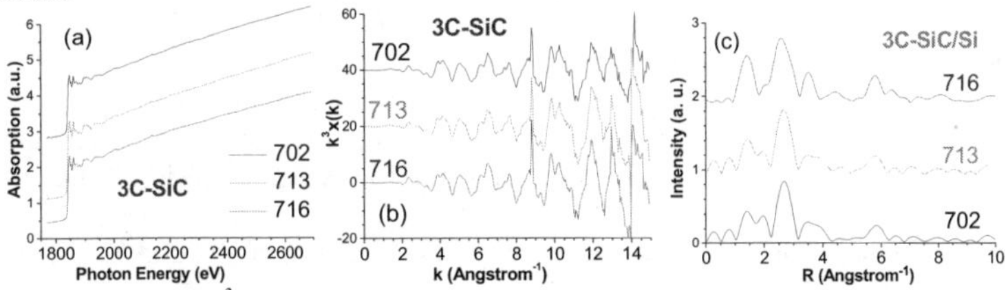

Fig. 3. (a) XAS raw spectra, (b) k^3-weighted $\chi(k)$ spectra transformed from (a) in energy scale to k-space scale with k^3-weigthed multiplication, and (c) Fourier Transform spectra by using program IFEFFIT, for three 3C-SiC/Si, 702 (3μm), 716 (5.4μm) and 713 (5.6μm).

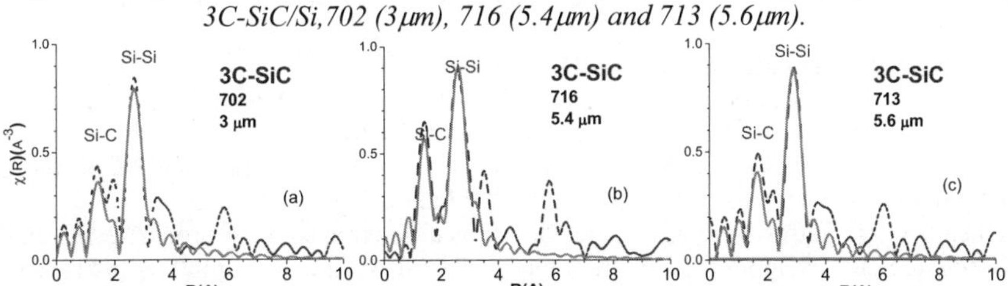

Fig. 4. Simulation fits on Fourier Transform spectra in Fig. 3 (c), for three 3C-SiC/Si samples.

Table 1. Simulation results on the Fourier transform Si K-edge EXAFS spectra {R(Si-C) is the bond length of Si-C, R2(Si-Si) is the bond length of Si-Si.}

Sample	Thickness (μm)	R(Si-C) (Å)	R2(Si-Si) (Å)
C702	3.0	1.87866	3.07046
C716	5.4	1.88270	3.07450
C713	5.6	1.88260	3.07440

Summary

We have performed a combined investigation of Raman scattering (both visible and UV) and synchrotron radiation extended X-ray absorption fine structure (EXAFS) on a series of CVD grown 3C-SiC thin films on Si (100) wafers. The Si-C and Si-Si bond lengths are obtained for experimental samples. Micro-visible and UV excitation Raman spectra were measured at room temperature. Further spectral line shape analyses from the spatial correlation modeling on the TO phonon and simulation on the LO phonon-plasmon coupling are under way, which can lead to the further comparison of the crystalline perfection corresponding to the EXAFS results, and the determination of free carrier concentrations for the 3C-SiC/Si samples. These results on the atomic bonding and Raman properties are useful to set up a base for further penetrating studies.

Acknowledgements

The work at National Taiwan University was supported by the National Science Council funds, NSC 98-2221-E-002-015-MY3 and NSC 98-3114-E-005-002-CC2. We would like to acknowledge an open-project fund support from the State Key Laboratory of Opto-Electronic Materials and Technologies (Sun Yatsen University), Guangzhou, China.

References

[1] J. A. Powell, P. Pirouz, and W. J. Choyke, in Z. C. Feng (Ed.), *Semiconductor Interfaces, Microstructures and Devices: Properties and Applications*, Institute of Physics Publishing, Bristol, 1993, pp. 257-293.

[2] Zhe Chuan Feng, in Z. C. Feng (Ed.), *SiC Power Materials – Devices and Applications*, Springer, Berlin, 2004, pp. 209-276.

[3] S. Nishino, J. A. Powell, and H. A. Will, Appl. Phys. Lett. **42** (1983) 460.

[4] Zhe Chuan Feng and Jian H. Zhao (Eds.), *Silicon Carbide: Materials, Processings and Devices*, Taylor & Francis Books Inc., New York, 2003.

[5] W. J. Choyke, H. Matsunami, and G. Pensl (Eds.), *Silicon Carbide: Recent Major Advances*, Springer, Berlin, 2004.

[6] V. Grivickas, G. Manolis, K. Gulbinas, K. Jarašiūnas and M. Kato, Appl. Phys. Lett. **95** (2009) 242110.

[7] F. Liu, C. Carraro, J. Chu, and R. Maboudian, J. Appl. Phys. **106** (2009) 013505.

[8] W.-Y. Chen, W.-L. Wang, J.-M. Liu, C.-C. Chen, J.-C. Hwang, C.-F. Huang, and L. Chang, J. Electrochem. Soc. **157** (2010) H377.

[9] R. Anzalone, M. Camarda, C. Locke, D. Alquier, A. Severino, M. Italia, D. Rodilosso, C. Tringali, A. La Magna, G. Foti, S. E. Saddow, F. La Via, and G. D'Arrigo, "Low Stress Heteroepitaxial 3C-SiC Films Characterized by Microstructure Fabrication and Finite Elements Analysis", J. Electrochem. Soc. **157** (2010) H438.

[10] G. Manolis, G. Zoulis, S. Juillaguet, J. Lorenzzi, G. Ferro, J. Camassel, and K. Jarašiūnas, Mater. Sci. Forum **645-648** (2010) 443.

[11] E. Saito, S. N. Filimonov and M. Suemitsu, "Growth Rate Anomaly in Ultralow-Pressure Chemical Vapor Deposition of 3C-SiC on Si(001) Using Monomethylsilane", Jap. J. Appl. Phys. **50** (2011) 010203.

[12] Z. C. Feng, Microelectronic Engineering **83** (2006) 165.

[13] M. Tallarida, D. Schmeisser, F. Zheng, and F. J. Himpsel, "X-ray absorption and photoemission spectroscopy of 3C- and 4H-SiC", Surface Science **600** (2006) 3879.

[14] F. Takano, W. Wang, H. Ofuchi, T. Ohshima, and H. Akinaga, "Structural and Magnetic Properties of Mn-Doped SiC", AIP Conference Proceedings **1199** (2010) 443.

[15] Z. C. Feng, W. J. Choyke and J. A. Powell, "Raman determination of layer stresses and strains for heterostructures and its application to the cubic 3C-SiC/Si system", J. Appl. Phys. **64** (1988) 6827.

[16] S. Rohmfeld, M. Hundhausen, L. Ley, C. A. Zorman, and M. Mehregany, "Quantitative evaluation of biaxial strain in epitaxial 3*C*-SiC layers on Si (100) substrates by Raman spectroscopy," J. Appl. Phys. **91** (2002) 1113.

[17] Y. L. Tu, Z. C. Feng, L. Y. Jang, and W. Lu, "Synchrotron Radiation X-ray Absorption Fine-structure and Raman Scattering Studies on 6H-SiC Materials", Proc. SPIE Vol. **7425** (2009) 74250Y.

[18] Y. L. Wu, Z. C. Feng, J. F. Lee, W. Tong, B. K. Wagner and I. Ferguson, "X-ray absorption and Raman study of GaN films grown on different substrates by different techniques", Thin Solid Films **518** (2010) 7775.

Materials Science Forum Vols. 717-720 (2012) pp 509-512
© (2012) Trans Tech Publications, Switzerland
doi:10.4028/www.scientific.net/MSF.717-720.509

4H-SiC wafers studied by X-ray absorption and Raman scattering

Qiang Xu[1,2,a], Hua Yang Sun[3,b], Cheng Chen[3,c], Ling-Yun Jang[4,d], RUSLI[5,e], Suwan P. Mendis[6,f], Chin Che Tin[6,g], Zhi Ren Qiu[3,h,*], Zhengyun Wu[1,I,+], Chee Wee Liu[2,7,j], and Zhe Chuan Feng[2,k,#]

[1]Department of Physics, Xiamen University, Xiamen, 361005 China
[2]Institute of Photonics & Optoelectronics, Department of Electrical Engineering, and Center for Emerging Material & Advanced Devices, National Taiwan University, Taipei, 106-17 Taiwan
[3]State Key Laboratory of Optoelectronic Materials and Technologies and School of Physics and Engineering, Sun Yat-Sen University, Guangzhou, 510275 China
[4]National Synchrotron Radiation Research Center, Hsinchu, 30076 Taiwan
[5]School of Electrical, Electronic Engineering, Nanyang Technological University, 639798 Singapore
[6]Department of Physics, Auburn University, Auburn, AL 36849, USA
[7]National Nano Device Laboratories, Science-based Industrial Park, Hsinchu, 30078 Taiwan

[a]xuqiang-1322@163.com, [b]sunhygwjl@163.com, [c]ynu2008@163.com, [d]lyjang@nsrrc.org.tw, [e]erusli@ntu.edu.sg, [f]mendisp@auburn.edu, [g]tinchin@auburn.edu, [h]stsqzr@mail.sysu.edu.cn, [i]zhywu@xmu.edu.cn, [j]cheeweeliu@gmail.com, [k]fengzc@cc.ee.ntu.edu.tw [*,+,#]corresponding author

Keywords: 4H-SiC, synchrotron radiation, X-ray absorption, Raman scattering, photoluminescence

Abstract. Synchrotron radiation X-ray absorption and UV 325 nm excitation Raman scattering-photoluminescence (PL) have been employed to investigate a series of 4H-SiC wafers, including bulk, epitaxial single or multiple layer structures grown by chemical vapor deposition. Significant results on the atomic bonding and PL-Raman properties are obtained from these comparative studies.

Introduction

Silicon carbide (SiC), a wide band gap semiconductor with an indirect band structure, is an attractive material for high-temperature and high-voltage power applications and optical sensors in the ultraviolet region [1-3]. There are a large number of polytypes for the SiC crystals. 4H- and 6H-SiC are of considerable technological interest due to their large band gaps and high electron mobility as well as commercial availability [1-3]. Among various SiC polytypes, 4H-SiC has the higher electron mobility and weaker anisotropy in electric transport properties, and therefore the hottest research focus in recent years [4-6]. In this paper, synchrotron radiation X-ray absorption spectroscopy (XAS), in particular, the extended X-ray absorption fine structure (EXAFS), and UV excitation Raman scattering-photoluminescence (PL) were employed to study a series of 4H-SiC samples.

Experimental

Experimental samples were prepared under different conditions, with different structure and doping concentration, and on different substrates. Some are bulk samples, some are thin films with different doping concentration, and some have multiple layer structures. Table 1 lists the sample information (from the grower) involved in this study.

Table 1. Sample Information of 4H-SiC

	Structure (on 4H-SiC substrate)	Resistivity [Ω-cm]	Doping concentration [x 10^{18} cm^{-3}]
4H01	N^- bare	0.021	6.4
4H05	$N^-/N/N^+$	0.016	10.4 / 11
4H07	P^-/P^+	11.3	0.0068
4H12	$N^+/N^-/P^-$/Si-doped	0.01	10.8 / 1.256 / 11.5
4HZ02	N-type	>0.05	1.45

Synchrotron Radiation X-ray Absorption Spectroscopy (XAS) Results and Analyses

Figure 1 shows XAS raw data from the Si K-edge for five 4H-SiC samples, taken at the National Synchrotron Radiation Research Center, Hsinchu, Taiwan, beamline 16A1. These absorption spectra were recorded by detecting the fluorescence from the sample, which is proportional to the X-ray absorption, as a function of incident photon energy in the 1,770-2,550 eV range. Here we focus on the extended X-ray absorption fine structure (EXAFS) oscillations region, from about tens of eV above the absorption edge to up to 2,550 eV photon energy. Data processing and analysis was performed, including the transform of the spectra from energy to k-space with weighted multiplication, subtracting the non-oscillatory background, Fourier-transform of the weighted oscillations in R space to obtain the approximate radial position of the coordination shells surrounding the Si absorbing atoms, sorting out and back transforming in the k-space the peak corresponding to a given coordination shell, and curve fitting of these characteristic oscillations [7].

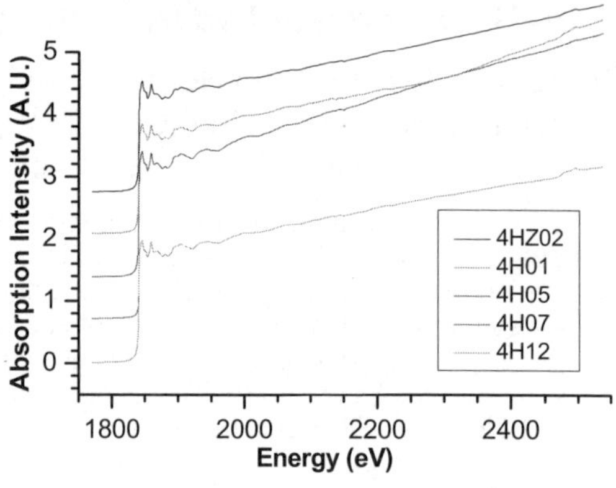

Figure 1. XAS raw data from the Si K-edge for five experimental 4H-SiC samples.

Figure 2 presents Fourier transformation spectra and fitting results for three 4H-SiC samples studied. From the fitting results, the shell information could be classified. The first peak, near 1.9 Ångstroms (Å), tells us the position of the reflective wave from the nearest atoms, i.e., Si-C bonding here. The second peak at ~3 Å indicates the second reflective wave from next nearest atoms, i.e., Si-Si bonding. Table 2 presents the fitting results of three 4H-SiC samples 4H05, 4H012 and 4HZ02. The Si-C bond length, R(Si-C), of 1.899 Å, 1.899 Å and 1.905 Å, and the Si-Si bond length, R(Si-Si), of 3.089 Å, 3.091 Å and 3.097 Å, are obtained for these three samples, respectively.

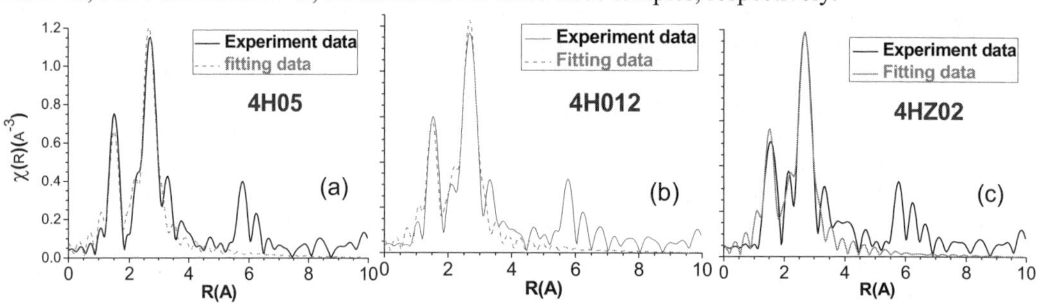

Figure 2. Fourier transformation spectra and fitting results of three 4H-SiC samples studied.

Table 2. EXAFS fitting results of 4H-SiC

Sample	R(Si-C) [Å]	R(Si-Si) [Å]
4H05	1.8968	3.0886
4H012	1.8988	3.0906
4HZ02	1.9047	3.0965

UV Excitation Raman-photoluminescence Results and Eiscussion

Figure 3 shows combined micro-photoluminescence (PL)-Raman spectra of two 4H-SiC samples, measured at room temperature (RT) and under UV excitation of 325 nm from a HeCd laser. PL peaks at ~3.17 eV and the Raman mode from the transverse optical (TO) phonon below but near 800 cm^{-1} are observed from these two samples (and also the other three, not shown). The Raman longitudinal

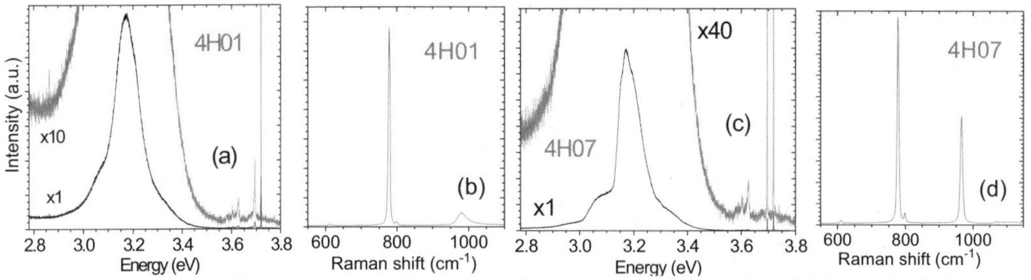

Figure 3. UV (325 nm) excited micro-PL-Raman spectra from two 4H-SiC samples.

optical (LO) phonon mode near 960 cm^{-1} is seen from the sample 4H07 with the top lower p-type doping layer. But for the n-type bare substrate 4H01 with higher doping level of 6.4 x 10^{18} cm^{-3}, a weak but broad band appears between 950-1040 cm^{-1}, indicating the LO phonon-plasma coupling.

Raman spectral line shape analysis can be applied to the TO phonon mode, from which useful information on the material properties is obtained. In an ideal semiconductor, the crystal lattice translation symmetry leads to plane wave phonon eigenstates. Due to the energy and momentum conservation, only **q**=0 phonons at the center of the Brillouin zone (Γ point) participate in the first-order Raman scattering process. However, disorder or finite-size effects may partially or completely relax the momentum conservation, leading to a downshift and broadening of the Raman peak. From a quantitative model, the spatial correlation model, the first-order Raman spectrum I(ω) can be expressed as [8,9]

$$I(\omega) \propto \int_0^1 \exp(-q^2L^2/4)\frac{d^3q}{[\omega-\omega(q)]^2+(\Gamma_0/2)^2}, \quad (1)$$

where q is expressed in units of 2π/a, a is the lattice constant, L is the correlation length in units of a, and Γ_0 is the natural or intrinsic line width. The dispersion relation ω(q) for optical phonons can be represented by two analytical forms: $\omega^2(q) = A + \{A^2\text{-}B[1\text{-}\cos(\pi q)]\}^{1/2}$, or $\omega(q) = A - Bq^2$, with A and B as adjustable parameters. Consequently, the calculated Raman spectrum, including the line width Γ(FWHM), can be obtained and fitted to the experimental data. Figure 4 (a) shows a typical line shape fit for the E_2(TO) mode from 4H01. Fitted parameters from this and other 4H-SiC samples are listed in Table 2.

Raman spectral line shape analysis can also lead to an optical determination of an electrical property of SiC, free carrier concentration, i.e., the doping level. When doped with impurities in SiC, the lattice vibration will be affected by free carriers through their interactions, leading to coupling of the LO-phonon and plasma (LOPC). Three mechanisms are responsible for this coupling: deformation potential, electro-optic and density-fluctuation mechanisms. In wide bandgap semiconductors such as SiC, GaN, GaP, etc., the deformation potential and the electro-optic mechanisms are dominant compared with the density fluctuation mechanism. A set of formulas was developed and applied to determine the free-carrier concentration in these wide bandgap semiconductors [8 and refs. therein]. The simulation process has been done for our experimental

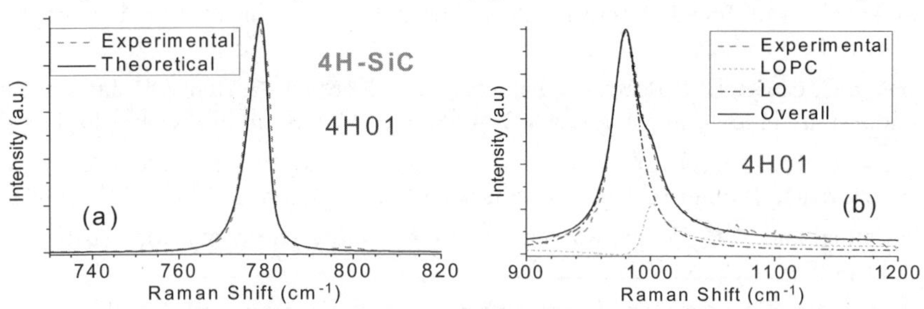

Figure 4. Theoretical fits on 4H-SiC #4H01: (a) spatial correlation model fits on E_2(TO) mode and (b) two mode fits on LO and LOPC phonon modes, respectively.

Table 3. Fitted parameters on TO and LOPC modes for samples 4H01, 4H05, 4H07, 4H12, 4HZ02

Sample no.	A [cm^{-1}]	B [cm^{-1}]	L [Å]	Γ_0 [cm^{-1}]	ω_P [cm^{-1}]	n [$x10^{18}$ cm^{-3}]
4H01	780.6	515	33	2.3	305	3.38
4H05	778.7	515	45	4.0	725	19.1
4H07	778.3	515	55	4.5	735	19.7
4H12	779.2	515	43	3.9	730	19.4
4HZ02	779.2	515	43	3.9	730	19.4

$\omega_p^{\;2} = 4\pi e^2 n / m^* \varepsilon$ (m^* is the electron effective mass) $m^*=0.48m_0$, $m_0=9.11x10^{-31}kg$

4H-SiC {details not shown due to the page limit}. Fig. 4 (b) presents the theoretical two mode fits on LO and LOPC phonon modes of 4H-SiC #4H01. The obtained values of plasma frequency, ω_p, and carrier concentration, n, for five experimental samples are given in the last two columns of Table 3.

Summary

We have performed a combined investigation of synchrotron radiation extended X-ray absorption fine structure (EXAFS) and UV 325 nm excitation Raman scattering-photoluminescence (PL) on a series of 4H-SiC wafers, including bulk, epitaxial single or multiple layer structures by chemical vapor deposition. The C-C and Si-C bond lengths and RT UV-PL emission band from the indirect energy gap are obtained for these 4H-SiC samples. Micro-UV Raman measurements and spectral line shape analyses from the spatial correlation modeling on the TO phonon and simulation on the LO-plasma phonon coupling led to the further comparison of the crystalline perfection corresponding to EXAFS, and the determination of free carrier concentrations for the 4H-SiC samples.

Acknowledgements

The work at National Taiwan University was supported by the National Science Council funds of NSC 98-2221-E-002-015-MY3 and NSC 98-3114-E-005-002-CC2, and by NTU Excellent Research Project (10R80908). We also acknowledge an open-project fund support from State Key laboratory of Opto-Electronic Material and Technologies (Sun Yatsen University), Guangzhou, China.

References

[1] Zhe Chuan Feng and Jian H. Zhao (Eds.), Silicon Carbide: Materials, Processing and Devices, Taylor & Francis Books Inc., New York, 2003.

[2] W. J. Choyke, H. Matsunami, G. Pensl (Eds.), Silicon Carbide: Recent Major Advances, Springer, Berlin, 2004.

[3] Zhe Chuan Feng (Ed.), SiC Power Materials–Devices and Applications, Springer, Berlin, 2004.

[4] T. Egilsson, I.G. Ivanov, N.T. Son, A. Henry, J.P. Bergman, and E. Janzén, Exciton and defect photoluminescence from SiC, in Zhe Chuan Feng and Jian H. Zhao (Eds.), in Silicon Carbide: Materials, Processings and Devices, Taylor & Francis Books, Inc., New York, 2003, pp. 81-120.

[5] Suk Chung, Ronen A. Berechman, Martha R. McCartney, and Marek Skowronski, Electronic structure analysis of threading screw dislocations in 4H–SiC using electron holography, J. Appl. Phys. 109 (2011) 034906.

[6] S. Leone, F.C. Beyer, H. Pedersen, S. Andersson, O. Kordina, A. Henry, E. Janzén, Chlorinated precursors study in low-temperature CVD of 4H-SiC, Thin Solid Films 519 (2011) 3074.

[7] Y. L. Tu, Z. C. Feng, L. Y. Jang, and W. Lu, Synchrotron Radiation X-ray Absorption Fine-structure and Raman Scattering Studies on 6H-SiC Materials, SPIE 7425 (2009) 74250Y.

[8] P. Parayanthal and F. H. Pollak, Raman scattering in alloy semiconductors: spatial correlation' model, Phys. Rev. Lett. 52 (1984) 1822.

[9] Z. C. Feng, Micro-Raman scattering and micro-photoluminescence of GaN thin films grown on sapphire by metalorganic chemical vapor deposition, Optical Engineering 41 (2002) 2022.

Materials Science Forum Vols. 717-720 (2012) pp 513-516

doi:10.4028/www.scientific.net/MSF.717-720.513

Mechanical Properties of Cubic SiC, GaN and AlN Thin Films

Jörg Pezoldt[1,a], Rolf Grieseler[2,b], Thorsten Schupp[3,c], Donat J. As[3,d] and Peter Schaaf[2,e]

[1] FG Nanotechnologie, Institut für Mikro- und Nanotechnologien MacroNano®, Technische Universität Ilmenau, Postfach 100565, 98684 Ilmenau, Germany

[2] FG Werkstoffe der Elektrotechnik, Institut für Mikro- und Nanotechnologien MacroNano®, Technische Universität Ilmenau, Postfach 100565, 98684 Ilmenau, Germany

[3]Department Physik, Universität Paderborn, Warburgerstraße 100, 33098 Paderborn, Germany

[a]joerg.pezoldt@tu-ilmenau.de, [b]rolf.grieseler@tu-ilmenau.de, [c]schupp@post.com, [d]d.as@uni-paderborn.de, [e]peter.schaaf@tu-ilmenau.de

Keywords: heteroepitaxy, mechanical properties, indentation modulus, nanoindentation, cubic III-Nitrides.

Abstract. Cubic polytypes of SiC, GaN and AlN were grown on silicon by molecular beam epitaxy. The mechanical properties of the epitaxial layers were investigated by nanoindentation. For 3C-SiC grown on Si(111) and Si(100) an extremal behaviour of the indentation modulus in dependence on the surface preparation with germanium prior to the carbonization was obtained. The maximum values of the indentation modulus for both Si orientations were achieved at 1 ML Ge.

Introduction

The knowledge and control of mechanical properties and residual stress are essential for micro- and nanoelectromechanical systems (MEMS and NEMS). MEMS and NEMS cover a wide application range with varying designs from cantilevers/bridges up to membranes. Due to their physical properties, SiC and group III-Nitrides, especially AlN, are preferable materials for MEMS and NEMS applications [1, 2]. Compared to non-piezoelectric materials, the piezoelectricity adds an additional energy dissipation path which might affect the vibrational properties and the quality factor of the MEMS device [3]. From this point of view, cubic modifications of these materials, as non piezoelectric materials, offer the advantage avoiding pronounced piezoelectric effects. The mechanical and dissipative properties are affected by the epitaxial layer quality, their dislocation densities and the residual stress. These quantities in turn depend on the substrate used and the growth technology. The mechanical properties of 3C-SiC are mainly studied on thick epitaxial layers, but advanced applications [4] require thin films exhibiting different mechanical properties [2]. In contrast to 3C-SiC, the growth of cubic III-nitrides on Si substrates is challenging and experimental studies of the mechanical properties of cubic III-N materials are rare [5] so that predominantly theoretical investigations can be found.

We report on the mechanical properties of cubic AlN and GaN grown by plasma assisted molecular beam epitaxial growth on Si(100) substrates, as well as on 3C-SiC grown on Ge modified Si(100) and Si(111).

Experimental

The III-nitride epitaxial growth was carried out on 3C-SiC/Si(100) templates. They were fabricated using UHV-CVD [6]. Reflection high energy electron diffraction (RHEED) of the (0,0) streak was used to control the growth of 3C-GaN(100) and 3C-AlN(100) on 3C-SiC(100)/Si(100) templates [7, 8]. The growth was realized under metal-rich conditions. The thickness of the 3C-AlN was between 100 and 200 nm, whereas the thicknesses of the 3C-GaN layers were 600 and 1000 nm.

The 3C-SiC(100) and 3C-SiC(111) heteroepitaxial layers were grown by solid source molecular beam epitaxy using a specific interface design technique based on the incorporation of Ge into the 3C-SiC/Si interface and into the near interface region [9]. The Ge precoverage was varied between 0 and 2 monolayers (ML) with respect to the silicon surface. The growth process was monitored by *in situ* RHEED and UV-Vis spectroscopic ellipsometry. The thickness of the epitaxial layers was between 100 and 200 nm.

Nanoindentation measurements were carried out to evaluate the mechanical properties of the cubic SiC, AlN and GaN epitaxial layers grown on silicon substrates. The experiments were performed using a Fischerscope® HM2000 equipped with a Vickers pyramid. The maximum load used in the nanoindentation experiments was 50 mN.

Results and Discussion

Measurement technique. In nanoindentation measurements the penetration depth should be equal to or less then one-tenth of the film thickness if the intrinsic properties of a thin film on a substrate are to be measured. For larger depths an effective composite modulus will be acquired, because with increasing penetration depth the indenter deforms not only the film, but also the substrate. It is therefore the value of the composite modulus which is obtained for larger penetration depth. This composite effect increases with increasing load and penetration depth. In addition to the standard load and unload measurement an alternative method can be used to test the mechanical properties of a composite material, here a thin epitaxial layer grown on a single crystalline substrate. The Enhanced Stiffness Procedure (ESP) uses a series of load and partial unload steps [10]. The measurement is carried out in a series of measurement cycles consisting of three steps. In the first step the load is increased to a certain value (pure load step). Subsequently, the load is reduced to a fraction of the maximum load (partial unload step). Afterwards, the load is increased to the previous maximum load and further increased to the next load value (pure load step). Within the measurement procedure each single entire load and partial unload step has a constant time step. Fig. 1 displays the dependence of the indentation depth on the load acquired during an ESP measurement cycle including all loading and unloading steps.

The obtained depth on load dependence was used to determine the Martens hardness and the composite modulus depth dependence of the heteroepitaxial layer – substrate system. An example is shown in Fig. 2. The indentation modulus was obtained for every reversal point of the indentation depth - load curve shown in Fig. 1, after the elastic modulus correction of the tip rounding [10]. The Martens hardness depth dependence was determined using the Martens hardness correction [10].

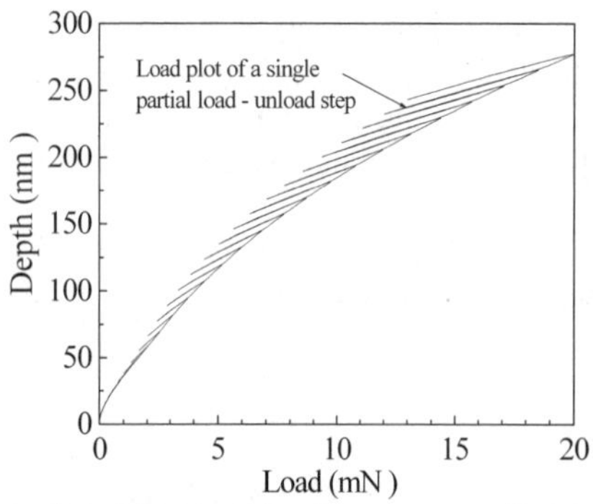

Fig.1 Indentation depth versus applied load of the ESP measurement procedure (163 nm 3C-SiC(100) on Si(100)).

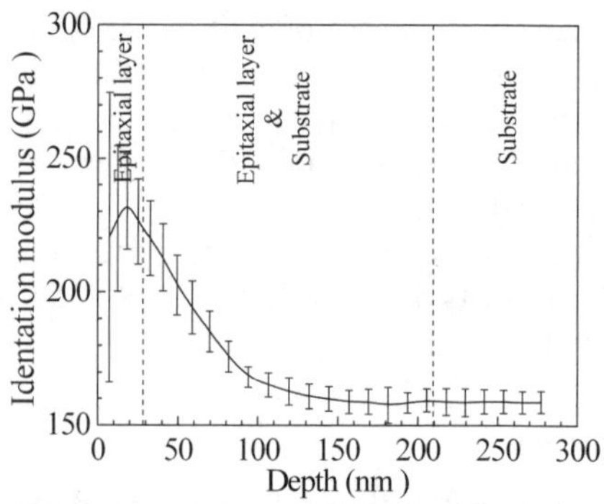

Fig.2 Depth dependence of the indentation modulus determined from indentation depth load curve (163 nm 3C-SiC(100) on Si(100)).

The composite modulus depth dependence shows three distinct regions. At a small indentation depth, below 20 % of the overall layer thickness, the indentation modulus is determined by the epitaxial layer. For intermediate depths a decrease of the indentation modulus from the layer to the

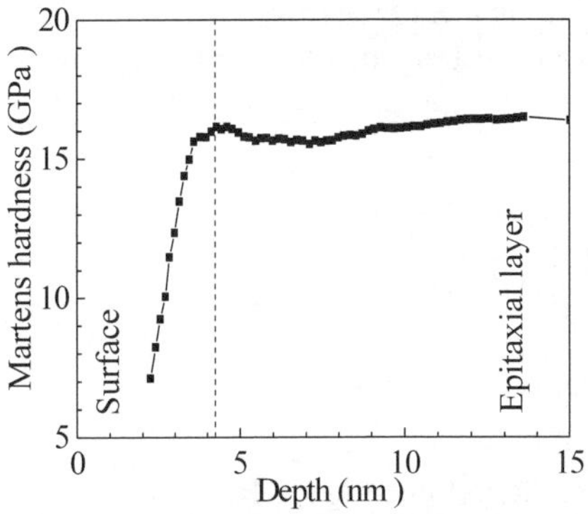

Fig. 3 Martens hardness versus depth for a 3C-SiC(111) layer grown on Si(111) for the depth region determined bv initial ESP measurements.

substrate values can be noticed. Thus, the epitaxial layer and the substrate contribute to the measured indentation modulus. This region extends into the substrate. In this case the indentation modulus or the hardness of the epitaxial layer can be determined using appropriate models for the composite hardness or modulus [11]. Investigations are underway to verify an appropriate model. At larger depth the elastic modulus is determined only by the properties of the substrate.

The ESP measurement procedure was used to identify the depth where the influence of the substrate material on the mechanical properties can be neglected. An example of the nanoindentation test is shown in Fig. 3, where the dependence of the Martens hardness on the indentation depth is given. It is evident that at depths exceeding 5 nm constant Martens hardness was obtained.

Measurement results. The results of the indentation tests for the 3C-SiC(100) on Si(100) and 3C-SiC(111) on Si(111) samples are shown in Figs. 4 and 5. In both cases the indentation modulus increases with increasing Ge coverage below 1ML Ge. At higher Ge coverages the indentation modulus decreases. The observed behaviour correlates with the dependencies of the quality factors and resonance frequencies of doubly clamped resonators fabricated from 3C-SiC heteroepitaxial layers grown on Ge modified Si(100) and Si(111) substrates [12, 13]. An increased indentation modulus can be caused by hardening due to grain boundaries and increased dislocation densities or by an improved material quality. In a previous study it was shown that the full width of the half maximum of the 3C-SiC(111) diffraction peak decreases and the terrace width increases if 3C-SiC is grown on Ge pretreated wafers [9]. Furthermore, the residual stress is reduced at Ge precoverages where the indentation modulus shows increased values [14-16]. Thus, the increased indentation modulus might not be caused by dislocation hardening, but could be related to an improvement of the crystalline structure of the 3C-SiC heteroepitaxial layers.

The results of the nanoindentation tests of 3C-GaN and 3C-AlN are summarized in Table 1. The values reported in Table 1 are averaged values, where at least five nanoindentation measurements were performed. The obtained values are in good agreement with the values obtained in [13] determined using doubly clamped resonator beams. The 3C-GaN(100) elastic modulus was found to be in the in the range between 235 and 250 MPa [13].

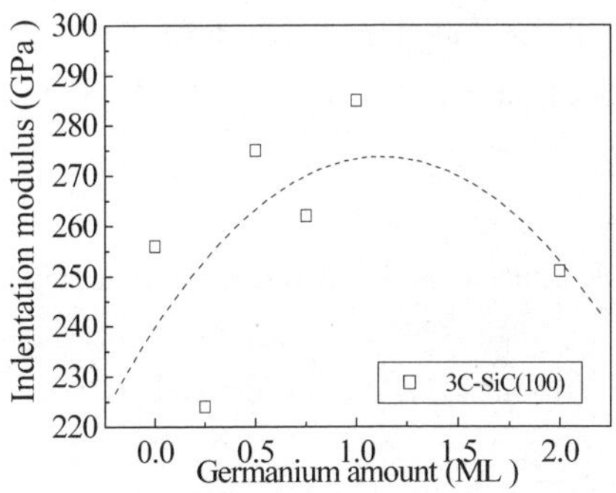

Fig. 4 Indentation modulus versus germanium precoverage prior to 3C-SiC(100) on Si(100) epitaxy. Dotted line represents the least square fit of the data.

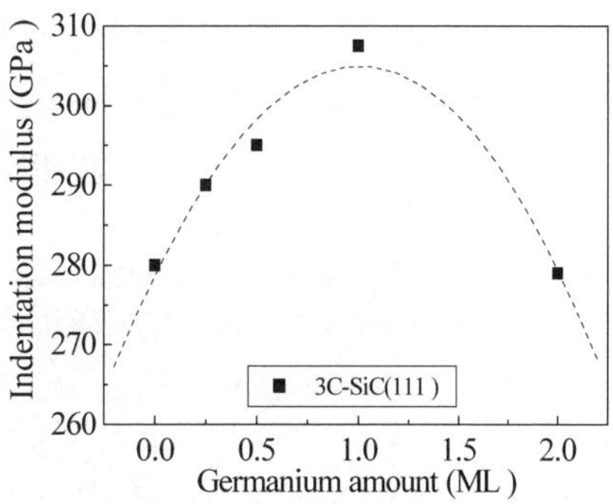

Fig. 5 Indentation modulus versus germanium precoverage prior to 3C-SiC(111) on Si(111) epitaxy. Dotted line represents the least square fit of the data

Table 1 Measured 3C-AlN and 3C-GaN mechanical properties

Material type	Substrate	Thickness [nm]	Hardness [GPa]	Indentation Modlus [GPa]	Compliance [µm/N]
3C-GaN(100)	3C-SiC(100)/Si(100)	600	11.3	220	48.3
3C-GaN(100)	3C-SiC(100)/Si(100)	1000	9.8	262	3.7
3C-AlN(100)	3C-SiC(100)/Si(100)	100	4.2	129	-
3C-AlN(100)	3C-SiC(100)	220	12.8	241	41.4
3C-AlN(111)	3C-SiC(111)/Si(111)	180	12.4	220	46.6

Summary

The mechanical properties of cubic SiC, AlN and GaN were determined using ESP. The obtained results correlate with data extracted from vibration frequencies of MEMS structures.

Acknowledgements. The work was supported by TMBWK contract B714-09065 and by the DFG graduate program GRK 1464 “Micro- and Nanostructures in Optoelectronics and Photonics”.

References

[1] M. Mehregany, C.A. Zorman, N. Rajan and C.H. Wu: Proc. IEEE Vol. 86 (1998), p. 1594

[2] V. Cimalla, J. Pezoldt and O. Ambacher: J. Phys. D Vol. 40 (2007), p. 6386

[3] K. Ishii and S. Tashiro: Jpn. J. Appl. Phys. Vol. 49 (2010), p. 09MD01

[4] X.L. Fang, M.H. Matheny, C.A. Zorman, M. Mehregany and M.L Roukes: Nano Lett. Vol. 10 (2010), p. 2891

[5] F. Niebelschütz, K. Brueckner, W. Jatal, E. Tschumak, D.J. As, M.A. Hein and J. Pezoldt: Phys. Status Solidi C Vol. 7 (2010), p. 116

[6] Ch. Förster, V. Cimalla, O. Ambacher and J. Pezoldt: Mater. Sci. Forum Vols. 483-485 (2005), p. 201

[7] T. Schupp, K. Lischka and D.J. As: J. Cryst. Growth Vol. 312 (2010), p. 1500

[8] J. Schörmann. S. Potthast, D.J. As and K. Lischka: Appl. Phys. Lett. Vol. 90 (2007), p. 041918

[9] J. Pezoldt, P. Masri, M.R. Laridjani, M. Averous, T. Wöhner, J.A. Schaefer, T. Stauden, G. Ecke, R. Pieterwas and L. Spiess: Mater. Sci. Forum Vols. 338-342 (2000), p. 289

[10] G. Bosch, http://www.helmut-fischer.com/globalfiles/DE_ESP_appreport_EN.pdf

[11] J. Mečnik, D. Minz, E.R. Weppelmann and M.V. Swain: J. Mater. Res. Vol. 12 (1997), p. 2475

[12] F. Niebelschütz, K. Brueckner, V. Cimalla, M.A. Hein and J. Pezoldt: Mater. Sci. Forum Vols. 615-617 (2009), p. 621

[13] F. Niebelschütz, W. Zhao, K. Brückner, K. Tonisch, M. Linß, M.A. Hein and J. Pezoldt: Mater. Sci. Forum Vols. 645-648 (2010), p. 861

[14] Ch. Zgheib, L.E. McNeil, M. Kazan, P. Masri, F.M. Morales, O. Ambacher and J. Pezoldt: Appl. Phys. Lett. Vol. 87 (2005), p. 041905

[15] Ch. Zgheib, L.E. McNeil, P. Masri, Ch. Förster, F.M. Morales, Th. Stauden, O. Ambacher and J. Pezoldt: Appl. Phys. Lett. Vol. 88 (2006), p. 211909

[16] J. Pezoldt, Th. Stauden, F. Niebelschütz, M.A. Alsioufy, R. Nader and P. Masri: Mater. Sci. Forum: Vols. 645-648 (2010), p. 159

Materials Science Forum Vols. 717-720 (2012) pp 517-520

doi:10.4028/www.scientific.net/MSF.717-720.517

Extended characterization of the stress fields in the heteroepitaxial growth of 3C-SiC on silicon for sensors and device applications

Massimo Camarda, Ruggero Anzalone, Nicolò Piluso, Andrea Severino, Andrea Canino, Francesco La Via and Antonino La Magna

Istituto per la Microelettronica e Microsistemi IMM-CNR, sezione di Catania, Stradale Primosole 50, 95121, Catania, Italy

massimo.camarda@imm.cnr.it

Keywords: stress, MEMS, finite element simulation

Abstract. Using several characterization techniques (μ-Raman, mechanical profilometer and microstructure deflections) together with a recent stress model [1] we study the heteroepitaxial growth of cubic silicon carbide on silicon (100). We find that the observed inconsistency between experimental results might be the result of defects generated on the silicon substrate during the carbonization process. In such a situation wafer curvature techniques do not allow the determination of the stress field in the grown films.

Introduction

Silicon Carbide (SiC) is recognized as an excellent material for high-power, high-frequency and high-temperature electronics due its outstanding electrical and thermal properties. Recently, due to its exceptional electrical, mechanical and chemical properties as compared with silicon (Si), SiC is also receiving added attention for its potential application in micro- and nano-electromechanical systems (MEMS and NEMS) [2,3]. In this respect, the heteroepitaxial growth of 3C-SiC on Silicon has the further advantage of a reduced production cost compared to the use of the hexagonal polytype. Unfortunately, due to high residual stresses which normally arise during the growth process, the use of hetero-epitaxial 3C-SiC/Si MEMS has been hindered. Indeed, in 3C-SiC/Si thin film growth, the defects-related stress leads to a bowing of the wafer and to the deflection of the microstructures, with important implications with regard to processing and device behaviour [4,5,6,7]. Stress measurement techniques are therefore an essential tool for both process development and process monitoring.

For these reasons we have analyzed the heteroepitaxial chemical vapor deposition (CVD) growth of single-crystal cubic silicon carbide (3C-SiC) on silicon (100) through optical profilometer, confocal μ-Raman, and microstructure deflections and compared the observed measurements with a recently developed stress theory that can incorporate a defect-related strain in the silicon substrate [1]. Using this theory we show that a possible explanation for the observed inconsistency between curvature techniques and the other analysis can be attributed to the presence of a defective region on the silicon substrate, generated during the carbonization step, which induces a compressive strain in the 3C-SiC/Si(100) system and results in an anomalous (downward) curvature of the wafer.

Experimental Details

A 3C-SiC single crystal film was hetero-epitaxially grown on a (100) nominally *on-axis* 2' silicon wafer. The growth was carried out in a home-made horizontal hot wall inductively heated low pressure CVD reactor by applying a conventional 2 step process (carbonization followed by standard growth; details of the growth process are presented in ref. [6]). Usually, the sample is rotating during the growth to reduce thickness variation and increase process uniformity. For the needs of this study, i.e., to analyze the stress evolution as function of the thickness, we have instead grown the epilayer without holder rotation. This results in a variation of the film thickness along the gas direction due to a depletion of the precursors. The thickness of the grown film ranges from ~2μm to ~3.2 μm, measured through Fourier Transform InfraRed spectroscopy (FTIR).

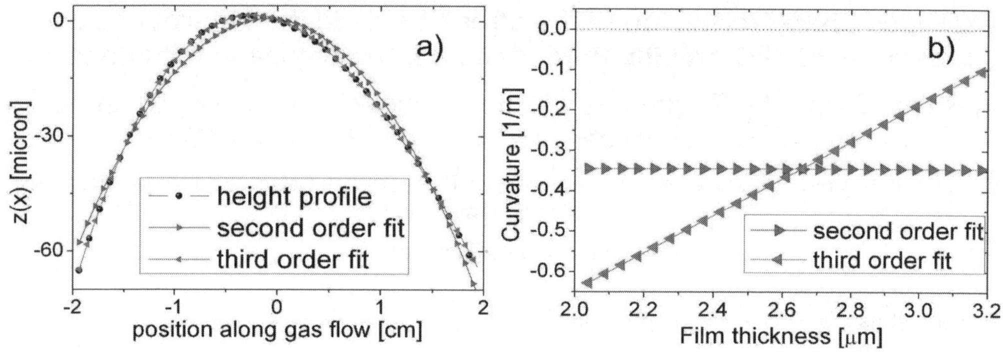

Fig.1 a) Height profile of a single wafer along the gas flow obtained through mechanical profilometer. b) Extracted curvatures as function of the film thickness assuming either a second order or a third order polynomial fit of the height profile in a)

Results and Discussion

Epiwafer Characterization. We begin by considering the planarized height profile along the gas flow obtained using a mechanical profilometer and subtracting the measured final profile from that of the initial silicon substrate (to avoid any bowing generated by the non-planar substrate). We fit this profile, *z(x)*, using either a second order or a third order polynomial function (Fig.1a). As can be seen the second order function does not correctly fit the experimental data. This is due to the local change in the system curvature induced by the thickness gradient. Using the fitted parameters we can extract the curvature (K_{macro}) of the system (being $K_{macro} = d^2z(x)/dx^2$) as a function of the thickness (Fig.1b). The curvature obtained through the third order fit shows a peculiar trend, with the absolute value of the curvature, $|K_{macro}(h_{film})|$, that decreases for increasing thicknesses (Fig.1b). This is in contrast with the "standard" behaviour for thin films (see e.g. refs. [8,9,10]) where $|K_{macro}(h_{film})|$ increases for increasing film thickness due to the increase of the stress associated with the film/substrate interface. The observed anomalous trend suggests the possibility that the stress generated by the film/substrate interface, which increases with the film thickness, is compensating another stress source, which is dominating the wafer curvature in the limit of thin films (in the considered case, this secondary stress, is dominating the curvature, i.e., it is imposing a downward curvature, for film thicknesses lower than ~3.5μm). This same behaviour, i.e., the decrease of $|K_{macro}(h_{film})|$ for increasing film thicknesses, has been observed also in other SiC/Si(100) samples grown in the same reactor.

To better elucidate the stress field profile in the system we have fabricated a set of microstructures (mainly planar rotators (PR) [11]) and analyzed the deflections of the free-standing microstructures as function of the thickness (Fig.2). The in-plane and out-of-plane deflection of the PR is related to the residual stress in the film (specifically to the uniform and gradient components of the residual stress function [1,7]). As can be seen in Fig.2a, the out-of-plane deflection quickly decreases for increasing thicknesses, this is, in turn, related to a variation of the averaged gradient stress along the film thickness, i.e., the stress function along the SiC thin film is non-linear [1].

Another important result is the sign of the in-plane rotation, which, being positive, indicates a state of tensile stress in the film [11]. This result is in contrast with the observed downward curvature of the wafer. The state of tensile stress has also been confirmed by Raman [1]. There is therefore an evident inconsistency between the downward curvature of the wafer, which is generally associated with compressive film stress [8,9,10], the evolution of the curvature as function of the film thickness on one side and the all showing a film under a state of tensile stress.

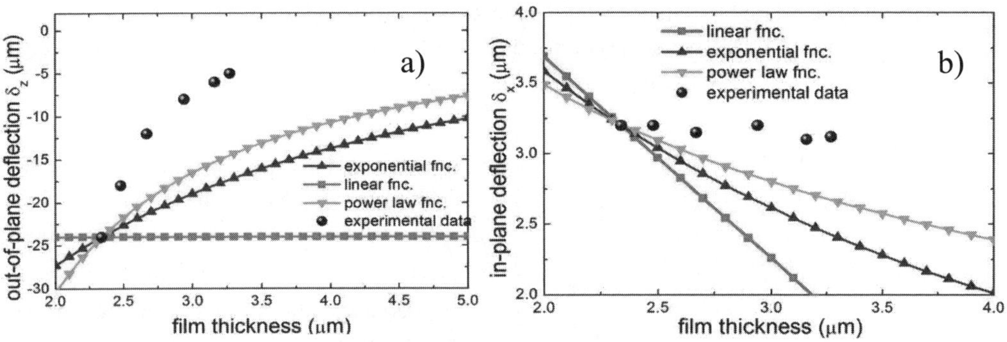

Fig2. Comparison between simulated and experimental out-of-plane (a) and in-plane (b) deflection for different possible residual stress functions. As can be seen, the linear approximation, the standard approximation used, does not correctly describe the deflections as a function of the film thickness. Calculation parameters are: Esi=200Gpa, E_{SiC}=420Gpa, υ_{Si}=0.18 υ_{SiC}=0.27, h_{Si}=525μm.

Theoretical framework. With the aim of solving these inconsistencies we analyze these results considering a recently developed stress theory that correlates global (wafer-scale) measurements to those obtained by means of local indicators (Raman and microstructure). This theory is based on the minimization of the total elastic energy with respect to all the allowed macro and micro displacements (for details of the model see ref. [1]). We begin by using this theory to investigate the best functional form for $\varepsilon_{def}(z)$, comparing the predicted in-plane and out-of-plane deflections of the model with the measured ones (Fig.2). As can be seen, the linear approximation, the standard functional form used for $\varepsilon_{def}(z)$, does not correctly fit the experimental data. The best functional form found is the power law function: $\varepsilon_{def}(z) = Az^B$ where A and B are fitting parameters. Furthermore, independently of the functional form used, the curvature of the wafer and the in-plane rotation of the PRs have been found to be tightly connected, i.e., they always indicate a consistent state of stress in the status, either compressive or tensile. This means that none of the functional forms considered allow for a decoupling of the macro/micro measurements and for an explanation of the experimental results.

To solve this inconsistency we need to critically reconsider the experimental results with respect to the probed regions of the SiC/Si system. Planar rotators, XRD and Raman all probe the stress/strain of the film *only*, whereas the wafer curvature technique is related to the *full* system (film and substrate). If the defective stress (i.e., the stress that generates the deformations of the system) is located only in the film, that is, if the substrate represents only an elastic constraint to the system, than all techniques will be consistent with each other, otherwise the wafer curvature measurement will be an indicator of a defects located in the substrate. To validate this idea we have theoretically calculated the macro and micro deflections assuming the presence of a region with a compressive stress in the substrate close to the hetero interface. As can be seen in Fig.3, this new defective stress profile generates a decoupling of the wafer curvature from the film stress status and, furthermore, describes the correct qualitative behaviour of the curvature as a function of the thickness (a decrease of $|K_{macro}(h_{film})|$ for increasing film thicknesses).

Conclusions

We have found a good agreement between the theoretical and experimental values of the wafer curvature and PR deflections as a function of the film thickness. We found that a possible explanation of the inconsistency between wafer curvature measurements, PR in-plane deflections and Raman could be related to defects generated in the silicon during the carbonization process. The origin of this strong compressive stress/strain, of the order of $\varepsilon_{def}(-2\mu m < z < 0) = -0.01$, located in the silicon near the SiC/Si interface, is currently under investigation.

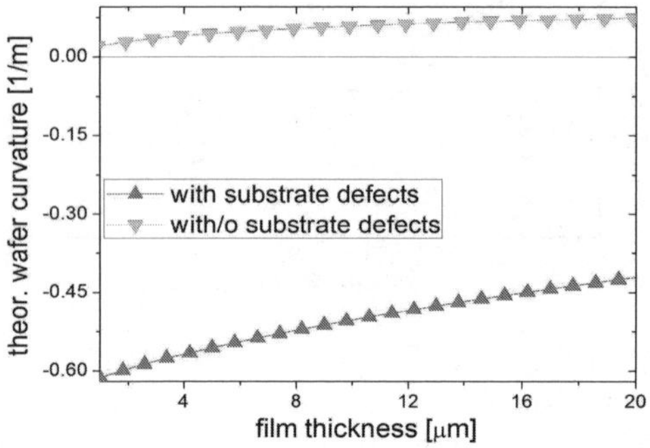

Fig3. Wafer curvature as function of film thickness considering defective $\varepsilon_{def}(z < 0) \neq 0$ and non-defective ($\varepsilon_{def}(z < 0) = 0$) substrates. In both cases considered the stress function in the film is a power law one.

References

[1] M. Camarda, R. Anzalone, A. Severino, N. Piluso, A. Canino, F. La Via and A. La Magna, arXiv:1110.4727v1

[2] X. L. Feng, M. H. Matheny, C. A. Zorman, M. Mehregany, M. L. Roukes, Nano Letters 10 2891 (2010)

[3] X. Fu , J. L. Dunning, C. A. Zorman and M. Mehregany, Sensors and Actuators A: Physical 119 (2005) 169

[4] A. Severino, C. Bongiorno, N. Piluso, M. Italia, M. Camarda, M. Mauceri, G. Condorelli, M. A. Di Stefano, B. Cafra, A. La Magna and F. La Via, Thin Solid Films 518 (2010) S165

[5] R. Anzalone, M. Camarda, C. Locke, D. Alquier, A. Severino, M. Italia, D. Rodilosso, C. Tringali, A. La Magna, G. Foti, S. E. Saddow, F. La Via, and G. D'Arrigo J. Electrochem. Soc. 157 (2010) H438-H442

[6] R. Anzalone, G. D'Arrigo, M. Camarda, C. Locke, S. E. Saddow, and F. La Via, Journal of Micromechanical Systems 20 (2011) 745

[7] R. Anzalone, M. Camarda, A. Canino, N. Piluso, F. La Via and G. D'Arrigo, Electrochemical and Solid State Letters 14 (2011) H161

[8] G. G. Stoney. Proc. R. Soc. Lond. A 172 (1909) 82

[9] C. A. Klein, Journal of Applied Physics 88 (2000) 5487

[10] L. B. Freund, J. A. Floro and E. Chason, Appl. Phys. Lett. 74 (1999) 1987

[11] B. P. van Drieënhuizen, J. F. L. Goosen, P. J. French and R. F. Wolffenbuttel, Sensors and Actuators A 37 (1993) 756

Materials Science Forum Vols. 717-720 (2012) pp 521-524

doi:10.4028/www.scientific.net/MSF.717-720.521

Stress evaluation on hetero-epitaxial 3C-SiC film on (100) Si substrates

R. Anzalone[1,*], M. Camarda[1], C. Locke[2], J. Carballo[2], N. Piluso[1], G. D'Arrigo[1], A. Severino[1], A.A. Volinsky[3], S.E. Saddow[2,4] and F. La Via[1]

[1]IMM-CNR, sezione di Catania, Stradale Primosole 50, 95121, Catania, Italy

[2]Dept. of Electrical Eng., Univ. of South Florida, Tampa, FL 33620, USA

[3]Dept. of Mechanical Eng., Univ. of South Florida Tampa, Florida 33620, USA

[4]Dept. of Molecular Pharmacology and Physiology, University of South Florida, Tampa, FL 33620, USA

[*]Corresponding author: ruggero.anzalone@imm.cnr.it

Keywords: 3C-SiC heteroepitaxy, growth rate, residual stress.

Abstract. SiC is a candidate material for micro- and nano-electromechanical systems (MEMS and NEMS). In order to understand the impact that the growth rate has on the residual stress of CVD-grown 3C-SiC hetero-epitaxial films on Si substrates, growth experiments were performed and the resulting stress was evaluated. The film thickness was held constant at ~2.5 μm independent of the growth rate so as to allow for direct comparison of films as a function of the growth rate. Supported by profilometry, Raman and micro-machined free-standing structures, this study shows that the growth rate is a fundamental parameter for the low-defect and the low-stress hetero-epitaxial growth process of 3C-SiC on Si substrates.

Introduction

Silicon Carbide (SiC) has long been recognized as an excellent material for high-power, high-frequency and high-temperature electronics due its outstanding electrical and thermal properties. SiC is now receiving added attention for its potential application in micro- and nano-electromechanical systems (MEMS and NEMS) due to its exceptional electrical, mechanical and chemical properties as compared with silicon (Si), which is currently the leading material in these technology areas.

Among the polytypes of SiC, cubic SiC (i.e., 3C-SiC) possesses unique properties, such as high electron drift velocity, which are more suitable for high-frequency power devices [1]. However, the most important property of 3C–SiC is that it can be grown on large diameter Si substrates, offering the possibility for low-cost batch processing, making SiC more attractive for sensors and device applications [2].

Unfortunately, due to high residual stresses (which normally arise during the growth process), the use of hetero-epitaxial 3C-SiC for Si-based MEMS fabrication techniques has been somewhat limited. The residual strain/stress field in a thin film determines the final wafer bow, which has important implications with regard to the processing, the epitaxial quality and the film's properties [3-4]. Therefore, it is necessary to reduce and control the residual stress in 3C-SiC films for high performance MEMS devices.

Experimental

For a better understanding of the impact that the growth rate has on the residual stress, wafer bow and film crystallinity of LPCVD-grown 3C-SiC heteroepitaxial films, three 3C-SiC films of the same thickness were grown at three different deposition rates on individual 50 mm diameter (100) Si wafers. A high-quality 3C-SiC epitaxial multi-step growth process was used to reduce the defect density and to improve the crystalline quality [5]. The process was adopted to grow three 3C-SiC films on (100) Si substrates, with a growth rate of 2.45, 3.21, 4 μm/h, respectively. The film thickness, measured with a FT-IR (Fourier Transform InfraRed) system at the wafer centre, is 2.4 μm, for all the samples grown.

Results and Discussion

The sample profile [z(x)] was observed from surface profiler measurements along the <110> direction (orthogonal to the wafer flat). The film profiles are shown in Fig. 1 for the three growth rates; the asymmetry of the measured data is due to the slight wafer tilt. From this analysis all three films are under compressive stress. The stress (σ) was calculated using a modified Stoney's equation that takes into account the elastic relation of the film to wafer bending [6]:

$$\sigma = \frac{M_{film}h}{\Delta R}\left(\frac{1+4mn+6mn^2 4mn^3+m^2n^4}{6mn(1+n)}\right), \quad (1)$$

where $m=M_{film}/M_{sub}$ are the biaxial modulus $(M=E/(1-\upsilon))$ of the film and substrate and $n=t/h$, with t and h the individual thicknesses of the substrate and film, respectively.

The values of Young's modulus (E) along the <100> axis and Poisson's ratio (υ) were taken to be 130 GPa and 0.279 for silicon [10] and 379 GPa and 0.44 for silicon carbide, respectively [7]. Average film and substrate thicknesses were measured by weighing the wafer before and after the growth. ΔR is the change in radius of curvature determined from $(\Delta R)=(1/R_2)-(1/R_1)$, where R_1 is the radius of curvature of the substrate prior to film deposition, while, R_2 is the radius of curvature after the film is deposited.

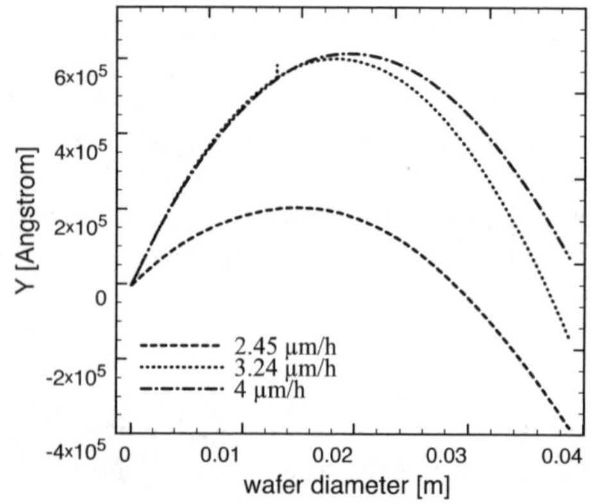

Fig. 1. Wafer profile for the three different growth rate values.

The curvature K of the samples [d^2z/dx^2] (not shown), can be calculated directly from the deformation profile $z(x)$ and gives information about the strain in the sample. The calculated curvature values are -0.08, -0.15 and -0.26 m^{-1} for the films grown at 2.45, 3.21, 4.00 µm/h, respectively. From these values it is possible to observe that the wafer curvature (the inverse of the curvature radius) decreases with decreasing growth rate. From the measured sample curvature, one can determine the average residual stress within the layer using Eq. 1.

The calculated stress increases with growth rate, from 0.31 GPa to 0.89 GPa for 3C-SiC films grown at 2.45 and 4.00 µm/h, respectively, (Fig. 2). From Stoney's equation a lower stress value corresponds to a lower growth rate, with all films under compressive stress.

The crystalline quality was also measured by Raman spectroscopy using an HR800 integrated system by Horiba Jobin Yvon in a back-scattering configuration. The excitation wavelength is supplied by a He-Ne laser at 632.8 nm. The Full Width at Half Maximum (FWHM) values of the TO vibration mode of 8.57 cm^{-1}, 9.01 cm^{-1} and 9.24 cm^{-1} were found for 3C-SiC films grown at 2.45, 3.21, and 4.00 µm/h, respectively (not shown). From the TO Raman mode peaks it is clear that the growth rate is strictly related to the growth parameters, with an evident increase in the film quality (in terms of the defect density) with the decreasing of growth rate. In Figure 3 is shown the TO and LO Raman shift for the three different growth rates. From this analysis it is possible to observe that the stress nature is tensile and the sample growth at 2.45 µm/h is farther from the theoretical stress free value (dotted line) than the other samples; thus the 2.45 micron/h sample is the *most stressed* one. This result is totally at odds with the stress calculated with the Stoney's equation and with the curvature measurement that shows a film under compressive stress.

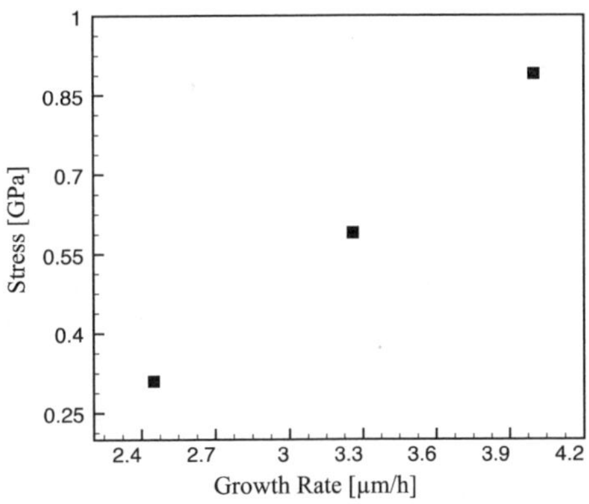

Fig. 2. Residual stress (σ) calculated by the modified Stoney's equation as a function of the growth rate.

To support the Raman analysis, on the three samples we micro-machined different free standing structures as stress probes. The planar rotating probe is a microstructure extremely sensitive as a stress probe, more than the free-standing cantilever, commonly used for stress analysis. In Figure 4 is shown a SEM image of the structure. From the rotation of the central arm it is clear that all the films are under tensile stress (inset, Fig. 4) [8].

As can be seen from the experimental data, there is an inconsistency between confocal Raman and micro-structure rotation, which both indicate a tensile residual film stress, while the convex wafer bow and the Stoney's equation would indicate a film under compressive stress. In order to explain the variation of wafer bow as a function of growth parameters many theories have been presented [9]. To try to understand this apparent disagreement between different experimental data we developed a new theory capable to include the stress field inside the substrate. In particular, we studied a systematic approach to correlate global (wafer-scale) measurements to those obtained by local indicators (as cantilevers or planar rotators) of the strain status. It is assuming that the source of this discrepancy is the initial stress in the substrate, related to the defects in the silicon substrate (voids, among others) [10-11].

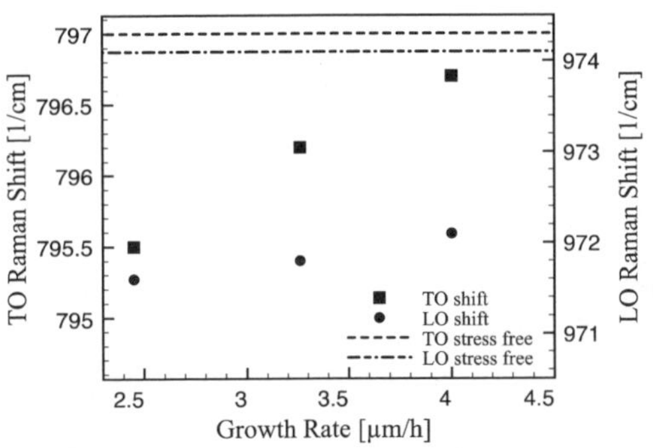

Fig. 3. TO and LO Raman shift as a function of film thickness for different growth rate values.

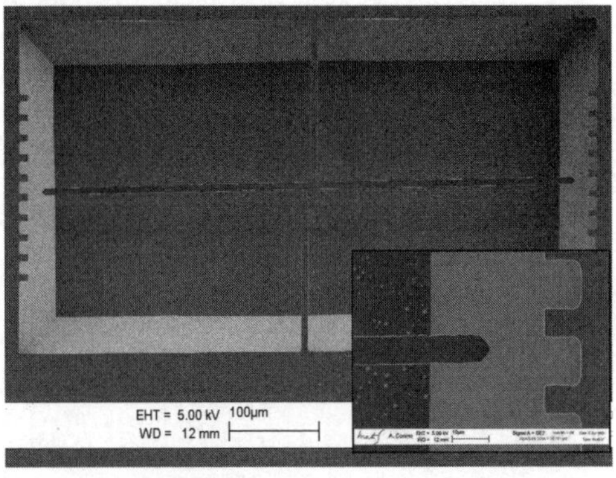

Fig. 4. SEM plan view of planar rotating probe. From the rotation of the main arm (inset) it is possible to observe that the sample is under tensile stress.

Summary

In this work, different residual stress analyses performed by modified Stoney's equation (through optical curvature measurement), Raman shift analysis and free-standing structures were presented. From these measurements was found an apparent disagreement about the nature of the stress. These discrepancies between the experimental data can be explained assuming a strong stress field located in the substrate and related to defects generated in the silicon during the growth process.

References

[1] W. E. Nelson, F. A. Halden and A. Rosengreen, *J. Appl. Phys*. 37 (1966) 33.

[2] P.M. Sarro, *Sens. Actuators A* 82 (2000) p. 210.

[3] D. Reddy, A.A. Volinsky, C. Frewin, C. Locke and S.E. Saddow, *Mat. Res. Soc. Symp. Proc.* 1049 (2008) AA3.6.

[4] R. Anzalone, M. Camarda, C. Locke, D. Alquier, A. Severino, M. Italia, D. Rodilosso, C. Tringali, A. La Magna, G. Foti, S. E. Saddow, F. La Via, and G. D'Arrigo, *J. Electrochem. Soc*. Vol. 157, no. 4, (2010) H438.

[5] C. L. Frewin, C. Locke, J. Wang, P. Spagnol and S. E. Saddow, *J. Cryst Growth*, Vol. 311, Issue 17 (2009) 4179-4182.

[6] Claude A. Klein, *J. Appl. Phys*. 88 (2000) 5487.

[7] K. Kunc, M. Balkanski and M. Nusimovici, *Phys. Stat. Sol.* (b) 72 (1975) 229.

[8] R. Anzalone, G. D'Arrigo, M. Camarda, C. Locke, S. E. Saddow, and F. La Via, *J. of Microelectromechanical Systems*, Vol. 20, No. 3 (2011) 745-752

[9] M. Zielinski, S. Ndiaye, T. Chassagne, S. Juillaguet, R. Lewandowska, M. Portail, A. Leycuras and J. Camassel, *Phys. Stat. Sol. A* 204 (2007) 981-986.

[10] M. Camarda, R. Anzalone, G. D'Arrigo, A. Severino, N. Piluso, A. Canino, F. La Via and A. La Magna, *Thin Solid Films*, accepted.

[11] M. Camarda, R. Anzalone, G. D'Arrigo, A. Severino, N. Piluso, A. Canino, F. La Via, and A. La Magna, arXiv:1110.4727v1

Materials Science Forum Vols. 717-720 (2012) pp 525-528

doi:10.4028/www.scientific.net/MSF.717-720.525

Micro-Raman analysis of a micromachined 3C-SiC cantilever

N. Piluso[1], R. Anzalone[1], M. Camarda[1], A. Severino[1], G. D'Arrigo[1], A. La Magna, F. La Via[1]

[1]Istituto per la Microelettronica e Microsistemi IMM-CNR, sezione di Catania, Stradale Primosole 50, 95121, Catania, Italy

Corresponding author: nicolo.piluso@imm.cnr.it

Keywords: micro-Raman, stress relaxation, microstructures, Finite element modeling

Abstract. In this work, Raman microscopy is used to study the stress distribution in a 3C-SiC cantilever. Also, a comparison between the strain distribution observed on the microstructure, using the shift of transverse optical mode in micro-Raman maps, with the values predicted using a recent analytic theory [1] has been done. Along the width of the cantilever is observed a reduction of stress ascribed to the etching processes that removes a thin layer of the interface between the 3C-SiC film and the substrate close to the edge of the microstructure. It is possible to show that this variation can be ascribed to a non-linearity of the strain field along the 3C-SiC film thickness. Also, helped by Finite Element Modeling, we determined the stress tensor along the cantilever. This result shows that, for a complete stress description of the cantilevers, it is necessary to take into account the role of diagonal and off-diagonal stress tensorial components.

Introduction

The study of stress in 3C-SiC free standing structure, such as a cantilever, represents a crucial step in order to optimize the manufacturing processes for Micro Electro-Mechanical Systems (MEMS) [2]. Hetero-epitaxy of 3C-SiC on Si substrates is still affected by a low crystalline quality of the epilayer and by the bending of the entire wafer processed. Defects, together with the SiC/Si interface, generate a stress field into the hetero-epitaxial system.

The residual stress is strongly affected by deposition conditions and subsequent fabrication processes. Stress measurement techniques, such as micro-Raman microscopy, are essential for both process development and process monitoring. A spatially resolved stress map (with a lateral resolution of 1 μm) on a micro-machined freestanding cantilever is obtainable by using back scattering Raman microscopy, a not destructive technique particularly useful for study of the local stress in micro-structures with the diameter of the laser spot less than 2 μm. This spot size allows an acquisition of Raman spectra with a sufficiently high spatial resolution, ensuring a local analysis and a study of the distribution of the stress field in comparison with the morphology of the structure.

Experimental Details

We collected Raman spectra using a HR800 integrated system by Horiba Jobin Yvon in a back-scattering configuration with a microscope coupled confocally to a 800 mm focal length spectrograph and a grating with 2400gr/mm. The excitation wavelength is supplied by a He-Ne laser (632.8 nm) that was focussed on the sample by a x100 objective with numerical aperture (NA) of 0.95. The power of the laser is 20mW and no heat effect during the measurements was observed. This is mainly due to the transparency of the film at the wavelength of the laser. The mean thickness of the epitaxial layers is about 3 μm, thus we probe the entire layer, since the 3C-SiC band gap is higher (about 520nm) than the laser excitation photon energy. The typical Raman signal of a 3C-SiC monocrystal epitaxial film consists of transverse (TO) and longitudinal (LO) optical phonon mode. We analyzed the TO mode because it is not affected by doping (in this case unintentionally doping) and it is suitable to probe the order of the crystal lattice, as well as its stress field. A shift of the TO Raman peak indicates the presence of stress in the film mainly due to the difference in the thermal expansion coefficients and to the lattice mismatch between the Si(001) substrate and the epitaxial film (3C-SiC). The description of the fabrication steps of the analyzed structures can be found in Ref. [2].

Results and Discussion

The cantilever analysed has a length of 200 μm and a width of 10 μm. In figure 1 is shown the TO Raman shift map of the cantilever (A) and linear Raman scans along the edges, the centre (B) and the width (C) of the micro-structure.

The unreleased and released regions show a mean value of 795.5cm^{-1} and 795.7cm^{-1}, respectively. This difference suggests the presence of an average residual tensile strain in the unreleased film, consistent with what is found through in-plane deflections of planar-rotators [3]. Although the linear scan performed along the cantilever (see Fig.1A and Fig.1B) does not reveal a particular trend of the Raman shift, where the TO peak is constant for the whole length of the cantilever, the Raman analysis performed along its width reveals a different Raman shift, and thus, a different stress reduction between the edges and the center of the micro-structure (Fig.1C). This is probably a result of the asymmetric etching processes (Inductively Coupled Plasma). Indeed, we have found that the fabrication process resulted in an under etching of the microstructure whose thickness is reduced close to the edges with a removal of highly defective material near the Si/3C-

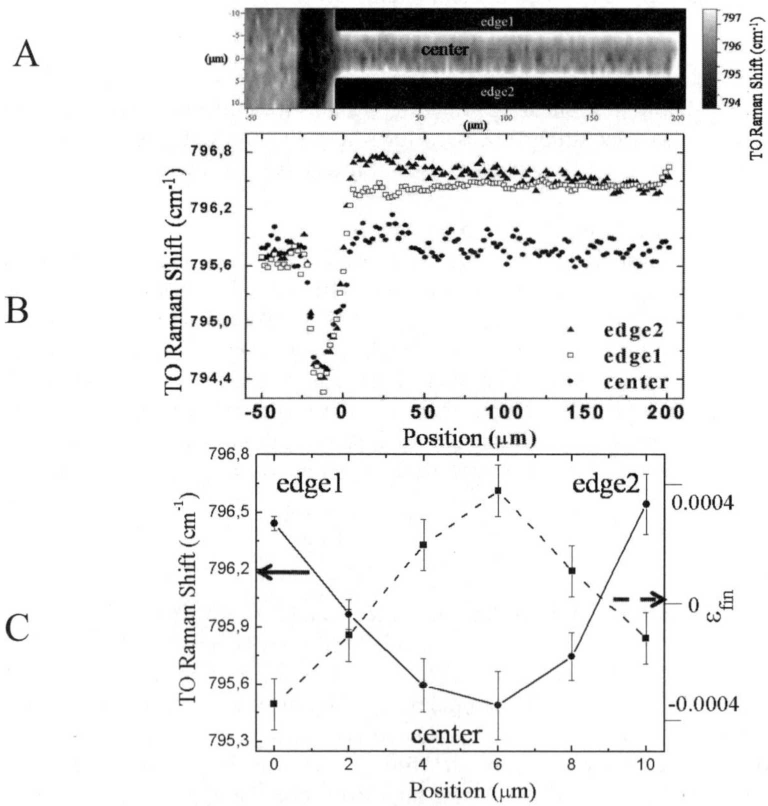

Fig. 1. (A) TO Raman shift map of the cantilever. **(B)** TO Raman scans performed along the edges and the center of the cantilever. **(C)** Trend of the TO Raman shift (left) and of the calculated (using ref. [8]) final strain along the width of the cantilever (right).

SiC interface. This film thickness variation was estimated to be around 120 nm by AFM height profilometry of the back side of the microstructure (not shown). Since the interface 3C-SiC/Si is characterized by a high density of defects [4] the removal of such a region close to the edges of the cantilever leads to a more efficient stress reduction with respect to the centre of the micro-structure.

To correlate the observed averaged final strain, $<\varepsilon(z)>$, to the initial defective strain we used a recently developed theory that correlates the defective, the residual and the final strain with the macro (wafer bow) and micro deflections [5]. It has been found good agreement between the theoretical and experimental values of the averaged final strain $<\varepsilon_{fin}>$ assuming a non-linear initial defective strain $\varepsilon_{def}(z)$ in free-standing 3C-SiC/Si(100) microstructures. The non-linearity of the initial defective strain is strongly motivated by the non-linearity of the density of extended defects (mainly stacking faults and twins) as a function of the film thickness observed in this hetero-epitaxy [6].

The largest Raman shift variation is localized on the undercut region (ranging between 1.3 – 2.3 cm^{-1}). Close to the undercut, the free standing structure shows a complex stress distribution. It has been simulated the free standing microstructures using FEM (finite element modeling) which allows us to determine the full stress tensor at every point of the microstructure (see figure 2A). Then, following the experimental analysis, we have calculated the stress tensor along a line starting from the anchored part and along the microstructure (see figure 2A). To demonstrate the limit of the relation between Raman shift and the residual stress determined in [7,8] we have separated the stress tensor into two components: the diagonal and off-diagonal parts, that is: $<\sigma_{diag}>=(\sigma_{11}+\sigma_{22}+\sigma_{33})/3$ and $<\sigma_{off\text{-}diag}>=(\sigma_{12}+\sigma_{23}+\sigma_{31})/3$.

As can be seen comparing figures 2B and 2C, the $<\sigma_{off\text{-}diag}>$ is non-zero close to the undercut. For this reason, to fully describe the stress field map in terms of the Raman-shift, it is necessary to modify the Raman-stress relation to correctly link the Raman-shift to the full stress tensor. This theoretical approach has been already used in [9,10] in the case of stressed silicon. Work is underway to implement the same theoretical framework in the case of microstructures in silicon carbide epitaxy.

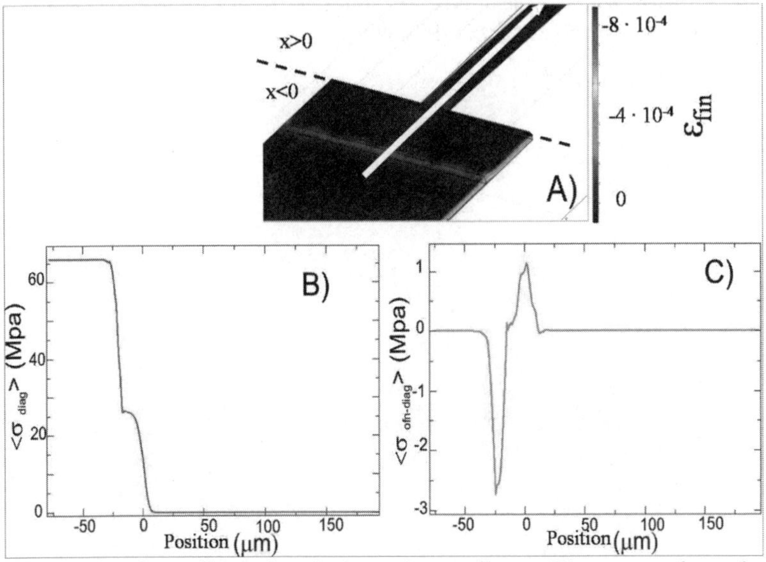

Fig. 2. (A) Three-dimensional view of a cantilever **(B)** $<\sigma_{diag}>$ along the arrow **(C)** $<\sigma_{off\text{-}diag}>$ along the arrow.

Conclusion

The residual stress related to the hetero-epitaxy itself is a crucial point for the development of the SiC base sensor and microsystems. Raman analysis seems to be a very suitable method for the stress investigation. Along the width of the cantilever is revealed a trend of stress reduction ascribed to the etching processes that removes a thin layer of the interface between the 3C-SiC film and the substrate close to the edge region. Such a trend of the strain relaxation with the thickness of the film can be explained by assuming a non-linear initial defective strain in free-standing 3C-SiC/Si(100) microstructures. Close to the undercut, the stress field cannot be considered as simple biaxial stress. For this reason FEM must be used to determine the form of the stress tensor at every point of the microstructure and to correctly link the observed Raman shift to the different components (diagonal and off-diagonal) of the stress tensor.

References

[1] M. Camarda, R. Anzalone, G. D'Arrigo, A. Severino, N. Piluso, A. Canino, F. La Via, and A. La Magna, arXiv:1110.4727v1

[2] R. Anzalone, M. Camarda, C. Locke, D. Alquier, A. Severino, M. Italia, D. Rodilosso, C. Tringali, A. La Magna, G. Foti, S. E. Saddow, F. La Via, and G. D'Arrigo, *J. Electrochem. Soc.,* 157 (4) (2010) H438

[3] R. Anzalone, G. D'Arrigo, M. Camarda, C. Locke, S.E. Saddow and F. La Via, *IEEE J. Microelectromechanical Systems* 20,(2011) 745

[4] A. Severino, C. Frewin, C. Bongiorno, S.E. Saddow and F. La Via, *Diamond Relat. Mater.* 18 (2009) 1440

[5] M. Camarda, N. Piluso, R. Anzalone, A. La Magna and F. La Via, *HeteroSiC 2011*, in press

[6] R. Anzalone, M. Camarda, A. Canino, N. Piluso, F. La Via, and G. D'Arrigo, *Electrochem. Solid-State Lett.* 14 (2011) H161

[7] D. Olego, M. Cardona and P. Vogl, *Phys. Rev. B*, Vol 25 (6) (1982) 3878

[8] S. Rohmfeld, M. Hundhausen, L. Ley, C.A. Zorman and M. Mehregany, *J. Appl. Phys.* 91 (3) (2002) 1113

[9] I. De Wolf, D. J. Howard, K. Maex and S. K. Jones, *J. Appl. Phys.* 79 (9) (1996) 7148

[10] I. De Wolf, *J. Raman Spectrosc.*, 30 (1999) 877

Materials Science Forum Vols. 717-720 (2012) pp 529-532

doi:10.4028/www.scientific.net/MSF.717-720.529

Single-Crystal SiC Resonators by Photoelectrochemical Etching

Mohammad M. Islam[1], Chih-Fang Huang[2], Feng Zhao[3]*

[1]Department of Electrical Engineering, University of South Carolina, Columbia, SC 29208 USA

[2]Institute of Electronics Engineering, National Tsing Hua University, Hsinchu 30013, Taiwan, R.O.C

[3]School of Engineering and Computer Science, Washington State University, WA 98686 USA

*feng.zhao@wsu.edu

Keywords: Single crystal, SiC, resonator, photoelectrochemical etching.

Abstract. In this paper, we report single-crystal 4H-SiC resonant structures fabricated by dopant-selective photoelectrochemical etching. The frequency response of the resonant beams was characterized by a dynamic scanning method using AFM with the beams excited by a piezoelectric actuator under atmosphere pressure and room temperature. The beam with a length of 35 μm shows mechanical resonance at 945 kHz. The Young's modulus of single-crystal SiC was derived from the measured resonant frequency. Single-crystal 4H-SiC resonators developed in this study fully exploit the excellent electrical, mechanical, and chemical properties of SiC, while dopant-selective photoelectrochemical etching technique significantly simplifies the fabrication process.

Introduction

SiC has unique material properties including wide bandgap, high thermal conductivity, high inertness to corrosive environments, high temperature stability and biocompatibility, which make it a promising material alternative to Si for developing MEMS operated in hostile environments [1]. In addition, SiC has larger E/ρ ratio between Young's modulus and density, leading to a higher resonant frequency and better quality factor than Si (provided that all other boundary conditions such as geometry, environment … are the same) for MEMS resonant devices used in frequency filtering in high performance communication transceivers, oscillators and high sensitivity sensing.

The superior chemical resistance of SiC to almost all chemicals except molten KOH makes the fabrication of suspended single-crystal SiC MEMS structures a challenging task. In the past, efforts have been focused on the fabrication of polycrystalline SiC films or heteroepitaxial 3C-SiC films on sacrificial Si or SOI substrate by surface micromachining techniques [2], with the MEMS electrical, mechanical, and chemical performance for high temperature and harsh environment limited by the substrate. The techniques to fabricate single-crystal SiC structures include bulk micromachining to etch through the nearly 300 μm SiC substrate from the backside using ICP etching process [3]. Recently a surface micromachining technique − dopant-selective photoelectrochemical (PEC) etching was developed [4] by the authors to fabricate single-crystal 4H-SiC MEMS with a good control of the MEMS structures and simplified fabrication process.

In this report, single-crystal 4H-SiC resonator structures were fabricated using PEC etching. The resonant response of the resonators was characterized using a dynamic method by AFM to demonstrate the high-frequency performance, with the Young's modulus of single-crystal 4H-SiC extracted from the measurements.

Fabrication

Single-crystal p-type 4H-SiC resonant structures were fabricated on an n-type substrate (1×10^{18} cm^{-3}). The p-type layer was grown by CVD with the nominal doping of 2×10^{17} cm^{-3} and a thickness of 100 nm, which determines the thickness of the microstructures. The lateral geometry and layout of the microstructures were defined by photolithography and Ni etch mask, followed by a reactive ion etch (RIE) in SF_6/O_2 plasma to remove the p-layer in unmasked region. The etching selectivity between SiC and Ni was 25:1. After RIE, the Ni mask was stripped off and another Ni layer of 100 nm was deposited on the backside of the sample. The resonant structures were released by PEC etching. The sample surface was exposed to diluted KOH (1% in DI water by weight) and UV light from a 100 W Hg arc lamp filtered with a 250-400 nm bandpass filter. A Pt wire was immersed in the solution and connected to the cathode while the Ni layer on the backside of the sample was connected to the anode of the voltage source. The applied voltage was 0.12 V, and the etch rate of 4H-SiC was found to be 2.2 μm/hour. The process flow and PEC etching setup is shown in Fig. 1. The mechanism about PEC etching of 4H-SiC was explained by the authors in reference [4].

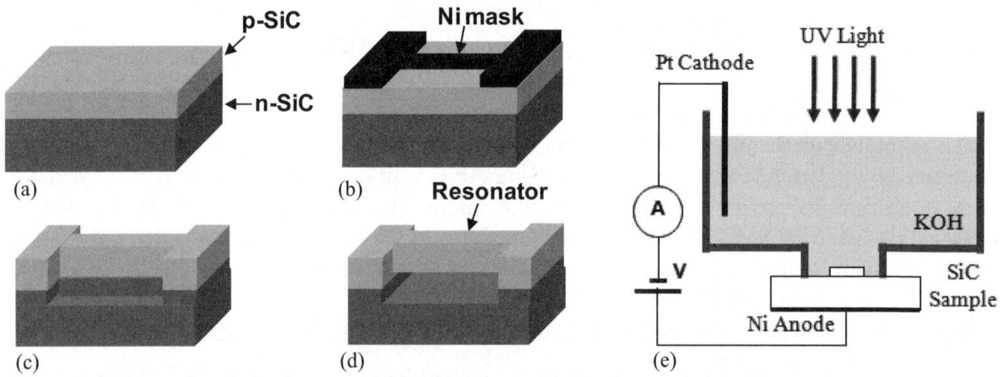

Figure 1: Schematic of single-crystal SiC resonator fabrication process. (a) starting material with p-SiC on n-SiC substrate; (b) after photolithography and Ni etch mask lift-off; (c) after SiC RIE etching and Ni mask removal; (d) after PEC etching with p-SiC resonator structure released; (e) PEC etching setup.

Results and Discussion

Fig. 2 shows the SEM images of two single-crystal 4H-SiC resonator structures released by PEC etching. The images show that PEC etching did not leave obvious damage to the suspended structures, and the straightness of the beams indicates the absence of stress gradient in the structures.

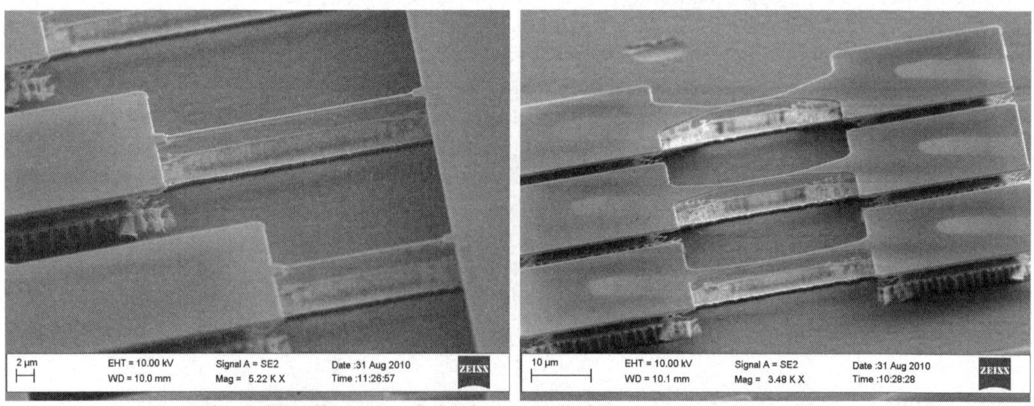

Fig. 2 SEM images of single-crystal SiC resonators structures.

The frequency response of the microstructures was characterized by a dynamic scanning measurement [5] using an AFM probe operated in intermitted contact mode. This measurement is based on the coupled interaction between the suspended beam structure and AFM probe, from which the dynamic response of the structure is obtained. The driving forces used to actuate the suspended resonant structure can be mechanical, thermal, piezoelectric, electrostatic, etc, and in our setup we used a piezoelectric actuator in direct contact to the backside of the sample with the frequency swept. Vibration response of the resonating beam was recorded by the deflection of the AFM cantilever which continuously scanned the center of the beam in intermittent contact mode. When the excitation was near the resonant frequency of the beam, the out-of-plane vibration of the beam was strongly coupled with the AFM probe, which is indicated by the additional displacement shown in the AFM image. All measurements were performed in air at room temperature. Fig 3 shows the resonant response of a 50 μm long doubly-clamped beam with the resonant frequency of 510 kHz and the quality factor of 18.2. The schematic of the coupling between the beam and AFM probe at frequencies below, at and above the resonant frequency is inserted in the figure.

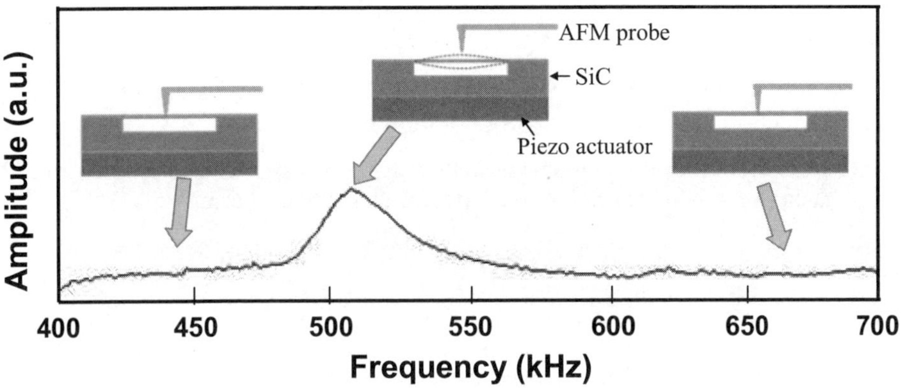

Fig. 3 Resonant response of a 50 μm long doubly-clamped beam obtained from the dynamic scanning measurement. The resonant frequency is f_0=510 kHz.

The resonant frequency of doubly-clamped beams with four different lengths was characterized and the results are summarized in Table 1 and plotted in Figure 4. The resonant frequency of the ideal beams (with a rectangular cross-section) is defined by the Young's modulus E, the mass density ρ and the structure's dimensions (length L and thickness t) by the following equation:

$$f_0 = 1.03 \cdot \sqrt{\frac{E}{\rho}} \cdot \frac{t}{L^2} \quad (1)$$

For 4H-SiC, ρ= 3.21 g/cm^3. The thickness of the beams equals the thickness of the p-type layer which is 0.1 μm. A Young's modulus of 480 GPa was extracted from the measured resonant frequency using Eq. (1). This value agrees well with other published data [1] and as expected, it is very close to that of 6H-SiC [6].

Table 1: Resonant frequency and quality factor of 1.5-μm wide resonant beams with various lengths.

Beam length (μm)	35	40	45	50
Resonant frequency (kHz)	945	800	670	510
Quality factor	63.0	36.4	26.8	18.2

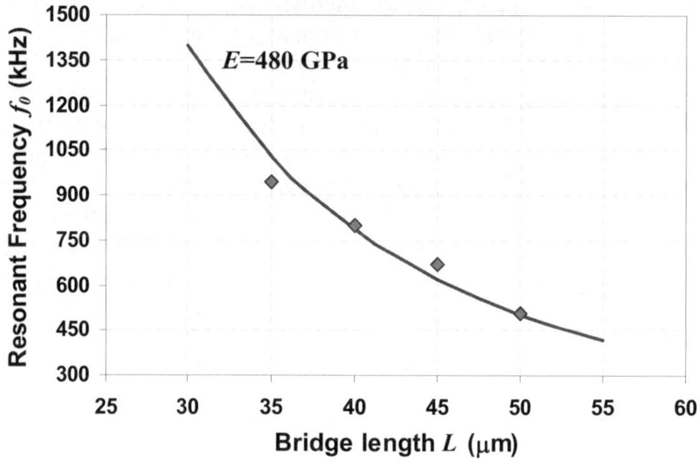

Figure 4 Resonant frequency of 1.5 μm wide doubly-clamped beams with various lengths. The Young's modulus of single-crystal 4H-SiC was extracted using Eq. (1).

Summary

In this paper, a well-controlled dopant-selective PEC etching was applied to fabricate resonant structures on single-crystal 4H-SiC, an ideal material to realize MEMS in high temperature and harsh environment applications. The selective etching process simplifies the fabrication of SiC MEMS, and is applicable for other SiC polytypes as well. SEM shows that there is no stress introduced in the resonant structures by the etching process. The frequency response of the resonant beams was characterized by a dynamic scanning measurement and the resonant frequency from different structures was obtained. The Young's modulus of single-crystal 4H-SiC was extracted from the measurements and the value agrees well with other published data. With this simplified and well-controlled etching technique, combined with the excellent material properties of single-crystal SiC, resonant devices for fast, high-resolution sensors and actuators used for harsh environments are expected to be achievable.

References

[1] V. Cimalla, J. Pezoldt, O. Ambacher, Group III nitride and SiC based MEMS and NEMS: materials properties, technology and applications, J. Phys. D: Appl. Phys. 40 (2007) 6386-6434.

[2] F. Liu, C. Carraro, A.P. Pisano, R. Maboudia, Growth and characterization of nitrogen-doped polycrystalline 3C-SiC thin films for harsh environment MEMS applications, J. Micromech. Microeng. 20 (2010) 035011.

[3] T.K. Hossain, S. MacLaren, J.M. Engel, C. Liu, I. Adesida, R.S. Okojie, The fabrication of suspended micromechanical structures from bulk 6H-SiC using an ICP-RIE system, J. Micromech. Microeng. 16 (2006) 751-756.

[4] F. Zhao, M.M. Islam, C.F. Huang, Photoelectrochemical etching to fabricate single-crystal SiC MEMS for harsh environments, Mater. Lett. 65 (2011) 409–412.

[5] B. Illic, S. Krylov, L.M. Bellan, H.G. Craighead, Dynamic characterization of nanoelectromechanical oscillators by atomic force microscopy, J. Appl. Phys. 101 (2007) 044308.

[6] M. Fukuhara and Y. Abe, High-temperature elastic moduli and internal frictions of α-SiC ceramic, J Mater. Sci. Lett. 12 (1993) 681-683.

Materials Science Forum Vols. 717-720 (2012) pp 533-536

doi:10.4028/www.scientific.net/MSF.717-720.533

Amorphous Silicon Carbide (*a*-SiC) Thin Square Membranes for Resonant Micromechanical Devices

Andrew C. Barnes[1], Christian A. Zorman[1]*, Philip X.-L. Feng[1]*

[1]Electrical Engineering and Computer Science, Case School of Engineering
Case Western Reserve University, Cleveland, OH 44106, USA
andrew.barnes@case.edu, *christian.zorman@case.edu, *philip.feng@case.edu
*Corresponding authors

Keywords: Amorphous Silicon Carbide (*a*-SiC), SiC Membrane, Micromechanical Resonator, Vibrational Modes, Nano/Microelectromechanical Systems (MEMS/MEMS)

Abstract. We report an initial experimental exploration of engineering very thin, suspended amorphous silicon carbide (*a*-SiC) membranes into vibrating micromechanical devices. We show that micromachined *a*-SiC thin square membranes can make interesting multiple-mode flexural resonators, with frequency spectra exhibiting *many* measurable resonant modes over a wide frequency range (100kHz–10MHz) in the low radio frequency (RF) bands. Initial demonstration and preliminary data suggest interesting and rich dynamical, nonlinear, and dissipative properties in these micromechanical resonances. Specifically, for instance, at room temperature ($T\approx300$K) and in moderate vacuum (*e.g.*, ~20mTorr), resonant modes of an *a*-SiC square membrane (thickness: $t\approx1.6\mu$m, size: 1mm×1mm) are observed in the ~100kHz–5MHz range, with measured quality factors (Q's) in the range of ~2,500–9,000.

Introduction

Amorphous silicon carbide (*a*-SiC) thin films have previously been developed and engineered as a low-κ dielectric for low-parasitic, high-speed interconnects in electronics for high-density CMOS integrated circuits in the ultra large scale integration (ULSI) era [1]. They have also been explored as emerging thin films for important optical applications such as electroluminescence and large area flat displays [2]. Recently we have been exploring the *micro- & nano-mechanical* behavior of *a*-SiC thin films for their potential in enabling new types of electromechanically coupled devices for miniature sensors and integrated transducers [3,4]. In exploring and engineering new types of nano/microelectromechanical systems (NEMS/MEMS), the unique electrical, mechanical, thermal, and optical properties of *a*-SiC are very attractive, which combine and add on to the great comparative advantages of low-cost, easier, and CMOS-compatible material processes. Very thin suspended membrane structures with thicknesses in the range of few tens of nanometers to a few microns are a generic type of building blocks for NEMS/MEMS. Conventional thin membrane MEMS make great sensors for detection of physical quantities such as fluidic pressure, adsorbate mass, contact and non-contact forces, radiation, temperature changes (or heat), *etc.*, all with ever increasing sensitivities as devices are scaled down to NEMS regime. Moreover, in such sensing applications, membrane-based physical sensors may also have certain compelling advantages, in comparison to their conventional counterparts in cantilevers, beams, and wires [5,6]. Lately, based on 3C-SiC membranes of 100–500nm in thickness, we have developed multiple-mode high-Q nanomechanical resonators [7].

In this work, we describe our initial experimental effort on developing some of the first vibrating micromechanical devices based on thin square *a*-SiC membranes. With the demonstration of *a*-SiC membrane resonators and their multiple-mode operations, this study may provide new ways of precisely measuring physical properties of such membranes, and also open possibilities for fabrication and optimization of new sensors, particularly for critical applications involving high temperatures and harsh environments to which SiC is especially suited.

SiC Membrane Device Fabrication

Figure 1a,b show the typical *a*-SiC membranes realized by using a bulk micromachining process. In brief, a double-side polished (100) silicon (Si) wafer is coated with an *a*-SiC thin film by plasma enhanced chemical vapor deposition (PECVD) on the top side and a low temperature oxide (SiO_2) layer is deposited by low pressure chemical vapor deposition (LPCVD) on the backside (Fig. 1c). A photoresist mask is then used to pattern windows and channels in the oxide (Fig. 1d). The windows are designed to form free standing membranes 1mm by 1mm in size and ~1.6μm thick. The backside channels (*i.e.*, gaps between adjacent membrane units) are designed to terminate on the <111> boundaries before etching through the substrate. The membranes may then be separated into individual chips without the need for mechanical saw dicing which would be destructive to the membranes. The Si wafer was then etched with KOH to form the membrane and dicing channels (Fig. 1e). However, due to insufficient corner compensation the substrate could be over etched resulting in the uneven bottom surface (Fig. 1b). The over etching of the bottom planes of the substrate should not affect the membrane as the edges of the membranes are well maintained. This process creates large aspect ratio (~625:1) membranes that may allow for unique properties that have not been investigated to date.

Figure 1: Amorphous SiC (*a*-SiC) square membranes and simplified device fabrication process. (a) Optical micrograph of the top side of an *a*-SiC membrane (center square, size: 1mm×1mm). (b) Micrograph from bottom side showing the etched Si substrate. (c), (d) & (e) Sequential cross sections showing the bulk micromachining process.

Experimental Apparatus and Techniques

We characterize the *a*-SiC membrane devices resonance behavior by using an experimental system (Fig. 2) that incorporates (i) an open-loop two-port frequency-sweeping electrical resonance measurement via a radio frequency (RF) network analyzer, and (ii) a laser interferometry for membrane's motion detection and for converting the interferometric signal into electrical readout through a sensitive photodetector.

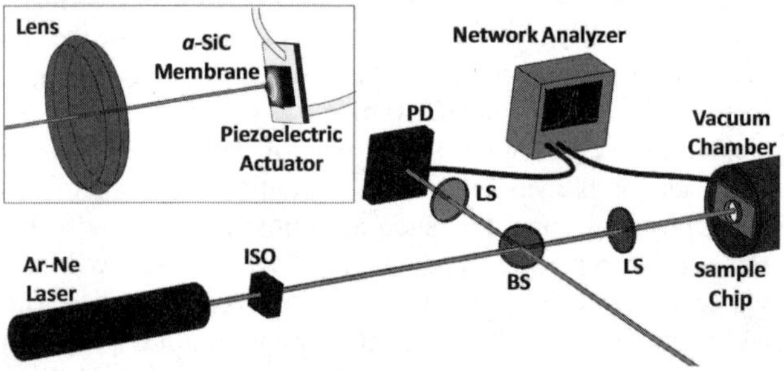

Figure 2: Schematic illustration of the experimental system for measurement of the *a*-SiC membrane resonance behavior, including a laser interferometry, in combination with a two-port electrical measurement via a radio frequency (RF) network analyzer. The Ar-Ne laser passes a laser beam through the isolator (ISO), then a beam splitter (BS). Laser beam is then focused onto the device membrane through an objective lens (LS). The interference signal is then focused onto a photodetector (PD). Inset: A close-in view illustrating a laser beam focused upon an *a*-SiC membrane (with a color-mapped mode shape) chip mounted on a piezoelectric actuator which is electrically accessible.

Sample chips consisting of *a*-SiC membranes are assembled in standard ceramic packages. A piezoelectric (disk) actuator is wire bonded within the package for piezoelectric actuation and excitation of the membranes.

The sample assembly is loaded inside a vacuum chamber (Fig. 2) to permit reflection of the laser off the membrane through an optical window. Electrical feedthroughs provide access to devices for two-port electrical measurement. Data are taken at room temperature, with strong response peaks emerging in vacuum (~20mTorr).

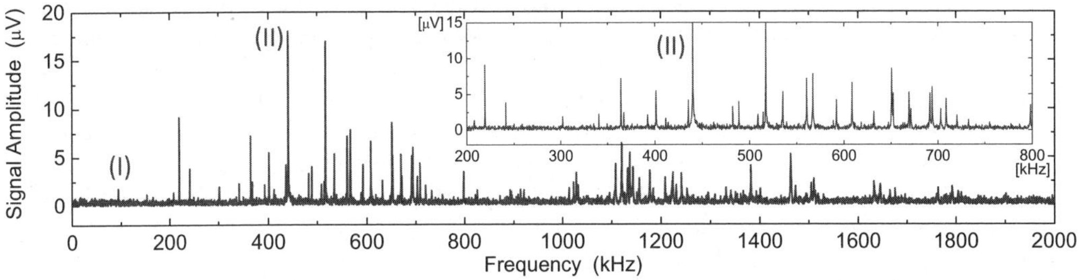

Figure 3: Wide band (from DC to 2MHz) frequency response and resonances measured from a 1mm×1mm×1.6μm *a*-SiC membrane in its multiple-mode operations, demonstrating many sharp peaks of micromechanical resonances. The detection scheme obtains very low parasitic background off resonances. The inset demonstrates a close-in and clearer view of the measured resonance peaks in a slightly narrower frequency range (200–800kHz). Details of the labeled peaks (I) and (II) are presented in Figure 4.

Experimental Results and Discussions

Figure 3 displays the measured frequency response in a relatively wide band (from DC to 2MHz) by sweeping-frequency transmission measurements in the two-port network analysis. These measurements yield many vibrational modes that are easily seen above the background (off-resonance) response (*e.g.*, the off resonance background is also the response level measured when the laser beam is focused off the suspended membrane structures). Of the most apparent resonance peaks, the lowest resonance frequency is at 94.55kHz, which is assumed to be the fundamental flexural mode of the membrane (labeled (I) in Fig. 3). The highest amplitude peak (labeled (II), Fig. 3) at 442.5 KHz is used to further refine and tune the focusing of the laser setup to obtain the overall greatest amplitude resonant response. Many of the larger peaks appear in pairs, often with repeatable frequency differences between accompanying or adjacent peaks. This presumably is from the frequency splitting due to slight difference in side lengths of the membranes (*i.e.*, the membranes are not ideal or perfect squares).

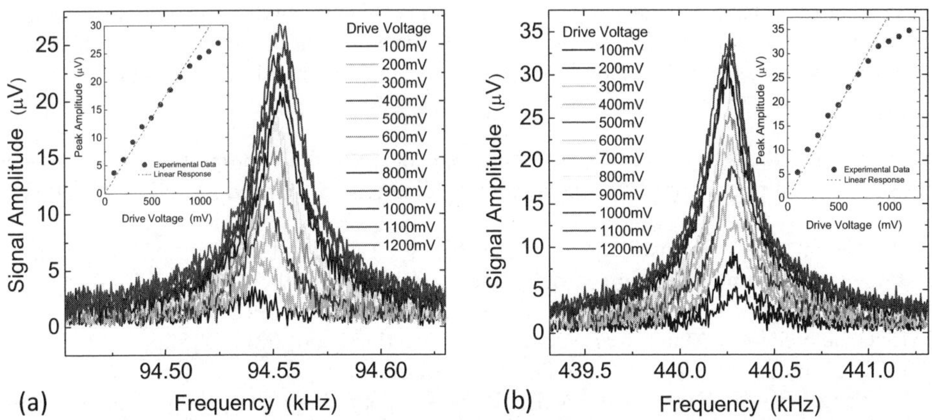

Figure 4: Examples of measured resonance data with details, from selected individual resonance peaks. (a) Measured response of the 94.55kHz resonance (peak (I) in Figure 3). (b) Measured 440.25kHz resonance (peak (II) in Figure 3). Insets (for both (a) and (b) panels): Amplitude of the resonance peak as a function of the drive (actuation) voltage.

Figure 4 demonstrates two examples of selected individual peaks and some technical details observed from the measurements. A narrow range frequency sweep, 94.45 to 94.65 kHz, was conducted to examine the lowest frequency peak obtained from the large scale sweep. Figure 4a shows the raw data from the sweep for drive amplitude varying from 100mV to 1200mV. It can be seen that the center frequency shifts from 94.54kHz to 94.56kHz as the amplitude of the input is increased. The data also demonstrate the nonlinearity of the amplitude of the peak with increasing drive. The inset in Fig. 4a shows the expected linear response of the membrane as the pink line and the results of the peak amplitude as measured. Between 800 and 900 mV the peak amplitude starts to diverge from the linear response. This suggests the onset of nonlinearity may be due to mechanical Duffing behavior as seen in previously reported 3C-SiC nanomembrane [7].

A second narrow range, 439.0–441.5kHz, sweep is performed on the peak labeled (II) in Fig. 3. This higher mode peak exhibits similar results as observed for peak (I). Figure 4b shows this peak is centered at 440.25KHz with peak amplitude of ~35μV for a drive voltage of 1200mV. This peak also demonstrates nonlinearity starting at a drive voltage of ~800mV.

Figure 5 shows the measured quality factors (Q's) of various large-signal modes (those whose peaks are high and Q's are reliably measured), as a function of the resonance frequency of the mode. While the overall trend confirms the well-known trade-off between resonance frequency f and Q – *i.e.*, Q decreases as resonance frequency is increased, it is very interesting that the Q does *not* decrease monotonously with increasing frequency. The measured data fall in the $f \times Q$ product contours of 1.35×10^9 and 3.5×10^9.

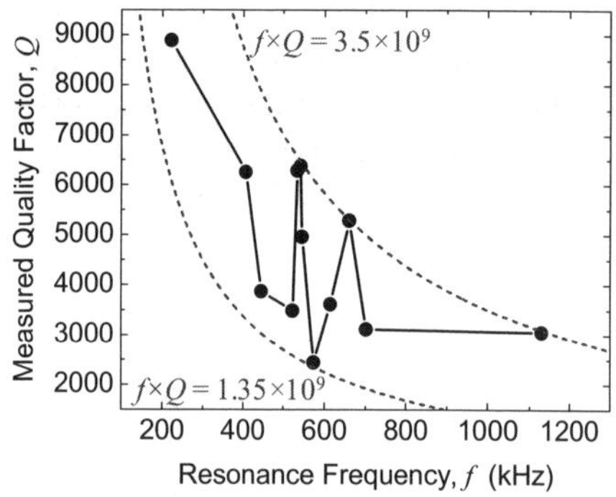

Figure 5: Measured quality factors (Q's) of the various resonance modes, as a function of the resonance frequency. The $f \times Q$ products fall in the range of ~$1.35–3.5\times10^9$.

Conclusion

In summary, in this work we have experimentally demonstrated interesting micromechanical resonators vibrating with many high-order flexural modes, based upon thin *a*-SiC suspended membranes that are realized by a convenient bulk machining process. Initial measurements show that these resonances have considerably high Q's at non-stringent conditions (room temperature and very moderate vacuum). The multiple modes and engineerable dissipations may lead to new high-performance physical sensors based on further elaborate engineering of these membranes.

References

[1] X.M. Li, T.K.S. Wong, D.J. Yang, Structural and electronic properties of low dielectric constant carbon rich amorphous silicon carbide, Diamond & Related Materials 12 (2003) 963-967.

[2] W.A. Nevin, H. Yamagishi, M. Yamaguchi, Y. Tawada, Emission of blue light from hydrogenated amorphous silicon carbide, Nature 368 (1994) 529-531.

[3] J.G. Du, N. Singh, J.B. Summers, C.A. Zorman, Development of PECVD SiC for MEMS using 3MS as the precursor, Mater. Res. Soc. Symp. Proc. 911 (2006) 0911-B05-28.

[4] J.G. Du, Ph.D. Thesis, Case Western Reserve University (2007).

[5] X.L. Feng, R. He, P. Yang, M.L. Roukes, Very high frequency silicon nanowire electromechanical resonators, Nano Letters 7 (2007) 1953-1959.

[6] X.L. Feng, M.H. Matheny, R.B. Karabalin, C.A. Zorman, M. Mehregany, M.L. Roukes, Silicon carbide (SiC) top-down nanowire electromechanical resonators, Digest of Tech. Papers, The 15th Int. Conf. on Solid-State Sensors, Actuators, & Microsystems (Transducers'09) (2009) 2246-2249.

[7] A.C. Barnes, R.C. Roberts, N.C. Tien, C.A. Zorman, P.X.L. Feng, Silicon carbide (SiC) membrane nanomechanical resonators with multiple vibrational modes, Digest of Tech. Papers, The 16th Int. Conf. on Solid-State Sensors, Actuators, & Microsystems (Transducers'11) (2011) 2614-2617.

Materials Science Forum Vols. 717-720 (2012) pp 537-540
© (2012) Trans Tech Publications, Switzerland
doi:10.4028/www.scientific.net/MSF.717-720.537

Amorphous Silicon Carbide as a Non-Biofouling Structural Material for Biomedical Microdevices

C. Zorman[1*], A. Eldridge[2], J. Du[1], M. Johnston[2], A. Dubnisheva[2], S. Manley[2], W. Fissell[2], A. Fleischman[2] and S. Roy[2,3]

[1]Electrical Engineering and Computer Science, Case Western Reserve Univ. Cleveland, OH, USA

[2]Biomedical Engineering, Lerner Research Institute, Cleveland Clinic Cleveland, OH, USA

[3]Bioengineering and Therapeutic Sciences, Univ. of California, San Francisco, CA, USA

[*]caz@case.edu

Keywords: Amorphous SiC, bulk micromachined diaphragms, MEMS, biocompatibility, poly(ethylene) glycol, Young's modulus

Abstract: This paper details the development of amorphous hydrogenated silicon carbide (a-SiC:H) films as structural material that is resistant to biofouling. The a-SiC:H films were deposited by PECVD and evaluated for their mechanical and anti-biofouling properties. It was found that the as-deposited films exhibited compressive residual stresses that could be converted to moderate tensile stresses upon a post deposition anneal. The amorphous films exhibited a much lower Young's modulus but similar burst stress when compared to polycrystalline 3C-SiC films of like thickness. The as-deposited a-SiC:H films were more resistant to biofouling than Si and SiO_2 surfaces. Coating the a-SiC:H films with polyethylene glycol (PEG) significantly improved the anti-fouling characteristics for extended periods.

Introduction

Implantable Si MEMS devices generally rely on packaging approaches that utilize polymer encapsulants (ie., PDMS) or ceramic/metallic enclosures. Conventional packaging can result in two primary drawbacks: (1) attenuation of signal/stimulus between the physiological environment and device; and, (2) increased size that detracts from the benefits of miniaturization. The ideal package would consist of a low form factor structure that is chemically inert, resistant to fouling, mechanically-durable and inherently part of the device structure. Although known for its use in extreme environment applications, SiC has recently been used as a structural material in biomedical microdevices due in part to its biocompatibility and compatibility with commonly used sterilization procedures [1]. This paper reports the development of a-SiC:H thin films for biomedical microdevice applications requiring a non-biofouling, corrosion resistant, mechanically-robust material that can serve both structural and packaging functions.

Experimental

Amorphous SiC films were deposited on (100) Si wafers in an UltraDep II^{TM} dual frequency, plasma enhanced chemical vapor deposition system. Trimethylsilane ($(CH_3)_3SiH$) was used as the precursor gas with helium serving as a carrier gas. Flow rates for trimethylsilane and helium were 38 sccm and 160 sccm, respectively. Depositions were performed at a substrate temperature of 350ºC, a low frequency (400 kHz) power of 100W and a high frequency (13.56 kHz) power of 400W. The deposition pressure was nominally 8 Torr. When desired, the films were annealed after deposition at 450ºC in a nitrogen-filed, atmospheric furnace. Wafer curvature measurements were used to determine the average residual stresses in select films.

To evaluate the Young's modulus, residual stress and burst stress of the a-SiC:H films, load-deflection testing was performed on micromachined diaphragms. For these studies, 2.5 μm thick, 1 x 1 mm^2 a-SiC:H diaphragms were fabricated by conventional Si bulk micromachining.

Load-deflection testing was performed on the micromachined diaphragms using a custom-built microscope-based interferometer specifically designed to accommodate pressurized diaphragms. Details about the apparatus and the technique used to determine the center deflection of a pressurized diaphragm can be found elsewhere [2]. The Young's modulus and residual stress of the diaphragm can be determined by fitting the following equation to a plot of center deflection versus applied pressure:

$$P = \frac{t}{a^2} W_O \left[C_1 \sigma_O + \frac{f(\nu)}{a^2} \frac{E}{1-\nu} W_O^2 \right] \tag{1}$$

where P is the applied differential pressure, W_0 is the center deflection, t is the diaphragm thickness, a is one-half of the diaphragm side length, $C_1 = 3.41$ [ref], σ_0 is the residual stress, E is the Young's modulus, ν is Poisson's ratio, and $f(\nu) = 1.37(1.446\text{-}0.427\nu)$ [ref]. For this work, a Poisson's ratio of 0.23 was used for a-SiC:H [ref].

Following load-deflection testing, burst testing was performed on select diaphragms by simply increasing the applied pressure up to the burst point and recording the maximum pressure achieved. A finite element model (FEM) of the diaphragm is then constructed using the results from the load-deflection testing as inputs. The load-deflection behavior of the model diaphragm is then extrapolated to the measured burst pressure. The stress distribution in the model diaphragm is then calculated by FEM at the burst pressure. Since the diaphragm is known to have burst at this pressure, it is assumed that failure occurred in the region of highest stress. This stress is assigned to be the burst stress.

To examine biofouling, the performance of uncoated and poly(ethylene) glycol (PEG) coated a-SiC:H samples were compared with uncoated and PEG-coated surfaces of (100) Si, and 100 nm--thick thermal SiO_2 films. PEG is among the most commonly used coating materials to suppress biofouling on Si [3]. For this study a 500 nm a-SiC:H film was deposited by PECVD using the conditions detailed above. PEG was covalently bonded to MEMS substrates using a technique detailed elsewhere [4]. Chips were initially soaked in an EtOH:H_2O solution, followed by a Piranha clean and immersion in a solution of 20 mmol trimethoxysilane-PEG, in acidified anhydrous toluene for 12+ hours and then rinsed with DI water. X-ray photoelectron spectroscopy (XPS) analysis was used to examine the surface of the samples before and after PEG coating, as well as after the DI water rinse, in order to determine the effectiveness of the process.

Biofouling studies were performed using cell media consisting of High Glucose DMEM, 1% L-glutamine, 1% penicillin/streptomycin, and 10% fetal bovine serum. Madin-Darby Canine Kidney cells were plated on the chips and polyester control wells at $3\text{x}10^5$cells/cm^2. Media was exchanged every 48 hours, and after 12 days, the media was replaced with 0 calcium media for half of the samples. At day 14, all samples were inspected using fluorescence microscopy.

Results

Wafer curvature measurements were performed on as-deposited and annealed wafers to calculate the average residual stress and thus determine the best processing conditions to prepare samples for load-deflection testing. Films were deposited to a nominal thickness of 1.5 μm on (100) Si substrates over 19 distinct runs. No post processing steps were performed for this sample set prior to measuring wafer curvature. The average residual stress was -135 ± -12 MPa. The compressive stress in the as-deposited films makes load-deflection testing of diaphragms very difficult, therefore a post deposition annealing

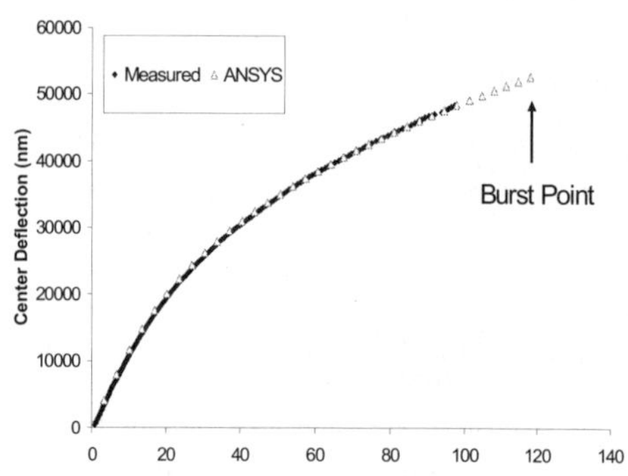

Fig. 1. Measured and modeled load deflection behavior of an a-SiC:H diaphragm.

step was performed to modify the residual stress. It was found that the residual stress changes to ~ 110 MPa upon annealing the a-SiC:H films for 30 min at 450°C. The moderate tensile stresses produced under these conditions are ideally suited for micromachined diaphragms, and therefore all samples for load-deflection testing were subjected to post-deposition annealing.

A typical plot of center-deflection versus applied pressure for the a-SiC:H films investigated in this study is shown in Fig. 1 above. Applying Eq. 1 to these data yield a Young's modulus of 75 ± 10 GPa and a residual stress of 202 ± 20 MPa for a sample set of 10 diaphragms. Fig. 1 also shows the load-deflection behavior for the model diaphragm created by FEM. There is excellent agreement between the experimental data and the FEM model for the full range of applied pressures. Extrapolation of the FEM results to the burst pressure is also shown in Fig. 1. From the FEM analysis, the burst stress for the a-SiC:H diaphragms is estimated to be nominally 800 MPa.

For the PEG-coating and biofouling studies, sample coupons were extracted from a single wafer that supported an a-SiC:H film with a compressive residual stress of -50 MPa. Figure 2 shows XPS peaks between 280 and 295 eV for as-deposited, PEG-coated and PEG-coated and water-treated a-SiC films. The C-O peak at 287eV in the PEG coated sample and the C-O shoulder at 287eV in the C-C peak in the DI water treated sample indicate both the presence and durability of the PEG coating on the a-SiC:H films.

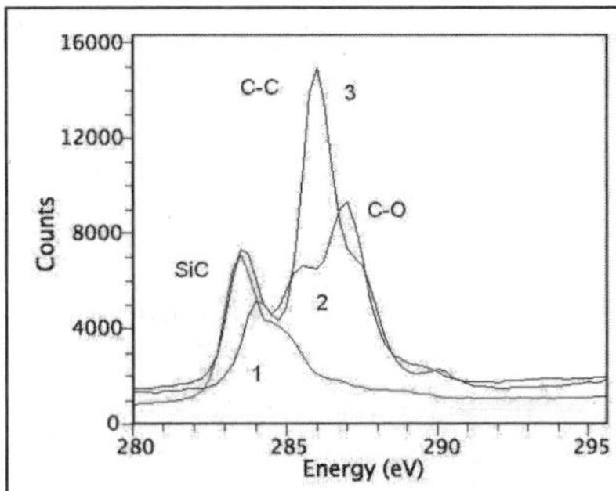

Fig 2. XPS analysis: (left) Carbon 1S peaks for ("1") uncoated, ("2") PEG coated, and ("3") H_2O exposed surfaces. The C-O bond peak indicates the presence of a PEG film on SiC.

Fig. 3 presents a series of immunofluorescence micrographs from the biofouling study for uncoated and PEG-coated Si, a-SiC:H and thermally-grown SiO_2. The images clearly show that cells readily attach to uncoated and PEG-coated Si and SiO_2 surfaces after 12 days of exposure. In contrast, cells resist attachment to the uncoated a-SiC:H surfaces while the PEG-coated a-SiC:H surfaces inhibit attachment altogether.

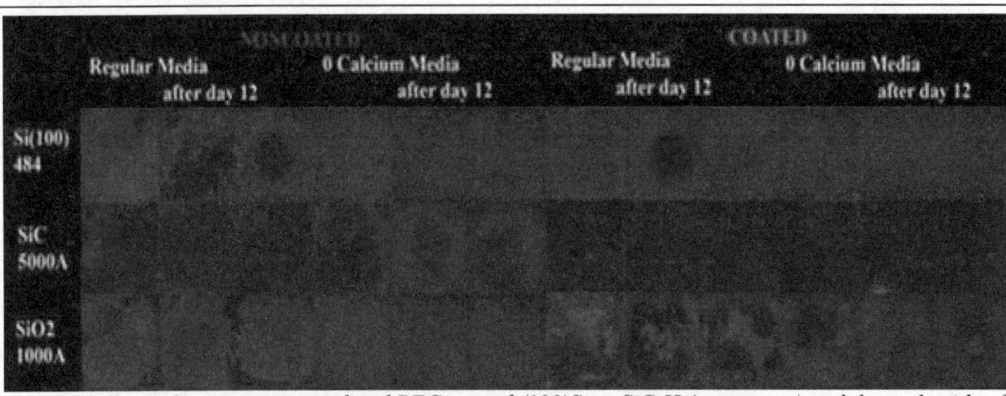

Fig. 3. Cell attachment to uncoated and PEG-coated (100)Si, a-SiC:H (center row) and thermal oxide of Madin-Darby Canine Kidney cells in regular and 0 calcium media after 12 days

FTIR analysis (Fig. 4) was performed on an unprocessed test coupon from the aforementioned biofouling experiment as well as on one of the diaphragm chips from the load-deflection experiment. FTIR analysis performed on the non-biofouling sample (σ = -50 MPa) exhibits peaks corresponding to $Si(CH_3)$ (1259/cm) and SiH_n. (2900/cm). In contrast, the mechanically-robust sample (σ = 110 MPa) exhibits a peak not seen in the nonfouling sample associated with the Si-H stretching mode (2089/cm). Both samples exhibit a Si-C bond peak at 810/cm characteristic of SiC.

Discussion and Conclusion

In terms of mechanical properties, the data reported herein suggest that a-SiC:H deposited by PECVD using trimethylsilane has the characteristics need for a MEMS structural/packaging material. A Young's modulus of 75 GPa is considerably lower that what is typically reported for polycrystalline and single crystalline 3C-SiC (350 to 400 GPa) [5], and moderately lower than what has been reported for a-SiC films deposited by PECVD using SiH_4 and CH_4 (129 GPa) [6]. The lower Young's modulus is advantageous in situations where the SiC film is to be used in conjunction with polymeric materials (ie., as a protective or diffusion barrier layer) which typically have Young's moduli in the low GPa range. A burst stress of 800 MPa compares quite favorably with that measured for polycrystalline 3C-SiC (920 MPa) [5].

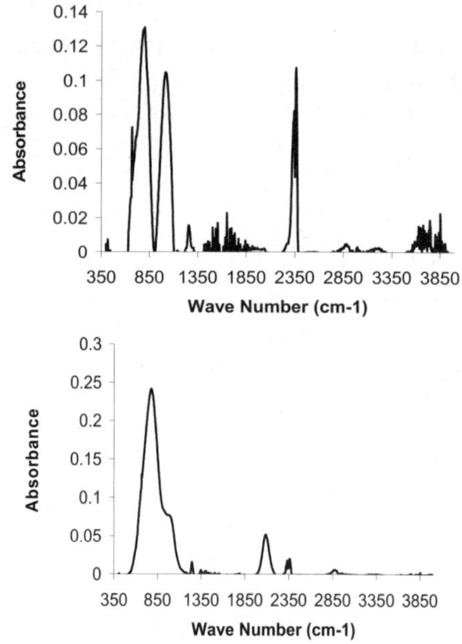

Fig 4. FTIR spectra: (top) non-biofouling SiC film, and (bottom) mechanically-robust SiC film.

The a-SiC:H films investigated in this project performed equally well in the biofouling study. Untreated SiC films resisted biofouling much better than their silicon derivative counterparts, although evidence of some fouling could be observed on the SiC surfaces. Unlike Si and SiO_2, SiC is known to be hydrophobic, and it is this property that is likely the key contributor to its antifouling characteristics. In terms of PEG-coated samples, the data not only indicate that a-SiC:H can successfully be coated with PEG, but also that the PEG coating on a-SiC:H is quite durable, having maintained its potency after 12 days whereas the same PEG coating on Si and SiO_2 samples have lost nearly all its effectiveness. Interpretation of the FTIR spectra is ongoing, but the data suggest that residual stress in the film could be related to the chemical bond structure.

Acknowledgements

This work was funded by NSF (0327674 and 0244048), the Dept. of Defense (W81XWH-05-2-0010), Case Alumni Assoc., and the Diabetes Assoc. of Greater Cleveland (DDRI).

References

[1] G. Kotzar, M. Freas, P. Abel, A. Fleischman, S. Roy, C. Zorman, J. Moran, J. Melzak, Biomat, 23, (2002) 2737-2750.

[2] J. Mitchell, C.A. Zorman, T. Kicher, S. Roy and M. Mehregany, J Aerospace Eng, 16 (2003) 46-54.

[3] M. Zhang, T. Desai, M. Ferrari, Biomat, 19 (1998) 953-960.

[4] T. Desai, D. Hansford, L. Leoni, M. Essenpreis, and M. Ferrari, Biosen Bioelect, 15 (2000) 453-462.

[5] S. Roy, C. A. Zorman, M. Mehregany, R. DeAnna, and C. Deeb, J Appl Phys, 99, (2006) 44108-1-13.

[6] R.J. Parro, M.C. Scardelletti, N.C. Varaljay, S. Zimmerman, and C.A. Zorman, Solid State Elect, 52, (2008) 1647-1651.

Materials Science Forum Vols. 717-720 (2012) pp 541-544
doi:10.4028/www.scientific.net/MSF.717-720.541

Seebeck Coefficient of Heavily Doped Polycrystalline 3C-SiC Deposited by LPCVD

Man I Lei[1,a] and Mehran Mehregany[2,b,*]

[1]Materials Science and Engineering Department, Case Western Reserve University, 10900 Euclid Avenue, Cleveland, OH 44106 USA

[2]Electrical Engineering and Computer Science, Case Western Reserve University, 10900 Euclid Avenue, Cleveland, OH 44106 USA

[a]leimani@gmail.com, [b]mehran@case.edu, *corresponding author

Keywords: polycrystalline silicon carbide, thermoelectric, Seebeck coefficient, thermopile

Abstract. The Seebeck coefficient of heavily-nitrogen-doped n-type polycrystalline 3C-SiC (n-SiC) and platinum (Pt) thin films has been measured from room temperature up to 300 °C by using a microfabricated test structure. At room temperature, the absolute Seebeck coefficient of the n-SiC is -10 μV/°C. With ambient temperature increase, the absolute Seebeck coefficient of the n-SiC is found to gradually increase, reaching -20 μV/°C at 300 °C.

Introduction

Silicon Carbide (SiC) is of interest in micro-electro-mechanical system (MEMS) for high temperature and harsh environment sensor applications due to SiC's excellent mechanical properties and high temperature stability [1, 2]. While the mechanical and electrical properties have been studied [3, 4], little information on the thermal properties of SiC thin films is available. Here, we have studied the Seebeck coefficient of a heavily-nitrogen-doped polycrystalline SiC thin film (n-SiC) deposited by low pressure chemical vapor deposition (LPCVD) from room temperature up to 300 °C.

To observe the Seebeck effect, a circuit composed of two materials is always used. The thermoelectric voltage (ΔV) generated due to the temperature difference (ΔT) occurring in the cold and hot ends of thermocouple junctions can be expressed by Eq. (1a), where n is the number of thermocouple junctions, and S_A and S_B are the Seebeck coefficients of materials A and B, respectively, with

$$\Delta V = nS_{A/B}\Delta T \tag{1a}$$

and $$S_{A/B} = S_A - S_B. \tag{1b}$$

In this study, a microfabricated test structure is used to measure the Seebeck coefficient of n-SiC and platinum (Pt) from room temperature to 300 °C. Then, the absolute Seebeck coefficient of n-SiC is compared with platinum, which acts as a reference element with well-established absolute Seebeck coefficient data [5, 6].

Design and Fabrication

The test structure depicted in Fig. 1 was designed for the Seebeck coefficient measurement. Each thermocouple is comprised of one 15 μm-wide, 930 nm-thick n-SiC leg and one 8 μm-wide, 300 nm-thick Pt leg. Both elements were 750 μm-long. There were a total of 16 thermocouples, with 32 junctions connected together in series to create the thermopile. The temperature sensors (Fig. 1) were made of the n-SiC material and used to measure the hot- and cold-junction temperatures. The cold junctions of the thermopile were connected to the substrate, whereas the hot junctions were suspended over a cavity to enhance thermal isolation. The ΔV and ΔT measured experimentally were used to calculate the Seebeck coefficient of $S_{Pt/SiC}$ based on Eq. (1a).

Fabrication of this test structure required 5 lithography steps. Starting with (100) Si wafers having 0.5 μm-thick of thermally grown SiO_2, a 1 μm-thick heavily-nitrogen-doped n-SiC film was deposited by LPCVD using SiH_2Cl_2 and C_2H_2 as precursor gases and NH_3 as the doping gas [7]. The n-SiC film was patterned using reactive ion etching (RIE) in a mixture of Freon® 23 (CHF_3) and sulfur hexafluoride (SF_6) in order to define the heater, thermocouple element and temperature sensors. Next, the n-SiC film was thermally oxidized at 1000 °C to create a 0.15 μm-thick SiO_2 layer in order to electrically isolate the n-SiC and the subsequent Pt layers. As a result, the n-SiC film thickness decreased to 930 nm-thick. Next, a 300 nm-thick film of Pt was deposited by RF sputtering, using a 20 nm-thick titanium (Ti) adhesion layer, to create the second thermocouple element. Contact windows were opened on contact pads and thermocouple junctions prior to metallization. For high temperature Ohmic contact with n-SiC, a stack of Ti(100nm)/$TaSi_2$(400nm)/Pt (200nm) metallization [8] was sputtered and patterned by lift-off. Release holes, exposing the substrate, were created by buffered oxide etching of the masking SiO2, and the suspended sensor structure was obtained by anisotropic etching in a tetramethylammonium hydroixde (TMAH) solution at 65°C to form the substrate cavity.

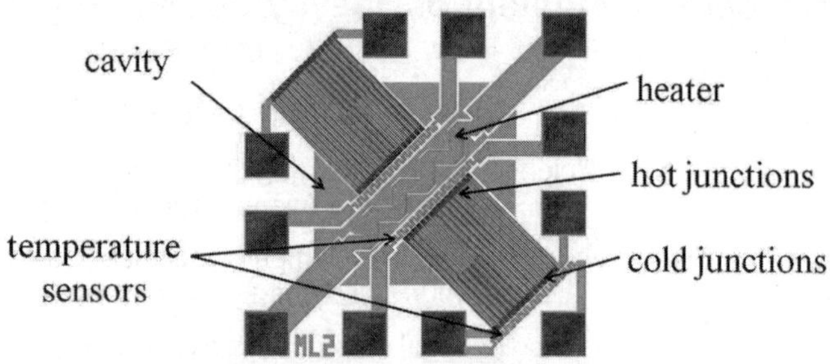

Fig. 1: Test structure for Seebeck coefficient measurement.

Results and Discussion

To measure the Seebeck coefficient of the Pt/n-SiC thermopile from room temperature to 300 °C, the test structure was mounted in a ceramic package, and gold wire bonding was used. The package was placed sensor-side down on a hotplate. The hotplate temperature represented the ambient temperature, while the n-SiC heater supplied the temperature difference between the hot- and cold-junctions. For room temperature measurement, the hotplate was off. A constant current of 2.0 mA was applied to the n-SiC heater, and the Pt/n-SiC thermopile produced 1.07 mV. Figure 2 presents the thermopile voltage response when 2.0 mA of constant current is applied to the n-SiC heater at 24 °C, 100 °C, 200 °C and 300 °C hotplate temperatures. From Fig. 2, the baseline voltage shifted up with increasing hotplate temperature. This baseline change was possibly caused by a small finite temperature difference that existed between the hot- and cold-junctions due to non-uniform heating within the test structure at elevated temperatures. The baseline shift is not likely to affect the measurement, since the n-SiC heater supplies a ΔT to the thermopile in addition to the ambient temperature, and the Seebeck coefficient only depends on ΔV measured by the thermopile in response to change of ΔT, but not the absolute temperature.

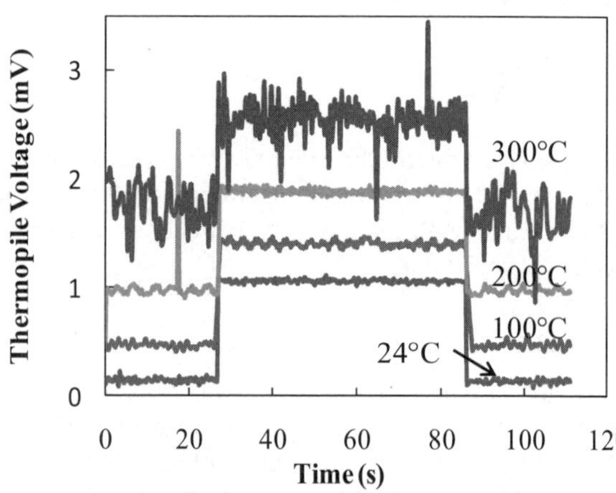

Fig. 2: Voltage generated by the thermopile at different temperatures in response to 2.0 mA of constant heater current.

As in Fig. 2, the thermopile voltage gets noiser as the temperature increases, and this noise is most likely cause by the thermal noise generated from the thermopile.

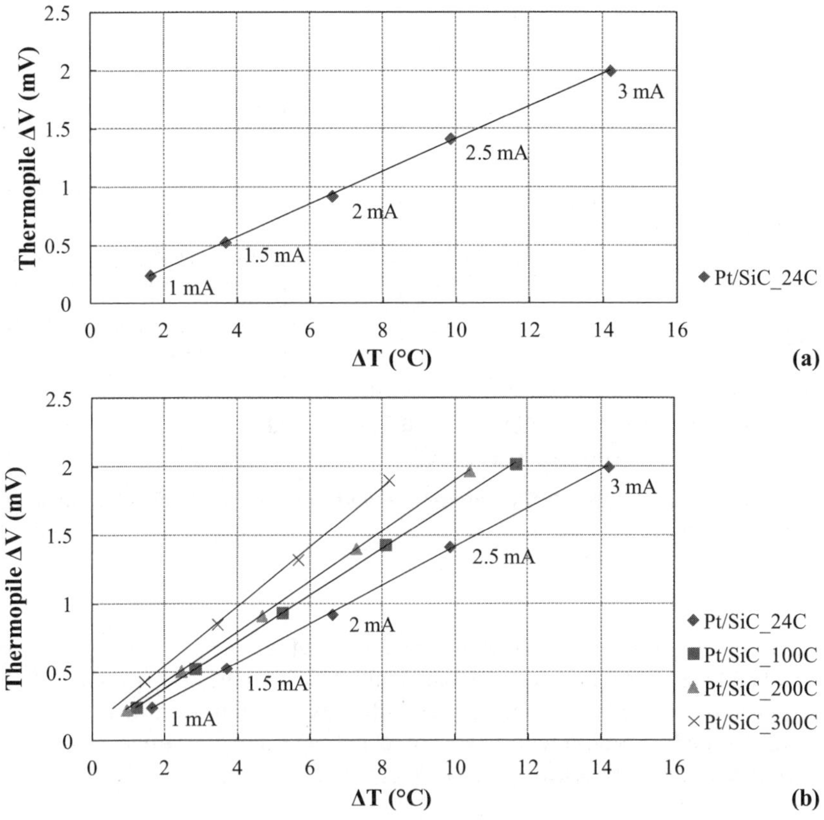

Fig. 3: The voltage generated by the thermopile is plotted against the temperature difference (ΔT) between the hot- and cold-junctions for different heater currents: **(a)** at room temperature; and **(b)** at elevated temperature up to 300 °C.

At room temperature, a constant current of 1.0 mA, 1.5 mA, 2.0 mA, 2.5 mA or 3.0 mA was applied to the n-SiC heater. From Fig. 3a, the amount of voltage generated by the thermopile is found to be linearly proportional to the ΔT between the hot- and cold-junction. The ΔT was measured by the temperature sensors (as in Fig. 1) located at the hot- and cold- junctions using the temperature coefficient of resistance technique. Then, the Seebeck coefficient of Pt/n-SiC was calculated to be 5 μV/°C at 24 °C based on Eq. (1a). The same measurement was performed at different hotplate temperatures as presented in Fig. 3b. With increasing temperature, the Seebeck coefficient of Pt/n-SiC increased to 7.5 μV/°C at 300 °C.

The Seebeck coefficient of Pt reported in the literature as a function of temperature is plotted in Fig. 4 [5]. These values were used to calculate the absolute Seebeck coefficients of n-SiC based on Eq. (1b). The Seebeck coefficient of the Pt/n-SiC thermopile and the absolute Seebeck coefficient of n-SiC as a function of temperature are also shown in Fig. 4. The Seebeck coefficient of n-SiC is -10 μV/°C at room temperature, increasing to -20 μV/°C at 300 °C.

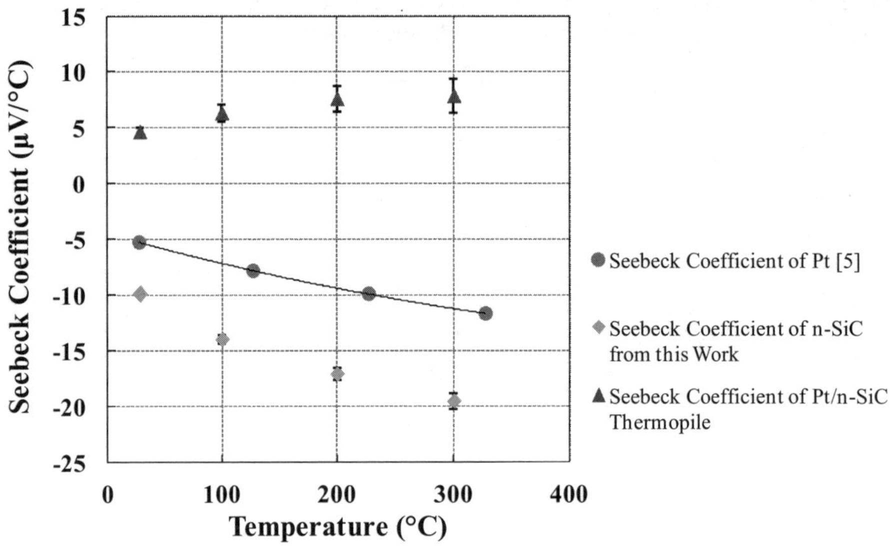

Fig. 4: Seebeck coefficient of the Pt/SiC thermopile and the absolute Seebeck coefficient of n-SiC at different temperatures.

Summary

A microfabricated test structure was developed and used to measure the combined Seebeck coefficient of Pt and n-SiC as a function of temperature. Based on these measurements, the absolute Seebeck coefficient of the studied n-type polycrystalline 3C-SiC thin film was calculated. At room temperature, the absolute Seebeck coefficient of the n-type polycrystalline 3C-SiC thin film is found to be -10 μV/°C, and it increases to -20 μV/°C at 300 °C.

References

[1] M. Mehregany, C. A. Zorman, N. Rajan and C. H. Wu, Silicon carbide MEMS for harsh environments, Proceedings of the IEEE 86 (1998) 1594-1609.

[2] N. G. Wright and A. B. Horsfall, SiC sensors: A review, J. Phys. D: Appl. Phys. 40 (2007) 6345.

[3] C. S. Roper, V. Radmilovic, R. T. Howe and R. Maboudian, Electrical and mechanical characterization of doped and annealed polycrystalline 3C-SiC thin films, J. Electrochem. Soc. 156 (2009) D5-D10.

[4] F. Liu, C. Carraro, A. A. Pisano and R. Maboudian, Growth and characterization of nitrogen-doped polycrystalline 3C-SiC thin films for harsh environment MEMS applications, J. Micromech. Microeng. 20 (2010) 035011.

[5] J. P. Moore and R. S. Graves, Absolute Seebeck coefficient of platinum from 80 to 340 K and the thermal and electrical conductivities of lead from 80 to 400 K, J. Appl. Phys. 44 (1973) 1174-1178.

[6] D. D. Pollock, Thermoelectric phenomena, in: Physical Properties of Materials for Engineers, CRC Press, Inc., Florida, 2002, pp. 229-264.

[7] X. Fu, J. L. Dunning, C. A. Zorman and M. Mehregany, Polycrystalline 3C-SiC thin films deposited by dual precursor LPCVD for MEMS applications, Sens. Actuators, A. 119 (2005) 169-176.

[8] S. Jin, X. Fu and M. Mehregany, Ohmic contacts on m-type polycrystalline silicon carbide with Ti/$TiSi_2$/Pt, in TRANSDUCERS, Solid-State Sensors, Actuators and Microsystems, 15th International Conference on 2009, Denvor, CO (2009) 1083-1086.

Materials Science Forum Vols. 717-720 (2012) pp 545-548
© (2012) Trans Tech Publications, Switzerland
doi:10.4028/www.scientific.net/MSF.717-720.545

OBIC measurements on avalanche diodes in 4H-SiC for the determination of impact ionization coefficients

D. M. Nguyen[1,a,*], C. Raynaud[1,b], M. Lazar[1,c], G. Pâques[2,d], S. Scharnholz[2,e], N. Dheilly[1,f], D. Tournier[1,g], D. Planson[1,h]

[1] Université de Lyon, INSA de Lyon, Ampère, UMR 5005, 20 avenue A. Einstein, F-69621 Villeurbanne Cedex, France

[2] French-German Research Institute of Saint Louis (ISL), 5 rue du Général Cassagou, F-68301 Saint Louis Cedex, France

[a]duy-minh.nguyen@insa-lyon.fr, [b]christophe.raynaud@insa-lyon.fr, [c]mihai.lazar@insa-lyon.fr, [d]gontran.paques@rwth-aachen.de, [e]sigo.scharnholz@isl.eu,[f]nicolas.dheilly@insa-lyon.fr, [g]dominique.tournier@insa-lyon.fr, [h]dominique.planson@insa-lyon.fr, *corresponding author

Keywords: 4H-SiC, Avalanche diode, OBIC, Impact ionization coefficients, Breakdown voltage, Device simulation.

Abstract. Optical Beam Induced Current (OBIC) measurements have been performed on 4H-SiC avalanche diodes with a very thin and highly doped avalanche region. The light source used in this study is an Ar-laser with a wavelength of 351 nm which results in a mixed carrier injection. From these measurements, impact ionization coefficients for 4H-SiC have been extracted in the electric field range from 3 to 4.8 MV/cm. In combination with ionization coefficients in our previous paper extracted from diodes with a low-doped avalanche region, we propose a set of parameters of impact ionization coefficients for 4H-SiC, applicable to a wide electric field range.

Introduction

The knowledge of impact ionization coefficients is essential for the accurate prediction of the avalanche breakdown voltage of power devices. 4H-SiC is increasingly preferable to 6H-SiC for power device applications due to its high mobility, high critical electric field, high thermal stability and especially crystal maturity. However, few articles on the impact ionization coefficients in 4H-SiC, particularly in high electric fields, have been published [1-3]. In this paper, we present extracted ionization coefficients for 4H-SiC in high electric fields from OBIC measurements on avalanche diodes with a thin and highly doped avalanche region.

Experimental

The avalanche diodes used in this study show an abrupt avalanche breakdown voltage of about 59 V which is presented in detail in a previous paper [4]. To perform OBIC measurements, optical windows have been opened on the anode metallizations by using the Secondary Ion Mass Spectrometry (SIMS) technique [5]. OBIC spectra have been measured on two diodes of the same substrate.

Fig. 1 shows the schematic structure of the diodes excited by laser light focused on the optical window. The light source used in this study is an Ar-laser with a wavelength of 351 nm which presents a penetration length in 4H-SiC of 35 μm [6]. This penetration length results in a mixed carrier injection for the diodes. This is taken into account in our calculations.

Results and Discussion

The OBIC spectra, measured on the diodes for reverse bias voltages ranging from zero volts to avalanche voltage, are presented in Fig. 2. An increase of the OBIC signal with reverse bias voltage (V_r) can clearly be observed over the optical window. From 0 V to 20 V, the OBIC signal is almost identical, which indicates that the space charge region of the diodes is depleted at 0 V. The increase of the OBIC signal starts at 30 V and continues until the breakdown voltage. From Fig. 2, we can deduce the multiplication coefficient (M), which is the ratio of the OBIC signal at a reverse voltage V_r to the OBIC signal at 0 V.

The measured multiplication coefficient (M) as a function of V_r is shown in Fig. 3. A high value (greater than 100) of the multiplication coefficient was achieved at the maximum applied voltage. The impact ionization coefficients are determined experimentally by fitting the measured M-V_r curve with the empirical law of impact ionization coefficients (Eq.1) suggested by Chynoweth [7]:

$$\alpha_{n,p} = a_{n,p} \times \exp\left(-\frac{b_{n,p}}{E}\right), \tag{1}$$

where $\alpha_{n,p}$ represent the ionization coefficients for electrons and holes, respectively, E is the electric field and $a_{n,p}$, $b_{n,p}$ are fitting parameters.

The fitted M-V_r curve is also shown in Fig. 3 and the extracted parameters of the electron and hole impact ionization coefficients in the electric field range of (3 - 4.8) MV/cm are given in Table 1. The minimum value of this field range corresponds to the beginning of the multiplication and the maximum to the breakdown field value. Simulated multiplication curves by Sentaurus [8] using coefficients reported in our previous paper [9] and literature data [1-3] are also presented in Fig. 3.

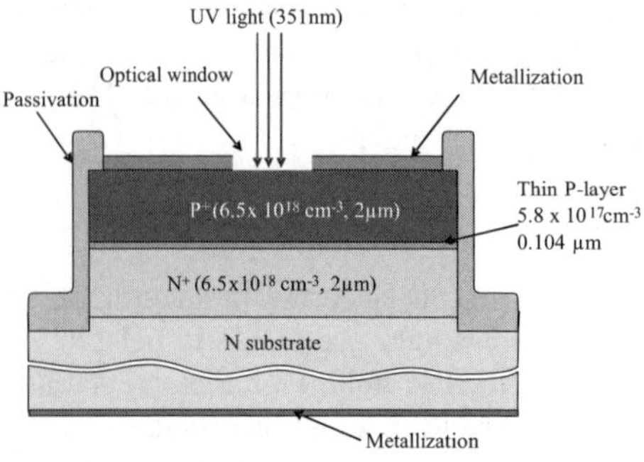

Fig. 1. Schematic structure of the diodes with an optical window.

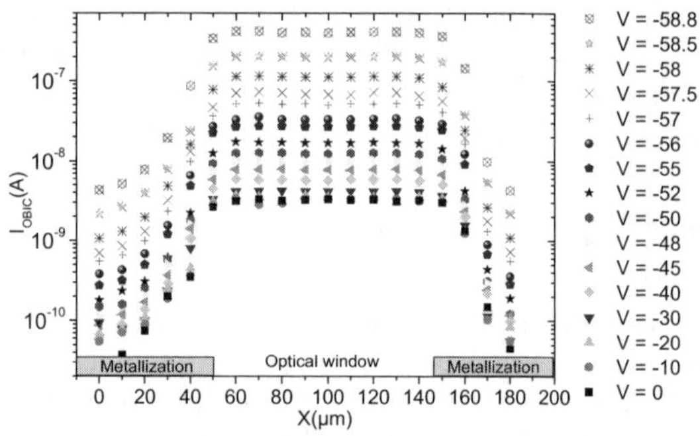

Fig. 2. OBIC spectra measured for numerous reverse bias voltages in volts. Same spectra are obtained for two diodes.

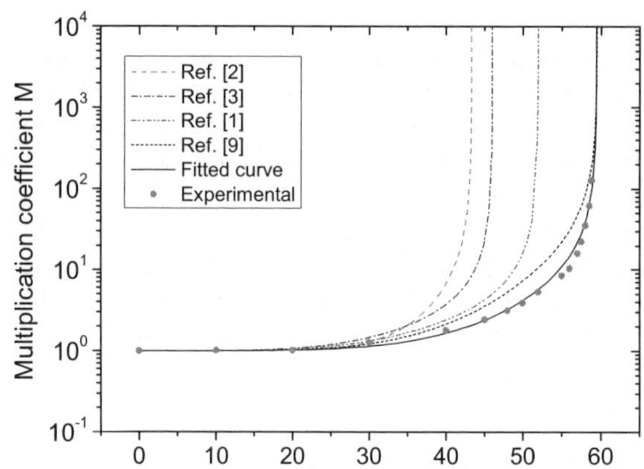

Fig. 3. Multiplication coefficient versus reverse bias voltage.

We can see that the results of simulation with the extracted coefficients in [9] and this study provide the same breakdown voltage but the shape of the multiplication curve is somewhat different. The simulated curve with the data in [9] starts to rise earlier and joins the simulated curve with the parameters deduced from this study in the avalanche region. A good agreement was obtained between the simulation with our parameters and the experimental result. Among all the coefficients in the literature, the ones published in [1] give the closest result to ours.

Fig. 4 represents the dependence of our extracted impact ionization coefficients on the electric field in the range from 3 to 4.8 MV/cm from this study and in the range from 1.5 to 2.7 MV/cm in [9] in comparison with reported data in the literature [1-3]. At high electric field values, the increase of ionization coefficients of electrons is faster than that of holes. In any case, the ionization coefficient of electrons is always much lower than that of the holes.

In combination with data reported in [9], we propose, in Table 1, a set of parameters of impact ionization coefficients for 4H-SiC, applicable to a wide electric field range with a threshold field of 3 MV/cm. This is the field value to switch the parameters between the two field ranges.

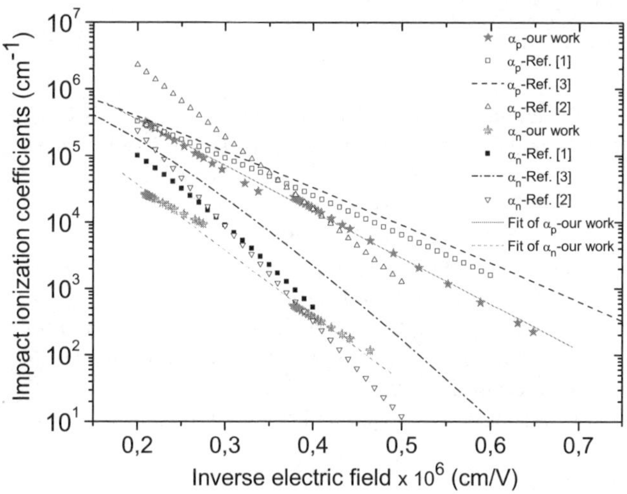

Fig. 4. Dependence of impact ionization coefficients on the electric field.

Table 1. Parameters of impact ionization coefficients for 4H-SiC applicable to a wide electric field range with a threshold field of 3 MV/cm.

4H-SiC	Parameters of impact ionization coefficients			
	a_p $(\times 10^7 cm^{-1})$	b_p $(\times 10^7 Vcm^{-1})$	a_n $(\times 10^6 cm^{-1})$	b_n $(\times 10^7 Vcm^{-1})$
E $\geq$ 3 MV/cm (3 – 4.8 MV/cm)	1.52	1.86	0.68	1.56
E < 3 MV/cm (1.5 – 2.7 MV/cm)	1.56	1.72	0.46	1.78

Table 2. Parameters of impact ionization coefficients for 4H-SiC used to fit the α_p and α_n in our work presented in Fig. 4.

4H-SiC	Parameters of impact ionization coefficients			
	a_p $(\times 10^7 cm^{-1})$	b_p $(\times 10^7 Vcm^{-1})$	a_n $(\times 10^7 cm^{-1})$	b_n $(\times 10^7 Vcm^{-1})$
E = 1.5 – 4.8 MV/cm	0.85	1.597	0.336	2.26

Another method is to use the same parameters of impact ionization coefficients for the full range of the examined electric field as presented in Table 2.

Conclusion

4H-SiC avalanche diodes with a very thin and highly doped avalanche region used in this study show an abrupt avalanche breakdown voltage of about 59 V. Optical windows have been opened on the anode metallizations and OBIC measurements have been performed on these diodes. From these measurements, impact ionization coefficients for 4H-SiC have been extracted in the electric field range of (3 - 4.8) MV/cm. In combination with ionization coefficients in our previous paper in the field range of (1.5 - 2.7) MV/cm extracted from diodes with a low-doped avalanche region, we propose a set of parameters of impact ionization coefficients for 4H-SiC, applicable to a wide electric field range.

Acknowledgements

This work was supported in part by ANR-VHVD-SiC project (France-ANR-08-BLAN-0191) and the authors would like to acknowledge Francois Jomard from the Université de Versailles-Saint-Quentin (UVSQ) for the SIMS analyses.

References

[1] A. O. Konstantinov, Q. Wahab, N. Nordel and U. Lindefelt, Appl. Phys. Lett. **71** (1997), p. 90.
[2] T. Hatakeyama, T. Watanabe, T. Shinohe, K. Kojima, K. Arai and N. Sano, Appl. Phys. Lett. **85** (2004), p.1380.
[3] W. S. Loh, B. K. Ng, J. S. Ng, S. I. Soloviev, H-Y. Cha, P. M. Sandvik, C. M. Johnson and J. P. R. David, IEEE T. Electron Dev. **55** (2008), p. 1984.
[4] D. M. Nguyen, G. Pâques, N. Dheilly, C. Raynaud, D. Tournier, J. P. Konrath, S. Scharnholz and D. Planson, Mater. Sci. Forum **679-680** (2011), p. 567.
[5] M. Lazar, F. Jomard, D. M. Nguyen, C. Raynaud, G. Pâques, S. Scharnholz, D. Tournier and D. Planson, "SIMS analyses applied to open an optical window in 4H-SiC devices for electro-optical measurements", these proceedings.
[6] S. G. Sdrihara, T. J. Eperjesi, R. P. Devaty and W. J. Choyke, Mater. Sci. Eng. B **61-62** (1999), p. 229.
[7] A. G. Chynoweth, Phys. Rev. **109** (1958), p.1537.
[8] Sentaurus, TCAD simulation tool by Synopsys Inc., ver. E-2010 (2010).
[9] D. M. Nguyen, C. Raynaud, N. Dheilly, M. Lazar, D. Tournier, P. Brosselard and D. Planson, Diam. Relat. Mater. **20** (2011), p. 395.

Materials Science Forum Vols. 717-720 (2012) pp 549-552

doi:10.4028/www.scientific.net/MSF.717-720.549

Radiation Hardness of Wide-Bandgap Materials as Exemplified by SiC Nuclear Radiation Detectors

A.M. Ivanov[1,a], A.V. Sadokhin[1,b], N.B. Strokan[1,c], A.A. Lebedev[1,d], and V.V. Kozlovski[2,e]

[1]Ioffe Physicotechnical Institute, Russian Academy of Sciences, 26 Politekhnicheskaya str., St. Petersburg, 194021 Russia

[2]St. Petersburg State Polytechnical University, 29 Polytekhnichaskaya, St. Petersburg, 195251, Russia

[a]alexandr.ivanov@mail.ioffe.ru, [b]aln9t7o9n@mail.ioffe.ru, [c]nikita.strokan@mail.ioffe.ru [d]shura.lebe@mail.ioffe.ru, [e]kozlovski@tuexph.stu.neva.ru

Keywords: Radiation Hardness, Carrier Localization, Polarization, Nuclear Radiation Detector

Abstract. The polarization effect characteristically occurs in detectors based on wide-bandgap materials at considerable concentrations of radiation defects. The appearance of an electromotive force in the bulk of a detector is due to the long-term capture of carriers at deep levels related to radiation centers. The kinetics and strength of the polarization field have been determined. The capture can be controlled by varying the detector temperature, with a compromise reached at the "optimal" temperature between the generation current and the position of the deepest of the levels whose contribution to the loss of charge via capture is negligible. It has been found that the depth of a level (related to the energy gap width) is close to 1/3, irrespective of the material. The optimal temperatures are specific to individual materials.

Introduction

The notion radiation hardness means that, despite a hard radiation fluence received by a detector, its parameters remain in a required range of values. Important factors are the nature and concentration of centers that appear in the material. These centers change the generation-recombination rate of carriers and the conductivity of the starting material.

Below, the relationship between the above processes is traced for the example of operation of heavily irradiated SiC nuclear radiation detectors. The association between the conductivity and the nature of carrier capture and accumulation of a charge by radiation defects is demonstrated. The optimal operating temperature is found, at which carrier capture by centers is precluded, but only a minimal undesirable rise in the generation current occurs.

Mechanism of Carrier Capture and Polarization Effect

SiC belongs to a class of wide-bandgap semiconductors. Consequently, an important role in changes in conductivity belongs to the self-compensation process [1]. It was noted already in the first studies that the conductivity of SiC decreases to the point of its becoming a semi-insulator (for example [2]). On the mechanism by which the nonequilibrium charge drifting in the detector is captured such a change means that the localization (trapping) dominates over the recombination. The carrier localization on deep levels (long-term capture) inevitably gives rise to "polarization" of the detector bulk. We observed this negative effect as the appearance of an "internal" electric field opposite to that created by the voltage applied to the detector [3–6]. The field appeared when a detector (4H-SiC surface barrier structure) with a considerable (1.2×10^{17} cm^{-3}) concentration of primary radiation defects was tested with α-particles.

Figure 1 (curve 1) illustrates the characteristic decay of the charge collection efficiency (CCE)

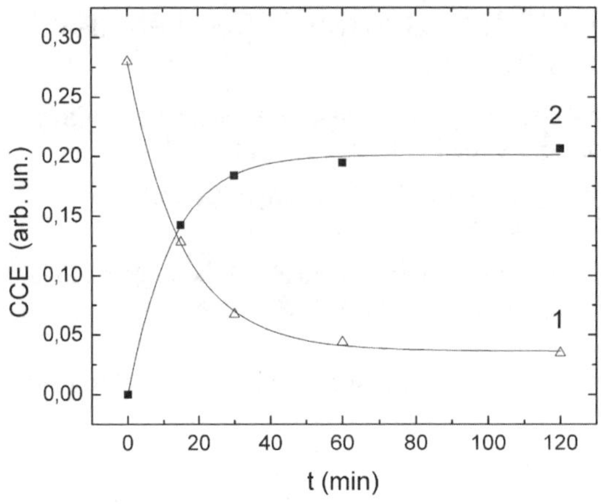

Fig. 1. (1) Decay of the CCE in the course of time and (2) rise in the opposite-polarity signal because of the polarization of the detector bulk.

with the time of testing. It is important that switching-off the external voltage source did not result in disappearance of the signal. However, the signal changed its polarity and its value increased in the course of time (curve 2). The effect is attributed to the appearance of an electromotive force because of the accumulation of the charge of captured nonequilibrium electrons and holes. Created by α-particles in pairs, they are separated by the external field and, being captured, form space charges of opposite signs. At room temperatures, the polarization effect can hinder use of detectors based on wide-bandgap materials.

Let us analyze the kinetics and strength of the polarization field F_p.

(1) First, values of F_p can be found from curve 1 for the run of the dependence of the CCE on the time of testing. We use the known Hecht formula [7] for describing the CCE. With the experimental values of the CCE available in Fig. 1, we find values of the argument of the function. We fully attribute the decrease in the CCE in the course of measurements to the appearance of the polarization electromotive force. Figure 2 (curve 1) shows thus obtained values of the polarization field, which are as high as $F_p \approx 75$ kV/cm.

(2) It is noteworthy that there is a positive feedback in the charge accumulation kinetics. The process is accelerated over the course of time because the CCE decreases as the field in the detector becomes weaker. On the whole, the process is represented by two equations that describe the increase in the charge density (σ) on plates of a capacitor equivalent to the detector and in the polarization field strength $F_p = \sigma/(\varepsilon\varepsilon_0)$ related to this charge

$$\frac{d\sigma}{dt} = Y(1 - CCE) - \frac{\sigma}{\gamma}. \qquad (1)$$

Here $\varepsilon_0 = 8.85\ 10^{-12}$ F/m and ε is the relative dielectric constant. The constant $\gamma = 12$ min describes the relaxation of the charge being captured. Under the conditions in which the data in Fig. 1 was obtained (5.4 MeV α-particles, rate of incidence onto the detector ~30 s^{-1}), the generation rate is $Y \approx 3.4\ 10^{-10}$ C s^{-1} cm^{-2}. Our interest here is in the solution to Eq. 1 with the initial conditions $\sigma = \sigma_0$ at t = 0. The solution makes it possible to find how σ (and, consequently, also the field F_p) depend on time by using several successive steps over the time intervals in Fig. 1. Accordingly, the values of the initial conditions, obtained as a solution to the equation in the preceding step, increase, too. The field strengths determined in this way are compared in Fig. 2 with F_p found directly from the experimental curve 1 in Fig. 1. As follows from Fig. 2, the values of F_p coincide as regards the run in the course of time. This confirms the validity of describing the kinetics by Eq. 1.

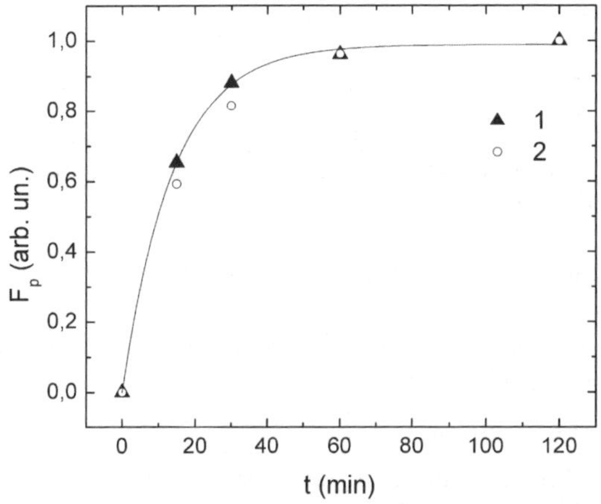

Fig. 2. Time dependence of the polarization field strength: (1) according to the data in Fig. 1 and (2) based on the charge accumulation kinetics. The maximum field strength for the data in Fig. 1 is 75 kV/cm. The solid line is a unification of the data (1) and (2).

Optimal Operating Temperature in a Single-level Model

Raising the temperature effectively diminishes the carrier capture. Unfortunately, the generation current grows simultaneously. Let us find the temperature T_0 at which carrier capture can be precluded and the generation current does not exceed a prescribed value.

We now consider for simplicity the case of a single level in the upper half of the energy gap lying at a depth E_m below the conduction band minimum. Then, capture means that a nonequilibrium carrier is localized at a level for a time exceeding the time of signal shaping by electronic circuits. Nonequilibrium holes, a component of pairs generated by radiation, do not recombine. Indeed, pairs are separated by the strong electric field in the operating zone of the detector. The recombination with holes from the valence band is not considered because of the sharp decrease in the conductivity of irradiated SiC. The quantities of our interest, T_0 and E_m, are related by two standard expressions [8]. The first of these describes the time constant θ of thermal excitation of an electron from the level E_m into the conduction band, and the second, the generation current density J_g. The expressions make it possible, with the given values for the generation current density J_g and time constant θ, to algebraically find the operating temperature T_0 and the E_m position

$$T_0 = \frac{E_g}{k \ln\left(\frac{A\theta B}{J_g}\right)}, \tag{2}$$

$$\frac{E_m}{E_g} = \frac{\ln(A\theta)}{\ln\left(\frac{A\theta B}{J_g}\right)}. \tag{3}$$

Here k is the Boltzmann constant; $A = \sigma_n v N_c$; $B = 0.5ew\sigma_n v M(N_c N_v)^{1/2}$. σ_n is the electron capture cross-section, v is the thermal velocity of electrons, N_c and N_v are the effective densities of states in the conduction and valence bands, e is the elementary charge, w is the width of the field region, and M is the E_m concentration.

Relation (3) shows that, in terms of the model under consideration, the position of the E_m level (normalized to the energy gap width E_g) is independent of E_g. Consequently, the fraction of the energy gap (E_m/E_g) in which carriers are excited from levels in a prescribed time θ is the same for different materials within this model. It is apparent that identical values of θ will be reached at

Table 1. Temperature T_0 and depth of the E_m level, which provide a generation current density of 1 nA cm^{-2} and a time constant θ of 10 μs for thermal release of a carrier from the level.

Parameter/material	Si	4H-SiC	Diamond
Energy gap width E_g [eV]	1.12	3.23	5.5
Depth of the E_m level [eV]	0.35	0.96	1.76
Temperature T_0 [K]	266	737	1245
E_m/E_g ratio	0.31	0.30	0.32

Note: the cross-section $\sigma_n = 10^{-15}$ cm^2 and concentration $M = 10^{15}$ cm^{-3} are taken for the E_m level. The densities of states, N_c and N_v, are proportional to $T^{3/2}$.

different temperatures, and the nature of a material is manifested in values of the temperature T_0 in accordance with Eq. 2.

Table 1 presents comparative data obtained not only for 4H-SiC, but also for Si and diamond as narrow-gap and wide-bandgap materials, respectively.

Summary

It was shown that three related processes are manifested in detectors based on wide-bandgap materials upon introduction of radiation defects. The deep conductivity compensation predetermines localization as a dominant kind of capture, which, in turn, gives rise to an electromotive force of polarization.

The situation was considered in which the carrier generation and capture are governed by radiation centers. A model is suggested in which the time during which a carrier is captured by a center and (simultaneously) the generation current density are set. For the simplest case of a single level, the optimal operating temperature and the depth of a level excluded from capture were determined (Table 1). A comparison of various materials demonstrated that only values of this temperature are strictly individual, whereas the level depth related to the energy gap width has a common value close to 1/3 (at the specified deep center parameters).

This result implies that comparison of the radiation hardness of materials is correct at optimal temperatures. For example, a polarization effect reducing the electric field in the detector appears in wide-bandgap materials at room temperatures. This diminishes CCE (in addition to making shorter the carrier lifetime). In an experiment, a compromise between the generation current and the positions of radiation centers excluded from carrier capture can be reached by choosing the temperature.

Acknowledgements

The authors are deeply grateful to M.E. Levinshtein for helpful discussions and valuable comments. The study was supported by a Grant of the President of the Russian Federation in Support of Leading Scientific Schools (NSh-3306.2010.2).

References

[1] G. Mandel, Phys. Rev. A 134 (1964) 1073-1079.
[2] A.A. Lebedev, A.I. Veinger, V.V. Kozlovski, D.V. Davydov, N.S. Savkina and A.M. Strelchuk, J. Appl. Phys. 88 (2000) 6265-6271.
[3] A.M. Ivanov, A.A. Lebedev and N.B. Strokan, Semiconductors 40 (2006) 1227-1231.
[4] A.M. Ivanov, N.B. Strokan, E.V. Bogdanova and A.A. Lebedev, Semiconductors 41 (2007) 115-119
[5] A.M. Ivanov, N.B. Strokan and A.A. Lebedev, Semiconductors 42 (2008) 1117-1121.
[6] A.M. Ivanov, N.B. Strokan and A.A. Lebedev, Nucl. Instrum. Methods A. 597 Issues 2-3 (2008) 203-206.
[7] K. Hecht, Zeitschrift Für Physik 77 (1932) 235–241.
[8] S. Sze, Physics of Semiconductor Devices, Wiley, New York, 1981.

Materials Science Forum Vols. 717-720 (2012) pp 553-556
© (2012) Trans Tech Publications, Switzerland
doi:10.4028/www.scientific.net/MSF.717-720.553

Terahertz electroluminescence of 6H-SiC natural SiC superlattice in Bloch oscillations regime

V.I. Sankin[a,*], A.V. Andrianov[1,b], A.G. Petrov[c], P.P. Shkrebiy and A.O. Zakhar`in[d] *corresponding author V.I. Sankin

A.F. Ioffe Physico-Technical Institute of RAS, 194021 St.-Petersburg

[a]sankin@mail.ioffe.ru, [b]alex.andrianov@mail.ioffe.ru, [c]petrov@mail.ioffe.ru, [d]Alex.Zaharin@mail.ioffe.ru, *corresponding author

Keywords: natural superlattice, miniband, Wannier-Stark localization, electric field, negative differential conduction, terahertz, impact ionization.

Abstract We report on efficient terahertz emission in the region of 1.5-2 THz from high electric field biased 6H-SiC structures with a natural superlattice at liquid helium temperatures. The shape of the emission spectrum, the linear dependence of its maximum on the bias and the characteristic field strength required to achieve the emission allow the emission to be attributed to steady-state Bloch oscillations of electrons in the SiC natural superlattice.

Introduction

One of the earliest projects of an electrically pumped THz emitter was based on the possibility of oscillatory motion of electrons in crystals at high field bias [1]. The Bloch oscillations (BO) with frequency (ν)

$$\nu = eFd/h \quad (1)$$

where e, h and d are the electron charge, the Planck constant and the crystal lattice period respectively, originate due to the acceleration of electrons in the electric field and the Bragg reflection of them at the Brillouin zone boundary. The threshold field of the BO regime, F_t, corresponds to the condition:

$$elF_t = E_1 \quad (2)$$

where l is the electron mean free path, and E_1 is the width of the allowed electron band (the conduction band in a bulk crystal or the first miniband in a supperlattice). Esaki and Tsu [2] pointed out that BO of electrons in artificial superlattices with narrow minibands can be observable even for modest fields.

In spite of numerous attempts, the THz radiation due to steady-state Bloch oscillations of electrons in the artificial superlattices was not observed up to now [3].

One interesting system demonstrating axial superperiodicity, but free of heterointerfaces, is a natural superlattice (NSL) in SiC crystals. This superperiodicity is self-organized in the main SiC crystal lattice and the NSL period $d = c/2$, where c is the period of the 6H-SiC crystal [4].

Electron transport studies on SiC structures with the natural superlattices [5] have demonstrated pronounced effects of the negative differential conductivity (NDC). The threshold fields of the NDC onset at 300 K were found to be $F_t \sim 110 \pm 20$ kV/cm, $F_t \sim 150 \pm 25$kV/cm and $F_t \sim 290 \pm 45$ kV/cm for the 8H, 6H and 4H polytypes, respectively [5], that is in good agreement with Eq. 2 taking into account the different widths of the first electron miniband in these polytypes (140, 260 and 500 meV for 8H-, 6H- and 4H-SiC, respectively [5]). These results provide a research guideline and well-grounded hope of achieving THz emission due to the Bloch oscillations in SiC NSL. In this paper we report on the first experimental observation of THz electroluminescence (1.5-2 THz spectral range) from 6H-SiC structures with a natural superlattice.

Experiment

The samples studied in this work were unipolar 6H-SiC n^+–n^-–n^+ diode structures. The n^- epitaxial layer (the base of the diode) was grown by the sublimation method on an on-axis 6H-SiC (0001) Lely substrate, which had $N_d - N_a \sim 2\text{x}10^{18}$ cm^{-3} and the thickness of 200 μm. The base had a donor concentration of 10^{15} cm^{-3} < $N_d - N_a$ < 10^{16} cm^{-3} and the thickness varied from 2 to 4 μm. The top n^+-layer with $N_d - N_a \sim 10^{20}$ cm^{-3} was fabricated on the n^- epitaxial layer by ion implantation of nitrogen with subsequent annealing at T =1500°C, t =15 min, in argon gas. The square mesas were made by dry etching after photolithographic procedure. The area of the square mesa was equaled to S = 3x10^{-5} cm^2. A sputtered and annealed (at 900°C, t = 2 min, in vacuum) nickel film with a thickness of 0.2 μm was used as contact to the n^+-layers. A common contact area was located on the upper surface of the substrate. Insulator layers were created by proton implantation on the mesa-periphery and on the upper surface of the substrate. The sample under test was fed with a train of 8 pulses (the negative polarity was applied to the top n^+-SiC layer), where each pulse in the train was 1 μs in duration with a 950 μs time interval between the pulses. The duty cycle of the train was 50% at a frequency of 75 Hz. Such a bias was used to minimize lattice heating effects.

The spectral measurements were performed in spectral region 0.15 – 10 THz with a spectral resolution of 0.6 meV using a step-scan Fourier spectrometer. The THz emission signal was measured using a liquid-helium-cooled Si bolometer and a lock-in amplifier. The geometry of the available structures permitted the observation of THz radiation only through the substrate. Polarization of the emission was not expected in this geometry of the experiment and, therefore, all measurements were made in the regime of integrated polarization.

Results and Discussion

An intense THz signal was detected at bias voltages exceeding 190-195 V (Fig. 1). The existence of such a threshold voltage for the THz emission can be explained by the impurity breakdown in the top n^+-SiC layer, which is required for injection of electrons into the NSL. We have determined that the spectrally integrated THz emission peak power is about 10 μW at 6.9 W of peak pumping power. The corresponding external quantum yield of the THz emission is about 0.01 photons/electron.

Fig. 1 demonstrates a set of THz electroluminescence (EL) spectra measured at different amplitudes of the bias voltage. It is seen that the THz emission spectrum consists of a single, symmetrical line and its maximum evidently shifts to higher frequencies with an increase in the bias voltage. The shift in the emission maximum is of order of 1.5 meV as the bias is varied from 200 to 255 V. As seen from Fig. 2, the shift of the THz emission peak versus the bias voltage can be well approximated by a linear law. It is important to note that the FWHM of the emission line is almost constant and equal to ~2.9 meV (0.7 THz) as the bias voltage varies from 200 to ~240 V (Fig. 2). Some increase in the spectral width of the emission line at voltages higher than 250 V is likely to be caused by the lattice heating effect.

The experimental data allow the observed THz EL to be attributed to spontaneous radiation resulting from electron Bloch oscillations in a SiC natural superlattice. Using Eq. 1 the electric field strength F required for the BO can be estimated. The estimations give the F to be of the order of 8.5 x 10^4 V/cm that is slightly less than the magnitude of F_t for the BO regime obtained from high-field transport measurement data at 300K on 6H-SiC (see Introduction). However, the agreement between the electric fields is quite reasonable if an increase of the electron scattering time at 7 K compared to its value at 300 K and a corresponding decrease of the threshold field of the BO regime are taken into account. The value of the electric field F implies that only a small part (about 15%) of the bias voltage drops on the base of the structure.

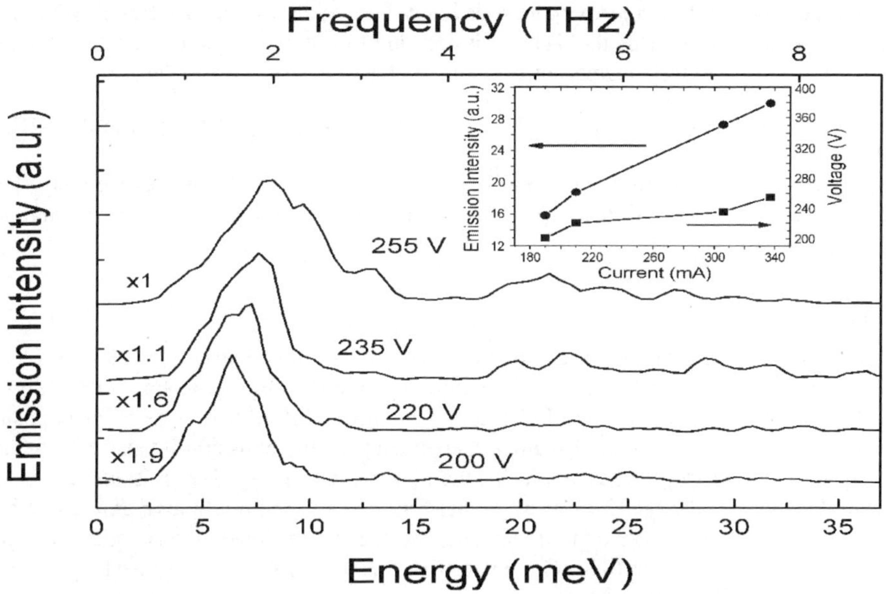

Fig. 1. Spectra of the THz emission from the 6H-SiC mesa structure at several bias voltages. The inset shows the I-V characteristic of the structure and the dependence of the spectrally integrated THz emission intensity on the current through the structure. Temperature T = 7 K.

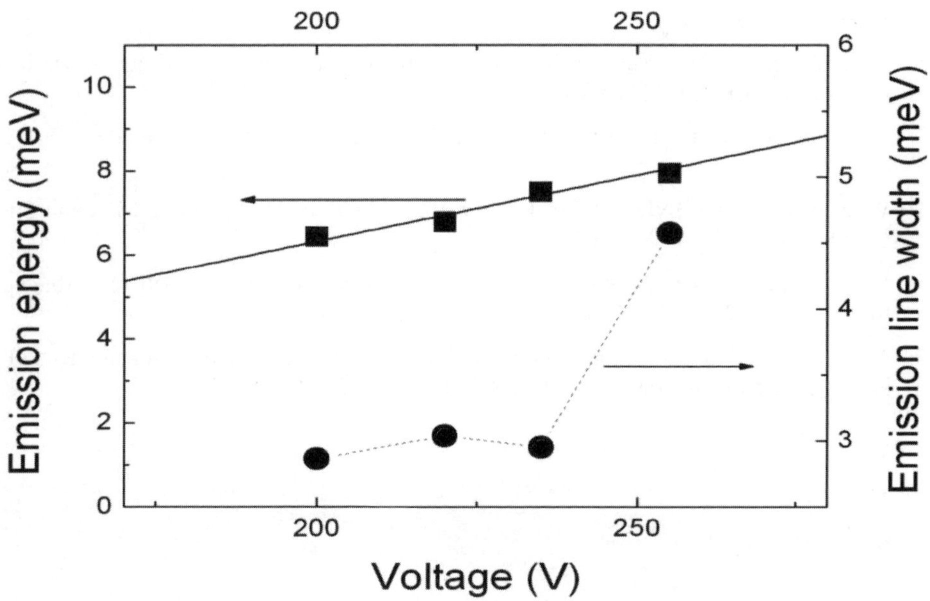

Fig. 2. Dependences of the peak position of the THz emission line (solid) and the emission line width (dotted) on the bias voltage.

It is necessary to add that the experimental geometry used was far from optimal. Therefore, in the case of the observation of emission from the side facet of the structure the expected external quantum yield of the THz emission should be higher by an order of magnitude.

Using the tight-binding approximation developed for such systems by Bouchard and Luban [6], the internal quantum yield of the THz emission due to optical transitions between Wannier-Stark levels in the natural superlattice of 6H-SiC was estimated. It is given by $\eta = \frac{L}{d} \times \frac{\tau_{rad}^{-1}}{\tau_{nonrad}^{-1}} \approx 2666.6 \times (1.8 \times 10^{5} s^{-1} / 1.3 \times 10^{10} s^{-1}) \approx 0.04$, where $L = 2$ μm is the NSL thickness, and τ_{rad} and τ_{nonrad} are the times for radiative and nonradiative transitions, respectively. This value is in reasonable agreement with the experiment if the non-optimality of the experimental geometry is taken into account.

Summary

The efficient THz EL from 6H-SiC structures with a natural supperlattice has been observed. The properties of the observed EL are: i) the single, symmetrical emission line with the maximum at 1.5-2 THz and the spectral width of 0.7 THz; ii) the linear shift of the emission peak position versus the bias electric field; iii) the almost linear dependence of the emission intensity on the current; iv) the characteristic electric field required for the observation of the THz emission and its correlation with the threshold of the NDC onset in the 6H-SiC NSL; v) the qualitative agreement between the results of the THz emission experiment and the theory of spontaneous radiative transitions of electrons in BO regime for 6H-SiC NSL testify that the emission is caused by the steady-state electron Bloch oscillations in the SiC NSL.

The discovered intensive THz EL with a variable frequency can find practical applications for electrically tunable THz emitters.

References

[1] C. Zener, Theory of the Electrical Breakdown of Solid Dielectrics, Proc. R. Soc. London Ser. A 145 (1934) 523-529.

[2] L. Esaki and R. Tsu, Superlattice and negative differential conductivity in semiconductors, IBM J. Res. Dev., 14 (1970) 61-65.

[3] P. Voisin, Wannier-Stark Effects in Semiconductor Superlattices, Ann. Phys. Fr., 22 (1997) 681-705.

[4] W. J. Choyke and Lyle Patrick, Exciton Recombination Radiation and Phonon Spectrum of 6H-SiC, Phys. Rev. 127 (1962) 1868-1877.

[5] V. I. Sankin, Wannier-Stark localization in the natural superlattice of silicon carbide polytypes, Semicond. 36 (2002) 717-739.

[6] A. M. Bouchard and M. Luban, Bloch oscillations and other dynamical phenomena of electrons in semiconductor superlattices, Phys. Rev. B 52 (1995) 5105-5123.

Materials Science Forum Vols. 717-720 (2012) pp 557-560

doi:10.4028/www.scientific.net/MSF.717-720.557

Emission enhancement of SiC/SiO_2 core/shell nanowires induced by the oxide shell

Filippo Fabbri[1,a], Francesca Rossi[1,b], Giovanni Attolini[1,c], Matteo Bosi[1,d,*], Giancarlo Salviati[1,e], Salvatore Iannotta[1,f], Lucrezia Aversa[2,g], Roberto Verucchi[2,h], Marco Nardi[2,3,i], Naoki Fukata[4,l], Benjamin Dierre[4,m], and Takashi Sekiguchi[5,n]

[1] IMEM-CNR Institute, Area delle Scienze 37/A, 43124 Parma (Italy)
[2] IMEM-CNR Institute, Via alla Cascata 56/C – Povo, 38123 Trento (Italy)
[3]Humboldt-Universitat zu Berlin, Institut fur Physik, Newtonstrasse 15, 12489 Berlin, Germany
[4] International Center for Materials Nanoarchitectonics, NIMS & PRESTO JST, 1-1 Namiki, Tsukuba, Ibaraki, 305-0044 (Japan)
[5] Nano device characterization group, Advanced Electronic Materials Center, NIMS, 1-1 Namiki, Tsukuba, Ibaraki, 305-0044 (Japan)

[a]fabbri@imem.cnr.it, [b]frossi@imem.cnr.it, [c]giovanni@imem.cnr.it [d]bosi@imem.cnr.it, [e]salviati@imem.cnr.it, [f]Iannotta@imem.cnr.it, [g]aversa@science.unitn.it,

[h]verucchi@science.unitn.it, [i]marco.nardi@physik.hu-berlin.de, [l]fukata.naoki@nims.go.jp,

[m]dierre.benjamin@nims.go.jp, [n]sekiguchi.takashi@nims.go.jp, *corresponding author

Keywords: Nanowire, characterization, emission, nanostructures

Abstract. In this work we report the enhancement of the 3C-SiC band edge luminescence induced by the SiO_2 shell in a SiC/SiO_2 core/shell nanowires (NWs) system. We demonstrate that the shell enhances the SiC near band edge luminescence and we argue the formation of a type-I quantum well between the SiC core and the SiO_2 shell, with the consequent injection of carriers from the larger band-gap shell to the narrower band-gap core.

Introduction

SiC is known to have poor optical emission; nevertheless considerable efforts have been made to increase its light-emission efficiency in order to adopt this material for optical devices. For instance, the use of porous or nanosized cubic SiC structures permits to significantly improve the SiC emission, but the effect was mainly due to surface states or quantum confinement [1-3].

In this work we report the enhancement of the 3C-SiC band edge luminescence induced by the SiO_2 shell in a SiC/SiO_2 core/shell nanowires (NWs) system. The optical emission was investigated using SEM-Cathodoluminescence (SEM-CL) spectroscopy. Raman spectroscopy is used to study the change in NW core/shell structures induced by HF etching.

High resolution Transmission Electron Microscopy (TEM), Energy Filtered High-Angle Annular Dark Field (HAADF) imaging spectroscopy and Energy Dispersive X-rays (EDX) are used to assess the structural properties of the NWs.

Experimental

β-SiC/SiO_2 core/shell NWs are grown by a CVD process on n-type Si (001) substrates, using carbon monoxide (CO) as the carbon source and nickel nitrate as the catalyst. The synthesis, performed at temperatures of 1100°C in an open-tube, is based on the carbothermal reduction. Silica was present as the native oxide on the substrate surface, where the reaction is catalyzed by Ni [3].

The morphology, structure and optical properties of the nanowires are characterized by SEM in a Field-emission Scanning Electron Microscope JEOL 6400F, by TEM in a Field-emission Transmission Electron Microscope JEOL 2200FS working at 200 kV, and by SEM-Cathodoluminescence in a Cambridge 360 Stereoscan microscope equipped with a Gatan MonoCL2 system. The as-grown nanowires were etched in HF:H_2O (1:3), to selectively remove the SiO_2 shell and to study NW luminescence dependence on the shell thickness. Micro-Raman experiments were carried out at room temperature with a 100X objective and a 532 nm excitation laser with an excitation power of 0.5 mW. The spectral resolution was about 0.2–0.3 cm^{-1}.

Results and Discussion

Fig. 1 shows the nanowires arranged into dense networks randomly oriented on the silicon surface. They are several microns long with diameters below 80 nm.

The core/shell geometry of the NWs and the SiC structure has been examined by TEM (Fig. 2a). HAADF imaging based on the atomic number contrast shows the crystalline core wrapped by the amorphous shell in a cylindrical geometry. EDX point analyses (Fig. 2b) confirms that the amorphous layer is SiO_x with x very close to 2. High resolution TEM studies show that in the inner core perfect 3C-SiC segments alternate with defective regions, where planar defects and in particular stacking faults on (111) planes perpendicular to the growth axis are formed. In these regions, stacking faults can induce very local segments of different polytypes (e.g. 2H, 4H, 6H).

Micro-Raman analysis on NWs as grown and after 30 seconds HF etch (Fig. 3) shows different peaks, centered at around 700, 741, 796 and 900 (broad band) cm^{-1}. The peak at 796 cm^{-1} is attributed to the 3C-SiC core TO band and does not change after etching. The intensity of the band at 700 cm^{-1} decreases after the etch, indicating that this feature is not related to the silicon carbide core. SiO_2 does not show a Raman peak at this frequency, so we propose that this feature could be possibly due to SiO_x or oxicarbide compounds.

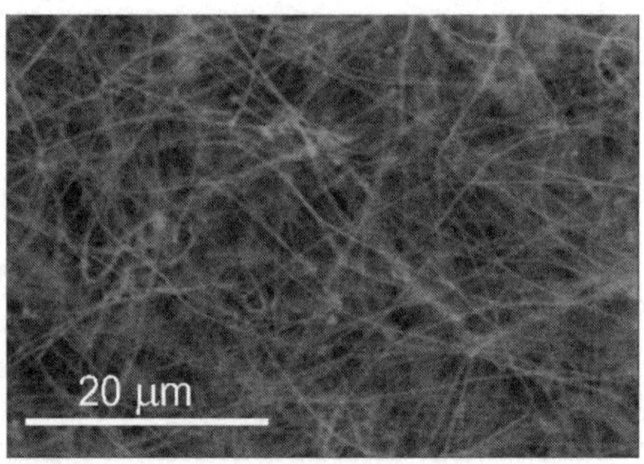

Fig. 1: SEM image of SiC/SiO_2 core/shell nanowires.

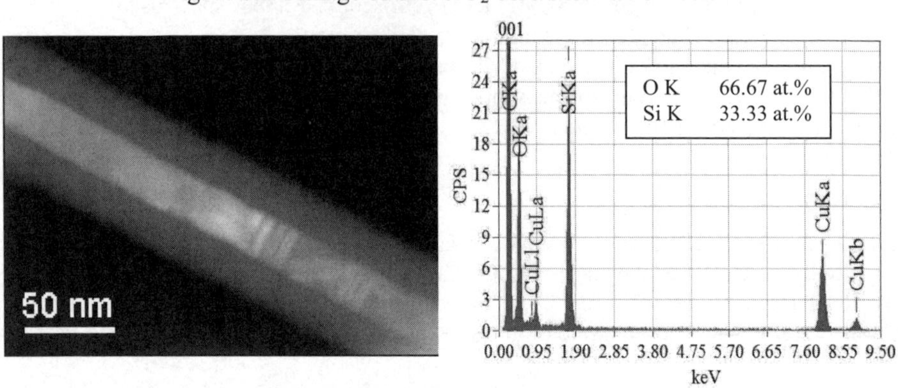

Fig. 2: (a, left) TEM image of SiC/SiO2 core/shell nanowires. (b, right) EDX point spectrum collected on the amorphous shell.

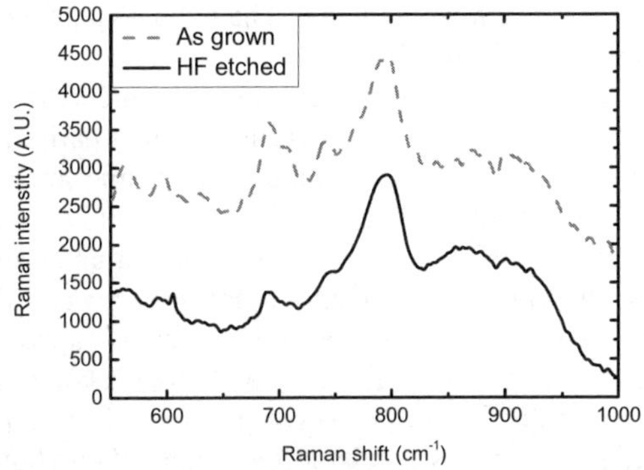

Fig. 3: Raman analysis of core/shell nanowire samples. The as grown spectrum was shifted by 2000 A.U. for clarity

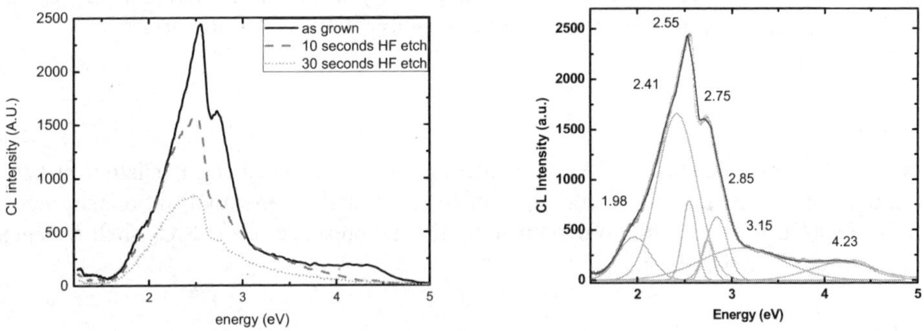

Fig. 4. Left: CL spectra on a single NW taken at different HF etching times. Right: deconvolution of CL spectra on as-grown NW.

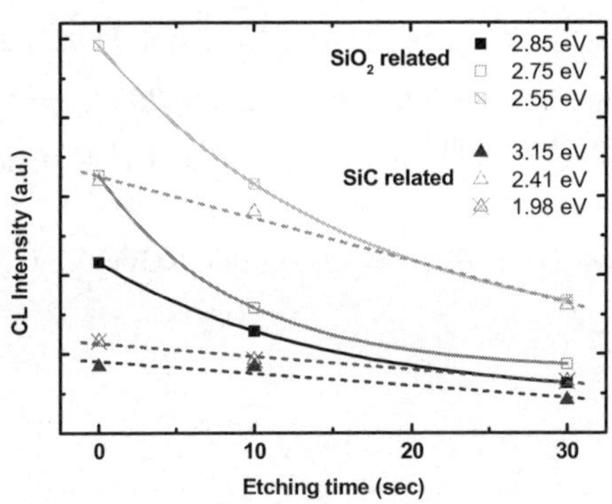

Fig. 5: CL intensity of each band as a function of the etching time.

CL spectroscopy was performed on single nanowires: a broad band with a complex spectral shape is detected, due to the superposition of multiple emissions from the SiC/SiO_2 core/shell nanostructure. Based on peak analysis by Gaussian deconvolution (Fig. 4, right) we were able to identify and assign the different emission components. The main peak, centred at 2.41 eV, is related to the near-band-edge emission (NBE) of the 3C-SiC core.

By varying the SiO_2 shell diameter we were able to experimentally verify the shell-induced enhancement of the crystalline SiC core luminescence: in order to partially remove the SiO_2 shell from the NW we performed the HF etch by varying the etching time (etch rate of about 50 nm per minute) and, subsequently, we measured the CL emission of the SiC/SiO_2 NW with different SiO_2 shell diameter after each etch process. Figure 4 shows the spectra taken on a single NW after HF etching of different durations, showing a significant CL spectra intensity decrease due to the shell removal. Figure 5 shows the CL intensity of each band as a function of the etching time, demonstrating the different decay behavior of the core and shell related emissions. All the SiO_2-related bands show an exponential decay, with time constants of 10–15 s related to the removal of about 50% of the shell thickness. As for SiC-related emissions, they show a linear decrease as a function of the etching time. Since the etching is selective and does not affect the core size, the decrease of the core luminescence must be related to the removal of the wrapping layer. We can conclude that in our system the amorphous shell is beneficial in enhancing the luminescence intensity of the crystalline core, preferentially the SiC NBE radiative recombination. Besides the effectiveness as a carrier injector region, this could be partly related to the fact that the shell can act as a passivation layer to reduce the non-radiative recombination related to surface states, as in the case of entirely crystalline core/shell systems.

Conclusions

We were able to grow core/shell SiC/SiO_2 nanowires and we investigated their lattice structure by Raman and TEM. By etching the outer SiO_2 shell in HF and by the analysis of the nanowires' luminescence by CL we were able to demonstrate that the presence of the SiO_2 shell enhances the SiC near band edge luminescence.

Acknowledgements

This work has been partly supported by Fondazione Cariparma by the BioNiMed project.

References

[1] L. S. Liao, X. M. Bao, Z. F. Yang and N. B. Min, Appl. Phys. Lett. 66 (1995) 2382.

[2] X. M. Liu and K. F. Yao, Nanotechnology 16 (2005) 2932.

[3] X. L. Wu, J. Y. Fan, T. Qiu, X. Yang, G. G. Siu and P. K. Chu, Phys. Rev. Lett. 94 (2005) 026102.

[4] G. Attolini, F. Rossi, M. Bosi, B. E. Watts and G. Salviati, J. Non-Cryst. Solids 354 (2008) 5227.

Materials Science Forum Vols. 717-720 (2012) pp 561-564

doi:10.4028/www.scientific.net/MSF.717-720.561

Theoretical study of thermoelectric properties of SiC nanowires

J. H. Choi[1, 2 a], M. Pala[1], L. Latu-Romain[2], and E. Bano[1]

[1]IMEP-LAHC/ Grenoble INP, MINATEC, 3, parvis Louis Néel-BP 257, 38016 Grenoble Cedex 1, France

[2]LTM/CNRS, 17 rue des Martyrs (CEA-LETI), 38054 Grenoble Cedex 9, France

[a]choiji@minatec.grenoble-inp.fr

Keywords: Thermoelectric properties, Figure of merit, Nanowires

Abstract. We present a theoretical study of thermoelectric properties of silicon carbide (SiC) nanowires as a function of the wire cross section at high temperature (500K), based on non-equilibrium classical molecular dynamics simulations for the lattice thermal transport and the non-equilibrium green's function for the electrical transport. Our calculations show that figure of merit (ZT) increases with decreasing cross section area: ZT of SiC nanowire at 2x2 nm^2 has a maximum value in the range of 0.65 - 0.89 at 500 K, which is 7 - 8 times larger than the maximum ZT of a SiC thin film (0.125 at 973 K). These results show that SiC may be a promising material for thermoelectric applications operating at high temperature.

Introduction

The combination of the distinctive physical and chemical properties of silicon carbide (SiC) and unique advantages of nanowires (NWs) make them outstanding building blocks for the fabrication of nano-electronic devices, which can operate in harsh environment. Our group has demonstrated that SiC NWs field effect transistors (FETs) have similar electrical behavior compared with Si NWs FETs in the ballistic regime [1]. However, compared with the study of electronic properties of SiC NWs [1,2,3], much less has been done for thermal properties of SiC NWs [4]. To the best of our knowledge, there is no report of experimental or theoretical results about the thermoelectric properties of SiC NWs.

Bulk SiC is a poor thermoelectric material due to its high thermal conductivity (300 $Wm^{-1}K^{-1}$) [5]. However, one-dimensional structures, such as NWs, are theoretically predicted by Hicks and Dresselhaus to have a larger ZT compared with the bulk counterpart, due to their novel band structure [6]. In particular, SiC is expected to be promising at a higher temperature region due to wide gap semiconductor characteristics. Therefore, these unique properties of SiC, if coupled with the advantages of NWs, will make it an outstanding candidate for thermoelectric devices, which operate at high temperature.

In this work, we first attempt to predict the thermoelectric properties of SiC NWs with different square cross sections in the ballistic regime at high temperature (500K).

Background and Computational Methods

Figure of Merit (ZT). The performance of a thermoelectric device is characterized by the figure of merit ZT (a dimensionless value), which can be calculated by Eq (1):

$$ZT = \frac{TS^2G}{\kappa}, \qquad (1)$$

where Z is a measure of a material's thermoelectric properties, T is the temperature [K], S is the Seebeck coefficient [V/K] , G is the electrical conductance [Ω^{-1}], and κ is the thermal conductance [W/K], which is the sum of the electron (κ_e) and phonon (lattice vibration) (κ_l) contributions.

In order to increase the ZT, G should be large to minimize joule heating losses and maximize the electron contribution for the thermoelectric effect. S also should be as large as possible to generate a large voltage difference between the two contacts even for a small temperature difference. In addition, it must have small κ to maintain the temperature gradient between two contacts [7]. However, the quantities S, G and κ are not mutually exclusive, but coupled with each other. Therefore, it is very difficult to control these variables independently.

Thermoelectric Power Factor (S^2G). The numerator of Z, which is (S^2G), is called the power factor. It quantifies the electrical performance of thermoelectric materials. The conductance of SiC NWs (in Eq. (2)) and the corresponding Seebeck coefficient (in Eq. (3)) were computed within the framework of the quantum-transport theory of Landauer–Büttiker [8]:

$$G = \frac{2q^2}{h}\int_{-\infty}^{\infty} T(E)(-\frac{df}{dE})dE \qquad (2)$$

$$S \equiv -\frac{\Delta V}{\Delta T}\bigg|_{I=0} = -\frac{1}{qT}\int_{-\infty}^{\infty}(E-E_F)T(E)(f_1(E)-f_2(E))dE \Big/ \int_{-\infty}^{\infty} T(E)(f_1(E)-f_2(E))dE \qquad (3)$$

where q is the electron charge, T is temperature, h is the Planck constant, $T(E)$ is the transmission coefficient, f_1 and f_2 are equilibrium Fermi–Dirac distributions for the contacts, and E_F is the Fermi level at both contacts.

Thermal Conductance (κ). The thermal properties of NWs are much different from those of the bulk material. The high surface to volume ratio of NWs may increase the boundary inelastic scatterings by the surface and hence dramatically decreases the thermal conductivity with respect to the bulk value.

Experimental and theoretical results show NW structure to be effective at reducing the lattice thermal conductivity of SiC. From experimental measurements, it is shown that the thermal conductivity of Si NWs is about 2 orders of magnitude smaller than that of bulk crystals [9]. SiC NWs present lower thermal conductivity (82 ± 6 $Wm^{-1}K^{-1}$) with diameter around 60 nm, compared to bulk SiC (300 $Wm^{-1}K^{-1}$) [10]. From theoretical calculations, N. Papanikolaou has estimated the lattice thermal conductivity of SiC NWs at high temperature (500 K) by using non-equilibrium classical molecular dynamics simulation [4]. According to their simulation results, SiC NWs with diameter from 2 to 4 nm show a thermal conductivity of 3 to 6 $Wm^{-1}K^{-1}$ at 500K. We implemented this simulated lattice thermal conductivity into the calculation of thermoelectric properties of SiC NWs.

The electron thermal conductance (κ_e) was calculated by means of the quantum-transport theory of Landauer–Büttiker, which can be calculated according to Eq (4):

$$\kappa_e = \frac{2}{hT}\int_{-\infty}^{\infty} T(E)(E-E_F)^2(-\frac{df}{dE})dE - TS^2G \qquad (4)$$

Results and Discussion

Fig. 1 shows the power factor (S^2G) and thermal conductance (κ, κ_e and κ_l) in squared cross-section SiC NWs as a function of the wire cross section. The power factor has its maximum value near the threshold voltage and increases with increasing wire cross section.

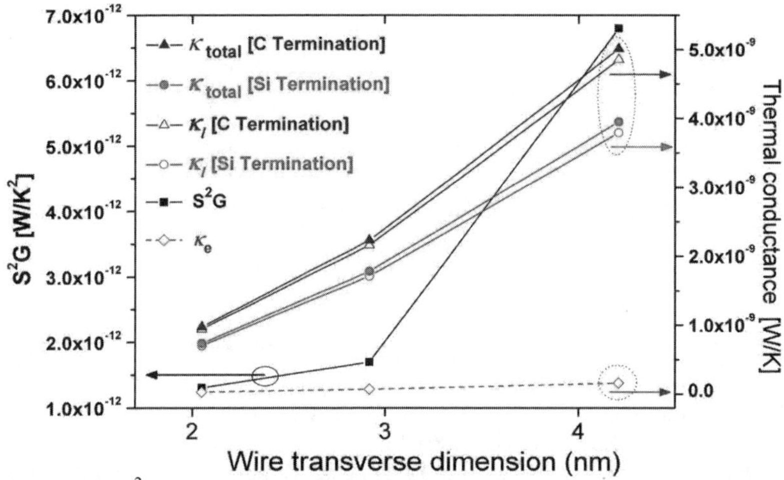

Figure 1. Power factor (S^2G) and thermal conductance as a function of the wire transverse dimension.

An interesting observation is that the smaller wire cross section increases the surface to volume ratio, therefore more phonons are diffusely reflected at the surface. As a result, thermal conductance was gradually reduced with decreasing cross-sectional area. The electron contribution for thermal conductivity is much lower than the phonon (lattice) contribution. The lattice thermal conductivity depends sensitively not only on the cross section of the nanowire but also on surface terminations (Si and C) of the NW. A Si terminated SiC NW has a lower thermal conductance than a C terminated SiC NW. This might be due to stronger scattering on a more disordered Si surface compared to the C one [4].

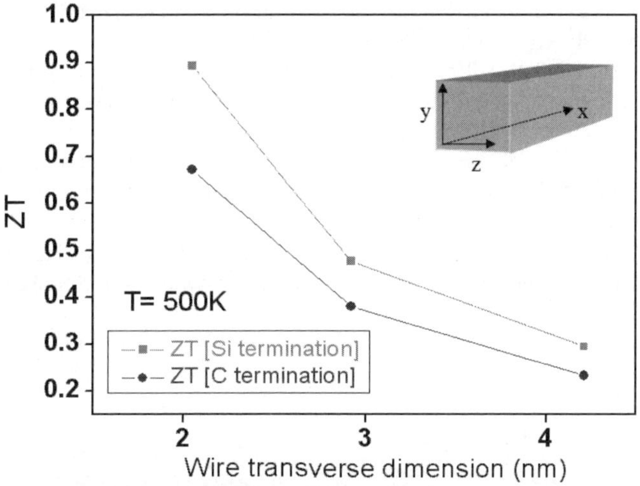

Figure 2. Figure of merit (ZT) in square SiC NWs as a function of the wire transverse dimension (inset: schematic structure of the simulated square NWs)

Thermal conductance decreases faster than a power factor with decreasing wire cross section. Hence, ZT gradually increases with decreasing wire cross section, as shown in Fig. 2. The ZT of a SiC NW at 500 K can reach values of up to 0.6 - 0.89 at 2x2 nm^2, which is 5 - 7 times larger than the maximum ZT value of a SiC thin film (0.125 at 973 K) [11]. This value is also comparable to the highest experimental ZT value of a Si NW (0.6) at room temperature [9]. This ZT enhancement mainly comes from a drastic reduction of thermal conductivity through the enhanced scattering of phonons, without significant reduction of the Seebeck coefficient (S) and electrical conductance (G).

Summary

We have investigated the size dependence of thermoelectric properties of SiC NWs. When cross sectional area of the NW decreases, the power factor decreases slowly while the thermal conductance reduces remarkably. Hence, ZT is gradually increasing with decreasing wire cross section. ZT reaches a value of 0.65 - 0.89 for a wire of cross section 2x2 nm^2, which is 5 - 7 times larger than the highest experimentally reported ZT value for a SiC thin film (0.125 at 973K). Based on simulated results, SiC NWs could be promising structures for thermoelectric applications operating at high temperature.

References

[1] K. Rogdakis, S. Poli, E. Bano, K. Zekentes, M. G. Pala, Phonon and surface roughness limited mobility of gate-all-around 3C-SiC and Si nanowire FETs, Nanotechnology 20 (2009) 295202.

[2] W. M. Zhou, F. Fang, Z. Y. Hou, L. J. Yan, Y. F. Zhang, Field-Effect Transistor Based on β-SiC Nanowire, IEEE Elec. Dev. Lett. 27 (2006) 463.

[3] Chan-Oh Jang, Tae-Hong Kim, Seung-Yong Lee, Dong-Joo Kim, Sang-Kwon Lee, Low resistance ohmic contacts to SiC nanowires and their applications to field-effect transistors, Nanotechnology 19 (2008) 345203.

[4] N Papanikolaou, Lattice thermal conductivity of SiC nanowires, J. Phys.: Condens. Matter 20 (2008) 135201.

[5] Ju Li, Lisa Porter, Sidney Yip, Atomistic modeling of finite-temperature properties of crystalline β-SiC, J. Nucl. Mater. 246 (1997) 53.

[6] L. D. Hicks, M. S. Dresselhaus, Thermoelectric figure of merit of a one-dimensional conductor, Phys. Rev. B 47 (1993) 16631.

[7] Allon I. Hochbaum, Peidong Yang, Semiconductor Nanowires for Energy Conversion, Chem. Rev. 110 (2010) 527.

[8] Raseong Kim, Supriyo Datta, Mark S. Lundstrom, Influence of dimensionality on thermoelectric device performance, J. Appl. Phys. 105 (2009) 034506.

[9] Allon I. Hochbaum, Renkun Chen, Raul Diaz Delgado, Wenjie Liang, Erik C. Garnett, Mark Najarian, Arun Majumdar, Peidong Yang, Enhanced thermoelectric performance of rough silicon nanowires, Nature 45 (2008) 163.

[10] K. M. Lee, T. Y. Choi, S. K. Lee, D. Poulikakos, Focused ion beam-assisted manipulation of single and double β-SiC nanowires and their thermal conductivity measurements by the four-point probe 3ω method, Nanotechnology 21 (2010) 125301.

[11] X. H. Wang, A. Yamamoto, K. Eguchi, H. Obara, T. Yoshida, Thermoelectric properties of SiC thick films deposited by thermal plasma physical vapor deposition, Sci. Technol. Adv. Mat. 4 (2003) 167.

Materials Science Forum Vols. 717-720 (2012) pp 565-568

doi:10.4028/www.scientific.net/MSF.717-720.565

GaAs Nanowires: A New Place to Explore Polytype Physics

Tawinan Cheiwchanchamnangij[1], Thomas Birkel[1], Walter R. L. Lambrecht[1,a,*], and Al. L. Efros[2]

[1] Department of Physics, Case Western Reserve University, Cleveland, OH 44106-7079 USA

[2] Center for Computational Materials Science, Naval Research Laboratory, Washington DC20375

[a]walter.lambrecht@case.edu,*corresponding author

Keywords: polytypism, GW calculations, nanowire, GaAs

Abstract. Recently, polytypism has been observed in nanowires in materials (e.g. GaAs) for which normally only one crystal structure is stable. Here we provide band structure parameters for wurzite and 4H GaAs and use them for modeling the nanowire electronic states. The band gap, crystal field splitting, and its strain dependence are calculated using the quasiparticle self-consistent *GW* method. The nanowire electronic states are obtained in the envelope function approximation within a simplified cylindrical model. The crystal field splitting of the wurtzite GaAs valence band is found to be 180 meV while in 4H-GaAs it is 69 meV, suggesting a downward bowing as a function of hexagonality. The conduction band minimum at Γ changes symmetry character under strain. We discuss the consequences for nanowires and determine the conditions under which a polarization reversal of photoluminescence can occur from mostly perpendicular to parallel to the wire.

Introduction

Polytypism was until recently observed only in a few semiconductor materials systems, most notably SiC, but also ZnS. Most other tetrahedrally bonded semiconductors have a clear preference for either the (cubic) zincblende (ZB) structure or the (hexagonal) wurtzite (WZ) structure. However, recently, the wurtzite structure was observed in nanowires of GaAs grown by catalytic liquid vapor solid growth methods [1-3]. Depending on the growth conditions, such nanowires can either have the WZ or ZB structure and it proved possible to change the structure along the length of the nanowire. In some sections of the nanowires, multiply twinned regions were observed. While it is not yet clear if these have the same regularity as polytypes, such as 4H and 6H, which switch from cubic to hexagonal stacking every layer or every other layer, it raises the question of what the electronic band structure would be of hypothetical bulk GaAs in these new polytypes. The band structure parameters such as the effective masses or more generally the Kohn-Luttinger type Hamiltonian of the valence band manifold, including the crystal field and spin-orbit splittings are required to model the electronic states in the nanowires within the envelope function approximation [4]. Recently, we presented a study of the wurtzite and zincblende GaAs band structure parameters as a function of strain [5]. Here, we extend this study to the 4H polytype and discuss briefly its application to nanowire electronic states.

Computational Method

Our first principles band structure calculations are performed within the quasiparticle self-consistent *GW* method (QS*GW*) [6,7]. Here *G* and *W* stand for the one-electron Green's function and screened Coulomb interaction. The non-local and energy dependent self-energy operator $\Sigma = iGW$ determines shifts of the one-electron eigenvalues obtained from a reference effective independent electron Hamiltonian H_0. Usually, the local density approximation (LDA) is used for the latter, and the corrections are obtained within perturbation theory. In the quasiparticle self-consistent version of the *GW* method, the H_0 already contains a non-local but energy independent exchange correlation potential, extracted from the previous iteration Σ. It thus uses the best possible H_0, in the sense that the self-energy corrections become as small as possible. In practice, this method was shown to provide accurate predictions of the quasiparticle band structure of many materials, in particular band gaps of most semiconductors agree to within about 0.1 eV with experiment [7]. The method is implemented in the framework of the full-potential (FP) linearized muffin-tin orbital (LMTO) method [8]. Details of the implementation can be found in Ref. [6] and of our particular application to WZ GaAs in Ref. [5].

Results for Bulk Electronic Band Structures

In this paper, we focus only on the details of the band structure near the band gap, which are relevant for the nanowire modeling. The band gaps of ZB, WZ and 4H are listed in Table 1. We include the symmetry labels of the valence band maximum (VBM) and conduction band minimum (CBM) to identify the particular band gap. The second row below the headings uses double group notation and includes spin-orbit coupling, while the others do not. Excellent agreement with experiment is obtained for ZB. We note that in WZ the lowest direct gap corresponds to the Γ_3 symmetry. This state corresponds to the L_1^c state of ZB in a band-folding scheme. These calculations were performed at the experimental lattice volume but the *c/a* and Ga-As bond length (along c) were optimized and found to be 1.650 and 2.455 Å, respectively, in WZ. In the 4H structure, the optimized c/2a is 1.6385 and the optimized bond lengths along the c-direction are 2.447 Å and 2.457 Å in cubic and hexagonal layers respectively. A small tensional strain, either isotropic in volume or uniaxial along c, will interchange the Γ_1^c and Γ_3^c levels. Relatedly, a transition from direct gap at Γ to indirect gap at L has been observed in ZB GaAs under strong compressive stress as occurs in shock wave experiments [9]. Which one is lowest at zero strain is also sensitive to k-point sampling. It is important to note that transitions from the VBM to the Γ_3^c are dipole forbidden. In contrast to some previous work [3,10], and agreement with others [11], we find a slight increase in band gap from ZB to WZ. The gap increase in 4H is almost exactly half of that of WZ and the CBM has Γ_1 symmetry. The band structure near the gap in 4H is shown in Fig. 1, while for WZ they are given in Ref. [5].

We now turn to the fine structure of the VBM manifold. The crystal field splitting parameter $\Delta_1=\Gamma_5^v-\Gamma_1^v$ is found to be 180 meV in WZ GaAs. This value is sensitive to the lattice constant and to using QSGW as opposed to LDA. In LDA the value is 212 meV. The reason for the overestimate is that because the gap is too small in LDA, the Γ_1^v and Γ_1^c states repel each other too strongly.

This crystal field splitting is fairly large. As a function of uniaxial strain along the c direction, ε_{zz}, it behaves as $\Delta_1=\Delta_1(0)+(3/2)D_3\varepsilon_{zz}$ with D_3=7.68 eV. For a compressive ε_{zz}=−0.018, the valence bands switch around. This is important for polarization of photoluminescence because the $\Gamma_5^v-\Gamma_1^c$ transition is allowed for $E \perp c$ while $\Gamma_1^v-\Gamma_1^c$ is allowed for $E \parallel c$. The strain required for a polarization reversal seems

Table 1: QSGW band gaps E_g and $\Delta_1 = \Gamma_5^v-\Gamma_1^v$ in GaAs polytypes.

structure	symmmetry	E_g (eV)	Δ_1 (meV)
ZB	$\Gamma_{15}^v-\Gamma_1^c$	1.595	0
ZB	$\Gamma_8^v-\Gamma_7^c$	1.503*	
ZB	$\Gamma_{15}^v-L_1^c$	1.884	
WZ	$\Gamma_5^v-\Gamma_1^c$	1.675	180
WZ	$\Gamma_5^v-\Gamma_3^c$	1.646	
4H	$\Gamma_5^v-\Gamma_1^c$	1.638	69

* This value should be compared to the experimental value of 1.519 eV.

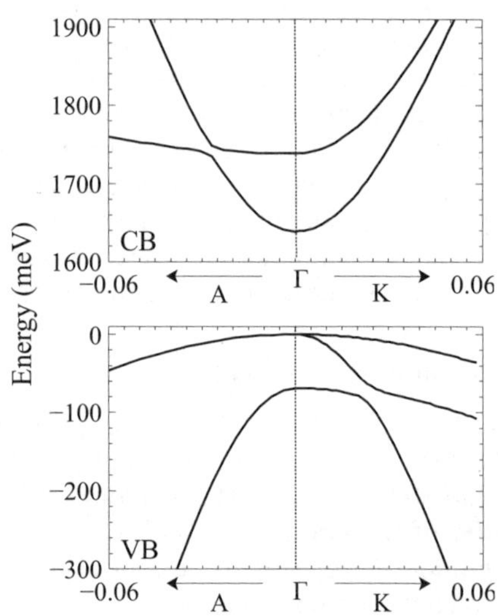

Figure 1: 4H GaAs QSGW band structure near the band gap.

large. In fact, by the time we have such a large compressive strain, the conduction band minimum has Γ_3 symmetry and the PL should be strongly suppressed because conduction electrons should thermalize to the lowest CBM.

In contrast to the simple expectation that the crystal field splitting would vary proportional to the degree of hexagonality (1/2 in 4H) we find that in the 4H polytype the crystal field splitting is significantly less than half that of WZ. Introducing a bowing parameter b through the equation, $\Delta_1(h)=h\Delta_1(\mathrm{WZ})+bh(1-h)$, we find b=−84 meV. This would predict a Δ_1 of 41 meV in 6H (h=1/3) and 29 meV in 8H (h=1/4). This suggests that even in somewhat random multiply twinned regions, the crystal field splitting could be strongly reduced. When spin-orbit coupling is included, the levels become

$$\Gamma_9 = \Delta_1 + \Delta_2$$
$$\Gamma_{7\pm} = \frac{\Delta_1 - \Delta_2}{2} \pm \sqrt{\left(\frac{\Delta_1 - \Delta_2}{2}\right)^2 + 2\Delta_3^2} \qquad (1)$$

with Δ_2=115 meV and Δ_3=113 meV the nearly isotropic spin-orbit coupling parameters, about 1/3 of the cubic spin-orbit coupling parameter. This leads to a splitting of the upper levels Γ_9-Γ_{7+}=100 meV, while the lower Γ_{7-}, the so-called crystal field split-off level lies about 450 meV below the Γ_9 in wurtzite. This justifies using a simple quasicubic Kohn-Luttinger model later in which the spin-orbit coupling is large and we focus on the upper J=3/2 states only but with splitting of the J_z=±3/2 (heavy hole) and J_z=±1/2 (light hole) states, Δ of about 100 meV. Keeping the same spin-orbit coupling parameters as for 4H leads to splittings Δ in 4H, 6H, and 8H of 45, 29, and 21 meV. A small amount of compressive uniaxial strain could further reduce these values to almost zero.

In Ref. [5] we have fully determined the Rashba-Sheka-Pikus Hamiltonian parameters for the VBM manifold in WZ and the Kohn-Luttinger (KL) parameters in ZB GaAs. Below, we will only use the conventional KL parameters γ_1=7.3, γ_2=2.1 and γ_3=3.0 which we determined for ZB [5].

Envelope Function Approximation Model of Nanowire Electronic States

The nanowire (NW) exciton states can be calculated within an effective mass approximation method [4]. The nanowire is assumed to be cylindrical and of radius R, and its axis is along the c-axis of WZ or the [111] axis of ZB. For radii of order 50 nm, which are large compared to the exciton Bohr radius (of order 5 nm), one assumes the weak confinement limit. The motion of the center of mass of the exciton is dominated by the hole, because it is much heavier than the electron and thus described by a wave function of the form:

$$\Psi^F(\rho,\phi,z) = \sum_{\mu=-3/2}^{3/2} u_\mu \chi_\mu^F(k_\perp \rho) e^{i(F-\mu)\phi} e^{ik_z z} \qquad (2)$$

where u_μ are the Bloch functions of the J=3/2 VBM manifold within a modified KL Hamiltonian. For simplicity the latter is taken axially symmetric (γ=γ_2=γ_3=2.5). In that case, the total angular momentum about the z-axis is conserved and defines a quantum number F. We focus on solutions independent of z and thus take k_z=0. Within the KL Hamiltonian there is a coupling between the μ=3/2 and μ=−1/2 bands and the μ=1/2 and μ=−3/2 envelope functions. For total angular momentum F=3/2, we can either have m=0 with μ=3/2 or m=2 with μ=−1/2. For a given energy E, the light hole and heavy hole band solutions have different $k_\perp$ and hence they cannot both simultaneously go to zero at ρ=R. Therefore the exciton eigenstates of the NW are mixtures of these two states. In ZB NW, the ground state of the holes corresponds to F=1/2. This is significant for polarization because the μ=±3/2 Bloch states only have perpendicular orbital character, X and Y, while the μ=±1/2 Bloch states have a Z component parallel to c. Another important point to note is that the perpendicular component of the electric field of the photon impinging on the NW is reduced by a factor $(\varepsilon_1+\varepsilon_2)/2\varepsilon_2$ where ε_1 is the dielectric constant inside the NW (ε_1 =13.2 for GaAs) and ε_2=1, the dielectric constant in the medium outside. Thus the polarization of the ZB NW PL is mostly parallel to the NW.

In WZ nanowires, the situation is different because there is a splitting Δ separating the heavy and light hole bands. Thus the ground state at T=0 is clearly the F=3/2 state which has purely perpendicular character. The important point is that this perpendicular polarization is disfavored by the dielectric constant factor. Thus at slightly elevated temperature, and for a small enough radius, the parallel component becomes of comparable magnitude to the perpendicular component and a switch in polarization can occur. Exploring this quantitatively, we find that this occurs only for sufficiently small Δ (below about 20 meV) and below a critical radius, which is larger for smaller Δ. The temperature at which the switch from mostly perpendicular to mostly parallel occurs depends on the splitting between the F=3/2 and F=1/2 exciton states. Strain can reduce Δ but an unrealistically strong compressive strain is required to reduce the value from 100 meV down to order 5-20 meV. However if the NW does not have pure WZ structure but has lower hexagonality, i.e. is multiply twinned, then our calculations above predict that Δ could easily be reduced to about 6 meV, the value required to observe a full polarization switch for 50 nm radius NW.

Conclusions

In this paper, we summarized the important band structure parameters of WZ and 4H GaAs relevant to nanowires. We find that the crystal field splitting as a function of hexagonality has a downward bowing, i.e. it decreases faster than linear and that this plays an important role in explaining the polarization reversal in GaAs nanowires as function of temperature [12].

References

[1] D. Spirkoska, J. Arbiol, A. Gustafsson, S. Conesa-Boj, F. Glas, I. Zardo, M. Heigoldt, M. H. Gass, A. L. Bleloch, S. Estrade, M. Kaniber, J. Rossler, F. Peiro, J. R. Morante, G. Abstreiter, L. Samuelson, and A. Fontcuberta i Morral, Phys. Rev. B 80 (2009) 245325.

[2] I. Zardo, S. Conesa-Boj, F. Peiro, J. R. Morante, J. Arbiol, E. Uccelli, G. Abstreiter, and A. Fontcuberta i Morral, Phys. Rev. B 80 (2009) 245324.

[3] M. Heiss, S. Conesa-Boj, J. Ren, H.-H. Tseng, A. Gali, A. Rudolph, E. Uccelli, F. Peiró, J. R. Morante, D. Schuh, E. Reiger, E. Kaxiras, J. Arbiol, and A. Fontcuberta i Morral, Phys. Rev. B 83 (2011) 045303.

[4] Peter C. Sercel and Kerry J. Vahala, Phys. Rev. B 42 (1990) 3690.

[5] Tawinan Cheiwchanchamnangij and Walter R. L. Lambrecht, Phys. Rev. B 84 (2011) 035203.

[6] T. Kotani, M. van Schilfgaarde, and S. V. Faleev, Phys. Rev. B 76 (2007) 165106.

[7] M. van Schilfgaarde, T. Kotani, and S. Faleev, Phys. Rev. Lett. 96 (2006) 226402.

[8] M. Methfessel, M. van Schilfgaarde, and R. A. Casali, in Electronic Structure and Physical Properties of Solids. The Use of the LMTO Method, Lecture Notes in Physics, edited by H. Dreyssé (Springer, Berlin, 2000), Vol. 535, p. 114.

[9] P. Grivickas, M. D. McCluskey, and Y. M. Gupta, Phys. Rev. B 80 (2009) 073201.

[10] A. De and C. E. Pryor, Phys. Rev. B 81 (2010) 155210.

[11] Z. Zanolli, F. Fuchs, J. Furthmüller, U. von Barth, and F. Bechstedt, Phys. Rev. B 75 (2007) 245121.

[12] D. Spirkoska, Al. L. Efros, W. R. Lambrecht, T. Cheiwchanchamnangij, A. Fontcuberta i Morral, and G. Abstreiter, submitted to Phys. Rev. B (2011).

Materials Science Forum Vols. 717-720 (2012) pp 569-572

doi:10.4028/www.scientific.net/MSF.717-720.569

The atomic step induced by off angle CMP influences the electrical properties of the SiC surface

Yayoi Tanaka[1,a], Takao Kanda[2], Kazuyuki Nagatoshi[3], Masamichi Yoshimura[4] and Osamu Eryu[1]

[1] Nagoya Institute of Technology, Gokiso-cho, Showa-ku, Nagoya, Aichi 466-8555, Japan

[2] POVALKOGYOCO.,Ltd., 30 Nakafuka Noda-cho Nakamura-ku, Nagoya, Aichi 453-0858, Japan

[3] FUJIMI INCORPORATED, 1-1, Chiryo-2, Nishibiwajima-cho, Kiyosu, Aichi 452-8502, Japan

[4] Toyota Technological Institute, 2-12-1 Hisakata, Tempaku-ku, Nagoya, Aichi 468-8511, Japan

[a]19516505@stn.nitech.ac.jp

Keywords: atomic step, CMP, LEISS, CAFM, SBD

Abstract. In this work, we made an off-axis surface of SiC with the chemical-mechanical polishing (CMP) method to examine the influence of an atomic step on the electronic properties of devices. Low energy ion scattering spectroscopy (LEISS) shows the Si-face and C-face with SiC structure on the polished SiC monocrystal. We considered that the atomic step on the surface influenced device properties, especially the leakage current of a Schottky diode. C atoms at the step and terrace, which generated the leakage current at the metal-semiconductor interface of a Schottky diode, have been clarified by conductive atomic force microscopy (CAFM).

Introduction

The CMP method realized the surface of SiC that was smooth at an atomic level in the monocrystalline state [1]. A parallel striped pattern with atomic steps and terraces was observed on the surface after CMP by the AFM method. The polarity of the atomic step ridge and terrace of the SiC surface is different because Si and C have different electronegativity. It is reported that the surface obtained by the CMP method leads to improved electrical properties of Schottky barrier diodes (SBD) and metal-oxide-semiconductor field-effect transistors (MOSFETs) [1,2]. However, the relationship between atomic steps and electrical properties of SiC devices is not clear. In this work, we made an off-axis surface of SiC using the CMP method to examine the influence of an atomic step on surface electronic properties. Electrical properties of the atomic step and terrace structure of a surface were observed by CAFM [3]. CAFM measurement evaluates the point electrical characteristics.

Experiment

SiC substrates of n-type 6H-SiC ($N_D - N_A \cong 1.23\times10^{18}$ cm^{-3}) (0001) Si face and (000-1) C face were used. The off angled surfaces were made with various off-axis angles toward [1-100] by the CMP method [1]. The lattice structure of the surface was evaluated using LEISS [4,5]. LEISS measurements were carried out with He^+ as the primary ion beam. The CAFM measurements were used to clarify the electrical characteristics of the atomic step and terrace structure on the SiC surface. The SBDs characteristics were evaluated by current-voltage (I-V) characteristics for samples of on-axis and 3.5 degrees off-axis C-face samples.

Results and Discussion

LEISS measurement was used to analyze crystallinity on the SiC surface layer. We prepared the SiC substrate with several off angles of: on-axis, 1, 1.5, 2 and 3.5 degrees off-axis. Figure 1 shows signal of Si yield for the surface when the azimuth angle of the incident beam was changed from -60 degrees to 60 degrees with fixed incident angle of 15 degrees and 55 degrees. This means that the depth of penetration into the crystal surface becomes shallow at low incident angle and the scattered ions include only the information from the surface of 1 nm and 3 nm in depth. As shown by the results in Fig. 1, from the surface were obtained signals that reflected single crystal depth of 3 nm from the surface. Similarly the results show that the off angle for the C-axis is made by CMP. Figure 1 shows that the CMP method produces a surface with single crystal structure for both on and off axis SiC substrates.

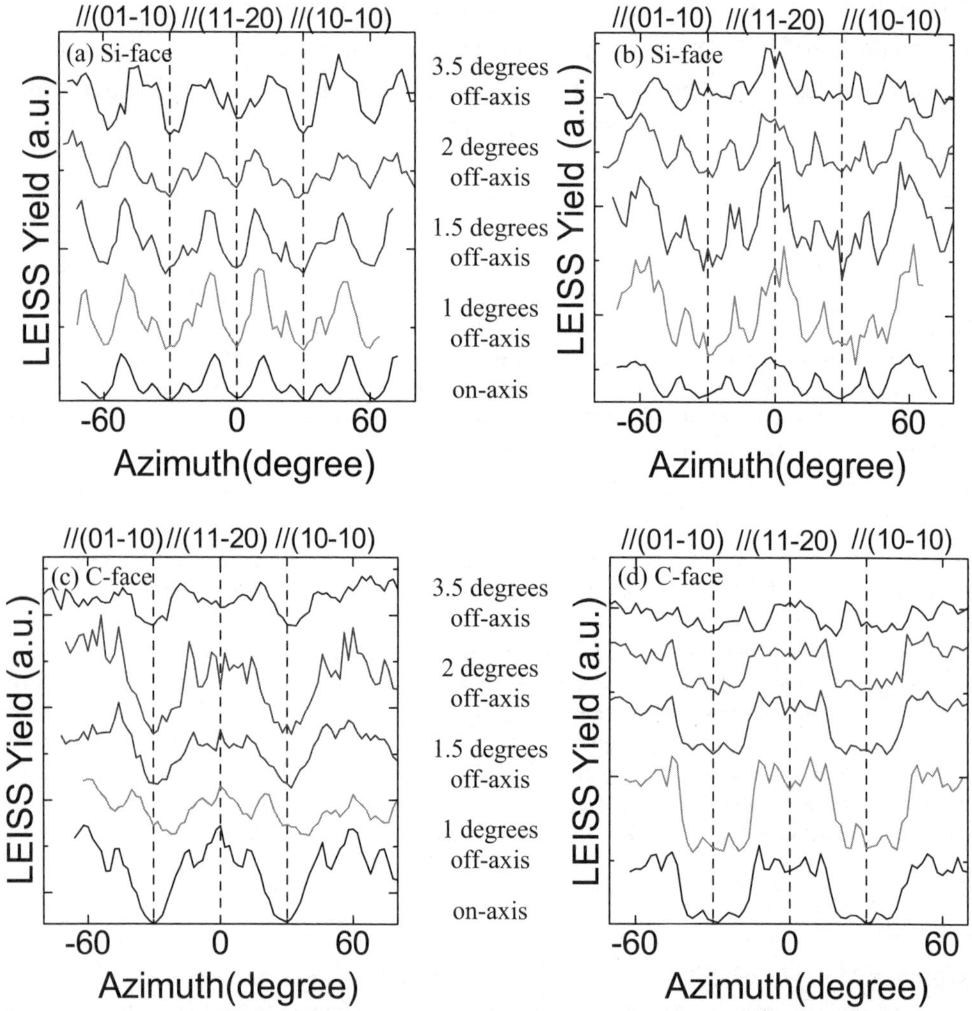

Fig. 1: Comparison of LEISS signals for SiC surface on-axis, 1 degrees off-axis, 1.5 degrees off-axis, 2 degrees off-axis and 3.5 degrees off-axis prepared by CMP. Signals from surface: (a), (c) 2～3 layers (b), (d) 10 layers.

Figure 2 shows a non contact dynamic mode AFM image of the on-axis Si-face. The symbol × indicates the positions of CAFM in Fig. 2 (a). A diamond coated probe with radius of ～100 nm was used for CAFM measurements. Figure 2 (b) shows an atomic step and terrace structure. The step height in Fig. 2 (b) is equivalent to the height of a SiC single atomic layer of 0.25 nm.

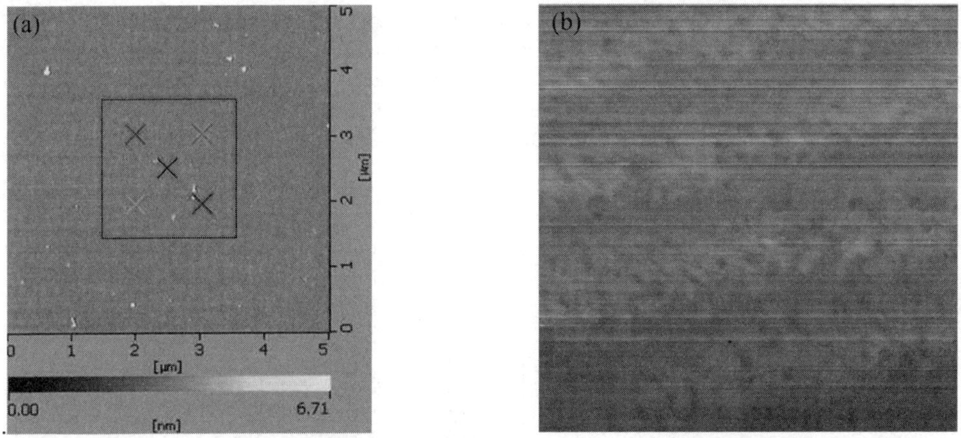

Fig. 2: (a) Topography signal AFM image of on-axis Si-face in 5 μ m×5 μ m and (b) in 1 μ m×1 μ m. (a) shows the points (×) for CAFM in an area of 2 μ m × 2 μ m.

Figure 3 shows I-V curves of SiC surfaces with different off axis orientations using the CAFM method. The width of the atomic step was 250 nm on the on-axis face and 4.1 nm for the 3.5 degrees off face [1]. The electrical properties were influenced by the distance of the atomic step on the Si-face, but little influenced on the C-face. The Si-face has terraces with Si atoms and steps with C atoms. C has a larger electronegativity than that of Si, so that C atoms are negatively charged [6]. We consider that the difference of electrical properties between the Si-face and C-face for CAFM measurement was caused by the negative charge of C atoms. A negative charge at the interface of the CAFM probe and SiC surface causes reduced band bending in a Schottky contact with reverse bias. The surface of off-axis has many atomic steps with C atoms compared with an on-axis Si-face. We think that the atomic steps with negatively charged C atoms cause current to leak abruptly at low voltage.

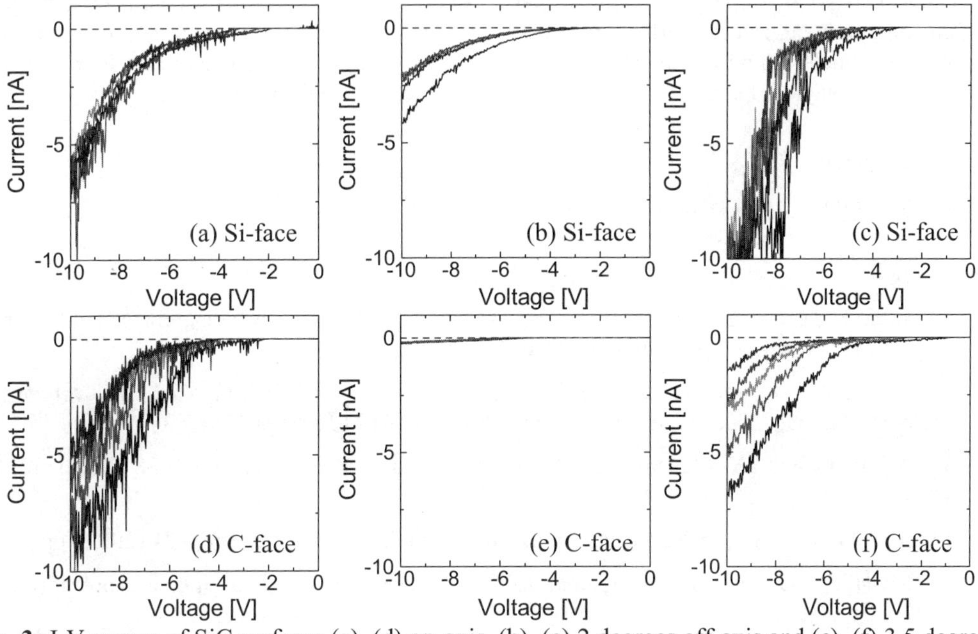

Fig. 3: I-V curves of SiC surfaces (a), (d) on-axis, (b), (e) 2 degrees off-axis and (c), (f) 3.5 degrees off-axis for 5 points in the range of 2 μ m where the I-V characteristics were measured.

Figure 4 shows reverse I-V characteristics of SBDs contacted with Au to the SiC surface on-axis and 3.5 degrees off-axis on the C-face. The breakdown voltage for on-axis was low compared to 3.5 degrees off-axis. The terraces of on-axis material are surface C atoms, so that the current leakage occurred at lower voltage. On the basis of the electrical properties as reported in Fig. 3 and Fig. 4, we conclude the origin of leakage current in SBDs are the C atoms on the SiC surface.

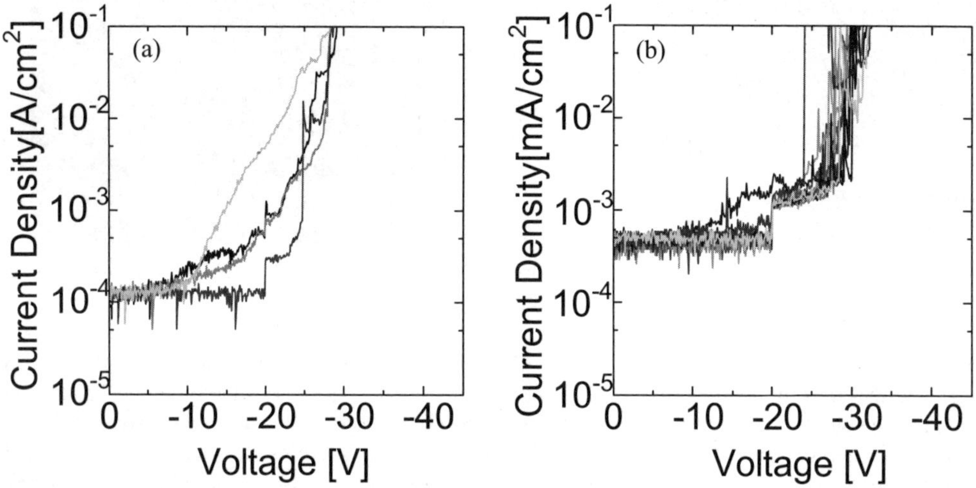

Fig. 4: Reverse I-V characteristics of SBDs on the SiC surface (a) on-axis and (b) 3.5 degrees off-axis on the C-face.

Conclusions

We made SiC substrate surfaces of different atomic step densities by CMP in order to clarify the influence of an atomic step on electronic properties of devices. Current-voltage characteristics by the CAFM measurement revealed effects indicating that the electronegativity of Si and C atoms is an important parameter for a SBD interface. The leakage current at a metal-SiC interface in reverse bias is concluded thought to be related with negatively charged surface C atoms.

Acknowledgments

We are thankful to Gemma Rius for measurement of CAFM.

References

[1] K. Hotta, K. Hirose, Y. Tanaka, K. Kawata and O. Eryu: Silicon Carbide and Related Materials 2007 Part 2, Mater. Sci. Forum Vols. **600-603** (2009) pp. 823-826

[2] T. Ohshima, S. Onoda, T. Kamada, K.Hotta, K. Kawata and O. Eryu: Silicon Carbide and Related Materials 2008, Mater. Sci. Forum Vols. **651-617** (2009) pp. 781-784

[3] I. T. Clark, G. Rius, Y. Matsuoka and M. Yoshimura: J. Vac. Sci. Technol. B **28** (2010) pp.1148 -1152

[4] O. Eryu, M. Mizuta: Rep. Res. Cent. Ion Beam Technol., Hosei Univ. Suppl **21** (2003) pp. 41-44

[5] M. Matsui, F. Uchida, K. Katsuyama, T. Tokunaga and M. Kojima: Jpn. J. Appl. Phys. **37**, (1998) pp. 2043–2050

[6] J. C. Phillips: Phys. Rev. Lett. **20** (1968) pp. 550-553

Materials Science Forum Vols. 717-720 (2012) pp 573-576
© (2012) Trans Tech Publications, Switzerland
doi:10.4028/www.scientific.net/MSF.717-720.573

Surface phase diagram of 4H-SiC {0001} step-terrace structures during Si-vapor etching in a TaC crucible

Shoji Ushio[1,a], Tatsuya Karaki[1], Kenta Hagiwara[1], Noboru Ohtani[1,2,b] and Tadaaki Kaneko[1,2,c]

[1]School of Science and Technology, Kwansei Gakuin University, 2-1, Gakuen, Sanda, Hyogo, 669-1337, Japan

[2]R&D Center for SiC Materials and Processes, Kwansei Gakuin University, 2-1, Gakuen, Sanda, Hyogo, 669-1337, Japan

[a]buj37401@kwansei.ac.jp, [b]ohtani.noboru@kwansei.ac.jp, [c]kaneko@kwansei.ac.jp

Keywords: 4H-SiC, step-terrace structure, etching, high temperature annealing

Abstract. Step-terrace structures were observed at on-axis/4° off 4H-SiC {0001} surfaces after Si-vapor etching. The Si-vapor etching is our original technique to planarize and etch the SiC surfaces by utilizing a TaC crucible in temperature ranging from 1600 to 2200 °C. The structures obtained after the Si-vapor etching obviously indicated temperature dependence. There were two types of step-terrace structures at on-axis surfaces. One is step bunched surfaces consisting of full unit cell height (= 1.0 nm) steps with {1-10n} facets at the step edges. They were observed at 4H-SiC (0001) at lower temperatures below 2000 °C. The other is smooth isotropic surfaces with half unit cell height (= 0.5 nm) steps and without any stable facets at the step edges. They were observed at 4H-SiC (0001) at higher temperatures above 2000 °C and in all temperature conditions (1600 - 2200 °C) at 4H-SiC (000-1). A similar tendency was also confirmed at 4° off 4H-SiC {0001} surfaces. From the comparison with 6H-SiC, macro step bunching (~10 nm height) was revealed to be a unique phenomenon on the 4H-SiC (0001) surface in the Si-vapor etching.

Introduction

Precise control of the surface step-terrace structures of single crystal 4H/6H-SiC has received special attention in almost every process for device fabrication, such as bulk/epitaxial growth and formation of oxide/SiC or metal/SiC interfaces. Furthermore, 4H/6H-SiC {0001} surfaces with well-ordered step-terrace structure have also become important as a template for the epitaxial growth of high quality graphene [1,2]. So far, the engineering technique to control the SiC surfaces has not been straightforward, mainly due to the necessity of high temperature etching and annealing processes under certain vapor phase conditions such as hydrogen and argon. In those processes, the consequent stoichiometric imbalance is induced by different vapor pressures from the SiC constituent elements, silicon and carbon, resulting in the degradation of the SiC surfaces [3,4]. Concerning the details of the step- terrace structure obtained by etching, step heights and shapes of step edges have been discussed with tilted angles from the {0001} basal plane and tilted directions [5-7]. However, for precise control of the SiC step-terrace structure, it is still unclear which condition leads to the structures with specific step heights and step-edge shapes in spite of a number of results using hydrogen etching techniques [8].

As a new method for planarizing and cleaning of SiC surfaces, Si-vapor etching in a semi-closed TaC crucible has revealed its high degree of availability for the SiC surface control and a successive epitaxial graphene growth [9,10]. In contrast to the conventional methods such as hydrogen etching, the Si-vapor etching requires no external gas introduction for surface reaction control but utilizes Si vapor pressure from residual solid Si inside the crucible with no temperature gradient. As a result of thermal flux balancing between sublimation from SiC surfaces and residual Si vapor pressure, the etching reaction takes place at 4H/6H-SiC {0001} while maintaining surface stoichiometry and uniform step-terrace structures up to 2200 °C. In this method, the etching rate is controlled simply by the processing temperature itself.

In this study, wide range thermal stabilities of SiC step-terrace structures in Si-vapor environment were investigated and a comprehensive understanding of SiC surface step dynamics was considered.

Experiments

The experiments in this study were carried out in an ultrahigh vacuum (UHV) system consisting of an all metal chamber with resistive heating apparatus (KGX2000 from Epiquest and Sunric, max. temperature: 2200 °C, ramp rate: ~1000 °C/min, background pressure: ~10^{-9} Torr). Heating treatments were performed simply by radiation from surrounding resistive heating meshes. Pieces of normally on-axis (< 0.2° off) and 4° off towards <11$\bar{2}$0> 4H-SiC {0001} substrates were placed on the bottom of a large diameter TaC crucible (carburized tantalum from Toyotanso, 6-inch multi-wafer available). Processing temperatures were measured by a pyrometer and thermocouples. Prior to the loading into the chamber, substrates were cleaned by a standard RCA etch and subsequent hydrofluoric acid treatment. In the UHV chamber, substrates firstly were degassed at about 800 °C for 1 hour and flushed to targeting temperatures by directly transferring the crucible into the heated region in the vacuum chamber. Si-vapor etching was performed at the targeting temperature for certain duration time. Atomic force microscopy (AFM) and scanning electron microscopy (SEM) measurements were utilized for observation of surface morphology after the Si-vapor etching.

Results and Discussion

Initial surface morphologies are altered by Si-vapor etching with increasing etched depth. Since the etched depth is directly determined by temperature and duration time, surfaces etched at lower temperature and for shorter annealing times retain some trails of initial surface morphologies, such as polishing scratches. However, once surfaces form some stable structures by Si-vapor etching with a sufficient etched depth, there is no distinguishable dependence of etched depth any more. Therefore, the following results are stable structures after Si-vapor etching at each temperature.

Fig.1 represents typical surface morphologies of 4H-SiC {0001} substrates after Si-vapor etching. The surfaces represent uniform step-terrace structures over the entire substrates after 1800 °C for 10 min and 2150 °C for 30 min. For the lower temperature on the on-axis Si-face Fig.1 (a), the etched surface has greater than full unit cell height steps (= 1.0 nm) with stable {1-10n} crystallographic facets at step edges. On the other hand, for the higher temperature Fig.1 (b), the surface has half unit cell height steps (= 0.5 nm) with no particular stable facet. A similar tendency was also observed at the surfaces of 4° off Si-face substrates, as shown in Fig.1 (c), (d). Macro step bunching (mainly ~10 nm height) was found in association with {1-10n} "zigzag" facets for a lower temperature range up to about 2100 °C. Higher temperature above 2100 °C leads to smooth isotropic surfaces on 4° off Si-face substrates. In contrast to the above mentioned Si-face, almost the same results appeared at each surface of on-axis and 4° off C-face substrates for the various temperatures (Fig.1 (e), (f)), which look like the surfaces for the high temperature range of the Si-face.

The structural tendency of 4H-SiC {0001} surfaces after Si-vapor etching is summarized as a function of temperature, as shown in Fig. 2. This is a kind of morphological surface phase diagram of 4H-SiC in Si-vapor etching. Here, macro step bunching structures obtained at on-axis 4H-SiC (0001) surfaces in low temperature range (below 1900 °C) have large step heights in multiples of full unit cell height. Thus, there are two types of stability of step-terrace structures after Si-vapor etching. One is the stability of full unit cell height steps with zigzag edges and the other is the stability of half unit cell height steps with smooth isotropic edges. The stability of full unit cell height steps has been observed commonly at vicinal 6H-SiC (0001) surfaces after the conventional hydrogen etching. In this case, the formation of full unit cell height steps is explained as the catching up of "fast"-etched half unit cell

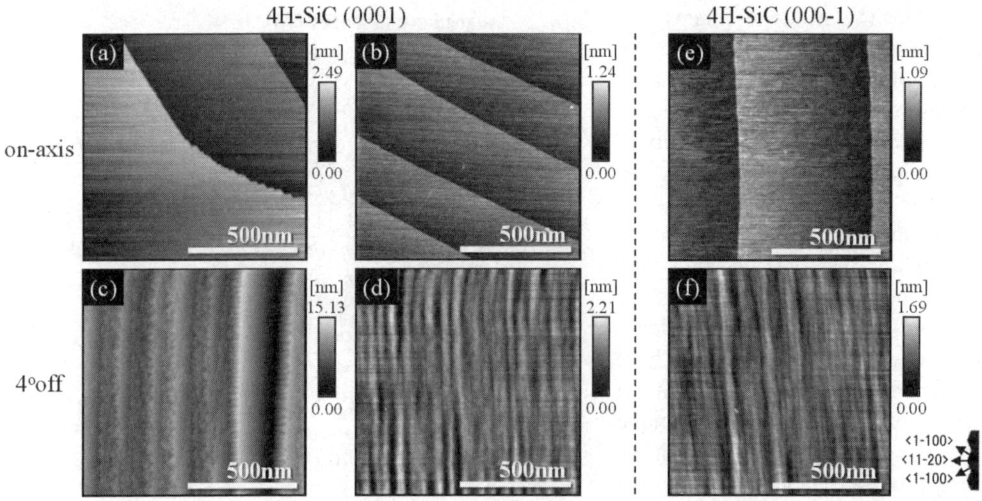

Fig.1 Step-terrace structures at on-axis 4H-SiC (0001) Si-face surfaces after Si-vapor etching at (a) 1800 °C for 10 min and (b) 2150 °C for 30 min. (c), (d) are 4° off 4H-SiC (0001) Si-face surfaces under the same conditions as (a),(b), respectively. (e), (f) are on-axis and 4 ° off 4H-SiC (000-1) C-face surfaces after Si-vapor etching at 1800 °C for 10 min.

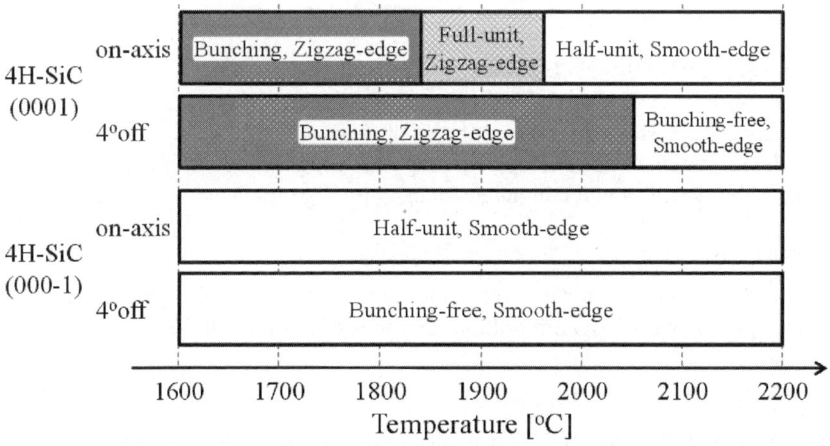

Fig.2 Surface phase diagram of 4H-SiC {0001} step-terrace structures after the Si-vapor etching as a function of temperature. Step-terrace structures are represented by the step heights and the shapes of step-edges.

height steps to the adjacent "slow"-etched half unit cell height steps along a certain <1-100> direction [5]. However, the stability of full unit cell height steps was observed even for the on-axis surface which has no specific surface inclination in this study. In fact, all step edges appearing on 4H-SiC (0001) surfaces in low temperature range formed zigzag {1-10n} facets, as seen in Fig.1 (a), (c). Therefore, the appearances of macro step bunching (large terrace surface) and zigzag-edge structures can be interpreted as the stability of the (0001) basal plane and {1-10n} facets, respectively. While those facets are stable in low temperature range, they are unstable at higher temperature above 2100 °C. From this viewpoint, the formation of macro step bunching on a 4° off Si-face surface is explained in Fig.3. As an interesting feature of the step bunching structure, zigzag step edges seem to line up in their vertex positions just beneath large terraces, as seen in Fig.3 (a). These aligned zigzag steps are assumed to contribute to the formation of the macro step bunching. Once the aligned zigzag steps are formed, etching is suppressed at the {1-10n} planes of bunched zigzag edge. Due to the decrease of etching rate, the step bunching and the terrace width are enlarged by catching up of lower steps and running away of upper steps, as illustrated in Fig.3 (b). Meanwhile, the absence of step bunching in high temperature range is considered to be attributed to the decrease of stability of aligned zigzag edge {1-10n} facets. Note that in the case of 4H-SiC (000-1) and 6H-SiC {0001}, no macro step bunching was obtained on 4° off surfaces and half-unit cell height steps were dominant for on-axis surfaces (not shown for 6H-SiC). These observations imply that the formation of macro step bunching is a unique phenomenon on 4H-SiC (0001) in the etching process. They also imply that the stability of {1-10n} planes on 4H-SiC (0001) is higher than that on 6H-SiC (0001). Especially in the case of the C-face for both 4H- and 6H-SiC, the concept of "fast" or "slow" -etched steps doesn't exist because surfaces have no full unit cell steps. Results in this study bring about the first systematic view in understanding the thermal stability of SiC step-terrace structures.

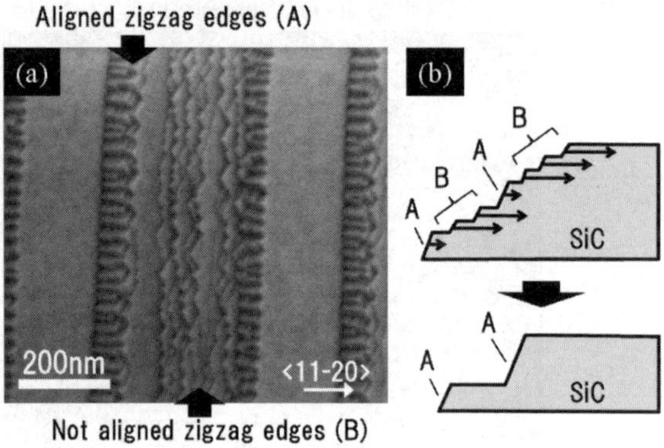

Fig.3 (a) SEM image of 4° off 4H-SiC (0001) Si-face surface after Si-vapor etching at 1800 ^{o}C for 2h. (b) Schematic illustration of the macro step bunching model.

Summary

Temperature dependence (1600 - 2200 ^{o}C) of Si-vapor etched step-terrace structures on 4H-SiC {0001} surfaces were discussed. There are two types of step-terrace structures in terms of the step heights and the shapes of the step edges. From the comparison of on-axis with 4° off 4H-SiC (0001) Si-face surfaces, the formation of macro step bunching appearing for the lower temperature range was interpreted by the stability of full unit cell height steps with {1-10n} facets at the step edges.

Acknowledgements

The authors thank Mr. S. Ashizaki for his continuous encouragement and support on this work. We gratefully acknowlege Toyota Motor Corporation, Toyotanso, Epiquest and Sunric.

References

[1] J. Robinson, X. Weng, K. Trumbull, R. Cavalero, M. Wetherington, E. Frantz, M. LaBella, Z. Hughes, M. Fanton, and D. Snyder, ACS Nano. **4** (2010) 153.

[2] T. Ohta, N. C. Bartelt, S. Nie, K. Thürmer, and G. L. Kellogg, Phys. Rev. B **81** (2010) 121411(R).

[3] O.J. Guy, M. Lodzinski, K.S. Teng, T.G.G. Maffeis, M. Tan, I. Blackwood, P.R. Dunstan, O. Al-Hartomy, S. P. Wilks, T. Wilby, N. Rimmer, D. Lewis, and J. Hopkins, Appl. Sur. Sci. **254** (2008) 8098.

[4] G. Younes, G. Ferro, C. Jacquier, J. Dazord, and Y. Monteil, Appl. Surf. Sci. **207** (2003) 200.

[5] S. Nakamura, T. Kimoto, H. Matsunami, S. Tanaka, N. Teraguchi, and Akira Suzuki, Appl. Phys. Lett. **76** (2000) 3412.

[6] S. Soubatch, S.E. Saddow, S.P. Rao, W.Y. Lee, M. Konuma, and U. Starke, Mater. Sci. Forum **483-485** (2005) 761.

[7] A. Nakajima, H. Yokoya, Y. Furukawa, and H. Yonezu, J. Appl. Phys. **97** (2005) 104919.

[8] F. J. Ferrer, E. Moreau, D. Vignaud, S. Godey, and X. Wallart, Semicond. Sci. Technol. **24** (2009) 125014.

[9] S. Ushio, A. Yoshii, N. Tamai, N. Ohtani, and T. Kaneko, Phys. Status Solidi C **8** (2011) 580.

[10] S. Ushio, A. Adachi, K. Matsuda, N. Ohtani, and T. Kaneko, Mater. Sci. Forum, **679-680** (2011) 777.

Materials Science Forum Vols. 717-720 (2012) pp 577-580
© (2012) Trans Tech Publications, Switzerland
doi:10.4028/www.scientific.net/MSF.717-720.577

Morphological instability of 4H-SiC (0001) basal plane surface during Si-vapor thermal etching

Shoji Ushio[1,a], Kohei Nakanishi[1], Noboru Ohtani[1,2,b] and Tadaaki Kaneko[1,2,c]

[1]School of Science and Technology, Kwansei Gakuin University, 2-1, Gakuen, Sanda, Hyogo, 669-1337, Japan

[2]R&D Center for SiC Materials and Processes, Kwansei Gakuin University, 2-1, Gakuen, Sanda, Hyogo, 669-1337, Japan

[a]buj37401@kwansei.ac.jp, [b]ohtani.noboru@kwansei.ac.jp, [c]kaneko@kwansei.ac.jp

Keywords: 4H-SiC, step-free surface, Si-vapor etching, morphological instability, dislocation

Abstract. The morphological instability appeared at step-free 4H-SiC (0001) basal plane surfaces was investigated. The step-free (0001) basal plane surfaces were fabricated at the bottom of inverted-mesa structure by the method combining a laser digging and Si-vapor thermal etching. By repeated Si-vapor etching treatments, randomly created crater and maze structures were cyclically appeared at the (0001) basal plane surfaces. On the other hand, in the presence of threading dislocations, the bottom surfaces of the inverted-mesa structure were dominated by SiC 2 bilayer steps produced from dislocation cores. For the use of extremely low off-axis 4H-SiC (0001) surfaces, the difficulty of controlling 4H-SiC (0001) terrace surfaces was revealed.

Introduction

Morphological instability of 4H-SiC (0001) surface is a fundamental and critical phenomenon to control homo-/hetero- epitaxial growth and fabricate power devices. In particular, surface step dynamics and kinetics plays a key role for etching and epitaxial growth of SiC. So far, instability studies on SiC surfaces have been focused mainly on step dynamics rather than terrace's, because of the urgent applications for SiC power devices where various kinds of high off-axis surfaces have been used for step-controlled epitaxy and etching. Recently, extremely low off-axis surface has been focused because of the use as the template for hetero epitaxy such as GaN [1,2] and for the growth of high quality epitaxial graphene. Step-free surface, which has been suggested as the extremely low off-axis surface, is able to be fabricated in two different methods. One is the method based on step-flow growth at the top of mesa structure. The other is the method based on etching which proceeds in a reverse step-flow manner at the bottom of inverted-mesa structure. In both cases, it is difficult to obtain large area step-free surfaces due to the complexity in controlling the flux balances of growth and etching species. We have been reported the method to fabricate 4H-SiC (0001) step-free surfaces at the bottom of inverted-mesa structures by employing Si-vapor thermal etching. This etching technique is achieved by high temperature annealing up to 2200 °C in Si-vapor environment, where thermally decomposed SiC surface is compensated by a Si-vapor flux from Si solid source in a same semi-closed container made of TaC. Furthermore, the epitaxial graphene growth using the step-free surface as an initial template has already been demonstrated by our group, where single-/bi-layer control is possible and the stability of the SiC basal plane is a critical factor for the control of epitaxial graphene layer thickness [3].

In this study, we aimed to investigate the thermal stability of 4H-SiC (0001) basal plane by utilizing locally fabricated step-free surface as a limiting case of low off-axis surface. We first report the observation of the morphological instability of step-free 4H-SiC (0001) basal plane surfaces during Si-vapor thermal etching in the temperature range of 1400 °C - 2100 °C. The morphological change at the bottom of inverted-mesa structures with threading dislocations was also discussed. The findings in this study are important not only for epitaxial graphene growth on 4H-SiC (0001), which proceeds at SiC surface step edges rather than at terrace surfaces, but also for SiC homo epitaxial growth. This is because degradation in SiC epitaxial layer such as the formation of step bunching directly depends on initial SiC surface morphologies. Especially in the case of use of low off-axis 4H-SiC (0001) surfaces, it was indicated that there is the necessity to consider the surface instability of (0001) basal plane in addition to the influence of dislocations on SiC surface step dynamics.

Experiments

4H-SiC (0001) normally on-axis substrates were used in this study. The step-free initial SiC surfaces were fabricated at the bottom of inverted-mesa structure by the method combining a direct laser digging and Si-vapor etching, as reported in ref. [3]. The original experimental setup consisting of a TaC crucible and an ultrahigh vacuum metal chamber was utilized for the Si-vapor etching technique. It enables us to heat samples in the crucible up to 2200 °C from degassed temperature ~ 800 °C within one minute. The diameter and the depth of the inverted-mesa structures were 50 - 200 μm and 10 - 30 μm, respectively. Following the formation of the inverted-mesa structure by the direct laser digging, steps at the bottom of the inverted-mesa structure were swept out by Si-vapor etching. Therefore, SiC step-free (0001) basal plane regions are able to be enlarged during the etching. After the formation of the step-free surface at the bottom of the inverted-mesa structure, additional Si-vapor etching was performed repeatedly for the same sample. Fig.1 shows the schematic illustration of the concept for the experimental process in this study. Surface morphologies were monitored by atomic force microscope (AFM), scanning electron microscope (SEM) and laser microscope.

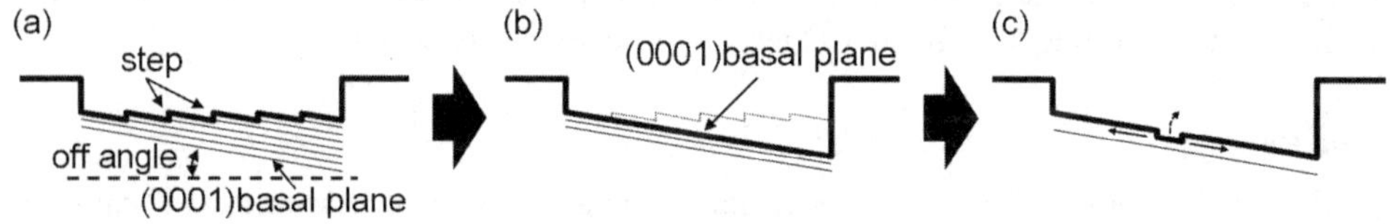

Fig.1. Schematic illustration of the experimental method in this study. (a) Fabrication of the inverted-mesa structure at a 4H-SiC (0001) surface. (b) Si-vapor etching for the formation of the step-free (0001) basal plane region at the bottom of the inverted-mesa structure. (c) Characteristic morphological instability was appeared at the bottom surface of the inverted-mesa structure by additional Si-vapor etching.

Results and Discussion

Fig.2 shows the typical surface morphologies appeared at the bottom of an inverted-mesa structure after several cycles of Si-vapor etching. Randomly created "crater" and "maze" structures were observed as shown in Fig.2 (a) and (b), respectively. A perfect step-free surface was also shown in Fig.2 (c). Note that these images were obtained from an identical area. As can be realized from those images, the surface morphology was altered drastically by Si-vapor etching. And more interestingly, the structural features (crater and maze structures) cyclically appeared with respect to each additional etching treatment at 1600 - 1900 °C. The depths of the crater structures in Fig. 2 (a) are in the range of 0.2 - 0.3 nm, which corresponds to SiC 1 bilayer. Stability of the top basal plane is considered to be broken by the randomly created craters. Note that an initial (0001) basal plane is always broken by the formation of these craters. On the other hand, the steps with the half unit cell height of 4H-SiC (= 0.5 nm) are dominant for the maze structures in Fig.2 (b). Stability of the bottom basal plane maintains until the whole maze structures are etched away. The conceivable etching process at 4H-SiC step-free surfaces is illustrated in Fig.3. It is a kind of "layer-by-layer" etching model. First, the 0.2 - 0.3 nm depth crater structures appear at the step-free (0001) basal plane randomly. Then the structures expand each area during Si-vapor etching. After the crater structures sufficiently expanded and connected with neighboring structures during Si-vapor etching, 0.5 nm depth areas appear in the crater structures. Once the 0.5 nm depth areas appeared, etching proceeds and these areas are expanded like a maze. Finally, the maze structures are swept away by Si-vapor etching, resulting in

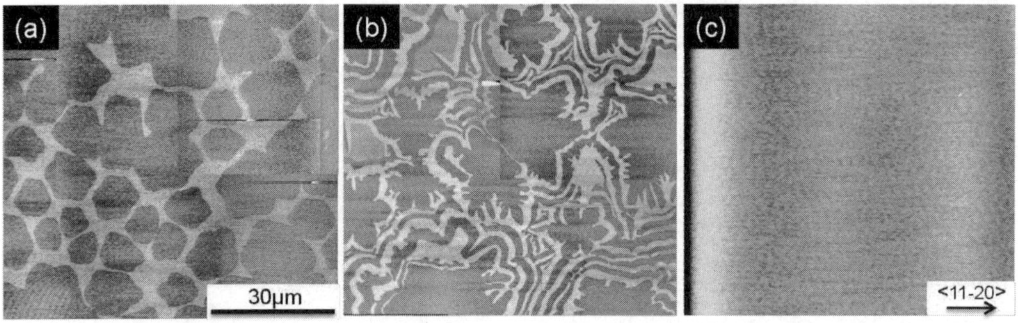

Fig.2. AFM topographic images of the identical step-free region at on-axis 4H-SiC (0001) (a) after Si-vapor etching at 1800 °C for 10 min at the end of the several cycles of Si-vapor etching at 1900 °C (SiC 1 bilayer steps are dominant), (b) after the additional Si-vapor etching at 1800 °C for 20 min (SiC 2 bilayer steps are dominant), and (c) after the additional Si-vapor etching at 2100 °C for 10min (perfect step-free surface) at the end of the several cycles of Si-vapor etching.

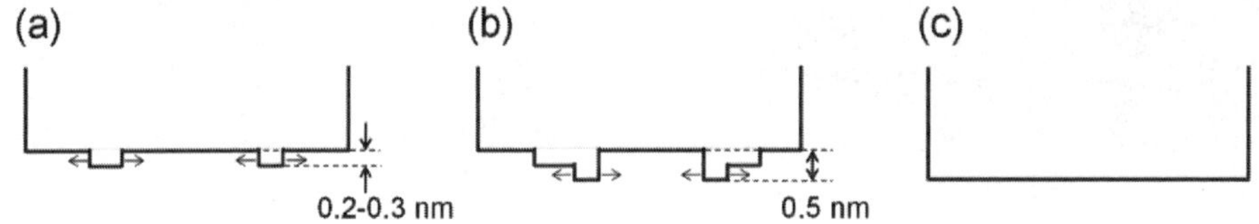

Fig.3. Cross-sectional etching model at a step-free 4H-SiC (0001) basal plane. (a) First, the crater structures with 0.2 - 0.3 nm depth are randomly created at the (0001) basal plane surface and are expanded each area. (b) Next, the maze structures with ~ 0.5 nm depth appear in the crater structures. (c) At last, the maze structures are etched away and the step-free (0001) basal plane appears once again.

the re-emergence of step-free surface. The order of etching rate in this process was estimated below 10 nm/min. Here, it is noteworthly that the morphological instability such as the formation of crater and maze structures was not observed in the case that the bottom surface of inverted-mesa structure is relatively small (< 50 μm diameter) and in high temperature etching conditions [as shown in Fig.2 (c) 2100 °C]. In the etching mechanism, the bottom surface of the crater structures seems to be a transitional surface condition from the initial basal plane to the next stable basal plane with 0.5 nm depth from the initial basal plane. In terms of shapes of structures, maze structures have more strong anisotropy than crater structures. It indicates that the bottom surfaces of crater structures have a different surface state from the initial step-free surface and the bottom surface of maze structures, such as different surface reconstructions.

As another case of surface morphology at the bottom of inverted-mesa structure after Si-vapor etching, two types of SiC step structures were observed, as shown in Fig.4. The common feature of them was homo-centric SiC 2 bilayer steps (~ 0.5 nm height) and these centers always appeared at same positions even after a number of Si-vapor etching treatments. They are classified according to shapes of SiC steps. One has a spiral shape of SiC steps [Fig.4 (b)], the other has a concentric circle shape of SiC steps [Fig.4 (c)-(f)]. Thus, the former comes from a threading screw dislocation (TSD) and the latter is considered to come from a threading edge dislocation (TED). This was also confirmed by a conventional molten KOH etching analysis. As seen in Fig.4, the shapes of steps from TED have temperature dependence. It shows strong anisotropic shapes in the low temperature range [Fig.4 (c)–(e): 1400-1800 °C], while an isotropic shape in the high temperature condition [Fig.4 (f): 2100 °C]. The depressed region at the TED core becomes more circular shape from triangle shape with increasing etching temperature. The triangle shapes of SiC steps reflect the stability of {1-10n} facets at the step edges. Although the dendritic shape shown in Fig.4 (d) is very strange and unique, it needs a more detailed discussion elsewhere. By contrast, the shape of steps from TSDs has no such clear temperature dependence. In both TSD and TED cases, SiC steps seem to be constantly produced from the dislocation cores during Si-vapor etching. So, threading dislocations act as the supply source of steps at the SiC surface without off-axis –induced steps in the etching process.

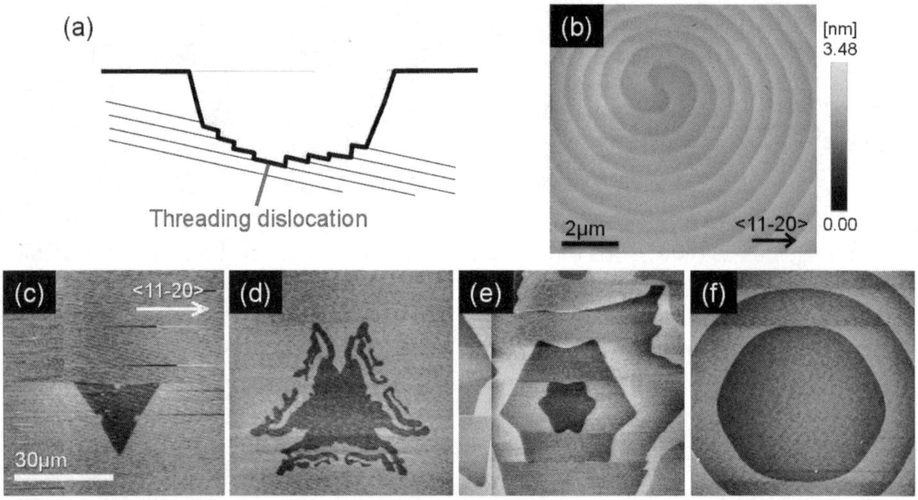

Fig.4. (a) Cross sectional illustration of the inverted-mesa structure which has a threading dislocation at the bottom. (b) AFM topographic image obtained from a region including a threading screw dislocation (TSD) after Si-vapor etching at 1900 °C for 8 hours. (c)-(f) AFM topographic images obtained from an identical region including a threading edge dislocation (TED). At the ends of several cycles of Si-vapor etching, (c) 1400 °C for 10 min, (d) 1600 °C for 1 hour, (e) 1800 °C for 10 min, (f) 2100 °C for 10 min Si-vapor etching were performed, respectively.

Summary

We investigated the morphological instability at step-free 4H-SiC (0001) surfaces during Si-vapor thermal etching. Randomly created crater and maze structures were observed by AFM measurement. A kind of layer-by-layer etching model was suggested. In contrast to the step-free surfaces, the morphological changes at the bottom of inverted-mesa structures with threading dislocations were also discussed. It revealed that dislocation cores act as the supply source of steps at the SiC surface in the etching process. The morphological instability and the influence of dislocations on SiC step dynamics at large (0001) basal plane surfaces, which were discussed in this report, are important to be considered for the future application of extremely low off-axis 4H-SiC (0001) surfaces.

Acknowledgement

The authors are grateful to Mr. S. Ashizaki for his encouragement and support on this study.

References

[1] J.A.Powell, P.G.Neudeck, A.J.Trunek, G.M.Beheim, L.G.Matus, R.W.Hoffman,Jr., J.Keys, Appl. Phys. Lett. **77** (2000) 1449.

[2] J.D.Caldwell, M.A.Mastro, K.D.Hobart, O.J.Glembocki, C.R.Eddy,Jr., N.D.Bassim, R.T.Holm, R.L.Henry, M.E.Twigg, F.Kub, P.G.Neudeck, A.J.Trunek, J.A.Powell, Appl. Phys. Lett. **88** (2006) 263509.

[3] S.Ushio, Y.Kutsuma, A.Yoshii, N.Tamai, N.Ohtani, T.Kaneko, Jpn. J. Appl. Phys. **50** (2011) 070104.

Materials Science Forum Vols. 717-720 (2012) pp 581-584

doi:10.4028/www.scientific.net/MSF.717-720.581

First-principles Analysis of Dissociative Absorption of HF Molecule at SiC Surface Step Edge

Kouji Inagaki[a], Bui Van Pho, Kazuto Yamauchi and Yoshitada Morikawa

Osaka University, 2-1 Yamada-oka Suita, Osaka, Japan

[a] inagaki@prec.eng.osaka-u.ac.jp

Keywords: Catalyst-referred etching, Hydrogen fluoride, First-principles calculation, Dissociative absorption, Reaction barrier.

Abstract. It has been reported that a SiC surface can be etched off in a concentrated aqueous solution of hydrogen fluoride by making contact with a Pt plate catalyst, named Catalyst-referred Etching (CARE) [1]. In this report, we present first-principles simulations on the initial stage of the etching process. Reaction barrier heights of dissociative absorption reactions of a hydrogen fluoride molecule breaking into a couple of kinds of back-bonds at a step edge of 3C-SiC(111) are analyzed.

1. Introduction

Fabrication of atomically flat SiC or GaN surfaces by a wet process turned out to be realized by means of a newly developed surface planarization technique named CARE (CAtalyst Referred Etching) [1,2]. In the method, a catalyst material with a plate shape and a target material, both immersed into an etching solution, are rubbed against each other so as to promote chemical etching reactions at the interface, which do not occur in any ordinary room environment. After the etching, the finished surfaces are atomically flat with almost no introduction of defects. This method, however, requires further improvement. In processing a SiC surface, there are two problems to be overcome: The removal rate is not sufficiently high for practical use, and expensive Pt catalysts and toxic HF solution are used in the process. To explore more effective, cheaper and safer new materials, it is important to understand the etching reaction at the atomic level. In this paper, we present first-principles molecular-dynamics simulations on the initial stage of the reaction. Taking into account experimental facts that (a) step-terrace surfaces can be obtained after the CARE process, (b) etching does not occur without HF, and (c) most of HF does not dissociate into ions but exists in molecular form, it is expected that (i) etching should occur at a step edge of the surface and (ii) an HF molecule should be an etchant of the reaction. Therefore we analyzed dissociative absorption of an HF molecule into a Si-C bond at the step edge of the SiC surface.

2. Calculation

We performed first-principles molecular-dynamics calculations using the "STATE" (Simulation Tool for Atom TEechnology) code [3,4]. The calculations are based on density functional theory [5,6] within the generalized gradient approximation proposed by Perdew [7]. Electron-ion interactions were described by ultrasoft pseudopotentials [8]. Wave functions and the augmented charge density were expanded by a plane-wave basis set with cutoff energies of 25 and 225 Ry, respectively. A 3×2 k-point mesh was used for surface Brillouin zone integration. To calculate reaction paths, the climbing-image nudged-elastic band method (CI-NEB) [10,11] is adopted. This method is based on an iterative optimization scheme to find a reaction path with reaction barriers from an initial guess of the reaction path. In CI-NEB, reaction path is expressed by a discrete sequence of atomic configurations, called replicas, in which the direction of the reaction is approximated by a finite difference of the replicas. They are optimized to have no gradient in the directions perpendicular to the reaction path and keep distances between replicas. Before starting optimization of a reaction path, the initial and the final configurations of the reaction are optimized by ordinary molecular dynamics calculations. An initial guess of the reaction path is then generated usually by simply interpolating the initial and the final configurations. When unphysical models with

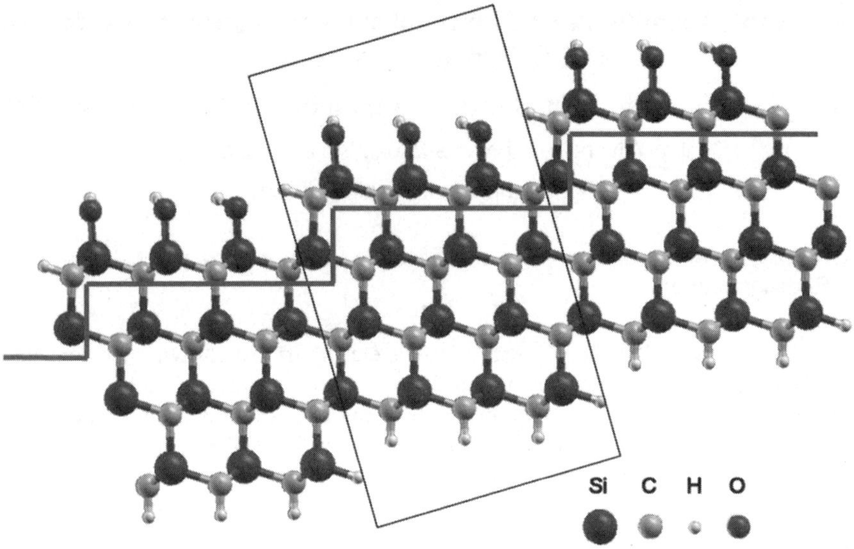

Fig. 1: Employed super cell model. Four bilayer 3C-SiC(111) surface with steps. Surface Si and C atoms at the step edge are terminated by an OH group and by a H atom, respectively. Two units in the direction parallel to the step (perpendicular to the indicated plane) are taken into account. The rectangle indicates the unit cell. The size of the super cell is 6.2Å×9.3Å×21.2Å.

short inter-atomic distances are generated, such configurations were rearranged by hand. After several optimization steps, the motion of the replica which has the highest energy is changed to move towards the higher energy direction along the reaction path. In this way, this replica reaches a saddle point and the others align along the valley of the potential surface. The optimization is continued until the residual forces acting on all the atoms at the saddle point become smaller than 10^{-2} Hartree/Bohr.

We employed a repeated slab model consisting of four bilayers of 3C-SiC (111) with steps, which is equivalent to the (221) Si-terminated surface, including two units along the step and with a rectangular supercell of 6.2 x 9.3 x 21.2 Å, as depicted in Fig.1. The vacuum region between slabs is about 8.35Å. Although there are many polytypes of SiC, the atomic configurations of the topmost bilayer around a surface step are all similar. Therefore we consider that the 3C-SiC surface model is sufficient for the purpose of this study. All dangling bonds of surface Si atoms are terminated by OH according to Ref. [9]. The termination type of C atoms at a step edge has not been reported yet. Our first-principles calculations of the formation energies of H-termination and OH-termination show that H-termination is more stable by 1.74 eV. H-termination is therefore employed in this study. The bottom of the slab is terminated by hydrogen atoms.

Using this model, we analyzed the reaction paths and the barrier heights of the dissociative absorption reactions of a HF molecule onto a SiC surface step. We investigated a couple of HF absorption sites as indicated in Fig. 2(a). The obtained final and transition states are shown in Fig. 2 (b)-(e). The first site is at the bond directing perpendicular to the step edge [back bond (BB)] and the other is at the bond lying along the step edge [side bond (SB)]. The energy differences between the initial and the final configurations, and the obtained reaction barrier heights are shown in Table 1. The interatomic distances between the atoms, which correspond to the dissociative absorption, in the transition states are summarized in Table 2. Both of the reaction barriers turn out to be higher than 2.0 eV. Since they are far larger than the hydrogen bond energy, these values are reliable. Nevertheless, the hydrogen bonds between the surface terminating atoms and the surrounding water molecules are neglected in this simulation. These values are quite large, thus such reaction does not occur in the room condition. These reactions might occur with certein catalytic or mechanical assistance in the CARE process. On the other hand, the total energies of the final states are quite different. This is due to a difference in mechanical effect: In the case of SB, the motion of the Si atom in the direction parallel to the step edge induces a large bulk strain, while, in the case of BB, the motion perpendicular to the step edge induces no bulk strain. This is consistent with the fact that the interatomic distances between the reacting atoms in the transition state in BB in Table 2 are longer than those in SB. A small but lower barrier height in the SB than that in the BB is inconsistent with this consideration. This small differnce is due to the artifical enhancement of the hydrogen bonds between OH terminations of the reacting Si and of the neghiboring Si in SB. In order to obtain more accurate energy, the neglected surrounding water molecules should be taken into account. The higher energy of the final state in SB cause a lower reaction barrier of the reverse reactions, thus the atomic configuration quickly return to the initial state in SB. Therefore, it can be concluded that the dominat process is BB.

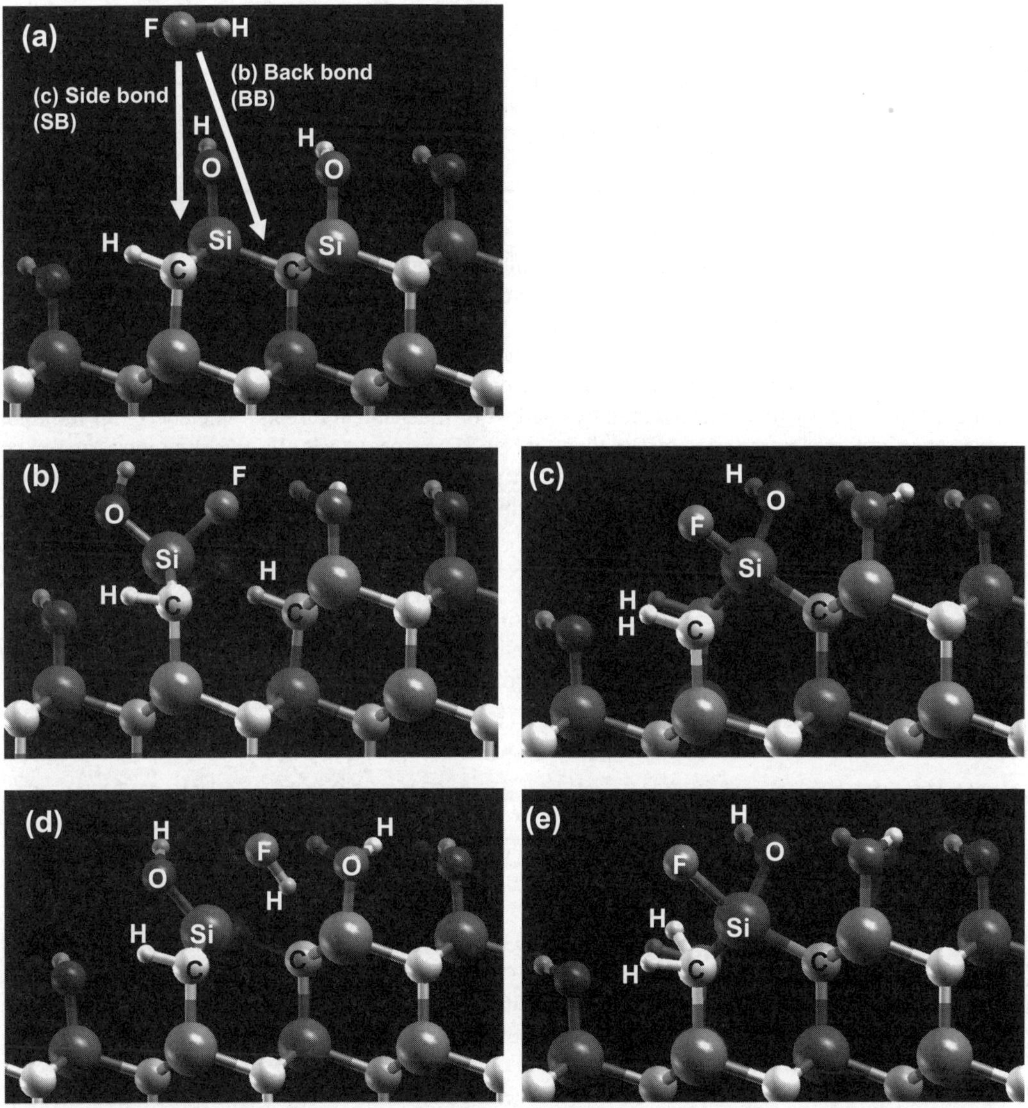

Fig. 2: Reaction paths of dissociative absorption of a HF molecule at a step edge of the SiC surface. Initial configuration (a). Final configurations: insertion into Si-C back bond (BB) (b) and that into side bond (SB) (c). (d) and (e) are the transition states of the reactions (b) and (c), respectively. Deeper atoms are expressed by darker spheres. In (c), two H atoms terminate the left C atom.

Table 1 Energy gain and barrier height for dissociative absorption of HF molecule into SiC back bonds at surface step edge.

Model	Energy difference (eV)	Barrier height (eV)
HF into SB	1.53	2.07
HF into BB	-0.37	2.14

Table 2 Interatomic distances between Si, C, H and F atoms in the transition states.

	SB Fig.1(d) (Å)	BB Fig.1 (e) (Å)	Initial (Å)	Final (Å)
H-F	1.53	1.92	0.95	-
Si-C	2.30	2.35	1.90	-
Si-F	1.91	2.26	-	1.67
C-H	1.15	1.70	-	1.11

Summary

Dissociative adsorption processes of a HF molecule into back bonds at a step edge of SiC (111) are investigated. The termination structure of a C atom at the step edge is determined as H termination. Reaction barriers and total energies after dissociative absorption of a HF molecule into SiC back bonds are calculated for a couple of different back bond sites. Barrier heights are larger than 2.0 eV in these cases, which indicates that these reactions do not occur in ordinary room conditions. It is considered to occur via reactions assisted by catalyst or mechanical effects. The energy of the final configuration is lower in the case of BB than that of SB by about 1.9 eV, since the introduction of bulk strain is larger in SB than in BB. Dissociative absorption of HF into the BB site should be dominant reaction in the initial stage of CARE etching.

Acknowledgments

This research was partially supported by a Grant-in-Aid (No. 22560110) and a grant of the Global COE Program “Center of Excellence for Atomically Controlled Fabrication Technology” from the Ministry of Education, Culture, Sports, Science and Technology.

References

[1] H. Hara, Y. Sano, K. Arima, K, Yagi, J. Murata, A. Kubota, H. Mimura, and K. Yamauchi, Sci. and Technol. of Adv. Mater. 8 (2007) 162.

[2] K. Arima, H. Hara, J. Murata, T. Ishida, R. Okamoto, K. Yagi, Y. Sano, H. Mimura, and K. Yamauchi, Appl. Phys. Lett. 90 (2007) 202106.

[3] Y. Morikawa, Phys. Rev. B 51 (1995) 14802.

[4] Y. Morikawa, Phys. Rev. B 63 (2001) 033405.

[5] P. Hohenberg and W. Kohn, Phys. Rev. 136 (1964) B864.

[6] W. Kohn and L. J. Sham, Phys. Rev. 140 (1965) A1133.

[7] J. P. Perdew, K. Burke, and M. Ernzerhof, Phys. Rev. Lett. 77 (1996) 3865.

[8] D. Vanderbilt, Phys. Rev. B 41 (1990) 7892.

[9] S. J. Schoell, J. Howgate, M. Hoeb, M. Auernhommer, J. A. Garrido, M. Stutzmann, M. S. Brandt, and I. D. Sharp, Appl. Phys. Lett. 98 (2001) 182106.

[10] G. Mills, H. Jonsson, and G. K. Schenter, Surf. Sci. 324 (1994) 305.

[11] G. Henkelman, B. P. Uberuaga, and H. Jonsson, J. Chem. Phys. 113 (2000) 9901.

Materials Science Forum Vols. 717-720 (2012) pp 585-588
© (2012) Trans Tech Publications, Switzerland
doi:10.4028/www.scientific.net/MSF.717-720.585

Characterization of Photoelectrochemical Properties of SiC as a Water Splitting Material

Tomonari Yasuda [1,a], Masashi Kato [1,b] and Masaya Ichimura[1,c]

[1] Nagoya Institute of Technology, Gokiso-cho, Showa-ku, Nagoya-shi, 466-8555, Japan

[a] cja16597@stn.nitech.ac.jp, [b]kato.masashi@nitech.ac.jp

[c] ichimura.masaya@nitech.ac.jp

Keywords: water splitting, hydrogen, photoelectrochemical properties

Abstract. Hydrogen attracts attention as an eco-friendly energy resource. The water splitting by semiconductor materials can generate hydrogen without CO_2 emission. However, the hydrogen conversion efficiency using conventional materials is not high enough, or the materials corrode easily. On the other hand, silicon carbide (SiC) is expected to be a water splitting material showing high conversion efficiency without corrosion. In this study, we characterized band edge potentials for 4H-, 6H- and 3C-SiC, and we revealed that they are capable of water splitting. We also estimated conversion efficiencies by photocurrent measurements in electrolytes for bulk 4H- and 6H-SiC.

Introduction

Hydrogen attracts attention as an eco-friendly energy resource. However, hydrogen is usually produced from fossil fuel, and, in this production, a green house gas, CO_2, is also generated. The water splitting by solar light illumination of semiconductor materials can generate hydrogen without CO_2 emission. The conversion efficiency or durability of this method is not high enough for practical use. For example, metal oxides are conventional water splitting materials without corrosion in electrolytes. However, because they have wide band gaps, the conversion efficiencies for these materials are only up to 1-2%. On the other hand, semiconductor materials other than metal oxides sometimes show high conversion efficiencies, but they are weak against corrosion. Thus, new water splitting materials, such as GaN, have been studied to obtain both high conversion efficiency and resistance to corrosion [1]. In general, a water splitting semiconductor material should meet three criteria: appropriate bandgap (1.6-2.2eV), good material durability and band edges straddling H_2O redox potentials. Silicon carbide (SiC) is chemically stable and can be made p-type with high crystal quality. P-type materials are generally more stable than n-type materials for water splitting, and thus p-type SiC could also be resistant to corrosion. Because of the wide band gap (> 2.9 eV), theoretical efficiencies reported in Ref. [2] for 4H- and 6H-SiC are not so high, ~1% and 2%. However, that for 3C-SiC is 8% because it has a band gap of 2.2 eV and can absorb visible light. Although many studies about electrochemical properties of SiC have been reported so far [3-6], reports on SiC as a water splitting material are rare [6,7]. In this study, the photoelectrochemical properties of both n- and p-type SiC as a water splitting material were characterized and the conversion efficiencies were also estimated.

Experiment

Samples employed in this study were bulk grown n- and p-type 4H-and 6H-SiC along with n-type 3C-SiC. The thickness and resistivity of the samples are listed in Table 1. We fabricated ohmic contacts on the backside of the samples. Then the samples were fixed by silicone resin on polycarbonate plates with only the sample surface exposed. A wire was bonded on the ohmic contact and connected to a potentiostat so that the sample acted as the working electrode in the three electrode cell. Pt was used as a counter electrode, and a saturated calomel electrode (SCE) was used as a reference electrode.

Current-voltage (*I-V*) characteristics were measured under dark and light conditions. The electrolyte was a 0.5mol/l Na_2SO_4 aqueous soliution and the light source is a solar simulator whose power is 780 mW/cm^2. The band edge potential was estimated by the capacitance-voltage (*C-V*)

measurement with a measurement frequency of 1 kHz in electrolytes. As electrolytes, aqueous solutions with HCl (pH 1.5), Na_2SO_4 (pH 7) and NH_4OH (pH 12.2) were employed. Time dependence of photocurrent under illumination by the solar simulator was measured in a 1mol/l H_2SO_4 aqueous solution with a potential of 0 V vs SCE. All the experiments were done at room temperature.

Results and Discussion

Figure 1 shows *I-V* characteristics for n-type 4H-SiC. In the dark condition, current was observed at negative potentials vs SCE, and rectifying characteristics were observed. Under light illumitnation, photocurernt is observed at 0 V and positive potentials vs SCE. Figure 2 shows *I-V* characteristics for p-type 4H-SiC. Rectifying characteristics were observed and polarity was opposite to that for n-type 4H-SiC. If we illuminate n-type SiC in electrolyte, it will be oxidized because photogenerated holes react with H_2O in the electrolyte as shown in the inset of Fig. 1. On the other hand, in the case of p-type SiC, photogenerated electrons react with H^+ ions as shown in inset of Fig. 2. As shown in Fig. 2, photocurrent from p-type 4H-SiC was observed under illumination below 0.5 V vs SCE. For 6H-SiC, similar results were obtained. In contrast, n-type 3C-SiC showed rectifying characteristics but little photocurrent at 0 V vs SCE under light illumination. Figure 3 shows a Mott-Schottky plot for n-type 4H-SiC. In this plot, the intercept with the x-axis corresponds to the flat band potential, V_{fb}. For this sample $V_{fb} = -1.84$ V. Using V_{fb} and the Fermi level estimated from donor or acceptor concentrations, we obtained the band edge potentials in various pH for the samples as shown in Fig. 4. The band edge potential for p-type 6H-SiC is not shown in this figure because the *C-V* measurement was not able to be performed for this sample due to high resistivity. In this figure, we also show the redox potentials of H^+ / H_2 and O_2 / H_2O as the top and bottom of the gray band. If the edge of the conduction band is located above the redox potential of H^+ / H^2 and the edge of the valence band is located below the redox potential of O_2 / H_2O, water splitting is possible. The conduction band edge of 3C-SiC is near to the redox potential of H^+ / H_2, while other polytypes clearly have band edge potentials capable of water splitting in all pH. Although, in principle, the band edge potential does not

Table 1 Thickness and resistivity of the samples

	n-4H	p-4H	n-6H	p-6H	n-3C
Thickness (μm)	350	346	260	400	303
Resistivity (Ωcm)	0.01-0.05	500-2500	0.1	1.5×10^{7}-1.0×10^{12}	1.0
Doping density (cm^{-3})	1×10^{19}	2×10^{17}	4×10^{19}	-	2×10^{19}

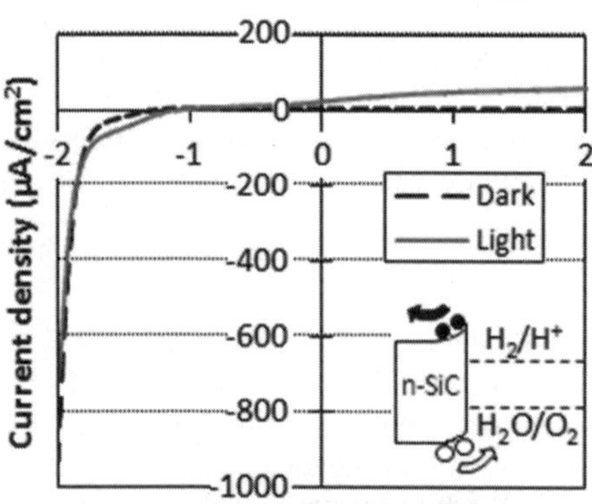

Fig. 1 *I-V* characteristics in a 0.5mol/l Na_2SO_4 aqueous solution for n-type 4H-SiC. The insert is a schematic of water splitting for n-type samples.

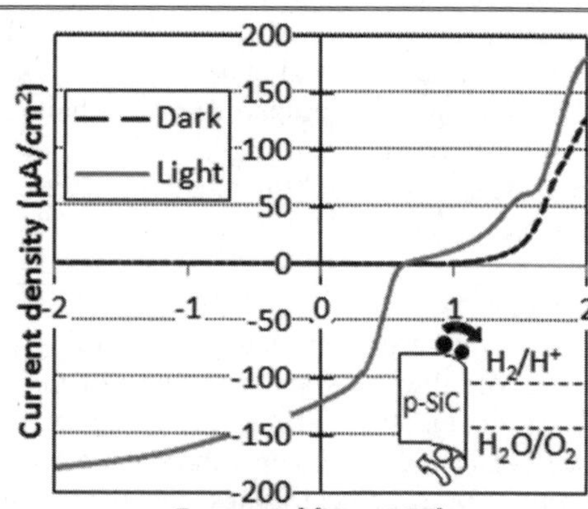

Fig. 2 *I-V* characteristics in a 0.5mol/l Na_2SO_4 aqueous solution for p-type 4H-SiC. The insert is a schematic of water splitting for p-type samples.

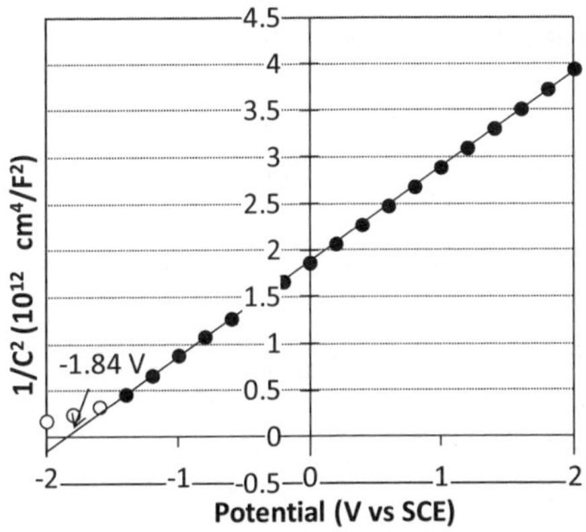

Fig. 3 Mott-Schottky plot for n-type 4H-SiC in pH 7 (0.5mol/l Na_2SO_4 aqueous solution). The intercept with the horizontal axis corresponds to the flat band potential.

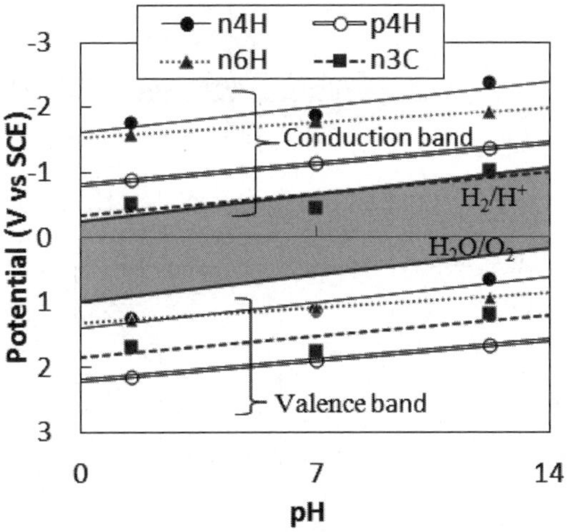

Fig. 4 Band edge potential of SiC versus pH. The redox potentials of H^+/H_2 and of O_2/H_2O are also indicated by the top and bottom lines of the gray band.

depend on the conduction type, the estimated band edge potential for n-type 4H-SiC is 1 eV above that of p-type 4H-SiC. The reason is still unclear and further investigation of the band edge potential is required.

Figure 5 shows the time dependence of photocurrents under light illumination for the n-type samples. In all the n-type samples, the photocurrents are very small and decrease gradually. Figure 6 shows photocurrents for the p-type samples. The photocurrents are larger than those for the n-type samples and do not significantly decrease with time. Small steps observed in photocurrent for the p-type 4H-SiC may be due to instability of the light source. We calculated the conversion efficiency η from

$$\eta\,(\%) = \frac{I \times 1.23}{L} \times 100 \tag{1}$$

where I is the photocurrent, L is the light intensity, and 1.23 V is the redox potential width between H^+ / H_2 and O_2 / H_2O. Table 2 lists the obtained conversion efficiencies for the samples. In this estimation, we use the stable photocurrents after the initial reduction for the n-type samples. The conversion efficiency for n-type 3C-SiC is not shown due to its very small photocurrent. The efficiencies for the n-type samples are much lower than those for the p-type samples. It is known that SiC is oxidized by light illumination in an electrolyte [7], and the resulting SiO_2 behaves as a resistant layer for photocurrent. Therefore, the small and decreasing photocurrent for the n-type samples would be caused by SiO_2 formation. Among all the samples, the highest conversion efficiency of 0.024% is obtained for p-type 4H-SiC.

We confirmed no corrosion for the p-type samples after the measurements of photocurrent by optical microscopy and a profilometer observation, and thus p-type SiC is promising for water splitting application. We should note that the efficiencies obtained in this study are calculated from photocurrent and not from measured volume of H_2. In addition, photocurrent was measured by the three electrodes system with the potential adjusted at 0 V vs SCE. If we use a two electrode system, the conversion efficiency may be different. In addition, the samples employed in this study were bulk ones, and thus carrier lifetimes should be short and doping densities are relatively high. These factors have negative influence on the efficiency. We are now trying to characterize epitaxial samples to improve efficiencies. The epitaxial samples have higher crystalline quality than the bulk samples and are expected to show higher efficiencies. Results for the epitaxial samples will be reported elsewhere.

Table 2 Conversion efficiencies estimated from photocurrent

Conduction type	4H	6H
n	6.3×10^{-5} %	4.7×10^{-5} %
p	0.024 %	0.011 %

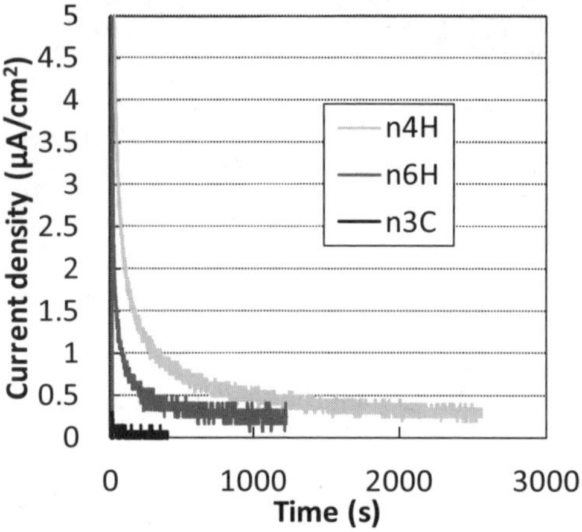

Fig. 5 Time dependence of photocurrent for the n-type samples.

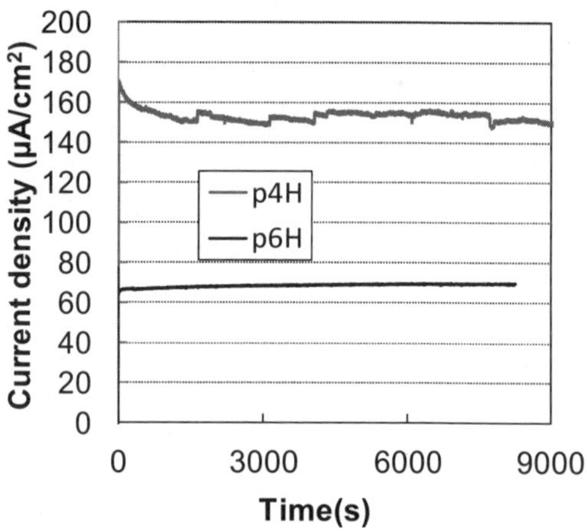

Fig. 6 Time dependence of photocurrent for the p-type samples.

Conclusions

We characterized photoelectrochemical properties of SiC as a water splitting material. Band edge potentials were estimated from *C-V* measurements for 4H-, 6H-SiC and 3C-SiC, and all the polytypes seem to have band edge potentials capable of water splitting. From photocurrent under light illumination in electrolytes, the conversion efficiencies were estimated to be 0.024% and 0.011% for p-type 4H- and 6H-SiC, respectively, and p-type SiC did not corrode under illumination in the electrolytes. These results suggest that p-type SiC is promising as a water splitting material.

Acknowledgements

We would like to thank Dr. T. Hatayama of Nara Institute of Science and Technology for sample preparation.

References

[1] K. Fujii, K. Ohkawa, Journal of the Electrochemical Society 153 (3) (2006) A468-A471

[2] S. G. Sridhara, T. J. Eperjesi, R. P. Devaty, W. J. Choyke, Materials Science and Enginerring B61-62 (1999) 229-233

[3] C. Schnabel, M. Worner, B. Gonzalez, I. Olmo, A. M. Braun, Electrochimica Acta 47 (2001) 719-727.

[4] J. V. Lagemaat, D. Vanmakelbergh, J. J. Kelly, Journal of Applied Physics 83 (1998) 6089-6094.

[5] D. H. V. Dorp, N. Heijnen, M. D. Vece, J. J. Kelly, Angewandte Chemie International Edition 48 (2009) 6085-6087.

[6] J. Akikusa, S. U. M. Khan, International Journal of Hydrogen Energy 27 (2002) 863-870

[7] I. Lauermann, R. Memming, D. Meissner, Journal of The Electrochemical Society, 144, January (1997) 73-80.

Materials Science Forum Vols. 717-720 (2012) pp 589-592
© (2012) Trans Tech Publications, Switzerland
doi:10.4028/www.scientific.net/MSF.717-720.589

Room temperature physical characterization of implanted 4H- and 6H-SiC

Konstantinos Zekentes[1,a,*], Katerina Tsagaraki[1,b], Maria Androulidaki[1,c], Maria Kayambaki[1,d] Antonis Stavrinidis[1,e] Hervé Peyre[2,f] and Jean Camassel[2,g]

[1]MRG-IESL/ FORTH, Vassilika Vouton, PO Box 1385 Heraklion, Greece

[2]GES, Université Montpellier 2, CC047, 2 place Eugène Bataillon, F-34095 Montpellier, France

[a]trifili@physics.uoc.gr, [b]ktsag@physics.uoc.gr, [c]pyrhnas@physics.uoc.gr, [d]mmic@physics.uoc.gr, [e]astav@physics.uoc.gr, [f]peyre@ges.univ-montp2.fr, [g]jean.camassel@ges.univ-montp2.fr, *corresponding author

Keywords. Ion implantation, X-Ray Diffraction, Room temperature photoluminescence, Electrochemical, C-V, mercury-probe.

Abstract. The purpose of the present study is to determine the appropriate physical characterization methods for evaluating material quality during the fabrication steps of a 4H-SiC Static Induction Transistor (SIT). The most important fabrication step, in terms of material quality, is the gate implantation and post-implantation annealing. Separate “witness” samples from the processed sample have been used for evaluating implantation and post-implantation annealing. Secondary Ion Mass Spectroscopy (SIMS), optical transmission, room-temperature photoluminescence (RTPL), High Resolution X-Ray diffraction (HRXRD) and C-V measurements with Hg-probe and electrochemical (ECV) cells have been employed in the frame of the present study. HRXRD and ECV have been proved particularly suitable for characterizing the implanted layers.

Introduction

Ion implantation is the unique method for selective area doping in the case of SiC. Many device geometries are based on ion implantation although very few of them use an ion-implanted layer as an emitter. The main reason is that the quality of these layers and especially that of p-type material is worse in comparison to epitaxial layers. Indeed, in the case of SiC-based transistors, ion implantation is mainly used for gate formation and edge periphery isolation purposes. Post-implantation annealing can’t result simultaneously in a high electrical activation of dopant, smooth surface and high carrier mobility value. A consequence of this fact is that a single sheet resistance measurement is not sufficient to ensure that implanted layers lead to proper device operation. Various studies using material characterization methods such us XRD [1, 2] and RTPL [2] have been reported. For doping profiling, capacitance measurements are used, based mainly on Hg probe, but this technique presents many limits on carrier delineation due to a poor contact capability and a limited dynamic range. Scanning capacitance microscopy has been successfully used to make a two dimensional profile of the implanted ions [3] but it is a time consuming method. Electrochemical C-V (ECV) profiling has been successfully applied for the doping profiling of n- and p-type 6H- and 4H-SiC epilayers and substrates [4] but not in a systematic way in the case of implanted layers.

The purpose of the present study is to determine the appropriate physical characterization methods for evaluating material quality following ion implantation and post-implantation annealing during the fabrication steps of a typical SiC device like the SIT.

Experimental

For the purposes of the present investigation, separate “witness” samples from the processed sample have been used for evaluating implantation and post-implantation annealing. Secondary Ion Mass Spectroscopy (SIMS), optical transmission in the visible and ultraviolet, RTPL, High Resolution

XRD (HRXRD) and C-V measurements with Hg-probe and electrochemical cells have been used in addition to the commonly employed Scanning Electron Microscopy (SEM) method. Different samples have been implanted and annealed in different experimental setups (implanters and furnaces) to offer the necessary variation for the material evaluation process. The investigated samples served as "witness" samples for the implantation steps of the n^+ contact and p^+ gate during the fabrication of 4H-SiC SITs [4]. Both 4H- and 6H-SiC samples have been used as witnesses in the implantation and post-implantation-annealing process steps.

Results and Discussion

Figs. 1 and 2 show the C-V profiles measured on highly doped n- and p-type SiC layers formed by N and Al implantation, respectively, as well as the corresponding SIMS profiles. There is a net difference between the SIMS and Hg-probe profiles below the top 50 nm of the nitrogen-implanted layer. A possible explanation would be a nitrogen outdiffusion during annealing but this does not seem probable, taking into account that the difference is so huge as well as that the surface has been covered with a C-cap and that the temperature of the annealing was relatively low (1600°C). Rather, the measuring limitations of the Hg-probe C-V profiling are likely the reason for this discrepancy. This is not the case for ECV profiling performed on the Al-implanted layer (Fig. 2), where the top 350 nm of the implanted layer could be profiled and the dopant activation could be determined. Note also on Fig. 2 that most of the Al atoms are active and there is no important outdiffusion effect as the profile is similar to that simulated with TRIM (not shown). The ECV technique can be easily applied during device fabrication once a top or back-side ohmic contact has been formed.

Fig. 3 shows the XRD θ-2θ scans for an as-grown, an as-implanted and an as-post-implantation-annealed 4H-SiC sample. The as-implanted sample exhibits multiple diffraction features at smaller

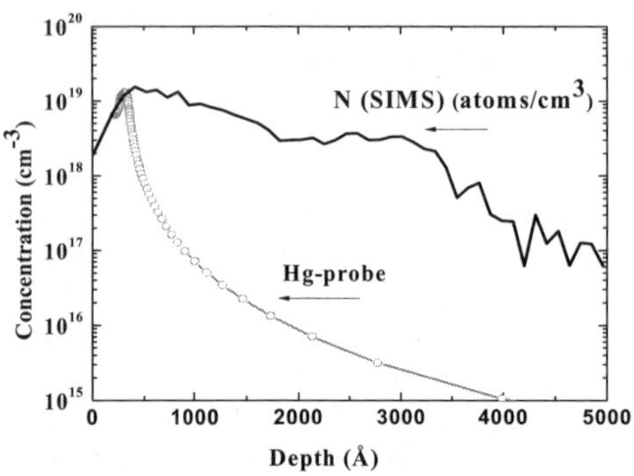

Fig. 1. SIMS (continuous line) and Hg-probe C-V profile (squares) measured on a N-implanted 4H-SiC sample. SIMS has been measured prior to annealing while the C-V profile afterwards.

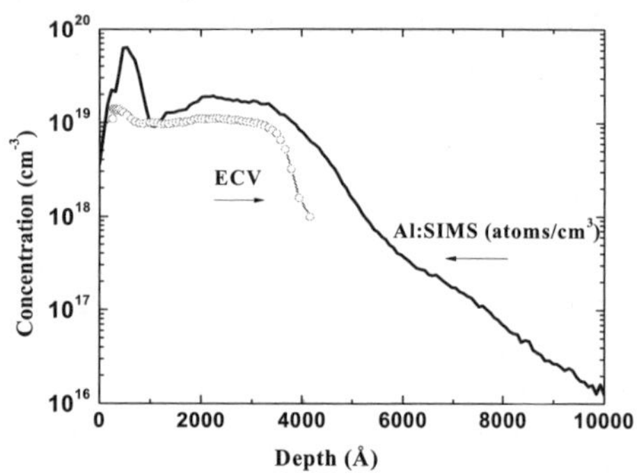

Fig. 2. SIMS (continuous line) and ECV (dotted line) profile from an Al-implanted 6H-SiC sample. Both measurements were performed after post-implantation annealing.

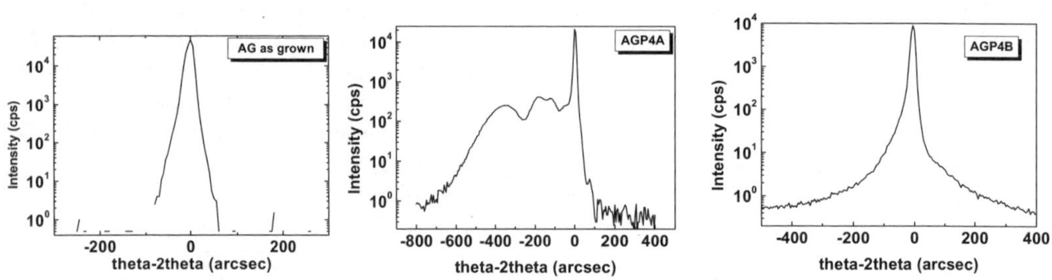

Fig. 3. θ-2θ XRD (0004) graphs measured under the same conditions for N-implanted 4H-SiC samples: as-grown (left), as-implanted (middle) and as-post-implantation-annealed (right).

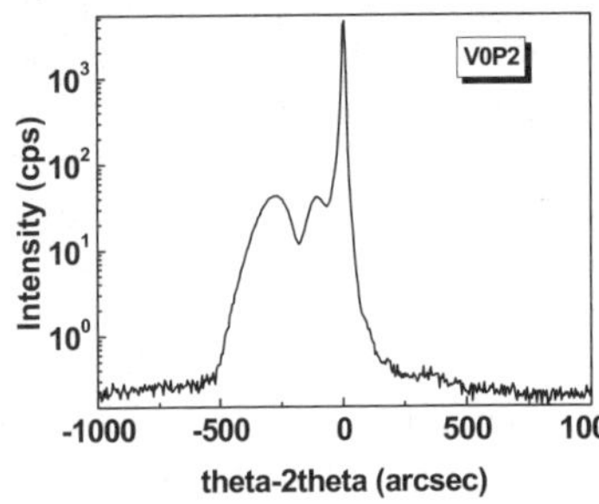

Fig. 4. θ-2θ XRD graph from an "as-N-implanted" 6H-SiC sample.

angles than the main SiC peak due to the multiple implantation steps performed to obtain a "box-profile". Each implantation step creates a disordered layer, for which the degree of disorder and the location from the sample surface depend on implantation dose and energy, respectively. In addition, the implanted ions are located in interstitial sites forming a new highly disordered lattice with a lattice constant higher than that of the SiC lattice since the density of the implanted ions is smaller than that of "matrix" material. For this reason, the additional features appear at smaller angles than the main SiC peak. After annealing, these multiple diffraction features disappeared indicating a complete crystal restoring and even an improvement of crystal quality in this particular case of nitrogen implantation. Similar results have been observed in ω-scan rocking curves. However, this was not the case for the Al-implanted samples where the XRD showed systematically that the post-implantation annealing didn't completely restore the crystal quality.

An interesting comparison between two "as-implanted", with nitrogen ions, samples at different implantation runs can be done based on Figs. 3 and 4. 4H-SiC sample AGP4A has been subjected to 7 consecutive implantation steps of increasing energy and decreasing dose while 6H-SiC sample V0P2 has been subjected to 5 implantation steps similar to the first 5 corresponding steps of sample AGP4A, i.e., the two highest implantation energy steps with a lower nitrogen dose are missing. This means that there are two "extra" layers in sample AG4A located deeper than the first five layers and with a lower implanted nitrogen ion density. The latter is equivalent to a higher lattice constant resulting in a higher angle difference from the main peak of the multiple features for the sample AGP4A. Indeed, three extra peaks, located at -350, -180 and -120 arc·sec from the main (0004) reflection, are observed for sample AGP4A in the θ-2θ scan, whereas for sample V0P2 two peaks, located at -270 and -105 arc·sec from the main (0006) reflection, are observed.

The RTPL is also a contactless method for characterizing material quality and, thus, suitable to apply during device fabrication. Fig. 5 shows typical RTPL spectra from the same 4H-SiC samples as in Fig. 3. All spectra exhibit an edge luminescence (EL) peak at 391nm and a green-yellow (GYL) band around 570 nm, which according to the bibliography is related to residual doping and particularly to boron. Note that the intensity ratio between the EL and the GYL peaks is higher for the as-post-implantation-annealed sample in comparison to the as-grown one. The as-implanted sample AGP4A presents an "attenuated" spectrum (Fig. 5 middle) in comparison to the as-grown one with a higher attenuation for the EL peak. This is observed for all as-implanted samples whatever the implanted species (Al, N) and/or the polytype (4H-, 6H-).

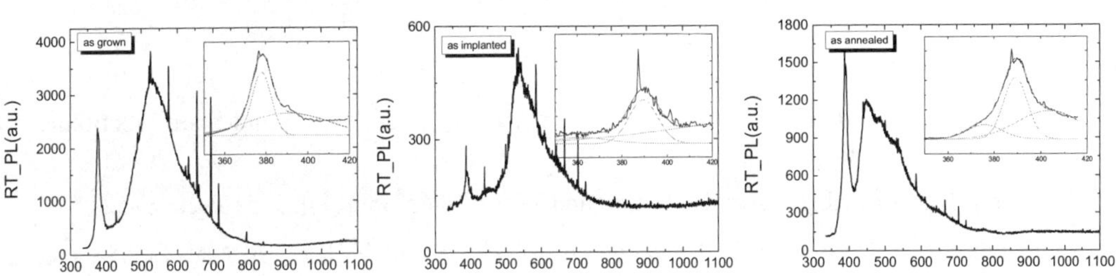

Fig. 5. RTPL spectra (linear scale) spectra measured under the same conditions for nitrogen-implanted 4H-SiC samples: as-grown (left), as-implanted (middle) and as-post-implantation-annealed (right). The peak energies are determined by de-convolution of experimental data by applying Gaussian fits (see insert).

Optical transmission measurements (Fig. 6) show that, in the case of the as-implanted samples, there is, indeed, a corresponding decrease of the transmission. Note that a cw He-Cd (325nm) laser has been used for RTPL measurements and the penetration depth for this laser is 4.3 and 7.5 μm for the 6H-SiC and 4H-SiC, respectively. So, most of the PL signal originates from the n^+ substrate and n^- epilayer for the 6H- and 4H-SiC samples, respectively, and not from the implanted layers. However, the absorption is higher in an implanted layer, often even after the post-implantation annealing, and the signal from the underneath layers is attenuated as shown in Figs. 5 and 6. Thus, in the optimum case, the post-implantation-annealed samples must exhibit a spectrum similar to the as-grown one without attenuation and especially the relative intensity between EL and GYL peaks should be the same.

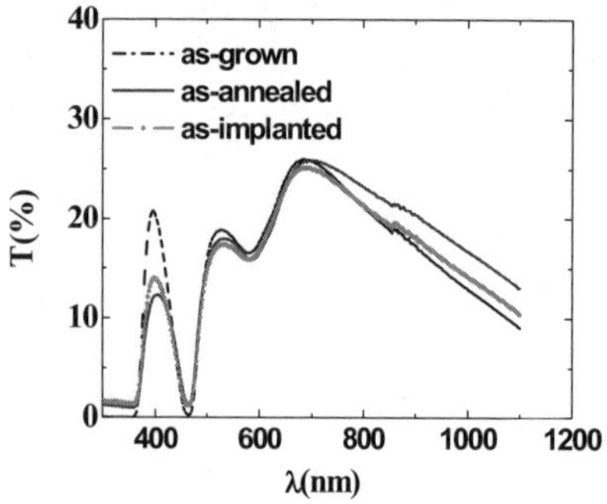

Fig. 6. Optical transmission spectra for the samples of Fig. 5.

In addition, in the case of nitrogen-implanted-and-annealed samples the broad GYL is "shifted" to lower wavelengths in comparison to the spectra from the as-grown samples and a deconvolution with Gaussians showed the appearance of an additional peak in the 420-450 nm wavelength range. A further study with a larger number of samples is necessary to connect this "shift" with the implanted ions in the SiC crystal.

RTPL spectra seem to confirm the findings from XRD scans that the post-nitrogen-implantation annealing restores the crystal quality and even improves it in comparison to the as-grown samples while this is not the case for Al-implanted samples.

Conclusions

The Electrochemical C-V technique seems appropriate for determining the doping profiling of implanted layers. HRXRD gives straightforward information on material quality after post-implantation annealing. RTPL results are indicative of material quality but need more experimental study to detect any effect on the recorded spectra from the dopant incorporation.

All these techniques are device-fabrication compatible but require the incorporation in the processed wafer of bare areas for physical characterization purposes.

Acknowledgments

This work has been supported by the projects *Anti-SiC* (Program SYNERGASIA, GSRT) and *LAST-POWER* (ENIAC project). K.Z. acknowledges also the support of the Office of Naval Research Global (ONRG) covering its participation to ICSCRM'11.

References

[1] K.B. Mulpuri, S.B. Qadri, J. Grun, C.K. Manka and M.C. Ridgway. Solid-State Electronics 50 (2006) pp.1035–1040

[2] Z.C. Feng, S.C. Lien, J.H. Zhao, X.W. Sun and W. Lu, Thin Solid Films 516 (2008) 5217–5222

[3] F. Giannazzo, P. Musumeci, L. Calcagno, A. Makhtari and V. Raineri, Mat. Sci. Semicond. Processing, Vol 4 (1-3) (2001) pp. 195-199

[4] M. Kayambaki and K. Zekentes, Mater. Sci. Forum Vols. 338-342 (2000) pp. 1061-1064

[5] K. Zekentes, M. Kayambaki and S. Mousset, Mater. Sci. Forum 457-460 (2004) pp. 681-684

[6] A. Stavrinidis, G. Konstantinidis, M. Kayambaki, F. Cayrel, D. Alquier, Z. Gao and K. Zekentes, these Proceedings

CHAPTER 4:

Graphene

Materials Science Forum Vols. 717-720 (2012) pp 595-600
© (2012) Trans Tech Publications, Switzerland
doi:10.4028/www.scientific.net/MSF.717-720.595

Electronic and Structural Properties of Turbostratic Epitaxial Graphene on the 6H-SiC (000-1) Surface

Thushari Jayasekera[1,a], K.W. Kim[2,b] and M. Buongiorno Nardelli[3,4,c]

[1]Department of Physics, Southern Illinois University, Carbondale, 62901 IL, USA.

[2]Department of Electrical and Computer Engineering, North Carolina State University, Raleigh 27695, NC, USA.

[3]Department of Physics, University of North Texas, Denton, 76203, TX, USA.

[4]Computer Science and Mathematics Division, Oak Ridge National Laboratory, Oak Ridge, Tennessee, 37831, USA.

[a]tjayasekera@physics.siu.edu, [b]kwk@ncsu.edu [c]mbnardelli@ncsu.edu

Keywords: epitaxial graphene, turbostratic, Moire patterns, SiC C-face

Abstract. We propose an atomistic model to study the interface properties of mis-oriented (turbostratic) epitaxial graphene on the 6H-SiC (000-1) surface. Using calculations from first principles, we compare the energetics and structural/electronic properties of AB and turbostratic stacking sequences within a model based on the Si adatom surface reconstruction. Our calculations show that the systems with AB and turbostratic sequences are very close in energy, demonstrating the possibility of observation of Moire patterns in epitaxial graphene on the C-face of 6H-SiC. The two-dimensional electron gas behavior is preserved in the epitaxial turbostratic graphene systems. However, there are deviations from the ideal turbostratic graphene.

Introduction

The possibility of designing lithographically patterned structures on epitaxial graphene has created an enormous interest with a high expectation for its potential device applications [1-4]. The progress towards epitaxial graphene in electronic applications will require an in-depth understanding of the electronic properties that take place at the interface. In this work, we investigate the electronic properties of epitaxial graphene on the C-face of 6H-SiC with AB and turbostratic stacking sequences.

Epitaxial growth of graphene on 6H-SiC is obtained by the sublimation of Si atoms from one of the polar surfaces of SiC in the (0001) direction. The growth of graphene on the Si-face of SiC is characterized by an electron-doped graphene layer followed by a buffer layer of carbon, which is also explained by the first principles density functional theory calculations (DFT) [5-7]. On the Si-terminated surface of SiC, a thermal annealing process gives a $\sqrt{3}$x$\sqrt{3}$ Cos 30^0 (R3) surface reconstruction followed by a 6$\sqrt{3}$ x 6$\sqrt{3}$ Cos 30^0 (6R3) surface reconstruction. Since the translational unit cell for the 6R3 system is large, most of the computational investigations have been carried out using the R3 model, irrespective of the 7% mismatch between the R3 cell of SiC and the graphene 2x2 supercell. Despite this mismatch, the computational calculations correctly produce the experimentally observed buffer layer and electron doped graphene layer on the Si-face of SiC. While the electronic properties of epitaxial graphene on the Si-face of SiC have been investigated by many theoretical and experimental groups, very few studies have been devoted to understanding the properties of epitaxial graphene on the C-face [8-9]. The results of the existing theoretical calculations are also not in full agreement with the experimental observations. The electronic and structural properties of epitaxial graphene on the C-face, unlike on the Si-face, are shown to be significantly dependant on the growth conditions. Hass *et al.* have reported a strongly interacting first C layer (buffer layer), whereas a weakly bound or decoupled graphene layer is reported by many other experiments [10,11]. A wide range of electronic structure properties also parallels these structural differences. De Heer *et al.* reported an electron-doped graphene layer on the C-face of SiC with the Dirac point approximately 300 meV below the Fermi level [12], while Emtsev *et al.* reported this value to be 200 meV [13]. Different growth conditions can also result in different graphitic stacking sequences in multi-layered samples. Turbostratic stacking is the most commonly observed, even though little has been done in understanding their electronic properties [14]. In this article, we study the electronic properties of bi-layer epitaxial graphene systems on SiC (000-1) with turbostratic and AB stacking sequences using first principles density functional theory calculations.

Theoretical Model and Calculation Details

One major difficulty of modeling turbostratic graphene structures on SiC is to combine their commensurate character in the supercell structures. Our atomistic model is based on the Si adatom surface reconstructed SiC (000-1) surface, which has been suggested by A. Seubert *et al.* [15] and later used for studying the epitaxial graphene on the C-face of SiC [8-9].

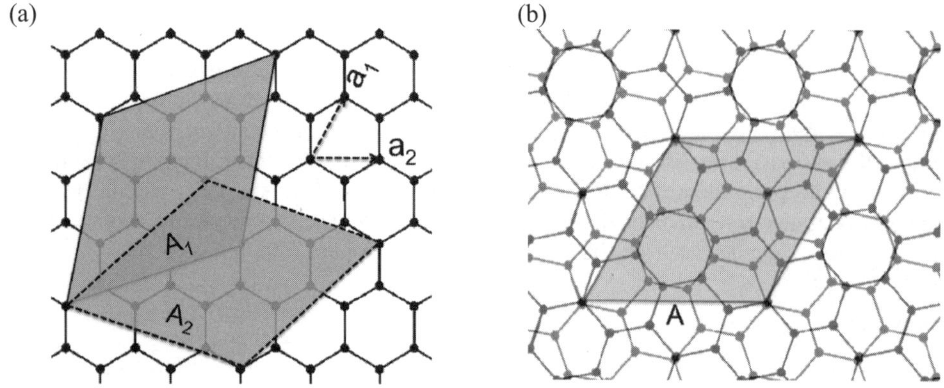

Fig. 1. (a) shows the two basis vectors A_1 and A_2 as mentioned in the text. By superimposing these two vectors onto A, an ideal Moire pattern structure is obtained with the $\sqrt{7} \times \sqrt{7}$ supercell periodicity as shown in Fig. 1 (b). Two basis vectors of the hexagonal lattice a_1 and a_2 are also shown in Fig. 1 (a).

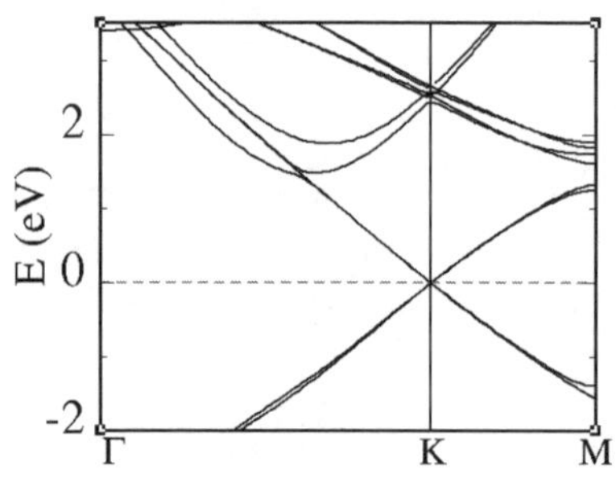

Fig. 2. Electronic band structure of ideal turbostratic graphene bilayers, which is depicted in Fig. 1(b)

This model follows the experimentally observed (3x3) surface reconstruction, which later evolves into a (2x2)c reconstruction. In the (2x2)c SiC unit cell, there are 4 C atoms open to the surface. In the model used by Seubert *et al.*, three out of the four surface C atoms are bonded to a Si adatom. The position of the Si adatom is chosen such that the surface energy is lowered. In this Si adatom model, two surface dangling bonds are present due to the unbound C atom ("C rest atom") and the Si adatom. This model has been earlier used to study the electronic and structural properties of epitaxial graphene (with Bernel stacking) on the C-face of SiC, in which a 5x5 graphene supercell is overlaid with a 4x4 SiC

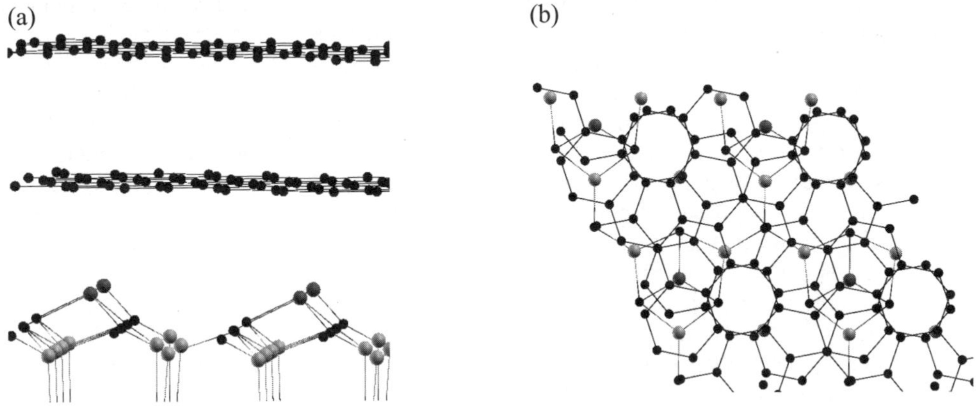

Fig. 3. (a) side and (b) top view of the atomistic arrangement used in this work: √7x√7 asymmetric graphene cells are overlaid on (2x2)c SiC supercell in the Si adatom model. We have denoted C atoms in solid black, Si atoms in ash and Si adatoms in dark-ash circles.

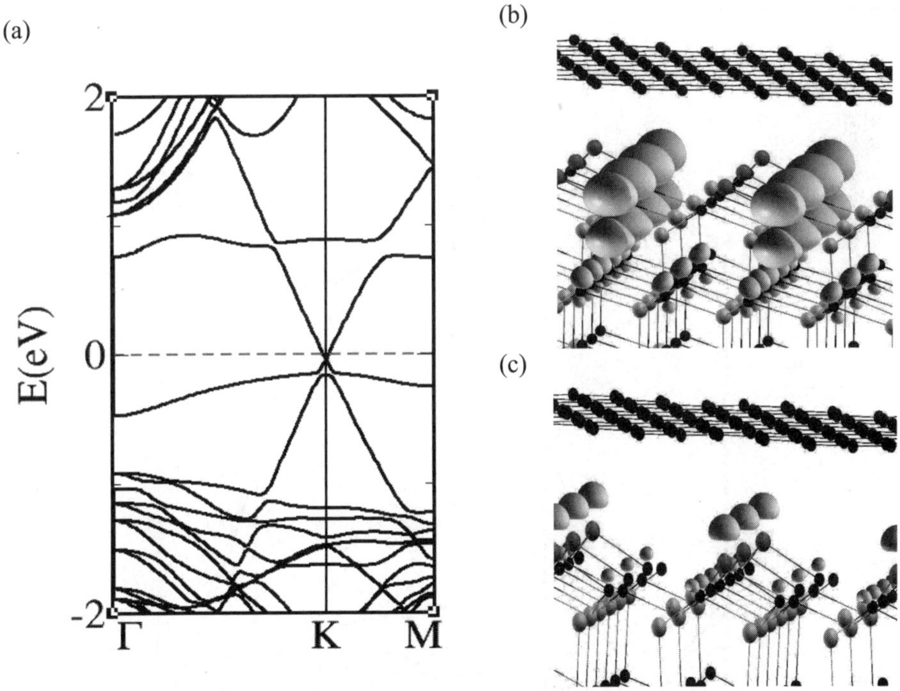

Fig. 4. (a) the electronic band structure of a single C layer on top of the SiC C-face, (b) the electronic states corresponding to Si adatoms which are empty and lie around 1 eV above the Fermi level, and (c) the filled electronic state near the Fermi level which is derived from the dangling electron on the C-rest atom.

supercell [8,9]. This model is, however, not suitable for studying the multi layer epitaxial graphene (MEG) with turbostratic stacking sequence. We utilize a method of graphene supercells without mirror symmetry as proposed by Kolmogorov and Crespi and later used by Latil *et al.* [16,17] along with the Si-adatom surface reconstruction to propose the following model. This model is suitable for studying all AA, AB and turbostratically stacked epitaxial graphene structures on the C-face of SiC. The smallest supercell for studying the turbostratically stacked graphene is defined by two basis vectors $A_1=2a_1+a_2$, and $A_2=-a_1+3a_2$ and contains 14 atoms ($\sqrt{7}x\sqrt{7}$ graphene cell), as shown in the Fig. 1. Fig. 2 shows the electronic band structure for the turbostratically stacked graphene bilayers, which shows the massless Dirac Fermion behavior. The $\sqrt{7}x\sqrt{7}$ graphene cell is commensurate with $(2x2)_C$ SiC supercells with a 4% residual mismatch. In order to keep the number of atoms to a computationally manageable size, our simulation supercell is constructed by superimposing a $\sqrt{7}x\sqrt{7}$ graphene layer on a (2x2)C SiC substrate, with a residual lattice mismatch of ~4%. In this geometry the substrate is stretched in order to match the graphene periodicity and avoid spurious electronic effects that would come from the application of a strain to the graphene layers. Given the relatively large band gap of SiC, the stretching of the substrate does not change significantly the electronic structure of the resulting system, as shown in Fig. 4.

The calculation cell consists of 6 SiC bilayers (each layer has 4 Si atoms and 4 C atoms). The Si-terminated surface is saturated with Hydrogen. A Si adatom surface reconstruction model is used for the C-terminated surface such that there is one Si adatom per supercell. The supercell has a $\sqrt{7}x\sqrt{7}$ graphene cell on top of the surface reconstructed C-terminated surface, which gives totally 67 atoms in the supercell. We have also considered bilayer graphene systems with both the AB and turbostratic stacking sequences.

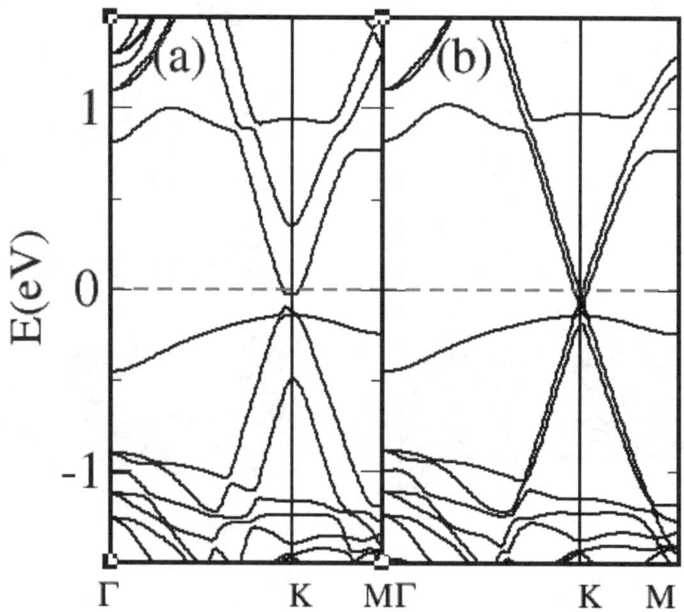

Fig. 5. The electronic band structures for (a) AB and (b) turbostratic bilayer graphene on the C-face of SiC.

All calculations were performed in the framework of density functional theory as it is implemented in the Quantum Espresso Package [18]. We have used a 12x12x1 Monkhorst-Pack Grid for Brilluoin zone integration, 12-15 Å to separate the slabs, 35 Ry energy cutoff for the plane wave basis expansion, and the Perdew-Burke, Ernzerhof (PBE) generalized gradient approximation (GGA) for the exchange correlation functional. The geometries have been optimized to forces less than 0.025 eV/Å. It is important to note that one common limitation of the DFT calculation is its inability to describe the long-range van der Waals (vdW) interactions, which is an essential component in layered materials like graphene. We have included vdW interactions in our calculations for the GGA exchange and correlation functional using an empirical potential of the form C_6R^{-6} added to the regular density functional energy as proposed by Grimme *et al.* [19] and recently implemented by Barone *et al.* [20] in the Quantum Espresso code.

Results

Fig. 4a shows the electronic band structure of a single carbon layer on top of the C-face of 6H-SiC within the prescribed model. It is clear that the first C layer has graphene properties; thus there is no signature of a buffer layer. The electronic states derived from the "C rest atoms" are filled and lie closer to the Fermi level as shown in Fig. 4c. Empty electronic states responsible for the dangling electrons in the Si adatom lie around 1 eV above the Fermi level (Fig. 4b). The average distance between the C layer and the Si adatom is around 2.9 Å. The graphene layer is shown to be slightly electron-doped. This is consistent with previous calculations. However, the amount of electron transfer is less in this model compared to the model we have used in our previous calculations [9].

We are particularly interested in understanding and comparing the structural and electronic properties in two C layers on the C-face of SiC in 2 types of stacking sequences: AB and turbostratic. The AB stacked system has an interlayer distance of 3.20 Å, which is a little lower, compared to the 3.28 Å interlayer distance in the turbostratic system. The total energies of these two systems are nearly equal, the difference being 5.4 meV per $\sqrt{7}$x$\sqrt{7}$ graphene supercell. There is a 0.1 eV gap in the Bernel stacked bilayer system. The turbostratic system is slightly electron-doped. There is a difference between the turbostratically-stacked graphene on SiC and ideal turbostratic graphene. The ideal turbostratic graphene shows a complete Dirac behavior (Fig. 2); however, the substrate induces a mismatch between the two Dirac cones (Fig. 5b). The mismatch is relatively small so that it is possible to observe the two-dimensional electron behavior in this three-dimensional system.

Summary

In summary, we have proposed an atomistic model for studying the turobostratically stacked multi-layer graphene. The first C layer behaves like graphene. Thus there is no buffer layer, which is a result of interactions between the C rest atoms and Si adatoms at the interface. The two-dimensional electron behavior is preserved in the turbostratically stacked graphene as the experiments suggested. The graphene is electron-doped and the total energy is comparable to the Bernal stacked graphene. It is possible to reproduce a variety of experimental observations such as AB and turbostratically stacked graphene on the C-face epitaxial graphene as the experiments reported. The atomistic model we presented in this article will be a starting point for further investigations of possible band engineering of the multi-layer epitaxial graphene on the C-face of SiC.

Acknowledgements

This work was supported, in part, by the DARPA/HRL, CERA, the FCRP FENA programs, and NSF. M.B.N. wishes to acknowledge partial support from the Office of Basic Energy Sciences, US Department of Energy at Oak Ridge National Laboratory, under Contract No. DE-AC05-00OR22725 with UT-Battelle, LLC. Calculations have been performed at NCCS-ORNL and HPC-NCSU.

References

[1] K. S. Novoselov, A. K. Geim, S. V. Morozov, D. Jiang, Y. Zhang, S. V. Dubonos, I. V. Grigorieva, and A. A. Firsov, *Science* **306** (2004) 666.

[2] Y. Zhang, Y. W. Tan, H. L. Stormer, and P. Kim, Nature (London) **438** (2005) 201.

[3] C. Berger, Z. Song, T. Li, X. Li, A. Y. Ogbazghi, R. Feng, Z. Dai, A. N. Marchenkov, E. H. Conrad, P. N. First, and W. A. deHeer, *J. Phys. Chem. B* **108** (2004) 19912.

[4] W. A. de Heer, C. Berger, X. Wu, P. N. First, E. H. Conrad, X. Li, T. Li, M. Sprinkle, J. Hass, M. L. Sadowski, P. Potenmski, and G. Martinez, *Solid State Commun.* **143** (2007) 92.

[5] A. Mattausch and O. Pankratov, *Phys. Rev. Lett.* **99** (2007) 076802.

[6] F. Varchon, R. Feng, J. Hass, X. Li, B. N. Nguyen, C. Naud, P. Mallet, J. Y. Veuillen, C. Berger, E. H. Conrad, and L. Magaud, *Phys. Rev. Lett.* **99** (2007) 126805.

[7] T. Jayasekera, B. D. Kong, K. W. Kim, and M. Buongiorno Nardelli, *Phys. Rev. Lett.* **104** (2010) 146801.

[8] L. Magaud, F. Hielbel, F. Varchon, P. Mallet, and J. Y. Veullin, *Phys. Rev. B* **79** (2009) 161405.

[9] T. Jayasekera, S. Xu, K.W. Kim, and M. Buongiorno Nardelli, *Phys. Rev. B* **84** (2011) 035442.

[10] J. Hass, W. A. de Heer, and E. H. Conrad, *J. Phys.: Condens. Matter* **20** (2008) 323202.

[11] C. Berger, X. Wu, P. N. First, E. H. Conrad, X. Li, M. Sprinkle, J. Hass, F. Varchon, L. Magaud, M. Sadowski, M. Potemski, G. Martinez, and W. A. de Heer, *Adv. Solid State Phys.* **47** (2008) 145.

[12] W. A. de Heer, C. Berger, X. Wu, M. Sprinkle, Y. Hu, M. Ruan, J. A. Storscio, P. N. First, R. Haddon, B. Piot, C. Faugeras. M. Potemski, and J. S. Moon, *J. Phys. D: Appl. Phys.* **43** (2010) 374007.

[13] K. V. Emtsev, F. Speck, Th. Seyller, L. Ley, and J. D. Riley, *Phys. Rev. B* **77** (2008) 155303.

[14] J. Hass, F. Varchon, J. E. Millian-Otoya, M. Sprinkle, N. Sharma, W. A. de Heer, C. Berger, P. N. First, L. Magaud, and E. H. Conrad, *Phys. Rev. Lett.* **100** (2008) 125504.

[15] A. Seubert, J. Bernhardt, M. Nerding, U. Starke, and K. Heinz, *Surf. Sci.* **45** (2000) 454.

[16] S. Latil, V. Meunier, and L. Henrard, *Phys. Rev. B* **76** (2007) 201402.

[17] A. N. Kolmogorov and V. H. Crespi, *Phys. Rev. B* **71** (2005) 235415.

[18] P. Giannozzi *et al.*, *J. Phys:. Condens Matter* **21** (2009) 395502.

[19] S. Grimme, *J. Comput. Chem.* **25** (2004) 1463.

[20] V. Barone, M. Casarin, D. Forrer, M. Pavone, M. Sambi, and A. Vittadini, *J. Comput. Chem.* **30** (2009) 934.

Materials Science Forum Vols. 717-720 (2012) pp 601-604
© (2012) Trans Tech Publications, Switzerland
doi:10.4028/www.scientific.net/MSF.717-720.601

Spatially graded graphitization on 4H-SiC (0001) with Si-sublimation gradient for high quality epitaxial graphene growth

Shoji Ushio[1,a], Yasunori Kutsuma[1], Arata Yoshii[1], Naoto Tamai[1,b], Noboru Ohtani[1,2,c] and Tadaaki Kaneko[1,2,d,]

[1]School of Science and Technology, Kwansei Gakuin University, 2-1, Gakuen, Sanda, Hyogo, 669-1337, Japan

[2]R&D Center for SiC Materials and Processes, Kwansei Gakuin University, 2-1, Gakuen, Sanda, Hyogo, 669-1337, Japan

[a]buj37401@kwansei.ac.jp, [b]tamai@kwansei.ac.jp, [c]ohtani.noboru@kwansei.ac.jp, [d]kaneko@kwansei.ac.jp

Keywords: epitaxial graphene, Si-face SiC, Raman spectroscopy, scanning electron microscopy

Abstract. We report a new approach to produce high quality epitaxial graphene based on the concept of controlling the Si sublimation rate from the SiC surface. By using a mask to suppress Si sublimation from the SiC surface in ultrahigh vacuum, epitaxial graphene growth on 4H-SiC (0001) was locally controlled. Spatially graded surface graphitization was confirmed in a SEM contrast from the outside unmasked region to the inside masked region. The contrast is discussed via Raman characterization as the surface composition changes from SiC to multilayer graphene. Results indicate two types of growth processes of epitaxial graphene on 4H-SiC (0001) step-terrace structures.

Introduction

Epitaxial graphene growth by thermal decomposition of hexagonal SiC {0001} surfaces is one of the most promising approaches to obtain a wafer scale graphene sheet with controlled thickness [1]. The growth process has been discussed as the preferential sublimation of Si [2]. In recent days, ambient pressure of argon is attracting attention for the fabrication of high quality graphene which has higher carrier mobility and pit less morphology [3]. The formation of this improved graphene is interpreted as the effect of a suppression of abrupt Si sublimation at higher temperature, where sufficient surface diffusion of carbon is expected [4]. As another important topic of the growth environment, Camara et al. have reported a graphite cap method to inhibit Si sublimation from the SiC surface [8]. It led to monolayer graphene ribbon on the C-face, on which control of the graphene thickness has been thought to be more difficult than for the Si-face. In addition, Si vapor pressure is also mentionable from the principle of the growth mechanism itself. Indeed, the initiation temperature of the graphene growth is shifted to higher temperature when increasing the Si vapor pressure in the growth environment [2]. From this standpoint, some experimental attempts to synthesize high quality epitaxial graphene have been performed [5,6]. Besides, it has been found out that the Si vapor environment has the function to etch graphene layers and SiC surfaces resulting in uniform step-terrace structures at the SiC surfaces [7].

In this study, we report a new approach to produce high quality epitaxial graphene based on the concept of controlling the Si sublimation rate from the SiC surface. In order to do that, a simple experimental configuration has been employed by putting a mask material onto the SiC substrate surface to be thermally graphitized. Due to the masking effect with a narrow opening between the mask and the substrate, Si sublimation from the substrate is more suppressed behind the mask. Thus, spatially graded surface graphitization takes place from the outside unmasked region to the inside masked region. As a result, compositional change from graphene to SiC arises on a single SiC (0001) surface, unlike the case of graphite cap reported on the C-face. The advantage of this method is in the controlled growth scheme of epitaxial graphene which proceeds in one lateral direction from the mask edge to the inside of the mask by suppressing Si sublimation. We also discuss the contribution of the Si-vapor environment on high quality graphene growth by utilizing a SiC substrate as a mask material. In this case, the Si-vapor arises and circulates between the mask and sample surfaces.

Experiments

The substrates used in this study were on-axis 4H-SiC (0001) Si-face wafer materials cut into 10×10 mm^2 or 5×10 mm^2 pieces. As a mask material for local suppression of Si sublimation, another 5×5 mm^2 piece of on-axis 4H-SiC substrate was directly placed on the surface of on-axis 4H-SiC (0001). The configuration was Si-face to Si-face. Prior to experiments, samples were prepared by a standard cleaning procedure ending up with dipping in dilute hydrofluoric acid and rinsing in deionized water. Heating of the samples was performed in an ultrahigh vacuum (UHV) system utilizing a TaC crucible in the temperature range from 800 °C to 2200 °C (1000 °C /min) by radiation from surrounding resistive heaters. Graphitization temperatures were 1800 - 2000 °C. To obtain the initial flatness at SiC surfaces, Si-vapor etching was performed before the graphitization annealing in some experiments. By utilizing the Si-vapor etching, initial SiC surfaces with polished scratches were modified to atomically flat surfaces with uniform step-terrace structures. Details of the Si-vapor etching and the annealing procedure in UHV were reported in our previous work [7]. Graphitized surfaces were characterized by atomic force microscopy (AFM), low-energy scanning electron microscopy (1keV-SEM) and Raman spectroscopy (Ar^+ laser, 488nm excitation).

Results and Discussion

The cross-sectional configuration of the experiments and an example of a wide area SEM top-view image obtained at 4H-SiC (0001) surface after masked graphitization are shown in Fig.1. The boundary line between the masked region and the unmasked region was deduced by the surface morphology and the size of the mask substrate. Note that the unmasked graphitized surface on which Si-vapor etching has not been performed before graphitization has many pits and some traces of polishing scratches, while the masked surface has apparently a pit less morphology. Although the initial flatness at SiC surfaces is important for the reproducibility of the separation between the sample and the mask, there is no any notable difference in surface morphology at the masked region between Si-vapor etched surfaces and no Si-vapor etched surfaces. This is considered as the effect of the suppression of abrupt Si sublimation and a surface re-arrangement including SiC regrowth and etching induced by circulating Si-vapor between the mask and the substrate. Here, it is also confirmed that the contrast in the SEM image is separable from the inside of the masked region to the outside

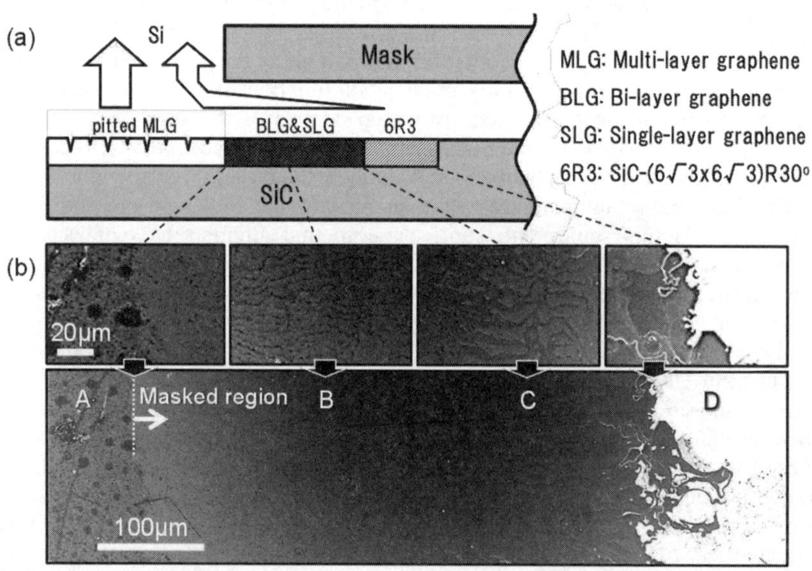

Fig.1 (a) Schematic illustration of the cross-sectional sample configuration. The surface compositional change from SiC to multi-layer graphene takes place at 4H-SiC (0001) surfaces by masked graphitization. (b) SEM top-view image of the gradually changing region at 4H-SiC (0001) by graphitization at 2000 °C for 1h (Si-vapor etching pretreatment was not performed).

unmasked region as the contrast gets darker. The length of the gradually changing region observed in SEM contrast increased as a function of temperature and duration time of heating. Recently, it was reported that darker contrast in an SEM image indicates thicker graphene in the case of exfoliated graphene put on oxidized Si [9], and the graphene area seem to be darker than the SiC area in the case of epitaxial graphene on C-face SiC [10]. Thus, it is proposed that the different contrast regions reflect the surface compositional change of 4H-SiC (0001), as represented in Fig.1. As an evidence for this proposal, Raman mapping images obtained from the surfaces at regions A and B in Fig. 1 (b) are shown in Fig.2 (a) and (b), respectively. From the relation between the number of graphene layers and the 2D full width at half maximum (FWHM) [11], the region of Fig. 2 (a) is considered to be a multi-layer (> 2ML) graphene region. On the other hand, the region of Fig. 2 (b) is considered as the boundary region between single- (2D-FWHM ~40 cm^{-1}) and bi-layer (~65 cm^{-1}) graphene. Especially, thicker graphene seems to be dominant at step-edges of the thermally decomposed SiC surface, where the SEM contrast is darker than terrace surfaces. However, there is another dark contrast region from the region B up to the brightest SiC region D. Considering the surface transition from SiC to graphene, there is a $(6\sqrt{3}\times6\sqrt{3})R30^{\circ}$ surface as a C-rich SiC surface reconstruction before the formation of graphene at Si-face SiC. Therefore, the other contrast region darker than that of the SiC region is deduced as the C-rich SiC-$(6\sqrt{3}\times6\sqrt{3})R30^{\circ}$ reconstructed surface. Interestingly, the boundary region C in Fig.1 (b) corresponding to the region between single-layer graphene and the $(6\sqrt{3}\times6\sqrt{3})R30^{\circ}$ surface exhibits various types of dendritic shapes (see in Fig.3 as another example). In the variation of annealing conditions (1800 – 2000 °C, 15 - 60 min) in this study, similar dendritic shapes appeared at surfaces of all samples. In Fig.3, dark contrast regions reflect the strong tendency of graphene growth to initiate from SiC step edges. Besides, graphene growth also proceeds from the innermost terrace surfaces at lower sides of steps. This type of graphene growth proceeds in the reverse direction from the growth from step edges, as mentioned in ref. [12]. Obviously, there is a difference in the expansion mechanism of graphene area between the growth from step-edges (type-A) and the growth from the innermost terrace surfaces at lower sides of steps (type-B). Typical examples of SEM and AFM images indicating them are shown in Fig.4. While the type-A graphene grows a belt-like shape along the step edges, the type-B graphene grows a net-like shape at the terrace surface. The difference is interpreted in terms of the number of SiC bilayers decomposing normal to the surface to supply the

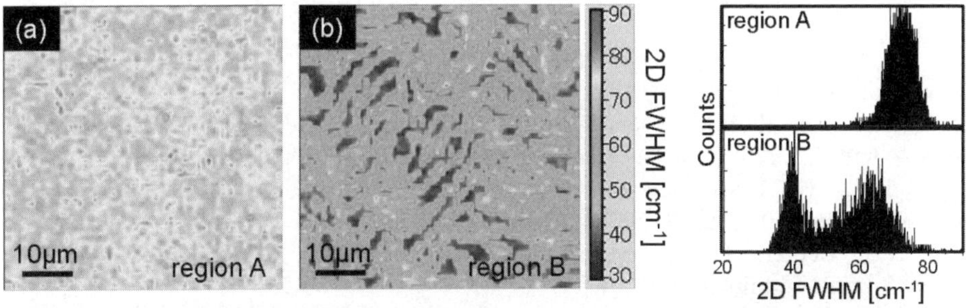

Fig.2 Raman mapping images of 2D FWHM and their frequency distributions obtained from the unmasked region A (a) and the masked region B (b) at the sample surface in Fig.1 (b).

Fig.3 SEM image of the boundary region between the 4H-SiC (0001) $(6\sqrt{3}\times6\sqrt{3})R30^{\circ}$ reconstructed surface "6R3" and the single-layer graphene "SLG", as represented region C in Fig.1 (b). Experimental condition of this sample was 1900 °C for 15 min. It exhibits the dendritic shape of epitaxial graphene as darker contrast at 4H-SiC (0001) surface.

amount of carbon atoms for graphene growth [13]. The formation of a single layer graphene requires the carbon contained in ~3.14 SiC bilayers [14]. Thus, from the shapes of graphene growth areas, it is presumed that the type-A graphene grows by carbon atoms supplied from more than 3 SiC bilayers, while the type-B graphene grows by carbon atoms supplied from less than 3 SiC bilayers. These two types of growth mechanism might be induced by the Si-vapor from the mask and the substrate, which is interpreted as the clear distinction from the growth in argon ambient and the graphite cap method.

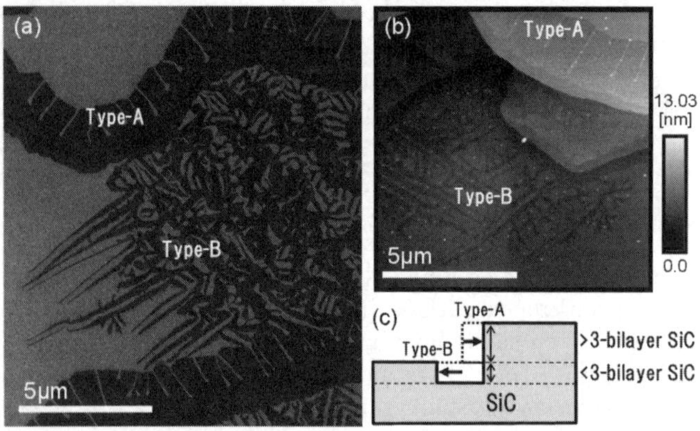

Fig.4 (a) Magnified SEM image in Fig.3. (b) AFM topographic image obtained from a similar region. (c) Cross-sectional growth model of epitaxial graphene at the SiC step-terrace surface. Two types of epitaxial graphene growth are shown in each figure.

Summary

Spatially graded surface graphitization was demonstrated at 4H-SiC (0001) by putting a mask on the surface. SEM and Raman mapping revealed compositional changing surface domains from SiC to multilayer graphene. Two types of graphene growth scheme at 4H-SiC (0001) step-terrace structure were revealed. One starts from step edges, and the other starts from the innermost terrace surfaces. Results indicate the growth mechanism in Si-vapor environment and create an expectation of high quality graphene growth in Si-vapor environment.

Acknowledgement

The authors are grateful to Mr. S. Ashizaki for his encouragement and support on this work.

References

[1] W. Lu, J.J. Boeckl, W.C. Mitchel, Appl. Phys. **43** (2010) 374004.
[2] R.M. Tromp and J.B. Hannon, Phys. Rev. Lett. **102** (2009) 106104.
[3] J.L. Tedesco, B. VanMil, R.L. Myers-Ward, J. Culbertson, G. Jernigan, P. Campbell, J.M. McCrate, S.A. Kitt, C.E ddy, Jr.,D.K. Gaskill: ECS Trans. **19** [5] (2009) 137.
[4] C. Virojanadara, M. Syväjarvi, R. Yakimova, L.I. Johansson, A.A. Zakharov, T. Balasubramanian, Phys. Rev. B **78** (2008) 245403.
[5] W.A. de Heer, C. Berger, M. Ruan, M. Sprinkle, X. Li, Y. Hu, B. Zhang, J. Hankinson, E.H. Conrad, PNAS Early Eddition (2011) 16899.
[6] X.Z. Yu, C.G. Hwang, C.M. Jozwiak, A. Köhl, A.K. Schmid, A. Lanzara, J. Electron Spectrosc. Relat. Phenom. **184** (2011) 100.
[7] S. Ushio, A. Adachi, K. Matsuda, N. Ohtani, T. Kaneko, Mater. Sci. Forum, **679-680** (2011) 777.
[8] N. Camara, J-R. Huntzinger, G. Rius, A. Tiberj, N. Mestres, F. Pérez-Murano, P. Godignon, J. Camassel, Phys. Rev. B **80** (2009) 125410.
[9] H. Hiura, H. Miyazaki, K. Tsukagoshi, Appl. Phys. Express **3** (2010) 095101.
[10] D.K. Gaskill, J.K. Hite, J.C. Culbertson, G.G. Jernigan, J.L. Tedesco, L.O. Nyakiti, V.D. Wheeler, R.L. Myers-Ward, N.Y. Garces, C.R. Eddy, Jr., Mater. Sci. Forum, **679-680** (2011) 789.
[11] D.S. Lee, C. Riedl, B. Krauss, K. von Klitzing, U. Starke, J.H. Smet, Nano Lett. **8** (2008) 4320.
[12] M.L. Bolen, S.E. Harrison, L.B. Biedermann, M.A. Capano, Phys. Rev. B **80** (2009) 115433.
[13] T. Ohta, N.C. Bartelt, S. Nie, K. Thrümer, G.L. Kellogg, Phys. Rev. B **81** (2010) 121411(R).
[14] J. Hass, W.A. de Heer, E.H. Conrad, J. Phys: Condens. Matter. **20** (2008) 323202.

Materials Science Forum Vols. 717-720 (2012) pp 605-608
© (2012) Trans Tech Publications, Switzerland
doi:10.4028/www.scientific.net/MSF.717-720.605

Control of epitaxial graphene thickness on 4H-SiC(0001) and buffer layer removal through hydrogen intercalation

J. Hassan[1,a*], C. Virojanadara[1,b], A. Meyer[2,c], I.G. Ivanov[1,d], J.I. Flege[2,e], S. Watcharinyanon[1,f], J. Falta[2,g], L.I. Johansson[1,h] and E. Janzén[1,i]

[1]Department of Physics, Chemistry and Biology, Linköping University, SE-581 83 Linköping, Sweden

[2]Institute of Solid State Physics, University of Bremen, 28359 Bremen, Germany

[a]jawul@ifm.liu.se, [b]chavi@ifm.liu.se, [c]meyer@ifp.uni-bremen.de, [d]iiv@ifm.liu.se, [e]flege@ifp.uni-bremen.de, [f]somwat@ifm.liu.se, [g]falta@ifp.uni-bremen.de, [h]lejoh@ifm.liu.se, [i]erj@ifm.liu.se

Keywords: epitaxial graphene, growth, hydrogen intercalation, LEEM, ARPES, mobility

Abstract. We report graphene thickness, uniformity and surface morphology dependence on the growth temperature and local variations in the off-cut of Si-face 4H-SiC on-axis substrates. The transformation of the buffer layer through hydrogen intercalation and the subsequent influence on the charge carrier mobility are also studied. A hot-wall CVD reactor was used for in-situ etching, graphene growth in vacuum and the hydrogen intercalation process. The number of graphene layers is found to be dependent on the growth temperature while the surface morphology also depends on the local off-cut of the substrate and results in a non-homogeneous surface. Additionally, the influence of dislocations on surface morphology and graphene thickness uniformity is also presented.

Introduction

High charge carrier mobility in graphene [1] makes it a suitable material for high frequency electronic devices. Epitaxial graphene on SiC has in recent years received considerable attention due to the fact that large-area graphene can directly be obtained on semi-insulating (SI) substrates that easily could be integrated into current electronic device technology. Several methods including sublimation of the SiC surface under vacuum [2], in Ar ambient [3] or CVD epitaxial growth under controlled flow of the carbon source [4] have been shown to produce high quality graphene layers on SiC substrates. However, there are several issues related to, e.g., the control of the number of graphene layers, large area thickness uniformity, and surface morphology which severely deteriorate the properties of graphene. Furthermore, the reproducibility of the growth process to achieve homogeneous thickness with high carrier mobility and efficient removal of the covalent interlayer (buffer layer) is also important. The electrical and optical properties of graphene are directly related to the number of graphene layers in a stack and its thickness uniformity. Therefore, a good control of the number of layers and the thickness homogeneity over a large area is required.

In this report we present a detailed study of how the growth temperature affects the number of graphene layers, the thickness uniformity, the surface morphology and graphene intrinsic properties. We also explored the transformation of the so-called buffer layer into a first graphene layer through hydrogen intercalation and the corresponding changes in the electrical properties of graphene.

Experimental Details

A hot-wall chemical vapor deposition (HWCVD) reactor was used to grow epitaxial graphene. The HWCVD reactor has the advantage of good control over a wide range of temperature and pressure with very high uniformity of temperature distribution, which is very crucial for achieving uniform graphene thickness over a large area. Also, in-situ etching, graphene growth, and hydrogen intercalation processes can be performed during a single sequence. Nominally on-axis 0.08^{o} SI Si-face 4H-SiC (0001) was used as the substrate. All samples were cut from the same substrate to avoid the influence of different structural quality and un-intentional off-cut variation. The growth

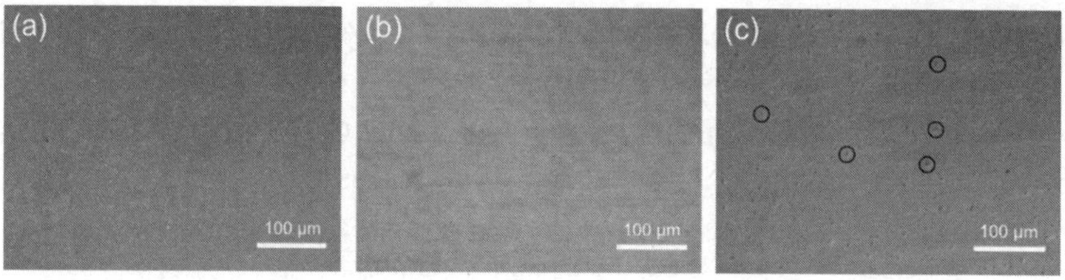

Fig. 1 Optical image taken from graphene sample grown at (a) 1400 (b) 1500 and (c) 1600 °C.

was performed at three different temperatures, 1400, 1500, and 1600 °C, at pressures of 5 x 10^{-5} mbar to 5 x 10^{-6} mbar for the same duration of 50 minutes. In-situ surface treatment of SiC substrates was performed in a mixture of hydrogen and silane at 1580 °C [5] prior to the graphene growth. Hydrogen intercalation under the graphene layers was done in the same reactor in the temperature range of 600-700 °C under hydrogen ambient at 500 mbar for 40 minutes.

The surface morphology was studied using optical microscopy while the detailed step structure on the surface was observed using atomic force microscopy (AFM) in tapping mode. Contactless measurements of the sheet resistance, carrier density and mobility in as-grown graphene samples and after hydrogen intercalation were monitored using micro-wave reflectivity measurements at room temperature. The direct observation of the π-band structure, the number of graphene layers and thickness uniformity were made using angle-resolved photoelectron spectroscopy (ARPES), low-energy electron diffraction (LEED) and low-energy electron microscopy (LEEM).

Results and Discussion

Optical analysis of graphene layers grown at three different temperatures, 1400, 1500 and 1600 °C, but for the same time interval reveals different surface morphologies (Fig. 1a-c) depending on the growth temperature. Step-bunching is observed in all samples; however, the overall surface morphology appears rougher with increasing growth temperature. Also, a high density of etch pits is observed for growth at the relatively high temperature of 1600 °C (Fig. 1c). Heavy step-bunching and defect selective etch pits are typical features of Si-face SiC substrates etched in a mixture of hydrogen and propane/silane even at low temperatures of about 1400 °C [5]. A detailed analysis of etched surfaces has shown that such pits are related to threading screw dislocations (TSDs) and micropipes in the SiC substrate [5]. Even though all of the samples were in-situ etched under the same conditions, after the growth of graphene, etch pits are only observed in the samples grown at 1600 °C while no such pits are observed in the samples grown at relatively low temperatures of 1400 and 1500 °C. The defect selective etch pits, produced during in-situ etching, vanished during in-vacuum annealing at temperatures up to 1500 °C while above this temperature the etch pits apparently continue to grow deep into the substrate. This observation shows that TSDs related defect selective etching can be controlled at relatively low growth temperature and hence surface morphology and graphene thickness uniformity can be improved.

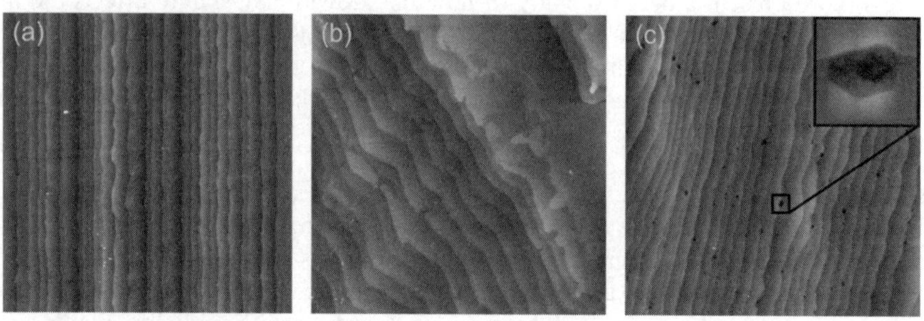

Fig. 2 (50x50 μm^2) tapping mode AFM image of graphene sample grown at (a) 1400 °C, RMS 4.6 nm, (b) 1500 °C, RMS 2.4 nm and (c) 1600 °C, RMS 3.5 nm. In-set in (c) shows high magnification image of a threading screw dislocation related etch pit.

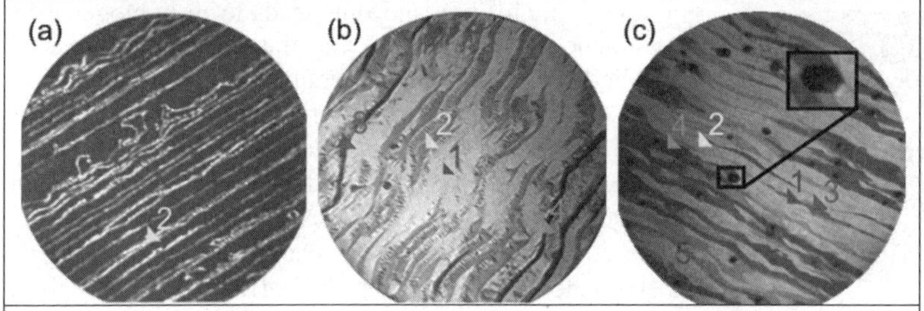

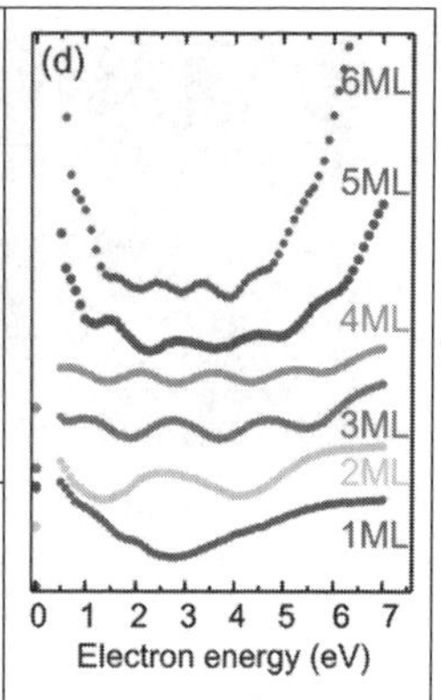

Fig. 3 LEEM images from as-grown sample at (a) 1400 °C, (b) 1500 °C, (c) 1600 °C at a field of view of 25 μm, in-set is an enlarged image of graphitic region related to TSD and (d) electron reflectivity curves corresponding to different number of graphene layers.

The AFM analysis of the samples confirms surface step-bunching in all cases with step heights in the range of 4-7 nm. Different values of surface roughness have been observed in the samples grown at different temperature and also within the same sample depending on the area selected for AFM measurement, which is an inherent property of SiC. However, the overall surface morphology is found to be dependent on the growth temperature with increased surface roughness at high growth temperature. Even though all samples were etched under the same conditions prior to graphene growth, the high surface roughness value in Fig. 2a is related to large basal plane bending of the substrate. SiC basal plane bending [6] related to the SiC bulk growth process is a well-known phenomenon and leads to different local off-cuts in the nominally on-axis substrate and results in large steps on the surface. Therefore, along with growth temperature, SiC basal plane bending present in the substrate also acts as an additional source of surface roughening and is evident from different surface morphologies observed in the same sample.

Another important observation was made through the comparison of surface step-structure of the SiC substrate after in-situ etching and after graphene growth either in vacuum or in Ar ambient in the temperature range of 1400-1600 °C. After in-situ etching, the surface is covered with fine micro-steps of 0.5-1 nm height originating from TSDs along with the large steps related to local off-cut and step-bunching [5]. The micro-step structure completely disappears after the growth of graphene in vacuum, and instead the surface is found to be covered only with large steps of 4-7 nm height depending on the local off-cut, growth temperature, and time. This phenomenon leads to wider terraces due to accumulation of micro-steps, however, at the cost of increased step-height. Similar behavior was observed in the samples grown in Ar ambient at 1600 °C and 500 mbar. This indicates that high temperature in-situ etching of the substrate in a mixture of hydrogen and propane/silane is a different process compared to thermal decomposition of the SiC surface under vacuum or Ar ambient and, therefore, leads to different surface morphology and step-structure.

As-grown graphene layers show mobilities in the range of 300-600 $cm^2V^{-1}s^{-1}$ and charge carrier density of 1-2 x 10^{13} cm^{-3} with n-type conductivity. These are typical values for epitaxial graphene on the Si-face of SiC reported in literature [7]. LEEM measurements and direct observation of π-band structure using ARPES reveal 1-2 and 1-4 layers of graphene for the samples grown at 1400 °C (Fig. 3a) and 1500 °C (Fig. 3b), respectively, with a uniform single layer on the terraces and two layers on the step-edges. Up to 6 graphene layers are observed in the samples grown at 1600 °C (Fig. 3c); however, the major area of the sample is covered with a single layer. Another major difference in the LEEM images taken from graphene layers grown at 1600 °C has been observed in the form of dark spots with graphite-like properties. Such dark spots are not observed in LEEM images taken from graphene samples grown at relatively low temperatures of 1400 and 1500 °C. The density and distribution of dark spots in LEEM images (Fig. 3c) are found to be comparable to the etch pits in optical (Fig. 1c) and AFM images (Fig. 2c) taken from the same sample, providing evidence that the dark spots are related to TSDs in the substrate intersecting the surface. Also, a high magnification LEEM image of a dark spot (inset in Fig. 3c) indeed shows a hexagonal structure, typical of TSD related etch pits (inset in Fig. 2c). This suggests that at high growth temperature, the Si sublimation or removal rate around TSDs and micropipes is high, which results

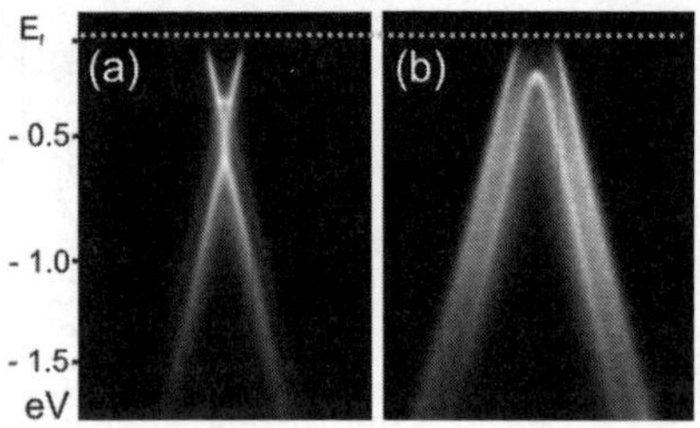

Fig. 4 π-band structure at the K-point of (a) as-grown graphene layer at 1500 °C (b) after hydrogen treatment.

in graphitic inclusions and thus leads to in-homogeneities in graphene thickness. However, this phenomenon can be controlled at low growth temperature (Fig. 3a and b).

The intrinsic properties of graphene on the Si-face of SiC are hindered by the so-called carbon buffer layer at the interface [8] which is covalently bonded to the SiC substrate and is believed to be responsible for high n-type carrier concentration and low charge carrier mobility in the graphene layer. However, the buffer layer can be lifted up by termination of the SiC surface with hydrogen through diffusion to the interface. The buffer layer transforms into a first graphene layer, and therefore the total number of graphene layers is increased by one. In as-grown samples (at 1400 and 1500 °C) two layers of graphene are mainly observed on the step edges while the rest of the surface is covered with a single layer (Fig. 4a). After treating the samples in hydrogen ambient at a pressure of 500 mbar and a temperature about 700 °C, the number of corresponding layers has increased to 2-3 as evident from the two branches exhibiting linear π-band dispersion around the K-point (Fig. 4b). In addition, while the as-grown graphene exhibits the Fermi level above the crossing point of the π-bands (Fig. 4a), after hydrogen intercalation the Fermi level is shifted below this crossing point indicating a p-type doping (Fig. 4b). However, hydrogen intercalation was not observed in the sample grown at 1600 °C, probably due to multiple layers of graphene. A considerable increase in mobility by a factor of 3 has been observed after hydrogen treatment with the best value of 1600 $cm^2V^{-1}s^{-1}$ in samples grown at 1500 °C.

Summary

High temperature growth leads to up to 6 graphene layers, high surface roughness and graphitization of defect selective etch pits related to TSDs. At low temperature growth (1400-1500 °C), a relatively smooth surface and uniform coverage with one graphene layer is observed while the step-edges are covered with two layers. Surface morphology is also found to be dependent on the local off-cut in on-axis substrates. The fine surface step-structure related to TSDs formed during in-situ etching completely disappears after graphene growth either in vacuum or in Ar. Hydrogen intercalation effectively transforms the buffer layer into the first graphene layer and significantly improves carrier mobility.

Acknowledgements

This work is supported by the European Science Foundation (ESF) under the EUROCORES program EuroGRAPHENE and European and Swedish Research Councils (VR) as well as the Swedish Government Strategic Research Area Grant in Materials Science.

References

[1] A. K. Geim and K. S. Novoselov, Nature Materials 6 (2007) 183.
[2] W. A. de Heer, C. Berger, X. S. Wu, P. N. First, E. H. Conrad, X. B. Li, T. B. Li, M. Sprinkle, J. Hass, M. L. Sadowski, M. Potemski, and G. Martinez, Solid State Communications 143 (2007) 92.
[3] C. Virojanadara, M. Syvajarvi, R. Yakimova, L. I. Johansson, A. A. Zakharov, and T. Balasubramanian, Physical Review B 78 (2008) 245403.
[4] A. Michon, S. Vezian, A. Ouerghi, M. Zielinski, T. Chassagne, and M. Portail, Applied Physics Letters 97 (2010) 171909.
[5] J. Hassan, J. P. Bergman, A. Henry, and E. Janzen, Journal of Crystal Growth 310 (2008) 4430.
[6] J. W. Lee, M. Skowronski, E. K. Sanchez, and G. Chung, Journal of Crystal Growth 310 (2008) 4126.
[7] G. G. Jernigan, B. L. VanMil, J. L. Tedesco, J. G. Tischler, E. R. Glaser, A. Davidson, P. M. Campbell, and D. K. Gaskill, Nano Letters 9 (2009) 2605.
[8] A. Mattausch and O. Pankratov, Physical Review Letters 99 (2007) 076802.

Materials Science Forum Vols. 717-720 (2012) pp 609-612

doi:10.4028/www.scientific.net/MSF.717-720.609

Graphene on Carbon-face SiC{0001} Surfaces Formed in a Disilane Environment

N. Srivastava[a], Guowei He[b], and R. M. Feenstra[c, *]

Dept. Physics, Carnegie Mellon University, Pittsburgh, PA, 15213, USA

[a]nishthas@andrew.cmu.edu, [b]guoweih@andrew.cmu.edu, [c]feenstra@cmu.edu, *corresponding author

Keywords: C-face, graphene, interface structure, low energy electron microscopy, disilane

Abstract. The formation of epitaxial graphene on SiC($000\bar{1}$) in a disilane environment is studied. The higher graphitization temperature required, compared to formation in vacuum, results in improved thickness uniformity of the graphene films. Some areas of the surface display unique electron reflectivity curves not seen in vacuum-prepared samples. Using selected area diffraction, these areas are found to have a graphene/SiC interface structure with a graphene-like buffer layer [analogous to what occurs on SiC(0001) surfaces].

Introduction

Epitaxial graphene on SiC{0001} has been intensively studied over the past five years as a potential means of producing *large area* graphene for electronic applications [1]. There are two inequivalent faces of SiC{0001} – the (0001) face, also known as the Si-face, and the ($000\bar{1}$) face or C-face. On both surfaces graphene can be formed by heating the SiC in vacuum, causing preferential sublimation of the Si atoms thereby leaving behind an excess of C atoms which self-assembles into graphene. Preparation is also possible in other environments including argon [2-4], or disilane [5]. The advantage of using these environments is that they permit independent control of the sample temperature and the Si partial pressure during the graphene formation, thus enabling a reduction in Si sublimation rate (at a given temperature) so that a higher formation temperature can be used, hence, improving the graphene quality.

The prior work involving graphene formation in argon has been most successful for the Si-face [2-4]. In fact, for the C-face, improvement in structural properties using an argon environment was *not* obtained, particularly for thin, monolayer (ML) graphene films [6,7]. The reason for this lack of success on the C-face occurs, we believe, is unintentional oxidation of the C-face surface in the presence of the Ar background [7]. With the goal of achieving uniform, thin graphene layers on the C-face, we have studied the graphene formation in a disilane environment. For disilane pressures around 10^{-5} Torr we find modest improvement in the graphene structural properties compared to vacuum formation. At higher pressure near 10^{-4} Torr, we obtain considerably better uniformity in the graphene thickness for thin ≈ML-thick films. We also find the presence of a new graphene structure on the surface compared to that observed in vacuum-prepared samples, which we identify as a *graphene-like buffer layer* bonded to the SiC, analogous to the $6\sqrt{3}\times6\sqrt{3}$-R30° (denoted $6\sqrt{3}$ for short) layer that acts as a buffer on the Si-face [8,9].

Experimental

Experiments are performed on nominally on-axis, n-type 6H-SiC or semi-insulating 4H-SiC wafers purchased from Cree, Inc., with no apparent differences between results for the two types of wafers. These wafers are cut into 1×1 cm^2 samples. Prior to graphene formation, the samples are cleaned in 5×10^{-5} Torr of disilane by heating to 850°C for 5 min, which is found to remove polishing damage in a similar manner as for H-etching [10]. To form the graphene, the samples are then heated in 10^{-6} – 10^{-4} Torr of disilane at 1200 - 1400°C for 10 to 30 min. Temperature is measured using a disappearing filament pyrometer [10]. Low-energy electron microscopy (LEEM) and diffraction

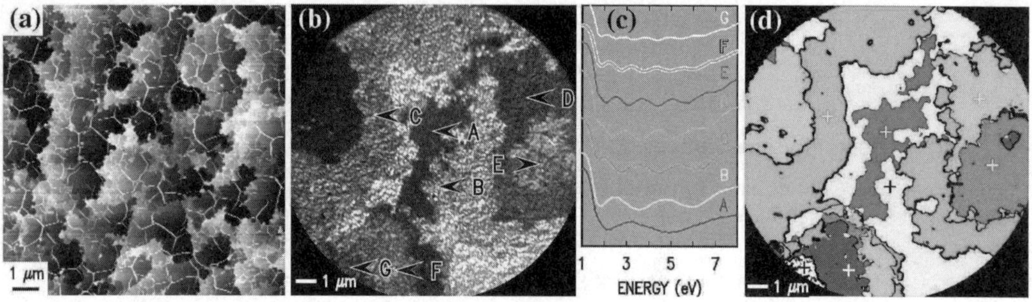

FIG 1. Results for graphene on the C-face prepared by annealing in 1×10^{-6} Torr of disilane at 1290°C for 15 min. (a) AFM image, displayed using gray scale range of 4 nm. (b) LEEM image acquired at electron beam energy of 2.3 eV. (c) Intensity of reflected electrons from the points marked in (b), as a function of electron beam energy. Curves A – G correspond to 2 – 8 ML of graphene, respectively. (d) Map of local graphene thickness, using the same shades as in (c). Small white or black crosses mark the same locations as in (b).

(LEED) are performed in an Elmitec LEEM III system. During LEEM, the sample and the electron gun are kept at a potential of -20 kV and LEEM images are acquired with electrons having energy, set by varying the bias on the sample, in the range of 0-10 eV. The intensity of the reflected electrons from different regions of the sample is plotted as a function of the beam energy, with the number of graphene MLs given by the number of local minima in the reflectivity curve [11]. From sequences of images acquired at energies varying by 0.1 eV, color-coded maps of the graphene thickness are generated using the method described in Ref. [4]. Surface morphology is measured in air by atomic force microscopy (AFM), using a Digital Instruments Nanoscope III in tapping mode.

Results

Results for graphene formation on the C-face at a disilane pressure of 1×10^{-6} Torr are shown in Fig. 1. The AFM image of Fig. 1(a) shows the surface morphology, whereas the LEEM results of Figs. 1(b) and 1(c) can be analyzed to produce a map of the local graphene thickness, Fig. 1(d). At this relatively low disilane pressure, there is a modest improvement in structural homogeneity compared to vacuum formation, similar to the situation found for Si-face graphene formed under low or high (10^{-6} – 10^{-4} Torr) disilane pressure [12]. The annealing temperature used for the sample of Fig. 1 is about 100°C higher than for a film with the same average thickness (4.2 ML) formed in vacuum, so this modest improvement in quality is expected. This graphene film displays a considerable range of thicknesses, similar to that found for vacuum preparation [7]. However, one notable difference between the results of Fig. 1 compared to our vacuum-prepared C-face graphene [7] is the network of raised (white) lines prominently seen in Fig. 1(a). These ridges arise from the different thermal expansion coefficients between the graphene and the SiC. As emphasized by Hass et al., the presence of such features is an indicator of a structurally ideal graphene film [1].

Results for a C-face film prepared under a higher pressure of disilane, 5×10^{-5} Torr, are shown in Fig. 2. A thinner film was formed in this case as compared with Fig. 1. The surface morphology measured by AFM, shown in Fig. 2(a), displays islands of graphene as revealed again by the presence of the raised (white) lines on the surface, surrounded by featureless areas that apparently contain no graphene. The electron reflectivity curves in Fig. 2(c) contain a number of new features compared to those in Fig. 1(c). Considering first the curves B, C, and D of Fig. 2(c), they contain 1, 1, or 2 minima in the reflectivity over the energy range 2.0 – 6.2 eV, as expected for 1, 1, or 2 ML of graphene, respectively, but curve B also contains a new minimum near 6.8 eV. This new feature is even more intense at surface areas that display a reflectivity curve completely different than those associated with graphene layers, as seen in curve A of Fig. 2(c).

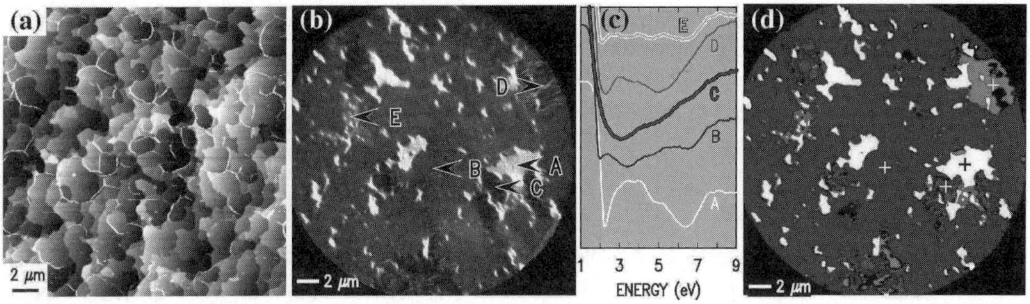

FIG 2. Results for graphene on the C-face prepared by annealing in 5×10^{-5} Torr of disilane at 1270°C for 15 min. (a) AFM image displayed using gray scale range of 3 nm. (b) LEEM image acquired at electron beam energy of 3.2 eV. (c) Intensity of reflected electrons from the points marked in (b), as a function of electron beam energy. Curve A corresponds to the oxidized buffer layer, B to 1 ML of graphene on that oxidized buffer, C to a ML of graphene on a non-oxidized buffer, D to 2 ML of graphene on non-oxidized buffer, and E to an ill-defined region. (d) Map of local graphene thickness, using the same shades as in (c). Small white or black crosses mark the same locations as in (b).

We associate this new reflectivity curve A of Fig. 2(c) with a graphene-like layer in contact with the SiC, i.e. the base or "buffer" layer of graphene. There are several reasons for this identification: (i) this curve appears only on samples with very thin (sub-ML) films of graphene films, consistent with it being the layer nearest the SiC, (ii) the reflectivity curve itself resembles that found for the $6\sqrt{3}$ buffer layer found on the Si-face of SiC [11], albeit with the extra minimum near 6.3 eV which we explain below is associated with oxidation of the underlying SiC, (iii) the reflectivity minimum near 6.3 eV also appears in the curve of some overlying graphene, i.e. curves B of Fig. 2(c) [albeit shifted up to 6.8 eV], consistent with our identification of this layer being the base layer, and (iv) electron diffraction of this layer reveals spots identical with those found on the overlying graphene layer, as further discussed below, demonstrating that this base layer does indeed have the graphene structure [13]. Regarding curve E of Fig. 2(c), this curve appears to arise from a combination of areas with characteristics as in curves A – D.

Broad-area low-energy electron diffraction (LEED) patterns of the sample of Fig. 2 reveal a $\sqrt{3}\times\sqrt{3}$-R30° arrangement of spots in addition to the usual graphene streaks, as shown in Ref. [13], and an intensity vs. energy analysis of those spots reveals that the SiC is covered with the well known silicate layer [14]. Those patterns were, however, acquired *ex situ* after the sample had been exposed to air for several days or more. LEED patterns acquired *in situ* immediately following the graphene formation still reveal the graphene streaks, but instead of the $\sqrt{3}\times\sqrt{3}$-R30° silicate pattern they show a complex pattern of satellite spots around the main SiC spots, somewhat similar to the $6\sqrt{3}\times6\sqrt{3}$-R30° pattern seen for the Si-face buffer layer [9]. These observations are consistent with our identification of the base graphene layer in the samples of Fig. 2 being an oxidized graphene-like buffer layer. The effect of the oxidation is to decouple the buffer layer from the SiC (hence making it into the first graphene layer), as previously observed on the Si-face using either hydrogenation [15] or oxidation [16]. Selected-area LEED patterns acquired from the sample of Fig. 2, after air exposure, reveal spots with precisely the graphene wavevector at *all* points over the surface, i.e. in particular for the areas in Fig. 2 displaying reflectivity curves like that of curve A in Fig. 2(b) [13]. This observation of the graphene structure for those areas confirms our identification of their structure as being a graphene-like buffer layer.

For vacuum-prepared graphene on the C-face, 2×2 and 3×3 LEED patterns are commonly observed, and we observe the same in our work [12]. These arrangements are found to persist at the graphene/SiC interface [17], and their presence has been associated with a weakly-bonded graphene layer on the SiC [9]. In contrast, the C-face buffer layer we observe here corresponds to a strongly-bonded graphene layer on the SiC, and indeed, this type of bonding arrangement was observed in one prior study [18]. We conclude that the structure of the interface between graphene and C-face SiC depends on the preparation conditions for the graphene. The buffer-layer structure found here is analogous to what occurs for Si-face SiC, and it likely provides a template for the formation of the graphene layers. Control of this interface structure thus provides a route for improved graphene formation on C-face SiC.

Summary

Graphitization of the C-face of SiC in a background of disilane leads to thinner and more uniform films as the pressure of the disilane is increased. Some regions show unique reflectivity curves in LEEM which we identify as a graphene-like buffer layer with the underlying SiC being oxidized.

Acknowledgements

Discussions with Luxmi and Patrick Mende are gratefully acknowledged. This work was supported by the National Science Foundation (grant DMR-0856240).

References

[1] J. Hass, W. A. de Heer, and E. H. Conrad: J. Phys.: Condens. Matter Vol. 20 (2008), p. 323202

[2] C. Virojanadara, M. Syväjarvi, R. Yakimova, L. I. Johansson, A. A. Zakharov, and T. Balasubramanian: Phys. Rev. B Vol. 78 (2008), p. 245403

[3] K. V. Emtsev et al.: Nat. Mat. Vol. 8 (2009), p. 203

[4] Luxmi, N. Srivastava, R. M. Feenstra, and P. J. Fisher: J. Vac. Sci. Technol. A Vol. 28 (2010), p. C5C1

[5] R. M. Tromp and J. B. Hannon: Phys. Rev. Lett. Vol. 102 (2009), p. 106104

[6] J. L. Tedesco, G. G. Jernigan, J. C. Culbertson, J. K. Hite, Y. Yang, K. M. Daniels, R. L. Myers-Ward, C. R. Eddy, Jr., J. A. Robinson, K. A. Trumball, M. T. Wetherington, P. M. Campbell, and D. K. Gaskill: Appl. Phys. Lett. Vol. 96 (2009), p. 222103

[7] Luxmi, N. Srivastava, G. He, R. M. Feenstra, and P. J. Fisher: Phys. Rev. B Vol. 82 (2010), p. 235406

[8] F. Varchon, R. Feng, J. Hass, X. Li, B. Ngoc Nguyen, C. Naud, P. Mallet, J-Y. Veuillen, C. Berger, E. H. Conrad, and L. Magaud: Phys. Rev. Lett. Vol. 99 (2007), p. 126805

[9] K. V. Emtsev, F. Speck, Th. Seyller, L. Ley, and J. D. Riley: Phys. Rev. B Vol. 77 (2008), p. 155303

[10] Luxmi, S. Nie, P. J. Fisher, R. M. Feenstra, G. Gu, and Y. Sun: J. Electron. Mater. Vol. 38 (2009), p. 718

[11] H. Hibino, H. Kageshima, F. Maeda, M. Nagase, Y. Kobayashi and H. Yamaguchi: Phys. Rev. B Vol. 77 (2008), p. 075413

[12] N. Srivastava, G. He, Luxmi, P. C. Mende, R. M. Feenstra, and Y. Sun: submitted to J. Phys. D: Appl. Phys.; arXiv:1109.1224

[13] N. Srivastava, G. He, and R. M. Feenstra: submitted to Phys. Rev. B; arXiv:1110.6562.

[14] U. Starke, J. Schardt, J. Bernhardt, and K. Heinz: J. Vac. Sci. Technol. A Vol. 17 (1999), p. 1688

[15] C. Riedl, C. Coletti, T. Iwasaki, A. A. Zakharov, and U. Starke: Phys. Rev. Lett. Vol. 103 (2009), p. 246804

[16] S. Oida, F. R. McFreely, J. B. Hannon, R. M. Tromp, M. Copel, Z. Chen, Y. Sun, D. B. Farmer, and J. Yurkas: Phys. Rev. B Vol. 82 (2010), p. 041411

[17] F. Hiebel, P. Mallet, F. Varchon, L. Magaud, and J-Y. Veuillen: Phys. Rev. B Vol. 78 (2008), p. 153412

[18] J. Hass, R. Feng, J. E. Millán-Otoya, X. Li, M. Sprinkle, P. N. First, W. A. de Heer, and E. H. Conrad: Phys. Rev. B Vol. 75 (2007), p. 214109

Materials Science Forum Vols. 717-720 (2012) pp 613-616

doi:10.4028/www.scientific.net/MSF.717-720.613

The registry of graphene layers grown on SiC(000-1)

L.I. Johansson[1,a], S. Watcharinyanon[1,b], A.A. Zakharov[2,c], R. Yakimova[1,d] and C. Virojanadara [1,e]

[1] IFM, Linkoping University, SE-58183 Linkoping, Sweden
[2] MAX-lab, Lund University, SE 22100 Lund, Sweden

[a]lij@ifm.liu.se, [b]somwat@ifm.liu.se, [c]Alexei.Zakharov@maxlab.lu.se, [d]roy@ifm.liu.se, [e]chavi@ifm.liu.se

Keywords: graphene, C-face SiC, ordered grains, LEEM, XPEEM, μ-LEED and ARPES

Abstract. Graphene samples were grown on the C-face of SiC, at high temperature in a furnace and an Ar ambient, and were investigated using LEEM, XPEEM, LEED, XPS and ARPES. Formation of fairly large grains (crystallographic domains) of graphene exhibiting sharp 1x1 patterns in μ-LEED was revealed and that different grains showed different azimuthal orientations. Selective area constant initial energy photoelectron angular distribution patterns recorded showed the same results, ordered grains and no rotational disorder between adjacent layers. A grain size of up to a few μm was obtained on some samples.

Introduction

Epitaxial graphene grown on the C-terminated SiC(000-1) surface has been claimed to be dramatically different [1-7] compared to graphene grown on the Si-terminated surface. A rotational disorder between adjacent graphene layers has been reported for C-face graphene. Whereas Si-face graphene exhibits sharp Low Energy Electron Diffraction (LEED) spots, the diffraction pattern from C-face graphene was smeared out into a strongly modulated diffraction ring [2,4,5,7]. This, together with Surface X-ray diffraction results [1,2,5], was interpreted to indicate that the graphene layers on the C-face stack in such a way that adjacent layers are rotated with respect to each other. The experimental results indicated some preferred rotation angles but a turbostratic structure was proposed [2, 4], and rotated interleaved layers were suggested. The rotational disorder produced moiré patterns in STM topographs [6], and was suggested to explain why epitaxial graphene films, tens of layers thick, show single layer electronic properties [2, 5], i.e., show a single π-band with the Dirac point at the Fermi energy. For Si–face graphene the structural quality was shown to improve dramatically upon furnace growth at a high temperature and in an argon [8,9] environment, compared to *in situ* preparation. Considerably larger graphene domains were obtained by furnace growth at high temperature. Although a recent study of C-face graphene did not show a similar improvement [7] after growth at 1600°C in a 1 atm argon environment, we wanted to find out if a higher growth temperature would give large enough grains/domains so single grains could be directly investigated also for C-face graphene.

We found [10] formation of fairly large grains (crystallographic domains) of graphene exhibiting sharp 1x1 spots in μ-LEED while adjacent grains were found to show different azimuthal orientations. The macro-LEED patterns were found to mimic earlier recorded [2,4,5,7] LEED patterns and indicated contributions from several grains of different azimuthal orientations. Selective area constant initial energy photoelectron angular distribution patterns collected showed the same result. When utilizing a small aperture (probing area) the six Dirac cones at the K-points of the Brillouin zone could be clearly resolved while several Dirac cone replicas from differently oriented grains were detected when a large aperture was utilized. These findings show clearly the existence of ordered multilayer graphene grains with different azimuthal orientations and not that adjacent graphene layers are rotationally disorder as earlier reported for C-face graphene.

Experimental

Graphene samples were grown on nominally on-axis C-face SiC and both n-type 6H and 4H substrates were used. High temperature sublimation in an inert gas environment was applied [8] and the temperature range 1800-2000°C, pressure range 500-850 mbar and average growth time of 15 min were used. The thickness, morphology and electronic structure of these graphene samples were investigated using Low-Energy Electron Microscopy (LEEM), X-ray Photo Electron Emission Microscopy (XPEEM), Photo Electron Spectroscopy (PES) and LEED on beam line I311 at the MAX laboratory. In the LEEM instrument it was possible to switch between energy filtered images of the surface and energy filtered photoelectron emission angular distribution patterns. Angle Resolved Photo Electron Spectroscopy (ARPES) experiments were performed also at beam line I4.

Results and Discussion

A LEEM image collected from a sample at a voltage of 2.7 eV and a field of view of 10 μm is shown in Fig. 1a). This sample was grown on an on-axis 4H-SiC substrate. Extracted reflectivity curves from the areas labeled 1 to 8, respectively, are plotted in Fig. 1b). Although clear domains

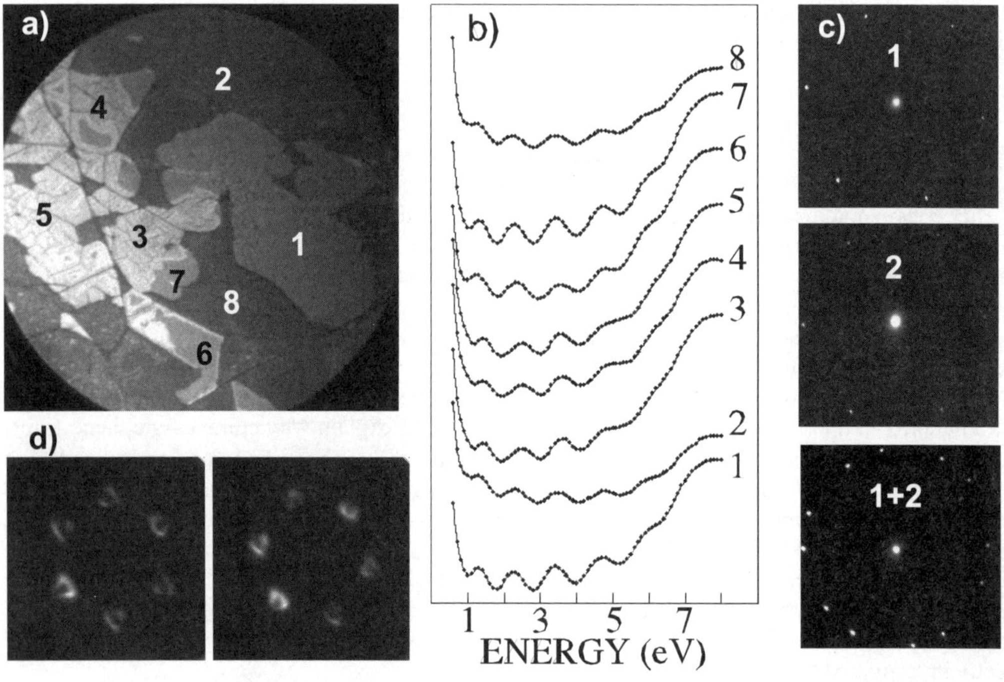

Fig. 1 a) LEEM image recorded from a sample at 2.7 eV and a field of view of 10 μm. b) Electron reflectivity curves extracted from the different areas labeled 1-8. c) The two upper panels show μ-LEED patterns collected at 45 eV using a probing area with a diameter of 400 nm from the positions labeled 1 and 2 in Fig. 1a). The lower panel shows the LEED pattern collected using a probing diameter of 5 μm at position 2. d) Shows two constant energy photoelectron angular distribution patterns (E_i,k_x,k_y) of the π-band collected using a probing area diameter of 800 nm, a photon energy of 45 eV and at an energy of about 2 eV below the Fermi energy. One Dirac cone centered at each of the six K-points in the Brillouin zone is clearly observed.

appear in the LEEM image the reflectivity curves show that most have the same number of graphene layers. Six minima are clearly observed in curves 1, 2, 6, 7 and 8 while curves 3, 4 and 5 appear to have only five minima. However, when looking in detail at these latter three curves one finds that there actually appears a weak minimum at around 6.5 eV also in these. The coverage may thus actually be 6 ML (ML= monolayer) over the whole area in Fig. 1a). Whether the whole surface has a 6ML coverage or some areas have 5 ML is, however, not important for the message we want to convey from this sample. The main message is contained in Figs. 1c) and d). Selected area LEED patterns collected using a probing area with a diameter of 400 nm from the positions labeled 1 and 2 in Fig. 1a) are shown in the two upper panels in Fig. 1c). It is obvious from these LEED patterns that a grain size larger than 400 nm has been achieved on this sample and that the stacking of the graphene layers in such that adjacent layers are not rotated with respect to each other. The third, bottom, panel in Fig. 1c) shows the LEED pattern collected using a 5 μm large probing area diameter at position 2. The sampled area then includes domains 2 and 1 and it is obvious that two grains of different azimuthal orientations are now probed since this LEED pattern just represents a superposition of the patterns shown in the upper two panels. To further emphasize that grains of multilayer graphene do form on the C-face and stack in such a way that no rotational disorder exists between adjacent layers, some constant energy photoelectron angular distribution patterns (E_i,k_x,k_y) of the π-band [11] were also recorded. Two such patterns collected using a probing area diameter of 800 nm, a photon energy of 45 eV and at an energy of about 2 eV below the Fermi energy are shown in Fig. 1d). Only six Dirac cones, which appear triangular at this initial energy [11], centered at the six K-points in the Brillouin zone are observed in these patterns. They were collected at two different locations outside the area shown in Fig. 1a) but where the number of graphene layers was also six. These patterns show clearly that well ordered grains of multilayer graphene larger than 800 nm has formed and that adjacent graphene layers are not rotated relative to each other.

The LEEM image in Fig. 2a) was recorded from a graphene sample grown on an on axis 6H substrate. The reflectivity curves, in Fig. 2b), show that on the areas labeled 1, 2, 3 and 4 respectively, 3, 1, 4 and 2 ML of graphene have formed. The XPEEM image in Fig. 2c) shows essentially the same area as the LEEM image in Fig. 2a). This XPEEM image shows the variation in work function over the surface. The light grey areas correspond to regions where no graphene has formed while the darker gray areas indicate graphene formation. Even on such a non-homogeneous surface the graphene formed on the different patches surprisingly contained fairly large well

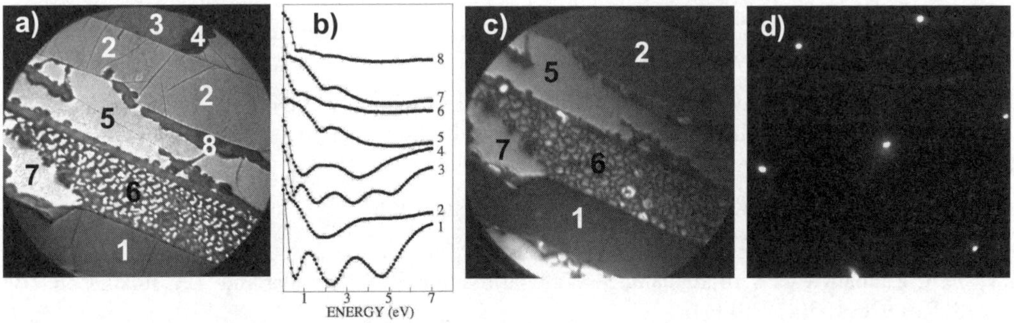

Fig. 2 a) LEEM image recorded from a sample at 0.8 eV and a field of view of 25 μm. b) Electron reflectivity curves extracted from the different areas labeled from 1 (bottom) to 8 (top curve). c) XPEEM image collected, from essentially the same area as in Fig. 2a), at an energy of 0.1 eV using 130 eV photons. d) μ-LEED pattern collected at 47 eV from the area labeled 1 using a probing area of 5 μm.

ordered grains. The μ-LEED pattern shown in Fig. 2d) was collected from the 3ML area labeled 1 using a probing area of 5 μm. This 1x1 pattern demonstrates again that fairly large multilayer graphene grains, >5 μm, form that show no trace of rotational disorder between adjacent layers.

Summary

The μ-LEED and constant energy photoelectron angular distribution patterns reported show unambiguously the same results, formation of ordered multilayer grains of graphene on the C-face of SiC with no rotational disorder between adjacent layers. It has earlier been claimed that adjacent graphene layers on the C-face are rotated with respect to each other, and may be interleaved. The reason we can observe ordered grains (crystallographic domains) of multilayer grapheme, we suggest, is that larger grains form at the higher growth temperature we have utilized, compared to earlier efforts [1-7] where lower growth temperatures have been applied, resulting in smaller grain sizes.

Acknowledgement

Support from the EU project ConceptGRAPHENE is gratefully acknowledged.

References

[1] J. Hass, R. Feng, J.E. Millan-Otoya, X. Li, M. Sprinkle, P.N. First, W.A. de Heer, E.H. Conrad and C. Berger, Phys. Rev. B **75** (2007) 214109.

[2] J. Hass, F. Varchon, J.E. Millan-Otoya, M. Sprinkle, N. Sharma, W.A. de Heer, C. Berger, P.N. First, L. Magaud and E.H. Conrad, Phys. Rev. Lett. **100** (2008) 125504.

[3] J Hass, W.A. de Heer and J. Conrad, J. Phys.: Condens. Matter **20** (2008) 323202.

[4] K.V. Emtsev, F. Speck, Th. Seyller, L. Ley and J.D. Riley, Phys. Rev. B **77** (2008) 155303.

[5] M Sprinkle, D. Siegel, Y. Hu, J. Hicks, A. Tejeda, A. Taleb-Ibrahimi, P. Le Fe`vre, F. Bertran, S. Vizzini, H. Enriquez, S. Chiang, P. Soukiassian, C. Berger, W. A. de Heer, A. Lanzara, and E. H. Conrad, Phys. Rev. Lett. **103** (2009) 226803.

[6] D.L. Miller, K.D. Kubista, G.M. Rutter, M. Ruan, W.A. de Heer, P.N. First and J.A. Stroscio, Phys. Rev. B **81** (2010) 125427.

[7] L. N. Srivastava, G. He, R.M. Feenstra and P.J. Fisher, Phys. Rev. B **82** (2010) 235406.

[8] C Virojanadara, M. Syväjarvi, R. Yakimova, L. I. Johansson, A.A. Zakharov, and T. Balasubramanian, Phys. Rev. B **78** (2008) 245403.

[9] K.V. Emtsev, A. Bostwick, K. Horn, J. Jobst, G.L. Kellogg, L. Ley, J.L. McChesney, T. Ohta, S.A. Reshanov, J. Röhrl, E. Rotenberg, A.K. Schmid, D. Waldmann, H.B. Weber and Th. Seyller, Nature Mater. **8** (2009) 203

[10] L.I. Johansson, S. Watcharinyanon, A.A. Zakharov, T. Iakimov, R. Yakimova and C. Virojanadara, Phys. Rev. B in press.

[11] A.A. Zakharov, C. Virojanadara, S. Watcharinyanon, R. Yakimova and L.I. Johansson, IBM J. Res. Dev. 55(4) (2011) 6.

Materials Science Forum Vols. 717-720 (2012) pp 617-620
© (2012) Trans Tech Publications, Switzerland
doi:10.4028/www.scientific.net/MSF.717-720.617

Large area quasi-free standing monolayer graphene on 3C-SiC(111)

U. Starke[1,a,*], C. Coletti[1,4,b], K.V. Emtsev[1,c],
A.A. Zakharov[2,d], T. Ouisse[3,e] and D. Chaussende[3,f]

[1]Max-Planck-Institut für Festkörperforschung, Heisenbergstr. 1, D-70569 Stuttgart, Germany

[2]MAX-lab, Lund University, Lund, S-22100, Sweden

[3]Laboratoire des Matériaux et du Génie Physique - CNRS UMR5628 – Grenoble INP, Minatec 3 parvis Louis Néel, BP 257, 38016 Grenoble, France

[4]present address: Center for Nanotechnology Innovation @NEST, Istituto Italiano di Tecnologia, Piazza San Silvestro 12, 56127 Pisa, Italy

[a]u.starke@fkf.mpg.de, [b]camilla.coletti@iit.it, [c]k.emtsev@fkf.mpg.de,
[d]alexei.zakharov@maxlab.lu.se, [e]thierry.ouisse@grenoble-inp.fr,
[f]didier.chaussende@grenoble-inp.fr, *corresponding author

Keywords: Graphene, Intercalation, Quasi-free standing graphene, 3C-SiC, ARPES, AFM, LEEM, LEED, XPS

Abstract. Large scale, homogeneous quasi-free standing monolayer graphene is obtained on a (111) oriented cubic SiC bulk crystal. The free standing monolayer was prepared on the 3C-SiC(111) surface by hydrogen intercalation of a $(6\sqrt{3}\times 6\sqrt{3})R30°$-reconstructed carbon monolayer, so-called zerolayer graphene, which had been grown in Ar atmosphere. The regular morphology of the surface, the complete chemical and structural decoupling of the graphene layer from the SiC substrate as well as the development of sharp monolayer π-bands are demonstrated. On the resulting sample, homogeneous graphene monolayer domains extend over areas of hundreds of square-micrometers.

Introduction

The growth of large scale, homogeneous graphene monolayers on the 4H and 6H polytypes of SiC by thermal decomposition of the topmost SiC layers has been investigated intensively [1,2]. In contrast, on cubic SiC substrates, namely the corresponding Si-face, i.e. 3C-SiC(111), just a few studies of graphene growth have been performed, yielding grain sizes of only a few micrometers and with limited homogeneity of the layers obtained [3-5]. The graphene layers presented so far on 3C-SiC(111) were grown on top of the carbon interface layer which couples by covalent bonds to the SiC substrate and by this coupling strongly influences the electronic structure of the graphene via doping [2]. In this paper, we present the preparation of homogenous, quasi-free standing monolayer graphene (QFMLG) on 3C-SiC(111) by growth in Ar atmosphere and subsequent hydrogen intercalation which yields domains extending over areas of hundreds of micrometers [6]. The morphological, structural and electronic properties are characterized.

Experimental Setup

The 3C-SiC(111) sample was cut from cubic SiC bulk material that had been grown on a 4H-SiC(0001) substrate by heteroepitaxy using the continuous feed-physical vapour transport (CF-PVT) method [7]. The 4H-SiC substrate was polished away after the growth process. Areas that were free of double positioning boundaries in the cubic crystal structure were selected and a sample of approximately 7 mm side length was obtained as shown in the photograph in Figure 1(a). Atomic force microscopy (AFM) was carried out in ambient environment. Graphene growth and hydrogen treatment were performed in a quartz glass reactor by induction heating (RF) of a graphite susceptor. The samples were characterized in ultra-high vacuum (UHV). The surface structure was analyzed by low-energy electron diffraction (LEED). The electronic structure was investigated by

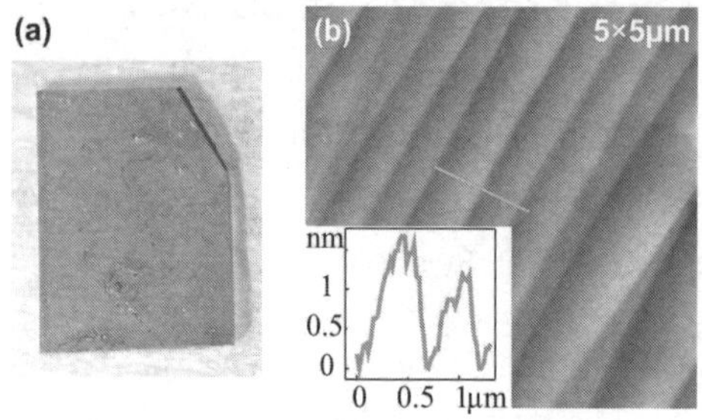

Figure 1: (a) Photograph of the 3C-SiC(111) sample (approx. 7mm on the long side) cut from the CF-PVT grown boule. (b) AFM micrograph for the as-grown 3C-SiC(111) sample. Inset: Height profile along the green line indicated in the main image.

X-ray photoelectron spectroscopy (XPS) with Mg K_α radiation (hν=1253.6 eV) and angle-resolved photoemission spectroscopy (ARPES) using the He II line (hν=40.8 eV) from a UV discharge source. Low-energy electron microscopy (LEEM) was performed using the LEEM III instrument from ELMITEC.

Homogeneous $(6\sqrt{3}\times 6\sqrt{3})$R30° Reconstruction on 3C-SiC(111)

The high structural quality on an atomic level of the as grown 3C-SiC(111) sample surface is demonstrated in Figure 1(b) by AFM. The micrograph shown reveals atomically flat terraces with step heights around 1 nm as drawn from the line profile plotted in the inset. The $(6\sqrt{3}\times 6\sqrt{3})$R30° reconstructed initial carbon layer was grown on these substrates by RF heating in an Ar atmosphere (800 °C, 10 min, 800 mBar) [8]. This layer represents the buffer layer or zerolayer graphene (ZLG) which evolves prior to normal epitaxial graphene growth on SiC(0001). While growth in UHV leads to an inhomogeneous surface with small islands of different graphene thickness [9] as displayed in Figure 2(a), the RF method [8] – similar to the case of 4H and 6H-substrates – yields flat homogeneous layers with terrace widths in the 2-4 μm regime and step heights of about 8 nm, cf. Figure 2(b). The LEED pattern of the RF grown ZLG shows the fractional order spots of the $(6\sqrt{3}\times 6\sqrt{3})$R30° reconstruction around the first order diffraction spots of the SiC(111) substrate. At an electron energy of 188 eV as displayed in Figure 2(c), the extinction of three of these six first order spots corroborates the complete absence of twinning domains in the SiC substrate [6]. XPS of the C 1s core level shows – besides emission from the SiC substrate – two interface components as revealed from a line shape analysis, see panel (d). The two components, S1 and S2, correspond to the carbon species in the ZLG covalently bound to SiC, and the sp^2 hybridized carbon bound planar within the ZLG layer, respectively [2,10].

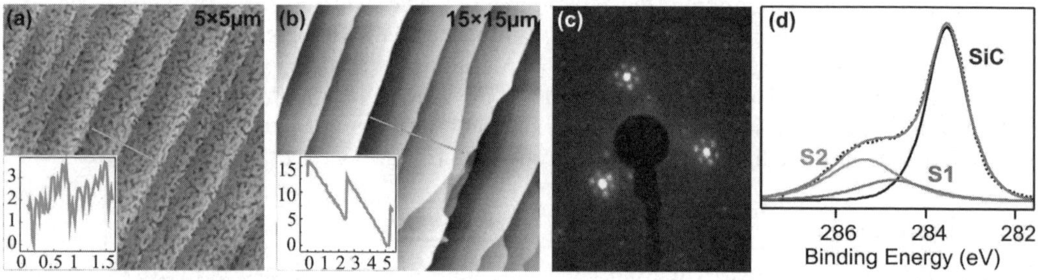

Figure 2: AFM micrographs of (a) a UHV grown ZLG and (b) an RF-grown ZLG on 3C-SiC(111). The insets display height profiles along the green lines indicated in the respective images (hor. axis: μm, vert. axis: nm). (c) LEED pattern of an RF-grown ZLG at 188 eV. (d) C1s core level spectrum for RF-grown ZLG. Spectral components obtained from a line shape analysis are indicated. S1 and S2 reflect carbon within the ZLG, S1 covalently bound to SiC, and S2 sp^2 hybridized and planar within the ZLG, SiC indicates emission from carbon in the bulk

Growth of QFMLG on 3C-SiC(111)

Subsequently, the sample was processed by annealing to around 850 °C in a molecular hydrogen flow at atmospheric pressure. While due to the covalent bonds between as grown ZLG and the SiC substrate the p_z-orbitals of the carbon atoms cannot form a delocalized π-system [2], after the hydrogen treatment sharp π-bands evolve as measured by ARPES and plotted in Figure 3(a). This indicates that the ZLG was turned into QFMLG by hydrogen intercalation [11]. In this state, the dangling bonds of the topmost Si atoms of the substrate are passivated by hydrogen and the carbon layer is decoupled from the SiC substrate as corroborated by the suppression of the reconstruction spots in LEED, cf. Figure 3(b), and by the disappearance of the interface components S1 and S2 in XPS, whereas a sharp emission line emerges from the graphene layer (indicated by "g"), as seen in Figure 3(c). However, in contrast to quasi-free standing graphene on hexagonal SiC crystals [11], the π-band dispersion of the QFMLG on 3C-SiC(111) indicates a notable *n*-type doping, see panel (a), which requires further investigation. LEEM images acquired on a large field of view (~57 μm) demonstrate an extremely homogeneous graphene thickness over the entire surface, as indicated by the lack of contrast in the grayscale as shown in panel (d). At a larger magnification (~23 μm), details of the surface morphology can be observed, cf. panel (e). A darker contrast, symptomatic of a different graphene thickness, emerges only at the step edges (see panel e) or in restricted domains, see panel (d). Electron reflectivity spectra extracted from energy series of such LEEM images can be used to count the number of graphene layers present [12] by counting the number of minima in the energy range between 0 and 6 eV. Such spectra evaluated for areas of different contrast in the region shown in panel (e) and plotted in panel (f) confirm that practically the entire surface is covered by monolayer graphene (areas with grey contrast in panel e, analyzed regions are labeled A

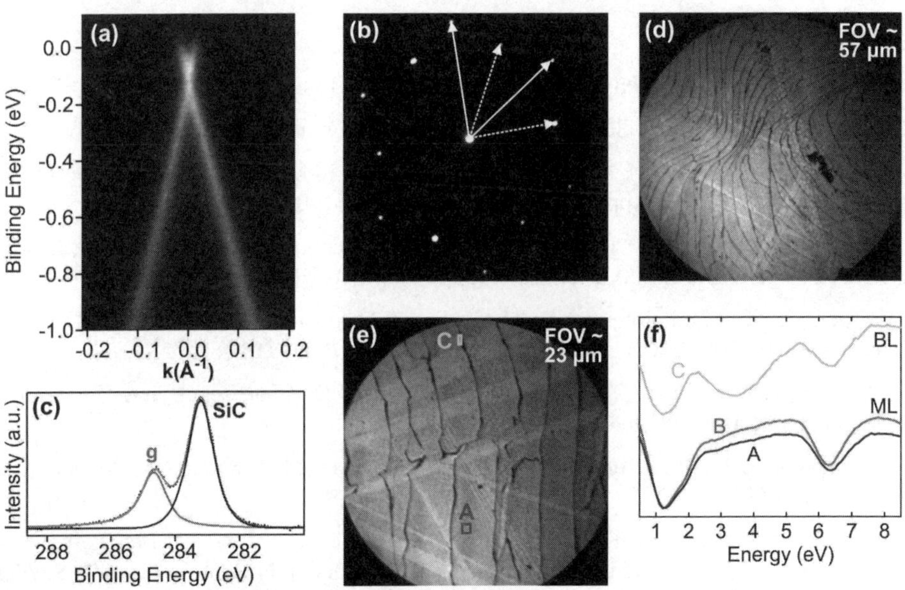

Figure 3: QFMLG on 3C-SiC(111) after hydrogen intercalation of RF-grown ZLG: (a) ARPES measurement of the π-band dispersion at the $\overline{\mathbf{K}}$-point of the graphene Brillouin zone, acquired perpendicular to the $\overline{\Gamma\mathbf{K}}$-direction. (b) Micro-LEED image taken at 98 eV. Arrows indicate first order diffraction spots for graphene (solid lines) and SiC (dashed lines). (c) C1s core level spectrum with fitted components indicated for graphene (g) and SiC bulk emission, respectively. (d) LEEM overview image (energy E=19 eV, ~57μm field of view ((FOV)). (e) LEEM micrograph (E=4 eV, ~ 23μm FOV). Representative regions are labeled with A, B and C. (f) Electron reflectivity spectra measured for regions A, B and C in panel (e). Curve C is shifted on the y-axis for better display.

and B) while bilayer graphene is only present at the darker step edges (region C). We note that, in contrast to as grown graphene on SiC(0001) [12], an extra minimum appears above 6 eV, whose origin is not yet clear at present. We observe areas of lighter gray contrast that are running along certain crystal orientations (region B). Nevertheless, they are clearly identified as monolayer graphene also by the LEEM intensities. Hence, the contrast difference is not due to the graphene properties. We speculate that it could be caused by defects in the substrate, which might be mediated by strain. LEEM analysis also reveals that the large atomically flat macro-terraces that are homogeneously covered by QFMLG run uninterrupted for hundreds of micrometers. This thickness homogeneity is highly remarkable considering that, until now, the maximum lateral dimension reported for homogenous graphene domains on SiC(111) was roughly 1 μm [3-5].

Summary

In conclusion, this work demonstrates that large area QFMLG can be produced on 3C-SiC(111) substrates. The high quality of the graphene layer obtained suggests that 3C-SiC(111) might be an appealing and cost effective platform for the future development of graphene technology.

Acknowledgements

C.C. acknowledges the Alexander von Humboldt Foundation for financial support. This research was partially funded by the European Community's Seventh Framework Programme (FP7/2007-2013) under grant agreement no. 226716. We are indebted to the staff at MAX-Lab (Lund, Sweden) for their advice and support.

References

[1] U. Starke and C. Riedl, J. Phys.: Cond. Mat. **21** (2009) 134016.

[2] C. Riedl, C. Coletti and U. Starke, J. Phys. D: Appl. Phys. **43** (2010) 374009.

[3] A. Ouerghi, R. Belkhou, M. Marangolo, M.G. Silly, S. El Moussaoui, M. Eddrief, L. Largeau, M. Portail and F. Sirotti, Appl. Phys. Lett. **97** (2010) 161905.

[4] A. Ouerghi, A. Kahouli, D. Lucot, M. Portail, L. Travers, J. Gierak, J. Penuelas, P. Jegou, A. Shukla, T. Chassagne and M. Zielinski, Appl. Phys. Lett. **96** (2010) 191910.

[5] A. Ouerghi, M. Marangolo, R. Belkhou, S. El Moussaoui, M. G. Silly, M. Eddrief, L. Largeau, M. Portail, B. Fain and, F. Sirotti, Phys. Rev. B **82** (2010) 125445.

[6] C. Coletti, K.V. Emtsev, A.A. Zakharov, T. Ouisse, D. Chaussende and U. Starke, Appl. Phys. Lett. **99** (2011) 081904.

[7] D. Chaussende, L. Latu-Romain, L. Auvray, M. Ucar, M. Pons and R. Madar, Mat. Sci. Forum **483-485** (2005) 225.

[8] K.V. Emtsev, A. Bostwick, K. Horn, J. Jobst, G.L. Kellogg, L. Ley, J.L. McChesney, T. Ohta, S.A. Reshanov, E. Rotenberg, A.K. Schmid, D. Waldmann, H.B. Weber and T. Seyller, Nat. Mater. **8** (2009) 203.

[9] C. Riedl, U. Starke, J. Bernhardt, M. Franke and K. Heinz, Phys. Rev. B **76** (2007) 245406.

[10] K.V. Emtsev, F. Speck, Th. Seyller, L. Ley and J.D. Riley, Phys. Rev. B **77** (2008) 155303.

[11] C. Riedl, C. Coletti, T. Iwasaki, A.A. Zakharov and U. Starke, Phys. Rev. Lett. **103** (2009) 246804.

[12] H. Hibino, H. Kageshima, F. Maeda, M. Nagase, Y. Kobayashi and H. Yamaguchi, Phys. Rev. B **77** (2008) 075413.

Materials Science Forum Vols. 717-720 (2012) pp 621-624
© (2012) Trans Tech Publications, Switzerland
doi:10.4028/www.scientific.net/MSF.717-720.621

CVD Growth of Graphene on 2" 3C-SiC/Si Templates: Influence of Substrate Orientation and Wafer Homogeneity

M. Portail[1,a], A. Michon[1], S. Vézian[1], D. Lefebvre[1], S. Chenot[1], A. Ouerghi [2], M. Zielinski[3], Th. Chassagne[3] and Y. Cordier[1]

[1]CNRS – CRHEA, Rue Bernard Gregory, 06560 Valbonne, France

[2]CNRS – LPN, Route de Nozay, 91460 Marcoussis, France

[3]NOVASiC, Savoie Technolac, Arche Bat4, 73375 Le Bourget du lac, France

[a]mpo@crhea.cnrs.fr

Keywords: 3C-SiC, graphene, propane - argon CVD.

Abstract. Structural and electrical properties of graphitic phases formed on 3C-SiC(111)/Si and 3C-SiC(100)/Si templates, using propane-argon gas mixtures under CVD environment, are presented. On 3C-SiC(111), the graphitic phase is graphene. It presents good electrical conductivity at the macroscopic scale. The opposite case is observed on 3C-SiC(100), for which the graphitic phase develops more rapidly but with a high degree of disorientation. The graphitization is homogeneous over the whole surface of a 2'' wafer. It can be coupled with the 3C-SiC growth and allows to grow, in a single process, Graphene on Silicon wafers.

Introduction

The fabrication of "Graphene on Silicon" (GOS) wafers is an interesting approach in order to take advantage of the electronic properties of graphene and to benefit from the existing facilities of Si technology. To achieve this goal, the use of 3C-SiC templates, grown on silicon substrates, could be a key point. Their production is currently well addressed by Chemical Vapour Deposition (CVD) and an important number of publications deals with graphitization of bulk hexagonal SiC. The possibility to form conjointly graphene/3C-SiC heterostructures within the same CVD equipment should give opportunity to promote GOS for industrial applications. This approach has been already proposed by different authors but, till now, the graphitization of 3C-SiC/Si templates has been demonstrated almost exclusively by thermal annealing performed under ultra high vacuum (UHV) conditions [1-3]. Furthermore, for two of these works [2, 3], the initial 3C-SiC templates were previously grown by CVD. This implies that such GOS samples are the result of two highly technologically different steps. On the other hand, one group has realised some GOS samples in a single process, using an intermediate 3C-SiC buffer, under high vacuum conditions [1]. But such conditions are unfavourable for growing high quality 3C-SiC (in comparison to what is achieved by CVD) films and could be detrimental for obtaining high quality graphene. Recently, the graphitization of hexagonal SiC substrates has been demonstrated at high pressures using pure argon [4, 5], propane-hydrogen [6] or propane-argon [7] atmospheres. The higher uniformity of the graphene films obtained by this way, as well as the fact that SiC epitaxy reactors are suitable equipment to operate at pressures of interest (from some mbar to one atmosphere), is a strong motivation for targeting graphene formation on silicon *via* a SiC buffer by CVD.

In order to contribute to this field, we have explored, in this work, some aspects of the graphitization of 3C-SiC epilayers using propane-argon mixtures under CVD environment. In particular, we have focused on the following questions: *(i)* are 3C-SiC(111) (silicon terminated faces) and 3C-SiC(100) surfaces subject to the same kind of graphitization? *(ii)* how do they compare to bulk 6H-SiC? *(iii)* how uniform is the graphitization across the wafer surface ?

Experimental Details

3C-SiC deposition and graphitization were done using a home made resistively heated hot wall CVD reactor [8]. During treatment, Si wafers are placed on a rotating sample holder in order to improve thickness uniformity. 3C-SiC epilayers were grown on 2'' highly resistive silicon substrates (ρ >10000 ohm.cm), oriented along (111) and (100) crystalline directions, using a classical two step deposition process using hydrogen as carrier gas and propane and silane as precursors (see details in [9]). 3C-SiC epilayers are unintentionally doped. Their typical thickness is about 1 μm.

After SiC growth and cooling down, the graphitization was performed at 1300°C under a mixture of argon and propane (0.028% propane/argon) during 30 minutes at a pressure ranging from 6 to 700mbar. When using 2'' Si substrates, both SiC growth and graphitization were conducted without any air exposure between the two stages. For certain cases, in order to directly compare the effect of template orientation for a given graphitization condition, the graphitization was carried out on pieces of previously grown 3C-SiC(111), 3C-SiC(100) epilayers and a 6H-SiC(0001) sample serving as reference. Structural characterizations were carried out by means of X ray photoemission spectroscopy (XPS) and low energy electron diffraction (LEED). Basic electrical properties of the surface after graphitization have been investigated by measuring I(V) curves using a Hg probe (the typical distance between contact points is about 2 mm).

Results and Discussion

Formation and Properties of Graphitic Phases. In the first part of this study, we have investigated the presence of a graphitic phase after treatment under propane-argon atmosphere. In order to apply the same graphitization condition to each type of substrate the process was run simultaneously on 3C-SiC(111)/Si, 3C-SiC(100)/Si and 6H-SiC(0001) samples. Fig. 1 presents XPS C1s core levels recorded after graphitization conducted at 1300°C during 30 minutes at 400mbar. For all samples, the C1s core levels are multi component signals.

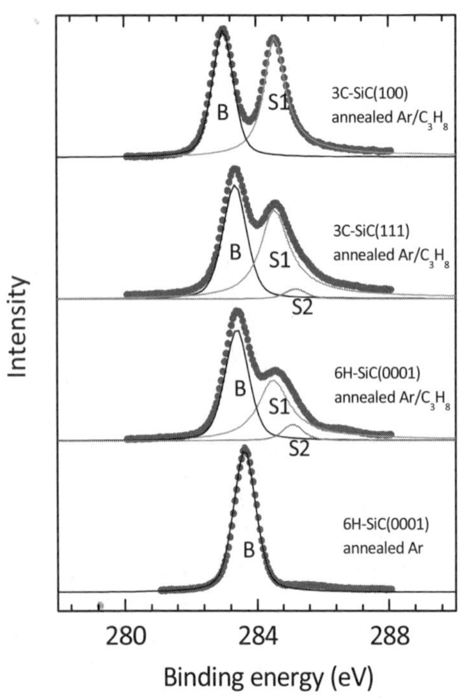

Fig. 1: C1s XPS core levels on 3C-SiC(100), 3C-SiC(111) and 6H-SiC(0001) samples graphitized at 1300°C – C_3H_8/Ar = 0.028% - 400 mbar – 30 minutes; comparison with 6H-SiC(0001) annealed under the same conditions of temperature and pressure but only under argon (lower spectrum); points = experiment – lines = fit

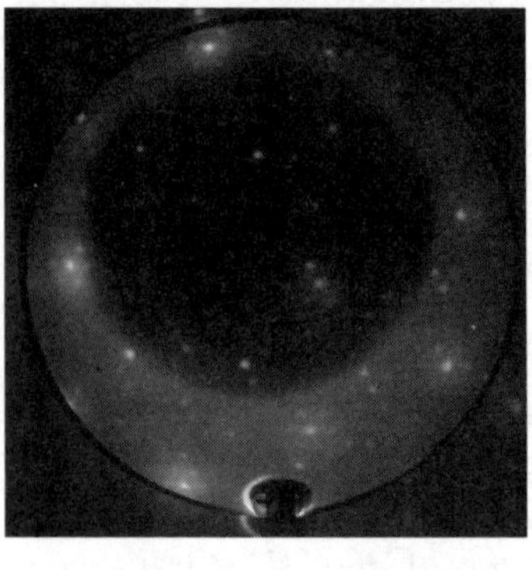

Fig. 2: LEED pattern on 3C-SiC(111), graphitized at 1300°C – C_3H_8/Ar = 0.028% - 400 mbar – 30 minutes

Basing on previous works dealing with graphene formed on 6H-SiC(0001) under UHV [10-11], the B component can be related to carbon atoms involved in Si-C bonding. The S1 component, located for all cases at 284.5eV, can be related to carbon atoms involved in sp^2 bonding. For 3C-SiC(111) and 6H-SiC(0001) samples, a third component is observed at a higher binding energy (BE) (S2 component - 285.2eV). Using a X ray synchrotron source coupled with XPS facilities, the previous cited works have revealed that a buffer layer, between the SiC and the graphene plane, induces such a C1s component having a fine structure. Even if the poorer spectral resolution used in the present work does not permit to evidence the fine structure, we can attribute the S2 component to the buffer contribution. This is well confirmed by LEED patterns which show a $(6\sqrt{3}\times 6\sqrt{3})R30°$ reconstruction for 3C-SiC(111) and 6H-SiC(0001) samples (see Fig. 2). Thus, the graphitic phases formed on these orientations can be attributed to graphene developing on the underlying SiC *via* the formation of an intermediate layer, similarly to what has been reported by thermal graphitization. In the case of the 3C-SiC(100) surface, the intensity of the sp^2 component is higher than on the two other surfaces showing a faster formation of the surface phase. Furthermore, only a diffuse background, without marked diffraction spots, is visible. This indicates a poor level of organisation of the graphitic phase on this surface orientation. Therefore, 3C-SiC(100) faces experience a different kind of graphitization which leads to the formation of a graphitic phase decoupled from the substrate (as observed on 6H-SiC(000-1) faces) but having a high degree of disorder.

The importance of adding a carbon precursor diluted with argon is illustrated in Fig. 1. The bottom spectrum has been recorded on 6H-SiC(0001) annealed under the same conditions of temperature, pressure and time as that of the other spectra in the figure but under pure argon atmosphere. It is clear that no graphitic phase has developed. This shows that the presence of carbon precursor is mandatory for graphitizing the SiC surface at low temperatures. The question of the growth mode must be considered. Indeed, without a carbon precursor, the surface does not experience any carbon enrichment, which indicates none, or negligible, surface evaporation. Therefore, the graphene formation should occur *via* a deposition mechanism. But, following this assumption, how can we justify the presence of the same buffer layer as that formed by thermal annealing? Hwang *et al.*, using a similar gas environment to that of the present study, have observed the formation of graphitic carbon on a sapphire substrate [7]. This leads to consider that the deposition regime is prominent in their case. Based on the intensity ratio of our C1s components, the graphene coverage in our study is about 4-5 monolayers, which is well below those reported by Hwang *et al.* We can suppose that we are operating in a different growth regime but further investigations are required to understand which driving forces are governing the graphitization. In particular, a detailed study of the kinematic of the graphitization is required and is in progress.

Electrical Behaviour. The second part of the study explores the uniformity of the electrical behaviour of the graphitic phases using I(V) measurements between two contacts at the surface. For this part, different graphitizations have been done under different pressure conditions. For all of them, the presence of the graphitic phase has been systematically verified by XPS prior to I(V) measurements. These last are displayed in Fig. 3. For 3C-SiC(111) faces, a transition from rectifier to ohmic behaviour has been observed, irrespective of the working pressure. This transition is readily visible through the significant increase of the current density as well as the symmetrisation of the I(V) curve with respect to the origin. Of course, such measurements do not give any information about the carrier mobility and do not allow to assess graphene quality. But they demonstrate that graphene coverage is suitable to allow electrical conductivity. It can be added that the surface roughness of the underlying 3C-SiC(111) epilayer does not limit the current flow. The rectifier/ohmic transition is not observed for 3C-SiC(100) faces. In that case, as mentioned before, the graphitic coverage, measured by XPS, is slightly higher than on 3C-SiC(111) faces, but for all working pressures, I(V) curves remain of rectifier type (low current density and non symmetric shape of the curve, typical of p-type epilayers). This highlights a strong difference of electrical continuity of the graphitic phases developed either on 3C-SiC(111) or 3C-SiC(100) faces.

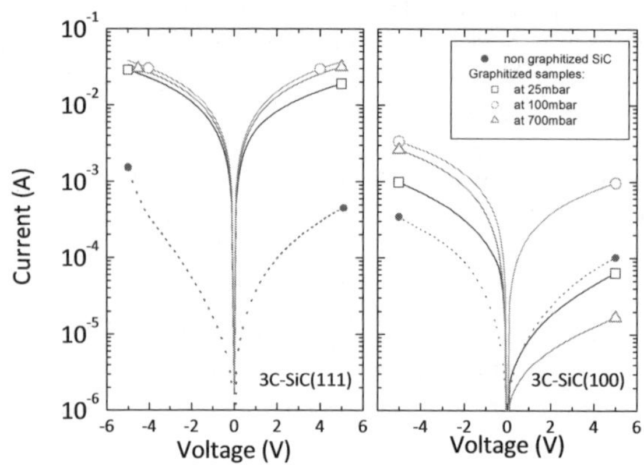

Fig. 3: I(V) curves of graphitized 3C-SiC epilayers showing the dependence on working pressure

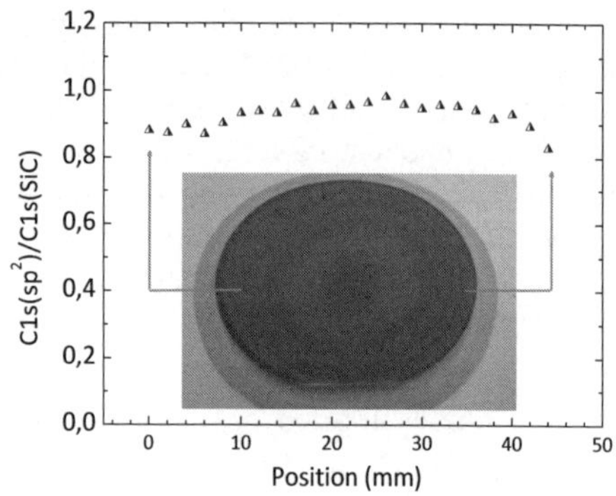

Fig. 4: Profile along wafer diameter of C1s(sp^2)/C1s(SiC) intensity ratio measured on graphitized 3C-SiC(111) (1300°C – 600 mbar-30 min)

Wafer Homogeneity. Finally, as our work is motivated by the idea of developing large scale graphene on silicon wafers, the uniformity of the graphitic coverage has been investigated by performing XPS mapping across the wafer surface. C1s core levels have been recorded every 2mm along a 3C-SiC(111) wafer diameter. C1s(sp^2)/C1s(SiC) intensity ratios have been determined for each position. As illustrated on Fig. 4, very low dispersion (standard deviation of 4%) of the intensity ratio profile has been observed. This demonstrates that, within our experimental set up, the graphitization stage is homogeneous on a whole 2'', allowing the realisation of wide scale GOS wafers.

Conclusion

Formation of graphene on 3C-SiC(111)/Si and 3C-SiC(100)/Si templates, using a propane-argon gas mixture, has been explored. Both 3C-SiC growth and graphitization can be done within the same apparatus, with a good homogeneity on a 2'' Si wafer, which is a good starting point for direct growth of Graphene on Silicon wafers using CVD. Graphitic phases are obtained on both orientations but major differences in terms of structural properties and electrical behaviour have been highlighted. The way the graphitic phase is developing on surface (*via* only matter deposition or *via* a balance between deposition and evaporation) remains unclear and is currently investigated.

References

[1] M. Suemitsu *et al.*, J. Phys. D Vol. 43 (2010), p. 374012
[2] V. Aristov *et al.*, Nano Lett. Vol. 10 (2010), p. 992
[3] A. Ouerghi *et al.*, Appl. Phys. Lett. Vol. 96 (2010), p. 191910
[4] C. Virojanadara *et al.*, Phys. Rev. B Vol. 78 (2008), p. 245403
[5] K. V. Emtsev *et al.*, Nat. Mat. Vol. 8 (2009), p. 203
[6] A. Michon *et al.*, Appl. Phys. Lett. Vol. 97 (2010), p. 171909
[7] J. Hwang *et al.*, J. Cryst. Growth Vol. 312 (2010), p. 3219
[8] A. Leycuras, Mat. Sci. Forum Vol. 338-342 (2000), p. 241
[9] M. Portail *et al.*, J. Appl. Phys. Vol. 105 (2009), p. 083505
[10] K. V. Emtsev *et al.*, Phys. Rev. B Vol. 77 (2008), p. 155303
[11] C. Riedl *et al.*, Phys. Rev. Lett. Vol. 103 (2009), p. 246804

Materials Science Forum Vols. 717-720 (2012) pp 625-628

doi:10.4028/www.scientific.net/MSF.717-720.625

Structural and Electrical Properties of Graphene Films Grown by Propane/Hydrogen CVD on 6H-SiC(0001)

A. Michon[1,a], E. Roudon[1], M. Portail[1], B. Jouault[2,b], S. Contreras[2], S. Chenot[1], Y. Cordier[1], D. Lefebvre[1], S. Vézian[1], M. Zielinski[3], T. Chassagne[3,c], and Jean Camassel[2]

[1] Centre de Recherche sur l'Hétéro-Epitaxie et ses Applications, UPR 10, CNRS, rue Bernard Gregory, 06560 VALBONNE, France

[2] Laboratoire Charles Coulomb, UMR 5221, CNRS-UM2, Université Montpellier 2, Place Eugène Bataillon, 34095 MONTPELLIER Cedex 05, France

[3] NOVASiC, Savoie Technolac, Arche Bât. 4, BP 267, 73375 LE BOURGET DU LAC Cedex, France

[a] am@crhea.cnrs.fr, [b] benoit.jouault@univ-montp2.fr, [c] tchassagne@novasic.com,

Keywords: Graphene, 6H-SiC, 3C-SiC/Si, CVD

Abstract. We have grown graphene on SiC(0001) using propane-hydrogen CVD. In this work, we present the effects of growth pressure and temperature on structural and electrical properties. Structural characterizations evidence the formation of graphene with in-plane rotational disorder, except for low growth pressure and high growth temperature which lead to the formation of a $(6\sqrt{3}\times6\sqrt{3})$-30° interface between graphene and SiC. Electrical properties of samples presenting different graphene/SiC stacking and interfaces are compared and discussed.

Introduction

The isolation of graphene and the first study of its electronic properties in 2004 have opened the way for new fundamental studies as well as for applications. While a lot of studies have focused on transport properties, several efforts have been devoted to graphene growth methods. Among these methods, SiC annealing (silicon sublimation and graphene formation from carbon of the substrate) has been widely studied due to its ability to form graphene on a large area semi-insulating substrate [1,2]. In contrast to the annealing method, few works attempted to obtain graphene from an external carbon source, possibly through a direct growth process. The first demonstrations have been done using ultra-high vacuum conditions (molecular beam epitaxy - MBE) in 2009 [3,4], and subsequently under vapor phase conditions (chemical vapor deposition - CVD) in 2010, using either argon [5] or hydrogen [6] as the carrier gas. Following these first works, similar methods have been studied using different growth conditions [7,8] or substrates [9]. One of the advantages of direct growth methods is a lower growth temperature with respect to the annealing method [4,6]. This allows to grow graphene under vapor phase on low cost and large area 3C-SiC/Si templates [6,7,10], in spite of the temperature limitation due to the melting point of silicon. In addition to the reduced growth temperature, a CVD process using hydrogen as the carrier gas (propane-hydrogen CVD) offers the possibility to control the graphene/SiC interface on the Si-face, and consequently graphene stacking [10-11].

In this work, we study structural and electrical properties of graphene films grown using propane-hydrogen CVD on 6H-SiC(0001) (Si-face). Low energy electron diffraction (LEED), atomic force microscopy (AFM) and X-ray photoemission (XPS) are used to study the influence of growth pressure and temperature on the graphene interface and morphology, while the transmission line method (TLM) and van der Pauw (vdP) configuration are used for electrical measurements. We discuss the effects of growth pressure and temperature on structural and electrical properties.

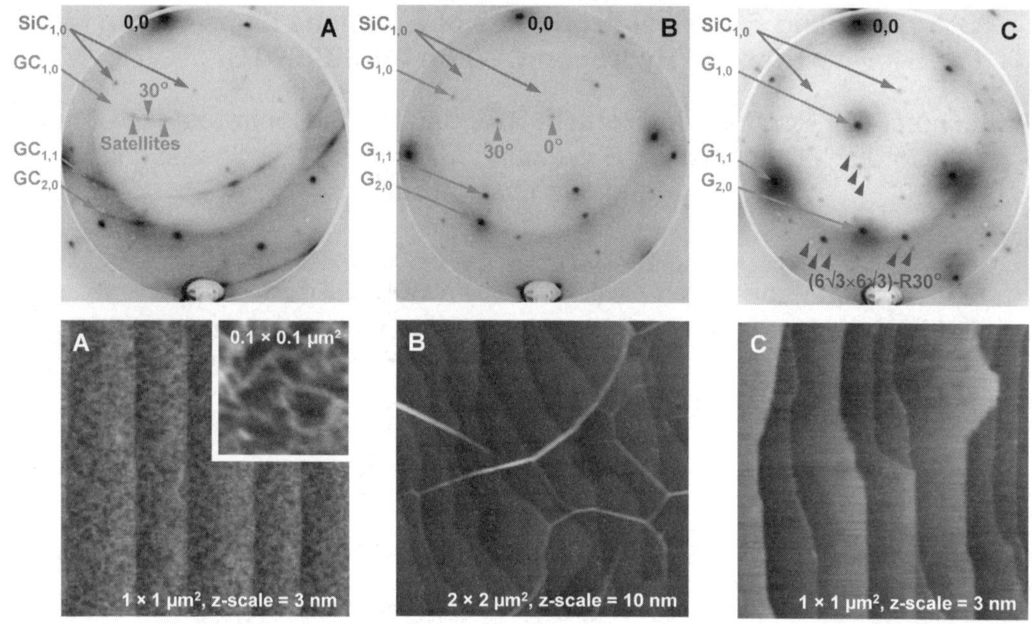

Figure 1. LEED patterns and AFM views of graphene films grown on 6H-SiC(0001) under different pressure and temperature (A: 1150°C/400 mbar, B: 1550°C/800 mbar, C: 1550°C/100 mbar). In LEED patterns, G and GC denote graphene and graphitic circle, respectively.

Growth and Technology

Graphene films are grown on semi-insulating on-axis 6H-SiC(0001) (Si-face) substrates in a horizontal hot-wall CVD reactor using hydrogen as the carrier gas and propane as the precursor. Details on substrate preparation and growth can be found in Ref. [6]. The growth temperature and pressure are 1150°C/400 mbar for sample A, 1550°C/800 mbar for sample B, and 1550°C/100 mbar for sample C. The propane flow is set to 5 (sample A and B) or 3.5 sccm (sample C) for 5 minutes. Growth conditions are summarized in Table 1.

Samples are studied using LEED, AFM and XPS. TLM and van der Pauw measurements are used for electrical characterizations. For that purpose, the technology begins with Ti/Au contact deposition using UV lithography, metallization, and a lift-off technique. Then, UV lithography and O_2 plasma etching are used to define the graphene ribbon between contacts. The graphene ribbon width is 125 µm, and the distances between contacts range from 4 to 160 µm. For the vdP measurements, the dimensions of the square patterns are $500{\times}500\ \mu m^2$. With respect to substrate atomic steps, TLM directions and the sides of the van der Pauw square patterns are roughly oriented at ±45°.

Effects of Pressure and Growth Temperature on Structural Properties

Figure 1 presents LEED patterns (45° configuration) observed on samples A, B, and C. For all samples, we can see spots corresponding to SiC and graphene. The observation of SiC spots for all samples indicates that the substrate is covered by a small number of graphene layers, of the order of 3 to 6 monolayers. This is consistent thickness estimations (see Table 1) from XPS measurements [12]. The sample C, grown at low pressure and high temperature, presents a characteristic $(6\sqrt{3}\times6\sqrt{3})$-R30° interface reconstruction which is commonly observed on the Si-face [1,2].

Accordingly, the graphene is oriented at 30° with respect to the SiC. Samples A and B present multiple graphene orientations without interface reconstruction. The LEED pattern of sample A, grown at low temperature, presents multiple graphene orientations with satellites, and furthermore, an elongation of the spots which tend to form graphitic circles. This is characteristic of in-plane rotational disorder (IRD), which is generally observed on the C-face [13,14], but which can be observed on the Si-face using propane-hydrogen CVD [6,10,11]. The LEED pattern of sample B, grown at high pressure and high temperature, presents spotty 30° and 0° graphene orientations. While the presence of 0° orientation has already been observed [6,11], the absence of elongated spots and of satellites makes this LEED pattern original.

Figure 1 also presents AFM views of the samples. Associated with the presence of IRD on sample A and B, we can see on the AFM views the presence of ridges [14]. Sample A (low growth temperature) presents a high density of small ridges defining small domains, while sample B (high growth temperature) presents large ridges, crossing SiC steps, and defining large domains. Sample C, which present a (6√3×6√3)-R30° interface reconstruction on the LEED pattern, does not present ridges on the AFM image. For comparison, samples grown with similar conditions to samples A, B, and C, but without propane flow present only parallel SiC atomic steps (not shown).

We have previously described the formation of IRD on the Si-face when using propane-hydrogen CVD [6] and the transition from IRD to the (6√3×6√3)-R30° interface with increasing growth temperature [11]. A small effect of the growth pressure is also evidenced [10] (preferential (6√3×6√3)-R30° interface formation at low pressure). The observations described here are consistent with these previous results. The observation of IRD for growth at high temperature and high pressure, shows that the formation of a (6√3×6√3)-R30° interface on the Si-face is an unusual case of propane-hydrogen CVD. In addition, we also evidence the variability in orientation and morphology of graphene films grown using propane-hydrogen CVD (Table 1). In the following, we will study the electrical properties of the graphene films presented above.

Table 1. Growth conditions, structural and electrical properties of samples A, B, and C.

Sample	A	B	C
Growth temperature	1150°C	1550°C	1550°C
Growth pressure	400 mbar	800 mbar	100 mbar
Structure (LEED)	IRD: 30°, satellites and circles	IRD: 0 and 30°	(6√3×6√3)-R30°
Morphology (AFM)	Small and dense ridges	Large and spaced ridges	SiC terraces
Thickness (XPS)	2 – 3 monolayers	3 – 5 monolayers	3 – 5 monolayers
Sheet resistance $R_\square$ (300 K)	4.8 – 6.0 kΩ/□	0.7 – 2.2 kΩ/□	5.8 – 5.9 / 3.3 – 3.4 kΩ/□
Contact resistivity ρ_C (300 K)	4 – 6 × 10^{-4} Ω.cm^2	1 – 24 × 10^{-4} Ω.cm^2	1 – 4 / 1 – 2 × 10^{-4} Ω.cm^2
Mobility μ (2 K)	-	≈ 200 cm^2/V.s	≈ 400 cm^2/V.s
Sheet density N (2 K)	-	3 × 10^{13} /cm^2	5 × 10^{12} /cm^2

Electrical Properties

Table 1 summarizes growth conditions and properties of samples A, B, and C. The average number of graphene layers was deduced from the intensity ratio on XPS spectra. For sample A, the sheet resistance ($R_\square$) ranges from 4.8 to 6.0 kΩ/□, while the specific contact resistivity (ρ_C) ranges from 4 to 6 × 10^{-4} Ω.cm^2. Sample B presents smaller and more dispersed values: $R_\square$ ranges from 0.7 to 2.2 kΩ/□, and ρ_C from 1 to 24 × 10^{-4} Ω.cm^2. Finally, electrical anisotropy is evidenced on sample C: for a set of TLM patterns, $R_\square$ ≈ 5.8 kΩ/□ and ρ_C ≈ 3.5 × 10^{-4} Ω.cm^2, and for the other set (perpendicular to the first), $R_\square$ ≈ 3.3 kΩ/□ and ρ_C ≈ 1.5 × 10^{-4} Ω.cm^2.

We have also performed magneto-transport measurements using the vdP configuration at 2 K on samples B and C (see table 1). We have measured on sample B (presenting IRD) the carrier density $n_s \approx 3 \times 10^{13}/\text{cm}^2$ and mobility $\mu \approx 200$ cm^2/V.s, and on sample C (presenting (6√3×6√3)-R30° interface), $n_s \approx 5 \times 10^{12}/\text{cm}^2$ and mobility $\mu \approx 400$ cm^2/V.s. These mobilities are lower than for the best graphene layers obtained using SiC annealing [1,15,16]. One could expect higher mobility on sample B than on sample C [17], but this is not the case.

As TLM patterns are oriented at ±45° with respect to SiC steps, SiC steps should have similar effects on any pattern. So, the anisotropy observed on sample C should correspond to the anisotropy of the hexagonal graphene lattice, which is preserved by the (6√3×6√3)-R30° interface and by the associated stacking of graphene layers. On the contrary, the IRD on sample A and B should average the angle-dependent electrical properties, leading to the absence of anisotropy. TLM curves (resistance *vs.* distance) from sample B frequently present a nonlinear behavior with an increase of the slope when increasing the distance between contacts. We explain this by the presence of large domains (as can be seen in the AFM image): for small distances between contacts, conduction should occur mainly through parallel domains (*i.e.*, some domains join the two contacts), and with increasing distance, conduction occurs through domains and ridges (*i.e.*, fewer domains join the two contacts simultaneously), ridges being more resistive. Sample A, due to its small domain size (less than the smaller distance between contacts), presents more linear TLM curves, with higher resistivity. The presence of large domains on sample B could also partly explain the large dispersion of $R_\square$.

Conclusion

We have grown graphene on SiC(0001) using propane-hydrogen CVD and studied the influence of growth parameters on structural and electrical properties. The growth method leads generally to the formation of graphene presenting IRD (multiple graphene orientations on LEED and ridge morphology on AFM views). Increasing growth temperature leads to a reduction of the ridge density together with a reduction of the sheet resistance. For low pressure and high temperature, graphene grows on a (6√3×6√3)-R30° interface. Electrical measurements can be correlated to structural properties, but further characterizations are required to confirm our first conclusions.

References

[1] C. Berger et al., Science 312 (2006) 1191-1196.
[2] K.V. Emtsev et al., Phys. Rev. B 77 (2008) 155303.
[3] E. Moreau et al., Phys. Stat. Sol. a 207 (2009) 300-303.
[4] A. Al-Temimy et al., Appl. Phys. Lett. 95 (2009) 231907.
[5] J. Hwang et al., J. Crystal Growth 312 (2010) 3219-3224.
[6] A. Michon et al., Appl. Phys. Lett. 97 (2010) 171909.
[7] M. Portail et al., Graphene 2011 (Bilbao), poster communication.
[8] J. Robinson et al., Graphene 2011 (Bilbao), oral communication.
[9] M.A. Fanton et al., ACS Nano. 5 (2011) 8062-8069.
[10] A. Michon et al., to be published in Mat. Sci. Forum (HeteroSiC-WASMPE 2011 proceedings).
[11] A. Michon et al., to be published in Phys. Stat. Sol. c (ISCS 2011 proceedings).
[12] Th. Seyller et al., Surf. Sci. 600 (2006), 3906-3911.
[13] J. Hass et al., Phys. Rev. Lett. 100 (2008), 125504.
[14] F. Varchon et al., Phys. Rev. B 77 (2008), 165415.
[15] J. Jobst et al., Phys. Rev. B 81 (2010) 195434.
[16] B. Jouault et al., Phys. Rev. B 82 (2010) 085438.
[17] J.L. Tedesco et al., Appl. Phys. Lett. 95 (2009) 122102.

Materials Science Forum Vols. 717-720 (2012) pp 629-632

doi:10.4028/www.scientific.net/MSF.717-720.629

Local Solid Phase Epitaxy of Few-Layer Graphene on Silicon Carbide

Enrique Escobedo-Cousin [a], Konstantin Vassilevski [b,*], Irina Nikitina [c], Nick Wright [d], Anthony O'Neill [e], Alton Horsfall [f], and Jonathan Goss [g]

School of Electrical, Electronic and Computer Engineering, Newcastle University, Newcastle upon Tyne, NE1 7RU, England, United Kingdom

[a] enrique.escobedo-cousin@ncl.ac.uk, [b] k.vasilevskiy@ncl.ac.uk, [c] irina.nikitina@ncl.ac.uk, [d] n.g.wright@ncl.ac.uk, [e] anthony.oneill@ncl.ac.uk, [f] a.b.horsfall@ncl.ac.uk, [g] j.p.goss@ncl.ac.uk, * corresponding author

Keywords: graphene, few-layer graphene, nickel silicide, Raman scattering.

Abstract. Patterned Few-Layers Graphene (FLG) films were grown by local solid phase epitaxy from nickel silicide supersaturated with carbon. The process was realised by annealing of thin Ni films deposited on the carbon-terminated surface of a 6H-SiC semi-insulating wafer followed by wet processing to remove the resulting nickel silicide. Raman spectroscopy was used to investigate both the formation and subsequent removal of nickel silicide during processing. Characterisation of the resulting FLG films was carried out by Raman spectroscopy and Atomic Force Microscopy (AFM). The thickness of the final FLG film estimated from the Raman spectra varied from 1 to 3 monolayers for initial Ni layers varying from 3 to 20 nm thick. AFM observations revealed process-induced surface roughening in FLG films. However, electrical conductivity measurements by Transfer Length Method (TLM) structures confirmed that roughness does not compromise the film sheet resistivity.

Introduction

Graphene, a single atomic layer of hexagonally arranged carbon atoms, is considered as the most likely material of choice for the post-silicon era because of its unique electrical properties [1]. It has been intensively studied and many different methods of graphene formation have been developed. However the two most commonly used methods are (i) exfoliation of highly oriented pyrolytic graphite, which allows to produce mono or bi-layer graphene films [2], and (ii) growth of few-layer graphene (FLG) films on hexagonal polytypes of SiC by silicon sublimation in vacuum [3] or argon [4]. Graphene formation on SiC surfaces yields large area films, suitable for further use in electronic devices, but this method requires annealing at temperatures usually exceeding 1500 °C to provide an efficient sublimation of Si atoms. The FLG thickness strongly depends on the process duration and temperature. Conversely, a controlled amount of silicon can be consumed from the SiC surface by the solid state chemical reaction between SiC and Ni. This reaction occurs at relatively low temperatures, starting at around 600 °C, where silicon sublimation does not occur. Since the solubility of carbon in nickel silicide does not exceed one percent, the excess carbon forms precipitates in relatively thick (of micron scale) nickel silicide films or highly textured graphitic flakes with basal planes parallel to the SiC surface in nanoscale films [5]. We have demonstrated recently that FLG films can be grown at the Ni_2Si/SiC interface and that the nickel silicide can be removed selectively [6]. This method potentially allows the precise control of FLG film thickness and device geometry, and is compatible with standard patterning techniques.

This paper presents results from patterned FLG films grown by local solid phase epitaxy on semi-insulating 6H-SiC substrates from a nickel silicide supersaturated with carbon.

Experimental

Commercial 6H-SiC semi-insulating wafers [7] polished at NOVASiC [8] were used in this study. FLG films were grown on the C-face with on-axis orientation. The samples were cleaned by degreasing with organic solvents followed by RCA cleaning and oxidation in dry oxygen at 1100 °C for 6 h. After oxide removal in buffered oxide etch (BOE), 3 – 20 nm thick nickel layers were deposited by e-beam evaporation at the base pressure of 1×10^{-6} Torr or by magnetron sputtering. Immediately prior to placing the samples in the vacuum chamber, they were immersed in BOE for 30 s followed by rinsing in DI water and drying with a nitrogen jet. Deposited metal films were patterned using a lift-off procedure. To form FLG films, the samples were annealed in a Jipelec JetFirst 200C rapid thermal processing system under high vacuum (10^{-4} mbar) at 1000-1150 °C for 60 - 600 s. Nickel silicide films were stripped by etching in a $HF:HNO_3:CH_3COOH$ (1:3:2) mixture for 2 min at room temperature.

Micro-Raman spectra were collected using a LabRAM HP Raman microscope fitted with an argon ion laser. The spectra were taken at room temperature in the confocal backscattering configuration using an excitation wavelength of 514.5 nm and a 100× objective with 0.9 numerical aperture, resulting in a spot size diameter of around 700 nm. Sample surface topography was studied by optical and atomic force microscopy (AFM) using an XE-150 AFM from Park Systems.

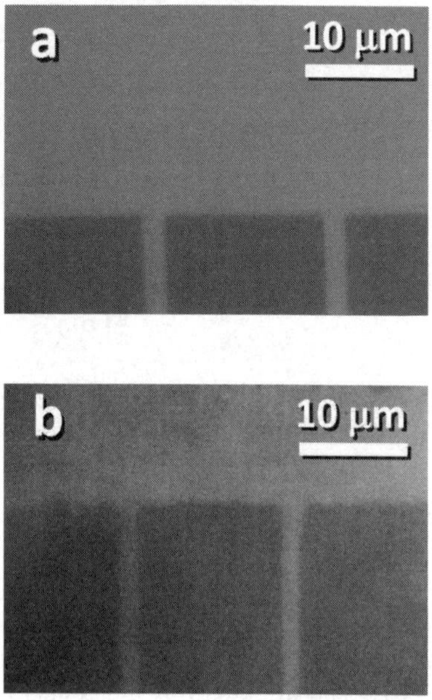

Fig. 1. Optical micrograph of patterned (a) Ni_2Si film (light area) grown on the C-face of a 6H-SiC on-axis substrate by annealing of 7 nm thick Ni at 1010°C for 200 s in vacuum and (b) FLG film (light area) after removal of Ni_2Si. The width of the strip lines is 2 µm.

Results and Discussion

Fig. 1 shows optical micrographs comparing the annealed film topography before and after the removal of nickel silicide. It is clearly seen that the continuous patterned FLG film remains on the SiC surface after Ni_2Si removal. Both he formation and the removal of Ni_2Si were verified using Raman spectroscopy. Fig. 2 compares the Raman spectra of a 20 nm thick Ni film as-deposited (thin, dashed line) with the same film after annealing at 1080 °C for 600 s (thick, solid line). The spectrum corresponding to the annealed film exhibits the characteristic peaks for Ni_2Si, i.e., increased intensity in the 100 cm^{-1} and 140 cm^{-1} bands compared with the as-deposited Ni film, which confirms the Ni_2Si formation. Similarly, the removal of Ni_2Si was also verified by Raman. The spectrum for the FLG film after the Ni_2Si etching is shown by the dotted line. It exhibits low intensity in the 100 cm^{-1} and 140 cm^{-1} bands which confirms the Ni_2Si removal.

The formation of FLG films during the solid-state chemical reaction between silicon carbide and the deposited nickel film was also clearly detected by Raman spectroscopy. Typical Raman spectra taken from the grown FLG films formed from initial layers of Ni with thicknesses of 3, 7 and 20 nm are shown in Fig. 3. The clearly resolved D, G, and 2D bands confirm the graphene formation: the G band at ~1580 cm^{-1} proves the presence of graphitic domains; the intense 2D band located at ~2707 cm^{-1} is attributed to long range ordering in FLG; and the D band at ~1350 cm^{-1} reflects a structural disorder in the films [9]. The minimum thickness of the initial nickel film required to produce a FLG film can be easily calculated using atomic densities of nickel, graphene and silicon carbide. It should also be taken into account that two atoms of nickel form Ni_2Si and release one carbon atom. Assuming that the Ni_2Si and FLG films are continuous and do not form islands and

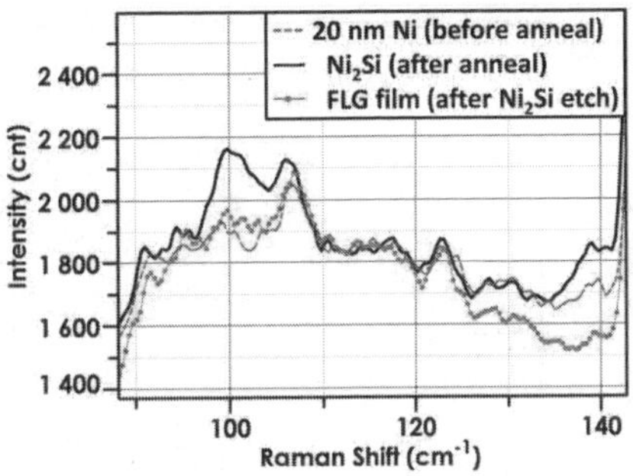

Fig. 2. Raman spectra of a 20 nm thick Ni film as deposited (thin, dashed line), the Ni_2Si formed following a 1080 °C anneal for 600 s in vacuum (thick, solid line), and the final FLG resulting from the Ni_2Si removal (dotted line).

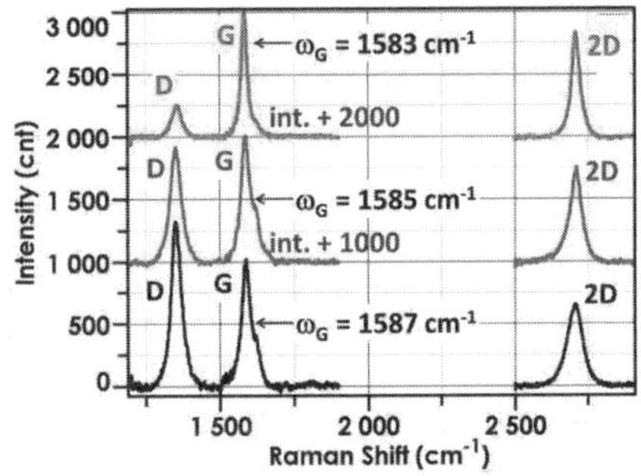

Fig. 3. Raman spectra from the FLG film grown on the C-face of a 6H-SiC on-axis substrate from an initial Ni layer thickness of 3 nm (bottom), 7 nm (middle) and 20 nm (top). Annealing was at 1150 °C in vacuum and then the Ni_2Si was removed. The SiC substrate signal was subtracted.

holes, this estimation gives the minimum required nickel film thickness of 0.84 nm for formation of a single graphene layer. As a result, the FLG films formed from initial 7 and 3 nm thick nickel are expected to contain no more than 8 and 3 graphene monolayers, respectively.

The thickness or number of monolayers (n) constituting each FLG film can be estimated either by the G peak position [10] or by the ratio of the integrated intensities of the G and 2D bands [11]. The spectra in Fig. 3 exhibit a shift of the G band towards higher frequencies with decreasing thickness of the initial Ni film. According to the relation $\omega_G = 1581.6+11/(1 + n^{1.6})$ [10], the number of monolayers found for FLG films with initial Ni layer thickness of 20, 7 and 3 nm were 3 (±1), 2 (±0.5) and 1 (±0.3), respectively.

The surface topography of the FLG films was investigated by AFM. Fig. 4 shows 2.5 × 2.5 μm^2 AFM scans of FLG films produced from initial Ni films of 3, 7 and 20 nm thickness, all annealed at 1080 °C. There is a clear roughening of the FLG surface with increasing thickness of the initial Ni layer, with the rms roughness growing from 2.5 nm in Fig. 4a (3 nm thick initial Ni) to 9 nm in Fig. 4c (20 nm thick initial Ni). Surface roughening occurs at high temperature during the Ni_2Si formation and can result in the creation of the island shaped structures observed in figures 4b and 4c, which may compromise the FLG film continuity.

In order to test the FLG film continuity, electrical conductivity measurements were carried out using transfer length method (TLM) structures with 4 μm wide stripes. The separation between Ni_2Si contact pads varied from 4 to 16 μm. Fig. 5 shows the FLG film resistance as a function of the distance for the roughest FLG film (produced from the 20 nm thick initial Ni, shown in Fig. 4c). Despite the high film roughness, the extracted resistance shows a linear dependence on contact separation, indicating that the film remains continuous. These results suggest that FLG films grown by this procedure can be used to fabricate FLG based electronic devices.

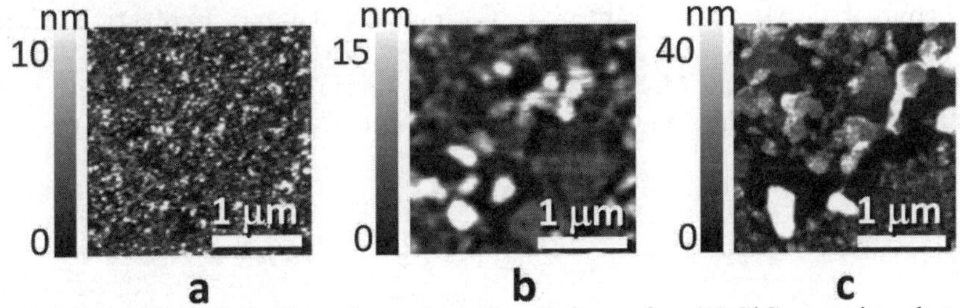

Fig. 4. AFM images of FLG films formed on the C-face of a 6H-SiC on-axis substrate from initial Ni film thicknesses of (a) 3 nm; (b) 7 nm and (c) 20 nm. The Ni layers were annealed at 1080 °C and then the Ni_2Si was removed.

Summary

Patterned Few Layers Graphene (FLG) films were grown by local solid phase epitaxy from nickel silicide supersaturated with carbon. The process was realised by annealing of patterned Ni films deposited on the carbon-terminated surface of semi-insulating 6H-SiC followed by wet processing to remove the resulting nickel silicide. Raman spectroscopy confirmed both the formation and subsequent removal of nickel silicide during processing. Electrical conductivity measurements using TLM structures confirmed the continuity of the FLG films despite the rough surface observed by AFM on films produced from thick initial Ni layers. Raman spectroscopy was also used to assess the FLG film thickness by monitoring the position of the G band in the Raman spectrum. It was observed that the thickness of the final FLG film increased from 1 to 3 monolayers for initial Ni layers varying from 3 to 20 nm thick. Due to its compatibility with conventional device processing schemes, FLG films grown by this procedure can be used to fabricate FLG based electronic devices.

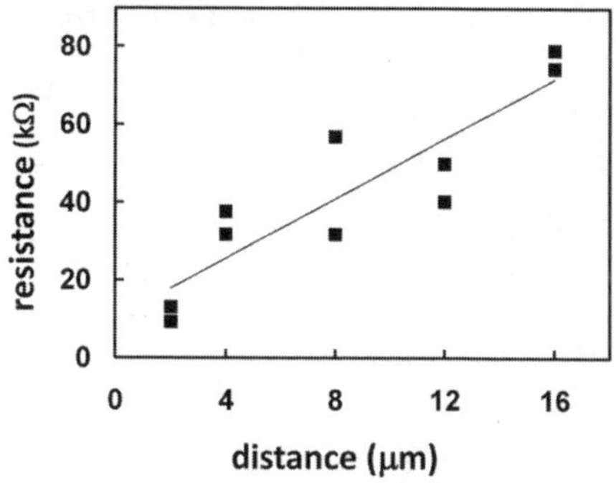

Fig. 5. Resistance of the FLG film measured on TLM structures. The FLG film was formed on the C-face of a 6H-SiC on-axis substrate from a 20 nm thick initial Ni layer, annealed at 1080 °C for 600 s.

Acknowledgment

This work was supported by the Leverhulme Trust (grant F/00 125/AN).

References

[1] A. K. Geim, Science 324 (2009) 1530-1534.

[2] K. S. Novoselov, A. K. Geim, S. V. Morozov, D. Jiang, Y. Zhang, S. V. Dubonos, I. V. Grigorieva and A. A. Firsov, Science 306 (2004) 666-669.

[3] C. Berger, Z. Song, T. Li, X. Li, A. Y. Ogbazghi, R. Feng, Z. Dai, A. N. Marchenkov, E. H. Conrad, P. N. First and W. A. de Heer, J. Phys. Chem. B 108 (2004) 19912-19916.

[4] K. V. Emtsev, A. Bostwick, K. Horn, J. Jobst, G. L. Kellogg, L. Ley, J. L. McChesney, T. Ohta, S. A. Reshanov, J. Rohrl, E. Rotenberg, A. K. Schmid, D. Waldmann, H. B. Weber and T. Seyller, Nat. Mater. 8 (2009) 203-207.

[5] A. Hähnel, V. Ischenko and J. Woltersdorf, Mat. Chem. and Phys. 110 (2-3), (2008), 303-310.

[6] K. Vassilevski, I. P. Nikitina, A. B. Horsfall, N. G. Wright and C. M. Johnson, Mat. Science Forum 645-648 (2010) 589-592.

[7] Information on http://www.tankeblue.com.

[8] Information on http://www.novasic.com.

[9] A. C. Ferrari, J. C. Meyer, V. Scardaci, C. Casiraghi, M. Lazzeri, F. Mauri, S. Piscanec, D. Jiang, K. S. Novoselov, S. Roth and A. K. Geim, Phys. Rev. Lett. 97 (2006) 187401.

[10] H. Wang, Y. Wang, X. Cao, M. Feng and G. Lan, J. Raman Spec. 40 (2009) 1791-1796.

[11] D. Graf, F. Molitor, K. Ensslin, C. Stampfer, A. Jungen, C. Hierold and L. Wirtz, Nano Lett. 7 (2007) 238-242.

Materials Science Forum Vols. 717-720 (2012) pp 633-636

doi:10.4028/www.scientific.net/MSF.717-720.633

Study of epitaxial graphene on non-polar 6H-SiC faces

B.K. Daas[1,a], K.Daniels[1], S.Shetu[1], T.S. Sudarshan[1], and MVS Chandrashekhar[1]

[1]Department of Electrical and Computer Engineering, University of South Carolina

301 S. Main St, Columbia, SC 29208, USA

[a]Daas@email.sc.edu

Keywords: Non-polar, 6H-SiC, Epitaxial graphene

Abstract. We present epitaxial graphene (EG) growth on non-polar a-plane and m-plane 6H-SiC faces where material characterization is compared with that known for EG grown on polar faces. Atomic force microscopy (AFM) surface morphology exhibits nanocrystalline graphite like features for non-polar faces, while the polar silicon face shows step like features. This differing behavior is attributed to the lack of a hexagonal template on the non-polar faces. Non-polar faces also exhibit greater disorder and red shift of all Raman peaks (D, G and 2D) with increasing temperature. This is attributed to decreasing stress with increasing temperature. These variations provide evidence of different EG growth mechanisms on non-polar and polar faces, likely due to differences in surface free energy. We also present differences between a-plane $(11\bar{2}0)$ EG and m-plane $(1\bar{1}00)$ EG in terms of morphology, thickness and Raman characteristics.

Introduction

Graphene, a two-dimensional (2D) form of honeycomb crystal structure, is the basic building block of other sp^2 carbon nanomaterials, such as nanographite sheets and carbon nanotubes that exhibit unusual electronic and optical properties [1]. Since its discovery, graphene has been produced using a mechanical exfoliation technique whereby graphene layers are peeled off layer by layer using scotch tape [2]. This technique is not suitable for large scale production due to poor yield and repeatability, lack of process control and small sample size. As an alternative, large-area epitaxial graphene (EG) is grown by thermal decomposition of polar c-plane (both Si and C-face) of 4H or 6H SiC in ultra high vacuum or argon environment at high temperature. In this technique, Si sublimes off from the SiC substrate, leaving behind carbon atoms which rearrange themselves into a graphene layer [3]. The growth mechanism of EG on SiC is still a current issue of research because it involves high temperatures with several competing process, such as silicon desorption, carbon diffusion, island nucleation, etc. For the polar c-plane Si face, EG growth is due to the step flow mechanism whereby adjacent steps react with different speeds and the released carbon produces characteristics fingers when a step of higher speed catches up with a slower moving step [4]. Though there is argument about the growth mechanism involved on the C-face [5,6], Tedesco et al. [6] have shown that argon mediated EG growth on the C-face is defect mediated, as defects are at a higher energy compared to the surface.

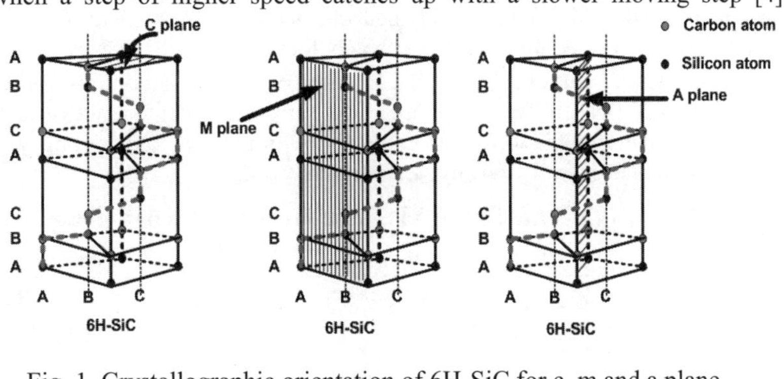

Fig. 1. Crystallographic orientation of 6H-SiC for c, m and a plane, respectively.

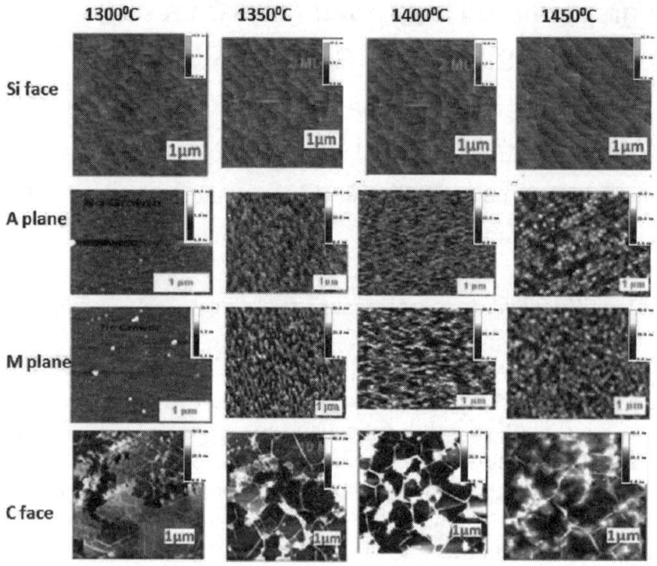

Fig. 2. AFM images of epitaxial graphene grown on a, m and c plane 6H-SiC substrates at three different growth temperatures 1350^{0}C, 1400^{0}C and 1450^{0}C, respectively.

To understand the graphene growth mechanism on SiC and to correlate differences between faces on c-plane SiC, it is essential to study EG growth on a and m plane, EG-a and EG-m (non-polar faces) SiC simultaneously and characterize it in detail. Fig. 1 shows the differences in crystallographic orientation of 6H-SiC a, m and c plane crystal. It is evident that non-polar faces (a and m planes) have more carbon atoms per unit area in a bilayer. These faces also lack a hexagonal template compared to the polar (c plane) Si-face. Difference in crystallographic orientation will be used for explaining graphene morphology and thickness. In this paper we study EG grown on non-polar n-type 6H-SiC, both on a-plane $(11\bar{2}0)$, EG-a, and m-plane $(1\bar{1}00)$, EG-m, faces, by the well known thermal decomposition technique in an inductively heated RF furnace at high temperature and high vacuum.

Experimental Details

Epitaxial growth of large-area graphene by thermal decomposition of commercial <0001> 4H and 6H SiC substrates at high temperature and vacuum [3] can produces EG a few monolayers (ML) to >50 ML thick, depending on growth conditions. In our experiments, EG was grown on a and m plane CMP polished 6H-SiC substrates, nitrogen doped ~10^{17}/cm^3 with rms roughness <0.75nm measured in a 2.5X2.5μm^2 window. The polar c plane Si face was commercially CMP polished with RMS roughness ~0.5nm while the C-face was optically polished with RMS roughness ~1nm. $1\text{cm}\times1\text{cm}$ samples were degreased using Trichloroethylene (TCE), acetone and methanol, respectively. They were then rinsed in DI water for three minutes. The samples were finally dipped in HF for two minutes to remove native oxide and rinsed with DI water before being blown dry with UHP Ar gas. They were then set in the crucible in an inductively heater furnace where high vacuum was maintained (<10^{-6} Torr) and baked out at 1000°C for 13 to 15 hours. We note that no H_2 etching was performed on the SiC surface before EG growth. The temperature was slowly raised to the growth temperature (1300-1450°C), using optimized conditions for c plane growth. In every growth, in addition to the a and m plane sample, c plane witness samples were included for comparison. All growths were performed for 60 minutes before cooling to 1000°C at a ramp rate of 7~8°C/min and eventually to room temperature.

Table 1. EG thickness using XPS both for polar and non-polar faces under same growth conditions.

EG grown at 1350^{0}C	Polar face		Non-polar face	
	Si	**C**	**m-plane**	**a-plane**
No of monolayer	2	9	4	3

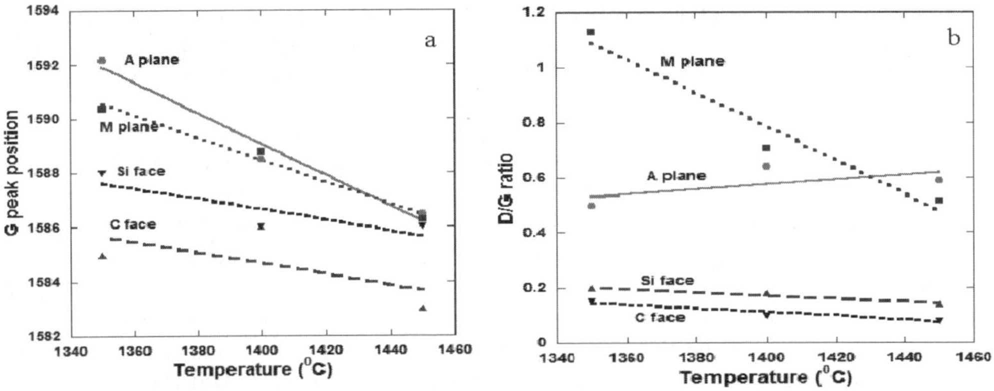

Fig. 3. Variation of a) G peak and b) D/G ratio of polar and non-polar faces as a function of growth temperature. Corresponding polar face data are also presented.

Results and Discussion

Atomic Force Microscopy. After growth, atomic force microscopy (AFM) measurements in tapping mode were carried out on grown EG-a, EG-m and EG-c to analyze the EG surface and RMS roughness was analyzed within a $2.5\times2.5\,\mu m^2$ window for all cases. Fig. 2 shows the AFM of EG-a, EG-m and EG-c grown at temperatures 1300°C to 1450°C. EG-c shows step like features on the silicon face [4] at all growth temperatures (rms roughness ~0.4nm at 1350°C), confirming growth by step flow that become more prominent with increasing growth temperature. Both EG-a and EG-m exhibit nanocrystalline graphite-like features, which also become more prominent with increasing growth temperature. RMS roughness is ~1nm at 1350°C growth temperature. There appears to be an underlying long-range periodicity that may be from steps on the surface. The poorer morphology of the EG-a and EG-m compared to EG-c is believed to originate from the crystallographic orientation of the substrate, which does not provide a hexagonal template. We restrict our discussion in this paper to a comparison of the material quality and do not comment on the growth mechanism.

Raman Spectroscopy. Raman spectroscopy was performed using a micro-Raman setup with laser excitation wavelength at 632nm with a spot size of ~2μm. Raman was used to observe G, D and 2D peaks which are due to in plane-vibrations, disorder and double resonance, respectively [7]. Raman analyses of EG-a and EG-m for different growth conditions are shown in Fig. 3, whereas with the same growth condition, our previous work shows EG-c with the ratios of intensities of D peak and G peak I_D/I_G <0.2, demonstrating the quality of graphene on a c plane [8]. For, EG-m and EG-a, the G (1586 to 1590cm^{-1}), D (1334 to 1348 cm^{-1}) and 2D peaks (2657 to 2678 cm^{-1}) confirm the presence of graphene [7,8].

All peaks are red shifted with increasing growth temperature, likely due to thermal stress relief. Fig. 3 shows a shift in the G peak and variation of the I_D/I_G ratio as a function of growth temperature, where corresponding polar face (Si and C face) data are presented for comparison. In Fig. 3(a) the C face shows a lower G peak red shift compared to the Si face, because of thicker layers grown with less stress. EG-a and EG-m exhibit greater red shifts in the G peak compared to the polar faces with EG-a showing the greatest shift of all faces. This observation, along with the lower D/G ratio (discussed later), which should increase the G-peak position, indicates greater strain in the EG-a surface (Fig. 3(a)) while the strain is lower in EG-m. The greater G-peak positions for both non-polar faces compared to the polar faces are largely due to the significantly higher disorder in these films [7].

The disorder in the graphitized layers is quantified by the relative peak intensities of Raman D and G peaks, (I_D/I_G) [7], called the disorder ratio. Fig. 3(b) shows a comparison of the disorder ratio as a function of growth temperature for both polar and nonpolar faces. Polar face EG has a lower disorder ratio (<0.2) than nonpolar face EG. For the polar faces, the Si face I_D/I_G is 0.1~0.2, while I_D/I_G <0.05 for the C face. The disorder ratio (I_D/I_G) for EG-m decreases from 1.1 to 0.5 as the growth temperature varied from 1350°C to 1450°C, indicating increasing grain size, while I_D/I_G for EG-a does not vary significantly. The reason for this difference is at present unclear, but we speculate that this difference arises from the differences in surface energy [9], leading to different growth mechanisms as evident by the nanocrystalline surface morphology through AFM. Both EG-m and EG-a exhibit much greater disorder than the polar Si-face (<0.1, optimized).

X-ray Photoelectron Spectroscopy. X-ray photoelectron spectroscopy (XPS) measurements were performed using a Kratos AXIS Ultra DLD XPS system with a Al Kα source. Using XPS we obtained the C1s and Si2p peaks, at normal and 70° beam incidence angles. The graphene C1s peak was normalized to the SiC substrate C1s peak, from which the thickness was determined as described elsewhere [10]. Our result indicates that both EG-m and EG-a layers are thicker than their corresponding polar c-plane Si-face sample while EG-m shows thicker growth than EG-a. These thickness differences (Table 1) can be explained from the crystallographic structure shown in Fig. 1. For a c plane, the Si face has one carbon atom at the first bi-layer of the single crystal, whereas for the m and a planes there are five and four, respectively (Fig. 1), to share with the adjacent unit cell. The key here is that we are counting the carbon atom density within a bi-layer, and not only the atoms at the surface. This number is the same on the C-and Si faces, as the arrangement on those faces is the same. The arrangement on the a and m planes is different, leading to a different density of carbon atoms/bi-layer. This indicates higher carbon atom density in a bi-layer of a and m plane compared to Si face. Higher carbon atom density at the surface provides the possibility of more carbon atoms to rearrange themselves as Si sublimes off during EG. This is reflected through the thickness analysis (shown in Table 1), where the c plane Si face has 2 ML whereas the a and m planes have 3 ML and 4 ML, respectively, at 1350°C growth temperature. This thickness difference may also be due to the defective nature of the non-polar EG layers, allowing more Si escape from the surface.

Conclusion

In summary, we have demonstrated EG growth on polar c-plane silicon and carbon faces while non-polar face EG was compared with the polar faces in surface morphology and graphitization quality. We believe that nanocrystalline graphite features on non-polar faces originate from the lack of a hexagonal template in the substrate. Raman measurements show that polar faces exhibit better graphitization quality than non-polar faces. Non-polar face graphitization is thicker than polar Si face as confirmed through XPS. Comparing the non-polar face graphitization, between a and m planes, it is evident that a-plane graphitization is better quality than m plane graphitization, as determined by Raman, while m-plane graphitization is thicker than a-plane. In this paper, we focus on comparing graphitization quality between polar and non-polar faces, rather than the growth mechanism. Further investigation is underway to clarify the growth mechanism on non-polar faces.

Acknowledgement

The authors would like to acknowledge the Southeastern Center for Electrical Engineering Education, and the National Science Foundation (NSF) ECCS-EPDT Grant #1029346 under the supervision of program director Rajinder Khosla for funding this work.

References

[1] A.H. Castro Neto et al., Rev. Mod. Phys. 81 (2009) 109
[2] A.K. Geim and K.S. Novoselov, Nat. Mater. 6 (2007) 183
[3] Jahan M. Dawlaty, et al., Appl. Phys. Lett. 93 (2008) 131905
[4] M. Hupalo, E. Conrad,and M. C. Tringides, arXiv:0809.3619v1
[5] Luxmi, N. Srivastava et al., Phys. Rev. B 82 (2010) 235406
[6] J. L. Tedesco, G. G. Jernigan, et al., Appl. Phys. Lett. 96 (2010) 222103
[7] A.C Ferrari and J. Robertson, Phys. Rev. B 61 (2000) 14095-14107
[8] B.K. Daas et al., J. Appl. Phys. 110 (2011) 113114
[9] Eric Pearson and William A Tiller, Journal of Crystal Growth 70 (1984) 33
[10] P.J. Cumpson, Surf. Interface Anal. 29 (2000) 403

Materials Science Forum Vols. 717-720 (2012) pp 637-640

doi:10.4028/www.scientific.net/MSF.717-720.637

Micro- and Nano-Scale Electrical Characterization of Epitaxial Graphene on Off-Axis 4H-SiC (0001)

F. Giannazzo [1,a,*], C. Bongiorno [1,b], S. Di Franco[1,c], E. Rimini[1,d], V. Raineri[1,e]

[1]CNR-IMM, Strada VIII 5, 95121 Catania, Italy.

[a] filippo.giannazzo@imm.cnr.it, [b] corrado.bongiorno@imm.cnr.it, [c]salvatore.difranco@imm.cnr.it, [d]emanuele.rimini@imm.cnr.it, [e]vito.raineri@imm.cnr.it, *corresponding author

Keywords: Epitaxial graphene, specific contact resistance, mobility, electron mean free path

Abstract. The electrical properties of epitaxial graphene (EG) grown on 8° off-axis 4H-SiC (0001) by annealing at 1600 °C in inert gas ambient (Ar) were studied. The sheet resistance of the EG layers (R_{sh}=740±50Ω/sq) and the specific contact resistance of Ni contacts to EG ($\rho_c \approx 6\times10^{-5}$ Ωcm^2) were evaluated on micrometer scale by measurements on transmission line model (TLM) structures. Si_3N_4 was evaluated as a gate dielectric, showing excellent coverage to EG and a limited effect on its conductivity. The high n-type doping (~10^{13} cm^{-2}) of EG, as well as the field effect mobility (μ) dependence on n were determined using top gated field effect transistors (FETs) with Si_3N_4 gate dielectric. Electron mean free path (l_{loc}) and mobility (μ_{loc}) were also locally determined, on submicrometer scale, by scanning probe microscopy, showing a broad distribution of μ_{loc} values, with the most probable value very similar to the macroscopic carrier mobility μ.

Introduction

Epitaxial graphene (EG) obtained by high temperature decomposition of hexagonal SiC (h-SiC) represents a promising material for the development of future high frequency electronics [1]. The electronic properties of this material have been mainly studied on EG grown on on-axis semi-insulating h-SiC, Si face [2,3] , due to the fact that single or few layers of graphene can be obtained in a controlled way by properly choosing the process temperature and the inert gas pressure [3]. However, in the perspective of future industrial applications, the growth of EG on off-axis 4H-SiC (0001) has been recently investigated, and the number of graphene layers as a function of the annealing temperature was determined in a wide temperature range, from 1600 to 2000 °C [4].

In this paper, a study of the electronic properties of EG grown on 8° off-axis 4H-SiC (0001) at 1600 °C is reported. The sheet resistance of EG and the specific contact resistance of Ni contacts to EG were evaluated on micrometer scale by transmission line model (TLM) measurements. The carrier density (n) as a function of the gate bias and the field effect mobility dependence on n were determined on top gated FETs with Si_3N_4 gate dielectric. Furthermore, electron mean free path and mobility were locally determined, on submicrometer scale, by scanning probe microscopy, showing a broad distribution with the most probable value very similar to the macroscopic carrier mobility.

Experimental Details

Highly doped 8° off-axis 4H-SiC (0001) with low doped (~10^{14} cm^{-3}) epi-layers on top were used as the substrates for graphene growth. Thermal decomposition of SiC was carried out in argon (Ar) ambient in an industrial furnace by Centrotherm Thermal Solutions at 1600 °C. After growth, samples for device fabrication were preliminarily characterized by nondestructive methods, i.e. Raman spectroscopy and tapping-mode atomic force microscopy (AFM) [5], to check graphene formation and its uniformity. AFM measurements of the step height in selectively etched EG at different sample positions, as well as high resolution transmission electron microscopy (HRTEM) on other SiC samples annealed in the same run, were carried out to determine the number of layers in EG [4]. The cross-comparison of all of these methods indicated that EG grown at 1600 °C is made up of three graphene layers on most of the area. TLM and long channel top gated FETs were fabricated as follows. The first step of device fabrication was lateral insulation of individual test structures by selective etching of EG around the effective area by O_2 plasma. Afterwards, metal pads of TLM structures, as well as source and drain contacts of FETs were obtained by lift-off of a Ni/Au bilayer. At this stage of device fabrication, the sheet resistance of EG and the specific contact resistance of Ni/EG interface were determined by TLM measurements. After that, the Si_3N_4 gate dielectric, ~40 nm thick, was deposited by plasma-enhanced chemical vapor deposition (PECVD). It was selectively etched by CHF_3 plasma on the TLM metal pads and on the source and drain contacts of FETs. Top gate contacts were finally obtained by Pt/Au lift-off.

Micro-scale Electrical Characterization

The sheet resistance R_{sh} of graphene layers as well as the specific contact resistance (ρ_c) of Ni contacts to EG was determined by measurements on TLM structures, consisting of a set of Ni pads having identical geometry (width W=200 μm and length L=100 μm) and different spacing d deposited on a laterally insulated rectangular graphene area. Fig.1(a) shows an optical microscopy image of two TLM structures. In Fig.1(b), the I-V curves measured between pads with different separations are reported, showing the Ohmic behavior of Ni/EG contacts. Finally, in Fig.1(c) the resistance R_T between two adjacent metal pads is reported as a function of the distance d. $R_{sh}=740\pm50\Omega/\square$ as well as the contact resistance $R_C=14\pm5\Omega$ were determined by fitting of the data with the linear relation $R_T=2R_C+R_{sh}d/W$. The specific contact resistance ρ_c (independent of pad geometry) can be obtained as $\rho_c= R_CWL_T$, with L_T being the transfer length, i.e. the effective length of graphene under the pads involved in current transfer from the contact to graphene and from graphene to the contact. According to the TLM theory, $L_T^2=(\rho_c/R_{sh}')$, where R_{sh}' is the sheet resistance of graphene under the metal [6]. Due to the charge transfer from metal to the few layers of graphene [7], R_{sh}' can be very different than R_{sh} in EG between the pads. As discussed in the following, EG exhibits an overall n-type doping ($\approx10^{13}$ cm^{-2}) induced by the presence of a buffer layer at the interface with SiC [5]. This doping is spread between the 3 graphene layers (higher in the layer closest to the buffer layer, lower in the topmost one). However, in graphene regions coated by Ni, the work-function difference between the metal and graphene ($W_{Ni}-W_{Gr}\approx1$eV) causes a positive charge transfer to EG [8], leading to a counter doping effect, especially in the topmost graphene layer. This results in an increase of the sheet resistance of EG under the Ni contact. Taking into account these issues, $R_{sh}'=1160\pm50\Omega/\square$, $L_T=2.4\pm0.8\mu m$ and $\rho_c=(6.7\pm4.8)\times10^{-5}$ Ωcm^2 were estimated.

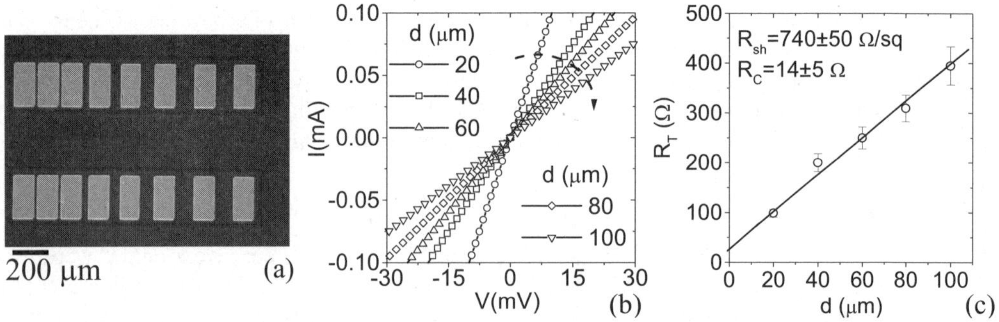

Fig.1 (a) Optical microscopy of two TLM structures. (b) I-V curves measured between pads with different spacings d. (c) Linear fit of the resistance R_T between two adjacent metal pads as a function of d, yielding the sheet resistance of EG and the contact resistance.

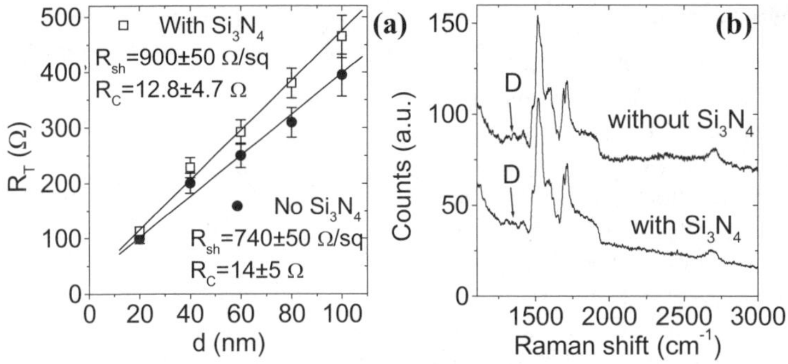

Fig.2 Measurements on TLM structures where the graphene region between metal pads is coated by 40 nm Si_3N_4 (squares) or not coated (dots) (a). Raman spectra collected in the same regions before and after Si_3N_4 deposition (b).

Fig.2(a) shows measurements on TLM structures with EG between metal pads coated by Si_3N_4. The comparison with measurements on structures without the Si_3N_4 layer demonstrates only a limited increase of EG sheet resistance (15-20% with respect to bare EG) due to dielectric deposition. Raman spectra collected in the same regions before and after Si_3N_4 deposition (Fig.2(b)) do not show a significant increase of the intensity in the defect related D band, confirming no structural degradation of EG. Carrier mobility in the EG layer was evaluated using top gated FETs with ~40 nm thick Si_3N_4 gate dielectric. Long channel devices (gate length L_G from 5 to 20 μm) were used to maximize the weight of the channel resistance with respect to contact resistance on the source-drain current I_D. A cross-section schematic of a FET is shown in Fig.3(a), and a top-view optical micrograph of a device with L_G=10 μm and W=100 μm in Fig.3(b). The drain current vs top-gate bias (I_D-V_{tg}) characteristic (Fig.3(c)) exhibit an ambipolar behavior. The minimum conductivity (neutrality point) is shifted to negative gate bias (V_{NP}≈-10 V), indicating n-type doping of EG. The carrier density (*n*) as a function of the gate bias (upper scale in Fig.3(c)) was calculated as $n=\varepsilon_0\varepsilon(V_{tg}-V_{NP})/(qt)$, where ε_0 and ε are the vacuum and Si_3N_4 relative dielectric constants, respectively, *q* the electron charge and *t* the Si_3N_4 film thickness. EG is highly n-type doped ($n\approx10^{13}$ cm^{-2}) at V_{tg}=0 V. This result is consistent with the commonly reported doping values in EG grown on the Si face of SiC and is due to the presence of a positively charged buffer layer at the interface with SiC [5]. The mobility μ vs carrier density in the electron conduction branch (Fig.3(d)) was obtained from the conductivity σ and the carrier density *n* in Fig.3(c), as $\mu=\sigma/(qn)$. $\mu\approx700$ $cm^2V^{-1}s^{-1}$

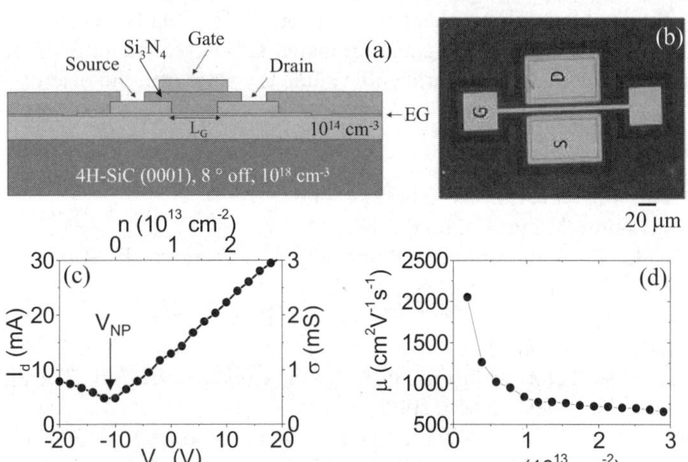

Fig.3 (a) Schematic of the cross-section and (b) optical micrograph of a FET with L_G=10 μm and W=100 μm. (c) Drain current vs top gate bias and carrier density. (d) Calculated mobility μ in the electron branch.

was obtained at V_{tg}=0V, whereas $\mu\approx2000$ $cm^2V^{-1}s^{-1}$ was obtained near the neutrality point.

Nano-scale Electrical Characterization

The mobility values extracted from measurements on FETs are average values from several μm^2 areas and are the result of the combination of different scattering sources inhomogeneously distributed on EG. To get a deeper understanding of the lateral homogeneity of the carrier mobility, the electron mean free path in EG was locally probed by a scanning probe method based on scanning capacitance microscopy [9,10,11].

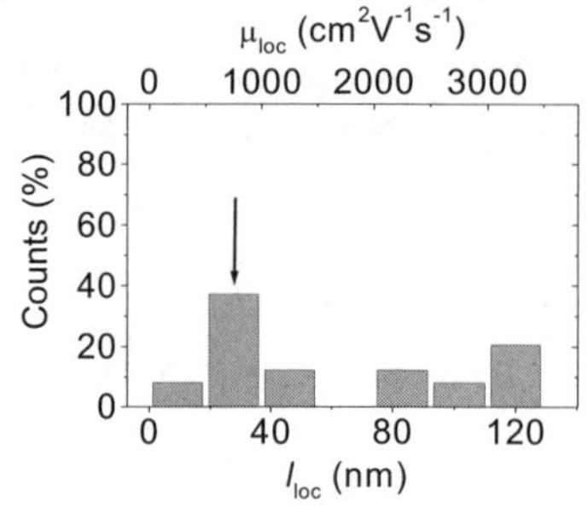

Fig.4 Histogram of the locally probed values of electron mean free path (lower horizontal scale) and mobility (upper horizontal scale) on 10×10 sample positions in 1μm^2 area on bare EG.

Measurements were performed in a bare EG region of 1μm^2 size located between two TLM pads. On that area the tip is displaced on a regular array of 10×10 positions spaced by ~100 nm and for each position the local electron mean free path (l_{loc}) was measured. In Fig.4, lower scale, the histogram of the probed l_{loc} values is reported. A broad distribution, ranging from ~10 to ~120 nm can be observed. Assuming an average carrier concentration $n \approx 10^{13}$ cm^{-2}, the distribution of the local carrier mobility (μ_{loc}) values (upper scale) is calculated from the distribution of l_{loc} values, using the relation $\mu_{loc}=ql_{loc}/\hbar(\pi n)^{-1/2}$ between mobility and mean free path. It is worth comparing the average μ extracted from long-channel FETs with this distribution, to understand how μ_{local} values combine to yield the macroscopic μ. Although this subject deserves further modeling, it can be observed that the most probable μ_{local} value (700±50 cm^{-2}V^{-1}s^{-1}) is very similar to the macroscopic carrier mobility obtained on a FET at V_{tg}=0 V.

Conclusion

Macroscopic measurements on TLM structures were used to determine the sheet resistance of EG on 4H-SiC (0001) and its modification after deposition of a dielectric overlayer. Si_3N_4 was evaluated as a gate dielectric for EG, showing excellent coverage to EG and a limited effect on its conductivity. The specific contact resistance of Ni contacts to EG was evaluated, taking into account the effect of charge transfer from metal to EG. The mobility dependence on the carrier density *n* in EG was obtained using long channel top gated FETs. Finally, electron mean free path and mobility were locally determined on submicrometer scale by scanning probe microscopy, showing a broad distribution with the most probable value very similar to the macroscopic mobility.

References

[1] Y.-M. Lin, et al., Science 327 (2010) 662.
[2] C. Virojanadara, et al., Phys. Rev. B 78 (2008) 245403.
[3] K. V. Emtsev, et al., Nature Mater. 8 (2009) 203.
[4] C. Vecchio, S. Sonde, C. Bongiorno, M. Rambach, R. Yakimova, E. Rimini, V. Raineri, F. Giannazzo, Nanoscale Research Letters 6 (2011) 269.
[5] S. Sonde, F. Giannazzo, V. Raineri, R. Yakimova, J.-R. Huntzinger, A. Tiberj, J. Camassel, Phys. Rev. B 80 (2009) 241406(R).
[6] D.K. Schroder, *Semiconductor Material and Device Characterization*, Third Edition, John Wiley and Sons, Hoboken, New Jersey, 2006.
[7] F. Xia, V. Perebeinos, Y.-M. Lin, Y. Wu, P. Avouris, Nature Nanotech. 6 (2011) 179-184.
[8] P. A. Khomyakov, et al., Phys. Rev. B 79 (2009) 195425.
[9] F. Giannazzo, S. Sonde, V. Raineri, E. Rimini, Nano Lett. 9 (2009) 23-29.
[10] F. Giannazzo, S. Sonde, V. Raineri, and E. Rimini, Appl. Phys. Lett. 95 (2009) 263109.
[11] S. Sonde, F. Giannazzo, C. Vecchio, R. Yakimova, E. Rimini, V. Raineri, Appl. Phys. Lett. 97 (2010) 132101.

Materials Science Forum Vols. 717-720 (2012) pp 641-644

doi:10.4028/www.scientific.net/MSF.717-720.641

Electrical Characterization of the Graphene-SiC Heterojunction

T.J. Anderson[1,a,*], K.D. Hobart[1,b], L.O. Nyakiti[2,c], V.D. Wheeler[2,d], R.L. Myers-Ward[1,e], J.D. Caldwell[1,f], F.J. Bezares[2,g], D.K. Gaskill[1,h], C.R. Eddy, Jr[1,i], F.J. Kub[1,j], G.G. Jernigan[1,k], M.J. Tadjer[3,l], and E.A. Imhoff[1,m]

[1]U.S. Naval Research Laboratory, 4555 Overlook Ave SW, Washington, DC 20375 USA

[2]American Society for Engineering Education, Washington, DC USA (residing at NRL)

[3]Universidad Politécnica de Madrid, Madrid 28040, Spain

[a]travis.anderson@nrl.navy.mil, [b]karl.hobart@nrl.navy.mil, [c]luke.nyakiti.ctr@nrl.navy.mil, [d]virginia.wheeler.ctr@nrl.navy.mil, [e]rachael.myers-ward@nrl.navy.mil, [f]joshua.caldwell@nrl.navy.mil, [g]francisco.bezares.ctr@nrl.navy.mil, [h]kurt.gaskill@nrl.navy.mil, [i]chip.eddy@nrl.navy.mil, [j]fritz.kub@nrl.navy.mil, [k]glenn.jernigan@nrl.navy.mil, [l]mtadjer@die.upm.es, [m]gene.imhoff@nrl.navy.mil, *corresponding author

Keywords: Graphene, p-SiC, n-SiC, heterojunction

Abstract. Graphene, a 2D material, has motivated significant research in the study of its in-plane charge carrier transport in order to understand and exploit its unique physical and electrical properties. The vertical graphene-semiconductor system, however, also presents opportunities for unique devices, yet there have been few attempts to understand the properties of carrier transport through the graphene sheet into an underlying substrate. In this work, we investigate the epitaxial graphene/4H-SiC system, studying both p and n-type SiC substrates with varying doping levels in order to better understand this vertical heterojunction.

Introduction

Graphene is an emerging 2D electrical and optical material. It is now well-known for its unique physical and in-plane electrical transport properties [1,2]. This material is obtained by mechanical exfoliation of highly ordered pyrolytic graphite (HOPG) [3], sublimation of Si from SiC substrates [4,5], or grown by chemical vapor deposition (CVD) on Cu or Ni surfaces [6,7]. In all of these cases, there is only van der Waals bonding between the graphene and the underlying surfaces. The unique ability to either grow graphene directly on a semiconductor (SiC) or transfer layers to alternative semiconductor surfaces presents opportunities for novel device structures. The low, uniform optical absorption properties also make graphene viable as a transparent contact. To date there have been few attempts to understand the properties of electron or hole transport through the graphene sheet into the underlying substrate. It has been shown in the literature, however, that based on a graphene work function of 4.5-5.0 eV, a Schottky contact is formed when graphene sheets are transferred to various semiconductors, such as Si, GaAs, and SiC [8-11]. Furthermore, highly sp^3-bonded, heavily doped nanocrystalline diamond has also been shown to form a Schottky contact when seeded and grown on 4H-SiC [12]. In this work, we present the electrical characterization and analysis of the vertical heterojunction formed between graphene and 4H-SiC.

Experimental

Epitaxial graphene (EG) was formed on the Si-face 4H-SiC substrates using Si sublimation techniques described elsewhere in the literature [4]. The SiC substrates were conductive with either 10 μm p-type ($N_A = 4\times10^{15}$ cm^{-3}) epitaxial layers on p^+ substrates offcut 8° towards [1-102] or 5 μm n-type ($N_D = 1\times10^{16}$, 1×10^{17}, 1×10^{18} cm^{-3}) epitaxial layers on n^+ substrates offcut 4° towards [1-102]. Atomic Force Microscopy (AFM) and Raman spectroscopy were performed to characterize the graphene and estimate the average thickness to be 2 layers based on attenuation of the SiC signal. Device fabrication was performed using an entirely etch-back process to avoid chemical modification of the graphene by organic solvents. Initially, a HF dip was performed to clean and passivate the surface, followed by a blanket e-beam metal deposition (Ti/Al 20/100 nm). The metal was then patterned by etching the aluminum in Transene Al Etch Type A with a photoresist mask, followed by a buffered oxide etch (BOE) of the Ti layer. The photoresist was stripped by a flood exposure and develop step to again avoid the use of organic solvents. Isolated graphene mesas were formed by O_2 plasma etching with the metal as the etch mask. A final lithography and wet etch step was used to further etch the metal to form contact pads. The mesa diameters ranged from 70-520 μm, with metal pads of diameter 20 μm smaller than the mesa. A device cross-section and optical image are shown in Figure 1.

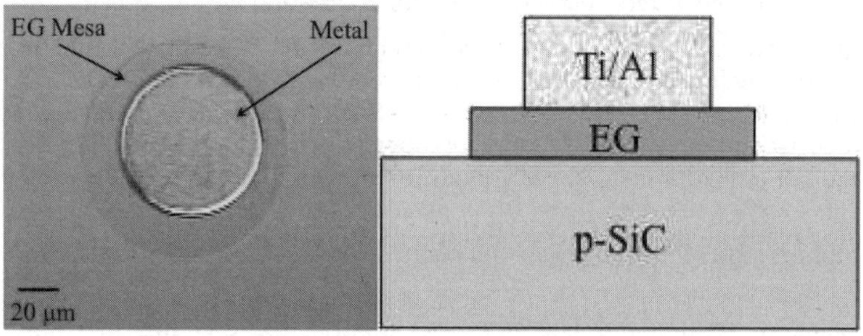

Figure 1. Optical image and device cross-section

Results

The lateral current flow through the graphene layer was probed by the transmission line method (TLM) and van der Pauw analysis of graphene mesas. The relevant data are summarized in Table 1. It was found that the graphene sheet resistance was highly dependent upon the orientation of the TLM pattern, consistent with scanning tunneling microscopy (STM) observations of conductance anisotropy in graphene sheets on SiC [13]. Measuring with the TLM pattern oriented parallel to the steps on the SiC surface, an average R_{SH} of 250 Ω/□was calculated. However, measuring with the pattern oriented perpendicular to the steps an average R_{SH} of 700 Ω/□ was calculated. Finally, measuring with the pattern oriented at a 45° angle to the steps resulted in the precise median value of 475 Ω/□ The average contact resistance for the metal-graphene contact was 1.5 Ω-mm, independent of orientation. Hall measurements on van der Pauw patterns confirmed the average R_{SH} value, as this measurement will always be at an angle to the steps, and indicated an electron mobility of 400 cm^2/V-s and N_{SH} of $1.5x10^{13}$ cm^{-2}. These values were observed in all samples, independent of substrate doping, and are typical of monolayer graphene on a buffer layer.

Table 1. TLM and Hall data with contacts oriented parallel, perpendicular, or at a 45° angle relative to the steps of the SiC surface.

	TLM orientation relative to steps on SiC surface		
	parallel	perpendicular	45° angle
R_{SH} (Ω/□)	253	712	481
R_C (Ω-mm)	2.09	1.37	1.16
μ_n (cm^2/V-s)	373	402	400
N_{SH} (cm^{-2})	$1.91x10^{13}$	$1.82x10^{13}$	$1.75x10^{13}$

Initially, the EG/p-SiC system was studied. A vertical I-V measurement, shown with temperature dependence in Figure 2, demonstrated rectifying characteristics with a large turn-on voltage offset of >1V. Applying thermionic emission theory to this curve, a barrier height of 1.3 eV and ideality factor of 2 was calculated. This is much higher than expected based on the relative electron affinities, and the ideality factor implies recombination, suggesting two-carrier transport. The barrier height increased by less than 0.3 eV over the temperature range 30-210 °C, which is much less than one would expect if thermionic emission dominated, and can be attributed purely to temperature-dependent changes in the Fermi level of the SiC. Furthermore, low temperature I-V measurements have shown carrier freeze-out at ~100K, consistent with the freeze-out temperature of Al-doped p-type SiC. Capacitance-voltage (CV) measurements confirmed the SiC to be p-type with a hole concentration of $4x10^{15}$ cm^{-3}.

To further probe this material system, comparable samples were fabricated with n-SiC epitaxial layers. The vertical IV measurement on n-SiC epi, shown in Figure 2, also demonstrated rectifying characteristics, though with much higher reverse leakage. Furthermore, the current was related to SiC doping, as the higher doped SiC layer demonstrated higher current, and a much stronger temperature dependence was observed. The barrier height was also calculated to be 1.3 eV, but the ideality factor was much lower, around 1.2. This implies that the transport mechanism is thermionic emission. C-V measurements (not shown) confirmed the SiC to be n-type with an electron concentration to be $1x10^{16}$ cm^{-3}.

Electroluminescence (EL) imaging of the EG/p-SiC device, shown in Figure 3, indicates emission around the edge of the graphene mesa in the forward bias region. This is evidence of minority carrier injection into the SiC from the graphene, confirming the 2-carrier transport mechanism hypothesized above. Under reverse bias conditions, the dominant current is electron injection from SiC into grapheme; thus there is negligible current observed on the EG/p-SiC system, while it is expected that there will be a significant leakage current in the EG/n-SiC system. Under forward bias, the dominant current is electron injection from graphene into SiC. EL was not observed in the EG/n-SiC system under any bias condition.

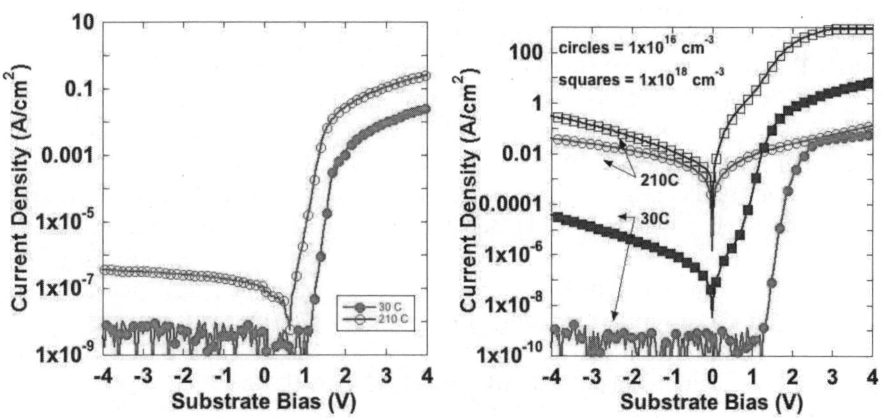

Figure 2. I-V curves from a vertical measurement of the EG/p-SiC (left) and EG/n-SiC (right) systems, with temperature dependence. Bias polarity is relative to the substrate.

Overall, these experimental observations are more consistent with a heterojunction rather than a Schottky-Barrier. In this case, an isotype (n-n) heterojunction is observed in n-SiC, leading to a majority carrier device with thermionic emission as the transport mechanism, demonstrating exponential forward I-V behavior, linear reverse I-V behavior, and a relationship between substrate doping and both forward and reverse current. Furthermore, an anisotype (p-n) heterojunction is observed in the p-SiC case, with the dominant current due to recombination and diffusion in the forward bias region, and reverse leakage due to thermal generation [14]. A complete derivation of the equations governing this system is beyond the scope of this work.

Conclusion

In conclusion, carrier transport in the direction normal to the graphene plane has been studied using the epitaxial graphene on SiC system. The SiC doping was varied between p and n-type, while the graphene was consistently n-type. This enabled the exploration of p-n and n-n junctions. It was determined that graphene on SiC forms a Type I heterojunction rather than a simple Schottky contact.

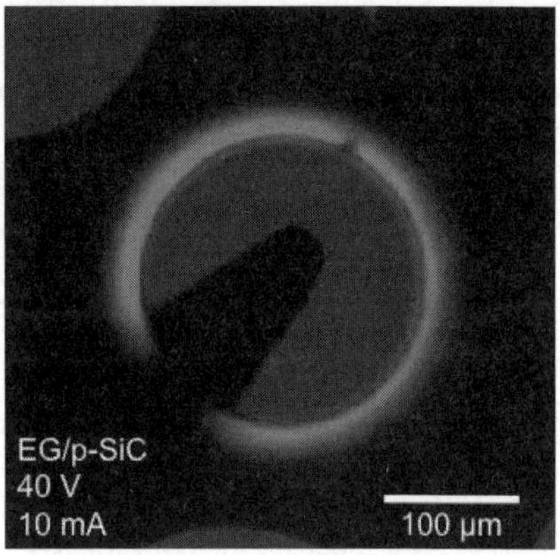

Figure 3. False color EL image of EG/p-SiC under high forward bias.

References

[1] P. Avouris, Z. Chen, and V. Perebeinos, Carbon-based electronics, Nat. Nanotech. 2 (2007) 605

[2] C. Berger, Z. Song, T. Li, X. Li, A.Y. Ogbazghi, R. Feng, Z. Dai, A.N. Marchenkov, E.H. Conrad, P.N. First, and W.A. de Heer, Ultrathin Epitaxial Graphite: 2D Electron Gas Properties and a Route toward Graphene-based Nanoelectronics, J. Phys. Chem. B 108 (2004) 19912

[3] A. Geim and A.H. MacDonald, Graphene: Exploring carbon flatland. Phys. Today 60, 35 (2007)

[4] J.L. Tedesco, B.L. VanMil, R.L. Myers-Ward, J.M. McCrate, S.A. Kitt, P.M. Campbell, G.G. Jernigan, J.C. Culbertson, C.R. Eddy, Jr, D.K. Gaskill, Hall effect mobility of epitaxial graphene grown on silicon carbide, Appl. Phys. Lett. 95 (2009) 122102

[5] P. Sutter, Epitaxial Graphene: How silicon leaves the scene, Nat. Mater. 8 (2009) 171

[6] S. Bae, H. Kim, Y. Lee, X. Xu, J.S. Park, Y. Zheng, J. Balakrishnan, T. Lei, H.R. Kim, Y.I. Song, Y.J. Kim, K.S. Kim, B. Ozyilmaz, J.H. Ahn, B.H. Hong, and S. Iijima, Roll-to-roll production of 30-inch graphene films for transparent electrodes. Nat. Nanotechnol. 5 (2010) 574

[7] Y. Lee, S. Bae, H. Jang, S. Jang, S.E. Zhu, S.H. Sim, Y.I. Song, B.H. Hong, and J.H. Ahn, Vertical Pillar-Superlattice Array and Graphene Hybrid Light Emitting Diodes, Nano Lett. 2010 (2010) 490

[8] S. Tongay, T. Schumann, and A.F. Hebard, Graphite based Schottky diodes formed on Si, GaAs, and 4H-SiC substrates, Appl. Phys. Lett. 95 (2009) 222103

[9] X. Li, H. Zhu, K. Wang, A. Cao, J. Wei, C. li, Y. Jia, Z. Li, X. Li, and D. Wu, Graphene-On-Silicon Schottky Junction Solar Cells, Adv. Mater. 22 (2010) 2743-2748

[10] Chun-Chung Chen, Mehmet Aykol, Chia-Chi Chang, A. F. J. Levi, and Stephen B. Cronin, Graphene-Silicon Diodes, Nano Lett. 2011 (2011) 1863

[11] S. Tongay, T. Schumann, X. Miao, B.R. Appleton, and A.F. Hebard, Tuning Schottky diodes at the many-layer-graphene/semiconductor interface by doping, Carbon 49 (2011) 2033

[12] M.J. Tadjer, K.D. Hobart, J.D. Caldwell, J.E. Butler, K.X. Liu, C.R. Eddy, Jr, D.K. Gaskill, K.K. Lew, B.L. VanMil, R.L. Myers-Ward, M.G. Ancona, F.J. Kub, and T.I. Feygelson, Nanocrystalline diamond films as UV-semitransparent Schottky contacts to 4H-SiC, Appl. Phys. Lett. 91 (2007) 163508

[13] M. Yakes, D. Gunlycke, J.L. Tedesco, P.M. Campbell, R.L. Myers-Ward, C.R. Eddy, Jr, D.K. Gaskill, P.E. Sheehan, and A.R. Laracuente, Conductance Anisotropy in Epitaxial Graphene Sheets Generated by Substrate Interactions, Nano Lett. 2010 (2010) 1559

[14] S. Sze, K. Ng. Physics of Semiconductor Devices, third ed. Wiley, 2007 (125-129)

Materials Science Forum Vols. 717-720 (2012) pp 645-648

doi:10.4028/www.scientific.net/MSF.717-720.645

Epitaxial Single-Layer Graphene on the SiC Substrate

Sergei Yu. Davydov[1, a] and Alexander A. Lebedev[1, b]

[1]Ioffe Physical-Technical Institute, Russian Academy of Sciences, Politekhnicheskaya ul. 26, St. Petersburg, 194021 Russia

[a]sergei_davydov@mail.ru, [b]shura.lebe@mail.ioffe.ru

Keywords: graphene, density of states, Green function.

Abstract. The analytical expression for the density-of-states (DOS) of single-layer graphene interacting with the SiC surface (epitaxial graphene) is obtained. The silicon carbide DOS is described within the scope of the Haldane-Anderson model. It is shown that due to the interaction with the substrate the gap of about 0.01-0.06 eV arises in the epitaxial graphene DOS. The estimation indicates that the electron charge of about (-10^{-3}) *e*/atom transfers from the substrate to graphene.

Introduction

It is well known now that the epitaxial graphene gains some extra electronic charge transferred from the supported substrate. To compensate this charge some special efforts have to be undertaken [1,2]. One way to neutralize graphene is the adsorption of acceptor species onto it. A corresponding theoretical model has been proposed in [3,4]. However, for the description of the SiC substrate the Anderson Hamiltonian has been used, which is more appropriate for metals. Here we take for this purpose the Haldane-Anderson model. Thus, general results obtained below can be applied to any semiconductor substrate.

Model

We begin with the consideration of the isolated single-level carbon atom adsorption on the SiC substrate. In the Haldane-Anderson model the substrate DOS $\rho_s(\omega)$ is given by the following expression: $\rho_s(\omega)=\rho_s$ for $|\omega|\geq E_g/2$ and $\rho_s(\omega)=0$ 0 for $|\omega|< E_g/2$, where ω is energy and E_g is the substrate energy gap. Such a DOS corresponds to the adatom's quasilevel half-width function $\Gamma(\omega)=\Gamma$ for $|\omega|\geq E_g/2$ and $\Gamma(\omega)=0$ for $|\omega|<E_g/2$, while for the adatom's quasilevel shift function $\Lambda(\omega)$ we obtain $\Lambda(\omega)=(\Gamma/\pi)\ln|(\omega-E_g/2)/(\omega+E_g/2)|$ [5].

Now we turn to the isolated carbon adatom Green function $g_a(\omega)=(\omega-\varepsilon-\Lambda(\omega)+i\Gamma(\omega))^{-1}$ and then to the single-layer graphene (SLG) adsorbed on the SiC surface. Assuming that the nearest-neighbor (n.n.) adatoms interact with the hopping amplitude t we find that the epitaxial graphene DOS is

$$\rho_{\pm}(k_x,k_y,\omega)=\pi^{-1}\Gamma(\omega)/[(\omega-\varepsilon_{\pm}(k_x,k_y,\omega))^2+\Gamma(\omega)^2], \qquad (1)$$

where $\varepsilon_{\pm}(k_x,k_y)=\varepsilon\pm tf(k_x,k_y)$ is the electron dispersion for the π^* and π bands correspondingly and

$$f^2(k_x,k_y)=3+2\cos(k_x a\sqrt{3})+4\cos(k_x a\sqrt{3}/2)\cos(3k_y a/2), \qquad (2)$$

is the periodic function of the wave vectors k_x, k_y of the Brillouin zone (BZ). In what follows we need the energy DOS function $\rho_{\pm}(\omega)$, which can be found by summation of the $\rho_{\pm}(k_x,k_y,\omega)$ over the BZ. To obtain the analytical results we use the simplified dispersion function $\varepsilon_{\pm}(\vec{q})=\varepsilon\pm(3ta/2)|\vec{q}|$, where $q=K-k$ and K is the wave vector of the Dirac point and a is the nearest neighbor separation between the graphene adatoms. This simplification permits to find the analytical expressions for the epitaxial graphene DOS $\rho_{\pm}^{c,v}(\overline{\Omega},\xi)$ and $\rho_{\pm}^{g}(\overline{\Omega},\xi)$, where $\xi=(3taq_B/2)$ (q_B is the cut-off vector), $\overline{\Omega}=\omega-\varepsilon-\Lambda(\omega)$, index + (−) corresponds to $\omega>\varepsilon$ ($\omega<\varepsilon$), indexes *c*, *v* denote the energy domains of the conduction and valence SiC bands, while g stands for the substrate gap energy interval.

Epitaxial Graphene DOS

Let us introduce dimensionless quantities: $\overline{\Omega}^*=2\overline{\Omega}/E_g$, $\Omega^*=\omega^*-\varepsilon^*$, $\omega^*=2\omega/E_g$, $\varepsilon^*=2\varepsilon/E_g$, $\xi^*=2\xi/E_g$, $\beta=2\Gamma/E_g$. Then

$$\overline{\rho}_{\pm}^{c,v}(\overline{\Omega}^*)\equiv\rho_{\pm}^{c,v}(\overline{\Omega}^*)\frac{E_g}{2}= =\frac{\beta}{\pi(\xi^*)^2}\left[\ln\frac{(\pm\xi^*-\overline{\Omega}^*)^2+\beta^2}{(\overline{\Omega}^*)^2+\beta^2}+\frac{2\overline{\Omega}^*}{\beta}\left(\tan^{-1}\frac{\overline{\Omega}^*}{\beta}-\tan^{-1}\frac{\overline{\Omega}^*\mp\xi^*}{\beta}\right)\right]. \quad (3)$$

Analysis of the $\overline{\rho}_{\pm}^{c,v}(\overline{\Omega},\xi)$ shows that the increase of the parameter Γ, i.e., the strength of the graphene-SiC interaction, leads to the decrease of the DOS maxima and makes DOS more smooth. It is easy to understand if one approximates $\rho_{\pm}^{c,v}(\overline{\Omega},\xi)$ near the maximum as $\Gamma/[(\omega-\omega^*)^2+\Gamma^2]$, where ω^* is the energy, corresponding to the maximum. More thorough analysis is given in Ref. [6].

As to the $\rho_{\pm}^{g}(\overline{\Omega},\xi)$, its dimensionless form can be written as

$$f_{\pm}^{g}\equiv\rho_{\pm}^{g}(\overline{\Omega})\frac{E_g}{2}=\frac{2\overline{\Omega}^*}{\pi(\xi^*)^2}\left(\tan^{-1}\frac{\overline{\Omega}^*}{s^*}-\tan^{-1}\frac{\overline{\Omega}^*\mp\xi^*}{s^*}\right), \quad (4)$$

where $s^*=0^+$, the minus and plus signs in the last term corresponding to energy $\omega^*>\varepsilon^*$ and $\omega^*<\varepsilon^*$.

Let us show the band gap appearance in the epitaxial graphene DOS. The arctangent can ony be $\pm(\pi/2)$ for positive and negative arguments, respectively. In order for the expression (4) to vanish, it is necessary and sufficient that the inequalities $\overline{\Omega}^*(\overline{\Omega}^*\mp\xi^*)>0$ should be satisfied. In the case $\omega^*>\varepsilon^*$ we obtain the conditions for the band gap appearance in the DOS in the form of the inequalities $\overline{\Omega}^*>\xi^*$ and $\overline{\Omega}^*<0$. In the case $\omega^*<\varepsilon^*$, these conditions are reduced to the inequalities $\overline{\Omega}^*>0$ and $\overline{\Omega}^*<-\xi^*$.

Now we estimate the problem parameters, beginning with the determination of the energy zero coinciding with the substrate midgap, i.e. with the energy $-(\chi+E_g/2)$ relative to vacuum, where χ is the energy of the SiC polytype electron affinity. For the 6H-SiC polytype, we have χ = 3.45 eV and E_g = 3 eV [7], so that $(\chi+E_g/2)$ = 4.95 eV. The work function ϕ_G of epitaxial graphene varies from 4.0 [2] to 5.1 eV [8]. Thus for the intrinsic 6H-SiC, ε ranges from − 0.15 to 0.95 eV, i.e. ε^* ranges from − 0.1 to 0.63. Let us set ϕ_G = 5.1 eV, i.e. ε^* = − 0.1. Hereafter, let β = 0.1 or 1. Since $t\sim3$, then $\xi=(3taq_B/2)\sim4.5(aq_B)$ eV, and we take $\xi^*=3$.

It has been found in [6] that for the value β = 0.1 and ε^* = − 0.1, ξ^*=3 we have three energy domains: (−1, − 0.1), (− 0.1, −0.094), and (− 0.094, 1). In the first domain (−1, − 0.1) the DOS (4) is given by

$$f_{-}^{g}=-\frac{2\overline{\Omega}^*}{(\xi^*)^2}. \quad (5)$$

In the second domain $(-0.1, -0.094)$ the DOS $f_+^g = 0$. In the third domain $(-0.09, 1)$, we have

$$f_+^g = \frac{2\overline{\Omega}^*}{(\xi^*)^2}. \tag{6}$$

At $\beta = 1$, we have the same three domains, but with -0.061 instead of -0.094. Accordingly, in the domains $(-1, -0.1)$, the DOS of epitaxial graphene is given by Eqs. (5) and (6), respectively. In the domain $(-0.061, 1)$, the DOS vanishes.

Thus, for $\varepsilon^* = -0.1$, $\xi^* = 3$, we obtained the band gaps 0.006 and 0.039 for $\beta = 1$ and $\beta = 0.1$, respectively, or passing onto dimensional quantities, $\Delta = \Delta^*(E_g/2) \approx 0.01$ and 0.06 eV. Certainly, these are very narrow band gaps. However, according to various experimental estimations (see, e.g., [9-12]), the band gap Δ varies from 0 to 0.3 eV. For example, in [13], it was shown that $\Delta \approx 0.03$ eV for the graphene layer formed on the Si face of silicon carbide, while $\Delta < 0.01$ eV is for the C face. Thus, the estimations in the present study are quite reasonable.

It can also be shown (see [6]) that with $|\varepsilon^*| << 1$ we have

$$\Delta^* = \pm \frac{2\beta}{\pi + 2\beta} \tag{7}$$

for $\varepsilon^* > 0$ and $\Delta^* < |\varepsilon^*|$, respectively. In the first case the band gap is located at the top of the substrate band gap, while in the second case it is at the bottom of the substrate band gap.

Charge Transfer

Graphene adatom's occupation number n_G is the sum of the contribution n_v from the π band energy domain overlapped with the SiC valence band and the contributions $n_\mp^g$ from the π and π^* bands overlapped with the SiC gap, which are

$$n_v = \int_{-W^*}^{-1} \overline{\rho}_-^v(\omega^*)d\omega^*, \quad n_-^g = \int_{-1}^{\varepsilon^*} f_-^g(\omega^*)d\omega^*, \quad n_+^g = \int_{\overline{\omega}^*}^{0} f_+^g(\omega^*)d\omega^*. \tag{8}$$

Here $-W^*$ is the dimensionless energy of the bottom edge of the graphene π band, $\overline{\omega}^* \approx -0.094$ at $\beta = 0.1$ and $\overline{\omega}^* \approx -0.061$ at $\beta = 1$, 0 is the energy of the Dirac point ε_D. To determine the contribution n_v, numerical calculations are required; the contributions $n_\mp^g$ can be obtained in an analytical form:

$$n_-^g = (\xi^*)^{-2}\{(1+\varepsilon^*)^2 + (2\beta/\pi)[2\ln 2 - (1-\varepsilon^*)\ln(1-\varepsilon^*) - (1+\varepsilon^*)\ln(1+\varepsilon^*)]\}, \tag{9}$$

$$n_+^g(0) = (\xi^*)^{-2}\{\overline{\omega}^*(2\varepsilon^* - \overline{\omega}^*) - (2\beta/\pi)[(1-\overline{\omega}^*)\ln(1-\overline{\omega}^*) + (1+\overline{\omega}^*)\ln(1+\overline{\omega}^*)]\}, \tag{10}$$

where the notation $n_+^g(0)$ emphasizes that the occupation numbers are calculated at the zero Fermi level (see below). Since for the isolated graphene $n_G^0 = 1$ and at low temperatures $n_v + n_-^g \approx 1$, the excess charge, accumulated on graphene per atom is $\delta q \approx -en_+^g$. Estimations for $\varepsilon^* = -0.1$, $\xi^* = 3$ give the surface charge density $\sigma_s = \delta q / S$ as $-0.4 \cdot 10^{-4}$ e/Å^2 at $\beta = 0.1$ and $-0.3 \cdot 10^{-3}$ e/Å^2 at $\beta = 1$ (S is the area per graphene atom). We recall that these estimations are made for the intrinsic semiconductors.

However, in [2] (see also references therein), it was shown that $\varepsilon_F - \varepsilon_D \approx 0.4$ eV. Assuming as before, $\varepsilon^* = -0.1$, we can conclude that $\varepsilon_F^* \approx 0.17$. Since in the case at hand $|\varepsilon^*| << 1$, $|\overline{\omega}^*| << 1$, and $|\varepsilon_F^*| << 1$, we can write approximately

$$n_-^g = (\xi^*)^{-2}\{1 + (4\beta\ln 2/\pi) + 2\varepsilon^* + [1 - (4\beta/\pi)](\varepsilon^*)^2\}, \tag{11}$$

$$n_+^g(0) = [\overline{\omega}^*/(\xi^*)^2]\{2\varepsilon^* - [(4\beta/\pi) + 1](\overline{\omega}^*)\}, \tag{12}$$

$$n_+^g(\varepsilon_F^*) = [(\overline{\omega}^* - \varepsilon_F^*)/(\xi^*)^2]\{2\varepsilon^* - [(4\beta/\pi) + 1](\overline{\omega}^* + \varepsilon_F^*)\}. \tag{13}$$

We also note that $n_-^g \approx 0.10$ and 0.19 per graphene atom at $\beta = 0.1$ and 1 respectively. Hence according tot he above assumption, $n_v \approx$ 0.90 and 0.81 at $\beta = 0.1$ and 1, respectively. In [2], the density of the graphene surface charge is estimated as $0.73 \cdot 10^{13} \text{cm}^{-2}$, which is on average higher than our estimates for $\varepsilon_F^* = 0$ and $\varepsilon_F^* \approx 0.17$ by a factor of 2-3 and ~5, respectively.

Let us now assume that the graphene monolayer is forrmed on the p-type substrate and $\varepsilon_F^* < \varepsilon^*$. Then $n_G = n_v + n_-^g(\varepsilon_F)$, where

$$n_-^g = \int_{-1}^{\varepsilon_F^*} f_-^g(\omega^*) d\omega^* . \tag{14}$$

Substituting (5) into (14), we obtain

$$\begin{aligned} n_-^g(\varepsilon_F^*) &= (\xi^*)^{-2}\{(1+\varepsilon_F^*)(1+2\varepsilon^* - \varepsilon_F^*) + \\ &+ (2\beta/\pi)[2\ln 2 - (1-\varepsilon_F^*)\ln(1-\varepsilon_F^*) - (1+\varepsilon_F^*)\ln(1+\varepsilon_F^*)]\} \end{aligned} . \tag{15}$$

At $\varepsilon_F^* = -0.27$ and $\varepsilon^* = -0.10$, we find $n_-^g \approx 0.10$ at $\beta = 0.1$ and $n_-^g \approx 0.18$ at $\beta = 1$, which, with the accuracy up to the second sign, almost does not differ from the case $\varepsilon_F^* = 0$. This result is explained by the fact that all events occur near the substrate midgap; hence, all parameters (except for ξ^*) appear small.

Summary

We propose here a simple model which permits to achieve the analytical description of the graphene-semiconductor system. The obtained results for the epitaxial graphene gap induced by the SiC substrate and the charge transfer value are in qualitative agreement with the known results found by other authors by numerical calculations. This is the proof of our model's capability.

Our numerical estimations are given for the graphene fabricated just on the 6H-SiC substrate. This is the case for the C-terminated face $(000\bar{1})$ of the substrate. In the case of the Si-terminated face (0001) the intermediate buffer layer between the single-layer graphene and SiC substrate is formed. We suppose, however, that our model can describe this latter situation also provided the corresponding energy parameters for the buffer layer are taken into account.

Acknowledgements

This study was supported by the Russian Foundation for Basic Research (project no. 11-02-0662-a); the program "Quantum Physics of Condensed Matter" of the Presidium of the Russian Academy of Sciences (contract 0.2.740.11.0108); the program "Development of the Scientific Potential of Higher Schools" (2009-2011) of the Ministry of Education and Science of the Russian Federation (project 2.1.1/10269).

References

[1] S.Yu. Davydov and A.A. Lebedev, Tech. Phys. Lett. **36** (2010) 856.
[2] C. Coletti, C. Riedel, D.S. Lee, B. Krauss, L. Patthey, K. von Klitzig, J.H. Smet and U. Starke, Phys. Rev. B **81** (2010) 235401.
[3] S.Yu. Davydov, Semiconductors **45** (2011) 618.
[4] S.Yu. Davydov. Tech. Phys. Lett. **37** (2011) 476.
[5] S.Yu. Davydov and S.V. Troshin: Phys. Solid State **49** (2007) 1583.
[6] S.Yu. Davydov. Semiconductors **45** (2011) 1070.
[7] S.Yu. Davydov. Semiconductors **41** (2007) 718.
[8] A. Mattausch and O. Pankratov, Phys.Rev. Lett. **99** (2007) 076802.
[9] L. Vitalis, C. Riedel, R. Jhmann, I. Brihuega, U. Starke, and K. Kern, Surf. Sci. **602** (2008) L127.
[10] T. Seyller, A. Bostwick, K.V. Emtsev, K. Horn, L. Ley, J.L. McChesney, T. Ohta, J.D. Riley, E. Rotenberg, and F. Speck, Phys. Stat. Sol. (b) **245** (2008) 1436.
[11] O. Pankratov, S. Hensel, and M. Bockstedte, Phys. Rev. B **82** (2010) 121416 (R).
[12] T. Jayasekera, S. Xu, K.W. Kim, and B. Nardelli. Phys. Rev. B **84** (2011) 035442.
[13] S. Kopylov, A. Tzalenchuk, S. Kubatkin, and V.I. Val'ko. Appl. Phys. Lett. **97** (2010) 112109.

Materials Science Forum Vols. 717-720 (2012) pp 649-652
© (2012) Trans Tech Publications, Switzerland
doi:10.4028/www.scientific.net/MSF.717-720.649

Decoupling the Graphene Buffer Layer from SiC(0001) via Interface Oxidation

Markus Ostler[1,a], Roland J. Koch[1], Florian Speck[1], Felix Fromm[1], Hendrik Vita[2], Martin Hundhausen[1], Karsten Horn[2] and Thomas Seyller[1,b]

[1]FAU Erlangen-Nürnberg, Technische Physik, Erwin-Rommel-Str.1, 91058 Erlangen, Germany

[2]Fritz-Haber-Institut der Max-Planck-Gesellschaft, Faradayweg 4-6, 14195 Berlin, Germany

[a]Markus.Ostler@physik.uni-erlangen.de, [b]Thomas.Seyller@physik.uni-erlangen.de

Keywords: graphene, oxidation, intercalation, ARPES, XPS, Raman spectroscopy, Si-face

Abstract. Epitaxial graphene (EG) grown on SiC(0001) resides on the so-called buffer layer. This carbon rich $(6\sqrt{3}\times6\sqrt{3})R30°$ reconstruction is covalently bound to the topmost silicon atoms of the SiC. Decoupling the graphene buffer layer from the SiC interface is a well studied topic since successful intercalation has been shown for hydrogen [1-3]. Recently, intercalation was also shown for oxygen [4, 5]. We present angle-resolved photoemission spectroscopy (ARPES), x-ray induced photoelectron spectroscopy (XPS) and Raman spectroscopy studies to determine the quality of oxygen intercalated buffer layer samples in terms of decoupling and integrity of the transformed graphene layer. The decoupling effect is demonstrated by ARPES measurements showing a graphene-like π band. XPS shows whether the oxidation takes place in the buffer layer or at the interface. Raman spectroscopy is well suited to investigate oxygen induced defects in graphene-like material.

Introduction

High quality epitaxial graphene (EG) can be grown by thermal decomposition of 6H-SiC at elevated temperatures in argon atmosphere [6]. The thickness of the graphene layer grown on the SiC(0001) surface, or Si-face, can be easily controlled to achieve monolayer graphene samples. This is not the case for EG grown on the C-face, where producing a monolayer sample can be very challenging. The disadvantage of EG grown on the Si-face is that it always resides on the buffer layer, which influences the graphene layer on top. This leads, for example, to strong n-type doping of monolayer EG on the Si-face. However, it has previously been shown that the buffer layer can be decoupled from the SiC substrate via hydrogenation of the interface [1-3]. This transforms the buffer layer into quasi-freestanding monolayer graphene on top of SiC(0001). In recent work, it was shown that the chemical modification of the interface can also be achieved via oxidation [4, 5]. This was shown for graphene on ruthenium [4] and for the buffer layer grown with disilane and ethylene on SiC [5].

Experimental

Buffer layer samples were grown by annealing of SiC(0001) at elevated temperatures in argon atmosphere as described elsewhere [7]. Oxygen treatment was carried out using two different approaches: (a) *in situ* annealing at higher temperatures (ca. 750 °C) and lower oxygen pressure (ca. 10^{-4} mbar), or (b) *ex situ* annealing at lower temperatures (ca. 270 °C) and high oxygen pressure (100-1000 mbar). For the *in situ* approach, the sample was heated to ca. 750 °C for 30 min under constant oxygen flow in ultra high vacuum (UHV) conditions. Oxygen was inserted through a leak valve and transported directly to the sample surface via a tube leading to a partial oxygen pressure at the sample which is about two orders of magnitude higher than the chamber pressure. With a chamber pressure of $5\cdot10^{-7}$ torr for 30 min, this results in a dose of $9\cdot10^{4}$ L (Langmuir). *Ex situ* oxygen treatment was carried out in a furnace which can be evacuated and then filled with pure oxygen. Best results, judging by ARPES spectra, were obtained at a temperature of 270 °C and 200 mbar oxygen atmosphere.

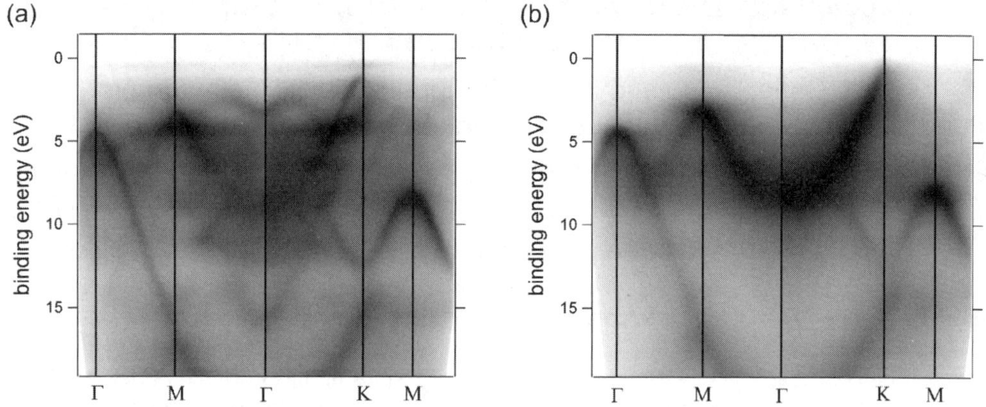

Fig. 1: Band structure measured by ARPES at hν = 65 eV of (a) pristine buffer layer and (b) after in situ oxygen treatment.

We investigated the oxygen treated buffer layer samples by means of ARPES, XPS and Raman spectroscopy. ARPES measurements were carried out at the synchrotron BESSY II with a toroidal analyzer at beamline U125-2/SGM and with a *Specs* PHOIBOS 100 analyzer at beamline UE56-2/PGM-2. XPS data were also taken with these two setups at BESSY II. Additional XPS measurements were carried out in our home laboratory with a *Specs* PHOIBOS 150 analyzer and a monochromated Al-K_α light source. Micro-Raman spectra were measured under ambient conditions using a *Jobin Yvon* T64000 triple spectrometer with a liquid N_2 cooled CCD detector. For excitation a frequency doubled Nd:YVO_4 laser with a wavelength of 532 nm was focused on the sample by a 100 × objective.

Results and Discussion

ARPES data of a pristine buffer layer sample measured using the toroidal analyzer are shown in Fig. 1a. The valence band dispersion, displayed along the $\overline{\Gamma KM}$ and $\overline{\Gamma M\Gamma}$ directions in the graphene Brillouin zone, exhibits typical features of a buffer layer sample, i.e. graphene-like σ bands and discontinuous π states around Γ [8]. In addition there is a weak graphene-like π band visible at the K point. This is probably due to slight overheating during sample preparation leading to a small fraction of the surface (< 5%) being covered by graphene. Fig. 1b shows the valence band dispersion after *in situ* oxygen treatment at a temperature of 750 °C. A prominent π band with a linear dispersion at the K point is clearly visible. This is typical of uncoupled graphene. However, for perfect uncoupled graphene one would expect a sharper π band. All features assigned to the buffer layer disappeared, leading to the suggestion that the buffer layer was transformed into graphene.

Fig. 2 shows XPS C1s core level spectra taken before (a) and after (b) oxygen treatment of the same sample at a photon energy of 384 eV. Deconvolution of the buffer layer spectrum into voigt lines yields three components. The most prominent component S2 at 285.8 eV binding energy can be attributed to carbon atoms of the buffer layer which are only bound in plane to other carbon atoms. The second component of the buffer layer S1 at 285.1 eV belongs to carbon atoms covalently bound to the topmost silicon atoms of the substrate [8]. After treatment, the buffer layer components are drastically weakened and a new asymmetric component at 285.3 eV arises, which can be identified as graphene. The data are best fitted by a Mahan line for the graphene component and three voigt lines for the remaining buffer layer and SiC bulk components. It is interesting to note that the buffer layer was not completely transformed. Treatment at temperatures above 800 °C resulted in significant etching of the graphene. Furthermore, there are no CO groups detectable in the XPS spectra, which would be expected above 288 eV. In the corresponding Si2p core level

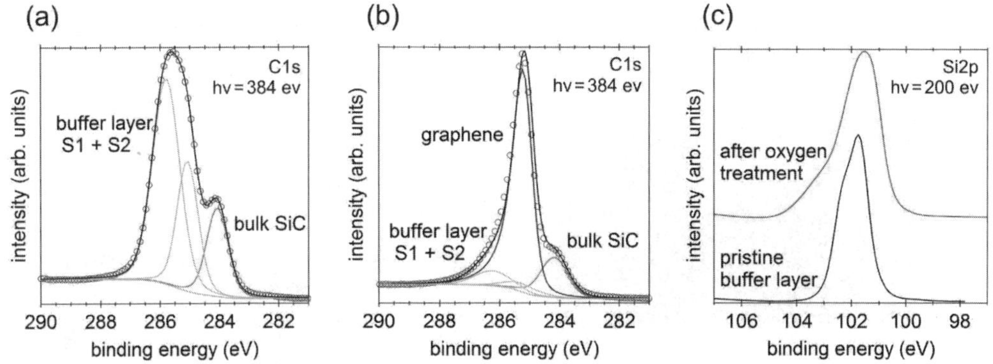

Fig. 2: C1s core level spectra of (a) pristine buffer layer, (b) after oxygen treatment, and (c) Si2p core level spectrum before and after oxygen treatment.

spectrum shown in Fig. 2c a broadening after oxygen treatment is clearly visible. Therefore, we suggest that the oxygen does not directly modify the buffer layer, but oxidizes the silicon at the interface and in this way substitutes the carbon atoms of the Si–C bond resulting in a decoupled graphene layer on top of a SiO_x interface.

Fig. 3a shows the Raman spectra as measured before and after an *in situ* oxygen treatment at a temperature of 600 °C. For better visibility the Raman signal of the bare SiC substrate is subtracted from the measurements in Fig. 3b. The most prominent feature is the defect induced D-peak at 1316 cm^{-1} followed by the G-peak at 1565 cm^{-1}. Assuming point like defects, the distance between defects L_D can be estimated from the intensity ratio I_D/I_G of the D- and G-peak [9] to be about 7 nm. The presence of the D' (1595 cm^{-1}) and D+D' (2896 cm^{-1}) peaks and the weakness of the 2D-Peak at 2629 cm^{-1} are typical of graphene with high defect density and were also seen in with argon ions bombarded exfoliated graphene [9]. The limitations of the oxygen treatment become obvious when the Raman spectrum is compared to a spectrum from a hydrogen intercalated buffer layer sample as shown in Fig. 3b with an offset. The hydrogen intercalated sample was prepared as described in ref. [2] and exhibits a rather small D-peak, a sharp G-peak and a prominent 2D-peak.

Oida et al. [5] proposed a low-temperature (250 °C), high-pressure oxidation process to minimize the introduction of defects to the graphene layer. In our experiments best results, judging by ARPES spectra, were obtained at a temperature of 270 °C and 200 mbar oxygen atmosphere.

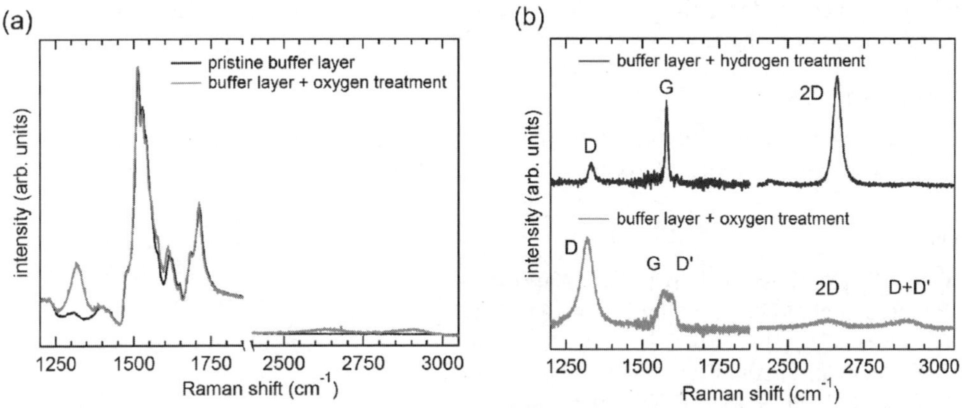

Fig. 3: Raman spectra (a) as measured before and after oxygen treatment and (b) SiC background corrected spectrum of oxygen treated sample. For comparison a spectrum of an hydrogen intercalated sample is also shown in (b) with an offset in intensity.

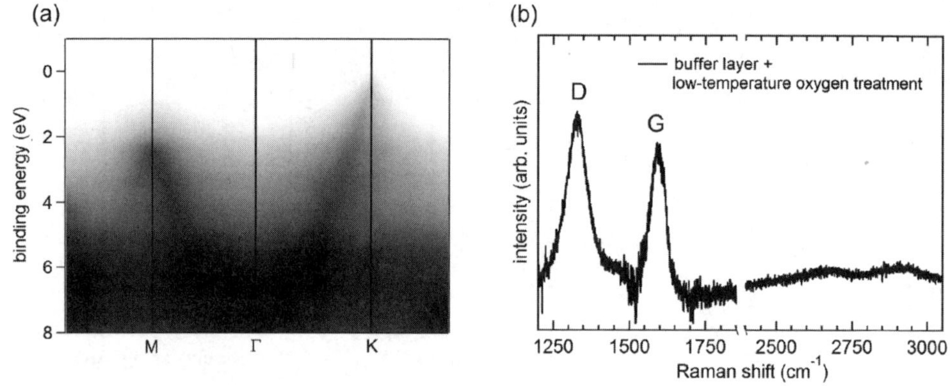

Fig. 4: ARPES spectrum of buffer layer after low-temperature oxygen treatment taken at hν = 120 eV (a) and corresponding Raman spectrum (b).

Fig. 4a shows corresponding ARPES data with a clearly visible graphene-like π band suggesting decoupling of the buffer layer. There is no improvement to the *in situ* process in terms of sharpness of the π band. XPS data (not shown) also supports the transformation of the buffer layer in graphene and the oxidation of the interface, as is the case for the *in situ* process. The Raman spectrum of the *ex situ* treated sample is shown in Fig. 4b. The intensity ratio I_D/I_G is smaller, but the full width at half-maximum (FWHM) of the D-peak is larger as for the *in situ* process. As described in ref. [9], one intensity ratio I_D/I_G corresponds to a higher and a lower defect density depending on the FWHM of the D-peak. Therefore, it is likely that the sample has a higher defect density, although the ratio I_D/I_G is smaller. Based on the bulk to surface intensity ratios in the C1s core level spectra Oida et al. [5] claim that the graphene layer is not chemically attacked by the oxygen at 250 °C. However, our ARPES and Raman data suggest that introduction of defects can not be excluded by XPS data alone. We saw no significant improvement of the low-temperature, high-pressure (*ex situ*) treatment over the high-temperature and low-pressure (*in situ*) treatment.

Conclusion

In summary we showed by means of ARPES and XPS that both presented oxygen treatments decouple the buffer layer by oxidation of the interface transforming the buffer layer into graphene. However, Raman spectroscopy showed that this was not possible without introducing a large amount of defects in the graphene probably due to etching by oxygen. Therefore, the oxygen treatment is inferior to the already well established hydrogen intercalation process.

Acknowledgements

This work was supported by the DFG through contract SE1087/9-1 and by the ESF in the framework of the EUROCORES program EuroGRAPHENE.

References

[1] C. Riedl et al., Phys. Rev. Lett. 103 (2009) 246804.
[2] F. Speck et al., Mater. Sci. Forum 645-648 (2010) 629-632.
[3] C. Virojanadara et al., Surf. Sci. 604 (2010) L4-L7.
[4] P. Sutter et al., J. Am. Chem. Soc. 132 (2010) 8175-8179.
[5] S. Oida et al., Phys. Rev. B 82 (2010) 041411.
[6] K. V. Emtsev et al., Nat. Mater. 8 (3) (2009) 203-207.
[7] M. Ostler et al., physica status solidi (b), 247 (2010) 2924–2926.
[8] K. V. Emtsev et al., Phys. Rev. B 77 (2008) 155303.
[9] L. G. Cançado et al., Nano Lett. 11 (2011) 3190-3196.

Materials Science Forum Vols. 717-720 (2012) pp 653-656

doi:10.4028/www.scientific.net/MSF.717-720.653

Studies of Li intercalation into epitaxial graphene on SiC(0001)

C. Virojanadara[1,a], S. Watcharinyanon[1], A. A. Zakharov[2], R. Yakimova[1] and L. I. Johansson[1]

[1]Department of Physics, Chemistry and Biology, Linköping University, S- 58183 Linköping, Sweden

[2]MAX-lab, Lund University, S-22100 Lund, Sweden

[a] chavi@ifm.liu.se

Keywords: graphene, intercalation, functionalization, doping, electronic structure

Abstract. Detailed studies of Li deposition on monolayer graphene grown on the Si-face SiC surface were performed using LEEM, μ-LEED, PES and ARPES. Li was found to intercalate directly after the deposition at room temperature. However, excess Li was also observed on the surface and found to form a compound with carbon atoms. This compound is suggested to give rise to a new (√3x√3) R30° surface reconstruction. After annealing the (√3x√3) R30° reconstruction vanished and only a (1x1) graphene diffraction pattern was visible. At the same time a severe change was observed in the graphene morphology, especially from the *ex-situ* grown graphene, i.e., extended areas of cracks/wrinkles were observed. These wrinkles/cracks did not disappear even after heating at temperature of 500-1000°C when no Li signal was detected.

Introduction

The recent realization of two-dimensional graphene sheets has raised enormous interest in trying to make use of their peculiar electronic structure. However, the graphene electronic properties are affected by elastic deformations, interlayer coupling and interactions with the substrate. Especially for a graphene sheet prepared epitaxially on a SiC(0001) substrate it is known that the first carbon layer, the so called "buffer layer", bonds covalently to the SiC substrate and generates unsaturated dangling bonds which result in a pinning of the Fermi level [1]. To find different ways than the earlier reports (H, Au [2-4]) to decouple the buffer layer from the substrate is therefore of great interest. Moreover, lithium-graphite intercalation compounds (Li-GIC's) are of interest for the development of rechargeable Li-ion batteries for which the lifetime is particularly crucial. It has been claimed that the graphitic anodes suffer severe surface structural disordering upon prolonged cycling in such batteries, which shortens their lifetime [5-7]. Graphene may be a better material than graphite for such applications. Therefore we pursued the detailed experimental investigation on the influence of lithium (Li) exposures on monolayer graphene grown on the silicon terminated SiC(0001) surface.

Experimental

Our graphene samples were grown on nominally on-axis 6H-SiC(0001) substrates. The growth of homogeneous monolayer graphene on the SiC substrate was performed either by heating the sample *in situ* at a temperature of 1300°C for a few minutes in vacuum or *ex situ* prepared at a temperature of 2000°C in an inductively heated furnace with 1 atm argon ambient for 30 min.[8]. The samples were kept at room temperature and exposed to a Li flux from a SAES getter source for a few minutes. The evolution of the Li intercalation, morphology and electronic structure with annealing temperature was monitored using Low Energy Electron Microscopy (LEEM), micro-Low Energy Electron Diffraction (μ-LEED), Photoemission (PES) and also Angle Resolved Photoemission (ARPES) on beam lines I311 and I4 at the MAX laboratory, Lund, Sweden [8-9].

Results and Discussion

Figs. 1(a)-(b) display LEEM images of an *in situ* prepared graphene layer before and after Li deposition at room temperature, respectively. Small islands were observed directly after Li deposition at room temperature, which were found to coalesce with time and also with heating. The electron reflectivity curves (I-V) collected before and after Li deposition show significant differences. In Fig. 1(c), one minimum is observed at energy of ~ 3 eV. This represents monolayer graphene before deposition [10-12]. The I-V spectrum collected after Li deposition is plotted in Fig. 1(d) and two minima are now observed at energies of around -0.4 and 5 eV, which suggests the presence of two monolayer graphene. It should be noted that the separation of the two minima in Fig. 1(d) appears significantly different compared to those displayed in Figs.1(e) and 1(f) where the signals were collected from the as grown 2 ML sample [11] and from a hydrogen-intercalated bi-layer sample [3], respectively. Slightly shifted minima were observed for the hydrogen intercalated sample but the separation between the minima remained similar to that for the *ex situ* grown bi-layer graphene sample, i.e., cf. Figs. 1(d) and (e). This is, however, not the case after Li deposition where a larger separation between the two minima is clearly observed. We suggest this to be caused by Li atoms penetrating through, and in between, the graphene and buffer layer and into the interface and thereby inducing changes in the electron diffraction path. A lowering of the work function of the sample was also clearly observed by LEEM.

A well ordered (1x1) graphene pattern with the additional $6\sqrt{3}\text{x}6\sqrt{3}$-R30° spots from the buffer layer is normally observed for monolayer graphene, as displayed in Fig. 2(a). Directly after Li deposition the $6\sqrt{3}$ spots can no longer be observed but instead a new reconstructed pattern similar to $\sqrt{3}\text{x}\sqrt{3}$-R30° could be detected as shown in Fig. 2(b). This observation agrees well with the predictions of an earlier theoretical study [13]. This may imply that Li atoms have occupied 1/3 of the π-bonds on the graphene surface, giving rise to this new reconstructed LEED pattern. However, this pattern faded away directly upon heating and then mainly the graphene spots could be observed, as illustrated in Fig. 2(c).

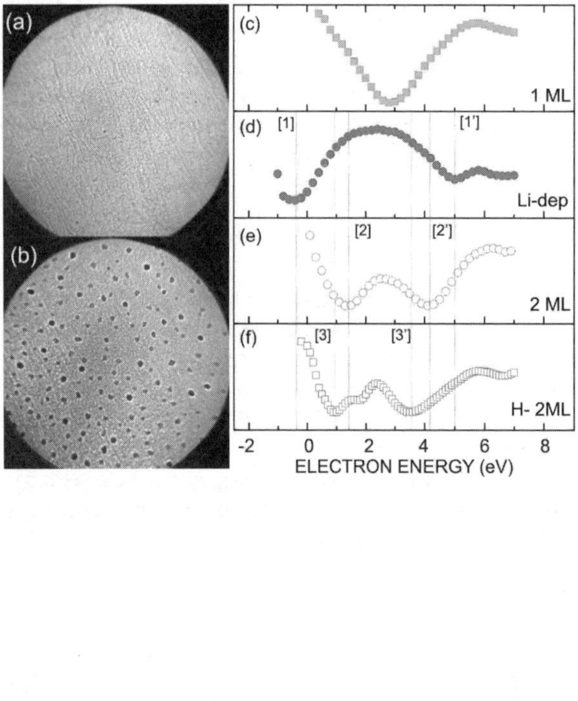

Fig. 1. (a) LEEM image of a monolayer of graphene grown on SiC(0001); the field of view (FOV) is 50 μm and the electron energy is -0.5 eV. (b) after Li deposition with same FOV and the electron energy -1.7 eV. Electron reflectivity curve recorded (c) from 1 ML graphene before Li deposition, (d) after Li deposition, (e) and (f) from a bi-layer *ex situ* grown sample and from a 2 ML sample obtained after hydrogen intercalation, respectively.

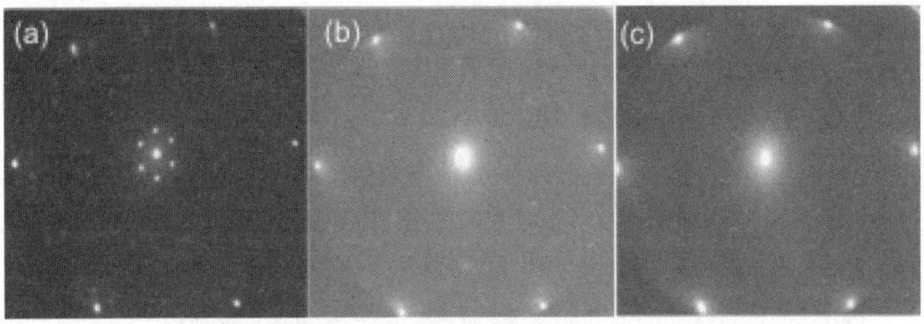

Fig. 2. μ-LEED images collected at Ekin =40 eV from (a) 1ML graphene, (b) after Li deposition and (c) after annealing to 325°C.

The ARPES results obtained from this sample also support the point of view that the Li atoms have penetrated the graphene and buffer layer and have intercalated, as demonstrated in Fig. 3. Before Li deposition the monolayer graphene sample showed a single π-band and a Dirac point at ca. 0.4 eV below the Fermi level as displayed in Fig. 3(a). Of particular interest is that after Li deposition, three π-bands are clearly observed, Fig. 3(b). It is, however, impossible to create tri-layer graphene from only a single sheet of graphene plus a buffer layer. Therefore we suggest that the three π-bands represent a mixture of bi- and mono-layer graphene but with different amount of doping, i.e. an unevenly distribution of Li in the surface region (as indicated by Fig. 1(b)). After heating the sample from 250° to 450°C only two sharp π-bands are observed suggesting that a more uniform intercalation and distribution of Li has been accomplished. In addition, the Dirac point energy (E_D) is after the deposition shifted towards higher binding energy by about 0.25-0.5 eV, depending on the Li coverage. This indicates, as expected, electron doping of the graphene instead of the hole doping observed upon hydrogenation.

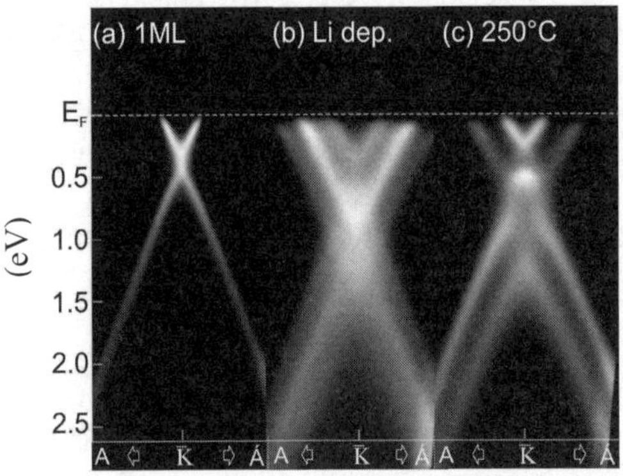

Fig.3. The π band around the K point recorded from (a) monolayer graphene, (b) after Li deposition and (c) after annealing at 250°C. Direction of scan is perpendicular to the Γ-K direction.

Effects of Li were also studied on the monolayer graphene sample prepared by the *ex-situ* method. The graphene sample prepared this way is known to produce a large and homogeneous monolayer graphene sample, as illustrated in Fig. 4(a). The large bright gray area in Fig. 4(a) represents a monolayer graphene and the darker islands represent 2 ML. The fine lines observed on the bright gray area were suggested to correspond to step bunching from the SiC substrate that occurred during the graphene growth procedure [11, 14]. A similar (√3x√3) R30° reconstructed surface was observed directly after Li deposition. However, after annealing at 180°C the morphology of the sample was surprisingly different from those observed using the *in situ* prepared graphene. A significant degradation of the graphene layer was clearly observed as illustrated in Fig. 4(b), where cracks/wrinkles were observed over the whole surface area. This observation may explain the reduction in efficiency of the fuel cell with usage.

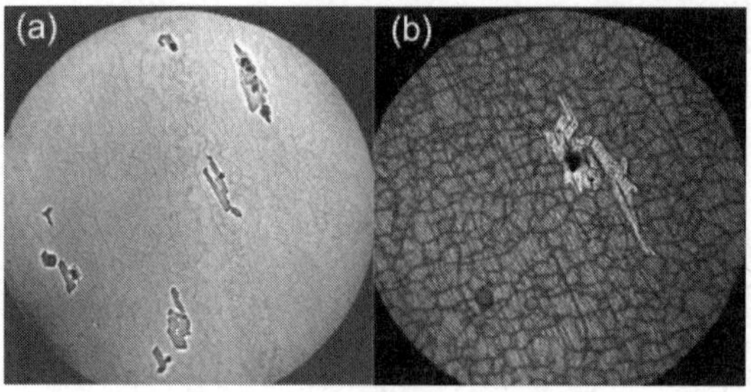

Fig. 4. LEEM images taken from (a) 1 ML graphene (FOV of 50μm and energy -0.38 eV) and (b) after Li deposition and annealing at 180°C (FOV, 20 μm and energy -0.02 eV).

Summary

We have demonstrated an intercalation process of Li at room temperature in monolayer graphene samples prepared epitaxially on SiC(0001). Our results show that Li atoms penetrate through the graphene as well as the carbon buffer layer and intercalate at the interface between SiC and the buffer layer. These Li atoms do at the same time also lift up part of the buffer layer from the substrate and transform it into a second graphene layer. The process starts immediately after deposition. Moreover, Li is suggested to initially bond with graphene π electrons and produce a new reconstructed surface and the (√3x√3) R30° diffraction pattern is observed. More uniform Li distribution and intercalation is obtained after annealing as observed from the two distinct π-bands appearing at the K-point. Li gives rise to electron doping of the graphene and results in a lowering of the Dirac point energy (E_D) by 0.25-0.5 eV depending on Li coverage. The Li intercalated samples were found to be stable up to a temperature of around 450°C. In addition, Li was also found to deteriorate the quality of the graphene layer, especially for the 1 ML graphene sheet prepared by the *ex situ* method. Wrinkles/cracks and defects were clearly seen to form. These wrinkles/cracks did not disappear even after heating at temperatures of 500-1000°C and no Li atoms remained on the substrate. The observed deterioration in quality of the graphene upon Li intercalation may be connected to an explanation why the efficiency in fuel cell reduces with usage.

References

[1]. C. Berger *et al.*, Adv. in Solid State Phys. 47 (2008) 145 and refs therein.
[2]. C. Riedl *et al.*, Phys. Rev. Lett. 103 (2009) 246804.
[3]. C. Virojanadara *et al.*, Surf. Sci. 604 (2010) L4.
[4]. I. Gierz *et al.*, Phys. Rev. B 81 (2010) 235408.
[5]. R. Kostecki *et al.*, J. Power Sources 119 (2003) 550.
[6]. L. J. Hardwick *et al*,. J. Electrochem. Soc. 155 (2008) A442.
[7]. V. A. Sethuraman *et al.*, J. Power Sources 195 (2010) 3655.
[8]. C. Virojanadara *et al.*, Phys, Rev. B 82 (2010) 205402.
[9]. C. Virojanadara *et al*,. New J. Phys.12 (2010) 125015.
[10]. H. Hibino *et al.,* Phys. Rev. B 77 (2008) 075413.
[11]. C. Virojanadara *et al*,. Phys. Rev. B 78 (2008) 245403.
[12]. C. Virojanadara *et al.*, J. Phys. D: Appl. Phys. 43 (2010) 374010.
[13]. M. Farjam *et al.*, Phys. Rev. B 79 (2009) 045417.
[14]. C. Virojanadara *et al.*, Surf. Sci. Lett. 603 (2009) L87.

Materials Science Forum Vols. 717-720 (2012) pp 657-660

doi:10.4028/www.scientific.net/MSF.717-720.657

Plasma-based chemical modification of epitaxial graphene

S.G. Walton[1,a,*], S.C. Hernández [1,b,4], M. Baraket [1,c,4], V.D. Wheeler[2,d,5], L.O. Nyakiti[2,e,5], R.L. Myers-Ward[2,f], C.R. Eddy Jr.[2,g], D. K. Gaskill[2,h]

[1]U.S. Naval Research Laboratory, Plasma Physics Division, Washington, DC 20375, USA

[2]Advanced SiC Epitaxial Research Laboratory, U.S. Naval Research Laboratory, Electronics Science and Technology Division, Washington, DC 20375, USA

[4]NRC/NRL Postdoctoral Research Associate

[5]ASEE/NRL Postdoctoral Research Associate

[a]scott.walton@nrl.navy.mil, [b]sandra.hangarter.ctr@nrl.navy.mil, [c]mira.baraket.ctr@nrl.navy.mil, [d]virginia.wheeler.ctr@nrl.navy.mil, [e]luke.nyakiti.ctr@nrl.navy.mil, [f]rachael.myers-ward@nrl.navy.mil, [g]chip.eddy@nrl.navy.mil, [h]kurt.gaskill@nrl.navy.mil, *corresponding author

Keywords: Epitaxial Graphene, Functionalization, Plasma, Oxygen

Abstract. In this work, the chemical modification of epitaxial graphene on SiC using electron beam generated plasmas produced in mixtures of argon and oxygen is demonstrated. The treatment imparts oxygen functional groups on the surface with concentrations ranging up to about 12 at.%, depending on treatment parameters. Surface characterization shows the incorporation of oxygen leads to a disruption of π-bonds, and an altering of bulk electrical properties.

Introduction

Graphene, the sp^2 bonded monolayer of carbon atoms, has attracted much attention for its unmatched properties and potential in a wide variety of device applications. Epitaxial graphene (EG), derived from the sublimation of Si from SiC substrates, has been demonstrated to produce a continuous single layer graphene sheet with exceptional properties which holds promise for large scale uniform production [1,2,3] and is of particular interest to the microelectronics industry. However, the use of graphene in devices depends strongly on the ability to manage its properties and on methods to engineer a band gap in this material [4]; chemical doping is one such approach [5,6]. The addition of oxygen, for example, can be used to regulate the conductivity of graphene and can induce the formation of a band gap [7].

There are a number of approaches to incorporating functional groups, either within the plane or on the surface of graphene. Of those, plasma processing represents a well-tested, flexible and scalable technique for materials modification. However, the high energy of ions incident to the substrate surface inherent to many plasma-based tools is of considerable concern in the processing of graphene. Indeed, typical ion energies in high-density, low-pressure systems are above 20 eV and the C-C bond strength is about 3.6 eV, while the low energy thresholds for vacancy creation and displacement in graphite are 7.44 eV [8] and 35 eV [9], respectively.

The Naval Research Laboratory has developed a novel plasma processing tool, which utilizes high-energy electron beams to produce the plasma. When high-energy (~ 2 keV) electron beams are used as the primary ionization source, the resultant plasmas have several unique characteristics when compared to conventional low-pressure plasma sources. Relevant to this work is the inherently low electron temperature, which leads to a low plasma potential and thus limits the energy at which ions impact adjacent surfaces. The electron temperature of the electron beam generated plasmas used in our system varies between about 0.4 and 1 eV [10]. Since the plasma potential is typically 4-5 times the value of the temperature (in volts), the corresponding maximum ion energy at adjacent surface is in the range of 2-5 eV. These values are particularly important in the processing of graphene since they are at or below the aforementioned energy thresholds of

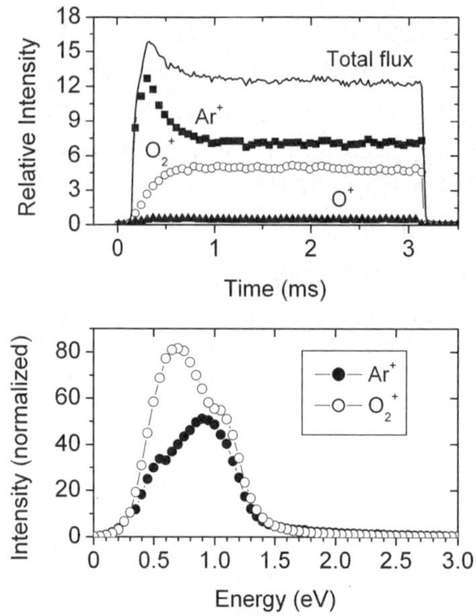

Fig. 1 (Top) Time dependent flux and (Bottom) energy distribution of ions in electron beam generated plasmas produced in Ar/O₂ gas mixtures (50 mTorr).

concern, thereby providing a "gentle" modification of the substrate. In previous work [11], we demonstrated the use of these plasmas to decorate the surface of graphene, produced via CVD on copper surfaces, with both fluorine and oxygen. In this work, we report the use of these plasmas to functionalize the surface of epitaxial graphene derived from 6H-SiC.

Experimental

The processing reactor and its operation have been described previously [11]. Briefly, pulsed, sheet-like, 2 keV electron beams were injected into an argon/oxygen background to produce a similarly shaped plasma. Electrodes for diagnostics or processing can be located adjacent to the sheet plasma. For this work, the duty factor was 10% and oxygen was introduced at 5% of the total flow rate.

Ion flux measurements were made using a mass-resolving, ion energy analyzer (Hiden Analytical). Details of the measurement technique can be found in [12].

For plasma functionalization, EG samples were grown by Si sublimation on the Si-face of 8 x 8 mm^2 semi-insulating 6H-SiC substrates (II-VI, Inc.) [3] and placed on a grounded substrate holder. Beam pulse widths were 2ms (20 ms period) and the total treatment time was 60 s, resulting in a total plasma exposure time of 6 s. Operating pressures were varied between 50 and 90 mTorr.

A Thermo Scientific K-Alpha X-ray photoelectron spectroscopy (XPS) system equipped with a monochromatic Al-source (1486.6 eV) was used for chemical analysis. Raman characterization was performed using an InVia Raman microscope (Renishaw) equipped with a 50x objective lens, a 514.5nm diode laser excitation on 1800-line grating, at a set source power of 20mW. Hall effect characteristics were measured at room temperature in a van der Pauw configuration, with a magnetic field of 2060 G and measurement currents of 1 and 50 μA.

Results and Discussion

Shown in Fig 1 are the time-dependent flux of ions and their energy distributions measured at an electrode located adjacent to the plasmas used for the processing of EG. At these pressures and flow rates, Ar^+ and O_2^+ ions will comprise the majority of ions, along with a small flux of O^+ ions. In addition, fluxes of O and photons are expected. This combination of species will provide not only reactive O but also deliver energy to the surface, which facilitates the formation of oxygen functional groups at the surface of graphene. Importantly, the energy of the ions is not high enough to cause any erosion of the grapheme [11]. Generally, an increase in operating pressure will increase the magnitude of the total flux of all species but it will not change the temporal behavior nor the magnitudes of the ion energies [13].

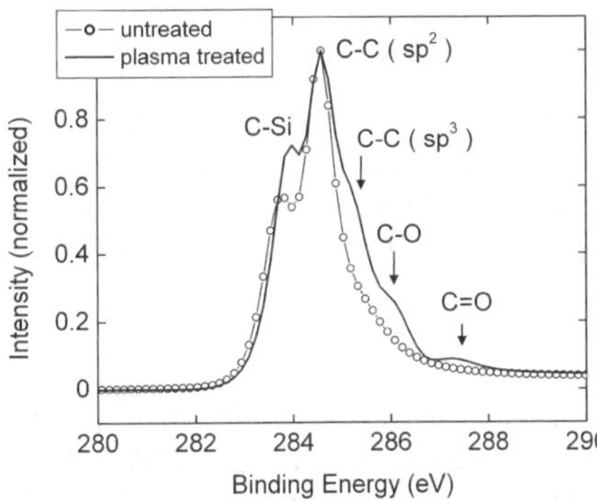

Fig. 2 XPS C1s spectra of untreated (open circles) and plasma treated (solid line) EG. Plasma treatment was performed at 90mT with an Ar/O_2 (5%) plasma.

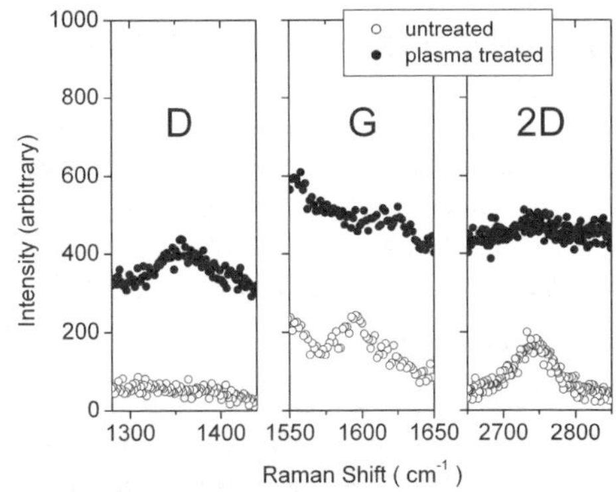

Fig. 3 Snapshots of Raman spectra of EG showing the D, G and 2D modes before (open circles) and after (solid circles) plasma treatment at 90mT.

Chemical and structural changes in EG, due to plasma processing, were tracked using XPS and Raman spectroscopy. As expected, the oxygen content increased with increasing operating pressure, from 3.6 at.% at 50mT to 11.7 at.% at 90 mTorr. A better understanding can be developed by comparing the high-resolution C1s spectra of graphene before and after processing (see Fig. 2). Unprocessed EG was de-convoluted into three components centered at about 283.7eV, 284.5eV and 285.3eV which correspond to (Si-C), (C-C), and the interfacial layer between the SiC and EG, respectively [14]. Following plasma treatment, the C1s spectrum broadened due to the presence of oxygen with peaks corresponding to C-O (286eV) and C=O (287eV). The incorporation of oxygen causes a disruption of sp^2 hybridization leading to the formation of sp^3 hybridization represented by a peak located at about 285.5 eV [11,15].

The details of the binding can be further understood by the Raman spectra, examples of which are shown in Fig. 3. Untreated EG show a weak G peak (≈1600 cm^{-1}) and a 2D peak (≈2730cm^{-1}), the latter is indicative of monolayer graphene [16]. The weak intensity of the G peak is attributed to the background contribution of SiC, which was not subtracted; the 2D peak was easily distinguishable from the background. The D to G ratios (I_D/I_G) of the untreated EG samples range from 0.1 to 0.5 demonstrating defect free domains on the order of 200-300nm, as determined from the equations developed by Tuinstra and Koeing [17,18]. After plasma treatment, the Raman spectra show an increase in the disorder-induced D peak and decreases in the G and 2D peaks.

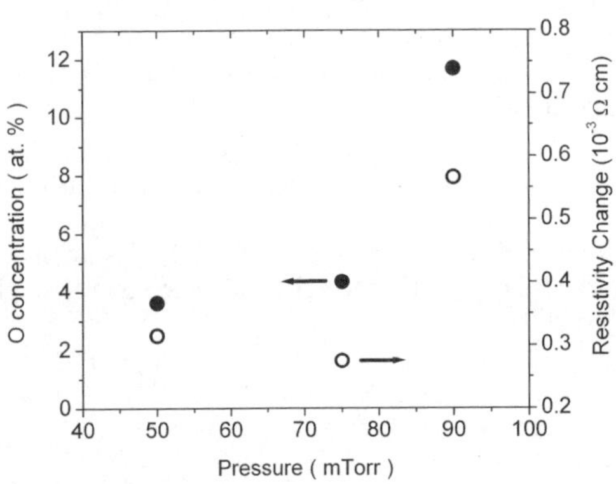

Fig. 4 Dependence of oxygen concentration (solid circles) and change in resistivity (open circles) at different processing pressures.

This addition of oxygen and the subsequent changes in graphene structure increase the resistivity of EG, as can be seen in Fig. 4. Untreated EG samples had an average resistivity of $8.1x10^{-5}$ Ωcm. EG samples with 3.6-4.3 at.% oxygen correspond to a 5-fold increase in resistivity whereas EG with 11.7at.% oxygen correspond to an 8-fold increase in resistivity. The C-O and C=O entities that are introduced serve as scattering sites for the flow of electrons and so the observed increasing resistivity with increasing oxygen content is to be expected.

Summary

Initial results using electron beam generated plasmas produced in Ar/O_2 to functionalize epitaxial graphene on silicon carbide with oxygen are reported. Plasma diagnostic results show a time-dependent ion flux that is characterized by energies less than 2 eV. Oxygen concentrations on the surface of epitaxial graphene can be increased by increasing the operating pressure. Under all treatment conditions, graphene remained, albeit with increased defects and sheet resistance as a function of increasing oxygen concentrations. It is expected that further control over the oxygen concentrations can be realized by controlling the exposure times.

Acknowledgements

This work is supported by the Naval Research Laboratory base program. MB and SCH acknowledge postdoctoral fellowship support from the National Research Council. VDW and LON acknowledge postdoctoral fellowship support from the American Society of Engineering Education.

References

[1] A. K. Geim, Graphene: Status and Prospects Science 324 (2009) 1530-1534.

[2] C. Berger, Z. Song, X. Li, X. Wu, N. Brown, C. Naud, D. Mayou, T. Li, J. Hass, A. N. Marchenkov, E. H. Conrad, P. N. First, and W. A. de Heer, Electronic Confinement and Coherence in Patterned Epitaxial Graphene, Science 312 (2006) 1191-1196.

[3] J. K. Hite, M. E. Twigg, J. L. Tedesco, A. L. Friedman, R. L. Myers-Ward, C. R. Eddy, Jr., and D. K. Gaskill, Conductance Anisotropy in Epitaxial Graphene Sheets Generated by Substrate Interactions, Nano Lett. 11 (2010) 1190-1194.

[4] D. A. Abanin, A. V. Shytov, and L. S. Levitov, Peierls-type instability and tunable band gap in functionalized graphene, Phys. Rev. Lett. 105 (2010) 086802-4.

[5] T. Ohta, A. Bostwick, T. Seyller, K. Horn, and E. Rotenberg, Controlling the Electronic Structure of Bilayer Graphene, Science 313 (2006) 951-954

[6] D. K. Samarakoon and X.-Q. Wang, Tunable band gap in hydrogenated bilayer graphene, Acs Nano 4 (2010) 4126-4130.

[7] M. Wu, C. Cao, and J. Z. Jiang, Light non-metallic atom (B, N, O and F) -doped graphene: a first-principles study, Nanotechnol. 21 (2010) 505202-6.

[8] J.R. Hahn, H. Kang, S. Song, and I.C. Jeon, Observation of charge enhancement induced by graphite atomic vacancy: A comparative STM and AFM study, Phys. Rev. B 53 (1996) R1725–R1728

[9] D. Marton, K.J. Boyd, T. Lytle and J.W. Rabalais, Near Threshold Ion Induced Defect Production in Graphite, Phys. Rev. B 48 (1993) 6757-6766.

[10] S.G. Walton, C. Muratore, D. Leonhardt, R.F. Fernsler, D.D. Blackwell, and R.A. Meger, Electron beam-generated plasmas for materials processing, Surf. Coat. Technol., 186 (1-2) (2004) 40-46.

[11] M. Baraket, S.G. Walton, E.H. Lock, J.T. Robinson, F.K. Perkins, Electron beam generated plasmas for the functionalization of graphene, Appl. Phys. Lett. 96 (2010) 231501-3.

[12] S.G. Walton, D. Leonhardt, and R.F. Fernsler, Time-resolved Diagnostics in a Pulsed, Electron Beam-Generated Plasma, IEEE Trans. Plasma Science, 33 (2005) 838-843.

[13] S.G. Walton,* E.H. Lock, A. Ni , M. Baraket, R.F. Fernsler, D.D. Pappas, K.E. Strawhecker, and A.A. Bujanda, Study of plasma-polyethylene interactions using electron beam generated plasmas produced in Ar/SF_6 mixtures, J. Appl. Polym. Sci. 117 (2010) 3515-3523.

[14] G.G. Jernigan, B.L. VanMil, J.L. Tedesco, J.G. Tischler, E.R. Glaser, A. Davidson III, P.M. Campbell, and D.K. Gaskill, Comparison of epitaxial graphene on Si-face and C-face 4H SiC formed by ultrahigh vacuum and RF furnace production, Nano Lett. 9 (2009) 2605-2609.

[15] N. M. Rodriguez, P. E. Anderson, A. Wootsch, U. Wild, R. Schlogl, and Z. Paal, XPS, EM, and Catalytic Studies of the Accumulation of Carbon on Pt Black, J. Catal. 197 (2001) 365-377.

[16] M. Hundhausen, R. Puesche, J. Roehrl, and L. Ley, Characterization of defects in silicon carbide by Raman spectroscopy, Phys. Stat. Sol. B-Basic 245 (2008) 1356-1368.

[17] A. C. Ferrari, B. Kleinsorge, G. Adamopoulos, J. Robertson, W. I. Milne, V. Stolojan, L. M. Brown, A. LiBassi, and B. K. Tanner, Determination of bonding in amorphous carbons by electron energy loss spectroscopy, Raman scattering and Xray reflectivity, J Non-Cryst. Solids 266 (2000) 765-768.

[18] A. C. Ferrari and J. Robertson, Interpretation of Raman spectra of disordered and amorphous carbon, Phys. Rev. B 61 (2000) 14095-14107.

Materials Science Forum Vols. 717-720 (2012) pp 661-664

doi:10.4028/www.scientific.net/MSF.717-720.661

Evidence of Electrochemical Graphene Functionalization by Raman Spectroscopy

Kevin M. Daniels[1, a, *], B. K. Daas[1], N. Srivastava[2], C. Williams[3], R. M. Feenstra[2], T.S. Sudarshan[1], and MVS Chandrashekhar[1]

[1]Department of Electrical Engineering, University of South Carolina, Columbia, SC 29208, USA

[2]Department of Physics, Carnegie Mellon University, Pittsburgh, PA 15213, USA

[3]Department of Chemical Engineering, University of South Carolina, Columbia, SC 29208, USA

[a]danielkm@email.sc.edu, *corresponding author

Keywords: Graphene, Electrochemistry, Functionalization, Raman

Abstract. Electrochemical functionalization of epitaxial graphene samples on Si-face 6H-SiC are presented in this work. Three semi-insulating 6H-SiC substrates cut from different boules with varying off cut angle (on axis, 0.5° and 1.0° degrees off axis in the [11-20] direction) were diced into 10mm x 10mm samples and quality epitaxial graphene (EG) grown on top. A home-built electrochemical cell was then used with current applied though a 10% H_2SO_4 solution, with a Pt wire and exposed graphene as the anode and cathode, respectively. Functionalization was determined using Raman spectroscopy and measured by an increase in I(D)/I(G) ratio, increase in fluorescence background and introduction of the C-H bond peak at ~2930 cm^{-1}. Components of the Raman spectra before and after functionalization of all samples used were analyzed to show a substrate dependent effect on functionalization with values such as I(D)/I(G) ratio and normalized fluorescence varying between the substrates.

Introduction

Epitaxial graphene has sparked much interest in the materials and device community, boasting such properties as high electrical conductivity, high thermal conductivity and high tensile strength due to the C-C bond, high room temperature mobility, quasi-ballistic transport of carriers, low noise [1-4] and ability to be produced on large areas on commercial SiC substrates [5]. Epitaxial graphene is produced by thermal desorption of Si from SiC in a vacuum [4] or in argon [6,7]. One of the unique properties of graphene is its zero-bandgap, making it a semi-metal. This zero-gap precludes the fabrication of traditional semiconductor devices due to excessive band-to-band leakage.

However, when hydrogen reacts with the π-bonds of graphene, the delocalized π-electron becomes localized at the C-H bond, decreasing the conductivity, changing the bond hybridization from sp^2 to sp^3 and increasing the C-C bond length ~7% from 1.42Å to 1.52Å [8]. This hydride of graphene is called *graphane*. The electron localization at the sp^3 bond leads to a decrease in conductivity, and opens up a bandgap varying from 0-3.5eV depending on the degree of hydrogenation [8]. This enables new applications in bandgap engineered electronics using carbon-based materials. Furthermore, graphane is as thermodynamically stable as comparable hydrocarbons, more stable than metal hydrides and more stable than graphene by ~0.15eV [8]. To form graphane, graphene must react with hydrogen. While bulk graphite has been observed to be chemically inert, graphene has shown enhanced reactivity with atoms such as fluorine [9] and, most importantly, hydrogen. The extent of the reactivity of graphene was quantified by an increase in the Raman D peak resulting in a higher I(D)/I(G) ratio after functionalization [10].

The difficulty with functionalizing graphene is the need for atomic hydrogen, as hydrogen gas H_2 is unreactive. Techniques implemented by other groups involve *in situ* development of atomic hydrogen by plasma-assistance [11] or hot filaments [12], as the H-H bond in hydrogen gas requires high energy/temperature to break [11]. With the limitations of current techniques to functionalize graphene presented, an alternative, electrochemical means is demonstrated in this paper. Electrochemistry offers the most controlled route to systematic functionalization, as the extent of the hydrogenation of graphene can be precisely controlled by changing the duration of the functionalization. Such control is not as easily achievable using previous techniques. Furthermore, through electrochemistry, reactions can be conducted at ambient conditions, as opposed to the harsh environments in other techniques. The convenience and controllability of electrochemical hydrogenation of graphene therefore provides a more realistic approach for a tunable bandgap in graphene. In this paper, atomic hydrogen is generated through an acidic electrolyte (H_2SO_4-sulfuric acid), giving free H^+ ions that can then be easily attracted to an appropriately biased negative graphene cathode for functionalization.

Experimental

Three 6H SiC semi-insulating, on-axis wafers were obtained from II-IV, Inc. for this study. The Si-face was chemical-mechanical polished (CMP). 10 x 10 mm^2 samples were diced from these wafers and thoroughly cleaned using a standard RCA clean (TCE, acetone, methanol) and HF to remove any native oxide. Growth of EG was performed by thermal decomposition of a SiC substrate *in vacuo* [13], $<10^{-5}$ Torr, using an RF furnace. Temperatures for growth ranged form 1250-1450°C, with quality of EG growth verified with atomic force microscopy (AFM) and Raman, showing I(D)/I(G) ratios < 0.2, demonstrating the high quality of the starting material [14,15]. EG layer thicknesses were estimated by x-ray photoelectron spectroscopy, Raman spectroscopy [16] and infrared transmission measurements [17]. For this study only Si-face EG growths were functionalized. EG grown on the non-polar faces of SiC, a and m-plane, which have been shown to have higher I(D)/I(G) ratio than EG grown on polar faces, were used to demonstrate that fluorescence is not a result of large I(D)/I(G) ratio but a result of functionalization.

Atomic hydrogen was generated using a home-built electrochemical setup with current applied though a 10% H_2SO_4 acid solution. A 99.6% Pt wire and exposed EG (approximately a 4 mm diameter circular area) were used as the anode and cathode, respectively. With this setup, we believe H^+ ions in the H_2SO_4 electrolyte are attracted to the exposed graphene. Oxidation occurs at the Pt anode during this process, and has been investigated and discussed in greater detail elsewhere [18,19]. Thus, in this investigation, we limit ourselves to the reduction of the graphene cathode. Functionalization was confirmed using graphene's and functionalized graphene's known Raman peaks [14], and changes in the fluorescence background in the Raman spectra [15].

Raman spectroscopy was performed using a micro-Raman setup with laser excitation wavelength at 632 nm with a spot size of ~2 μm. The Raman system was calibrated using the known Si peak at 520.7 cm^{-1}. Reference blank substrate spectra were scaled appropriately and subtracted from the EG spectra to show only the graphene and functionalized graphene peaks [11,15]. All the spectra shown in this paper are difference Raman spectra obtained in this manner. Raman was used as an indication of possible functionalization and reversal by the behavior of the D, G and 2D peaks, which correspond to the disorder-induced peak, in-plane vibrations and double resonant [16], respectively, as well as a C-H bond peak introduced at ~2930 cm^{-1} [12]. The D peak is expected to increase as a function of hydrogenation [10, 11] and further support sp^3 hybridization [8, 10, 11].

Substrate dependence was investigated by taking sets of three or more samples from each of the three wafers and functionalizing them using the described experiment. D peak position, the ratio of the D and G peak intensities or I(D)/I(G) ratio, and normalized fluorescence, slope(m)/I(G), where slope(m) is defined as the absolute value of the rate of change of the baseline over the entire Raman spectra before SiC substrate subtraction, were recorded from the pristine EG and functionalized graphene. Samples grown on the a and m planes of SiC, not functionalized, were used to show that the resulting "disorder" shown in the functionalized graphene, which signifies functionalization, was significantly different than that of damaged or nanocystalline graphite [15].

Results

Fig. 1 shows that the I(D)/I(G) ratio has little effect on the D-peak position in unfuctionalized graphene and non-polar face grown graphene, but functionalized samples are red-shifted with respect to their pristine state, with substantially larger I(D)/I(G) ratios. In Fig. 2, when comparing the D-peak position and the normalized florescence slope, the trends in the converted and unconverted graphene are entirely different with much higher slopes seen in the converted samples. Fig. 3 shows a similar trend as Fig. 2, comparing I(D)/I(G) with the normalized fluorescence. As expected, fluorescence is only seen in the functionalized samples as it is only present in hydrocarbons, not in pure carbon. The effects of functionalization are shown to be substrate dependent in Table 1.

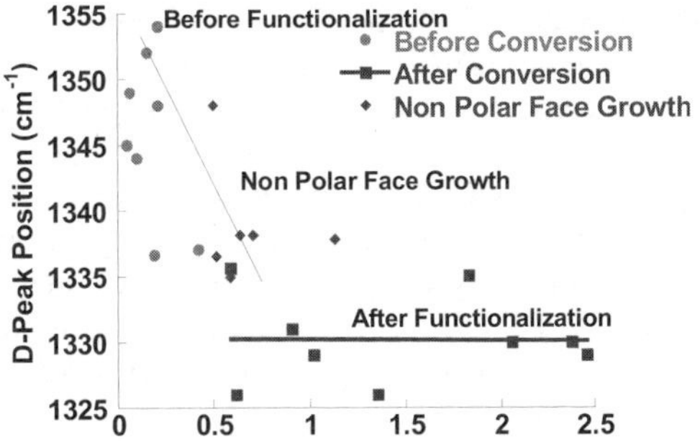

Figure 1: D-peak position compared to the I(D)/I(G) ratio of the Raman spectra of pristine EG, functionalized graphene and nanocrystalline graphene (non-polar face a/m plane growth). Functionalized graphene shows significant red shift in the D-peak position and I(D)/I(G) ratios as high as 2.5.

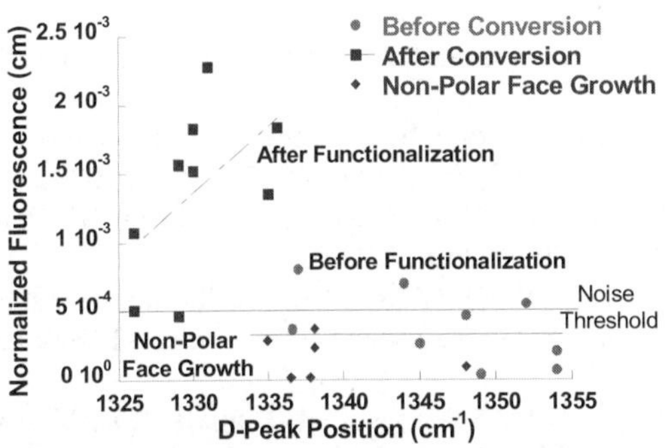

Figure 2: Normalized fluorescence, slope(m)/I(G) compared to D-peak position of the Raman spectra of pristine EG, functionalized graphene and nanocrystalline graphene (non-polar face a/m plane growth). Increasing normalized fluorescence corresponds with blue-shifting of D-peak position among functionalized graphene. Pristine EG and nanocrystalline normalized fluorescence remain below noise threshold.

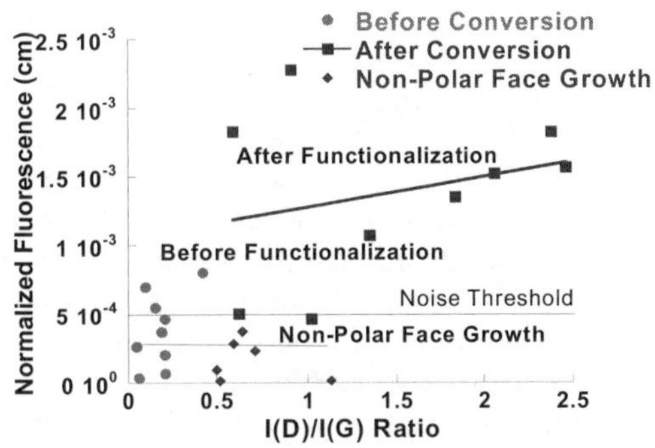

Figure 3: Normalized fluorescence, slope(m)/I(G) compared to I(D)/I(G) ratio of the Raman spectra of pristine EG, functionalized graphene and nanocrystalline graphene (non-polar face a/m plane growth). Increasing I(D)/I(G) ratio matches increasing normalized fluorescence while pristine EG and nanocrystalline remains below the noise threshold.

Table 1: Average Raman Parameters from Each Substrate before and after Functionalization

Substrate	D-peak Position Before (cm^{-1})	D-peak Position After (cm^{-1})	I(D)/I(G) ratio Before	I(D)/I(G) ratio After	Normalized Fluorescence Before (cm)	Normalized Fluorescence After (cm)
SI (1° off)	1348	1330.25	0.21	1.91	3.66×10^{-04}	1.44×10^{-03}
SI2 (on-axis)	1344	1332	0.17	1.32	4.24×10^{-04}	1.89×10^{-03}
SI3 (0.5° off)	1346.5	1327.5	0.13	0.82	3.65×10^{-04}	4.88×10^{-04}

Conclusion

An electrochemical process was developed with the purpose of functionalizing graphene grown on SiC substrates. This was done to try to induce a bandgap in the material. Raman spectroscopy was used to characterize the material before and after functionalization. The Raman spectra showed an increase in a sharp D peak, demonstrating possible surface functionalization, introduction of a peak at 2930cm^{-1}, indicating formation of C-H bonds, and D and 2D peak shifts from pristine EG to functionalized graphene, indicating likely sp^3 hybridization. These claims were distinguished from lattice damage by the presence of the fluorescence background in the functionalized graphene in addition to the lack of fluorescence found in the "disordered" EG grown on a and m-planes of SiC despite having comparably high I(D)/I(G) ratios to functionalized graphene on the Si-face.

By functionalizing samples from three different substrates, substrate dependence was observed based on varying I(D)/I(G) ratios and normalized fluorescence between each substrate with evidence pointing to differences between off-cut angle in the [11-20] direction of SiC as well as variation in off-cut angle in other directions. While evidence of hydrogenation has been shown by Raman, further study is needed to say graphane has been formed.

References

[1] W.A de Heer, C. Berger, X. Wu, P. N. First, E. H. Conrad, X. Li, T. Li, M. Sprinkle, J. Hass, M. L. Sadowski, M. Potemski, G. Martinez, Solid State Commun. 143 (2007) 92-100

[2] A.K. Geim, K.S. Novoselov, Nat. Mater. 6 (2007) 183-191

[3] A.H. Castro Neto, F. Guiena, N.M.R. Peres, K.S. Novoselov, A.K. Geim., Rev. Mod. Phys. 81 (2009) 109-162

[4] J.L. Tedesco, B.L. VanMil, R.L. Myers-Ward, J.M. McCrate, S.A. Kitt, P.M. Campbell, G.G. Jernigan, J.C. Culbertson, C.R. Eddy, D.K. Gaskill., Appl. Phys. Lett. 95 (2009) 122102

[5] G.G. Jernigan, B.L. VanMil, J.L. Tedesco, J.G. Tischler, E.R. Glaser, A. Davidson III, P.M. Campbell, D.K. Gaskill., Nano Lett., Vol. 9, No. 7 (2009) 2605-2609

[6] K.V. Emtsev, A. Bostwick, K. Horn, J. Jobst, G.L. Kellogg, L. Ley, J.L. McChesney, T. Ohta, S.A. Reshanov, J. Röhrl, E. Rotenberg, A.K. Schmid, D. Waldmann, H.B. Weber, T. Seyller, Nature Mater. 8 (2009) 203

[7] J.L. Tedesco, G.G. Jernigan, J.C. Culbertson, J.K. Hite, Y.Yang, K.M. Daniels, R.L. Myers-Ward, C.R. Eddy, J.A. Robinson, K.A. Trumbull, M.T. Wetherington, P.M. Campbell, and D.K. Gaskill, Appl. Phys. Lett. 96 (2010) 222103

[8] Jorge O. Sofo, Ajay S. Chaudhari, Greg D. Barber, Phys. Rev. B. 75 (2007) 153401

[9] D.W. Boukhvalov and M.I. Katsnelson, J.Phys.: Condens. Matter 21 (2009) 344205

[10] R. Sharma, J. H. Baik, C. J. Perera, M. S. Strano, Nano Lett. 10 (2010) 398-405

[11] D.C. Elias, R.R. Nair, T.M.G. Mohiuddin, S.V. Morozov, P. Blake, M.P. Halsall, A.C. Ferrari, D.W. Boukhvalov, M.I. Katsnelson, A.K. Geim, K.S. Novoselov, Science 30, Vol. 323, No. 5914 (2009) 610-613

[12] C. Riedl, C. Coletti, T. Iwasaki, A. A. Zakharov, U. Starke, Phys Rev Lett 103 (2009) 246804

[13] I. Chowdhury, M.V.S. Chandrashekhar, P.B. Klein, J.D. Caldwell, T.S. Sudarshan, J. Cryst. Growth,Vol. 316 (2011) 60-66

[14] C. Srinivasan and R. Saraswathi, Curr. Sci., Vol. 97, No. 3 (2009) 302-303

[15] A.C. Ferrari and J. Robertson, Phys. Rev. B 61 (2000) 14095-14107

[16] S. Shirvaraman, M.V.S. Chandrashekhar, J.J. Boeckl, M.G. Spencer. J. Electron. Mater Vol. 38, No. 6 (2009) 725-730

[17] J. M. Dawlaty, S. Shirvaraman, J. Strait, P. George, M.V.S. Chandrashekhar, F. Rana, M.G. Spencer, D. Veksler, Y. Chen, Appl. Phys. Lett. 93 (2008) 131905

[18] J. Rand. Electroanal. Chem, 35 (1972) 209-218

[19] K. Ota, S. Nishigori, N. Kamiya, J. Electroanal. Chem. 257 (1988) 205-215

Materials Science Forum Vols. 717-720 (2012) pp 665-668

doi:10.4028/www.scientific.net/MSF.717-720.665

Molecular Gas Adsorption Induced Carrier Transport Studies of Epitaxial Graphene using IR Reflection Spectroscopy

B.K. Daas[1,a], W.K. Nomani[1], K.M. Daniels[1], T.S. Sudarshan[1], Goutam Koley[1] and MVS Chandrashekhar[1]

[1]Department of Electrical Engineering,University of South Carolina, Columbia SC 29208,USA

Daas@email.sc.edu[a]

Keywords: Epitaxial Graphene, Molecular Adsorption, IR reflection, Scattering.

Abstract: We investigate molecular adsorption doping by electron withdrawing NO_2 and electron donating NH_3 on epitaxial graphene grown on C-face SiC substrates. Amperometric measurements show conductance changes upon introduction of molecular adsorbents on epitaxial graphene. Conductance changes are a trade-off between carrier concentration and scattering, and manifest at direct current and optical frequencies. We therefore investigate changes in the infrared (IR) reflection spectra to correlate these two frequency domains, as reflectance changes are due to a change of epitaxial graphene (EG) surface conductance. We match theory with experimental IR data and extract changes in carrier concentration and scattering due to gas adsorption. Finally, we separate the intraband and interband scattering contributions to the electronic transport under gas adsorption. The results indicate that, under gas adsorption, the influence of interband scattering cannot be neglected, even at DC.

Introduction

Graphene, a two-dimensional (2D) honeycomb carbon crystal, the basic building block of other sp^2 carbon nanomaterials, such as nanographite sheets and carbon nanotubes, exhibits unusual electronic and optical properties [1-3]. Graphene has a unique electronic band structure, with the conduction band touching the valance band at the K and K' points in the Brillouin zone. This gives graphene its characteristic linear Dirac-Fermion energy dispersion ($E=hk\,v_F$), where E is the energy, h is Planck's constant, k is the wave vector and v_F ($\sim10^6$m/s) is the Fermi velocity. Therefore electrons in the ideal graphene sheet exhibit very high electron mobility [4]. This truly 2D structure with high tunable semi-metallic conductivity has made graphene an ideal candidate for gaseous and vapor sensing applications. The sensitivity of exfoliated graphene to NH_3 and NO_2 [5] and that of epitaxial graphene [6] have been established by molecular adsorption doping. Due to adsorption of impurities on the epitaxial graphene (EG) surface, the electron concentration can either increase or decrease depending on whether the adsorbent is electron donating or withdrawing. These adsorbents can also influence carrier scattering, all of which change the surface conductivity. In our earlier paper [7], we established a change of conductivity due to molecular adsorption doping by DC measurements. In this paper we emphasize carrier transport studies through Fourier transform infrared (FTIR) measurement for molecular adsorption doping. Theoretically, we account for the influence of surface impurities on the conductivity of EG by molecular gas adsorption [5] using the Boltzmann picture. In our previous work [7], we have shown evidence of a surface plasmon polariton at the EG/SiC interface through IR reflection spectra. Here we investigate the carrier transport by FTIR reflection spectroscopy along with a mathematical model to extract various transport parameters, including the influence of interband scattering, which is non-negligible at IR frequencies and high surface impurity concentrations.

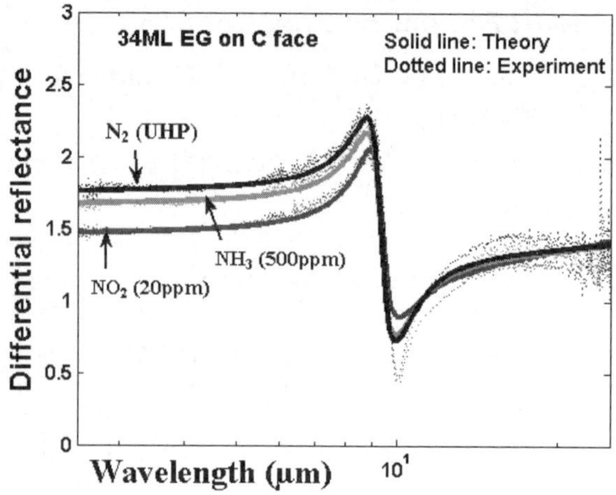

Fig. 1. IR reflection measurement on 34 ML EG on C face in N_2, NH_3 and NO_2 environments (SiC reference).

Experiment

Epitaxial growth of large-area graphene by thermal decomposition of commercial <0001> 4H and 6H SiC substrates at high temperature and vacuum has been demonstrated [8]. This produces EG a few monolayers (ML) to >50 ML thick, depending on growth conditions. In our experiments, EG was grown on the C-face of commercial n^+ 8° off axis 4H-SiC substrates, nitrogen doped at ~$10^{19}/cm^3$. 1cmx1cm samples were degreased using trichloroethylene, acetone and methanol, respectively. They were then rinsed in DI water for three minutes. The samples were finally dipped in HF for two minutes to remove native oxide and rinsed with DI water before being blown dry. They were then set in the crucible in an inductively heated furnace where high vacuum was maintained (<10^{-6} Torr) and baked out at 1000°C for 13 to 15 hours. The temperature was slowly raised to the growth temperature (1250-1400°C). All growths were performed for 60 minutes before cooling to 1000°C at a ramp rate of 7~8°C/min and eventually to room temperature. Slow temperature ramps were utilized to minimize thermal stress on the samples.

After growth, AFM (atomic force microscopy) was used to investigate the EG surface morphology. The morphology is similar to that observed by others [4,7-8]. Raman measurements were carried out on EG on both carbon (C) and silicon (Si) faces. Micro-Raman spectroscopy using a 632nm laser shows the G peak (~1590cm^{-1}), D peak (~1350cm^{-1}) and 2D peak (~2700cm^{-1}) characteristic of EG [9]. The ratio of intensities of the D-peak to G-peak, $I_D/I_G \leq 0.2$, demonstrates the high quality of our graphene [9]. X-ray photoelectron spectroscopy (XPS) measurements were done to obtain the thickness in monolayer's (ML) on EG. The thickness extracted by XPS was consistent with our FTIR measurements [10].

Results and Discussion

NO_2 is a strong oxidizer with electron withdrawing capability, and is expected to decrease electron carrier concentration on the EG surface, while the converse is true for NH_3 as it is electron donating. However, our measurements are based on EG grown on C-face SiC, which is 34 ML thick, much greater than the 1ML screening length in graphene. Thus, this layer is expected to be close to electrically neutral, i.e., the Fermi level is close to the Dirac point (K and K' in the Brillouin zone). Therefore, the carrier concentration will increase in both cases regardless of whether the adsorbate is electron withdrawing or electron donating. Polar molecules [6] change EG conductivity by a) inducing carriers in the EG and b) increasing scattering, i.e., decreasing mobility. Surface carrier concentration in EG is given by

$$n_s = \int_0^\infty D(E) f(E - E_F) dE \qquad (1)$$

where D(E) is the density of states, the Fermi energy function is $f(E - E_F)$, and E_F is the Fermi level. Scattering (τ) includes both intra and interband scattering [11]. Increase or decrease of conductivity is a tradeoff between carrier concentration and scattering. Thus, it is critical to investigate the role of E_F and τ in response to gas adsorption.

Fourier transform infrared reflection (FTIR) measurements were carried out with a blank SiC substrate, cut from the same wafer as the grown samples, as the reference. All reflectance measurement presented here are the differential reflectance with respect to the SiC substrate. The number of monolayers (ML), E_F and τ were extracted by fitting measurement to theory [7]. A more detailed analysis of this data allowed the extraction of the influence of interband scattering as well, which is non-negligible at IR frequencies and high surface impurity concentrations. Fig. 1 shows the IR reflection spectra for ~34ML EG in N_2, NH_3 and NO_2 environments, where we find that reflectance (indicative of conductivity, with higher conductivity leading to higher reflectance) changes for different gases due to adsorption of surface impurities. Reflectivity decreases both in NO_2 and NH_3 environments compared to N_2 with NO_2, showing greater decrease than NH_3. We extract ML, E_F and intra- and inter-band scattering times by matching experimental data with the theory. We also modified our earlier proposed mathematical model [7] to account for impurity adsorption, where we match optical conductivity [12] at high frequency with conductivity in the Random Phase Approximation (RPA) to extract surface impurity concentration, n_i, by

$$\sigma_{T=0}^{RPA}=\frac{e^2}{\pi h}\left[\frac{n_s}{n_i G[4r_s/(2-\pi r_s)]}+\frac{n_i F[4r_s/(2-\pi r_s)]}{4n_s}\right] \tag{2}$$

where h is Planck's constant,

$$r_s=\frac{e^2}{4\pi\varepsilon_0\varepsilon_{SiC}v_F\hbar} \tag{3}$$

where e is the electron charge $1.6\text{x}10^{-19}$C , v_F is the Fermi velocity $1.1\text{x}10^6$ m/s, ε_{SiC} is the dielectric constant of SiC, which has different values for high frequency (ε_{SiC}~6.5) and low frequency (ε_{SiC}~9.52) regimes, and the G and F functions are defined [12] as

$$G(x)=\frac{x^2}{8}\int_0^{2\pi}\frac{\sin^2\theta}{(\sin\frac{\theta}{2}+x)^2}d\theta \quad \text{and} \quad F(x)=\frac{x^2}{8}\int_0^{2\pi}\frac{(1-\cos\theta)^2}{(\sin\frac{\theta}{2}+x)^2}d\theta \tag{4}$$

Considering two limiting values of the SiC dielectric constant (high frequency~6.5 and low frequency~9.52), two different values of r_s (high frequency~0.31 and low frequency~0.21) were calculated. For the high frequency r_s =0.31 was used for the conductivity,

$$\sigma_{T=0}^{RPA}[highfrequency]=\frac{e^2}{\pi h}\left[\frac{n_s}{n_i G[4r_s/(2-\pi r_s)]}\right], \tag{5}$$

which is matched with the optical conductivity to extract n_i. A similar procedure was used for the low frequency side, where r_s ~0.21 arises from ε_{SiC}~9.52, using:

$$\sigma_{T=0}^{RPA}[lowfrequency]=\frac{e^2}{\pi h}\left[\frac{n_i F[4r_s/(2-\pi r_s)]}{4n_s}\right]. \tag{6}$$

Values of n_i extracted at both these frequency regimes were consistent. For further confirmation, we calculate intra- and inter-band scattering from n_i using equations presented elsewhere [12] and the results were found to be consistent with our extracted data within the experimental error limit.

Table 1: Our extracted parameters provide values of the Fermi level, impurity density and scattering times in three different gas environments when IR reflection experiment data matches with theory.

Gas	Fermi level (meV)	Impurity (cm^{-2})	Intraband scattering time (s)	Interband scattering time(s)
N_2	25	2×10^{11}	2.8 to 9×10^{-14}	6 to 2.7×10^{-14}
NH_3	30	6×10^{12}	9e-15 to 6×10^{-15}	2 to 1.6×10^{-15}
NO_2	35	2×10^{13}	2e-15 to 9×10^{-16}	5 to 2.6×10^{-16}

Table 1 shows the extracted carrier transport parameters as a function of gas adsorbents. With N_2 gas, Fermi level is ~25meV, close to neutral because our EG is thick (34ML). As N_2 is an inert gas and should not contribute any impurity on the EG surface. Our extracted parameters indicates a surface impurity concentration of $2x10^{11}cm^{-2}$, fairly consistent with an ex-situ sample that has not had any degassing or other processing performed on it. For NH_3 and NO_2, the surface impurity concentration is higher than N_2 because of the nature of the gas interaction with the carriers on the EG surface. In both cases, the intra and interband scattering times shorten, due to more scattering from the adsorbed impurity. Overall conductivity decreases both for NH_3 and NO_2 compared to the N_2 environment. With NO_2, overall conductivity decreases more than for NH_3 despite a strong increase in carrier concentration, indicating that the EG surface has greater affinity for NO_2 adsorption, leading to much greater carrier scattering at the adsorption surface.

In summary, we find that the infrared carrier transport in EG can be described only by accounting for the influence of interband scattering. In doing so, we find a self-consistent explanation of carrier transport in EG, where carrier scattering (interband and intraband) increases in response to surface impurities, along with a change in carrier concentration caused by charge transfer between the molecules and EG. This leads to an increase/decrease in overall conductivity/reflectivity depending on the tradeoff between scattering and carrier concentration.

Acknowledgement

The authors would like to acknowledge the Southeastern Center for Electrical Engineering Education, and the National Science Foundation (NSF) ECCS-EPDT Grant #1029346 under the supervision of program director Rajinder Khosla for funding this work.

References

[1] A.H. Castro Neto, et al., Rev. Mod. Phys. 81 (2009) 109
[2] K. S. Novoselov, et al., Nature-London 438 (2005) 197-201
[3] P.R. Wallace, Phys. Rev. 71 (1947) 622
[4] A.K. Geim and K.S. Novoselov, Nat. Mater. 6 (2007) 183
[5] Y. Dan, Y. Lu, N.J. Kybert, Z. Luo and A.T.C. Johnson, Nano Lett. 9 (2009) 1472
[6] Mohammad Qazi, et al., Appl. Phys. Express 3 (2010) 075101
[7] B.K. Daas, et al., submitted to Journal of Applied Physics arXiv:1010.4080v1
[8] C. Berger, et al., J. Phys. Chem. B 108 (2004) 19912
[9] A.C. Ferrari, et al., Phys. Rev. Lett. 7 (2006) 238
[10] Shriram Shivaraman, et al., J. Electron. Mater. 38 (2009) 6
[11] Jahan M. Dawlaty, et al., Appl. Phys. Lett. 93 (2008) 131905
[12] S.Y. Liu, X.L. Lei, Norman and J.M. Horing: arXiv:0707.0559v3 [cond-matt.mtrl-sci]

Materials Science Forum Vols. 717-720 (2012) pp 669-674

doi:10.4028/www.scientific.net/MSF.717-720.669

High Performance RF FETs Using High-k Dielectrics on Wafer-scale Quasi-Free-Standing Epitaxial Graphene

Joshua A. Robinson[1,2,a], Matthew J. Hollander[1,2,3,b], Michael LaBella III[2], Kathleen Trumbull[2], Mike Zhu[2,3], Randall Cavalero[4], and David Snyder[1,2]

[1]181 Materials Research Institute, 230 Innovation Blvd., Department of Materials Science and Engineering, Pennsylvania State University, University Park, PA 16802, USA

[2]181 Materials Research Institute, Materials Research Institute, 230 Innovation Blvd., Electro-Optics Center, Pennsylvania State University, University Park, PA 16802, USA

[3]169 Materials Research Institute, 230 Innovation Blvd., Department of Electrical Engineering, Pennsylvania State University, University Park, PA 16802, USA

[a]jrobinson@psu.edu, [b]mjh423@psu.edu

Keywords: Epitaxial graphene, hydrogen intercalation, RF, transistor, FET, high-k

Abstract. We explore the effect of processing on graphene/metal ohmic contact resistance, the integration of high-κ dielectric seeds and overlayers on carrier transport in epitaxial graphene, and directly demonstrate the importance of buffer elimination at the graphene/silicon carbide (0001) interface for high frequency applications. We present a robust method for forming high quality ohmic contacts to graphene, which improves the contact resistance by nearly six thousand times compared to untreated metal/graphene interfaces. Optimal specific contact resistance for treated Ti/Au contacts is found to average < 100 Ohm-μm. Additionally, we introduce a novel seeding technique for depositing dielectrics by atomic layer deposition that utilizes direct deposition of high-κ seed layers and can lead to an increase in Hall mobility up to 70% from as-grown. Finally, we demonstrate that buffer elimination at the graphene/silicon carbide (0001) interface results in excellent high frequency performance of graphene transistors with current gain cutoff frequency > 130 GHz at 75 nm gate lengths.

Introduction

Owing to its excellent intrinsic properties [1,2], graphene has the potential for operating in the terahertz regime [3]. In fact, recent reports have shown the potential for exfoliated [4], chemical vapor deposited [5], and epitaxial graphene [6] to exceed frequency performance of Si-MOSFET technology. Graphene derived from silicon carbide (SiC) – referred to as epitaxial graphene – has proven to be an excellent material system for high frequency electronic applications [6]. In order to achieve state-of-the-art graphene technologies, one must address several key features in material and device development: graphene carrier transport, Ohmic contact resistance, dielectric integration, and device scaling.

Experimental Methods

Epitaxial graphene is synthesized on 15x15 mm squares and 100 mm wafers via silicon sublimation from the silicon face of SiC (0001) at 1625 °C, for 15 min at 1 torr [7]. Following synthesis, silicon face epitaxial graphene (EG_{Si}) is characterized via Raman spectroscopy, x-ray photoelectron spectroscopy (XPS), and Lehighton non-contact mobility and sheet resistance (LEI, Inc). Raman confirms the successful growth of monolayer EG_{Si} on the SiC (0001) terrace and bilayer EG_{Si} on the terrace step edge [7**Error! Bookmark not defined.**,8,9]. Subsequently, samples undergo exposure to molecular hydrogen (H_2) at 600 – 1200 °C, 600 Torr for 30 – 120 min (referred to as hydrogen intercalation). Following hydrogen intercalation, peak fitting of the 2D Raman spectra indicates the presence of bilayer quasi-free standing EG_{Si} (QF-EG_{Si}) on the SiC (0001) terrace center, and multilayer QF-EG_{Si} on the terrace step edge.

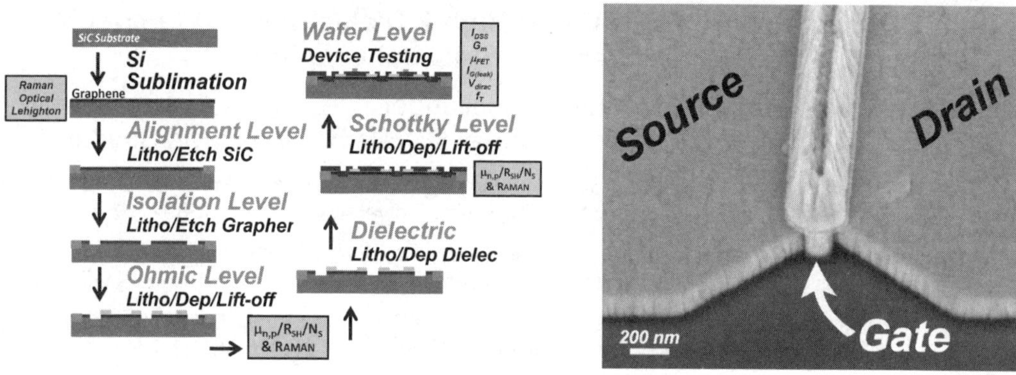

(a) (b)

Figure 1: (a) Generalized device fabrication schematic illustrating the various steps to achieve scaled graphene transistors. (b) Scanning electron micrograph of 75 nm gate length transistor.

Figure 1 demonstrates the fabrication process. Device processing utilizes standard photolithographic techniques. First, an oxygen plasma etch is utilized to isolate the channel and eliminate any shunt conductance outside the active region. Following graphene isolation, samples are patterned for Ti/Au (10/50 nm) ohmic metal deposition. Integration of a gate dielectric with graphene is comprised of various steps, with atomic layer deposition (ALD) and electron beam physical vapor deposition (EBPVD) being utilized to prepare either EBPVD or EBPVD seeded ALD films. EBPVD seeded ALD dielectrics utilize a ~2 – 3 nm seed layer of SiO_2, Al_2O_3, or HfO_2 deposited via non-reactive EBPVD at < 10^{-6} Torr. Immediately following EBPVD seeding, ~8 – 10 nm of Al_2O_3 or HfO_2 is deposited via ALD (Cambridge Nanotech, Inc "Savannah") to complete the gate stack. Source/drain spacings and gate lengths of various combinations were evaluated in this work. Table 1 lists the device dimensions investigated in this work.

Table 1: Device dimensions investigated in this work

Gate Width [μm]	Gate Length [nm]	Source/Drain Spacing [nm]
10	75	150
10	250	325
10	750	750

Results & Discussion

Hydrogen intercalation of EG_{Si} results in elimination of the graphene/SiC (0001) buffer layer. Evidenced by XPS, the S1 and S2 components of the C1s spectra present in as-grown graphene are eliminated upon full hydrogen intercalation (Fig. 2a), indicating successful passivation of the graphene/SiC (0001) buffer layer. The shift in the position of the SiC peak with hydrogen intercalation (Fig. 2a) is not fully understood at this time, although it has been previously reported in the literature. Additionally, we find that Raman spectroscopy directly correlates with the degree of hydrogen intercalation, showing an increase in the graphene signal strength after complete hydrogen intercalation. As grown EG_{Si} is n-type in nature due to the band alignment between graphene and SiC (0001) [10], with typical carrier mobility and density averaging 800 – 900 cm^2/Vs and 3 – 6 x 10^{12} cm^{-2}, respectively. Once total buffer elimination is achieved, all Hall structures exhibit p-type behavior as the majority carrier type is converted from electrons to holes. Additionally, there is a nearly 200% increase in the majority carrier concentration (now holes) to 1.3 x 10^{13} carriers/cm^2 [11]. Despite this increase in carrier concentration, carrier mobility is found to increase by 200 – 300% resulting in an average of 2050 cm^2/Vs after hydrogen intercalation at 1050 °C for 60 min (Figure 2b). The increase in Hall mobility indicates an enhancement in transport properties with hydrogen intercalation.

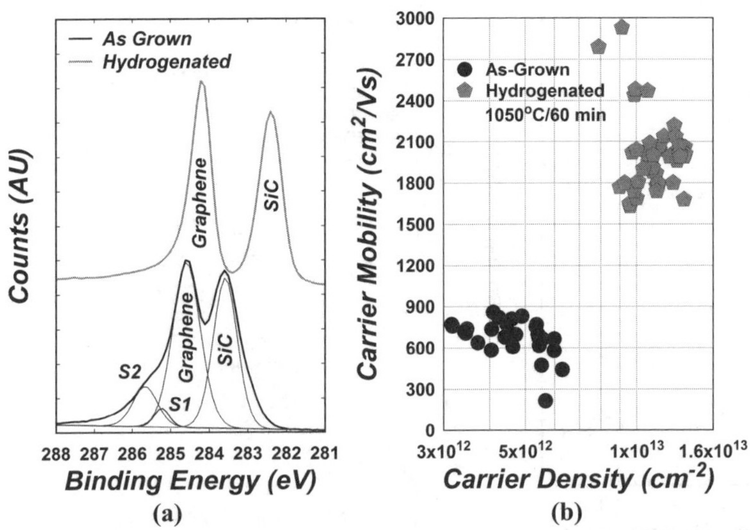

Figure 2: (a) XPS, and (b) Hall mobility and carrier concentration of as-grown and hydrogenated graphene. Note that elimination of S1 and S2 in XPS indicates buffer elimination, which correlates with change in carrier type from n to p-type, as well as with enhanced carrier transport.

The contact resistance (R_c) is highly sensitive to interfacial reactions or device processing conditions that could lead to interfacial contamination. Thus it is an ideal way to evaluate the quality of the metal/graphene electronic interface. We find that the utilization of an oxygen plasma-etch serves well to remove residual resist prior to metal deposition when implemented properly and can lead to a significant improvement in R_c. Specifically, utilization of a low power O_2 plasma treatment ("ash") prior to metal deposition provides superior contact resistances compared to un-treated or post-processed contacts [12].

Although there is degradation in the structure of the graphene (evidenced by the increase in Raman D/G ratio) even under the most benign conditions, R_c is shown to improve with O_2 plasma treatment as the treatment time increases from 30 – 90 sec to an average value of 200 Ohm-μm. Upon reaching 120 sec, R_c begins to degrade and is correlated with severe deterioration in the Raman 2D peak intensity, suggesting that the bulk of the sp^2-bonding expected in graphene has been destroyed. Following the low power O_2 plasma etch to remove photoresist residue, post-processing heat treatments of the contacts further improve contact resistance by a factor of five to an average of 35 Ohm-μm.

In order to evaluate the effect of dielectric integration on carrier transport, Hall effect measurements are made *prior to* and *following* deposition of the gate dielectric on samples that have not undergone hydrogen intercalation. Prior to deposition of a dielectric, as-grown samples exhibit Hall mobility (μ_{Hall}) values of 700 – 1100 $cm^2/V{\cdot}s$ and carrier concentration (n_s) values of 5 – $8x10^{12}$ cm^{-2} (n-type), typical of Si-face EG. Following deposition of the gate dielectric, there is significant variation in the carrier mobility based on the seeding process. All dielectrics are found to cause an increase in n_s; however, not all dielectrics cause a decrease in μ_{Hall}. Figure 3a shows that high-κ seeded dielectrics deposited by EBPVD and oxide seeded ALD result in an increase in n_s *and* an increase in μ_{Hall}. Conversely, oxidized metal seeded dielectrics deposited by ALD lead to decreased μ_{Hall} with increased n_s. The measured increase of μ_{Hall} with an oxide seed is attributed not only to dielectric screening, signifying a reduction in scattering by remote charged impurities, but also to the removal of the metal to metal-oxide phase transition that occurs when metal seeds are utilized in the device fabrication process [13,14].

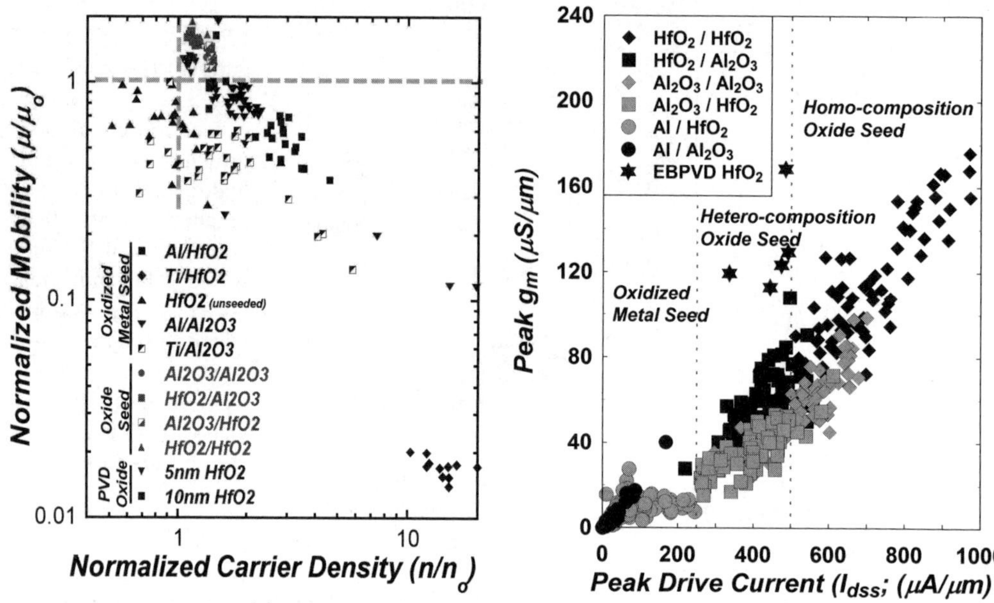

(a) (b)

Figure 3: Evaluation of the normalized Hall mobility (μ/μ_o) and charge carrier density (n/n_o) of the underlying graphene (a), where μ_o and n_o represent as-grown mobility and carrier density respectively, indicates that use of an oxidized metal seed results in an average 30 – 50% decrease in carrier mobility, while the use of a direct deposited oxide seed can improve Hall mobility by as much as 70% [14]. Additionally, evaluation of transconductance and sautration current (b) verifies that oxidized metal seeded dielectrics result in poor FET performance, while homogeneous high-κ seeded O-ALD gate dielectrics produce the highest performance GFETs, attributed to increased dielectric screening.

Additionally, Figure 3b plots peak transconductance ($g_{m,max}$) and peak drive current, I_{dss} (I_{ds} @ V_{gs}=3V, V_{ds}=1V), extracted from I_{ds}-V_{gs} sweeps. GFETs utilizing oxidized metal seeded ALD gates (Fig. 3; left-most section) exhibit average I_{dss} < 250 μA/μm and average $g_{m,max}$ < 30 μS/μm and represent the lowest performance dielectric in this work. Low I_{dss} and $g_{m,max}$ for oxidized metal seeded ALD is attributed to reduced mobility relative to high-κ seeded gates (Fig. 3a). On the other hand, GFETs utilizing high-κ EBPVD or oxide seeded heterogeneous ALD gates show a three to four times increase in I_{dss} to 250 – 500 μA/μm and $g_{m,max}$ to 40 – 80 μS/μm, with 5 nm EBPVD HfO_2 gated GFETS displaying $g_{m,max}$ up to 170 μS/μm. Homogeneous oxide seeded ALD gated GFETs exhibit exceptional FET characteristics with I_{dss}= 550 – 1000 μA/μm and $g_{m,max}$ = 75 – 180 μS/μm depending on gate composition. The increase in performance for GFETs utilizing high-κ EBPVD and high-κ oxide seeded ALD dielectrics is attributed to a reduction of remote charged impurity scattering through dielectric screening, leading to increased carrier mobility. The superior performance of homogeneous oxide seeded ALD gated GFETs relative to heterogeneous oxide seeded ALD is attributed to removal of the seed/overlayer interface which is thought to negatively affect carrier transport in the channel through increased interface roughness scattering [14].

In addition to enhanced DC performance, the frequency response of EG_{Si}-based transistors benefits from the process of improved contacts, utilization of oxide seeded ALD processes, and graphene hydrogen intercalation. The combination of enhanced transport properties and limited device parasitics lead to excellent *extrinsic* RF performance with *extrinsic* small signal current gain cutoff frequencies of up to 20 GHz achievable at a gate length (L_g) of 250 nm. By scaling source/drain distance and gate length, we are able to achieve *average intrinsic* current gain cut-off frequencies of 15, 46, and 134 GHz for 750, 250, and 75 nm gate lengths respectively (Figure 4), where the intrinsic frequency response is found to scale with $1/L_g$. Intrinsic small signal current gain is extracted using short and open deembed structures to remove pad parasitics from the as measured graphene devices as well as any other parasitic resistances or capactiacnes.

Although the *intrinsic* performance is comparable to other high performance graphene FETs reported in the literature, the *extrinsic* device performance indicates that significant work must continue to reduce access resistance, improve the oxide/graphene interface, and enhance gate/channel electrostatic coupling to close the gap between reported *intrinsic* f_T and measured *extrinsic* f_T before graphene transistors can truly be considered for ultra-high frequency operation.

Summary

We have considered several factors for the development of epitaxial graphene transistors including contact resistance, hydrogen intercalation, integration of high-k dielectrics with EG, and device scaling. Through continued improvement of these items, we have achieved high performance extrinsic and intrinsic device operation, with anticipated improvements in device parasitics to further enhance epitaxial graphene technology.

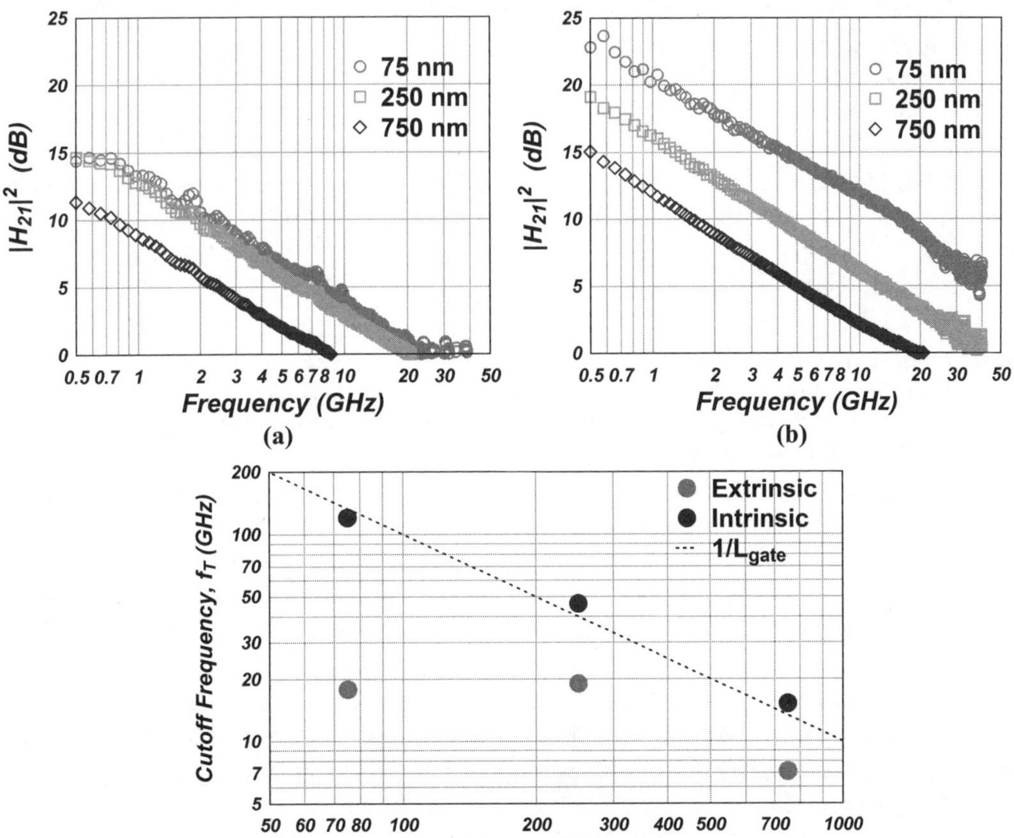

Figure 4: Representative extrinsic (a) and intrinsic (b) small signal current gain as a function of frequency for hydrogen intercalated graphene FETs. Average small signal current gain cut-off frequencies (c) utilizing source/drain scaling in conjunction with gate scaling to achieve $1/L_g$ frequency response behavior, where L_g represents gate length.

References

[1] A. K. Geim, K. S. Novoselov, The rise of graphene, Nature Materials, 6 (2007) 183-191.

[2] K. S. Novoselov, A. K. Geim, S. V. Morozov, D. Jiang, Y. Zhang, S.V. Dubonos, I. V. Grigorieva, A. Firsov, Electric field effect in atomically thin carbon films, Science. 306 (2004) 666-669.

[3] L. Liao, J. W. Bai, R. Cheng, Y. C. Lin, S. Jiang, Y. Q. Qu, Y. Huang, X. F. Duan, Top-gated graphene nanoribbon transistors with ultrathin high-k dielectrics, Nano Lett. 10 (2010) 3952-3956.

[4] L. Liao, Y. C. Lin, M. Q. Bao, R. Cheng, J. W. Bai, Y. Liu, Y. Q. Qu, K. L. Wang, Y. Huang, X. F. Duan, High-speed graphene transistors with a self-aligned nanowire gate, Nature 467 (2010) 305-308.

[5] Y. Q. Wu, Y. M. Lin, A. A. Bol, K. A. Jenkins, F. N. Xia, D. B. Farmer, Y. Zhu, P. Avouris, High-frequency, scaled graphene transistors on diamond-like carbon, Nature 472 (2011) 74-78.

[6] Y. M. Lin, C. Dimitrakopoulos, K. A. Jenkins, D. B. Farmer, H. Y. Chiu, A. Grill, P. Avouris, 100-GHz transistors from wafer-scale epitaxial graphene, Science 327 (2010) 662-662.

[7] J.A. Robinson, M. Wetherington, J.L. Tedesco, P.M. Campbell, X. Weng, J. Stitt, M. Fanton, E. Frantz, D. Snyder, B.L. VanMil, G.G. Jernigan, R.L. Meyers-Ward, C.R. Eddy, D.K. Gaskill, Correlating Raman spectral signatures with carrier mobility in epitaxial graphene: a guide to achieving high mobility on the wafer scale, Nano Lett. 9 (2009) 2873-2876.

[8] J. Röhrl, M. Hundhausen, K. V. Emtsev, T. Seyller, R. Graupner, L. Ley, Raman spectra of epitaxial graphene on SiC(0001), Appl. Phys. Lett. 92 (2008) 201918.

[9] D. Graf, F. Molitor, K. Ensslin, C. Stampfer, A. Jungen, C. Hierold, L. Wirtz, Spatially resolved Raman spectroscopy of single- and few-layer graphene, Nano Lett. 7 (2007) 238-242.

[10] K. Emtsev, A. Bostwick, K. Horn, J. Jobst, G. Kellogg, L. Ley, J. McChesney, T. Ohta, S. Reshanov, J. Röhrl, E. Rotenberg, A. Schmid, A. Waldmann, H. Weber, T. Seyller, Towards wafer-size graphene layers by atmospheric pressure graphitization of silicon carbide, Nature Materials 8 (2009) 203-207.

[11] J.A. Robinson, M.J. Hollander, M. LaBella, III, K.A. Trumbull, R. Cavalero, and D.W. Snyder, Epitaxial graphene transistors: enhancing performance via hydrogen intercalation, Nano Lett. 11 (2011) 3875-3880.

[12] J.A. Robinson, M. LaBella, M. Zhu, M.J. Hollander, R. Kasarda, Z. Hughes, K.A. Trumbull, R. Cavalero, D. Snyder, Contacting graphene, App. Phys. Lett. 98 (2011) 053103.

[13] J.A. Robinson, M. Labella, K.A. Trumball, X.J. Weng, R. Cavelero, T. Daniels, Z. Hughes, M.J. Hollander, M. Fanton, D. Snyder, Epitaxial graphene materials integration: effects of dielectric overlayers on structural and electronic properties, ACS Nano 4 (2010) 2667-2672.

[14] M.J. Hollander, M. LaBella, Z.R. Hughes, M. Zhu, K.A. Trumbull, R. Cavalero, D.W. Snyder, X. Wang, E. Hwang, S. Datta, and J.A. Robinson, Enhanced transport and transistor performance with oxide seeded high-k gate dielectrics on wafer-scale epitaxial graphene, Nano Lett. 11 (2011) 3601-3607.

Materials Science Forum Vols. 717-720 (2012) pp 675-678

doi:10.4028/www.scientific.net/MSF.717-720.675

Gated Epitaxial Graphene Devices

Daniel Waldmann[1,a], Johannes Jobst[1,b], Florian Speck[2,c], Thomas Seyller[2,d], Michael Krieger[1,e,*] and Heiko B. Weber[1,f]

[1]Lehrstuhl für Angewandte Physik, University of Erlangen-Nuremberg, Staudtstr. 7 / A3, D-91058 Erlangen, Germany

[2]Lehrstuhl für Technische Physik, University of Erlangen-Nuremberg, Erwin-Rommel-Str. 1, D-91058 Erlangen, Germany

[a]daniel.waldmann@physik.uni-erlangen.de, [b]johannes.jobst@physik.uni-erlangen.de, [c]florian.speck@physik.uni-erlangen.de, [d]thomas.seyller@physik.uni-erlangen.de, [e]michael.krieger@physik.uni-erlangen.de, [f]heiko.weber@physik.uni-erlangen.de, *corresponding author

Keywords: Bottom gate, epitaxial graphene, buried gate

Abstract. A bottom gate scheme is presented to tune the charge density of epitaxial graphene via a gate voltage while leaving the surface open for further manipulation or investigation. Depending on the doping concentration of the buried gate layer, the temperature and illumination, the bottom gate structure can be operated in two regimes with distinct capacitances. A model is proposed, which quantitatively describes the gate operation. The model is verified by a control experiment with an illuminated gate structure using UV light. Using UV illumination the Schottky capacitor (SC) regime, which provides improved gate efficiency, can be used even at low temperatures.

Introduction

Similar to exfoliated graphene [1], epitaxial graphene grown on silicon carbide (SiC) by thermal decomposition exhibits exceptional properties. In both materials the charge carriers are governed by the Dirac-Weyl Hamiltonian. Moreover, epitaxial graphene can be fabricated on the wafer scale [2], an important aspect with respect to industrial applications. However, for sensor devices and especially for surface science experiments, a bottom gate would be desirable in order to tune the charge carrier density but keeping the graphene layer accessible. This can be realized by combining graphene and SiC semiconductor technology [3], which is the aim of this work.

Experimental

Commercial on-axis semi-insulating 6H-SiC (0001) wafers were used as base material for the devices. Prior to the graphene fabrication, a buried gate layer was implanted at T_{impl} = 500°C by using nitrogen (N) ions at implantation energies E = 1 MeV (samples 1 and 2) or E = 2 MeV (sample 3) resulting in an implantation depth of 1.0 µm or 1.4 µm, respectively. The dose was varied from $D = 7 \times 10^{12}$ cm^{-2} to 3×10^{14} cm^{-2}. A vertical conductive channel implanted through a shadow mask (N box profile, [N] = 10^{19} cm^{-3}) serves as an electrical contact to the buried gate (Fig. 1a,b). Thermal annealing was carried out at T = 1550°C for 15 min followed by hydrogen etching for 90 s. Then, two types of graphene layers were grown on different samples: mono-layer graphene (MLG, sample 1) [2], which is separated from the SiC substrate by the so-called buffer layer, and quasi free-standing monolayer graphene (QFMLG, sample 2 and 3) obtained after conversion of the buffer layer by hydrogen intercalation [4, 5]. The high quality of the graphene layers was verified using Raman spectroscopy. Finally, graphene Hall bar structures were fabricated by means of electron beam lithography. The samples were analyzed by measuring the resistivity and the Hall coefficient as a function of the magnetic field and the temperature.

Results and Discussion

Fig. 2 displays the room temperature resistivity ρ (filled symbols) and charge carrier density n (open symbols) as a function of the gate voltage V_g applied to the bottom gate for the MLG (squares) as well as for the QFMLG (circles) structure. Surprisingly, the MLG sample reveals no visible gate effect in the whole voltage range V_g = (0 ... 100) V. This is explained by the effective

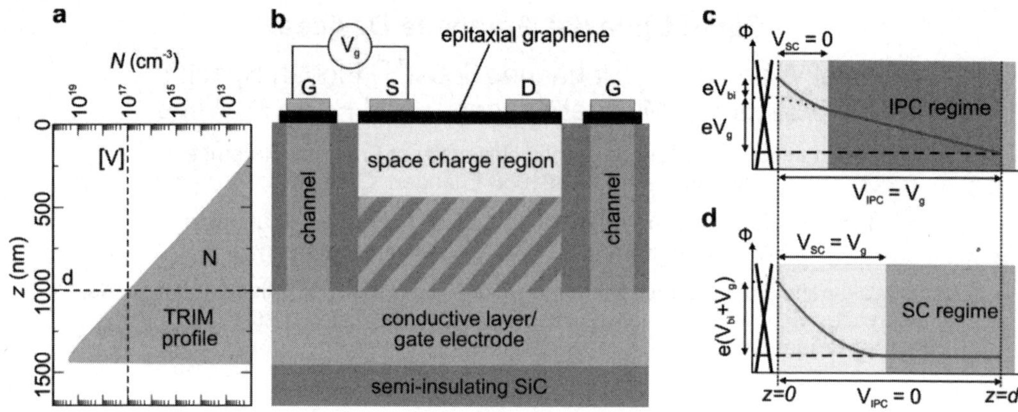

Fig. 1. Scheme of the bottom-gated graphene device. The conductive layer (orange) is created by implantation of nitrogen (a, b). The SiC is conducting (orange) where the implanted dose exceeds [V] and insulating (grey) elsewhere. Below the interface between epitaxial graphene (black) and SiC, a space charge region (yellow) is formed. The graphene is contacted through source (S) and drain (D) titanium/gold contacts; the gate contact (G) is placed on top of vertical conductive channels connected to the implanted gate layer. (c,d) Band diagrams of the SiC conduction band (red curve) in the IPC regime (c) and in the SC regime (d); the graphene at $z = 0$ is depicted as the Dirac cone. (color online only)

shielding of the electric field by the dangling bond states at the SiC/buffer layer interface [4]. By passivation of these dangling bonds with hydrogen and thus producing QFMLG, the electric coupling of the buried gate to the graphene layer is recovered. Driving the gate voltage V_g from 0 V to 90 V, the conductivity of the graphene layer can be controlled from hole to electron conduction; charge neutrality is observed at $V_g \approx 80$ V, where n diverges due to the vanishing Hall coefficient. In the transition region (Fig. 2, dashed line) both holes and electrons contribute to the conductivity.

The gate operation can be distinguished into two regimes, which are characterized by two distinct capacitances: when the implanted N concentration or the temperature is low, the insulating layer between the buried gate and the QFMLG has negligible conductivity (implanted plate capacitor (IPC) regime, Fig. 1c). The second regime (Schottky capacitor (SC) regime) occurs, when the insulating layer becomes sufficiently conductive (Fig. 1d). The latter regime is reached at temperatures above the transition temperature T_t (Fig. 3, black circles; remark: the capacitance of the gate structure was determined in 'DC-mode' by measuring the charge carrier density at $V_g = 0$ V and $V_g = 3$ V and calculating $C_{meas} = [p(V_g = 3\text{ V}) - p(V_g = 0\text{ V})] / 3\text{ V}$). The two regimes as well as the transition can be understood by calculating the temperature dependent voltage drop of V_g along the semiconductor between buried gate and graphene as follows. At the graphene/SiC interface, a Schottky contact with a space charge region is formed. Thus, the gate voltage V_g divides into two parts:

$$V_g = V_{SC}(T) + V_{IPC}(T). \tag{1}$$

$V_{SC}(T)$ and $V_{IPC}(T)$ are the temperature dependent voltages that drop across the Schottky contact and the insulating SiC layer between graphene and implanted gate, respectively. $V_{IPC}(T)$ can be calculated from the Schottky leakage current $I_{SC}(T)$ and the resistivity of the SiC ρ_{SiC} and is given by

$$V_{IPC}(T) = \frac{d}{A}\rho_{SiC}(T) I_{SC}(T) \quad (d\text{: distance of buried gate from graphene, } A\text{: gate area}), \tag{2}$$

where

$$I_{SC}(T) = I_0 T^2 \exp\left(-\frac{\Phi_{b,eff}}{k_B T}\right)\left[\exp\left(\frac{qV_{SC}}{k_B T}\right) - 1\right] \tag{3}$$

with the effective Schottky barrier height $\Phi_{b,eff}$ [6]. The resistivity of highly V compensated SiC can be approximated by $\rho_{SiC}(T) = \rho_0 \exp(E_{V,acc} / k_B T)$ with the activation energy of the V acceptor $E_{V,acc} = 0.7$ eV [7]. $V_{SC}(T)$ is determined by solving eq. (1) combined with eq. (2) and (3) numerically for a given V_g and T. The overall capacitance C is composed of the contribution from the charge stored in the implanted plate capacitor and from the space charge of the Schottky contact

$$C = \frac{e}{V_g}\left[\frac{\varepsilon_{SiC}\varepsilon_0}{ed} V_{IPC} + (w - w_{min})N_d\right] \quad (4)$$

with the width of the space charge region $w = \sqrt{2\varepsilon_{SiC}\varepsilon_0(V_{SC} + V_{bi})/(eN_d)}$, $w_{min} = w(V_{SC} = 0)$, the built-in voltage V_{bi} of the graphene/SiC Schottky contact and the net doping concentration N_d in the space charge region. All parameters except N_d are experimentally determined from independent measurements: $\Phi_{b,eff} = 0.14$ eV and $I_0 = 1.1 \times 10^{-3}$ nA/K^2 are taken from the leakage current $I_g(V_g,T)$, ρ_0 is obtained from I_g measured in forward direction of the Schottky contact at $T = 295$ K, and V_{bi} is measured by X-ray photo electron spectroscopy (XPS) [8]. $N_d = 3.1 \times 10^{16}$ cm^{-3} was used to fit eq. (4) to the experimental data points (Fig. 3) with excellent agreement even in the transition regime around $T \approx 275$ K. The value of N_d is very reasonable, since the corresponding layer is V-doped, highly compensated and exhibits an unknown concentration of residual implantation damage induced defects. For $T << T_t$, a capacitance of $C_{IPC} \approx 8.5$ pF is found. For $T > T_t$, the capacitance C_{SC} increases to more than 30 pF. Therefore, the best gate coupling is achieved with the sample being in the SC regime. Figure 1c and 1d displays the band scheme of the gate structure as a function of the vertical coordinate z for the IPC regime and the SC regime, respectively. The SC regime (Fig. 1d) represents a normal Schottky contact. In the IPC regime (Fig. 1c), however, the applied gate voltage V_g drops completely across the insulating layer and the band bending in the space charge region of the Schottky contact is governed by the built-in voltage V_{bi} only.

As a further validation of the model, the sample was illuminated with ultraviolet (UV) light with a photon energy larger than the SiC bandgap. As a consequence, free charge carriers are created in the SiC by absorption of photons and the resistivity of the insulating layer is lowered similar to the case of elevated temperatures. Figure 3 displays the measured capacitance under UV illumination (green squares). Even for $T < T_t$, the high value of the capacitance C_{SC} of the SC regime is observed in agreement with the expectation. It is remarkable that the sample remains in the SC regime at low temperatures, even when the UV light is switched off. Figure 4 displays the response of the source-drain current I_{SD} on a change of V_g from 0 V to 3 V at $t = 15$ s for $T = 35$ K (black, dashed curve) and $T = 100$ K (red, solid curve) with the sample being illuminated with UV light. In both cases, the sample is in the SC regime due to the UV light and, as a consequence, the same gate response $\Delta I_{SD} = 1$ μA is observed. At $t = 150$ s, the UV light is switched off. At $T = 100$ K, I_{SD} increases indicating the transition to the IPC regime and thus a reduction of the gate coupling. At $T = 35$ K, however, no change on I_{SD} is observed and the sample remains in the SC regime. This behaviour is explained by the extremely high resistivity of the insulating layer at $T << T_t$ without UV light (> 10^{13} Ω), which prevents the stored charge in the space charge region from escaping. This effect opens the possibility of low temperature measurements in the favourable SC regime without the disturbing influence of the UV light.

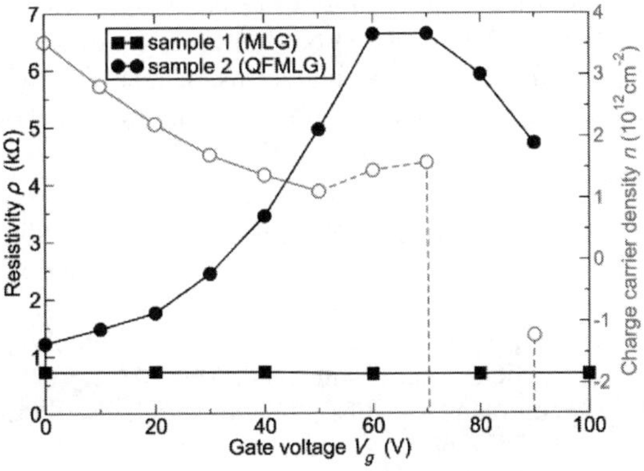

Fig. 2. Resistivity ρ (full symbols) and charge carrier density n (open symbols) as a function of gate voltage V_g taken on the MLG sample (squares) and on the QFMLG sample (circles). In the transition region (dashed line), both holes and electrons contribute to the conductivity.

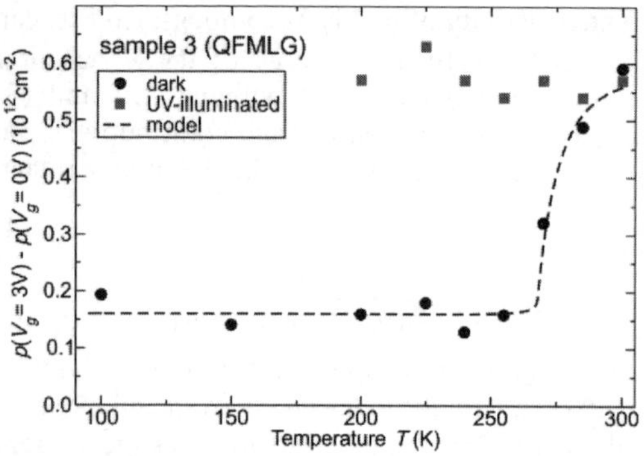

Fig. 3. Induced charge carrier concentration in the graphene as a function of the temperature between V_g = 0V and 3 V measured in the dark (circles) or under UV illumination (squares). For T > 280 K and under UV illumination, the sample is in the SC regime, whereas for T < 270 K the low capacitance of the IPC regime is revealed.

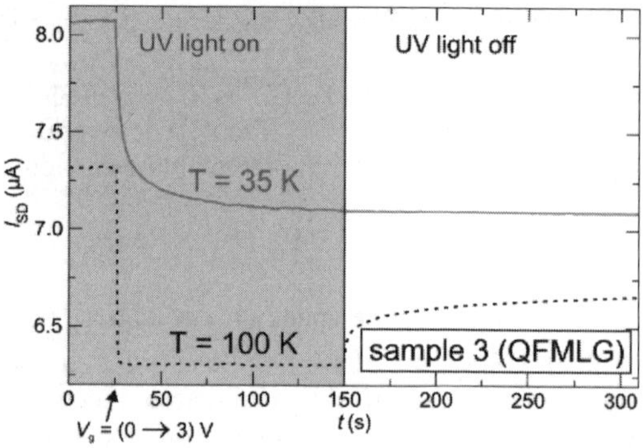

Fig. 4. Gate response of the source-drain current I_{SD} at T = 35 K (solid, red line) and T = 100 K (dashed, black line). At t = 0, the sample is in the SC regime due to the illumination with UV light. At t = 25 s the gate voltage V_g is changed from 0 V to 3 V and at t = 150 s the UV is switched off. At T = 100 K the sample undergoes a slow transition back to the IPC regime whereas at T = 35 K it remains in the SC regime.

Summary and Conclusion

A bottom-gate scheme for epitaxial graphene on SiC is presented. Using hydrogen intercalated quasi-free standing epitaxial graphene, it is demonstrated that the conductivity can be tuned from hole to electron conduction. Two gate regimes have been found: The IPC regime reveals the smaller gate capacitance, which is determined by the distance of the implanted buried gate from the graphene layer. The SC regime, however, is observed, when the insulating layer in between becomes conductive and the gate capacitance is governed from the space charge region of the graphene/SiC Schottky contact only. This requires high temperatures $T > T_t$ or illumination with UV light. A model is presented, which describes both regimes as well as the transition region quantitatively. Using UV light at low temperatures and applying a gate voltage to the bottom gate, the state of the gate (SC regime) can be kept even after switching off the UV light. Altogether, the properties of the presented bottom gate structure can be tailored over a wide range by appropriate choice of the doping concentrations. The surface of the graphene is left open and thus any surface treatment and measurement on a working device is possible.

References

[1] K. S. Novoselov, A. K. Geim, S. V. Morozov, D. Jiang, M. I. Katsnelson, I. V. Grigorieva, S. V. Dubonos, A. A. Firsov, Nature 438 (2005) 197-200.

[2] K. V. Emtsev, A. Bostwick, K. Horn, J. Jobst, G. L. Kellogg, L. Ley, J. L. McChesney, T. Ohta, S. A. Reshanov, J. Röhr, E. Rotenberg, A. K. Schmid, D. Waldmann, H. B. Weber, Th. Seyller, Nature Mater. 8 (2009) 203-207.

[3] D. Waldmann, J. Jobst, F. Speck, Th. Seyller, M. Krieger, H. B. Weber, Nature Mater. 10 (2011) 357-360.

[4] C. Riedl, C. Coletti, T. Iwasaki, A.A. Zakharov, U. Starke, Phys. Rev. Lett. 103 (2009) 246804.

[5] F. Speck, J. Jobst, F. Fromm, M. Ostler, D. Waldmann, H. B. Weber, Th. Seyller, Mater. Sci. Forum 645-648 (2010) 629-632.

[6] E. H. Rhoderick, J. Phys. D 5 (1972) 1920.

[7] J. R. Jenny, J. Skowronski, W. C. Mitchel, H. M. Hobgood, R. C. Glass, G. Augustine, R. H. Hopkins, Appl. Phys. Lett. 68 (1996) 1963.

[8] Th. Seyller, K. V. Emtsev, F. Speck, K. Y. Gao, L. Ley, Appl. Phys. Lett. 88 (2006) 242103.

Materials Science Forum Vols. 717-720 (2012) pp 679-682

doi:10.4028/www.scientific.net/MSF.717-720.679

Suppression of Hole Current in Graphene Transistors with n-type Doped SiC Source/Drain Regions

Yuichi Nagahisa[1,a] and Eisuke Tokumitsu[1,b]

[1]Precision and Intelligence Lab., Tokyo institute of Technology

4259-R2-19 Nagatsuta, Midori-ku, Yokohama 226-8503, Japan

[a]y.nagahisa@neuro.pi.titech.ac.jp, [b]tokumitu@neuro.pi.titech.ac.jp

Keywords: graphene, transistor, on/off ratio, ambipolar behavior

Abstract. To achieve graphene channel transistors which have a high on/off drain current ratio and unipolar behavior of drain current – gate voltage (I_D-V_G) characteristics, we fabricated and characterized top gated graphene channel transistors with n-type doped SiC source/drain regions. The graphene layer was formed on a SiC substrate by high temperature annealing in vacuum, and Al_2O_3 was used as a gate insulator. For the graphene channel transistor with heavily doped n-SiC source/drain regions (doping concentration N_D=4.5x10^{19}cm^{-3}) and a 4~6 monolayer (ML) graphene channel, ambipolar behavior was observed. On the other hand, when N_D was reduced to 4.5x10^{18}cm^{-3} and a thin graphene layer was used, the suppression of hole current in the I_D-V_G curve was observed.

Introduction

Graphene has attracted great interest as a candidate electronic material for the post silicon era, because it has several excellent and unique electrical properties such as extraordinary high carrier mobility. However, due to the lack of a bandgap, most graphene channel transistors ever reported have a small on/off drain current ratio. The main reason for the poor on/off ratio is the large off current. The lack of bandgap causes ambipolar behavior of the drain current – gate voltage (I_D-V_G) characteristics. In addition, even when the Fermi level is at the Dirac point, some thermally excited carriers (holes and electrons) in the graphene channel make the off current large. These problems limit the application of graphene channel transistor as a switching device. To increase the on/off drain current ratio, many attempts at opening the band gap of graphene have been reported. There are mainly two methods to open the band gap of graphene: one is breaking the inversion symmetry in bilayer graphenes [1], and the other is fabrication of graphene nanoribbon [2]. However the band gap of graphene obtained by these methods is not large enough to achieve high on/off drain current ratio, and to obtain unipolar behavior.

In this work, we focus on the suppression of ambipolar behavior of graphene channel transistors by using semiconductor source/drain regions [3,4,5]. For instance, Zhu and Woo [3] suggested that the ambipolar behavior can be suppressed by using heavily doped n^+-Si as source/drain regions which act as the barriers against the hole injection from the graphene channel. Their simulation showed that a large on/off drain current ratio of about 10^3 will be achievable, and that the graphene transistor will show a unipolar I_D-V_G characteristic, both of which were caused by suppression of the hole drain current. However, fabrication of graphene channel transistors on a Si substrate with n^+-Si source/drain regions is not an easy task. On the other hand, controlled graphene layers can be formed on a SiC substrate on a wafer scale by annealing the SiC substrate at high temperature in vacuum. Hence, in this work, we have fabricated graphene channel transistors on a SiC substrate with n-type doped SiC source/drain regions and evaluated their electrical properties.

Experimental Procedure

In this work, we fabricated and characterized graphene channel transistors as shown in Figs. 1(a) and (b). The fabrication process flow is shown in Fig. 2. Transistors were fabricated on an 8°-off 4H-SiC(0001) p substrate with a p-type epi-layer (acceptor concentration was 1.1x10^{16} cm^{-3}). Before

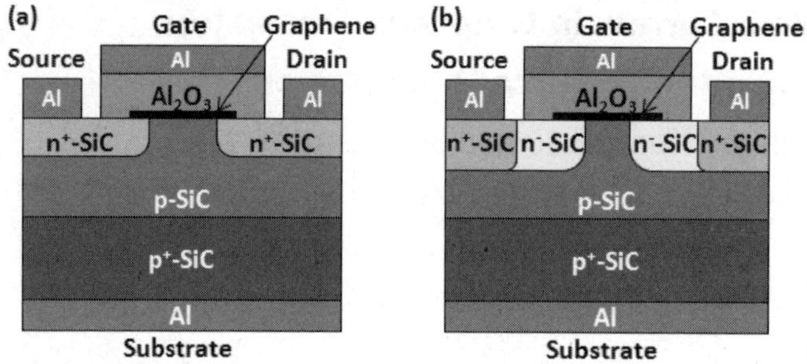

Fig. 1. The diagrams (a) and (b) show the transistor structures fabricated in this work. The doping concentrations of n^+-SiC and n^--SiC regions are $4.5\times10^{19}cm^{-3}$ and $4.5\times10^{18}cm^{-3}$ respectively.

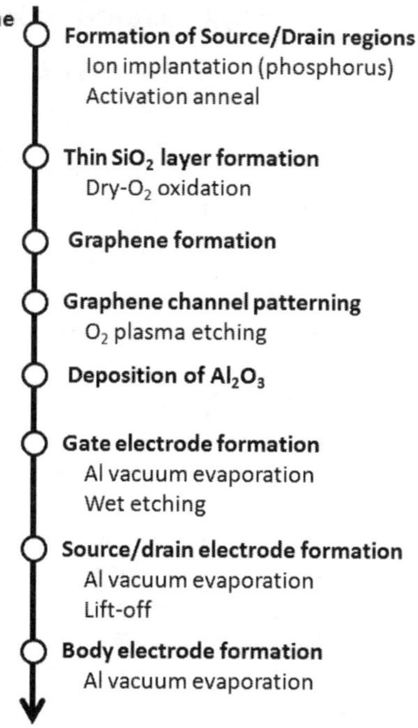

Fig. 2. Process flow of transistor fabrication.

graphene formation process, we formed a thin SiO_2 layer on the SiC substrate by rapid thermal anneal in dry-O_2 ambient at 1150°C for 30 min. The SiO_2 layer was formed to protect the surface of the SiC substrate as the substrate temperature rises. This process is necessary to form graphene by vacuum anneal [6]. Graphene layers were formed by annealing the SiC substrate at 1300°C or 1400°C under $5x10^{-5}$ Torr.

Graphene layer formation on the SiC substrate was confirmed by observation of the G-peak using Raman spectroscopy and we estimated the number of graphene layers by considering the attenuation of the peak intensity due to the SiC substrate, as reported by Shivaraman et al. [7]. In addition, to confirm the graphene formation, the cross sectional transmission electron micrograph (TEM) was observed. An Al_2O_3 film, which was deposited by metalorganic chemical vapor deposition (MOCVD) at 330°C using triethylaluminium (TEA) and O_2 as precursors, was used as the gate insulator. The thickness of the Al_2O_3 layer was controlled to be around 70 nm.

For characterizing electrical property of the graphene transistor, we measured the I_D-V_G characteristic by using a semiconductor parameter analyzer (Agilent Technologies, HP4145C). The measurements were performed at room temperature while connecting the source and substrate body electrodes to the ground.

Results and Discussion

We first confirmed the formation of graphene at both temperatures of 1300°C and 1400°C. Fig. 3 shows the cross section TEM image of graphene grown on SiC at 1400°C and the Al_2O_3 film deposited to protect the graphene surface. It is confirmed that the number of graphene layers is 4~6 ML. The estimated number of graphene layers by Raman spectroscopy measurement tended to be fewer than the actual number. The estimated layer thickness of the graphene layer formed at 1400°C is 2~4ML,

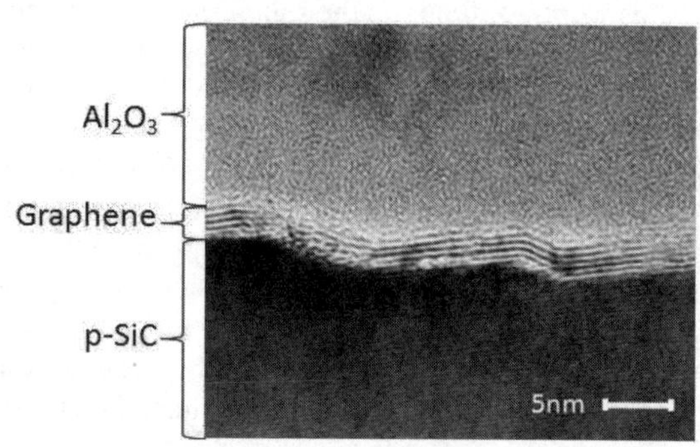

Fig. 3. TEM cross section of graphene grown on SiC substrate.

Table 1. Fabrication conditions of graphene channel transistors

Transistor Number	Temperature of Graphene Formation :T_G	Source/Drain regions	Thickness of Al_2O_3 Gate Insulator
#1	1400°C	n^+-SiC: N_D=4.5x10^{19}cm^{-3}	52nm
#2	1400°C	n^--SiC: N_D=4.5x10^{18}cm^{-3}	58nm
#3	1300°C	n^+-SiC: N_D=4.5x10^{19}cm^{-3}	87nm
#4	1300°C	n^--SiC: N_D=4.5x10^{18}cm^{-3}	87nm

and that of graphene formed at 1300°C is 0~1ML. Although the actual number of layers of graphene formed at 1300°C was not directly observed by TEM, it was estimated to be fewer than the number of layers of graphene formed at 1400°C.

Next, we fabricate graphene channel transistors for four different sets of conditions as shown in Table. 1. We changed two paramaters, T_G : temperature of graphene formation, and the doping concentration of source/drain regions by fabricating two types of transistor as shown in Fig. 1(a),(b). The I_D-V_G characteristics of the fabricated graphene channel transistors are shown in Fig. 4. The gate voltage was first swept from the negative to positive voltages, and then returned to the negative voltage. The drain voltages to measure the I_D-V_G characteristics for transistors #2 and #4 were set at 5V to obtain enough drain current, because transistors #2 and #4 have high contact resistance due to the low doping concentration of the source/drain regions. Each I_D-V_G characteristic shows large hysteresis, and the origin of the hysteresis is probably because of the poor properties of the Al_2O_3/graphene interface and/or Al_2O_3 layer. Figs. 4(a), 4(b), and 4(c) show typical ambipolar behavior of the graphene channel transistor in spite of the use of n-SiC source/drain regions. There is a single minimal value in a single sweep of V_G due to the reduction of the carrier concentration, which

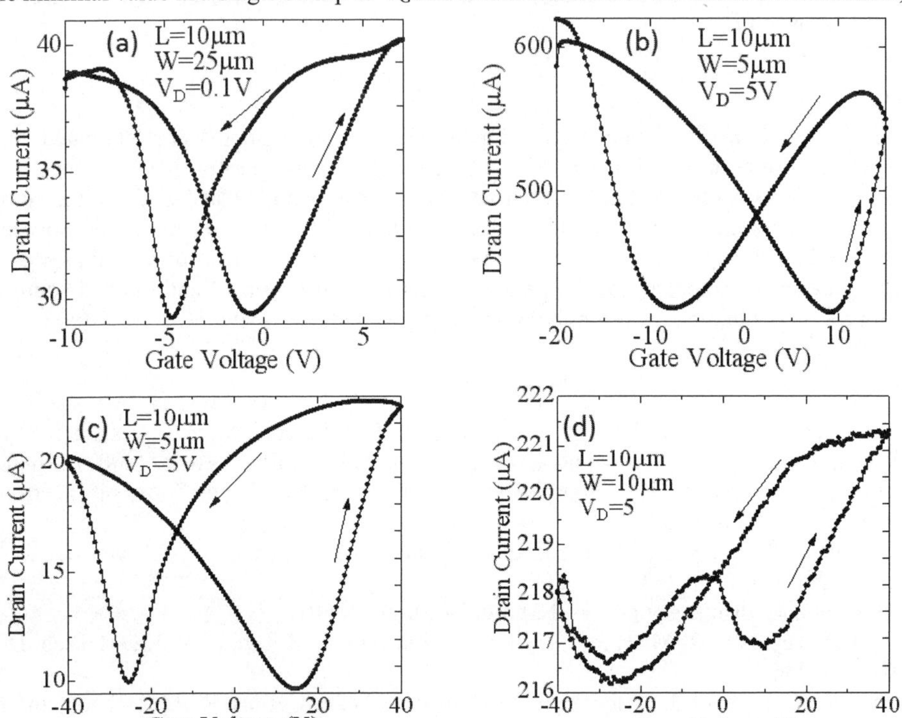

Fig. 4. The I_D-V_G characteristics of graphene channel transistors we fabricated.
(a)I_D-V_G characteristic of #1 (b)I_D-V_G characteristic of #2
(c)I_D-V_G characteristic of #3 (d)I_D-V_G characteristic of #4

is caused when the Fermi level is located at the Dirac point of graphene. When the positive bias is applied, the electron current flows, whereas the hole current is dominant when the negative bias was applied. On the other hand, Fig. 4(d) shows a different I_D-V_G characteristic. In the Fig. 4(d), the drain current in the negative bias region is suppressed, which suggests the suppression of the hole current. The I_D-V_G curve of transistor #4 is clearly different from the I_D-V_G curves of the other transistors on the negative bias side, where the hole current is dominant. Although the I_D-V_G characteristic shown in Fig. 4(d) shows a certain noise, such suppression effect was observed reproducibly. So the suppression effect observed here is likely to be caused by the n^--SiC source/drain regions.

When we first fabricated transistor #1 as shown in Fig.1(a), the transistor showed ambipolar behavior as shown in Fig. 4(a), which indicates that n^+-SiC does not suppress the hole current. There are two possible reasons. One is that heavily doped n^+-SiC causes tunneling current at the graphene/n^+-SiC junction, so that n^+-SiC does not act as a barrier against hole injection from the graphene channel. The other is that the large number of graphene layers, 4~6 ML, screens the gate electric field so that the field effect due to the gate field does not extend to the graphene/n^+-SiC junction. To confirm these considerations, we fabricated transistors #2, #3 and #4. Although #2 and #3 showed ambipolar behavior, #4 showed suppression of the hole current by n^--SiC and partial unipolar behavior. This indicates that both of these two reasons contribute to the ambipolar behavior of the graphene channel transistor with n-type doped source/drain regions.

However, the suppression effect against hole current of transistor #4 is still not enough. Although the hole current of transistor #4 is slightly suppressed, the off current of the transistor is still large at the negative gate voltage region. So the transistor shows low on/off ratio (less than 1.1). In this work, we confirmed that the number of graphene layers and the doping concentration of the source/drain regions were important parameters to suppress the hole current. In addition, interface properties of Al_2O_3/graphene and graphene/SiC are also important. Further optimization and study are necessary to improve the on/off ratio.

Summary

Graphene channel transistors with n-type doped source/drain regions were fabricated. For the graphene channel transistor with the n^+-SiC source/drain regions and 4~6 ML graphene channel, ambipolar behavior was observed in I_D-V_G curve. On the other hand, it was found that the ambipolar behavior was suppressed for the transistor with a thinner graphene channel and n^--SiC source/drain regions. Therefore, to suppress the ambipolar behavior, it is important to form a thinner graphene channel to prevent the screening of the gate electric field and to form lightly doped source/drain regions to prevent the tunneling current between graphene and the n-type SiC source/drain.

References

[1] E. V. Castro, K. S. Novoselov, S. V. Morozov, N. M. R. Peres, J. M. B. Lopes dos Santos, J. Nilsson, F. Guiea, A. K. Geim, and A. H. Castro Neto, Phys. Rev. Lett. 99 (2007) 216802.

[2] G. Liang, N. Neophytos, D. Nikonov, and M. Lundstrom, IEEE Transactions on Electron Devices, 54, 4 (2007) pp 677-682.

[3] J. Zhu and J. Woo, ESSDERC- Proceedings of the 37th European Solid-State Device Research Conference, (2007) pp 243-246.

[4] E. Sano and T. Otsuji, Appl. Phys. Express 2 (2009) 061601.

[5] K. Majumdar, K. V. R. M. Murali, N. Bhat, F. Xia, and Y. M. Lin, IEEE IEDM Tech. Dig., (2010) pp 736-739.

[6] W. Lu, W. C. Mitchel, J. J. Boeckl, T. R. Crenshaw, W. E. Collins, R. P.H . Chang, and L. C. Feldman, Journal of Electronic Materials, 38, 6 (2009) pp 731-736.

[7] S. Shivaraman, M. V.S . Chandrashekhar, J. J. Boeckl and M. G. Spencer, Journal of Electronic Materials 38, 6 (2009) pp 725-730.

Materials Science Forum Vols. 717-720 (2012) pp 683-686
© (2012) Trans Tech Publications, Switzerland
doi:10.4028/www.scientific.net/MSF.717-720.683

T- and Y-Branched Three-Terminal Junction Graphene Devices

Jörg Pezoldt[1,a], Robert Göckeritz[1,2,b], Bernd Hähnlein[1,c],Benjamin Händel[1,d], and Frank Schwierz[3,e]

[1]FG Nanotechnologie, Institut für Mikro- und Nanotechnologien MacroNano®, Technische Universität Ilmenau, Postfach 100565, 98684 Ilmenau, Germany

[2]FG Nanostrukturierte Materialien, Institut für Physik, Martin-Luther-Universität Halle-Wittenberg, Von-Danckelmann-Platz 3, 06120 Halle (Saale), Germany

[3]FG Festkörperelektronik, Institut für Mikro- und Nanotechnologien MacroNano®, Technische Universität Ilmenau, Postfach 100565, 98684 Ilmenau, Germany

[a]joerg.pezoldt@tu-ilmenau.de, [b]robert.goeckeritz@physik.uni-halle.de, [c]bernd.haehnlein@tu-ilmenau.de, [d]benjamin.haendel@tu-ilmenau.de, [e]frank.schwierz@tu-ilmenau.de,

Keywords: three-terminal junction, graphene, heteroepitaxy.

Abstract. Heteroepitaxial graphene on semiinsulating silicon carbide was used to fabricate nanoelectronic devices. T- and Y-branched graphene three-terminal junction devices were realized. Room temperature electrical measurements demonstrate pronounced nonlinear electrical properties of the devices. Voltage rectification at room temperature was observed. Increasing branch width reduces the curvature of the voltage rectification response curve of the three-terminal junctions.

Introduction

The progress in lateral structuring in combination with low dimensional structures exhibiting high charge carrier mobility allows the fabrication of new types of devices. They can be divided into two classes. The first, exploits tunnel transport, such as quantum dot devices, single electron transistors, tunneling and resonant tunneling devices [1]. The second is based on ballistic or diffusive-ballistic transport effects, like T- or Y-shaped junctions operating as ballistic rectifiers or logic gates [2]. The T- or Y-shaped junctions belong to the device group of three-terminal junctions (TTJ) or three-terminal ballistic junctions. They are exploiting specific effects in the nonlinear response regime of charge carrier transport appearing in two dimensional charge carrier gases confined laterally in channels [3, 4]. The main property of this device type is a negative voltage response of the central branch of the TTJ if finite voltages are applied to the left and the right branch of a symmetric TTJ. In the ideal device the voltage response of the central branch is equal to the half of the voltage difference applied between the left and the right branch of the TTJ. In case of pure ohmic behavior at the central branch of the TTJ the voltage would be zero and independent on the applied push-pull voltage [2]. TTJ as elementary building blocks have been used to realize different types of electronic devices, for example, frequency doublers and mixers, logic gates, latch and detectors [2, 5, 6]. The relative simple fabrication requirements and implementation abilities into post CMOS circuits make them attractive for future low power high speed electronics.

TTJ devices were mainly demonstrated on III-V heterostructures [2, 5, 6], Recently, voltage rectification was demonstrated on exfoliated graphene [7] and epitaxial graphene [8, 9]. In this report voltage rectification of two TTJ device types fabricated on epitaxial graphene grown on semiinsulating silicon carbide will be shown. The impact of the device design on the TTJ properties will be demonstrated.

Experimental

Epitaxial graphene was formed on Si-face on-axis semiinsulating 4H-SiC. The graphene growth was carried out in a high temperature furnace at 1800°C in an argon atmosphere at atmospheric pressure. Prior to graphitization procedure the samples were pretreated using a direct capping technique with a graphite cover placed directly onto the SiC surface while annealing at 1800 °C for 1 min [10]. This preparation step was essential to achieve stepped SiC morphology. The graphene on SiC was grown using a heating rate of 10 K/min at the same Ar ambient conditions without a subsequent hydrogen intercalation. Raman spectroscopy and atomic force microscopy measurements confirmed the presence of few-layer graphene on top of the stepped SiC terraces with a width of a few microns. The layer number varied between 2 and 4 layers across the sample and within the devices. The hall mobility of the graphene was determined to be between 800 and 1000 cm2 V-1 s-1.

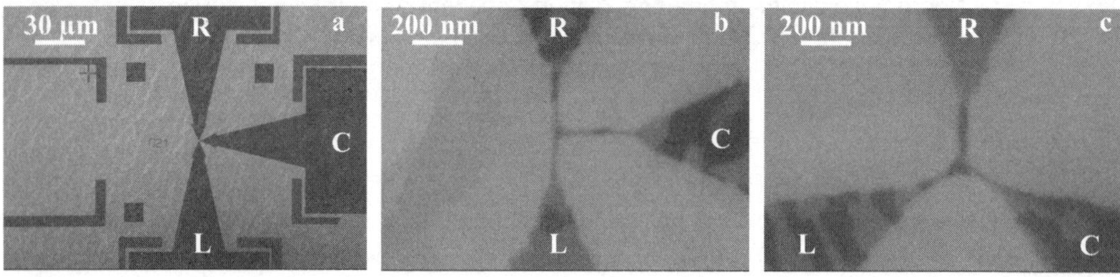

Fig. 1 Scanning electron microscopy images of the all graphene TTJ devices: (a) device layout, (b) T-branch TTJ, (c) Y-branch TTJ. R, L and C denote for right, left and centre TTJ terminal.

The devices were fabricated using electron beam lithography and a negative tone HSQ based resist. The graphene layer was structured using oxygen plasma in an electron cyclotron resonance plasma etching equipment. The fabricated TTJ structures were all-carbon structures, i.e. beside the active device parts the interconnections and the pads were also directly patterned into the epitaxial graphene. Metals were only used for the alignment marks. The HSQ was removed by dipping the samples into a buffered oxide etcher consisting of diluted HF. A typical device layout is shown in Fig. 1. Fig. 1a displays scanning electron microscopy (SEM) images of the device layout consisting of the TTJ structure, interconnects and pads. The active device part is located between the arrows-like interconnect structures. Figs. 1b and 1c display enlarged SEM images of the Y- and T-branched TTJ structures, respectively.

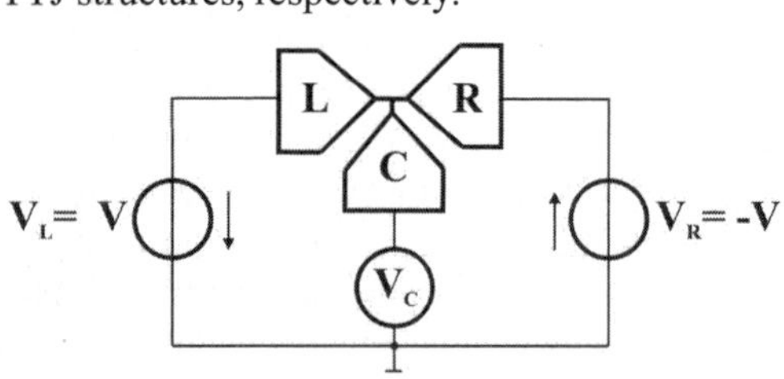

Fig. 2 Measurement set-up

The TTJ devices were measured at room temperature using a three point setup given in Fig. 2. The parasitic resistances were eliminated using separate force and sense circuits. For the characterisation of the nonlinear electrical properties, the push-pull measurement technique was used. In this configuration a voltage VL = V and VR = -V or vice versa is applied to the terminals on the left (L) and on the right (R) side of the TTJ structure. The voltage response is measured at the centre terminal (C) with a high impedance voltmeter.

Results and discussion

The dependence of the output voltage VC measured at the centre terminal C on the applied push-pull voltage V of the T- and Y-branch TTJ devices shown in Figs. 1b and 1c are given in Figs. 3a and 3b, respectively. For both device types a negative curvature of the VC versus V curve was observed. This response behaviour of the centre terminal voltage is called voltage rectification [3]. It can be described by extending the Landauer-Büttiger formalism [11] to the nonlinear regime by including the energy dependence of the transmissivity, the actual distribution function of the charge carrier and quantum mechanical scattering effects [3]. Another approach to explain the nonlinear effect was based on intervalley electron transfer [12] or surface charges [13]. Independent on the theoretical approach used to describe the nonlinear response of the TTJ the form of the bell shaped down-bending curve can be described by the following expression:

$$V_C = k\, V^2 + O(V^4), \tag{1}$$

with the curvature k and the residual function O(V4) [3]. The fitting curves using expression (1) describe the experimental curve reasonable well. They are shown together with the experimental curves in Fig. 3. At higher push-pull voltages the response curve shows a linear dependence. The related asymptotic curves are given in Fig. 3. By extrapolating the linear fit of the high voltage region to VC = 0 an onset voltage Von can be defined. For both types of devices Von exceeds 0.5 V with respect to the position of the maximum. Such large values can not be explained by the nonlinear ballistic model [14] alone. Therefore, other models have to be developed to describe the observed behaviour.

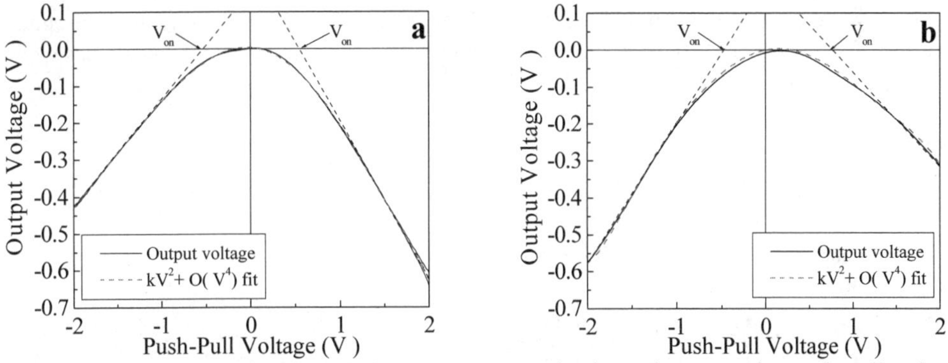

Fig. 3 The centre terminal voltage VC response curve in dependence on the push-pull voltage V: (a) T-branched TTJ (horizontal branch length LH = 395 nm, horizontal branch width WH = 15 nm), (b) Y-branched TTJ (branch length LY = 210 nm, branch width WY = 210 nm, angle between bran-ches αY = 120°.

Both voltage responses exhibit an asymmetric behaviour which can be assigned to geometrical asymmetries in the TTJ structure and to thickness inhomogeneities of the epitaxial graphene layer. For example, the width of the branches show slight differences due to different electron beam writing directions and line edge roughness (Fig. 1). The graphene thickness inhomogeneities are visible in Fig. 1 due to the changing shade of grey in the device structure. According to [3]

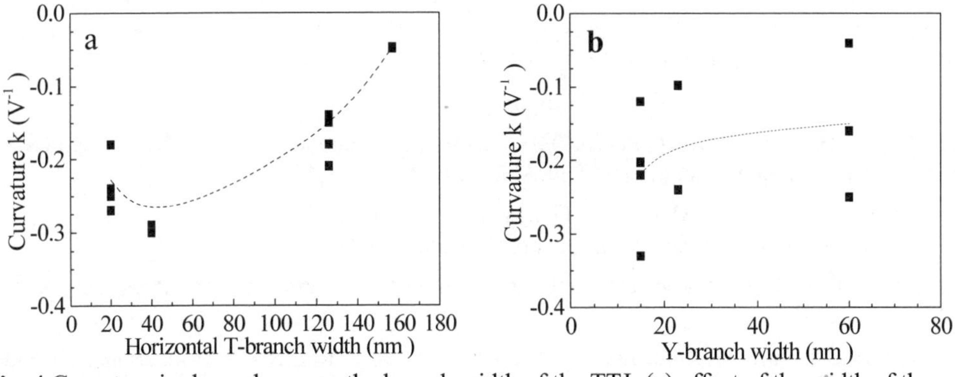

Fig. 4 Curvature in dependence on the branch width of the TTJ: (a) effect of the width of the horizontal branch W_H in the T-branch TTJ, (b) effect of the width of the Y-branch in the Y-TTJ.

symmetry breaking effects in the device structure are changing the symmetry of the electron transmission between the left and the central, and the right and the central branch. As a consequence the output voltage shows an asymmetric behaviour.

In Fig. 4 the dependence of the curvature k of the T- branch (Fig. 4a) and Y-branch (Fig 4b) TTJ as a function of the width of the horizontal branch in the T-TTJ and the branch width of the Y-TTJ is given, respectively. The dependence of the absolute value of k versus the branch width for T-TTJ shows a maximum around 40 nm, where as the dependence of the Y-TTJ shows a monotonically decreasing behaviour. On the one hand side, k it is directly connected to the transport mechanism in the device reflecting the amount of ballistic transport in the TTJ [12 - 14]. Thus the increasing absolute value of k is connected to a more pronounced ballistic transport in the TTJ. This can be expected if the width and the length of the structures are shrinking [14]. On the other hand, within the limits of the surface charge model the increase of the absolute k value can be explained by an increasing voltage drop at the anode [15]. The extremum in Fig. 4a cannot be explained within the framework of the mentioned theories.

Summary

Voltage rectification was demonstrated for T- and Y-branched TTJ devices fabricated on epitaxial graphene on Si-face 4H-SiC. It was shown that shrinking dimensions led to an increase of the volage rectification effect.

References

[1] D. Goldhaber-Gordon, M.S. Montemerlo, J.C. Love, G.J. Opiteck and J.C. Ellenbogen: Proc. IEEE Vol. 85 (1997), p. 521

[2] I. Worschech, D. Hartmann, S. Reitzenstein and A. Forchel: J. Phys.: Condens. Mater. Vol. 17 (2005), p. R775

[3] H.Q. Xu: Appl. Phys. Lett. Vol. 78 (2001), p. 2064

[4] H.Q. Xu: Appl. Phys. Lett. Vol. 80 (2002), p. 853

[5] I. Iniguez-de-la-Torre, T. Gonzalez, D. Pardo, C. Gardes, Y. Roelens, S. Bollart, A. Curuchet, C. Gaquiere and J. Mateos: Semicond. Sci. Technol. Vol. 25 (2010), p. 125013

[6] J. Sun, D. Wallin, I Maximov and H.Q. Xu: Electr, Dev. Lett. Vol.29 (2008), p 540

[7] A. Jacobson, I. Shorubalko, L. Maag, U. Sennhauser and K. Ensslin: Appl. Phys. Let. Vol. 97 (2010), p. 032110

[8] R. Göckeritz, K. Tonisch, W. Jatal, L. Hiller, F. Schwierz and J. Pezoldt: Adv. Mater. Res. Vol. 324 (2011), p. 427

[9] R. Göckeritz, J. Pezoldt and F. Schwierz: Appl. Phys. Lett. Vol. 99 (2011), p. 173111

[10]R. Göckeritz, D. Schmidt, M. Beleites, G. Seifert, S. Krischok, M. Himmerlich and J. Pezoldt: Mater. Sci. Forum Vol. 679-680 (2011), p. 785

[11]M. Büttiger: Phys. Rev. Lett. Vol. 57 (1986), p. 1761

[12]J. Mateos, B.G. Vasallo, D. Pardo, T. Gonzalez, E. Pichonat, J.-S. Galloo, S. Bollart, Y. Roelens and A. Cappy: Electr. Dev. Lett. Vol. 25 (2004), p. 235

[13]I. Iniguez-de-la-Torre, J. Mateos, T. Gonzalez, D. Pardo, J.S. Galloo, S. Bollaert, Y. Roelens and A. Cappy: Semicond. Sci. Technol. Vol. 22 (2007), p. 663

[14]H. Irie, Q. Diduck, M. Margala, R. Sobolewski and M.J. Feldmann: Appl. Phys. Lett. Vol. 93 (2008), p. 053502

[15]I. Iniguez-de-la-Torre, T. Gonzalez, D. Pardo, C. Gardes, Y. Roelens, S. Bollaert and J. Mateos: J. Appl. Phys. Vol. 105 (2009), p. 094504

Materials Science Forum Vols. 717-720 (2012) pp 687-690
© (2012) Trans Tech Publications, Switzerland
doi:10.4028/www.scientific.net/MSF.717-720.687

Development of FETs and resistive devices based on epitaxially grown single layer graphene on SiC for highly sensitive gas detection

R. Pearce[1, a,*], R. Yakimova[1, b], J. Eriksson[1, c] L. Hultman[1, d], M. Andersson[1, e] and A. Lloyd Spetz[1, f]

[1]Department of Physics, Chemistry and Biology, Linköping University, 58183 Linköping, Sweden

[a]ruth@ifm.liu.se, [b]roy@ifm.liu.se, [c]jenser@ifm.liu.se, dlarhu@ifm.liu.se, [e]mikan@ifm.liu.se, [f]spetz@ifm.liu.se. *corresponding author

Keywords: epitaxial graphene, gra-FET, NO_2 sensor, work function

Abstract. Epitaxially grown single layer graphene on silicon carbide (SiC) resistive sensors were characterised for NO_2 response at room and elevated temperatures, with an n-p type transition observed with increasing NO_2 concentration for all sensors. The concentration of NO_2 required to cause this transition varied for different graphene samples and is attributed to varying degrees of substrate induced Fermi-level (E_F) pinning above the Dirac point. The work function of a single layer device increased steadily with increasing NO_2 concentration indicating no change in reaction mechanism for high and low concentrations despite a change in sensor response direction. Epitaxially grown graphene device preparation is challenging due to poor adhesion of the graphene layer to the substrate. A field effect transistor (FET) device is presented which does not require wire bonding to contacts on graphene.

Introduction

Graphene is a one atom thick, honeycomb layer of carbon, and is particularly interesting as a gas sensor material. Single layer graphene has every atom on the surface so every atom is exposed to gas. Graphene has few crystal defects which leads to low noise devices and also exhibits high metallic conductivity, even when charge carrier concentration is low. Thus, changes in number of even a few charge carriers can cause notable changes in conductivity and sensor signal. Epitaxial graphene on SiC has clear advantages as a sensor material over graphene produced by, e.g., exfoliation from highly ordered pyrolytic graphite or by reduction of graphene oxide such as uniformity of coverage and control over the number of graphene layers, and therefore, electronic properties. Epitaxial graphene is n-type due to charge donation from the substrate; however, in an air ambient, electron withdrawal due to adsorbed oxygen reduces the concentration of electronic carriers, pinning E_F near to the Dirac point. Resistive epitaxial graphene sensors have already been demonstrated to be very sensitive gas sensors, with single layer devices showing a much greater sensitivity than multi-layer devices [1, 2]. A gate electrode can be used to bias the graphene layer so that E_F lies at the Dirac point where gas response is expected to be greatest. Graphene on SiC FETs utilize top gate technology; however, this device design is not suitable for sensing as gases cannot diffuse through the dielectric layer. The incorporation of a graphene layer as the channel in a gas sensitive, suspended gate field effect transistor device would maximise the potential of epitaxial graphene as a feasible gas sensing layer, by enabling further control over Fermi level pinning.

Experimental

Single layer graphene on the Si-face of on-axis, 4H-SiC (0001) was fabricated by the sublimation of Si by annealing in Ar at 2000°C. Fig. 1 shows the results of typical attempts to wire bond Ti/Au contacts on graphene on SiC resulting in either detachment of small areas of the contact and underlying graphene from the SiC substrate or detachment of the entire contact and underlying graphene due to the poor adhesion of graphene to the buffer-layer. Resistive sensor devices were fabricated by adhesion of gold bond wires to the graphene surface with silver nanoparticle glue (Epoteck E3018). The devices and a Pt100 temperature sensor were glued to a heater suspended above the header to enable rapid changes in temperature. The devices were exposed to NO_2 in a

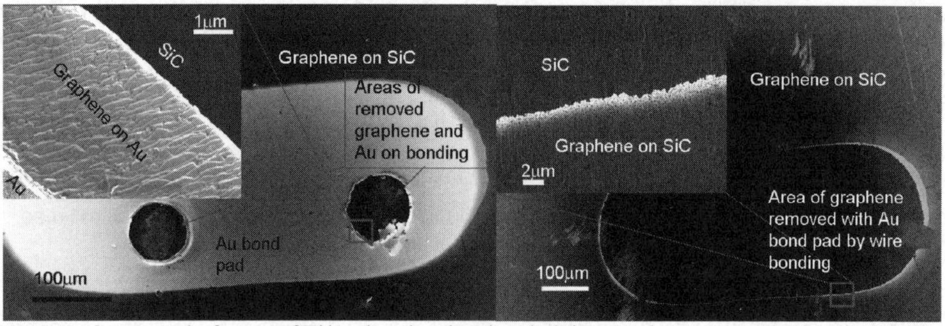

Fig. 1: Left: removal of areas of Ti/Au bond pad and underlying graphene upon wire bonding. Right: removal of whole Ti/Au bond pad and underlying graphene upon wire bonding.

carrier gas of N_2 and 20% O_2 whilst held in a stainless steel flow cell with a head space of ~1 cm^3. The gases were mixed with an in-house built gasmixer system comprising Bronkhorst flow controllers controlled with Lab VIEW. The flow rate during gas exposure testing was maintained at 200mlmin^{-1}. The resistance of the devices was monitored using a Keithley SourceMeter. Work function measurements were carried out with a Kelvin Probe with a 0.5mm gold tip (KP Instruments).

Graphene Characterisation

The graphene devices were characterised by atomic force microscopy (AFM) and Kelvin probe force microscopy (KPFM) which shows the number of graphene layers (Fig. 2) [3]. The statistical potential distribution reveals homogeneous single layer graphene which follows the SiC step edges. Some wrinkles are visible in the graphene of device 1. More defined SiC step edges with corresponding potential drops at the step edges are observed for the graphene of device 2.

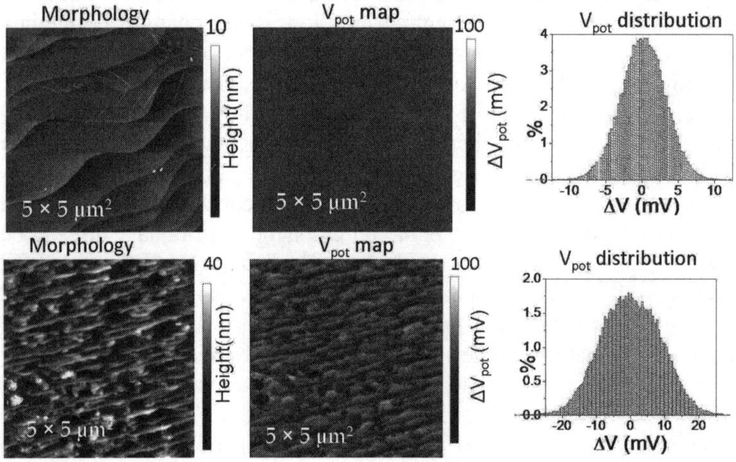

Fig 2: Typical AFM and potential maps of the graphene/SiC sensor devices. The average potential distributions are shown in the histograms to the right, with both device 1 (top) and device 2 (lower) showing single layer graphene coverage demonstrated by the homogeneous surface potential maps.

Sensitivity to Gases

Graphene on SiC is n-type due to electron donation from the substrate / buffer layer. The response of device 1 to ppb concentrations of NO_2 is typical of an n-type conductor (Fig. 3a) The recovery times are significantly longer at lower NO_2 concentrations, suggesting NO_2 first adsorbs at defect or higher energy binding sites. At higher NO_2 concentrations the device undergoes an n-p transition due to electron withdrawal (hole doping) by the adsorbed NO_2 (Fig. 3c). Device 2 also undergoes an n-p transition, however, at lower NO_2 concentrations than device 1 (Fig. 3b). It is theorised that one electron is transferred from the graphene to each adsorbed molecule of NO_2 [4]. The n-p transition occurs at differing NO_2 concentrations for the sensors; however, due to long response times it is difficult to estimate the exact NO_2 concentration required to cause the transition.

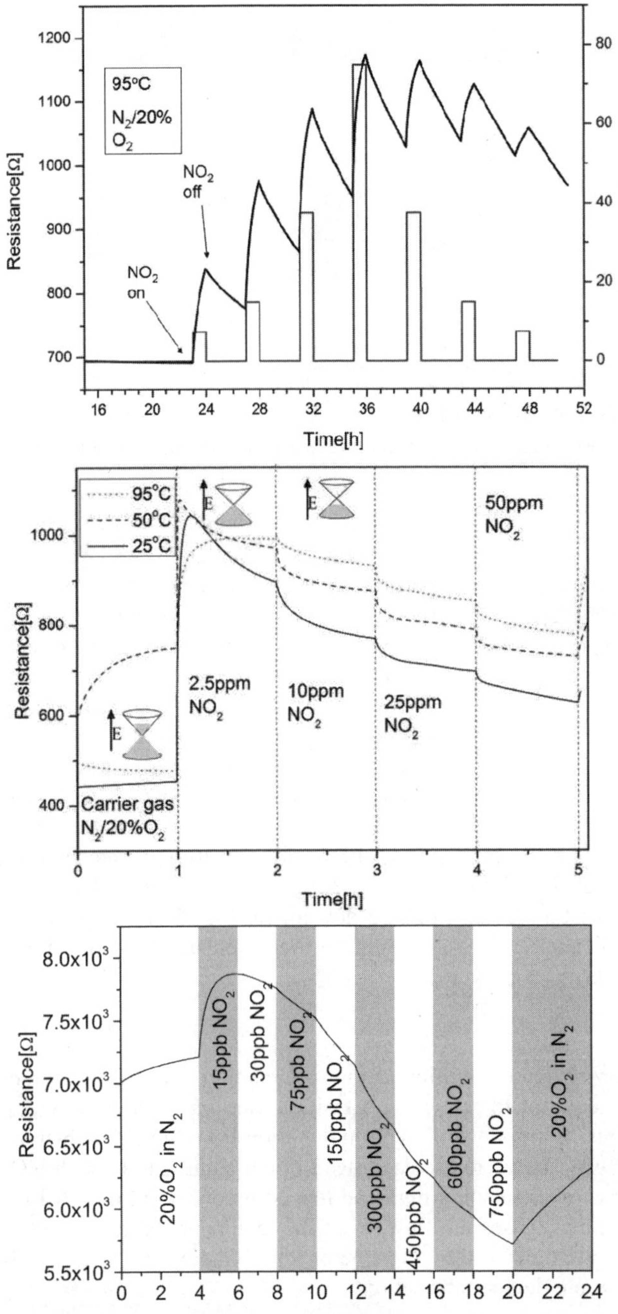

Fig. 3a: Response of device 1 to pulses of ppb concentration NO_2 in a N_2/20% O_2 background at 95°C, **b**: Response of device 1 to ppm concentrations of NO_2 showing an n-p type transition at all tested temperatures (the schematic band diagrams depict the Fermi level at each gas concentration), **c**: n-p transition of device 2 upon stepwise exposure to ppb concentrations of NO_2 showing an n-p transition at a lower NO_2 concentration.

Work Function Measurements

Device 1 was cleaned by flowing N_2 with 20% O_2 over the sensor at a high source drain current until adsorbed gases were desorbed and the sensor response was stable. The device was then left to stabilise with a low source drain current (0.01mA). When stable, the device resistance was measured. The work function was measured with a Kelvin probe (KP instruments) with no applied source drain current. The device was left to stabilise for 24 hours at each NO_2 concentration before resistance and work function were recorded. Oxygen chemisorbed on the graphene surface is thought to increase the resistance of the n-type graphene layer due to the removal of electrons. NO_2 also binds strongly on the surface, acting as an electron acceptor and increasing the resistance of the graphene device when n-type and reducing the resistance of the device when p-type. The work function changes during gas adsorption due to a shift in E_F energy which is brought about by chemical doping of the graphene by the adsorbed gas species. The work function (Fig. 4) increases steadily with NO_2 exposure, indicating that the reaction mechanism does not change at high and low NO_2 concentrations. The n-p type transition, indicated by the change in direction of sensor response, is observed to occur between 15 and 30ppb. Despite the very low NO_2 surface coverage expected at ppb concentrations the work function increases by ~0.5eV on adsorption of 300ppb NO_2.

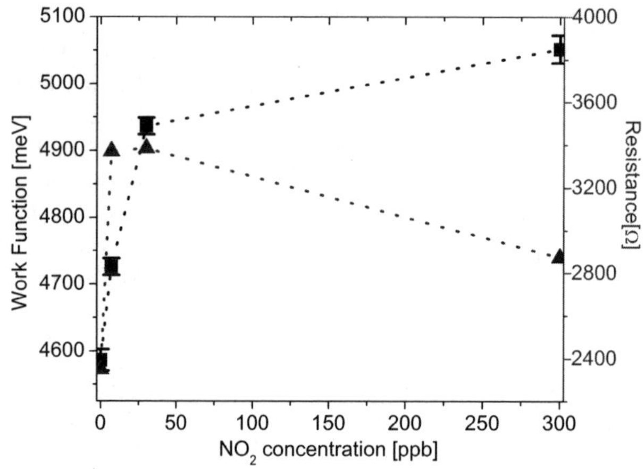

Fig. 4: Change in resistance (squares) and work function (triangles) of device 1 as a function of NO_2 conc.

New Designs for FET Devices

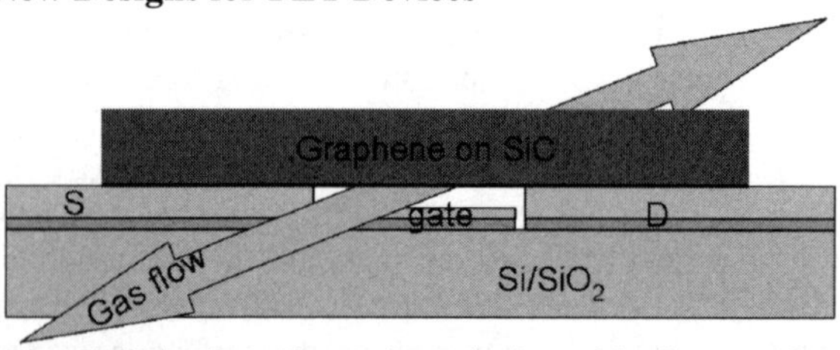

Fig. 5: Side view of transistor device with all patterning and bonding carried out on the Si/SiO_2 part to preserve the properties of the graphene on SiC layer.

Whilst resistive epitaxial graphene sensors have been demonstrated to be very sensitive [1, 2] the incorporation of a graphene layer as the channel in a gas sensitive field effect transistor device would enable control over E_F pinning in the graphene channel by applying a gate field. This control over the position of E_F could achieve a yet more sensitive and reproducible sensor response. The largest gas response is expected when E_F lies at the Dirac point where the density of states is zero. Graphene/SiC FETs are usually top gated with a dielectric deposited atop a graphene channel. To allow for gas interaction with the epitaxially grown graphene, a new type of field effect transistor device must be developed. Fig. 5 shows a device which enables biasing of the graphene channel and gas sorption on the graphene channel without requiring wire bonding to contacts on the graphene layer [5].

Summary

Single layer graphene on SiC resistive sensor devices exhibited an n-p transition upon adsorption of NO_2 with the amount of adsorbed NO_2 required to cause the transition varying between devices. This variation is thought to be due to differing amounts of electron donation from the SiC substrate. The work function of a graphene device was shown to increase steadily but significantly with NO_2 exposure, indicating no change in response mechanism at high and low NO_2 concentrations. A FET device with a suspended gate which allows gases to flow over the graphene layer has been developed. The device does not require patterning of the graphene or wire bonding to contact pads on the graphene, thus allowing for very simple device fabrication.

References

[1] R. Pearce, T. Iakimov, M. Andersson *et al.*, Sensors and Actuators B: Chemical 155 (2011) 451-455

[2] R. Pearce, T. Iakimov, M. Andersson *et al.*, in Proceedings of the International IEEE Conference on Sensors, Waikoloa Hawaii US 1-4 November (2010) 898

[3] T. Filleter, K. V. Emtsev, Th. Seyller, and R. Bennewitz, Appl. Phys. Lett. 93 (2008) 133117

[4] T. O. Wehling, K. S. Novoselov, S. V. Morozov *et al.*, Nano Letters 8 (2007) 173-177

[5] R. Pearce, M. Andersson, L. Hultman, *et al.*, Patent app. no. PCT/SE2011/05056 (2011)

Materials Science Forum Vols. 717-720 (2012) pp 691-694
© (2012) Trans Tech Publications, Switzerland
doi:10.4028/www.scientific.net/MSF.717-720.691

Temperature Dependent Chemical Sensitivity of Epitaxial Graphene

V.K. Nagareddy[1,a,*], D.K. Gaskill[2,b], J.L. Tedesco[2], R.L. Myers-Ward[2,c], C.R. Eddy Jr[2,d], J.P. Goss[1,e], N.G. Wright[1,f], and A.B. Horsfall[1,g]

[1]School of Electrical, Electronic & Computer Engineering, Newcastle University, NE1 7RU, UK

[2]U.S. Naval Research Laboratory, 4555 Overlook Avenue SW, Washington D.C. 20375, USA

[a,*]v.k.nagareddy@ncl.ac.uk, [b]kurt.gaskill@nrl.navy.mil, [c]rachael.myers-ward@nrl.navy.mil, [d]chip.eddy@nrl.navy.mil, [e]j.p.goss@ncl.ac.uk, [f]n.g.wright@ncl.ac.uk, [g]alton.horsfall@ncl.ac.uk
*corresponding author

Keywords: Epitaxial Graphene, Chemical Sensing, Polar Protic, Polar Aprotic, Dipole Moment.

Abstract. We investigated the chemical sensing mechanism of epitaxial graphene grown on 6H-SiC (0001) to different polar solvents and their behavior at higher temperatures. We show that at 300 K the sensitivity of the graphene sensor increases with the dipole moment of a solvent and decreases significantly as the temperature increased to 42.5 K. Using electrical measurements, we also show that graphene can effectively discriminate between polar protic and polar aprotic solvents with the shift in device electrical resistance at 300 K.

Introduction

Graphene, a two dimensional form of carbon, has been demonstrated as a promising sensing material because of its unique physiochemical properties such as ultrahigh surface-to-volume ratio, excellent thermal and electrical conductivity and extraordinary carrier mobility at room temperature [1]. Previous studies have shown that graphene can be extremely sensitive to various gases with a single molecule detection limit for certain gas species [2]. However, a substantial proportion of these studies were performed either on graphene oxide or mechanically exfoliated graphene, which are not suitable for large scale industrial production [3]. When considered for practical applications, epitaxial graphene grown on SiC is desirable, because of its ability to produce wafer size films directly on insulating SiC substrates. Moreover, the chemical sensing behaviour of graphene to various organic solvent vapors has been overlooked in the literature. The effect of temperature on the chemical sensing mechanism of graphene has not been discussed to date. Since military and aerospace applications require operation of devices over a range of temperatures, addressing the sensing mechanism of graphene with temperature is also of great interest.

In this work we show that the epitaxial graphene grown on 6H-SiC (0001) is sensitive to a variety of chemicals including polar protic (hydrogen peroxide (H_2O_2), De-Ionised (DI) water (H_2O)) and polar aprotic (Di-methyl acetamide (DMA), Di-methyl sulphoxide (DMSO), Acetic anhydride and N-methyl-2-pyrrolidone, (NMP)) solvents. We show that the sensitivity of our graphene sensor increases with the increase in dipole moment (μ). We also analysed the temperature dependent chemical sensing behavior of graphene to these solvents, which showed a considerable decrease in magnitude of response with the increasing temperature due to the decrease in equilibrium concentration of analyte molecules on the graphene surface.

Experimental

The epitaxial graphene films were synthesized on semi-insulating, 6H-SiC (0001) substrates in a commercial chemical vapour deposition system by thermal decomposition of SiC at 1500°C in a vacuum of about 10^{-5} mbar. The sample preparation procedure, growth morphology and Raman characterization of the graphene films are described elsewhere [4, 5]. The inset of Fig. 1 shows the prototype sensor structure used in this study, where seventy devices were fabricated on a 5×5mm^2 sample. These devices were electrically isolated by lithographically patterning the graphene film into mesas by means of SF_6 based reactive ion etching. Ti/Au (10nm/300nm) metal films were then deposited on the patterned graphene using e-beam evaporation, followed by lift-off in warm acetone (~50°C). In order to eliminate any photo-resist contamination and its possible impact on the electrical and sensing properties of the graphene, the sample was annealed in an N_2/H_2 atmosphere at 400°C for 120 sec using a rapid thermal processing system prior to the measurements. The ohmic nature of the contacts was verified from the linear I-V characteristics (Fig. 1) as expected for a graphene-metal based system. The baseline resistance of all the devices on the sample at 300 K varied between 20 kΩ to 30 kΩ. To avoid the current variation between devices, all results reported in this work are extracted from the same device.

For chemical sensing experiments, two *polar protic* solvents (H_2O_2 and H_2O) and four *polar aprotic* solvents (NMP, DMSO, DMA and acetic anhydride) were used. For each chemical, the change in electrical resistance of the device was recorded upon exposure. The device was exposed to analyte vapours by bringing a cotton swab soaked with the target analyte approximately 1cm away from the device under test. A dry cotton swab was used as a control reference, which registered as no detectable change in response. The analyte exposure started at t = 0 sec, and at t = 120 sec the response of the device was measured, having allowed sufficient time for the analyte to adsorb on to the graphene surface. Subsequently, the cotton swab was removed from the proximity of the device and the sample was cleaned with DI water and ambient annealed at 150°C to recover the sensor's initial state before re-measuring the device. Repeated exposure and annealing cycles showed no 'poisoning' effects of these chemicals on the device performance. This was repeated five times to obtain an average response for each analyte tested. All sensing experiments reported here were performed under ambient conditions at the temperatures between 300 K and 425 K using a Cascade Microtech probe station and a Keithley 4200 semiconductor parameter analyser.

Results and Discussion

Fig. 1 shows the I-V characteristics of the graphene sensor to various polar solvents at 300 K. Upon exposure to the aprotic solvents, the sensor showed a decrease in resistance, whereas, for the protic solvents resistance increased. This shows that graphene is responsive to both solvent types, which can be uniquely determined by the nature of the response. Fig. 2 shows the magnitude of change in resistance for five different polar solvents tested in this work as a function of their dipole moment over a range from 2.01 D to 4.09 D (here D represents Debyes, which is a common measurement unit for bond dipole moments). The sensitivity of the sensor was determined by $S=(R-R_0)/R_0$, where R_0 and R are resistances of the device before and after the chemical exposure respectively.

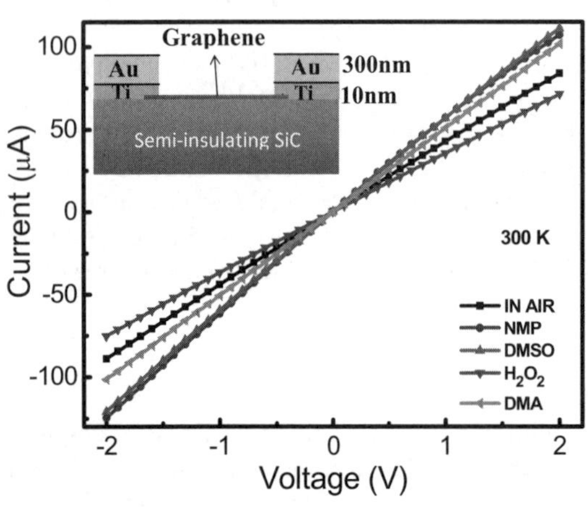

Fig. 1. I-V characteristics of the sensor exposed to polar protic and aprotic chemicals. Inset: device structure of the sensor used in this study

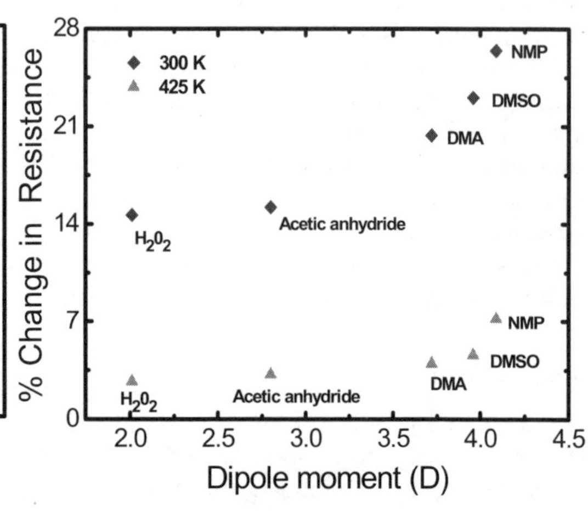

Fig. 2. Relationship between the dipole moment of a solvent and the sensitivity of the sensor.

We observed an increase in the magnitude of sensor response with increasing dipole moment. In order to understand this physical mechanism of the response of the resistivity of graphene to polar chemicals, the impact of the electronic interaction between analyte molecules and the graphene surface is considered.

Previous studies on as-grown graphene on SiC have shown that graphene is n-doped due to the interaction with the substrate, leaving the graphene film with an electron rich surface [6]. Upon exposure to polar chemicals, negative surface charges on graphene are expected to attract the positive-charged poles of molecules, which strongly adsorb to the graphene. This strong interaction leads to charge transfer between the adsorbed analyte and graphene, shifting the Fermi level in the graphene above or below the Dirac point [7]. Such type of adsorbate induced doping or surface transfer doping by analyte molecules was experimentally observed before [7], which can affect the charge carrier density of the graphene film. The nature of the doping and the magnitude of the response depend on the polarity and dipole moment of the analyte being used. For instance, at 300 K, aprotic NMP with a high dipole moment of 4.09 D showed the strongest response with 27% decrease in resistance, while protic H_2O_2 with a low dipole moment of 2.01 D showed the lowest response with 15% increase in resistance. The increase and decrease in resistance change are attributed to the electron donating and withdrawing behavior of these chemicals, in which the charge carrier density of the graphene layer is significantly affected as a result of chemical doping. For example, aprotic solvents with strong electron donating groups such as carbonyl (C=O) and sulphoxide (S=O) exert an electron pumping effect into the graphene during exposure, resulting in larger reduction in resistance. However, protic solvents with attached hydroxyl groups (-OH groups) exert an electron withdrawing effect and leads to an increase in resistance.

In addition to the dipole moment, the vapour pressure of a solvent can also affect the magnitude of the resistance change [8]. NMP has the lowest vapour pressure of 0.040 kPa of the all six tested solvents and water has the highest vapour pressure of 3.18 kPa. Hence, larger numbers of NMP vapour molecules are expected to interact with the graphene surface, contributing to the highest magnitude change in the resistance at 300 K, in comparison to water as shown in the Fig. 3. The observed response to all the solvent vapours was rapid and consistent with the dipole moment, and clearly showed discriminating behavior between protic and aprotic solvents. While the change in resistance for NMP and H_2O saturated within 50 sec of exposure, DMSO, DMF, DMA, and H_2O_2 showed changes in resistance even at t = 120 sec. However, the recovery of the base line after discontinuing the exposure of the solvent vapour showed a low desorption rate in comparison to very strong adsorption, consistent with previous studies [2]. Nevertheless, the base line can be fully retrieved by following the steps outlined in the experimental section.

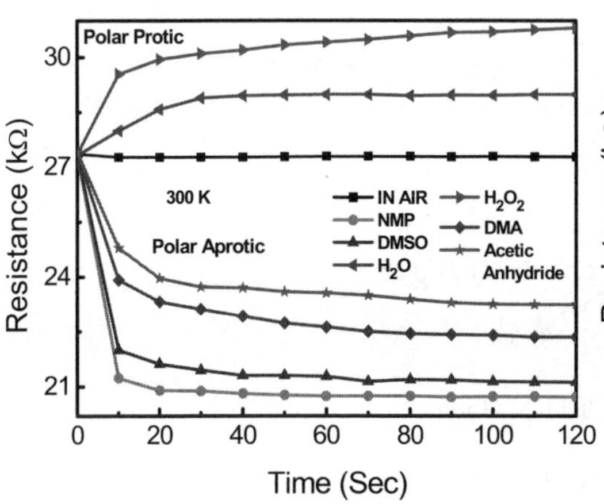

Fig. 3. Sensing behaviour of graphene to polar protic and aprotic solvents at 300 K.

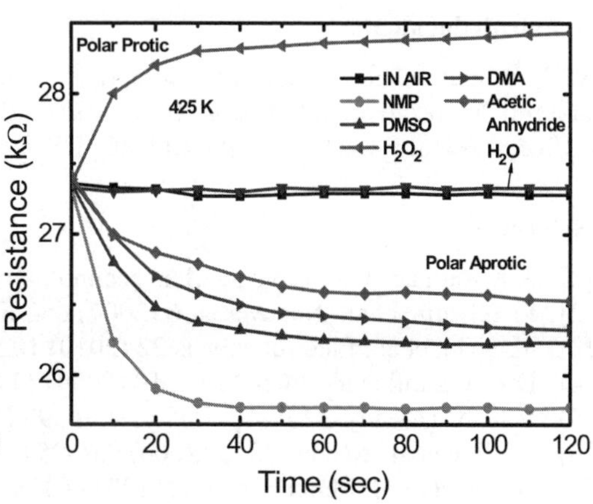

Fig. 4. Sensing behaviour of graphene to polar protic and aprotic solvents at 425 K.

To understand the temperature dependent sensitivity of graphene to these chemicals, sensing measurements were performed at temperatures between 300 K and 425 K. As shown in the Fig. 2, a significant decrease in the magnitude of response with respect to the solvent's dipole moment was observed due to the weaker analyte-graphene interactions at higher temperatures, which resulted in negligible change to the I-V characteristics. Fig. 4 shows the discriminating behavior of protic and aprotic solvents at 425 K, in which all the chemicals have shown a very weak response (less than 5%) except NMP which still showed ~7% change. This is possibly related to its heavy molecular weight. In the case of water, no change in resistance was detected above 100°C as expected. Moreover, with every 25 K rise in temperature, a rapid decrease in sensitivity was recorded for all the solvent types (Fig. 5). The observed decrease in sensitivity with temperature is consistent with the reported studies on carbon nanotube chemical sensors [9]. This decrease was attributed to the rise in vaporization rate of solvents with the increasing temperature, which decreases the equilibrium concentration of analyte molecules on the graphene surface. Although the sensitivities decreased with the increasing temperature, the response rate and the recovery time have improved substantially for all the chemicals.

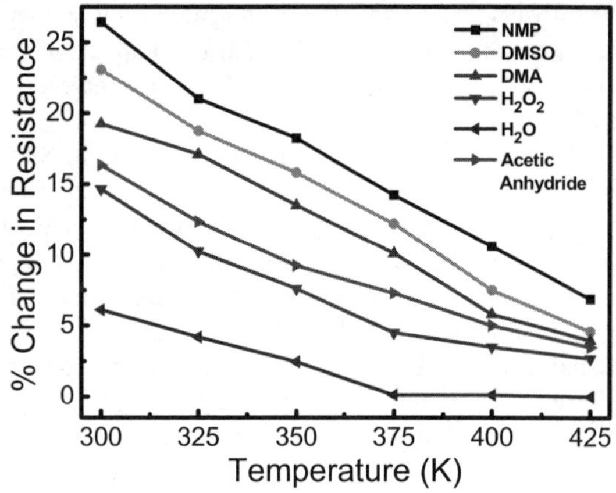

Fig. 5. Temperature dependent chemical sensitivity of epitaxial graphene to protic and aprotic solvents.

Summary

In summary, using the change in device electrical characteristics, we have demonstrated the feasibility of epitaxial graphene as an organic solvent sensor. This sensor can uniquely discriminate between polar protic and polar aprotic solvents at 300 K but this effect is limited at elevated temperatures. For all six solvents tested here, we observed an increase in sensitivity with the increase in the solvent's dipole moment at 300 K. At high temperatures, recovery rate of the sensor increased substantially at an expense of sensitivity.

Acknowledgements

VKN is grateful for studentship support from BAE Systems through the DHPA scheme. JLT is grateful for postdoctoral support from ASEE. Portions of this work were supported by the U.S. Office of Naval Research and the EPSRC, UK

References

[1] K.R Ratinac et al., Environ. Sci. Technol. 44 (2010) 1167-1176.
[2] F. Schedin, et al., Nat. Mater. 6 (2007) 652-655.
[3] Y. Shao, et al., Electroanalysis 22 (2010) 1027-1036.
[4] D.K. Gaskill et al., ECS Trans. 19 (2009) 117-124.
[5] V.K. Nagareddy et al., Appl. Phys. Lett. 99 (2011) 073506.
[6] T. Ohta et al., Science. 313 (2006) 951-954.
[7] H. Liu et al., J. Mater. Chem. 21 (2011) 3335-3345.
[8] J. Choi et al., Macro Mol. Res. 19 (2011) 980-983.
[9] N. L. Lala et al., Sensors 9 (2011) 86-101.

Keyword Index

Q

R

S

T

Author Index

T